Lecture Notes of the Institute for Computer Sciences, Social Informatics and Telecommunications Engineering 394

More information about this series at http://www.springer.com/series/8197

Jinbo Xiong · Shaoen Wu · Changgen Peng ·
Youliang Tian (Eds.)

Mobile Multimedia Communications

14th EAI International Conference, Mobimedia 2021
Virtual Event, July 23–25, 2021
Proceedings

Editors
Jinbo Xiong
Fujian Normal University
Fuzhou, China

Shaoen Wu
Illinois State University
Normal, IL, USA

Changgen Peng
Guizhou University
Guiyang, China

Youliang Tian
Guizhou University
Guiyang, China

ISSN 1867-8211 ISSN 1867-822X (electronic)
Lecture Notes of the Institute for Computer Sciences, Social Informatics and Telecommunications Engineering
ISBN 978-3-030-89813-7 ISBN 978-3-030-89814-4 (eBook)
https://doi.org/10.1007/978-3-030-89814-4

This Springer imprint is published by the registered company Springer Nature Switzerland AG
The registered company address is: Gewerbestrasse 11, 6330 Cham, Switzerland

Preface

We are delighted to introduce the proceedings of the 2021 European Alliance for Innovation (EAI) International Conference on Mobile Multimedia Communications (EAI Mobimedia 2021). This international conference facilitates technical discourse on deploying multimedia services and applications in mobile environments, which usually requires an interdisciplinary approach where multimedia, networking, and physical layer issues are jointly addressed. Within this scope, Mobimedia is intended to provide a unique international forum for researchers from industry and academia, working on multimedia coding, mobile communications, multimedia security and privacy, and networking, to explore and exchange ideas in new technologies, applications, and standards.

The technical program of Mobimedia 2021 consisted of 66 full papers, which were presented in eight technical sessions at the main conference: Session 1 - Internet of Things and Wireless Communications; Session 2 - Communication Strategy Optimization and Task Scheduling; Session 3 - Cryptography Security and Privacy Protection; Session 4 - Privacy Computing Technology; Session 5 - Cyberspace Security and Access Control; Session 6 - Neural Networks and Feature Learning; Session 7 - Task Classification and Prediction; and Session 8 - Object Recognition and Detection. Aside from the high-quality technical paper presentations, the technical program also featured three keynote speeches and five invited talks. The three keynote speeches were given by Shiwen Mao from Auburn University, USA, Shaoen Wu from Illinois State University, USA, and Yan Zhang from the University of Oslo, Norway. The five invited talks were presented by Fenghua Li from the Institute of Information Engineering, China, Zheng Qin from Hunan University, China, Xinyi Huang from Fujian Normal University, China, Jin Li from Guangzhou University, China, and Chao Shen from Xi'an Jiaotong University, China.

Coordination with the steering chairs, Imrich Chlamtac, Yang Qing, Dapeng Wu, and Yun Lin was essential for the success of the conference. We sincerely appreciate their constant support and guidance. It was a great pleasure to work with such an excellent organizing committee team for their hard work in organizing and supporting the conference. In particular, we are grateful to the Technical Program Committee who completed the peer-review process of technical papers and made a high-quality technical program. We would also like to express our gratitude to the local chairs, Youliang Tian and Yongbin Qin, for their meticulous organization, arrangement, and coordination of the conference. We are also grateful to Conference Manager Natasha Onofrei for her support and all the authors who submitted their papers to the Mobimedia 2021 conference.

We strongly believe that the EAI Mobimedia conference provides a good forum for all researchers, developers, and practitioners to discuss all science and technology aspects that are relevant to mobile multimedia communications. We also expect that the future conferences will be as successful and stimulating as Mobimedia 2021, as indicated by the contributions presented in this volume.

September 2021

Fenghua Li
Jianfeng Ma
Shiwen Mao
Changgen Peng
Youliang Tian
Shaoen Wu
Jinbo Xiong

Organization

Steering Committee

Chair

Imrich Chlamtac	University of Trento, Italy

Steering Committee Co-chair

Honggang Wang	University of Massachusetts Dartmouth, USA

Steering Committee

Chonggang Wang	InterDigital Communications, USA
Yonggang Wen	Nanyang Technological University, Singapore
Wei Wang	San Diego State University, USA
Shaoen Wu	Illinois State University, USA
Qing Yang	University of North Texas, USA
Shiwen Mao	Auburn University, USA
Dapeng Wu	Chongqing University of Posts and Telecommunications, China
Yun Lin	Harbin Engineering University, China

Organizing Committee

General Chair

Jianfeng Ma	Xidian University, China

General Co-chairs

Shiwen Mao	Auburn University, USA
Fenghua Li	Institute of Information Engineering, China

Technical Program Committee Chair and Co-chair

Jinbo Xiong	Fujian Normal University, China
Shaoen Wu	Illinois State University, USA
Changgen Peng	Guizhou University, China
Dianhui Chu	Harbin Institute of Technology, China

Sponsorship and Exhibit Chairs

Liang Wan	Guizhou University, China
Hongfa Ding	Guizhou University of Finance and Economics, China

Local Chairs

Youliang Tian	Guizhou University, China
Yongbin Qin	Guizhou University, China

Workshops Chairs

Lei Chen	Georgia Southern University, USA
Wenjia Li	New York Institute of Technology, USA

Publicity and Social Media Chair

Xuhui Chen	Kent State University, USA

Publications Chairs

Wei Wu	Fujian Normal University, China
Youliang Tian	Guizhou University, China
Yun Lin	Harbin Engineering University, China

Web Chair

Biao Jin	Fujian Normal University, China

Posters and PhD Track Chairs

Qing Yang	University of North Texas, USA
Weitian Tong	Eastern Michigan University, USA

Panels Chairs

Zuobin Ying	Nanyang Technological University, China
Weinan Gao	Georgia Southern University, USA

Demos Chairs

Siqi Ma	University of Queensland, Australia
Changqing Luo	Virginia Commonwealth University, USA

Tutorials Chairs

Hongtao Xie	University of Science and Technology of China, China
Mohammad Ali	Amirkabir University of Technology, Iran

Technical Program Committee

Anan Liu	Tianjin University, China
Atef Mohamad	Georgia Southern University, USA
Biao Jin	Fujian Normal University, China
Bin Xiao	Chongqing University of Posts and Telecommunications, China
Bowen Zhao	Singapore Management University, Singapore
Changgen Peng	Guizhou University, China
Changguang Wang	Heibei Normal University, China
Changqing Luo	Virginia Commonwealth University, USA
Cheng Zhang	Waseda University, Japan
Chihua Chen	Fuzhou University, China
Chunjie Cao	Hainan University, China
Dapeng Wu	Chongqing University of Posts and Telecommunications, China
Dianhui Chu	Harbin Institute of Technology, China
Farhan Siddiqui	Dickinson College, USA
Guodong Wang	Massachusetts College of Liberal Arts, USA
Hai Liu	Guizhou University, China
Hongfa Ding	Guizhou University, China
Hongtao Li	Shanxi Normal University, China
Hongtao Xie	University of Science and Technology of China, China
Jeonghwa Lee	Shippensburg University of Pennsylvania, USA
Jian Yu	Tianjin University, China
Jianhua Li	UniCloud Australia, Australia
Jiawen Kang	Nanyang Technological University, Singapore
Jiayin Li	Fuzhou University, China
Jin Wang	University of Massachusetts Dartmouth, USA
Jin Xu	University of Electronic Science and Technology of China, China
Jingda Guo	University of North Texas, USA
Jingjing Guo	Xidian University, China
Jun Zhang	Hefei University of Technology, China
Kang Chen	Southern Illinois University, USA
Kashinath Basu	Oxford Brookes University, UK
Lanxiang Chen	Fujian Normal University, China
Lei Chen	Georgia Southern University, USA
Liangming Wang	Jiangsu University, China
Lina Pu	University of Southern Mississippi, USA
Liu Cui	West Chester University of Pennsylvania, USA
Mike Wittie	Montana State University, USA
Ming Yang	Kennesaw State University in Georgia, USA
Mingwei Lin	Fujian Normal University, China

Mohammad Ali	Amirkabir University of Technology, Iran
Pengfei Wu	National University of Singapore, Singapore
Qi Chen	University of North Texas, USA
Qi Jiang	Xidian University, China
Qi Li	Nanjing University of Posts and Telecommunications, China
Qingchen Zhang	St. Francis Xavier University, Canada
Qingsong Zhao	Nanjing Agricultural University, China
Qingzhong Liu	Sam Houston State University, USA
Rami Haddad	Georgia Southern University, USA
Rui Wang	Civil Aviation University of China, China
Ruxin Dai	University of Wisconsin–River Falls, USA
Shaojing Fu	National University of Defense Technology, China
Siqi Ma	University of Queensland, Australia
Tao Feng	Lanzhou University of Technology, China
Tao Xiang	Chongqing University, China
Tao Zhang	Xidian University, China
Teng Wang	Xi'an University of Posts and Telecommunications, China
Weidong Yang	Henan Polytechnic University, China
Weinan Gao	Georgia Southern University, USA
Weitian Tong	Eastern Michigan University, USA
Wenjia Li	New York Institute of Technology, USA
Wonyong Yoon	Dong-A University, South Korea
Xiaotian Zhou	Shandong University, China
Ximeng Liu	Fuzhou University, China
Xing Liu	Wuhan University of Technology, China
Xingsi Xue	Fujian University of Technology, China
Xinhong Hei	Xi'an University of Technology, China
Xiwei Wang	Northeastern Illinois University, USA
Xu Yuan	University of Louisiana at Lafayette, USA
Yanxiao Zhao	Virginia Commonwealth University, USA
Yinbin Miao	Xidian University, China
Yongbin Qin	Guizhou University, China
Youliang Tian	Guizhou University, China
Yuanbo Guo	University of Information Engineering, China
Yuansong Qiao	Athlone Institute of Technology, Ireland
Yulong Shen	Xidian University, China
Yun Lin	Harbin Engineering University, China
Zeping Li	Guizhou University, China
Zhigang Yang	Chongqing University of Arts and Sciences, China
Zhu Hui	Xidian University, China
Zuobin Ying	Nanyang Technological University, Singapore

Co-sponsor

Science and Technology on Communication Information Security Control Laboratory, China Academy of Engineering (CAE)

Contents

Communication Strategy Optimization and Task Scheduling

Cryptography Security and Privacy Protection

Neural Networks and Feature Learning

Task Classification and Prediction

Internet of Things and Wireless Communications

Physical-Layer Network Coding in 6G Enabled Marine Internet of Things

Zhuoran Cai[1](✉) and Yun Lin[2]

[1] School of OPTO-Electronic Information Science and Technology, Yantai University, Yantai, People's Republic of China
caizhuoran@ytu.edu.cn

[2] College of Information and Communication, Harbin Engineering University, Harbin, People's Republic of China
linyun@hrbeu.edu.cn

Abstract. The ocean is a treasure trove of human resources, a blue gem on the earth, and a battlefield where soldiers meet. With the continuous advancement of science and technology, informationalized and modern marine research and the rational development of marine resources have attracted the attention of countries all over the world. Marine development must be accompanied by operations such as acquisition, transmission and processing of marine information. The Marine Internet of Things is the main carrier of ocean information acquisition and transmission, and also an important part of the integrated sky, ground and sea network explicitly involved in the 6G communication network. The underwater transmission of 6G Marine Internet of Things partly adopts underwater acoustic communication. The biggest problem facing underwater acoustic communication is the strong interference and strong fading of underwater acoustic channels. Considering the complexity of underwater acoustic communication environment and the asymmetry of channels, asymmetric two-way relay underwater acoustic communication system model is set up based on the ray model. In order to solve the link asymmetric problems in shallow underwater acoustic communication, a physical-layer network coding scheme based on asymmetric modulation is proposed.

Keywords: Shallow underwater acoustic communication · Asymmetric channel · Physical-layer Network Coding · Asymmetric modulation · Bit error rate

1 Introduction

Driven by economic and military needs, countries around the world are committed to developing marine resources. The vision of the sixth-generation mobile communication system (6G) clearly proposes the formation of an integrated sky, ground and sea network. Among them, the ocean part is composed of the Marine Internet of Things as a part of 6G [1–3]. The transmission of underwater information in the Marine Internet of Things adopts the method of underwater acoustic communication. Underwater acoustic

J. Xiong et al. (Eds.): MobiMedia 2021, LNICST 394, pp. 3–21, 2021.
https://doi.org/10.1007/978-3-030-89814-4_1

communication is a key technology in the development of marine resources, the measurement of marine environmental data, the search and rescue of underwater targets, the detection of submarine targets, and military early warning [1, 4–7]. Due to various random factors such as underwater communication transmission media, terrain conditions, and natural environment, it is very different from land communication. It is generally used to call the continental shelf sea area within a depth of 200 m as a shallow sea area. In the shallow sea, the transmission of sound waves will always reflect when touching the bottom and the sea surface. The underwater acoustic channel has the characteristics of significant multipath effect, large propagation delay, and severe propagation loss, which greatly limit the capacity of underwater acoustic communication. Due to the flow of water, the activities of ships or other underwater equipment, the movement of marine life, and other factors, the position of underwater terminals such as underwater sensors is not fixed under water and the underwater environment changes rapidly. Therefore, the time-varying characteristics of the underwater acoustic channel are obvious.

Underwater information transmission can be divided into wired transmission and wireless transmission. Wired transmission is achieved by using underwater or optical cables [8–13]. The main advantages of wired transmission are stable signals and strong anti-interference ability. However, underwater cables are not only expensive, but the investment in laying cables on the seabed is also huge, and the construction is extremely difficult. Once the cables are laid, it is difficult to move. These factors greatly limit the development and application of underwater wired communications. While the underwater wireless information has become an important method of underwater communication because of its small investment, convenience and flexibility, and no spatial location limitation. Owing to the unique medium of seawater and the complexity of the seabed environment, the transmission distance of electromagnetic waves or light waves widely used in water surface communications and land communications in water is very short, and sound waves can be transmitted farther in the ocean than the above two forms. Since the electromagnetic signals severely attenuated in an underwater transmission process, the underwater wireless communication cannot use electromagnetic waves as carriers like terrestrial wireless communication. At this stage, sound waves are the main information transmission carriers used in underwater wireless communication. The underwater acoustic channel is very complicated and has serious time-space-frequency change characteristics, multipath effect, strong fading and strong noise, and the available bandwidth is extremely narrow, which seriously affects the underwater acoustic communication performance [14].

Restricted by the characteristics of underwater acoustic channels, underwater acoustic communication faces challenges in terms of low transmission rate, high bit error rate and short transmission distance. In order to overcome these challenges, a technology with unique advantages in improving communication efficiency - Physical-Layer Network Coding (PNC) has attracted the attention of experts and scholars. The idea of physical layer network coding originated from Network Coding (NC). In 2000, R. Ahlswede et al. proposed the concept of network coding in [15], which allows routers to process the transmitted information. At this time, the use of network coding can improve the throughput of the entire network. The gain in communication efficiency brought about by network coding has aroused the interest of researchers. At present, people have reached a certain level of research on network coding. In 2006, S. Zhang and others proposed physical layer network coding technology based on the transmission characteristics of electromagnetic waves, breaking through the limitations of traditional point-to-point systems [16, 17]. PNC allows two or more sources to send information at the same time. Due to the broadcast characteristics of wireless media, the information of different sources is superimposed in space. PNC regards this natural superposition as the sum of different signals, and then uses a series of signal processing and mapping methods to obtain network-encoded signals. In the typical application scenario of the physical layer network coding - Two-Way Relay Channel (TWRC), PNC can increase the system throughput by 100% compared with the traditional point-to-point scheme, and increase the system throughput by 50% compared with the NC scheme in theory.

The advantages of physical layer network coding in improving the throughput of the communication network are consistent with the inefficiency of underwater acoustic communication. But up to now, most of the research on the application of physical layer network coding in TWRC is based on the premise that all channel states of the uplink and downlink are the same, that is, the channel is symmetric. However, in practical applications, due to the influence of many factors such as the communication distance and obstacles, the channel state of each channel will be somewhat different, so the communication channel of the Two-Way Relay Channel is not ideal. Therefore, when the channel conditions are asymmetric in TWRC, design an effective asymmetric channel physical layer network coding scheme, which has certain reference significance for the practical application of physical layer network coding in actual communication systems. Under the channel environment of underwater acoustic communication, with the flow of water, the activities of ships or other underwater equipment, the movement of marine life, etc., the position of underwater terminals such as underwater sensors, submarines and underwater unmanned aerial vehicles under water is even more unlikely to be fixed. The asymmetry of the channel is also an important feature of underwater acoustic communication, so it is of great significance to study the asymmetric physical layer network coding and its application in underwater acoustic communication.

2 Asymmetric Shallow Seawater Acoustic Channel Model

2.1 Point-To-Point Ray Model

In this study, the "ray model" is used to describe and analyze the shallow sea acoustic channel. The underwater acoustic ray model can obtain the channel's impulse response based on the reflection and attenuation of the multipath signal. Authors in [16] give a schematic diagram of the ray model, as shown in Fig. 1.

The energy of the sound wave is transmitted through the ray. If the path of the ray travels differently, there will be different arrival times and phases. The representation of the multipath signal can be obtained according to the attenuation and the delay of the path.

Signal attenuation mainly considers three main forms of attenuation. First, extended attenuation due to the expansion of the wavefront during propagation. In this study, we assume that the wavefront of an acoustic wave expands spherically in sea water, then the amplitude of the acoustic wave decays with the reciprocal of the propagation distance.

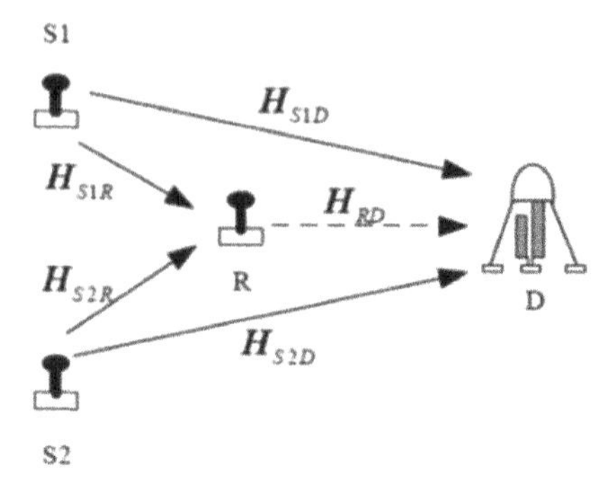

Fig. 1. Shallow sea acoustic ray model

Second, the reduction in sound intensity due to the absorption surface is called absorption attenuation, which is related to the distance and the working frequency of the sound wave. Third, the loss caused by the reflection of the interface during the transmission of the sound wave is called reflection attenuation.

There is a large gap between the theoretical value of absorption attenuation and the actual value, so it is generally expressed by an empirical formula [17]:

$$AL = r * 10\ lg\ \alpha \tag{1}$$

Where r is the propagation distance and α is the absorption coefficient.

$$\alpha(f) = \frac{0.102f^2}{1+f^2} + \frac{40.7f^2}{4100+f^2}(dB/km) \tag{2}$$

Reflection attenuation needs to consider the two reflection surfaces of the sea floor and the sea surface. The bottom of the shallow sea can be regarded as uniform and smooth fine sand or silt, so the reflection coefficient of the bottom is close to 1. In order to simplify the model parameters, the bottom reflection coefficient is assumed to be $|r_b| = 0.9$. The reflection coefficient of the sea surface can be calculated by the Bechmann-Spezzichino model [18]:

$$|r_s| = \sqrt{\frac{1+\left(\frac{f}{f_1}\right)^2}{1+\left(\frac{f}{f_2}\right)^2}}, f_2 = 378w^{-2}, f_1 = \sqrt{10}f_2 \tag{3}$$

Where f is the working frequency of the acoustic wave signal in kHz; w is the wind speed in knots (1 knots = 1.852 km/h = 0.514 m/s).

The propagation speed of sound waves in seawater changes with the difference of the seabed environment. It can also be regarded as constant in a certain period of time, so the speed of sound in the shallow sea can be regarded as a constant, where the average speed of the sound wave is 1500 m/s. The delay of each path can be calculated according to the distance and sound speed of each path.

After obtaining the delay and attenuation of the multipath signal, the received signal can be expressed as [16]:

$$r(t) = \alpha \frac{e^{jw(t-t_D)}}{D} + \alpha \sum_{n=1}^{\infty} \begin{bmatrix} \frac{R_{SS_n}}{SS_n} e^{jw(t-t_{SS_n})} + \frac{R_{SB_n}}{SB_n} e^{jw(t-t_{SB_n})} + \\ \frac{R_{BS_n}}{BS_n} e^{jw(t-t_{BS_n})} + \frac{R_{BB_n}}{BB_n} e^{jw(t-t_{BB_n})} \end{bmatrix} \quad (4)$$

2.2 Asymmetric Two-Way Relay Underwater Acoustic Channel Model

Establish the asymmetric two-way relay underwater acoustic channel ray model, as shown in Fig. 2. Here nodes M and N are no longer horizontally symmetrical about relay nodes, where a_1: distance between node M and the seabed; a_2: distance between node N and the seabed; b: distance between relay node R and the seabed; L_1: horizontal distance between node M and relay node R; L_2: the horizontal distance between node N and relay node R; h: the distance from the seabed to the sea surface. For the sake of clarity, only the direct wave and the reflected wave number 1 are given here.

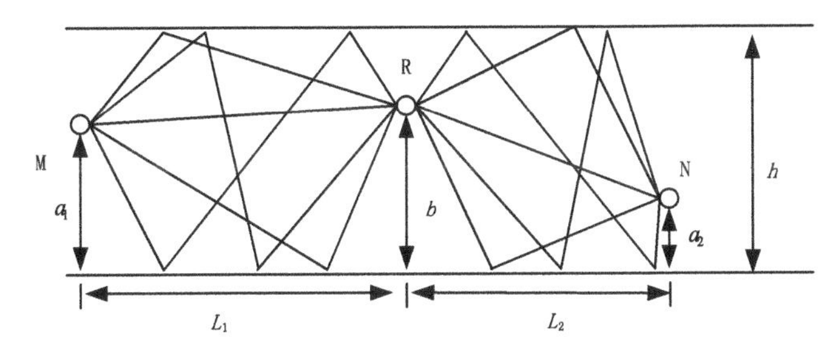

Fig. 2. Asymmetric two-way relay underwater acoustic channel ray model

It can be seen intuitively from Fig. 2 that because the two nodes M and N are not symmetrical about the relay node, the acoustic wave signals from the two nodes have completely different paths to the relay node, so the signal attenuation and delay are also different.

3 Existing Physical Layer Network Coding Scheme

3.1 Physical Layer Network Coding Scheme

If the system uses the traditional physical layer network coding scheme, the two users use ideally synchronized (time, frequency and phase) carriers $Re\left[e^{j(2\pi f_c t+\theta)}\right]$. The superimposed signal of a data packet (N symbols) period received by the relay node R is

$$y_R(t) = \sum_{i=1}^{n} \sum_{n=1}^{2} \left\{ \sqrt{g_{nR}} h_{nR} x_n^i p(t - iT_s) Re\left[e^{j(2\pi f_c t+\theta)}\right] \right\} + w_R(t) \quad (5)$$

where for x_i^n, $i = 1, 2$ represents the i-th symbol of the $\mathbf{x}_n$ of the data packet of BPSK modulated information of users M and N respectively. T_s is the symbol period and $W_{R(t)}$

is the additive noise. $p(t - iT_s) = rect(t - iT_s) = u[t - (i - 1)T_s] - u[t - iT_s]$ denotes rectangular pulse shaping function.

A data packet baseband signal vector which is down-converted from the carrier frequency and low-pass filtered can be expressed as follow:

$$y_R = \sum_{n=1}^{2} \sqrt{g_{nR}} h_{nR} x_n + w_R \tag{6}$$

Here $w_R = \boldsymbol{w}_{Rc} + j\boldsymbol{w}_{Rs}$ is the Additive White Gaussian Noise vector at the relay R, and the variance of each dimension is $N_0/2$. In order to obtain the XOR integrated data packet s_R, it is necessary to calculate the log-likelihood ratio of each symbol and make a decision. The log-likelihood ratio of the i-th symbol is,

$$\Lambda^j = log\left(\frac{\sum_{c^j = \pm(\sqrt{g_1 R} h_1 R + \sqrt{g_2 R} h_2 R)} exp\left(\frac{-\|y_R^i - c^i\|^2}{N_0}\right)}{\sum_{c^j = \pm(\sqrt{g_1 R} h_1 R - \sqrt{g_2 R} h_2 R)} exp\left(\frac{-\|y_R^i - c^i\|^2}{N_0}\right)}\right) \tag{7}$$

The decision formula is

$$\Lambda^j > 0\left(s_R^j = 0\right), \Lambda^{j,} < 0\left(s_R^j = 1\right) \tag{8}$$

The integrated data packet $s_R = \boldsymbol{s}_1 \oplus \boldsymbol{s}_2$ modulated by BPSK is broadcast to user M and user N. In the downlink phase, the baseband signals received by two users are

$$\boldsymbol{y}_n = \sqrt{g_{Rn}} h_{Rn} \boldsymbol{x}_R + \boldsymbol{w}_n, \; n = 1, 2 \tag{9}$$

After BPSK demodulation, the user can decode the other's information according to their own information

$$\widehat{s_n} = \widehat{s_R} \oplus s_{3-n} = 1, 2 \tag{10}$$

3.2 Bi-orthogonal Physical Layer Network Coding Scheme

In this scheme, two users use orthogonal carriers instead of ideal synchronization carriers to transmit data packets. Without loss of generality, assume that user M and user N respectively use $Re\left[e^{j(2\pi f_c t + \theta)}\right]$ and $Re\left[e^{j\left(2\pi f_c t + \theta - \frac{\pi}{2}\right)}\right] = Im\left[e^{j(2\pi f_c t + \theta)}\right]$ as carriers. The phase difference between the two carriers is π/2. So the superimposed signal received at the relay is

$$y_R(t) = \sum_{i=1}^{N} \sum_{n=1}^{2} \left\{\sqrt{g_{nR}} h_{nR} x_n^i Re\left[e^{j\left(2\pi f_c t + \theta - \frac{(n-1)\pi}{2}\right)}\right] p(t - iT_s)\right\} + w_R(t) \tag{11}$$

According to (11), it can be seen that ${x_1}^i$ and ${x_2}^i$ can be regarded as the in-phase component and the quadrature component of a symbol period, respectively, similar to the conventional QPSK modulation symbol. Therefore, the relay R can obtain the estimated

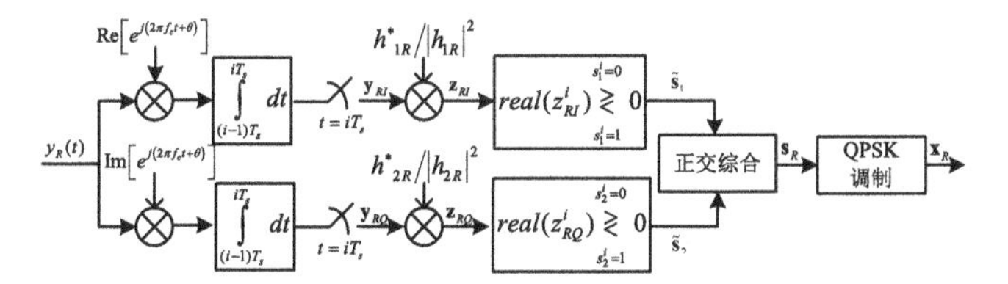

Fig. 3. Block diagram of relay R processing l

values of the information vectors s1 and s2 by using a correlation detector in the two branches respectively. The specific block diagram is shown in Fig. 3.

Taking the in-phase component as an example, after multiplying the carrier $Re\left[e^{j(2\pi f_c t+\theta)}\right]$, rounding and sampling can obtain the baseband signal of the in-phase branch:

$$y_{RI} = \sqrt{g_{1R}} h_{1R} \boldsymbol{x}_1 + \boldsymbol{w}_{Rc} \tag{12}$$

Then multiply $h^*{}_{1R}/|h_{1R}|^2$ and then obtain the estimated vector $\widetilde{s_1}$ through maximum ratio combination and hard decision, similarly can obtain $\widetilde{s_2}$

$$y_{RQ} = \sqrt{g_{2R}} h_{2R} \boldsymbol{x}_2 + \boldsymbol{w}_{Rs} \tag{13}$$

Then according to Fig. 3 we can get the comprehensive information

$$s_R = \frac{(\tilde{s}_1 + j\tilde{s}_2)}{\sqrt{2}} \tag{14}$$

Multiplying $1/\sqrt{2}$ here is to ensure that the transmission energy of relay R is 1. Compared to XOR synthesis and linear synthesis, (11) is called orthogonal synthesis. Therefore, the method of using orthogonal carriers for the two users and orthogonal synthesis at the relay is called bi-orthogonal physical layer network coding.

In the downlink phase, the integrated data packet s_R is modulated by QPSK and broadcast to two users. The data packet after QPSK modulation is

$$\boldsymbol{x}_R = \frac{\left[(1 - 2\tilde{s}_1) + j(1 - 2\tilde{s}_2)\right]}{\sqrt{2}} \tag{15}$$

The data that user M and user N want to obtain are in the quadrature and in-phase portions of the broadcast signal, respectively. Taking user M as an example, when receiving the signal from relay R, user M uses the same detector as the quadrature branch of the traditional QPSK correlation detector, multiplies the carrier $Re\left[e^{j\left(2\pi f_c t+\theta-\frac{\pi}{2}\right)}\right] = Im\left[e^{j(2\pi f_c t+\theta)}\right]$ first, and then round and sample to obtain the baseband signal.

$$y_1 = \sqrt{\frac{g_{R1}}{2}} h_{R1} \mathrm{Im}(\mathbf{x}_R) + \boldsymbol{w}_{1s} \tag{16}$$

where $\boldsymbol{w}_{1s}$ is the orthogonal component vector of the AWGN vector at user M.

Due to the orthogonal synthesis on the relay R, the in-phase part of the broadcast signal has no effect on the orthogonal component that the user M wants. Then multiply $h_{RI}{}^{*}/|h_{RI}|^2$ and make a hard decision to get the information $\tilde{s}_2$ sent by user N. Since the information $\hat{s}_1$ that user N wants to obtain is in the in-phase part of the broadcast signal, user N is multiplied by the carrier $Re\left[e^{j(2\pi f_c t+\theta)}\right]$. We can get the baseband signal

$$y_2 = \sqrt{\frac{g_{R2}}{2}} h_{R2} Re(\mathbf{x}_R) + \boldsymbol{w}_{2c} \tag{17}$$

where $\boldsymbol{w}_{2c}$ is the in-phase component vector of the AWGN vector at user N.

Since two users use orthogonal carriers and the relay uses orthogonal synthesis, the BER of s_1 and s_2 can be calculated separately. Taking $\boldsymbol{s}_1$ as an example, in the uplink stage, according to (9), s_1 is obtained by estimating the in-phase branch of the correlation detector at the relay R. Then we can get the error probability of s_1 in the upstream stage as

$$P_{e\_s_1\_u} = \frac{1}{2}\left[1 - \sqrt{\frac{\left(\frac{2g_{1R}}{N_0}\right)}{\left(1 + \frac{2g_{1R}}{N_0}\right)}}\right] \tag{18}$$

In the downlink phase, user N detects s_1 through (14). The error probability of s_1 in the downlink phase is

$$P_{e\_s_1\_d} = \frac{1}{2}\left[1 - \sqrt{\frac{\left(\frac{g_{R2}}{N_0}\right)}{\left(1 + \frac{g_{R2}}{N_0}\right)}}\right] \tag{19}$$

If and only if there is a mistake in the uplink or downlink, user N will receive the error bit, so the bit error rate of s_1 is

$$\mathrm{BER}_{s_1} = \frac{1}{2}\left\{1 - \sqrt{\frac{\frac{2g_{1R}g_{R2}}{N_0^2}}{\left[\left(1 + \frac{2g_{1R}}{N_0}\right)\left(1 + \frac{g_{R2}}{N_0}\right)\right]}}\right\} \tag{20}$$

Similarly, according to formulas (10) and (13), the bit error rate of s_2 can be obtained

$$\mathrm{BER}_{s_2} = \frac{1}{2}\left\{1 - \sqrt{\frac{\frac{2g_{2R}g_{R1}}{N_0^2}}{\left[\left(1 + \frac{2g_{2R}}{N_0}\right)\left(1 + \frac{g_{R1}}{N_0}\right)\right]}}\right\} \tag{21}$$

Finally, the bit error rate of the BQ-PNC system is as follow

$$\mathrm{BER} = \frac{1}{2}\left[1 - \frac{1}{2}\sqrt{\frac{\frac{2g_{1R}g_{R2}}{N_0^2}}{\left(1 + \frac{2g_{1R}}{N_0}\right)\left(1 + \frac{g_{R2}}{N_0}\right)}} - \frac{1}{2}\sqrt{\frac{\frac{2g_{2R}g_{R1}}{N_0^2}}{\left(1 + \frac{2g_{2R}}{N_0}\right)\left(1 + \frac{g_{R1}}{N_0}\right)}}\right] \tag{22}$$

4 Underwater Asymmetric Physical Layer Network Coding Scheme

Aiming at the phenomenon that the data transmission efficiency is affected by the weak link quality in the asymmetric two-way relay underwater acoustic channel model, the channel gain of the strong link quality cannot be fully utilized, a new modulation scheme is proposed. The scheme adopts different modulation methods for asymmetric two link nodes. The nodes with weak link quality use low-order modulation methods and the nodes with strong link quality use relatively high-order modulation methods. The resulting superimposed signal is broadcast by designing a corresponding code mapping scheme. The specific block diagram is shown as Fig. 4, in which the modulation methods of M and N are different.

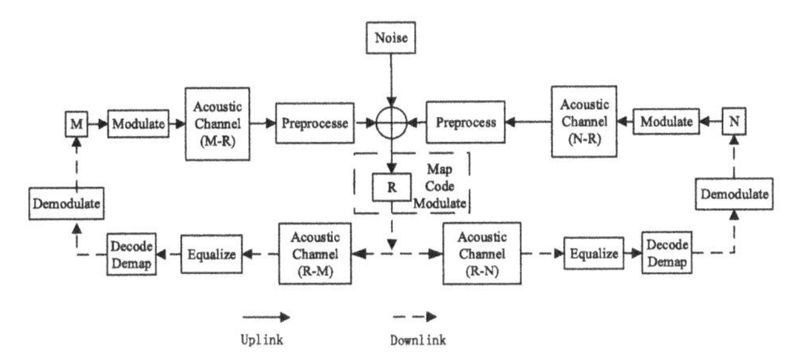

Fig. 4. Asymmetric two-way relay underwater acoustic channel ray model

In the bidirectional relay channel model, the bit information sent by node M is s_1, and the bit information sent by node N is s_2. The original bit information undergoes corresponding modulation to obtain modulation information. The modulation information is transmitted through the channel and finally reaches the relay node under the influence of noise. The superimposed signal received by the relay can be expressed as

$$r = h_1 S(M_1(s_1)) + h_2 S(M_2(s_2)) + n \tag{23}$$

where M_1 and M_2 represent the modulation modes of users M and N; h_1, h_2 represent the channel parameters from users on both sides to the relay; n is underwater acoustic channel noise. The channel parameters h_1 and h_2 will lead to the superimposed signal constellation to rotate because of the asymmetry of the channel. Therefore, it is necessary to perform channel compensation on the uplink signal, which can be called signal preprocessing.

$$r = \frac{1}{h_1} h_1 S(M_1(s_1)) + \frac{1}{h_2} h_2 S(M_2(s_2)) + n = S(M_1(s_1)) + S(M_2(s_2)) + n \tag{24}$$

Then the relay node R encodes the received superimposed signal through the corresponding encoding scheme. In the downlink stage, the encoded signal is modulated and broadcast to M and N, after the corresponding equalization to remove the influence of the channel, and then through the corresponding decoding and demodulation scheme, using its own data to obtain the information sent by the user of the other party.

4.1 Channel Compensation Preprocessing

Due to the asymmetry of the channel parameters h_1 and h_2, it is necessary to perform a pre-processing operation of channel fading compensation on the uplink signal. Compensating channel fading often adopts three technologies: channel equalization, diversity reception and channel coding, which can be used either individually or in combination. Channel equalization can compensate for the problem of intersymbol interference caused by multipath effects in time division channels. An equalizer is equivalent to configuring a filter that compensates and corrects system characteristics. Time-domain equalization is often used in digital communications. According to whether the receiver's decision result should be feedback to the equalizer to adjust the parameters, it can be divided into two categories: nonlinear equalizer and linear equalizer. For the underwater acoustic communication system, channel equalization is selected to preprocess the channel. Because of the characteristics of time-varying and randomness in shallow sea acoustic channels, the MMSE equalizer [19] is still used here for channel compensation. That is, a set of training sequences is inserted into the data at the transmitting terminal, and the receiving terminal uses the training sequence to estimate channel information and uses the channel information to correct the influence of the underwater acoustic channel. The criterion of the MMSE is to minimize the mean square error of the received signal and the signal at the transmitting terminal. $\boldsymbol{h}_{MR} = [h_{MR}(1)\ldots h_{MR}(K)]_{1\times K}$ represents the vector of the channel coefficient from user M to relay node R, where K represents the number of multipath signals (K can be obtained from the ray model). For convenience, it is assumed that the delay of each path signal compared to the previous path signal is one chip length. Therefore, the signal received by user M after passing through the underwater acoustic channel can be obtained:

$$r_1(i) = \boldsymbol{H}_{MR}\boldsymbol{S}(M_1(s_1)) + \boldsymbol{n} \tag{25}$$

Equation (26) is the channel state matrix

$$\boldsymbol{H}_{MR} \triangleq \begin{bmatrix} h_{MR}(1) & 0 & \cdots & 0 \\ \vdots & h_{MR}(0) & \ddots & \vdots \\ h_{MR}(K) & \vdots & \ddots & 0 \\ 0 & h_{MR}(K) & \ldots & h_{MR}(1) \\ \vdots & \vdots & \ddots & \vdots \\ 0 & 0 & \cdots & h_{MR}(K) \end{bmatrix}_{N\times N} \tag{26}$$

Using the estimated value of channel coefficients $\hat{h}_{MR}(i), i = 1, 2, \cdots, K$ in MMSE equalizer to construct a channel state estimation matrix $\hat{\boldsymbol{H}}_{MR}$. The MMSE channel estimation method aims that $E\left[\left|\hat{\boldsymbol{H}}_{MR}-\boldsymbol{H}_{MR}\right|^2\right]$ is the smallest.

$$\hat{\boldsymbol{H}}_{MR} \triangleq \begin{bmatrix} \hat{h}_{MR}(1) & 0 & \cdots & 0 \\ \vdots & \hat{h}_{MR}(0) & \ddots & \vdots \\ \hat{h}_{MR}(K) & \vdots & \ddots & 0 \\ 0 & \hat{h}_{MR}(K) & \cdots & \hat{h}_{MR}(1) \\ \vdots & \vdots & \ddots & \vdots \\ 0 & 0 & \cdots & \hat{h}_{MR}(K) \end{bmatrix}_{N\times N} \tag{27}$$

Build MMSE equalizer under the condition that noise and signal are not related

$$\boldsymbol{H}^{\dagger}=\left(\left(\hat{\boldsymbol{H}}_{MR}\right)*\hat{\boldsymbol{H}}_{MR}+\sigma_n^2\boldsymbol{I}\right)^{-1}\left(\hat{\boldsymbol{H}}_{MR}\right) \tag{28}$$

where σ_n^2 is noise variance.

Use (28) to balance the channel influence and get the modulated signal sent by the transmitter

$$\hat{\mathbf{S}}(M_1(s_1))=\boldsymbol{H}^{\dagger}*\boldsymbol{r}_1 \tag{29}$$

In the MMSE estimation of channel coefficients, the influence of noise is considered, then the mean square error of the obtained estimation matrix and the actual channel matrix is small.

4.2 Asymmetric Physical Layer Network Coding Scheme

BPSK-QPSK Asymmetric Physical Layer Network Coding Scheme

In the asymmetric two-way relay system model, if user M uses BPSK modulation, and user N uses QPSK modulation. The information bit of user M is $m_1 \in \{0, 1\}$, and after modulation is $s_1 \in \{+1, -1\}$; the information bit of user N is $m_2 \in \{00, 01, 11, 10\}$, and after modulation is $s_2 \in \left\{\exp(j(\frac{\pi}{4})), \exp(j(\frac{3\pi}{4})), \exp\left(j\left(\frac{5\pi}{4}\right)\right), \exp(j(\frac{7\pi}{4}))\right\}$. However, the channels of the two users' information to reach the relay are different, and channel compensation is needed for the two signals. Turn s_1 and s_2 into in-phase components and quadrature components.

$$s_1=\sqrt{P_{bpsk}}\alpha_{bpsk},\quad s_2=\sqrt{\frac{P_{qpsk}}{\chi_{qpsk}}}\left(\alpha_{qpsk}+j\beta_{qpsk}\right) \tag{30}$$

After preprocessing by the equalizer, the ideal superimposed signal received by the relay is

$$r_{\text{sup}}=I+jQ+n=\sqrt{P_{bpsk}}\alpha_{bpsk}+\sqrt{\frac{P_{qpsk}}{\chi_{qpsk}}}\left(\alpha_{qpsk}+j\beta_{qpsk}\right)+n$$

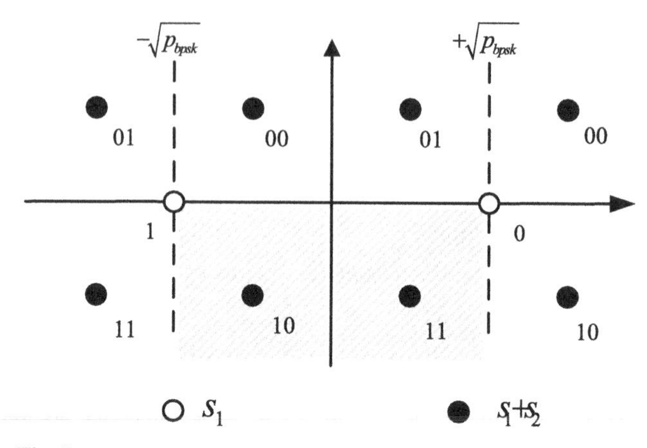

Fig. 5. BPSK-QPSK modulated superimposed signal constellation

$$= \left(\sqrt{P_{bpsk}}\alpha_{bpsk} + \sqrt{\frac{P_{qpsk}}{\chi_{qpsk}}}\alpha_{qpsk}\right) + j\left(\sqrt{\frac{P_{qpsk}}{\chi_{qpsk}}}\beta_{qpsk}\right) + n \tag{31}$$

The constellation of the superimposed signal is shown in Fig. 5. The white constellation point in Fig. 4 represents the signal constellation point of user M after BPSK modulation. The black constellation point represents the constellation point of the superimposed signal received at the relay after user M undergoes BPSK modulation and user N undergoes QPSK modulation.

BPSK-QPSK Modulated Relay Coding Mapping Scheme

Similarly, for the asymmetrically modulated superimposed signal received at the relay, coding and mapping are required. In Fig. 5, there are a total of eight superimposed information, indicating the four signal relationships of two users, so 2 bits are required for coding. The coding method is to XOR the real part information of the M user information with the user N, while the imaginary part remains unchanged. That is, when the real part of the constellation point of the superimposed signal is at $\left[-\sqrt{P_{bpsk}}, \sqrt{P_{bpsk}}\right]$, it is represented by the symbol 1. The virtual part of the constellation point of the superimposed signal is indicated by 10 when it corresponds to the positive semi-axis area, and is indicated by 11 when it falls within negative half-axis area. When the real part of the constellation point of the superimposed signal is outside the area of $\left[-\sqrt{P_{bpsk}}, \sqrt{P_{bpsk}}\right]$, it is represented by the symbol 0. The virtual part of the constellation point of the superimposed signal is indicated by 00 when it corresponds to the positive semi-axis area, and is indicated by 01 when it falls within negative semi-axis area. As shown in Fig. 5, the black constellation point's coding result is 11.

According to the above design, the coding and mapping rules of the relay can be obtained as (32).

$$\hat{s} = \begin{cases} (+1,-1)s_1, s_2 \in \{(-1) \cap (-1,-1), (+1) \cap (+1,-1)\} \\ (-1,+1)s_1, s_2 \in \{(-1) \cap (+1,+1), (+1) \cap (-1,+1)\} \\ (+1,+1)s_1, s_2 \in \{(-1) \cap (-1,+1), (+1) \cap (+1,+1)\} \\ (-1,-1)s_1, s_2 \in \{(-1) \cap (+1,-1), (+1) \cap (-1,-1)\} \end{cases} \tag{32}$$

BPSK-QPSK Modulated Relay Decoding Demapping Scheme

In the downlink stage, relay R broadcasts the encoded 2 bits after QPSK modulation to two users, and the information has been equalized by the equalizer. For the coding and mapping rules of (32), corresponding decoding and demapping design can be carried out, that is, the counterparty information is obtained according to the coding information of the received superimposed signal and its own information.

Decoding and demapping scheme at user M: Knowing the BPSK modulated signal of user M, use the BPSK modulated signal and the real part of the received signal to XOR to get the first information of user N, and the second information of user N is the imaginary part of the received signal.

Decoding and demapping scheme at user N: Knowing the QPSK modulated signal of user N, the information of user M can be obtained by XORing the real part of QPSK modulated signal with the real part of the received signal.

Then after the above decoding and demapping process, the receiving terminal can obtain the data information of the other user through demodulation.

QPSK-8PSK Asymmetric Physical Layer Network Coding Scheme

Here we discuss another asymmetric modulation scheme. In the asymmetric two-way relay system model, user M uses the QPSK modulation method, the information bit of the user M is $m_1 \in \{00, 01, 11, 10\}$ which is $s_1 \in \left\{\exp\left(j\left(\frac{\pi}{4}\right)\right), \exp\left(j\left(\frac{3\pi}{4}\right)\right), \exp\left(j\left(\frac{5\pi}{4}\right)\right), \exp\left(j\left(\frac{7\pi}{4}\right)\right)\right\}$ after QPSK modulation. User N uses the 8PSK modulation method, the information bit of the user N is $m_2 \in \{000, 001, 010, 011, 100, 101, 110, 111\}$, which is $s_2 \in \exp\left(j\left(\frac{(2m+1)\pi}{8}\right)\right)$ after QPSK modulation, in which m is the decimal of m_2. However, the information of the two users reaches the relay through different channels, so it is necessary to perform channel compensation on the two signals. After preprocessing, the superimposed signal received by the relay node is:

$$r_{\text{sup}} = I + jQ + n = \sqrt{\frac{P_{qpsk}}{\chi_{qpsk}}}\left(\alpha_{qpsk} + j\beta_{qpsk}\right) + \sqrt{P_{8psk}}\exp\left(j\frac{2m+1}{8}\pi\right) + n \tag{33}$$

Here $\alpha_{qpsk}, \beta_{qpsk} \in \{\pm 1\}$ represents the in-phase and quadrature components of the QPSK modulated signal of user M, and P_{8psk}, P_{qpsk} represents the transmission power of users M and N, respectively. The 8PSK modulation adopted by user N is to modulate 8 bits of information on 8 phases on a circle with radius P_{8psk}. Ideally, the constellation of the superimposed signal is shown in Fig. 6. In Fig. 6, the white constellation points represent the signal constellation points of the user M after QPSK modulation, and the black constellation point represents the constellation point of the superimposed signal received by the user N and the user M after QPSK modulation at the relay.

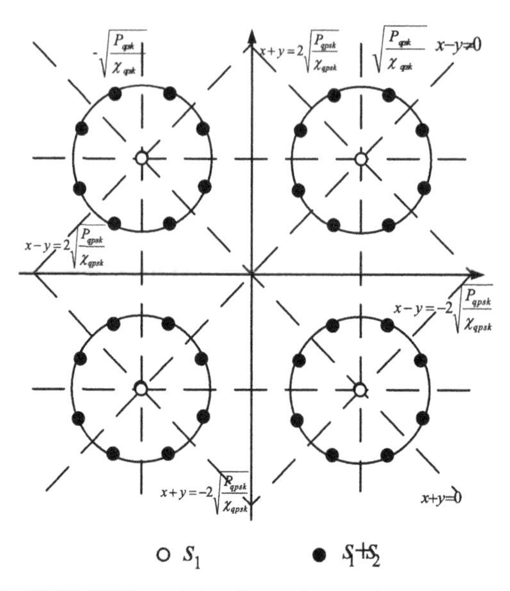

Fig. 6. QPSK-8PSK modulated superimposed signal constellation

QPSK-8PSK Modulation Coding Mapping Scheme

In this scheme, the asymmetrically modulated superimposed signal received at the relay needs to be coded and mapped. There are 32 superimposed information in Fig. 6, which represents the eight kinds of asymmetrically modulated signal relationship between two users, so it takes 3 bits to encode for these relationships. The coding method here is: the first two bits of the encoded information are XORed by the information of user M and the first two bits of information of user N, and the third bit of information is determined by the 8PSK modulation form of user N's information. If the absolute value of the real part of the modulation signal is greater than the absolute value of the imaginary part, the third bit is 0, otherwise if the absolute value of the real part of the modulation signal is less than the absolute value of the imaginary part, the third bit is 1. Intuitively compare the original QPSK and the original 8PSK: when their in-phase components have the same sign and the quadrature components have the same sign, it is represented by 00, and if the absolute value of the original 8PSK real part is greater than the imaginary part, the code is 000, otherwise the code is 001; when their in-phase components have the same sign and the quadrature components have the opposite sign, it is represented by 01, and if the absolute value of the original 8PSK real part is greater than the imaginary part, the code is 010, otherwise the code is 011; when their in-phase components have the opposite sign and the quadrature components have the same sign, it is represented by 10, and if the absolute value of the original 8PSK real part is greater than the imaginary part, the code is 100, otherwise the code is 101; when their in-phase components have the opposite sign and the quadrature components have the opposite sign, it is represented by 11, and if the absolute value of the original 8PSK real part is greater than the imaginary part, the code is 110, otherwise the code is 111; The coding and mapping rules at the relay are as shown in (34).

$$\hat{s} = \begin{cases} (011) s_1, s_2 \in \begin{Bmatrix} (+1,+1) \cap \left(\frac{13\pi}{8}\right), (+1,-1) \cap \left(\frac{3\pi}{8}\right), \\ (-1,-1) \cap \left(\frac{5\pi}{8}\right), (-1,+1) \cap \left(\frac{11\pi}{8}\right) \end{Bmatrix} \\ (001) s_1, s_2 \in \begin{Bmatrix} (+1,+1) \cap \left(\frac{15\pi}{8}\right), (+1,-1) \cap \left(\frac{\pi}{8}\right), \\ (-1,-1) \cap \left(\frac{7\pi}{8}\right), (-1,+1) \cap \left(\frac{9\pi}{8}\right) \end{Bmatrix} \\ (101) s_1, s_2 \in \begin{Bmatrix} (+1,+1) \cap \left(\frac{5\pi}{8}\right), (+1,-1) \cap \left(\frac{11\pi}{8}\right), \\ (-1,-1) \cap \left(\frac{13\pi}{8}\right), (-1,+1) \cap \left(\frac{3\pi}{8}\right) \end{Bmatrix} \\ (100) s_1, s_2 \in \begin{Bmatrix} (+1,+1) \cap \left(\frac{7\pi}{8}\right), (+1,-1) \cap \left(\frac{9\pi}{8}\right), \\ (-1,-1) \cap \left(\frac{15\pi}{8}\right), (-1,+1) \cap \left(\frac{\pi}{8}\right) \end{Bmatrix} \\ (001)' s_1, s_2 \in \begin{Bmatrix} (+1,+1) \cap \left(\frac{3\pi}{8}\right), (+1,-1) \cap \left(\frac{13\pi}{8}\right), \\ (-1,-1) \cap \left(\frac{11\pi}{8}\right), (-1,+1) \cap \left(\frac{5\pi}{8}\right) \end{Bmatrix} \\ (000) s_1, s_2 \in \begin{Bmatrix} (+1,+1) \cap \left(\frac{\pi}{8}\right), (+1,-1) \cap \left(\frac{15\pi}{8}\right), \\ (-1,-1) \cap \left(\frac{9\pi}{8}\right), (-1,+1) \cap \left(\frac{7\pi}{8}\right) \end{Bmatrix} \\ (110) s_1, s_2 \in \begin{Bmatrix} (+1,+1) \cap \left(\frac{11\pi}{8}\right), (+1,-1) \cap \left(\frac{5\pi}{8}\right), \\ (-1,-1) \cap \left(\frac{3\pi}{8}\right), (-1,+1) \cap \left(\frac{13\pi}{8}\right) \end{Bmatrix} \\ (111) s_1, s_2 \in \begin{Bmatrix} (+1,+1) \cap \left(\frac{9\pi}{8}\right), (+1,-1) \cap \left(\frac{7\pi}{8}\right), \\ (-1,-1) \cap \left(\frac{\pi}{8}\right), (-1,+1) \cap \left(\frac{15\pi}{8}\right) \end{Bmatrix} \end{cases} \quad (34)$$

QPSK-8PSK Modulated Relay Decoding Demapping Scheme

Here, 3 bit encoding is used, so in the downlink stage, relay R broadcasts the encoded information using 8PSK modulation to two users. The information is also processed by the equalizer to compensate the downlink channel attenuation. According to the coding and mapping rules of (34), corresponding decoding and demapping rules are designed.

Decoding and demapping scheme at user M: Knowing the QPSK modulated signal of user M, using the QPSK modulated signal to XOR with the received signal to get the first two bits of information of user N, and the third information of the user N is determined by the third information of the received information. If the third bit of the received information is 0, the absolute value of the real part of the original 8PSK is greater than the absolute value of the imaginary part. We can refer to the first two bits of information to determine the third bit of N; if the third bit of information in the received message is 1, the absolute value of the real part of the original 8PSK is smaller than the absolute value of the imaginary part. Similarly, the third bit of N information can be determined.

Decoding and demapping scheme at user N: Knowing the 8PSK modulated signal of user N, retain the symbols of the real and imaginary parts, and converts it into a unit signal, that is, exp{(j(π/8)), (j(3π/8)), (j(5π/8)), (j(7π/8)), (j(9π/8)), (j(11π/8)), (j(13π/8)), (j(15π/8))} is converted into{$1+j$, $1+j$, $-1+j$, $-1+j$, $-1-j$, $-1-j$, $1-j$, $1-j$}, then XOR it with the received signal to obtain the information of the user M. So the receiver can get the other user's data information through demodulation.

5 Simulation and Analysis

Firstly, the parameters of the shallow sea environment are assumed. The working frequency of sound wave is $f = 8$ kHz; the speed of sound wave propagation in seawater $c = 1500$ m/s; sea surface wind speed $s = 8$ m/s; seabed reflection coefficient is 0.9; each frame data length is 128 bits. In the previous analysis of the underwater acoustic channel model, we define a1: distance between node M and the seabed; a2: distance from node N to the seabed; b: distance from relay node R to the seabed; L1: horizontal distance from node M to relay node R; L2: horizontal distance from node N to relay node R; h: the distance from the bottom to the sea surface. In order to verify the performance in an asymmetric underwater acoustic channel, consider the asymmetric case: in the underwater acoustic channel, the uplink and downlink stages' parameters are $a_1 = 20$ m, $a_2 = 40$ m, $b = 60$ m, $L_1 = 1000$ m, $L_2 = 800$ m, $h = 100$ m. That is, the user M and the user N are always asymmetric about the relay node R, and the two nodes maintain a fixed position during the uplink stage and the downlink stage. BPSK-QPSK asymmetric modulation and QPSK-8PSK asymmetric modulation are used respectively.

The two schemes of BPSK-QPSK asymmetric modulation and QPSK-8PSK asymmetric modulation are simulated in the shallow seawater asymmetric bidirectional relay channel, and their BER performances are observed respectively. The BER curve is shown in Fig. 7, which is obtained by statistically averaging 1000 independent simulation experiments.

The simulation results in Fig. 7 show that the asymmetric modulation scheme can obtain a BER curve similar to the symmetric modulation scheme, which can verify the feasibility of the asymmetric modulation scheme. It can be seen that the BPSK-QPSK asymmetric modulation has better BER performance than the QPSK-8PSK asymmetric modulation scheme. When the BER reaches 10^{-3}, the BPSK-QPSK asymmetric modulation scheme differs from the QPSK-8PSK asymmetric modulation by approximately 3 dB in SNR.

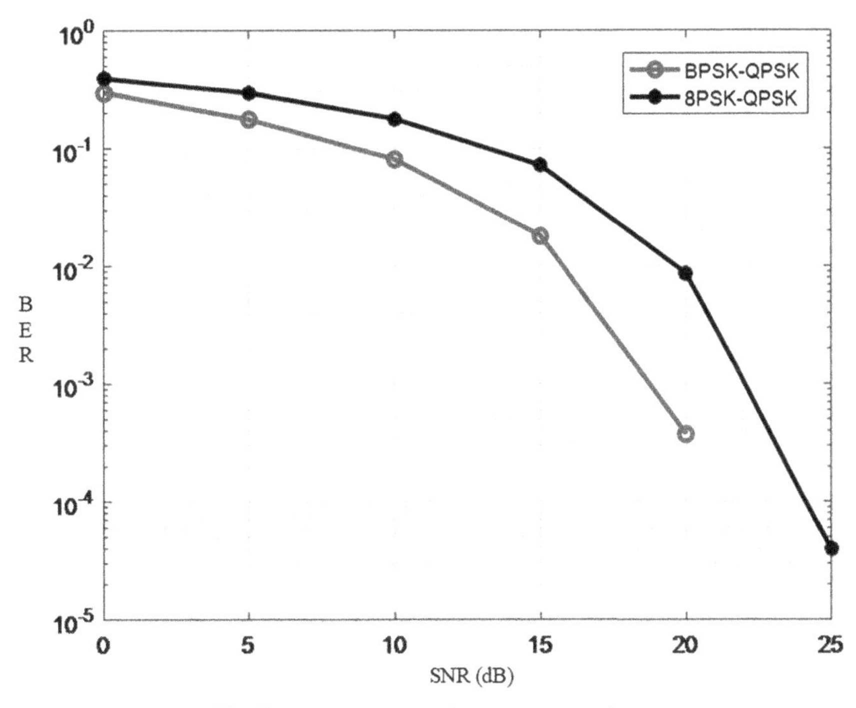

Fig. 7. Bit error rate performance comparison

6 Conclusion

In the 6G Marine Internet of Things, for the situation that the asymmetry of the underwater acoustic channel will cause the data transmission efficiency to be affected by the weak link quality, making the channel gain of the strong link quality not fully utilized, a new modulation scheme, asymmetric modulation has been proposed. After preprocessing the superimposed signal, the relay mapping schemes and demapping schemes under different modulation schemes are designed, and simulation experiments are carried out according to the designed scheme to verify the feasibility of the asymmetric modulation scheme. Comparing the performance of the two schemes under asymmetric conditions, the BER performance of BPSK-QPSK asymmetric modulation is better than that of QPSK-8PSK asymmetric modulation scheme. When the system reaches a stable state. Modern wireless communication systems have different requirements for different multimedia services. In underwater actual combat, real-time voice information transmission is required. At this time, a scheme with higher data transmission efficiency is selected. While actual underwater resource detection and other activities will involve the transmission of more video and image information, more attention should be paid to the reliability of data transmission. Considering the different requirements of different services, we can find the balance between BER performance and communication efficiency according to the actual situation. In an asymmetric underwater communication environment, the use of an asymmetric modulation scheme can make the data transmission efficiency as small as possible due to the weak link quality, so that the communication system can make full use of the strong link quality channel gain, and then consider the business needs to select the appropriate asymmetric modulation scheme.

References

1. Morozs, N., Mitchell, P.D., Diamant, R.: Scalable adaptive networking for the internet of underwater things. IEEE Internet Things J. **7**(10), 10023–10037 (2020). https://doi.org/10.1109/JIOT.2020.2988621
2. Qi, Q., Chen, X.: Wireless powered massive access for cellular internet of things with imperfect SIC and nonlinear EH. IEEE Internet Things J. **6**(2), 3110–3120 (2019)
3. Li, B., Rong, Y.: AF MIMO relay systems with wireless powered relay node and direct link. IEEE Trans. Commun. **66**(4), 1508–1519 (2018)
4. Zhai, D., Chen, H., Lin, Z., Li, Y., Vucetic, B.: Accumulate then transmit: multiuser scheduling in full-duplex wireless-powered IoT systems. IEEE Internet Things J. **5**(4), 2753–2767 (2018)
5. Li, C., Yang, H.J., Sun, F., Cioffi, J.M., Yang, L.: Multiuser overhearing for cooperative two-way multiantenna relays. IEEE Trans. Veh. Technol. **65**(5), 3796–3802 (2016)
6. Ma, Y., Chen, H., Lin, Z., Li, Y., Vucetic, B.: Distributed and optimal resource allocation for power beacon-assisted wireless-powered communications. IEEE Trans. Commun. **63**(10), 3569–3583 (2015)
7. Zhou, Y., Diamant, R.: A parallel decoding approach for mitigating near-far interference in internet of underwater things. IEEE Internet Things J. **7**(10), 9747–9759 (2020). https://doi.org/10.1109/JIOT.2020.2988246
8. Sun, X., Li, Y., Wang, N., Li, Z., Liu, M., Gui, G.: Towards self-adaptive selection of kernel functions for support vector regression in IoT based marine data prediction. IEEE Internet Things J. **7**(10), 9943–9952 (2020). https://doi.org/10.1109/JIOT.2020.2988050
9. Luccio, D.D., et al.: Coastal marine data crowdsourcing using the internet of floating things: improving the results of a water quality model. IEEE Access **8**, 101209–101223 (2020). https://doi.org/10.1109/ACCESS.2020.2996778
10. Wang, Q., Li, J., Qi, Q., Zhou, P., Wu, D.O.: A game theoretic routing protocol for 3D underwater acoustic sensor networks. IEEE Internet Things J. **7**(10), 9846–9857 (2020). https://doi.org/10.1109/JIOT.2020.2988503
11. Jouhari, M., Ibrahimi, K., Tembine, H., Ben-Othman, J.: Underwater wireless sensor networks: a survey on enabling technologies, localization protocols, and internet of underwater things. IEEE Access **7**, 96879–96899 (2019). https://doi.org/10.1109/ACCESS.2019.2928876
12. Abdel-Basset, M., Mohamed, R., Elhoseny, M., Bashir, A.K., Jolfaei, A., Kumar, N.: Energy-aware marine predators algorithm for task scheduling in IoT-based fog computing applications. IEEE Trans. Indust. Inf. **17**(7), 5068–5076 (2021). https://doi.org/10.1109/TII.2020.3001067
13. Qi, Q., Chen, X., Ng, D.W.K.: Robust beamforming for NOMA-based cellular massive IoT with SWIPT. IEEE Trans. Signal Process. **68**, 211–224 (2020)
14. Zhang, S., Liew, S.C., Lam, P.: Hot topic: physical-layer network coding. Proc. ACM Mobicom. **2006**, 358–365 (2006)
15. Popovski, P., Yomo, H.: The anti-packets can increase the achievable throughput of a wireless multi-hop network. IEEE International Conference on Communications, pp. 3885–3890 (2006)
16. Zielinski, A., Young-Hoon, Y., Lixue, W.: Performance analysis of digital acoustic communication in a shallow water channel. IEEE J. Oceanic Eng. **20**(4), 293–299 (1995)

17. Xiong, J., Ma, R., Chen, L., et al.: A personalized privacy protection framework for mobile crowdsensing in IIoT. IEEE Trans. Industr. Inf. **16**(6), 4231–4241 (2020)
18. Xiong, J., Zhao, M., Bhuiyan, M., et al.: An AI-enabled three-party game framework for guaranteed data privacy in mobile edge crowdsensing of IoT. IEEE Trans. Indust. Inf. **17**(2), 922–933 (2021)
19. Xiong, J., Ren, J., Chen, L., et al.: Enhancing privacy and availability for data clustering in intelligent electrical service of IoT. IEEE Internet Things J. **6**(2), 1530–1540 (2019)

A New Load Balance Scheme for Heterogeneous Entities in Cloud Network Convergence

Jiaji Liu, Zhiwei Zhang(✉), Wangzhe Xu, Xinghui Zhu, and Xuewen Dong

School of Computer Science and Technology, Xidian University, Xi'an Shaanxi 710071, China
zwzhang@xidian.edu.cn

Abstract. For future Internet and next-generation network, the cloud networking convergence is one of the most popular research directions, and it has attracted widespread attention from academia as well as industry. Network adapting cloud and network cloudification are two dimensions in cloud network convergence that can break the closeness and independence between cloud and network. However, the techniques related to the network adapting cloud and network cloudification unavoidably introduce more heterogeneous devices, services and users. That disables the existing load balance schemes which are almost proposed for data centers in cloud computing environments, where the entities are typically standard hardware and software modules. As a result, the overhead and cost of load balance shcemes would be raised significantly in the progress of cloud network convergence. Therefore, in this paper, to make the most usage of heterogeneous entities and encourage the development of future Internet as well as next-generation networking, we propose a model and the requirements of load balance for heterogeneous entities in the convergence of cloud and network, then we present a concrete load balance scheme. Finally, we discuss the abilities and applications of our proposed model and scheme.

Keywords: Load balance · Cloud network convergence · Heterogeneous entities · Next-generation network · Private cloud computing

1 Introduction

Nowadays, the convergence of cloud and network is the development trend for future Internet and next-generation networking. Among the existing researches and applications in the convergence of cloud and network, two techniques, the network adapting cloud and the network cloudification, are playing key roles in realizing the convergence. However, these two techniques introduce heterogeneousness unavoidably. It is well known that the more different entities, the more complex to deal with. Furthermore, the previous load balance schemes have been designed for data centers in cloud computing environments or communication network operators where the entities are typically standard hardware and software modules. Therefore, few of the previous cloud and network load balance schemes are still feasible in convergence environments.

J. Xiong et al. (Eds.): MobiMedia 2021, LNICST 394, pp. 22–32, 2021.
https://doi.org/10.1007/978-3-030-89814-4_2

1.1 Related Works

Many methods have been proposed to reduce energy consumptions in both academic and industrial fields. At the hardware level, there are techniques such as optimized memory and dynamic frequency conversion [1]. At the software level, the most important method to reduce energy consumption is resource scheduling, which can achieve load balancing while ensuring that energy consumption is as low as possible.

Load scheduling algorithms can be divided into two categories: one is the load balancing algorithm implemented on the physical machine with tasks as the scheduling unit [2, 3]; another is the load balancing algorithm implemented on the physical machine with resources as the scheduling unit [4]. Besides, many researchers have conducted research on energy conservation. Online migration of virtual machines is an important method for server consolidation. However, migration has many negative effects, such as service interruption, network congestion and additional migration costs [5]. Therefore, it is also important to improve resource utilization and reduce energy consumption by migrating virtual machines. The static integration method [6–8] gives a mapping plan for virtual machines and physical servers, minimizing the number of servers or the overall cost. The dynamic integration method [9, 10] dynamically reconfigures the cluster through virtual machine migration to run on fewer nodes.

Our work in this paper is to analyze the status of services in the cloud and combine the ideas of load balancing and server integration in the environment of cloud networking integration to schedule resources in the cloud under different conditions. Therefore, the cloud network integration environment can be seen as a private cloud computing platform with a group of physical computers, the software-defined network and other technologies are used to manage the communication among the physical and virtual machines.

2 Contributions

In this paper, we focus on the problem of how to minimize the overhead and cost of load balance in cloud network convergence environment when taking the heterogeneousness into account. Our main contribution is that we propose the model and requirements of load balance for heterogeneous entities in the environment of cloud network convergence, and we present a concrete load balance scheme based on the proposed model.

2.1 Organization

In Sect. 2 we present the model and requirements of load balance for heterogeneous entities in cloud network convergence, then we describe the architecture of the load balance system in Sect. 3, with which we propose a new concrete load balance scheme in Sect. 4. In Sect. 5, we analyze and discuss the proposed scheme. Finally, Sect. 6 concludes this paper and discusses our future work on load balance in the environments of future Internet and next-generation networking.

3 The Model and Requirements of Load Balance for Heterogeneous Entities

In this section, we present the model and requirements of load balance for heterogeneous entities.

3.1 Entities Model

In an enterprise's private cloud, assume there are N physical machines (PM), and they can be represented as

$$\{\{PMa1, ..., PMan\}, \{PMb1, ..., PMbn\}, ..., \{PMx1, ..., PMxn\}\}, \tag{1}$$

where every physical machine is classified into one category represented as a set. In each set, machines have the same configuration. The existing researches [11] have found that power consumption and CPU utilization have a linear relationship. Therefore, the previous work uses the average power consumption measured when the load of a physical machine is 80% to 90%, called PPM. In short, PPM means the power consumption of PM. Based on the classification set of physical machines, the power consumption is divided into sets

$$\{\{PPMa2, ..., PPMa\mathrm{n}\}, \{PPMb1, ..., PPMbn\}, ..., \{PPMx1, .., PPMxn\}\}. \tag{2}$$

The previous work uses the average performance efficiency of physical machines measured when the load of a physical machine is 80% to 90%, called FPM. In short, FPM means the performance efficiency of PM.

$$\{\{FPMa2, ..., FPMan\}, \{FPMb1, ..., FPMbn\}, ..., \{FPMx1, ..., FPMxn\}\}. \tag{3}$$

Note that we should arrange the values in each set from small to large. Besides, we classify the services or requests. Assume there are N services (requests) in a private cloud, and they can be represented as

$$\{\{U\mathrm{a1},...,U\mathrm{an}\},\{ U\mathrm{b1},..,U\mathrm{bn}\},...,\{ U\mathrm{x1},...,U\mathrm{xn}\} \} . \tag{4}$$

Note that the sets in expression (4) and the sets in expression (1) have a one-to-one mapping relationship. Figure 1 illustrates the one-to-one correspondence between request sets and physical machine sets. Furthermore, there is a many-to-many relationship between requests and physical in the corresponding sets.

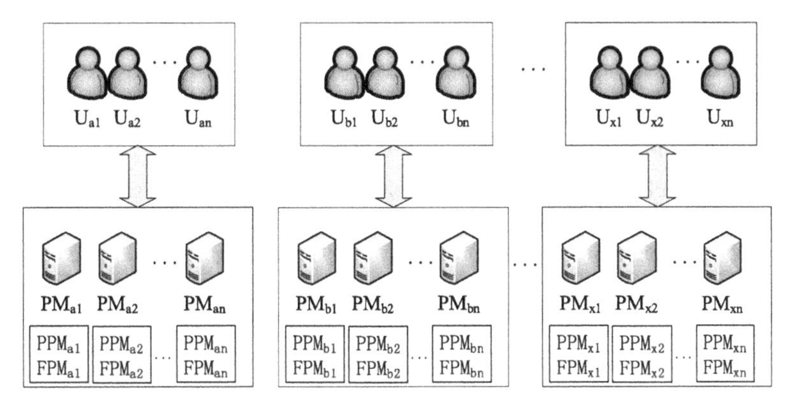

Fig. 1. Correspondence between requests and physical machines.

3.2 Load Balance Requirements

In private clouds, the actions of users, services and other entities could present kinds of regulations according to some contexts.

3.2.1 Service State Transition

In order to quantify the periodic and regular changes in services or requests, we introduce the strategy of service state transition (SST) in this section, in which we divide the service cycle into several time segments. Typically, a private cloud data center can also be divided into three states: idle (T0), normal (T1) and busy (T2). Change between states is determined by the results of historical value and cloud data center monitoring module, shown as Fig. 2.

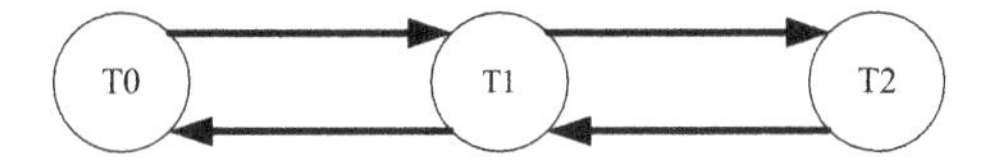

Fig. 2. Service state transition.

3.2.2 Scheduling Strategy

We apply resource scheduling method in SST to realize the load balance of cloud data center resources and the least energy consumption in each state. The core of scheduling strategy is a load balancing resource scheduling method and the underlying technology of load balancing resource scheduling method is resource migration. Considering green computing, virtual machines should be migrated to the lowest energy-consuming server. The technique of live or online migration is being used for efficient resource allocation in cloud computing. In order to select the virtual machine to be migrated, various threshold-based methods are introduced in. These thresholds are set to determine whether the server is over-utilized or under-utilized. In this paper, with the idea of Maximum Correlation

Policy (MC), we propose a new virtual machine migration strategy according to the previous SST.

3.2.3 The Minimum Energy Consumption Strategy

In this paper, we characterize the energy consumption of different servers as a fixed value according to the load value. We introduce the idea of server integration into our proposed SST strategy. So that, the problem of server integration can be saved by two steps: (1) choose the right service for scheduling and turn off the right physical machines; (2) choose the right physical machines for integrated services.

4 Design of Load Balance System Architecture

In this section, we present the system architecture of the resource scheduling model.

4.1 Service Request Learner

When users submit their service requests, service request learner (SRL) records the number of times each service accesses the cloud data center on a time scale, so as to judge the service status and the overall status of the data center. Further, SRL can judge the status of the data center in advance with the learned periodic data values, so that relevant strategies can be selected faster.

4.2 Local Resource Monitor

The local resource monitor (LRM) collects information about allocated VMS and containers running on physical machines. The LRM can save the resource utilization history of its host, and the local resource monitor monitors physical machine resource usage in real-time. At the last, LRM will transmit its data to the data center physical machine.

4.3 Global Resource Scheduler

The global resource scheduler (GRS), shown as Fig. 3, works at the central physical machine. The GRS is responsible for detecting the current load conditions (peak and nonpeak) by using the host utilization information stored in the LRM. The GRS also stores the load threshold, energy consumption value, and performance value of each physical machine. The communications between LRM and GRS are enabled by the sharing of common data structures and by using regular information polling.

4.4 Physical Machine

In the cloud, PM provides the hardware infrastructure for creating virtual machines, it can also be used to hold multiple containers. In other words, PM is a physical platform for virtual machines (VM) and containers. So, in this paper, we see PM as the basic building block in architecture.

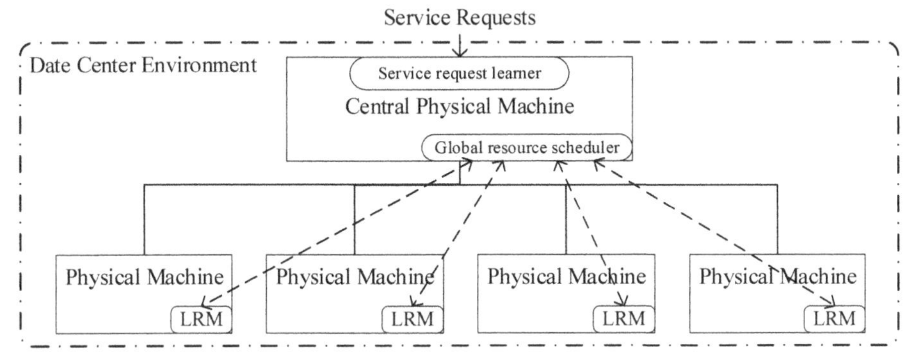

Fig. 3. Model Architecture Diagram.

5 Load Balance Scheme in Cloud Network Convergence

In this section, we present the details of the model and the design process.

5.1 From Idle State to Normal State

When SRL finds that the service volume of the entire system begins to exceed the set threshold, or when GRS finds that system resource usage starts to rise to a certain threshold the system state changes from T0 to T1. Thresholds are generated by periodic SRL and GRS records. When data center status changes from idle (T0) to normal (T1). In each PM cluster, the currently active PM set is called *SL*, and the currently dormant PM set is called *Si*. *Rsi* is used as the resource threshold of the set *SL*.

LRM gets the number of instances in each PM and gets the resource usage of each PM, called *Rpu*. LRM obtains the remaining resources of each PM, called *Rpi*. The energy consumption of each PM is represented by *PPM*. The performance efficiency of each PM is represented by *FPM*. When a service or request enters the data center, match its corresponding virtual machine instance, called VM. The amount of resources used by VM is called *RVm*.

When data center status changes from idle (T0) to normal (T1). And when a request corresponds to an instance on an inactive PM in the cluster. Moreover, the resources of active physical machines in the current cluster meet the resource requirements of the instance, the instance can be migrated to the active cluster. the problem of the location of instance can be saved by two steps: (1) selection of the active physical machine (2) choose the right new physical machine. For the selection of active PM, we introduce the idea of improving the minimum number of connections as the criterion. For each PM in the cluster, LRM obtains the amount of resource used on the PM, called *RVmi*; the total amount of resources in PM, called *Rpi*; the number of VM instances in PM, called *Gi*. Based on *(Rvmi/Rpi) *Gi*, select the PM with the smallest value,

$$Min(RVMi/RPi) * Gi \ (i = 1, 2, 3, ..., n). \tag{5}$$

Figure 4 illustrates the situation when the service corresponding instance is on the PM, this PM can be in a dormant cluster or in an active cluster.

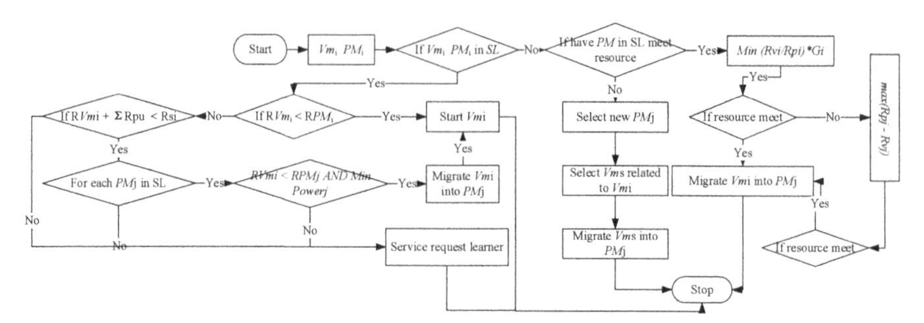

Fig. 4. Flowchart for VM instance in sleeping PM cluster.

5.2 From Normal State to Busy State

When SRL detects the status of the cloud data center from normal (T1) to busy (T2) through historical data, or when GRS detects that the used resources of the data center are approaching the threshold resource value, we judge that the status of the data center has changed. We have done the following to balance the load while reducing energy consumption as much as possible. (1) Detect whether the VM carried by the active PM in each existing cluster can be migrated. (2) If no PM is detected, add the dormant PM to the active cluster and perform load balancing. (3) If a PM is detected, the data center can shut down the existing PM and select a new PM cluster to ensure the load while reducing energy consumption.

We judge whether the instance in PM can be migrated through the dual-threshold method and prediction method. The level corresponding to each instance is recorded in GRS, and the instances that handle special services have a higher level which cannot be easily moved. SRL records the number of times a user accesses the cloud center by time. Only instances whose threshold is lower than the threshold set in SRL and GRS can be scheduled. LRM records the resource consumption of an instance on a time scale. According to the recorded value of LRM, it is found that the corresponding value of the instance will be higher than the threshold after a short period of time, so the instance will not be migrated.

When all instances on the PM can be migrated, a new PM collection will be obtained. The new PM needs to meet the following,

$$\sum PPMi < \sum PPMj + Power, \tag{6}$$

$$\sum Ri > \sum Rj + \nabla Rt * \alpha + Rp * \beta. \tag{7}$$

Table 1. Description of the parameter in the formula (6) and (7).

Parameter	Description
$\sum PPMi$	Total energy consumption value generated by the selected new PM set
$\sum PPMj$	Total energy consumption generated by the selected PM set by step (1)
Power	The value of the minimum energy consumption in the dormant PM
$\sum Ri$	The total resource value generated by the selected new PM set
$\sum Rj$	The total resource generated by the selected PM set by step (1)
∇Rt	Maximum used resource difference between in T1 state and in T2 state
α	When GRS has no recorded value, it is set to 0. Otherwise, it is set to 1
β	When GRS has no recorded value, it is set to 1. Otherwise, it is set to 0

Figure 5 illustrates the process of changing the cloud data center from T1 state to T2 state, where the instance migration strategy is a migration strategy based on the relevance idea mentioned above.

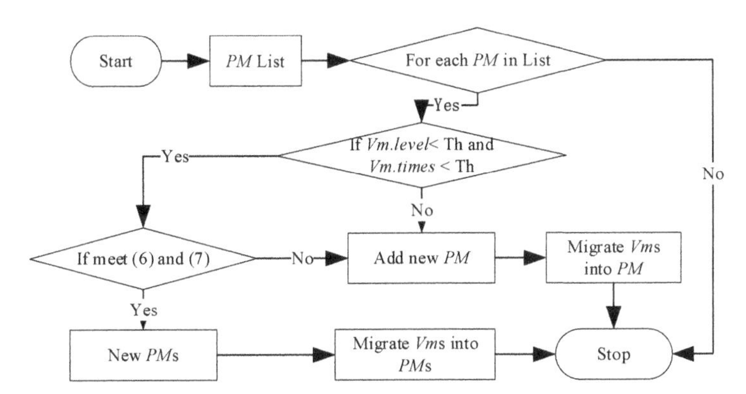

Fig. 5. Flowchart for the cluster status from T1 to T2.

5.3 From Busy State to Normal State or Form Normal State to Idle State

When SRL detects the status of the cloud data center from busy (T2) to normal (T1) or from normal (T1) to idle (T0) through historical data, or when GRS detects that the used resources of the data center below the threshold resource value, we judge that the status of the data center has changed. In the process of state transition, as business processing decreases, server idle resources increase, and servers can be integrated by five steps: (1) LRM obtains the used resource *Ru* and idle resource *Rd* of the active PM in the cluster; (2) eliminate PMs that cannot be integrated through the dual-threshold method; (3) close the instances that are no longer in use, and close the PM when all instances are closed; (4) find PM pairs with *Ru* less than *Rd*. Otherwise, find PM with *Ru* less than the sum of

$Rd * \alpha$;(5) migrate all instances in the PM to other PM, and use the improved minimum number of connections to ensure load balance during migration, and the resources used by other PMs must not exceed the threshold. Then put the corresponding PM to sleep. Figure 6 illustrates the processing flow of reducing energy consumption by combining server integration ideas when the cluster status changes.

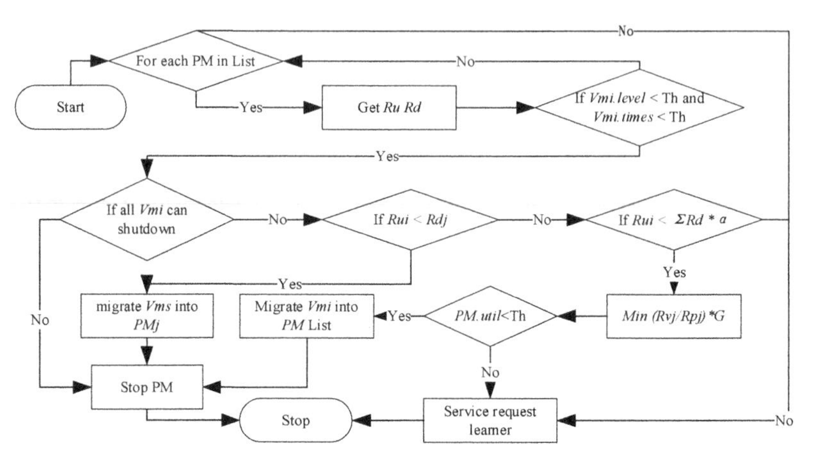

Fig. 6. Flowchart for the cluster status from T2 to T1 or T1 to T0.

To this end, the above content fully demonstrates the process of resource scheduling in the data center according to state changes in a cycle of time.

6 Result and Discussion

In order to analyze the performance of the service state-based scheduling strategy proposed in this paper, the simulation experiment of this paper is carried out in the *CloudSim-Toolkit* [12] environment, and the *bindCloudletToVm* method in the *DatacenterBroker* class is rewritten to map the services and instances accordingly. Figure 7 illustrates that compared with the ordinary private cloud data center, our proposed strategy can shorten processing time under the same workload.

We simulated a data center comprising 30, 50 and 100 VMs respectively. Each VM is modeled to have one CPU core with the performance equivalent to 1,000 MIPS, 2 GB of RAM and 10,000 Mb storage. Figure 7(a) illustrates that when the number of VMs is 30, the total service time of Our Balance Schedule (OBS) is 13.1% better than that of the None Balanced Scheme (NBS); when the number of VMs reaches 50, the total service time of Our Balance Schedule (OBS) is 16.8% better than the None Balanced Scheme (NBS). Notably, when the number of VMs is about 100, the total service time of OBS is 20.5% better than that of NBS.

An improved instance migration strategy based on correlation is introduced in the instance migration, and instance scheduling is carried out according to the correlation between services, which reduces the scheduling overhead to the greatest extent and

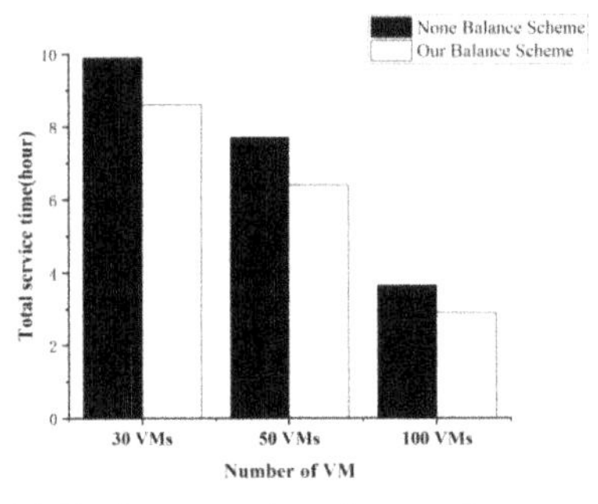

(a) Task processing time variance.

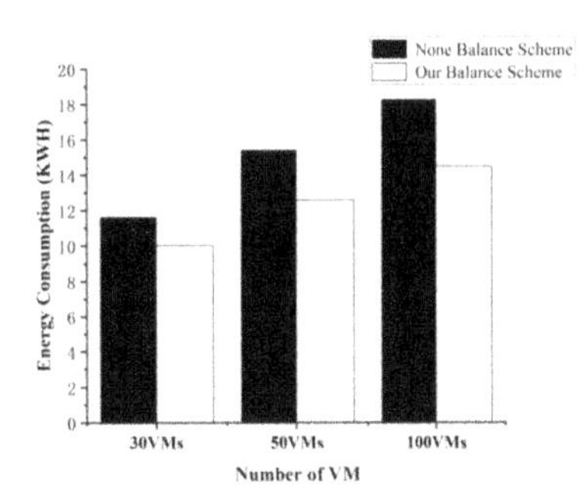

(b) Power consumption variance.

Fig. 7. Illustration of task processing time variance and power consumption variance

effectively reduces the number of instance migrations in the integrated cloud center. Figure 7(b) illustrates the change in data center energy consumption as the number of VMs changes.

7 Conclusion

Cloud networking convergence is the development trend of future Internet and next-generation networking. In this paper, we propose the model and requirements of load balance for heterogeneous entities in the convergence of cloud and network, and we present a concrete load balance scheme with a set of detailed algorithms. With the improvement of cloud network convergence, there should be paid more attention in the study of load balance.

Acknowledgment. This work is supported by National Key R&D Program of China (No.2017YFB1400700), Key R&D Program of Hainan Provincial (ZDYF2019202) and the Fundamental Research Funds for the Central Universities (JB210301).

References

1. Mezmaz, M., Melab, N., Kessaci, Y., et al.: A parallel bi-objective hybrid metaheuristic for energy-aware scheduling for cloud computing systems. J. Parallel Distrib. Comput. **71**(11), 1497–1508 (2011)
2. Panda, S.K., Jana, P.K.: Load balanced task scheduling for cloud computing: a probabilistic approach. Knowl. Inf. Syst. **61**(3), 1607–1631 (2019)
3. Singh, R.M., Paul, S., Kumar, A.: Task scheduling in cloud computing. Int. J. Comp. Sci. Inf. Technol. **5**(6), 7940–7944 (2014)
4. Nashaat, H., Ashry, N., Rizk, R.: Smart elastic scheduling algorithm for virtual machine migration in cloud computing. J. Supercomput. **75**(7), 3842–3865 (2019)
5. Bari, M.F., Zhani, M.F., Zhang, Q., et al.: CQNCR: optimal VM migration planning in cloud data centers. In: 2014 IFIP Networking Conference, pp,:1–9. IEEE (2014)

6. Zhang, S., Qian, Z., Luo, Z., et al.: Burstiness-aware resource reservation for server consolidation in computing clouds. IEEE Trans. Parallel Distrib. Syst. **27**(4), 964–977 (2015)
7. He, L., Zou, D., Zhang, Z., et al.: Developing resource consolidation frameworks for moldable virtual machines in clouds. Futur. Gener. Comput. Syst. **32**, 69–81 (2014)
8. Mohamadi Bahram Abadi, R., Rahmani, A.M., Alizadeh, S.H.: Server consolidation techniques in virtualized data centers of cloud environments: a systematic literature review. Software Pract. Exp. **48**(9), 1688–1726 (2018)
9. Biswas, J., Ray, M., Sondur, S., et al.: Coordinated power management in data center networks. Sustain. Comput. Inf. Syst. **22**, 1–12 (2019)
10. Choudhary, A., Govil, M.C., Singh, G., et al.: A critical survey of live virtual machine migration techniques. J. Cloud Comput. **6**(1), 1–41 (2017)
11. Kusic, D., Kephart, J.O., Hanson, J.E., et al.: Power and performance management of virtualized computing environments via lookahead control. Clust. Comput. **12**(1), 1–15 (2009)
12. Calheiros, R.N., Ranjan, R., De Rose, C.A.F., et al.: Cloudsim: a novel framework for modeling and simulation of cloud computing infrastructures and services. arXiv preprint arXiv: 0903.2525 (2009)

Enhancing Wi-Fi Device Authentication Protocol Leveraging Channel State Information

Bing Chen[1,2,3], Yubo Song[1,3](✉), Tianqi Wu[1,3], Tianyu Zheng[1,3], Hongyuan Chen[1,3], Junbo Wang[2,3], and Tao Li[1,3]

[1] School of Cyber Science and Engineering, Key Laboratory of Computer Network Technology of Jiangsu Province, Southeast University, Nanjing, China
[2] School of Information Science and Engineering, Southeast University, Nanjing, China
[3] Purple Mountain Laboratories, Nanjing, China
songyubo@seu.edu.cn

Abstract. Wi-Fi device authentication is crucial for defending against impersonation attacks and information forgery attacks. Most of the existing authentication technologies rely on complex cryptographic algorithms. However, they cannot be supported well on the devices with limited hardware resources. A fine-grained device authentication technology based on channel state information (CSI) provides a non-cryptographic method, which uses the fingerprint extracted from CSI for authentication since CSI can uniquely identify the device in a limited time. But maintaining a fingerprint database for fingerprint matching is a challenging work. Firstly, the fingerprints extracted from the CSI are time-sensitive, which means that the fingerprint database must be updated in real time; Secondly, the authentication device may collect false fingerprints under the attack of identity-based attackers, which means that the authenticity of the fingerprint used to update the database must be checked. In this paper, we propose an enhancing Wi-Fi device authentication protocol based on CSI to implement the fingerprint database update and the device authentication. We provide a viable method of database update and an authentication algorithm based on Local Outlier Factor (LOF). We also present a complete authentication process. In addition, we evaluate the performance of our CSI-based authentication algorithm and database updating method. The experiments showed that the accuracy of the authentication algorithm is up to 97.1% and our database updating method can help the system maintain high accuracy.

Keywords: Wi-Fi device authentication · Channel state information (CSI) · Local outlier factor (LOF) · Wireless physical fingerprintfing

J. Xiong et al. (Eds.): MobiMedia 2021, LNICST 394, pp. 33–46, 2021.
https://doi.org/10.1007/978-3-030-89814-4_3

1 Introduction

Nowadays, Wi-Fi has become one of the most important association technologies. However, there are serious security problems in Wi-Fi association. Attackers can get other devices' identity information through wireless sniffing, and then use the information to disguise themselves as legitimate devices [7,8]. Attackers can steal confidential data or attack the internal websites after getting authorization [1], or they can control other devices by sending spurious instructions [14]. Therefore, the authentication of Wi-Fi devices is indispensable. 802.11i provides some cryptographic-based device authentication method, but has proven security weaknesses [2,3,12]. What's more, a part of Wi-Fi devices does not have enough hardware resources to support these authentication schemes, which brings challenges to the usage of cryptographic-based device authentication technology.

In recent years, people tried to use wireless channel characteristics for device authentication. One of the methods is to extract fingerprints that can identify the device from CSI and use fingerprint matching to verify devices' identity [4,8,13]. The principle of this method is that under the influence of multipath and environmental fading, the amplitude and phase of each sub-carrier contains unique spatial information, which means devices at different locations have different CSI. Some existing CSI-based device authentication schemes can do identify matching with high accuracy [5,6,9–11]. [12] provides an identification scheme and authentication protocol in detail, and it also gives a framework to defend against attackers. As we all know, device authentication is a long-term work, and the channel state will change over time. Thus, the CSI of the device will also change and cause the current CSI and the fingerprints in database to not match, and the legal device will be rejected. Maintaining a valid fingerprint database in the authentication device during the long-term work is significant for device authentication. However, all the authentication system given above do not provide a solution for fingerprint database update.

In this paper, we propose an enhancing Wi-Fi device authentication protocol. We focus our attention on two application scenarios, including the authentication in access phase and the authentication in association phase. In two scenarios, we provide different fingerprint database updating method and authentication processes. Based on the fact that CSI can uniquely identify devices [7], our protocol extracts fingerprints from CSI and uses them to verify the identity of unknown devices by fingerprint matching.

The main contributions of this paper are as follows:

- We propose an enhancing Wi-Fi device authentication protocol based on CSI.
- We provide a method of fingerprint database update, which can update the fingerprint database effectively in both access phase and association phase. It can help to maintain a low false rejection rate during long-term work, and it can effectively detect the identity-based attackers on the other hand.
- We provide a complete Wi-Fi device authentication framework, which covers the authentication process and the fingerprint database updating method in both access and association phase.

This paper is organized as follows: Sect. 2 gives an overview of the authentication framework. Section 3 provides an authentication method based on CSI; Sect. 4 gives a detailed description of our fingerprint database updating method; In Sect. 5, we introduce the enhancing Wi-Fi device authentication protocol. Section 6 gives the evaluation. Section 7 concludes the paper.

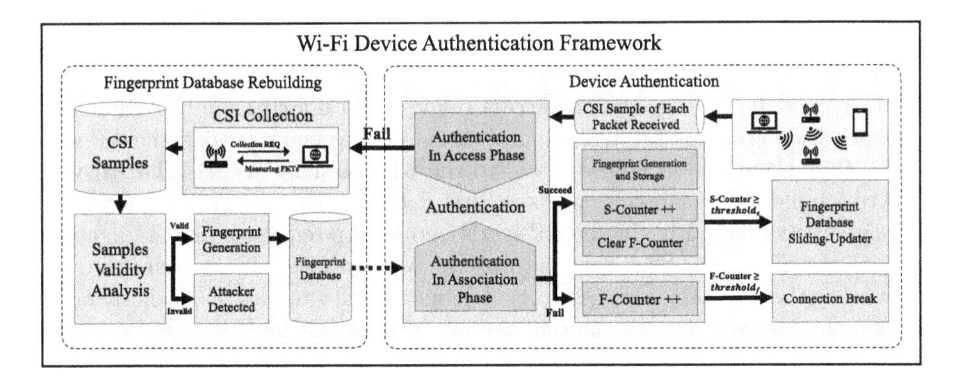

Fig. 1. The framework of our Wi-Fi device authentication system

2 Framework Overview

We design a Wi-Fi device authentication framework to provide device authentication service in both access phase and association phase, as shown in Fig. 1. The framework is mainly implemented on the Wi-Fi Access Point (AP). Some working steps that require signaling interaction between the AP and the working station (STA) can be completed by installing our authentication protocol on the both devices.

It can be seen in Fig. 1 that the Wi-Fi device authentication framework is mainly composed of two parts, including the device authentication and the fingerprint database rebuilding. But according to the logical function, the framework should be divided into two parts including device authentication and fingerprint database update. The device authentication mainly completes two tasks. The first task is to verify the identity of the STA in access phase and decide whether to allow access. The second task is to check the identity of data frames sent by the STA during the association phase. The third task is to extract fingerprints from the received data packets and use them to update the fingerprint database. The fingerprint database update is mainly responsible for the rebuilding and the sliding-updating of the fingerprint database.

We describe the whole authentication framework by simulating the device authentication process in two application scenarios, including the access phase and the association phase.

2.1 Authentication in Access Phase

According to the access process of the 802.11 device, the STA needs to send an authentication request to the AP. The AP first extracts CSI sample from the request frame, and then sends it to the authentication module. At the same time, the fingerprints belonging to this STA in the fingerprint database will be sent to the authentication module (shown with the dotted line), and the module will check if the CSI sample is an outlier in the fingerprints based on LOF. If the sample is not an outlier, it means that the access request is indeed from the STA, and the AP accepts the access request and informs the STA that the authentication is successful. If the sample is checked to be an outlier, the AP will return an authentication failure signal to the STA to deny access. The AP will start to rebuild the fingerprint database after detecting the authentication failure event in access phase. Firstly, the AP sends a collection request named Collection REQ to the STA. The STA then returns the Measuring PKTs, which are used to measure the CSI. After the AP collects enough CSI samples, it sends them to the samples validity analysis module to detect whether the CSI samples really comes from the STA. If the sample set is valid, it used to generate fingerprints and rebuild the fingerprint database. Otherwise, the fingerprint database will not be changed.

2.2 Authentication in Association Phase

For each data packet from the STA, the AP extracts the CSI and sends it to the authentication module. The authentication process of the module is the same as the authentication in access phase, but some follow-up processing should be finished according to different authenticating results. It can be seen from Fig. 1 that after successful authentication, the AP uses the CSI samples to generate a new fingerprint and store it in a buffer. The AP maintains a counter called S-Counter to record the number of data packets that have been successfully authenticated since the last update of the fingerprint database. If the value is greater than or equal to the $threshold_s$, it will update the fingerprint database with the fingerprints in the buffer. We provide a fingerprint database sliding-updater to maintain the database. Meanwhile, The AP also maintains a counter called F-Counter. It records the number of the frames that continuously fail the authentication. When the authentication fails, the F-Counter is updated, and the AP determines whether the connection should continue according to whether the $threshold_f$ is reached.

3 Authentication Leveraging CSI Fingerprint

In this section, we focus on the device authentication based on CSI fingerprint. We first explain the basic idea of how to do authentication using CSI fingerprint, and then we describe the algorithm of device authentication.

3.1 Basic Idea

CSI includes the amplitude and phase of all sub-carriers used to transmit data. These values will vary with factors such as channel fading and the multipath effect. We know that the channel between two Wi-Fi devices are unique, which means that the CSI measured by the two devices is also unique. In the light of this idea, the AP can uniquely mark the device by extracting and recording the CSI of the STA. Figure 2 shows the amplitude image of the CSI obtained from three devices placed in different positions. We can see that the CSI images of the same device are concentrated, while the CSI images between different devices are very different.

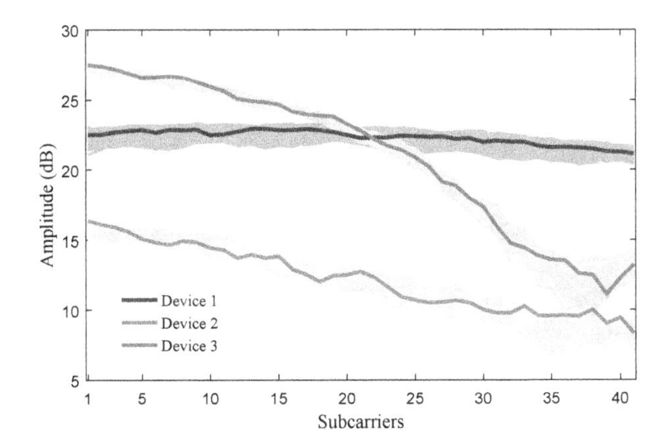

Fig. 2. CSI samples of three devices placed in different locations

In the realistic authentication work, we need to match an unknown CSI sample with the local fingerprint set, and determine the result of the authentication based on the matching result. Inspired by the feature of the CSI images in Fig. 2, we consider using the Local Outlier Factor (LOF) algorithm to do CSI matching. We use the samples in the local fingerprint set as the reference sample group, and calculate the LOF of the unknown sample in the reference sample group to determine whether it is an outlier, thereby obtaining the matching result.

We found in the test that there are often values that far exceed the expected numerical fluctuation range, as shown in Fig. 3(a). They do not contain any information of device identity and will reduce the accuracy of system. Therefore, we use the Hampel Identifier to exclude these outliers. In addition, the CSI samples of the same device will change under noise interference. Even though the CSI images of the same device still have the same trend, but the dispersion increases. The local outlier factor describes the difference between samples based on the Euclidean distance. It means the noise will cause the reference sample group become discrete and reduce the probability of abnormal CSI being detected. Therefore, we smooth the CSI samples in the time domain.

3.2 Algorithm Description

Fingerprint Generation. The Fingerprint generation algorithm consists of two stages, including the outlier elimination and the smoothing.

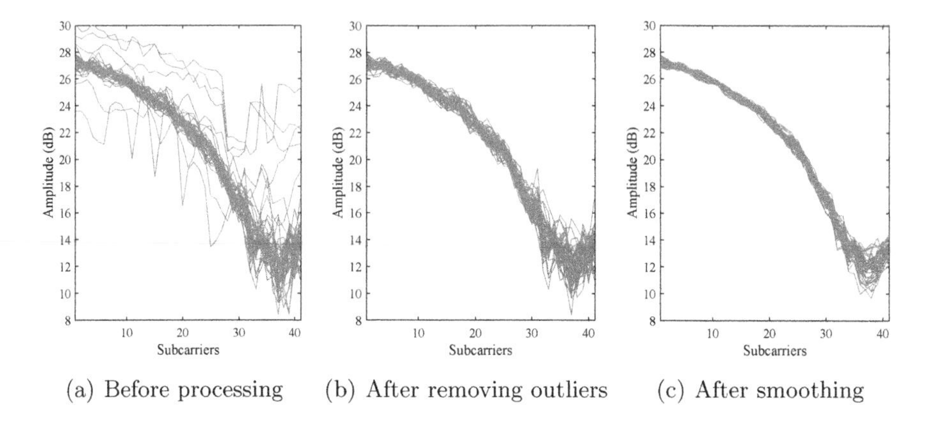

(a) Before processing (b) After removing outliers (c) After smoothing

Fig. 3. CSI samples in different processing stage

Outlier Elimination. We perform outlier detection on each subcarrier individually. Let the k-th CSI sample from device D be $C_D^k = \{C_{D,1}^k, ..., C_{D,N}^k\}(k = 1, ..., K)$, where N is the number of subcarrier and K is the number of samples obtained from device D. We arrange the amplitude samples of n-th subcarrier in chronological order, which can be presented as $L_{D,n} = (C_{D,n}^1, ..., C_{D,n}^K)$. We select the median of the window composed of $C_{D,n}^I$ and its $2l$ surrounding samples. We use the absolute deviation of each sample to estimate the standard deviation of the samples in the window. The standard deviation is shown as follows:

$$\sigma_{D,n}^I = \frac{1}{\gamma} median\left(\mid C_{D,n}^i - C_{D,n}^I \mid\right), \tag{1}$$

where $C_{D,n}^I$ indicates the median of the window , $C_{D,n}^i$ is the sample in the window. We use $\gamma = \sqrt{2} erfinv(0.5)$, where $erfinv$ on behalf of the inverse error function.

We then perfrom the following numerical substitution on each amplitude sample to exclude the outlier:

$$C_{D,n}^i = \begin{cases} C_{D,n}^i & \mid C_{D,n}^i - C_{D,n}^I \mid \leq \eta\sigma_{D,n}^I \\ C_{D,n}^I & \mid C_{D,n}^i - C_{D,n}^I \mid > \eta\sigma_{D,n}^I, \end{cases} \tag{2}$$

where η is a threshold used to judge whether the sample is outlier.

We repeat the above work on each subcarrier. Figure 3(b) shows the CSI amplitude images after removing the outliers.

Smoothing. We smooth the amplitude of each sub-carrier in the time domain to reduce this effect. The smoothing process is described as follows:

$$\widetilde{C}_{D,n}^{k} = \frac{1}{w} \sum_{max(0,k-\lfloor \frac{w}{2} \rfloor)}^{min(K,k+\lfloor \frac{w-1}{2} \rfloor)} C_{D,n}^{k}, \tag{3}$$

where w indicates the length of smoothing window. Figure 3(c) shows the CSI amplitude images after smoothing.

Fingerprint Matching. In the sample space S_D composed of the local fingerprints of Device D and the new sample with $D's$ identity, we define the distance between the k-th and the r-th sample as:

$$d_{k,r} = \left(\sum_{n=1}^{N} \left(C_{D,n}^{k} - C_{D,n}^{r} \right)^2 \right)^{\frac{1}{2}}. \tag{4}$$

We define $d_p(k)$ as the distance between the k-th sample and the p-th farthest sample. Also, we define the p-distance neighborhood of the k-th sample as the set of samples whose distance from it is less than or equal to $d_p(k)$, and denote it by $N_p(k)$. In addition, we define the p-reachable distance from the k-th and the r-th sample as:

$$rech_p(k, r) = max(d_p(r), d_{k,r}). \tag{5}$$

Therefore, the local reachable density of the k-th sample can be expressed as:

$$lrd_p(k) = \frac{\mid N_p(k) \mid}{\sum_{r \in N_p(k)} rech_p(k, r)}. \tag{6}$$

Finally, we get the local outlier factor of the k-th sample:

$$LOF_p(k) = \frac{\sum_{r \in N_p(k)} lrd_p(r)}{\mid N_p(k) \mid \cdot lrd_p(k)}. \tag{7}$$

The Fig. 1 tells us that if the k-th sample does not belong to device D, then it will be far away from the surrounding samples, and the local reachable density of it should be small. On the contrary, if the samples around the k-th sample are clustered together, the local reachability density of them will be very large. It is reflected on the LOF that if the k-th sample belongs to device D, the LOF of it is approximately 1; if it does not belong to device D, the LOF is a larger value. Thus, We do the following judgement to verify the identity:

$$\begin{cases} LOF_p(u) \leq mean\left(LOF_p(f)\right) + 10 * std\left(LOF_p(f)\right) & success; \\ LOF_p(u) > mean\left(LOF_p(f)\right) + 10 * std\left(LOF_p(f)\right) & failure, \end{cases} \tag{8}$$

where u indicates the new CSI sample with D's identity, f represents D's fingerprint in local database, $mean\left(LOF_p(f)\right)$ indicates the mean of $LOF_p(f)$ and $std\left(LOF_p(f)\right)$ indicates the standard deviation.

The Fig. 4 shows the LOF obtained from two different devices. We use 44 CSI samples obtained from a device to generate a fingerprint database, and collected 44 CSI samples of this device and another device at an adjacent time. Blue bar represents the LOF of the fingerprints in the library, red bar represents the LOF of the samples collected from the same device, and yellow bar is the LOF of the CSI samples from another device. The blue dashed line shows the threshold, and it can be seen that only 1 sample was wrongly judged.

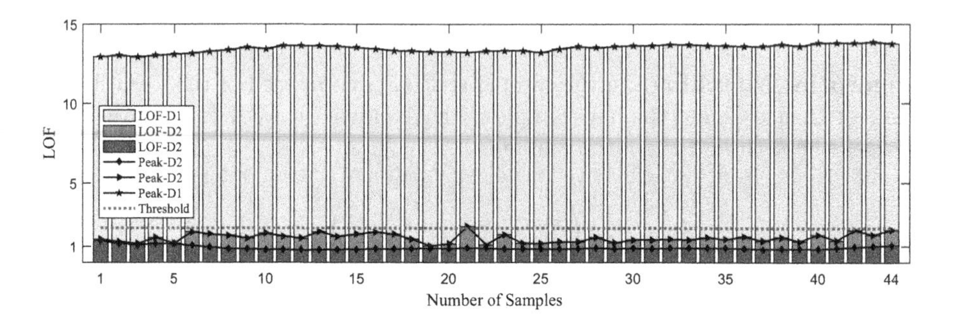

Fig. 4. The LOF of the samples collected from two devices

4 CSI Fingerprint Update

In this section, we will focus on the method of the fingerprint database update. Fingerprint database update is important because if the fingerprints in the database become outmoded, any access request sent from STA will be refused because the CSI of the STA is different with the fingerprints in AP's database. The updating method of fingerprint database presented in this paper can improve this problem without compromising the safety of the system. We will describe this method from two application scenarios.

4.1 CSI Fingerprint Update in Access Phase

In the initial state, the STA is not connected. At this time, the fingerprints of the STA stored in the AP may or may not be able to match with the current CSI. If the matching succeeds, the access work can be finished without maintaining the fingerprint database. Therefore, we will only discuss the updating method in the case of matching fails.

We discuss the algorithm in two situations: 1) the STA requests access; and 2) the attacker requests access.

The STA Requests Access. When the STA requests access, the AP uses an invalid fingerprint for authentication, which causes the authentication to fail. The AP then sends the CSI collection request, and the STA sends CSI measurement packets after receiving the request. If there is no attacker or the attacker does not do any response, the master device collects a true and effective CSI sample set. It then uses the sample set to generate fingerprints and overwrite the original data in the fingerprint database. If the attacker exists and sends measurement frames at the same time, the CSI sample set collected by the AP will contain two distinct group of samples. We use the mean of samples as the center, Fig. 2 shows that when the samples come from one device, all the samples are concentrated in the narrow band at the center. When the samples come from two different devices, it can be expected that most of the samples are far away from the center. We use the standard deviation of samples to quantify this dispersion:

$$\sigma = \left[\frac{1}{K} \sum_{k=1}^{K} (C_D^k - C_D)^2 \right]^{\frac{1}{2}}, \tag{9}$$

where C_D is the mean of all samples.

We set a threshold for estimate. If the standard deviation exceeds the threshold, the sample set is judged to be invalid and the fingerprint database will not be maintained. Regardless of whether the fingerprint database is successfully maintained, the child device will continue to request access. If the update is successful, the STA can be successfully accessed in the next request. Otherwise, the STA must wait until the attacker stops attacking before it can successfully access.

The Attacker Requests Access. When an attacker requests access, authentication will fail and CSI collection will start. In our working environment, the STA being impersonated will also receive the measurement request and send the measuring packets to the AP. Similarly, we consider two possibilities. If the attacker does not send measuring packets, the AP will update the fingerprint database with the CSI of the real device, and the attacker still cannot be accessed. If the attacker sends measuring packets, it cannot pass the sample validity check, and thus cannot construct a false fingerprint database in the AP.

4.2 CSI Fingerprint Update in Association Phase

We assume that the STA has been successfully connected and starts to transmit data. The successful access of the STA means that the fingerprints in the AP is valid, which means that we can successfully authenticate the data sent by the STA. The key for updating the fingerprint database in association phase is that we will store the CSI samples of data packets and use them to generate a new fingerprint set to update the fingerprint database (as described in Sect. 2). We propose a simple fingerprint database sliding-updater, which uses the new fingerprints to replace the $threshold_f$ earliest generated fingerprints in the database (just like a sliding-window). Through the updater, we ensure that the fingerprints in database and the CSI samples obtained from packets received recently are always highly correlated.

However, the real-time update of the fingerprint database is based on the active communication between the AP and the STA. When the STA is in sleeping mode, keeping the CSI collection will cause greater power consumption and increase the network burden, so this is not recommended. After a device returns to be active, the fingerprints stored in the authenticator may become invalid. In order to reestablish the fingerprint database, we disconnect after $threshold_f$ consecutive data packets fail the authentication (see Sect. 2), at this time the update of the fingerprint database will return to the first situation.

In addition, we must also consider the existence of the attacker. When an attacker uses a forged identity to send a data packet, the AP will get the failed authentication result and discard the packet. If the AP continuously receives $threshold_f$ packets from the attacker, the connection is broken. We can find that the attacker cannot attack the AP effectively during this whole process.

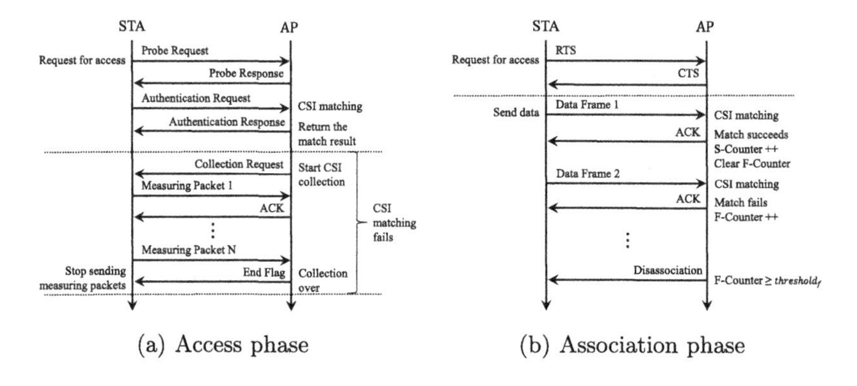

(a) Access phase (b) Association phase

Fig. 5. The procedure of CSI-based authentication protocol

5 The CSI-based Authentication Protocol

It can be seen from the previous section that the association and cooperation between the AP and the STA are the key to do device authentication and fingerprint database update in both the access and the association phase. Therefore, we present an enhancing Wi-Fi device authentication protocol in this section, which is designed based on the existing Wi-Fi association process.

5.1 The Authentication Procedure in Access Phase

This subsection shows the authentication procedure in access phase. As shown in Fig. 5(a), the process includes an authentication stage and a CSI collection stage, where the CSI collection stage is triggered only when the identity authentication fails. The steps are shown as follows:

- STA sends Probe Request to request access to the network. AP responds with Probe Response after listening to the request, indicating that it can provide access service and is ready for authentication.

- STA then sends Authentication Request for the authentication. AP extracts CSI from the control frame and performs CSI matching. Then AP returns Authentication Response to inform STA of the authentication result.
- If the matching is successful, the access authentication ends. If the match fails, AP sends Collection Request to STA, requesting to start CSI collection.
- After receiving the request, STA starts to send Measuring Packets with a fixed data length of 24 bytes. After receiving the data packets, AP responds with ACK to tell STA to continue or stop.

5.2 The Continuous Authentication in Association Phase

Figure 5(b) shows the continuous authentication in association phase. In Wi-Fi association, CSMA/CA is used to avoid the collisions among packets. The STA exchanges RTS/CTS with the AP before starting to transmit data to inform other devices to remain silent. After exchanging RTS/CTS, the AP performs CSI matching on each received data frame, and judges whether the current connection is valid according to the matching situation, and decides whether to continue association or not (as shown in Sect. 2).

- STA sends RTS to tell AP that it is going to send data. AP responds with CTS after receiving the request, indicating that it is ready receiving and clearing the channel.
- STA starts to send Data Frames. When AP receives a Data Frame, it responds with ACK. AP extracts the CSI of the frame and performs CSI matching. If the match is successful, the S-Counter is increased by 1, and the F-Counter is cleared. If the match fails, the F-Counter is increased by 1.
- If F-Counter reaches $threshold_f$, AP sends Disassociation to STA to disconnect.

6 Performance Evaluation

In this section, we evaluate the performance of two parts of our authentication framework. The first part is the authentication algorithm based on LOF. The CSI samples used in this process is obtained within a short time interval. The second part is the fingerprint database sliding-updater. To evaluate the performance of our sliding-updater, we simulate the long-term working scenario by collecting CSI within a long time interval.

In our experiments, the STAs were fixedly placed in different locations in the laboratory. They transmitted frames at a rate of 100 pkt/sec in a 20 MHz wireless channel. The AP was also placed in a fixed location and listened to the frames sent by STAs. Since the AP will received irrelevant frames from other unknown Wi-Fi devices, we set the MAC of each STA and filter the frames by MAC checking. For each valid frame received, the AP got CSI information from it and saved.

To evaluate the performance of the fingerprint database sliding-updater, we first used a fixed fingerprint database to do authentication. Then we updated the fingerprint database during authentication to see the improvement of performance.

We use three metrics: false rejection rate (FRR), false acceptance rate (FAR) and accuracy rate (ACC) to quantify the performance of the system. FRR is the probability of matching failure for samples from legal device. FAR is the probability of successful matching for samples from attackers. ACC is equal to the proportion of correctly matching samples in the total sample set. We use the samples collected in the same time on the same device. After successively authenticating 100 samples, we obtain the authentication accuracy of the 100 samples as the accuracy of the current system.

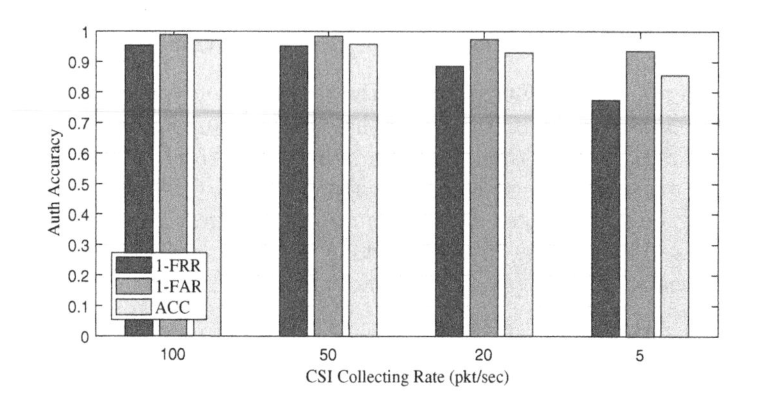

Fig. 6. The value of each metrics with different CSI collecting rate

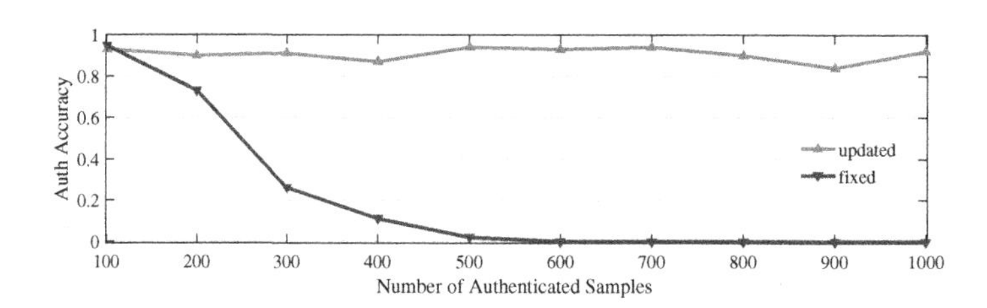

Fig. 7. The accuracy of the authentication systerm tested with updated and fixed fingerprint database

In the experiments, we got more than 96,000 CSI samples using four devices (ETTUS USRP B210) in our laboratory. In the first phase, we evaluated our authentication algorithm based on LOF. We got the three metrics in four different packet sending rates. We set the size of the local fingerprint database to 100, and did authentication on more than 1,000 CSI samples. The Fig. 6 shows the test results. To observe the results more clearly, we use $1 - FRR$ and $1 - FAR$ to represent the test results. The results show that when the data packet rate is at its maximum, the three indicators all reach their maximum values, with the maximum values being 95.4%, 98.8% and 97.1%. It can be seen from it that when the data packet rate decreases, FRR and FAR increase accordingly. The accuracy of authentication also decreases, but it remains above 85%. This is because when the packet rate decreases, the time difference between each CSI sample increases, so the correlation between samples will decrease. In particularly, we can see that FRR is growing faster than FAR in the Fig. 6.

In the second stage, we evaluated the performance of the fingerprint database sliding-updater. The two sets of data shown in the Fig. 7 are the accuracy of system with updating the database and not updating. We can see that when the system uses the sliding-updater to continuously update the fingerprint database during the test, the accuracy of the system is maintained above 80%, and the highest accuracy is 94%. When the fingerprint database is fixed, the accuracy of the system decreases rapidly during the authentication, and it is almost 0 after authenticating the 600th CSI sample.

We obtained good results in both experiments. However, the high accuracy of the system may be due to the over-fitting of the data and the stability of the test environment considering that our experiment has limited data and single test environment.

7 Conclusion

In this paper, we propose an enhancing Wi-Fi device authentication protocol. We also give a complete Wi-Fi device authentication framework. The framework mainly contains two parts, including the device authentication and the fingerprint database update. Our protocol works on two application scenarios, including the authentication in access phase and the authentication in association phase. We provide different fingerprint database updating method and authentication processes in two scenarios. In the first scenario, we reestablish the fingerprint database after authenticating fails with new CSI samples. We use the standard deviation of the new CSI sample set to decide whether there is an attacker. In the second scenario, we use the database sliding-updater to update the fingerprints. We also provide a fingerprint generating method and an authentication method based on LOF. The evaluation shows that our authentication method has a good performance as well as the sliding-updater.

Acknowledgment. This work is supported by Frontiers Science Center for Mobile Information Communication and Security, Southeast University, Nanjing, China. This work is also supported by Zhishan Youth Scholar Program Of SEU, Nanjing, China.

References

1. Alshudukhi, J.S., Mohammed, B.A., Al-Mekhlafi, Z.G.: An efficient conditional privacy-preserving authentication scheme for the prevention of side-channel attacks in vehicular ad hoc networks. IEEE Access **8**, 226624–226636 (2020). https://doi.org/10.1109/ACCESS.2020.3045940
2. Chatterjee, U., Sadhukhan, R., Mukhopadhyay, D., Subhra Chakraborty, R., Mahata, D., M.Prabhu, M.: Stupify: A hardware countermeasure of kracks in wpa2 using physically unclonable functions. In: Companion Proceedings of the Web Conference 2020, WWW 2020, pp. 217–221. Association for Computing Machinery, New York (2020). https://doi.org/10.1145/3366424.3383545

3. Elhigazi, A., Razak, S.A., Hamdan, M., Mohammed, B., Abaker, I., Elsafi, A.: Authentication flooding dos attack detection and prevention in 802.11. In: 2020 IEEE Student Conference on Research and Development (SCOReD), pp. 325–329, September 2020. https://doi.org/10.1109/SCOReD50371.2020.9250990
4. Liao, R., Wen, H., Pan, F., Song, H., Xu, A., Jiang, Y.: A novel physical layer authentication method with convolutional neural network. In: 2019 IEEE International Conference on Artificial Intelligence and Computer Applications (ICAICA), pp. 231–235, March 2019. https://doi.org/10.1109/ICAICA.2019.8873460
5. Liu, H., Wang, Y., Liu, J., Yang, J., Chen, Y.: Practical user authentication leveraging channel state information (csi). In: Proceedings of the 9th ACM Symposium on Information, Computer and Communications Security, ASIA CCS 2014, pp. 389–400. Association for Computing Machinery, New York (2014). https://doi.org/10.1145/2590296.2590321
6. Liu, M., Mukherjee, A., Zhang, Z., Liu, X.: Tbas: Enhancing wi-fi authentication by actively eliciting channel state information. In: 2016 13th Annual IEEE International Conference on Sensing, Communication, and Networking (SECON), pp. 1–9, June 2016. https://doi.org/10.1109/SAHCN.2016.7733021
7. Liu, S.: Mac spoofing attack detection based on physical layer characteristics in wireless networks. In: 2019 IEEE International Conference on Computational Electromagnetics (ICCEM), pp. 1–3, March 2019. https://doi.org/10.1109/COMPEM.2019.8779180
8. Madani, P., Vlajic, N., Sadeghpour, S.: Mac-layer spoofing detection and prevention in iot systems: Randomized moving target approach. In: Proceedings of the 2020 Joint Workshop on CPS&IoT Security and Privacy, CPSIOTSEC 2020, pp. 71–80. Association for Computing Machinery, New York (2020). https://doi.org/10.1145/3411498.3419968
9. Rocamora, J.M., Ho, I.W.H., Mak, M.W.: Fingerprint quality classification for csi-based indoor positioning systems. In: Proceedings of the ACM MobiHoc Workshop on Pervasive Systems in the IoT Era, PERSIST-IoT 2019, pp. 31–36. Association for Computing Machinery, New York (2019). https://doi.org/10.1145/3331052.3332475
10. St. Germain, K., Kragh, F.: Multi-transmitter physical layer authentication using channel state information and deep learning. In: 2020 14th International Conference on Signal Processing and Communication Systems (ICSPCS), pp. 1–8, December 2020. https://doi.org/10.1109/ICSPCS50536.2020.9310034
11. St. Germain, K., Kragh, F.: Physical-layer authentication using channel state information and machine learning. In: 2020 14th International Conference on Signal Processing and Communication Systems (ICSPCS), pp. 1–8, December 2020. https://doi.org/10.1109/ICSPCS50536.2020.9310070
12. Troya, A.S., Astudillo, J.J., Romero, C.G., Sáenz, F.G., Díaz, J.: Vulnerability detection in 802.11i wireless networks through link layer analysis. In: 2014 IEEE Latin-America Conference on Communications (LATINCOM), pp. 1–6, November 2014. https://doi.org/10.1109/LATINCOM.2014.7041875
13. Wang, Q., Li, H., Zhao, D., Chen, Z., Ye, S., Cai, J.: Deep neural networks for csi-based authentication. IEEE Access **7**, 123026–123034 (2019). https://doi.org/10.1109/ACCESS.2019.2938533
14. Yan, C., Ge, J.: Synchronous control of master-slave manipulator system under deception attacks. In: 2020 Chinese Control And Decision Conference (CCDC), pp. 1778–1782 August 2020. https://doi.org/10.1109/CCDC49329.2020.9164635

Reverse Analysis Method of Unknown Protocol Syntax in Mobile Multimedia Communications

Yichuan Wang, Binbin Bai, Zhigang Liu, Xinhong Hei(✉), and Han Yu

Laboratory for Network Computing and Security Technology,
Xi'an University of Technology, Shannxi, China
heixinhong@xaut.edu.cn

Abstract. With the development of mobile multimedia communication, the frequency of a few protocols and special communication protocols increases rapidly, and the probability of network attack events is frightening. This has caused people financial losses and psychological panic. At present, the commonly used protocol recognition tools use a single and targeted method, which has a lot of limitations. Based on the existing algorithms, this paper presents a new protocol feature extraction method, which greatly improves the efficiency of feature extraction. In this paper, the idea of Apriori algorithm is improved, and the feature string is searched under the idea of composite features of CFI algorithm. Combining the advantages of previous algorithms, a more efficient OFS (Optimal Feature Strings) algorithm is proposed, which can perform better when facing the feature extraction problem of bit stream protocol. Then it compares with the existing algorithms, tests the superiority of the new algorithm in running time, and further illustrates the accuracy and correctness of the algorithm.

Keywords: Protocol recognition · Feature extraction · Composite feature · Communication security

1 Introduction

With the rapid development of China's social economy, China's science and technology level has been improved as a whole. As the product of information society, mobile multimedia communication technology is widely used in all walks of life. By using the advantages and values of mobile multimedia communication technology, it has made great contributions to the rapid development of social information in China [1]. As multimedia communication technology involves many fields, such as service, military, finance and administration, it plays an important role in modern social production activities. Therefore, this paper analyzes the specific application and development trend of multimedia communication technology [2,3], so as to improve the role of multimedia communication technology in promoting the development of modernization and make it more in line with the needs of social development goals.

J. Xiong et al. (Eds.): MobiMedia 2021, LNICST 394, pp. 47–62, 2021.
https://doi.org/10.1007/978-3-030-89814-4_4

As the most widely used modern technology in the field of science and technology, multimedia communication technology occupies an important development position [4]. Due to the rapid development of multimedia communication technology, it directly affects the development and openness of modern society. Therefore, strengthening the application of multimedia communication technology and making clear the development trend of multimedia communication technology are conducive to promoting the rapid development of science and technology society in China. Multimedia communication technology is mainly through the media to process the information and data, and display the data in the form of files, audio or pictures. It is an effective method to realize the rapid dissemination of information and interactive processing [5].

In recent years, a variety of network security incidents have frequently appeared in the public's vision [6,7]. The endless malicious network attacks have brought a lot of economic losses and psychological panic to the people. Network security has a great impact on the people, enterprises and countries. At this time, reasonable and effective network security supervision is of great significance. In order to better regulate network security, it is necessary to identify and analyze the unknown protocols in the network [9].

The three basic elements of the protocol are semantics, syntax and timing [10]. The inference of protocol message format and the determination of its field content belong to the content of protocol syntax analysis. The analysis and extraction of protocol syntax is the basis of protocol analysis and identification. It needs to analyze the control statement of protocol message, and extract the semantics of protocol based on data mining and sequence ratio method. The purpose of protocol syntax rule inference is to build a logical model of protocol syntax, focusing on the inherent logical relationship between protocol messages [11]. How the protocol interacts must follow certain syntax rules.

By analyzing the role of network protocol specification in the field of network supervision, we can obtain the network traffic information in the target network [12]. By classifying the traffic generated by these protocols, the network usage can be identified, the network expansion plan can be formulated, and the bandwidth of specific protocols can be controlled. Protocol analysis can help analyze network vulnerabilities, or provide useful information for firewalls and intrusion detection and defense systems, so as to discover and prevent previously unknown attacks.

With the rapid development of computer network and communication technology, the rise of various network services based on data transmission has enriched people's life. At the same time, the network security issues related to privacy and sensitive data have been paid more and more attention. Usually, security protocols run in a very complex development environment [13]. In order to reveal each vulnerability of communication protocol accurately and effectively through formal analysis and reasoning of security protocol model, it is necessary to cover every detail in modeling [14]. However, this is difficult to achieve, and when the security vulnerability is not considered in the model, it is difficult to improve the protocol.

Bitstream protocol format analyzer works at the bottom of network environment. By analyzing and processing the bitstream protocol data captured in the vehicle, the content of these data can be analyzed in real time, and then the protocol format can be analyzed [15]. Further ensure the safety of vehicles and the privacy of passengers. The current network protocol analysis method is to analyze a large number of protocol frames, and the data frame itself is relatively complex, and the algorithm will run for a long time [16]. How to optimize the algorithm is a research direction that needs to be studied continuously.

The rest of this paper is arranged as follows. The first part introduces the development of unknown protocols in network security [17,18]. The second part introduces the related work we have done in unknown protocol parsing. The third part is about the Basic knowledge reserve. In the fourth part, we propose a new protocol format analysis algorithm. In the fifth part, we analyze the performance of the new algorithm from several aspects and compare it with other algorithms. Finally, let's summarize our work.

2 Related Work

In the aspect of protocol feature extraction, pattern recognition and data mining have some applications. For example, Wang Xufang based on the existing BM algorithm and analyzed the common pattern matching algorithm [19]. Data mining, such as unsupervised, supervised and semi supervised learning methods to study. The algorithm used in this paper is optimized by the main idea of association rule mining algorithm in data mining.

Protocol informatics project published by Beddoe, referred to as PI project [20], is used for protocol reverse analysis, but its idea is also applicable to protocol identification. In bioinformatics, it is necessary to search for specific genes that produce proteins from DNA [21]. Similar to this, many network applications will identify certain protocols through some specific domains. Therefore, PI project introduces sequence alignment algorithm to match these similar fields. The purpose of sequence alignment algorithm is to find a way of permutation, combination or insertion to make the two sequences most similar after splitting and reorganizing the sequence [22]. That is, the two string sequences are arranged up and down, cut off or insert spaces in some positions, and then compare their string matching on each block in turn, so as to find out an arrangement method that makes the two sequences most similar. Similarly, for many protocols, there are some similar signature codes or feature fields [23]. Therefore, many researchers use this feature to match these feature fields in the way similar to pattern matching, so as to judge and identify a certain protocol. Good results have been achieved in some specific protocols [30]. This method belongs to the customized feature extraction method, but there may be some missing reports, because there may be a special case, that is, some packets using the protocol do not have the feature field, and if the protocol has update iteration, it is necessary to re study the protocol extraction features.

In addition to the above methods for extracting special fields, there are also methods for feature extraction using flow statistics. For example, Andrew

Moore [24] extracted 248 features for network flows, including packet level and flow-i Even, TCP protocol acknowledgement number and sequence number and some flag bits, but in order to calculate all these characteristics, a lot of computing resources are needed, and there will be some invalid features, which are not helpful for classification. Therefore, Nguyen [25] screened these features to a certain extent, and selected 20 distinguishing features, which greatly improved the computational efficiency.

For example, Chen Liang [26] and others have proposed a feature extraction method based on Chi square statistics, which optimizes feature extraction by introducing chi square statistics in statistical theory. By analyzing the difference between the target protocol traffic and the total traffic or other traffic, the validity of the feature is defined, and the optimal feature is selected, and then the eDonkey protocol is used to optimize the feature extraction. Experiments are carried out on the protocol, and good results are achieved in the recognition of the protocol, but this is a customized feature extraction method.

Chu Huilin [27] and others used the ReliefF algorithm to filter the irrelevant features in the traffic characteristics to obtain the optimal feature subset, and then combined with genetic algorithm and support vector machine to optimize the model parameters of support vector machine, so as to achieve better classification effect. In addition, Ma Yongli et al. [28] used correlation feature selection and genetic search methods to filter out some useless features from all traffic attribute features, so as to obtain a new feature subset. All features of the subset are relatively representative traffic attribute features, so better classification effect can be obtained. However, these two methods are not universal enough, and each protocol needs to be analyzed separately.

Yang Feihu [29] optimized ReliefF algorithm and combined ReliefF algorithm with mutual information method. Firstly, the ReliefF algorithm is used to remove some classification independent features, and then the mutual information method is used to remove redundant features. According to the classification results, the threshold is adjusted and iterated. Finally, through a number of comparative experiments, it is proved that the optimized algorithm achieves better results in classification accuracy and feature dimension reduction effect. However, because of the idea of wrapper, the time complexity of the algorithm becomes higher.

In 2016, Guan Lei et al. [31] introduced big data analysis technology into the field of network security situation awareness, and proposed a multi-functional security situation assessment platform. The platform provides many functions that need to be used in the situation assessment process, which has a good reference significance for the design and implementation of the subsequent situation assessment platform.

In 2016, Zhang Shuwen [32] and others proposed a hierarchical network security situation awareness model based on information fusion. The model uses D-S evidence theory for data fusion, and evaluates situation parameters through neural network technology. Finally, experiments show that the model can solve the problem of strong subjectivity in parameter setting in the field of situation awareness.

In 2018, zjfan et al. [33] focused on network security situation prediction, and proposed a prediction method based on spatio-temporal analysis. The method predicted the network security situation from two dimensions of time and space. Finally, experiments show that this method can more accurately predict the change trend of future network security compared with other state aware prediction algorithms.

3 Protocol Feature Extraction Algorithm

3.1 Algorithm Flow

Algorithm Related Definitions. In order to better illustrate the algorithm, some concepts are introduced here.

Definition 1 minimum support: A user-defined reasonable threshold used to measure the size of support, which represents the minimum statistical significance of the data, and can be expressed as Min_Sup .

Definition 2 frequent substring: Suppose there are M data frame messages with the length of $L1$ bit sequence. If there is a substring β with length of $L2(L1 \geqslant L2)$, if β appears in K data frames, the probability of β occurrence is $p(K/M)$. If the probability of a string occurrence is greater than or equal to Min_Sup, then it is called frequent substring [3].

$$Seq = \{\beta | P(\beta) \geqslant Min\_Sup\} \tag{1}$$

Define 3 minimum frequent substring length: A user-defined value. The length of a frequent substring must meet the minimum frequent substring length. Otherwise, it will be filtered. The value is expressed as Min_Len.

Definition 4 protocol features: If a certain frequent substring β appears frequently at a certain or multiple positions in the protocol data frame, it is considered that the frequent substring may be the protocol feature of the protocol, expressed as Fi [3].

Algorithm Data Initialization. The algorithm data is initialized in five steps:

(1) Input support threshold Min_Sup, traverse the data set, find out the data with the longest length in the data set, and record the length as Max_len;
(2) Use a one-dimensional vector, $localVector$, and use Max_len initializes it, and all elements in it are initialized to 0 by default;
(3) Go through all the data frames of the data set, and record whether the element of each position of each data is '0'. If it is '0', then add one to the location of the $localVector$. For example, if the data[i] in the i-th position is equal to '0', then the $localVector[i]$ plus one;
(4) Traverse the vector $localVector$ once and calculate the support of each position. If the support of a position $sup \geqslant Min\_Sup$ or $sup \leqslant 1 - Min\_Sup$ (assuming$Min_Sup > 0.5$), it means that the position may exist in a feature string, otherwise, the position cannot exist in a feature string.

After calculating the support of each location, two important definitions need to be introduced:

Define 5 bad characters: If a location is not supported within the range described in step (4), the location is considered a bad character and is represented as Ci.

Define 6 ideal strings: A substring that appears between two adjacent bad characters in the $localVector$ and is represented as Pi.

For Definition 6, it is clear that a data frame collection has only one bad character $C1$, then the substring from 0 to $C1$ of the $localVector$ (containing 0 characters but not $C1$) is considered an ideal string. Similarly, a substring from $C1$ to the end of the $localVector$ (containing the end character but not the $C1$ character) is also an ideal string. You can pass the minimum frequent substring length Min_Len partially filters the ideal string.

(5) At this point, through the location of the $localVector$ and bad characters, it is easy to get all the ideal strings, and record all the ideal strings in a set $prunSet$.

Data Reprocessing. After data preprocessing in the previous step, a $prunSet$ has been obtained that contains all possible locations where the signature string may appear. However, since the data in the $prunSet$ is calculated from the frequency at each location, such data must contain a large amount of redundant data. That is, the calculation of each location can get a range, but the range is too large, which is not friendly for subsequent specific operations to find the feature string. The reason for this is not difficult to find. Obviously, because the frequency statistics for each location ignore the continuous attributes of the string, the result is a wide range. It is not difficult to imagine that reprocessing using the continuous properties of a string is a good way. See the following steps for specific operation.

(1) Traverses through each data Str in the $prunSet$, creating a one-dimensional vector $localVector$ with the length of Str and setting it to 0.
(2) Traverse the dataset $dataSet$ to intercept strings of data of the same length and location as Str. Depending on the minimum frequent substring length Min_Len of Definition 3 make a slice and determine. If they are equal, the $localVector[i]$ plus one for the slicing position i. If not, do not operate.

The updated $prunSet$ can then be obtained by referring to the third, fourth, and fifth steps of the preprocessing operation. At this point, the data processing operation has been completed.

3.2 Algorithm Flow

The algorithm is described in Table 1. The flow chart of the algorithm is shown in Fig. 1.

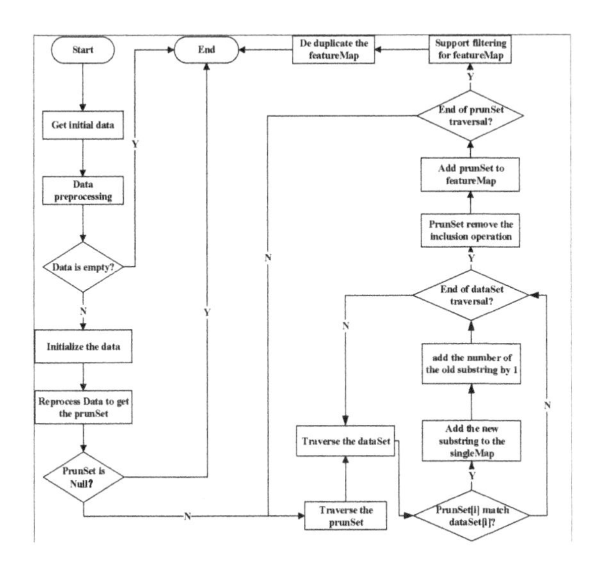

Fig. 1. Flow chart of OFS algorithm.

Table 1. Description of OFS algorithm.

Input	The ideal string set of data preprocessing is *prunSet*; the data frame set is *dataSet*; the minimum support *Min_Sup*;
Output	Frequent itemset *featureMap*
1	Judge whether *prunSet* is empty;
2	If it is empty, the algorithm ends; otherwise, the *prunSet* is reprocessed;
3	Judge whether the *prunSet* is empty. If it is, the algorithm ends. Otherwise, traverse the *prunSet*
4	Traverse each piece of data in the *prunSet* in the data frame set *dataSet*;
5	From each data frame in the *dataSet*, the strings with the same position and length as the data are intercepted and compared. If any substring matches successfully, it will be added to *singleMap*;
6	Whether the *dataSet* traversal is finished. If not, return to step 4. Otherwise, the data in *singleMap* will be removed and included;
7	Add the data in *singleMap* to *featureMap*;
8	Judge whether the traversal of *prunSet* is finished. If not, return to step 3. Otherwise, filter the support of *featureMap*;
9	De duplicate *featureMap*;
10	At the end, the *featureMap* is output as the final frequent itemset;

From Table 1, it can be seen that after data processing, OFS algorithm is simpler, but its efficiency is not low. This is because, after data preprocessing, the initial ideal string is obtained, and the minimum frequent substring length is used when data is processed, which is often accompanied by a continuous nature, so this operation can further narrow the ideal string from preprocessing, and it often excludes a large amount of useless information. This makes the subsequent operation very efficient. In contrast, data preprocessing only uses fixed locations, and the ideal string computed does not utilize the continuity property, so the ideal string obtained has a large range and contains a lot of useless information. However, both data preprocessing and data reprocessing are indispensable, and it is these two key operations that greatly improve the efficiency of the algorithm.

Get Itemset Procedure. As you can see from the process, $prunSet$ store all the ideal set of strings, so the process of getting frequent substrings is naturally also from the ideal set. Assuming that an ideal string is "010000100010010101#20" and the string intercepted from a data set in the data frame collection $dataSet$ is "01000010101#20", you can see that the two strings are only different at 29 positions. Then, from this data, you can get two substrings of "010000100#20" and "10010101#30". At this time, you put them into $singleMap$, and then intercept and compare all data frames of $dataSet$, the ideal set of substrings can be obtained. A $singleMap$ of all ideal strings can be obtained by acquiring all ideal strings.

Remove Inclusion Operation. After getting the itemset of the ideal string, you get a $singleMap$ belonging to the ideal string. The $singleMap$ needs to be stripped of inclusions at this point, considering three scenarios.

After Inclusion: In an ideal string, the following two substrings appear: "0001001011#302", "01001011#304". Obviously, the substring at position 304 is a true suffix of the substring at position 302, which is called After inclusion.

Pre Inclusion: In an ideal string, there are two substrings as follows: "1111 111110100#361", "11111111101#361". Obviously, the latter is a true prefix of the former, then this case is called pre inclusion.

Mutual Inclusion: In an ideal string, the following two substrings appear: "0100010010#156", "0001001010#158". Obviously, if a true prefix of the latter is a true suffix of the former, the case is called mutual inclusion.

After inclusion will lead to the number of substring statistics error, resulting in frequent substring missing. Because they are counted individually in the $singleMap$. In an extreme case, "0001001011" appears in the first 50% of the data frame set dataset, while "01001011" appears in the last 50% of the data frame set. If $Minsup$ is 0.7, then both substrings cannot be used as frequent substrings. However, the string "01001011#304" is obviously a characteristic string, because it actually appears in 100% of the data. Therefore, when dealing with such cases, it is necessary to add the number of times of the string "0001001011" in the *singleMap* to the string “01001011#304”, so that the statistics are complete. Similarly, for post inclusion, you need to add the number of times a longer substring has in the *singleMap* to another substring. For mutual inclusion, we need to add location information to the mutual inclusion part of two strings to form a new substring, and add the times of both in *singleMap* to the new substring. Before processing these three situations, the *singleMap* is copied to a tmp *SingleMap*. Whether the number of times is increased or the new string is added, the *singleMap* needs to be updated after processing.

Get Frequent Substring. After all the ideal strings in the *prunSet* get item set and remove inclusion operations, each substring and corresponding times in the *singleMap* of each ideal string will be added to the *featureMap*.

After that, the support of each substring in the *featureMap* is calculated, and all the substrings with support less than *Min_sup* are deleted.

At this time, it is possible that the substrings in the *featureMap* are duplicated due to the inclusion conditions mentioned above. For example, in an extreme case, if the supports of strings “1111111110100” and “11111111101” are greater than the minimum support, then both of them will not be removed. But obviously, the same position of the string, only need to leave a longer. So at this time, delete the short string. It can be seen that when it is not necessary to remove the inclusion as mentioned above, the judgment is divided into several situations. The processing method is relatively simple. You only need to judge whether there is a substring included in another substring. If so, you just need to delete the shorter string, such as “0001001011#302” , and “01001011#304” , and just delete “01001011#304”.

At this point, the final frequent itemset of the data frame set dataset has been obtained.

However, the generation of association rules still follows the analysis method of association rules of Apriori algorithm.

3.3 Algorithm Evaluation

Evaluating an algorithm needs to be judged from many perspectives. The most common means are to calculate the time and space complexity of the algorithm.

Time Complexity. Suppose the data frame collection *dataSet* has n data frames and the average length of the data frame is m. Then first iterate through the *dataSet* to initialize the vector *localVector* with $O(mn)$ time complexity. The ideal set of strings, *prunSet*, is obtained through the *localVector* with $O(m)$ time complexity. Then all the ideal strings in the *prunSet* add up to no more than m, and each ideal string in the *prunSet* is compared with the dataset to get a substring, which has an $O(mn)$ time complexity. Overall, the final time complexity of the algorithm is $O(mn)$. This also demonstrates the superiority of the new algorithm.

Spatial Complexity. All operations are $localVectors$ based on the initial data preprocessing, so all subsequent operations will no longer exceed the $localVector$, so the spatial complexity of the algorithm is $O(m)$.

4 Experimental Results and Analysis

The details of seven sets of data frame sets are introduced. Seven sets of data frames are different protocol files. Among them, DNS protocol file size is 9384 kb, HTTP protocol file size is 43642 kb, HTTP2 protocol file size is 46337 kb, ICMP Protocol file size is 750 kb, ICMP2 protocol file size is 6071 kb, OICQ protocol file size is 20119 kb, SSDP protocol file size is 6839 kb. The detailed data set is shown in Table 2. The specific size and running time of the two algorithms are detailed in Table 3 (the data in the table are arranged in ascending order of file size). The seven groups of data are protocol data intercepted from Wireshark. The running time of the two algorithms comes from the execution time of the console program.

Table 2. Protocol data set.

Protocol type	Total number of data frames (Pieces)	Total data frame size (KB)
ARP-like protocol	30000	2639
DNS-like protocol	113000	9384
HTTP-like protocol	230000	43642
HTTP2-like protocol	238000	46337
ICMP-like protocol	10000	750
ICMP2-like protocol	31000	6071
OICQ-like protocol	267000	20119
SSDP-like protocol	108000	6839
TCP-like protocol	30000	2595
UDP-like protocol	30000	2565
Train Set	240000	20834

Table 3. Running time comparison of CFI algorithm and OFS algorithm

File size (KB)	750	6071	6839	9384	20119	43642	46337
CFI time (s)	9.2	160.2	2198.4	77.0	391.7	508.9	583.5
OFS time(s)	0.4	1.3	0.5	1.7	3.8	16.6	18.2

In Table 3, due to space constraints, only one decimal place is reserved according to the rounding principle. We can see the superiority of OFS algorithm. In the SSDP protocol file, the CFI algorithm duration is 2198.4 S. Considering the overall situation, the running time data of SSDP is regarded as a bad data. Then we can draw a broken line diagram of the two algorithms, and we can see the difference between them more clearly, as shown in Fig. 2.

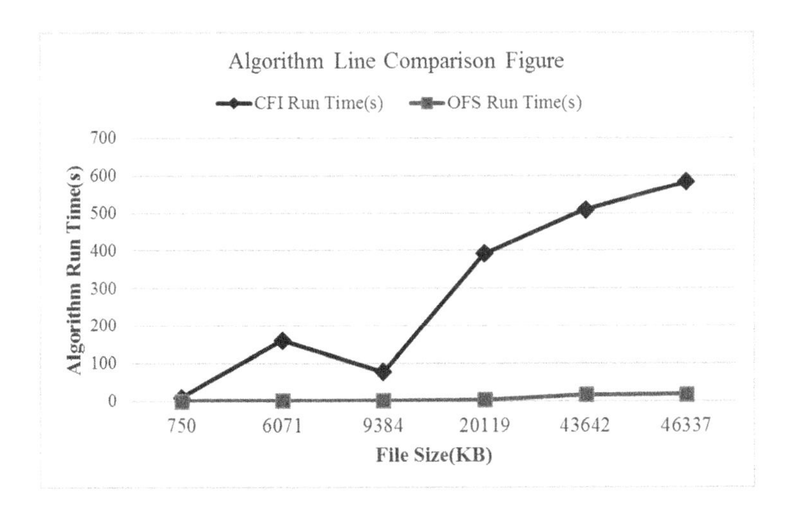

Fig. 2. Broken line comparison between CFI algorithm and OFS algorithm.

In the broken line comparison diagram in Figs. 4 and 5, it can be clearly seen that the running time of OFS algorithm is only 34.9 s even if the http2 protocol frame set with the largest dataset has 46337 kb data, which fully illustrates the advantages of OFS algorithm.

After feature extraction of the input data, the extracted feature string will be saved. The result of the feature string is shown in Fig. 3.

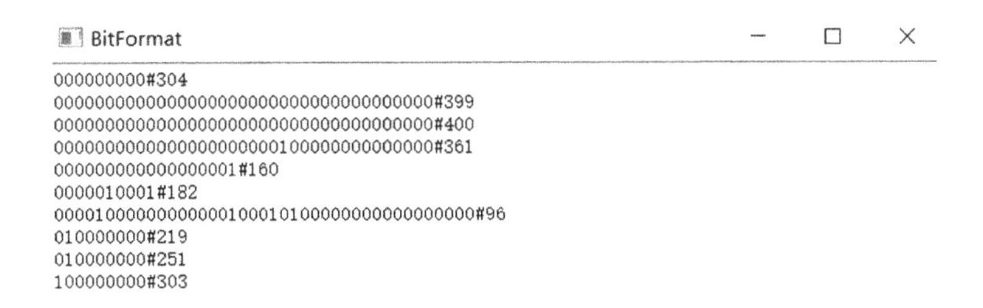

Fig. 3. Display of feature extraction results.

From the Fig. 3, we can clearly see the result of protocol features. The left side of the protocol features is the feature string, and the right side is the feature string position. The picture shows that the method can achieve good results.

The OFS algorithm is embedded into the unknown protocol syntax reverse analysis system, and different similarity is output according to different support degrees. The results are shown in Fig. 4. When six protocols are selected, ICMP is the result when the support degree is 0.60, ICMP2 is the result when the support degree is 0.65, ICMP3 is the result when the support degree is 0.70, and ICMP4 is the result when the support degree is 0.75. It can be seen from the figure that with the increase of support, the similarity increases gradually.

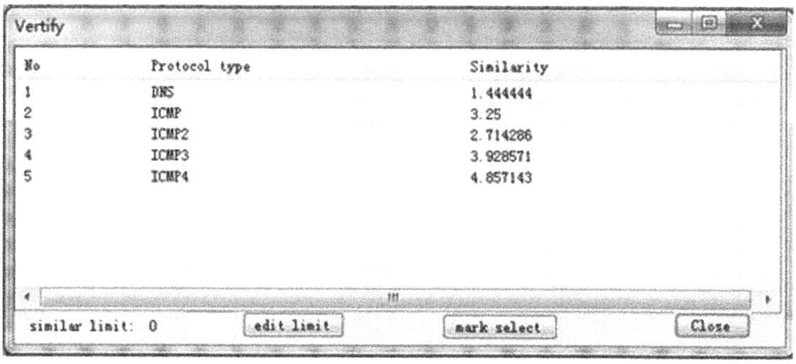

Vertify

No	Protocol type	Similarity
1	DNS	1.444444
2	ICMP	3.25
3	ICMP2	2.714286
4	ICMP3	3.928571
5	ICMP4	4.857143

similar limit: 0 edit limit mark select Close

Fig. 4. Comparison of the results of different supports.

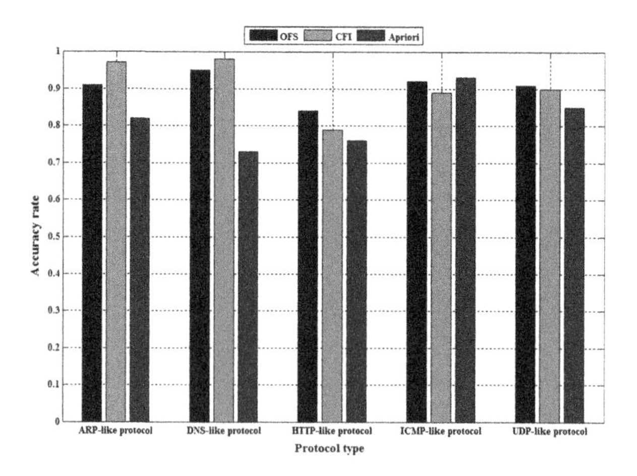

Fig. 5. Protocol accuracy comparison figure.

The Fig. 5 above shows the comparison of protocol accuracy. From the figure, we can see that although the running time of ofs is greatly reduced, its accuracy rate is basically not affected, which is equivalent to that of CFI algorithm. Therefore, it has certain advantages to apply the algorithm to the identification of unknown protocols.

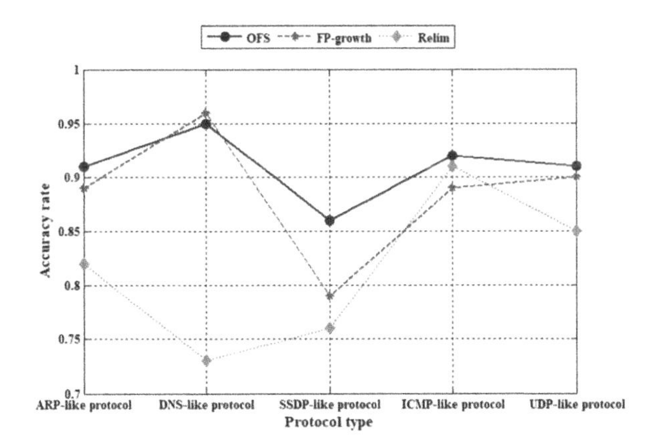

Fig. 6. Protocol accuracy comparison.

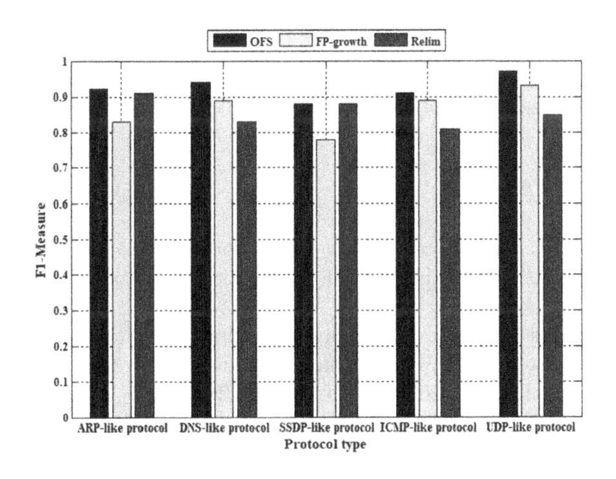

Fig. 7. Protocol F1-measure comparison.

Figures 6 and 7 are the experimental charts of accuracy and F1 value comparison of OFS algorithm with FP-growth algorithm and Relim algorithm. From the accuracy comparison chart of Fig. 6, it can be seen that the accuracy of OFS algorithm is more stable than other algorithms. From the F1 value comparison chart of Fig. 7, it can be seen that the performance evaluation of OFS algorithm is slightly higher than FP-growth algorithm and relim algorithm, about 5% higher than FP-growth algorithm.

5 Conclusion

In this paper, the bitstream data frame set is taken as the research object. The main work is to extract the feature string from the data frame set, and put the feature string into the database. If there is a new data frame, the data frame will be compared with the features in the library, and finally the protocol of the data frame can be identified. In this paper, the Apriori algorithm is improved, and the CFI algorithm is used to find the feature string. Combining the advantages of the previous algorithms, a more efficient OFS algorithm is proposed, which has better performance in the face of feature extraction of bitstream protocol. In this paper, based on OFS algorithm, the frequent itemsets of the data frame set are extracted more efficiently, and the strong association rules of the items in the frequent substring itemsets are analyzed by the Apriori algorithm, and the frequent substrings with low recognition degree are removed, and the more representative frequent subsequence sets are stored in the agreement feature database. Experimental results show that OFS has a good effect on protocol reverse analysis, and greatly speeds up the efficiency of the algorithm based on the original CFI algorithm.

Acknowledgment. This research work is supposed by the National Joint Funds of China (U20B2050), National Key R&D Program of China(2018YFB1201500), National Natural Science Founds of China (62072368, 61773313, 61702411), National Natural Science Founds of Shaanxi (2017JQ6020, 2016JQ6041), Key Research and Development Program of Shaanxi Province (2020GY-039, 2017ZDXMGY-098, 2019TD-014).

References

1. Zhou, Z., Zhao, L.: Cloud computing model for big data processing and performance optimization of multimedia communication. Comput. Commun. **160**, 326–332 (2020)
2. Jiang, Y.: Wireless resource management mechanism with green communication for multimedia streaming. Multimedia Tools and Applications **78**(7), 8699–8710 (2018). https://doi.org/10.1007/s11042-018-6149-4
3. Hei, X., Bai, B., Wang, Y., et al.: Feature extraction optimization for bitstream communication protocol format reverse analysis. In: 2019 18th IEEE International Conference on Trust, Security and Privacy Iin Computing and Communications/13th IEEE International Conference On Big Data Science And Engineering (TrustCom/BigDataSE), pp. 662–669 IEEE (2019)
4. Dai Y.F.: Research and application of apriori algorithm in analysis of students' achievement. In: Information Engineering Research Institute, USA. Proceedings of 2012 Third International Conference on Theoretical and Mathematical Foundations of Computer Science (ICTMF 2012), vol. 38, pp. 894–899. Information Engineering Research Institute (2012)

5. Yang, H., Li, P., Zhu, Q., et al.: The application layer protocol identification method based on semisupervised learning. IEEE **6**, 115–120 (2011)
6. Yuan, Z., Xue, Y., Dong, Y.: Havesting unique characteristics in packet sequences for effective application classification. In: Proceedings of the 1st IEEE Conference on Communications and Network Security, pp. 341–349. IEEE Communication Society, Los Alamitos (2013)
7. Mohsin, A.H., Bakar, K.A., Zainal, A.: Optimal control overhead based multi-metric routing for MANET. Wirel. Netw. (2), 1–17 (2017)
8. Naseem, M., Kumar, C.: Queue-based multiple path load balancing routing protocol for MANETs. Int. J. Commun. Syst. **30**(6), e3141 (2017)
9. Pradittasnee, L., Camtepe, S., Tian, Y.C.: Efficient route update and maintenance for reliable routing in large-scale sensor networks. IEEE Trans. Ind. Inf. **13**, 144–156 (2016)
10. Zheng, J.: Research on a cluster system for binary data frames of wireless sensor network. Cluster Comput. **19**(2), 783–791 (2016)
11. Tao, L.: Packet signature mining for application identification using an improved apriori algorithm. Nanjing University of Science and Technology, Shanghai University of Finance and Economics. Proceedings of 2015 IEEE International Conference on Progress in Informatics and Computing (PIC 2015 V1). Nanjing University of Science and Technology, Shanghai University of Finance and Economics, IEEE Beijing Section, pp. 664–668 (2015)
12. Lin, Y., Nie, Z., Ma, H.: Structural damage detection with automatic feature-extraction through deep learning. Comput. Aided Civ. Infrastruct. Eng. **32**(12), 1025–1046 (2017)
13. Song, Z., Wu, B.: Anomaly detection based on feature extraction of unknown protocol payload format. In: 2020 IEEE 5th Information Technology and Mechatronics Engineering Conference (ITOEC), pp. 704–709. IEEE (2020)
14. Sengupta, S., Chowdhary, A., Sabur, A., et al.: A survey of moving target defenses for network security. IEEE Commun. Surv. Tutorials **22**, 1909–1941 (2020)
15. Jeong, C., Ahn, M., Lee, H., Jung, Y.: Automatic classification of transformed protocols using deep learning. In: Park, J.H., Shen, H., Sung, Y., Tian, H. (eds.) PDCAT 2018. CCIS, vol. 931, pp. 153–158. Springer, Singapore (2019). https://doi.org/10.1007/978-981-13-5907-1_16
16. Tang, H., Xiao, B., Li, W., Wang, G.: Pixel convolutional neural network for multi-focus image fusion. Inf. Sci. **433**, 125–141 (2018)
17. Xiao, B., Wang, K., Bi, X., Li, W., Han, J.: 2D-LBP: an enhanced local binary feature for texture image classification. IEEE Trans. Circuits Syst. Video Technol. **29**(9), 2796–2808 (2019)
18. Shen, Z., Lee, P.P., Shu, J., et al.: Encoding-aware data placement for efficient degraded reads in XOR-coded storage systems: algorithms and evaluation. IEEE Trans. Parallel Distrib. Syst. **29**(12), 2757–2770 (2018)
19. Cheng, Y., Wang, F., Jiang, H., et al.: A communication-reduced and computation-balanced framework for fast graph computation. Front. Comput. Sci. Chin. **12**(5), 887–907 (2018)
20. Lin, B., Guo, W., Xiong, N., et al.: A pretreatment workflow scheduling approach for big data applications in multicloud environments. IEEE Trans. Netw. Serv. Manage. **13**(3), 581–594 (2016)
21. Wang, J., Zhang, X., Lin, Y., et al.: Event-triggered dissipative control for networked stochastic systems under non-uniform sampling. Inf. Sci. **447**, 216–228 (2018)

22. Zou, J., Dong, L., Wu, W.: New algorithms for the unbalanced generalized birthday problem. IET Inf. Sec. **12**, 527–533 (2018). https://doi.org/10.1049/iet-ifs.2017.0495
23. Rawat, D.B., Garuba, M., Chen, L., et al.: On the security of information dissemination in the internet-of-vehicles. Tsinghua Sci. Technol. **22**(4), 437–445 (2017)
24. Radhakrishna, V., Aljawarneh, S., Kumar, P.V., et al.: A novel fuzzy gaussian-based dissimilarity measure for discovering similarity temporal association patterns. Soft Comput. **22**(6), 1903–1919 (2018)
25. Wang, C., Zheng, X.: Application of improved time series apriori algorithm by frequent itemsets in association rule data mining based on temporal constraint. Evol. Intell. **13**(1), 39–49 (2020)
26. Cui, W., Kannan, J., Wang, H.J.: Discoverer: automatic protocol reverse engineering from network traces. In: Proceedings of the 16th USENIX Security Symposium, pp. 199–212. IEEE (2007)
27. Trifilo, A., Burschka, S., Biersack, E:. Traffic to protocol reverse engineering. In: Proceedings of the Second IEEE International Conference on Computational Intelligence for Security and Defense Applications, pp. 1–8. IEEE (2009)
28. Sabokrou, M., Fayyaz, M., Fathy, M., et al.: Deep-anomaly: fully convolutional neural network for fast anomaly detection in crowded scenes. Comput. Vis. Image Understand. **172**, 88–97 (2018)
29. Giuseppe, S., Massimiliano, G., Antonio, M., et al.: A CNN-based fusion method for feature extraction from sentinel data. Remote Sens. **10**(2), 236 (2018)
30. Kumar, B., Anand, D.K., Anjanappa, M., et al.: Feature extraction and validation within a flexible manufacturing protocol. Knowl. Based Syst. **6**(3), 130–140 (1993)
31. Yi, L., Hua, P., Zhen-Hua, Z.: Protocol recognition feature extraction algorithm of high frequency communication signals based on wavelet de-noising. J. Inf. Eng. Univ. **13**, 438–442 (2012)
32. Ding-Ding, F., Rong-Feng, Z., An-Min, Z.: Protocol feature extraction and anomaly detection based on flow characteristics of industrial control system. Mod. Comput. (2019)
33. Lin., C.C., Wang, C.N., et al.: Combined image enhancement, feature extraction, and classification protocol to improve detection and diagnosis of rotator-cuff tears on MR imaging. Magn. Reson. Med. Sci. Mrms Official J. J. Soc. Magn. Reson. Med. **13**, 155–166 (2014)
34. Xiong, J., Bi, R., Zhao, M., Guo, J., Yang, Q.: Edge-assisted privacy-preserving raw data sharing framework for connected autonomous vehicles. IEEE Wirel. Commun. **27**(3), 24–30 (2020)
35. Tian, Y., Wang, Z., Xiong, J., Ma, J.: A blockchain-based secure key management scheme with trustworthiness in DWSNs. IEEE Trans. Ind. Inf. **16**(9), 6193–6202 (2020)
36. Xiong, J., et al.: A personalized privacy protection framework for mobile crowdsensing in IoT. IEEE Trans. Ind. Inf. **16**(6), 4231–4241 (2020)
37. Xiong, J., Zhao, M., Bhuiyan, M.Z.A., Chen, L., Tian, Y.: An AI-enabled three-party game framework for guaranteed data privacy in mobile edge crowdsensing of IoT. IEEE Trans. Ind. Inf. **17**(2), 922–933 (2021)
38. Xiong, J., et al.: Enhancing privacy and availability for data clustering in intelligent electrical service of IoT. IEEE Internet Things J. **6**(2), 1530–1540 (2019)

Key Technologies of Space-Air-Ground Integrated Network: A Comprehensive Review

Chuanfeng Wei[1,2,3,4(✉)], Yuan Zhang[2,3,4], Ruyan Wang[2,3,4], Dapeng Wu[2,3,4], and Zhidu Li[2,3,4]

[1] Macro Net Communication Co., Ltd., CASC, Beijing , China
[2] School of Communication and Information Engineering, Chongqing University of Posts and Telecommunications, Chongqing, China
[3] Key Laboratory of Optical Communication and Networks, Chongqing, China
[4] Key Laboratory of Ubiquitous Sensing and Networking, Chongqing, China

Abstract. Facing the urgent needs of wide area intelligent network and global random access, the independent development of terrestrial cellular communication system and satellite communication system will face great challenges in the future. Space-air-ground integrated network is considered to be potential in integrating the space-based network and terrestrial network to realize unified and efficient resource scheduling and network management. In this paper, the architecture, functional requirements, challenges and key technologies of the space-air-ground integrated network are reviewed. It is expected that the paper is able to provide insightful guidelines on the research of the space-air-ground integrated network.

Keywords: Space-air-ground integrated network · Software defined network · Mobile edge computing · Network slicing · Deep reinforcement learning

1 Introduction

With the development of science and technology in conjunction with the continuous expansion of human production and activity space, the emerging network technologies represented by Internet of Things (IoT) will gradually become the main body of the future network demands [1–3]. Compared with the communication needs of ordinary people, IoT communication would greatly expand in terms of space range and communication content so that a variety of IoT devices and services would cover a wider area such as mountains, deserts, oceans, deep underground, air, and space [4]. Owing to the advances in wireless network in

This work was supported in part by the National Natural Science Foundation of China under grants 61871062 and 61771082, and in part by Natural Science Foundation of Chongqing under grant cstc2020jcyj-zdxmX0024, and in part by University Innovation Research Group of Chongqing under grant CXQT20017.

J. Xiong et al. (Eds.): MobiMedia 2021, LNICST 394, pp. 63–80, 2021.
https://doi.org/10.1007/978-3-030-89814-4_5

recent years, 5G network technology has provided more flexible services, larger capacity, and higher efficiency for new network applications, such as virtual reality, autonomous driving, and smart cities, and has entered the deployment and actual commercial use phases [5]. As for IoT applications, 5G network has specifically devised two significant service scenarios, namely ultra-reliable and low latency communication (uRLLC) and massive machine type communication (mMTC) [6–8].

The technologies promoted by 5G network technology, like narrow band IoT (NB-IoT), beamforming, uplink/downlink decoupling, are able to solve key technical issues in IoT, such as the wide area coverage, energy consumption, and massive connections [9]. However, large-scale 5G network deployment would cause huge costs which consist of the infrastructure cost brought by intensive base station deployment and backhaul network construction, as well as the installation, rent and maintenance cost for optical fiber cable. [10] Meanwhile, it is hard for ground-based networks to cover extremely remote areas, oceans, deep underground, air, and even deep space. Therefore, 5G ground-based network technology is difficult to meet the ubiquitous communication need due to the greatly expanded network space [11]. Furthermore, the demand for multi-dimensional comprehensive information resources in future information services will gradually increase since the efficient operation of services in the fields of national strategic security, disaster prevention and mitigation, aerospace and navigation, education and medical care, environmental monitoring, and traffic management all rely on the comprehensive application of multi-dimensional information such as air, space, and ground [12]. In this context, constructing space-air-ground integrated network, deeply integrated space-based network, air-based network and ground-based network, so as to give full play to the functions from different network dimensions is able to break the barriers of data sharing between independent network systems and realize wide-area full coverage and network interconnection, which will further trigger an unprecedented information revolution.

Space-air-ground integrated network is a large-scale heterogeneous network that can collect and process network resources, which has the characteristics of wide coverage, multiple coexisted protocols, highly dynamic nodes, variable topology and high throughput for providing seamless wireless access services to the world. Space-air-ground integrated network needs to be intelligent to enhance network performance and management efficiency urgently due to its own heterogeneity and other characteristics. To adapt to 5G ecosystem, integrated systems, especially satellite networks, have to provide wireless communication services in a more flexible, agile and cost-effective manner [13].

This paper reviews the basic architecture, network requirements and the faced challenges of space-air-ground integrated network. Additionally, the application and advantages of mobile edge computing (MEC), software defined network (SDN), network slicing and deep reinforcement learning (DRL) in space-air-ground integrated network are discussed in detail. It is expected that the paper is able to provide insightful guidelines on the research of the space-air-ground integrated network.

The reminder of this paper is organized as follows. Section 2 introduces the architecture of the space-air-ground integrated network. In Sect. 3, existing issues and challenges of the space-air-ground integrated network are discussed. In Sect. 4, key technologies related to the space-air-ground integrated network are introduced and discussed in detail. Section 5 finally concludes the paper.

2 Space-Air-Ground Integrated Network

2.1 Network Architecture

Space-air-ground integrated network is based on ground-based network, supplemented and extended by space-based network and air-based network, providing various network applications in a wide area with ubiquitous, intelligent, collaborative and efficient information assurance infrastructure, whose architecture is shown in Fig. 1. Specifically, ground-based network, mainly consisting of ground internet and mobile communication network, is responsible for network services in service-intensive areas. Air-based network is composed of high-altitude communication platforms and unmanned aerial vehicle (UAV) ad hoc networks, which has functions of coverage enhancement, enabling edge service and flexible network reconfiguration. Space-based network is composed of various satellite systems to form space-based backbone networks and space-based access networks, realizing functions such as global coverage, ubiquitous connection and broadband access. Space-air-ground integrated network is able to utilize various comprehensive resources effectively through the in-depth integration of multi-dimensional networks to perform intelligent network control and information processing, so as to cope with network services with different requirements, achieving the objective of network integration, functional service and application customization, where space-based network is the core enabling technology to build the ubiquitous space-air-ground integrated network.

Space-based network contains satellites, constellations and corresponding ground infrastructure, like terrestrial stations and network operation and control centers, in which satellites and constellations have different characteristics and work in different orbits. According to the height above the ground, these satellites can be divided into three categories: geostationary earth orbit (GEO) satellites, medium earth orbit (MEO) satellites and low earth orbit (LEO) satellites. On the other hand, they can also be divided into narrowband and broadband satellite networks according to the channel bandwidth.

- Narrowband satellite network mainly refers to MEO/LEO satellite systems, such as the Iridium and Globalstar satellite systems, which mainly provide voice and low-rate data services for global users.
- Broadband satellite network can transmit large amounts of data due to its wide frequency band, where broadband is a general term in fixed or wireless communications. Broadband satellite system usually uses the Ka frequency

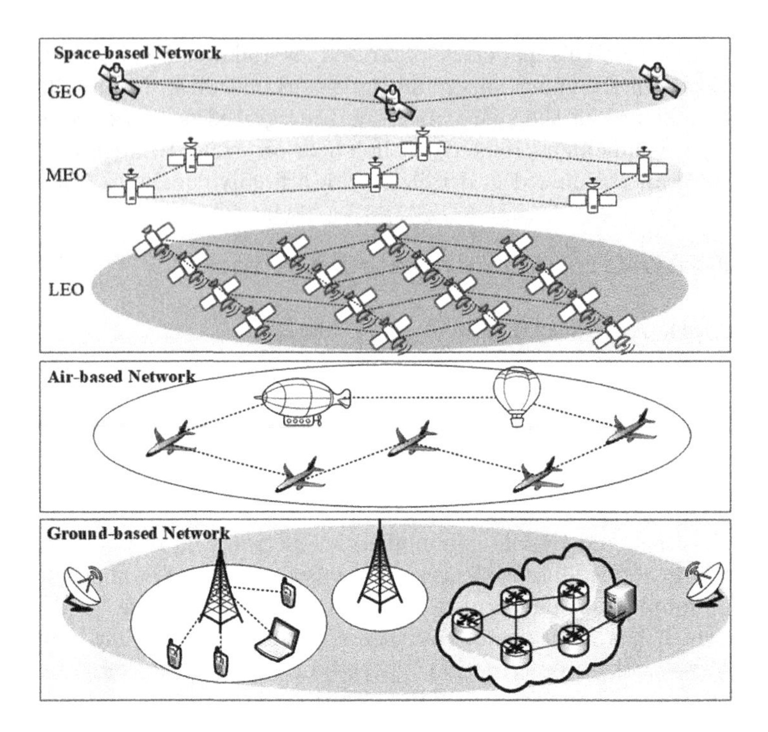

Fig. 1. Space-air-ground integrated network architecture.

band with a data transmission rate of up to 10 Gbps [14] to provide services for high-speed internet access and multimedia information. In addition, the data transmission rate of broadband satellite system is expected to reach 1000 Gbps by 2020 [15].

- Multi-layer satellite network is mainly formed by satellite networks in different orbits [16], which is a kind of practical next-generation satellite network architectures. GEO, MEO and LEO satellite systems could build a typical, tridimensional and multi-layer satellite network through inter-satellite link and inter-layer link (ILL).

Air-based network uses space aircraft as carriers for information acquisition, transmission and processing. UAVs, airships and balloons are the main infrastructures that constitute high-altitude and low-altitude platforms, providing broadband wireless communication complementary to the ground-based network [17]. Compared with the base stations in the ground-based network, the space-based network has the characteristics of low cost, easy deployment and large coverage area to provide high-speed wireless access services in the area covered by it.

Ground-Based Network. Ground-based network is mainly composed of ground mobile cellular networks, mobile ad hoc networks [18], worldwide interpretability for microwave access (WiMAX) network [19], and wireless local area network (WLAN). Ground-based network can provide users with high-speed data transmission, however, whose coverage is limited, especially in rural and remote areas.

Comparison of Various Networks. The space-based network can provide seamless coverage to the global, but owns higher propagation delay. On the other hand, ground-based network infrastructure is vulnerable to natural disasters or man-made damage although it has lower transmission delay. Space-based network has advantages in terms of transmission delay and network coverage, but its capacity and wireless links are limited and unstable, respectively, which should be taken into consideration when deploying such a network [20].

2.2 Network Requirements

Compared with traditional satellite networks, space-air-ground integrated network should own more comprehensive capabilities, not only traditional communication capability, but also computing, artificial intelligence (AI) and security capabilities.

(1) Requirements on communication capability
Communication capability is the basic requirement for the future development of 6G space-air-ground integrated network. Owing to this, space-air-ground integrated network allows terminal travelling speed to exceed 1000 km/h, transmission rate to be greater than 1 Gbps, transmission delay to be less than 10 ms, and spectrum efficiency to be more than 4 times higher than traditional satellite communication systems, respectively.

(2) Requirements on computing capability
In the future, network nodes such as constellation satellites and high-altitude platforms would be an important part of the 6G computing network. It is necessary to provide a high-performance computing platform under resource-constrained conditions for supporting a computing rate of up to tens of trillions per second and the ability to coordinate and migrate computing resource dynamically with the goal of creating a high-performance, low-latency, and large-bandwidth service environment for wide-area business and network management and adapting to the characteristics of rapid changes in network topology and limited satellite load resources.

(3) Requirements on AI capability
AI capability is the core capability of space-air-ground integrated communication system, including five capabilities: perception, learning, reasoning, prediction, and decision-making. Perception capability is able to perceive transmission requirements and security risks through ubiquitous information collection followed by the identification of new services and new threats using self-learning

and self-evolution capabilities. Reasoning capability can be adopted to reason and predict possible future service changes and network events through the big data technology. Eventually, use various technologies comprehensively to make decisions to provide users with personalized services and jointly optimize and schedule system resources.

(4) Requirements on security capability
The security capability of space-air-ground integrated network incorporates the security of data transmission and network behavior with the objective to solve the security and privacy issues caused by the wide participation of various subjects in heterogeneous complex networks. Space-air-ground integrated network needs to form a growing endogenous security mechanism to deal with dynamically changing network security threats.

3 Exiting Issues and Challenges

Space-air-ground integrated network has the characteristics of heterogeneity, highly dynamic of space nodes, highly dynamic of network topology, extremely large space-time spans, limited resources of space nodes and vulnerable attribute of satellite broadcast transmission links, which puts forward higher requirements for the design of network architecture, satellite-to-ground integrated communication standards, and inter-satellite networking protocols.

(1) Highly dynamic of space nodes
The nodes of the traditional ground-based network are relatively fixed, while high-speed relative movements between various satellites and the ground-based network exist in space-air-ground integrated network. In case of considering the access and networking of nodes such as high-altitude platforms, airplanes and low-altitude UAVs, their moving characteristics are usually less regular and the impact will be greater. One of the main effects brought by the highly dynamic of nodes is the serious Doppler frequency offset. At the same time, the communication links are prone to high interruption rate and high symbol error rate. These transmission characteristics put forward higher requirements for the design of space-air-ground integrated communication standards.

(2) Highly dynamic of network topology
In space-air-ground integrated communication system, the network consists of nodes at different levels such as satellites, high-altitude platforms, mid-to-low-altitude suspension/aircraft and ground equipment, and has a tridimensional architecture that is completely different from traditional ground-based cellular communication networks. The movements of nodes in the network will also cause the network topology to exhibit highly dynamic characteristics.

On the one hand, highly dynamic of network topology will cause link changes, making it difficult to transmit data through fixed traffic. On the other hand, the network protocols face the contradiction between asymmetric links, link quality changes and highly reliable transmission feedback control under different changes such as multi-hop and relay, which will cause low transmission efficiency at the application layer and even fail to guarantee data transmission quality.

(3) Extremely large space-time spans
Due to the long distance and fast speed of altitude platform, the links between network nodes far exceed the ground-based network in space-time spans. On the one hand, the transmission signal between nodes has a large attenuation loss and is easily affected by a series of factors such as orbital changes, elevation angles, sun glints, atmospheric scattering, rain attenuation, and blocking, resulting in weak received signals and various interferences, which would be more serious in the high frequency band. On the other hand, the influence of long-distance transmission and space environment will also cause high delay and large jitter in the communication process, resulting the contradiction between timely adjustment and long delay in the feedback adaptation mechanism, which makes it difficult for conventional error control methods to work.

(4) Heterogeneous network interconnection
Space-air-ground integrated communication system is composed of a variety of heterogeneous networks, and the environment and characteristics of each network vary greatly. For the satellite space-based network, the satellite signal and transmission characteristics of different orbital positions are quite different; for the ground-based network, the different communication environment conditions such as air, space and ground are also very different. In order to achieve the integration, compatibility and amalgamation of air, space, and ground-based networks, it is necessary to have uniformity and similarity in the design of waveform systems and communication standards. The aforementioned heterogeneous network interconnection characteristics will inevitably make this goal more difficult.

(5) Limited load resources
The communication satellites are systems with limited power resources. On the one hand, due to the rocket launch capability, the weight and size of communication satellites are limited, which restricts the size of solar panels and the available power resources. On the other hand, the power consumption of the payload is greatly limited by the satellites thermal radiation capability. The larger the area of thermal radiator is, the stronger the heat dissipation capacity is, and the higher power consumption can be supported. However, due to the limitation of the outline size of the satellite fairing and the influence of the satellite antenna, the size of the thermal radiator cannot be increased without limit. As the demand for satellite communications develops toward higher peak rates and more connections, the contradiction between on-board power resource constraints and increased transmission power and on-board processing capabilities will further intensify.

(6) Wide area transmission security
Space-air-ground integrated network uses satellites, high-altitude platforms and other means to achieve wide area coverage of users. However, the wireless channel of satellite communication has the characteristics of openness and broadcasting, which makes the information transmission channel uncontrollable and the wireless link more susceptible to threats such as man-made interference, attacks, eavesdropping and replay. Therefore, the development of space-air-ground integrated

networks needs to solve the transmission security challenges under the conditions of wide area coverage.

4 Key Technology

4.1 Mobile Edge Computing

As one of the important technologies of mobile networks, MEC is playing an increasingly important role in the process of current network development. MEC can be regarded as a cloud service platform running at the edge of the network, which can support the deployment of service processing and resource scheduling functions, improve service performance and users experience and reduce the transmitted data of backhaul links and bandwidth pressure of the core network to a certain extent [21].

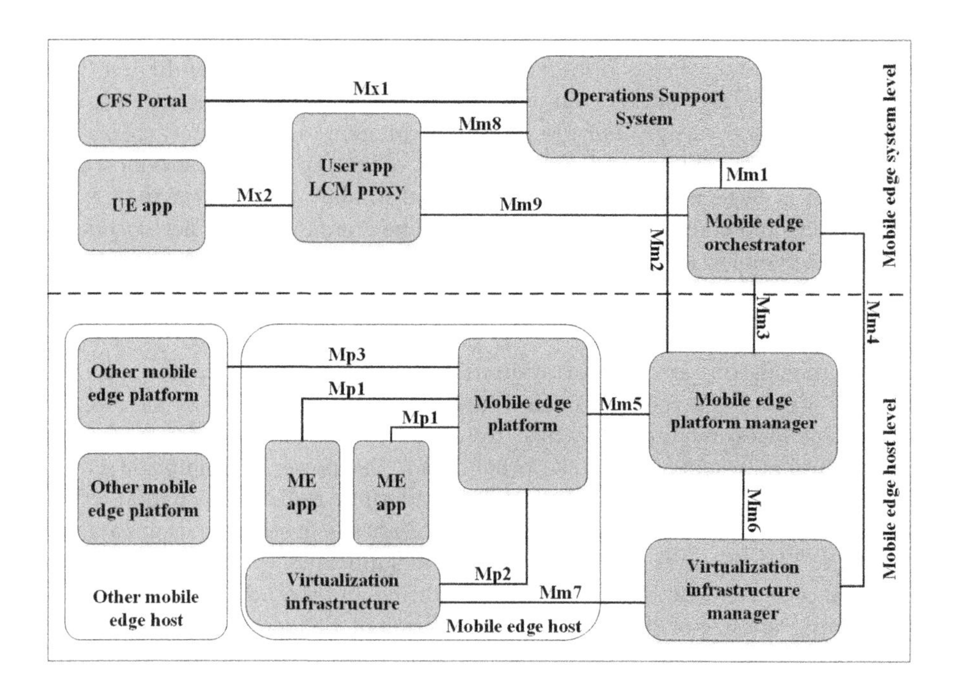

Fig. 2. MEC reference architecture.

An MEC reference system architecture is depicted in Fig. 2, which was proposed by the European Telecommunication Standards Institute (ETSI) [22]. It can be seen from Fig. 2 that MEC can serve users directly through user terminal applications or customer facing service (CFS) portal. Both the user terminal and the CFS portal interact with the MEC system through MEC system-level management.

Mobile edge users can instantiate, terminate or relocate related applications and services through the user application lifecycle management agent managed by the MEC system. Then, the operation support system (OSS) decides whether to approve the related request, and the approved request will be forwarded to the mobile edge orchestrator which is the core function of MEC system-level management and responsible for maintaining an overall view of available computing/caching/network resources and MEC services. The mobile edge orchestrator allocates virtualized MEC resources to the application to be launched based on application requirements, like latency. In addition, the orchestrator can flexibly extend the available resources down/up to the already running applications.

The MEC system-level management is interconnected with the MEC server-level management that constitutes the mobile edge platform and the virtualization platform manager. The former is responsible for the management of application life cycle, application rules and service authorization, traffic rules, etc.; the latter is responsible for allocating, managing and releasing the virtualized computing and caching resources provided by the virtualized infrastructure in the MEC server. The MEC server, an important part of the reference architecture, represents virtualized resources and carries MEC applications that run as virtual machines on the virtualized infrastructure. The advantages of introducing MEC from the perspective of user terminals are to further satisfy users QoS requirements and reduce terminal energy consumption. From the perspective of operators, it is mainly to further reduce core network traffic and improve network scalability and security. The introduction of edge computing can effectively deal with some of the challenges faced in space-air-ground integrated network, which has become an important development trend in the future. These advantages are discussed as follows separately.

(1) Real-time QoS guarantee
With the rapid development of smart terminals, their performance is constantly improved and perfected, but most smart terminals still lack sufficient performance to complete real-time use cases with predefined QoS requirements. Offloading the computing tasks to cloud servers through satellites with computing offloading can effectively meet the computing requirements of terminal devices with limited capacity. However, the rapid development of various emerging service applications, such as Internet of Things, autonomous driving, 4 K/8 K video transmission, etc., has put forward new requirements such as ultra-low latency and high QoS guarantees for space-air-ground integrated network. Through edge computing, the users computing tasks can be directly processed at the edge server without being transmitted to the remote cloud server, which greatly reduces the processing delay of the task and improves the users QoE. Compared with cloud computing, the offloading solution based on edge computing can effectively reduce delay when processing IoT data and guarantee the QoS requirements of users [23].

(2) Energy consumption optimization
Energy consumption is one of the most important parameters when considering mobile devices. Although the processing power of smart terminals is steadily

improving, battery life has not improved at the desired rate. With the development of various computationally intensive applications, executing these applications on the device itself will result in very high energy consumption. In this case, although offloading the computing tasks to the cloud server can reduce the computing energy consumption of the mobile devices to process the computing tasks to a certain extent, the transmission energy consumption of the mobile devices for the transmission tasks will also increase due to the long distance between the cloud server and mobile terminals. Therefore, the introduction of edge computing in space-air-ground integrated network and offloading computing tasks to the edge of the network closer to users will help to further reduce the energy consumption of mobile devices. Energy consumption of applications, like face recognition and augmented reality, is analyzed in [24], which shows that compared with cloud servers, offloading tasks to edge servers can effectively reduce the energy consumption of device terminals. Therefore, the introduction of edge computing is very necessary for mobile devices with limited battery energy in space-air-ground integrated network.

(3) Core network traffic scheduling
The core network has limited bandwidth and is vulnerable to congestion. According to the latest forecast report released by Cisco, by 2021, the total amount of global equipment will reach 75 billion, and mobile traffic will exceed 24.3 EB/month [25]. Thus, operators face huge challenges in managing accumulated data traffic with different sizes and characteristics. In the traditional satellite-ground collaborative network, the traffic generated by mobile devices accesses the core network through satellites or other access devices and further accesses the cloud server. Assume that these services can be satisfied at the edge of space-air-ground integrated network, the burden on the core network can be greatly reduced and bandwidth utilization can be optimized. This transformation of the network prevents billions of devices at the edge from consuming the limited bandwidth of the core network which makes that the services that the core network is responsible for become manageable in scale and be simplified. Therefore, the introduction of edge computing can effectively solve the congestion problem of the core network and data center.

(4) Scalability
The number of terminal devices is expected to reach trillions within a few years. Therefore, the scalability issue is one of the major challenges facing space-air-ground integrated network [26]. To support these real-time changing dynamic requirements, the cloud can be scaled accordingly. However, transmitting large amounts of data to the cloud server could cause congestion in the data center. Moreover, it is more difficult for operators to work due to the constantly changing data traffic generated by terminal equipment. In this case, the centralized structure of cloud computing cannot provide a scalable environment for data and applications. Introducing edge computing in space-air-ground integrated network and distributing services and applications in the form of virtual machines (VM) on edge servers for copying them can greatly improve the scalability of the entire system [27]. The corresponding service can be copied to another nearby edge

server and the request can be further processed when the edge server becomes crowded and cannot satisfy the incoming request. In addition, edge computing can preprocess data at the edge of space-air-ground integrated network, which can greatly reduce the traffic forwarded to the cloud server and the scalability burden of the cloud.

(5) Security and reliability
In space-air-ground integrated network, if all data is transmitted back to the main server, the operation process and data are extremely vulnerable to attacks. Distributed edge computing will allocate data processing work among different data centers and devices. Thus, the attacker cannot affect the entire network by attacking one device. If the data is stored and analyzed locally, the security team can easily monitor it, thereby greatly improving the security of the entire system. Furthermore, compared with cloud computing, edge computing provides better reliability. Edge computing servers are deployed closer to users, so the possibility of network interruption is greatly reduced.

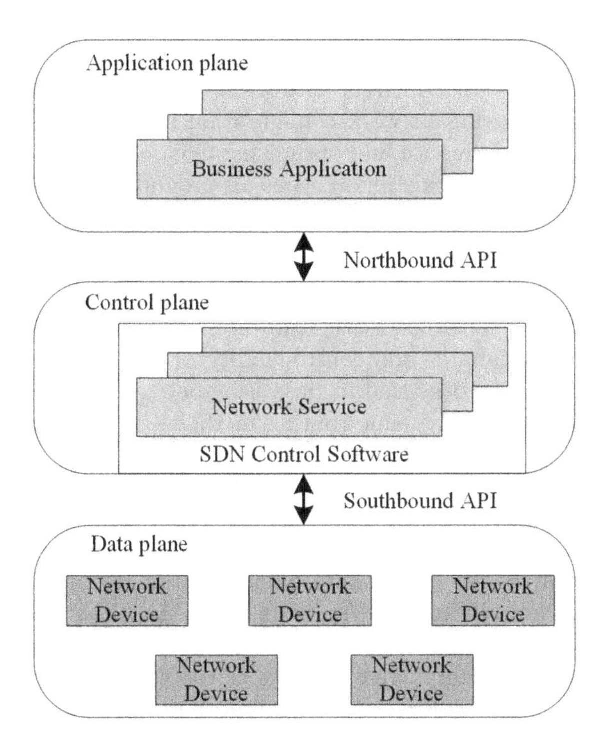

Fig. 3. SDN architecture.

4.2 Software Defined Network

As an emerging network architecture, SDN separates the control plane and data forwarding plane of traditional switching equipment, uses software technology to abstract the underlying network infrastructure, manages and allocates network resources flexibly on demand through the unified and programmable application program interface. The entire SDN architecture is composed of three logical planes: application plane, control plane and data plane, where data transmission and information exchange between logical planes are realized through application program interface, whose architecture is shown in Fig. 3. The devices in SDN have a unified logical structure. Under the action of the controller, these devices only need to execute data forwarding instructions, which can eliminate the network heterogeneity caused by different hardware devices. Using software concepts, SDN can redefine the network architecture to support the new requirements of the ecosystem in the future integrated network. In integrated systems, especially satellite networks, SDN can break through many limitations on network operations and providing end-to-end services, which has become a key factor in promoting technological innovation and network evolution. The introduction of SDN into space-air-ground integrated network can centrally manage heterogeneous networks with different network architectures and communication protocols, thereby reducing network configuration and management overhead, improving network performance and optimizing QoS.

In view of the problems existing in space-air-ground integrated network, the advantages of applying SDN technology to space-air-ground integrated network are as follows:

(1) Flexible routing strategy
In traditional satellite networks, static snapshot routing methods are generally used to ensure the reliability and controllability of the network, but the high dynamics of satellite networks lead to unsatisfactory requirements such as load balancing. Although some dynamic routing methods based on distributed link information collection can meet some of the above requirements, they make it impossible for satellites to easily obtain a global view of the network and can only achieve local optimal routing. In the SDN-based architecture, the control plane obtains a real-time global view of the network state through the network state information uploaded by the data plane and the communication between the controllers to centrally manage the satellite nodes of the data plane, which makes it possible to provide more flexible global routing calculations and routing strategies in the face of highly dynamic network topologies, such as load balancing, multicast path correction, and node failure management. In addition, the control plane with a global view of the network can effectively update the global configuration of the network when the space information network needs to be expanded or updated, allowing newly launched satellites to seamlessly access the existing space information network. On the other hand, the control plane can make timely adjustments to the network configuration, assign neighboring nodes to be responsible for the coverage area and network tasks of the failed node and replace the failed node when a satellite node fails [28].

(2) Convenient network configuration
Considering limited resources, small memory and low central processing unit (CPU) processing power in space information network with the increase of the number of space-based network applications, the amount and complexity of tasks that need to be handled by the on-board payload have also increased greatly, which makes the configuration of space information network very difficult compared with ground-based networks. The core idea of the SDN architecture, the separation of control and forwarding simplifies the processing function of the satellite node, eases this problem to a large extent. What the satellite node of the data plane needs to do is to receive and execute various configuration information distributed by the control plane and feedback its own network state information to the control plane. Complex network configuration and control functions, as well as functions like collecting information from the data plane and building a global view of the network, are handed over to the controller in the control plane of the space-ground double backbone to complete.

(3) Better compatibility
SDN architecture has a unified data exchange standard and programming interface, which can manage the entire network equipment in a unified manner when the network is heterogeneous. The flow table in SDN architecture abstracts the two-layer forwarding table and the three-layer routing table, integrates network configuration information at all levels and is able to simultaneously process various protocols coexisting in the space information network, like delay tolerant network (DTN) etc., so as to well solve the problem of heterogeneity of space-based network protocols.

(4) Lower hardware cost
In traditional satellite networks, satellite nodes have to complete complicated processing tasks which are often the most complicated and expensive part. After adopting the SDN-based architecture, the data plane satellite node is just a simple network forwarding device, which simplifies the architecture of the satellite function, effectively reduce the satellite design and production costs, simplify the complexity of satellite management and make space information network more flexible and controllable. In addition, the control plane structure and inter-satellite link forwarding mode of the space-ground double backbone also reduces the number of required terrestrial stations and the investment in infrastructure.

4.3 Network Slicing

Network slicing was first proposed as a key technology in 5G mobile communication networks and one of the important features of 5G network architecture. Applying slicing technology to space-air-ground integrated network can expand the application range of 5G network slicing and improve overall network performance.

The network slicing technology in 5G networks allows operators who share the same infrastructure to configure the network for the slicing and define specific functions. However, for different application scenarios, it has different network bandwidth and node computing processing capabilities, and can be flexibly and dynamically create and repeal slices based on the operators strategy. Applying network slicing technology to space-air-ground integrated network slicing can build mutually independent virtual logical networks on general equipment, provide customized network functions for a variety of different services, meet users QoS requirements, and improve the scalability and the flexibility of network resource. For applications with relatively high delay requirements, the ground-based network is adopted. At the same time, the broadcast application signals can be transmitted through satellites, which greatly improves the utilization of network resources. Space-air-ground integrated network slicing is composed of physical networks and virtual networks that provide different services for different applications, as shown in Fig. 4 [29].

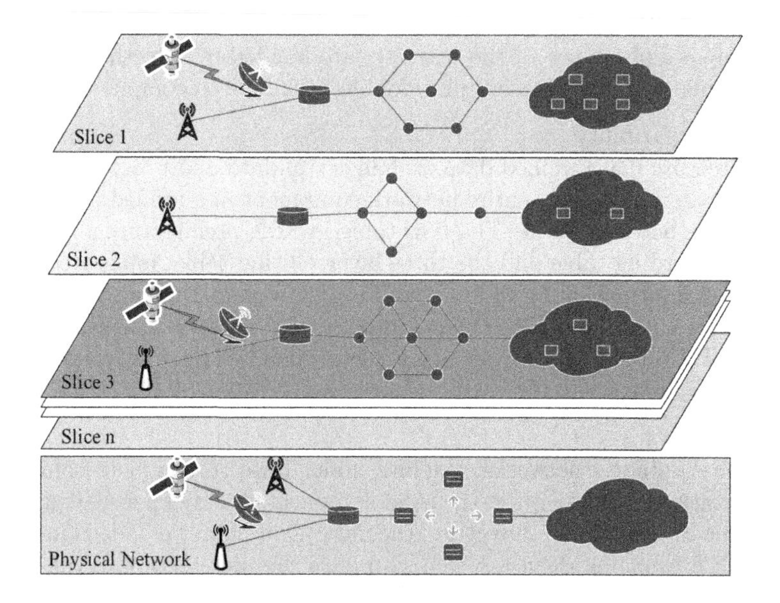

Fig. 4. Network slicing diagram.

The realization of converged network slicing is mainly based on satellite-ground segment virtualization, virtual network embedding (VNE), service function chaining (SFC), etc.

The satellite communication system is composed of space segment, ground segment and user segment. The ground segment includes satellite ground gateway station, ground satellite control centers, tracking, telemetry and command stations. The gateway station is the interface between the satellite system and the ground public network, which can be used by ground users to enter and exit the satellite system to form a link. The ground segment is the management segment of the satellite network, completing the connection between the satellite network and the ground-based network, to allocate resources and charge. Similar to VM, the virtualization of the satellite-ground segment is mainly to virtualize the functions of the satellite-ground segment equipment such as gateways, satellite control centers, tracking, telemetry and command stations, and realize their functions on general equipment rather than specific equipment. Run the same virtual machine on different devices to achieve the same functions. The satellite-ground segment virtualization process firstly converts the satellite signal from the analog signal to digital signal through the receiver, and then transmits it to the cloud server through the network protocol to virtualize the function.

VNE needs to solve the problem of mapping between virtual network functions and physical networks. The computing power required by virtual network nodes and the bandwidth requirements required by virtual network links are mapped to physical network devices through a certain function. How to establish a suitable mapping that meets the virtual network requirements is an important research problem.

SFC links different virtual network function blocks according to application requirements to construct a complete function link. The traffic and requests in the virtual network fluctuate continuously over time, and algorithms are required to continuously optimize according to the dynamic virtual network environment to balance the stability and effectiveness of the virtual function link. In addition, for some applications that require relatively high latency, it is necessary to allocate resources in advance according to the prediction of the virtual network environment.

The virtualization of the satellite-ground segment, VNE and SFC have jointly laid the foundation for converged network slicing.

4.4 Deep Reinforcement Learning

As one of the most important research directions in the field of machine learning, RL has had a significant impact on the development of artificial intelligence in the past 20 years [30]. RL is a learning process in which the agent can make decisions regularly, observe the results, and then automatically adjust its strategy to achieve the optimal strategy. Although the convergence of this learning process has been proven, it usually takes a lot of time to explore and obtain knowledge of the entire system to ensure convergence to the optimal strategy. Therefore, naive RL is not suitable for large-scale and high-complexity network environment such as space-air-ground integrated network. In order to overcome the limitations of RL, DRL has received more and more attention as a new breakthrough technology. Different from the traditional RL enumerating the mapping relationship between the environmental state and the optimal strategy action through the Q table, DRL uses deep neural network (DNN) to replace the Q table, which can theoretically fit the complex mapping relationship of any characteristic, thereby improving the learning rate and the performance of the RL algorithm. In practice, deep learning has become the theoretical support for emerging industries such as robot control, computer vision, voice recognition, and natural language processing. In the field of communication and networking, deep learning has been used as an emerging tool to effectively solve various problems and challenges [31]. For the future networks represented by the space-air-ground integrated network, it contains a variety of heterogeneous and complex network slices or elements, such as IoT devices, mobile users, UAV base stations, and low-orbit satellite nodes, etc. These heterogeneous network entities need to make various decisions on different space-time scales in a centralized or distributed manner, like network and spectrum selective access, data rate and transmit power control, base station and satellite handover, to achieve the maximization of different network optimization goals including throughput and the minimization of network energy consumption [32]. In a highly dynamic and uncertain network environment, most decision-making problems can be modeled as Markov decision process (MDP). Although MDP can be solved in theory by dynamic programming, heuristic algorithm, and RL technology, considering the large-scale and complex modern networks, technologies like dynamic programming and heuristic algorithm would be difficult to be used due to the large amount of calculation. Therefore, DRL has developed into a core solution to overcome this challenge [33]. Applying DRL method to space-air-ground integrated network has the following advantages.

(1) DRL is able to solve network optimization problems in complex environments. Non-convex and complex problems can be optimized without complete and accurate network information by deploying DRL algorithms on network controllers, like base stations or core network controllers.
(2) DRL allows network entities to establish knowledge of the network environment during the learning process without the need to preset channel models or user mobility modes. For example, in space-air-ground integrated network, the network can monitor user distribution or network environment changes in real time using DRL, and gradually learn the optimal base station selection, channel selection, handover, caching and offloading decisions, without having to be based on abstract or inaccurate environmental model.
(3) DRL greatly improves the response speed, especially in complex problems with large state and action space. Therefore, in a large-scale network represented by space-air-ground integrated network, DRL allows the network controller to dynamically control a large number of mobile users and heterogeneous devices based on relatively real-time environmental information.
(4) Some other problems in space-air-ground integrated network, such as cyber-physical attacks, interference management and data offloading, can be modeled as game theory related problems. DRL has recently been used as an effective tool to solve some complex game theory problems, like finding the Nash equilibrium without complete information.

5 Conclusion

This paper provides an overview on the space-air-ground integrated network. Network capability requirements, challenges, as well as the potential key technologies are discussed. In specific, MEC can reduce the data transmission of backhaul links and the bandwidth pressure on the core network, such that service performance and user experience are improved. SDN is able to manage the network resources centrally, thereby reducing network configuration and management overheads. Network slicing can provide customized network functions for a variety of different services, such that network resource scalability and networking flexibility can be guaranteed. DRL can is suitable for important tasks such as service orchestration, resource scheduling, network access, and mobility management in the future space-air-ground integrated network.

References

1. Chettri, L., Bera, R.: A comprehensive survey on internet of things (IoT) toward 5G wireless systems. IEEE Internet Things J. **7**(1), 16–32 (2020)
2. Xiong, J., et al.: A personalized privacy protection framework for mobile crowdsensing in IoT. IEEE Trans. Ind. Inf. **16**(6), 4231–4241 (2020)
3. Xiong, J., et al.: Enhancing privacy and availability for data clustering in intelligent electrical service of IoT. IEEE Internet Things J. **6**(2), 1530–1540 (2019)
4. Cirillo, F., Gmez, D., Diez, L., Elicegui Maestro, I., Gilbert, T.B.J., Akhavan, R.: Smart city IoT services creation through large-scale collaboration. IEEE Internet Things J. **7**(6), 5267–5275 (2020)
5. Huang, S., Zeng, Z., Ota, K., Dong, M., Wang, T., Xiong, N.: An intelligent collaboration trust interconnections system for mobile information control in ubiquitous 5G networks. IEEE Trans. Netw. Sci. Eng. **8**, 347–365 (2020). https://doi.org/10.1109/TNSE.2020.3038454
6. Li, J., Zhang, X.: Deep Reinforcement Learning-Based Joint Scheduling of eMBB and URLLC in 5G Networks. IEEE Wireless Communications Letters **9**(9), 1543–1546 (2020)
7. Xiong, J., Zhao, M., Bhuiyan, M.Z.A., Chen, L., Tian, Y.: An AI-enabled three-party game framework for guaranteed data privacy in mobile edge crowdsensing of IoT. IEEE Trans. Ind. Inf. **17**(2), 922–933 (2021). https://doi.org/10.1109/TII.2019.2957130
8. Pokhrel, S.R., Ding, J., Park, J., Park, O.-S., Choi, J.: Towards enabling critical mMTC: a review of urllc within mMTC. IEEE Access **8**, 131796–131813 (2020)
9. Wan, S., Hu, J., Chen, C., Jolfaei, A., Mumtaz, S., Pei, Q.: Fair-hierarchical scheduling for diversified services in space, air and ground for 6G-dense internet of things. IEEE Trans. Netw. Sci. Eng, (2020). https://doi.org/10.1109/TNSE.2020.3035616
10. Ye, J., Dang, S., Shihada, B., Alouini, M.S.: Space-air-ground integrated networks: outage performance analysis. IEEE Trans. Wirel. Commun. **19**(12), 7897–7912 (2020)
11. Niu, Z., Shen, X.S., Zhang, Q., Tang, Y.: Space-air-ground integrated vehicular network for connected and automated vehicles: challenges and solutions. Intell. Converged Netw. **1**(2), 142–169 (2020)
12. Sharma, S.R.S., Vishwakarma, N., Madhukumar, A.: HAPS-based relaying for integrated space-air-ground networks with hybrid FSO/RF communication: a performance analysis. IEEE Trans. Aerosp. Electron. Syst. **57**(3), 1581–1599 (2021). https://doi.org/10.1109/TAES.2021.3050663
13. Jiang, C., Li, Z.: Decreasing big data application latency in satellite link by caching and peer selection. IEEE Trans. Netw. Sci. Eng. **7**(4), 2555–2565 (2020)
14. Farserotu, J., Prasad, R.: A survey of future broadband multimedia satellite systems, issues and trends. IEEE Commun. Mag. **38**(3), 128–133 (2000)

15. Gharanjik, A., M.R., B.S., Arapoglou, P., Ottersten, B.: Multiple gateway transmit diversity in Q/V band feeder links. IEEE Trans. Commun. **63**(3), 916–926 (2015)
16. Nishiyama, H., Tada, Y., Kato, N., et al.: Toward optimized traffic distribution for efficient network capacity utilization in two-layered satellite. IEEE Trans. Veh. Technol. **62**(3), 1303–1313 (2013)
17. Chandrasekharan, S., Gomez, K., Al-Houran, A., et al.: Designing and implementing future aerial communication networks. IEEE Commun. Mag. **54**(5), 26–34 (2016)
18. Anjum, S., Noor, R.M., Anisi, M.H.: Review on MANET based communication for search and rescue operations. Wirel. Pers. Commun. **94**(1), 31–52 (2017)
19. Rooyen, M.V., Odendaal, J.W., Joubert, J.: High-gain directional antenna for WLAN and WiMAX applications. IEEE Antennas Wirel. Propag. Lett. **16**, 492–495 (2017)
20. Zhang, N., Zhang, S., Yang, P., et al.: Software defined space-air-ground integrated vehicular networks: challenges and solutions. IEEE Commun. Mag. **55**(7), 101–109 (2017)
21. Taleb, T., Samdanis, K., Mada, B., et al.: On multi-access edge computing: a survey of the emerging 5G network edge architecture and orchestration. IEEE Commun. Surv. Tutorials **19**(3), 1657–1681 (2017)
22. Xie, R., Lian, X., Jia, Q.: Survey on computation offloading in mobile edge computing. J. Commun. **39**(11), 138–155 (2018)
23. Shukla, R.M., Munir, A.: A computation offloading scheme leveraging parameter tuning for real-time IoT devices. In: IEEE International Symposium Nanoelectronic Information Systems pp. 208–209 (2016)
24. Ha, K., Pillai, P., Lewis, G., et al.: The impact of mobile multimedia applications on data center consolidation. In: IEEE International Conference on Cloud Engineering, pp. 166–176 (2013)
25. CISCO: Cisco visual networking index: global mobile data traffic forecast update 2016–2021 white paper (2017)
26. Wang, W., Tong, Y., Li, L., et al.: Near optimal timing and frequency offset estimation for 5G integrated LEO satellite communication system. IEEE Access **7**, 113298–113310 (2019)
27. Vaquero, L.M., Rodero-merino, L.: Finding your way in the fog: towards a comprehensive definition of fog computing. Comput. Commun. Rev. **44**(5), 27–32 (2014)
28. Ren, C., et al.: Enhancing traffic engineering performance and flow manageability in hybrid SDN. In: IEEE GLOBECOM (2016)
29. Ordonez-Lucena, J., et al.: Network slicing for 5G with SDN/NFV: concepts, architectures, and challenges. IEEE Commun. Mag. **55**(5), 80–87 (2017)
30. Sutton, R., Barto, A.: Reinforcement Learning: An Introduction. The MIT Press, Cambridge (1998)
31. Luong, N., Hoang, D., Gong, S., et al.: Applications of deep reinforcement learning in communications and networking: a survey. IEEE Commun. Surv. Tutorials **21**(4), 3133–3174 (2019)
32. Liu, J., Shi, Y., et al.: Space-air-ground integrated network: a survey. IEEE Commun. Surv. Tutorials **20**(4), 2714–2741 (2018)
33. Kato, N., Fadlullah, Z.M., Tang, F., et al.: Optimizing space-air-ground integrated networks by artificial intelligence. IEEE Wirel. Commun. **26**(4), 140–147 (2019)

Decentralized Certificate Management for Network Function Virtualization (NFV) Implementation in 5G Networks

Junzhi Yan(✉), Bo Yang, Li Su, Shen He, and Ning Dong

China Mobile Research Institute, Beijing, China
{yanjunzhi,yangbo,suli,heshen,dongning}@chinamobile.com

Abstract. The certificate cost and certificate management complexity increase when PKI is leveraged into Network Function Virtualization (NFV), a significant enabling technology for 5G networks. The expected security of PKI cannot be met because the certificate revocation inquiry is unavailable during the intranet implementation in the operator's core network. This paper analyses the issues and challenges during the NFV implementation, and proposes a blockchain based decentralized NFV certificate management mechanism. During instantiation, the Virtual Network Functions (VNF) instance generates certificates according to the certificate profile provided in the VNF package. The certificates submitted to the decentralized certificate management system by the instance will be validated by corresponding participants. The certificates will be recorded into the ledger after validation and consensus, and then it will be trusted by the participants. The performance analysis shows the transaction efficiency is non-critical, and the transaction delay of seconds is acceptable in this decentralized system. The delay of the certificate inquiry is critical, and it can be fulfilled by the decentralized deployment of inquiry nodes.

Keywords: Blockchain · NFV · Certificate management · PKI

1 Introduction

Network Function Virtualization (NFV), featured as decoupling software from hardware, flexible network function deployment, and dynamic operation, is a significant enabling technology for 5G networks. In NFV, network functions are implemented by vendors in software components known as Virtual Network Functions (VNFs), which are deployed on cloud infrastructure or massively distributed servers instead of dedicated hardware [1].

The architectural framework of NFV defined by the European Telecommunication Standardization Institute (ETSI) is depicted in Fig. 1. It enabled the execution and deployment of VNF on NFV infrastructure comprising a pool of network, storage, and computing resources. The NFV infrastructure is usually a decentralized cloud infrastructure

J. Xiong et al. (Eds.): MobiMedia 2021, LNICST 394, pp. 81–93, 2021.
https://doi.org/10.1007/978-3-030-89814-4_6

in which servers are distributed over various locations. ETSI defines network functions, including VNFs. The operation, deployment, and execution of network services and VNFs in NFV infrastructure are controlled by an orchestration and management system, whose performance is steered by NFV descriptors [1, 2].

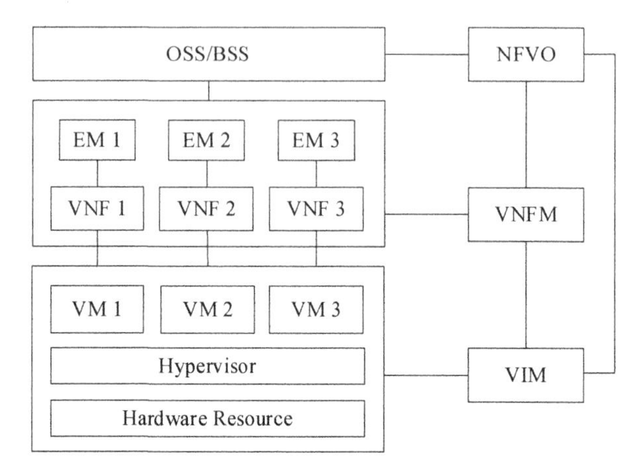

Fig. 1. Architectural framework of NFV defined by ETSI [2]

Typically, NFV is capable of overcoming certain 5G challenges, such as, reducing the energy cost by maximizing the resource usage, scaling and mobilizing VNFs from one resource to another, ensuring VNFs performance operations [3]. A VNF is a virtualisation of a network function in a legacy non-virtualised network. In 5G networks, Network Functions (NFs) are defined in 3GPP TS 23.501 [4].

PKI certificates are widely used by the VNF, MANO (Management and Orchestration), OSS/BSS/EM (Operation Support Systems, Business Support System, Element Management) in NFV. These certificates are used for authentication and secure communication. The NFs in 5G networks use TLS protocol to connect each other [5]. However, some issues and challenges arise during the deployment of the NFV. These issues and challenges are related with the certificate cost, across-domain trust, CRL/OCSP (Certificate Revocation List/Online Certificate Status Protocol) services, certificate validation and certificate maintenance. The essence of some issues is the lack of trust amongst the multiple participants in the NFV deployment. Blockchain featured as decentralization and tamper resistant may benefit PKI technology [6]. The blockchain based decentralized PKI is a significant trend for PKI technology [7], which could be used to facilitate the certificate management of NFV.

The main contribution of this paper is the blockchain based decentralized NFV certificate management system, which aims to solve the issues in NFV implementation in telecommunication operator's network. Section 2 discusses the related researches. The issues and challenges aroused during the NFV implementation is discussed in Sect. 3. Section 4 provides the framework of decentralized NFV certificate management mechanism, and the certificate management method. The performance is analyzed in Sect. 5. The conclusion is in Sect. 6.

2 Related Works

ETSI has published series of NFV standards, of which ETSI GS NFV 002 defines the architectural framework [2], ETSI GS NFV 001 [8] provides a list of use cases and examples of target network functions for virtualization,ETSI GR NFV SEC 005 [9] analyses the certificate management using traditional PKI.

An overview of enabling technologies like NFV and SDN for 5G was provided in [10]. It described the 5G slicing notion and the prominent challenges associated with it, and highlighted a few challenges for ensuring an envisaged 5G networking system.

The concept of Network Slicing (NS) as a service was presented for providing customized services in [11]. An outline of sophisticated standardization views of NS was provided in [12]. There are crucial challenges in NS faced by telecommunication operators such as difficulties in attaining end-to-end NS, stable migration, interoperability, and roaming. The work highlighted that base station virtualization and wireless resource sharing to formulate appropriate requirements and create standardized slices should be emphasized. 5G architecture design was presented for NS and building NFV, Software Defined Network (SDN) technologies in [13]. It emphasized schemes which provide effective substrate resource utilization for NS. In [14], the performance deterioration issue of virtualized access points occurring due to NFV implementation was addressed and an overcoming approach was presented. In [15], a framework for mobile network virtualization comprising three planes, namely control, cognitive, and data planes, was presented. A blockchain-based secure key management scheme was proposed in [16] to address the security of key management caused by a non-trusted base station, it was suited to improve the trustworthiness of the base station. The incentive mechanism combining edge computing was addressed in [17].

Some typical researches focusing on the decentralized PKI could be found in [7, 18–22]. A blockchain based PKI framework in mobile networks was proposed in [7]. It focused on the problems when traditional PKI is leveraged into mobile networks. The system was constituted by submission nodes, validator nodes, inquiry nodes. It provided some scenarios and application cases in mobile networks. The optimizations for certificate storage in blockchain based PKI system was analyzed in [18]. The provided methods aimed to improve the storage efficiency of specific nodes in blockchain based PKI system. Research in [19] focused on the trust among multiple CAs using blockchain, and provided some use cases in mobile networks.

The implementation of Yakubov et al. [20] used the standard X.509v3 certificate with an addition to the extension fields to indicate its location in the blockchain. The smart contract of each CA contained one list with all issued certificates and another list for revoked certificates. BlockPKI [21] required multiple CAs to perform a complete domain validation from different vantage points for an increased resilience to compromise and hijacking, scale to a high number of CAs by using an efficient multi-signature scheme, and provided a framework for paying multiple CAs automatically. SCPKI [22] worked on Ethereum blockchain, and used an entity or authority in the system to verify another entity's identity. It could be used to detect rogue certificates when they are published.

Standard development organizations such as ISO/IEC, ITU-T have begun to study and standardize blockchain based PKI and certificate management technology. These works are focusing on the profile and the mechanism of blockchain recorded certificates. However, these normative works are still under development.

3 Issues and Challenges

In NFV, there are mainly three kinds of use cases for the use of certificates [9], i.e., VNF certificate use case, MANO certificate use case, and OSS/BSS/EM certificate use case. The VNF certificate use case will be discussed in this paper. The other two use cases are similar.

A VNF component instance (VNFCI) needs one or more certificates provisioned to attest its identity to the VNFM or EM to establish a secure connection between them. In NFV implementation, the number of VNF certificates is far more than that in the other two use cases. The management of VNF certificates will be discussed in this paper. However, the certificates in the other two use cases could use the same method as VNF certificates.

By using traditional solutions, each instance of VNF could enroll certificates to CA/RA directly, or by a delegator such as VNFM [9]. However, the issues and challenges are as follows:

- Cost of certificates

  VNFs are implemented with one or more VNF components. While a VNF component instance composed of various VNFCIs could have multiple logical identities, each of which is represented by a certificate, to communicate with different peers [9]. As a result, there will be a huge number of certificates required for the VNFs in 5G networks. It will be costly to use certificates issued by commercial CAs. The telecommunication operators prefer to use their own CA, vendor's CA or designated CA to provide certificate service due to the cost. This may cause the problem of trust across CA domains.
- Trust across CA domains

  A VNFCI may communicate with another VNFCI in another telecommunication operator's network. These two VNFCIs may be configured with the certificates issued by different CAs. There are several traditional methods to deal with multiple CAs, including trusted root list, cross certification, bridge CA, each with its own pros and cons as illustrated as following.

  - Trusted root list: It relies on the list maintained by the relying party. Certificates issued by the roots which are not in the list will not be trusted. It will be costly to update the list.
  - Cross certification: It's suitable for a small amount of CAs. If there are a large amount of CAs, the cross relationship will make a complex structure. Moreover, the usage of certificate policies will be limited after multiple mappings.
  - Bridge CA: The bridge CA will connect multiple CAs. The certificate chain would be longer, and the validation would be much more complex and expensive.

- CRL/OCSP unavailable due to intranet implementation
  The 5G network functions are deployed in the telecommunication operator's core network with no connection to the Internet, which means CRL/OCSP are unavailable. Moreover, the telecommunication operator's core network is usually divided into different security domains. These security domains are isolated physically or logically. The entity in one security domain cannot communicate with the entities in another security domain directly. In practice, the telecommunication operator's CA/RA service, including CRL/OCSP services, are implemented in different security domains from the 5G NFs in the core network. This means the VNFCIs in the 5G core network cannot access the CRL/OCSP services provided by the telecommunication operator's CA. Unless, the telecommunication operator deploys the CRL/OCSP services in each security domain, which is a complex and costly work.
- Certificate validation
  Before a VNFCI gets a certificate issued by the CA/RA, the VNFCI will be validated by the CA/RA. The subject field in the certificate may be an IP address, FQDN, or other unique identifiers, and these information is related with the deployment. One VNFCI could have several functionalities and several logical interfaces, and it could have several identities for different functionalities and interfaces. It is impossible for the CA/RA to validate the subject field, since the CA/RA does not get the information related with the deployment or the identities of the VNFCI. In practice, these is an endorsement for the subject field. The endorsement is provided by some designated administrators. In accordance with the use cases, there will be kinds of certificates endorsed by different administrators. In such case, the deep cooperation between the CA/RA and the administrators is significant and it makes the certificate management complicated.
- Certificate maintenance
  Each certificate has a validation period, which means it may expire. The certificate to be expired needs to be renewed. Or else it will not be trusted by the relying party. There was once some mobile service became unavailable because of the expired certificates. In 5G networks, there will be more than thousands of VNF certificates. It has to be ensured each certificate be renewed before it expires, and be revoked when it is insecure or the VNFCI is terminated. In practice, the certificate renew could be optional, since the VNFCI with an expired certificate could be terminated, and a new VNFCI could be instantiated with a new certificate.

The essence of the above issues is the lack of trust amongst the multiple participants (such as vendors, administrators of the telecommunication operator, CA/RAs) during the NFV implementation. A secured information sharing and trusted endorsement method is necessary to solve the issues. The blockchain is featured as decentralization and tamper-resistant. The endorsement and consensus mechanisms in blockchain help to make the information submitted to the participants in the blockchain system be trusted. It provides a decentralized way to solve the issues of the NFV certificate management.

4 Decentralized NFV Certificate Management

4.1 Framework

A blockchain based PKI framework was proposed in [7]. It consists submission nodes, validator nodes, and inquiry nodes, while the submission node is used to submit certificates to the blockchain based PKI system, the validator node is the node to verify the received requests and generate new blocks, and the inquiry node works to provide certificate and status inquiry service. In this VNF certificate management scenario, the framework for VNF certificate management is shown in Fig. 2.

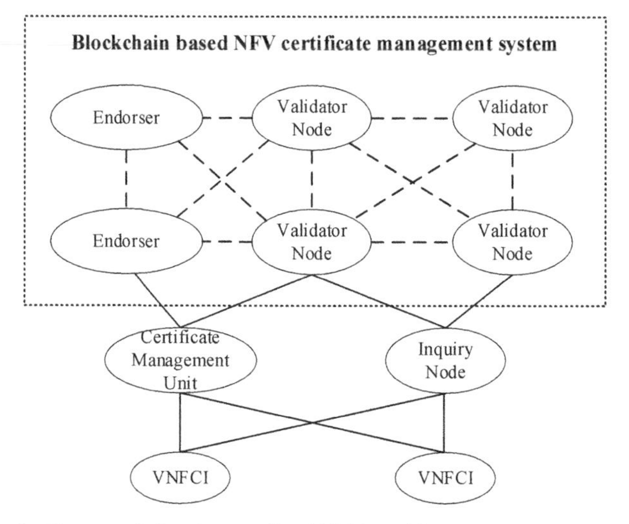

Fig. 2. Framework for decentralized NFV certificate management system

The VNFCI is the owner of the certificate.

The Certificate Management Unit (CMU) works as a client to submit certificates and related information into the blockchain based NFV certificate management system. The CMU could be a function in NFV architecture, e.g., located in VNFM, and it also could be independent to the NFV architecture.

The endorser is the node to endorse the identity in the submitted certificates. Only the endorsed certificates and requests could be processed by the validator nodes.

The validator node is the node to verify the received requests and generate new blocks. It validates the certificates and request according to the policies. The validator nodes are held by vendors, operators, and CAs. One node could act as both an endorser and a validator.

The inquiry node provides certificate inquiry services. It needs to receive new blocks, but do not need to participate into the generation of new blocks. The inquiry nodes are held and deployed by any party which is capable to access the blockchain based certificate management system.

4.2 Certificate Enrollment

During instantiation, VNFCI needs to enroll certificates to communicate with other VNFCI or MANO/OSS/BSS/EM. The certificate could be a certificate issued by CA/RA as descripted in [9]. It could also be a self-signed certificate generated by the VNFCI. The self-signed certificate management will be discussed in this paper. The certificate profile is provided to the VNFCI during the VNF configuration. While the management of certificates issued by traditional CAs is also supported, which is similar to the self-signed certificates.

The VNF configuration is based on parameterization captured at design time, included in the VNF package, and complemented during the VNF instantiation. Before a VNF is installed, the VNF package will be on-boarded by NFVO. The VNF package includes a component of VNFD (Virtualised Network Function Descriptor), which is a deployment template describing a VNF in terms of deployment and operational behavior requirements [23]. The VNFM accesses to the VNFD, and configures the certificate profile during the VNF instantiation. The VNFCI enrolls a certificate as follows and the message flow is shown in Fig. 3.

0. The VNFM generates the certificate profile and initial credential for each VNFCI which are included in the VNFD, sends the certificate profile and token for each VNFCI to the CMU.

The VNF parameters describing the certificate profile in the VNFD can be declared to be configurable during the VNF design phase, and further be configured by the VNFM during or after the VNF instantiation [24]. The certificate profile declares the information used to generate the certificate, such as the subject, key usage, basic constraint [7, 25].

The subject field identifies the entity associated with the public key stored in the subject public key field, and contains a distinguished name. The distinguished name may be an FQDN, a serial number, or other kinds of names, according to the operator's policy. It is suggested to include the operator's information in the distinguished name field, so as to identify the HPLMN (Home Public Land Mobile Network) in roaming scenarios. Multiply names could be addressed in the SAN (Subject Alternative Name) field [25]. The address of the inquiry node could be included in the extension field of the certificate.

The VNFM sends the certificate profile and a token to CMU. The token and information in the certificate profile will be used to validate the submitted VNFCI certificates. For the sake of simplicity, we use a token which is the value of multiple hash operations on the initial credential. The initial credential is kept as a secret by the VNFCI. Denote the initial credential by x, the token by y. Then we have

$$\mathrm{H}\left(\mathrm{H}(\ldots \mathrm{H}(x))\right) = y \tag{1}$$

y is the value of multiple times (e.g., n times) hash operations of x.

1. The VNFCI generates a self-signed certificate, and submits the certificate publish request to the CMU.

The public-private key pair used to generate a self-signed certificate is generated using the methods addressed in [9]. The VNFCI generates the certificate using the information and certificate profile provided in the VNFD, and then generates the authentication credential based on the initial credential. The authentication credential is the value (denoted by y_1) of multiple hash operations (e.g., n-1 times) on the initial credential (x), of which the hash value equals the token (y). The VNFCI submits certificate publish request to the CMU, while the request consists the certificate and the authentication credential.

Message flows for VNFCI certificate enrollment:

0. VNFM: Configures the certificate profile in VNFD, generates an initial credential and a token
   VNFM → VNFCI:
   {certificate profile, initial credential x}
   VNFM → CMU:
   {certificate profile, token y}

1. VNFCI: Generates a certificate using the certificate profile
   VNFCI → CMU: certificate publish request
   {certificate, authentication credential y_1*}*

2. CMU: verifies and signs the request, update the token
   The identity in the certificate is correct
   Authentication credential is valid, update the token into y_1
   CMU → Blockchain based system: certificate publish request
   {certificate, signature}

3. Endorsers: verify the request
   Signature is correct
   Validator nodes: record the certificate and status into a newly generated block after consensus

Fig. 3. Message flows for VNFCI certificate enrollment

2. The CMU verifies the certificate publish request, signs the request and transmits it to the blockchain based certificate management system.

The CMU verifies the certificate in the request to ensure it is consistent with the certificate profile, and the information contained in the certificate is correct (e.g., the information in the subject field is valid). The authentication credential is verified to ensure it is consistent with the token. Then the CMU signs the request and submits it to the blockchain based certificate management system. The token could only be used once so as to prevent replay attacks. Thus, CMU updates the token from y into y_1. The one-time token makes it possible for the VNFCI to enroll multiple certificates.

3. The endorsers in the blockchain system verifies the request, endorses the verified request. The certificate in the endorsed request will be recorded into the ledger by the validator nodes.

The endorsers verify the signature of the certificate publish request. After verification, the endorsers sign the request with their private keys. The validator nodes record the certificates in the endorsed requests into the ledger after consensus. The endorsement policy is made and configured by the participants.

4.3 Certificate Revocation

A VNFCI certificate needs to be revoked, when it is insecure or the VNFCI is terminated. At first, the VNFCI generates and submits a certificate revocation request to the CMU. Or, the CMU generates the certificate revocation request according to the policy. The request contains the certificate or its identifier, and then it is signed by the CMU.

The CMU submits the certificate revocation request to the blockchain based certificate management system. The endorsers and validator nodes verify the request and then update the status of the certificate as "revoked" in the ledger.

4.4 Certificate Renewal

The certificate to be expired needs to be renewed. The certificate renewal request is initiated by the VNFCI. The CMU could indicate the VNFCI to initiate a certificate renewal process.

The VNFCI generates the certificate renewal request and submits it to the CMU. The request contains the certificate to be renewed or its identifier, the new certificate, and the signature signed by the private key corresponding to the certificate to be renewed. The CMU submits the certificate renewal request to the blockchain based certificate management system. The endorsers and validator nodes verify the request and then record the new certificate into the ledge, and update the status of the former certificate as "revoked" in the ledger.

4.5 Certificate Inquiry

Ideally, the NFV infrustructure of all the telecommunication operators use the blockchain based certificate management solution. However, in practice, some operators may use blockchain based solution while others use traditional PKI solution. The certificate inquiry is discussed as follows in non-roaming scenario and roaming scenario, in which the VPLMN (Visited Public Land Mobile Network) uses the blockchain based solution.

1. Non-roaming scenario
   When a VNFCI receives a certificate from another VNFCI, it inquires the certificate and its status from the inquiry node of the blockchain based certificate management system, the inquiry node finds the inquired certificate and its status, and feedbacks them to the relying party. The relying party verifies the certificate and its status to ensure the certificate is valid.

2. Roaming scenario
   Figure 4 depicts a simplified certificate inquiry architecture in the case of local break out scenario which was defined in [4]. It shows an example of local break out scenario. Usually, each operator only trusts its own system, including the NFV certificate management system. In this case, the VPLMN uses the blockchain based solution, HPLMN 1 uses the traditional PKI solution, and HPLMN 2 uses the blockchain based solution which is independent to the VPLMN.

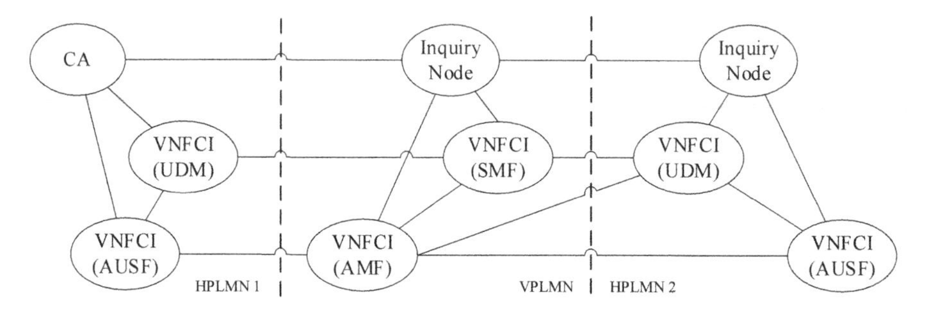

Fig. 4. Certificate inquiry architecture for roaming 5G system

The inquiry node in the VPLMN connects the CRL/OCSP servers used by HPLMN 1, and the inquiry node in NFV certificate management system of HPLMN 2. When a VNFCI in the VPLMN receives a certificate from the VNFCI of other PLMN, it connects the inquiry node in VPLMN for the status of the certificate. The certificate in this blockchain based solution contains the operator's information or the address of the inquiry node. The certificate in the traditional PKI solution contains the CRL/OCSP address. As a result, the inquiry node in the VPLMN connects the CRL/OCSP server in HPLMN 1, and the inquiry node in HPLMN 2, according to the information included in the certificate. The inquiry node in the VPLMN inquiries the certificate status and feedback the status to the VNFCI in the VPLMN.

5 Performance Consideration

5.1 Transaction Efficiency of Certificate Management

The certificate management requests including certificate enrollment, renewal and revocation are blockchain transactions, which means the transaction validation is needed and new blocks are generated. Transaction throughput, which is the number of transactions could be processed in a given time period, determines the efficiency of a blockchain based system. High transaction throughput means more requests could be processed in a given time period.

In NFV scenario, the certificate enrollment happens during instantiation. It happens once or no more than several times for each VNFCI. Usually, the validity of a certificate is 1-year long. However, it could be configured according to the operator's policy. The longer is the validity, the less certificate renewal is needed. Each certificate can only be revoked once. As a result, the certificate for each VNFCI needs no more than two transactions (certificate enrollment/renewal, and certificate revocation) per year on average. Even if there are millions of VNFCIs in the operator's network, about tens of transactions happen per minute. The transaction efficiency is non-critical in this decentralized system.

5.2 Transaction Delay

Each certificate management request may result in a new record in the ledger. Transaction delay means the time from the certificate management request submitted to the blockchain based system to the time that the request be processed and recorded into a new block or be rejected. It usually takes minutes to instantiate a VNFCI, so the transaction delay of seconds is acceptable. The most commonly used blockchain framework such as Fabric and Etherum support the transaction delay of seconds. Both of them could be used to implement the decentralized NFV certificate management system.

5.3 Performance of Certificate Inquiry

In the traditional PKI system, the certificate status is inquired by using CRL or OCSP, which is centralized service provided by CA. In blockchain based NFV certificate management system, each node capable to access the ledger could provide certificate status inquiry service. This makes the inquiry service be decentralized. When an inquiry node is deployed on the edge of the operator's core network and Internet, it could provide local certificate inquiry service for the entities in the core network. This may greatly enhance the availability and efficiency of certificate status inquiry service.

5.4 Other Considerations

Some other considerations about the cost, trust across domains, compatibility are as follows:

- Cost: There is no need for the operator and vendor to deploy and maintain a CA infrastructure for the NFV implementation, so the cost is reduced.
- Trust across domains: The nodes in the decentralized system consist operators, vendors, traditional CAs, which may be from different trust domains. The endorsement and consensus mechanisms make all the records in the ledger be trusted by the multiple participants from different domains according to the policy. It makes the trust between different trust domains be available.
- Compatibility: X.509v3 certificate [26] is supported in blockchain based certificate management system, so as to be compatible with the traditional PKI system and applications.

6 Conclusion

Decentralized PKI is a significant direction for PKI technology. This paper analyses the issues and challenges related to the certificate management aroused during the NFV implementation in the 5G networks, and proposes a blockchain based decentralized NFV certificate management mechanism. The system could establish the trust among the participants in the NFV implementation, such as vendors, operators, even traditional CAs. It could ease the work load of the certificate management, reduce the cost to deploy and maintain the CA, and make certificate status inquiry available in the 5G core network. The analysis shows that the performance of transaction efficiency is non-critical in the blockchain based decentralized system. The high performance of the certificate inquiry could be facilitated by the decentralized deployment of inquiry nodes. This work could also facilitate the certificate usage in other scenarios in the telecommunication operator's networks.

References

1. Srinivasan, S.: A literature review of network function virtualization (NFV) in 5G networks. Int. J. Comput. Trends Technol. **68**(10), 49–55 (2020)
2. ETSI GS NFV 002: Network Functions Virtualisation (NFV); Architectural Framework (2014)
3. Ordonez-Lucena, J., Ameigeiras, P., Lopez, D., et al.: Network slicing for 5G with SDN/NFV: concepts, architectures, and challenges. IEEE Commun. Mag. **55**(5), 80–87 (2017)
4. 3GPP TS 23.501: System Architecture for the 5G System (2019)
5. 3GPP TS 33.501: Security architecture and procedures for 5G system (2019)
6. Hepp, T., Spaeh, F., Schoenhals, A., et al.: Exploring potentials and challenges of blockchain-based public key infrastructures. In: 2019 IEEE Conference on Computer Communications Workshops (IEEE INFOCOM 2019), pp. 847–852. IEEE (2019)
7. Yan, J., Hang, X., Yang, B., et al.: Blockchain based PKI and certificates management in mobile networks. In: IEEE 19th International Conference on Trust, Security and Privacy in Computing and Communications (TrustCom 2020), Los Alamitos, California, pp. 1764–1770. IEEE Computer Society (2021)
8. ETSI GS NFV 001: Network Functions Virtualisation (NFV); Use Cases (2013).
9. ETSI GR NFV-SEC 005: Network Functions Virtualisation (NFV); Trust; Report on Certificate Management (2019)
10. Yousaf, F.Z., Bredel, M., Schaller, S., Schneider, F.: NFV and SDN—Key technology enablers for 5G networks. IEEE J. Sel. Areas Commun. **35**(11), 2468–2478 (2017)
11. Zhou, X., Li, R., Chen, T., et al.: Network slicing as a service: enabling enterprises' own software-defined cellular networks. IEEE Commun. Mag. **54**(7), 146–153 (2016)
12. Kim, D., Kim, S.: Network slicing as enablers for 5G services: state of the art and challenges for the mobile industry. Telecommun. Syst. **71**(3), 517–527 (2019)
13. Yousaf, F.Z., Gramaglia, M., Friderikos, et al.: Network slicing with flexible mobility and QoS/QoE support for 5G networks. In: IEEE International Conference on Communications Workshops (ICC Workshops 2017), pp. 1195–1201. IEEE (2017)
14. Wang, X., Xu, C., Zhao, G., et al.: Tuna: an efficient and practical scheme for wireless access points in 5G networks virtualization. IEEE Commun. Lett. **22**(4), 748–751 (2017)
15. Feng, Z., Qiu, C., Feng, Z., et al.: An effective approach to 5G: wireless network virtualization. IEEE Commun. Mag. **53**(12), 53–59 (2015)

16. Tian, Y., Wang, Z., Xiong, J., et al.: A blockchain-based secure key management scheme with trustworthiness in DWSNs. IEEE Trans. Industr. Inf. **16**(9), 6193–6202 (2020)
17. Xiong, J., Chen, X., Yang, Q., et al.: A task-oriented user selection incentive mechanism in edge-aided mobile crowdsensing. IEEE Trans. Network Sci. Eng. **7**(4), 2347–2360 (2020)
18. Yan, J., Yang, B., Su, L., He, S.: Storage optimization for certificates in blockchain based PKI system. In: Xu, Ke., Zhu, J., Song, X., Lu, Z. (eds.) CBCC 2020. CCIS, vol. 1305, pp. 116–125. Springer, Singapore (2021). https://doi.org/10.1007/978-981-33-6478-3_8
19. Yan, J., Peng, J., Zuo, M., et al.: Blockchain based PKI certificate system. Telecom Eng. Tech. Stand. **2017**(11), 16–20 (2017)
20. Yakubov, A., Shbair, W.M., Wallbom, A., et al.: A blockchain-based PKI management framework. In: 2018 IEEE/IFIP Network Operations and Management Symposium (NOMS 2018), pp. 1–6. IEEE (2018)
21. Dykcik, L., Chuat, L., Szalachowski, P., et al.: BlockPKI: an automated, resilient, and transparent public-key infrastructure. In: 2018 IEEE International Conference on Data Mining Workshops (ICDMW 2018), pp. 105–114. IEEE (2018)
22. Al-Bassam, M.: SCPKI: A smart contract based PKI and identity system. In: Proceedings of the ACM Workshop on Blockchain, Cryptocurrencies and Contracts, pp. 35–40. ACM (2017)
23. ETSI GS NFV-IFA 011: network functions virtualisation (NFV) release 4; management and orchestration; VNF descriptor and packaging specification, (2020)
24. ETSI GS NFV-IFA 008: network functions virtualisation (NFV); management and orchestration; Ve-Vnfm reference point - interface and information model specification (2019)
25. IETF RFC 5280: Internet X.509 public key infrastructure certificate and certificate revocation list (CRL) profile (2008)
26. ITU-T X.509. The directory: public-key and attribute certificate frameworks (2016)

Stochastic Performance Analysis of Edge-Device Collaboration

Zhidu Li[1,2,3](✉), Xue Jiang[1,2,3], Ruyan Wang[1,2,3], and Dapeng Wu[1,2,3]

[1] School of Communication and Information Engineering, Chongqing University of Posts and Telecommunications, Chongqing, China
lizd@cqupt.edu.cn

[2] Key Laboratory of Optical Communication and Networks, Chongqing, China

[3] Key Laboratory of Ubiquitous Sensing and Networking, Chongqing, China

Abstract. As a new computing paradigm, edge-device collaboration is able to reduce the task execution delay for the devices with limited energy. However, deterministic delay guarantee is unavailable due to the random task arrivals and time-varying channel fading. This paper proposes to study the task execution delay of edge-device collaboration from the probabilistic view of point. Specifically, an optimization problem is formulated with objective of minimizing the proportion of tasks that are not served in time. Then, a general delay violation probability for stochastic system is derived, based on which, the task execution delay violation probability at local device and that at the edge server are obtained respectively. Thereafter, the optimization problem can be transferred into a tractable one where task offloading proportion and task offloading rate can be jointly optimized. The effectiveness of the proposed scheme is finally validated by extensive numerical simulations.

Keywords: Edge-device collaboration · Stochastic delay guarantee · Task offloading · Resource allocation

1 Introduction

With the rapid development of 5G techniques, communication and computing at the edge has become a new trend in the current network [1–3]. Edge-device collaboration is able to makes full use of the advantages of local computing and fast edge computing, such that various types of tasks can be executed efficiently. Therefore, edge-device collaboration reveals its huge potential in different fields,

This work was supported in part by the National Natural Science Foundation of China under grants 61901078, 61871062 and 61771082, and in part by Natural Science Foundation of Chongqing under grant cstc2020jcyj-zdxmX0024, and in part by the Science and Technology Research Program of Chongqing Municipal Education Commission under grant KJQN201900609, and in part by University Innovation Research Group of Chongqing under grant CXQT20017.

J. Xiong et al. (Eds.): MobiMedia 2021, LNICST 394, pp. 94–106, 2021.
https://doi.org/10.1007/978-3-030-89814-4_7

such as industrial Internet, smart city, intelligent transportation and intelligent medical treatment [4–6].

In addition to advantages, there are some challenges in practical edge-device collaboration. On one hand, the processing capability of a device is limited, which may not meet the requirements of delay sensitive tasks, especially when task load is heavy [7]. On the other hand, though edge server has relatively high processing capability, the limited wireless bandwidth becomes the bottleneck of edge computing [8]. Thus, how to assign tasks to the device and edge server to execute is critical for an efficient edge-device collaboration system. However, as task arrivals and wireless channel are time-varying, task offloading may fail sometime, which brings additional difficulty in the optimization of edge-device collaboration [9].

In the literature, edge-device collaboration has been widely concerned. The existing research on edge-device collaboration mainly considers the optimization of the overall performance of the system, such as energy efficiency, computational efficiency and resource utilization. In [10], computational efficiency maximization problem of wireless mobile edge computing (MEC) networks was studied under partial and binary computing offloading modes, where an energy harvesting model was considered at the device side. In [11], task offloading and resource allocation among multiple users served by a base station was studied. With consideration of limited resources, mobility of UE and task delay requirements, a scheme was proposed to obtained the trade-off between service delay and energy consumption. Work [12] designed a game theory based deep reinforcement learning (DRL) algorithm to maximize the energy utility of a network where users may refuse to disclose their network bandwidth and preference information. In order to realize reliable computing offloading of delay sensitive services in Internet of Vehicle, a scheme was proposed for an edge-device collaboration system [13]. In [14], a two-stage joint optimization model was studied to maximize the energy efficiency of an edge-device collaboration system. Work [15] also considered the system reliability and used a double time scale mechanism to minimize the energy consumption of users. However, there are some shortcakes in the existing researches of edge-device collaboration. Firstly, existing studies assume that the task arrivals are deterministic. Hence, the schemes proposed by those studies are not appropriate to the scenario where tasks are randomly generated. Secondly, perfect instantaneous channel information is assumed to be available in those works, which are not reasonable in practical network. Thirdly, existing schemes are based on time block, meaning that a policy is optimized and carried out in a short time, which brings heavy overheads to the system. In summary, how to guarantee the stochastic delay performance of task in an edge-device collaboration is still an open problem.

Motivated by this, we focus on an edge-device collaboration system where the randomness of task arrivals and that of wireless channel are both taken into account. We aim to serve as many tasks as possible under a given delay requirement. The formulated optimization problem is regard to the task allocation and offloading rate configuration. The violation probability of task execution is derived at the local device side and edge server respectively. Thereafter, a two-dimension search approach is proposed to obtain the optimal edge-device collaboration decision. Numerical results verify that the proposed scheme is effective and the task execution delay can be guaranteed from the statistical point of view.

The remaining of this paper is organized as follows. Section 2 introduces the system model. In Sect. 3, performance analysis as well as system optimization are conducted. In Sect. 4, numerical results were presented, compared and discussed. Finally, Sect. 5 concludes the paper.

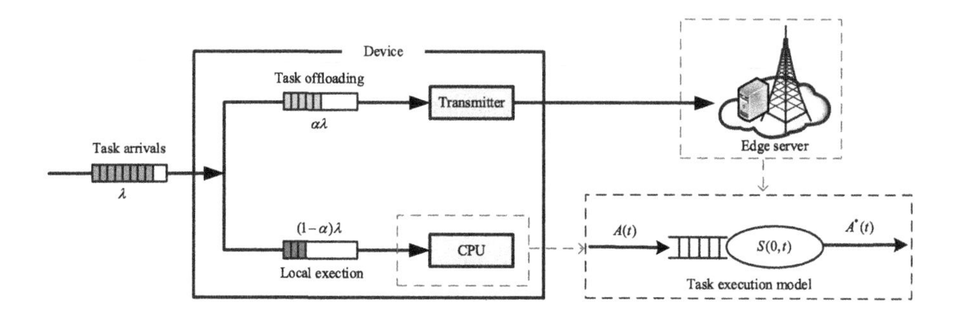

Fig. 1. Task execution in edge-device collaboration manner.

2 System Model

As depicted in Fig. 1, we consider a single-server-single-device edge-device collaboration scenario where the random arrival tasks can be executed by the local device directly or uploaded to the edge server to execute based on first-in-first-out (FIFO) manner. The task arrival process at the device side is assumed to follow Poisson distribution with average rate λ. The packet length of each task is constant and denoted by L. Due to the limited energy of device and the time-varying channel fading, in order to reduce the task execution delay, some tasks should be executed locally and some should be offloaded to the edge server. Here, we use α $(0 \leq \alpha \leq 1)$ to denote the edge-device collaboration parameter, representing the proportion of task offloaded to the edge server. According to the superposition of Poisson process, the task arrival process at the device side and that at the edge server side both follow Poisson distribution with average rates $(1-\alpha)\lambda$ and $\alpha\lambda$ respectively.

In regard to task offloading process, the small-scale fading gain of the channel, denoted by $h(t)$, is assumed to follow Rayleigh distribution with envelope probability density function

$$g_{|h|}(x) = 2xe^{-x^2}. \tag{1}$$

Let p denote the transmission power of the device, l denote the distance between the device and the edge server, ζ denote the path loss constant, φ denote the path loss factor, W denote the uplink bandwidth, N_0 denote the power spectral density of white Gaussian noise. The instantaneous channel capacity for task offloading at time t hold as

$$C(t) = W\log_2\left(1 + \frac{p\zeta l^{-\varphi}|h(t)|^2}{N_0 W}\right), \tag{2}$$

Additionally, as instantaneous channel state is usually unavailable to the local device, it is practical for the device to upload tasks with a fixed rate. Typically, a task is considered to be transmitted successfully if $R \le C(t)$, otherwise, the task offloading fails. Hence, with a given task offloading rate R, the instantaneous channel fading parameter for successful task offloading should hold as follows

$$\begin{gathered} R \le W\log_2\left(1 + \frac{p\zeta l^{-\varphi}|h(t)|^2}{N_0 W}\right) \\ \Updownarrow \\ |h(t)| \ge \sqrt{\frac{N_0 W}{p\zeta l^{-\varphi}}\left(2^{\frac{R}{W}} - 1\right)} \triangleq \delta, \end{gathered} \tag{3}$$

where δ denotes the threshold of channel fading gain, beyond which the task offloading is successful. As a result, the probability of successful task offloading P^{ON} and that of failure task offloading P^{OFF} can be obtained as

$$P^{\mathrm{ON}} = \Pr\{|h(t)| \ge \delta\} = \int_{\delta}^{\infty} 2xe^{-x^2}dx = e^{-\delta^2}, \tag{4}$$

$$P^{\mathrm{OFF}} = \Pr\{|h(t)| < \delta\} = 1 - P^{\mathrm{ON}} = 1 - e^{-\delta^2}. \tag{5}$$

In this paper, we use $D(t(i))$ to denote the execution delay of the i-th task arriving at time t. Besides, the maximum tolerance delay is denoted by d. As mentioned before, deterministic delay guarantee is unavailable to the considered system. Hence, the aim of this paper is to minimize the proportion of tasks that are not served under the delay requirement. Let N denote the number of tasks. The optimization problem can be formulated as follows

$$\begin{gathered} \min_{R,\alpha} \lim_{N\to\infty} \frac{\sum_{i=1}^{N} I_{\{D(t(i))>d\}}}{N}, \\ \text{s.t.}\ \ 0 \le \alpha \le 1 \\ R > 0 \end{gathered} \tag{6}$$

where $I_{\{D(t(i))>d\}}$ is the indicator function. If the execution delay of the ith task exceeds the maximum tolerable delay d, there holds $I_{\{D(t(i))>d\}} = 1$, otherwise,

$I_{\{D(t(i))>d\}} = 0$. The challenge of solving problem (7) lies in characterization of the proportion of tasks that do not meet the delay requirement. In what follows, we use deal with the optimization problem with the help of network calculus theory.

3 Performance Analysis

As the task arrival process is Poisson distributed and the wireless channel is Rayleigh distributed, the task arrival process and task execution process are both as independently and identically distributed (i.i.d). Hence, if the considered system is stable, the delay violation probability suffered by each task is identical. Consequently, we have

$$\begin{aligned}
&\min_{R,\alpha} \lim_{N\to\infty} \frac{\sum_{i=1}^{N} I_{\{D(t(i))>d\}}}{N} \\
&= \min_{R,\alpha} \lim_{N\to\infty} \frac{\sum_{i=1}^{\alpha N} I_{\{D_e(t(i))>d\}} + \sum_{i=1}^{(1-\alpha)N} I_{\{D_l(t(i))>d\}}}{N} \\
&= \alpha \Pr\{D_e > d\} + (1-\alpha)\Pr\{D_l > d\}
\end{aligned} \tag{7}$$

Here, $\Pr\{D_e > d\}$ and $\Pr\{D_l > d\}$ denote the delay violation probabilities of the tasks executed by the edge server and device respectively. Therefore, problem (7) can be transferred to the following problem

$$\begin{aligned}
&\min_{R,\alpha} \ \alpha \Pr\{D_e > d\} + (1-\alpha)\Pr\{D_l > d\} \\
&\text{s.t.} \ \ 0 \le \alpha \le 1 \\
&\qquad R > 0
\end{aligned} \tag{8}$$

To deal with problem (8), we need to derive the delay violation probability of the tasks executed at the device and at the edge server respectively.

3.1 General Delay Violation Probability

In this section, we first derive a general delay violation probability for a system with both i.i.d arrival and i.i.d service processes. Let $A(s,t)$ and $A^*(s,t)$ denote the cumulative arrivals and cumulative departures of a system during time $(s,t]$. And let $S(s,t)$ denote the cumulative service capacity during time $(s,t]$. According to network calculus theory [16], we have

$$A^*(0,t) = \inf_{0\le s\le t} \{A(0,s) + S(s,t)\}. \tag{9}$$

As for any $d \geq 0$, there always holds $\{D(t) > d\} \subseteq \{A(0,t) > A^*(0,t+d)\}$, we then have

$$\begin{aligned} &\Pr\{D > d\} \leq \Pr\{A(0,t) - A^*(0,t+d) > 0\} \\ =& \Pr\{A(0,t) - \inf_{0 \leq s \leq t+d} \{A(0,s) + S(0,t+d-s)\} > 0\} \\ =& \Pr\{\sup_{0 \leq s \leq t} \{A(s,t)\} - S(0,t)\} > S(0,d)\}. \end{aligned} \tag{10}$$

We then let

$$\begin{aligned} V_s &= e^{\theta(A(t-s,t) - S(t-s,t))} \\ Y_k &= A(k-1,k) \\ Z_k &= S(k-1,k) \end{aligned} \tag{11}$$

There holds

$$\begin{aligned} V_{s+1} &= e^{\theta(A(t-s-1,t) - S(t-s-1,t))} \\ &= e^{\theta \sum_{k=t-s}^{t} (Y_k - Z_k)} \\ &= V_s e^{\theta(Y_{t-s} - Z_{t-s})}. \end{aligned} \tag{12}$$

Because the task arrival and system service processes are independent of each other and both have independent and identically distributed increments, there holds

$$\begin{aligned} &\mathbb{E}\left[V_{s+1} \left| V_1, V_2, \cdots, V_s \right.\right] \\ &= \mathbb{E}\left[V_{s+1} \left| Y_t, Y_{t-1}, \cdots, Y_{t-s+1}, Z_t, Z_{t-1}, \cdots, Z_{t-s+1} \right.\right] \\ &= \mathbb{E}\left[V_s e^{\theta(Y_{t-s} - Z_{t-s})} \left| Y_t, Y_{t-1}, \cdots, Y_{t-s+1} \right.\right] \\ &= \mathbb{E}\left[V_s \left| Y_t, Y_{t-1}, \cdots, Y_{t-s+1} \right.\right] \mathbb{E}\left[e^{\theta Y_{t-s}}\right] \mathbb{E}\left[e^{-\theta Z_{t-s}}\right] \\ &= V_s \mathbb{E}\left[e^{\theta A(0,1)}\right] \mathbb{E}\left[e^{-\theta S(0,1)}\right] \\ &\overset{(a)}{\leq} V_s, \end{aligned} \tag{13}$$

Here, step (a) holds since the considered system is stable. Hence, $V_1, V_2, \cdots, V_s$ constitute a nonnegative supermartingale [17,18], we can then obtain the general delay violation probability as

$$\begin{aligned} &\Pr\{D(t) > d\} \\ &= \Pr\{\sup_{0 \leq s \leq t} \{e^{A(s,t) - S(s,t)}\} > e^{S(0,d)}\} \\ &\leq \Pr\{\sup_{1 \leq s \leq t} \{V_{t-s}\} > e^{S(0,d)}\} \\ &\leq \Pr\{V_1 > e^{S(0,d)}\} \\ &\overset{(a)}{\leq} \mathbb{E}[e^{-\theta S(0,d)}] \mathbb{E}\left[e^{\theta A(0,1)}\right] \mathbb{E}[e^{-\theta S(0,1)}] \\ &\overset{(b)}{\leq} \mathbb{E}[e^{-\theta S(0,d)}] \end{aligned} \tag{14}$$

Here step (a) holds based on Chernoff boundary. In step (b), we apply the system stability. More specific, the following equation holds for a stable system.

$$\mathbb{E}[e^{\theta A(0,1)}] \mathbb{E}[e^{-\theta S(0,1)}] \leq 1 \tag{15}$$

where θ is a free parameter. According to (14), delay violation probability decreases as θ increases. Hence, optimal θ can be obtained as

$$\theta^{\text{opt}} = \max\{\theta : \mathbb{E}[e^{\theta A(0,1)}]\mathbb{E}[e^{-\theta S(0,1)}] \leq 1\} \tag{16}$$

Based on (14) and (16), the delay violation probability of the tasks executed at the device and at the edge server can be further derived respectively.

3.2 Local Execution Delay Performance

Let λ_l denote the task arrival rate for the local execution, and there holds $\lambda_l = (1-\alpha)\lambda$. Besides, let $A_l(s,t)$ and $S_l(s,t)$ denote the cumulative arrivals and cumulative service capacity of at the device side during time $(s,t]$. We then have

$$S_l(s,t) = \frac{f_l}{k}(t-s), \tag{17}$$

where f_l denotes the CPU frequency of the device and k denotes the required cycles for executing one bit of each task. Therefore, the delay violation probability at the device side holds as

$$\begin{aligned}\Pr\{D_l > d\} =& \Pr\{D_l(t(i)) > d\} \\ \leq& \mathbb{E}[e^{-\theta_l S_l(0,d)}] \\ =& e^{-\theta_l \frac{f_l}{k} d}\end{aligned} \tag{18}$$

Here, as the task arrival process is Poisson distributed, we have

$$\begin{aligned}&\mathbb{E}\left[e^{\theta_l A_l(0,1)}\right] \\ &= \sum_{n=0}^{\infty} \mathbb{E}\left[e^{\theta_l A_l(0,1)} \left| N(t) = n\right.\right] \Pr\{N(t) = n\}] \\ &= \sum_{n=0}^{\infty} \mathbb{E}\left[e^{\theta_l A_l(0,1)} \left| N(t) = n\right.\right] e^{-\lambda_l} \frac{\lambda_l^n}{n!} \\ &= \sum_{n=0}^{\infty} \mathbb{E}\left[e^{\theta_l n L}\right] e^{-\lambda_l} \frac{\lambda_l^n}{n!} \\ &= e^{-\lambda_l} \sum_{n=0}^{\infty} \frac{\left(\lambda_l e^{\theta_l L}\right)^n}{n!} \\ &= e^{(1-\alpha)\lambda\left(e^{\theta_l L}-1\right)}.\end{aligned} \tag{19}$$

Thus, the optimal free parameter θ_l holds as

$$\begin{aligned}\theta_l^{\text{opt}} &= \max\{\theta_l : \mathbb{E}[e^{\theta_l A_l(0,1)}]\mathbb{E}[e^{-\theta_l S_l(0,1)}] \leq 1\} \\ &= \max\{\theta_l : \frac{(1-\alpha)\lambda(e^{\theta_l L}-1)}{\theta_l} \leq \frac{f_l}{k}\}\end{aligned} \tag{20}$$

3.3 Edge Execution Delay Performance

As the task processing capacity of the edge server is much more powerful than that of the device, we only consider the task offloading delay as the task execution delay at the edge server. Let λ_e denote the task arrival rate for the local execution, and there holds $\lambda_e = \alpha\lambda$. Besides, let $A_e(s,t)$ and $S_e(s,t)$ denote the cumulative arrivals and cumulative service capacity of at the edge server during time $(s,t]$. As tasks can only be offloaded successfully when $R < C(t)$, we have

$$S_e(t-1,t)\begin{cases} = R, & R \leq C(t) \\ = 0, & R > C(t) \end{cases} \tag{21}$$

Therefore, the delay violation probability at the edge server holds as

$$\begin{aligned} \Pr\{D_e > d\} &= \Pr\{D_e(t(i)) > d\} \\ &\leq \mathbb{E}[e^{-\theta_e S_e(0,d)}] \\ &= (e^{-\theta_e R} P^{\mathrm{ON}} + P^{\mathrm{OFF}})^d \end{aligned}, \tag{22}$$

where P^{ON} and P^{OFF} can be obtained from (4) and (5) respectively.

Similar to (19), we also have

$$\mathbb{E}\left[e^{\theta_e A_e(0,1)}\right] = e^{\alpha\lambda\left(e^{\theta_e L}-1\right)}. \tag{23}$$

Thus, the optimal free parameter θ_e holds as

$$\begin{aligned} \theta_e^{\mathrm{opt}} &= \max\{\theta_e : \mathbb{E}[e^{\theta_e A_e(0,1)}]\mathbb{E}[e^{-\theta_e S_e(0,1)}] \leq 1\} \\ &= \max\{\theta_e : \alpha\lambda(e^{\theta_e L} - 1) \leq -\ln(e^{-\theta_e R} P^{\mathrm{ON}} + P^{\mathrm{OFF}})\} \end{aligned} \tag{24}$$

3.4 Problem Solution

From (18) and (22), problem (8) can be further transferred into a tractable one

$$\begin{aligned} \min_{R,\alpha} \ & \alpha(e^{-\theta_e R} P^{\mathrm{ON}} + P^{\mathrm{OFF}})^d + (1-\alpha)(e^{-\theta_l \frac{f_l}{k} d}) \\ \text{s.t.} \ & 0 \leq \alpha \leq 1 \\ & R > 0 \end{aligned}. \tag{25}$$

The above problem can be solved through two-dimension search. Note that the system optimization is from the statistical view of point, which means it is applied to long-term edge-device collaboration. Thus, it is reasonable to apply two-dimension search though such approach has a high computation complexity.

4 Results

Simulation results are provided and discussed in this section. And we use MATLAB to carried out the simulation experiments. Different activation factors and contention intensity are considered in the simulation. The simulation parameters are set as Table 1.

The value of the average transmission rate is related to the user's transmission rate, and the relationship is shown in Fig. 2. And it depicts that there is an

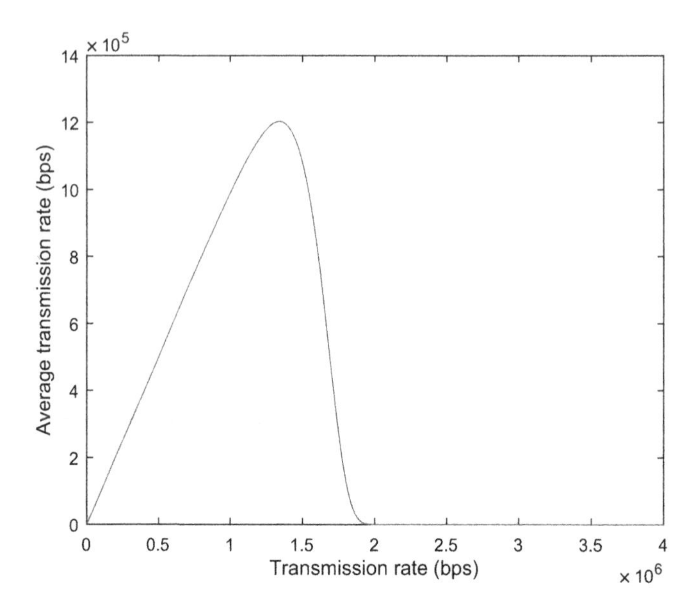

Fig. 2. Relationship between average transmission rate and user's transmission rate.

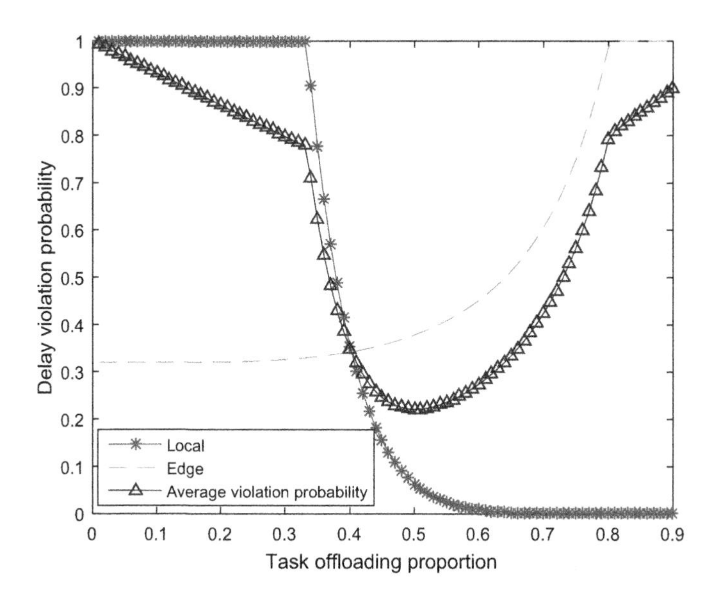

Fig. 3. Relationship between violation probability and task offloading proportion.

Table 1. Simulation parameters

Symbol	Value
Wireless channel bandwidth W	100 kHz
Noise power N_0	1×10^{-17} W
Transmission power of device p	0.1 W
Distance between device and edge server l	10 m
Task arrival rate λ	15 packets/s
Size of each task package	1×10^5 bits
CPU frequency of device f	1×10^9 cycles/s
Number of cycles required for 1 bit data processing k	1×10^3 cycles
Maximum tolerable delay d	0.5 s
Path loss	$10^{-3}l^{-3}$

optimal transmission data rate $R = R^{opt}$ to maximize the average transmission rate of the channel $\bar{R}$. From Fig. 2, we can get $R^{opt}= 1.34 \times 10^6$ bit/s.

Figure 3 shows the relationship between local and offloading delay violation probabilities and task offloading proportion. For local computing, the local delay violation probability decreases with the increase of task offloading proportion. The delay violation probability of edge computing increases with the increase of task offloading proportion. This is because with the increase of task offloading proportion, more tasks are allocated to the edge server for processing, which increases the load of wireless channel and is difficult to guarantee the delay requirement. On the contrary, with the increase of task offloading proportion, the number of tasks left for local processing gradually decreases, and the delay violation probability of local devices becomes smaller and smaller under the condition of the same computing frequency. It can be seen from Fig. 3 that there is an optimal task offloading proportion of 0.50 to minimize the average delay violation probability.

Figure 4 is the relationship between the offloading violation probability and the delay requirement at three transmission rates. It can be seen from the figure that with the increase of delay demand, the offloading violation probability decreases exponentially, and the delay guarantee reliability also increases. Moreover, when the transmission rate is smaller, the offloading violation probability is smaller. This is because the smaller the transmission rate, the less likely it is to exceed the capacity of the channel, thus the greater the probability of transmission success and the smaller the probability of offloading violation.

Figure 5 shows the influence of task offloading rate on the probability of offloading violation under different delay requirements. When the time delay requirement is 0.6 s, the violation probability under different offloading rates is the lowest among them. The reason is that the lower the delay requirement, the less the violation of task delay occurs, that is, the smaller the probability of offloading failure. It can be seen from the figure that there are optimal task offloading rates under different delay requirements, and the delay violation probability is the minimum.

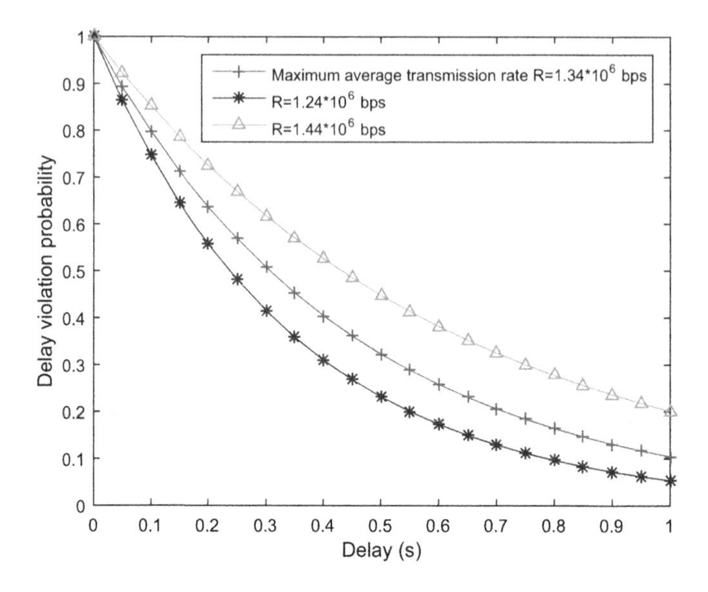

Fig. 4. Relationship between delay violation probability and delay demand.

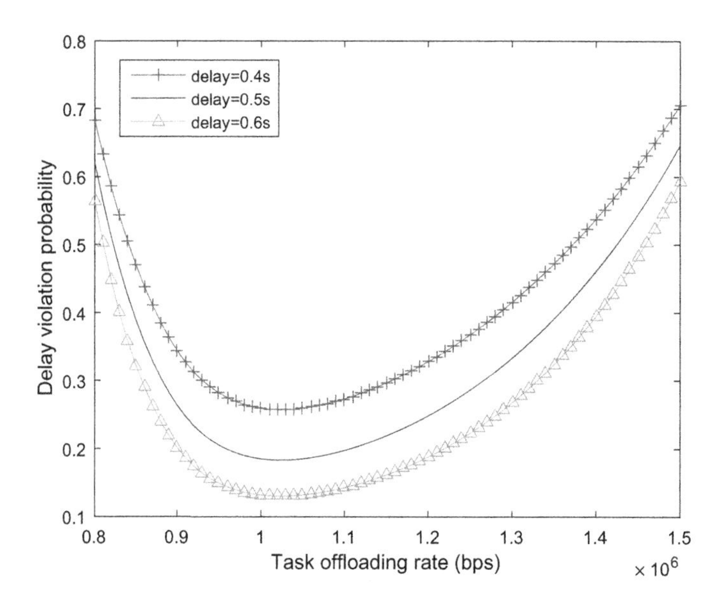

Fig. 5. The relationship between delay violation probability and task offloading rate.

5 Conclusions

In this paper, stochastic delay guarantee was studied in an edge-device collaboration system. The task execution delay violation probability at the local device and that at the edge server were derived respectively. The task allocation and task offloading rate were jointly optimized to minimize the proportion of tasks that could not be served in time. The effectiveness of the proposed scheme was finally validated by extensive numerical simulations. Since this paper only considers a single-node-single-server scenario, the future research will pay more attention to a multi-node-multi-server scenario, where edge-device collaboration is more complex.

References

1. Kong, X., et al.: Mobile edge collaboration optimization for wearable internet of things: a network representation-based framework. IEEE Trans. Ind. Inf. **17**, 5050–5058 (2020). https://doi.org/10.1109/TII.2020.3016037
2. Chen, Z., Li, D., Cuo, J.: Analysis and implementation of a qos-aware routing protocol based on a heterogeneous network. Int. J. Performability Eng. **16**(8), 1235–1244 (2020)
3. Xiong, J., Chen, X., Yang, Q., Chen, L., Yao, Z.: A task-oriented user selection incentive mechanism in edge-aided mobile crowdsensing. IEEE Trans. Netw. Sci. Eng. **7**(4), 2347–2360 (2020)
4. Liu, Y., Peng, M., Shou, G., Chen, Y., Chen, S.: Toward edge intelligence: multiaccess edge computing for 5G and internet of things. IEEE Internet Things J. **7**(8), 6722–6747 (2020)
5. Taleb, T., Samdanis, K., Mada, B., Flinck, H., Dutta, S., Sabella, D.: On multi-access edge computing: a survey of the emerging 5G network edge cloud architecture and orchestration. IEEE Commun. Surv. Tutorials **19**(3), 1657–1681 (2017)
6. Xiong, J., Bi, R., Zhao, M., Guo, J., Yang, Q.: Edge-assisted privacy-preserving raw data sharing framework for connected autonomous vehicles. IEEE Wirel. Commun. **27**(3), 24–30 (2020)
7. Li, Y., Wang, X., Gan, X., Jin, H., Fu, L., Wang, X.: Learning-aided computation offloading for trusted collaborative mobile edge computing. IEEE Trans. Mob. Comput. **19**(12), 2833–2849 (2020)
8. Gao, W., Zhao, Z., Yu, Z., Min, G., Yang, M., Huang, W.: Edge-computing-based channel allocation for deadline-driven IoT networks. IEEE Trans. Ind. Inf. **16**(10), 6693–6702 (2020)
9. Li, S., Lin, S., Cai, L., Li, W., Zhu, G.: Joint resource allocation and computation offloading with time-varying fading channel in vehicular edge computing. IEEE Trans. Veh. Technol. **69**(3), 3384–3398 (2020)
10. Zhou, F., Hu, R.Q.: Computation efficiency maximization in wireless-powered mobile edge computing networks. IEEE Trans. Wirel. Commun. **19**, 3170–3184 (2020)
11. Zhan, W., Luo, C., Min, G., Wang, C., Zhu, Q., Duan, H.: Mobility-aware multi-user offloading optimization for mobile edge computing. IEEE Trans. Veh. Technol. **69**(3), 3341–3356 (2020)
12. Zhan, Y., Guo, S., Li, P., Zhang, J.: A deep reinforcement learning based offloading game in edge computing. IEEE Trans. Comput. **69**(6), 883–893 (2020)

13. Hou, X., et al.: Reliable computation offloading for edge-computing-enabled software-defined IoV. IEEE Internet Things J. **7**(8), 7097–7111 (2020)
14. Cao, T., et al.: Reliable and efficient multimedia service optimization for edge computing-based 5G networks: game theoretic approaches. IEEE Trans. Netw. Serv. Manage. **17**(3), 1610–1625 (2020)
15. Liu, C., Bennis, M., Debbah, M., Poor, H.V.: Dynamic task offloading and resource allocation for ultra-reliable low-latency edge computing. IEEE Trans. Commun. **67**(6), 4132–4150 (2019)
16. Jiang, Y.: Network calculus and queueing theory: two sides of one coin. In: ICST Conference on Performance Evaluation Methodologies and Tools, ICST, pp. 37–48 (2009)
17. Ding, J., Zhu, X., Chen, X.: State equations in stochastic process algebra models. IEEE Access **7**, 61195–61203 (2019)
18. Kingman, J.F.C.: A martingale inequality in the theory of queues. Math. Proc. Cambridge Philos. Soc. **60**(2), 359–361 (1964)

Hybrid Human-Artificial Intelligence Enabled Edge Caching Based on Interest Evolution

Zhidu Li[1,2,3](✉), Fuxiang Li[1,2,3], Dapeng Wu[1,2,3], Honggang Wang[4], and Ruyan Wang[1,2,3]

[1] School of Communication and Information Engineering, Chongqing University of Posts and Telecommunications, Chongqing, China
lizd@cqupt.edu.cn

[2] Key Laboratory of Optical Communication and Networks, Chongqing, China

[3] Key Laboratory of Ubiquitous Sensing and Networking, Chongqing, China

[4] Electrical and Computer Engineering Department, University of Massachusetts Dartmouth, Dartmouth, USA

Abstract. How to cache appropriable contents for users from huge amount of candidates is a challenge in edge caching network. To address this challenge, this paper studies an edge caching scheme based on user interest, where an interest extraction and evolution network is developed. Specifically, the input features are first classified and embedding. The user interest is then mined and modeled according to the user historical behaviors with the gated recurrent unit network. Thereafter, the user interest evolution process is studied by analyzing the impact of the previous interests on the current interest through an attention mechanism. The group interest model is further studied by merging user interest evolution and social relationships among contents, based on which edge caching scheme is obtained. The effectiveness of the proposed scheme is finally validated by extensive experiments with a real-world dataset. The analysis in this paper sheds new light on edge content caching from user interest evolution perspective.

Keywords: Edge caching · User interest evolution · Group interest · Cache hit rate · User hit rate

This work was supported in part by the National Natural Science Foundation of China under grants 61901078, 61871062 and 61771082, and in part by Natural Science Foundation of Chongqing under grant cstc2020jcyj-zdxmX0024, and in part by the Science and Technology Research Program of Chongqing Municipal Education Commission under grant KJQN201900609, and in part by University Innovation Research Group of Chongqing under grant CXQT20017.

J. Xiong et al. (Eds.): MobiMedia 2021, LNICST 394, pp. 107–122, 2021.
https://doi.org/10.1007/978-3-030-89814-4_8

1 Introduction

With the widespread the rapid development of communication techniques, watching videos online has become one of the most popular entertainments [1–3]. However, due to the huge population and variety of contents, traditional cloud-based caching may not meet the service of quality for users. As a result, edge caching technique is developed to enable popular contents cached at an edge server close to users [4,5].

Due to limited caching capacity, how to cache contents efficiently is a fundamental issue of edge caching. As a consensus, user requests are highly related to user interest. Hence, an accurate prediction on user interest is able to guarantee an effective content caching decision, such that reducing the waiting time of content requests. User interest, however, is difficult to predict due to its time-varying property. For instance, a user is interested in the comedy movie at time t_1 while his/her interest tends towards the romance one at time t_2, as shown in Fig. 1. Consequently, dynamic user interest mining is critical in for edge caching.

In the literature, edge caching are usually based on content popularity with assumption of Zipf distribution [6]. In [7], an integrated content distribution framework was proposed cache contents based on contextual information. In [8], deep learning was applied to predict and cache popular contents in the edge server. In [9], content offloading and caching were jointly optimized by using reinforcement learning. Works [6–9] mainly designed caching schemes in terms of communication optimization, while user interest was not taken into account. Recently, some researches try to design caching policy based on recommendation system approach [10,11]. Such approach performs well in content feature characterization and user interest mining [12–16]. In [14], classical matrix factorization of the recommender system and convolutional neural network were combined to predict user interest, based on which contents are selected to cache at the edge server. In [15], the idea of soft cache was proposed to improve the cache hit rate. In detail, similar contents were recommended to a user before relaying the requested contents from the cloud if the user request was not meet at the edge server. To characterize the temporal features of content popularity, work [16] proposed a advanced long and short-term memory network. However, the above mentioned works only studied the caching performance from the edge server, while user hit rate was omitted. Additionally, how to employ the time-varying user interest to improve the edge caching performance is still an open issue.

Motivated by this, we study a hybrid human-artificial intelligence enabled edge caching scheme in a edge caching network. A cloud-edge-end collaboration framework is first constructed to improve the cache hit rate and user hit rate at the same time. An interest extraction and evolution network is then proposed the analyze the individual interest on the contents. Specifically, the gated recurrent unit (GRU) is employed to capture the user interest over a period of time. Additionally, attention mechanism is applied to analyze the impacts of user historical behaviors on the future behavior. Moreover, an intelligent based content prediction scheme and a socially-aware content prediction scheme are merged into a group interest model, based on which, the edge caching can be finally decided. The effectiveness of the proposed caching scheme is further validated compared to four baseline schemes.

The rest of the paper is organized as follows. Section 2 introduces the system model as well as performance metrics. Section 3 proposes the IEEN model for user interest mining. In Sect. 4, content caching scheme is studied. In Sect. 5, extensive experiment results are presented and discussed. Finally, Sect. 6 concludes the paper.

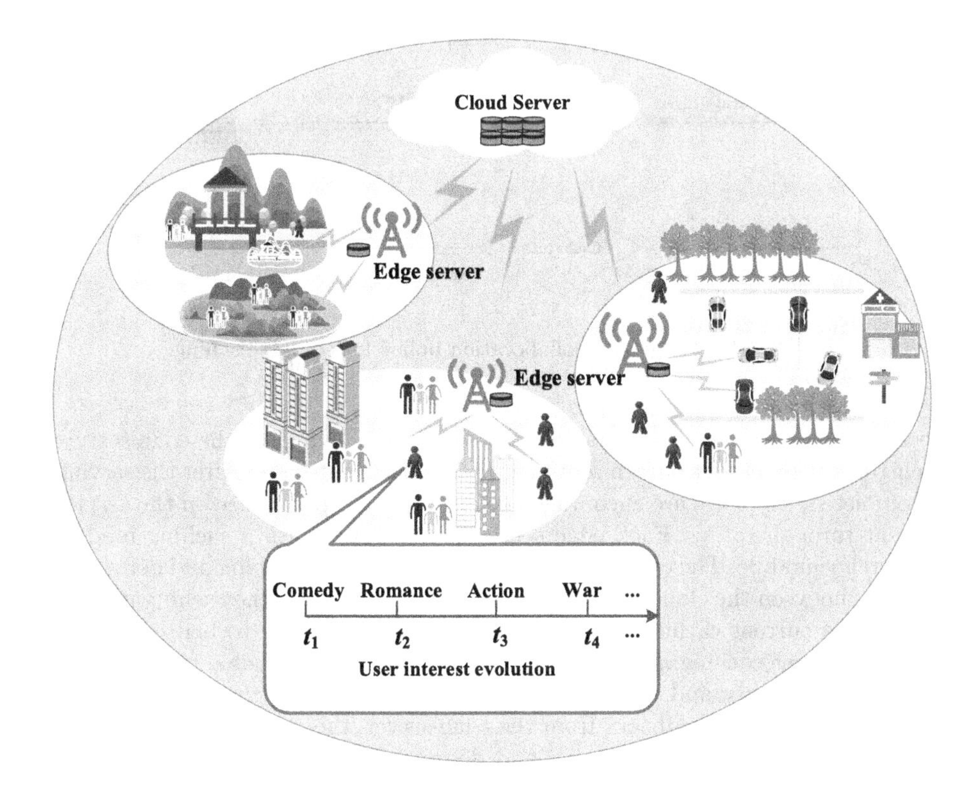

Fig. 1. Network model.

2 System Model

As depicted in Fig. 1, the considered network aims to provide services to users located in different types of areas wherein the users may have heterogeneous interests. A cloud server is deployed and assumed to store all of the contents that may be requested by the users. The contents stored in the cloud is denoted by set $\mathcal{F} = \{f_1, f_2, \cdots, f_C\}$, where f_c denotes the c-th content and C denotes the number of contents. Besides, the edge servers are deployed in each area to cache contents from the cloud server. In this regard, user requests can be responded more quickly if they are hit by the edge servers.

In this paper, we propose a cloud-edge-end collaboration policy to deal with such interest evolution problem and improve the edge caching performance. As

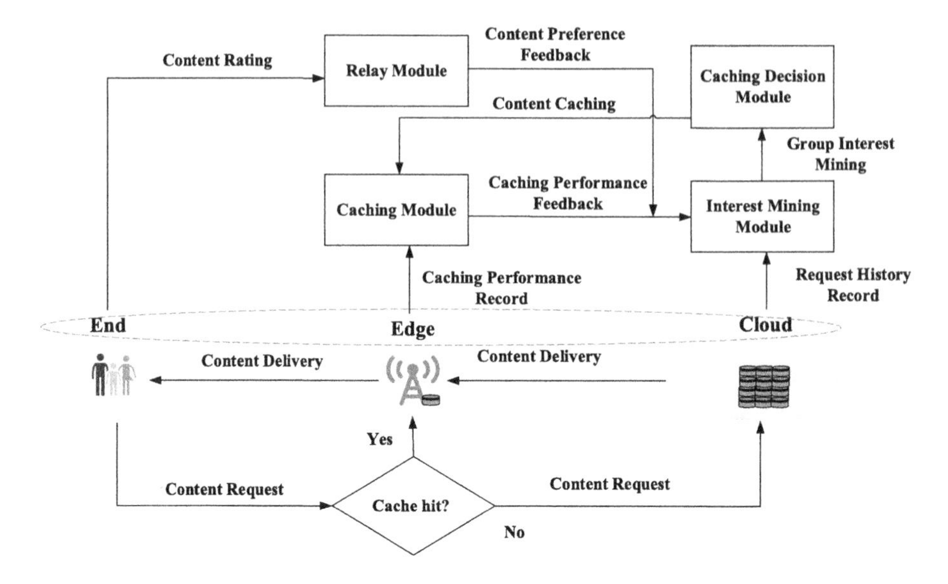

Fig. 2. Cloud-edge-end collaboration policy for content caching.

depicted in Fig. 2, the end users are responsible for sending to the edge servers their content requests that can represent their current interests. After the serving is completed, the users are encouraged to express their preference on the service content through rating. Each edge server is configured with a caching module and a relay module. The caching module is responsible for caching and delivering contents between the cloud server and the users. In addition, the caching module records the current caching performance and sends feedback to help the cloud server to adjust caching decision. The relay module is responsible for collecting user request history and content preference and forwarding them to the cloud server. Based on the feedbacks from the edge server, the cloud server mines the law of individual interest evolution and designs group interest model to make caching decision.

In practical, an edge caching decision duration should be larger than the reading or watching time of a content, since a user has a higher probability to stay within the service coverage of an edge sever. Consequently, a user may request multiple contents and a content may be requested by multiple users within an edge caching decision duration. Without loss of generality, we focus on an edge server network with caching capacity Φ. The users within the service coverage of such edge server is denoted by set $\mathcal{U} = \{U_1, U_2, \cdots, U_N\}$, where N denotes the number of users. Besides, we use binary variable $\alpha_{n,c}$ to indicate the user request. If content f_c is requested by user U_n, $\alpha_{n,c} = 1$, otherwise $\alpha_{n,c} = 0$. Similarly, binary variable $\beta_c = 1$ indicates content f_c is cached by the edge server, and otherwise $\beta_c = 0$. From the perspective of service provider, an effective caching scheme should make sure 1) as many cached contents as possible can be requested by users, and 2) as many user requests as possible can be hit at the same time, such that a high profit can be acquired. In this sense, we introduce two metrics to evaluate the caching performance, which are cache hit rate (CHR) and user hit rate (UHR).

The CHR is defined as the proportion of the cached contents that are requested by users during an edge caching decision duration, there holds

$$\text{CHR} = \frac{\sum\limits_{f_c \in \mathcal{F}} \min\{\sum\limits_{U_n \in \mathcal{U}} \alpha_{n,c}, 1\}\beta_c}{\Phi}. \tag{1}$$

The UHR is defined as the proportion of users whose requests can be hit by the edge server, there holds

$$\text{UHR} = \frac{\sum\limits_{U_n \in \mathcal{U}} \min\{\sum\limits_{f_c \in \mathcal{F}} \alpha_{n,c}\beta_c, 1\}}{\sum\limits_{U_n \in \mathcal{U}} \min\{\sum\limits_{f_c \in \mathcal{F}} \alpha_{n,c}, 1\}}. \tag{2}$$

Therefore, we can formulate two optimization problems about edge caching scheme in terms of CHR and UHR respectively.

$$\begin{aligned} \mathbf{P1} \quad & \max_{\boldsymbol{\beta}} \ \text{CHR} = \frac{\sum\limits_{f_c \in \mathcal{F}} \min\{\sum\limits_{U_n \in \mathcal{U}} \alpha_{n,c}, 1\}\beta_c}{\Phi} \\ \text{s.t.} \quad & \text{C1}: \sum_{f_m \in \mathcal{M}} \beta_m \leq \Phi \end{aligned} \tag{3}$$

$$\begin{aligned} \mathbf{P2} \quad & \max_{\boldsymbol{\beta}} \ \text{UHR} = \frac{\sum\limits_{U_n \in \mathcal{U}} \min\{\sum\limits_{f_c \in \mathcal{F}} \alpha_{n,c}\beta_c, 1\}}{\sum\limits_{U_n \in \mathcal{U}} \min\{\sum\limits_{f_c \in \mathcal{F}} \alpha_{n,c}, 1\}} \\ \text{s.t.} \quad & \text{C1}: \sum_{f_m \in \mathcal{M}} \beta_m \leq \Phi \end{aligned} \tag{4}$$

where $\boldsymbol{\beta} = \{\beta_1, \beta_2, \cdots, \beta_C\}$ denotes the content caching decision. In problems P1 and P2, constraint C1 means that the number of contents selected to cache should not exceed the edge caching capacity. As mentioned before, user interest is time-varying and cannot be characterized by a model-based approach. Thus, the user requests $\boldsymbol{\alpha} = \{\alpha_{n,c} : 1 \leq n \leq N, 1 \leq c \leq C\}$ cannot be ascertain exactly in advance, which means maximizing CHR or UHR is infeasible. Moreover, it is unknown that whether there exits conflict between CHR and UHR. In this paper, data-based idea is resorted to design a hybrid human-artificial intelligence enabled content caching scheme where user interest evolution and group interest are both taken into account.

3 Individual Interest Evolution

In this section, an interest extraction and evolution network (IEEN) is designed to analyze the interest evolution process. As depicted in Fig. 3, our proposed

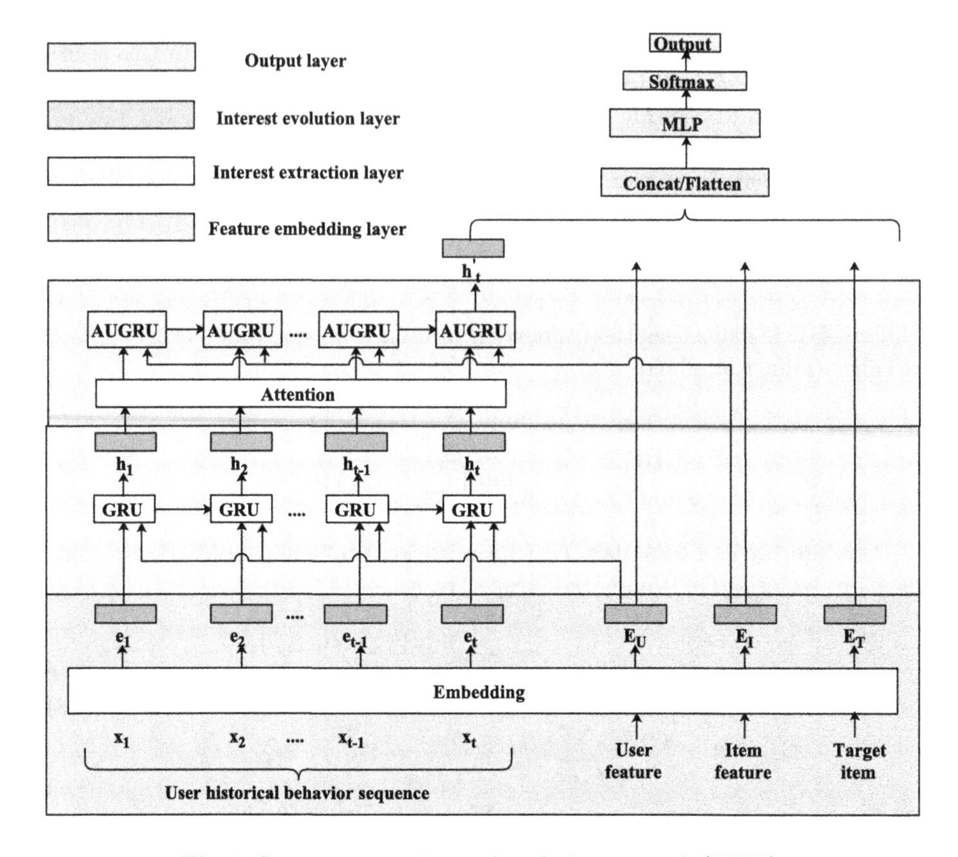

Fig. 3. Interest extraction and evolution network (IEEN).

IEEN model consists of four layers, i.e., feature embedding layer, interest extraction layer, interest evolution layer, and output layer. The feature embedding layer is responsible for converting the sparse high-dimensional input features into the dense low-dimensional ones, such that the features are more easily represented. The interest extraction layer is responsible for learning the behaviors of a user according to the input dense low-dimensional features. In particular, GRU is resorted to extract user interest from the historical behavior sequence preliminarily. The interest evolution layer is responsible for analyzing the relationship among the historical interests and then deducing the future interest for a user. In the output layer, user interest on a content is obtained through feature merging with a full-connective network.

3.1 Feature Embedding

In order to guarantee high prediction accuracy of user interest, the input features should be well represented by IEEN. In this paper, we choose four fields as the input of IEEN model, which are user behavior sequence, target item, item feature, and user feature. The user behavior sequence includes a list of historical IDs of items browsed by the user, which can be expressed as:

$$\mathbf{X}_n = [x_1, x_2, x_3, ..., x_{t_0}, ..., x_t], \tag{5}$$

where $\mathbf{X}_n \in \mathbb{R}^{1\times t}$ represents the behavior sequence of the i-th user, x_{t_0} denotes the ID of the t_0-th item browsed by the user, and t denotes the length of the user behavior sequence.

Besides, target feature is defined as the items that are requested by the users, such as music, video and etc. Item feature usually includes item attributes, item category, description information, etc. User feature includes the user gender, age, occupation, residence information and etc. Among the item features and user features, item category, gender, age and etc. all belong to category features that can be expressed as sparse feature vectors with the one-hot encoding scheme. However, gradient update will be degraded if the sparse category features are directly sent to the interest extraction layer, which increases the model training complexity significantly. Hence, we apply feature embedding to convert the high-dimensional sparse vectors into low-dimensional dense vectors. Note that the user features, e.g., gender, age, and etc., are statistically stable within a large time scale, the model training time can consequently be reduced by pre-training the embedded user features. Moreover, feature embedding extracts the connection between different feature vectors, which enhances the model memory ability.

3.2 Interest Extraction

As the time-varying user behaviors can be represented by a sequence, we resort to the GRU model to extract the user interest from the user behavior sequence. Compared with recurrent neural network (RNN) and long short-term memory network (LSTM), GRU can retain more information of the long-term sequence. Additionally, the structure of the GRU model is simpler and the training speed is faster [17]. The GRU model consists of the update gate and reset gate. The update gate decides how much of the previous information in the user behavior sequence to remember and to be passed along to the future. The reset gate determines how much of the previous information should be omitted. The GRU model can be expressed as follows:

$$\mathbf{z_t} = \sigma\left(\mathbf{W}^z(\mathbf{e}_t + \mathbf{E}_U) + \mathbf{N}^z\mathbf{h}_{t-1} + \mathbf{b}^z\right), \tag{6}$$

$$\mathbf{r_t} = \sigma\left(\mathbf{W}^r(\mathbf{e}_t + \mathbf{E}_U) + \mathbf{N}^r\mathbf{h}_{t-1} + \mathbf{b}^{\mathrm{r}}\right), \tag{7}$$

$$\tilde{\mathbf{h}}_t = \tanh\left(\mathbf{W}^h(\mathbf{e}_t + \mathbf{E}_U) + \mathbf{N}^h(\mathbf{r_t} \odot \mathbf{h}_{t-1}) + \mathbf{b}^h\right), \tag{8}$$

$$\mathbf{h}_t = (\mathbf{1} - \mathbf{z_t}) \odot \mathbf{h}_{t-1} + \mathbf{z_t} \odot \tilde{\mathbf{h}}_t, \tag{9}$$

where $\mathbf{z_t}$, $\mathbf{r_t}$, $\tilde{\mathbf{h}}_t$, $\mathbf{h}_t$ denotes the update gate, the reset gate, candidate hidden state vector, the hidden state vector of the current time step, respectively, σ is the sigmoid activation function, $\mathbf{W}^z, \mathbf{W}^r, \mathbf{W}^h$ and $\mathbf{N}^z, \mathbf{N}^r, \mathbf{N}^h$ are the training parameters, $\mathbf{e}(t)$ denotes the embedding vector of x_t, $\mathbf{E}_U$ denotes the embedding of user feature, $\mathbf{b}^z, \mathbf{b}^r, \mathbf{b}^h$ denote biases, $\odot$ denotes element-wise multiplication.

3.3 User Interest Evolution

As user interest is time-varying and highly related among different time, historical user interest should be considered while predicting the future user interest. Intuitively, different previous interests play different roles in the final prediction. In order to find out the impacts of different previous interests on the interest evolution process, an attention mechanism is introduced. The attention mechanism can automatically learn weights and extract important user interest features from a long user behavior sequence. The output of the attention network holds as [18]

$$\begin{aligned}\mathbf{a}(t) &= \mathrm{softmax}\left(\frac{\mathbf{Q}\mathbf{K}^{\mathrm{T}}}{\sqrt{d_k}}\right)\mathbf{V} \\ &= \mathrm{softmax}\left(\frac{\mathbf{Q}\mathbf{h}_t^{\mathrm{T}}}{\sqrt{d_k}}\right)\mathbf{h}_t \\ &= \mathrm{softmax}\left(\frac{\mathbf{Q}\left\|\mathbf{h}_t\right\|}{\sqrt{d_k}}\right)\end{aligned} \tag{10}$$

where $\mathbf{Q} = [\mathbf{h}_1, \mathbf{h}_2, ...\mathbf{h}_{t_0}, ...\mathbf{h}_t]$ denotes the output sequence of the interest extraction layer, $\mathbf{h}_{t_0}$ denotes the output of t_0-th GRU in the interest extraction layer, $\mathbf{K} = \mathbf{V}$ denote the output of t-th GRU in the interest extraction layer, d_k represents the normalization parameter. From (10), the attention weight of use interest can be obtained. Note that in the long-term sequence, there may exist interest shift caused by other reasons, the attention network can ignore these small deviations and mines the main characteristics from the long-term sequence.

As user interest evolutes with time, we thus apply GRU with attention update gate(AUGRU) to learn the relationship between user interest of different time. The input of the GRU model includes the output attention weight $\mathbf{a}(t)$ and the item features $\mathbf{E}_I$. We use attention weight $\mathbf{a}(t)$ to update the gate of GRU.

$$\mathbf{z}_\mathbf{t}^{'} = \mathbf{a}\left(\mathbf{t}\right) * \mathbf{z}_\mathbf{t} \tag{11}$$

$$\mathbf{h}_\mathbf{t}^{'} = \left(\mathbf{1} - \mathbf{z}_\mathbf{t}^{'}\right) \odot \mathbf{h}_{\mathbf{t}-1}^{'} + \mathbf{z}_\mathbf{t}^{'} \odot \tilde{\mathbf{h}}_\mathbf{t}^{'} \tag{12}$$

The hidden state vector of the current time step $\mathbf{h}_t'$ can be trained and obtained similarly as the one in interest extraction layer.

3.4 Output Layer

In the output layer, the user preference of a given content can be analyzed by a full-connective network with the predicted future individual interest vector $\mathbf{h}_t'$, the user feature $\mathbf{E}_U$, the item features $\mathbf{E}_I$ and the target item information $\mathbf{E}_T$. The user preference on a content is formulated as a classification problem. Hence, cross-entropy loss function is employed for training the whole IEEN.

$$\mathrm{Loss} = -\frac{1}{M}\sum_{(\mathbf{v},y)\in\mathbf{D}}(y\log \mathrm{IEEN}(\mathbf{v}) + (1-y)\log(1-\mathrm{IEEN}(\mathbf{v})), \tag{13}$$

where $\mathbf{D}$ denotes the training data set, M denotes the amount of training data, $\mathbf{v}$ denotes the input of IEEN including $\mathbf{X}_n$, $\mathbf{E}_U$, $\mathbf{E}_I$ and $\mathbf{E}_T$, IEEN$(\cdot)$ denotes the output of IEEN, $y \in \{0, 1\}$ denotes the label of content preference.

4 Hybrid Human-Artificial Intelligence Caching Scheme

For a real-world scenario, caching contents in terms of individual user interest directly may degrade the UHR, since a large amount of contents may be cached for a small amount of users that have high interest on lots of contents. As a result, edge caching should consider the common interest of each user. However, as common interest may not be the strongest interest of a user, caching contents in terms of common interest may lead to the CHR degradation if individual interest is omitted. In this section, we propose a hybrid human-artificial intelligence based group interest model which can guarantee high UHR and CHR at them same time.

Firstly, the obtained IEEN model is applied to predict the interest of each user on each content $\{p_{n,c} : 1 \leq n \leq N, 1 \leq c \leq C\}$. Then, interest evolutionary attention is resorted to quantify the group interest on each content, there holds

$$p_c^{\text{att}} = \text{softmax}(\frac{\sum_{t_0=1}^{t} \sum_{i=1}^{N} a_n(t_0) \odot p_{n,c}}{N}), \tag{14}$$

According to (14), we can sort the contents in a descend order in terms of p_c^{att}. The attention-based content set is denoted by $\mathcal{F}^{\text{att}}$.

In addition to the attention based approach that scores contents through user interest, the contents can also be scored based on the social relationship among contents. Specifically, association analysis method is applied to achieve the frequent itemsets. Let $\mathcal{X} \subseteq \mathcal{F}$ and $\mathcal{Y} \subseteq \mathcal{F}$ denote two different sets of the contents that are watched by users before. The support of $\mathcal{X}$ to $\mathcal{Y}$ can be represented as

$$\text{Support}(\mathcal{X} \rightarrow \mathcal{Y}) = \frac{||\mathcal{X} \cup \mathcal{Y}||}{N}, \tag{15}$$

where $||X \cup Y||$ denotes the number of users that watched the all the contents from set $\mathcal{X} \cup \mathcal{Y}$ before. By setting the support threshold, the frequent itemsets of items can be obtained as $\mathcal{Z}$. The prediction value of the new content f_c can be characterized with the confidence between $\mathcal{Z}$ and f_c, i.e.,

$$p_c^{\text{aa}} = \text{Confidence}(\mathcal{Z} \rightarrow f_c) = \frac{\text{Support}(\mathcal{Z} \cup f_c)}{\text{Support}(\mathcal{Z})}. \tag{16}$$

According to (16), we can sort the contents in a descend order in terms of p_c^{aa}. The content set based on association analysis is denoted by $\mathcal{F}^{\text{aa}}$.

By merging the content set obtained based on user interest and that based on social relationship of contents, the content caching scheme $\boldsymbol{\beta}$ can be obtained as

$$\beta_c = \begin{cases} 1, & \text{if } f_c \text{ belongs to the } m\text{-th } (m \le \Phi) \text{ element of } (\mathcal{F}^{\text{att}} \cap \mathcal{F}^{\text{aa}}) \\ 0, & \text{otherwise} \end{cases} \quad (17)$$

5 Experiment Results

In this section, we carry out extensive experiments to validate the effectiveness of the proposed caching scheme. In particular, a well-known real-world dataset, i.e., MovieLens, is applied for experiments [19]. The dataset includes more than 1 million iteration records between more than 6,000 users and 3,000 movies. Additionally, the proposed caching scheme is compared with four caching schemes in terms of cache hit rate and user hit rate. The four baseline schemes are introduced as follows.

- Caching based on attention weight (IEEN-A): Individual interest is first predicted by IEEN, based on which contents are cached based on the attention weights. Note that in this scheme, social relationship among contents are not considered.
- Caching based on association analysis (AA): In this scheme, contents are cached based on frequent itemsets.
- Cache based on popularity: In this scheme, the most watched contents are selected to cache.
- Cache based on user interest: Individual interest values are first predicted by IEEN, and then the contents with the highest scores are selected to cache.

In the data processing stage, the category features, such as gender, age, occupation, etc., are one-hot coded. For continuous features, we normalize them to speedup the model convergence. In order to verify the accuracy of the obtained IEEN model, the contents of each user are sorted by timestamp, and the last 10 contents of each user is selected as the test set. The rest of the data is used as the training set.

From the dataset, the ratings of movies from different users are divided into 5 levels with values $\{1, 2, 3, 4, 5\}$. The user distribution at each rating level is depicted in Fig. 4. It is observed that about half of the movies are rated from 1 to 3, and the remaining half are rated from 4 to 5. Hence, we can model the data label with binary value. In specific, if the rating of a content from a user is 4 or 5, the data label is considered to be positive, i.e., the user is interested in such content. Otherwise, the data label is considered to be negative, i.e., the user is not interested in such content. Based on the dataset and corresponding labels, the user interest evolution prediction can be carried out by the proposed IEEN model. The prediction results of a random selected user is depicted as Fig. 5. It is verified that the predicted interest at different time coincides with the positive labels.

Figure 6 depicts the cache hit rate varying with edge caching capacity under the scenario with 1000 users. It is observed that the proposed scheme outperforms the other baseline schemes especially when edge capacity is small. Hence, it is

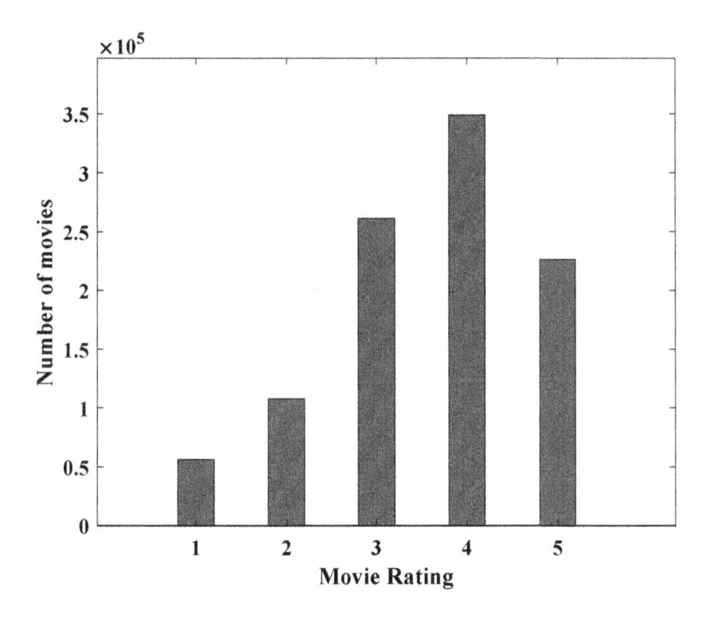

Fig. 4. Distribution of movie ratings in MovieLens dataset.

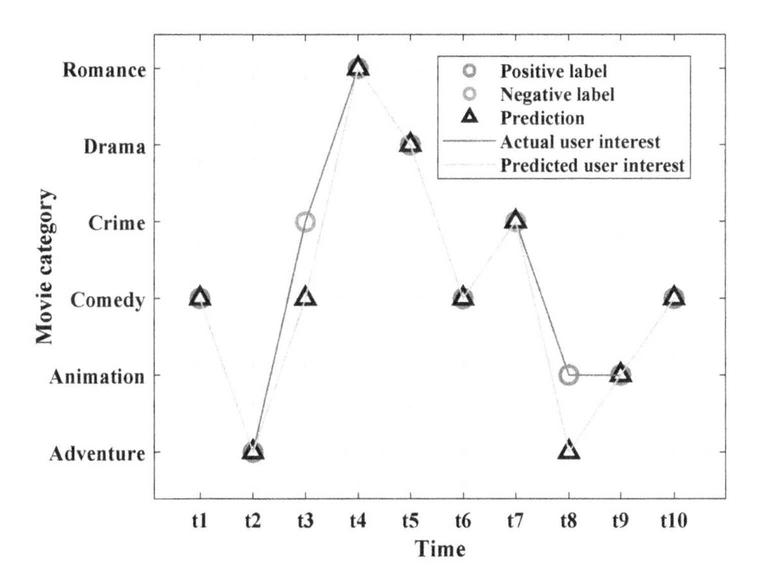

Fig. 5. User interest evolution process.

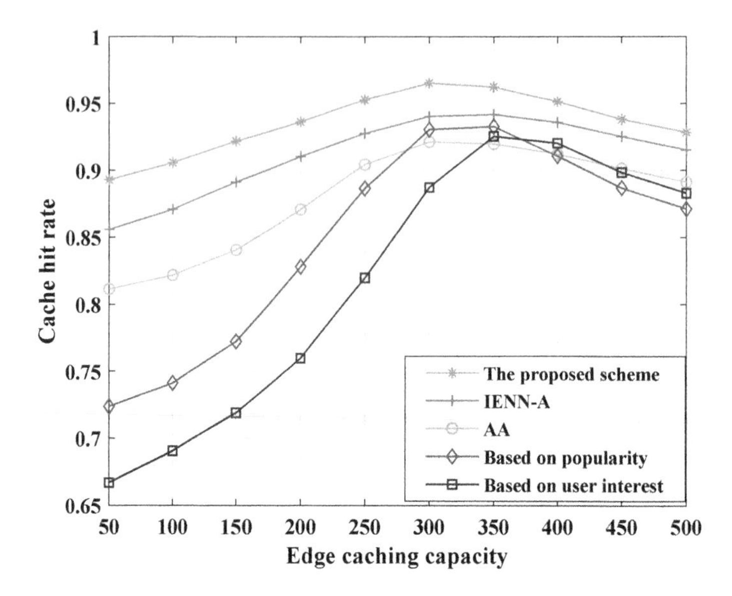

Fig. 6. Cache hit rate vs. edge caching capacity.

verified that merging user interest and social relationship of contents is able to improve the cache hit rate. Moreover, it is also found that there exists optimal caching capacity for each scheme, beyond which, more and more contents that are not interesting for users are cached. This phenomenon implies that caching capacity should be carefully designed from the profit view of point.

Figure 7 depicts the impact of cache capacity on user hit rate under the scenario with 1000 users. It is observed that the proposed scheme can guarantee more users' requests than the other baseline schemes. Additionally, user hit rate increases with the caching capacity when caching capacity is small. This is because a larger caching capacity can cache more contents that are interesting for some users but the interest values are smaller than that from other users. Additionally, it is found that when the caching capacity is large enough (e.g., $\Phi = 400$ for the proposed scheme), the user hit rate becomes invariant, which implies the system is statistically stable in this case.

Figure 8 depicts the impact of number of users on the cache hit rate under the scenario with caching capacity $\Phi = 300$. It is observed that the cache hit rate increases with the number of users. Besides, the proposed scheme achieves highest cache hit rate compared to the other four schemes especially when the number of users is small. In contrast, the scheme only based on individual interest suffers the worst cache hit rate. Such observations validate that group interest is useful in content caching and the proposed scheme is able to characterize the group interest accurately.

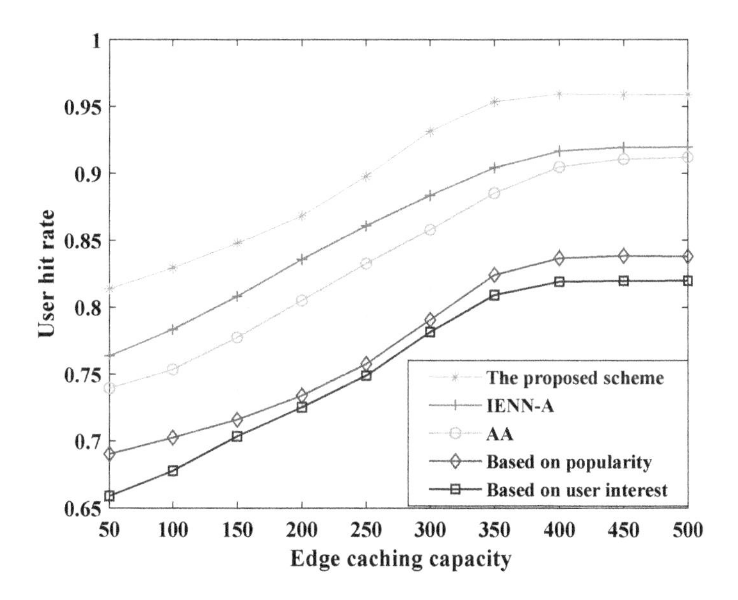

Fig. 7. User hit rate vs. edge caching capacity.

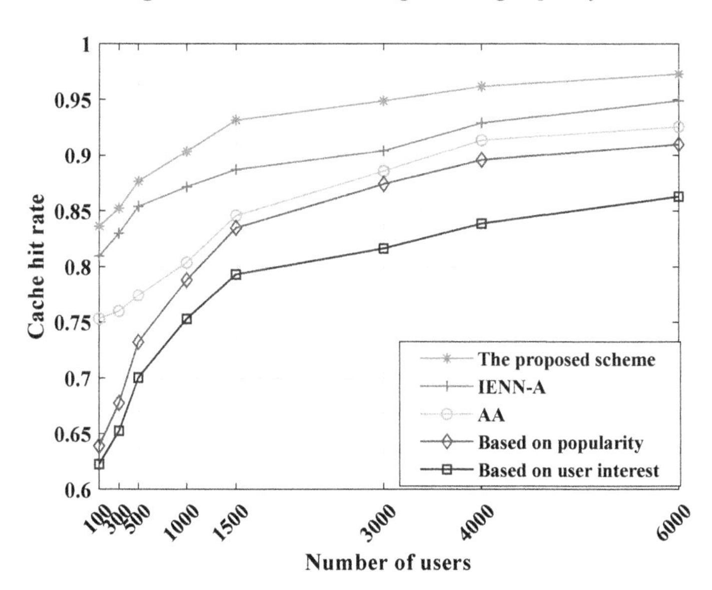

Fig. 8. Cache hit rate vs. the number of users.

Figure 9 depicts the user hit rate varying with the number of users under the scenario with caching capacity $\Phi = 300$. It is observed that as the number of users increases, the user hit rate first increases and then decreases. This is because when the number of users is small, user interest is hard to characterize. At this

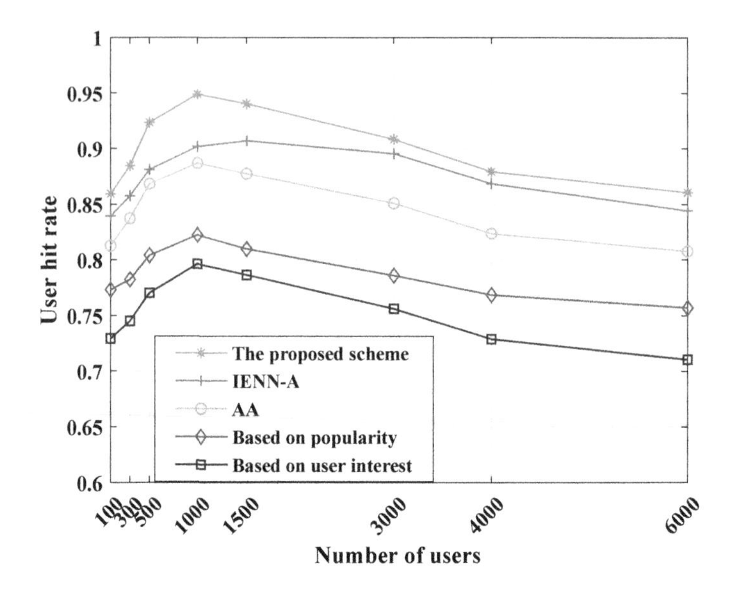

Fig. 9. User hit rate vs. the number of users.

case, increasing users is helpful to capture the common interest of users. However, as the number of users is large, increasing users may bring more interference on common interest capture, which degrades the user hit rate.

Figure 10 depicts the impact of historical behavior sequence length on the cache hit rate. The longer user historical behavior sequence are available, the

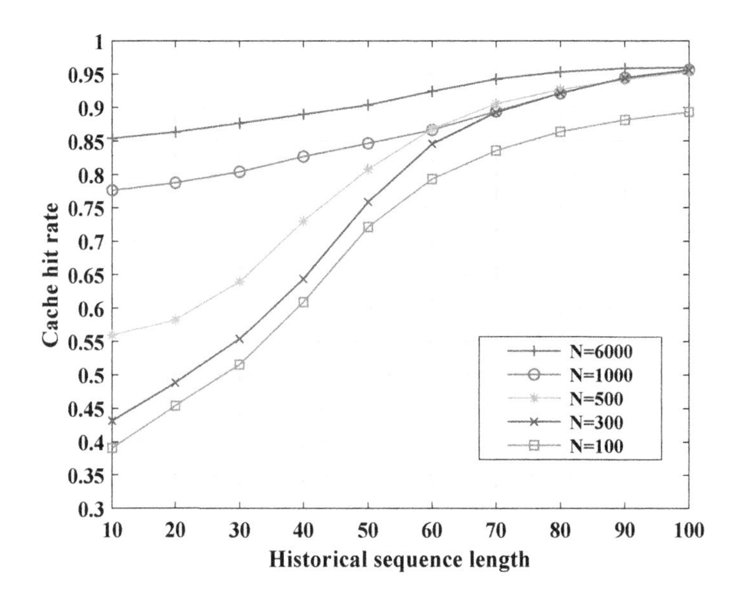

Fig. 10. Cache hit rate v.s. historical sequence length

higher cache hit rate can be obtained, since in this case the user interest evolution process are more dominant. Additionally, the scenario with more users can guarantee a higher cache hit rate. This is because increasing users can enriches features for model training.

6 Conclusions

In this paper, a hybrid human-artificial intelligence enabled edge caching scheme was studied in an edge caching network. The cache hit rate and user hit rate are first modeled. An interest extraction and evolution network was then constructed with consideration of user historical behaviors, content features and user features. Furthermore, a group interest model was proposed by merging weights from the intelligent attention approach and socially association analysis approach. Experiments revealed the impacts of edge caching capacity, the number of users as well as historical sequence length on the caching performance in terms of cache hit rate and user hit rate. Additionally, experiments validated that the proposed scheme performed better than the other baseline schemes with the help of a real-world dataset.

References

1. Sun, Y., Liu, J., Wang, J., Cao, Y., Kato, N.: When machine learning meets privacy in 6G: a survey. IEEE Commun. Surv. Tutorials **22**(4), 2694–2724 (2020)
2. Xiong, J., et al.: Enhancing privacy and availability for data clustering in intelligent electrical service of IoT. IEEE Internet Things J. **6**(2), 1530–1540 (2019)
3. Xiong, J., Chen, X., Yang, Q., Chen, L., Yao, Z.: A task-oriented user selection incentive mechanism in edge-aided mobile crowdsensing. IEEE Trans. Netw. Sci. Eng. **7**(4), 2347–2360 (2020)
4. Abuhadra, R., Hamdaoui, B.: Proactive in-network caching for mobile on-demand video streaming. In: IEEE International Conference on Communications (ICC), Kansas City, pp. 1–6 (2018)
5. Doan, T.V., Pajevic, L., Bajpai, V., Ott, J.: Tracing the path to youtube: a quantification of path lengths and latencies toward content caches. IEEE Commun. Mag. **57**(1), 80–86 (2019)
6. Jiang, W., Feng, G., Qin, S.: Optimal cooperative content caching and delivery policy for heterogeneous cellular networks. IEEE Trans. Mob. Comput. **16**(5), 1382–1393 (2017)
7. Xing, H., Song, W.: Collaborative content distribution in 5G mobile networks with edge caching. In: IEEE International Conference on Communications (ICC), Shanghai, China, pp. 1–6 (2019)
8. Liu, W., Zhang, J., Liang, Z., Peng, L., Cai, J.: Content popularity prediction and caching for ICN: a deep learning approach with SDN. IEEE Access **6**, 5075–5089 (2018)
9. Yang, Z., Liu, Y., Chen, Y., Tyson, G.: Deep reinforcement learning in cache-aided MEC networks. In: IEEE International Conference on Communications (ICC), Shanghai, China, pp. 1–6 (2019)

10. Zhang, S., Yao, L., Sun, A., Tay, Y.: Deep learning based recommender system: a survey and new perspectives. ACM Comput. Surv. (CSUR) **52**(1), 1–38 (2019)
11. Dara, S., Chowdary, C.R., Kumar, C.: A survey on group recommender systems. J. Intell. Inf. Syst. **54**(2), 271–295 (2019)
12. Wang, S., Cao, L., Wang, Y.: A survey on session-based recommender systems. arXiv preprint arXiv:1406.1078 (2019)
13. Wang, S., Hu, L., Wang, Y., Cao, L., Sheng, Q., Orgun, M.: Sequential recommender systems challenges, progress and prospects. arXiv preprint arXiv:2001.04830 (2019)
14. Yin, Y., Chen, L., Xu, Y., Wan, J., Zhang, H., Mai, Z.: QoS prediction for service recommendation with deep feature learning in edge computing environment. Mob. Netw. Appl. 1–11 (2019)
15. Costantini, M., Spyropoulos, T., Giannakas, T., Sermpezis, P.: Approximation guarantees for the joint optimization of caching and recommendation. In: ICC 2020–2020 IEEE International Conference on Communications (ICC), Dublin, Ireland, pp. 1–7 (2020)
16. Zhang, C., et al.: Toward edge-assisted video content intelligent caching with long short-term memory learning. IEEE Access **7**, 152832–152846 (2019)
17. Cho, K., et al.: Learning phrase representations using RNN encoder-decoder for statistical machine translation. arXiv preprint arXiv:1406.1078 (2014)
18. Vaswani, A., et al.: Attention is all you need. Presented at the advances in neural information processing systems. (2017)
19. Harper, F.M., Konstan, J.A.: The movielens datasets: history and context. ACM Trans. Interact. Intell. Syst. **5**(4), 1–19 (2015)

A New Way of Mobile Energy Trading

Chaoyue Tan[1], Yuling Chen[1,2](✉), Xiaojun Ren[1,2], and Changgen Peng[1]

[1] State Key Laboratory of Public Big Data, College of Computer Science and Technology, Guizhou University, Guiyang, China
Ylchen3@gzu.edu.cn

[2] Block Chain Laboratory of Agricultural Vegetables, Weifang University of Science and Technology, Shouguang 262700, China

Abstract. Current blockchain-based energy trading models raise serious concerns regarding the high and capped transaction latency and expensive service charges. In this study, a mobile energy trading scheme based on lightning network is presented. The focal point of the scheme lies in transfer of value occurs off-blockchain, which addresses the problem of transaction latency. Micropayment channels create a communication between two parties to update balances constantly, deferring what is broadcast to the blockchain in a single transaction netting out the total balance between those two parties. The found security is guaranteed by committing funds into a multisignature address. Additionally, a security analysis is conducted within the context of the proposed model to identify potential vulnerabilities.

Keywords: Blockchain · Lightning network · Mobile energy trading · Off-blockchain · Transaction latency

1 Introduction

In the mobile energy transaction scenario, a microgrid or distributed generation (DG) with small-scale and decentralized transaction characteristics can participate in the electricity trading market as a seller or buyer of electricity [1]. With the rapid expansion of the volume of distributed energy transactions, issues such as privacy leakage and low efficiency of transactions have been increasingly prominent. Due to the distributed structure and decentralization of the blockchain, it is similar to distributed energy transactions. The combination of energy trading provides a new development direction for energy trading [2]. Chen et al. [3] discussed the use of energy blockchain in the energy field and proposed development proposals based on the feature of energy transition trend and blockchain technology. Wang et al. [4] designed a blockchain-based EV charging pile sharing platform, and improved the transaction mechanism and process by using smart contracts. Zhang et al. [5] proposed a blockchain electric vehicle charging model which including three-layer distribution algorithm for optimal scheduling of electric vehicle electrical changing station, also verify the applicability of the model to distributed grid layout.

J. Xiong et al. (Eds.): MobiMedia 2021, LNICST 394, pp. 123–132, 2021.
https://doi.org/10.1007/978-3-030-89814-4_9

At present, the consensus schemes and network architecture of existing solutions are decentralized. Due to the lack of synergy and high efficiency among various chains, the large-scale application of new energy and the use of marketization are greatly restricted [6]. In 2016, the storage capacity of each block has approached 1 MB, which means that some transactions could not be packaged in the block in time and energy nodes need more time to confirm these transactions [7]. With the development of decentralized applications, there is an urgent need to modify the code of blockchain to improve the processing capacity and ensure the scalability of blockchain [8], e.g. Side Chain [9] and Block Slicing [10].

Lightning network is a decentralized trading network, which has been proven that it has strong adaptability to high-frequency transaction scenarios, can improve its scalability and execute massive transactions in a real-time way [11]. The author in reference [6] studied the development status of lightning network and its possible non-monetary uses, meanwhile created a new business model based on lightning applications (LAPPs), micro-channel payment and small transactions. The author in reference [12] discusses the requirements that need to be fulfilled to properly support micropayment in IoT, and further the extent to which different blockchain technologies can fulfill those requirements. As well as proves that the performance of Lightning network is superior to traditional blockchain solutions. Based on blockchain, lightning network and smart contract technology, the author in reference [13] proposed a charging pile sharing economy model, which points out that builds a blockchain-enabled energy trading platform is possible.

Generally, according to the backdrop discussed above, in this paper, we adopt lightning network to construct an off-chain bidirectional payment channel, which enables the buyer node to pay to seller nodes directly, and propose a mobile energy trading scheme. And then, we utilize multi-signature and a series of decrementing timelocks to ensure the security of the payment channel. Finally, we conduct a comprehensive experimental evaluation to evaluate the trading performance. Experimental evaluations show the effectiveness of the proposed mobile energy trading scheme by comparison with other schemes. The main contributions of this paper are as follows:

1. We propose a mobile energy trading scheme based on lightning network that can improve trading efficiency remarkably and decrease service charge by off-chain transactions.
2. Our scheme combines Lightning Network and energy transaction, which gives a new way to small high-frequency energy trading.

The rest of the paper is organized as follows. Section 2 gives some preliminaries. Section 3 gives the details of our proposed scheme. Section 4 provides some experimental results and evaluation analysis. Finally, Sect. 5 concludes the paper.

2 Preliminaries

2.1 Lightning Network

There are two core concepts of lightning network: Recoverable Sequence Maturity Contract (RSMC) and Hashed Time Lock Contract (HTLC). Assuming that there exists a micro-payment channel between two entities, before execute the trading operation, these entities will pre-deposit funds in the micro-payment channel, and then the funds allocation scheme will be confirmed by both entities in every subsequent transaction, the old version is signed for cancellation operation. When a transaction finished, the final transaction result with signatures of the entities will be broadcast to the blockchain network, and then the funds will be pay to the wallet address of both parties after the final confirmation.

There are two types of creation payment channel: direct payment channel and indirect payment channel. The direct payment channel means that energy nodes of a transaction apply for a multi-signed address on the blockchain and pre-deposit funds. The indirect payment channel means that energy nodes of a transaction build a payment channel with the help of middle nodes which have direct payment channel. The update of the payment channel refers to that both parties modify and record the new proportion of funds in the multi-signature address according to the transaction situation, and use their private keys to confirm. After each update of the proportion of funds, energy nodes will generate a new private key for the next record confirmation. When the transaction is over, the two parties will broadcast the final transaction results to the block chain for capital settlement, then closing the trading channel.

2.2 Multiple Short Signature Scheme [14]

Initialization: Key generation center (KGC) sets k as the security parameter, G_1 and G_2 are cyclic groups of order Q which is a large prime, P is a generator of G_1, e:$G_1 \times G_1 \rightarrow G_2$ is a safe bilinear mapping. Select the random number s as the master key of the system and calculate the public key $P_{pub} = sP$, use it as the main public key of the system, then select two collision-resistant Hash functions $H_1 : \{0, 1)^* \rightarrow Z_q^*$, $H_2 : \{0, 1)^*$. KGC public system parameters: $\{q, P, e, G_1, G_2, P_{pub}H_1, H_2\}$. And keep the system master key s secret.

Key Generation: Users register with KGC, which assigns each user a uniquely identify ID and randomly selects a key to generate a tag $K_{ID} \in Z_q^*$, Generate the user private key $x_{ID} = sH_1(ID,\ K_{ID}, s)$, Calculate the public key $y_{ID} = x_{ID}P$.

Multiple Signature: Multiple signatures for M users $L = \{U_1, U_2, \ldots U_M\}$, Their identity sets are $L_{ID} = \{ID_1, ID_2, \ldots ID_M\}$, The specific signature steps are as follows:

a. U_1 signs the message m and calculates $h = H_2(m)$, $S_1 = ID_{1xID_1h}$, send identify ID_1, message m, partial signature S_1 to the next user U_2.
b. After received message, U_2 firstly verifies validity of signature, and then sign.

   ① compute $h = H_2(m)$.
   ② Verify whether $e(S_1, P) = e(h, ID_{1yID_1})$ is true, if not, verification failed and return FALSE.
   ③ compute $S_2 = S_1 + ID_{2xID_2}, h$, send ID_1, ID_2, m and Part of the signature S_2 to next user U_3.

c. Multiple signature verification: user verifies the multiple signature *SM* of the given message *m*, and the correctness of the signature algorithm is verified as follows:

$$e(S_M, P) = e(\sum_{i=1}^{M} ID_i x_{ID_i}, h, P) = e(h \sum_{i=1}^{M} ID_i y_{ID_I})$$

3 The Proposed Scheme

3.1 System Model

The proposed scheme mainly includes two different entities: users and charging piles. Each user chooses different states, charge or discharge, according to current energy status. Charging piles include energy providers, community shared charging piles, shopping mall charging piles, etc. The parameters involved in our proposed scheme are shown in Table 1:

Table 1. Parameters

Description	Name
User	U_i
Distributed generation	*DG*
Charging piles	*CP*
Retail price	*S_EP*
Electricity sales	*ES*
Purchase tariff	*P_EP*
Electricity Purchasing	*EP*
Volume of electricity sold	*C_ES*
Volume of electricity Purchasing	*C_EP*

The main structure of our scheme is shown in Fig. 1. All participants, distributed generations (*DGs*), the charging piles (*CPs*), users, need to register before they trade so they can query some information of themselves and other registrants. When completed registration process, the electricity sales quotation *S_EP* and the electricity sales ES are submitted to the blockchain in each transaction cycle, while the user submits the electricity purchase quotation *P_EP* and the electricity purchase *EP*. The smart contract matches the quotations of both parties according to the quotation matching mechanism defined by the system, and announces transaction information, including the current selling price, buying price, and transaction price. Successfully matched transaction parties will issue transaction settlement certificates through the Lightning Network, including transaction electricity sales *C_ES* and transaction electricity purchase fees *C_EP*, as transaction vouchers, which will be uploaded to the blockchain after the transaction is completed for redistribution of the capital ratio and returned to the accounts of both parties.

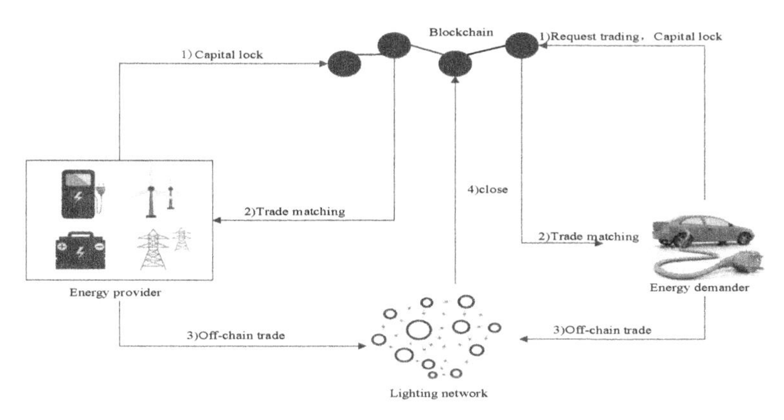

Fig. 1. System model of our proposed scheme

3.2 Registration

This part is divided into user registration and charging pile registration. 1) User registration: users login to ethereum through mobile terminals such as mobile phones to register. After registration, each user gets a unique *ID*. 2) Charging pile registration: the company to which the charging pile belongs will uniformly register each charging pile, and the smart contract will calculate according to the power of the charging pile and lock the corresponding funds on the blockchain. Similarly, each charging pile also has a corresponding identity Identification *ID*. The detail process of registration shows is Table 2.

Table 2 . Registration steps

Registration:	
Step1	KGC assign a unique identity ID to the user or the charging pile, select the private key randomly to generate the tag K_{ID}
Step2	Generate the user private key $x_{ID} = sH_1(ID, K_{ID}, s)$
Step3	Calculate the public key $y_{ID} = x_{ID}P$
Step4	*KGC* send the corresponding key to the user

3.3 Transaction Application

After registration, the user can log in to the Ethernet to initiate a transaction request and retrieve whether there is a transaction channel between both parties. If exists, the smart contract locks the corresponding funds on the blockchain according to the charge or discharge amount provided by the user. If does not exist, the user chooses whether to build a payment channel with the energy provider according to his own wishes. If the user chooses to build a payment channel, the payment channel is built according to the smart contract shown in Table 3, and if not, the best payment channel is matched for both parties in the transaction.

Table 3. Smart contract

Contract:	
Step1	Building *Founding Transaction (FT)*: pre-deposit funds in multi-signed address User & Sale, privacy key sign;
Step2	$U_i$building *Commitment Transaction1a (C1a):* The transaction contains two outputs that are used to redeem the funds they have locked up; First output: *User2 & Sale*, set *sequence number = 50*, Used to control the return of funds to both sides of the transaction wallet address time, when $C1a$ is triggered, the funds locked in FT will be returned to the Sale wallet address on a delayed basis; Second output: *Sale,* when $C1a$ is triggered,the funds locked in FT will be returned to the Sale wallet address immediately; * The execution condition of this transaction is that User quits the transaction before the transaction ends;
Step3	*DG* or *CP* constructs *Commitment Transaction1b (C1b);*
Step4	*DG* or *CP* sign *C1a then send to* U_i, U_i*sign C1b* and send to *DG* or *CP;*
Step5	Broadcast *Founding Transaction (FT)* to blockchain;

3.4 Off-Chain Transaction

When there is no payment channel between the two parties, the Lightning Network matches the best payment channel. After the matching is successful, the two parties will conduct the transaction through the intermediate node. The successful matching is that the transaction receiver sends the secret *R* to the transaction initiator within the specified time. That is, to verify the authenticity of the intermediate node, both parties to the transaction will pass a secret *R* through the intermediate node within a specified time, and start the transaction after successful verification. The transaction process is shown in Fig. 2:

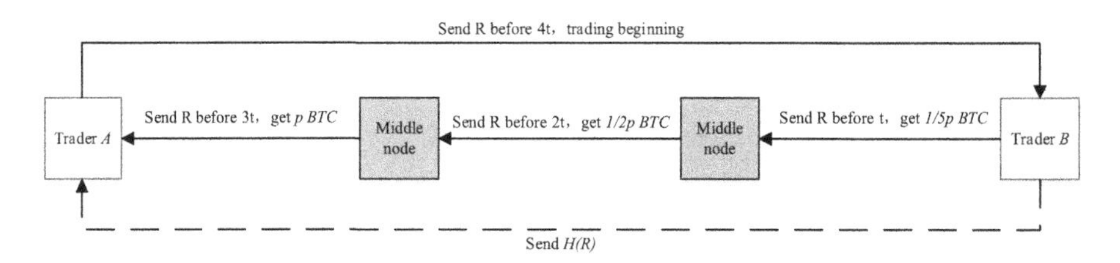

Fig. 2. Off-chain transaction

Price matching mechanism. This mechanism sorts the seller's quotations from low to high, and sorts the buyers' quotations from high to low. When the quotations are the same, they are sorted according to the time of submission. Among them, the optimal selling price (*OSR*) is the seller's lowest price, and the optimal buying price (*OBR*) is the buyer's highest price. When *OBR* is greater than *OSR*, the transaction can be conducted. Suppose that the lowest quotation of the seller is P_{min}, the highest quotation of the buyer is P_{max}, and the transaction price is $P_{fin} = \frac{P_{min}+P_{max}}{2}$. Each time the matching is completed, the buyer and the seller can re-adjust the price according to their own conditions for the next matching.

4 Analysis

4.1 Security Analysis

Anti-tampering attacks. The transaction mode of this agreement is to record each transaction and use the private key to sign for confirmation, upload the final transaction result to the blockchain for proportional distribution of funds. In each Transaction update, the private key recorded last time will be sent to the other party to construct a Breach Remedy Transaction, and attackers will upload the outdated Commitment Transaction record to the blockchain to gain benefits. As shown in Fig. 3, when a malicious attacker broadcasts an out-of-date Transaction certificate to the blockchain, the other party will immediately know and sign a Breach Remedy Transaction 1A with the private key and broadcast it to the blockchain. Due to the sequence number = 50 in the out-of-date Transaction certificate, it will delay getting its own funds for 50 blocks. An honest trader will be broadcast to the blockchain and immediately receive all the money.

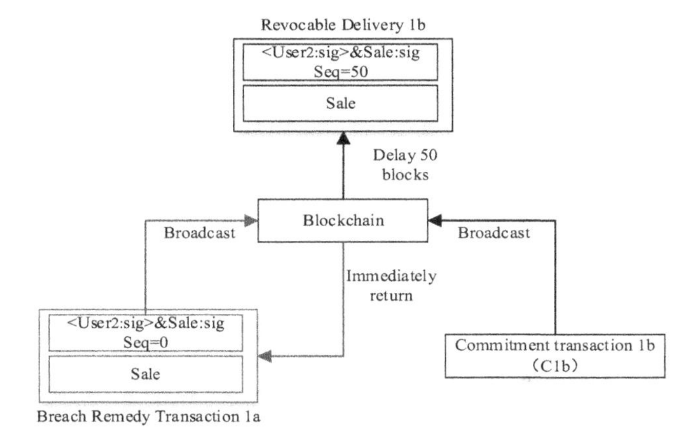

Fig. 3. Anti-tampering attacks

4.2 Transaction Performance Analysis

The size of a block in the Bitcoin blockchain is about 1M, and each transaction is about 250 bytes, which means there are about 4194 transactions in one block, so the transaction throughput of the blockchain is about 7 transactions per second. In addition, the transaction fee in the blockchain is charged by bytes, so the transaction fee is about from 0.0001 to 0.0005 BTC for one transaction [15]. Each transaction needs to wait for confirmation of 6 blocks before it is confirmed to be valid. If the transaction is small high-frequency transaction like energy trading, the waiting time will be longer, because the miners give priority to transactions with higher fees according to the level of transaction fees. When the two parties have a payment channel, they can conduct transactions directly through the payment channel without any commission. If there is no payment channel between the two parties, the transfer through the intermediate node only needs to pay a small fee to the intermediate node. However, when several small transactions are carried out on the blockchain, the fee will be much higher than that of the Lightning network (Table 4).

Table 4 . Trading contrast

Trading scheme	Trading cost	Trade confirmation speed	Handling capacity
Blockchain-based scheme	0.0001–0.0005BTC	60 min	About seven strokes per second
Our scheme	Petty handling charge	instant confirmation	The more trades, the faster

Table 5 shows the comparison of the mobile energy trading protocol in the proposed scheme and the existing protocol. Comparing from the Transaction performance, fairness of transaction, and key security. Here, "√" satisfies the performance and " ×" dissatisfies the performance.

Table 5. Trading contrast

Trading scheme	Transaction performance	Fairness of transaction	Defense MITM attack
Ayman et al. [16]	About seven strokes per second	√	×
Nurzhan et al. [17]	uncertain	√	×
Our scheme	Petty handling charge	√	√

5 Conclusion

In this paper, we have explored the opportunities brought by lightning network to empower energy trading based on the existing literature in the field, and builds a mobile energy trading solution based on lightning network. Our scheme states out a new way of energy trading. Particularly, we have first provided a bidirectional payment channels for small high-frequency transactions like energy trading. We have then analyzed in detail the limitation of blockchain for energy trading and expound the whole process of off-chain trading. Through the comprehensive analyze, we have proved that our scheme can reduce the block chain overload caused by small high-frequency transactions, moreover it avoids high transaction fees caused by mining and long waiting time for transaction confirmation, meanwhile, it greatly improves trading performance. In addition, due to the existence of Multi-signature, Hash Time Lock and Sequence Number, the security of this scheme is guaranteed. Bidirectional payment channels improve scalability of blockchain but it has limitations. Fee policies of intermediate nodes may influence the whole network, our next work is exploring how to balance fee of intermediate.

Acknowledgement. This work is financially supported by the National Natural Science Foundation of China under Grant No. 61962009. In part by the Major Scientific and Technological Special Project of GuiZhou Province under Grant No. 20183001. and in part by the Open Funding of GuiZhou Provincial Key Laboratory of Public Big Data under Grant No. 2018BDKFJJ003 and No. 2019BDKFJJ011.

References

1. Wang, J., Zhou, N., Wang, Q., et al.: Elecricity direct transaction mode and stratey in microgrid based on blockchain and continuous double auction mechanism. Proc. CSEE **38**(17), 5072-5084+5304 (2018)
2. Xu, J., Ma, L.: Application of block chain technology in distributed energy trading. Electr. Power Autom. Equipment **40**(08), 17-22+30 (2020)
3. Chen, Y., Zhao, Q., Gong, Y., et al.: Discussion on electric vehicle charging transaction based on block chain technology. Electr. Power Eng. Technol. **39**(06), 2–7 (2020)
4. Wang, G., Yang, J., Wang, S., et al.: Distributed optimization of power grid considering EV transfer scheduling and block-chain data storage. Autom. Electr. Power Syst. **43**(08), 110–116 (2019)
5. Zhang, X., Liu, C., Chai, K.K., Poslad, S.: A privacy-preserving consensus mechanism for an electric vehicle charging scheme. J. Network Comput. Appl. **174** (2021)
6. Ren, A., Feng, L., Cheong, S., et al.: Optimal fee structure for efficient lightning networks. In: 2018 IEEE 24th International Conference on Parallel and Distributed Systems (ICPADS). IEEE (2018)
7. Yu, H., Zhang, Z., Liu, J.: Research on scaling technology of bitcoin blockchain. J. Comput. Res. Dev. **54**(10), 2390–2403 (2017)
8. Harris, B.: Bitcoin and Lightning Network on Raspberry Pi. Apress, Berkeley, CA (2019)
9. Singh, A., Click, K., Parizi, R.M., Zhang, Q., Dehghantanha, A., Choo, K.-K.: Sidechain technologies in blockchain networks: an examination and state-of-the-art review. J. Network Comput. Appl. **149**, 102471 (2020)
10. Erdin, E., Cebe, M., Akkaya, K., Solak, S., Bulut, E., Uluagac, S.: A Bitcoin payment network with reduced transaction fees and confirmation times. Comput. Networks **172**, 107098 (2020)
11. Stasi, G., Avallone, S., Canonico, R., et al.: routing payments on the lightning network. In: 2018 IEEE International Conference on Internet of Things (iThings) and IEEE Green Computing and Communications (GreenCom) and IEEE Cyber, Physical and Social Computing (CPSCom) and IEEE Smart Data (SmartData). IEEE (2018)
12. Robert, J., Kubler, S., Ghatpande, S.: Enhanced Lightning Network (off-chain)-based micropayment in IoT ecosystems. Future Gener. Comput. Syst. **112**, 283–296 (2020)
13. Linghai, Q., Xue, L., Bing, Q., et al.: Shared economy model of charging pile based on block chain ecosystem. Electr. Power Constr. **38**(09), 1–7 (2017)

14. Liming, Z., Lanlan, C., Qing, Z.: Short multi-signature scheme for distributed approval workflow. Appl. Res. Comput. **37**(02), 521–525 (2020)
15. Fang, L., Zhuoran, L., He, Z.: Research on the progress in cross-chain technology of blockchains. J. Software **30**(6), 1649–1660 (2019)
16. Esmat, A., de Vos, M., Ghiassi-Farrokhfal, Y., Palensky, P., Epema, D.: A novel decentralized platform for peer-to-peer energy trading market with blockchain technology. Appl. Energy **282**, 116123 (2021). https://doi.org/10.1016/j.apenergy.2020.116123
17. Nurzhan, Z., Davor, S.: Security and privacy in decentralized energy trading through multi-signatures, blockchain and anonymous messaging streams. IEEE Trans. Depend. Secure Comput. **15**(1), 840–852 (2016)

A SCATC-Based Blockchain Index Structure

Xiaogang Xing[1], Yuling Chen[1,4(✉)], Tao Li[1], Yang Xin[2,3], and Hongwei Sun[4]

[1] State Key Laboratory of Public Big Data, College of Computer Science and Technology, Guizhou University, Guiyang 550025, China
Ylchen3@gzu.edu.cn

[2] School of Cyberspace Security, Beijing University of Posts and Telecommunications, Beijing 100876, China

[3] National Engineering Laboratory for Disaster Backup Recovery, Beijing 100876, China

[4] Blockchain Laboratory of Agricultural Vegetables, Weifang University of Science and Technology, WeiFang 262700, China

Abstract. Blockchain technology has the characteristics of decentralization and tamper resistance, which can store data safely and reduce the cost of trust effectively. However, the existing blockchain system has weak performance in data management, and only supports traversal queries with transaction hashes as keywords. The query method based on the account transaction trace chain (ATTC) improves the query efficiency of historical transactions of the account. However, the efficiency of querying accounts with longer transaction chains has not been effectively improved. Given the inefficiency and single method of the ATTC index in the query, we propose a subchain-based account transaction chain (SCATC) index structure. First, the account transaction chain is divided into subchains, and the last block of each subchain is connected by a hash pointer. The block-by-block query mode in ATTC is converted to the subchain-by-subchain query mode, which shortens the query path; then, the query algorithm is given for the SCATC index structure. Simulation analysis shows that the SCATC index structure significantly improves query efficiency.

Keywords: Blockchain · Query optimization · Hash pointer · Subchain

1 Introduction

In 2008, Bitcoin was proposed by Satoshi Nakamoto in "Bitcoin: A Peer-to-Peer Electronic Cash System" [1], marking the emergence of blockchain technology. Blockchain is a distributed database technology that has the characteristics of decentralization, traceability, tamper-proof, collective maintenance, etc. [2]. The emergence of this technology solves a series of problems such as high cost, low efficiency, and low trust brought by centralized institutions [3]. Level-DB is the mainstream database in the blockchain system, which is based on the storage structure of the LSM tree. This leads to the lower reading performance of the blockchain [4]. Besides, Level-DB only supports simple Key-Value queries, not relational queries [5]. When querying transactions, users can

J. Xiong et al. (Eds.): MobiMedia 2021, LNICST 394, pp. 133–141, 2021.
https://doi.org/10.1007/978-3-030-89814-4_10

only traverse in block order, which further reduces query efficiency [6]. The blockchain system only supports related queries with transaction hashes as keywords and does not query with account hashes as keywords. The query method is single.

To quickly query account historical transactions, an index structure that supports querying account transaction chains was proposed in the Education Certificate Blockchain (ECBC) [7]. This index structure has the characteristics of low latency and high throughput. You et al. [8] designed a hybrid index mechanism that supports blockchain transaction traceability based on the Ethereum state tree. In this mechanism, a hash pointer is embedded in the account transaction, which points to the block where the previous transaction. Through the pointer, the Account Transaction Trace Chain (ATTC) can be quickly traced. The query method based on ATTC improves the query efficiency of account transactions, but for some active accounts with longer transaction chain length, a longer chain still needs to be traversed. Besides, users do not always want to find all the historical transactions of an account, and it is still difficult to find target transactions in massive account data. In this regard, we improve the query scheme based on ATTC and propose a subchain-based account transaction chain (SCATC) index structure, which solves the shortcomings of the ATTC index structure in the query effectively.

The main contributions of this paper are as follows:

1. We divide the transaction chain into subchains and connect different subchains with hash pointers to shorten the query path when querying early historical transactions. This solution is not a query mode that uses space for time. While reducing the time complexity, the space complexity does not increase significantly.
2. We design a query algorithm for the SCATC index structure. The simulation results show that the SCATC-based query is more efficient when querying the early transactions of accounts.

The paper is organized as follows. Section 2 of this article introduces the related work of blockchain in the data query. Section 3 introduces the index structure based on SCATC and the query algorithm given in detail. Section 4 is efficiency analysis and simulation experiment. The full text is summarized in Sect. 5.

2 Related Works

To improve the efficiency of blockchain in data retrieval, Morishima et al. [9] propose to accelerate blockchain search through GPU using the higher computing power of GPU. Utilizing the feature that blockchain data does not need to be updated or deleted, an array-based Patricia tree structure is introduced, which is suitable for GPU processing. To study the identity verification and range query issues in the hybrid storage blockchain, Zhang et al. [10] used a unique gas cost model to design an authentication data structure GEM2-tree that can be effectively maintained by the blockchain. It not only saves gas consumption in smart contracts but also effectively supports identity verification queries. Aiming at the inefficient query of the ElasticChain [11] model on the blockchain, Jia et al. [12] propose an ElasticQM (elastic query model) query method based on the model. In the user layer, the model catches the user's first query result to improve the efficiency of the second query. In the data layer, the B-tree is combined with the Merkle tree to construct the blockchain data storage structure of the B-M tree. This storage structure improves the query efficiency of the internal data of the block. Jiao et al. [13] propose a blockchain database system framework, which realizes the application of data management on the blockchain. Combining red-black trees with Merkle trees, they propose a tamper-resistance index based on hash pointers. Through the index can realize the fast positioning of the data in the block. Zheng et al. [14] divide the data attributes on the blockchain into discrete attributes and continuous attributes and proposed a MHerkle tree index structure for different attributes, which supports range query. Ren et al. [15] introduce a DCOMB (Dual Combination Bloom filter) scheme, which converts the computing power used for Bitcoin mining into the computing power for data query. DCOMB has higher random read performance and lower error rate than COMB (Combination Bloom filter). The encrypted signature tree data structure of the Merkel Block Space Index (BSI) [6] modifies the Merkle KD-tree to support fast Spatio-temporal query processing. In Ethereum, when a user initiates a transaction, the system checks the status of the account. Wan et al. [16] built a Merkle Patricia tree account storage structure GMPT (Group Merkel Patricia Tree) to speed up the query of account status. However, GMPT does not support fast queries of historical transactions. For this, an index directory BKV (B-Key-Value) is constructed in combination with the B-tree index [17].

3 SCATC Index Structure

3.1 Index Design

Given ATTC's shortcomings in retrieval, we improve it based on the index structure. In ATTC, the transactions of accounts in different blocks are connected by hash pointers. The hash pointers here are called the first hash pointer (FHP).

In the SCATC index structure, account transaction chain is divided into subchains. Every $k(k > 1)$ block is divided into a subchain, and each subchain has a subchain number. The index structure of SCATC is shown in Fig. 1.

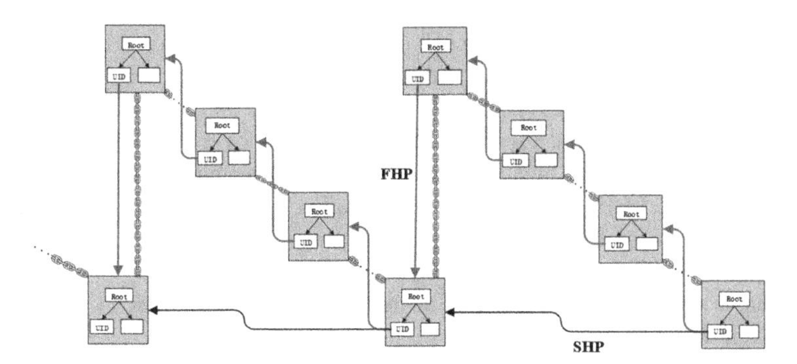

Fig. 1. SCATC index structure

Each transaction of the account will identify the location of the transaction when it enters the chain. For example, $Account_{n,k}$(*Account* is the account name, $n \geq 1$ and $k \geq 1$, n and k are both positive integers) means that the account is in the k th block in the nth subchain of the transaction chain. In ATTC, every time a user participates in a transaction of k blocks, another hash pointer is added to the account branch leaf node in the block $Account_{n,k}$ pointing to the block $Account_{n-1,k}$. The hash pointer connecting the blocks at the last block of the two subchains is second hash pointer (SHP).

Within each block body, each leaf node of the Merkle tree represents an account, and both FHP and SHP are stored in the pointer variable defined in the leaf node. For new blocks on the chain, the system will detect each account whose status has changed, and update the transaction chain and subchain for each account. The specific steps for dividing subchains are as follows:

Step 1: Firstly, determine whether the account is a new user one by one. If so, set the transaction subchain number of the account transaction chain and the block serial number in the subchain to 1. If not, go to *step 2*.

Step 2: Determine whether the block sequence number of the sub-chain where the previous block in the account transaction chain is located is less than $k - 1$. If so, the subchain number of the new transaction is the same as the previous block, and the sequence number of the block in the sub-chain is increased by 1.

Step 3: Determine whether the block sequence number in the subchain where the previous block in the transaction chain is equal to $k - 1$, if so, the subchain number of the new transaction is the same as the previous block, and the block sequence number in the subchain is k. At the same time, the secondary hash pointer SHP is added to the account branch node of the new block to point to the k th block of the previous subchain.

Step 4: Determine whether the block sequence number in the subchain where the previous block is located is equal to k. If so, the subchain number in the new transaction is increased by 1, and the block sequence number is 1. This block is the initial block of the new subchain in the transaction chain.

3.2 Algorithm Design

When inquiring about historical transactions, users can directly access the k th block of the previous subchain from the k th block of the latest subchain according to the SHP until the target subchain. Then traverse the blocks in the target subchain to obtain the transaction. Before the query reaches the target subchain, only one block is visited in all subchains except the latest subchain. The block-by-block traversal query method is transformed into a subchain-by-subchain query, which shortens the access path in the search process. The FHP in SCATC is not embedded in the transaction but embedded in the leaf nodes of the Merkle tree. When querying early historical transactions, the system will directly filter the user's recent transaction data.

To achieve rapid retrieval of data, we design the query algorithm as shown below for the SCATC index.

Algorithm 1 SCATC query algorithm

```
Input: Target account subchain
Output: Account transaction
1: TargetAccount_data=[]
2: p = LatestBlock.data
3: if p.Subchain_BlockNum<k:
4:     for i in range(LatestBlock, LatestSubchain_FirstBlock,-1):
5:         q= i.data
6:         for j in q:
7:             if j.Account==Target_Account:
8:                 TargetAccount_data.append(j)
9: for i in range(LatestSubchain_kBlock,TargetSubchain_ kBlock,-k):
10:    q= i.data
11:    for j in q:
12:        if j.Account==Target_Account:
13:            TargetAccount_data.append(j)
14: for i in range(TargetSubchain_kBlock, TargetSubchain_FisrtBlock,-1):
15:    q= i.data
16:    for j in q:
17:        if j.Account==Target_Account:
18:            TargetAccount_data.append(j)
19: return TargetAccount_data
```

The algorithm first creates a list *TargetAccount_data* to save the data of the target accounts that have been accessed. Lines 2–8 of the algorithm visit the latest block in the transaction chain. If the sequence number of the block is less than k, traverse from the latest block to the first block in the subchain. Lines 9–13 of the algorithm, according to the hash pointer in the k th block, access the k th block of the previous subchain until the k th block of the target subchain. During this process, only one block is visited in each subchain. Lines 14–18 of the algorithm traverse all the blocks in the target subchain.

4 Experiment and Analysis

4.1 Efficiency Analysis

The length of the subchain affects the scope and efficiency of the query. Assuming that the transaction chain length of the current target account is s, and the number of blocks in each subchain is $k(k > 1)$. When the transaction chain length s is determined, the number of subchains n and k are inversely proportional.

$$n = \frac{s}{k} \tag{1}$$

When k increases, the number of block accesses in the subchain will increase, and the query range will increase. The number n of subchains will continue to decrease with the increase of k, because when the query proceeds to the target subchain, other subchains only access the last block, which reduces the number of irrelevant blocks that need to be visited when locating the target subchain.

The ATTC-based query method requires access to the complete transaction chain; the number of irrelevant blocks accessed is t_1.

$$t_1 = n(k - 1) \tag{2}$$

In the SCATC-based query method, the number of blocks to be accessed in the initial query subchain is t_2.

$$t_2 = \frac{s}{k} + k - 1 \tag{3}$$

The number of blocks in the irrelevant subchain accessed is t_3.

$$t_3 = n - 1 \tag{4}$$

With the increasing number of users' transactions, n tends to increase monotonically. Equations (2) and (4) can be regarded as a linear function of t to n. In Eq. (2), the coefficient of the independent variable n is $k(k > 1)$, and Eq. (4) where the coefficient of the independent variable is 1. With the growth of n, the number of irrelevant blocks that need to be accessed increases rapidly based on the ATTC query method. The SCATC-based query method has a slower growth rate, and the larger the n, the more obvious the advantage of the SCATC-based query method.

4.2 Simulation Experiment

The simulation environment is a host computer, where the CPU is Intel(R) Core(TM) i7-5500U, 12 GB memory, and the 64-bit operating system Windows10 Professional Edition. The SCATC index structure is written and implemented in python language. The blockchain requires each full node to maintain a complete ledger, so the data retrieval of the simulation is performed locally.

The simulation compares the query efficiency of ATTC and SCATC query methods under different transaction chain lengths. Set the subchain length k to 10. The length of the transaction chain is set to 1000–6000 blocks, and the corresponding number of subchains is 100–600. The simulation experiments are divided into six groups according to different transaction chain lengths, and each group of simulations is repeated eight times. To better highlight the effect of simulation comparison, each query is tested with the initial subchain. Both query methods start from the latest block forward, so the query time in SCATC includes the time to locate the subchain. The simulation experimental data obtained are shown in Tables 1 and 2.

Table 1. ATTC

Number of blocks	Query time/Ms								AVG	Mean deviation
1000	21	25	13	13	19	20	12	23	18.3	33.4
2000	32	42	30	35	61	56	48	43	43.4	69.8
3000	52	55	47	42	56	39	50	49	48.8	36.4
4000	46	54	73	88	56	69	75	61	65.3	88.0
5000	85	99	88	65	73	79	92	87	83.5	67.0
6000	74	109	85	77	83	95	88	101	89.0	76.0

Table 2. SCATC

Number of blocks	Query time/Ms								AVG	Mean deviation
1000	15	12	16	15	15	17	13	13	14.5	11.0
2000	17	14	15	14	12	15	12	16	14.4	11.0
3000	16	18	14	13	15	17	16	13	15.3	12.0
4000	15	13	15	16	14	13	15	16	14.6	7.8
5000	17	17	17	13	16	16	19	18	16.6	9.8
6000	10	12	16	12	18	21	17	19	17.4	11.8

The average value of each subchain of simulation experimental data of ATTC and SCATC is plotted as a line chart shown in Fig. 2. As the length of the transaction chain continues to grow, the query time based on the ATTC query method is constantly increasing. However, the query method based on SCATC has not changed significantly in query efficiency as the length of the transaction chain continues to increase.

For active users in the blockchain system, the length of the transaction chain has increased at a faster rate. From a theoretical analysis, whether it is based on ATTC or SCATC query methods, as the transaction chain grows, the length of the transaction chain that needs to be traversed will be longer, and the query efficiency will show a downward trend. However, after the SCATC index structure divides the transaction chain into subchains, it greatly reduces the number of visits to irrelevant blocks. The limited length of the transaction chain cannot cause a significant change in SCATC's query efficiency.

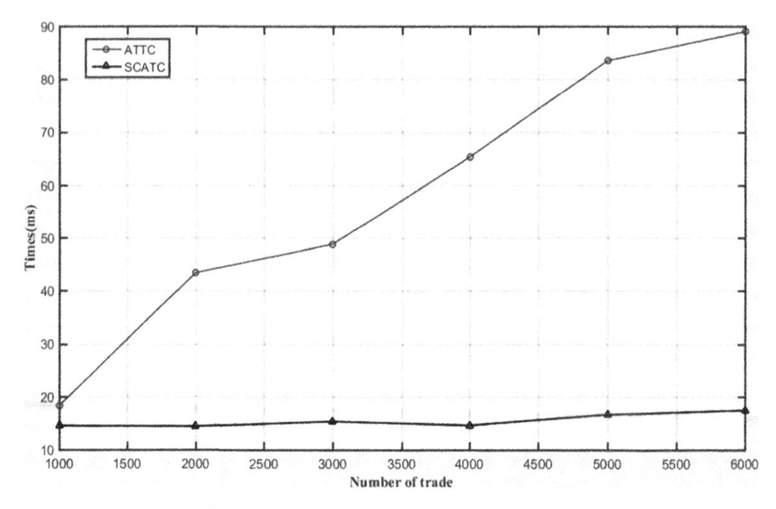

Fig. 2. Comparison of query efficiency

5 Conclusions

We improve the query efficiency of the ATTC index structure and proposes a SCATC index structure that supports querying account subchain data. We divide the transaction chain into subchains, add hash pointers to the account branch nodes of the block at the last block of each subchain, and each subchain is connected by hash pointers. Through this pointer, the query mode of traversing the transaction chain is converted to the subchain query mode, which effectively reduces the access to irrelevant block data and reduces the computational overhead. Besides, we also design a query algorithm for the SCATC index. Simulation experiments and analysis show that the index structure based on SCATC can improve the query efficiency of account transactions effectively.

Acknowledgement. This work is financially supported by the National Natural Science Foundation of China under Grant No. 61962009, Major Scientific and Technological Special Project of Guizhou Province under Grant No. 20183001, Foundation of Guizhou Provincial Key Laboratory of Public Big Data No.2018BDKFJJ005, in part by Talent project of Guizhou Big Data Academy. Guizhou Provincial Key Laboratory of Public Big Data.([2018]01) .

References

1. Nakamoto, S.: Bitcoin: A Peer-to-Peer Electronic Cash System[EB\OL] (2008). https://bitcoin.org/bitcoin.pdf.
2. He, P., Yu, G., Zhang, Y.F., Bao, Y.B.: Survey on blockchain technology and its application prospect. Comput. Sci. **44**(4), 1–7 (2017). (in Chinese with English abstract)
3. Yuan, Y., Wang, F.Y.: Blockchain: The state of the art and future trends. Acta Automatica Sinica 42(4), 481–494 (2016) (in Chinese with English abstract).[doi:https://doi.org/10.16383/j.aas.2016.c160158]

4. Wang, H.J., Dai, B.R., Li, C., Zhang, S.H.: Query optimization model for blockchain applications. Comput. Eng. Appl. **55**(22), 34-39+171 (2019)
5. Wang, Q.G., He, P., Nie, T.Z., Shen, D.R., Yu, G.: Overview of data storage and query technology in blockchain systems. Comput. Sci. **45**(12), 12–18 (2018)
6. Qu, Q., Nurgaliev, I., Muzammal, M., et al.: On spatio-temporal blockchain query processing. Futur. Gener. Comput. Syst. **98**, 208–218 (2019)
7. Yuqin, X., Zhao, S., Kong, L., Zheng, Y., Zhang, S., Li, Q.: ECBC: a high performance educational certificate blockchain with efficient query. In: Van Hung, D., Kapur, D. (eds.) Theoretical Aspects of Computing – ICTAC 2017, pp. 288–304. Springer International Publishing, Cham (2017). https://doi.org/10.1007/978-3-319-67729-3_17
8. You, Y., Kong, L.J., Xiao, Z.S., Zheng, Y.Q., Li, Q.Z.: Hybrid indexing scheme supporting blockchain transaction tracing. Comput. Intergrat. Manufact. Syst. **25**(04), 978–984 (2019)
9. Morishima, S., Matsutani, H.: Accelerating blockchain search of full nodes using GPUs. pp. 244–248 (2018)
10. Zhang, C., Xu, C., Xu, J., et al.: GEM^2-Tree: a gas-efficient structure for authenticated range queries in blockchain. In: 35th IEEE International Conference on Data Engineering (ICDE '19). IEEE (2019)
11. Jia, D., Xin, J., Wang, Z., Guo, W., Wang, G.: ElasticChain: support very large blockchain by reducing data redundancy. In: Cai, Yi., Ishikawa, Y., Jianliang, Xu. (eds.) Web and Big Data: Second International Joint Conference, APWeb-WAIM 2018, Macau, China, July 23-25, 2018, Proceedings, Part II, pp. 440–454. Springer International Publishing, Cham (2018). https://doi.org/10.1007/978-3-319-96893-3_33
12. Jia, D.Y., Xin, J.C., Wang, Z.Q., Guo, W., Wang, G.R.: Efficient query model for scalable storage capacity blockchain system. J. Software **30**(09), 2655–2670 (2019)
13. Jiao, T., et al.: Blockchain database: a database that can be queried and tamper-proof. J. Software **30**(09), 2671–2685 (2019)
14. Zheng, H.H., Shen, D.R., Nie, T.Z., Kou, Y.: Query optimization of blockchain system for hybrid indexing. Comput. Sci. **47**(10), 301–308 (2020)
15. Ren, Y., et al.: Data query mechanism based on hash computing power of blockchain in Internet of Things. Sensors **20**(1), 207 (2019)
16. Wan, L.: A query optimization method of blockchain electronic transaction based on group account. In: Atiquzzaman, M., Yen, N., Zheng, Xu. (eds.) Big Data Analytics for Cyber-Physical System in Smart City: BDCPS 2020, 28-29 December 2020, Shanghai, China, pp. 1358–1364. Springer Singapore, Singapore (2021). https://doi.org/10.1007/978-981-33-4572-0_196
17. Wan, L.: An optimization method for blockchain electronic transaction queries based on indexing technology. In: Atiquzzaman, M., Yen, N., Zheng, Xu. (eds.) Big Data Analytics for Cyber-Physical System in Smart City: BDCPS 2020, 28-29 December 2020, Shanghai, China, pp. 1273–1281. Springer Singapore, Singapore (2021). https://doi.org/10.1007/978-981-33-4572-0_183

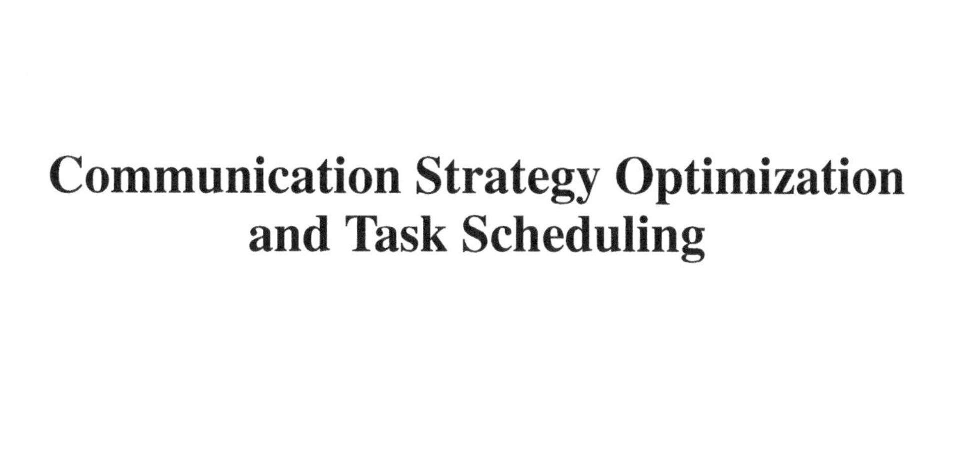

Communication Strategy Optimization and Task Scheduling

Comprehensive Valuation of Environmental Cost Based on Entropy Weight Method

Yihui Dong[1](✉), Jingchao Li[1], Jiayu Han[1], Zihao Zhu[2,3], and Kaixun Zhou[2,3]

[1] Shanghai Dianji University, Shanghai, China
[2] University of Shanghai for Science and Technology, Shanghai, China
[3] Shanghai University, Shanghai, China

Abstract. In the past few decades, economic decisions often exert an imperceptible influence on the biosphere. In order to get the real land use valuation of both small community projects and large national projects, environmental costs can be calculated by establishing a comprehensive valuation model. As fluctuations in the main indicators such as land area and biodiversity can affect the value of land use, we calculate the cost-benefit ratio with the comprehensive valuation model which is based on entropy weight method, and take both environmental costs indicators and economic factors into consideration. By using the model, we analyze the cases of Hong Kong Island and find out land indicator has the greatest impact on environmental costs. In general, if all the weights are variable, the deviation between the calculated indicator and the practical indicator will be narrowed. Land use project planners and managers can get recommendations from this model.

Keywords: Environmental costs · Entropy weight method · Comprehensive valuation model

1 Introduction

1.1 Background and Problem Analysis

Throughout the ages, ecosystem services are directly or indirectly beneficial for human life as natural processes, which provide four categories of services for us ——supporting services, provisioning services, regulating services and cultural services. However, people all over the world are potentially limiting or removing ecosystem services by altering it. In order to further illustrate the alteration, we can divide it into two parts ——local small-scale changes and large-scale projects. Although it seems insignificant to the total ability of the biosphere's functioning potential, they can damage the diversity of species and give rise to environmental degradation.

Actually, the impact of, or account for changes to, ecosystem services are not considered by most land use projects. And many negative changes like polluted rivers, poor air quality, hazardous waste sites, poorly treated waste water, climate changes are often not included in the plan. So, we constructed a model with high enough fidelity of the mathematical modelling and analyze to further manifest and elaborate our solutions.

J. Xiong et al. (Eds.): MobiMedia 2021, LNICST 394, pp. 145–156, 2021.
https://doi.org/10.1007/978-3-030-89814-4_11

To work out what is the cost of environmental degradation, we broke it down to 5 tasks to analyze this problem. What we need to do is shown below (Fig. 1) (Table 1):

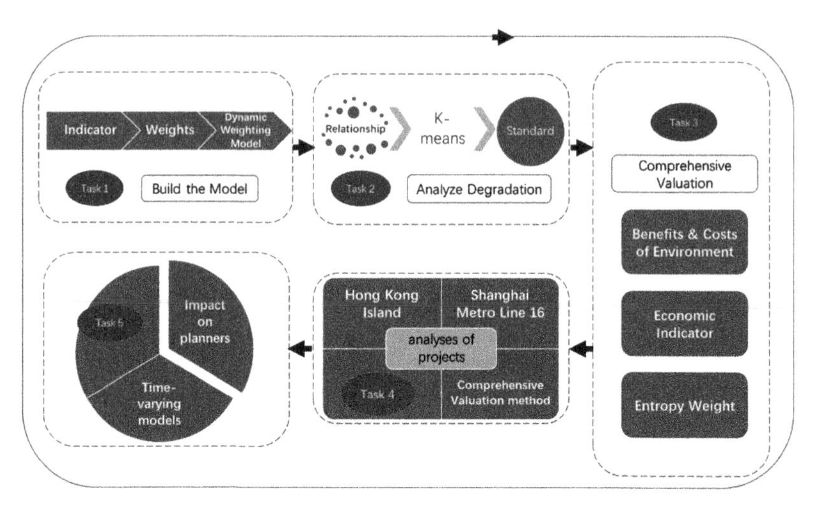

Fig. 1. Flow chart of this paper

2 Symbols and Definitions

Table 1. Symbols and Definitions

Symbols	Definitions
FA	Forest Area (% of land area)
AL	Arable land (% of land area)
TPA	Terrestrial Protected Areas (% of land area)
TPS	Threatened Plant Species
TBS	Threatened Bird Species
CDE	Carbon Dioxide Emissions (metric tons per capita)
AP	PM2.5 Air Pollution (micrograms per cubic meter)
IW	Improved Water (% of total population served with improved water)
IWW	Industrial Water Withdrawal (% of total water withdrawal)
FW	Freshwater Withdrawal
AQI	Air quality Indicator
WI	Water Indicator
LI	Land Indicator

(*continued*)

Table 1. (*continued*)

Symbols	Definitions
BI	Biodiversity Indicator
EWM	Entropy Weight Method
EWI	Entropy Weight Index
SWM	Subjective Weight Method
SWI	Subjective Weight Index
EI	Economic Indicator
EBL	Economic Benefits from Land (one hectare is a unit)
ERWR	Expenditure of the Restoration of Water Resource
ERAE	Expenditure of the Restoration of Atmospheric Environment
EMB	Expenditure of the Maintenance of Biodiversity
CE	Comprehensive Evaluation

3 The Model

3.1 Comprehensive Evaluation

We select the following indicators: environmental costs indicator and economic indicator. For environmental costs indicator, we take the factors of the environment itself and social factors into consideration. Our model includes four overall Indicators —— AQI, LI, BI, WI, which include ten small indicators. The 10 indicators include forest area, arable land, terrestrial protected areas, threatened plant species, threatened bird species, CO2 Emissions, PM2.5 air pollution, improved water, industrial water withdrawal, freshwater withdrawal (Fig 2).

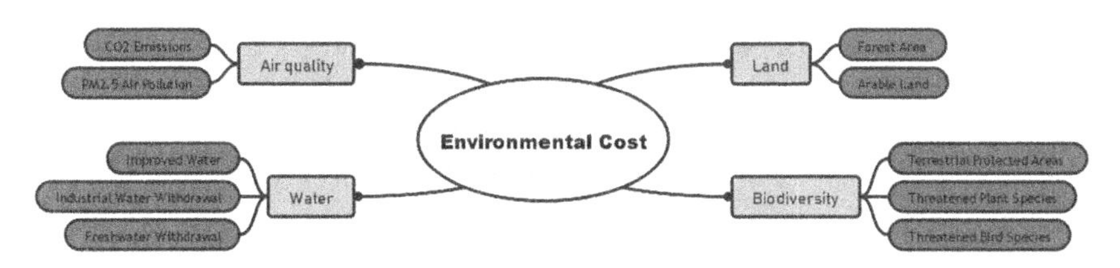

Fig. 2. All indicators

Dynamic weighting model consists of entropy weight method (EWM) and subjective weight method (SWM). Then, we get EWI and WEI respectively by using EWM and SWM. Finally, we get ECI, which is the combination of two methods. Through our ECI, we can put a value on the environmental costs of land use development projects.

p_i is the variance ratio of each indicator for every five years. As the great difference of the development of each country or region in different period, we should calculate the weight of each period from 1960 to 2015——5 years is a unit——to get a more scientific ratio every 5 years, and to analyze concretely, we should analyze the actual situation of every country or region. w_i is the weight of each Indicator

$$Land\ Index = \sum_{i=1}^{2} p_{i*}w_i \tag{1}$$

$$Biodiversity\ Index = \sum_{i=3}^{5} p_{i*}w_i \tag{2}$$

$$Air\ Quality\ Index = \sum_{i=6}^{7} p_{i*}w_i \tag{3}$$

$$Water\ Index = \sum_{i=8}^{10} p_{i*}w_i \tag{4}$$

3.1.1 Dynamic Weighting Models

With the evaluation indicators defined above, we further determine the weights of these indicators, resulting in the combination of primary indicators. Recalling on the Entropy Weight Method (EWM), we will carry out the standardized treatment, making the optimal and worst value of each variables after alternation be 1 and 0, respectively. The evaluation indexes are $X_1, X_2, X_3, \ldots, X_k$, where $X_i = \{x_{i1}, x_{i2}, \ldots, x_{in}\}$. Among there, For the sake of the cost-type indicators, the environmental costs is proportional to the value of the indicator. Nevertheless, in terms of the gain-type indicators, The higher the value, the lower the value of land development in the country. Thus, we have.

$$Y_{ij} = \frac{x_{ij} - \min(x_i)}{\max(x_i) - \min(x_i)}\ j = 1, 2, \ldots, \mathrm{n} \tag{5}$$

where Y_{ij} is the standardized value of each evaluation indicator of each country, $\max(x_i)$ and $\min(x_i)$ are the maximum and minimum value of the evaluation indicator X_i.

$$\max(x_i) = \max\{x_{i1}, x_{i2}, \ldots, x_{in}\},\ \min(x_i) = \min\{x_{i1}, x_{i2}, \ldots, x_{in}\}$$

After standardization, then we introduce.

$$T_{ij} = \frac{Y_{ij}}{\sum_{j=1}^{n} Y_{ij}} \tag{6}$$

According to the concepts of self-information and entropy in the information theory, we can calculate the information entropy.

$$E_i = -\ln(n)^{-1}\sum_{j=1}^{n} T_{ij}\ln T_{ij} \tag{7}$$

On the basis of the information entropy, we will further compute the weight of each evaluation indicator we defined before.

$$w_i = \frac{1 - E_i}{m - \sum_i E_i} \mathrm{i} = 1, 2, \ldots, \mathrm{m} \tag{8}$$

Subsequently, we can derive the four comprehensive evaluation indicators: air quality index, land index, water index and biodiversity index. Here after this paper will be abbreviated as LI, BI, AQI and WI respectively. On the basis of those calculated weights, we have

$$\begin{aligned} LI &= w_1p_1 + w_2p_2 \\ BI &= w_3p_3 + w_4p_4 + w_5p_5 \\ AQI &= w_6p_6 + w_7p_7 \\ WI &= w_8p_8 + w_9p_9 + w_{10}p_{10} \end{aligned} \tag{9}$$

$p_{nj}(n = 0 \sim 10)$

This is the variation of five years. Hence, we get EWI:

$$EWI = k_1 * LI + k_2 * BI + k_3 * AQI + k_4 * WI \tag{10}$$

According our method and model, it will cause environmental costs if the value of EWI is positive; it will generate environmental costs if the value of EWI is negative.

We assign weights (k_1, k_2, k_3, k_4,) for four indicators (LI, BI, AQI, WI,).

Environmental costs and profit are difficult to be quantized. To avoid the impact of the evaluation of land-use value, we need to define a subjective index to analyze the comprehensive land-use value. A series of human factors can exert impact on the weight of index. For instance, the plans and the thoughts of developers, the time of developing, the emotion of residents, the activities of residents, the interventions of government, etc. Subjective weight are adopted, thus we can get SWI (subjective weight index).

To compensate the deviation of the two methods, we adopt the combined index as the ultimate index. The calculating formula is shown below:

$$ECI = a{*}EWI{+}b * SWI(\mathrm{a} + \mathrm{b} = 1) \tag{11}$$

EWI in this equation accounts for main part.

Dynamic weighting model consists of entropy weight method (EWM) and subjective weight method (SWM). Then, we get EWI and WEI respectively by using EWM and SWM. Finally, we get ECI, which is the combination of two methods. Through our ECI, we can put a value on the environmental costs of land use development projects.

3.2 Environmental Degradation in Project Costs

3.2.1 Relationship between Indicators and Environmental Costs

After carefully analyzing the relevant information, we determine the variation tendency between Indicators and environmental costs are shown in the figure below (Table 2):

Table 2. Variation tendency between Indicators and environmental costs

Indicator	Variation Tendency(Index)	Indicator(Environment cost)
FA	↑	↓
AL	↑	↓
TPA	↑	↓
TPS	↑	↑
TBS	↑	↑
CDE	↑	↑
AP	↑	↑
IW	↑	↓
IWW	↑	↑
FW	↑	↑

3.2.2 The Determination of ECI Standard

To scientifically analyze the ECI, we define a set of standard for it. Symbol '−' means the profit of environment. Symbol '+' means the cost of environment. We also divide the degree into low, medium, high.

To determine the standard of environmental costs, we divide all countries into three categories with the K-means clustering algorithm: low environmental costs, medium environmental costs and high environmental costs. Therefore, we randomly selected 36 countries to determine the standard. we calculate its various indices. The K-means clustering algorithm allows us to divide 36 countries into three groups. Then use the index of the country with the lowest and highest indicator as the boundary. It will generate environmental profit if the value of EWI is negative. To scientifically analyze the ECI, we define a set of standard for it. The environmental costs and each standard Indicator are as follows (Fig. 3):

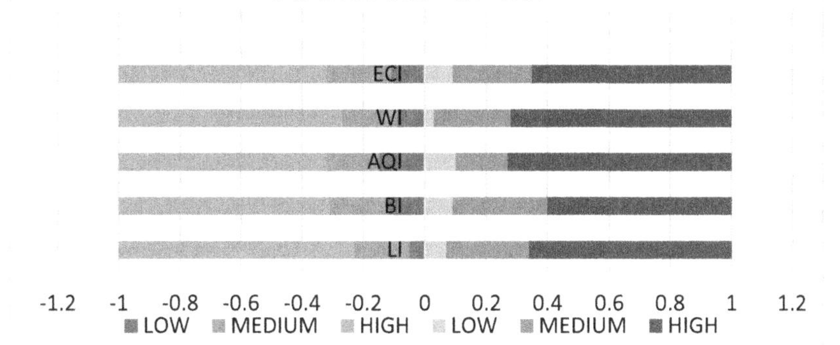

Fig. 3. Standard of ECI

The figure shows that it will generate high environmental profit if the value of ECI is more than −0.40, but less than −0.34. And LI has the most fast growing speed of environmental profit. It begins to generate high environmental profit when the value reaches around −0.23. We regard the overall economic factor as earnings if the economic indicator is positive, and we regard the overall economic factor as economic cost if the economic indicator is negative.

3.3 Comprehensive Valuation of Project

3.3.1 The Determination of Economic Indicator

The way of the selection of economic indicator is similar to ECI. Then, we get four Indicators: Economic benefits from land (one hectare is a unit), Expenditure of the restoration of water resource, Expenditure of the restoration of atmospheric environment, Expenditure of the maintenance of biodiversity.

We define the equation as follows:

$$EI = EBL - (ERWR + ERAE + EMB) \tag{12}$$

Finally, we get the economic Indicator.

3.3.2 Comprehensive Valuation Method

To compensate the deviation of the two methods, we adopt the combined Indicator as the ultimate Indicator. The calculating formula is shown below:

$$CE = c * EI - d * ECI(\mathrm{c} + \mathrm{d} = 1) \tag{13}$$

Reiteration: the plus-minus of ECI and EI is the same as above.

3.4 Comprehensive Valuation of Hong Kong Island

According to our analysis in Sect. 3.1.1, we get the weights of all the indicators. The weights of all the indicators is shown as below (Table 3):

Table 3. The weights of all the indicators

Indicator	Weights	Indicator	Weights
Land	0.3632	FA	0.8067
		AL	0.1933
Biodiversity	0.2199	TPA	0.2609
		TPS	0.4081
		TBS	0.3310
Air Quality	0.2873	CO2 Emissions	0.2437
		AP	0.7563
Water	0.1296	IW	0.6873
		IWW	0.1319
		FW	0.1806

And we also calculate the percentage of each indicator. The percentages of LI, BI, AQI and WI are 36%, 22%, 29% and 13%, respectively. As the weights of LI is greater than other three Indicators, we select LI to test and verify the weights we calculated before.

We divide the development period of Hong Kong into two 20-years-long period —— 1997 is the time node —— before 1997 and after 1997. The reason is that Hong Kong became British colony in 1842 and returned to China in 1997 (Fig. 4).

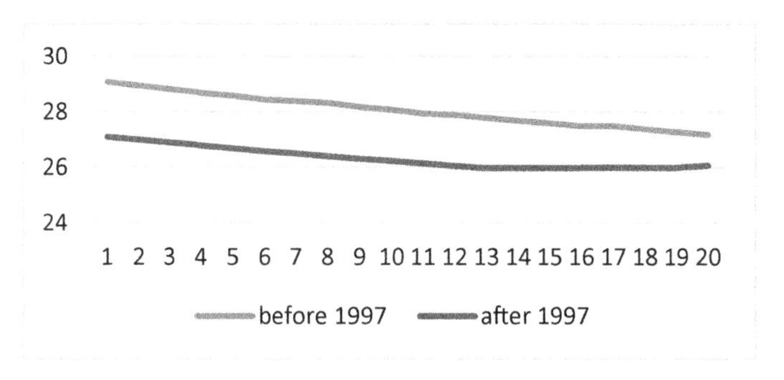

Fig. 4. The change rate of forest area

The line chart shows the change rate of forest area in Hong Kong. The blue curve represents the change rate of forest area before 1997; the orange curve represents the change rate of forest area after 1997. It is obvious that the change rate did not stop decreasing until 2010. (the white circle is 2010) Then, the line chart became smooth from 2010 to 2017 (Fig. 5).

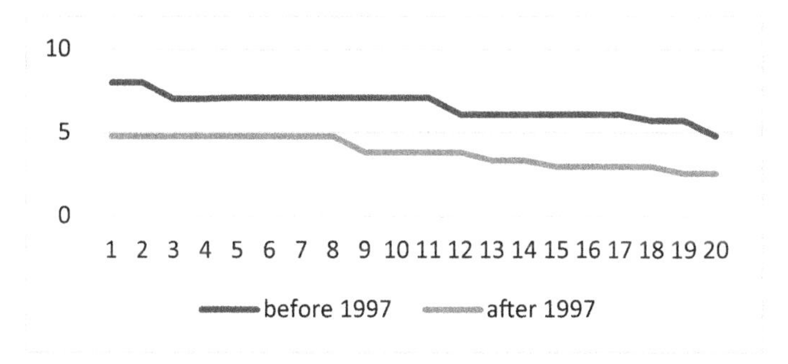

Fig. 5. The change rate of arable land

The line chart manifests the change rate of arable land in Hong Kong. The green curve represents the change rate of arable land before 1997; the brown curve represents the change rate of arable land after 1997. The change rate was keep decreasing all this time (Fig. 6).

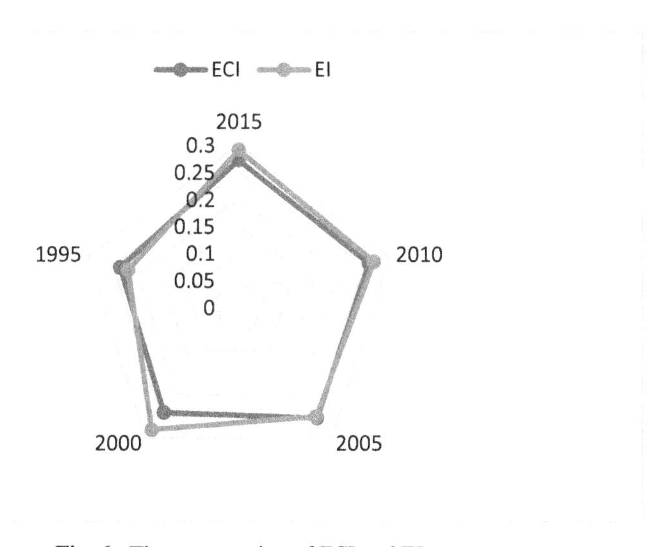

Fig. 6. The comparation of ECI and EI

According to our calculation, both ECI and EI were increasing from 1995 to 2015. Besides, the date of ECI was close to EI. Our comprehensive valuation also shows that the environmental problems are taken more seriously in the process of developing land for HKI projects. During this process, FA and AL were declining in a sharp way. The developers exchanged environmental costs for short-term economic profit. In the long run, economic profit cannot offset environmental cost when the weight of ECI is increasing in our comprehensive valuation. As the data in 1995, EI was less than ECI. Through our comprehensive valuation, EI should be greater than ECI in the next decade if we want to assure the sustainable development of land.

4 Impact of Modeling on Land-Use Project Planners and Managers

As many factors could lead to environmental costs, land-use project planners should consider the indicators we proposed in the process of land development and focus on the indicators with great weights. These great indicators occupy a dominant position in the environmental costs of land development. At the same time, the influence of social factors cannot be ignored. For example, the damage of environment can cause dissatisfaction among residents and terrible reputational cost of developers. In general, people will feel relaxed if we live in protected environment. The satisfied residents can promote the stability of society (Table 4).

Table 4. The Weights of all the indicators in China

Indicator	Weights	Indicator	Weights
Land	0.2338	FA	0.4682
		AL	0.5318
Biodiversity	0.3522	TPA	0.3022
		TPS	0.2979
		TBS	0.3999
Air Quality	0.2732	CO2 Emissions	0.5837
		AP	0.4163
Water	0.1408	IW	0.2545
		IWW	0.4223
		FW	0.3232

According to Fig. 10, BI tend to have the greatest weight for large-scale national land development projects in China. Therefore, In the process of land development for large national projects, the environmental costs brought by biodiversity is the most important part. We should avoid developing land in terrestrial protected areas, some areas with threatened plant species, and the other areas with endangered birds.

AQI also occupies a great weight. It will cause more environmental costs if air quality is not up to the standard in some areas. For example, the great carbon dioxide emissions with terrible PM2.5 in a specific small area can lead to more environmental costs.

In the process of land development for small national projects, forest area Indicator and arable land Indicator play the most indispensable roles through all the Indicators. It would cause huge environmental costs if we develop land in this area. Besides, it would accelerate the trend of environmental degradation.

Once thinking about these Indicators, especially the Indicators with great weights, we should take effective measures to reduce environmental costs and to increase revenue. In the long run, it can reduce the environmental damage caused by land development and mitigate the adverse effects of environmental degradation.

From what we have discussed above, the economic cost, the impact of environmental costs, the environment itself, and social factors should be taken into account when we are assessing the value of land development and utilization.

5 Time-Varying Models

As time goes by, all indicators will change with the land policy and development strategy of the country. Besides, some social factors such as population, living condition, and social stability also be taken into account. As the factors we discussed above, the weights of each Indicator should be variable. For instance, social factors have significant impact on the weights of ECI and EI in our comprehensive valuation. So, the developers will pay more attention to ECI when social factors are changing greatly. For example, on the case of HKI, the developers improved the weight of ECI in the comprehensive valuation of land development after 1997. After that, environmental degradation was alleviated by the improvement of EI. Actually, the weights of LI, BI, WI and AQI can be changed with social factors. In general, the deviation between calculated Indicators and practical Indicators would be narrowed if all the weights are variable.

6 Sensitivity Analysis

All Indicators will change with many factors such as the scale of project, social stability, natural disasters, etc. Based on our analysis and outcome above, we can determine some of Indicators which have high weights and get accurate value through the combination of EWM and SWM. The indicators with low weights cannot exert huge impact on our final comprehensive valuation of environmental costs even if there are some inevitable errors in it. For instance, on the case of HKI, FW changed 0.09 while the final ECI changed 0.002, which means some Indicators with low weights play slight roles in our model. Therefore, our model is stable because it is not easily affected by errors.

References

1. Costanza, R., et al.: The history of ecosystem services in economic theory and practice: from early notions to markets and payment schemes. Ecol. Econ. **69**(6), 1209–1218 (2010)
2. Land-use planning. https://en.wikipedia.org/wiki/Land-use_planning
3. Environmental protection. https://en.wikipedia.org/wiki/Environmental_protection
4. Ecosystem services. https://en.wikipedia.org/wiki/Ecosystem_services
5. land-use planning.https://planningtank.com/planning-techniques/land-use-planning
6. Land degradation and desertification
7. https://www.who.int/globalchange/ecosystems/desert/en/
8. Data of water index, land index and so on. https://cn.knoema.com/, https://data.worldbank.org/?year_high_desc=true

9. Mylonakis, J., Tahinakis, P.: The use of accounting information systems in the evaluation of environmental costs: a cost-benefit analysis model proposal. Int. J. Energy Res. **30**(11), 915–928 (2010)
10. Atkinson, G., et al.: Environmental cost-benefit analysis. Ann. Rev. Environ. Resour. **33**(1), 317–344 (2008)
11. Tan, Y., Zuo, J., Chen, J.: An environmental cost value model based on dynamic neural network prediction. In: Journal of Physics Conference Series, vol. 1325, pp. 012090 (2019)
12. Wencheng, Z., Bin, S., Economics, S.O., et al.: Environmental cost of China's exports: based on emissions intensity of value-added exports. J. Quant. Tech. Econ. (2017)
13. Yue-An, L., Jun, F.U., Yong, L., et al.: Environmental cost analysis and evaluation study of civil construction equipment in full life cycle. J. Zhejiang Sci-Tech Univ. (2014)
14. Nuță, F.M., Crețu, C.M., Nuță, A.C.: Environmental cost accounting – assessing the environmental responsibility effort. Proceedings, **7**(1). (2012)
15. Zhu, X., Chiong, R., Liu, K., Ren, M.: Dilemma of introducing a green product: Impacts of cost learning and environmental regulation. Appl. Math. Model. **92**, 829–847 (2020)

A QoE Evaluation and Adaptation Method for Multi-player Online Games

Zaijian Wang[1,4](✉), Wei Guo[1], Le Zhao[1], Weidong Yang[2], Shiwen Mao[3], and Zhipeng Li[1]

[1] The School of Physics and Electronic Information, Anhui Normal University, Wuhu 241002, Anhui, China
wangzaijian@ustc.edu

[2] The Key Laboratory of Grain Information Processing and Control at Henan University of Technology, Ministry of Education, Zhengzhou 450001, Henan, China
Yangweidong@haut.edu.cn

[3] Department of Electrical and Computer Engineering, Auburn University, Auburn, AL 36849-5201, USA
smao@ieee.org

[4] Anhui Province Key Laboratory of Optoelectric Materials Science and Technology, School of Physics and electronic information, Anhui Normal University, Wuhu 241002, China

Abstract. To evaluate the performance of multi-player online games (MPOGs) (e.g., the online Chinese poker game), this paper presents a novel Effective Quality of Experience (EQoE) concept that considers the interaction among multiple players. A novel QoE evaluation and adaptation method is proposed utilizing EQoE. Specifically, the proposed method assigns different weights to different players according to their roles in the multi-player online poker game. Some players with a lower weight are willing to reduce their own QoE to save network resources for players with a higher weight in order to keep the game going under poor network conditions. Our simulation study demonstrates the feasibility of the proposed QoE evaluation and adaptation method.

Keywords: Multi-Player Online Game (MPOG) · Resource allocation · Quality of Experience (QoE) · Mean Opinion Score (MOS) · 5G wireless

This work was supported in part by the National Natural Science Foundation of China (No. 61401004); Anhui Province Natural Science Foundation of China (No. 2008085MF222); Open fund of the Key Laboratory of Grain Information Processing and Control (No. KFJJ-2018-205) from the Key Laboratory of Grain Information Processing and Control (Henan University of Technology), The Ministry of Education; Key research projects of Anhui Provincial Department of Education of China (No. KJ2019A0490 and KJ2019A0491); and the NSF under Grant ECCS-1923717.

J. Xiong et al. (Eds.): MobiMedia 2021, LNICST 394, pp. 157–171, 2021.
https://doi.org/10.1007/978-3-030-89814-4_12

1 Introduction

As an important metric to evaluate users' satisfaction with perceived quality, Quality of Experience (QoE) is becoming one of the key metrics in the design of 5G/future networks [1], where a new era of personalized services are emerging that emphasize service experience and users' QoE [2]. QoE has been widely used to describe the overall level of satisfaction of service quality from user's perspective, and is a vital factor to evaluate the success of network traffic over future network systems (such as the 5G Tactile Internet).

Among many emerging multimedia applications, Multi-Player Online Game (MPOG) is nowadays one of the most popular applications, which usually is of large-scale over wide geographical locations and keeps attracting a large number of players. In 2010, the number of gamers has reached 20 million worldwide [3], and the PC online game market worldwide reaches $44.2 billion USD in 2020 (according to statistics.com). The MPOG is a mixture that consists of a number of different subgenres, which all have various, but stringent responsiveness requirements [3]. Different MPOGs can tolerate different performance thresholds, for example, 100 ms for First Person Shooter (FPS), 500 ms for Role Playing Game (RPG), and 1000 ms for Real Time Strategy (RTS) [4].

1.1 Motivation

It has been recognized that ensuring an acceptable QoE for all players is a fundamental criterion to sustain the economic success of the MPOG industry, which has been growing at a rapid pace and has attracted enormous attention in recent years [4]. Different from individual Quality of Service (QoS) metrics, the impact of various system factors on QoE depends on the ongoing interaction among users in the multiuser scenario [5].

Going beyond conversational single-user services, MPOGs are offered in different circumstances. For example, Destiny is a blend of shooter game and massive multi-player online game. It has attracted dozens of millions of players so far. In the game world, MPOG provides a platform for hundreds of thousands of players to participate in a game simultaneously. By dividing the game world into linked mini regions that are then distributed between servers, the players in each region are fetched by the server(s) hosting the region. Since game event handling, non-player character (NPC) control, physical hardware restrictions (such as CPU speed and the amount of memory that is available), and persistence in the game world management, MPOG's servers are usually heavily loaded [6].

Most existing research efforts on QoE evaluation are limited to the single-user scenario, which rarely consider issues or provide enhancements for multiuser scenarios. Considering fast development of MPOGs that should be offered with an acceptable QoE to all the concurrent players [6], this paper is focused on the challenging problem of QoE evaluation for MPOGs.

1.2 Challenges and Existing Solutions

A major challenge here is how to provide effective QoE evaluation, since QoE is influenced by both subjective and objective factors. The former includes bit rate, jitter, delay, and so forth. The latter involves user profile, emotion, education, surroundings, etc., which are related to human subjective factors and are hard to measure and quantify. There exists many prior works on QoE evaluation to address this challenging problem. For example, the authors in [7] proposed a continuous QoE prediction engine, which was driven by three QoE-aware inputs: a QoE memory descriptor that accounts for recency, rebuffering-aware information, and an objective measure of the perceptual video quality. In [8], Mok, Chang, and Li aimed to construct a predictive model by incorporating supervised learning algorithms, in which multiclass Naïve Bayes classifiers were applied to train a model. In a recent work [9], the authors proposed a variety of recurrent dynamic neural networks that conduct continuous-time subjective QoE prediction. A study was presented in another recent work [10] on subjective and objective quality assessment of compressed 4K ultra-high-definition videos in an immersive viewing environment. However, such existing research efforts on QoE evaluation are usually limited to the single-user scenario, which rarely consider the unique challenges arising in multiuser scenarios. Therefore, such approaches may not be effectively extended for MPOGs.

Unlike QoS, QoE is a subjective assessment of media quality of users and takes into account both technical parameters and usage context variables. System factors, context, and user, along with user behavior, will all influence a user's QoE, which in turn will influence the user's behavior and the current state of the user [5]. In practice, variations in service quality is unavoidable. For example, the same user may change her QoE requirements for the same traffic condition but in different scenes. Therefore, it is a non-trivial challenge for network operators and service providers, which are interested in seeking effective solutions to improve the overall network performance and revenue. The evaluation of QoE for the multiuser scenario is still an evolving, open problem, which requires far more works to be done in this domain. In this paper, we aim to tackle this challenging problem with a properly designed QoE evaluation method, which takes into account the interactions among multi-players to offer uninterrupted services in future network systems such as the 5G Tactile Internet.

1.3 Methodology and Innovations

Without loss of generality, we use interactive multi-player online poker game as an example, for which QoE modeling is a vital design factor. However, owing to the individual application types and user preferences, different players often have different QoE requirements even with the same data rate in the same game; even the same player may have different QoE requirements in different game scenes. Therefore, a multi-level QoE model should be maintained for the game players at runtime.

Since the players play the online poker game without needing 3D graphic rendering hardware or software, game streaming often consumes very little bandwidth. For a specific player, the throughput does not need to be overly increased, because the player's QoE cannot be further improved when the throughput is sufficiently large. In the interactive multi-player online poker game, players often have different levels of priority depending on their roles in the game. As illustrated in Fig. 1, some players can adjust their QoE requirements according to the

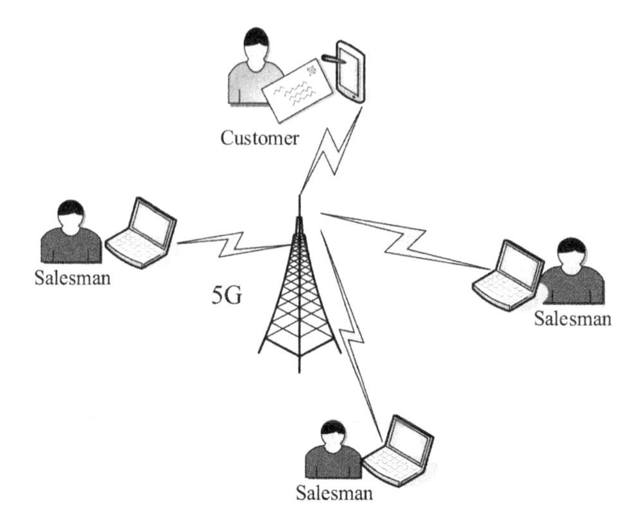

Fig. 1. An example scenario of the online Chinese poker game.

dynamic conditions to play the game with an acceptable QoE. Some players are more willing to neglect a decrease in network resources to maintain an optimal communication frequency for an enjoyable game. In the example shown in Fig. 1, a few salesman are playing the game to social with their customers. Because the online poker game will be terminated when any of the players decides to quit due to poor QoE, the salesmen are more willing to keep playing the game by decreasing their QoE level to provide more network resources to the customers.

Considering that different players usually have different roles in the game, due to different purposes such as business, social, team building, etc., we can make appropriate network selections accordingly in order to minimize the effect of poor network conditions for better game interactivity and an enjoyable user experience. Because some players are willing to save or provide network resources for some other players with a higher priority, by decreasing their own QoE level in order to continue the game, it is possible to maintain the quality of the game by relaxing some network constraints for such players.

Many existing methods are focused on individual player's perception of service quality, while ignoring the mutual influence of multiple players. We propose to consider the special characteristics of the multi-player online poker game, which influence the player's QoE accordingly. A new QoE evaluation method should reduce the impact of poor and unstable network conditions on the interactivity of the game, and maintain an enjoyable game for players even with poor or unstable network capabilities.

In this particle, our main objectives and contributions are: (i) to propose an "Effective QoE" (EQoE) concept by considering the interactions among multi-players; (ii) based on the concept of EQoE, we propose a novel QoE evaluation method considering the interactions among multi-players.

1.4 Organization

The remainder of this article is organized as follows. We discuss research challenges and existing solutions in Sect. 2. Our proposed QoE evaluation method is presented in Sect. 3. We then present a case study to demonstrate the performance of the proposed new approach in Sect. 4. Finally, we discuss open problems and future research directions in this problem area in Sect. 5, and conclude this paper in Sect. 6.

2 Research Challenges and Existing Solutions to Address These Challenges

Compared to the existing solutions that mainly implement QoE prediction for an individual user [9,10], some of the new challenges for QoE evaluation method for MPOGs include:

- It is non-trivial, if not impossible, to quantify the impact of various factors that influence end users' QoE.
- Devices need to have high processing capabilities to meet the high CPU demand imposed by online games.
- Since hundreds of thousands of players may be served (and hence constrained by) a single server, the players tend to flock to one area, causing the game related traffic to reach an unacceptable level, and eventually causing a low QoE for the players [6].
- How to model the activities of multiple players based on effective simulation models [11]?
- How to model the impact of various subjective and objective factors on the QoE?
- How to identify new, effective Key Performance Indicators (KPIs), which can truthfully represent the QoE?
- How to model players' preferences? The flocking of players to a hot-spot may be triggered either by players preference for a particular region of the game, or preference of players to achieve a bonus scores to improve their profiles [6]. All such factors should be considered in the QoE model.
- How to employ an integrated empirical and theoretical approach towards finding the structural patterns in gaming behaviors of both humans and artificial agents [11]?

Researchers have been making great effort to tackle some of the above issues. For example, in [12], the authors proposed a multiuser MAC scheduling scheme for the Wireless Virtual Reality service in a 5G MIMO-OFDM system, aiming to maximize the number of simultaneous Wireless Virtual Reality clients while guaranteeing their three ultra-high (3UH) QoE requirements. This scheme was composed of three novel functions, including video frame differentiation and delay-based weight calculation, spatial-frequency user selection based on maximum aggregate delay-capacity utility (ADCU), and link adaptation with a dynamic block-error-rate (BLER) target. In [13], the authors assessed the impact of several system factors, in the context of multi-party video-conferencing, on (i) user behavior, especially the interaction of participants, and (ii) the QoE considering user factors. Nightingale, et al. in [14] focused on providing a QoE prediction model for ultra-high-definition (UHD) video flows in emerging 5G networks. To improve QoE fairness and utilization of network resources, the authors in [15] presented an optimization framework based on sigmoidal programming to maximize the sigmoidal utility functions subject to network constraints. In this framework, the network bandwidth allocation problem for video traffic was formulated as a network utility maximization problem, which was a nonconvex optimization problem, and then solved with sigmoidal programming.

Although some advances having been made, the existing studies have not considered the weight effects of players participating in the game. Most of these QoE evaluation methods are focused on the single-user scenario, but rarely consider issues or provide enhancements for multi-player scenarios. Such approach are hard to be extended for effective QoE evaluation for MPOGs.

3 Proposed QoE Evaluation/Adaptation Method

3.1 System Framework

In the illustrative example shown in Fig. 1, there are four players participating in the online Chinese poker game, with strong interactivity among the multi-players. The players have different priorities depending on their roles in the game. However, a higher priority player will still be forced out of the Chinese poker game if any of the four players quits the game (i.e., the game will be terminated when the first player quits the game).

Generally, the player with a higher priority has more control of running the poker game. For example, the salesmen are more willing to play the poker game even if they have a low QoE (to social with the customer), while the customer wants to continue the game only if his/her QoE is satisfactory. Meanwhile, the individual QoE should be considered since different players have different QoEs. When any of the multi-players has a poor QoE, the corresponding player will quit from this game. In this case, the poker game will be interrupted even if the others with a higher priority are enjoying a satisfactory QoE.

In order to provide more network resources to the player with a higher priority, some other players should be willing to downgrade their QoE requirements due to other considerations (i.e., if they wish to play the game longer). In such cases, the players with a lower priority may still maintain a high Mean Opinion Score (MOS) value as estimated by typical QoE prediction methods.

Since some low-priority players can tolerate QoE variations within a certain range, we aim to design an effective QoE evaluation method from the user demand-centric perspective in terms of classical MOS for the multi-player scenario. We focus only on QoE evaluation, and assume that the priorities can be assigned with an existing method such as deep learning [17,18]. An agent located at the Base Station (BS) collects information from all players, and analyzes individual QoE levels and priorities in a particular context. After executing the proposed evaluation method, the agent can make an appropriate network selection according to the player's priority and network resource requirement. The main idea is that the proposed method may degrade some low-priority player's QoE momentarily, but the overall QoE the game will be maintained.

3.2 QoE Assessment

To represent the satisfaction for different players, this paper adopts MOS as a qualitative measure to characterize QoE since MOS has been recognized as the most popular descriptor of perceived multiple media quality. MOS is defined in ITU-T P.800, P.910 and P.920, which generally consists of five QoE levels. From high to low, the value of MOS represents "Best," "Excellent," "Good," "Poor," and "Bad" for users' QoE, respectively. "Bad" means that the perceived quality drops to an unacceptable level, while "Best" indicates that the user has enjoyed the most satisfactory experience [16].

Depending on whether the player is directly involved in the assessment, existing QoE prediction methods can be categorized into subjective methods and objective methods. The former is a direct means to obtain QoE by asking players. Therefore, it is expensive and ineffective to measure. The latter utilizes numerical quality metrics to approximate user-perceived service quality.

Since the data rate is one of the most important factors in determining QoE [7], throughput is sometimes used to represent the QoE a player can obtain for simplicity. In this paper, we adopt the QoE assessment method proposed in paper [16] for individual QoE assessment. Based on the method in [16], we utilize a discrete MOS model with five limited grades to evaluate each of the players, and consider a MOS of 2 as the minimum required rating to keep the game running. Assume there are N players. The relationship between data rate and MOS can be written as a bounded logarithmic relationship function as follows [16]:

$$\mathrm{MoS}(\theta_n) = \begin{cases} 5, \text{ for } \theta_n > \theta_n^4 \\ 4, \text{ for } 3 < a\log(\frac{\theta_n}{b}) \leq 4 \\ 3, \text{ for } 2 < a\log(\frac{\theta_n}{b}) \leq 3 \\ 2, \text{ for } 1 < a\log(\frac{\theta_n}{b}) \leq 2 \\ 1, \text{ for } \theta_n < \theta_n^1, \end{cases} \text{ for } n \in \{1, 2, ..., N\}, \tag{1}$$

where

$$\begin{cases} a = \frac{3.5}{\log(\theta_n^4/\theta_n^1)} \\ b = \theta_n^1 \left(\frac{\theta_n^1}{\theta_n^4}\right)^{\frac{1}{3.5}} \\ 0 \leq \theta_n^1 < \theta_n^4, \end{cases} \text{ for } n \in \{1, 2, ..., N\}. \tag{2}$$

In this model, θ_n denotes the average throughput of user n, and a, b, θ_n^1, and θ_n^4 are model parameters, where θ_n^1 represents the required minimum throughput and θ_n^4 is the recommended throughput.

3.3 The Concept of Effective QoE

For interactive multi-player online poker games, it is difficult to maintain a good game quality by considering only the individual QoE levels. When any of the players decides to quit due to an unacceptable QoE level, the online poker game will be stopped even if the other players have a satisfactory QoE. Therefore, we not only take account into the individual QoEs, but also consider the group QoE. By levering the network resources among the players, the game can be maintained at a high quality of QoE.

To effectively allocate network resources among the players, a concept of Effective QoE (EQoE) is proposed in this paper, which is different from the individual QoE. The formulation of EQoE is given in the following.

$$\mathrm{QoE}_e = \mathrm{QoE}_m + \sum_{n=1, n \neq m}^{N} \lambda_n (\mathrm{QoE}_n - \mathrm{MoS}_L) \tag{3}$$

$$\sum_{n=1}^{N} \lambda_n = 1, \tag{4}$$

where QoE_e represents EQoE, N is the number of players in the multi-player game, the nth player has the corresponding QoE value QoE_n obtained as in (1), λ_n denotes the player weight of the nth player, $\lambda_m = \max_n\{\lambda_n\}$, QoE_m is the QoE value of the player with weight λ_m, $\mathrm{MoS}_L = \theta_n^1$ represents the lowest acceptable QoE value, which is the minimal acceptable data rate of the player.

By computing (3), EQoE can be obtained for the poker game. According to the value of EQoE, we can adjust the network resource allocation to maintain the game at a good quality level. When $\mathrm{QoE}_e > \mathrm{QoE}_m$, there is extra QoE resource for the player with the highest weight in this case, and thus we can adjust the resource allocation to decrease other players' QoE level until $\mathrm{QoE}_e = \mathrm{QoE}_m$ one by one according to the ascending order of player weight. Otherwise, there is no extra QoE resource for the player with the highest weight in this case.

The player weights can be obtained by applying some existing methods (e.g., see reference [17] for details) or historical records of the players. This paper focuses on the design of a novel QoE evaluation method for multi-player online poker game.

3.4 The QoE Evaluation/Adaptation Method

Base on the above concept and model of EQoE, we propose a novel QoE evaluation method for the online Chinese poker game, in which the network resources are dynamically adjusted according to the network condition and player QoE levels until no extra QoE resource remains. The procedure of the proposed method is shown in Fig. 2. A detailed description of the scheme is given in the following.

Step 1: The MoS of each player n can be calculated based on an existing QoE evaluation method for individual QoE, as proposed in [16] (e.g., using (1)).

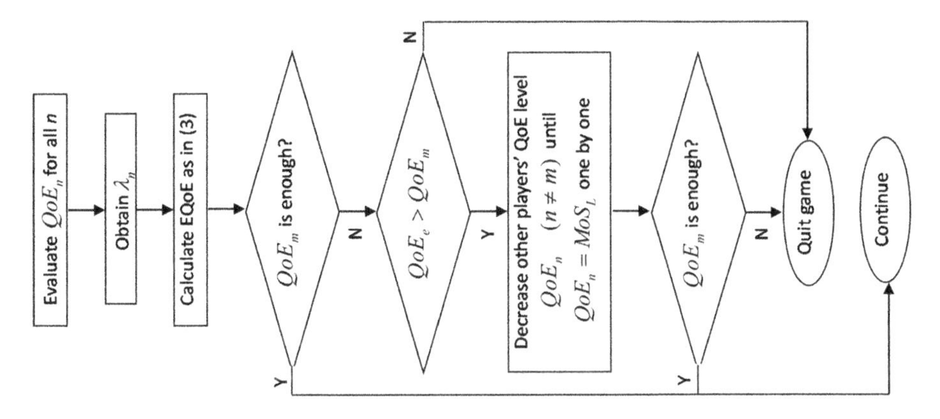

Fig. 2. Procedure of the proposed QoE prediction and adaptation method.

Step 2: Player weight is obtained by applying the deep learning method proposed in [17] or the historical records of each player.
Step 3: EQoE is obtained as in (3).
Step 4: If the player with the highest weight has enough QoE resource to play the poker game, then the agent goes to Step 11.
Step 5: According to the result of Step 3, the agent decides whether to adjust network resource allocation or not.
Step 6: If $\text{QoE}_e > \text{QoE}_m$, the agent has the capacity to adjust the network resource allocation, by decreasing the QoEs of the players with a lower weight to release network resource for the player with the highest weight or a higher priority. Then the agent goes to Step 8.
Step 7: If $\text{QoE}_e \leq \text{QoE}_m$, all of players will quit the current poker game. The agent goes to Step 12.
Step 8: By decreasing the QoE level of players with a lower weight until $\text{QoE}_n = \text{MoS}_L$ one by one according to ascending order of player weights, the agent allocates more network resources to the player with the highest weight.
Step 9: If QoE_m is satisfactory, the agent goes to Step 11.
Step 10: If QoE_m is not satisfactory, the agent goes to Step 12.
Step 11: Players can continual to play the game.
Step 12: Players need to quit the game.

Unlike the existing method [16], the proposed QoE evaluation and adaptation method takes into account that the players have different priorities depending on their roles in the game, and exploits the characteristics of the online Chinese poker game to guarantee the quality of game with the proposed EQoE concept.

4 Case Study

In this section, we evaluate the effectiveness of our proposed method using the online Chinese poker game as an example. Each game is typically played by 4 players, while a server can support a large number of games (and thus players) simultaneously. In the simulations of the typical online Chinese poker game, we assume 128 kbps as the required minimal throughput and 512 kbps as the recommended throughput as used for Skype. Therefore we have $\theta_n^1 = 128$ kbps and $\theta_n^4 = 512$ kbps.

In order to evaluate the effectiveness of our proposed method, we use the traditional method in [7] as a baseline scheme. The simulations are executed using the Matlab platform, and each result is the average of 50 runs. We compared the utilization of throughput in the proposed method with that of traditional QoE prediction method. It is assumed that the user's QoE and weight follow the Gaussian distribution in the range of [128 kbps, 512 kbps]. During the Chinese poker game, there are also some other real-time traffic with a higher priority (such as banking by phone, e-health services) transmitted, which also follows the Gaussian distribution in the range of [0, 50 Mbps].

When there is background traffic with higher priority, the available network throughput for the game players will be reduced. The traffic with higher priority will be protected by decreasing the game player's QoE levels temporarily with the proposed method. However, the traditional method will interrupt some players to provide network resources for the high-priority traffic. Therefore, we can see that the proposed method has obviously achieved a better performance than the traditional method in terms of throughput utilization and the number of players supported, as shown in Fig. 3, Fig. 4, Fig. 5, and Fig. 6.

Specifically, Fig. 3 shows that the proposed method has achieved higher throughput utilization than the traditional method under the same influence from the higher priority background traffic. This is because the traditional method does not take into account that all the 4 players of an online Chinese poker game will be interrupted when any of the players is interrupted due to insufficient throughput. Unlike the traditional method, the 4 players of the online Chinese poker game are assigned with different weights in the proposed scheme. Some players with a lower weight are willing to reduce their QoE level to save network resources for players with higher weights, in order to continue playing the game.

Figure 4 shows that the proposed method can accommodate more players than the traditional method. During the period of time when the high-priority background traffic is heavy, the number of players still remains high with the proposed method. However, the number of players is much smaller with the traditional method under the same background traffic pattern. This is also due to the specific design of the proposed scheme that introduce different priorities to the game players and exploit the priorities in network resource allocation.

Figure 5 presents the average throughputs of player with different weights with the proposed method. It can be seen that the throughputs are very different, since the proposed method maintains players' QoE according to their weights. When the background traffic with a higher priority is heavy, the average throughput of player with higher weight keeps stable. However, the average throughput of the player with a lower weight is much lower than that of the

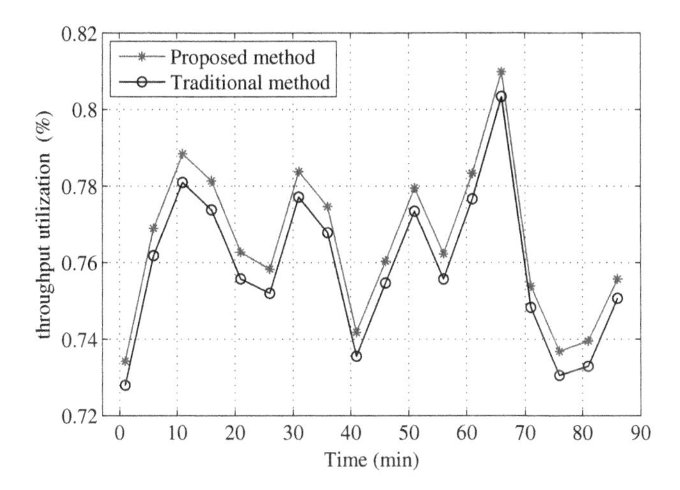

Fig. 3. Comparison of two methods in terms of throughput utilization.

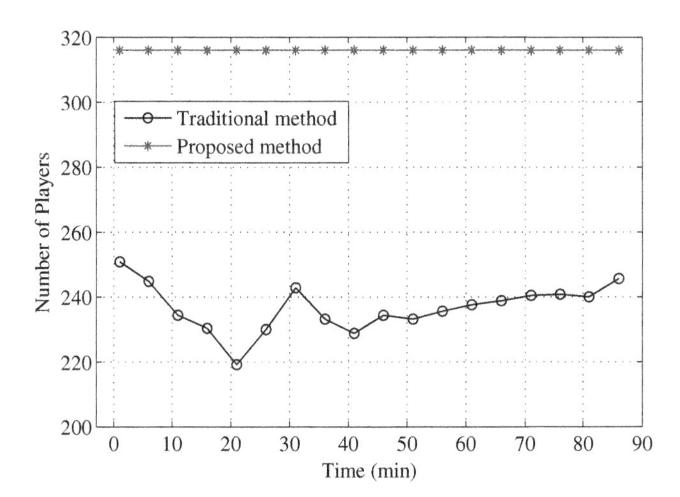

Fig. 4. Comparison of two methods in terms of the number of concurrent players supported.

player with a higher weight, and it changes with the requirements of the background traffic with higher priority. The average throughput of the player with a lower weight is decreased to better accommodate the higher priority traffic in order to keep the game going under poor network conditions. In Fig. 6, we can see that the average throughput of the player with a higher weight remains high, even though the high-priority background traffic is continuously increased, while the average throughput of the player with a lower weight decreases with the increased background traffic.

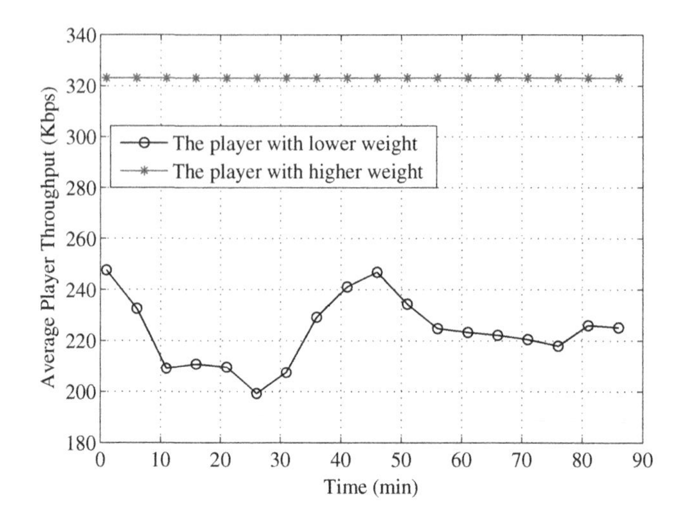

Fig. 5. Comparison of the average throughput of players with different weights over time.

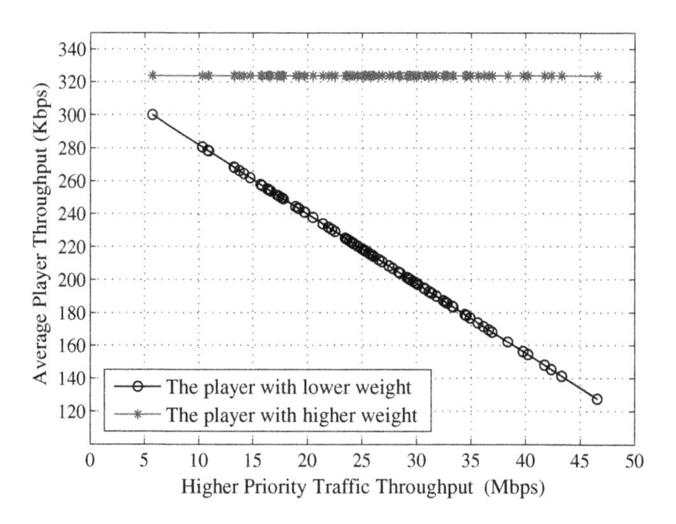

Fig. 6. Comparison of the average throughput of players wit different weights under increased high-priority background traffic.

Unlike the existing works that consider only individual QoE, our proposed method takes into account the fact that different players have different roles and weights in the multi-player online poker game, where some players with a lower weight are willing to reduce their QoEs to save network resources for players with a higher weight to keep the game going under poor network conditions.

5 Future Directions and Open Problems

QoE-related research and applications are still in the infancy stage, and only limited work has been done in the field. QoE-related technologies are lacking on a systemic and in-depth research. There are many obstacles in the development of QoE-related applications. In 5G and beyond network systems, many new online multimedia services and traffic types are emerging. Some potential challenges in the fields of QoE research are provided as follows.

- Subjective factors: QoE is influenced by a lot of subjective factors, such as user's mood, specific preference, attention, expectations, service ease of use, context, etc. Until today, there are also a lot of unknown factors. Some factors mentioned above still need further study. It is difficult to find kernel factors to describe QoE since a lot of subjective factors are highly sophisticated. Meanwhile, it is a challenge to effectively quantize and model these factors.
- Nonlinear fusion: QoE is affected by a variety of factors from different fields. The knowledge of factors is generally obtained from heterogeneous, autonomous sources, which utilize different technologies to acquire. It is very difficult to obtain the analytical relationship between QoE and the influencing factors. To mine potential and useful information, nonlinear fusion plays a vital role in the fields of QoE modeling. A new fusion method should be designed, which would solve the above problems.
- Dynamic QoE prediction: Because the user's QoE often evolves with time, space, context, knowledge, circumstance, and scenarios, the observed factors are often transient. Therefore, a prediction method should be capable of predicting future trends and should analyze and model dynamic factors together with dynamic requirements simultaneously.
- Real-time, online QoE assessment: With the explosion of data volume, traditional QoE prediction systems may not be able to handle the data effectively. In many applications, it is impossible to obtain the eventual data scale since the data is accumulated over time. Furthermore, different factors have different degrees of importance for different analytical goals. The QoE assessment method needs to consider the computational and storage loads of big data. From the perspective of data-processing platforms, an effective and efficient processing approach is critical for real-time online QoE prediction. Meanwhile, the QoE assessment method should also be focused on the complexity and scalability issues encountered in the development of QoE applications in future research.
- User-centric QoE methods: The QoE assessment methods should be capable of discovering users' latent intention from limited observed data. It should consider user profiles, social networks, and behaviors to model the user's intention.
- Security and privacy: Since QoE involves a lot of individual data, security and privacy should be considered, which still remain an open problem. For analyzing user's QoE, different data owners should warrant that analyzers have different access rights. The QoE method should avoid the violation of users'

privacy. Meanwhile, a QoE safety mechanism should be designed, including effective cryptography approaches, safety management, access control, and safe communications.
- Modeling: It is still a challenging task to model the various influencing factors since most of such factors are immeasurable. Meanwhile, it is difficulty to effectively observe the factors which are related to social psychology and cognitive science. The numerical presentation of QoE is still challenging since the QoE model could be different from time to time, impacted by both the network side and human side factors [19].
- The utility function: both the parametric QoE model and the personalized QoE model have several parameters calibrating the presentation of QoE, and thus, depending on the choice of parameters, the utility function may not be concave at all, making it hard to optimize in resource allocation [19].

We also note that the research on QoE evaluation for multiple players is still in its infancy. Heterogeneous network conditions and dynamic requirements make the multi-user QoE evaluation problem even more challenging. As discussed above, these and similar problems are subject to our future research.

6 Conclusions

This paper proposed a new QoE evaluation and adaptation method for MOPGs (especially, the online Chinese poker game), which considers the fact that different players have different weights when playing the game. Some player with a lower weight are willing to reduce their own QoE level to provide more network resources for players with a higher weight, in order to keep the game going under poor network conditions. We presented an EQoE concept and model by considering the interactions among multiple players. Simulation results reveal that the proposed method could effectively improve the game performance with respect to throughput utilization and the number of players supported, as compared to a traditional baseline method.

References

1. He, Z., Mao, S., Jiang, T.: A survey of QoE driven video streaming over cognitive radio networks. IEEE Network **29**(6), 20–25 (2015)
2. Xu, Y., Mao, S.: A survey of mobile cloud computing for rich media applications. IEEE Wireless Commun. **20**(3), 46–53 (2013)
3. Dhib, E., Boussetta, K., Zangar, N., Tabbane, N.: Modeling cloud gaming experience for massively multiplayer online Games. In: Proceedings of IEEE CCNC 2016, Las Vegas, NV, pp. 381–386, January 2016
4. Gao, C., Shen, H., Babar, M.A.: Concealing jitter in multi-player online games through predictive behaviour modeling. In: Proceedings of 20th IEEE International Conference on Computer Supported Cooperative Work in Design, Nanchang, China, pp. 62–67, May 2016

5. Schmitt, M., Redi, J., Bulterman, D., Cesar, P.S.: Towards individual QoE for multiparty videoconferencing. IEEE Trans. Multimedia **20**(7), 1781–1795 (2018)
6. Saeed, A., Olsen, R.L., Pedersen, J.M.: Optimizing the Loads of multi-player online game servers using Markov chains. In: Proceedings ICCCN 2015, Las Vegas, NV, pp. 1–5, August 2015
7. Bampis, C.G., Li, Z., Bovik, A.C.: Continuous prediction of streaming video QoE using dynamic networks. IEEE Signal Process. Lett. **24**(7), 1083–1087 (2017)
8. Mok, R.K.P., Chang, R.K.C., Li, W.: Detecting low-quality workers in QoE crowdtesting: a worker behavior-based approach. IEEE Trans. Multimedia **19**(3), 530–543 (2017)
9. Bampis, C.G., Li, Z., Katsavounidis, I., Bovik, A.C.: Recurrent and dynamic models for predicting streaming video Quality of Experience. IEEE Trans. Image Process. **27**(7), 3316–3331 (2018)
10. Cheon, M., Lee, J.-S.: Subjective and objective quality assessment of compressed 4K UHD videos for immersive experience. IEEE Trans. Circuits Syst. Video Technol. **28**(7), 1467–1480 (2018)
11. Schatten, M., Duric, B.O.: A social network analysis of a massively multi-player online role playing game. In: Proceedings of 4th International Conference on Modeling Simulation, Jeju Island, South Korea, pp. 37–42, November 2015
12. Huang, M., Zhang, X.: MAC scheduling for multiuser wireless virtual reality in 5G MIMO-OFDM systems. In: Proceedings of IEEE ICC 2018 Workshops, Kansas City, MO, pp. 1–6, May 2018
13. Amiri, M., Al Osman, H., Shirmohammadi, S.: Game-aware and SDN-assisted bandwidth allocation for data center networks. In: Proceedings of 2018 IEEE Conference on Multimedia Information Processing and Retrieval, Miami, FL, pp. 86–91, April 2018
14. Nightingale, J., Salva-Garcia, P., Calero, J.M.A., Wang, Q.: 5G-QoE: QoE modelling for Ultra-HD video streaming in 5G networks. IEEE Trans. Broadcasting **64**(2), 621–634 (2018)
15. Hemmati, M., McCormick, B., Shirmohammadi, S.: QoE-aware bandwidth allocation for video traffic using sigmoidal programming. IEEE Multimedia Mag. **24**(4), 80–90 (2017)
16. Shao, H., Zhao, H., Sun, Y., Zhang, J., Xu, Y.: QoE-aware downlink user-cell association in small cell networks: a transfer-matching game theoretic solution with peer effects. IEEE Access J. **4**, 10029–10041 (2016)
17. Vega, M.T., Mocanu, D.C., Famaey, J., Stavrou, S., Liotta, A.: Deep learning for quality assessment in live video streaming. IEEE Signal Process. Lett. **24**(6), 736–740 (2017)
18. Sun, Y., Peng, M., Zhou, Y., Huang, Y., Mao, S.: Application of machine learning in wireless networks: key technologies and open issues. IEEE Commun. Surv. Tutor. **12**(4), 3072–3108 (2019)
19. Xiao, K., Mao, S., Tugnait, J.K.: Robust QoE-driven DASH over OFDMA networks. IEEE Trans. Multimedia **22**(2), 474–486 (2020)

Traffic Characteristic and Overhead Analysis for Wireless Networks Based on Stochastic Network Calculus

Lishui Chen[1], Dapeng Wu[2], and Zhidu Li[2](✉)

[1] Science and Technology on Communication Networks Laboratory, The 54th Research Institute of CETC, Shijiazhuang 050081, Hebei, People's Republic of China
[2] School of Communication and Information Engineering, Chongqing University of Posts and Telecommunications, Chongqing, China

Abstract. With the rapid development of communication technology, impacts of various service characteristics on the performance of wireless networks draw numerous studies. However, few from such studies take into account the impacts of the overheads which always arise due to traffic influx. Stochastic network calculus is a newly but promising queuing theory, and its concept has proved to be a good tool for performance analysis in the field of communication networks. In this paper, we investigate the impacts of packet size and their overheads on wireless network performance. A wireless network model that comprises data channels and control channels is introduced. An optimization problem is then formulated with the objective to minimize the transmission rates under different probabilistic delay constraints in both data and control channels. Thereafter, we apply stochastic network calculus to solve the optimization problem. Finally, numerical results of minimum transmission rate, overhead arrival rate and packet loss probability are presented, wherein the impacts of packet size and their overheads are analyzed and discussed.

Keywords: Performance analysis · Packet size · Overheads · Probabilistic delay constraint · Stochastic network calculus

1 Introduction

With the development of communication technology, various mobile data services, such as video service, instant messaging service and so on, are loaded in high speed wireless networks, which have greatly enriched our daily lives [1–3]. Mobile terminals are now good alternatives to home television for watching videos and personal computers for electronic messages. Sometimes, the network seems quite affluent for users to enjoy a good time, but sometimes it seems too choky to be tolerated. This phenomenon is relevant to the traffic characteristics and the capacity of channels [4].

J. Xiong et al. (Eds.): MobiMedia 2021, LNICST 394, pp. 172–184, 2021.
https://doi.org/10.1007/978-3-030-89814-4_13

Recently, numerous studies have paid attention to the impacts of traffic characteristics on the performance of wireless networks. For example, work [5] shown that the data burst from delay-sensitive uplink services can have significant impacts on the network performance, and work [6] proposed a detailed performance assessment of VoIP traffic by carrying out experimental trials across a real LTE-A environment. To the best of our knowledge, few studies take into account the traffic characteristics and their overheads which may arise due to the traffic influx. However, these overheads do have significant impacts on the performance of wireless networks especially when data service is extracted from the real mobile networks. For example, WeChat service (a kind of instant messaging service) may cause congestions in wireless networks because of large amounts of control signaling. Therefore, comprehensive study on different traffic characteristics and their corresponding overheads is of great importance for the research of wireless resource allocation and quality of service guarantee provisioning.

Stochastic network calculus is one branch of network calculus which is an emerging queuing theory first proposed by Cruz in 1991 [7,8]. Nowadays, stochastic network calculus has evolved into a powerful and promising tool in analyzing network performance especially in delay and backlog analysis [9,10]. Compared with traditional queuing theories, stochastic network calculus deals with problem by giving probabilistic bound instead of deterministic mean value. Hence, it has great advantage on performance analysis in the situation where deterministic characteristics cannot be ascertained or the performance guarantee is flexible. Furthermore, stochastic network calculus has been widely used in many fields, such as performance analysis of cognitive radio [11,12], energy consumption analysis in wireless networks [13–15] just to mention a few.

Hence, this paper applies stochastic network calculus to study the traffic delay performance by taking the traffic characteristics and control overheads into account. Data channels and control channels are first modeled and an optimization problem is then formulated with the objective to minimize the transmission rates under different probabilistic delay constraints in both data and control channels. Finally, numerical results of minimum transmission rate, overhead arrival rate and packet loss probability are presented and discussed.

The remainder of this paper is organized as follows. Section 2 introduces the system model, where some assumptions are given. After that, an optimization problem is formulated, with the objective to minimize the transmission rates under different probabilistic delay constraints. Section 3 demonstrates the derivation of the relationship between the minimum transmission rate and the probabilistic delay constraint using stochastic network calculus. The solution of the optimization problem is also achieved at the end of this section. Then, numerical results about the impacts of packets size and their overheads are presented and discussed in Sect. 4. Finally, Sect. 5 concludes the paper.

2 System Model

In this paper, we consider a wireless network model depicted in Fig. 1, control channels, data channels as well as data flows and overheads are all taken into account. The data traffic and overheads are all transmitted as packets. Furthermore, all packets obey the scheduling rule of first in first out (FIFO). The data channels provide a constant total transmission rate of C_1 to transmit both data packets and their extra data overheads. Data overheads are the extra information which are generated and transmitted with data packets. The control channels provide a constant total transmission rate of C_2 to transmit the control overheads. In this paper, we only consider the control overheads which mainly arise whenever a session is setup to transmit data packets.

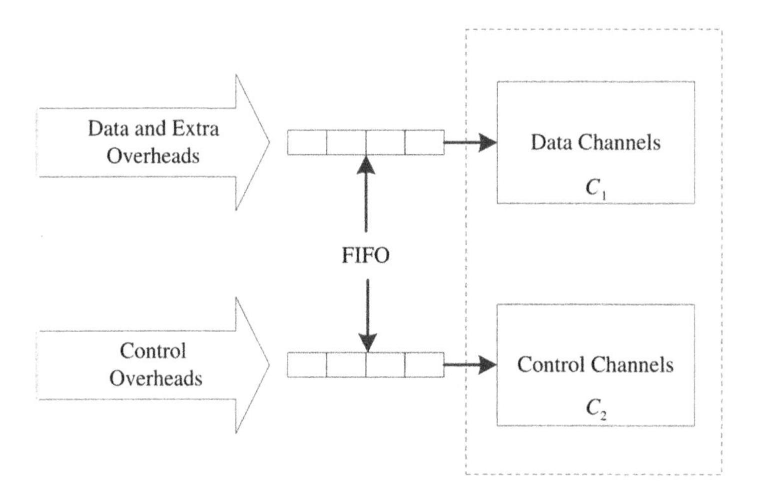

Fig. 1. System model

There are three steps for a successful data packet transmission: firstly, whenever there is a data packet to be transmitted, the system will generate a control overhead and transmit it through the control channels to the target system to setup a session; subsequently, the system generate a data overhead and pack it into the data packet; lastly, the data packet and its data overhead await their time slot to get transmitted through the data channels. For ease of expression and with the focus on the impacts of packet size and overheads, the transmission channels are assumed to be error free and only timeout packets are discarded. Therefore, a data packet is successfully transmitted if and only if the packet and its corresponding overheads are both successfully transmitted under a probabilistic delay constraint.

Transmission delay in the data channels and control channels are denoted by $D_1(t)$ and $D_2(t)$ respectively. Moreover, the transmission in the data channels has a delay requirement, represented by a probabilistic delay constraint (t_1, p_0), which means the probability of the delay $D_1(t)$ exceeding a delay time t_1 should be bounded by p_0, i.e. $\Pr\{D_1(t) > t_1\} \leq p_0$. Similarly, the transmission in control channels also has a probabilistic delay constraint $\Pr\{D_2(t) > t_2\} \leq p_0$.

For analyzing the impacts of packet size and overheads on the minimum transmission rate and packet loss probability under a probabilistic delay constraint, it is necessary to find the relationship between the minimum transmission rate and the probabilistic delay constraint. In addition, it is also required that the upper bound of the mean traffic arrival rate should not be larger than the transmission rate for the sake of ensuring stability in the system. Therefore, if the probabilistic delay constraint is given, an optimization problem which minimizes the transmission rates under probabilistic delay constraints in the data channels and the control channels respectively can be expressed as:

$$\begin{aligned} &\min \mathrm{C}_1(D_1(t)) \text{ and } \min \mathrm{C}_2(D_2(t)) \\ &\text{s.t. } \rho_1(\theta_1) \leq \mathrm{C}_1, \Pr\{D_1(t) > t_1\} \leq p_0, \\ &\quad\ \ \rho_2(\theta_2) \leq \mathrm{C}_2, \Pr\{D_2(t) > t_2\} \leq p_0 \end{aligned} \tag{1}$$

where $\rho(\theta_1)$ denotes the upper bound of mean traffic arrival rate for the data channels, $\rho(\theta_2)$ denotes the upper bound of mean arrival rate for the control channels, θ_1 and θ_2 are both optimization parameters to be used to determine the minimum of C_1 and C_2 respectively.

3 Performance Analysis

3.1 Stochastic Network Calculus Basics

As highlighted earlier, stochastic network calculus is a newly developed queuing theory which has been widely used in performance analysis. There are two basic concepts in stochastic network calculus: stochastic arrival curve (SAC) and stochastic service curve (SSC), which are used to describe the arrival process of input traffic and the service process of server respectively. The concept and proof of the theorem can be found in [9]. For convenience, we provide basic definitions here for our subsequent analysis.

Definition 1 (Stochastic Arrival Curve). A flow $A(t)$ is said to have a stochastic arrival curve $\alpha(t)$ with the bounding function $f(x)$, if for all $t \geq s \geq 0$ and all $x \geq 0$, there holds

$$\Pr\{\sup_{0 \leq s \leq t} \{A(s,t) - \alpha(t,s)\} > x\} \leq f(x) \tag{2}$$

Here,$A(s,t)$ denotes the cumulative amount of packets during the time period $(s,t]$, and $A(t) = A(0,t)$.

Definition 2 (Stochastic Service Curve). A system is said to provide a stochastic service curve $\beta(t)$ with the bounding function $g(x)$, if for all $t \geq s \geq 0$ and all $x \geq 0$, there holds

$$\Pr\{A \otimes \beta(t) - A^*(t) > x\} \leq g(x) \tag{3}$$

Here, $\otimes$ is the operation of minimum plus (min; +) convolution, $A \otimes \beta(t) \equiv \inf_{0 \leq s \leq t} \{A(s,t) + \beta(s)\}$. And $A^*(t)$ denotes the cumulative amount of the output packets by time t.

Theorem 1 (Probabilistic Delay Bound). In a given system, if the input has a stochastic arrival curve as $\alpha(t)$ with the bounding function $f(x)$, and the system provides to the input a stochastic service curve as $\beta(t)$ with the bounding function $g(x)$. Then for all $t \geq 0$ and all $x \geq 0$, the delay $D(t)$ is bounded by:

$$\Pr\{D(t) > h(\alpha + x, \beta)\} < f \otimes g(x) \tag{4}$$

where $h(\alpha+x, \beta) = \sup_{s \geq 0}\{\inf\{\tau \geq 0 : \alpha(s)+x \leq \beta(s+\tau)\}\}$ denotes the maximum horizontal distance between $\alpha(t)+x$ and $\beta(t)$. Here, $h(\alpha+x, \beta)$ denotes the delay time t, the relationship between $h(\alpha + x, \beta)$ and transmission rate C is:

$$h(\alpha + x, \beta) = t = \frac{t}{C} \tag{5}$$

3.2 Solution of the Optimization Problem

As the system model describes, a data overhead and a control overhead arise if and only if there is a data packet to be transmitted. Therefore, the numbers of arrival of data packets and control overheads as well as data overheads are the same.

Also, suppose the traffic flow is Poisson distribution with the expectation of λ. Using the knowledge of stochastic network calculus, the arrival curves with the bounding functions for both data channels and control channels are deduced respectively as follows [16].

Arrival curve for the data channels with the bounding function:

$$\begin{cases} \alpha_1(t) = \frac{\lambda t}{\theta_1}(e^{\theta_1(L+\sigma_1)} - 1) \\ f_1(x) = e^{-\theta_1 x} \end{cases} \tag{6}$$

where, L and σ_1 denote data packet size and data overhead size respectively, and θ_1 is optimization parameters mentioned earlier.

Arrival curve for the control channels with the bounding function:

$$\begin{cases} \alpha_2(t) = \frac{\lambda t}{\theta_2}(e^{\theta_2 \sigma_2} - 1) \\ f_2(x) = e^{-\theta_2 x} \end{cases} \tag{7}$$

where σ_2 denotes control overhead size and θ_2 is optimization parameters.

The service curve in the data channels with the bounding function is:

$$\begin{cases} \beta_1(t) = C_1 t \\ g_1(x) = 0 \end{cases} \tag{8}$$

Similarly, service curve in the control channels with the bounding function is:

$$\begin{cases} \beta_2(t) = C_2 t \\ g_2(x) = 0 \end{cases} \tag{9}$$

Let $t_1 = h(\alpha_1 + x, \beta_1)$, according to (4) and (5), the probabilistic delay bounding function for the data channels is derived as follows:

$$\Pr\{D_1(t) > t_1\} = \Pr(D_1(t) > h(\alpha_1 + x, \beta_1)) \leq f \otimes g(C_1 t_2) = e^{-\theta_1 C_1 t_1} \tag{10}$$

Similarly, the probabilistic delay bounding function for the control channels is derived as:

$$\Pr\{D_2(t) > t_2\} \leq f \otimes g(C_2 t_2) = e^{-\theta_2 C_2 t_2} \tag{11}$$

Hence, the optimization problem in (1) is clear, there holds

$$\begin{cases} \rho_1(\theta_1) = \frac{\lambda}{\theta_1}(e^{\theta_1(L+\sigma_1)} - 1) \leq C_1 \\ \Pr\{D_1(t) > t_1\} \leq e^{-\theta_1 C_1 t_1} = p_0 \\ \rho_2(\theta_2) = \frac{\lambda}{\theta_2}(e^{\theta_2\sigma_2} - 1) \leq C_2 \\ \Pr\{D_2(t) > t_2\} \leq e^{-\theta_2 C_2 t_2} = p_0 \end{cases} \tag{12}$$

By simplifying (12), we can get:

$$\begin{cases} C_1 \geq \frac{(L+\sigma_1)\log(1/p_0)}{t_1 \log(\frac{\log(1/p_0)}{\lambda t_1}+1)} \\ C_2 \geq \frac{\sigma_2 \log(1/p_0)}{t_2 \log(\frac{\log(1/p_0)}{\lambda t_2}+1)} \end{cases} \tag{13}$$

where the condition of equality holds if and only if

$$\begin{cases} \theta_1 = \frac{\log(\frac{\log(1/p_0)}{\lambda t_1}+1)}{L+\sigma_1} \\ \theta_2 = \frac{\log(\frac{\log(1/p_0)}{\lambda t_2}+1)}{\sigma_2} \end{cases} \tag{14}$$

Therefore, the optimization problem in (1) is solved. The minimum transmission rates for both data channels and control channels are achieved, there holds:

$$\begin{cases} C_{1,\min} = \frac{(L+\sigma_1)\log(1/p_0)}{t_1 \log(\frac{\log(1/p_0)}{r t_1/L}+1)} \\ C_{2,\min} = \frac{\sigma_2 \log(1/p_0)}{t_2 \log(\frac{\log(1/p_0)}{r t_2/L}+1)} \end{cases} \tag{15}$$

where $r = \lambda L$, denotes the mean traffic arrival rate.

4 Numerical Results and Analysis

In the preceding sections, the arrival processes of data packets and their overheads as well as service process are introduced and modeled. The considered model is applicable to different wireless network systems, such as Wi-Fi, LTE and NR. Afterwards, the relationships between the minimum transmission rate and the probabilistic delay constraint in both data channels and control channels are derived based on the theory of stochastic network calculus.

In this section, two types of mobile data services are considered, which are video service and instant messaging service (representing service with large size packets and service with small size packets respectively). The size of video packet is set to 5 Mbits whereas instant messaging (IM) packet is set to 5 kbits. Overheads of data packets for the data channels are set to 1 kbits while for the control channels are set to 0.5 kbits (i.e., $\sigma_1 = 1$ kbits,$\sigma_2 = 0.5$ kbits).

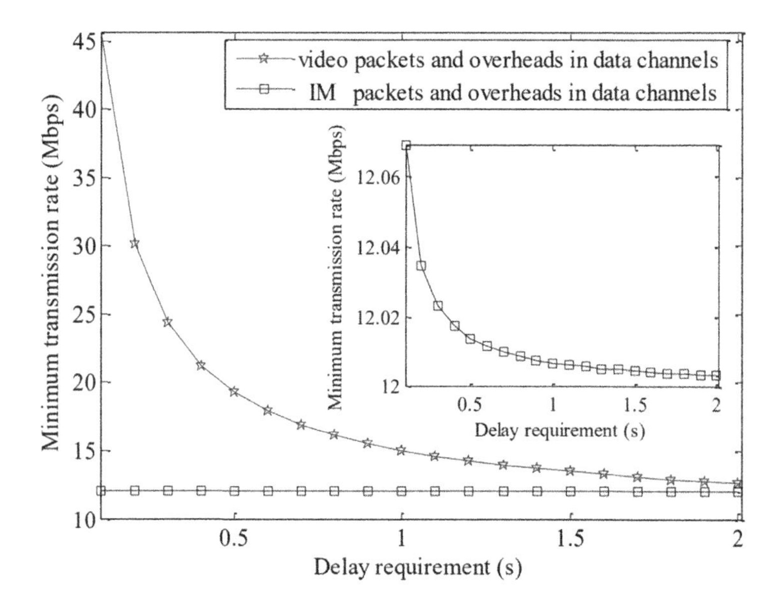

Fig. 2. Minimum transmission rate-delay requirement curves in data channels

4.1 Impacts of Packet Size on the Minimum Transmission Rate and Overhead Arrival Rate

In this subsection, impact of packet size on the minimum transmission rate is investigated. Probabilistic delay constraint is set to 0.1 (i.e. $p_0 = 0.1$) and traffic arrival rate is assumed to be 10 Mbps (i.e. $r = 10^7$bps). The minimum transmission rate requirement for the two traffics in the data channels and control channels are shown in Fig. 2 and Fig. 3 respectively.

Firstly, it is intuitive that every delay requirement is mapping to unique minimum transmission rate, which implies the optimization problem in (1) has unique solution. In addition, Fig. 2 and Fig. 3 also show that: the minimum transmission rate decreases as the delay requirement increases and will converge to a certain value when delay requirement increases largely enough (e.g. larger than 1.8 s). This is because larger delay requirement means looser delay constraint, which needs lower minimum transmission rate to guarantee it.

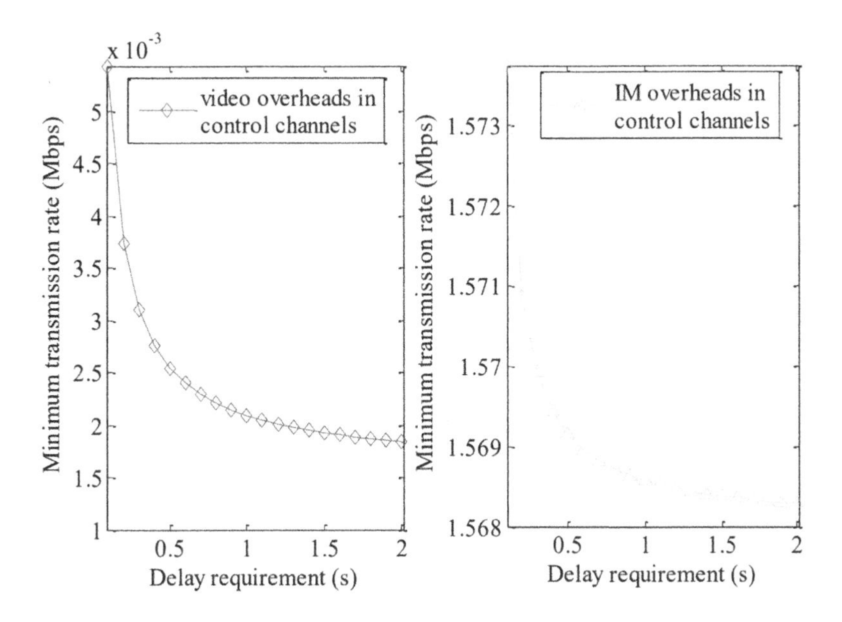

Fig. 3. Minimum transmission rate-delay requirement in control channels

Secondly, as demonstrated in Fig. 2, the minimum transmission rate for video packets changes more aggressively than the one for IM packets as the delay requirement changes. Besides, comparing video packets with IM packets, though they have the same traffic arrival rate, we found that higher transmission rate is needed to transmit larger packets while delay requirement is small (e.g. smaller than 1.5 s). For instant, when delay requirement is limited to 0.1 s, the minimum transmission rate for video packets to guarantee the probability delay constraint reaches to 45 Mbps, which is too high. We intuit that larger packets may have more stochastic characteristics in real arrival rate when other conditions are the same. It can also be mathematically explained that the upper bound of mean arrival traffic is an increasing function for packet size L when delay requirement is small according to (6).

Thirdly, as illustrated in Fig. 3, the minimum transmission rate for video control overheads is much lower than the one for IM control overheads while arrival rates of the both traffics are the same. This is because the arrival rate of video control overheads are much less than the one of IM overheads according to the expression of arrival rate for the control channels given as $r\sigma_2/L$, which are also demonstrated in Fig. 4.

Lastly, as Fig. 4 shows, the overhead arrival rates of video traffic is quite lower than the ones of IM traffic, which means the system generate less overheads while transmitting larger packet traffic.

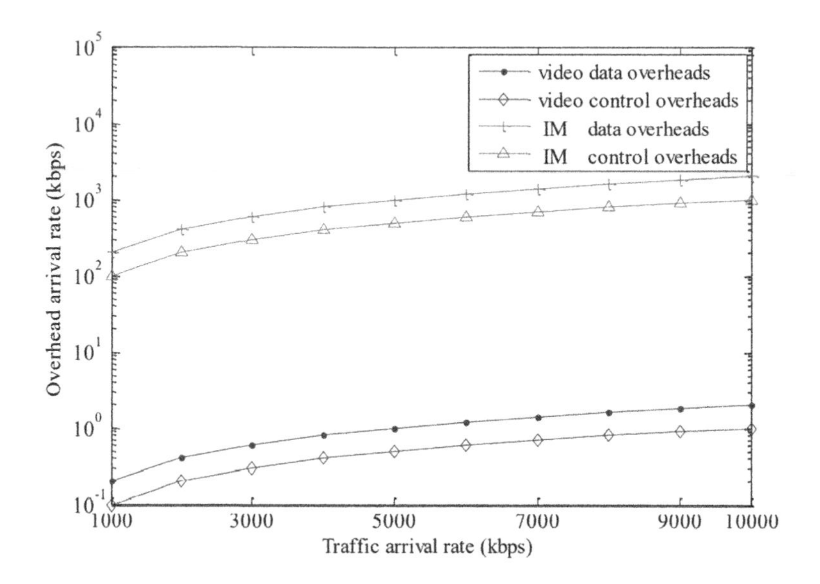

Fig. 4. Overhead arrival rate-traffic arrival rate for the two channels

Hence, packet size impacts greatly on the minimum transmission rate and the overhead arrival rate. The traffic of larger packets is more stochastic in real arrival rate. But large packets cause less overhead than small packets in both the data channels and the control channels when the arrival rates for the two traffics are the same.

4.2 Impacts of Control Overheads on Packet Loss Probability

In this subsection, we mainly focus on the impacts of control overheads on packet loss probability. As highlighted in the system model, a successful packet transmission holds if and only if the data packet and its corresponding overheads are transmitted under a probabilistic delay constraint. Otherwise, the timeout data packets will be discarded. Thus, packet loss probability and the probabilistic delay bound are equivalent in this paper.

For ease of analyzing, the network system is supposed to use OFDM technology with time slot of 0.5 ms. In every time slot, there are 7 OFDM symbols in time domain and 100 resource blocks (RB) in frequency domain. And there

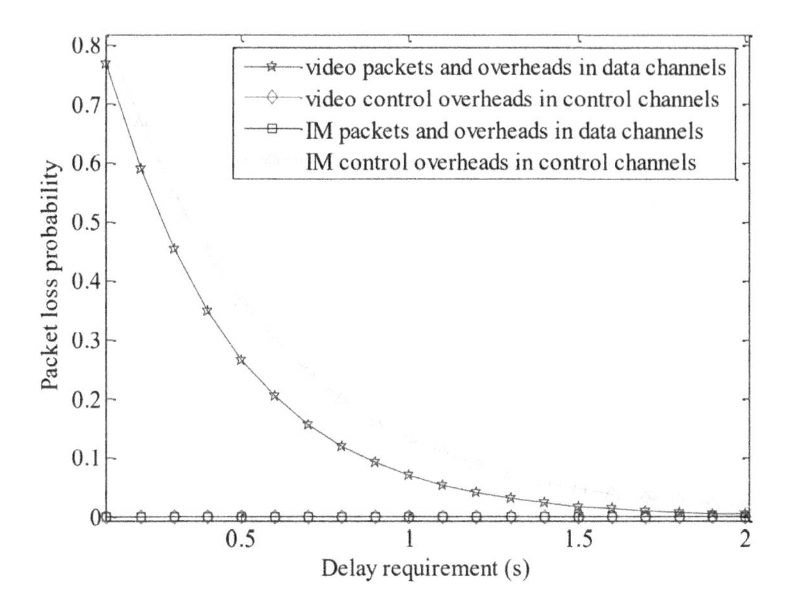

Fig. 5. Impact of control overheads on packet loss probability

are 12 sub-carriers in every RB. Besides, 16 QAM, 1/3 channel coding, single antenna are used as modulation, coding and transmission mode respectively. Then, the system channels can transmit 11.2 kbits flows per slot, which means the peak transmission rate is 22.4 Mbps. However, fixed overheads exist in RBs, such as PDCCH/PHICH/PSS overheads and so on, which are independent with the input traffic and provide other functions for the channels. These overheads are supposed to occupy 25.53% of RBs while the control overheads generated due to input packets in this paper are supposed to occupy 4.47% of RBs. Therefore, the peak transmission rate in the data channels is 15.68 Mbps, and in the control channels is 1.001 Mbps. Besides, the average arrival rate of data traffic is assumed to 10 Mbps. The packet loss probability-delay requirement curves of the four cases are shown in Fig. 5.

As Fig. 5 shows, while transmission rate and traffic arrival rate are definite, packet loss probability can be reduced at the expense of raising the delay requirement, which means the delay constraint turns loose. All curves converge to 0 when the delay requirement is large enough.

Moreover, an interesting phenomenon can be observed from Fig. 5. Packet loss probabilities of video control overheads and IM packets are close to 0 no matter what delay requirement is. However, packet loss probabilities of video packets and IM control overheads are greater than 0 when the delay requirement is not large enough. This indicates that packet loss probability for video service depends on the capacity of the data channels, while the one for IM service relies on the capacity of control channels. Now, we focus on the IM traffic. This is because the upper bound of arrival rate for the data channels is nearly 12.06 Mbps according to (6) and (14), which is quite less than the capacity of the data channels. On the other hand, the arrival rate of IM control overheads is large because the average arrival rate for control channels denoted by $r\sigma_2/L$ and its upper bound reaches up to 1.0005 Mbps according to (7) and (14), which may cause packet loss under a probabilistic delay constraint. Meanwhile, the packet loss probability in the control channels is much higher than the one in data channels. Thus, insufficient capacity of the control channels is the main factor that causes packet loss due to time out in this configuration while transmitting IM service. In contrast, the main factor of packet loss for video packet traffic is the insufficient capacity of the data channels and the impact of control overhead is negligible. Similar analysis for video traffic is omitted for saving space but could be inferred from IM analysis above.

In summary, we can verify that the results of our analysis in previous subsection that large packet traffic has less control overheads compare to small packet traffic if other conditions are the same. Even though the capacity of data channels is sufficient for traffic transmission, packet loss may occur due to timeout because of insufficient capacity of the control channels.

4.3 Impacts of Data Overheads on Packet Loss Probability

In this subsection, impacts of data overheads are discussed. And here, capacity of control channels is assumed to be infinite, which means packet loss does not occur in the control channels. As a result, whether a packet is transmitted successfully or not depends on the capacity of the data channels. Two types of packets are considered and transmission rate in data channels is set to 15.68 Mbps. For achieving the objective, three cases of traffic arrival rate are taken into account, e.g., case 1, case 2 and case 3. For case 1, mean arrival rate is set to 10 Mbps; for case 2, 13.06 Mbps; for case 3, 13.07 Mbps. Packet loss probability- delay requirement curves for these three cases are presented in Fig. 6.

It is obvious that the change of packet loss probability for IM packets is much more aggressively than the change of video packets as the traffic arrival rate changes. This implies that packet loss occurs more often as traffic arrival rate increases while transmitting small packet traffics. Typically, IM packets can be transmitted under a probabilistic delay constraint if mean traffic arrival rate is 13.06 Mbps but cannot be transmitted if the arrival rate reaches 13.07 Mbps. In contrast, the packet loss probabilities of video packets for the two arrival rates are similar.

The above phenomenon can be explained by centering on the data overheads in the data channels. Because overheads exist, the real arrival rate for the data channels is the sum of traffic arrival rate and data overhead arrival rate. Since the size of data overheads is fixed, according to the expression $r\sigma_1/L$, the actual mean arrival rate for the data channels is $(r + r\sigma_1/L)$. Note that, if the arrival rate of video packets is 13.06 Mbps or 13.07 Mbps, the actual mean arrival rates for data channels remain almost the same and less than the capacity of data channels. However, if the arrival rates of IM packets are the same as that of video packets, actual mean arrival rates reach up to 15.672 Mbps and 15.684 Mbps

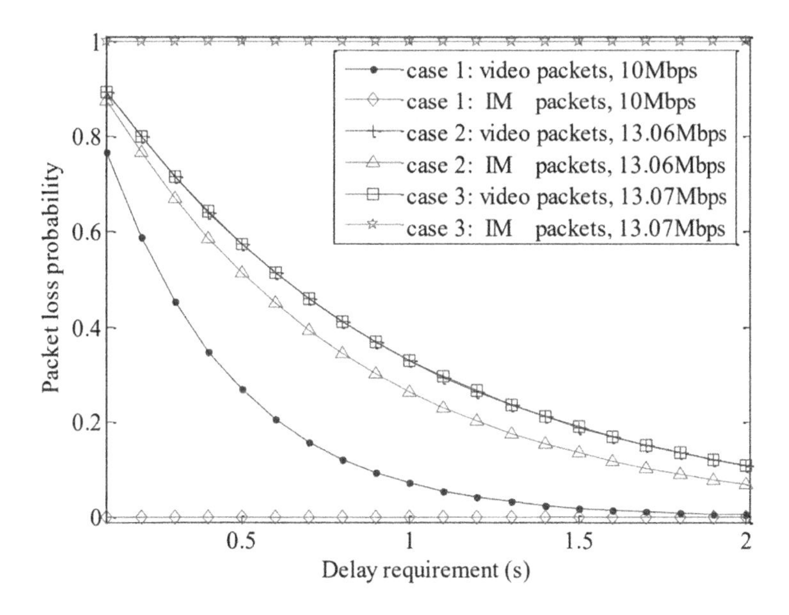

Fig. 6. Impact of control overheads on packet loss probability

respectively. And the later rate is larger than the channel capacity, which leads to packet loss due to timeout. In addition, IM packets can be transmitted under a probabilistic delay constraint, even if the actual mean arrival rate is quite close to but not beyond the channel capacity (e.g. 15.672 Mbps). This also verifies that small packet traffic is less stochastic than large packet traffic in arrival rate as mentioned earlier.

Therefore, we conclude that: compared with large packet traffic, small packet traffic generates more data overheads. And data overheads impact greatly on packet loss probability due to time out especially while transmitting small packet traffic. In addition, it also indicates that small packet traffic lowers the capacity of the data channels more severely.

5 Conclusions

In this paper, we use stochastic network calculus to investigate the impacts of traffic characteristics and overheads on wireless network performances. We formulate an optimization problem and solve it to find the minimum transmission rate which satisfies a given probabilistic delay constraint. Different types of traffics and their overheads are taken into account. The results of our analysis prove that packet size has great impacts on the minimum transmission rate as well as the arrival rate of overheads in both data channels and control channels. Meanwhile, both data overheads and control overheads have also great impacts on packet loss due to timeout, especially while transmitting small packet traffics.

Moreover, the idea of using stochastic network calculus to transform the complex traffic flows into linear flows and analyze the network performances with probabilistic bounds is generally awesome in performance analysis for different complex systems in wireless communications.

References

1. Xiong, J., Bi, R., Zhao, M., Guo, J., Yang, Q.: Edge-assisted privacy-preserving raw data sharing framework for connected autonomous vehicles. IEEE Wirel. Commun. **27**(3), 24–30 (2020)
2. Chen, Z., Li, D., Cuo, J.: Analysis and implementation of a QoS-aware routing protocol based on a heterogeneous network. Int. J. Performability Eng. **16**(8), 1235–1244 (2020)
3. Xiong, J., Chen, X., Yang, Q., Chen, L., Yao, Z.: A task-oriented user selection incentive mechanism in edge-aided mobile crowdsensing. IEEE Trans. Netw. Sci. Eng. **7**(4), 2347–2360 (2020)
4. Fernandes, D., Ferreira, A.G., Abrishambaf, R., Mendes, J., Cabral, J.: A low traffic overhead transmission power control for wireless body area networks. IEEE Sens. J. **18**(3), 1301–1313 (2018)
5. Li, M., Chen, H.: Energy-efficient traffic regulation and scheduling for video streaming services over LTE-A networks. IEEE Trans. Mob. Comput. **18**(2), 334–347 (2019)
6. Di Mauro, M., Liotta, A.: An experimental evaluation and characterization of VoIP over an LTE-A network. IEEE Trans. Netw. Serv. Manag. **17**(3), 1626–1639 (2020)
7. Cruz, R.L.: A calculus for network delay. I. Network elements in isolation. IEEE Trans. Inf. Theory **37**(1), 114–131 (1991)
8. Cruz, R.L.: A calculus for network delay. II. Network analysis. IEEE Trans. Inf. Theory **37**(1), 132–141 (1991). https://doi.org/10.1109/18.61110
9. Jiang, Y., Liu, Y.: Stochastic Network Calculus. Springer, London (2008). https://doi.org/10.1007/978-1-84800-127-5
10. Fidler, M.: Survey of deterministic and stochastic service curve models in the network calculus. IEEE Commun. Surv. Tutor. **12**(1), 59–86 (2010)
11. Zhao, L., Pop, P., Craciunas, S.S.: Worst-case latency analysis for IEEE 802.1Qbv time sensitive networks using network calculus. IEEE Access **6**, 41803–41815 (2018)
12. Mohammadpour, E., Stai, E., Le Boudec, J.: Improved credit bounds for the credit-based shaper in time-sensitive networking. IEEE Netw. Lett. **1**(3), 136–139 (2019)
13. Li, Z., Gao, Y., Li, P., Salihu, B.A., Sang, L., Yang, D.: Throughput analysis of an energy harvesting multichannel system under delay and energy storage constraints. IEEE Trans. Veh. Technol. **66**(9), 7818–7832 (2017)
14. Lin, Z., Chen, W.: Energy-efficient pushing with content consumption constraints: a network calculus approach. IEEE Trans. Green Commun. Netw. **4**(1), 301–314 (2020)
15. Cui, Q., et al.: Big data analytics and network calculus enabling intelligent management of autonomous vehicles in a smart city. IEEE Internet Things J. **6**(2), 2021–2034 (2019)
16. Jiang., Y.: A note on applying stochastic network calculus. In: Proceedings of SIGCOMM, vol. 10, pp. 16–20 (2010)

Entity Relationship Modeling for Enterprise Data Space Construction Driven by a Dynamic Detecting Probe

Ye Tao[1](✉), Shuaitong Guo[1], Ruichun Hou[2], Xiangqian Ding[2], and Dianhui Chu[3]

[1] College of Information Science and Technology, Qingdao University of Science and Technology, Qingdao, China
ye.tao@qust.edu.cn, forfree@mails.qust.edu.cn

[2] College of Information Science and Engineering, Ocean University of China, Qingdao, China
houruichun@ouc.edu.cn, dingxq1995@vip.sina.com

[3] School of Computer Science and Technology, Harbin Institute of Technology (Weihai), Weihai, China
chudh@hit.edu.cn

Abstract. To solve the problem of integrating and fusing scattered and heterogeneous data in the process of enterprise data space construction, we propose a novel entity association relationship modeling approach driven by dynamic detecting probes. By deploying acquisition units between the business logic layer and data access layer of different applications and dynamically collecting key information such as global data structure, related data and access logs, the entity association model for enterprise data space is constructed from three levels: schema, instance, and log. At the schema association level, a multidimensional similarity discrimination algorithm combined with semantic analysis is used to achieve the rapid fusion of similar entities; at the instance association level, a combination of feature vector-based similarity analysis and deep learning is used to complete the association matching of different entities for structured data such as numeric and character data and unstructured data such as long text data; at the log association level, the association between different entities and attributes is established by analyzing the equivalence relationships in the data access logs. In addition, to address the uncertainty problem in the association construction process, a fuzzy logic-based inference model is applied to obtain the final entity association construction scheme.

Keywords: Entity association · Data space · Fuzzy logic · Dynamic detecting probe

1 Introduction

In recent years, a low level of information sharing and a disconnect between information and business processes and applications have become common in enterprise business systems, which can easily lead to the formation of information silos within the enterprise

J. Xiong et al. (Eds.): MobiMedia 2021, LNICST 394, pp. 185–196, 2021.
https://doi.org/10.1007/978-3-030-89814-4_14

[1]. In particular, industrial software companies need substantial technical data support to deliver programmatic industrial processes and technologies, which requires not only a solution to information silos within the enterprise to achieve data sharing but also the integration of data with many different industrial enterprises [2, 3]. Therefore, to integrate internal or external data, some enterprises began to build data space systems, trying to integrate ERP, SCM, MES and other industrial software to eliminate information silos.

Building a data space system mainly requires the accurate establishment of associations between entities, which for this paper means integrating heterogeneous data from multiple source databases into a comprehensive enterprise data space through entity matching. Current research results mainly focus on discovering associations between entities or attributes through the semantic matching of dictionaries or semantic libraries, using data representation or content similarity judgments [4] to calculate the probability of similarity between data. Many of these methods have poor generalizability, slow response and low accuracy when attempting to discover the existence of associations from large amounts of data.

In this paper, we propose a new approach to discover entity association relationships in big data. First, this approach obtains business logic information and database data through dynamic probes deployed between the business logic layer and data access layer of different systems. Then, it portrays the similarity degree among entities in three dimensions, schema, instance and log, and gives the similarity values between entities in these different dimensions. Finally, based on the fuzzy logic inference method [5, 6], the similarity values between entities in different dimensions are converted into normalized values that can be uniformly measured to obtain the best matching results of entity association.

2 Related Work

In academic research, entity association is mainly divided into two types: schema matching [7] and instance analysis [8]. Schema matching extracts structural features from data sources as metadata and analyzes them to achieve association matching between data with fewer resources; matching based on instance analysis analyzes the data itself to obtain matching information, which usually consumes more resources but can obtain more accurate and comprehensive analysis results.

For schema matching academic research, in [9], He et al. used Structured Query Language (SQL) to extract features such as the name of the database, name of the schema, and type of column as metadata sets from each selected dataset. Then, they joined all metadata from each dataset into a metadata database. Finally, the correlation between the metadata was calculated by different methods to establish an association between the source data.

In academic research on instance analysis, the preprocessing that mines associations between data includes categorized data. For example, for the data conflict problem in data fusion, in [10], the conflict can be divided into two categories, uncertain conflict and contradictory conflict, and then the duplicate data of the same representation are fused, thus solving problems such as the possible conflict between different values for the same attribute.

Academics are also studying the integration of deep learning with logs. Mohanty et al. [11] cleaned the web log files collected by the IoT, built user profiles, saved similar information, and proposed a recommendation system based on rough fuzzy clustering to recommend e-commerce shopping sites to users. We offer a proposal for extracting the data association information in logs, this approach builds on the feature that logs contain association information between data.

The constantly increasing amount of data accumulating in the development of enterprises leads to an increasing size and number of categories of data, and methods such as schema matching, instance analysis, and log mining to analyze data from a single dimension may have problems such as not making full use of the diversity of data or incomplete analysis. Addressing the above issues, this paper analyzes the data from multiple dimensions by integrating schemas, instances, and logs to make full use of the diversity of data to establish entity associations.

3 Our Customized Framework

The entity association model in Fig. 1 shows the mapping relationship between multiple sources of data from different departments in the enterprise business system and the data space. According to the multidimensional analysis framework proposed in this paper, normalized similarity values between data that can be compared are obtained to establish the association relationship between entities. As shown in Fig. 1, R1 indicates a similarity value of 1 between its associated entities a_{13} and n_{11}.

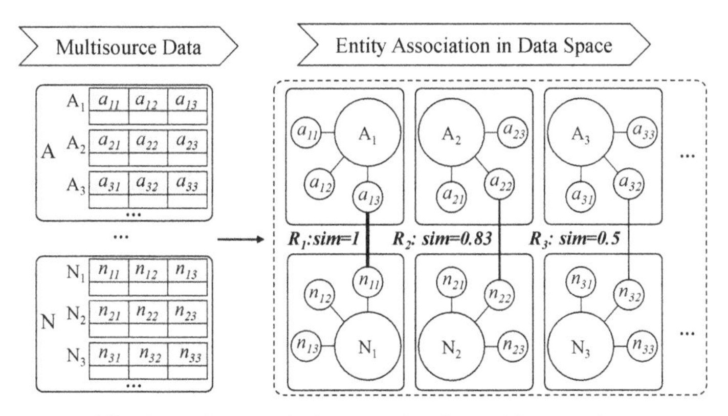

Fig. 1. Entity association mapping for multisource data

In the middle of the business logic layer and data access layer of each business system, such as ERP, CRM, and SCM, we deploy probes to obtain data. Then, the business logic layer of the data is stored as logs, and the rest of the data are stored in a relational database. To overcome the problems of large size and a variety of data types, the model pre-classifies the data based on their characteristics and nature, which improves the data processing and increases the accuracy of matching between entities. The structure and content of the data are divided into two categories, schema and instance, while logs as a carrier of business logic are grouped into a separate category. The similarity values between the data are analyzed and calculated in three dimensions: schema, instance and log. The schema matching analysis includes both attribute names and constraints, and the instance analysis is divided into three analysis methods according to data type: numeric, character and long text. Based on the attribute association information contained in SQL, the log analysis calculates the similarity values between the data. Finally, based on the fuzzy logic analyzer, a normalization calculation is performed based on similar values for the data in different dimensions to obtain the effective association values in the data space. The corresponding schema is shown in Fig. 2.

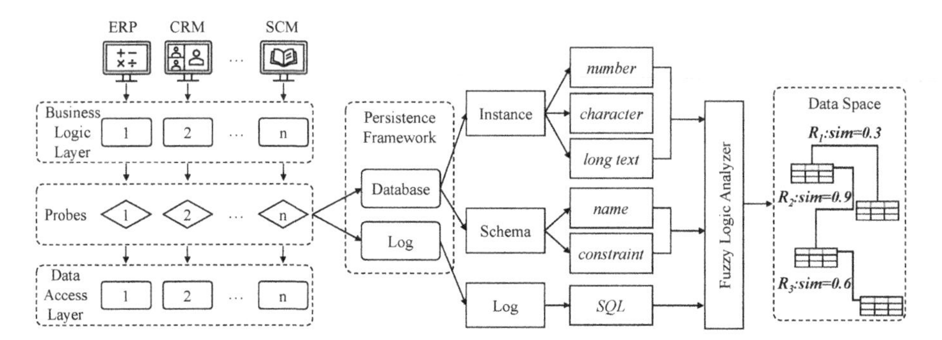

Fig. 2. A logical framework for multisource data analysis

3.1 Schema Similarity Model

Many different databases are developed by database designers to fit application scenarios, naming conventions, and other factors, but database designs generally contain table and field names, table structures, and data types. As such, the attribute names and constraints of the schema information in the database are extracted as the analysis content of the schema similarity model to measure the similarity between the data.

Name of Attribute. Attribute name analysis is divided into two types: plain text similarity and text semantic similarity analysis. The text similarity between attribute names is calculated by the edit distance algorithm, and text semantic similarity is calculated through a semantic library.

Edit distance is a way of quantifying how similar two strings are; it takes two words, w_1 and w_2, and finds the minimum number of operations required to convert w_1 to w_2. The plain text similarity value is defined according to the minimum number of edits.

$$S_{plain}(w_1, w_2) = 1 - \frac{D(w_1, w_2)}{Max(l_1, l_2)} \tag{1}$$

where l_1 and l_2 are the character lengths of w_1 and w_2, and D is the edit distance of w_1 and w_2.

Different expressions may be used for the description of the same entity. For example, if the information of an upstream company is recorded in the enterprise database, its attribute name can be named *CompanyID* and *SupplierID* based on different scenarios. To address the fact that plain text analysis cannot resolve the semantics between words, a semantic-based similarity analysis method is proposed. In particular, a tree semantic hierarchy is established for the attribute names, as shown in Fig. 3, and the similarity between words is calculated by the corresponding positions of the attribute names in the tree diagram.

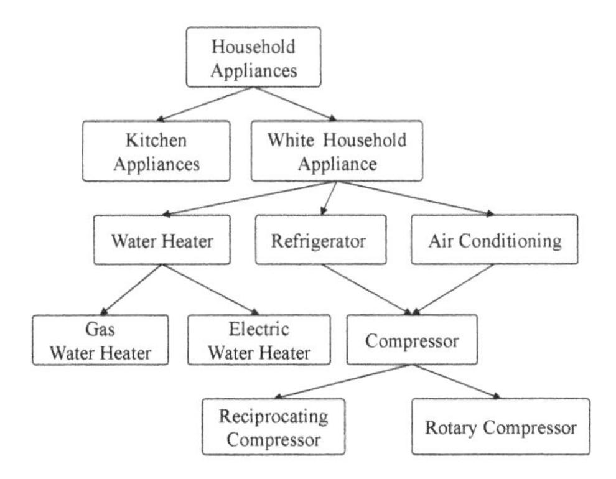

Fig. 3. Attribute name tree semantic hierarchy diagram

Therefore, the formula for calculating the semantic-based similarity is:

$$S_{sema}(w_1, w_2) = \frac{2H}{N_1 + N_2 + 2H} \tag{2}$$

where N_1 and N_2 are the shortest paths from words w_1 and w_2 to the nearest common parent word w, respectively, and H denotes the shortest path from w to the root node.

S_{name} is defined as the maximum of the plain text similarity and the semantic similarity of the text.

$$S_{name} = Max\left(S_{plain}, S_{sema}\right) \tag{3}$$

Constraint. Designers follow certain principles when programming columns in a database, such as the appropriate data type, whether it is empty, etc. The representative constraints selected from these rules can be used to explore the similarity between columns. We extracted the following constraints as features: type of each column, if the column is a primary or foreign key or not, if the column has constraint of null or not null, if the column has comments (Table 1).

Assume that the two columns requiring constraint similarity discrimination are A and B, and a_i and b_i are the values of the i-th candidate constraint corresponding to the attributes of the two columns, respectively, such that:

$$v_i = \begin{cases} 1 & a_i = b_i \\ 0 & otherwise \end{cases}, \quad i = 1, 2, \ldots, n \tag{4}$$

Table 1. Constraint features

$i = 1$	$i = 2$	$i = 3$	$i = 4$	$i = 5$
Type of column	Null	Primary key	Foreign key	Comments

where n is the number of candidate constraints; then, the attribute constraint similarity between column A and column B is:

$$S_{cons} = \frac{\sum_{i} v_i}{n} \tag{5}$$

Schema Similarity. S_{schema} includes attribute names and constraint analysis similar values by weighting.

$$S_{schema} = \alpha \cdot S_{name} + (1 - \alpha) \cdot S_{cons} \; (\alpha \in [0, 1]) \tag{6}$$

3.2 Instance Similarity Model

Since there are similarity trends in datasets representing similar entities, such as value intervals, and keywords. It is obvious that data categories are distinctive features of a dataset. Establishing differentiated feature extraction schemes for different classes of datasets can improve the accuracy of data association matching. Generally, if the data categories are different, there is no similar relationship.

Table 2. Data type categorization

Data type	Members
exact numeric data type	SMALLINT, MEDIUMINT, INT, BIGINT
approximate numeric data type	FLOAT, DOUBLE, DECIMAL
string data types	CHAR, VARCHAR, BLOB, TEXT

According to the different data types, instance analysis can be divided into the following three types: numeric, character, and long text. The numeric type refers to the exact numeric data type and the approximate numeric data types in Table 2. The string data types are divided into two categories, character and long text, according to the length of the text. After classifying and clustering the data, the similarities between the data are analyzed according to the process shown in Fig. 4.

Number. Number is scalar; considering the similarity between columns from the perspective of numerical distribution, the mean, median, mode, standard deviation, maximum, and minimum values are selected as the feature vector elements. The feature vector corresponding to each column is calculated, substituted into the cosine similarity formula, and the result is used as the numerical similarity value.

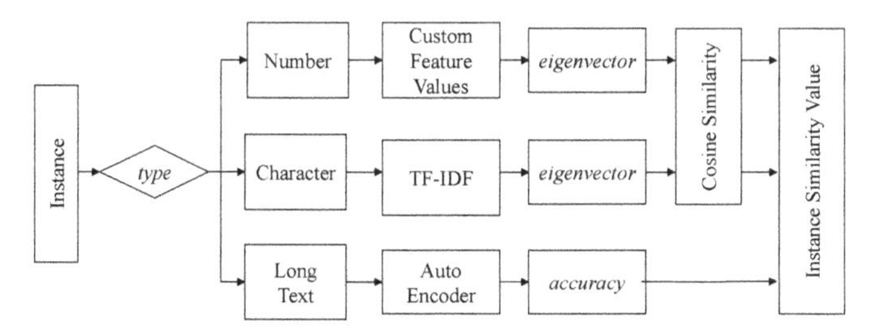

Fig. 4. Instance analyzer

Character. Character is short textual content, and it uses term frequency-inverse document frequency as the similarity calculation algorithm. First, the content of the columns that need to determine similarity is combined as a separate dataset. Then, the vectors for each column are found. Finally, the feature vectors are substituted into the cosine similarity formula to calculate the similarity value.

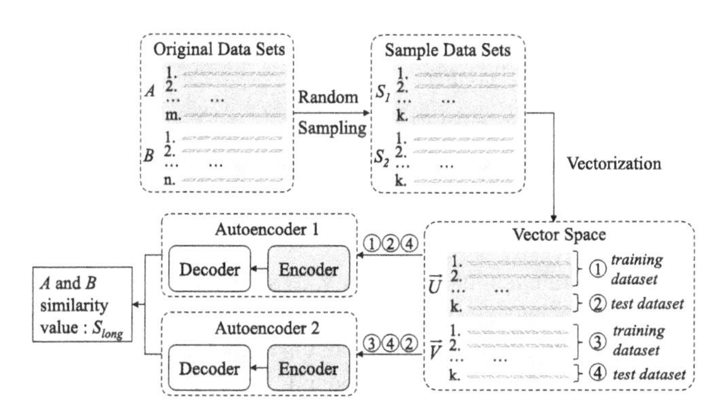

Fig. 5. The long text analysis process

Long Text. Long text is long text content, where the records in the columns are mapped as vectors, a model is built using an autoencoder, and the similarity values between columns are calculated based on the model. Assuming that A and B are two columns in the database, and they share the long text data type (Fig. 5), the overfitting problem of the model due to the large difference in the number of datasets is solved by randomly selecting k records in columns A and B as the sample data sets S_1 and S_2. Since vectors are required as input for the autoencoder, the text in the sample data sets is transformed into vectors $\vec{U}$ and $\vec{V}$. Then, the vectors are divided into a training set and test set, the autoencoder model is built using the training set, and the similarity of columns A and B is calculated according to the accuracy of the test set.

Algorithm 1. Long text similarity calculation method

$Input: x, y, \omega$
$Output: \lambda, \theta$
1: $a\_train, a\_test \leftarrow train\_test\_split(x)$
2: $b\_test \leftarrow y$
3: $a\_num, b\_num \leftarrow len(a\_test), len(b\_test)$
4: $\text{input} \leftarrow a\_train$
5: $encoded = Dense(input)$
6: $decoded = Dense(encoded)$
7: $autoencoded = Model(input, decoded)$
8: $a\_test\_predict = autoencoded(a\_test)$
9: $b\_test\_predict = autoencoded(b\_test)$
10: $For\ a, b\ in\ a\_test, a\_test\_predict$
11: $\quad s\_a\_num{+}{+} \leftarrow similarity(a, b) \geq \omega$
12: $\lambda \leftarrow \dfrac{num}{s\_a\_num}$
13: $For\ a, b\ in\ \text{b}\_test, b\_test\_predict$
14: $\quad s\_b\_num{+}{+} \leftarrow similarity(a, b) \geq \omega$
15: $\theta \leftarrow \dfrac{num}{s\_b\_num}$

The autoencoder model calculates similarity, as shown in Algorithm 1. For input, x is divided into a training set and a test set according to a custom scale, y is used as the test set, and ω is the custom text similarity threshold. On output, λ and θ is the percentage of the test set evaluated as similar. For Autoencoder 1, x and y for the input in Algorithm 1 are $\vec{U}$ in vector space and the test dataset of $\vec{V}$ in vector space; the output is λ_1, θ_1. For Autoencoder 2, x and y for the input in Algorithm 1 are $\vec{V}$ in vector space and test dataset of $\vec{U}$ in vector space, the output is λ_2 and θ_2. According to the results obtained from the Autoencoder, S_{long} represents two columns of similar values:

$$S_{long} = Min\left(\frac{\theta_1}{\lambda_1}, \frac{\theta_2}{\lambda_2}, 1\right) \tag{7}$$

3.3 Log Similarity Model

The business logic layer in the layered architecture mainly packages the attributes and behaviors of entities. Although the representation of entities varies across different business logics, similar entities have similar attributes and behaviors. The SQL commands recorded in the logs contain correlation relationships between columns, which can be used as a basis of analysis for measuring column similarity. The column-to-column similarity can be obtained by counting the number of equivalence relations in the log file.

Assuming that A and B are columns in the database, the log similarity value of columns A, B is calculated by:

$$S_{log} = \frac{N_{ab}}{N_a + N_b} \tag{8}$$

where a and b are names of columns A and B, N_a and N_b are the number of SQL commands containing a and b in the log and N_{ab} is the number of SQL commands containing both a and b in the log.

3.4 Fuzzy Logic Similarity

The similarity values obtained from the above calculation by schema, instance, and log similarity models are processed using fuzzy logic for standardization.

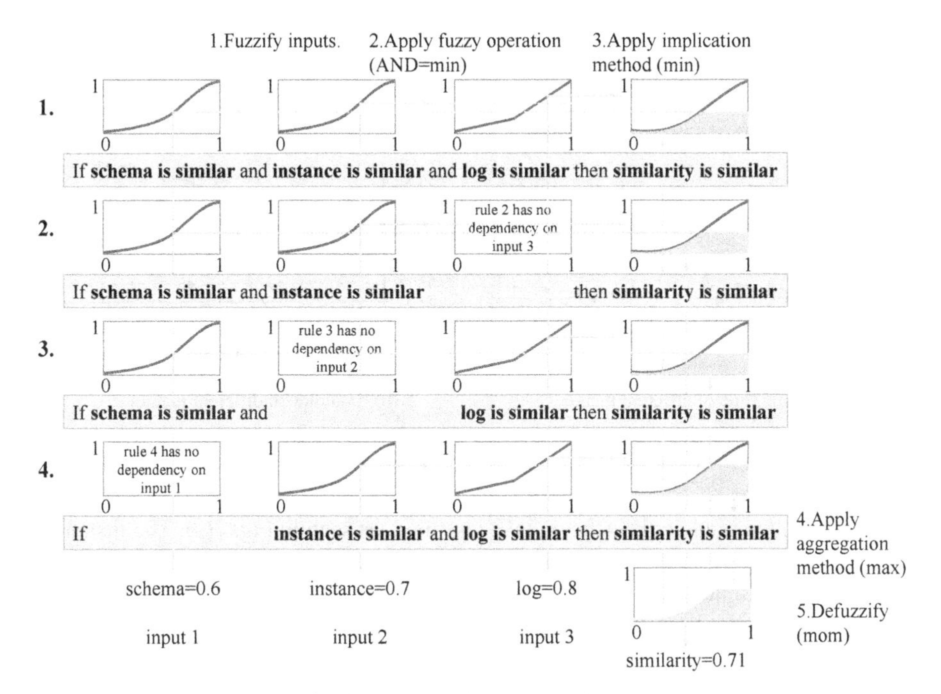

Fig. 6. Fuzzy logic instance diagram

While A and B are the columns in the database, the similarity values obtained from the above three-dimensional analysis are substituted into the affiliation function to obtain the affiliation values. The values that meet the fuzzy rules are aggregated according to the rules and defuzzified to obtain a normalized measure of column-to-column similarity. In Fig. 6, for example, A and B have similar values of 0.6, 0.7, and 0.8 in the schema, instance, and log dimensions. Through a series of fuzzy operations, the similarity value between A and B is 0.71.

4 Experiment

To verify the feasibility of the proposed framework, this paper uses data from all business systems of a company and stores them in a unified manner. The hardware environment for the experiments is an Intel(R) Xeon(R) Silver 4210 CPU @ 2.20 GHz, 64 GB RAM, RTX2080Ti*4. The results are the average of three replicated experiments. The dataset consists of Haier and public data set available on the Internet. It mainly includes the following categories: product data, enterprise operation data, value chain data, and external data [12].

4.1 Comparison of Experiments for the Different Solutions of Data Space Entity Association

A certain number of columns are randomly selected as samples from all data, the number of rows in each column is different, and experiments are conducted using schema matching (see Subsect. 3.1), instance analysis (see Subsect. 3.2), and the fuzzy logic-based model proposed in this paper to compare them in two ways: running time and accuracy.

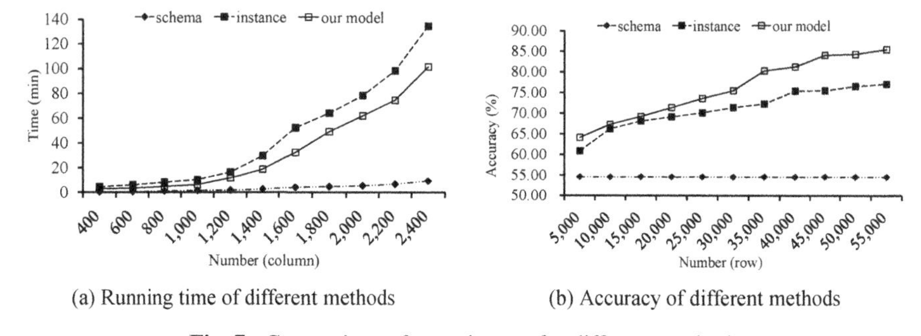

(a) Running time of different methods

(b) Accuracy of different methods

Fig. 7. Comparison of experiments for different methods

Experiments based on the schema take less time, as seen in Fig. 7(a). The instance-based method takes significantly more time for the same amount of data due to the comprehensive content analysis, while the method proposed in this paper includes instance analysis but takes less time than the instance-based method because the data are analyzed in categories during the instance analysis.

The accuracy of the schema-based method is found to remain unchanged when the volume of data is changed, as seen in Fig. 7(b), because the elements analyzed are constant; with an increase in the volume of data, the accuracy of the left two analysis methods generally tends to increase, but the method proposed in this paper maintains the highest accuracy in each experiment since data is preclassified.

As shown in Fig. 7, the proposed method in this paper can obtain a high accuracy rate in a short time with a moderate amount of data.

5 Conclusion

This paper proposes a hybrid data matching model based on schema, instances, and logs. The model consists of four main components: the front probe to acquire the analysis data, the analysis data, the three-dimensional outputs, and the normalized metric based on fuzzy logic. Experimental results show that the model provided in this paper has better results in terms of accuracy and efficient handling of mass data compared to previous single matching methods based on schema or instances. For further research, the focus is on how to establish a mapping relationship between data and weights and on establishing a guidance scheme for weight assignment to better address the impact of the randomness of multisource heterogeneous data on the accuracy of the results.

Acknowledgement. This work was supported in part by the National Key R&D Program of China under Grant 2018YFB1702902, and in part by the Shandong Province Colleges and Universities Young Talents Initiation Program under Grant 2019KJN047.

References

1. Li, G.: Research on innovation of enterprise management accounting informatization platform based on intelligent finance. In: Proceedings of the 1st International Symposium on Economic Development and Management Innovation (EDMI 2019), pp. 286–291. Atlantis Press, Paris (2019). https://doi.org/10.2991/edmi-19.2019.47
2. Nakamura, E.F., Loureiro, A.A.F., Frery, A.C.: Information fusion for wireless sensor networks: methods, models, and classifications. ACM Comput. Surv. **39** (2007). https://doi.org/10.1145/1267070.1267073
3. Boström, H., et al.: On the definition of information fusion as a field of research. IKI Tech. Reports. 1–8 (2007)
4. Alwan, A., Nordin, A., Alzeber, M., Zaid, A.: A survey of schema matching research using database schemas and instances. Int. J. Adv. Comput. Sci. Appl. 8 (2017). https://doi.org/10.14569/ijacsa.2017.081014
5. Li, X., Wen, H., Hu, Y., Jiang, L.: A novel beta parameter based fuzzy-logic controller for photovoltaic MPPT application. Renew. Energy **130**, 416–427 (2019). https://doi.org/10.1016/j.renene.2018.06.071
6. Roumila, Z., Rekioua, D., Rekioua, T.: Energy management based fuzzy logic controller of hybrid system wind/photovoltaic/diesel with storage battery. Int. J. Hydrogen Energy **42**, 19525–19535 (2017). https://doi.org/10.1016/j.ijhydene.2017.06.006
7. Tan, W., Mapforce, A.: Approximation Algorithms for Schema-Mapping Discovery. 42 (2017)
8. Wu, J., Pan, S., Zhu, X., Zhang, C., Wu, X.: Multi-instance learning with discriminative bag mapping. IEEE Trans. Knowl. Data Eng. **30**, 1065–1080 (2018). https://doi.org/10.1109/TKDE.2017.2788430
9. Gomes Dos Reis, D., Ladeira, M., Holanda, M., De Carvalho Victorino, M.: Large database schema matching using data mining techniques. In: IEEE International Conference on Data Mining Workshops, ICDMW, 2018-November, pp. 523–530 (2019). https://doi.org/10.1109/ICDMW.2018.00083
10. Bakhtouchi, A.: Data reconciliation and fusion methods: a survey. Appl. Comput. Informatics. 1–7 (2019). https://doi.org/10.1016/j.aci.2019.07.001

11. Mohanty, S.N., Rejina Parvin, J., Vinoth Kumar, K., Ramya, K.C., Sheeba Rani, S., Lakshmanaprabu, S.K.: Optimal rough fuzzy clustering for user profile ontology based web page recommendation analysis. J. Intell. Fuzzy Syst. **37**, 205–216 (2019). https://doi.org/10.3233/JIFS-179078
12. https://github.com/forgstfree/Industrial-Datasets. Accessed 06 May 2021

Software-Defined Task Scheduling Strategy Towards Edge Computing

Yue Guo, Junfeng Hou, Heng Wang, Changjin Li(✉), Hongjun Zhang, Guangxu Zhou, and Xudong Zhao

Xuchang Cigarette Factory, China Tobacco Henan Industrial Co., Ltd., Xuchang 461000, Henan, China

Abstract. Data processing has posed new challenges to transmission bandwidth and computing load under the existing cloud computing architectures. In this paper, a software-defined task cooperative scheduling structure is proposed towards edge computing. Specifically, a clustering algorithm based on two-way selection between idle users and overloaded users is firstly designed, which combines the historical information of idle users and the interest similarity of overloaded users to form the stable cooperative clusters. Then, a sub-task partitioning algorithm based on the optimal delay is presented to achieve the overall optimal delay with the guarantee that sub-tasks are completed simultaneously. Numerical results show that the proposed strategy is not only able to save the data transmission bandwidth significantly, but also achieve the optimal delay while ensuring the stability of cooperative clusters.

Keywords: Edge computing · Task scheduling · Software-defined

1 Introduction

Cloud computing can greatly improve the efficiency of data processing by offloading computation-intensive tasks to cloud servers with rich storage and computing resources [1]. However, the geographical and logical long-distance characteristics of users and cloud servers may lead to higher delay and error probability in communication. In order to solve this problem, mobile edge computing (MEC) is considered as a potential solution [2, 3]. MEC can reduce task completion delay effectively through migrating computing tasks from the cloud data center to the devices at the edge of network, like domiciliary WiFi routers and gateways. However, the thousands of billions access users and sensor nodes and the diversity of emerging task types in the future network scenario would result in the restriction on the tasks migrated to mobile edge servers due to MEC resources. Edge servers with more idle resources contribute to reduce computing delay, but may suffer from long communication delay due to unstable factors of communication links.

Researchers have proposed numerous solutions to this problem. Kong et al. [4], based on the characteristics of moving edge calculation, established a fine-grained task division model of directed acyclic graph, analyzed the problem of minimum energy

J. Xiong et al. (Eds.): MobiMedia 2021, LNICST 394, pp. 197–211, 2021.
https://doi.org/10.1007/978-3-030-89814-4_15

consumption and proposed a genetic algorithm to find the optimal solution, and finally obtained the migration decision results of sub-tasks. Considering the general three-layer fog computing network architecture and the mobility patterns of users, Wang et al. [5] jointly optimized the offloading decision and resource allocation, which was modeled as a mixed integer nonlinear programming problem. Then, the GCFSA selection algorithm and distributed resource optimization algorithm was proposed to solve the task offloading and resource allocation, respectively. Aiming to minimize delay with the constraints on power, a one-dimensional searching algorithm was proposed in [6] to obtain the optimal task scheduling strategy, where Markov decision process was utilized to solve the inevitable dual-time scale stochastic optimization problem in task scheduling strategy. In the process of pursuing high task execution efficiency, these methods only pay attention to the scheduling of tasks among servers and the allocation of resources among users, and lack the development of potential computing resources at the user level. The exponential increase in the number of users leads to a sharp increase in base station load. However, not every user equipment needs to offload tasks to edge servers. Reasonably arranging idle users and exploring the huge computing resources at the user level can effectively reduce the load on the edge server side and avoid uplink congestion. Therefore, how to reasonably distribute and schedule tasks according to the users' social relationship and resource distribution is the key issue in this paper.

In response to the above problems, this paper proposes a user cooperative computing task scheduling strategy based on the software-defined network framework in edge computing. Firstly, a cooperative cluster is established for each overloaded user, and idle users select a cluster to join according to the strength of the social relationship with the overloaded users in each cluster. Then, in-cluster tasks is scheduled, where resource weights of idle users are ranked followed by determining the task scheduling order according to the resource ranking, and an optimization equation is established. Finally, select the appropriate cooperative computing user through traversing the number of users who have joined the cooperative computing and calculate the task segmentation vector.

2 System Architecture and Network Model

The system architecture of this paper is divided into two layers, i.e. the terminal equipment layer at the bottom and the edge layer at the top, as shown in Fig. 1. The terminal equipment layer mainly includes the smart terminals used in people's daily life, such as mobile phones, tablet. Some computation-intensive tasks generated by the aforementioned smart terminals cannot be accomplished locally. Therefore, it is necessary to formulate a reasonable scheduling strategy for them to migrate the tasks to the suitable place for completing the tasks while meeting the task delay requirements. The other equipment of the terminal equipment layer are composed of idle terminal users who have no task to process. At certain moments, their CPU is in an idle state, which can provide computing resources for end users whose CPU is overloaded.

The edge layer mainly consists of macro base stations equipped with edge servers which contains SDN control module, user information management module (UIMM) and user fairness scheduling module (UFSM). UIMM is responsible for recording the basic information of users in the coverage area, mainly including the remaining amount of resources of idle users, the attribute information of the requested tasks, the social attributes of all user terminals and the spectrum resource situation in the coverage area. On the other hand, UFSM is responsible for executing the corresponding algorithm under the task requirements of the corresponding terminals according to the statistical information collected by UIMM to obtain the scheduling strategy achieving the optimal target indicator in the entire area. SDN control module forwards the scheduling policy results of all tasks in this area to the corresponding bottom terminals through the control plane according to the results of UFSM operation.

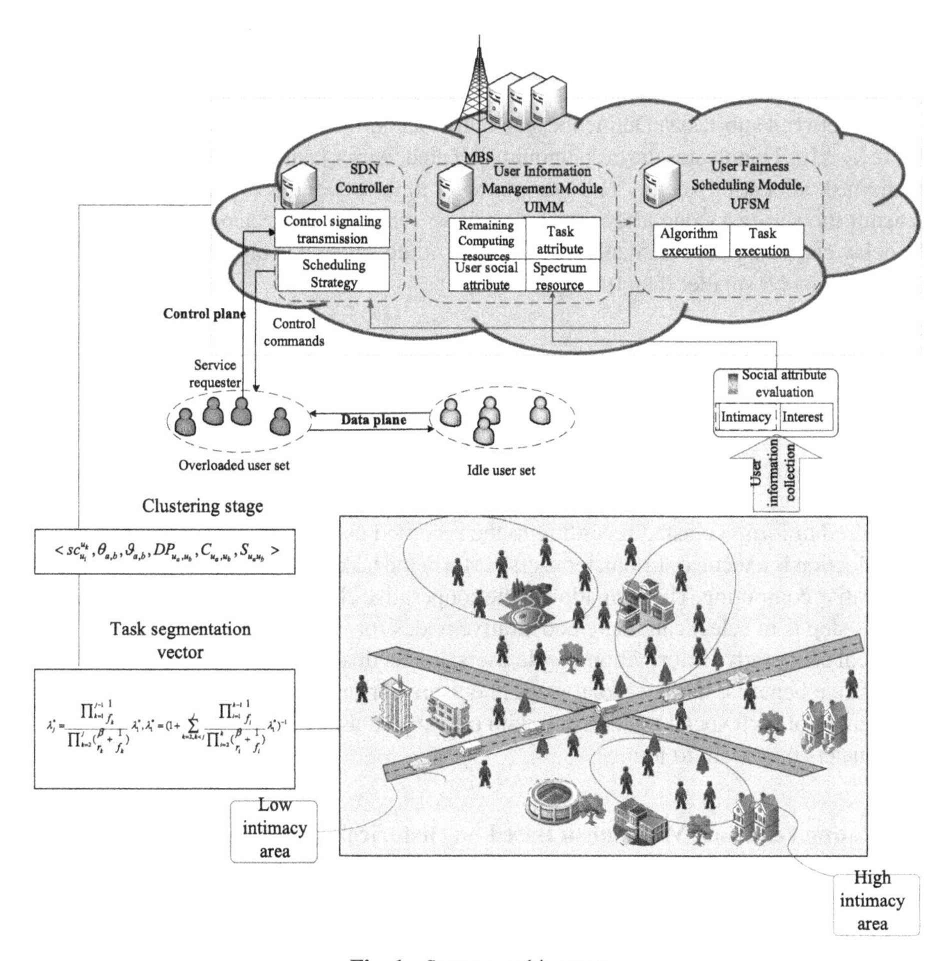

Fig. 1. System architecture

The terminal device layer is composed of multiple single-antenna users in this paper. Multiple end users are not only the generator of the tasks, but also the participants to assist in the execution of tasks. Denote the sets of terminals generating tasks and executing tasks as $M = \{1, 2, ..., m\}$ and $N = \{1, 2, ..., n\}$, respectively. The terminals generating tasks can select appropriate multiple idle users since a user is allowed to communicate with multiple users, and offload tasks to the selected users for execution through the wireless communication links.

In the scenario designed in this paper, servers execute a clustering algorithm based on the historical information of overloaded users and idle users. Thus, terminals with similar social attributes would be clustered into a cluster to perform the subsequent in-cluster task scheduling process. Assume that the overloaded user i and N_i idle users are clustered into a cluster, user i has a computation-intensive task to be completed, denoted as $T_i(s_i, c_i, t_i)$, where s_i indicates the amount of input data of this task, c_i is the number of CPU cycles required to complete this task, and t_i is the task delay constraint. Note that the above information is also recorded in UIMM module of edge servers. Suppose that tasks can be divided into multiple sub-tasks and assigned to multiple idle users. For simplicity, it is assumed that the task granularity has arbitrary precision and there is no overlap between sub-tasks. Define the size of the data offloaded to the idle user j as $\lambda_j s_i$, where $\lambda_j \in [0, 1]$ indicates the task amount allocated to user j, and let $\lambda = [\lambda_1, \lambda_2, ..., \lambda_j]$ be the task allocation vector. This paper uses a time-sequential offloading method to transmit the sub-tasks one after another as these sub-tasks are independent. Note that the delay caused by the backhaul links can be ignored since the output data size of each sub-task is much smaller than the input data.

3 Collaborative Cluster Formation Algorithm Based on Social Attributes

An in-cell user clustering algorithm is introduced in this section. Specifically, the willingness, revenue, and similarity of both idle users and overloaded users are quantitatively considered to form a cluster according to the recorded user information, and then two-way selection is executed. In-cluster users perform the tasks of overloaded users through cooperative computing. The formation of the cooperative cluster is divided into two steps: the first step is to select candidate cooperative nodes for overloaded users based on the historical cooperative information of idle users, the feedback evaluation mechanism and cooperative benefits; in the second step, idle users perform inverse selection based on the strength of their social relationship with overloaded users, and voluntarily select the user clusters they want to join.

3.1 Rating Feedback Mechanism Based on Historical Information

It is the basic demand of overloaded users to select suitable and high-reputation users for auxiliary computation. This section evaluates the reputation of users based on the service history information of each user, and on this basis, proposes a rating feedback mechanism, which is used as the criterion for users to choose to join the cooperative cluster.

Suppose user i in the cell has assisted other users in completing tasks in history, and record the total z_i times assists as $F = \{hc_1, hc_2, ..., hc_{z_i}\}$. Every time a user has been assisted, the assisted user will conduct a service evaluation on user i, and the evaluation feedback score is recorded as sc_i^k. The higher the score, the better the service quality provided by the user, and the corresponding score is directly uploaded to the UIMM module.

The level of the evaluation score depends on the task completion efficiency, which is manifested in the difference between the delay of completing the task and the original required delay of the task. This paper quantifies the evaluation score in the open interval of (0, 1). Specifically, it is quantified as 0.5 points when the actual task completion delay is equal to the task delay constraint; the score is required to increase from 0.5 to less than 1 when user i helps k complete the task at a faster speed; otherwise, user i would get a relative low score of 0–0.5. In order to meet the above requirements, the specific evaluation formula is shown in formula (1):

$$sc_{u_i}^{u_k} = \frac{2^{\frac{(t_k - to_{u_i,u_k})}{t_k}}}{2} \tag{1}$$

Where $sc_{u_i}^{u_k}$ represents the score obtained by user i assisting user k in completing the task, to_{u_i,u_k} represents the time it takes for user i to assist user k to complete the task, t_k represents the delay constraint of task k. Because some idle users have strong social relations and willingness to cooperate, but often have less resources to provide, they may get a low score. In order to weaken the impact of this situation on the user's score, this paper updates the user's score as shown in Eq. (2):

$$SCO_{u_i}^{u_k} = \frac{\sum_{u_i \in uc_i} sc_{u_i}^{u_k} \cdot \frac{c_{u_i}^{u_k}}{c_i}}{\sum_{u_i \in uc_i} sc_{u_i}^{u_k}} \tag{2}$$

Where uc_i represents the set of users assisted by user i. At the same time, in order to dynamically update the trust score of each user, the evaluation score of each time is processed by an exponential attenuation mechanism. The specific operation is shown in formula (3):

$$SCO_{u_i}^{u_k *} = \sum_{i=1}^{k} e^{1-i} \cdot SCO_{u_i}^{u_k} \tag{3}$$

To sum up, this mechanism can ensure that the score is true and effective to feedback the service status that users can provide.

3.2 Evaluation of Collaboration Benefits

In this paper, the parametric collaboration rate of return is defined to measure the revenue brought by users when they provide computing resources. The specific quantization process is as follows.

For a group of users a and b, if a is an overloaded user, that is, its own computing resources cannot meet the task requirements, and b is an idle user, then the task distribution between them can be divided into two parts: 1) User b is relatively idle; 2) The user is also overloaded. The computational resource relationship corresponding to the above two cases is as follows:

$$\theta_{a,b} = \begin{cases} \frac{f_b^l - \frac{c_b}{t_b}}{\frac{c_a}{t_a} - f_a^l}, & \frac{c_a}{t_a} > f_a^l \text{ and } \frac{c_b}{t_b} < f_b^l \\ 0, & \frac{c_a}{t_a} > f_a^l \text{ and } \frac{c_b}{t_b} > f_b^l \end{cases} \tag{4}$$

When $\theta_{a,b}$ is 0, it means that user b cannot provide effective assistance to user a; when $\theta_{a,b} > 1$, it means that user b's resources can meet the resources that user a lacks. When $0 < \theta_{a,b} < 1$, it means that user b alone cannot meet the resource needs of user a, and the assistance of more users is needed at this time.

In addition, the communication range between users is also another factor that measures the cost of collaborative computing. If two users are not within the communication range of each other, no matter how appropriate their computing resources are, they cannot provide assistance to each other. According to the requirements of D2D communication, a reliable connection can be established only when the distance between users a and b meets certain requirements. Therefore, parameter $\vartheta_{a,b}$ is defined in this paper to measure whether reliable communication is conducted between users. The specific definition is shown in formula (5):

$$\vartheta_{a,b} = sel(\gamma_{u_a,u_b}, \gamma_{u_a}^{th}) \tag{5}$$

Among them, $sel(a, b)$ is a selection function, when the parameter $a \geq b$, take a, and when $a < b$, take 0. Where γ_{u_a,u_b} represents the SNR formula of the channel between users a and b, which depends on the additive white Gaussian noise interference and the interference from the multiplexed cellular users to it. The formula is shown in formula (6):

$$\gamma_{u_a,u_b} = \frac{\left|h_{u_a,u_b}\right|^2 P_{u_a}}{\left|h_{m,u_a}\right|^2 P_m + \sigma^2} \tag{6}$$

$\gamma_{u_a}^{th}$ represents the SNR threshold that meets the requirements of the user's direct communication range. The specific calculation formula is shown in Eq. (7).

$$\gamma_{u_a,u_b}^{th} = 2^{\frac{s_a}{W\left(t_a - \frac{c_a - t_a f_a^l}{f_b^l - \frac{s_b}{t_b}}\right)}} - 1 \tag{7}$$

To sum up, $\vartheta_{a,b}$ describes whether direct communication can be conducted between two user terminals. When they can communicate with each other, the larger $\vartheta_{a,b}$ indicates that the link is more conducive to communication. According to parameter $\vartheta_{a,b}$ and the collaborative yield rate between users, this paper defines the cost of collaborative computing between users as shown in Eq. (8):

$$DP_{u_a,u_b} = \frac{\vartheta_{a,b}}{\theta_{a,b}} \tag{8}$$

It can be seen from the above formula that in the case of meeting the requirements of direct communication, the higher the rate of return of collaboration, the lower the cost of collaboration between users.

3.3 Social Attribute Driven Bidirectional Selection Clustering Algorithm

The social relationship between users can be measured from two aspects. One is the social relationship affected by intimacy, and the other is the social relationship affected by interest similarity. This section will measure the strength of social relationships between users from these two aspects.

Intimacy Relationship

The definition of intimacy $C_{a,b}$ represents the strength of the social relationship between user terminals a and b under the influence of intimacy. The larger the value, the stronger the intimacy between them, the more likely they are to trust each other and assist in calculations. Intimacy $C_{a,b}$ is affected by different scene factors. The degree to which users with different social identities are affected by different factors shows obvious differences. In this paper, the intimacy between users is defined as the sum of the product of weight factor and intimacy factor. The weight factor reflects the degree of influence of the scene on the user's intimacy, specifically the ratio of the average interaction interval of the user to time T in the time period T, as shown in Eq. (9):

$$Q^{i}_{u_a,u_b} = \frac{\sum_{p=1}^{\mathrm{ct}} \mathrm{STA}_{p+1} - \mathrm{EDT}_p}{T \cdot \mathrm{ct}} \tag{9}$$

Where $Q^{i}_{u_a,u_b}$ represents the weight factor of the i-th influencing factor between users a and b, and represents the importance of the influencing factor. ct represents the number of interactions between two users in time period T, STA_{p+1} represents the start time of $p+1$ interaction, and EDT_p represents the end time of p interaction.

The total intimacy between users a and b can be calculated as shown in Eq. (10):

$$C_{u_a,u_b} = \sum_{i=1}^{n} Q^{i}_{u_a,u_b} \cdot \zeta^{i}_{u_a,u_b} \tag{10}$$

Where $\zeta^{i}_{u_a,u_b}$ represents the closeness of users a and b corresponding to influencing factor i. In this paper, interaction frequency is used to measure intimacy under this factor. The interaction frequency between each pair of user terminals can be quantified, and the specific formula is shown in (11):

$$\zeta^{i}_{u_a,u_b} = \frac{\alpha_{u_a,u_b}}{\sum_{b \in U} \alpha_{u_a,u_b}} \times \frac{\eta_{u_a,u_b}}{\sum_{b \in U} \eta_{u_a,u_b}} \tag{11}$$

Where α_{u_a,u_b} represents the number of content sharing between users a and b, and η_{u_a,u_b} represents the average time for content sharing between users a and b. In summary, the total intimacy between users a and b is recorded as:

$$C_{u_a,u_b} = \sum_{i=1}^{n} Q^i_{u_a,u_b} \cdot \zeta^i_{u_a,u_b}$$
$$= \sum_{i=1}^{n} \frac{\sum_{p=1}^{ct} \mathrm{STA}_{p+1} - \mathrm{EDT}_p}{T \cdot \mathrm{ct}} \cdot \frac{\alpha_{u_a,u_b}}{\sum_{b \in U} \alpha_{u_a,u_b}} \times \frac{\eta_{u_a,u_b}}{\sum_{b \in U} \eta_{u_a,u_b}} \tag{12}$$

Interest Similarity Perception

Assuming that there are m users in this area, denoted as $U = \{u_1, u_2, \ldots, u_m\}$, there are a total of v points of interest. Since users' interest in different locations is affected by time period, in order to better measure the degree of interest, this paper divides the time period of a day into several time periods of corresponding size, denoted as $\{\chi_1, \chi_2, \ldots, \chi_w\}$, and there are w time periods in total. Each user's interest in the corresponding place at time χ_w is denoted as $I^v_{u_i}(\chi_w)$. This value is determined by the number of visits and the length of the time period between the corresponding time and the previous time. The formula is as (13):

$$I^v_{u_i}(\chi_w) = \frac{\chi_w - \chi_{w-1}}{\phi_w} \tag{13}$$

where χ_w and χ_{w-1} denote the length of time between the current moment and the previous statistical moment, respectively, and ϕ_w denotes the number of visits to the location within the time range. The interest matrix M is defined as a $v \times w$ matrix. Each row of the matrix corresponds to a different location, and each column represents a different time period.

After obtaining the interest matrix of all users, the similarity of interest of different users in the places of interest can be calculated based on the interest matrix. Firstly, the overall error of the degree of interest between two end users is calculated according to the definition of covariance, as shown in Eq. (14):

$$\mathrm{cov}(u_a, u_b) = \frac{\sum_{i=1}^{w} I^v_{u_a}(\chi_i) \times I^v_{u_b}(\chi_i)}{w} - \frac{\sum_{i=1}^{w} I^v_{u_a}(\chi_i) \sum_{i=1}^{w} I^v_{u_b}(\chi_i)}{w^2} \tag{14}$$

Since different individual users manifest variability for different locations of interest, in order to measure the similarity of interest among different users, it is necessary to know the fluctuation of the difference of each user's interest in different locations. Calculate the square of the difference between the interest degree of each end-user at different points of interest and the average interest degree, and take the average of the squares to measure the degree of fluctuation of interest of different users in each interest point. The formula is shown in (15):

$$D(u_a) = \sqrt{\frac{1}{v} \cdot \sum_{i=1}^{v} (I^v_{u_a}(\chi_i) - \sum_{i=1}^{v} I^v_{u_a}(\chi_i)/v)^2} \tag{15}$$

Finally, combining formulas (14) and (15), the interest similarity is derived by calculating the ratio of the difference in interest between users and the product of the fluctuation of users' respective interests, and the formula is shown in Eq. (16):

$$\rho(u_a, u_b) = \frac{\frac{\sum_{i=1}^{w} I_{u_a}^{v}(\chi_i) \times I_{u_b}^{v}(\chi_i)}{w} - \frac{\sum_{i=1}^{w} I_{u_a}^{v}(\chi_i) \sum_{i=1}^{w} I_{u_b}^{v}(\chi_i)}{w^2}}{\sqrt{\frac{1}{v} \cdot \sum_{i=1}^{v} (I_{u_a}^{v}(\chi_i) - \sum_{i=1}^{v} I_{u_a}^{v}(\chi_i)/v)^2} \times \sqrt{\frac{1}{v} \cdot \sum_{i=1}^{v} (I_{u_b}^{v}(\chi_i) - \sum_{i=1}^{v} I_{u_b}^{v}(\chi_i)/v)^2}} \tag{16}$$

$$|\rho(u_a, u_b)| \leq 1$$

From the above equation, the interest similarity takes values within the range of absolute values less than or equal to 1 when $\rho(u_a, u_b) = 0$ indicates that users a, b have almost no points of common interest.

In summary, the social relationship between users in the edge computing network needs to be considered from two aspects. The higher the social intensity between users, the higher the probability that both parties will perform assisted computing. Therefore, considering the intimacy relationship and interest similarity between user terminals, the strength of social relationship between them is shown in Eq. (17), where τ denotes the weighting factor.

$$S_{u_a u_b} = \tau \cdot C_{u_a, u_b} + (1 - \tau) \cdot \rho(u_a, u_b) \tag{17}$$

From Eqs. (12) and (16), it can be seen that in different scenarios, the strength of social relationships can be determined by a combination of several factors such as intimacy relationship and interest similarity. Therefore, according to the weight proportion of various influencing factors, the importance degree of the two types of social relations in different scenarios can be represented, i.e., the weight factor τ is shown in Eq. (18), where Q_{u_a,u_b}^{i} is the weight factor of the i-th influencing factor between users a, b, and $\rho(u_a, u_b)$ is the interest similarity of users a, b.

$$\tau = \frac{\sum_{i=1}^{n} Q_{u_a,u_b}^{i}}{\sum_{i=1}^{n} Q_{u_a,u_b}^{i} + \sum_{a=1}^{N} \sum_{b=1}^{N} |\rho(u_a, u_b)|} \tag{18}$$

4 Collaborative Scheduling Strategy in Cluster

The optimization problem can be established as shown in Eq. (19):

$$p1 : \min_{\boldsymbol{\lambda}} T(\boldsymbol{\lambda})$$
$$s.t. \sum_{j=1}^{N_i} \lambda_j = 1 \ (a) \tag{19}$$
$$0 \leq \lambda_j \leq 1 \ (b)$$

The two constraints ensure that there is no overlap in the offloading of subtasks. The delay for each subtask is a non-convex problem due to the sequential transfer of tasks. In this paper, a heuristic algorithm is proposed to achieve a suboptimal solution of the non-convex problem. The specific process is divided into three steps.

4.1 Idle User Selection

In this paper, we mainly consider the link quality and computation rate, and make the selection by the collaboration factor that will define the candidates. The collaboration factor is defined as:

$$w_j = \frac{\beta s_i}{r_{i,j}} + \frac{c_i}{f_j} \tag{20}$$

This represents the delay of offloading the total task to user j.

4.2 Task Scheduling

Intuitively, minimizing the maximum value of a set of variables is balancing these variables. The purpose of task scheduling is to balance the execution delay of each selected idle user. Therefore, the scheduling policy is to balance the latency of each selected user. To balance the latency between idle users is shown in Eq. (21):

$$\frac{c_i \lambda_1}{f_1} = (\frac{\beta s_i}{r_{i,1}} + \frac{c_i}{f_2})\lambda_2$$
$$\frac{c_i \lambda_2}{f_2} = (\frac{\beta s_i}{r_{i,2}} + \frac{c_i}{f_3})\lambda_3$$
$$\ldots,$$
$$\frac{c_i \lambda_{j-1}}{f_{j-1}} = (\frac{\beta s_i}{r_{i,j-1}} + \frac{c_i}{f_j})\lambda_j \tag{21}$$

According to the set of Eqs. (21), for a given j idle users, the optimal task assignment decision is shown in Eq. (22):

$$\lambda_j^* = \frac{\prod_{k=1}^{j-1} \frac{1}{f_k}}{\prod_{k=2}^{j} (\frac{\beta}{r_k} + \frac{1}{f_k})} \lambda_1^* \tag{22}$$

$$\lambda_1^* = (1 + \sum_{k=2,k<j}^{j} \frac{\prod_{i=1}^{k-1} \frac{1}{f_i}}{\prod_{i=2}^{k} (\frac{\beta}{r_i} + \frac{1}{f_i})})^{-1} \quad (23)$$

More idle users can help reduce the computational latency of the overall task, but can lead to a higher risk of offload failure. Therefore, we must find the optimal number of selected idle users by traversing from 1 to j. The number of users that minimizes the overall task latency is the optimal number. The corresponding task assignment strategy is the optimal decision. Table 4.3 summarizes the heuristic algorithm, which has a worst-case complexity of $O(j^3)$.

5 Numerical Results and Analysis

The performance of the proposed algorithm is evaluated by using MATLAB in this section. In the simulation environment, a single-cell multi-user model is adopted to measure the collaboration scores of users according to their historical information; the channel model is established as a Gaussian channel to measure the efficiency of task migration between users, thereby calculating the benefits of collaboration between users. Finally, the implementation of the clustering algorithm is verified according to the trajectory information and social information of the participants of the InfoCom06 conference, where the specific track information includes the start and end time of interaction between users and the number of interactions. Meanwhile, users' social relationship information can also be obtained from their activity trajectories to measure their closeness and similarity of interests. In terms of parameters, the configuration of the transmission link is based on the LTE system, and the channel bandwidth is 10 MHz. To simplify the problem, the noise spectral density is set as -174 dBm/Hz. The user's CPU rate ranges from 10×10^9 cycles per second to 100×10^9 cycles per second. And the task data size is set to 50–100 mbits to represent computationally intensive tasks. The main simulation parameters are shown in Table 1.

Table 1. Simulation parameter settings

Parameter	Parameter value
Transmission bandwidth	5 MHz
D2D user transmission power	15 dbm
User history information score	[0,1]
Noise spectral density	-174 dbm/Hz
Number of edge users	[10,100]
CPU cycles required for the task	$[1,50] \times 10^8$ cycles
User computing power	$[10,100] \times 10^9$ cycles/s
Task data size	[50,100] MBits
Task delay	[0,100] ms
Channel fading factor	10^{-4}

The proposed MDGTSM heuristic algorithm is compared with two typical task scheduling mechanisms. The first is based on the cooperative task computing mechanism of nodes in the fog wireless access network [10] (Cooperative Task Computing, CTC). The CTC algorithm splits the subtasks to multiple candidate terminals in parallel, and meets the ultra-low latency requirements by combining the settlement capacity of multiple powerful fog wireless access nodes and the near-field communication at the edge; The second is a task scheduling algorithm based on the task calculation time prediction algorithm [11] (Prediction of Tasks Computation Time Algorithm, PCA), which uses principal component analysis (PCA) and reduces the expected time of calculation (ETC) matrix to greatly increase the delay and reduce the computation and complexity. Moreover, it is also compared with two baseline schemes: (1) one-to-one offloading scheme: offload the whole task to the user with the lowest weight among idle users; (2) all offloading scheme: all idle users are involved Task execution.

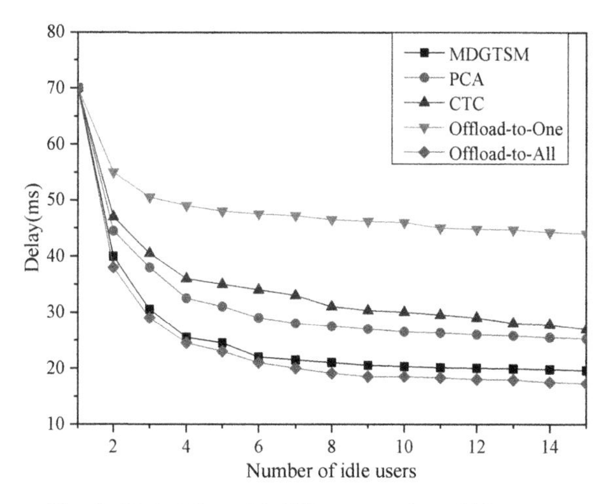

Fig. 2. Task delay with different number of idle users

Figure 2 shows the relationship between task delay and the number of idle users with different algorithms. As shown in the figure, the algorithm proposed in this paper achieves lower latency than the CTC algorithm and the PCA algorithm. Specifically, the algorithm proposed in this paper has similar performance to all offloading schemes, while the CTC algorithm has the highest latency.

Figure 3 depicts the relationship between the number of idle users and the probability of unloading failure. We can see that the offloading failure probability of the MDGTSM algorithm proposed in this paper is very close to that of the PCA algorithm, both of which are less than the failure probability of all offloading schemes but higher than that of one-to-one offloading. The CTC algorithm shows poor performance in this regard, because it tends to allocate more resource blocks for users with lower data rates.

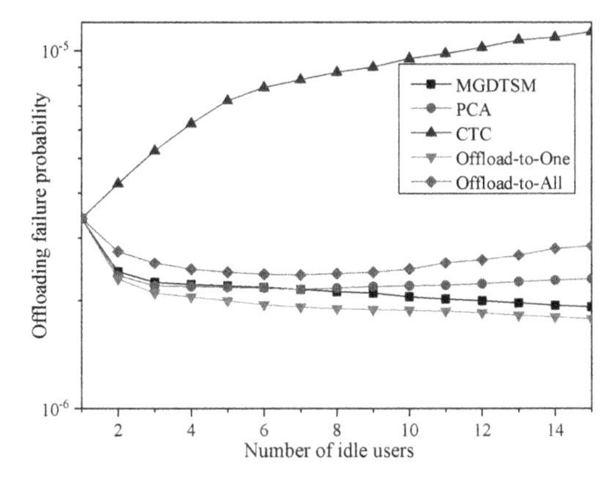

Fig. 3. Offloading failure probability with different number of idle users

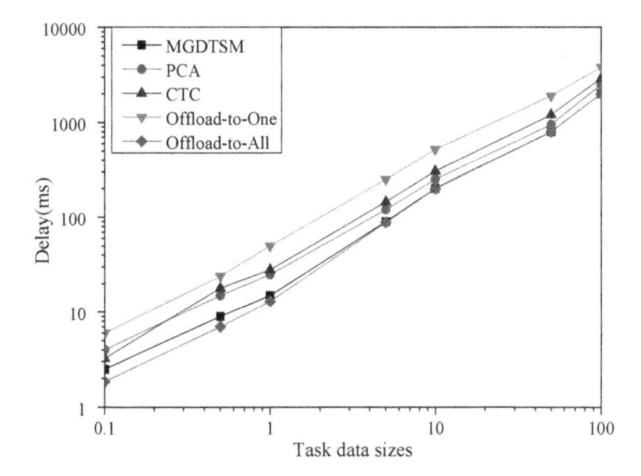

Fig. 4. Task delay with different task data sizes

Figures 4 and 5 describe the performance of the algorithm with different task data sizes with the total number of idle users set to 15. Obviously, the waiting time and the probability of offloading failure increase as the task data size increases. Specifically, as shown in Fig. 4, the MGDTSM algorithm and the PCA algorithm have better performance than the CTC algorithm. The CTC algorithm has the longest time delay among the three algorithms, while the MGDTSM and the PCA algorithm have similar waiting time delay and are both close to that of all offloading schemes. In Fig. 5, the CTC algorithm shows the worst offloading failure probability, while the performance of the other algorithms is very close to each other, especially at 50 Mbits.

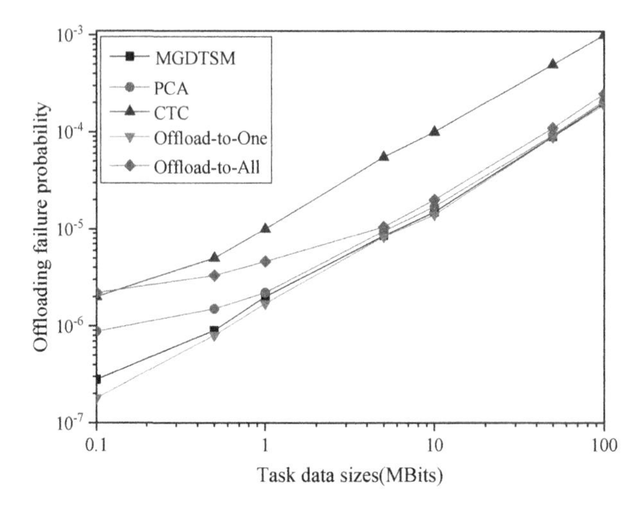

Fig. 5. Offloading failure probability with different task data sizes

6 Conclusion

In this paper, a task scheduling strategy for edge computing is presented. Firstly, the overloaded user makes positive selection according to the idle user's historical collaboration record and resource distribution. At the same time, idle users will also make reverse selection according to the social attributes of overloaded users to form a stable collaboration cluster. Secondly, the overload user in the cluster selects the appropriate user for unloading according to the link state and calculation rate of the idle user. The task is divided into subtasks which are offloaded in a reasonable sequence to ensure that the whole task is completed with low latency. Finally, the results show that the proposed strategy can effectively reduce the task completion delay, avoid the channel congestion, and optimize the service quality.

References

1. Armbrust, M., Fox, A., Griffith, R., et al.: A view of cloud computing. Commun. ACM **53**(4), 50–58 (2010)
2. Zhang, Q., Fitzek, F.H.P.: Mission critical IoT communication in 5g. In: Atanasovski, V., Leon-Garcia, A. (eds.) FABULOUS 2015. LNICSSITE, vol. 159, pp. 35–41. Springer, Cham (2015). https://doi.org/10.1007/978-3-319-27072-2_5
3. Weisong, S., Jie, C., Quan, Z., et al.: Edge computing: vision and challenges. IEEE Internet Things J. **3**(5), 637–646 (2016)
4. Kong, Y., Zhang, Y., Wang, Y., et al.: Energy saving strategy for task migration based on genetic algorithm. In: 2018 International Conference on Networking and Network Applications (NaNA), pp. 330–336. IEEE Press, Piscataway (2018)
5. Dongyu, W., Zhaolin, L., Xiaoxiang, W., et al.: Mobility-aware task offloading and migration schemes in fog computing networks. IEEE Access **7**, 43356–43368 (2019)
6. Zeng Deze, G., Lin, G.S., et al.: Joint optimization of task scheduling and image placement in fog computing supported software-defined embedded system. IEEE Trans. Comput. **65**(12), 3702–3712 (2016)

7. Chen, X., Lei, J., Wenzhong, L., et al.: Efficient multi-user computation offloading for mobile-edge cloud computing. IEEE/ACM Trans. Netw. **24**(5), 2795–2808 (2015)
8. Wang, C., Yu, F.R., Liang, C., et al.: Joint computation offloading and interference management in wireless cellular networks with mobile edge computing. IEEE Trans. Vehicul. Technol. **66**(8), 7432–7445 (2017)
9. Wang, X., Zhang, Y., Leung, V.C.M., et al.: D2D big data: content deliveries over wireless device-to-device sharing in large scale mobile networks. IEEE Wireless Commun. **25**(1), 32–38 (2018)
10. Chiu, T.-C., Chung, W.-H., Pang, A.-C., et al.: Ultra-low latency service provision in 5g fog-radio access networks. In: 2016 IEEE 27th Annual International Symposium on Personal. Indoor, and Mobile Radio Communications (PIMRC), pp. 1–6. IEEE Press, Piscataway (2016)
11. Al-Maytami, B.A., Fan, P., Hussain, A., et al.: A task scheduling algorithm with improved makespan based on prediction of tasks computation time algorithm for cloud computing. IEEE Access **7**, 160916–160926 (2019)

Intelligent Reflecting Surface-Assisted Full-Duplex UAV-Based Mobile Relay Communication

Mengqiu Chai[1], Shengjie Zhao[1(✉)], Yuan Liu[2], Fuqiang Ding[3], Hai Sun[3], and Runqing Jiang[3]

[1] School of Software Engineering, Tongji University, Shanghai 201804, China
{mengqiuchai,shengjiezhao}@tongji.edu.cn
[2] College of Electronic and Information Engineering, Tongji University, Shanghai 201804, China
tjyuanliu@tongji.edu.cn
[3] Shanghai Ideal Information Industry (Group) Co., Ltd, Shanghai, China

Abstract. Intelligent reflecting surface (IRS), a promising technology, can intelligently reflect the signals and improve the propagation environment, which makes it further achieve spectrum efficiency for wireless systems in the future. In this paper, we investigate an IRS-assisted full-duplex (FD) unmanned aerial vehicle (UAV) relay system with a source node and a destination node. Due to the obstacles, the source node can not communicate with the UAV-based mobile relay. The IRS on the wall of a building can reflect the signals from the source node to the relay. We aim to maximize the average rate of the system by jointly optimizing the phase shifter of the IRS, the trajectory of the UAV, and the transmit power of the UAV-based relay. On account of the non-convexity of the formulated problem, we propose a novel algorithm based on iteration that divides the problem into three sub-problems by the block coordinate descent and applies the successive convex approximation (SCA) method for obtaining an approximate optimal solution. Simulation results demonstrate the efficiency of our algorithm.

Keywords: Intelligent reflecting surface · Full-duplex relay · Phase control · Trajectory optimization

1 Introduction

With the incredible increase of the number of wireless devices in the forthcoming fifth-generation networks, the consumption of the network and the spectrum efficiency remain critical issues in the practical application [1]. As a transformative technology, intelligent reflecting surface (IRS) has been proposed to achieve spectrum efficiency with the lower hardware cost [2]. IRS is constituted of an array of IRS passive units that can independently incur a specific phase shift on the

J. Xiong et al. (Eds.): MobiMedia 2021, LNICST 394, pp. 212–223, 2021.
https://doi.org/10.1007/978-3-030-89814-4_16

incident signals [3]. Therefore, IRS can collaboratively improve the propagation environment and ameliorate the quality of the communication with low energy consumption, which makes IRS arouse widespread concern in communication applications [4]. In [5], the authors proposed a practical phase shift model of IRS which can capture the change of phase-dependent amplitude. A tutorial of new challenges when IRS is integrated into wireless networks has been provided in [6]. In [7], an IRS-assisted non-orthogonal multiple access (NOMA) system has been investigated, and a novel algorithm has been proposed to maximize the system throughput over the channel assignment, reflection coefficients, decoding order of NOMA users, power allocation. The asymptotic max-min signal-to-interference-plus-noise ratio (SINR) of an IRS-assisted multiple-input single-output (MISO) system has been studied, and the simulation results verify that IRS outperforms half-duplex relay [8]. The coverage analysis, the probability of signal-to-noise ratio (SNR) gain, and the delay outage rate of IRS-aided communications system have been investigated [9].

Due to the low energy consumption, flexible deployment, on-demand mobile features, unmanned aerial vehicles (UAVs) have been widely used in public and civil domains [10,11]. UAV can not only be connected to the communication network as a new type of mobile equipment but also can mount a flying base station and relay to guarantee the requirement of more complex communication requirement [12,13]. In [14], the authors investigated the energy efficiency in a single UAV-aided relaying system of two static ground nodes and defined an energy-efficiency metric. The outage probability (OP) of an amplify-and-forward (AF) UAV relaying network has been studied, and the mobile device, the UAV's trajectories, and transmit power of UAV can be optimized [15]. The energy harvesting practicality of a UAV-assisted relaying system has been studied in [16]. In [17], the authors have investigated the analytical expressions of outage probability, average bit error rate (BER), and the average capacity of an IRS-assisted half-duplex (HD)-UAV relaying system. However, most of the studies in UAV-based relay systems are in HD mode.

Considering the dense buildings in cities, it is difficult for direct communications in practical conditions, such as the link between two ground nodes with crowded buildings around. So we introduce the UAV and IRS technology into a system where an IRS-assisted FD UAV-based relay in deployed to maximize the average achievable rate. Different from the existing research results [17], we try to maximize the average achievable rate of the system by jointly optimizing the dynamic UAV's trajectory, transmit power of the UAV, and the phase shifter of IRS with the relay working in FD mode. Because multiple variables are coupled together, the problem is non-convex and hard to resolve. We propose an iterative algorithm that can solve the problem efficiently. Simulation results verify the efficiency of our algorithm.

The rest of this paper is organized as follows. Section 2 describes the system model and the formulated problem. In Sect. 3, we propose a joint optimization algorithm with the UAV's trajectory, transmit power of the UAV, and the phase shifter of IRS. Section 4 and Sect. 5 respectively represent the numerical result and conclusions.

2 System Model and Problem Formulation

2.1 System Model

As shown in Fig. 1, we consider an IRS-assisted FD UAV-based relaying system with one ground source node (S), an IRS installed on the building near S, a dynamic UAV-based relay station, one ground destination node (D). Due to the obstacles, S can not communicate with D and UAV-based relay. Therefore, IRS is introduced into the scenario to assist the communication. The IRS is equipped on the building near S to reflect the signals received from S to the UAV. The UAV-based relay works in the FD mode. To be specific, the UAV can receive and send data at the same time. A three-dimensional (3-D) Cartesian coordinate system is considered for the scenario. Therefore, the horizontal coordinates of S, D are denoted as $w_S = [x_S, y_S]^T$, $w_D = [x_D, y_D]^T$ respectively. Besides, the UAV flies from the initial point w_0 to the final point w_f at a fixed height H_U. For tractability, we divide the UAV's flight time T into M time slots, i.e., $T = M\tau$, where τ denotes the slot length. The coordinate changes between two time slots can be regarded as constant. Therefore, the UAV's horizontal coordinates changing with time can be given as $w_U[m] = [x_U[m], y_U[m]]^T$, $m \in \mathcal{M} = 1, 2, ..., M$. Accordingly, the flying constraints of the UAV can be given as

$$w_U[1] = w_0, \tag{1a}$$

$$||w_U[m] - w_U[m-1]|| \leq D_{\max}, m = 2, \cdots M \tag{1b}$$

$$w_U[M] = w_f, \tag{1c}$$

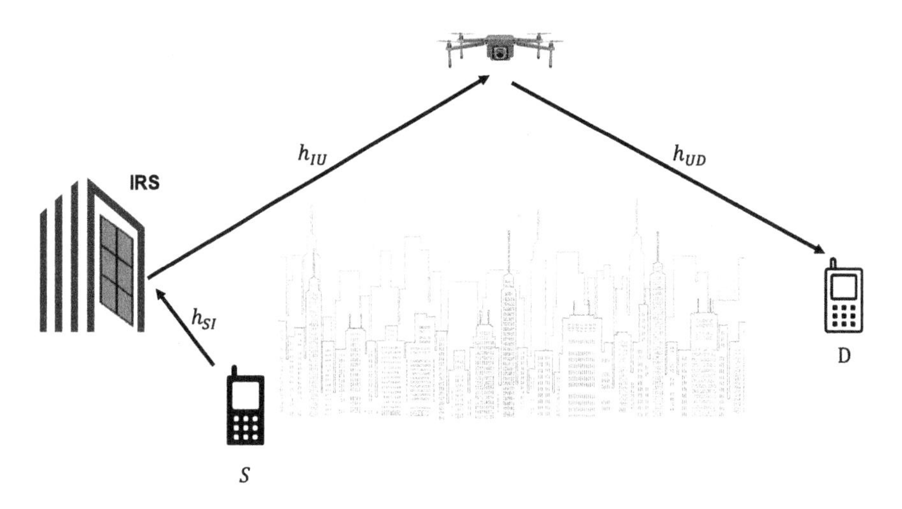

Fig. 1. An IRS-assisted DF UAV-based relaying system

where $D_{\max} = V_{\max}\tau$ denotes the maximum distance that the UAV can move at the maximum speed $V_{\max}$ in one time slot.

Moreover, S, D, and the UAV are all equipped with a single omni-directional antenna. The IRS is equipped with a uniform linear array (ULA) of N elements. A diagonal matrix $\Theta[m] = diag\{e^{j\theta_1[m]}, e^{j\theta_2[m]}, \ldots, e^{j\theta_N[m]}\}$, $\theta_n[m] \in [0, 2\pi)$, $n = 1, 2, \ldots, N$, represents the phase shifter matrix of the IRS in the mth time slot. The coordinate and the height of the first element at the IRS are denoted as $w_I = [x_I, y_I]^T$ and H_I. We suppose the first element can be regarded as the reference point of the IRS. Therefore, we use $w_I = [x_I, y_I]^T$ and H_I to denote the coordinate and height of the IRS.

2.2 Transmission Process

With the obstacles around, the UAV-based relay can not receive the signals from S directly, while the UAV can only receive the signals from S through IRS reflectively. We assume that all channel links are line-of-sight (LoS) channels, and the Doppler effect owing to the movement of UAV can be compensated [18]. The channel coefficient between S and the IRS, $\boldsymbol{h}_{SI} \in \mathcal{C}^{N\times 1}$, can be given as

$$\boldsymbol{h}_{SI} = \sqrt{\gamma d_{SI}^{-2}}[1, e^{-j\frac{2\pi}{\lambda_0}d_0\phi_{SI}}, \ldots, e^{-j\frac{2\pi}{\lambda_0}(N-1)d_0\phi_{SI}}]^T, \tag{2}$$

where γ denotes the channel gain at the reference distance $d_{ref} = 1\text{m}$, $d_{SI} = \sqrt{||\boldsymbol{w}_I - \boldsymbol{w}_S||^2 + H_I^2}$ is the distance between S and IRS, λ_0 is the carrier wavelength, d_0 is the antenna separation, $\phi_{SI} = \frac{x_I - x_S}{d_{SI}}$ denotes the cosine of the angle of arrival (AoA) of the signal from S to the IRS.

Similarly, the channel coefficient of the link between IRS and UAV, $\boldsymbol{h}_{IU} \in \mathcal{C}^{N\times 1}$, in the mth time slot can be given as

$$\boldsymbol{h}_{IU}[m] = \sqrt{\gamma d_{IU}^{-2}[m]}[1, e^{-j\frac{2\pi}{\lambda_0}d_0\phi_{IU}[m]}, \ldots, e^{-j\frac{2\pi}{\lambda_0}(N-1)d_0\phi_{IU}[m]}]^T, \tag{3}$$

where $d_{IU}[m] = \sqrt{||\boldsymbol{w}_U[m] - \boldsymbol{w}_I||^2 + (H_U - H_I)^2}$ denotes the distance between IRS and UAV, $\phi_{IU}[m] = \frac{x_U[m] - x_I}{d_{IU}[m]}$ is the cosine of the angle of departure (AoD) of the signal from IRS and UAV in the mth time slot. It is assumed that a large enough data buffer is equipped on the UAV, which can guarantee the data storage requirement for the decode-and-forward (DF) relay. Therefore, the achievable rate from IRS to UAV can be given as

$$R_S[m] = \log_2\left(1 + \frac{P_S\gamma^2|f[m]|^2}{d_{IU}^2[m]d_{SI}^2\left(\frac{P[m]}{\beta} + \sigma^2\right)}\right), \tag{4}$$

where $f[m] = \sum_{n=1}^{N} e^{j\theta_n[m] + j\frac{2\pi}{\lambda_0}(n-1)d_0(\phi_{IU}[m] - \phi_{SI})}$, P_S denotes the transmit power of S, $0 \le \frac{1}{\beta} \le 1$ denotes the self-interference coefficient of FD UAV-based relay, σ denotes the additive white Gaussian noise (AWGN) power. Accordingly, in the mth time slot, the transmission rate from DF UAV-based relay to D can be given as

$$R_D[m] = \log_2\left(1 + \frac{P[m]\gamma}{d_{UD}^2[m]\sigma^2}\right), \tag{5}$$

where $d_{UD}[m] = \sqrt{||\boldsymbol{w}_U[m] - \boldsymbol{w}_D||^2 + H_U^2}$, $P[m]$ denotes the transmit power of UAV. The average transmission rate of the system R is provided as follow

$$R = \frac{1}{M}\sum_{m=1}^{M} \min\{R_S[m], R_D[m]\}. \tag{6}$$

Due to the limitation of transmit power at UAV, the related power constraints of UAV can be given as

$$\frac{1}{M}\sum_{m=1}^{M} P[m] \leq \bar{P}, \tag{7a}$$

$$0 \leq P[m] \leq P_{\max}, m \in M, \tag{7b}$$

where $\bar{P}$ denotes the average transmit power and $P_{\max}$ denotes the peak transmit power of the UAV.

2.3 Problem Formulation

The goal of this paper is to maximize the average rate under the UAV's trajectory constraints (1) and the transmit power constraints (7) with respect to (w.r.t) the variables of UAV's trajectories $\boldsymbol{W}$, the transmit power of the UAV $\boldsymbol{P}$, and the phase shifter of the IRS $\boldsymbol{\Phi}$. Accordingly, the optimal problem can be given as

$$(\mathbf{P}0): \max_{\boldsymbol{W},\boldsymbol{P},\boldsymbol{\Phi}} \quad R \tag{8a}$$

$$\text{s.t.} \quad (1), (7). \tag{8b}$$

($\mathbf{P}0$) is not a convex problem and hard to solve. Therefore, we propose a novel algorithm based on iteration to solve ($\mathbf{P}0$) in the next section.

3 Proposed Algorithm

In this section, we propose an efficient algorithm to solve ($\mathbf{P}0$). We divide ($\mathbf{P}0$) into three sub-problems and use the block coordinate descent (BCD) method to solve one sub-problem at a time. For tractability, we introduce a variable ξ, and ($\mathbf{P}0$) can be transformed as

$$(\mathbf{P}1): \max_{\boldsymbol{W},\boldsymbol{P},\boldsymbol{\Phi}} \quad \xi \tag{9a}$$

$$\text{s.t.} \quad (1), (7), \tag{9b}$$

$$\frac{1}{M}\sum_{m=1}^{M} \min\{R_S[m], R_D[m]\} \geq \xi. \tag{9c}$$

3.1 Optimization of Phase Shifter

First, we optimize the phase shifter $\boldsymbol{\Phi}$ of IRS with given UAV's trajectories $\boldsymbol{W}$ and transmit power $\boldsymbol{P}$. The maximum value of $f[m]$ is N. Therefore, the optimal phase shifter of each elements in the mth time slot is

$$\theta_n^{\text{op}}[m] = \frac{2\pi}{\lambda_0}(n-1)d_0(\phi_{SI} - \phi_{IU}[m]). \tag{10}$$

Taking (10) into $R_S[m]$, $R_S[m]$ can be rewritten as

$$R_S^{\text{op}}[m] = \log_2\left(1 + \frac{P_S\gamma^2 N^2}{d_{IU}^2[m]d_{SI}^2(\frac{P[m]}{\beta} + \sigma^2)}\right). \tag{11}$$

3.2 Optimization of Trajectory

With the optimal phase shifter $\boldsymbol{\Phi}$ and specific transmit power $\boldsymbol{P}$, the sub-problem that optimizes the trajectories of the UAV $\boldsymbol{W}$ is studied in this section. Let $\eta_S[m] = \frac{P_S\gamma^2N^2}{d_{SI}^2(\frac{P[m]}{\beta}+\sigma^2)}$, $\eta_D[m] = \frac{P[m]\gamma}{\sigma^2}$. First, with the variables $\eta_S[m]$ and $\eta_D[m]$, we can respectively rewrite $R_S[m]$ and $R_D[m]$ as

$$G_S[m] = \log_2\left(1 + \frac{\eta_S[m]}{||\boldsymbol{w}_U[m] - \boldsymbol{w}_I||^2 + (H_U - H_I)^2}\right), \tag{12}$$

$$G_D[m] = \log_2\left(1 + \frac{\eta_D[m]}{||\boldsymbol{w}_U[m] - \boldsymbol{w}_D||^2 + H_U^2}\right). \tag{13}$$

Therefore, (**P**1) can be reformulated as (**P**2) can be transformed into

$$(\mathbf{P}2): \max_{\boldsymbol{W}} \ \xi \tag{14a}$$

$$\text{s.t.} \ \ (1), \tag{14b}$$

$$\frac{1}{M}\sum_{m=1}^{M} \min\{G_S[m], G_D[m]\} \geq \xi. \tag{14c}$$

Because the constraint (14c) is non-convex w.r.t $\boldsymbol{W}$, which makes (**P**2) hard to solve. It is worth noting that $G_S[m]$ is convex w.r.t the term $||\boldsymbol{w}_U[m] - \boldsymbol{w}_I||^2$. Consequently, by applying the first-order Taylor expansion of $G_S[m]$ at $||\boldsymbol{w}_U[m] - \boldsymbol{w}_I||^2$ in the kth iteration, the lower-bound of $G_S[m]$, denoted as $G_S^{\text{lb}}[m]$, can be written as

$$\begin{aligned} G_S[m] \geq & \log_2\left(1 + \frac{\eta_S[m]}{||\boldsymbol{w}_U^k[m] - \boldsymbol{w}_I||^2 + (H_U - H_I)^2}\right) \\ & - \zeta_S^k[m]\left(||\boldsymbol{w}_U[m] - \boldsymbol{w}_I||^2 - ||\boldsymbol{w}_U^k[m] - \boldsymbol{w}_I||^2\right) \triangleq G_S^{\text{lb}}[m], \end{aligned} \tag{15}$$

where

$$\zeta_S^k[m] = \frac{\eta_S[m]}{\ln 2(\eta_S[m] + ||\boldsymbol{w}_U^k[m] - \boldsymbol{w}_I||^2 + (H_U - H_I)^2)(||\boldsymbol{w}_U^k[m] - \boldsymbol{w}_I||^2 + (H_U - H_I)^2)}. \tag{16}$$

In the same way, in the kth iteration, the lower-bound of $R_D[m]$, denoted as $G_D^{\mathrm{lb}}[m]$, can be obtained as

$$G_D[m] \geq \log_2\left(1+\frac{\eta_D[m]}{||\boldsymbol{w}_U^k[m]-\boldsymbol{w}_D||^2+H_U^2}\right) \\ -\zeta_D^k[m]\left(||\boldsymbol{w}_U[m]-\boldsymbol{w}_D||^2-||\boldsymbol{w}_U^k[m]-\boldsymbol{w}_D||^2\right) \triangleq G_D^{\mathrm{lb}}[m], \tag{17}$$

where

$$\zeta_D^k[m] = \frac{\eta_D[m]}{\ln 2(\eta_D[m]+||\boldsymbol{w}_U^k[m]-\boldsymbol{w}_D||^2+H_U^2)(||\boldsymbol{w}_U^k[m]-\boldsymbol{w}_D||^2+H_U^2)}. \tag{18}$$

$G_S^{\mathrm{lb}}[m]$ and $G_D^{\mathrm{lb}}[m]$ are concave w.r.t $\boldsymbol{W}$. Taking $G_S^{\mathrm{lb}}[m]$ and $G_D^{\mathrm{lb}}[m]$ into (**P2**) and rewrite it into a convex problem as

$$(\mathbf{P3}): \max_{\boldsymbol{W}} \ \xi \tag{19a}$$

$$\text{s.t.} \ (1), \tag{19b}$$

$$\frac{1}{M}\sum_{m=1}^{M}\min\{G_S^{\mathrm{lb}}[m], G_D^{\mathrm{lb}}[m]\} \geq \xi, \tag{19c}$$

which can be solved easily.

3.3 Optimization of Transmit Power

By fixing the phase shifter of IRS $\boldsymbol{\Phi}$ and the UAV's trajectories $\boldsymbol{W}$, the UAV's transmit power $\boldsymbol{P}$ can be optimized in this section. For tractability, we define

$$\tau_S[m] = \frac{P_S\gamma^2N^2}{d_{SI}^2(||\boldsymbol{w}_U[m]-\boldsymbol{w}_I||^2+(H_U-H_I)^2)}, \tag{20}$$

$$\tau_D[m] = \frac{\gamma}{\sigma^2(||\boldsymbol{w}_U[m]-\boldsymbol{w}_D||^2+H_U^2)}. \tag{21}$$

Based on (20) and (21), we can respectively rewrite $R_S[m]$ and $R_D[m]$ as

$$L_S[m] = \log_2\left(1+\frac{\tau_S[m]}{\frac{P[m]}{\beta}+\sigma^2}\right), \tag{22}$$

$$L_D[m] = \log\left(1+\tau_D[m]P[m]\right). \tag{23}$$

Therefore, the subproblem of optimizing the transmit power at UAV can be expressed as

$$(\mathbf{P4}): \max_{\boldsymbol{P}} \ \xi \tag{24a}$$

$$\text{s.t.} \ (7), \tag{24b}$$

$$\frac{1}{M}\sum_{m=1}^{M}\min\{L_S[m], L_D[m]\} \geq \xi. \tag{24c}$$

$L_S[m]$ in constraints (24c) is convex w.r.t $\boldsymbol{P}$, which makes (**P**4) hard to solve. Like the operation in the optimization of the UAV's trajectories, an approximate optimal solution can be obtained based on the successive convex optimization (SCA) method. We use the first-order Taylor expansion of $L_S[m]$ at $P^k[m]$ in the kth iteration. $L_S[m]$ can be given as

$$L_S[m] \geq \log_2\left(1 + \frac{\tau_S[m]}{\frac{P^k[m]}{\beta} + \sigma^2}\right) - \mu_S^k[m]\left(P[m] - P^k[m]\right) \triangleq L_S^{\rm lb}[m], \quad (25)$$

where

$$\mu_S^k[m] = \frac{\tau_S[m]/\beta}{\ln 2(\tau_S[m] + P^k[m]/\beta + \sigma^2)(P^k[m]/\beta + \sigma^2)}. \quad (26)$$

Therefore, (**P**4) can be rewritten into a convex problem as

$$(\mathbf{P}5): \max_{\boldsymbol{P}} \ \xi \quad (27a)$$

$$\text{s.t.} \ \ (7), \quad (27b)$$

$$\frac{1}{M}\sum_{m=1}^{M} \min\{L_S^{\rm lb}[m], L_D[m]\} \geq \xi, \quad (27c)$$

3.4 Overall Algorithm

In summary, the optimal transmission rate R^* is obtained by the value of the IRS phase shifter $\boldsymbol{\Phi}^*$, the UAV's trajectories $\boldsymbol{W}^*$ and the UAV's transmit power $\boldsymbol{P}^*$ by Algorithm 1.

4 Simulation Result

In this section, numerical results are provided to evaluate the performance of the proposed algorithm in the IRS-assisted FD UAV-based mobile relaying communication. The coordinates of S, D, IRS, w_0, w_f are $[0, -200]$ m, $[-150, -50]$

Algorithm 1. Overall Algorithm

Input: $\boldsymbol{W}^{(0)}$, $\boldsymbol{P}^{(0)}$, $\boldsymbol{\Phi}^{(0)}$, ϵ, $k = 0$
Output: $\boldsymbol{W}^*$, $\boldsymbol{P}^*$, $\boldsymbol{\Phi}^*$, R^*
1: **do**
2: $k \leftarrow k + 1$
3: obtain $\boldsymbol{W}^{(k)}$ with given $\boldsymbol{P}^{(k-1)}$ by (**P**3)
4: obtain $\boldsymbol{\Phi}^{(k)}$ with given $\boldsymbol{W}^{(k)}$ by (10)
5: obtain $\boldsymbol{P}^{(k)}$ with given $\boldsymbol{W}^{(k)}$ by (**P**5)
6: obtain $R^{(k)}$ with given $\boldsymbol{W}^{(k)}$, $\boldsymbol{\Phi}^{(k)}$ and $\boldsymbol{P}^{(k)}$ by (6)
7: **while** $\frac{R^{(k)} - R^{(k-1)}}{R^{(k)}} > \epsilon$
8: $\boldsymbol{W}^* \leftarrow \boldsymbol{W}^k$
9: $\boldsymbol{\Phi}^* \leftarrow \boldsymbol{\Phi}^k$
10: $\boldsymbol{P}^* \leftarrow \boldsymbol{P}^k$
11: $R^* \leftarrow R^k$

m, [0, −100] m, [−200, 200] m, [200, 200] m, respectively. The height of IRS and UAV are $H_I = 10$ m and $H_U = 30$ m. The remaining parameters inspired by [19] are summarized in Table 1.

Table 1. Simulation parameters

Parameters	Example values
Maximum speed of the UAV, $V_{\max}$	20 m/s
Time slot length, τ	1 s
Noise variance, σ^2	−110 dBm
Reference channel gain at 1 m, γ	−20 dB
Self-interference cancellation coefficient, β	100 dB
Antenna separation d_0	$\lambda_0/2$
Average transmission power of UAV, $\bar{P}$	0.1 W
Peak transmission power of UAV, $P_{\max}$	0.4 W
Transmission power of S, P_S	0.1 W
Tolerance, ϵ	10^{-4}

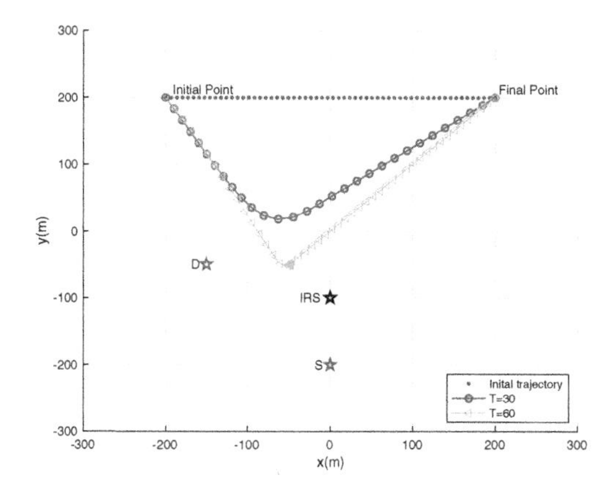

Fig. 2. Trajectories of the UAV with different T.

Figure 2 shows the trajectories of UAV in different flight time T with $N = 256$. It is obvious that for different flight time, the UAV's trajectory is obviously different. When T is equal to 60 s, the UAV first flies to the optimal position and hovers for a period of time before flying to the final point. In terms of $T = 30$ s, the UAV does not fly to a similar optimal position hovering due to the limitation of time. It flies as close to D and IRS as possible and then flies to the final point. It is worth noting that the UAV tends to come close to the midpoint of D and IRS because UAV can not communicate with S directly in our scenario.

Figure 3 illustrates the average rates with different N in the iteration. As the number of iterations increases, the average rates quickly stabilize at stable values. Furthermore, with the number of IRS elements N increasing, the average rate increases significantly. With N increasing from 64 to 256, almost 25% of the transmission rate increases. It shows that the assistance of the IRS not only establish a communication link between S and UAV, but also can effectively improve the transmission performance of the system by appropriately increasing the number of IRS elements.

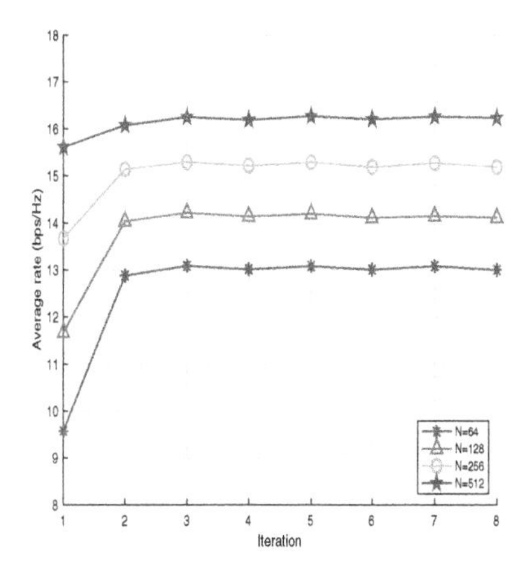

Fig. 3. Average rate with different N of the IRS versus Iteration

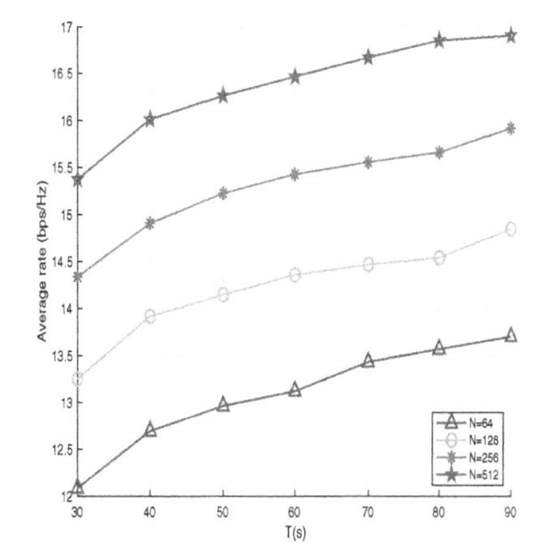

Fig. 4. Average rate with different N of the IRS versus T

Figure 4 compares the average rate of the different N under several values of flying time T. It is clear that as T increasing, the average rates increase significantly. It is because the UAV has more time to hover around the optimal point of a greater T. Moreover, the influence of the number of elements N also has a significant impact on system performance. It is straightforward that the average rate increases with the increase of N, which inspires us to use an IRS-assisted system to enhance the transmission performance of the communication system.

5 Conclusion

In this paper, we investigated an IRS-assisted DF UAV-based relay system. We jointly optimized the UAV's trajectory, the phase shifter of IRS, and transmit power of the UAV to maximize the average rate of the system. The non-convexity of the formulated problem results in solving the problem directly intractable. Therefore, we proposed an iterative algorithm on the basis of block coordinate descent and the SCA method. Numerical results show that the proposed scheme can significantly improve the transmission rate, which can solve the transmission problem in crowded cities in the future. Besides, the 3D trajectory design is left as the future work.

Acknowledgments. This work is supported in part by the Shanghai Aerospace Science and Technology (SAST) Innovation Fund under Grant SAST2019-092.

References

1. Boccardi, F., Heath, R.W., Lozano, A., Marzetta, T.L., Popovski, P.: Five disruptive technology directions for 5G. IEEE Commun. Mag. **52**(2), 74–80 (2014). https://doi.org/10.1109/MCOM.2014.6736746
2. Wu, Q., Zhang, R.: Intelligent reflecting surface enhanced wireless network via joint active and passive beamforming. IEEE Trans. Wirel. Commun. **18**(11), 5394–5409 (2019). https://doi.org/10.1109/TWC.2019.2936025
3. Basar, E., Di Renzo, M., De Rosny, J., Debbah, M., Alouini, M., Zhang, R.: Wireless communications through reconfigurable intelligent surfaces. IEEE Access **7**, 116753–116773 (2019). https://doi.org/10.1109/ACCESS.2019.2935192
4. Gong, S., et al.: Toward smart wireless communications via intelligent reflecting surfaces: a contemporary survey. IEEE Commun. Surv. Tutor. **22**(4), 2283–2314 (2020). https://doi.org/10.1109/COMST.2020.3004197
5. Abeywickrama, S., Zhang, R., Wu, Q., Yuen, C.: Intelligent reflecting surface: practical phase shift model and beamforming optimization. IEEE Trans. Commun. **68**(9), 5849–5863 (2020). https://doi.org/10.1109/TCOMM.2020.3001125
6. Wu, Q., Zhang, S., Zheng, B., You, C., Zhang, R.: Intelligent reflecting surface aided wireless communications: a tutorial. IEEE Trans. Commun. **69**, 3313–3351 (2021). https://doi.org/10.1109/TCOMM.2021.3051897
7. Zuo, J., Liu, Y., Qin, Z., Al-Dhahir, N.: Resource allocation in intelligent reflecting surface assisted NOMA systems. IEEE Trans. Commun. **68**(11), 7170–7183 (2020). https://doi.org/10.1109/TCOMM.2020.3016742
8. Nadeem, Q.U.A., Kammoun, A., Chaaban, A., Debbah, M., Alouini, M.S.: Asymptotic max-min SINR analysis of reconfigurable intelligent surface assisted miso systems. IEEE Trans. Wirel. Commun. **19**(12), 7748–7764 (2020). https://doi.org/10.1109/TWC.2020.2986438
9. Yang, L., Yang, Y., Hasna, M.O., Alouini, M.S.: Coverage, probability of SNR gain, and DOR analysis of RIS-aided communication systems. IEEE Wirel. Commun. Lett. **9**(8), 1268–1272 (2020). https://doi.org/10.1109/LWC.2020.2987798
10. Gupta, L., Jain, R., Vaszkun, G.: Survey of important issues in UAV communication networks. IEEE Commun. Surv. Tutor. **18**(2), 1123–1152 (2016). https://doi.org/10.1109/COMST.2015.2495297
11. Zhang, C., Fu, W.: Optimal model for patrols of UAVs in power grid under time constraints. Int. J. Perform. Eng. **17**(1), 103 (2021)
12. Fotouhi, A., et al.: Survey on UAV cellular communications: practical aspects, standardization advancements, regulation, and security challenges. IEEE Commun. Surv. Tutor. **21**(4), 3417–3442 (2019). https://doi.org/10.1109/COMST.2019.2906228
13. Wu, L., Wang, W.: Resource allocation optimization of UAVs-enabled air-ground collaborative emergency network in disaster area. Int. J. Perform. Eng. **15**(8), 2133 (2019). https://doi.org/10.23940/ijpe.19.08.p13.21332144

14. Choi, D.H., Kim, S.H., Sung, D.K.: Energy-efficient maneuvering and communication of a single UAV-based relay. IEEE Trans. Aerosp. Electron. Syst. **50**(3), 2320–2327 (2014). https://doi.org/10.1109/TAES.2013.130074
15. Zhang, S., Zhang, H., He, Q., Bian, K., Song, L.: Joint trajectory and power optimization for UAV relay networks. IEEE Commun. Lett. **22**(1), 161–164 (2018). https://doi.org/10.1109/LCOMM.2017.2763135
16. Yang, L., Chen, J., Hasna, M.O., Yang, H.: Outage performance of UAV-assisted relaying systems with RF energy harvesting. IEEE Commun. Lett. **22**(12), 2471–2474 (2018). https://doi.org/10.1109/LCOMM.2018.2876869
17. Yang, L., Meng, F., Zhang, J., Hasna, M.O., Renzo, M.D.: On the performance of RIS-assisted dual-hop UAV communication systems. IEEE Trans. Veh. Technol. **69**(9), 10385–10390 (2020). https://doi.org/10.1109/TVT.2020.3004598
18. Zhang, S., Zhang, H., Di, B., Song, L.: Joint trajectory and power optimization for UAV sensing over cellular networks. IEEE Commun. Lett. **22**(11), 2382–2385 (2018). https://doi.org/10.1109/LCOMM.2018.2868075
19. Li, S., Duo, B., Yuan, X., Liang, Y., Di Renzo, M.: Reconfigurable intelligent surface assisted UAV communication: joint trajectory design and passive beamforming. IEEE Wirel. Commun. Lett. **9**(5), 716–720 (2020)

Edge Cache Resource Allocation Strategy with Incentive Mechanism

Shengli Zhang[1], Xiaoguang Xu[2], Jiaxin Wei[2], Guangxu Zhou[2(✉)], Yue Guo[2], Junfeng Hou[2], and Zhunfeng Li[2]

[1] China Tobacco Henan Industrial Co., Ltd., Zhengzhou 450001, Henan, China
[2] Xuchang Cigarette Factory, China Tobacco Henan Industrial Co., Ltd., Xuchang 461000, Henan, China

Abstract. Edge computing technology can effectively alleviate the problems of network congestion and poor user experience quality in current mobile social networks. In this paper, an edge cache resource allocation strategy for mobile social networks based on incentive mechanism is presented. The incentive cache model is modeled as a Stackelberg game, and the utility functions of the macro base station and the social interest group are constructed respectively. The optimal unit cache resource price strategy for the macro base station and the optimal cache resource demand strategy for the social interest groups are obtained by using distributed iterative algorithm. The results show that the proposed mechanism has lower average content transmission delay and higher user experience quality.

Keywords: Mobile social network · Edge computing · Incentive mechanism · Cache allocation

1 Research Background

The introduction of edge cache [1, 2] technology into mobile social network [3] can effectively alleviate the network congestion and poor user experience quality in current mobile social networks [4]. Given the above problems, domestic and foreign researchers have carried out extensive and in-depth research. To improve the utilization rate of resources and the network capacity, literature [5] proposed a caching scheme with social awareness and payment incentives. Data caching was modeled as a socially aware payment game to motivate end users to cache content for other end users, and the cost function was constructed using social relation to minimize the cost of network access to content. To encourage users to share cached resources and improve the social welfare of cellular networks, literature [6] proposed a D2D caching strategy with joint incentive scheme, categorizing users according to their preference. The base station gave users incentives based on the size of cache resources shared by users, and contract theory was used to build optimization problems to motivate users to share cache resources and maximize base station utility. However, the literature [5] and [6] only considered motivating users to participate in content caching to allocate their cache resources, but not considering to

J. Xiong et al. (Eds.): MobiMedia 2021, LNICST 394, pp. 224–237, 2021.
https://doi.org/10.1007/978-3-030-89814-4_17

encourage selfish base stations to participate in content caching. Literature [7] proposed a cache resource allocation strategy to motivate selfish base stations to participate in content caching. The content server used the cache resources of the base station to provide services to users. The base station makes profits by renting cache resources to the content server, and the interaction between the base station and multiple content service providers is modeled as a Stackelberg game to maximize the utility of the base station and the content server ultimately. Literature [8] put forward an edge cache incentive mechanism based on contract theory. The Internet service providers rented out their resources to content service providers for profit, and the content server cached its most popular content in the rented base station to provide better service to users. Considering the different demands of users for service quality, the incentive mechanism is designed based on contract theory to maximize the utility of Internet service providers. The literature [7] and [8] only considered motivating selfish base stations to participate in content caching, but not considering motivating selfish users to participate in content caching. At present, there are few documents that simultaneously encourage selfish users and selfish base stations to participate in content caching.

Therefore, this paper proposes a mobile social network edge cache resource allocation strategy based on incentive mechanism. The macro base station gives corresponding incentives based on the contribution of social interest groups to cache process. Social interest groups pay incentives to base stations according to the number of cache resources given by base stations. The interaction between the macro base station and the social interest groups is modeled as a Stackelberg game problem, and the utility functions of macro base station and social interest groups are constructed based on unit cache resource price, cache resource demand, the contribution of social interest group. Using distributed iteration method, the optimal unit cache resource price of macro base station and the optimal cache resource demand of social interest groups are obtained.

2 System Model

2.1 Network Model

Consider a cellular network scenario in which all small base stations and end users in a macro base station have caching capabilities. The network is mainly composed of content servers, macro base stations, small base stations, and end users, as shown in Fig. 1.

Content Server: stores all the content that the user needs, but is far away from the end user. When the end user acquires the content from the content server, it will produce a long delay, which will lead to poor user experience quality.

Small base station: as an edge cache node, under the control of the macro base station, the content is cached for users who purchased its cache resources. Assuming that the coverage radius of the micro base station is r_s, it follows a homogeneous Poisson process with an intensity of λ_s in space [9], and the intensity λ_s represents the number of micro base stations per unit area. All micro base stations within the coverage of the macro base station have the same storage capacity, and the amount of content cached for each social interest group is at most C_{sn}.

End user: both as edge cache nodes and as content requestors. Assuming that the D2D communication radius of the end user is r_u, it follows a homogeneous Poisson process with an intensity of λ_u in space, the intensity of λ_u represents the number of end users in a unit area, and the maximum number of content cached for each social interest group is C_n. $N = \{1, \ldots, i, \ldots, N\}$ represents all social interest groups within the coverage of the macro base station.

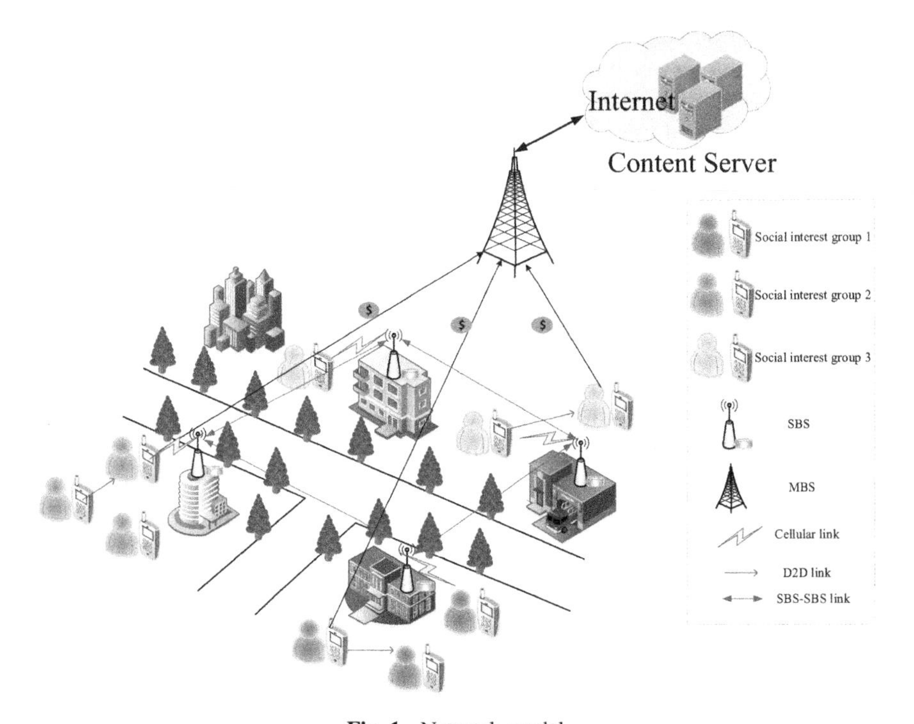

Fig. 1. Network model

In this cellular scenario, users can obtain the content in one of the following ways:

1. If the requested content is found in an adjacent end user, the user will first retrieve the requested content from the adjacent end user through the D2D link. And assume that the delay in retrieving content from adjacent end users is d_{D2D}.
2. If the content requested by the user cannot be found in an adjacent end user, the user obtains the requested content from the connected small base station through a cellular link. It is assumed that the delay of obtaining content from the micro-base station is d_{SBS}.
3. If the content requested by the user cannot be found in the connected small base station, the user obtains the requested content from the adjacent small base station through the connected small base station. It is assumed that the delay of obtaining content from the adjacent small base station is also d_{SBS}.
4. If the user cannot obtain the requested content through the above three methods, the user obtains the requested content from the content server through the connected

small base station. And assume that the delay in retrieving content from the content server is d.

2.2 Incentive Mechanism Model

When a large number of users at the same time obtain content from the base station, easy to cause network congestion, to alleviate the network congestion, the base station will encourage users to participate in content caching and sharing, according to the user to ease network congestion of interest groups in social contribution to give different incentives, improving user participation in mobile social network content caching and share the enthusiasm. The contribution of a social interest group can be measured by the number of shared cache resources provided by users in the group and the activity of users. The more shared cache resources, the greater the amount of cached content, the higher the activity of users, the higher the probability of meeting other users, and the higher the probability of successful content sharing. If Acer stands within range of the number of users with Shared cache resources for Q_u, according to the Shared cache resources in social interest group n provide Q_u and all users in the group's average active $\overline{V_n}$, determine the contribution of social interest group n, and do the normalized processing, the resulting social interest group contribution dg_n of n are as follows:

$$dg_n = \frac{2}{\pi} \arctan(\frac{Q_n}{Q_u} + \overline{V}_n) \tag{1}$$

2.3 Encounter Probability and Content Request Probability Model

The cache location includes the micro-base station and the end user within the coverage range of the Acer Station. The cache resources of all micro-base stations and the end user are managed by the Acer Station. The total cache resources managed by the Acer Station is Q, where $Q = Q_s + Q_u$. Each social interest groups based on their needs, and the utility macro base station stood to determine their cache resource demand, let $s = \{s_1, s_2, \cdots, s_n\}$ said Acer stand all social interest groups within a range of the cache capacity requirements of combination, where $0 \leq s_n \leq Q$ and $s_n = s_{sn} + s_{un}$, $s_{sn} = \alpha s_n$ said cache capacity, from the small base station, $s_{un} = (1-\alpha)s_n$ said the cache capacity from the adjacent user equipment. Assuming that the social interest group rents the small base station in the Acer station cell with the same probability, the micro-base station rented by the social interest group n can be modeled as a sparse homogeneous Poisson process of intensity $\lambda_s s_{sn}$, then the probability of users in the social interest group n within the coverage range of their rented small base station can be expressed as:

$$p_{ns} = 1 - \exp(-\lambda_s s_{sn} \pi r_s^2) \tag{2}$$

Similarly, the end users rented by the social interest group n can be modeled as a sparse homogeneous Poisson process of intensity $\lambda_u s_{un}$, and the probability of users in the social interest group n meeting their leased end users can be expressed as

$$p_{ns} = 1 - \exp(-\lambda_s s_{sn} \pi r_s^2) \tag{3}$$

Within different social interest groups, users will request different kinds of content and the popularity of their content will vary. Assumptions in the social interest group n, the user sets the contents of the request $F = \{F_1, F_2 \cdots F_m\}$, and assuming that each content of the same size. The content sets requested by users in the social interest group are arranged in descending order of popularity. The higher the popularity of the content at the top, that is, the content F_1 has the highest popularity, while content F_m has the lowest popularity. Users will request content F_f in the content set independently with probability, and it is assumed that $p_{n,f}$ follows the Zipf distribution with parameter γ [10]. That is:

$$p_{n,f} = \frac{f^{-\gamma}}{\sum_{i=1}^{m} i^{-\gamma}}, f = 1, 2, \cdots, m. \tag{4}$$

Where γ represents the content popularity parameter in the social interest group n.

3 Cache Resource Allocation Policy

In this part, we first model the cache resource allocation problem of the macro base station as a Stackelberg game model, then use the distributed iteration method to obtain the optimal unit cache resource price of the macro base station and the optimal cache resource demand of the social interest group, and finally get the edge cache resource allocation algorithm based on the incentive mechanism.

3.1 Stackelberg Game Model

The interaction between the macro base station and the social interest group during the allocation of the macro base station cache resources conforms to the relationship commonly used between two conflicting entities in game theory. Specifically, in the mobile social network, the interaction relationship between the macro base station and the social interest group conforms to the hierarchical relationship in the Stackelberg game, and there are two types of participants in this model: leaders and followers [11].

Leader: In this game model, the macro base station is regarded as the leader, and the leader is responsible for the pricing strategy of the cache resources it manages. Leaders can satisfy the needs of users in social interest groups by providing cache resources to social interest groups. At the same time, leaders can influence the demand of cache resources of social interest groups through pricing strategies to maximize their own utility.

Followers: Treat each social interest group as a follower. Followers will adjust their caching resource demand strategy according to the price of caching resource set by the leader and the total utility obtained by users in the social interest group.

Policy: For macro base station, the policy is to publish the price p per unit cache resource. For the social interest group, the strategy is to formulate the cache resource demand strategy $s = \{s_1, s_2, \cdots, s_n\}$, where s_n represents the size of the cache space purchased by the interest community n at the macro base station, and it satisfies $0 \leq s_n \leq Q$. The size of cache space purchased by all social interest groups on the macro base station should meet $\sum_{n=1}^{N} s_n \leq Q$.

The game between the macro base station and the Social Interest Group is divided into two stages. In the first stage, the macro base station first develops the unit cache resource price strategy p, and publishes this price strategy to all social interest groups. The social interest groups determine their own cache resource demand strategy s according to the received price strategy. In the second stage, after the macro base station learns the caching resource demand strategy of the social interest group, it readjusts the pricing strategy to maximize its utility.

3.2 The Utility Function

For the macro base station and the Social Interest Group, the utility function reflects the players' satisfaction with their choice of strategy. The proof of the existence of Nash equilibrium points can show that the non-cooperative game between social interest groups can reach Nash equilibrium, and any social interest groups will not gain if they want to change their own strategies.

1. Social interest group utility function

Social interest group utility function according to the definition of social interest groups, social interest groups in the same user will be asked to similar content and is willing to share the social interest groups to buy cache resource cost, namely the social interest group of the utility function for the effectiveness of its end users in the group, the sum of its utility function through the benefits and costs of two parts, the utility function U_n of social interest group n can be expressed as:

$$U_n = R_n(s_n) - C_n(s_n) \tag{5}$$

Where $R_n(s_n)$ and $C_n(s_n)$ represent the benefits and expenses of the social interest group respectively.

The content access delay of social interest group n is related to the size of leased cache resources s_n, which is an increase function of s_n. The incentive that social interest group n receives from the base station is related to its contribution degree dg_n. The higher the contribution degree, the more incentive it receives. Then, the income function $R_n(s_n)$ of social interest group n can be expressed as:

$$\begin{aligned} R_n(s_n) = {} & \eta_1 \cdot (d - d_{SBS}) \cdot p_{ns} \cdot \sum\nolimits_{f=1}^{C_{sn}} p_{n,f} + dg_n p_1 \\ & + \eta_1 \cdot (d - d_{D2D}) \cdot p_{nu} \cdot \sum\nolimits_{f=C_{sn}+1}^{C_n} p_{n,f} \end{aligned} \tag{6}$$

Where η_1 represents the revenue factor of delay saving, p_1 represents the unit contribution cost of the base station to the social interest group, $d - d_{SBS}$ represents the delay saving of acquiring content from the small base station, and $d - d_{D2D}$ represents the delay saving of acquiring content from adjacent end users.

For the social interest group n, its cost is the cost that the social interest group uses to motivate the base station to give cache resources, then the cost function $C_n(s_n)$ of the social interest group n can be expressed as:

$$C_n(s_n) = ps_n \tag{7}$$

Based on the benefit function and cost function of social interest group, the utility function U_n of social interest group n is finally obtained as follows:

$$U_n = \eta_1 \cdot (d - d_{SBS}) \cdot p_{ns} \cdot \sum_{f=1}^{C_{sn}} p_{n,f} + dg_n p_1 - ps_n + \eta_1 \cdot (d - d_{D2D}) \cdot p_{nu} \cdot \sum_{f=C_{sn}+1}^{C_n} p_{n,f} \quad (8)$$

According to the price strategy p of the current macro base station, each social interest group n adjusts its own cache resource demand strategy s_n and finally achieves the optimal cache resource demand strategy s_n^*. By maximizing its own utility function, the optimal cache resource demand strategy s_n^* can be obtained as follows:

$$s_n^* = \arg\ \max U_n(p^*, s_n, s_{-n}^*) \quad (9)$$

Where p^* represents the optimal price strategy selected by the macro base station, s_{-n}^* represents the optimal cache resource demand strategy selected by other social interest groups.

2. The utility function of macro base station

The macro base station needs to build and maintain the caching capability of the edge nodes, so the caching service provided for the social interest group is not free, and the cost is compensated by the incentives given by the social interest group. The utility function of the macro base station can be expressed as:

$$U_0 = R(p) - C(s) \quad (10)$$

Where $R(p)$ and $C(s)$ respectively represent the revenue and expense of the macro base station.

The revenue of the macro base station is the incentive fee given to it by the social interest group to obtain cache resources, which is an increasing function of the unit cache resource price p, then the revenue function $R(p)$ of the macro base station can be expressed as:

$$R(p) = \eta \sum_{n=1}^{N} ps_n \quad (11)$$

The cost of a macro base station is the cost of maintaining the cache capacity of the small base station and the cost of motivating social interest group. The cost function $C(s)$ of the macro base station can be expressed as:

$$C(s) = \begin{cases} \frac{1}{Q_s - \beta \sum_{i=1}^{n} s_{si}} + \sum_{i=1}^{n} dg_i p_1, & Q_s \geq \sum_{i=1}^{n} s_{si} \\ \infty, & Q_s < \sum_{i=1}^{n} s_{si} \end{cases} \quad (12)$$

Where Q_s represents the buffer capacity of all small base stations within the coverage of the macro base station, and β represents the unit cost of maintaining the buffer capacity of the small base station.

Based on the revenue and cost functions of the macro base station, the utility function U_0 of the macro base station can be obtained as:

$$U_0 = \eta \sum_{n=1}^{N} ps_n - \frac{1}{Q_s - \beta \sum_{i=1}^{n} s_i} - \sum_{i=1}^{n} dg_i p_1 \tag{13}$$

The macro base station needs to formulate an appropriate unit buffer resource price p^* to ensure its higher revenue. By maximizing the utility function U_0, the optimal unit cache resource price p^* can be obtained as:

$$p^* = \arg\ \max U_0(p, s^*) \tag{14}$$

Where s^* represents the optimal demand strategy for cache resources in the interest community.

3.3 Problem Solving and Resource Allocation Algorithm

This section will solve the Stackelberg game problem and give an edge cache resource allocation algorithm based on the incentive mechanism. Using the distributed iterative method, the strategy is continuously adjusted through multiple iterations to obtain the Nash equilibrium solution of the Stackelberg game.

Suppose that at time t, the macro base station announces the unit cache resource price $p(t)$ to all social interest groups. After obtaining the price strategy of the macro base station, the social interest group will adjust its cache resource demand at the macro base station according to its own needs and utility. Make it satisfy the Nash equilibrium solution. The change rate of the cache resource demand of the social interest group is proportional to the first-order partial derivative of its utility function. The social interest group needs to go through iterations in the non-cooperative game process, and change its cache resource demand strategy many times to achieve Nash equilibrium. In the iteration cycle $\Delta\tau$, the cache resource demand strategy iteration equation of social interest group n can be expressed as:

$$U_n(\tau + 1) = U_n(\tau) + \mu \dot{s}_n \tag{15}$$

Where $\mu > 0$ represents the iterative step size of the social interest group cache resource demand strategy, and $\dot{s}_n$ represents the first-order partial derivative of the utility function, as follows:

$$\dot{s}_n = \frac{\partial U_n(p, s_n, s_{-n})}{\partial s_n} \tag{16}$$

Since the utility function of the social interest group is a concave function, after multiple iterations, it can be ensured that the social interest group can converge to the Nash equilibrium point of the game.

After reaching the Nash equilibrium between the social interest groups, the macro base station adjusts its own unit cache resource price strategy to maximize its utility according to the cache resource demand strategy of the social interest group. The price iteration formula is as follows:

$$p(t+1) = p(t) + \theta \frac{\partial U_0(p(t), s(t))}{\partial p(t)} \tag{17}$$

Where θ represents the iteration step size of the unit cache resource price strategy of the macro base station.

The first-order partial derivative of the utility function of the macro base station can be calculated by a small change parameter ε, the formula is as follows:

$$\frac{\partial U_0(p(t), s(t))}{\partial p(t)} \approx \frac{U_0(\cdots, p(t)+\varepsilon, \cdots) - U_0(\cdots, p(t)-\varepsilon, \cdots)}{2\varepsilon} \tag{18}$$

Before the cache resource demand strategy of the social interest group reaches the Nash equilibrium, the unit cache resource price of the macro base station must remain unchanged until all the social interest groups reach the optimal cache resource demand strategy. The time used during this period is the macro base station's iteration cycle Δt, therefore, the iteration cycle Δt of the macro base station includes multiple iteration cycles $\Delta \tau$ of the social interest group. After many iterations, the macro base station and the social interest group will obtain the optimal price strategy p^* and the optimal cache resource demand strategy s_n^* respectively, that is, they both satisfy the Nash equilibrium (p^*, s_n^*). Under this equilibrium, any party of the game participants can not achieve higher revenue if it changes its strategy alone.

Intuitively, The entire iteration process is as follows:

1. Unit cache resource price strategy adjustment of macro base station: The macro base station adjusts its unit cache resource price strategy according to Eq. (17) and Eq. (18) at every moment t, and announces this strategy to the social interest group.
2. Cache resource demand strategy adjustment of social interest group: After receiving the new price strategy, the social interest group adjusts its own cache resource demand strategy according to Eq. (15) and Eq. (16) until the revenue of the social interest group reaches the maximum value. In this situation, all social interest groups have reached the Nash equilibrium.
3. If the revenue of the macro base station reaches the maximum value at this time, the iteration will end. Otherwise, at the next moment $t+1$, the macro base station returns to process 1 to continue to adjust the unit cache resource price strategy according to the cache resource demand strategy of the social interest group.

Based on the unit cache resource price of the macro base station, the cache resource demand of the social interest group, and the contribution of the social interest group, we can obtain the revenue function of the macro base station and the social interest group. By analyzing the Stackelberg game process between the macro base station and the social interest group and using distributed iterative algorithm, we can obtain the optimal price of the macro base station and the optimal cache resource strategy of the social interest group.

4 Experiment

4.1 Parameters

In this section, we will analyze the performance of the proposed SBSUC under the MATLAB platform. The macro base station manages the cache capacity of all social interest groups and small base stations. We assume that the cache capacity managed by the macro base station is $Q = 2$, and the coverage radius r_s of the small base station is 0.5 km. The D2D communication radius r_u between users is 0.1 km. The average activity of the three social interest groups are $\overline{V}_1$=0.4, $\overline{V}_2$=0.6, $\overline{V}_3$=0.8 respectively. We assume that the total number of popular content of each social interest group is the same as $m = 100$. The average transmission delay from the content server to the end user is 98 ms, and the average transmission delay from micro base stations to users is 10ms, and the average transmission delay from users to users is 3 ms. The initial caching resource strategy for each social interest group is 0.05, and the initial unit cache resource price strategy of macro base station is 0.1. The iteration step size μ of the social interest group cache resource demand strategy and the macro base station price strategy are both set to 0.1, and the convergence accuracy is $\varepsilon_1 = 10^{-3}$. We compare SBSUC with IUC [12], which is a resource allocation algorithm for incentivizing users to participate in content caching, and IBSC [13], which is a resource allocation algorithm for incentivizing base stations to participate in content caching.

4.2 Main Results

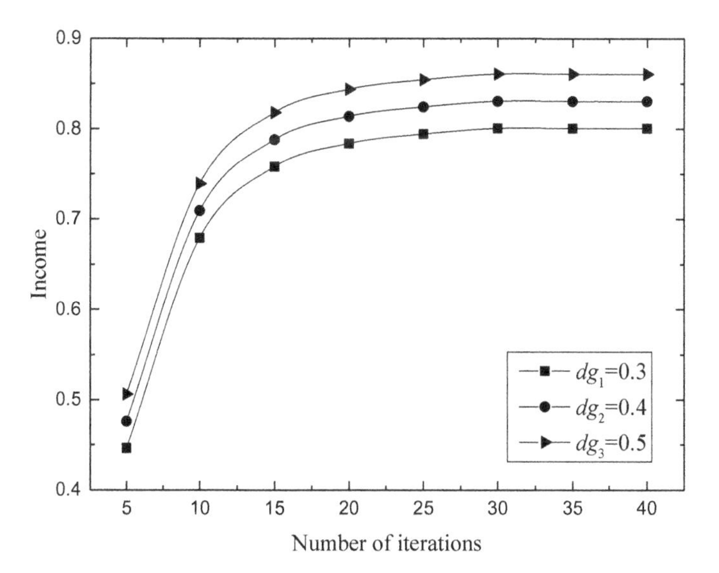

Fig. 2. The revenue in different contributions

Figure 2 shows when the content popularity parameter value of the three social interest groups is $r = 0.6$ and the macro base station is at the best unit cache resource price, as the number of iterations increases, the revenue changes in different contributions dg_n. It can be seen from the figure that as the number of iterations increases, the revenue of the social interest group gradually increases, and finally when the Nash equilibrium between the social interest groups is reached, the revenue tends to be stable. Among them, the revenue of the social interest group $dg_3 = 0.5$ is greater than the social interest group $dg_2 = 0.4$ and greater than the social interest group $dg_1 = 0.3$. This is because the greater the degree of contribution, the more incentives the macro base station gives, and therefore the higher the revenue of the social interest group.

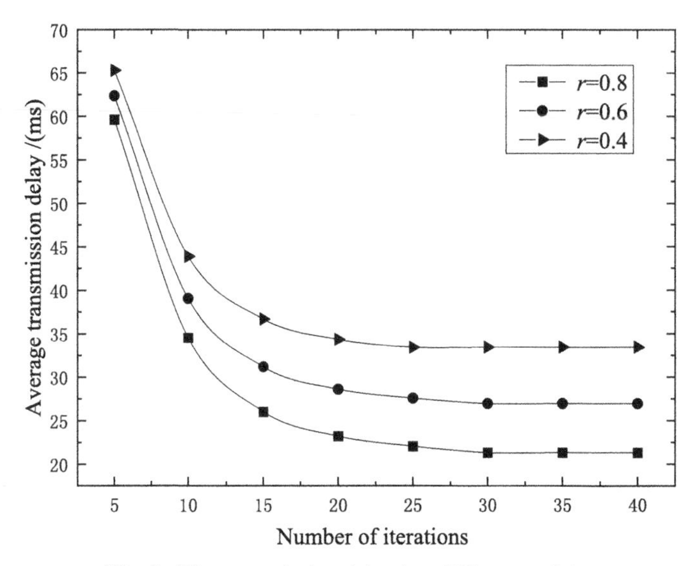

Fig. 3. The transmission delay in a different activity

Figure 3 shows that the contribution of the three social interest groups is the same $dg_n = 0.3$. At the same time, when the macro base station caches the price in the best unit, the average transmission delay of the users in the respective groups in the iterative process of the social interest group is changed. It can be seen from the figure that for social interest groups with different popularity parameter values, as the number of iterations increases, the average transmission delay for users in the group to obtain content gradually decreases and tends to stabilize. Among them, the popularity parameter the value of $r = 0.8$ delay in the social interest groups is less than the $r = 0.6$ delay in the social interest groups is less than $r = 0.4$ the delay in the social interest groups, because the popularity of parameter values r greater, popular content distribution is concentrated, the higher the contents of the cache hit rate, and therefore The lower the average time delay for users to access the content.

Figure 4 shows the change curve of the utility of the three social interest groups in the iterative process. It can be seen from the figure that as the number of iterations increases, the utility functions of the three social interest groups gradually increase and become stable. Among them, the utility of social interest group 3 is greater than that of social interest group 2 than that of social interest group 1, because of the content

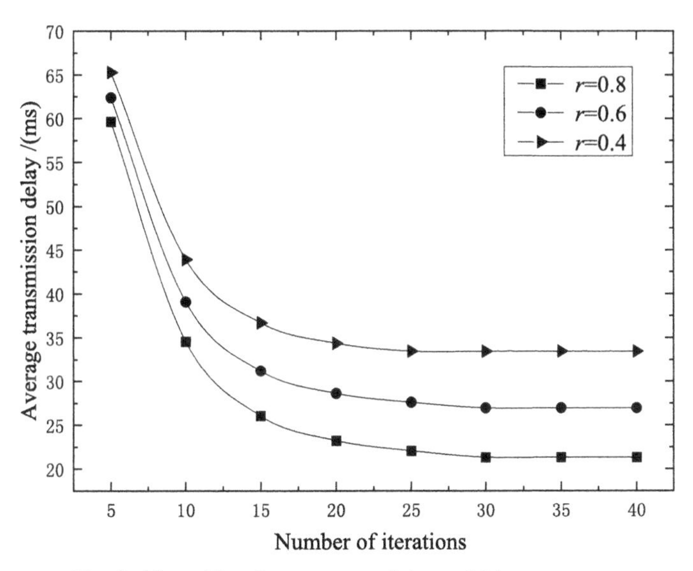

Fig. 4. The utility change curve of the social interest group

popularity parameter $r3 > r2 > r1$. The greater the popularity parameter value, the more concentrated the content distribution accessed by users in the group, the greater the buffer hit rate, the lower the average transmission delay, and the greater the utility. At the same time, the contribution degree $dg_3 > dg_2 > dg_1$, the greater the contribution degree of the social interest group, the more incentive the macro base station gives, so the greater the utility. Ultimately makes utility $SI3 > SI2 > SI1$.

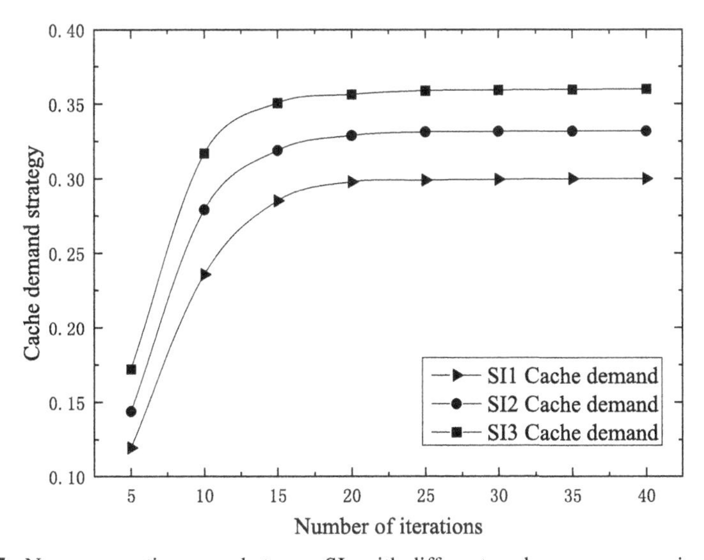

Fig. 5. Non-cooperative game between SIs with different cache resource requirements

Figure 5 shows the macro base station cache resource unit published price, with the increase in the number of iterations, 3 cooperatives pay non-cooperative game competition process between interest groups. As can be seen from the figure that increases as the number of iterations, 3 cache resources demand a social interest groups gradually increased, when the number of iterations reaches 20 or so times, 3 cache resources demand a stable social interest groups, that is, the non-cooperative game between social interest groups reaches Nash equilibrium. Where the social interest groups 3 cache resource demand is the greatest, because of its content popularity parameters biggest, popular content distribution is relatively more concentrated, higher cache hit rate, and therefore hire more cache resources can get more revenue.

Figure 6 shows the comparison of the IBSUC algorithm proposed in this chapter with the other two algorithms in terms of average transmission delay performance as the number of iterations increases. It can be seen from the figure that as the number of iterations increases, the average transmission delay of the three algorithms gradually decreases, and finally stabilizes. The performance of the proposed algorithm is better than that of the other two algorithms. This is because the proposed algorithm considers that the base station and the user are encouraged to participate in content caching at the same time. The user can obtain the requested content from neighboring users or the base station, which can reduce the number of users to the base station. And the average transmission delay from the base station to the content server to obtain the content. Compared with the IUC algorithm that only encourages users to participate in content caching and the IBSC algorithm that only encourages base stations to participate in content caching, the proposed algorithm can obtain more buffer capacity and therefore has a lower average transmission delay.

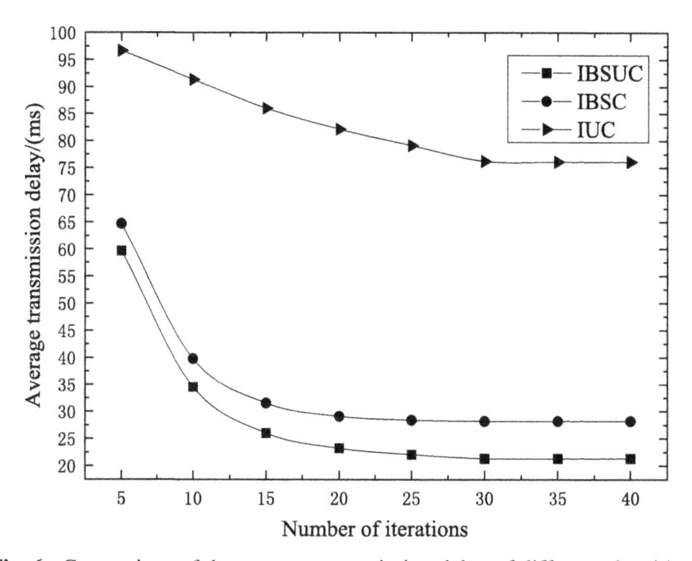

Fig. 6. Comparison of the average transmission delay of different algorithms

5 Conclusion

This paper proposes an incentive-based resource allocation strategy for edge caches in mobile social networks. Considering an edge cache resource allocation scenario within the coverage of a macro base station, the social interest groups purchase cache resources from the macro base station to cache popular content in the group. The incentive cache model is modeled as a Stackelberg game, and the utility functions of the macro base station and the social interest group are constructed respectively. The distributed iterative method is used to obtain the optimal unit cache resource price strategy of the macro base station and the optimal cache resource demand strategy of the social interest groups. The experimental results show that the proposed mechanism has lower average content transmission delay and higher user experience quality.

References

1. Saputra, Y.M., Hoang, D.T., Nguyen, D.N., et al.: A Novel mobile edge network architecture with joint caching-delivering and horizontal cooperation. IEEE Trans. Mob. Comput. **20**(1), 19–31 (2021)
2. Xia, X., Chen, F., He, Q., et al.: Online collaborative data caching in edge computing. IEEE Trans. Parallel Distrib. Syst. **32**(2), 281–294 (2021)
3. Bi, X., Qiu, T., Qu, W., et al.: Dynamically transient social community detection for mobile social networks. IEEE Internet Things J. **8**(3), 1282–1293 (2021)
4. Xu, Q., Su, Z., Zhang K.: HMM based cache pollution attack detection for edge computing enabled mobile social networks. In: 2019 IEEE International Conference on Communications (ICC), Shanghai, China, pp. 1–5. IEEE Press (2019)
5. Zhi, W., Zhu, K., Zhang, Y., et al.: Hierarchically social-aware incentivized caching for D2D communications. In: 2016 IEEE 22nd International Conference on Parallel and Distributed Systems (ICPADS), Wuhan, China, pp. 1521–9097. IEEE Press (2016)
6. Shuo, W., Xing, Z., Lin, W., et al.: Joint design of device to device caching strategy and incentive scheme in mobile edge networks. IET Commun. **12**(14), 1728–1736 (2018)
7. Zhao, K., Zhang, S., Zhang, N., et al.: Incentive mechanism for cached-enabled small cell sharing: a Stackelberg game approach. In: 2017 IEEE Global Communications Conference (GLOBECOM), Singapore, pp. 1–6. IEEE Press (2017)
8. Liu, T., Li, J., Shu, F., et al.: Incentive mechanism design for two-layer wireless edge caching networks using contract theory. IEEE Trans. Serv. Comput. **10**(11), 1 (2018)
9. Haenggi, M., Andrews, J.G., Baccelli, F., et al.: Stochastic geometry and random graphs for the analysis and design of wireless networks. IEEE J. Sel. Areas Commun. **27**(7), 1029–1046 (2009)
10. Shanmugam, K., Golrezaei, N., Dimakis, A.G., et al.: Femtocaching: wireless content delivery through distributed caching helpers. IEEE Trans. Inf. Theory **59**(12), 8402–8413 (2013)
11. Li, J., Chen, H., Chen, Y., et al.: Pricing and resource allocation via game theory for a small-cell video caching system. IEEE J. Sel. Areas Commun. **34**(8), 2115–2129 (2016)
12. Chen, Z., Liu, Y., Zhou, B., et al.: Caching incentive design in wireless D2D networks: a Stackelberg game approach. In: 2016 IEEE International Conference on Communications (ICC), Kuala Lumpur, Malaysia, pp. 1938–1883. IEEE Press (2016)
13. Ge, H., Jiang, Y., Bennis, M., et al.: Edge caching resource allocation in fog radio access networks: An Incentive Mechanism Based Approach. In: 2019 IEEE International Conference on Communications Workshops (ICC Workshops), Shanghai, China, pp. 1–6. IEEE Press (2019)

Energy-Efficient Optimization for Joint Design of Power Allocation and Beamforming in Downlink mmWave Communication Systems with NOMA

Jiali Cai[1(✉)], Xiangbin Yu[1,2], Xu Huang[1], and Cuimin Pan[1]

[1] College of Electronic and Information Engineering, Nanjing University of Aeronautics and Astronautics, Nanjing 210016, China

[2] Key Laboratory of Wireless Sensor Network and Communication, Shanghai Institute of Microsystem and Information Technology, Chinese Academy of Sciences, Shanghai, China

Abstract. In this paper, we develop two energy-efficient joint power allocation (PA) and beamforming (BF) design schemes for a two-user downlink millimeter-wave system with non-orthogonal multiple access under imperfect channel state information. The optimization problem is formulated by considering the maximum power and minimum rate constraints. Specifically, by means of block coordinate descent algorithm, the problem is transformed into two sub-problems: PA problem and BF problem. First, we derive a closed-form optimal PA solution for each iteration of concave-convex procedure so as to obtain a suboptimal PA scheme for the fixed BF. With this PA scheme, two suboptimal BF schemes are proposed based on successive convex approximation and one-dimensional search method, respectively. Simulation results validate the rational of two proposed schemes and show that the former scheme can achieve higher energy efficiency while the latter has small performance loss but lower complexity.

Keywords: Millimeter-wave · Non-orthogonal multiple access · Imperfect channel state information · Energy efficiency · Power allocation · Beamforming

1 Introduction

With the rapid development of mobile communications, spectrum resource shortage brings great challenges to existing technologies. Thus, a promising multiple

Supported by the Fundamental Research Funds for the Central Universities of NUAA (No. kfjj20200414), Natural Science Foundation of Jiangsu Province in China (No. BK20181289) and Open Research Fund Key Laboratory of Wireless Sensor Network and Communication of Chinese Academy of Sciences (2017006).

J. Xiong et al. (Eds.): MobiMedia 2021, LNICST 394, pp. 238–250, 2021.
https://doi.org/10.1007/978-3-030-89814-4_18

access technique, non-orthogonal multiple access (NOMA) has been proposed. NOMA adopts non-orthogonal transmission at the sender, actively introduces interference information, and realizes correct demodulation at the receiver by successive interference cancellation [1]. Besides, due to the large and available bandwidth, millimeter-wave (mmWave) communication is expected to support high data rates.

Therefore, in order to take advantage of both technologies, the studies about the combination of mmWave and NOMA have received more attention [2–8]. In [2] and [3], a joint beamforming (BF) and power allocation (PA) problem was formulated to maximize the sum rate for uplink and downlink mmWave-NOMA system respectively under perfect channel state information (CSI). Different from the studies above that focused on the spectral efficiency, [4] explored the energy-efficient PA problem for uplink mmWave-NOMA system, while [5] investigated the similar problem for downlink system. [6] studied mmWave-NOMA uplink systems and considered the fairness of the energy efficiency (EE) optimization problem. Combining with the mmWave channel estimation, the authors firstly used the analog BF scheme based on discrete Fourier transform codebook, and then put forward to a two-loop iterative algorithm for joint PA and digital BF design scheme of resource allocation. However, they didn't consider about the joint optimization problem since the analog BF vectors were selected by codebooks in the papers. In [7], a joint PA and BF scheme was developed so as to maximize the EE for downlink mmWave-NOMA system. And in [8], the authors proposed two beamwidth control methods and analyzed the energy-efficient digital precoder design by adopting a NOMA user scheduling algorithm for downlink mmWave-NOMA system.

Most of the existing works assume the perfect CSI is available, however, it can not be guaranteed in practice. Hence, we propose two schemes of joint PA and BF design for maximizing EE with imperfect CSI in the downlink mmWave-NOMA system. The main contributions in this paper are summarized as follows:

- With imperfect CSI, the EE maximization problem for two-user downlink mmWave-NOMA is formulated. The original optimization problem, which is difficult to solve directly, can be decomposed into two sub-problems. One is the PA problem and the other is the BF design problem. Then an efficient iterative algorithm via the block coordinate descent (BCD) method is developed.
- For the fixed BF, we address the PA problem by means of concave-convex procedure (CCCP). Moreover, the closed-form optimal solution for each iteration of CCCP is derived with the aid of Lambert W function. For the fixed PA, we adopt successive convex approximation (SCA) to obtain a suboptimal solution, and derive another low-complexity suboptimal solution by one-dimensional (1D) search method. The simulation shows that the former can achieve better EE performance, and the latter has better computational efficiency.

Notations: $(\cdot)^T$ and $(\cdot)^H$ stand for the transpose, conjugate transpose, respectively. $|\cdot|$ and $\|\cdot\|$ are absolute value and value of two-norm, respectively. Vectors and matrices are respectively represented by boldface lower and upper case symbols, $\mathbf{I}_N$ is an $N \times N$ identity matrix. $\mathcal{CN}(\mathbf{0}, \mathbf{R})$ denotes the complex Gaussian distribution with zero-mean and covariance matrix $\mathbf{R}$.

2 System Model and Problem Formulation

2.1 System Model

Considering a two-user downlink mmWave-NOMA system similar to that in [7], we assume that the base station (BS) is equipped with a radio frequency (RF) chain and N-element antenna array to support two single-antenna users. Each antenna branch of the BS has a phase shifter and a low-noise amplifier (LNA). All LNAs have the same scaling factor, therefore, the N dimensional BF vector $\mathbf{w}$ has constant-modulus (CM) elements, namely $|[\mathbf{w}]_n| = 1/\sqrt{N}, n = 1, 2, ..., N$. The mmWave channel between the BS and i-th user ($i = 1, 2$) with perfect CSI can be modeled as $\mathbf{h}_i = \sum_{l=1}^{L_i} \lambda_{i,l} \mathbf{a}(N, \theta_{i,l})$, where L_i, $\lambda_{i,l}$ and $\theta_{i,l}$ indicate the number of channel paths, complex gain and the angle of departure (AoD) of the l-th path between the BS and User-i, respectively, and $\mathbf{a}(N, \theta) = \left[e^{j\pi 0 \cos\theta}, e^{j\pi 1 \cos\theta}, ..., e^{j\pi (N-1) \cos\theta}\right]^T$ is the steering vector function [2–5,7]. In view of the difficulty of obtaining perfect CSI, we consider imperfect CSI at the BS. According to [9], the mmWave channel for the i-th user with imperfect CSI can be expressed as

$$\mathbf{h}_i = \hat{\mathbf{h}}_i + \sqrt{1-\rho}\tilde{\mathbf{h}}_i, \tag{1}$$

where $\hat{\mathbf{h}}_i$ denotes the estimated channel vector, ρ denotes the CSI accuracies of the non-line-of-sight (NLOS) components, $\tilde{\mathbf{h}}_i \sim \mathcal{CN}(\mathbf{0}, \mathbf{I}_N)$. Therefore, the received signal of User-i can be expressed as

$$\begin{aligned} \mathbf{y}_i = & \hat{\mathbf{h}}_i^H \mathbf{w} \sqrt{p_1} x_1 + \hat{\mathbf{h}}_i^H \mathbf{w} \sqrt{p_2} x_2 \\ & + \sqrt{1-\rho}\tilde{\mathbf{h}}_i^H \mathbf{w} \sqrt{p_1} x_1 + \sqrt{1-\rho}\tilde{\mathbf{h}}_i^H \mathbf{w} \sqrt{p_2} x_2 + \mathbf{n}_i, \end{aligned} \tag{2}$$

where x_i is the transmission signal of User-i, p_i denotes the transmission power and $\mathbf{n}_i$ denotes the Gaussian white noise with $\mathcal{CN}(\mathbf{0}, \sigma^2 \mathbf{I}_N)$.

2.2 Problem Formulation

For the above system, we need to consider two cases of decoding orders:

Case-1: The signal of User-1 x_1 is decoded first by regarding the signal of User-2 x_2 as interference, then x_2 is decoded after removing x_1. In this case, the achievable rates of two users are given as

$$\begin{cases} R_1^{(1)} = \log_2\left(1 + \frac{\hat{c}_1 p_1}{\hat{c}_1 p_2 + q(p_1+p_2)+\sigma^2}\right) \\ R_2^{(1)} = \log_2\left(1 + \frac{\hat{c}_2 p_2}{q(p_1+p_2)+\sigma^2}\right) \end{cases}, \tag{3}$$

where $q = 1 - \rho$ and $\hat{c}_i = \left|\hat{\mathbf{h}}_i^H \mathbf{w}\right|^2, i = 1, 2$, in which the implicit assumption is $\hat{c}_1 \leq \hat{c}_2$.

Case-2: The signal of User-2 x_2 is decoded first. In this case, the achievable rates of two users are given as

$$\begin{cases} R_1^{(2)} = \log_2\left(1 + \frac{\hat{c}_1 p_1}{q(p_1+p_2)+\sigma^2}\right) \\ R_2^{(2)} = \log_2\left(1 + \frac{\hat{c}_2 p_2}{\hat{c}_2 p_1 + q(p_1+p_2)+\sigma^2}\right) \end{cases}. \tag{4}$$

Similarly, the implicit assumption is $\hat{c}_1 \geq \hat{c}_2$.

Therefore, the EE optimization problem in both cases with imperfect CSI is formulated as

$$\begin{aligned} \max_{p_1, p_2, \mathbf{w}} \quad & \eta_{EE}^{(j)} = \frac{R_1^{(j)} + R_2^{(j)}}{\xi\left(p_1 + p_2\right) + P_C} \\ \text{s.t.} \quad & R_i^{(j)} \geq r_i, i = 1, 2, \\ & p_1 + p_2 \leq P_{\max}, \\ & \left|[\mathbf{w}]_n\right| = \frac{1}{\sqrt{N}}, n = 1, 2, ..., N, \end{aligned} \tag{5}$$

where $\eta_{EE}^{(j)}$ is the system EE in Case-j, r_i is the minimum rate of User-i, $P_{\max}$ is the maximum transmission power at the BS and ξ denotes the coefficient of LNAs. P_C is the fixed circuit power consumption, which can be expressed as $P_C = P_{BB} + P_{RF} + NP_{PS} + NP_{LNA}$, in which $P_{BB}, P_{RF}, P_{PS}, P_{LNA}$ are the power consumption of the baseband, the RF chain, the phase shifter and LNA, respectively.

3 Joint PA and BF Design Schemes

This section presents two schemes of joint PA and BF design. Due to the limited space, we only consider about the optimization problem in Case-2 under imperfect CSI, and that in Case-1 can be solved by using the similar algorithms. The superscript (j) in $\eta_{EE}^{(j)}$ and $R_i^{(j)}$ is omitted for the sake of discussion. Obviously, due to the non-concave objective function as well as the coupled optimization variables, the optimization problem (5) is difficult to solve directly. Therefore, we introduce a new variable $P = p_1 + p_2$, so that p_2 can be replaced as $P - p_1$, then (5) can be transformed into

$$\begin{aligned} \max_{p_1, P, \mathbf{w}} \quad & \eta_{EE} = \frac{\log_2\left(\left(1 + \frac{\hat{c}_1 p_1}{qP+\sigma^2}\right)\left(1 + \frac{\hat{c}_2 (P-p_1)}{\hat{c}_2 p_1 + qP + \sigma^2}\right)\right)}{\xi P + P_C} \\ \text{s.t.} \quad & \hat{c}_2(P - p_1) \geq \phi_2\left(\hat{c}_2 p_1 + qP + \sigma^2\right), \\ & \hat{c}_1 p_1 \geq \phi_1\left(qP + \sigma^2\right), \\ & 0 \leq P \leq P_{\max}, \\ & \left|[\mathbf{w}]_n\right| = \frac{1}{\sqrt{N}}, n = 1, 2, ..., N, \\ & \hat{c}_1 \geq \hat{c}_2, \end{aligned} \tag{6}$$

where $\phi_i = 2^{r_i} - 1, i = 1, 2$. According to the first derivative of the objective function in (6) with respect to p_1 and $\hat{c}_1 \geq \hat{c}_2$, we can get $\partial \eta_{EE} / \partial p_1 \geq 0$, i.e., η_{EE} is monotonically increasing with respect to p_1. Thus, the optimal p_1 is its upper bound, which can be obtained by the rate constraint of User-2:

$$p_1^* = \frac{\hat{c}_2 P - \phi_2 \left(qP + \sigma^2\right)}{\left(\phi_2 + 1\right) \hat{c}_2}. \tag{7}$$

With p_1^*, p_2 can be rewritten as $P - p_1^*$ and (6) can be reduced to the following problem with respect to $\{P, \mathbf{w}\}$:

$$\begin{aligned} \max_{P, \mathbf{w}} \quad & J = \frac{\log_2 \left(\frac{(\hat{c}_1 \hat{c}_2 + q(\phi_2+1)\hat{c}_2 - \phi_2 \hat{c}_1 q)P + \sigma^2((\phi_2+1)\hat{c}_2 - \phi_2 \hat{c}_1)}{(qP+\sigma^2)\hat{c}_2} \right)}{\xi P + P_C} \\ \text{s.t.} \quad & \hat{c}_1 \frac{\hat{c}_2 P - \phi_2 \left(qP + \sigma^2\right)}{\left(\phi_2 + 1\right) \hat{c}_2} \geq \phi_1 \left(qP + \sigma^2\right), \\ & 0 \leq P \leq P_{\max}, \\ & \left|[\mathbf{w}]_n\right| = \frac{1}{\sqrt{N}}, n = 1, 2, ..., N, \\ & \hat{c}_1 \geq \hat{c}_2, \end{aligned} \tag{8}$$

We can observe from (8) that when one of P and $\mathbf{w}$ is fixed, the resultant problem can be efficiently solved. Therefore, we adopt the BCD algorithm to solve this optimization problem. In particular, we decompose the problem (8) into PA problem and BF design problem, and then solve the PA problem for given $\mathbf{w}$, and solve the BF design problem for given P.

3.1 Solution of PA Problem

For any feasible $\mathbf{w}$, the PA problem in (8) can be formulated as

$$\begin{aligned} \max_{P} \quad & J \\ \text{s.t.} \quad & P_{\min} \leq P \leq P_{\max}, \end{aligned} \tag{9}$$

where $P_{\min} = \frac{\phi_1(\phi_2+1)\hat{c}_2\sigma^2 + \hat{c}_1\phi_2\sigma^2}{\hat{c}_1\hat{c}_2 - \hat{c}_1\phi_2 q - \phi_1(\phi_2+1)\hat{c}_2 q}$, which is obtained by the rate constraint of User-1. Besides, the numerator of J can be rewritten as $f_1(P) - f_2(P)$, where $f_1(P) = \log_2\left(\left(\hat{c}_1 + q(\phi_2+1) - \frac{\hat{c}_1\phi_2 q}{\hat{c}_2}\right)P + \sigma^2\left((\phi_2+1) - \frac{\phi_2\hat{c}_1}{\hat{c}_2}\right)\right)$, $f_2(P) = \log_2\left(qP + \sigma^2\right)$. Since $f_1(P)$ and $f_2(P)$ are convex, we can use CCCP to solve the difference of convex problem. By adopting first-order Taylor expansion at the initial point P_0, $f_2(P)$ is linearized as

$$f_2(P) = \log_2\left(qP_0 + \sigma^2\right) + \frac{q}{\ln 2\left(qP_0 + \sigma^2\right)}\left(P - P_0\right). \tag{10}$$

With (10), the optimization problem (9) for given P_0 can be transformed into

$$\begin{aligned} \max_{P} \quad & J_1 = \frac{\log_2(aP + b) - cP - d}{\xi P + P_C} \\ \text{s.t.} \quad & P_{\min} \leq P \leq P_{\max}, \end{aligned} \tag{11}$$

where

$$\begin{cases} a = \hat{c}_2\hat{c}_1 + q(\phi_2+1)\hat{c}_2 - \phi_2\hat{c}_1 q \\ b = \sigma^2\left((\phi_2+1)\hat{c}_2 - \phi_2\hat{c}_1\right) \\ c = \frac{q}{\ln 2(qP_0+\sigma^2)} \\ d = \log_2\left(\left(qP_0+\sigma^2\right)\hat{c}_2\right) - \frac{q}{\ln 2(qP_0+\sigma^2)}P_0 \end{cases}. \quad (12)$$

With the aid of Lambert W function $W(\cdot)$ [10], the solution of $\partial J_1/\partial P = 0$ is expressed as

$$\tilde{P} = \frac{1}{a}\left(\frac{\frac{aP_C}{\xi} - b}{W\left(\frac{\frac{aP_C}{\xi} - b}{\exp\left(\left(d - \frac{cP_C}{\xi}\right)\ln 2 + 1\right)}\right)} - b\right). \quad (13)$$

Then, the optimal solution of (12) is given by

$$\begin{cases} P^* = \min\left\{P_{\max}, \max\left\{P_{\min}, \tilde{P}\right\}\right\} \\ p_1^* = \frac{\hat{c}_2 P^* - \phi_2\left(qP^*+\sigma^2\right)}{(\phi_2+1)\hat{c}_2}, p_2^* = P^* - p_1^* \end{cases}. \quad (14)$$

Therefore, we calculate (14) iteratively by CCCP and obtain the converged solution as a suboptimal solution of (9). The algorithm for PA is summarized as **Algorithm** 1.

Algorithm 1. CCCP-based PA Algorithm

1: **Initialize**: tolerances $\epsilon_1 > 0$, the number of iterations $q = 0$ and initial point $P^{(q)}$.

2: **repeat**
3: Update $P_0 = P^{(q)}$.
4: $q = q + 1$.
5: Compute $P^{(q)}, p_1^{(q)}$ according to (14).
6: **until** $\left|P^{(q)} - P_0\right| \le \epsilon_1$
7: **Output**: $\hat{P}^* = P^{(q)}, \hat{p}_1^* = p_1^{(q)}, \hat{p}_2^* = P^* - p_1^*$.

3.2 Solution of BF Design Problem

For a given P, the BF problem in (8) can be expressed as

$$\begin{aligned} \max_{\mathbf{w}} \quad & J_2(\mathbf{w}) = \left|\hat{\mathbf{h}}_1^H\mathbf{w}\right|^2 P - \phi_2\left(qP+\sigma^2\right)\frac{\left|\hat{\mathbf{h}}_1^H\mathbf{w}\right|^2}{\left|\hat{\mathbf{h}}_2^H\mathbf{w}\right|^2} \\ \text{s.t.} \quad & \left|\hat{\mathbf{h}}_2^H\mathbf{w}\right|^2 \le \left|\hat{\mathbf{h}}_1^H\mathbf{w}\right|^2, \\ & \left|[\mathbf{w}]_n\right| = \frac{1}{\sqrt{N}}, n = 1, 2, ..., N. \end{aligned} \quad (15)$$

Let $\mathbf{H}_i = \hat{\mathbf{h}}_i \hat{\mathbf{h}}_i^H, i = 1, 2$, and introduce new auxiliary variables t_1, t_2, then the BF problem can be relaxed to the following problem:

$$\begin{aligned} \max_{\mathbf{w}, t_1, t_2} \quad & t_1 \\ \text{s.t.} \quad & 1 \leq \frac{\mathbf{w}^H \mathbf{H}_1 \mathbf{w}}{\mathbf{w}^H \mathbf{H}_2 \mathbf{w}} \leq t_2 \\ & \mathbf{w}^H \mathbf{H}_1 \mathbf{w} \geq \frac{t_1 + \phi_2 \left(qP + \sigma^2\right) t_2}{P} \\ & \left|[\mathbf{w}]_n\right| \leq \frac{1}{\sqrt{N}}, n = 1, 2, ..., N. \end{aligned} \tag{16}$$

In what follows, we adopt SCA method to address the problem (16). Accordingly, we need to solve the following convex optimization problem at the l-th iteration of SCA:

$$\begin{aligned} \max_{\mathbf{w}, t_1, t_2} \quad & t_1 \\ \text{s.t.} \quad & 2\text{Re}\left\{\mathbf{w}_0^H \mathbf{H}_2 \mathbf{w}\right\} - \mathbf{w}_0^H \mathbf{H}_2 \mathbf{w} \geq \frac{\mathbf{w}^H \mathbf{H}_1 \mathbf{w}}{t_2}, \\ & 2\text{Re}\left\{\mathbf{w}_0^H \mathbf{H}_1 \mathbf{w}\right\} - \mathbf{w}_0^H \mathbf{H}_1 \mathbf{w} \geq \mathbf{w}^H \mathbf{H}_2 \mathbf{w}, \\ & 2\text{Re}\left\{\mathbf{w}_0^H \mathbf{H}_1 \mathbf{w}\right\} - \mathbf{w}_0^H \mathbf{H}_1 \mathbf{w} \geq \frac{t_1 + \phi_2 \left(qP + \sigma^2\right) t_2}{P}, \\ & \left|[\mathbf{w}]_n\right| \leq \frac{1}{\sqrt{N}}, n = 1, 2, ..., N, \end{aligned} \tag{17}$$

where $\mathbf{w}_0 = \mathbf{w}^{(l-1)}$ denotes the value of $\mathbf{w}$ at the $(l-1)$-th iteration of SCA. Hence, CVX tool is used to obtain a suboptimal solution $\mathbf{w}^*$ of (15) [11]. To satisfy the CM constraint, we make the CM normalization for each element of $\mathbf{w}^*$ as follows:

$$[\mathbf{w}^*]_n = \frac{[\mathbf{w}^*]_n}{\sqrt{N} \left|[\mathbf{w}^*]_n\right|}, n = 1, ..., N. \tag{18}$$

As a result, we summarize the first joint design scheme in **Algorithm** 2. Furthermore, due to the poor calculation efficiency of CVX, we propose another scheme to address the BF design problem. When p_1, p_2 is known, the BF problem decomposed from the optimization problem (6) is reduced to

$$\mathcal{L}(\mathbf{w}) = \left(\frac{P}{qP + \sigma^2} - \frac{p_2}{\hat{c}_2 p_1 + qP + \sigma^2}\right) \hat{c}_1 + \frac{p_2}{\hat{c}_2 p_1 + qP + \sigma^2} \hat{c}_2. \tag{19}$$

Let $\alpha = \frac{p_2}{\hat{c}_2 p_1 + qP + \sigma^2}, \beta = \alpha \frac{qP + \sigma^2}{P}$, then $\mathcal{L}(\mathbf{w})$ can be rewritten as

$$\mathcal{L}(\mathbf{w}) = \frac{P}{qP + \sigma^2} \left[\left(1 - \beta\right) \hat{c}_1 + \beta \hat{c}_2\right]. \tag{20}$$

Algorithm 2. Joint PA and BF Algorithm with SCA

Initialize: tolerances $\epsilon_2, \epsilon_3 > 0$, the number of iterations $t = 0$, and $\{P^{(t)}, \mathbf{w}^{(t)}\}$.
repeat
 $t = t + 1$.
 Initialize the number of iterations $l = 0$ and initial point $\mathbf{w}^{(l)} = \mathbf{w}^{(t)}$.
 repeat
  $l = l + 1$.
  Calculate the optimal solution $\mathbf{w}^*$ by (17), (18).
  Update $\mathbf{w}^{(l)} = \mathbf{w}^*$.
 until $\left\|\mathbf{w}^{(l)} - \mathbf{w}^{(l-1)}\right\| \leq \epsilon_2$
 Update $\mathbf{w}^{(t)} = \mathbf{w}^{(l)}$.
 Calculate P^* by **Algorithm 1**.
 Update $P^{(t)} = P^*$.
until $\left\|\mathbf{w}^{(t)} - \mathbf{w}^{(t-1)}\right\| + \left|P^{(t)} - P^{(t-1)}\right| \leq \epsilon_3$
Output: $\hat{\mathbf{w}}^* = \mathbf{w}^{(t)}, \hat{P}^* = P^{(t)}, \hat{p}_1^*, \hat{p}_2^*$.

Since $\hat{c}_i = \mathbf{w}^H \hat{\mathbf{h}}_i \hat{\mathbf{h}}_i^H \mathbf{w}, i = 1, 2$, without considering the constraint of minimum rate, the BF problem can be formulated as

$$\begin{aligned} \max_{\mathbf{w}} \quad & J_3(\mathbf{w}) = \mathbf{w}^H \left[\sqrt{(1-\beta)}\hat{\mathbf{h}}_1\hat{\mathbf{h}}_1^H + \sqrt{\beta}\hat{\mathbf{h}}_2\hat{\mathbf{h}}_2^H\right]\mathbf{w} \\ \text{s.t.} \quad & \left|\hat{\mathbf{h}}_2^H\mathbf{w}\right|^2 \leq \left|\hat{\mathbf{h}}_1^H\mathbf{w}\right|^2, \\ & \left|[\mathbf{w}]_n\right| = \frac{1}{\sqrt{N}}, n = 1, 2, ..., N. \end{aligned} \tag{21}$$

To achieve the maximum $J_3(\mathbf{w})$, we first introduce the **Lemma** 1.

Lemma 1. *Given* $a_1, a_2 > 0$, *the following inequality holds*

$$a_1 |b_1|^2 + a_2 |b_2|^2 \geq |\sqrt{a_1}b_1 + \sqrt{a_2}b_2|^2 /2 \tag{22}$$

Proof. Since $a_1, a_2 > 0$, we can get

$$|\sqrt{a_1}b_1 - \sqrt{a_2}b_2|^2 \geq 0. \tag{23}$$

Expanding (23), we can obtain

$$a_1 |b_1|^2 + a_2 |b_2|^2 \geq 2\sqrt{a_1 a_2}\mathrm{Re}(b_1 b_2^*). \tag{24}$$

Then we have

$$\begin{aligned} 2\left(a_1 |b_1|^2 + a_2 |b_2|^2\right) &\geq a_1 |b_1|^2 + a_2 |b_2|^2 + 2\sqrt{a_1 a_2}\mathrm{Re}(b_1 b_2^*) \\ &= |\sqrt{a_1}b_1 + \sqrt{a_2}b_2|^2 \end{aligned} \tag{25}$$

Therefore the inequality (22) holds.

By using the **Lemma** 1, the problem (21) can be transformed into

$$\begin{aligned} \max_{\mathbf{w}} \quad & J_4(\mathbf{w}) = \left|\sqrt{(1-\beta)}\hat{\mathbf{h}}_1^H\mathbf{w} + \sqrt{\beta}\hat{\mathbf{h}}_2^H\mathbf{w}\right| \\ \text{s.t.} \quad & \left|\hat{\mathbf{h}}_2^H\mathbf{w}\right|^2 \leq \left|\hat{\mathbf{h}}_1^H\mathbf{w}\right|^2, \\ & \left|[\mathbf{w}]_n\right| = \frac{1}{\sqrt{N}}, n = 1, 2, ..., N. \end{aligned} \tag{26}$$

And a suboptimal solution can be obtained as

$$\mathbf{w}^* = \frac{1}{\sqrt{N}} \exp\left(j\angle\left\{\sqrt{(1-\beta)}\hat{\mathbf{h}}_1 + \sqrt{\beta}\hat{\mathbf{h}}_2\right\}\right). \tag{27}$$

Due to $\alpha \leq \frac{P}{qP+\sigma^2}$, we can obtain β belongs to [0,1]. Moreover, in order to find the optimal solution β^* corresponds to the maximum objective function of the original problem under the given p_1, p_2, we use the 1D search method, namely $\beta^* = \arg \max_{\beta\in[0,1]} \mathcal{L}(\mathbf{w})$. In particular, the BF scheme in [7] is a special case of this proposed scheme when $\rho = 1$.

Based on the analysis above, the second suboptimal joint design scheme is summarized as **Algorithm** 3.

Algorithm 3. Joint PA and BF Algorithm with 1D search

Initialize: tolerances $\epsilon_3 > 0$, the number of iterations $t = 0$, and $\{P^{(0)}, \mathbf{w}^{(0)}\}$, calculate $p_1^{(0)}, p_2^{(0)}$ by (14)
repeat
  $t = t + 1$.
  Calculate $\beta^* = \arg \max_{\beta\in[0,1]} \mathcal{L}(\mathbf{w})$
  Calculate $\mathbf{w}^*$ with β^* by (22).
  Update $\mathbf{w}^{(t)} = \mathbf{w}^*$.
  Calculate P^*, p_1^*, p_2^* by **Algorithm 1**.
  Update $P^{(t)} = P^*, p_1^{(t)} = p_1^*, p_2^{(t)} = p_2^*$.
until $\left\|\mathbf{w}^{(t)} - \mathbf{w}^{(t-1)}\right\| + \left|P^{(t)} - P^{(t-1)}\right| \leq \epsilon_3$
Output: $\hat{\mathbf{w}}^* = \mathbf{w}^{(t)}, \hat{P}^* = P^{(t)}, \hat{p}_1^*, \hat{p}_2^*$.

3.3 Complexity Analysis

We can find that **Algorithm** 2 solves the problem $L_1(L_2 + L_3)$ times, where L_1, L_2, L_3 denote the number of iterations of BCD, SCA and CCCP, respectively. Besides, CVX is used to solve (17) and its complexity is $\mathcal{O}\left(L_2(2N+2)^{3.5}\log(1/\varepsilon)\right)$, where $\mathcal{O}(\cdot)$ represents the big-O notation and ε is the solution accuracy [12]. Then the computational complexity is $\mathcal{O}\left(L_1\left(L_2(2N+2)^{3.5}\log(1/\varepsilon) + L_3\right)\right)$. By contrast, the complexity of **Algorithm** 3 is $\mathcal{O}\left(L_1(L_4 + L_3)\right)$, where L_4 depends on the step size of the 1D search method. Obviously, **Algorithm** 2 is more complicated than **Algorithm** 3.

4 Simulation Results

In this section, the performance for the mmWave-NOMA system with imperfect CSI is evaluated by computer simulation. For the simulation setup, it is assumed that the channel condition of User-1 is better than that of User-2. For the mmWave channel, we set $L_i = 4, i = 1, 2$, and the first channel path is line-of-sight path, where $|\lambda_{1,1}| = 1$, $|\lambda_{2,1}| = 0.3$, $\cos(\theta_{1,1}) = -0.25$, $\cos(\theta_{2,1}) = 0.4$ and the rest are NLOS paths, where $\lambda_{i,l}(l = 2, 3, 4) \in \mathcal{CN}(0, 10^{-15/10})$ and AoDs are uniformly distributed over $[0, 2\pi]$. The tolerances are $\epsilon_1 = \epsilon_2 = \epsilon_3 = 10^{-4}$. Other parameters are set as $r_1 = r_2 = r = 1$ bit/s/Hz, $N = 32$, $P_{BB} = 200$mW, $P_{RF} = 160$ mW, $P_{PS} = 20$ mW, $P_{LNA} = 40$ mW, $\xi = 1/0.38$, $\sigma^2 = 1$ mW [7], where the computer used is equipped with Intel Core i5-6300HQ 2.30 GHz and 4 G RAM.

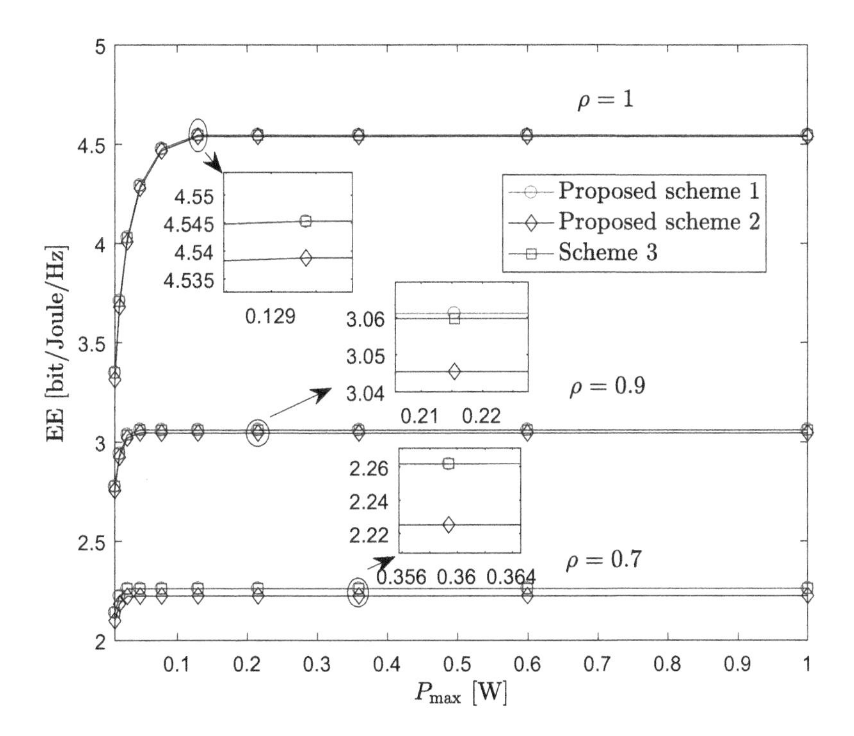

Fig. 1. Effect of imperfect CSI on the system EE under different schemes.

Figure 1 shows the system EE with different joint design schemes, including the BF design of "Proposed scheme 1" based on SCA and that of "Proposed scheme 2" based on 1D search method and "Scheme 3" that searches for the phase $\delta_n (n = 1, .., N)$ of the n-th entry of $\mathbf{w}$ over $[0, 2\pi]$ corresponds to maximum $J_3(\mathbf{w})$, where $\mathbf{w} = \frac{1}{\sqrt{N}} \left[e^{j\delta_0}, ..., e^{j\delta_N}\right]$. Moreover, "Scheme 3" uses N times

1D search which improves accuracy at the cost of efficiency. Obviously, the performance of both schemes is close to that of "Scheme 3", and "Proposed scheme 1" outperforms "Proposed scheme 2", while "Scheme 3" costs the most run time, followed by "Proposed scheme 1", and "Proposed scheme 2" has the lowest complexity. Specially, the average run time of three schemes when $\rho = 0.9$ is 5.1801 s, 0.4404 s and 17.2136 s, respectively and the result shows "Scheme 3" has the highest complexity because it requires N times 1D search method, and complexity of "Proposed scheme 1" is much more than that of "Proposed scheme 2" because CVX has poor computational efficiency, which results in longer run time. Besides, Fig. 1 presents the impact of imperfect CSI on the system EE, where $\rho \in \{1, 0.9, 0.7\}$. As can be observed, the system EE decreases along with the decrease of ρ, which concludes that the accuracy of channel estimation affects the EE significantly. This is because the channel estimation error results in the increase of interference when decoding the user's signals.

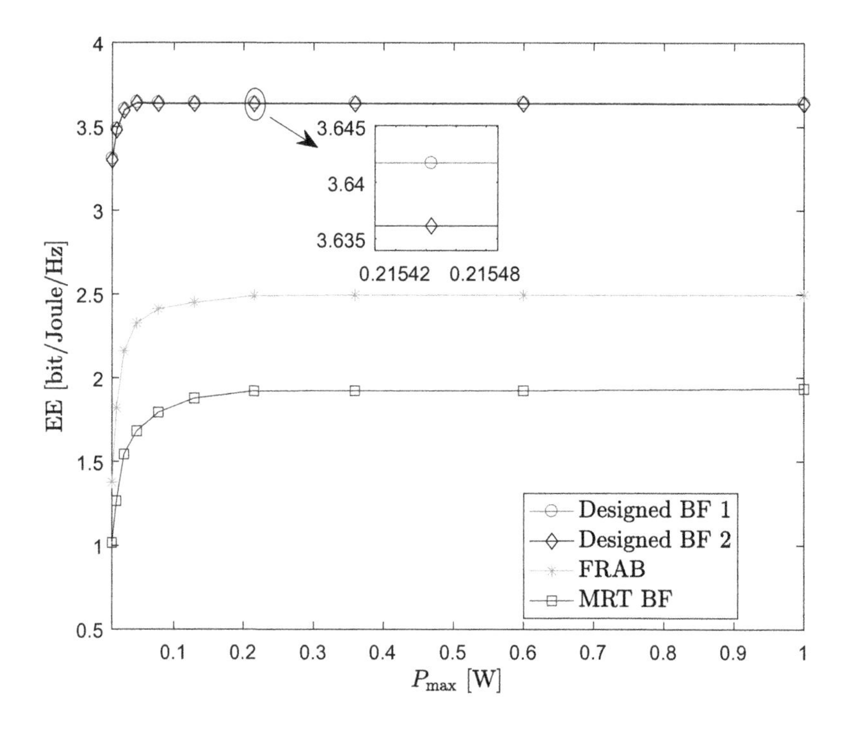

Fig. 2. EE performance of the mmWave-NOMA system under different BF schemes.

Figure 2 evaluates the EE performance with our proposed schemes and the finite resolution analog beamforming (FRAB) in [13], and maximal-ratio transmit (MRT) BF when $\rho = 0.95$, $r = 0.1$ bit/s/Hz, where "Designed BF 1" refers to SCA-based BF design, "Designed BF 2" refers to BF design based on 1D search, "FRAB" and "MRT BF" solve the PA problem under given BF. From Fig. 2, the proposed schemes have better performance than the other two schemes, since the BF vectors of "FRAB" scheme and "MRT BF" scheme are independent and not optimized with PA jointly, which verifies the effectiveness of two proposed schemes.

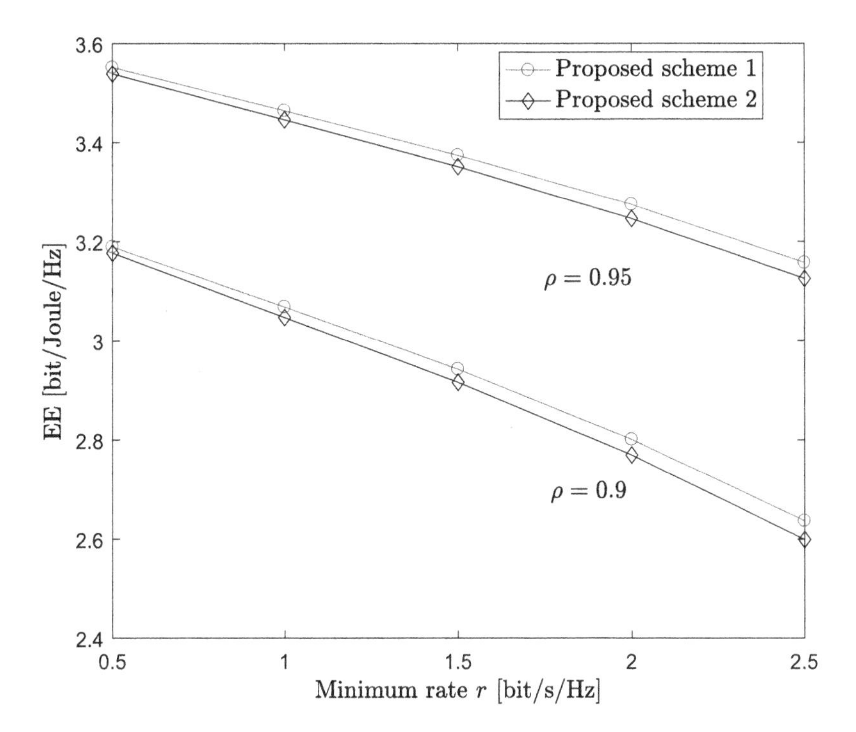

Fig. 3. EE comparison under different rate constraints.

Figure 3 plots the EE versus rate constraints with two schemes when $\rho \in \{0.95, 0.9\}$ and $P_{\max} = 1$ W. As shown in Fig. 3, the system EE declines with the minimum rates. And the higher the minimum rate, the more EE is reduced. This is because "Proposed scheme 2" is formulated without considering the rate constraint. The higher the minimum rate is, the more performance loss it will bring. Besides, the gap between two schemes is bigger with the increase of minimum rate, since the rate constraint has more effect on "Proposed scheme 2".

5 Conclusion

This paper studies on the energy-efficient joint PA and BF design problem for downlink mmWave-NOMA system under imperfect CSI. The basic idea is to decompose the optimization problem into the PA problem and BF design problem, and then solve them in succession by BCD. Specifically, for the PA problem, we derive the closed-form solution of CCCP by Lambert W function. For the BF design problem, we obtain a suboptimal solution by SCA and also derive a suboptimal beamforming utilizing the 1D search method. In the simulation, we analyze the effect of estimation error and rate constraints on the EE, respectively. The results prove the effectiveness of two proposed schemes. Moreover, the suboptimal scheme utilizing 1D search method can implement EE performance closing to that based on SCA and has better computational efficiency.

References

1. Dai, L., Wang, B., Ding, Z., Wang, Z., Chen, S., Hanzo, L.: A survey of non-orthogonal multiple access for 5G. IEEE Commun. Surv. Tutor. **20**(3), 2294–2323 (2018). Thirdquarter
2. Zhu, L., Zhang, J., Xiao, Z., Cao, X., Wu, D.O., Xia, X.G.: Joint power control and beamforming for uplink non-orthogonal multiple access in 5G millimeter-wave communications. IEEE Trans. Wirel. Commun. **17**(9), 6177–6189 (2018)
3. Xiao, Z., Zhu, L., Choi, J., Xia, P., Xia, X.: Joint power allocation and beamforming for non-orthogonal multiple access (NOMA) in 5G millimeter wave communications. IEEE Trans. Wirel. Commun. **17**(5), 2961–2974 (2018)
4. Zeng, M., Hao, W., Dobre, O.A., Poor, H.V.: Energy-efficient power allocation in uplink mmWave massive MIMO with NOMA. IEEE Trans. Veh. Technol. **68**(3), 3000–3004 (2019)
5. Hao, W., Zeng, M., Chu, Z., Yang, S.: Energy-efficient power allocation in millimeter wave massive MIMO with non-orthogonal multiple access. IEEE Wirel. Commun. Lett. **6**(6), 782–785 (2017)
6. Hao, W., et al.: Codebook-based max-min energy-efficient resource allocation for uplink mmWave MIMO-NOMA systems. IEEE Trans. Commun. **67**(12), 8303–8314 (2019)
7. Yu, X., Dang, X., Wen, B., Leung, S., Xu, F.: Energy-efficient power allocation for millimeter-wave system with non-orthogonal multiple access and beamforming. IEEE Trans. Veh. Technol. **68**(8), 7877–7889 (2019)
8. Wei, Z., Ng, D.W.K., Yuan, J.: NOMA for hybrid mmWave communication systems with beamwidth control. IEEE J. Sel. Top. Sig. Process. **13**(3), 567–583 (2019)
9. Guo, J., Yu, Q., Meng, W., Xiang, W.: Energy-efficient hybrid precoder with adaptive overlapped subarrays for large-array mmWave systems. IEEE Trans. Wirel. Commun. **19**(3), 1484–1502 (2020)
10. Hoorfar, A.: Inequalities on the Lambert W function and hyperpower function. J. Inequal. Pure Appl. Math. **9**(2), 5–9 (2013)
11. Grant, M., Boyd, S.: CVX: MATLAB software for disciplined convex programming, Version 2.2, January 2020. http://cvxr.com/cvx
12. Ben-Tal, A., Nemirovski, A.: Lectures on Modern Convex Optimization: Analysis, Algorithms, and Engineering Applications. SIAM Series Optimization, vol. 2, November 2001
13. Dai, L., Wang, B., Peng, M., Chen, S.: Hybrid precoding-based millimeter-wave massive MIMO-NOMA with simultaneous wireless information and power transfer. IEEE J. Sel. Areas Commun. **37**(1), 131–141 (2019)

Cryptography Security and Privacy Protection

A Reliable Voice Perceptual Hash Authentication Algorithm

Li Li, Yang Li(✉), Zizhen Wang, Xuemei Li, and Guozhen Shi

Beijing Electronic Science and Technology Institute, Beijing 100070, China

Abstract. In order to protect the authenticity and integrity of voice, this paper proposes a reliable voice perceptual hash authentication algorithm. This algorithm considers the dynamic characteristics of voice signals and proposes a scheme to construct a perceptual feature matrix containing voice static and dynamic characteristics based on the Mel frequency inverted spectrum coefficient and its first- and second-order difference parameters. In the process of perceptual hash chain construction, the feature matrix is degraded by two-dimensional non-negative matrix factorization method, and the result of decomposition is quantified to construct the perceptual hash chain. In the process of voice authentication, hamming distance method is used to measure the distance between perceptual hash chains. Experiments show that the algorithm has good distinguishability and robustness, and the voice perceptual hash authentication can be carried out accurately and reliably.

Keywords: Perceptual hash · Non-negative matrix factorization · MFCC

1 Introduction

Because of its convenience, voice is more suitable for communication in the case of inconvenient text input, as an indispensable communication method in modern society, its security problems can not be ignored. Due to the diversity of audio editing tools and simplification of operation, voice authenticity authentication and integrity protection are particularly important. Traditional crypto-hash authentication technology because of its high sensitivity to content changes, has not been applicable to voice content authentication, while voice perceptual hash authentication technology has a good distinguishability and robustness, can be well certified voice content, so as to complete the protection of voice authenticity and integrity.

Perceptual hash technology uniquely maps multimedia digital represents with the same perceptual content to a digital digest, was first applied in the field of image authentication [1–4], which completes image authentication and recognition by short summary of image perceptual information and summary-based comparison matching. It is now widely used in image authentication, image retrieval, voice content authentication and tamper detection and other fields [5–7]. The common method of voice perceptual hash feature extraction is to project a set of signals into another domain, with the main methods being wavelet transform [8, 9], discrete cosine transformation [10], short-term Fourier

J. Xiong et al. (Eds.): MobiMedia 2021, LNICST 394, pp. 253–263, 2021.
https://doi.org/10.1007/978-3-030-89814-4_19

transformation [11], etc. Then select feature values in the transformation domain, such as Mel-Frequency Cepstral Coefficient (MFCC), short-term zero-pass rate, short-term energy, etc. The extracted eigenvalues are analyzed, the perceptual hash chain is constructed, and the voice perceptual hash authentication is completed. By analyzing the extracted eigenvalues, perceptual hash chains are constructed, and use them to complete voice perceptual hash authentication. The paper [12] proposes a scheme for constructing a perceptual hash chain using short-term energy and short-term zero-pass rate to perceive audio content in different formats, with good robustness and security, and good computational efficiency. In paper [13], MFCC and LPCC feature parameters are fused to construct feature matrix, and the complexity of the feature matrix is reduced by using two-dimensional non-negative matrix factorization, effectively improve the robustness of hash authentication, but the matrix block decomposition method is lacking. The paper [14] uses MFCC parameters as perceptual features, which has good robustness, but the security of the algorithm depends on pseudo-random sequences, and the robustness to resampling and other operations is poor. The paper [15] presents 2 voice perceptual hash authentication algorithms based on voice spectral graph and pseudo-harmonic model, the former which can meet the higher demand for real-time performance, and the latter, which has relatively poor operational efficiency and differentiation but has better robustness.

In this paper, combining the auditory characteristics of human ear and fully considering the dynamic characteristics of voice signal, a reliable voice perceptual hash authentication algorithm is proposed: the algorithm extracts the MFCC feature parameters of voice and calculates its first- and second-order differences parameters, constructs the perceived feature matrix using these three sets of feature parameters, the feature matrix is degraded by two-dimensional non-negative matrix factorization, so as to construct the perceptual hash chain, and the normalized Hamming distance is used as a certification standard for perceptive hash authentication. Finally, the distinguishability and robustness of the algorithm are verified by experiments.

2 Voice Perceptual Hash Construction Algorithm

The voice perceptual hash construct in this paper is based on MFCC feature extraction, and the construction process is shown in Fig. 1.

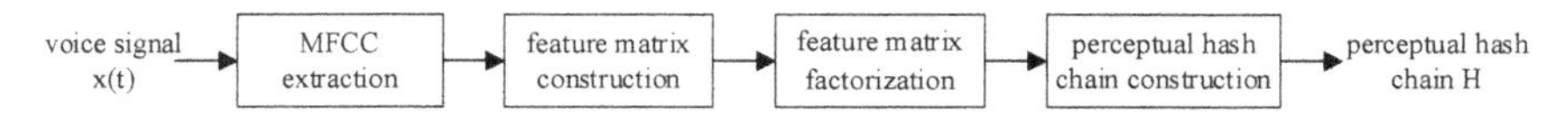

Fig. 1. Voice perceptual hash construction

2.1 MFCC Extraction

Mel-Frequency Cepstral Coefficient, MFCC is one of the important characteristic parameters used in voice recognition technology, and its physical meaning is the distribution of energy in the signal spectrum in different frequency bands. The MFCC is designed based on the auditory properties of the human ear, as a result of the frequency of sound heard by the human ear is not linear, the growth of the Mel-frequency is consistent with the auditory characteristics of the human ear, the actual frequency is linearly distributed below 1000 Hz, and increases logarithmically above 1000 Hz, the specific relationship between them is shown in formula (1):

$$f_{mel}(f) = 2595\lg(1 + \frac{f}{700}) \tag{1}$$

Among them, the f is the actual frequency, f_{mel} is the Mel-frequency.

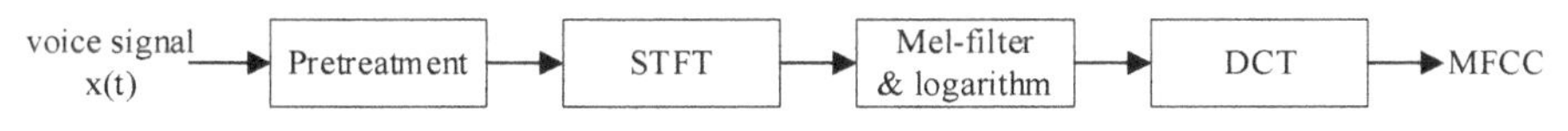

Fig. 2. MFCC extraction process

The specific process of MFCC extraction, as shown in Fig. 2, is divided into the following steps:

1> Pretreatment. Windowing and framing of the input voice signal, add Hanning window to the voice signal, and set frame length to 2048, frame shift to 512 for framing.

2> Short-term Fourier transform, STFT. Use fast Fourier transform (FFT) to calculate the results obtained in the previous step, to get spectrum of each frame to get the spectrum distribution information. This process of frame-by-frame fast Fourier transform is called short time Fourier transform. The FFT formula is as follows:

$$X_a(k) = \sum_{n=0}^{N-1} x(t)e^{-j2\pi k/N}, 0 \le k \le N \tag{2}$$

Among them, the $x(t)$ is the input voice signal, N is the number of Fourier transform points, and a is the frame index number.

3> Mel-filter and logarithm. The amplitude spectrum $|X_a(k)|$ is obtained by taking the mode of the spectrum $X_a(k)$ obtained in the previous step, and then the power spectrum $|X_a(k)|^2$ is obtained by squaring it. The power spectrum is passed through a group of M Mel-filter banks $H_m(k)$, and the dot product operation is performed and the logarithm is taken. In this paper, the number of Mel-filters M is set to 128, and the logarithm of the output of each filter is calculated as shown in formula (3):

$$S(m) = \ln(\sum_{k=0}^{N-1} |X_a(k)|^2 H_m(k)), 0 \le m \le M \tag{3}$$

Among them, the $S(m)$ is the logarithmic energy.

4> Discrete Cosine Transform, DCT. The MFCC coefficient is obtained by DCT:

$$MFCC(n) = \sum_{m=0}^{N-1} S(m)\cos(\frac{\pi n(m - 0.5)}{M}), n = 1, 2, \ldots L \tag{4}$$

Among them, the L is the order of MFCC. Taking the logarithmic energy $S(m)$ into DCT, the parameters of L-order MFCC are obtained. The purpose of DCT is to change the data distribution and separate the redundant data. Because most of the signal data will be concentrated in the low frequency region after DCT, only the low order coefficients of MFCC after DCT are needed. In this algorithm, $L = 13$ is set, the standard MFCC coefficients of order 13 are taken.

2.2 Feature Matrix Construction

The standard MFCC parameters can only reflect the static characteristics of voice, and the dynamic characteristics of voice can be described by the difference function of these static features, combining dynamic and static characteristics can effectively improve the recognition performance of the system. Therefore, in addition to the above 13-order standard MFCC feature parameters, the perceptual feature matrix constructed in this paper also includes the first- and second-order difference parameters of the 13th-order MFCC parameters used to describe dynamic characteristics in order to improve recognition performance. After framing a voice signal, the total number of frames obtained is m, then the perceptual feature parameters include $13 \times$ m-dimensional standard MFCC parameters, $13 \times$ m-dimensional first-order difference parameters and $13 \times$ m-dimensional second-order difference parameters, the perceptual feature matrix construction is shown in Fig. 3.

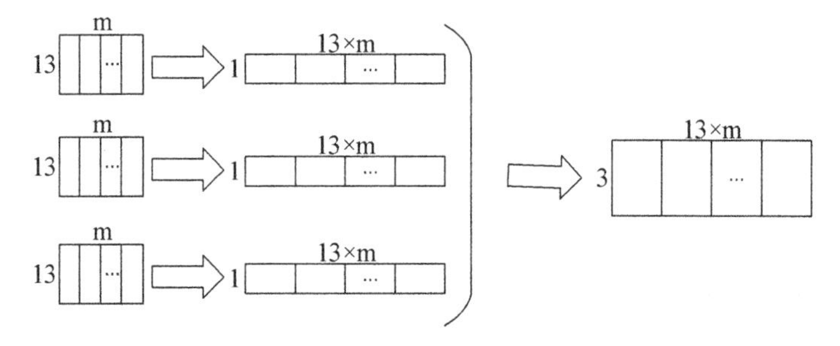

Fig. 3. Feature matrix construction process

The MFCC parameter and its first-order second-order differential parameters are $13 \times$ m dimensional matrixes, each of which tiles it by column, obtains 3 row matrixes which is $1 \times$ 13m-dimensional, and then stacks the 3 row matrixes into a matrix, the 1st is standard MFCC characteristic parameter, the 2nd is its first-order differences parameters, and the 3rd is its second-order differences parameters. Then we have a $3 \times$ 13m dimensional feature matrix which can be used for the construction of voice perceptual hash chain.

2.3 Feature Matrix Factorization

In this paper, use two-dimensional non-negative matrix factorization (2DNMF) to degraded reduction the feature matrix by using the non-negative matrix factorization

(NMF) method twice. Non-negative matrix decomposition exerts non-negative constraints on the data matrix, so that only addition combinations are allowed in the factorization process [16], resulting in the whole data being superimposed by part without positive or negative offsetting, and achieving the effect of the part expressing the whole, which coincides with the perceptual basis of human brain, and has the characteristics of fast convergence and small storage space. 2DNMF was first proposed by the paper [17] and applied to the field of image processing. The specific process of this method is shown in Fig. 4.

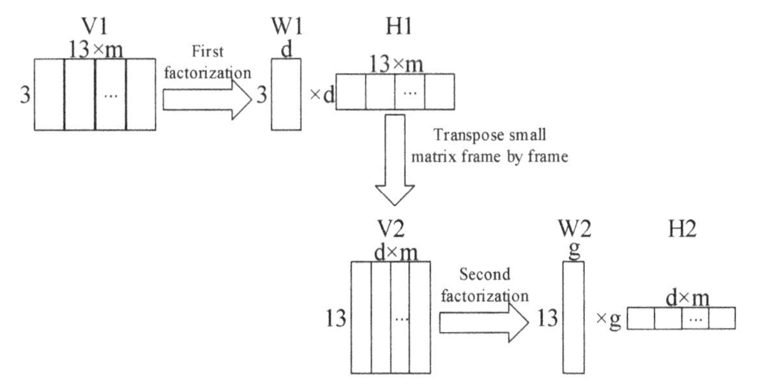

Fig. 4. Feature matrix factorization

The feature matrix constructed in Sect. 2.2 is denoted as $V1_{3\times 13m}$, which is decomposed into $W1_{3\times d}$ and $W1_{3\times d}$ by NMF:

$$V1_{3\times 13m} \approx W1_{3\times d}H1_{d\times 13m} \tag{5}$$

Among them, d is the decomposition rank, also called decomposition order. Take $H1_{d\times 13m}$, transpose each $d \times 13$ matrix frame by frame according to the number of total frames m, get $V2_{13\times dm}$, and then continue the matrix decomposition, that is

$$V2_{13\times dm} \approx W2_{13\times g}H2_{g\times dm} \tag{6}$$

Among them, g is the decomposition rank.

In the above decomposition process, the values of d and g need to satisfy $d < 3$, $g < 13$. In order to facilitate the subsequent calculation, the algorithm in this paper sets $d = 1$, $g = 5$, that is, after the decomposition of formula (5) and (6), the coefficient matrix is $H2_{5\times m}$, which is recorded as the result matrix C.

2.4 Perceptual Hash Chain Construction

The dimension reduced result matrix is used as the basis to construct the perceptual hash chain. Sum matrix C column by column according to the following formula (7) to get the element sum of each frame of the corresponding voice signal, which is recorded as $S(a)$:

$$S(a) = \sum_{i=1}^{g=5} C_{ia}, 1 \leq a \leq m \tag{7}$$

Among them, i is the row index of the matrix and a is the frame index. The average value of $S(a)$ is calculated as $\overline{S}$, and the voice perceptual hash chain H is constructed according to the following formula:

$$h_a = \begin{cases} 1, S(a) > \overline{S} \\ 0, other \end{cases}, 1 \leq a \leq m \tag{8}$$

Finally, the voice perceptual hash chain is $H = [h_1 h_2 ... h_m]$.

3 Perceptual Hash Authentication

In this algorithm, the authentication of voice is essentially the authentication of the perceptual hash chains of voice signals. In the certification process, the similarity of the two perceptual hash chains is measured by normalizing Hamming distance, and the normalized Hamming distance is defined as follows:

$$MFCC(n) = \sum_{m=0}^{N-1} S(m) \cos(\frac{\pi n(m - 0.5)}{M}), n = 1, 2, ...L \tag{9}$$

Among them, H^x and H^y are perceptual hash chains of two segments of voice respectively, L is the length of perceptual hash chain.

Normalized Hamming distance can also be defined as Bit Error Ratio (BER), and the hypothetical test using BER describes the perceptual hash chain authentication as follows: for two voice segments $x(t)$ and $y(t)$, construct the perceptual hash chain H^x and H^y separately, if the voice segments' perceptual content is same, then $D(H^x, H^y) \leq \tau$; if the voice segments' perceptual content is not same, then $D(H^x, H^y) > \tau$. Among them, τ is authentication threshold. Therefore, the perceptual hash chain authentication criteria are as follows: set the authentication threshold τ, compare the perceptual hash chains' mathematical distance D, if the mathematical distance less than the threshold, then the corresponding two voice segments are considered to be the same as the perceptual content, and the authentication is passed; otherwise, the authentication is not passed.

The specific authentication process is shown in Fig. 5, $x(t)$ is the original voice signal and $y(t)$ is the voice signal to be authenticated.

To evaluate the above authentication algorithm, define the false accept rate (FAR). FAR refers to the percentage that is incorrectly accepted, i.e., the percentage misjudged as passed by the perceptual hash chain that should not be passed:

$$\gamma_\tau = P(D \leq \tau) \tag{10}$$

Among them, γ_τ is FAR, $P(\bullet)$ is probability, τ is authentication threshold. The smaller τ, the lower the FAR, the better the distinguishability of the perceptual hash, and the higher the authentication accuracy.

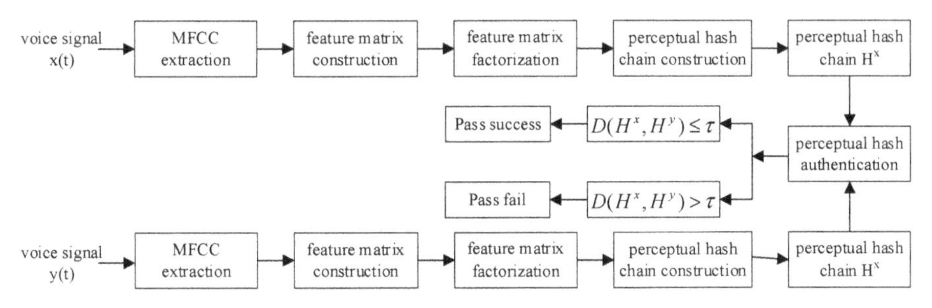

Fig. 5. Perceptual hash authentication process

4 Experimental Results and Analysis

In order to verify the distinguishability and robustness of the algorithm, the results of simulation experiments are carried out and analyzed. The hardware platform used in the experiment is Intel(R) Core(TM) i5-8300H CPU @ 2.30 GHz, the software environment is Python-3.6.9, and the audio data used in the experiment is 13388 audio files in wav format in THCHS-30 (Tsinghua University 30 h Chinese voice library). The main parameters of the experiment are set as follows: the frame length is 2048 and the frame shift is 512.

4.1 Distinguishability

Distinguishability is used to evaluate the reliability of algorithms to distinguish between different voice content by different or identical people. The BER for perceptual hash values for different content audio basically obeys the normal distribution. The audio data used in the experiment included 60 different speaker, 2 different voice segments for each speaker, and ensured that all voice contents are different, taking a total of 120 different audio files as test data for this section, as follows.

Construct the perceptual hash chain of these 120 audio segments separately, and the normalized Hamming distance D between each two hash chains is calculated by the formula (9), which is 7140 sets of data in total, the probability distribution histogram is shown in Fig. 6.

Assuming that the result conforms to a normal distribution, its mathematical expectations μ_D is 0.4497, its standard deviation σ_D is 0.0523, and theoretically its FAR is the probability integral of the normal distribution:

$$\gamma_\tau = P(D \le \tau) = \int_0^\tau \frac{1}{\sqrt{2\pi}\sigma} e^{\frac{-(x-\mu)^2}{2\sigma^2}} dx \tag{11}$$

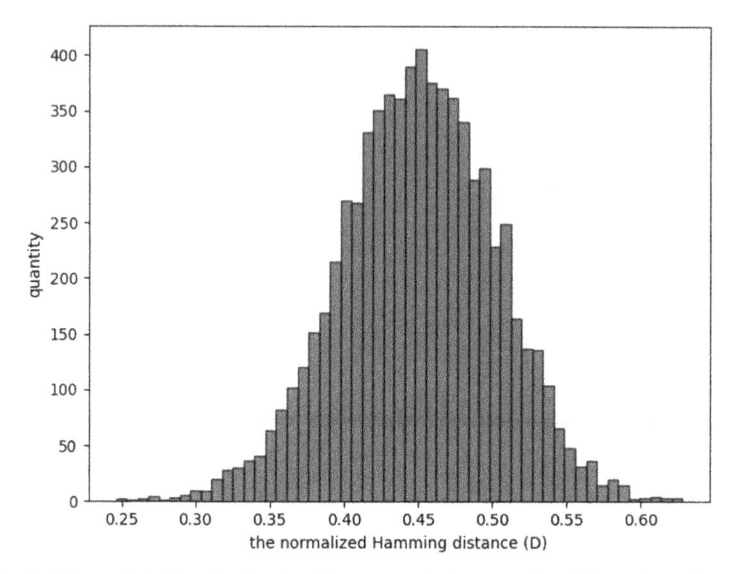

Fig. 6. The probability distribution histogram of the normalized Hamming distance

FAR reflects the distinguishability of the algorithm, the higher the threshold, the larger the FAR, the lower the distinguishability and the weaker the anti-collision ability of the algorithm. Figure 7 shows the comparison between the theoretical FAR and the experimental FAR. It can be seen from Fig. 7(a) that the two curves basically coincide, which proves that the experimental results are in line with the expectation and belong to normal distribution. Figure 7(b) enlarges the local part of the two curves. When the value τ is less than 0.3, the FAR is at a low level. At this time, the algorithm can be almost completely distinguished, so the threshold τ can be set to a value less than 0.3.

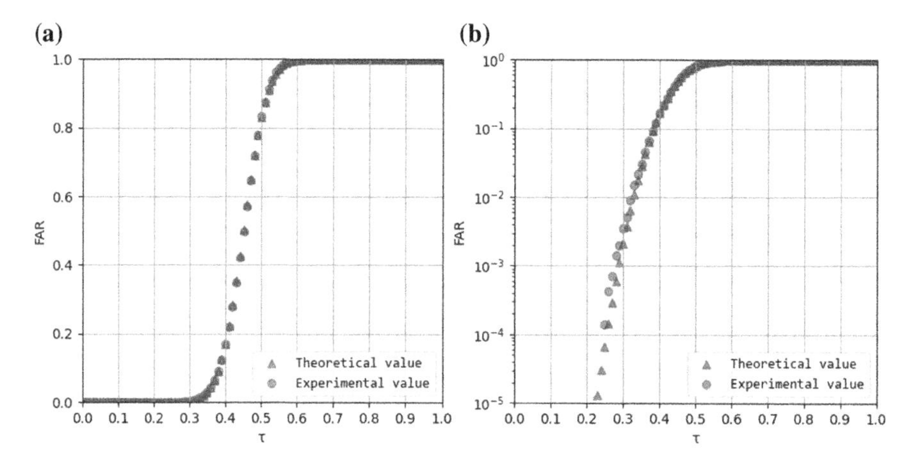

Fig. 7. The FAR

As shown in Fig. 7, the theoretical values of the FAR of this algorithm are 0.0021 when τ is 0.3, 6.717×10^{-5} when τ is 0.25 and 1.143×10^{-11} when τ is 0.2. That is, when there are enough voice sample data, the threshold value is set to 0.2, and about one of every 10^{11} voice segments will be wrongly recognized, which indicates that the algorithm in this paper has strong anti-collision ability and good distinguishability. However, compared with the literature [18, 19], there is still a large room for improvement.

4.2 Robustness

In order to verify the robustness of the algorithm proposed in this paper, that is, the same voice can still be effectively certified reliability after the content is maintained, this section uses some actions that do not change the voice content to interfere with the test audio, as shown in Table 1.

Table 1. Maintain content operations

The type of action	How it works	Symbol abbreviation
Turn down the volume	Reduce the volume by 50%	V.↓
Turn up the volume	Increase the volume by 50%	V.↑
Resampling	The sample rate is reduced to 0.8×10^4 and then to 1.6×10^4	R.8 → R.16
Echo	Overlays produce echoes	E.
Noise	50 dB Gauss noise	N.
Butterworth filtering	6th order Butterworth low pass filter, cutoff frequency 3400	B.W
FIR filtering	6th order FIR low pass filter, cutoff frequency 3400	F.I.R

In the experiment, all 13,388 audio test files are respectively operated as shown in Table 1, each test file generates 7 corresponding audio files, constructing the voice perceptual hash chain after the operation and the original voice perceptual hash chain respectively, and calculating the Hamming distance, which is worth 7 sets of BER averages, as shown in Table 2 and compared to the paper [12, 18, 19].

As can be seen from the data listed in Table 2, the algorithm proposed in this paper has a maximum value of 0.2450 for the audio BER mean with the same perceptual content, and the gap with paper [12] is small, indicating that the algorithm in this paper has good robustness. And by comparing with the data of paper [18, 19], for the operation of noise and low-pass filtering, the BER mean of this algorithm is obviously lower, which shows that the algorithm has better robustness in combating these operations.

Table 2. BER average

Action	This paper	Paper [12] I	Paper [12] II	Paper [12] III	Paper [12] IV	Paper [12] V	Paper [18]	Paper [19]
V.↓	0.0310	0.0202	0.0173	0.0346	0.1580	0.1097	0.0004	0.0002
V.↑	0.0821	0.0088	0.0095	0.0188	0.1013	0.1055	0.0116	0.0173
R.8 → R.16	0.0879	0.0841	0.0143	0.0230	0.0906	0.0508	0.0002	0.0026
E.	0.2450	0.1947	0.2002	0.1831	0.2327	0.2324	–	0.1137
N.	0.0260	0.0488	0.1564	0.1285	0.1834	0.0731	0.0581	–
B.W	0.0813	0.1450	0.1641	0.1449	0.1996	0.2109	0.1057	0.1412
F.I.R	0.0488	0.1551	0.1755	0.1611	0.2187	0.2056	0.1214	0.1520

The above experimental results show that the algorithm proposed in this paper has a high accuracy of voice perceptual hash authentication, and has good distinguishability and robustness. It can accurately and reliably authenticate the voice that has been operated by content hold. The disadvantage is that the FAR of this algorithm has not been reduced to the most ideal level.

5 Summary and Prospect

In this paper, a reliable voice perceptual hash authentication algorithm is proposed, which uses the 2DNMF method to degrade the MFCC feature matrix, and finally constructs the perceptual hash chain according to the decomposition results, and the perceptual hash authentication is carried out by normalizing hamming distance measurement. The experimental results show that the algorithm has good distinguishability and robustness, and has better robustness for low-pass filtering, noise and other disturbances. The next research direction is to refine the authentication results, analyze the relationship between the change of perceptual hash chain and content tampering, and detect and locate the tampering of the test voice after authentication.

References

1. Tang, Z., Zhang, X., Huang, L., et al.: Robust image hashing using ring-based entropies. Sig. Process. **93**(7), 2061–2069 (2013)
2. Niu, X., Jiao, Y.: An overview of perceptual hashing. Acta Electron. Sin. **07**, 1405–1411 (2008)
3. Li, Z., Zhu, M., Chen, Z.: Object tracking algorithm based on perception hash technology. J. Image Graph. **20**(006), 795–804 (2015)
4. Zhang, W., Kong, X., You, X.: Secure and robust image perceptual hashing. J. SE Univ. (Nat. Sci. Ed.) **000**(S1), 188–192 (2007)
5. Qin, C., Sun, M., Chang, C.C.: Perceptual hashing for color images based on hybrid extraction of structural features. Sig. Process. **142**, 194–205 (2018)

6. Sabahi, F., Ahmad, M.O., Swamy, M.N.S.: Content-based image retrieval using perceptual image hashing and hopfield neural network. In: 2018 IEEE 61st International Midwest Symposium on Circuits and Systems (MWSCAS), pp. 352–355. IEEE (2018)
7. Zhang, Q., Qiao, S., Huang, Y., et al.: A high-performance voice perceptual hashing authentication algorithm based on discrete wavelet transform and measurement matrix. Multimedia Tools Appl. **77**(16), 21653–21669 (2018)
8. Saikia, N., Bora, P.K.: Perceptual hash function for scalable video. Int. J. Inf. Secur. **13**(1), 81–93 (2014)
9. Yang, M., Tang, G., Liu, X., et al.: Low-light image enhancement based on Retinex theory and dual-tree complex wavelet transform. Optoelectron. Lett. **14**(6), 470–475 (2018)
10. Li, J., Wang, H., Jing, Y.: Audio perceptual hashing based on NMF and MDCT coefficients. Chin. J. Electron. **24**(3), 579–588 (2015)
11. Ramalingam, A., Krishnan, S.: Gaussian mixture modeling of short-time Fourier transform features for audio fingerprinting. IEEE Trans. Inf. Forensics Secur. **1**(4), 457–463 (2006)
12. Zhang, Q., Qiao, S., Zhang, T., Huang, Y.: Perceptual hashing authentication algorithm for multi-format audio based on energy to zero ratio. J. Huazhong Univ. Sci. Technol. (Nat. Sci. Ed.) **45**(09), 33–38 (2017)
13. Huang, Y., Zhang, Q., Yuan, Z., Yang, Z.: The hash algorithm of voice perception based on the integration of adaptive MFCC and LPCC. J. Huazhong Univ. Sci. Technol. (Nat. Sci. Ed.) **43**(02), 124–128 (2015)
14. Li, J., Wu, T., Wang, H.: Perceptual hashing based on correlation coefficient of MFCC for voice authentication. J. Beijing Univ. Posts Telecommun. **38**, 89 (2015)
15. Zhang, T.: Research on voice perceptual hashing authentication method and its application in mobile terminal. MS thesis. Lanzhou University of Technology (2018)
16. Bao, C., Bai, Z.: Voice enhancement based on nonnegative matrix factorization: an overview. J. Sig. Process. **36**(6), 791–803 (2020)
17. Zhang, D., Chen, S., Zhou, Z.-H.: Two-dimensional non-negative matrix factorization for face representation and recognition. In: Zhao, W., Gong, S., Tang, X. (eds.) AMFG 2005. LNCS, vol. 3723, pp. 350–363. Springer, Heidelberg (2005). https://doi.org/10.1007/11564386_27
18. Zhang, Y., Mi, B., Zhou, L., Zhang, T.: Speech perception hash authentication algorithm based on short-term autocorrelation. Radio Eng. **49**(10), 899–904 (2019)
19. Zhang, D.: Research on ciphertext voice content authentication and tampering recovery method based on perceptual hash in cloud environment. Lanzhou Univ. Technol. (2020)

PAID: Privacy-Preserving Incentive Mechanism Based on Truth Discovery for Mobile Crowdsensing

Tao Wan[1(✉)], Shixin Yue[1], and Weichuan Liao[2]

[1] School of Information Engineer, East China Jiaotong University, Nanchang 330013, China
[2] School of Science, East China Jiaotong University, Nanchang 330013, China

Abstract. Incentive mechanisms are an essential method to encourage users to participate in the mobile crowdsensing task. However, most incentive mechanisms based on data quality do not consider users' security and privacy protection. To overcome the above problems, we propose a privacy protection incentive mechanism based on truth discovery, named PAID. Specifically, we use the secure truth discovery scheme to calculate ground truth and the weight of users' data while protecting their privacy. Besides, to ensure the accuracy of the MCS results, a data eligibility assessment is proposed to remove unreliable user data before the truth discovery scheme. Finally, we distribute rewards to users based on their data quality. The analysis and evaluation demonstrate the security and effectiveness of our PAID.

Keywords: Incentive mechanism · Truth discovery · Mobile crowdsensing · Privacy-preserving

1 Introduction

As more and more sensors are integrated into human-carried mobile devices, such as GPS locators, gyroscopes, environmental sensors, and accelerometers, they can collect various types of data. Therefore, the MCS system [1,2] can utilize the sensors equipped in mobile devices to collect sensing data and complete various sensing tasks [3], such as navigation service, traffic monitoring, indoor positioning, and environmental monitoring. In general, the MCS system consists of three entities: a task requester, a sensing server, and participating users. The task requester publishes sensing tasks and pays awards for sensing results. The server recruits users according to the sensing task, processes the data from users, and sends the results to the task publisher. Users collect sensing data based on the requirements of the sensing task and get rewards.

In the practical MCS system, the sensing data collected by users are not always unreliable [4] due to various factors (such as poor sensor quality, lack of

J. Xiong et al. (Eds.): MobiMedia 2021, LNICST 394, pp. 264–277, 2021.
https://doi.org/10.1007/978-3-030-89814-4_20

effort, background noise). Therefore, the final result may be inaccurate if we treat the data provided by each user equally (e.g., averaging). To solve this problem, truth discovery [5,6] has been widely concerned by industry and academia. But one problem with these methods is that users have to be online to interact with the server. Therefore, if we design a truth discovery scheme that allows users to exit, the MCS system can get stronger robustness.

The proper functioning of the truth discovery requires enough users and high-quality sensing data. Generally, the MCS system utilizes an incentive mechanism [7] to motivate sufficient users to participate in sensing tasks. However, because of monetary incentives, malicious users attempt to earn rewards with little or no effort. Consequently, the evaluation of data quality is critical to the MCS system. To improve data quality, users who provide incorrect data can be removed before sensing data aggregated [8]. And the MCS system can output more accurate aggregation results.

Although the incentive mechanism has been improved a lot, users' privacy protection remains inadequate. When users submit sensing data, their sensitive or private information [9] may be leaked, including identity privacy, location privacy, and data privacy. And privacy disclosure [10] will reduce users' willingness to participate in sensing tasks. Recently, some researchers have designed the incentive mechanism scheme of privacy protection [11,12]. In [8], an incentive method is proposed to protect the user's identity and data privacy. Still, the user's sensing data will be submitted to the task publisher regardless of the privacy of the sensing data.

To address these issues, we propose a privacy-preserving incentive mechanism based on truth discovery, called PAID. In our PAID, the task publisher set data constraints, such as time, location, budget, and sensing data. If the user does not collect the sensing data at the required time and location or sensing data is not in the qualified range, we believe that the user's sensing data is not credible (i.e., unqualified). After removing the unqualified user's data, the qualified user's sensing data will be submitted to the server to calculate the ground truth and weight. We also design a secure truth discovery scheme, which can still work when some users drop out. Moreover, our truth discovery can ensure that other parties cannot obtain users' sensing data except users themselves. Finally, we calculate every user's data quality according to the weight and distribute the reward.

In summary, the main contributions of this paper are as follows:

- We propose a method to judge whether the data is in the qualified range. And this method will not disclose users' data and the qualified interval in the implementation process.
- We design a security truth discovery scheme, which can compute ground truth and users' weight. In this scheme, any party can not get the user's sensing data except himself. And the method can allow users to drop out.
- Our PAID accomplishes reward distribution according to data quality. The data quality is calculated by the weight.

2 Problem Statement

In this section, we introduce the background of truth discovery and our design goals.

2.1 Truth Discovery

Truth discovery [13] is widely used in the MCS system to solve the conflicts between sensing data collected from multiple sources. Although the methods of estimating weights and calculating ground truth are different, their general processes are similar. Specifically, truth discovery initializes a random ground truth and then iteratively updates the weight and ground truth until convergence.

Weight Update: Suppose that the ground truth of the object is fixed. If the user's sensing data is close to the ground truth, a higher weight should be assigned to the user. The weight w_i of each user u_i can be iteratively updated as follows:

$$w_i = log\left(\frac{\sum_{i'=1}^{|U|} d_{ist}(x_{i'}, x^*)}{d_{ist}(x_i, x^*)}\right) \tag{1}$$

where $d_{ist}(x_i, x^*)$ is a distance function, and $d_{ist}(x_i, x^*) = (x_i - x^*)^2$. We use U to represent the set of users, and $|U|$ is the number of users in the set U. The sensing data collected by the user u_i is denoted as x_i, which i is the number of u_i. And x^* is the estimated ground truth.

Truth Update: Similarly, we assume that the weight w_i of each user u_i is fixed. Then we can calculate the ground truth x^* as follows.

$$x^* = \frac{\sum_{i=1}^{|U|} w_i \cdot x_i}{\sum_{i=1}^{|U|} w_i} \tag{2}$$

The final ground truth x^* is obtained by iteratively running the weight update and the truth update when the convergence condition is satisfied.

2.2 Design Goals

In this section, we introduce the design goals of our PAID, which are divided into privacy and security goals and property goals.

The privacy goals can protect the user's private data, and the security goals can avoid malicious attacks. The details are as follows.

- **Privacy goals.** PAID can protect user's location privacy, data privacy, and identity privacy. Specifically, the location and sensing data of a user can not be obtained by any other parties except the user himself. And users' real identities would not be disclosed when performing a sensing task.

- **Security goals.** In our PAID, users can avoid the denial of payment attack (DoP) of TP. The server $\mathcal{S}$ cannot initiate an inference attack (IA) on users. The server $\mathcal{S}$ can resist the data pollution attack (DPA) launched by malicious users. And our PAID guarantees fairness by resisting the Sybil attack (SA).

Our PAID also requires the following property goals.

- **Eligibility.** If users' data do not meet the eligibility requirements, they cannot pass the eligibility assessment. In other words, the sensing data adopted by our PAID must be eligible.
- **Zero-knowledge.** When the server $\mathcal{S}$ assesses whether users' data meets the eligibility requirements, it cannot obtain the content of users' private data.

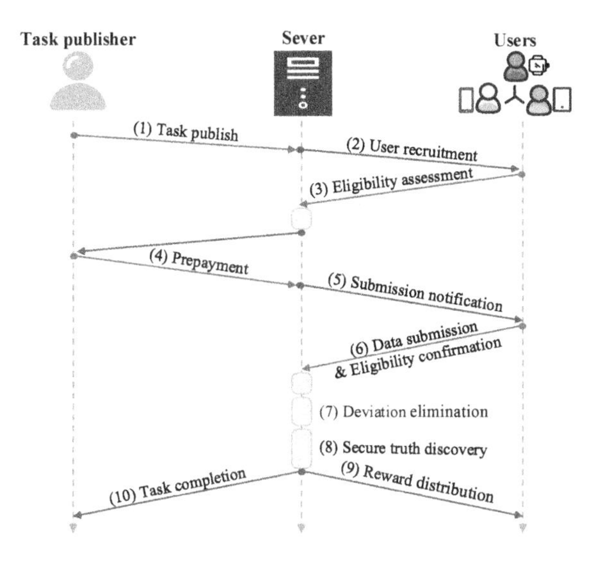

Fig. 1. System model of PAID.

3 Preliminaries

In this section, we review the cryptographic primitives used in our PAID.

3.1 Secret Sharing

We use Shamir's t-out-of-N secret sharing protocol, which can split each user's secret s into N shares, where any t shares can be used to reconstruct s. Still, it is impossible to get any information about s if the shares obtained by attackers are less than t.

Assume that some integers can be identified with distinct elements in a finite field $\mathcal{F}$, where $\mathcal{F}$ is parameterized with a size of $l > 2^k$ (which k is the security parameter). These integers can represent all users' IDs, and we use a symbol U to denote the set of users' IDs. Then the Shamir's secret sharing protocol consists of two steps as below.

- **Shamir.share**$(s, t, U) \rightarrow \{(u_i, s_i)\}_{u_i \in U}$. The inputs of the sharing algorithm are a secret s, a threshold $t \leq |U|$, and a set U of N field elements denoting the users' ID, where $|U| = N$. It outputs a set of shares s_i, each of which is associated with its corresponding the user u_i.
- **Shamir.recon**$(\{(u_i, s_i)\}_{u_i \in \mathcal{M}}, t) \rightarrow s$. The inputs of the reconstruction algorithm are the shares corresponding to a subset $\mathcal{M} \subseteq U$ and a threshold t, where $t \leq |\mathcal{M}|$, and it outputs the secret s.

3.2 Key Agreement

We utilize Diffie-Hellman key agreement called SIGMA [14] in our PAID to generate a session key between two users. Typically, SIGMA is described in three parts as follows.

- **KA.param**$(k) \rightarrow (\mathbb{G}, g, q, H)$. The algorithm's input is a security parameter k. It samples a group $\mathbb{G}$ of prime order q, along with a generator g and a hash function H, where H is set as SHA-256 for practicability in our model.
- **KA.gen**$(\mathbb{G}, g, q, H) \rightarrow (x, g^x)$. The algorithm's inputs are a group $\mathbb{G}$ of prime order q, along with a generator g and a hash function H. It samples a random $x \leftarrow Z_q$ and g^x, where x and g^x will be marked as the secret key $\mathtt{SK}_i$ and the public key $\mathtt{PK}_i$ in the following sections.
- **KA.agree**$(sign_j(g^{x_i}, g^{x_j}), MAC_k(u_j), x_i, g^{x_j}, i, j) \rightarrow s_{i,j}$. The algorithm's inputs are the user u_i's secret key x_i, the user u_j's public key g^{x_j}, signed signature $sign_j(g^{x_i}, g^{x_j})$ and $MAC_{k_v}(u_j)$ from the user u_j, where k_v is used as the MAC key. It outputs a session key $s_{i,j}$ between user u_i and user u_j. For simplicity, we use **KA.agree**$(x_i, g^{x_j}) \rightarrow s_{i,j}$ to represent the above process in the following sections.

3.3 Paillier Cryptosystem

The Paillier Cryptosystem is a probabilistic public key Cryptosystem. It consists of three parts as follows.

- **Paillier.gen**$(N, g) \rightarrow (sk_p, pk_p)$. The key distribution algorithm inputs are a number N and $g \leftarrow Z^*_{N^2}$, where N is the product of two large primes p, q. It outputs a secret key sk_p and a public key pk_p, where pk_p is computed by (N, g), and $sk_p = lcm(p-1, q-1)$.
- **Paillier.enc**$(m, pk_p) \rightarrow c$. The encryption algorithm inputs are a plaintext m (which $m < N$) and a public key pk_p. It outputs a ciphertext c.
- **Paillier.dec**$(c, sk_p) \rightarrow m$. The decryption algorithm inputs are a ciphertext c (which $c < N^2$) and a secret key sk_p. It outputs a plaintext m.

The Paillier cryptosystem has the property of homomorphic addition.

$$E_{pk}(a+b) = E_{pk}(a) \cdot E_{pk}(b) \pmod{N^2}, \tag{3}$$

We assume that E is an encryption function.

4 Technical Intuition

In this section, we first introduce how the interval judgment scheme can judge users' data eligibility under protecting users' privacy. Then, we notice that truth discovery mainly involves the aggregation of multiple users' data in a secure manner. Therefore, we require that the server $\mathcal{S}$ only get the sum of users' input, not content. And we propose a double-masking scheme to achieve this goal. Finally, we introduce the process of secure truth discovery.

4.1 Interval Judgment Scheme for Privacy Protection

In our PAID, we use the interval judgment scheme [15] based on the Paillier cryptosystem to determine the sensing data eligibility. Every user u_i provides a sensing data x_i, and the server provides a continuous integer interval $[y_1, y_2]$ ($y_1, y_2 \leftarrow Z^*$). The server $\mathcal{S}$ can judge whether the user u_i's sensing data x_i meets the interval range $[y_1, y_2]$ without knowing the data x_i. The user u_i also cannot obtain any information about the integer interval. The scheme is divided into four steps as follows.

- The user u_i gets $(pk_p, sk_p) \leftarrow \textbf{Paillier.gen}(N, g)$. Then u_i computes $E(x_i)$ using pk_p and sends it to $\mathcal{S}$.
- The server $\mathcal{S}$ picks two random numbers k, b $(k, b \leftarrow Z^*)$ to construct a monotone increasing (or decreasing) function $f(x_i) = kx_i + b$. Then the server $\mathcal{S}$ computes $f(y_1), f(y_2), c = E(x_i)^k E(b) = E(kx + b)$, and sends them to u_i.
- After receiving the information from the server $\mathcal{S}$, the user u_i gets $f(x_i) \leftarrow \textbf{Paillier.dec}(c, sk)$, then compares the size of $f(y_1)$, $f(y_2)$, and $f(x_i)$. Next, send the message to the server $\mathcal{S}$.
- After receiving the message from u_i, the server $\mathcal{S}$ judges whether $f(y_1) < f(x_d) < f(y_2)$. If so, we can know $x_i \in [y_1, y_2]$ because of the monotonicity of the function $f(x_i) = kx_i + b$, i.e., the user u_i passes the data eligibility assessment. Otherwise, the user u_i fails to pass the eligibility assessment of the server $\mathcal{S}$.

For simplicity, We formulate the above process as an interval judgment function denoted by $ins(x_i, y_1, y_2)$. If the user u_i passes the eligibility assessment of the server $\mathcal{S}$, $ins(x_i, y_1, y_2) = 1$, otherwise $ins(x_i, y_1, y_2) = 0$.

4.2 One-Masking to Protect Security

Assume that all users are represented in sequence as integers $1, \ldots, n$. And any pair of users (u_i, u_j), $i < j$ agree on a random value $r_{i,j}$. Let's add $r_{i,j}$ to the user u_i's data x_i and subtract $r_{i,j}$ from the user u_j's data x_j to mask all users' raw data. In other words, each user u_i computes as follows.

$$y_i = x_i + \sum_{u_j \in U : i < j} r_{i,j} - \sum_{u_j \in U : i > j} r_{j,i} \pmod{R}, \tag{4}$$

where we assume x_i and $\sum_{u_j \in U} r_{i,j}$ is in Z_R with order R for simplicity.

Then, each user u_i submits y_i to the server $\mathcal{S}$, and $\mathcal{S}$ computes:

$$\begin{aligned} z &= \sum_{u_i \in U} y_i \\ &= \sum_{u_i \in U} \left(x_i + \sum_{u_j \in U: i<j} r_{i,j} - \sum_{u_j \in U: i>j} r_{j,i} \right) \\ &= \sum_{u_i \in U} x_i \pmod{R}. \end{aligned} \tag{5}$$

However, this approach has two shortcomings. The first one is that every user u_i needs to exchange the value $r_{i,j}$ with all other users, which will result in quadratic communication overhead ($|U|^2$) if done naively. The second one is that the protocol will fail if any user u_i drops out since the server can't eliminate the value $r_{i,j}$ associated with u_i in the final aggregated results z.

4.3 Double-Masking to Protect Security

To solve these security problems, we introduce a double-masking scheme [16].

Every user u_i can get a session key $r_{i,j}$ with other user u_j by engaging the Diffie-Hellman key agreement after the server $\mathcal{S}$ broadcasting all of the Diffie-Hellman public keys.

We use the threshold secret sharing scheme to solve the issue that users are not allowed to drop out. Every user u_i can send his secret's shares to other users. Once some users cannot submit data in time, other users can recover masks associated with these users by submitting shares of these users' secrets to $\mathcal{S}$, as long as the number of dropped users is less than t (i.e., threshold of Shamir's secret sharing).

However, there is a problem that may lead to users' data leaked to $\mathcal{S}$. There is a scenario where a user u_i is very slow to send data to the server $\mathcal{S}$. The server $\mathcal{S}$ considers that the user u_i has dropped and asks for their shares of the user u_i's secret from all other users. Then, the server $\mathcal{S}$ receives the delayed data y_i after recovering u_i's mask. At this time, the server $\mathcal{S}$ can remove all the masks $r_{i,j}$ and get the plaintext x_i.

To improve the scheme, we introduce an additional random seed $\mathtt{n}_i$ to mask the data. Specifically, each user u_i selects a random seed $\mathtt{n}_i$ on the round of generating $r_{i,j}$, then creates and distributes shares of $\mathtt{n}_i$ to all other users during the secret sharing round. Now, users calculate y_i as follows:

$$\begin{aligned} y_i = & x_i + \mathbf{PRG}(\mathtt{n}_i) + \sum_{u_j \in U: i<j} \mathbf{PRG}(r_{i,j}) \\ & - \sum_{u_j \in U: i>j} \mathbf{PRG}(r_{j,i}) \pmod{R}. \end{aligned} \tag{6}$$

Note that an honest user will never reveal both kinds of shares of the same user to the server $\mathcal{S}$. During the recovery round, the server $\mathcal{S}$ can request either a share of $r_{i,j}$ or a share of $\mathtt{n}_i$ from each surviving user u_j. After gathering at least t shares of $r_{i,j}$ for all dropped users and t shares of $\mathtt{n}_i$ for all surviving users, the server $\mathcal{S}$ can eliminate the remaining masks to reveal the sum.

4.4 Secure Truth Discovery

In the secure truth discovery scheme [6], data exchange is between users and the server $\mathcal{S}$. The user u_i needs to collect sensing data x_i, perform the double-masking scheme to mask the raw input data, and then send the masked input data to $\mathcal{S}$. The server $\mathcal{S}$ receives masked input data from each user u_i and aggregates the input data of online users. The main process can be summarized as follows.

Part 1 (Key Generation). A trusted third party creates three key pairs for each user u_i signature, session key, and noise value. Then, each user u_i generates shares of $\mathtt{n}_i$ using secret sharing protocol and sends the encrypted information to $\mathcal{S}$.

Part 2 (Masking Data). Each user u_i uses the double-masking scheme to mask his input data and sends it to $\mathcal{S}$.

Part 3 (Unmasking). After receiving the masking data, the server $\mathcal{S}$ performs a summation operation to obtain the sensing data aggregation result of surviving users. For dropped users, the server $\mathcal{S}$ restores their noise using the secret sharing protocol then eliminates the impact on the aggregation results.

Part 4 (Computing Ground Truth and Weight). After the server $\mathcal{S}$ gets the aggregation result, the server $\mathcal{S}$ iteratively calculates the ground truth x^* and weight w_i of every user u_i according to Formula 1 and Formula 2 until convergence. And the server $\mathcal{S}$ initializes a random ground truth x^* in the first calculation.

5 Our Proposed Scheme

In this section, we introduce the process of our model. For convenience, we introduce a simple case. We set up a sensing task $\mathcal{T}$ to collect the temperature of urban roads in the evening. There are range requirements for time, location, and sensing data (i.e., temperature). To be more precise, the time range is required to be 5–8 pm on February 3rd, the location range is required to be 12.45–12.55 E and 41.79–41.99 N, and the temperature requirement is 10–15°C. In our PAID, we consider the range requirement as the data eligibility requirement $\mathcal{E}$. The data $\mathcal{D}_i$ ($\mathcal{D}_i = (x_i, \tau_i, \hat{\iota_i}, \tilde{\iota_i})$) collected by a user u_i meet the eligibility requirements $\mathcal{E}$, meaning that $10 \le x_i \le 15, 5 \le \tau_i \le 8, 12.45 \le \hat{\iota_i} \le 12.55, 41.79 \le \tilde{\iota_i} \le 41.99$. Figure 1 shows the flow of our PAID. And the specific steps are as follows.

Step 1 (Task Publish). The task publisher TP initializes a public key $pk_{\mathcal{T}}$ and a private key $sk_{\mathcal{T}}$, a reward control parameter π (π is a decimal number), a task budget B, the number of users N, and eligibility requirements $\mathcal{E}$ for a sensing task $\mathcal{T}$. The public key $pk_{\mathcal{T}}$ is used to encrypt the information that the server $\mathcal{S}$ needs to send to the TP, and the TP decrypts the ciphertext using the private key $sk_{\mathcal{T}}$. Then the TP sends the information $\{\mathcal{T}, pk_{\mathcal{T}}, \pi, N, B, \mathcal{E}\}$ to $\mathcal{S}$ as a task request.

Step 2 (User Recruitment). The server $\mathcal{S}$ broadcasts the sensing task information $\{\mathcal{T}, \pi, N, B\}$ and recruits N users who request to participate in the sensing task. Then $\mathcal{S}$ generates a key pair $\{\mathtt{PK}_{\mathcal{S}}^i, \mathtt{SK}_{\mathcal{S}}^i\}$ using the key agreement scheme for every user u_i and sends $\mathtt{PK}_{\mathcal{S}}^i$ to u_i.

Step 3 (Eligibility Assessment). Each user u_i confirms whether $c_i \leq \frac{B-\pi}{N}$, where c_i denotes the sensing cost of u_i, and the posted lowest reward is denoted as $\frac{B-\pi}{N}$. If $c_i \leq \frac{B-\pi}{N}$, u_i starts the sensing task and collects the data $\mathcal{D}_i$. The user u_i then generates a key pair $\{\mathtt{PK}_i, \mathtt{SK}_i\}$ using the key agreement scheme and computes a session key $k_i \leftarrow \mathbf{KA}.\mathbf{agree}(\mathtt{SK}_i, \mathtt{PK}_{\mathcal{S}}^i)$ as u_i's anonymous identity information. Then the user u_i performs the interval judgment scheme $ins(\mathcal{D}_i, \mathcal{E})$ and sends the public key $\mathtt{PK}_i$ to $\mathcal{S}$. Specifically, $ins(\mathcal{D}_i, \mathcal{E})$ is divided into $ins(x_i, \mathcal{E})$, $ins(\tau_i, \mathcal{E})$, $ins(\hat{\iota}_i, \mathcal{E})$, $ins(\tilde{\iota}_i, \mathcal{E})$.

Step 4 (Prepayment). After recruiting N eligible users, the server $\mathcal{S}$ requests TP to prepay a budget reward B for the sensing task $\mathcal{T}$ to prevent the denial of payment attack. And the server $\mathcal{S}$ calculates the session key $k_i \leftarrow \mathbf{KA}.\mathbf{agree}(\mathtt{SK}_{\mathcal{S}}^i, \mathtt{PK}_i)$ with the eligible user u_i.

Step 5 (Submission Notification). After getting the budget reward B, the server $\mathcal{S}$ informs the eligible user u_i $(1 \leq i \leq N)$ to submit data.

Step 6 (Data Submission & Eligibility Confirmation). After receiving the submission notification, each user u_i performs double masking scheme to mask the sensing data x_i and get y_i, at the same time, execute eligibility confirmation $ins(\mathcal{D}_i, \mathcal{E})$ to prevent malicious users from modifying data. Then u_i encrypts the data y_i using the symmetric encryption algorithm and sends the ciphertext $\mathbf{SEnc}(y_i, k_i)$ to $\mathcal{S}$. The session key k_i is the key of symmetric encryption.

Step 7 (Secure Truth Discovery). The server $\mathcal{S}$ computes the surviving user u_i's weight w_i and the ground truth x^* of the sensing object utilizing the truth discovery algorithm. The detailed algorithm process will be introduced later.

Step 8 (Reward Distribution). The server $\mathcal{S}$ calculates the sensing data quality $q_i = \frac{w_i}{\sum_{i=1}^{m} w_i}$ of u_i, where $\sum_{i=1}^{m} q_i = 1$, m is the number of online users. Then $\mathcal{S}$ pays a monetary reward $p_i = \frac{B}{m} + \pi \cdot (q_i - \bar{q})$ for u_i, where $\pi \cdot (q_i - \bar{q})$ denotes the payment parameter, $m \leq N$, and $1 \leq i \leq m$.

Step 9 (Task Completion). The server $\mathcal{S}$ encrypts the ground truth x^* using $pk_{\mathcal{T}}$ and sends $\mathbf{Enc}(x^*, pk_{\mathcal{T}})$ to TP. And the TP can decrypt the data using $sk_{\mathcal{T}}$, i.e., $x^* = \mathbf{Dec}(\mathbf{Enc}(x^*, pk_{\mathcal{T}}), sk_{\mathcal{T}})$.

6 Analysis

In this section, we introduce property analysis, privacy analysis, and security analysis to illustrate the feasibility of our PAID.

6.1 Property Analysis

In this section, we introduce eligibility, zero-knowledge of our PAID.

Theorem 1 (Eligibility). If the data $\mathcal{D}_i$ ($\mathcal{D}_i = (x_i, \tau_i, \hat{\iota}_i, \tilde{\iota}_i)$) collected by users do not meet the eligibility requirement $\mathcal{E}$, these users cannot pass the eligibility assessment.

Proof. We assume that the user's data are denoted as s, and the eligibility requirement interval is $[a, b]$. The user gets ciphertext $E(s)$ using homomorphic encryption. Then $\mathcal{S}$ picks different random k, b, and constructs a monotone increasing (or decreasing) function $f(x) = kx + b$. Then $\mathcal{S}$ computes $f(a)$, $f(b)$, and $c = E(s)^k E(b) = E(ks + b)$. When receiving $f(a)$, $f(b)$, c from $\mathcal{S}$, the user decrypts c to get $f(s)$ and compare the sizes of $f(a)$, $f(b)$, $f(s)$. Because the user does not know the monotonicity of the function, it is impossible to determine the size relationship among the three numbers. Therefore, if the user's data is not qualified, then it cannot pass the qualification judgment.

Theorem 2 (Zero-knowledge). The server $\mathcal{S}$ can determine whether the user's data meets the eligibility requirements, but it cannot know the user's specific data content.

Proof. Similar to the description in Theorem 1, we assume that the user's data is s, and the server $\mathcal{S}$ can receive the user's homomorphic encrypted ciphertext $E(s)$. Since the Paillier Cryptosystem is indistinguishable under the chosen plaintext attack, a malicious user has no way to recover the plaintext s. The server $\mathcal{S}$ may be curious about each user's data, but it cannot obtain each user's data s without knowing the secret key.

6.2 Privacy Analysis

In this section, we demonstrate the protection of user sensing data, location, and identity privacy in our PAID.

Theorem 3 (Data and location privacy protection). In addition to the user himself, other parties cannot obtain the user's sensing data and location data.

Proof. In PAID, the objects that steal users' data and location privacy are mainly the server $\mathcal{S}$ and external attackers. Specifically, the server $\mathcal{S}$ may obtain users' sensing data and location privacy in eligibility assessment and truth discovery. External attackers steal data and location privacy by eavesdropping on the communication between the server $\mathcal{S}$ and users.

According to Theorem 2, we can know that our PAID is zero-knowledge, so the server $\mathcal{S}$ cannot learn users' sensing data and location data in the eligibility assessment. In truth discovery, users' sensing data is sent to $\mathcal{S}$ after double-masking. However, the server $\mathcal{S}$ can't recover users' raw sensing data by double-masking sensing data. Furthermore, before the communication between the user u_i and $\mathcal{S}$, the data is encrypted by AES symmetric encryption function $\mathbf{SEnc}(y_i, k_i)$. Therefore, as long as $\mathbf{SEnc}(y_i, k_i)$ is secure, external attackers cannot steal the data y_i by eavesdropping communication.

Theorem 4 (Identity privacy protection). When users participate in a sensing task, they use an anonymous identity rather than their real identity. Therefore, any PPT adversary cannot distinguish the users' identities.

Proof. In PAID, the anonymous identity of a user u_i is represented by $k_i \leftarrow \mathbf{KA.agree}(\mathtt{SK}_i, \mathtt{PK}_{\mathcal{S}}^i)$, and the real identity of u_i is $\mathtt{SK}_i$ where $\mathtt{SK}_i = x_i \leftarrow Z_q$, and $\mathtt{PK}_{\mathcal{S}}^i = g^{x_S^i}$ ($\mathtt{PK}_{\mathcal{S}}^i$ is a token assigned by $\mathcal{S}$). The user u_i uses an anonymous identity k_i rather than a real identity $\mathtt{SK}_i$ to participate in a sensing task. Because of the DDH problem, the PPT adversary cannot get the real identity $\mathtt{SK}_i$ of the user u_i by the anonymous identity k_i. We omit the detailed proof, and interested readers can learn more details in the literature [14].

6.3 Security Analysis

In this section, we describe the attacks our PAID can resist, including *Denial of Payment attack* (DoP), *Inference attack* (IA), *Data pollution attack* (DPA), and *Sybil attack* (SA).

(1) *Resistance to denial of payment attack* (DoP). We use the prepayment mechanism in our PAID. At the beginning of a sensing task, the task publisher TP pays the monetary rewards of users to $\mathcal{S}$ in advance. If a malicious TP refuses to pay the monetary reward after receiving the data, $\mathcal{S}$ can pay the reward to users according to the reward distribution formula. Therefore, the TP cannot refuse to pay users the reward.
(2) *Resistance to inference attack* (IA). The server $\mathcal{S}$ cannot initiate an inference attack against users' data since our PAID is zero-knowledge.
(3) *Resistance to Data pollution attack* (DPA). Our PAID introduces eligibility assessment, and the unqualified data submitted by users are not used in the truth discovery algorithm. Therefore, our PAID can resist the Data pollution attack (DPA).
(4) *Resistance to Sybil attack* (SA). The anonymous identity k_i of a user u_i needs the information $\mathtt{PK}_i$ provided by the user and the token $\mathtt{PK}_{\mathcal{S}}^i$ assigned by $\mathcal{S}$. Each user can only obtain one token from $\mathcal{S}$, then get the anonymous identity k_i using the key agreement algorithm. Hence, untrusted users cannot forge vast fake identities to launch the Sybil attack (SA).

Table 1. Performance comparison between PAID and related work

Protocol	Computational overhead	Communication overhead
PAID	$4M_{N^2}$	3
Protocol 2 in [17]	$8M_{N^2}$	6

7 Performance Evaluation

In this section, we analyze the computational and communication overhead in the eligibility assessment. And Table 1 shows the performance comparison between our PAID and related work.

7.1 Computational Overhead

Since we use the Paillier homomorphic encryption in eligibility assessment, we use the modular exponentiation as the computational overhead indicator and ignore other operations. For convenience, the modular exponentiation in Paillier homomorphic encryption is denoted as M_{N^2}. The server $\mathcal{S}$ requires two encryptions, and users perform one encryption and one decryption. Therefore, the computational overhead of the eligibility assessment is $4M_{N^2}$.

7.2 Communication Overhead

Typically, we measure the communication overhead by communication rounds in secure multiparty computation. In our eligibility assessment, the interaction between server and user is 3 rounds.

8 Related Work

Truth discovery is an effective technology that can calculate the ground truth and users' quality from conflicting sensing data. Li et al. [13] Proposed a general truth discovery scheme, but privacy protection is not in their work scope. To protect users' privacy data, Miao et al. [18] proposed the first privacy-preserving truth discovery scheme using the Paillier cryptosystem, but the computational and communication costs are huge. Later, some works [19] improve the communication cost and privacy protection of truth discovery. However, these works do not take into account the failure of the MCS system caused by users' exit. And most existing works do not combine the incentive mechanism.

Another previous work related to this paper is the incentive mechanism in the MCS system. Some works [20] utilize the game theory model, such as the auction model, to implement incentive mechanisms but do not consider users' privacy leakage. In [12], the author designs privacy protection in the incentive mechanism. However, these works do not include the assessment of users who provide unqualified data in advance. Zhao et al. [8] presented an incentive mechanism model to evaluate the reliability of users' data while protecting data privacy. However, the users' sensing data needs to be submitted to the task publisher, so the sensing data privacy protection is still insufficient.

Different from existing work, we design an incentive mechanism based on truth discovery, which can remove unqualified users in advance. The incentive mechanism ensures that enough users participate in the sensing task and improve truth discovery accuracy.

9 Conclusion

In this paper, we propose a privacy-preserving incentive mechanism based on truth discovery in the MCS system. Specifically, we design an eligibility assessment scheme to estimate whether the data submitted by users are qualified. Next, the truth discovery scheme calculates the ground truth and the weight of each user using these qualified sensing data. Then we quantify the data quality of users by weight and distribute the rewards. Besides, we also demonstrate that PAID meets eligibility, zero-knowledge. And the analysis shows that our PAID can resist the Denial of Payment attack, Inference attack, Data pollution attack, and Sybil attack. In future work, we will demonstrate the efficiency of our model through experiments.

Acknowledgments. This work was supported by the National Nature Science Foundation of China (No. 61962022 and No. 62062034), the Key Research and Development Plan of Jiangxi Province (No. 20192BBE50077), and the Postgraduate Innovation Special Fund Project of Jiangxi Province (No. YC2020-S365).

References

1. Xiong, J., Zhao, M., Bhuiyan, M.Z.A., Chen, L., Tian, Y.: An AI-enabled three-party game framework for guaranteed data privacy in mobile edge crowdsensing of IoT. IEEE Trans. Industr. Inf. **17**(2), 922–933 (2019)
2. Wei, X., Sun, B., Cui, J.: Task replica assignment in mobile self-organized crowdsensing. Int. J. Performability Eng. **16**(1), 152–162 (2020)
3. Xiong, J., Chen, X., Yang, Q., Chen, L., Yao, Z.: A task-oriented user selection incentive mechanism in edge-aided mobile crowdsensing. IEEE Trans. Netw. Sci. Eng. **7**(4), 2347–2360 (2020)
4. Zhang, S., Li, H., Dai, Y., Li, J., He, M., Lu, R.: Verifiable outsourcing computation for matrix multiplication with improved efficiency and applicability. IEEE Internet Things J. **5**(6), 5076–5088 (2018)
5. Ouyang, R.W., Srivastava, M., Toniolo, A., Norman, T.J.: Truth discovery in crowdsourced detection of spatial events. IEEE Trans. Knowl. Data Eng. **28**(4), 1047–1060 (2015)
6. Xu, G., Li, H., Liu, S., Wen, M., Lu, R.: Efficient and privacy-preserving truth discovery in mobile crowd sensing systems. IEEE Trans. Veh. Technol. **68**(4), 3854–3865 (2019)

7. Jin, H., Su, L., Chen, D., Nahrstedt, K., Xu, J.: Quality of information aware incentive mechanisms for mobile crowd sensing systems. In: Proceedings of the 16th ACM International Symposium on Mobile Ad Hoc Networking and Computing, pp. 167–176 (2015)
8. Zhao, B., Tang, S., Liu, X., Zhang, X.: PACE: privacy-preserving and quality-aware incentive mechanism for mobile crowdsensing. IEEE Trans. Mob. Comput. **20**(5), 1924–1939 (2020)
9. Dharminder, D., Mishra, D.: LCPPA: lattice-based conditional privacy preserving authentication in vehicular communication. Trans. Emerg. Telecommun. Technol. **31**(2), e3810 (2020)
10. Xiong, J., et al.: A personalized privacy protection framework for mobile crowdsensing in IIoT. IEEE Trans. Industr. Inf. **16**(6), 4231–4241 (2019)
11. Zhao, B., Tang, S., Liu, X., Zhang, X., Chen, W.N.: IronM: privacy-preserving reliability estimation of heterogeneous data for mobile crowdsensing. IEEE Internet Things J. **7**(6), 5159–5170 (2020)
12. Wang, Z., Li, J., Hu, J., Ren, J., Li, Z., Li, Y.: Towards privacy-preserving incentive for mobile crowdsensing under an untrusted platform. In: IEEE INFOCOM 2019-IEEE Conference on Computer Communications, pp. 2053–2061. IEEE (2019)
13. Li, Q., Li, Y., Gao, J., Zhao, B., Fan, W., Han, J.: Resolving conflicts in heterogeneous data by truth discovery and source reliability estimation. In: Proceedings of the 2014 ACM SIGMOD International Conference on Management of Data, pp. 1187–1198 (2014)
14. Krawczyk, H.: SIGMA: the 'SIGn-and-MAc' approach to authenticated Diffie-Hellman and its use in the IKE protocols. In: Boneh, D. (ed.) CRYPTO 2003. LNCS, vol. 2729, pp. 400–425. Springer, Heidelberg (2003). https://doi.org/10.1007/978-3-540-45146-4_24
15. Chen, Z., Li, S., Chen, L., Huang, Q., Zhang, W.: Fully privacy-preserving determination of point-range relationship. SCIENTIA SINICA Informationis **48**(2), 187–204 (2018)
16. Bonawitz, K., et al.: Practical secure aggregation for privacy-preserving machine learning. In: proceedings of the 2017 ACM SIGSAC Conference on Computer and Communications Security, pp. 1175–1191 (2017)
17. Guo, Y., Zhou, S., Dou, J., Li, S., Wang, D.: Efficient privacy-preserving interval computation and its applications. Chin. J. Comput. **40**(39), 1–17 (2016)
18. Miao, C., et al.: Cloud-enabled privacy-preserving truth discovery in crowd sensing systems. In: Proceedings of the 13th ACM Conference on Embedded Networked Sensor Systems, pp. 183–196 (2015)
19. Zhang, C., Zhu, L., Xu, C., Liu, X., Sharif, K.: Reliable and privacy-preserving truth discovery for mobile crowdsensing systems. IEEE Trans. Dependable Secure Comput. (2019)
20. Zhang, X., Yang, Z., Zhou, Z., Cai, H., Chen, L., Li, X.: Free market of crowdsourcing: incentive mechanism design for mobile sensing. IEEE Trans. Parallel Distrib. Syst. **25**(12), 3190–3200 (2014)

An Efficient Certificateless Cloud Data Integrity Detection Scheme for Ecological Data

Yong Xie(✉), Muhammad Israr, Zhengliang Jiang, Pengfei Su, Ruoli Zhao, and Ruijiang Ma

Department of Computer Technology and Application, Qinghai University, Xining, China

Abstract. With the development of society, the ecological and environmental problems can not be ignored. To solve the ecological problems facing, a large amount of ecological data is indispensable. It is impractical to store a large amount of ecological data on a local server, but cloud storage can store it well. However, users are faced with the problems of untrustworthy cloud storage providers and vulnerable data. In order to solve related problems, this paper proposes an efficient certificateless cloud data integrity detection scheme for ecological data. The scheme adopts certificateless form to realize the audit, which can efficiently complete the user's audit requirements. The security analysis shows that this scheme can resist type I and type II adversary attacks. The computation and communication cost analysis results show that the efficiency advantage of this scheme is obvious.

Keywords: Integrity detection · Certificateless · Ecological data · Cloud data

1 Introduction

Eco-environmental issues refer to the global environmental pollution and ecological destruction caused by improper human behavior in the process of industrialization, which poses various realistic threats to human survival and development. For example, global climate change, biodiversity reduction, land desertification, ozone layer destruction and so on. The current ecological and environmental problems affecting the world affect the development of the world to a certain extent. Ecological and environmental issues cannot be ignored. With the development of Internet information technology, big data and other technologies have provided strong support for winning the battle against the ecological environment. But in this process, ecological data will increase massively. At this time,

The work was supported in part by the National Natural Science Foundation of China (61862052), and the Science and Technology Foundation of Qinghai Province (2019-ZJ-7065).

J. Xiong et al. (Eds.): MobiMedia 2021, LNICST 394, pp. 278–290, 2021.
https://doi.org/10.1007/978-3-030-89814-4_21

storing data on a local server is no longer an excellent method, and cloud storage can solve this problem well. Cloud storage technology allows users to outsource large amounts of data and store them in the cloud, thereby reducing user storage space management and computing costs. However, data security is also greatly threatened at this time. Untrusted cloud storage providers may tamper with or delete data for profit, or they may lose data due to hardware failure. In summary, it is imperative to ensure that the data stored in the cloud server is complete. An efficient certificate-free cloud data integrity detection scheme for ecological data is proposed in this paper, which can solve this problem. This paper uses a certificateless form, which solve key escrow issues and certificate management issues in the traditional public key cryptosystem and achieves high efficiency on this basis.

The rest of this paper is as follows: Section 2 introduces related work. Section 3 introduces the background. Section 4 introduces the system model. Section 5 introduces the plan. Section 6 proves the correctness of the scheme. Section 7 proves the security of the scheme. Section 8 analyzes the computation cost and communication cost of the scheme. Finally, Sect. 9 is the conclusion of the full text.

2 Related Work

It is not friendly to transmit data to the local server for verification. Ateniese et al. [1] defines provable data ownership (PDP), which promotes data integrity verification on cloud servers. In this scheme, a third-party auditor (TPA) was introduced for public review. Then Ateniese et al. [2] proposed a new scheme for the dynamically set scene, but it was unable to implement the insert operation. Many PDP schemes [3–7] were proposed later. However, schemes [2–7] are designed based on the traditional public key cryptosystem. In these schemes, the certificate authority generates the user's public-private key pair, and the generated certificates are managed by the certificate authority, which may cause certificate management problems.

Many years ago, Shamir et al. [8] proposed identity-based encryption technology. In this scenario, the public key is the identity of the owner, so the certificate is no longer needed. Taking advantage of identity-based encryption technology, many schemes [9–11] have been proposed. Work [9] proposed a new identity-based public audit scheme for aggregated signature, which solves the problem that storage users need to issue certificates before uploading data, which incurs huge costs. Work [10] adopts an effective certificate-based public key setting key management scheme, which combines identity based aggregate signature and public verification to construct a provable data integrity protocol, which reduces the audit time of a single TPA task. Work [11] proposes an identity-based audit scheme for multiple cloud environments. However, the above schemes have key escrow problem. The user's private key is completely generated by KGC. A malicious KGC may impersonate the user and may leak the user's private information, which is a security challenge.

The key escrow problem can be solved in a certificateless way. Work [12] proposed the concept of Certificateless Public Key Cryptography (CLPKC). In the scheme, the user's private key consists of two parts, one part is generated by the user, and the other part is generated by KGC. CLPKC solves the certificate management and key escrow problems in the traditional public key cryptosystem. Subsequently, the scheme [13] was also proposed, but this scheme is vulnerable to attacks by opponents, and some values can be used by opponents to replace the user's public key.

3 Background

3.1 Bilinear Pairings

G_1 is a cyclic group of prime q, where g_1 and g_2 are generators of G_1. There is a bilinear pairing operation $e : G_1 \times G_1 \rightarrow G_2$. The bilinear pair operation satisfies the following three properties.

Bilinear: If there are $a, b \in Z_q^*$ and $g_1, g_2 \in G_1$, then $e(g_1, g_2)^{ab} = e(g_1^a, g_2^b)$.

Non-degeneracy: If there is $P, Q \in G_1$, then $e(P, Q) \neq 1$.

Computability: For $P, Q \in G_1$, $e(P, Q)$ can be calculated by polynomial time algorithm.

In the security proof, we will use the following computation assumptions to construct.

Discrete Logarithm (DL) Problem: given $P, X \in G_1$, select an element $x \in Z_q^*$, the formula $X = xP$ exists.

Computational Diffie-Hellman (CDH) Problem: given $X = xp, Y = yP \in G_1$, select an element $R = xyP$, where $x \in Z_q^*$ and $y \in Z_q^*$ are unknown (Fig. 1).

4 System Model

The system model of this scheme consists of four parts, namely KGC, third-party auditors, cloud storage providers and group users.

The functions of each part are as follows:

Key generation center (KGC): KGC is responsible for generating the system master key, public parameters, and some public and private key pairs for group users and administrators. It is completely credible.

Third party auditor (TPA): Cloud data can be verified by TPA for completeness. And TPA cannot obtain any information in this process. It is completely credible.

Cloud storage provider (CSP): CSP is semi-trusted, and it may try to deceive data owners by forging integrity evidence. It provides sufficient storage space and retrieval functions.

Group users (Users): Users include a group manager and other users. Managers can register and track other users. Registered users can access and update data.

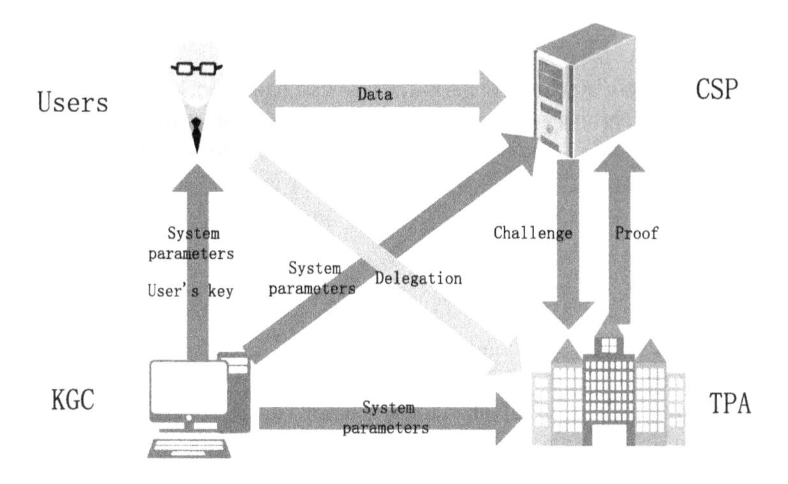

Fig. 1. System model

5 Scheme

The audit scenario for this paper consists of the following six phases.

5.1 Setup

In this section, KGC performs the following steps to generate system parameters.

Enter the security parameter η, and the system parameters are generated by KGC. G_1 is a cyclic group of prime q, where g_1 and g_2 are generators of G_1. There is a bilinear pairing operation $e : G_1 \times G_1 \rightarrow G_2$ and two hash functions $H_1 : \{0,1\}^* \rightarrow Z_q^*$, $H_2 : \{0,1\}^* \rightarrow G_1$ and $H_3 : \{0,1\}^* \rightarrow Z_q^*$. KGC chooses $x \in_R Z_q^*$ as the system master key and calculates the system public key $P_k = {g_1}^x$. The public parameter is $PP = \{G_1, G_2, g_1, g_2, q, H_1, H_2, P_k\}$. KGC discloses PP.

5.2 KeyGen

In this section, KGC and the user perform the following steps in order to generate the user's key.

The user identity information is id_o, and the user randomly selects $x_o \in_R Z_q^*$ and calculates $P_o = {g_1}^{x_o}$. The user sends $\{P_o, id_o\}$ to KGC through a secure channel. KGC randomly selects $t_o \in_R Z_q^*$ and then calculates $T_o = {g_1}^{t_o}$, $h_o = H_1(id_o, T_o, P_o)$ and $s_o = t_o + xh_o \ mod\ q$. KGC sends $\{s_o, T_o\}$ to users through a secure channel. The user uses $\{s_o, x_o\}$ as his private key and $\{P_o, T_o\}$ as his public key and publishes $\{P_o, T_o\}$. The user calculates $\alpha_o = H_3(id_o, P_o, T_o, P_k)$ and makes it public.

5.3 TagGen

This part is executed by the user. Assuming that the data block to be executed is $\{d_1, d_2, \cdots, d_n\}$, the user will generate a tag for each data block. The specific process is as follows

The user calculates $C_i = (H_2(i){g_2}^{d_i})^{s_o + \alpha_o x_o}$, where $i \in \{1, 2, \cdots, n\}$. The user uses $\{i, d_i, C_i\}$ as a tag for data block d_i, and then sends $\{i, d_i, C_i\}$ to CSP.

5.4 ChalGen

In this part, TPA sends a challenge message to the CSP after receiving the audit request sent by the user.

TPA randomly selects J blocks of data from $\{d_1, d_2, \cdots, d_n\}$, among which $J \in \{1, 2, \cdots, n\}$. TPA randomly selects $\tau_j \in_R Z_q^*$ for any $d_j \in \{d_1, d_2, \cdots, d_J\}$, and then sends $\{j, \tau_j\}_{j \in J}$ to CSP.

5.5 ProofGen

In this part, After receiving the challenge message sent by the TPA, the CSP calculates the evidence, and then sends the generated evidence to the TPA.

CSP calculates $C = \prod_{j=1}^{J} (C_j)^{\tau_j}$, where $d = \sum_{j=1}^{J} \tau_j d_j$. CAP sends $\{C, d\}$ to TPA.

5.6 ProofVerif

In this section, TPA will verify the evidence after receiving the evidence sent by the CSP.

The TPA verifies whether the formula $e(C, g_1) = e((\prod_{j=1}^{J} (H_2(j))^{\tau_j})g_2^d, T_o P_k^{h_o} P_o^{\alpha_o})$ is established. If it is established, it means that the data is correctly saved by the CSP; otherwise, the data may be tampered with or lost.

6 Correctness Analysis

Theorem 1: Confirm that the TPA verifies that the integrity part of the data is correct.

Proof: If the formula $e(C, g_1) = e((\prod_{j=1}^{J} (H_2(i))^{\tau_j})g_2^d, T_o P_k^{h_o} P_o^{\alpha_o})$ holds, it means that the TPA verification information is partly correct. The formula is derived as follows.

$$
\begin{aligned}
e(C,g_1) &= e(\prod_{j=1}^{J}(C_j)^{\tau_j},g_1)\\
&= e(\prod_{j=1}^{J}((H_2(j)g_2{}^{d_j})^{s_o+\alpha_o x_o})^{\tau_j},g_1)\\
&= e(\prod_{j=1}^{J}(H_2(j)^{\tau_j}g_2^{d_j\tau_j}),g_1^{s_o+\alpha_o x_o})\\
&= e((\prod_{j=1}^{J}(H_2(j))^{\tau_j})g_2^{\sum_{j=1}^{J}d_j\tau_j},T_o P_k^{h_o}P_o^{\alpha_o})\\
&= e((\prod_{j=1}^{J}(H_2(j))^{\tau_j})g_2^{d},T_o P_k^{h_o}P_o^{\alpha_o})
\end{aligned}
$$

7 Security Proof

First of all, the security model used is based on work [14].

7.1 Security Mode

This model has two types of opponents, namely Type I opponent $\mathcal{A}_1$ and Type II opponent $\mathcal{A}_2$. $\mathcal{A}_1$ can replace any user public key but cannot obtain the master key. $\mathcal{A}_2$ can obtain the master key but cannot replace any user's public key. The game between $\mathcal{A} \in \{\mathcal{A}_1, \mathcal{A}_2\}$ and challenger $\mathcal{C}$ formally defines the security of the proposed scheme. After the game starts, $\mathcal{C}$ executes the Setup algorithm to generate system parameters and transmits them to $\mathcal{A}$. Then, $\mathcal{A}$ performs the following query on $\mathcal{C}$.

$\mathcal{C}$ runs related operations to generate private key and public key $\{P_u, T_u\}$. $\mathcal{C}$ then sends it to $\mathcal{A}$.

PartialPrivateKey(id_u): $\mathcal{C}$ sends the partial private key $\{s_u, T_u\}$ to $\mathcal{A}$.

PublicKeyReplacement(id_u, $\{P_u, T_u\}'$): $\mathcal{C}$ replaces $\{P_u, T_u\}$ with $\{P_u, T_u\}'$.

SecretValue(id_u): $\mathcal{C}$ sends x_u to $\mathcal{A}$, where x_u is the secret value.

TagGen(id_u, d_j): $\mathcal{C}$ gets tags through TagGen algorithm and gives them to $\mathcal{A}$.

$\mathcal{A}$ falsifies the proof $\{C^*, d^*\}$ relative to J^*. The conditions for $\mathcal{A}$ to win are as follows.

(1) $ProofVerif(\alpha_o{}^*, P_o{}^*, T_o{}^*, P_k{}^*, h_o{}^*) \rightarrow True$
(2) If $\mathcal{A}$ is an I-type opponent, there will be no query about PartialPrivate Key($id_o{}^*$). If $\mathcal{A}$ is a Type II opponent, there will be no query about SecretValue($id_o{}^*$) in the game.
(3) There is no query related to TagGen($id_o{}^*$, $d_{i_j^*}{}^*$) in the game, where $i_j^* \in J^*$

Define the probability of $\mathcal{A}$ winning in the game as $Succ_{CLPA}^{cma}(\mathcal{A})$. For any $\mathcal{A} \in \{\mathcal{A}_1, \mathcal{A}_2\}$ if $Succ_{CLPA}^{cma}(\mathcal{A})$ can be ignored, the proposed scheme is secure against two types of opponents.

7.2 Security Analysis

This part proves through several lemmas and theorems that the scheme in this article can prevent the two kinds of opponents proposed. In the security proof, H_1 and H_2 are regarded as random predictions [15].

Lemma 1. The premise that the scheme in this paper is secure for Type I opponents is that the CDH problem is difficult to solve.

Proof. Define $\mathcal{A}_1$ as a Type I opponent. In the game, its probability ϵ cannot be ignored. The concept of the security proof in this paper is that if $\mathcal{A}_1$ exists, then an attacker can be generated through $\mathcal{A}_1$, and the attacker can crack the CDH problem. Given $(P, Y = P^a, \Theta = P^b)$, $\mathcal{C}$ sets $Y \rightarrow P_k$, chooses id_o as the challenge identity, and then transmits system parameters to $\mathcal{A}_1$. Then, $\mathcal{A}_1$ can execute the following query on $\mathcal{C}$.

$H_1(id_u, T_u, P_u)$: $\mathcal{C}$ keeps an initially empty list L_{H_1}. $\mathcal{C}$ judges whether $\{id_u, T_u, P_u, h_u\}$ is in L_{H_1}. If it exists, $\mathcal{C}$ transmits h_u to $\mathcal{A}_1$. Otherwise, $\mathcal{C}$ randomly selects $h_u \in Z_q^*$ and stores $\{id_u, T_u, P_u, h_u\}$ in L_{H_1}. $\mathcal{C}$ transmits h_u to $\mathcal{A}_1$.

$H_2(i, c_i, z_i, Z_i)$: $\mathcal{C}$ maintains an initially empty list L_{H_2}. $\mathcal{C}$ judges whether $\{i, c_i, z_i, Z_i\}$ is in L_{H_2}. If it exists, $\mathcal{C}$ transmits $H_2(i) = Z_i$ to $\mathcal{A}_1$. Otherwise, $\mathcal{C}$ randomly selects $b_i \in \{0, 1\}$ and $Pr[b_i = 0] = \lambda$, where $\lambda \in (0, 1)$. $\mathcal{C}$ chooses $z_i \in Z_q^*$. If $b_i = 0$, then $\mathcal{C}$ calculates $Z_i = z_i \cdot P$. Otherwise, $\mathcal{C}$ calculates $Z_i = z_i \cdot \Theta$. $\mathcal{C}$ stores $\{i, c_i, z_i, Z_i\}$ in the list L_{H_2}, and gives $H_2(i) = Z_i$ to $\mathcal{A}_1$.

$H_3(id_o, P_o, T_o, P_k)$: $\mathcal{C}$ maintains an initially empty list L_{H_3}. $\mathcal{C}$ checks whether $\{id_o, P_o, T_o, P_k, \alpha\}$ is in L_{H_3}. If it exists, $\mathcal{C}$ sends α to $\mathcal{A}_1$. Otherwise, $\mathcal{C}$ randomly selects a random value $\alpha \in Z_q^*$ and stores $\{id_o, P_o, T_o, P_k, \alpha\}$. $\mathcal{C}$ sends α to $\mathcal{A}_1$.

CreateUser(id_u): $\mathcal{C}$ keeps an initially empty list L_k. $\mathcal{C}$ judges whether $\{id_u, s_u, x_u, P_u, T_u\}$ is in L_k. If it is, then $\mathcal{C}$ gives $\{P_u, T_u\}$ to $\mathcal{A}_1$. If it is not and $id_o = id_u$, then $\mathcal{C}$ generates some random numbers $t_u, x_u \in Z_q^*$, sets $\perp \rightarrow s_u$, calculates $T_u = {g_1}^{t_u}$, $P_u = {g_1}^{x_u}$ and puts $\{id_u, s_u, x_u, P_u, T_u\}$ in L_k. If $\{id_u, s_u, x_u, P_u, T_u\}$ is not in L_k and $id_o! = id_u$, $\mathcal{C}$ generates some random numbers s_u, w_u, x_u, calculates $P_u = {g_1}^{x_u}$, $T_u = {g_1}^{s_u} / {g_1}^{w_u}$ and puts $\{id_u, s_u, x_u, P_u, T_u\}$ and $\{id_u, P_u, T_u, w_u\}$ in L_k and L_{H_1}. Then, $\mathcal{C}$ sends $\{P_u, T_u\}$ to $\mathcal{A}_1$.

PartialPrivateKey(id_u): $\mathcal{C}$ Check whether the formula $id_u = id_o$. If so, $\mathcal{C}$ interrupts the process. If not, $\mathcal{C}$ finds $\{id_u, s_u, x_u, P_u, T_u\}$ in L_k, and then sends $\{s_u, T_u\}$ to $\mathcal{A}_1$.

PublicKeyReplacement $(id_u, \{P_u, T_u\}^{'})$: $\mathcal{C}$ Look for $\{id_u, s_u, x_u, P_u, T_u\}$ in L_k, and replace $\{P_u, T_u\}$ with $\{P_u, T_u\}^{'}$.

SecretValue(id_u): $\mathcal{C}$ looks for $\{id_u, s_u, x_u, P_u, T_u\}$ in L_k, and sends x_u to $\mathcal{A}_1$.

TagGen$(id_u,\ d_j)$: After $\mathcal{C}$ receives the query of id_u and data d_i, $\mathcal{C}$ judges whether id_u and id_o are the same.. If they are equal, $\mathcal{C}$ finds the list L_k of $\{id_u, s_u, x_u, P_u, T_u\}$ and gets τ_i. Otherwise, $\mathcal{C}$ aborts.

Finally, $\mathcal{A}_1$ uses id_o^* to forge proof that $\{C^*, d^*\}$, $\{C^*, d^*\}$ and J^* are related. If $id_o! = id_o^*$ or $d_{i_j^*}{}^* = 0(i_j^* \in J^*)$, then $\mathcal{C}$ ends. Otherwise, $\mathcal{C}$ searches for $\{id_o, s_o, x_o, P_o, T_o\}$ and $\{i_j^*, c_{i_j^*}, z_{i_j^*}, Z_{i_j^*}\}$ in L_k and L_{H_2}, where $i_j^* \in J^*$. $\mathcal{C}$ gets

$$e(C^*, g_1) = e((\prod_{i_j^*=1}^{J^*} (H_2(i_j^*))^{\tau_{i_j^*}})g_2^{d^*}, T_o P_k^{h_o} P_o^{\alpha_o})$$

According to the bifurcation lemma [15], $\{C^{*'}, d^*\}$ can be obtained by choosing different H_1 and H_3. Where $\{C^{*'}, d^*\}$ satisfies

$$e(C^{*'}, g_1) = e((\prod_{i_j^*=1}^{J^*} (H_2(i_j^*))^{\tau_{i_j^*}})g_2^{d^*}, T_o P_k^{h_o^{'}} P_o^{\alpha_o^{'}})$$

From the above two equations. $\mathcal{C}$ gets

$$e((C^* - C^{*'}), g_1) = e(P_k^{h_o - h_o^{'}} P_o^{\alpha_o - \alpha_o^{'}}(\prod_{i_j^*=1}^{J^*} (H_2(i_j^*))^{\tau_{i_j^*}})g_2^{d^* t_o}, g_1)$$

Then, $\mathcal{C}$ outputs $\varPhi^{-1}$ and ω as answers to the CDH problem, where $\varPhi^{-1} \in Z_q^*$ and $\varPhi^{-1}.\varPhi \equiv 1 \ mod \ q$. $\varPhi = P_k^{h_o - h_o^{'}} P_o^{\alpha_o - \alpha_o^{'}}(\prod_{i_j^*=1}^{J^*} (H_2(i_j^*))^{\tau_{i_j^*}})$ and $\omega = (C^* - C^{*'})$.

We analyze the possibility of CDH problem being solved by $\mathcal{C}$. The following events need to be considered.

E_1: $\mathcal{C}$ is not interrupted.
E_2: $\mathcal{A}_1$ outputs the legal proof of J^* with $d_{i_j^*}{}^* = 1(i_j^* \in J^*)$.
E_3: $id_o^* = id_o$ is established.

It is not easy to obtain $Pr[E_1] \geq (1 - 1/q_h)^{q_{tag}}$, $Pr[E_2|E_1] \geq (1 - \lambda)^{q_{tag}}$ and $Pr[E_3|E_2 \wedge E_1] \geq 1/q_h$, where q_h and $q_t ag$ represent the number of H_1 queries and TagGen queries. Let E_{suc} prove that the CDH problem can be cracked by $\mathcal{C}$.

Then we have $E_{suc} = E_1 \wedge E_2 \wedge E_3$ and

$$\begin{aligned} Pr[E_1 \wedge E_2 \wedge E_3] &= Pr[E_1]Pr[E_2|E_1]Pr[E_3|E_1 \wedge E_2] \\ &\geq (1 - 1/q_h)(1 - \lambda)^{q_{tag}}(1 - 1/q_h)^{q_{cert}}\epsilon \end{aligned}$$

Since ϵ is not negligible, there is no attacker like $\mathcal{C}$ who can crack the CDH problem. Therefore, if the CDH problem is difficult to solve, the proposed scheme is secure for Type I opponents.

Lemma 2. The premise that the scheme in this paper is secure for Type II opponents is that the CDH problem is difficult to solve.

Proof. Assume that $\mathcal{A}_2$ is a Type II opponent, and its probability ϵ in the game cannot be ignored. If there is such an adversary, an attacker can be defined, and the attacker can crack the CDH problem. Given $(P, Y = P^a, \Theta = P^b)$, $\mathcal{C}$ chooses a random value $s \in Z_q^*$, calculates $P_k = {g_1}^s$, and then selects id_o as the challenge identity. $\mathcal{C}$ gives the key s and PP to $\mathcal{A}_2$. $\mathcal{C}$ solves $\mathcal{A}_2$'s H_1 queries, H_2 queries, H_3 queries and TagGen queries. The process is the same as the proof of Lemma 1. The other query steps of $\mathcal{A}_2$ are as follows.

CreateUser(id_u): $\mathcal{C}$ maintains an initially empty list L_k. $\mathcal{C}$ judges whether $\{id_u, s_u, x_u, P_u, T_u\}$ is in L_k. If it is, then $\mathcal{C}$ gives $\{P_u, T_u\}$ to $\mathcal{A}_2$. If it is not and $id_o = id_u$, then $\mathcal{C}$ generates some random numbers $t_u \in Z_q^*$, sets $\perp \rightarrow x_u$ and $Y \rightarrow P_u$, calculates $T_u = {g_1}^{t_u}$, $h_u = H_1(id_u, T_u, P_u)$, $s_u = t_o + sh_u \ mod\ q$ and puts $\{id_u, s_u, x_u, P_u, T_u\}$ in L_k. If $\{id_u, s_u, x_u, P_u, T_u\}$ is not in L_k and $id_o! = id_u$, $\mathcal{C}$ generates some random numbers s_u, w_u, x_u, calculates $P_u = {g_1}^{x_u}$, $T_u = {g_1}^{t_u}$, $h_u = H_1(id_u, T_u, P_u)$, $s_u = t_o + sh_u \ mod\ q$ and puts $\{id_u, s_u, x_u, P_u, T_u\}$ and $\{id_u, P_u, T_u, w_u\}$ in L_k and L_{H_1}. Then, $\mathcal{C}$ transmits $\{P_u, T_u\}$ to $\mathcal{A}_2$.

PartialPrivateKey(id_u): $\mathcal{C}$ looks for $\{id_u, s_u, x_u, P_u, T_u\}$ in L_k, and then sends $\{s_u, T_u\}$ to $\mathcal{A}_2$.

SecretValue(id_u): $\mathcal{C}$ Check whether the formula $id_u = id_o$. If so, $\mathcal{C}$ interrupts the game. If not, $\mathcal{C}$ finds $\{id_u, s_u, x_u, P_u, T_u\}$ in L_k, and sends x_u to $\mathcal{A}_2$.

Finally, $\mathcal{A}_2$ uses id_o^* to forge proof that $\{C^*, d^*\}$, $\{C^*, d^*\}$ and J^* are related. If $id_o! = id_o^*$, then $\mathcal{C}$ is interrupted. Otherwise, $\mathcal{C}$ finds $\{id_o, s_o, x_o, P_o, T_o\}$ and $\{i_j^*, c_{i_j^*}, z_{i_j^*}, Z_{i_j^*}\}$ in L_k and L_{H_2}, where $i_j^* \in J^*$. $\mathcal{C}$ gets

$$e(C^*, g_1) = e(P_k^{h_o} P_o^{\alpha_o} (\prod_{i_j^*=1}^{J^*} (H_2(i_j^*))^{\tau_{i_j^*}}) Y^{d^*}, g_2)$$

Then, $\mathcal{C}$ outputs Φ^{-1} and ω as answers to the CDH problem, where $\Phi^{-1} \in Z_q^*$ and $\Phi^{-1}.\Phi \equiv 1\ mod\ q$. $\Phi = P_k^{h_o} P_o^{\alpha_o} (\prod_{i_j^*=1}^{J^*} (H_2(i_j^*))^{\tau_{i_j^*}})$ and $\omega = (C^*)$.

We analyze the possibility of CDH problem being solved by $\mathcal{C}$. The following events need to be considered.

E_1: $\mathcal{C}$ is not interrupted.
E_2: $\mathcal{A}_1$ outputs the legal proof of J^* with ${d_{i_j^*}}^* = 1(i_j^* \in J^*)$.
E_3: $id_o^* = id_o$ is established.

It is not easy to obtain $Pr[E_1] \geq (1 - 1/q_h)^{q_{tag}}$, $Pr[E_2|E_1] \geq (1 - \lambda)^{q_{tag}}$ and $Pr[E_3|E_2 \wedge E_1] \geq 1/q_h$, where q_h and q_tag represent the number of H_1 queries and TagGen queries. Let E_{suc} prove that the CDH problem can be cracked by $\mathcal{C}$.

Then we have $E_{suc} = E_1 \wedge E_2 \wedge E_3$ and

$$\begin{aligned} Pr[E_1 \wedge E_2 \wedge E_3] &= Pr[E_1] Pr[E_2|E_1] Pr[E_3|E_1 \wedge E_2] \\ &\geq (1 - 1/q_h)(1 - \lambda)^{q_{tag}} (1 - 1/q_h)^{q_{cert}} \epsilon \end{aligned}$$

Since ϵ is not negligible, there is no attacker like $\mathcal{C}$ who can crack the CDH problem. Therefore, if the CDH problem is difficult to solve, the proposed scheme is secure for Type II opponents.

Theorem 1 is derived from Lemma 1 and Lemma 2. Theorem 1 the premise that the scheme in this paper is secure for both Type I and Type II opponents is that the CDH problem is difficult to solve.

8 Performance

We analyzed the performance of the scheme, related experimental environment and data reference [16]. In order to better demonstrate the advantages of the scheme proposed in this paper, we compare the scheme of this paper, scheme [14] and scheme [17] (For the sake of comparison, we assume that $s = 2\ and\ m = 3$ in the scheme [17].). Comparison method reference scheme [14]. The symbols used in this section are as follows Table 1.

Table 1. Run time of operations (millisecond)

Symbols	Definition	Time
T_H	The time spent in a hash operation mapped to the point	5.493
T_p	The time spent in a bilinear pairing operation	5.427
T_{mul}	The time it takes to perform a point multiplication operation	2.165
T_E	The time spent in an exponentiation operation	0.339
T_{add}	The time it takes to perform a point addition operation	0.013
T_h	The time spent in a regular hash operation	0.007
T_m	The time spent in a modular multiplication operation	0.001

8.1 Computation Cost

In order to compare the process more clearly. For the scheme, we only compare the three parts TagGen, ProofGen and ProofVerify. These three parts are the core of the scheme.

Regarding our scheme, part TagGen of it contains $2n$ exponentiation operations (EO), n regular hash operations (RHO) and n modular multiplication operations (MMO). The computation time of part TagGen is $2nT_E + nT_h + nT_m = 0.686n$ (ms). Part ProofGen contains J EO and $(J-1)$ MMO. The computation time of part ProofGen is $JT_E + (J-1)T_m = 0.340J - 0.001$ (ms). Part ProofVerify contains $(J+3)$ EO, J RHO and $(J+2)$ MMO. The computation time of part ProofVerify is $(J+3)T_E + (J)T_h + (J+2)T_m = 0.347J + 1.019$ (ms). The total time of our scheme is $(2n + 2J + 3)T_E + (n + J)T_h + (n + 2J + 1)T_m = 0.686n + 0.687J + 1.018$ (ms). The comparison results of other schemes are similar to this scheme, and the results are shown in Table 2.

Table 2. The comparison of communication cost

Schemes	Costs	Total
Work [14]	TagGen: $nT_H + 2nT_{mul} + nT_{add}$ ProofGen: $(J+1)T_{mul} + (J-1)T_{add}$ ProofVerify: $(J+1)T_H + (J+3)T_{mul} + (J+2)T_{add}$	$9.836n + 9.849J + 14.166$ (ms)
Work [17]	TagGen: $3nT_E + 4nT_m + nT_h$ ProofGen: $JT_E + (J-1)T_m$ ProofVerify: $(J+3)T_E + JT_h + (J+1)T_m$	$1.028n + 0.687J + 1.017$ (ms)
Our scheme	TagGen: $2nT_E + nT_h + nT_m$ ProofGen: $JT_E + (J-1)T_m$ ProofVerify:$(J+3)T_E + (J)T_h + (J+2)T_m$	$0.686n + 0.687J + 1.018$ (ms)

Tip: n represents the number of all data blocks.
s represents the number of data blocks that need to be verified ($s \leq n$).

It can be seen from Table 2 that s is less than or equal to n, so if you replace s with n for the data in the table and combined with practical applications, you can find $(1.373n+1.018)$ (ms) $\leq (1.715n+1.017)$ (ms) $\leq (19.685n+14.166)$ (ms). In summary, the advantages of the proposed scheme are obvious.

8.2 Communication Cost

The environmental reference of communication overhead [14], so the length of the elements in G_1 and Z_q^* are 1024 and 160 bits, respectively. Assume that n is 32 bits. n represents the amount of the data sub-block.

For scheme [14], in the communication process, what needs to be sent is the challenge $\{j, w_j\}_{j \in J}$ and the proof $\{R, S, \omega\}$, where $j \in \{1, 2, \cdots, n\}$, $\omega \in Z_q^*$, $R, S \in G_1$ and $d \in Z_q^*$. The result is $(32+160)J + 1024 + 1024 + 160 = 192J + 2208$ *bits*.

For scheme [17], in the communication process, what needs to be sent is the challenge $\{auth, \{AID\}_{CEMR_{PK}}, (j, v_j)\}$ and the proof $\{\sigma, u\}$, where $j \in \{1, 2, \cdots, n\}$, $\{AID\}_{CEMR_{PK}}, auth, v_j \in Z_q^*$, $\sigma \in G_1$ and $u \in Z_q^*$. Therefore, the communication cost is $(32+160)J + 160 + 160 + 1024 + 160 = 192J + 1504$ *bits*.

For our scheme, in the communication process, what needs to be sent is the challenge $\{j, \tau_j\}_{j \in J}$ and the proof $\{C, d\}$, where $j \in \{1, 2, \cdots, n\}$, $\tau_j \in Z_q^*$, $C \in G_1$ and $d \in Z_q^*$. Therefore, the communication cost is $(32+160)J + 1024 + 160 = 192J + 1184$ *bits*.

In summary, formula $192J + 1184$ *bits* $\leq 192J + 1504$ *bits* $\leq 192J + 2208$ *bits* holds, so the communication consumption in this paper is lower in comparison. The communication efficiency of the scheme in this paper has advantages.

9 Conclusion

This paper proposes an efficient certificateless cloud data integrity detection scheme for ecological data, which solves the problems of cloud storage service providers' untrustworthiness and storage data loss faced by ecological data cloud storage. This scheme uses a certificate-free form to avoid certificate management and key escrow problems. The security analysis results show that this scheme is secure. The analysis of computational consumption and communication consumption shows that this scheme has obvious efficiency advantages. Future research considers to further improve efficiency and enhance functions on the basis of this scheme.

References

1. Ateniese, G.: Provable data possession at untrusted stores. In: Proceedings of the 14th ACM Conference on Computer and Communications Security, pp. 598–609 (2007)
2. Ateniese, G., Di Pietro, R., Mancini, L.V., Tsudik, G.: Scalable and efficient provable data possession. In: Proceedings of the 4th International Conference on Security and Privacy in Communication Netowrks, pp. 1–10 (2008)
3. Ateniese, G., et al.: Remote data checking using provable data possession. ACM Trans. Inf. Syst. Secur. (TISSEC) **14**(1), 1–34 (2011)
4. Chris Erway, C., Küpçü, A., Papamanthou, C., Tamassia, R.: Dynamic provable data possession. ACM Trans. Inf. Syst. Secur. (TISSEC) **17**(4), 1–29 (2015)
5. Sebé, F., Domingo-Ferrer, J., Martinez-Balleste, A., Deswarte, Y., Quisquater, J.-J.: Efficient remote data possession checking in critical information infrastructures. IEEE Trans. Knowl. Data Eng. **20**(8), 1034–1038 (2008)
6. Wang, Q., Wang, C., Ren, K., Lou, W., Li, J.: Enabling public auditability and data dynamics for storage security in cloud computing. IEEE Trans. Parallel Distrib. Syst. **22**(5), 847–859 (2010)
7. Zhu, Y., Hongxin, H., Ahn, G.-J., Mengyang, Yu.: Cooperative provable data possession for integrity verification in multicloud storage. IEEE Trans. Parallel Distrib. Syst. **23**(12), 2231–2244 (2012)
8. Shamir, A.: Identity-Based cryptosystems and signature schemes. In: Blakley, G.R., Chaum, D. (eds.) CRYPTO 1984. LNCS, vol. 196, pp. 47–53. Springer, Heidelberg (1985). https://doi.org/10.1007/3-540-39568-7_5
9. Wang, H., Qianhong, W., Qin, B., Domingo-Ferrer, J.: Identity-based remote data possession checking in public clouds. IET Inf. Secur. **8**(2), 114–121 (2013)
10. Tan, S., Jia, Y.: NaEPASC: a novel and efficient public auditing scheme for cloud data. J. Zhejiang University SCIENCE C **15**(9), 794–804 (2014)
11. Wang, H.: Identity-based distributed provable data possession in multicloud storage. IEEE Trans. Serv. Comput. **8**(2), 328–340 (2014)
12. Al-Riyami, S.S., Paterson, K.G.: Certificateless public key cryptography. In: Laih, C.-S. (ed.) ASIACRYPT 2003. LNCS, vol. 2894, pp. 452–473. Springer, Heidelberg (2003). https://doi.org/10.1007/978-3-540-40061-5_29
13. Wang, B., Li, B., Li, H., Li, F.: Certificateless public auditing for data integrity in the cloud. In: 2013 IEEE Conference on Communications and Network Security (CNS), pp. 136–144. IEEE (2013)

14. He, D., Kumar, N., Wang, H., Wang, L., Choo, K.-K.R.: Privacy-preserving certificateless provable data possession scheme for big data storage on cloud. Appl. Math. Comput. **314**, 31–43 (2017)
15. Pointcheval, D., Stern, J.: Security arguments for digital signatures and blind signatures. J. Cryptol. **13**(3), 361–396 (2000)
16. Wu, L., Wang, J., Choo, K.-K.R., He, D.: Secure key agreement and key protection for mobile device user authentication. IEEE Trans. Inf. Forensics Secur. **14**(2), 319–330 (2018)
17. Zhou, L., Fu, A., Feng, J., Zhou, C.: An efficient and secure data integrity auditing scheme with traceability for cloud-based EMR. In: ICC 2020–2020 IEEE International Conference on Communications (ICC), pp. 1–6. IEEE (2020)

A Novel Privacy-Preserving Selective Data Aggregation with Revocation for Fog-Assisted IoT

Jianhong Zhang[1(✉)], Luo Ran[1], Dequan Xu[3], Jing Wang[2], Pei Liu[2], and Changgen Peng[3]

[1] School of Electrical and Computer Engineering, North China University of Technology, Beijing 100144, China
[2] Finance and Tax Innovation Department of JD Group, Beijing 100176, China
[3] Guizhou Provincial Key Laboratory of Public Big Data, Guizhou University, Guiyang 550025, Guizhou, China

Abstract. Internet-of-Things (IoT) can provide more convenient and intelligent services for our daily life by IoT devices collecting data. Fog computing enables ubiquitous perception, seamless connectivity, and real-time processing for B5G cellular IoT applications by making use of the advantages which fog nodes are deployed on the edge of the network, closer to data sources. Collecting sensor data by combining fog computing and the Internet of Things can enhance the security and efficiency of the B5G network in a low-cost way, which is very important for building a stable B5G network. Most existing data aggregation systems cannot support the aggregation of specific data types, which means that existing data aggregation systems have limitations in real-world applications. To solve these problems, in this work, we propose a novel privacy-preserving Selective Data Aggregation scheme with revocation for the fog-assisted IoT to address selective aggregation of privacy-preserving data and revocation of the application *App* by using homomorphic encryption and searchable encryption technique. The proposed scheme achieves not only privacy protection of data content, but also indistinguishability of data types. In the meanwhile, it enables that application *App* can simultaneously extract different types of data. Finally, security analysis show that the proposed scheme can achieve the corresponding security goals.

Keywords: Fog computing · Internet of Things · Data privacy · Data integrity · Selective data aggregation

1 Introduction

The appearance of the Internet of Things is one of the more remarkable phenomena of recent years. The Internet of Things refers to entities interconnected between heterogeneous entities, which may be sensors, devices, people or anything that requests or provides services. The IoT is changing the daily lives.

J. Xiong et al. (Eds.): MobiMedia 2021, LNICST 394, pp. 291–303, 2021.
https://doi.org/10.1007/978-3-030-89814-4_22

Through IoT devices which collect sensing data, IoT can provide real-time intelligent decision to better traffic conditions and forecast the weather. Up now, the rapid deployment of commercial 5G cellular networks offers a range of benefits to the Internet of Things that 4G or other technologies cannot offer.

The ultra-reliability and low latency of 5G will make self-driving cars, smart energy grids, factory automation and other demanding applications a reality. However, to realize these applications, a massive number of connected IoT devices need to be deployed, which will generate a large amount of data. It is reported that there will be 41 billion IoT devices by 2027. The data growth rate has been explosive due to consumer adoption and demand. If IoT devices with insufficient security design are connected to the 5G network, such massive growth in data traffic and connected IoT devices means more vulnerabilities, threats, and attacks resulting in catastrophic damages on financial markets and people's daily life. In addition, the diversity of the deployed nodes and access mechanisms at the edge of networks may result in some novel security challenges since the 5G networks have moved from centralized, hardware-based switching to distributed, software-defined digital routing. These problems pose an important challenge how to securely store, communicate, and compute these volumes of data.

To address these problems, fog computing paradigm [1] is proposed to enhance the IoT applications and to satisfy ultra-low delays requirement in 5G networks [2,3]. Due to being closer to where data is created and acted upon, fog computing makes some real-time and heterogeneous IoT applications feasible and practical [3]. Although fog computing overcomes the limitations of IoT devices and enables us to design a more capable architecture, it still unavoidably faces many security and privacy issues. As a non-trivial extension of cloud, some security and privacy issues in the context of cloud computing [4–7,10–28], still exist in the fog computing. Compared with the traditional Internet of things, the fog-assisted IoT confronts more complex network environment and network architecture. In addition to the traditional gateways for one application and fixed data sources, fog nodes need to collect data from multiple data sources and provide new aggregation services (selective data aggregation) for different applications, with different data types as intermediaries. For example, fog nodes (such as cellular base stations and roadside units) collect both the patient's physical condition data (such as heart rate and pulse rate) and road condition and traffic data (such as speed and traffic flow) to support disease monitoring applications and traffic sensing applications. It can not only provide real-time medical services to individual or community, but also improve the ability of healthcare organizations to monitor, track and control certain diseases on some regions. For a fog node, in order to conduct these different types of data, it first needs to distinguish different data types, and then performs data processing on the same type of data. In data processing procedure, both the data type and data content should be protected since the data type also leaks the privacy in an implied way [8,9], especially in the situations where the data sources come from the electronic bracelet. Obviously, traditional data aggregation techniques do not satisfy the kind of selective data processing. However, this kind of selective data aggregation is one of the most important operations in the statistics of data aggregation and data analysis. It is able to be used to analyze the difference of data traffic among different time slots in a certain App application. Thus, it is significant to study this kind of data aggregation technology.

To achieve selective data aggregation construction, in this work, we proposed a novel privacy-preserving selective data aggregation scheme with revocation for fog-assisted IoT by combining homomorphic encryption and signature. It can not only cope with both the data privacy and the data integrity, but also achieve the revocation of application Apps. And then we also analyze the security of the proposed scheme, the result show that our proposed scheme can achieve data privacy and data integrity. Finally, the proposed scheme can achieve better performance by experimental testing.

The rest of this paper is organized as follows. Section 2 give related background knowledge. Section 3 gives the detailed construction of the proposed scheme. Then, Sect. 4 presents security analysis of the proposed scheme. Section 5 analyzes the simulation results. Finally, Sect. 5 concludes this paper.

2 Preliminaries

In this part, we first give our network architecture, threat model, and identify our design goals. And then some related basic primitives are introduced.

2.1 Network Architecture

Take full advantage of cloud computing and fog computing, our IoT network architecture is a three-tier architecture, cloud layer, fog node layer and terminals layer. It is composed of four types of entities: a trust authority, a group of heterogeneous IoT devices, the deployed fog nodes at the network edges, and some application App which is run on cloud platform. Their detailed architecture is shown in Fig. 1.

1. **Trust authority (TA):** It is a trusted third party, and responsible to initialize the system and generate key materials for the other entities.
2. **IoT device:** They are terminal devices with embedded sensors and communication module, and can periodically gather and submit their sensing data to application App via the fog node. In general, IoT devices may be some fixed monitoring sensors, electronic bracelet, moving vehicles and so on, and provide various sensing data according to the detailed application requirements.
3. **Fog nodes:** They are deployed at the local network edges and serve as the middle-ware between IoT devices and application App in cloud. supporting latency data response) and storage capability (e.g. storing some data for data process).
4. **Application App:** The application App is some kinds of softwares. It can gather the sensing data from IoT devices via the fog nodes, and conduct data analysis according to specific requirements. And then some decision can be made or some system performance can be improved by analytic results.

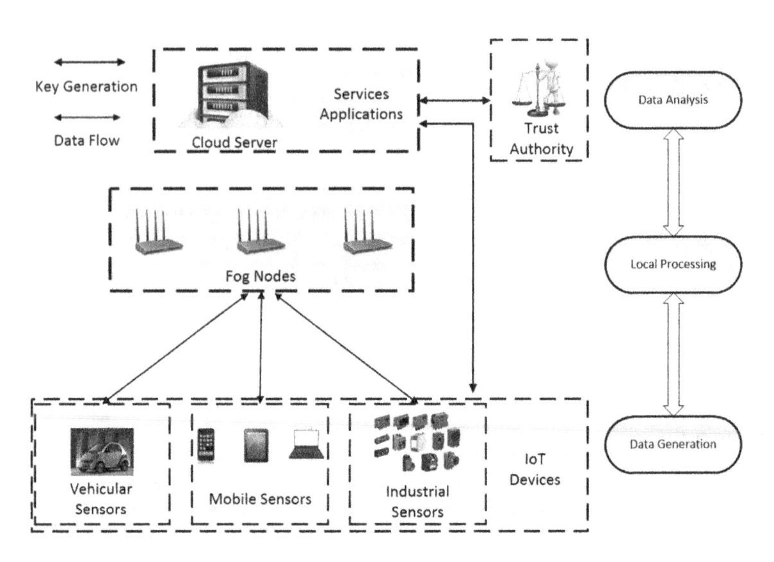

Fig. 1. Three-tried architecture of cloud and fog-based IoT network

2.2 Threat Model and Security Requeriments

In our network architecture, the trusted authority (TA) is a fully trusted-entity, the IoT devices and application Apps are honest and do not collude with the fog nodes, fog nodes are the honest-but-curious entity which follows protocol but will try to learn as much information as possible, without actively "cheating". And they should satisfy the following security requirements.

1. **Data Privacy:** The data privacy involves two aspects: data type privacy and sensing data privacy. Data type privacy means that, given sensing data, fog nodes can not determine its data type. Sensing data privacy indicates that sensing data and the aggregated result should satisfy confidentiality for fog node, and cannot be leaked in the data aggregation process.
2. **Data Integrity:** It means that the attackers can not forge and tamper the sensing data and the aggregated result.
3. **Indistinguishability:** It means that fog node cannot distinguish Whether $w_i = w_j$ under the condition that the ciphertexts $c_i = E(w_i)$ and $c_j = E(w_j)$ of data types w_i and w_j are given.

2.3 Design Goals

To construct privacy-preserving selective data aggregation scheme for fog-assist IoT networks, our goals are given as follows:

1. Security: The proposed scheme should satisfy data privacy and data integrity. Data privacy ensures that the data type and data content are confidential; and data integrity ensure that the sensing data and the aggregated result can not be tampered and modified by the attackers.

2. Efficiency: It means that all entities should each entity produces as low computational cost and communication cost as possible. And entities interact with each other as little as possible. Complex computation should be offloaded to fog node.
3. Easy Deployment: The scheme should ensure each entity to be easily deployed. Namely, the fog-assist network architecture should provide different applications, and the resource-limited IoT devices can expediently perform their key management.
4. Revocation: When an *App* application is revoked, fog node can recognize the service request from the IoT devices and delete the corresponding ciphertext.

2.4 Modified Paillier Homomorphic Cryptosystem

Paillier encryption is a kind of public key encryption scheme based on composite residuos classes. The security of the scheme is based on the difficulty to factor a big composite number N. The detail is given as follows: Let $(N, g, h = g^\theta \mod N^2)$ be public key, where $N = p \cdot q$, $p = 2p' + 1, q = 2p' + 1$, p', q' are two prime numbers, $g = -a^{2N} \mod N^2$, $a \in Z_{N^2}, \theta \in [1, N^2/2]$ are two random numbers. And the order of g is $2p'q'$. Let m be a encrypted plaintext, the ciphertext C is computed as follows: randomly select a number $r \in [1, N/4]$ to compute

$$(C_1 = g^r \mod N^2, C_2 = h^r(1 + m \cdot N) \mod N^2$$

To decrypt the ciphertext C, the plaintext m can be recovered by the key θ:

$$m = L(C_2/T_1^\theta \mod N^2)$$

where $L(x) = \frac{x-1 \mod N^2}{N}$. Additionally, if p, q are known, then $\lambda = 2p'q'$ can be obtained, thus,

$$C_2^\lambda = g^{\lambda \cdot r}(1 + m\lambda N) = (1 + m\lambda N)$$

if $gcd(N, \lambda) = 1$, then $m = L(C_2^\lambda) \cdot \lambda^{-1}$.

3 The Proposed Scheme

In this section, we proposed a novel privacy-preserving selective data aggregation scheme with revocation by combining Modified Paillier cryptosystem, search encryption and Lagrange interpolating polynomial technique. It is comprised of five phases: *System setup*, *data generation*, *fog-assisted selective aggregation*, *data-reading and verification*, and *revocation*. For the sake of easy explanation, the data format of the report which is generated by IoT is (id_i, τ_i, m_i, T_i), where id_i is the identifier of IoT device D_i, m_i is the data content which IoT device generates, τ_i is data type of m_i and T is the time slot of the reported data.

3.1 System Setup

In this phase, TA needs to run **Key Generation** algorithm, **Key Distribution** algorithm and **Selective Aggregation Initialization** algorithm to build system parameters and assign secret key for each App_j and IoT device D_i, respectively.

Key Generation. To bootstrap the entire system parameters, The trust authority (TA) takes a security parameter λ_1 as input and outputs the Modified Paillier crytosystem parameters $(n, g = \theta^2 \bmod n^2, h = g^x)$, where $x \in [1, (n^2)/2]$ is private key of TA, and $\theta \in Z_{n^2}$ is a random number. And then TA uses another security parameter λ_2 to produce two cyclic groups $\mathbb{G}_1$ and $\mathbb{G}_2$ with the same order p. Let ρ be a generator of group $\mathbb{G}_1$ and $e : \mathbb{G}_1 \times \mathbb{G}_1 \rightarrow \mathbb{G}_2$ be a bilinear map. H is a hash function. And TA randomly chooses $x_1, x_2, f_1, t \in Z_p$ and secretly constructs a polynomial

$$f(x) = x_1 + f_1 x$$

Finally, the public parameters PK is published as follow:

$$PK = (n, g, h, \mathbb{G}_1, \mathbb{G}_2, e, p, \rho, \rho_1, \rho_2, H, EK)$$

where $\rho_1 = \rho^{x_1}, \rho_2 = \rho^{x_2}$ and $EK = \rho^{f(t)/x_1}$.

Key Distribution. For each application service App_i, TA picks a number $t_i \in Z_p$ at random and builds a subset $\Theta_i = t_i \cup t$. And then it makes use of the Lagrange interpolation theorem to calculate App_j's secret key $SK_i = (SK_{i1}, SK_{i2})$ as follows:

$$SK_{i1} = \rho_2^{f(t_i)\cdot\triangle_{t_i,\Theta_i}(0)}, SK_{i2} = \rho_2{}^{x_1\cdot\Delta_{t,\Theta_i}(0)}$$

And add (ID_i^{App}, t_i, SK_i) to the list $\mathcal{K}$, where ID_i^{App} is the identifier of App_i. And TA selects a large prime $P \in [1, n/4]$ which satisfies that $(1, P, P^2, \cdots, P^d)$ is a super-increasing sequence, and randomly picks $\tau_i \in \lambda(n^2)/8$ to set App_i's public key as $h_i^{App} = g^{\tau_i}$. At last, (ID_i^{App}, h_i^{App}) should be published.

For all IoT devices $D_i, i = 1, \cdots, d$, if they enjoy the services of App_j application, they must register to App_j. Therefore, App_j firstly selects a random number $\beta \in Z_p$ as secret key, and then it distributes β to all IoT devices via a secure channel. Finally, (id_i, ρ_2^{β}) is published, where id_i is the identifier of IoT device D_i.

Selective Aggregation Initialization. For an application service App_j, it is assumed to have κ kinds of data types and $\kappa < d$. Let $\{w_1, \cdots, w_\kappa\}$ denote its data types. For $i = 1$ to κ, App_j randomly picks $\xi_i \in Z_p$ to compute the following ciphertext on data type w_i:

$$T_{i1} = \rho_1^{\xi_i}, T_{i2} = H(ID_j^{App}||w_i)^{\xi_i}, T_{i3} = SK_{j2}^{\xi_i}, T_{i4} = SK_{j1}^{\xi_i}$$

Finally, it sends κ ciphertexts $\{T_{i1}, T_{i2}, T_{i3}, T_{i4}\}_{i=1,\cdots,\kappa}$ and the large prime P to fog nodes for selective aggregation operations.

3.2 Data Generation

To report its sensing data (id_i, m_i, T) to the application App_j at time period T, an IoT device D_i needs to calculate two ciphertexts: one is to produce the cihpertext of data type, the other is to produce the ciphertext of the sensing data.

Data Type Encryption. For each IoT device, if its sensing data falls into data type w_l of application App_j, then the IoT device D_i picks two random numbers $r_{l1}, r_{l2} \in Z_p$ to calculate the ciphertext C_l of $w_l||ID_j^{App}$ as

$$C_{l1} = \rho_2^{r_{l2}} \cdot H(w_l||ID_j^{App})^{r_{l1}}, C_{l2} = \rho_1^{r_{l1}}, C_{l3} = EK^{r_{l2}}, C_{l4} = \rho^{r_{l2}}$$

Sensing Data Encryption. To encrypt the sensing data m_i for App_j, the IoT device D_i chooses two random numbers $r_i', \hat{r}_i \in [1, n/4]$ to compute a Paillier ciphertext

$$c_i = (c_{i1}, c_{i2}) = (g^{r_i'} \mod n^2, (h_j^{App})^{r_i'}(1 + m_i \cdot n) \mod n^2)$$
$$\sigma_i = (\sigma_{i1}, \sigma_{i2}) = ((H(ID_j^{App}||T)^{\hat{r}_i} \cdot \rho_1^{m_i})^{\beta}, \rho_2^{\hat{r}_i})$$

Finally, the IoT device D_i broadcasts $(id_i, ID_j^{App}, c_i, C_l, \sigma_i)$ to the fog node, where $C_l = (C_{l1}, C_{l2}, C_{l3}, C_{l4})$.

3.3 Fog-Assisted Selective Aggregation

After time period T, fog node receives m reports which are from the IoT devices D_i $(i = 1, \cdots, \pi)$, where π denotes the number of IoT devices. And then it executes the following two sub-phases:

Data Types Selection. To select data type, fog node executes Algorithm 1 to select data types, and produces data aggregation of the same type of all sensing reports. Finally, the aggregated result (s_{j1}, s_{j2}) of each data types $w_i, (i = 1, \cdots, \kappa)$ is returned.

If the sensing report does not satisfy Eq. (1), it means that the sensing data type is matched with all data types of w_i; Then it is dropped.

$$e(C_{l1}, T_{j1}) \stackrel{?}{=} e(C_{l2}, T_{j2}) \cdot e(C_{l3}, T_{j3}) \cdot e(C_{l4}, T_{i4}) \quad (1)$$

Content Aggregation. According to all sensing reports and the above data type selection, fog node can achieves all sensing reports aggregation by the following process:

$$CT = (CT_1, CT_2) = (\prod_{i \in \{1, \cdots, \kappa\}} s_{i1}^{P^i}, \prod_{i \in \{1, \cdots, \kappa\}} s_{i2}^{P^i})$$

For the sensing reports' signatures, they can be aggregated into

$$\sigma = (\sigma_1, \sigma_2) = (\prod_{i \in \{1, \cdots, m\}} \sigma_{i1}, \prod_{i \in \{1, \cdots, m\}} \sigma_{i2})$$

The reason that all signatures can be aggregated is that all IoT devices share a secret key β.

```
Input: (T_i)={T_i1, T_i2, T_i3, T_i4}, i = 1, ..., κ and
       {C_i1, C_i2, C_i3, C_i4}, i = 1, ..., m
Output: (s_j1, s_j2), j = 1, ..., κ
1  for (j = 1; j ≤ κ; j++) do
2      s_j1 = 1;
3      s_j2 = 1;
4      for (l = 1; l ≤ m; l++) do
5          if e(C_l1, T_j1) =? e(C_l2, T_j2) · e(C_l3, T_j3) · e(C_l4, T_i4) then
6              s_j1 = s_j1 · c_l1  mod n^2;
7              s_j2 = s_j2 · c_l2  mod n^2;
8          end
9      end
10
11 end
12 return (s_j1, s_j2), j = 1, ..., κ;
```

Algorithm 1: Data type selection algorithm

At last, the fog node forwards the selection aggregation results (CT, σ) to application App_j.

3.4 Data Reading and Verification

After receiving the aggregated results at time period T, application App_j can execute data reading and verification by its secret keys.

First, App_j uses its secret key x to decrypt the aggregated results of each data type by the following steps:

1. It computes
$$M = \frac{CT_2}{CT_1^x} = 1 + (\sum_{i=1}^{\kappa} M_i P^i)n \mod n^2$$
where M_i denotes the aggregated result of all sensing reports with data type w_i.
2. Then, it recovers $\sum_{i=1}^{\kappa} M_i P^i = \frac{M-1 \mod n^2}{n}$, And execute Algorithm 2 to recover the aggregated result of each data type.
3. Finally, The signature of aggregated result can also be verified by the following equation
$$e(\sigma_1, \rho_2) = (e(H(Id_j^{App} \| T), \sigma_2) e(\rho_1, \sigma_2)^{\sum_{i=1}^{\kappa} M_i})^{\beta}$$

Input: $M = M_1 + M_2P + \cdots + M_{\kappa-1}P^{\kappa-1}$, a super-increasing sequence $(1, P, P^2, \cdots, P^{\kappa-1})$ with $M_i < P - 1$
Output: $(M_1, \cdots, M_\kappa)$
1 set $\Phi_{\kappa-1} = M$;
2 **for** $(j = \kappa; j > 1; j - -)$ **do**
3 $\quad \Phi_{j-2} = \Phi_{j-1} \mod P^{j-1}$;
4 $\quad M_j = \frac{\Phi_{j-1} - \Phi_{j-2}}{P^{j-1}}$;
5 **end**
6 $M_1 = \Phi_0$;
7 return $(M_1, M_2, \cdots, M_\kappa)$;

Algorithm 2: Recover the aggregated report of all data types

3.5 Revocation

For achieving revocation, we introduce Revocation List (RL) to design a lightweight mechanism for application revocation. If an application App_j is were taken off the shelves and suspended, TA produces a revocation token $RvT_j = (SK_{j1}^k, SK_{j2}^k)$, where $k \in Z_p$ is a random number, and then it added RvT_j to RL. Note RL is broadcasted to all fog nodes.

In selective aggregation initialization phase, upon receiving $\{T_{i1}, T_{i2}, T_{i3}, T_{i4}\}$, fog node makes use of revocation token RvT_j in the RL and checks the following relation

$$e(T_{i3}, SK_{j1}^k) \stackrel{?}{=} e(SK_{j2}^r, T_{i4}) \tag{2}$$

If there exists a revocation token which make Eq. (2) true, it outputs 1; if Eq. (2) is false for all the revocation tokens in RL, it outputs 0.

4 Security Analysis

In the following, we first discuss the correction of data type selection, then demonstrate the security of the proposed scheme in terms of privacy and confidentiality of sensing data.

Theorem 1. For a sensing report, if $(C_{l1}, C_{l2}, C_{l3}, C_{l4})$ is the ciphertext of its data type w_l, then it must satisfy the verification equation Eq. (1).

Proof. Since the generated cihpertext of w_l by application App_j is $\{T_{l1}, T_{l2}, T_{l3}, T_{l4}\}$, then we have

$$\begin{aligned} & e(C_{l1}, T_{l1}) \\ &= e(\rho_2^{r_2} H(w_l)^{r_1}, \rho_1^{\xi_l}) = e(\rho_2^{r_2}, \rho_1^{\xi_l}) e(H(w_l)^{r_1}, \rho_1^{\xi_l}) \end{aligned}$$

$$e(C_{l2}, T_{l2}) = e(\rho_1^{r_1}, H(w_l)^{\xi_l})$$

$$e(C_{l3},T_{l3}) = e(\rho^{\frac{f(t)\cdot r_2}{\alpha_1}}, \rho_2^{\alpha_1\cdot\xi_l\cdot\Delta_{t,\Theta_i}(0)}) = e(\rho,\rho_2)^{r_2\xi_l f(t)\cdot\Delta_{t,\Theta_i}(0)}$$

$$e(C_{l4},T_{l4}) = e(\rho^{r_2}, \rho_2^{\xi_l f(t_i)\Delta_{t_i,\Theta_i}(0)}) = e(\rho, \rho_2^{r_2\xi_l f(t_i)\Delta_{t_i,\Theta_i}(0)})$$

According to Lagrange interpolation formula, we can know

$$\begin{aligned} & e(C_{l3},T_{l3})e(C_{l4},T_{l4}) \\ = & e(\rho,\rho_2)^{r_2\xi_l(f(t_i)\Delta_{t_i,\Theta_i}(0)+f(t)\Delta_{t,\Theta_i}(0))}) \\ = & e(\rho,\rho_2)^{\alpha_1 r_2\xi_l} = e(\rho_1,\rho_2)^{r_2\xi_l} \end{aligned}$$

Thus, we have $e(C_{l1},T_{l1}) = e(C_{l2},T_{l2})e(C_{l3},T_{l3})e(C_{l4},T_{l4})$ □

In the following, we will discuss data privacy and data integrity.

Sensing Data Privacy. In our three-tier architecture, to ensure the privacy of the transmitted data between fog node and IoT device, we adopt Paillier public encryption algorithm to encrypt the sensing report m_i from IoT devices. Because fog node does not know the corresponding private key, it makes that fog node cannot decrypt the corresponding ciphertext. In addition, although fog node can obtain the ciphertext $(g^{r_i'}, h_j^{App} = g^{\tau_i}, (h_j^{App})^{r_i'}(1+m_i\cdot n) \mod n^2)$, it is impossible to extract m_i from the ciphertext since to extract m_i needs that an adversary must know $(h_j^{App})^{r_i'}$. However, given $(g^{r_i'}, h_j^{App} = g^{\tau_i})$, to obtain $(h_j^{App})^{r_i'}$ is equivalent to solving computational Diffie-Hellman problem. Obviously, it is in contradiction with the difficulty of solving computational Diffie-Hellman assumption. Thus, sensing data's privacy is preserved.

For data type w_i, to ensure the privacy and secure matching of data type, we use secure encryption scheme with keyword search to encrypt data type, and make use of the search ability on the encrypted data to achieve the matching of data type. It appears in the form of the ciphertext $\{C_{l1}, C_{l2}, C_{l3}, C_{l4}\}$. Because the adopted searchable encryption scheme is semantically secure against an adaptive chosen keyword attack, Fog node can obtain nothing about data type w_i from $\{C_{l1}, C_{l2}, C_{l3}, C_{l4}\}$ in data type selection phase. In addition, given two ciphertexts C_l and C_l' of two data types, Fog node cannot also distinguish whether these two ciphertexts correspond to the same data type. For these two ciphertexts, they has the following formats:

$$C_{l1} = \rho_2^{r_{l2}}\cdot H(w_l||ID_j^{App})^{r_{l1}}, C_{l2} = \rho_1^{r_{l1}}, C_{l3} = EK^{r_{l2}}, C_{l4} = \rho^{r_{l2}}$$
$$C_{l1}' = \rho_2^{r_{l2}'}\cdot H(w_l'||ID_j^{App})^{r_{l1}'}, C_{l2}' = \rho_1^{r_{l1}'}, C_{l3}' = EK^{r_{l2}'}, C_{l4}' = \rho^{r_{l2}'}$$

where $r_{l1}, r_{l2}, r_{l1}', r_{l2}'$ are random numbers. To obtain the relation between the ciphertexts C_l' and C_l, an adversary must obtain $\rho_2^{r_{l2}-r_{l2}'}$ since given $\rho_2^{r_{l2}-r_{l2}'}$, an adversary can determine the relation of C_l and C_l. The reason is that the size of data type space is usually polynomial or low-entropy distribution, and an adversary can check whether the equation

$$e(\frac{C_{l1}/C'_{l1}}{\rho_2^{r_{l2}-r'_{l2}}}, \rho_1) = e(H(w_l||ID_j^{App}), \rho_1^{r_{l2}-r'_{l2}})$$

holds by exhaustive search attack. However, given $(\rho, \rho_2 = \rho^{x_2}, \rho^{r_{l2}-r'_{l2}})$, it is equivalent to solving the Computational Diffie-Hellman problem to obtain $\rho_2^{r_{l2}-r'_{12}}$. Obviously, it is inconsistent with the difficulty of solving the Computational Diffie-Hellman problem.

In summary, our scheme can preserve privacy of the content and achieve indistinguishability of data type.

Data Integrity. In content aggregation phase, fog node can aggregate all ciphertexts $c_i, i = 1, \cdots, m$ into a ciphertext CT by using homomorphism of Paillier encryption scheme, but fog node can not tamper/modify the IoT devices' aggregation results since linear homomorphic signature scheme is required in data verification phase. In our proposed scheme, the improved Paillier encryption only supports Additive homomorphic property, and the adopted homomorphic signature scheme only supports multiplicative homomorphic property. It enable that a tampered sensing report m'_i can not pass the verification of signature in data verification phase. Thus, our proposed scheme can achieve sensing data integrity.

Revocation. After an application App_i is revoked, TA needs to publish the corresponding revocation token $RvT_i = (SK^r_{i1}, SK^r_{i2})$ and add it in revocation list RL. Fog node deletes all the ciphertext of data types which correspond to the suspended application App_i by checking the relation according to the revocation token in the updated RL.

$$e(T_{i3}, SK^k_{j1}) \stackrel{?}{=} e(SK^r_{j2}, T_{i4}) \quad (3)$$

5 Conclusion

In this paper, we have proposed a privacy preserving selective data aggregation scheme with revocation for fog-assisted IoT. Then, our scheme has been proposed and designed particularly based on the features of fog computing and IoT, to guarantee data privacy and data integrity distributed fog nodes, and multiple application services. In the future work, we focus on selective forwarding attack: the attackers only selectively aggregate part of the data and forward the incomplete results to the application service.

Acknowledgement. This work is supported in part by The Natural Science Foundation of Beijing (No. 4212019), National Natural Science Foundation of China (No. 62172005), Guangxi Key Laboratory of Cryptography and Information Security (No. GCIS201808) and Foundation of Guizhou Provincial Key Laboratory of Public Big Data (No. 2019BDKF JJ012).

References

1. Dastjerdi, A.V., Buyya, R.: Fog computing: helping the internet of things realize its potential. IEEE Comput. **49**(8), 112–116 (2016)
2. Morocho, M., Lee, H., Lim, W.: Machine Learning for 5G/B5G mobile and wireless communications: potential, limitations, and future directions article. IEEE Access **7**, 137184–137206 (2020)
3. Chiang, M., Zhang, T.: Fog and IoT: an overview of research opportunities. IEEE Internet Things J. **3**(6), 854–864 (2016)
4. Takabi, H., Joshi, J.B., Ahn, G.J.: Security and privacy challenges in cloud computing environments. IEEE Secur. Priv. **8**, 271–350 (2010)
5. Huang, C., Liu, D., Shen, S.: Reliable and privacy-preserving selective data aggregation for fog-based IoT. In: 2018 IEEE International Conference on Communications (ICC), pp. 1–6 (2018)
6. Boneh, D., Gentry, C., Lynn, B., Shacham, H.: Aggregate and verifiably encrypted signatures from bilinear maps. In: Biham, E. (ed.) EUROCRYPT 2003. LNCS, vol. 2656, pp. 416–432. Springer, Heidelberg (2003). https://doi.org/10.1007/3-540-39200-9_26
7. Zhang, J., Zhu, J., Zhang, N.: An improved privacy-preserving collaborative filtering recommendation algorithm. In: Proceedings of Asia Conference on Information Systems, pp. 277–288 (2014)
8. Alghamdi, W.Y., Wu, H., Kanhere, S.S.: Reliable and secure end-to-end data aggregation using secret sharing in WSNs. In: IEEE WCNC, pp. 1–6 (2017)
9. Qian, J., Qiu, F., Wu, F., Ruan, N., Chen, G., Tang, S.: Privacy-preserving selective aggregation of online user behavior data. IEEE Trans. Comput. **66**(2), 326–338 (2017)
10. Hu, H., Lu, R., Zhang, Z., Shao, J.: REPLACE: a reliable trust-based platoon service recommendation scheme in VANET. IEEE Trans. Veh. Tech. **66**(2), 1786–1797 (2017)
11. Wang, K., Shao, Y., Shu, L., Zhu, C., Zhang, Y.: Mobile big data fault-tolerant processing for eHealth networks. IEEE Netw. **30**(1), 36–42 (2016)
12. Zhang, H., Qiu, Y., Long, K., Karagiannidis, G.K., Wang, X., Nallanathan, A.: Resource allocation in NOMA based fog radio access networks. IEEE Wirel. Commun. **25**(3), 110–115 (2018)
13. Zhang, J., Zhang, Q., Ji, S.: A fog-assisted privacy-preserving task allocation in crowdsourcing. IEEE Internet Things J. **7**(9), 8331–8342 (2020)
14. Zhang, J., Bai, W., Wang, Y.: Non-interactive ID-based proxy re-signature scheme for IoT based on mobile edge computing. IEEE Access **7**, 37865–37875 (2019)
15. Liu, X., Deng, R.H., Choo, K.K.R., Weng, J.: An efficient privacy-preserving outsourced calculation toolkit with multiple keys. IEEE Trans. Inf. Forensics Secur. **11**(11), 2401–2414 (2016)
16. Lu, R., Heung, K., Lashkari, A.H., Ghorbani, A.A.: A lightweight privacy-preserving data aggregation scheme for fog computing-enhanced IoT. IEEE Access **5**, 3302–3312 (2017)

17. Shen, H., Zhang, M., Shen, J.: Efficient privacy-preserving cube-data aggregation scheme for smart grids. IEEE Trans. Inf. Forensics Secur. **12**(6), 1369–1381 (2017)
18. Choubin, M., Taherpour, A., Rahmani, M.: Collaborative data aggregation using multiple antennas sensors and fusion centre with energy harvesting capability in WSN. IET Commun. **13**(13), 1971–1979 (2019)
19. Rezaeibagha, F., Yi, M., Huang, K., Chen, L.: Secure and efficient data aggregation for IoT monitoring systems. IEEE Internet Things J. (2020). https://doi.org/10.1109/JIOT.2020.3042204
20. Yan, O., Liu, A., Xiong, N., Wang, T.: An effective early message ahead join adaptive data aggregation scheme for sustainable IoT. IEEE Trans. Netw. Sci. Eng. (2020). https://doi.org/10.1109/TNSE.2020.3033938
21. Boneh, D., Goh, E.-J., Nissim, K.: Evaluating 2-DNF formulas on ciphertexts. In: Kilian, J. (ed.) TCC 2005. LNCS, vol. 3378, pp. 325–341. Springer, Heidelberg (2005). https://doi.org/10.1007/978-3-540-30576-7_18
22. Zhou, J., Cao, Z., Dong, X., Lin, X.: Security and privacy in cloud-assisted wireless wearable communications: Challenges, solutions, and future directions. IEEE Wireless Commun. **22**(2), 136–144 (2015)
23. Yi, X., Bouguettaya, A., Georgakopoulos, D., Song, A., Willemson, J.: Privacy protection for wireless medical sensor data. IEEE Trans. Dependable Sec. Comput. **13**(3), 369–380 (2016)
24. Bao, H., Lu, R.: A new differentially private data aggregation with fault tolerance for smart grid communications. IEEE Internet Things J. 2(3), 248–258 (2015)
25. Tang, W., Ren, J., Deng, K., Zhang, Y.: Secure data aggregation of lightweight e-healthcare IoT devices with fair incentives. IEEE Internet Things J. **6**(5), 8714–8726 (2019)
26. Xiong, J., et al.: Enhancing privacy and availability for data clustering in intelligent electrical service of IoT. IEEE Internet Things J. **6**(2), 1530–1540 (2019)
27. Xiong, J., et al.: A personalized privacy protection framework for mobile crowdsensing in IIoT. IEEE Trans. Industr. Inf. **16**(6), 4231–4241 (2020)
28. Xiong, J., Chen, X., Yang, Q., Chen, L., Yao, Z.: A task-oriented user selection incentive mechanism in edge-aided mobile crowdsensing. IEEE Trans. Netw. Sci. Eng. **7**(4), 2347–2360 (2020)

A Novel Location Privacy-Preserving Task Allocation Scheme for Spatial Crowdsourcing

Xuelun Huang[1], Shaojing Fu[1,2(✉)], Yuchuan Luo[1], and Liu Lin[1]

[1] College of Computer, National University of Defense Technology, Changsha, China
[2] Sate Key Laboratory of Cryptology, Beijing, China

Abstract. With the increasing popularity of big data and sharing economics, spatial crowdsourcing as a new computing paradigm has attracted the attention of both academia and industry. Task allocation is one of the indispensable processes in spatial crowdsourcing, but how to allocate tasks efficiently while protecting location privacy of tasks and workers is a tough problem. Most of the existing works focus on the selection of the workers privately. Few of them present solutions for secure problems in task delivery. To address this problem, we propose a novel privacy protection scheme that not only protects the location privacy of workers and tasks but also enables secure delivery of tasks with very little overhead. We first use the paillier homomorphic cryptosystem to protect the privacy of workers and tasks, then calculate travel information securely. Finally, let workers restore the tasks' location. In our scheme, only workers who meet the requirements can get the exact location of tasks. In addition, we prove the security of our method under the semi-honest model. Extensive experiments on real-world data sets demonstrate that our scheme achieves practical performance in terms of computational overhead and travel cost.

Keywords: Spatial crowdsourcing · Privacy-preserving · Homomorphic cryptosystem · Task allocation

1 Introduction

Crowdsourcing has gradually attracted the attention of all walks of life since Jeff Howe, a reporter for the Wired magazine, proposed it in 2006 [5]. Jeff Howe defines crowdsourcing as a company or organization posting problems on the network to collect better solutions. Nowadays, many crowdsourcing platforms (e.g., Amazon Mechanical MTurk1, TaskRabbit2) have been established to provide various kinds of crowdsourcing services. Mobile crowdsensing and spatial crowdsourcing also emerge as the times. Both of them require the participation of a large number of users, reduces costs, and leverages the advantages of the network to accumulate the resource of the public under different knowledge backgrounds. However, mobile crowdsensing focuses more on the use of mobile

J. Xiong et al. (Eds.): MobiMedia 2021, LNICST 394, pp. 304–322, 2021.
https://doi.org/10.1007/978-3-030-89814-4_23

devices for data perception, collection, and analysis, which does not pay attention to how tasks are allocated. Many mobile crowdsensing tasks are distributed and utilized through crowdsourcing. Therefore, solving task allocation problem in spatial crowdsourcing is also helpful to mobile crowdsensing.

With the rapid development of mobile internet technology, spatial crowdsourcing has also become popular. Spatial crowdsourcing plays a critical role in various fields, such as news, tourism, intelligence, disaster response, and urban planning [17]. Take real-time traffic condition monitoring as an example, it affects people's daily travel and lifestyle at all times. By obtaining the spatial distribution information of users at different times and corresponding various sensor data, such software can analyze and speculate real-time traffic conditions. When users use this software, they passively become crowdsourcing workers and share their spatiotemporal information and sensor data. In contrast to general crowdsourcing, spatial crowdsourcing adds more location requirements that need workers to reach the designated location to complete the task.

At the same time, people have paid attention to information security, hoping not only to enjoy the convenience of emerging techniques but also to protect their private information. When using crowdsourcing, the crowdsourcing platform needs the information of tasks and workers to perform task allocation, which usually contains a lot of private information. If it leaks out private information, disastrous consequences will spring out. For example, when users are enjoying the taxi service, they need to tell drivers where they are and where they want to go, but users do not want the platform to know. Because location data of users may indicate their home addresses, lifestyle, and other sensitive information. Attackers would know the real-time location of users once they grasped these privacies. These security risks depress the availability of crowdsourcing and may let some people refuse to use crowdsourcing. Thus, it's significant to allocate tasks efficiently at the premise of protecting privacy. To solve this problem, many feasible solutions have been proposed. [13] used homomorphic encryption and Yao's garbled circuits to achieve a secure task distribution. But it only protects the privacy of workers, without considering the privacy of tasks. However, task information will indirectly reveal the workers' information. [21] designed a grid-based position protection method for task distribution. But the distribution process involves heavy encryption and decryption operations, which is not efficient for practice. [23] proposed a novel spatial crowdsourcing framework without trusted third parties but providing differential privacy guarantees. But workers need to set up an acceptable location in advance, which requires a relatively sizeable storage space. [22] designed a data aggregation protocol based on k-anonymity, but it cannot well resist malicious deception of workers.

Most of the works focus on how to allocate tasks more securely and efficiently but ignore the next step after task allocation: how to deliver tasks to workers securely and efficiently. The major contributions of this article are as follows:

1. We propose a scheme for spatial crowdsourcing task allocation. Based on the two-server model and an additively homomorphic cryptographic cryptosystem, the proposed scheme protects the location privacy of workers and tasks without involving any online trusted third party (TTP).

2. We put forward a novel method to deliver tasks and calculate the travel information securely. This method ensures that only workers who meet the requirements can get the location information of tasks and also allows workers to reconstruct tasks' location with practical efficiency.
3. Security analysis in our paper indicates that our scheme can protect the location information about the works and tasks, as well as the data access patterns. Experimental results demonstrate that our scheme achieves practical performance in terms of computational overhead and travel cost.

We organize the rest of the paper as follows. Section 2 presents preliminaries, and Sect. 3 gives the problem formulation. Section 4 gives the proposed scheme. We present the security and performance analyses in Sect. 5 and introduce related work in Sect. 6. In Sect. 7, we conclude our work.

2 Preliminaries

In this section, we review the concepts and general procedures of the Paillier cryptosystem and spatial crowdsourcing, then introduce notations of this article.

2.1 Paillier Cryptosystem

The Paillier cryptosystem is a probabilistic public key encryption system invented by Paillier in 1999 [11]. The encryption algorithm is a homomorphic public key encryption system that satisfies addition and number multiplication homomorphism. Firstly, randomly select two large prime numbers p and q, which satisfy $gcd(pq,\ (p-1)(q-1)) = 1$. Then calculate $n = pq$ and $\lambda = lcm(p-1,\ q-1)$. And randomly select an positive integer g ($g \in Z^*_{n^2}$), which is less than n^2. Define $L(x) = \frac{(x-1)}{n}$, and there exists $\mu = (L(\ g^\lambda\ mod\ n^2))^{-1} mod\ n$. The public key PK is (n, g), the private key SK is (λ, μ). In encryption, randomly select a number $r \in Z_n$ and calculate the ciphertext $c = g^m \cdot r^n\ mod\ n^2$, where m is the original message. In decryption, calculate the plaintext $m = \frac{L(c^\lambda\ mod n^2)}{L(g^\lambda\ mod n^2)}\ mod\ n$.

We can express the homomorphic properties as : $c_1 \cdot c_2 = E[m_1, r_1] \cdot E[m_2, r_2] = g^{m_1+m_2}(r_1 \cdot r_2)^n mod\ n^2$, $D[c_1 \cdot c_2] = D[E[m_1, r_1]E[m_2, r_2] mod\ n^2] = m_1 + m_2\ mod\ n$. Here, $r_1, r_2 \in Z_n$ is a random number; $m_1, m_2 \in Z_n$ is plaintext; c_1, c_2 is the ciphertext of m_1, m_2.

2.2 Spatial Crowdsourcing

Spatial crowdsourcing applications are already very common in our daily life, such as Gigwalk, Easyshift, and Fieldagent, *etc.* It has a wide range of applications, but when people enjoy the happiness and convenience brought by these software, people inadvertently reveal a lot of their private information.

Spatial crowdsourcing often includes three parties, requesters, workers, and crowdsourcing platforms. First, requesters publish tasks on the crowdsourcing platform, then the crowdsourcing platform allocates tasks according to tasks' locations and workers' locations. And then, workers accept and complete tasks. There are two modes of spatial crowdsourcing task allocation [18]. One of them is the Server Assigned Tasks (SAT), which is a platform server-centric model. In this mode, the platform server assigns a nearby task to each worker after receiving the locations of all workers. Therefore, it is possible to assign nearby tasks to each worker when maximizing the overall number of tasks assigned. However, sending workers' location to the server may cause privacy threats. The other one is Worker Selected Tasks (WST), which is a user-centric model. Platform servers often issue space-aware tasks. Online workers can choose any spatial task without consulting with the server. Users submit less personal information, which can increase the participation of mobile users. However, some spatial tasks may never be assigned, and other tasks are assigned redundantly. And may not form a global optimal allocation (Table 1).

2.3 Notations

Table 2 presents several symbols for better readability.

Table 1. Notations

Notation	Meaning
w_i, t_i	Worker i, task i
(PK, SK)	Public key, Secret key
x_{w_i}, y_{w_i}	Coordinates of w_i, $0 < \mathrm{i} \leq \mathrm{m}$
x_{t_i}, y_{t_i}	Coordinates of t_i, $0 < \mathrm{i} \leq \mathrm{n}$
$D[\cdot]$	Decryption operation
$E[x_{w_i}], E[y_{w_i}]$	Encrypted coordinates of w_i
$E[dx_{w_i,t_i}], E[dy_{w_i,t_i}]$	The encrypted difference between w_i and t_i in the x, y direction
Dis_{w_i,t_i}	Distance between w_i and t_i
C_t, C_w	Task set, Worker set
sort	A function implements sorting from small to large
available	A function filtering out workers who are not currently available

3 Problem Formulation

3.1 System Model

Our system model is shown in Fig. 1. In the setting of our privacy-preserving scheme, we have four entities:

- **Requester :** A task requester is a user who first publishes tasks to CSP. The requester then waits for CSP to assign workers. Then wait for the workers to finish the task. For instance, in the taxi service, the taxi passenger is the requester, and he waits for the platform to assign him a driver.

- **Worker :** A task worker is a user who receives tasks. He decides whether to accept tasks according to his own will. When he accepts a task, he will finish the task quickly and efficiently. For instance, in the taxi service, the taxi driver is the worker, and he waits for the platform to allocate passengers to him.
- **CSP (Crowdsourcing Service Provider) :** Upon receiving a task, CSP cooperates with S to calculate the travel information. Also responsible for generating lists of candidate workers and interacting with workers.
- **S (Server) :** S is responsible for key generation and distribution. Assist CSP to complete travel information calculation. Also responsible for interacting with workers.

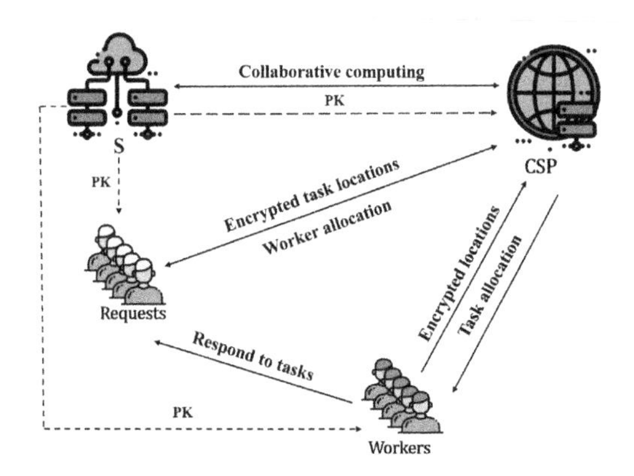

Fig. 1. System model

3.2 Threat Model

The threat model for each entity is set as follows:

- **Users:** (requesters and workers) are considered as fully trusted in the scheme where they could execute the operations properly and protect their locations and secret keys. Users wouldn't leak their locations actively or passively to the other entity, *e.g.* CSP or S.
- **CSP & S** are considered as *honest – but – curious* in the sense that they could execute the designed operations honestly. We also assume that they cannot collude with each other and wouldn't launch active attacks, such as collusion with users, pretending to be a requester (or a worker).

The assumption for CSP and S is reasonable (e.g. [1]), because most of the cloud service providers in the market are well-established IT companies and they understand the importance of reputation [8,9]. Active attacks are straightforward to detect and may damage their reputation once caught. Collusion between them is highly unlikely as it may damage their reputation and affect their revenues.

4 The Proposed Scheme

4.1 Overview

Our system model is shown in Fig. 1. The scheme consists of four parts, namely **KeyGen**, **DataEnc**, **TaskAllocation**, **TaskDelivery**.

- **KeyGen** $\rightarrow (PK, SK)$: S generate PK and SK for the Paillier Cryptosystem. Sending PK to CSP, requests, and workers, SK to CSP.
- **DataEnc** $\rightarrow (E[x_i], E[y_i])$: Workers encrypt their locations, and requesters encrypt the locations of tasks. Then they all send the encrypted locations to CSP.
- **TaskAllocation** $\rightarrow List(t, w)$: CSP receives requesters' tasks, then cooperates with S to calculate the travel message between tasks and workers, generating candidate worker lists based on this.
- **TaskDelivery** $\rightarrow (tan_{w_i,t_i}, signx_{w_i,t_i}, signy_{w_i,t_i})$: CSP notifies workers in the candidate worker list in turn until the task is assigned. Workers receive travel information about tasks after they accept tasks.

4.2 Scheme Details

4.2.1 Task Allocation

1) Distance Calculation. After receiving the encrypted tasks' locations and workers' locations, the CSP cooperates with S to calculate the distance between tasks and workers. The calculation procedures are shown in algorithm1.

First, CSP uses the homomorphic encryption property of the Paillier cryptosystem to calculate the coordinate distance values. Through step 1 completed by CSP, we can easily see that $E[dx_{w_i,t_i}] = E[x_{w_i} - x_{t_i}], E[dy_{w_i,t_i}] = E[y_{w_i} - y_{t_i}]$.

In order not to let S get any information about the location, then choose two random numbers $r_1, r_2 \in Z_N$, adding them to the result and send the result to S. Upon receiving a, b, c, d from CSP, because S has the decryption key SK, it can decrypt and calculate the corresponding squared distance, that is $ans1 = E[(dx_{w_i,t_i} + r_1) \cdot (dy_{w_i,t_i} + r_2)]$, $ans2 = E[(dy_{w_i,t_i} + r_1) \cdot (dy_{w_i,t_i} + r_2)]$.

So S cannot get the real distance value. CSP can use the properties of homomorphic encryption to remove the influence of $r1$ and $r2$ in the results. $Ans3$ and $ans4$ are the squared distances in the x and y directions of the real distance between workers and tasks. The decryption result of $ans3 \cdot ans4$ is the distance squared value. Same as before, CSP selects a random number to prevent S speculation and finally can get Dis_{w_i,t_i}.

2) Candidates List Generation.
Each worker and task can use Algorithm 1 to calculate the squared distance between the encrypted locations. Each requester submits the maximum acceptable distance D when submitting the task location. Only when the distance between workers and tasks is within the range of the maximum acceptable distance D, workers are eligible for the task. Therefore, by ranking the distance

Algorithm 1: Distance Calculation

Input: $E[x_{w_i}], E[y_{w_i}], E[x_{t_i}], E[y_{t_i}]$
Output: Dis_{w_i,t_i}
1. CSP:
    $E[dx_{w_i,t_i}] \leftarrow E[x_{w_i}] \cdot E[x_{t_i}]^{-1}$, $E[dy_{w_i,t_i}] \leftarrow E[y_{w_i}] \cdot E[y_{t_i}]^{-1}$;
    Pick two random numbers r_1, r_2;
    $a \leftarrow E[dx_{w_i,t_i}] \cdot E[r_1]$; $b \leftarrow E[dx_{w_i,t_i}] \cdot E[r_2]$;
    $c \leftarrow E[dy_{w_i,t_i}] \cdot E[r_1]$; $d \leftarrow E[dy_{w_i,t_i}] \cdot E[r_2]$;
    send a, b, c, d to S;
2. S:
    $a' \leftarrow D[a]$; $b' \leftarrow D[b]$; $c' \leftarrow D[c]$; $d' \leftarrow D[d]$;
    $ans1 \leftarrow E[a' \cdot b']$; $ans2 \leftarrow E[c' \cdot d']$, send $ans1, ans2$ to CSP;
3. CSP:
    Pick one random number r_3;
    $ans3 \leftarrow ans1 \cdot E[r_1 \times r_2]^{-1} \cdot E[dx_{w_i,t_i}]^{-r_1} \cdot E[dx_{w_i,t_i}]^{-r_2}$;
    $ans4 \leftarrow ans2 \cdot E[r_1 \times r_2]^{-1} \cdot E[dy_{w_i,t_i}]^{-r_1} \cdot E[dy_{w_i,t_i}]^{-r_2}$;
    $ans5 \leftarrow ans3 \cdot ans4 \cdot E[r_3]$, send $ans5$ to S;
4. S:
    Receive $ans5$ from CSP;
    $dis \leftarrow D[ans5]$, send dis to CSP;
5. CSP:
    $Dis_{w_i,t_i} \leftarrow dis - r_3$;

between tasks and workers, CSP can generate a list of candidate workers. To facilitate subsequent calculations, CSP records the distance values between workers and tasks in workers' database and tasks' database, respectively. We should note it as it randomly generates the random numbers mentioned in the algorithm in each cycle.

3) Task Allocation.
We devise efficient task allocation algorithms for both SAT and WTS. After the previous algorithm process, each task has a corresponding list of candidate workers. Therefore, the server allocation can directly notify workers in order according to the candidate worker list until the task is assigned. The procedures are shown in Algorithm 2.

In the SAT, the maximum number of candidate workers $maxNt$ is set to prevent too many workers who meet the requirements from affecting efficiency. Use counter to mark the number of selected workers and $t.d_{task}$ is the maximum distance D that the task t can accept. After the processing of **sort** and **available**, CSP can get an optional and orderly set of workers SC_w. After many cycles, each task can get a list of optional workers. It is worth noting that a worker may exist in multiple candidate worker lists at the same time. When a worker is invited to a task, the worker will decide whether to accept the task. And we assume that the workers in this article can only accept one task at a time, and each worker is busy and idle. When the worker accepts the task, his status will become busy, and the worker in the other candidate list will be invalid.

```
Algorithm 2: Task Allocation - SAT
Input: C_t, C_w, maxNt
Output: List_w = (t, w)
for each task t in C_t do
    counter = 0;
    SC_w = available ( sort ( C_w ) );
    for each worker w in SC_w do
        if w.d[t] > t.d_task then
            Break;
        else if counter == maxNt then
            Break;
        else
            List ← (t, w);
            ++counter;

    end
end
return List;
```

The worker mode is that the worker receives all the distance of tasks he satisfies, then chooses the task subjectively. But what criteria the worker uses to select the task based on the candidate task list is not the focus of this article, and this affects the subsequent implementation of the algorithm, so we just briefly introduce this algorithm.

In the WTS, $maxNw$ is the maximum number of tasks that the worker can select, and $w.d_{worker}$ is the maximum distance that the worker w can accept. After going through the previous algorithm, some workers may have some optional tasks in the database. Then these workers can choose tasks, which means that they must decide according to their preferences. For workers with optional tasks, after implementing Algorithm 3, they can get a list of candidate tasks ordered from near to far.

4.2.2 Task Delivery

1) Pre-processing.
The Paillier cryptosystem is carried out on positive integers, but the latitude and longitude coordinates are mostly floating-point. It also involves positive and negative numbers. Therefore, we need to convert the latitude and longitude to positive integers before encryption, then convert it to the original latitude and longitude after decrypting. We exploit the modular arithmetic properties of the Paillier scheme. We represent only integers between $(-n/3,\ n/3)$. Since n is a

Algorithm 3: Task Allocation - WTS

```
Input: C_w, maxNw
Output: List_t = (t, w)
for each worker w in C_w do
    counter = 0;
    SC_t = available ( sort ( C_w.t ) );
    for each task t in SCt do
        if t.d[w] > w.d_worker then
            Break;
        else if counter == maxNw then
            Break;
        else
            List ← (t, w);
            ++counter ;
    end
end
return List;
```

very large number, the longitude range is 0–180° and the latitude range is 0–90°, they are included in the range. Paillier homomorphic arithmetic works modulo n. We take the convention that a number $x < n/3$ is positive and that a number $x > 2n/3$ is negative. The range $n/3 < x < 2n/3$ allows for overflow detection. Representing floating-point numbers as integers is a harder task. Here we use a variant of fixed-precision arithmetic. In fixed precision, we encode by multiplying every float by a large number (*e.g.* $1e6$) and rounding the resulting product. We decode by dividing by that number. There are many other conversions, the specific details can be found in Sect. 5.

2) Travel Angle Calculation.
Before CSP notifies workers according to the list of candidate workers, it also needs to work with S to calculate the travel angle of workers. In this way, workers can combine their locations and travel angle to restore the locations of tasks. After pre-processing, the Paillier cryptosystem can encrypt and decrypt any real numbers in the range. Algorithm 4 shows the calculation process of workers' travel angle.

Algorithm 4: Travel Angle Calculation

Input: $E[x_{w_i}], E[y_{w_i}], E[x_{t_i}], E[y_{t_i}]$
Output: $tan_{w_i,t_i}, sign_{w_i,t_i}$
1. CSP:
    $E[dx_{w_i,t_i}] \leftarrow E[x_{w_i}] * E[x_{t_i}]^{-1}$; $E[dy_{w_i,t_i}] \leftarrow E[y_{w_i}] * E[y_{t_i}]^{-1}$;
    Pick random numbers r_4, r_5;
    $e \leftarrow E[dx_{w_i,t_i}]^{r_4}$; $f \leftarrow E[dy_{w_i,t_i}]^{r_5}$;
    send e, f to S;
2. S:
    $e' \leftarrow D[e]$; $f' \leftarrow D[f]$;
    $tan_{w_i,t_i} \leftarrow f'/e'$;
    if $e' > 0$, $signx_{w_i,t_i} = 1$, else $signx_{w_i,t_i} = -1$;
    if $f' > 0$, $signy_{w_i,t_i} = 1$, else $signy_{w_i,t_i} = -1$;
    Send $tan_{w_i,t_i}, signx_{w_i,t_i}$ to CSP;
3. CSP:
    if $r_4 < 0$, change $signx_{w_i,t_i}(2 \leftarrow 1, 1 \leftarrow 2)$;
    $tanr_{w_i,t_i} = tan_{w_i,t_i} * r_4/r_5$;
    if $tanr_{w_i,t_i} > 0$, $signy_{w_i,t_i} = signx_{w_i,t_i}$, else
$signy_{w_i,t_i} = -signx_{w_i,t_i}$;

Similar to Algorithm 1 at the beginning, we need to get the distance between the worker and the task first. The distance values calculated in Algorithm 1 have been recorded in the worker database, so we can directly call the distance result value according to the worker ID in the candidate worker list. The same is to prevent S from getting any valid information from the intermediate results, and the effect of the random number r_4, r_5 needs to be added, so e, f is calculated and sent to S. S decrypts the received information, calculates the division value tan, and sets the sign value according to the rules. It is not difficult to know that the result corresponding to tan_{w_i,t_i} is $dy_{w_i,t_i}/dx_{w_i,t_i}$, which is mathematically explained by the tan value of the trigonometric function. It should be noted that when the longitudes of the two points are the same, the denominator is 0. Although this situation does not occur frequently, to implement the scheme smoothly, we use a very small value instead of 0 in the code.

Finally, the CSP modifies the $signx_{w_i,t_i}, signy_{w_i,t_i}$ value according to the positive and negative values of r_4, r_5. And it is the next information to be sent to the workers, which is also stored in the workers' database.

3) Task Location Calculation.
After receiving messages $Dis_{w_i,t_i}, tan_{w_i,t_i}, signx_{w_i,t_i}, signy_{w_i,t_i}$, workers can estimate the location of the task.

According to $tan_{w_i,t_i} = \frac{dy_{w_i,t_i}}{dx_{w_i,t_i}} = \frac{y_{w_i} - y_{t_i}}{x_{w_i} - x_{t_i}}$.

We can get

$$dx_{w_i,t_i} = x_{w_i} - x_{t_i} = \pm\sqrt{\frac{Dis_{w_i,t_i}}{1+tan^2_{w_i,t_i}}}$$
$$x_{w_i} = x_{t_i} \pm \sqrt{\frac{Dis_{w_i,t_i}}{1+tan^2_{w_i,t_i}}} \tag{1}$$

$$dy_{w_i,t_i} = y_{w_i} - y_{t_i} = \pm tan_{w_i,t_i} \cdot \sqrt{\frac{Dis_{w_i,t_i}}{1+tan^2_{w_i,t_i}}}$$
$$y_{w_i} = y_{t_i} \pm tan_{w_i,t_i} \cdot \sqrt{\frac{Dis_{w_i,t_i}}{1+tan^2_{w_i,t_i}}} \tag{2}$$

The sign in the Eq. (1), (2) is determined by $signy_{w_i,t_i}$, $signx_{w_i,t_i}$ respectively. When $signx_{w_i,t_i} = 1$, the worker takes the positive sign, and when $sign_{w_i,t_i} = -1$, the worker takes the negative sign. So does $signy_{w_i,t_i}$.

5 Security and Performance Analysis

5.1 Security Analysis

We can summarize the security goal of our scheme as Theorem 1, 2, 3, followed by the proofs. Before providing the rigorous proofs to the privacy, we introduce the semantic security in Paillier homomorphic cryptosystems [3] and the security definition of the protocol under the semi-honest model in advance.

Definition 1 (semantic security in Paillier cryptosystem).

$$Pr\{c \leftarrow [m_1]\} - Pr\{c \leftarrow [m_2]\} \leq negl(\lambda) \tag{3}$$

In Eq. (3), m_1 and m_2 represent two plaintexts, c is the ciphertext of m_1 encrypted by paillier cryptosystem. $Pr\{c \leftarrow [m_1]\}$ is the probability that an attacker judges the message is m_1 after he observes c. $Pr\{c \leftarrow [m_2]\}$ is the probability that an attacker judges the message is m_2 after he observes c. $negl(\lambda)$ is a negligible polynomial. It means that an attacker can not distinguish m_1 from m_1 and m_2 after he observes c in Paillier cryptosystem.

Definition 2 (security in the semi-honest model [2]). Suppose a_i is the input of party P_i, $\prod_i(\pi)$ is the execution image of P_i, and b_i is the output of P_i computed from protocol π. If $\prod_i(\pi)$ can be simulated from a_i and b_i, then π is secure. In other words, distribution of the simulated image is computationally indistinguishable from $\prod_i(\pi)$.

Theorem 1. The *Distance Calculation* (*DC*) protocol described in Algorithm 1 is secure under the semi-honest model.

Proof. Here, let the execution image of CSP be denoted by $\prod_{CSP}(DC) = \{(ans1, ans2), (dis)\}$, where $dis = Dis_{w_i,t_i} + r_3$. Note that r_3 is a random number in Z_N. We assume $\prod_{CSP}^{S}(DC)$ means the simulated image of CSP, and $\prod_{CSP}^{S}(DC) = \{(ans1^s, ans2^s), (dis^s)\}$ where all the elements are randomly generated from Z_N. Since paillier cryptosystem is semantically secure, $(ans1, ans2)$ are computationally indistinguishable from $(ans1^s, ans2^s)$. Meanwhile, dis^s is randomly chosen from Z_N, dis is computationally indistinguishable from dis^s. Based on the above, we can draw a conclusion that $\prod_{CSP}(DC)$ is computationally indistinguishable from $\prod_{CSP}^{S}(DC)$.

Similarly, we can prove $\prod_{S}(DC)$ is computationally indistinguishable from $\prod_{S}^{S}(DC)$. Thus, combining the above analysis, we can confirm that DC protocol is sure under the semi-honest model.

Theorem 2. The *Travel Angle Calculation* (TAC) protocol described in Algorithm 4 is secure under the semi-honest model.

Proof. Here, let the execution image of S be denoted by $\prod_{S}(TAC) = \{e, f\}$. We assume $\prod_{S}^{S}(TAC)$ means the simulated image of S, and $\prod_{S}^{S}(TAC) = \{e^s, f^s\}$ where r_4, r_5 are randomly generated from Z_N. Since paillier cryptosystem is semantically secure, (e, f) are computationally indistinguishable from (e^s, f^s). Based on the above, we can draw a conclusion that $\prod_{S}(TAC)$ is computationally indistinguishable from $\prod_{S}^{S}(TAC)$.

Follow that familiar way, the execution image of CSP in TAC protocol is $\prod_{CSP}(TAC) = \{tan_{w_i,t_i}, signx_{w_i,t_i}, signy_{w_i,t_i}\}$. Where $signx_{w_i,t_i}, signy_{w_i,t_i}$ can regard as a random number in $\{1, -1\}$. tan_{w_i,t_i} is plain text with r_5/r_4, r_4 and r_5 are randomly generated from Z_N. So we can draw a conclusion that $\prod_{CSP}(TAC)$ is computationally indistinguishable from $\prod_{CSP}^{S}(TAC)$. Thus, combining the above analysis, we can confirm that TAC protocol is sure under the semi-honest model.

Theorem 3. The location privacy and the data access patterns are not be disclosed to CSP and S in our scheme. That is, CSP and S cannot infer the real location of workers or tasks from the historical records.

Proof. For location privacy, CSP gets $\{E[x_{w_i}], E[y_{w_i}]\}_{w_i \in C_w}$, $\{E[x_{t_i}], E[y_{t_i}]\}_{t_i \in C_t}$, $\{Dis_{w_i,t_i}\}_{w_i \in List_w}$, $\{signx_{w_i,t_i}\}_{w_i \in List_w}$, and $\{signy_{w_i,t_i}\}_{w_i \in List_w}$ in the whole process. S gets $\{tan_{w_i,t_i}\}_{w_i \in List_w}$ in the entire process. If CSP accidentally knows the real location of a worker, CSP can only infer the task location to which the worker is assigned. And can only guess that workers in the same candidate list is closer to the worker. If S accidentally learns the real location of a worker, S could hardly guess anything. If CSP or S know the location of a task, the situation is similar. So our scheme does not disclose location privacy to CSP and S.

From the perspective of S, S is semi-honest, and he calculates the received data according to the rules. All the data S obtains are added with random number $\{r_i\}_{1\leq i\leq 5}$ as a mask. S has no information to speculate on random numbers, so he cannot know any location information $\{(x_{w_i}, y_{w_i})\}_{w_i \in C_w}$ and $\{(x_{t_i}, y_{t_i})\}_{t_i \in C_t}$.

From the perspective of CSP, each task can get a list of candidate workers. Each worker will get $tan_{w_i,t_i}, sign_{w_i,t_i}$ of the corresponding task. According to the analysis of Theorem 1 & 2, CSP can only speculate that these candidate workers $\langle w_1, w_2, \cdots, w_k\rangle$ are located closer. But it is not possible to know the real range of its location $\{x_{w_i}, y_{w_i}\}_{1\leq i\leq k}$ and $\{x_{t_i}, y_{t_i}\}_{1\leq i\leq k}$. Therefore, the data access mode is not exposed to S and CSP.

5.2 Performance Analysis

In this section, we evaluate the proposed scheme and compare it with existing schemes.

5.2.1 Experimental Setup

In practice, take mobile taxi service as an example, the requester is a user who needs a taxi, and the worker corresponds to a taxi driver, each task usually needs only one worker. The characteristic of this application is that a worker needs to respond in time after each task is released, so it adopts the server distribution model.

Dataset: We conduct experiments on real datasets from the New York taxi website[1]. The data set is from New York taxis, and each month contains about one million pieces of information with real geographic location (using the 2015 data given on the website, the data after 2015 is not marked with latitude and longitude). First, remove duplicated coordinates, then randomly divide the data set into two parts, namely the worker and task locations. Each task requires the worker (representing the taxi driver) to arrive at the location of the user (passenger) to pick up the passenger and deliver it to the destination.

Baseline Approaches: We compare with 3 baseline approaches: (1) the method of our scheme without encryption protection. (2) PriRadar [21] designs a location protection method that maps the locations of workers and tasks to the grid. (3) our scheme.

Evaluation Metrics: We evaluate the effect of our scheme on running time, calculation error, and travel cost. The error is measured by the difference between the task coordinates restored by the worker and the real coordinates. We measure travel cost based on the distance between the last worker in the candidate worker list and the task location (represents the maximum travel cost). We should note in advance that each experimental result is an average result repeated at least 20 times.

[1] https://www1.nyc.gov/site/tlc/about/tlc-trip-record-data.page.

Setting: All the algorithms were implemented in python 3.4, including the implementation of Paillier cryptosystem[2]. We evaluated all the experiments on window 10, with Intel Core i7 at 2.8 GHz and 16 GB RAM.

5.2.2 Theoretical Analysis

As seen from Table 2, different features realized by different schemes, only our solution can satisfy all features.

Table 2. Notations

Features	DPSC [15]	DPGSC [14]	EDSC [7]	HEEDP [4]	ours
Location privacy-preserving	✓	✓	✓	✓	✓
Proctect workers and requesters	✗	✗	✓	N/A	✓
Server is untrusted	✓	✓	✗	N/A	✓
Don't need a trusted third party	✗	✗	✓	✓	✓
Using cryptographic approach	✗	✗	✓	✓	✓
Don't fake worker locations	✗	✗	✓	✓	✓

5.2.3 Experimental Results

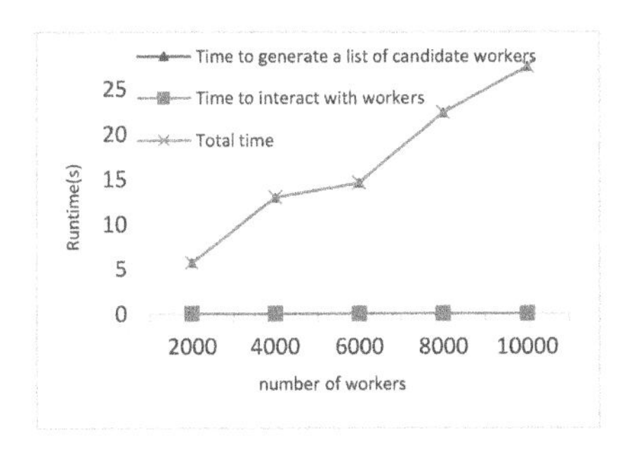

Fig. 2. Runtime

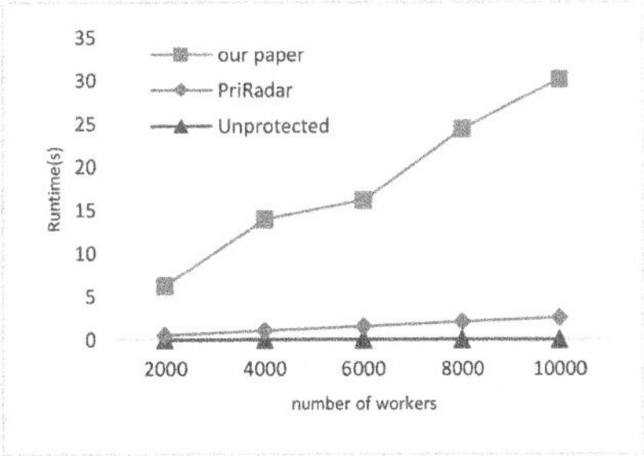

Fig. 3. Time to generate candidate workers under different tasks

Runtime. We evaluate the runtime under different numbers of workers and different schemes. Figure 2 indicates the time spent on the program under different numbers of workers. We can see that the time to generate the list of candidate workers increases continuously with the increase in the number of workers. Because each worker needs to calculate the distance from tasks' locations, and it is based on encrypted data. When interacting with workers, the time for workers to calculate the task location is not affected by the number of workers. On the one hand, because the calculations required by workers locally are very easy.

[2] https://python-paillier.readthedocs.io/en/develop/index.html.

On the other hand, because the candidate worker list sets a maximum number of workers to improve efficiency, So even if the amount of data is large, the calculation time is short. It can be found that we leave all or most of the calculations to the cloud platform, and local workers only need to perform simple calculations to get the real location of the task. Figure 3 shows the time for the two schemes to generate a list of candidate workers under different numbers of workers. We can see that the time for generating the candidate worker list in this paper is much higher. PriRadar didn't describe how to project the location on the grid in detail, so we can not know the runtime. But in the previous article, we theoretically analyzed the computational overhead. Our scheme requires fewer exponentiation operations than PriRadar [21] and has more advantages. Compared with the scheme without protection, it can be seen that the inevitable disadvantage of encryption technology is that it consumes a lot of running time. However, when the number of workers is 2,000, the average time is 6 s. When the number of workers is 10,000, the average time is almost half a minute, acceptable for a certain area. It is more reasonable to have hundreds of taxi drivers at the same time, but 10,000 drivers to be in a small area at the same time are not realistic. In reality, 10,000 drivers can be divided into small areas according to their locations. Then assign it again. It is very easy to complete.

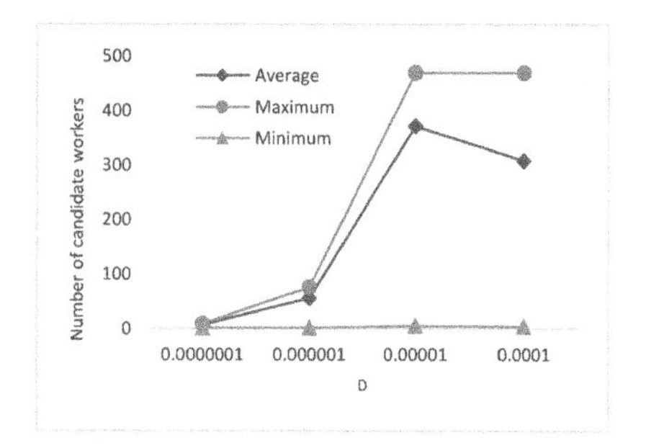

Fig. 4. Number of candidate workers under different D

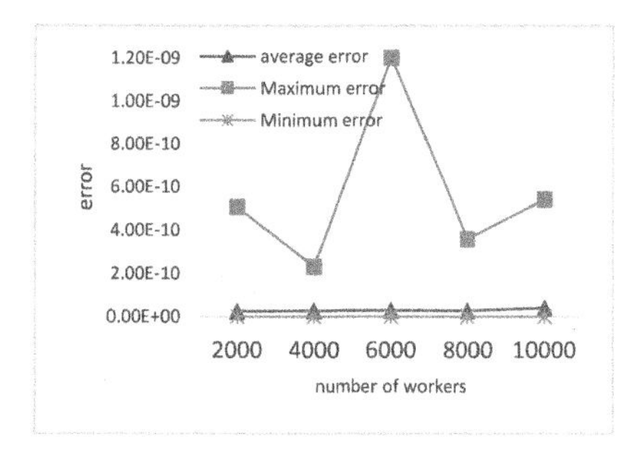

Fig. 5. Longitude error

Number of Candidate Workers. Figure 4 shows the number of candidate workers generated under different maximum acceptable distance D when the maximum number of candidate workers is not set. We can find that as D increases, the number of candidate workers also gradually increases. When $D = 0.00001$, the average number of workers is close to 400. When $D = 0.0001$, the average number of workers is higher than 300. We can see that for a task when D gradually increases, the number of qualified workers also increases. But only one worker is enough to complete a task, too many candidate workers may increase the running time. So set the maximum number of candidate workers is necessary.

Error Calculation. Figure 5 and Fig. 6 show the error value between the task location and the real task location recovered by the worker according to the information sent by CSP under different numbers of workers, Fig. 5 is longitude error

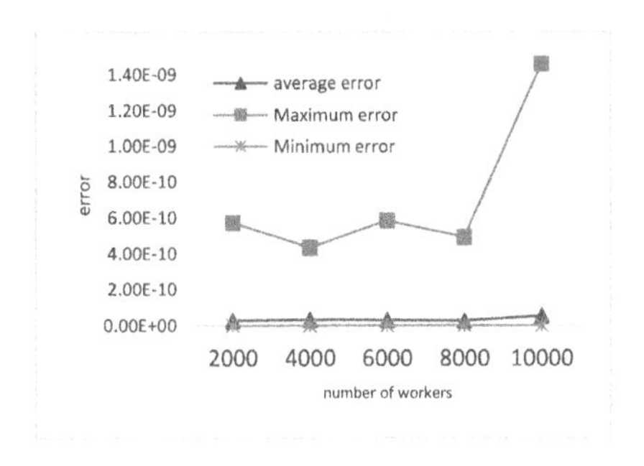

Fig. 6. Latitude error

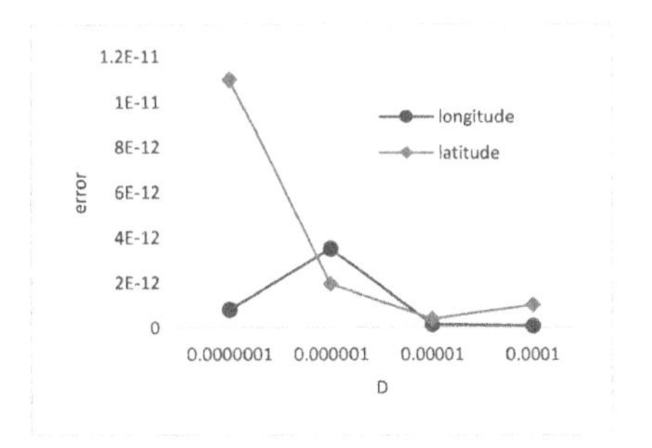

Fig. 7. Error under different D

and Fig. 6 is latitude error. It is found that the error is not significantly related to the number of workers, the minimum error is very close to 0. The average error is basically in the 10^{-10}, and the maximum error reaches the 10^{-9}. The amount of error is already tiny. Fig. 7 indicates the latitude and longitude error under different maximum acceptable distances D. We can find that there is no particularly obvious rule. The value is basically in the order of 10^{-11} to 10^{-12}. We can see that the change of D does not cause a large error change. The error is tiny. The calculations in the workers' local are easy, fast, and highly accurate.

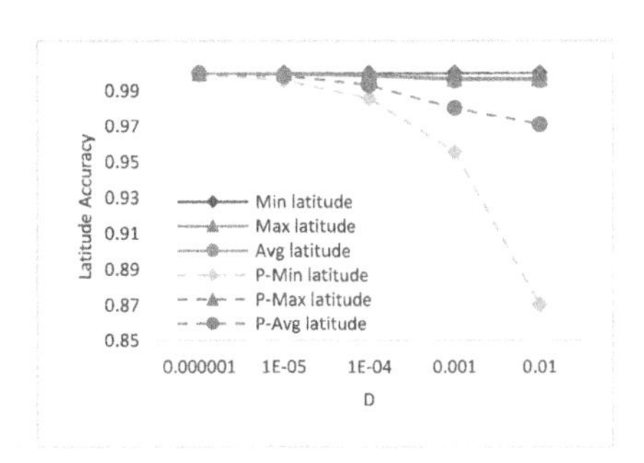

Fig. 8. Latitude travel cost of task allocation under different D

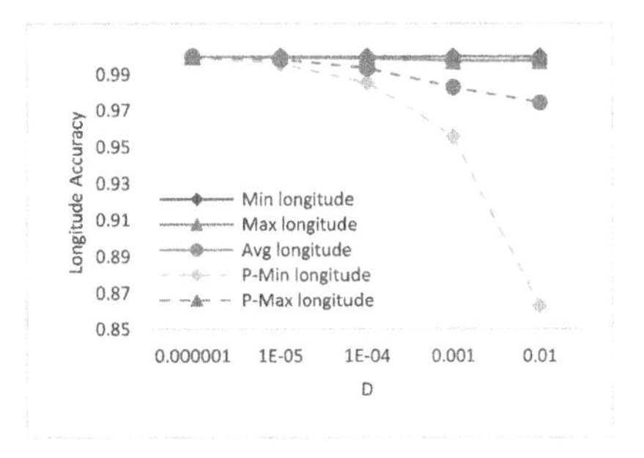

Fig. 9. Longitude travel cost of task allocation under different D

Travel Cost. Figure 8 and Fig. 9 indicate the latitude and longitude travel cost of the two schemes at different maximum acceptable distances D, calculated by the last worker in the list of candidate workers. It can be seen that with the continuous increase of the maximum acceptable distance D, the travel cost is increased, especially the PriRadar. However, our scheme has no particularly obvious impact, and the travel cost is lower. Figure 10 and Fig. 11 indicate the latitude and longitude travel distances under different workers. It can be seen that our travel distance is much smaller than in PriRadar [21]. The distance between the worker and the task is shorter, then when the worker receives the task, he is more willing to receive the task, and can quickly respond and complete the task. It also indicates that the travel cost of the task allocation of our scheme is less.

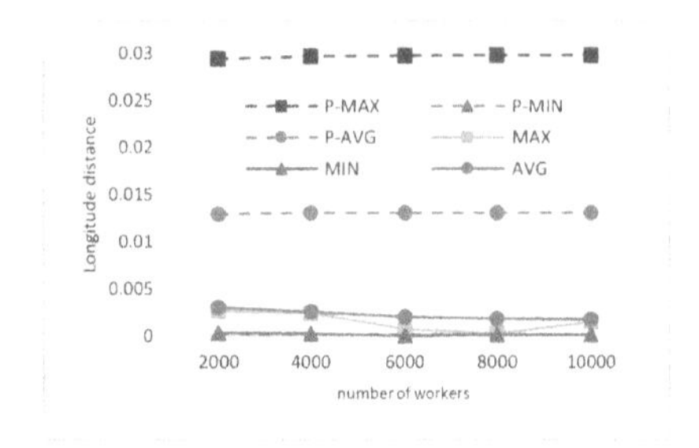

Fig. 10. Latitude travel distance under different tasks

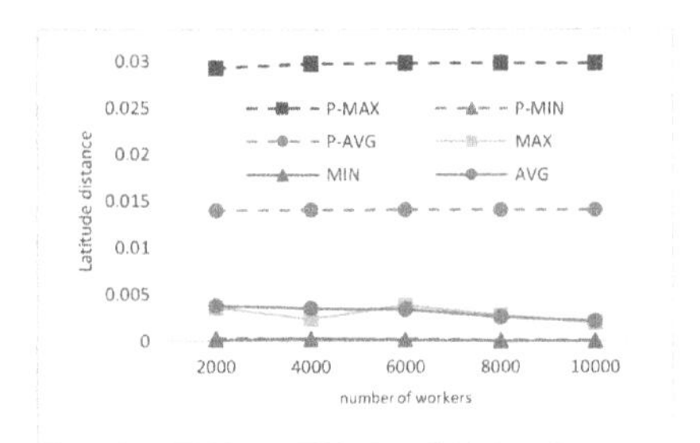

Fig. 11. Longitude travel distance under different tasks

In summary, our overall running time is acceptable and requires fewer exponentiation operations than PriRadar [21]. The maximum number of candidate workers set has well suppressed the increase in the amount of calculation. The calculation required locally by the worker is easy and the calculation error is negligible. The travel cost of our scheme is significantly less than [21], which indicates the superiority of our scheme.

6 Related Work

The privacy-preserving task allocation in spatial crowdsourcing has been an active area of research in recent years. Early schemes usually need a trusted third party [15], or a trusted data processing agency [6], but often cannot be realized in reality. At the same time, it may cause unnecessary delays in task allocation. When the staff wants to update their location, online TTP needs to publish new statistical location data, which may result in higher communication overhead [19]. And they usually only focus on protecting the location privacy of workers [12,15], and assume that the task location is public. However, the task location should also be protected, because workers who receive the task are often near the task location, and the requester often releases the task near its location [16].

The more mainstream solutions also have their advantages and disadvantages. In [10,20,22], k-anonymity and dummy users are leveraged to hide the location in a cloak region or a group of location data to achieve location privacy. But spatial anonymity cannot resist background knowledge attacks. Differential privacy protection and encryption technology can resist background knowledge attacks and have a top-level of privacy protection. [23] proposed a novel spatial crowdsourcing framework without using a trusted third party but providing a DP guarantee. Workers no longer submit their locations but choose their acceptable task set from public task location space, realizing the protection of workers' locations. But workers need to set up an acceptable location in advance, which requires a relatively sizeable storage space. The amount of noise is difficult to control, which can easily lead to reduced data availability. For example, [13] proposed an efficient protocol to securely compute the worker travel cost and select minimum cost worker in the encrypted domain, which reveals nothing about location privacy. But encryption technology usually has a relatively large operating overhead. [21] devised a grid-based location protection method, which can protect the locations of workers and tasks while keeping the distance-aware information on the protected locations. And they leveraged both attribute-based encryption and symmetric-key encryption to establish secure channels through servers, which ensures that the task is delivered securely and accurately by any untrusted server. But the delivery process involves numerous encryption and decryption operations, which was not efficient enough.

7 Conclusion

Aiming at the privacy protection of task allocation in spatial crowdsourcing, we proposed a task allocation scheme based on encryption protection. We use a two-server model and adopting paillier homomorphic cryptosystem to protect the location privacy of workers and tasks. And our scheme doesn't need any online TTP to involve. We especially realize the delivery of tasks' locations, which allows workers to achieve effectively privacy protection with a tiny computational overhead locally. At the same time, its advantages are proved by experiments on real data. Our scheme achieves practical performance in terms of computational overhead and travel cost and the running time is acceptable. For further work, we are going to focus on how to reduce the computing time of encryption protection technology, and how to further combine different kinds of protecting the task to improve efficiency.

Acknowledgment. This work is supported by the National Key Research and Development Program of China (No. 2018YFB0204301), National Nature Science Foundation of China (No. 62072466, No. U1811462), and the NUDT Grants (No. ZK19-38).

References

1. Bugiel, S., Nurnberger, S., Sadeghi, A.R., Schneider, T.: Twin clouds: an architecture for secure cloud computing (2011)
2. Goldreich, O.: Foundations of cryptography. II: basic applications 2 (2004). https://doi.org/10.1017/CBO9780511721656
3. Goldwasser, S.: The knowledge complexity of interactive proof system. SIAM J. Comput. **18**(1), 186–208 (1989)
4. Haiping, H., Tianhe, G., Ping, C., Reza, M., Tao, C.: Secure two-party distance computation protocol based on privacy homomorphism and scalar product in wireless sensor networks. Tsinghua Sci. Technol. (2016)
5. Howe, J.: The rise of crowdsourcing. Wired **14**(6), 176–183 (2006)
6. Kazemi, L., Shahabi, C.: A privacy-aware framework for participatory sensing. ACM SIGKDD Explor. Newsl. **13**(1), 43 (2011)
7. Liu, B., Chen, L., Zhu, X., Zhang, Y., Zhang, C., Qiu, W.: Protecting location privacy in spatial crowdsourcing using encrypted data. In: EDBT, pp. 478–481. OpenProceedings.org (2017)

8. Liu, L., Chen, R., Liu, X., Su, J., Qiao, L.: Towards practical privacy-preserving decision tree training and evaluation in the cloud. IEEE Trans. Inf. Forensics Secur. **15**, 2914–2929 (2020)
9. Miao, C., Jiang, W., Su, L., Li, Y., Guo, S.: Cloud-enabled privacy-preserving truth discovery in crowd sensing systems. In: ACM Conference on Embedded Networked Sensor Systems (2015)
10. Niu, B., Li, Q., Zhu, X., Cao, G., Hui, L.: Achieving K-anonymity in privacy-aware location-based services. In: IEEE INFOCOM 2014 - IEEE Conference on Computer Communications (2014)
11. Paillier, P.: Public-key cryptosystems based on composite degree residuosity classes. In: International Conference on Advances in Cryptology-Eurocrypt (1999)
12. Pournajaf, L., Li, X., Sunderam, V., Goryczka, S.: Spatial task assignment for crowd sensing with cloaked locations. In: IEEE International Conference on Mobile Data Management (2014)
13. Shen, Y., Huang, L., Li, L., Lu, X.: Towards preserving worker location privacy in spatial crowdsourcing. In: IEEE Global Communications Conference (2015)
14. To, H., Ghinita, G., Fan, L., Shahabi, C.: Differentially private location protection for worker datasets in spatial crowdsourcing. IEEE Trans. Mob. Comput. **16**(4), 934–949 (2017)
15. To, H., Ghinita, G., Shahabi, C.: A framework for protecting worker location privacy in spatial crowdsourcing. Proc. VLDB Endow. **7**(10), 919–930 (2014)
16. To, H., Shahabi, C.: Location privacy in spatial crowdsourcing. In: Gkoulalas-Divanis, A., Bettini, C. (eds.) Handbook of Mobile Data Privacy, pp. 167–194. Springer, Cham (2018). https://doi.org/10.1007/978-3-319-98161-1_7
17. To, H., Shahabi, C., Kazemi, L.: A server-assigned spatial crowdsourcing framework. ACM Trans. Spat. Algorithms Syst. **1**(1), 1–28 (2015)
18. Wang, Z., Hu, J., Jing, Z., Yang, D., Chen, H., Qian, W.: Pay on-demand: dynamic incentive and task selection for location-dependent mobile crowdsensing systems. In: IEEE International Conference on Distributed Computing Systems (2018)
19. Xiao, Y., Xiong, L.: Protecting locations with differential privacy under temporal correlations, pp. 1298–1309 (2015). https://doi.org/10.1145/2810103.2813640
20. Yang, D., Xi, F., Xue, G.: Truthful incentive mechanisms for K-anonymity location privacy. In: Infocom. IEEE (2013)
21. Yuan, D., Li, Q., Lia, G., Wang, Q., Ren, K.: PriRadar: a privacy-preserving framework for spatial crowdsourcing. IEEE Trans. Inf. Forensics Secur., 1 (2019)
22. Zhai, D., et al.: Towards secure and truthful task assignment in spatial crowdsourcing. World Wide Web **22**(5), 2017–2040 (2018). https://doi.org/10.1007/s11280-018-0638-2
23. Zhang, L., Xiong, P., Ren, W., Zhu, T.: A differentially private method for crowdsourcing data submission. Concurrency Comput. Pract. Exp. 31, **e5100** (2018)

Efficient RLWE-Based Multi-key Fully Homomorphic Encryption Without Key-Switching

Xiaoliang Che[1,2], Yao Liu[2], Shangwen Zheng[2], Longfei Liu[1,2], Tanping Zhou[1](✉), Xiaoyuan Yang[1,2], and Xu An Wang[1]

[1] College of Cryptographic Engineering, Engineering University of PAP, Xi'an 710086, Shaanxi, China

[2] Key Laboratory of Network and Information Security of the PAP, Xi'an 710086, Shaanxi, China

Abstract. The previous leveled BGV-type MKFHE schemes (e.g. CZW17, LZY$^+$19) based on the standard RLWE assumption are implemented by using key-switching and modulus-switching techniques. However, the frequent usage of key-switching causes the low efficiency of homomorphic multiplication operation. The CDKS19 scheme proposed two new simpler and faster relinearization algorithms, which supported the homomorphic computation with certain circuit depth. However, the construction that satisfies the fully homomorphic computation was not designed, and its relinearization performance can be further optimized.

In this paper, a more efficient leveled BGV-type MKFHE scheme without key-switching is constructed. Firstly, the generation method of evaluation key is improved, and two optimized generation algorithms of relinearization key are proposed. Secondly, following the relinearization algorithm framework of CDKS19, two efficient relinearization algorithms are proposed. The new algorithms are much faster to re-linearize the product of ciphertexts. Finally, using the optimized relinearization algorithms to replace the key-switching technology logically, and combining the modulus-switching technology, an efficient leveled MKFHE is constructed.

The results show that our MKFHE scheme is IND-CPA secure based on the standard RLWE assumption, and supports any parties dynamically join the homomorphic computation at any time. Moreover, the time complexity of relinearization and decryption is less than that of CDKS19. So it is a leveled BGV-type MKFHE scheme with more efficient homomorphic computation.

Keywords: Multi-key fully homomorphic encryption · Key-switching technique · Relinearization algorithm · The time complexity

1 Introduction

The typical FHE schemes can only support homomorphic computation of ciphertext for a single party, that is, all ciphertexts participating in computation correspond to the

J. Xiong et al. (Eds.): MobiMedia 2021, LNICST 394, pp. 323–342, 2021.
https://doi.org/10.1007/978-3-030-89814-4_24

one secret key. However, in many scenarios, it is usually necessary to calculate the data uploaded to the cloud by multi parties in the network. In order to meet this practical demand, multi-key fully homomorphic encryption (MKFHE) [1] has been proposed. It can realize the effective integration of multi-party data under the condition of protecting data security, which has important research meaning and high application value. One of the most appealing applications of MKFHE is to construct on-the-fly multiparty computation (MPC) protocols [2, 3].

1.1 Background

Multi-key fully homomorphic encryption can solve the problem of homomorphic operation between ciphertexts of different parties, and the results of the operation can be jointly decrypted by the secret keys. MKFHE can realize the multi-party data security calculation, which is mainly divided into four types: NTRU-type MKFHE, GSW-type MKFHE, TFHE-type MKFHE, and BGV-type MKFHE.

In 2012, López-Alt et al. first proposed the NTRU-type MKFHE based on the NTRU cryptosystem [4], which was optimized later in DHS16 [5]. In PKC2017, Chongchitmate et al. gave a basic framework for constructing MKFHE with circuit privacy characteristics and proposed an MKFHE scheme CO17 [6] that can protect the circuit privacy. However, the security of NTRU-type MKFHE is based on a new and somewhat non-standard assumption on RLWE. Based on the prime cyclotomic polynomial ring, Che et al. constructed an efficient NTRU-type MKFHE scheme CZL$^+$20 [7] by using the ciphertext dimension extension technology, which eliminates the key-switching process in the relinearization and effectively reduces the key size. But the ciphertext size of the scheme increases.

In CRYPTO2015, Clear and McGoldrick proposed the first GSW-type MKFHE scheme CM15 based on LWE problem [8], which proposes a transformation model from FHE to MKFHE. This transformation model is widely adopted by most MKFHE schemes based on LWE or RLWE problems. In EUROCRYPT 2016, Mukherjee and Wichs presented a construction of MKFHE scheme MW16 [9] based on LWE that simplifies the scheme of CM15 and admits a simple 1-round threshold decryption protocol. Based on this threshold MKFHE, they successfully constructed a 2-round MPC protocol upon it in the common random string (CRS) model. The schemes CM15 and MW16 need to determine all the involved parties before the homomorphic computation and do not allow any new party to join in, which is called single-hop MKFHE [10]. In TCC2016, Peikert and Shiehian proposed a notion of multi-hop MKFHE PS16 [10], in which the calculated ciphertexts can be used in further homomorphic computations involving additional parties. That is, any parties can dynamically join the homomorphic computation at any time. However, the disadvantage is that the number of parties is limited. In CRYPTO2016, A similar notion named fully dynamic MKFHE BP16 [11] was proposed by Brakerski and Perlman. A slight difference is that the bound of the number of parties does not need to be input during the setup procedure in fully dynamic MKFHE. The length of extended ciphertext only increases linearly with the number of parties. However, in the process of homomorphic computation, the scheme needs to use the parties' joint public key to run the bootstrap process, so the efficiency of ciphertext computation is low.

In ASIACRYPT2016, Chillotti et al. constructed the fully homomorphic scheme CGGI16 [12] based on a variant of GSW13 [13] on the T = (0,1] ring. In the scheme, the external product of TGSW ciphertext (matrix) and TLWE ciphertext (vector) is used to replace the product of TGSW ciphertext (matrix) and TGSW ciphertext (matrix). Therefore, the addition operation on polynomial exponent is more efficient, such that the time of the bootstrap process and the size of the bootstrap key are greatly reduced. In ASIACRYPT2017, Chilotti et al. optimized the accumulation process in the CGGI16 scheme and proposed CGGI17 [14], which reduced the bootstrap time to 13ms. In ASIACRYPT2019, Chen et al. designed an efficient ciphertext expansion algorithm based on CGGI17, realized the efficient expansion evaluation key, and proposed an MKFHE scheme CCS19 [15]. The ciphertext length of the scheme increases linearly with the number of parties. And also, they compiled an MKFHE software library MKTFHE, which has important guiding significance for the application of MKFHE schemes. However, these TFHE-type MKFHE schemes do not support the packaging technique, thereby resulting in a large expansion rate similar to TFHE.

In TCC2017, Chen et al. proposed the first BGV-type multi-hop MKFHE scheme CZW17 [16]. They used the GSW-type expansion algorithm to encrypt the secret key to generate the joint evaluation key of the party set. CZW17 supports the ciphertext packaging technology based on the Chinese Remainder Theory (CRT) and can be used to construct a 2-round MPC protocol. In 2019, Li et al. [17] put forward a nested ciphertext extension method, which reduces the size of the evaluation key and the ciphertext. In 2019, Chen et al. optimized the relinearization process and constructed an efficient MKFHE scheme [18], that is called the CDKS19 scheme. Because of its efficient homomorphic computation, it is applied to the neural network to perform the privacy computation. Our work is focused on the BGV-type MKFHE scheme.

1.2 Our Contributions

Fully homomorphic encryption schemes [e.g. 13, 19, 20] are the very attractive cryptography primitive, but they are limited to a single party. The multi-key FHE scheme can realize the homomorphic operation of multiple parties, but its efficiency of homomorphic operation is lower than that of single party FHE scheme. The RLWE-based FHE scheme has high security and good operation speed, so it has numerous theoretical and practical applications. For example, the CKKS17 [21] scheme has been widely applied for its efficient homomorphic operation. However, Li and Micciancio [22] found the security problems existing in CKKS17 and gave remedial measures. Recently, Cheon et al. published an announcement [23] and fixed the security loopholes of CKKS17. Another example is the CDKS19 [18] scheme, which is applied to the privacy-preserving operation of neural networks for its fast relinearization. Our work is to learn from the advantages of BGV-type FHE, and further improve them to design a more efficient leveled MKFHE scheme.

(1) Aiming at the CDKS19 scheme, we analyze the shortcomings of its specific relinearization algorithm. By reducing the public key and adding the auxiliary key, we improve the generation method of the evaluation key and propose two efficient generation algorithms of the relinearization key.

(2) Using the optimized generation algorithms of relinearization key, we modify the relinearization algorithm following the CDSK19 framework, so that the size of evaluation key and relinearization key is greatly reduced, and the efficiency of relinearization is higher.
(3) According to the optimized relinearization algorithms, we design a leveled MKFHE scheme combining with the modulus-switching technology. The scheme does not need to perform the key-switching process, so it is more efficient in the homomorphic computation process.

1.3 Overview of Our Construction

The relinearization idea of CDKS19 scheme is as follows. For the party $1 \leq i, j \leq k$, the method combines the i-th evaluation key $\mathbf{D}_i = [\mathbf{d}_{i,0}|\mathbf{d}_{i,1}|\mathbf{d}_{i,2}] \in R_q^{d\times 3}$ with the j-th public key $\mathbf{b}_j \approx -s_j \cdot \mathbf{a}(\text{mod}q) \in R_q^d$ to generate the relinearization key $\mathbf{K}_{i,j} \in R_q^{d\times 3}$ such that $\mathbf{K}_{i,j} \cdot (1, s_i, s_j) \approx s_i s_j \cdot \mathbf{g}(\text{mod}q)$. For the set with k parties, taking the homomorphic multiplication of parties i and j as an example, the product of their ciphertexts is $\overline{\mathbf{ct}} = (\mathbf{c}_i \otimes \mathbf{c}_j) = \{c_{i,j}\}_{0\leq i,j\leq k}$, and the corresponding joint secret key for decryption is $\mathbf{s} = (1, s_1, s_2, \cdots, s_k) \in \chi^{k+1}$. The purpose of relinearization is to make $\langle\overline{\mathbf{ct}}, \mathbf{s}\otimes\mathbf{s}\rangle = \langle\mathbf{c}_{\text{new}}, \mathbf{s}\rangle$ hold, where $\mathbf{c}_{\text{new}} \in R_q^{k+1}$ is a new ciphertext. Only two parties' keys are used in the relinearization, and it does not need to perform the RGSW ciphertext expansion algorithm in CZW17 [16] and LZY$^+$19 [17] schemes, so the computation efficiency is high. However, the size of evaluation key $\mathbf{D}_i \in R_q^{d\times 3}$ is still large, and the relinearization key $\mathbf{K}_{i,j}$ of evaluation key is also large.

In this paper, we choose the appropriate polynomial public parameter $a \leftarrow R_q$ to reduce the public key size, that is $b_j \approx -s_j \cdot a(\text{ mod } q) \in R_q$. By improving the generation method of evaluation key, we reduce the size of evaluation key $\mathbf{D}_i$ such that $\mathbf{K}_{i,j} \in R_q^3$, which greatly simplifies the relinearization key and improves the generation efficiency of evaluation key. The first method is to select another random vector $\mathbf{a} \leftarrow U(R_q^d)$, generate the party's auxiliary key $\mathbf{p}_i = s_i\mathbf{a} + \mathbf{e} \in R_q^d$, and then output the evaluation key $\mathbf{D}_i = [\mathbf{d}_0|d_1] \in R_q^{d+1}$. Combining with the j-th public key $b_j \approx -s_j \cdot a(\text{mod}q) \in R_q$, we calculate $[k_{i,j,0}|k_{i,j,1}] = \mathbf{g}^{-1}(b_j) \cdot [\mathbf{d}_0|\mathbf{a}]$ and $k_{i,j,2} = d_1$. Finally, we obtain the relinearization key $\mathbf{K}_{i,j} = [k_{i,j,0}|k_{i,j,1}|k_{i,j,2}] \in R_q^3$. The second method is to select a modulus P and the random vector $\mathbf{d} \leftarrow U(R_{P\cdot q}^d)$, generate the party's auxiliary key $\mathbf{p}_i = s_i\mathbf{d} + \mathbf{e} \in R_{P\cdot q}^d$. We calculate $\mathbf{d}_0 = -\mathbf{p}_i + \mathbf{e}_1 + r_i \cdot \mathbf{g}(\text{mod}P \cdot q)$ and $d_1 = r_i \cdot a + e_2 + P \cdot \mu(\text{mod}P \cdot q)$, and then obtain the evaluation key $\mathbf{D}_i = [\mathbf{d}_0|d_1] \in R_q^{d+1}$. By calculating $[k_{i,j,0}|k_{i,j,1}] = \mathbf{g}^{-1}(b_j) \cdot [\mathbf{d}_0|\mathbf{d}]$ and $k_{i,j,2} = d_1$, we finally get $\mathbf{K}_{i,j} = [k_{i,j,0}|k_{i,j,1}|k_{i,j,2}] \in R_{P\cdot q}^3$. Through the above two methods, by reducing the using times of function $\mathbf{g}$, we decrease the size of evaluation key and relinearization key, so that to improve the generation efficiency of evaluation key. Furthermore, in Sect. 3, two optimized relinearization algorithms are proposed. Combined with the modulus switching technology, an efficient leveled MKFHE scheme is designed in Sect. 4.

2 Preliminaries

2.1 Basic Notation

We denote vectors in bold, e.g. $\mathbf{a}$, and matrices in upper-case bold, e.g. $\mathbf{A}$. We denote by $\langle \mathbf{u}, \mathbf{v} \rangle$·the usual dot product of two vectors $\mathbf{u}, \mathbf{v}$. For a security parameter λ and a positive integer m, let $\Phi_m(X)$ denote the m-th cyclotomic polynomial with the degree $n = \phi(m)$, where $\phi(\cdot)$ is the Euler's function. We work over rings $R = \mathbb{Z}[X]/\Phi_m$ and $R_q = R/qR$ for a prime integer $q = q(\lambda)$. Addition and multiplication in these rings are done component-wise in their coefficients, and the coefficients in R_q are reduced in $[-q/2, q/2)$(except for $q = 2$). Let $\psi = \psi(\lambda)$ be a B-bound error distribution over R whose coefficients are in the range $[-B, B]$. For a probability distribution D, $x \leftarrow D$ denotes that x is sampled from D, and $x \leftarrow U(D)$ denotes that x is sampled uniformly from D. For $a \in R$, we use $\|a\|_\infty = \max_{0 \le i \le n-1} |a_i|$ to denote the standard l_∞-norm and use $\|a\|_1 = \sum_{i=0}^{n-1} |a_j|$ to denote the standard l_1-norm.

2.2 Leveled Multi-key FHE

We now introduce the cryptographic definition of a leveled multi-key FHE, which is defined in CZW17.

Definition 1 (Multi-key FHE) [16]. Let $\mathcal{C}$ be a class of circuits. A leveled multi-key FHE scheme $\mathcal{E} = (Setup, KeyGen, Enc, Eval, Dec)$ is described as follows:

- Setup$(1^\lambda, 1^K, 1^L)$: Given the security parameter λ, the circuit depth L, and the number of distinct users K that can be tolerated in an evaluation, outputs the public parameters pp.
- KeyGen(pp): Given the public parameters pp, derives and outputs a public key pk_i, a secret key sk_i, and the evaluation keys evk_i of party $i (i = 1, \ldots, K)$.
- Enc(pk_i, μ): Given a public key pk_i and message μ, outputs a ciphertext ct_i.
- Dec$((sk_{i_1}, sk_{i_2}, \ldots, sk_{i_k}), ct_S)$: Given a ciphertext ct_S corresponding to a set of users $S = \{i_1, i_2, \ldots, i_k\} \subseteq [K]$, and their secret keys $sk_S = \{sk_{i_1}, sk_{i_2}, \ldots, sk_{i_k}\}$, outputs the message μ.
- Eval$(\mathcal{C}, (ct_{S_1}, pk_{S_1}, evk_{S_1}), \ldots, (ct_{S_t}, pk_{S_t}, evk_{S_t}))$: On input a Boolean circuit $\mathcal{C}$ along with t tuples $(ct_{S_i}, pk_{S_i}, evk_{S_i})_{i=1,\ldots,t}$, each tuple comprises of a ciphertext ct_{S_i} corresponding to a user set S_i, a set of public keys $pk_{S_i} = \{pk_j, \forall j \in S_i\}$, and the evaluation keys evk_{S_i}, outputs a ciphertext ct_S corresponding to a set of secret keys indexed by $S = \cup_{i=1}^{t} S_i \subseteq [K]$.

2.3 RLWE Problem

Refers [24] and [25] for explaining the RLWE problem in more details. We use this discrete distribution as the RLWE error distribution. Here we define the RLWE distribution and decisional problem associated with it. Let $R^\vee$ be the dual fractional ideal of R and write $R_q = R^\vee/qR^\vee$. For a positive integer modulus $q \ge 2$, $s \in R_q^\vee$, and an error distribution ψ, we define $A_{N,q,\psi}(s)$ as the RLWE distribution obtained by sampling $a \leftarrow R_q$

uniformly at random, $e \leftarrow \psi$ and returning $(a, a \cdot s + e) \in R_q \times R_q^\vee$. The (decision) ring learning with errors, denoted by $\mathrm{RLWE}_{N,q,\chi}(D)$, is a problem to distinguish arbitrarily many independent samples chosen according to $A_{N,q,\chi}(s)$ for a random choice of s sampled from the distribution $s \leftarrow \chi$ over $R^\vee$ from the same number of uniformly random and independent samples from $R_q \times R_q^\vee$.

2.4 Two Techniques

Gadget Decomposition [18]. Let $\mathbf{g} = (\partial_i) \in Z^d$ be a gadget vector and q an integer. The gadget decomposition, denoted by $\mathbf{g}^{-1}$, is a function from R_q to R_2^d which transforms an element $a \in R_q$ into a vector $\mathbf{u} = (u_0, u_1, \cdots, u_{d-1}) \in R^d$ of small polynomials such that $a = \sum_{i=0}^{d-1} \partial_i \cdot u_i (\bmod q)$. The gadget decomposition technique is widely used in the construction of HE schemes, such as bit decomposition [13, 26], base decomposition [12, 14], and RNS-based decomposition [27]. Our implementation exploits the bit decomposition for efficiency.

Modulus-Switching [19, 28]. Since the error involved in the ciphertext grows with homomorphic operations, modulus switching which can change the inner modulus q_{l+1} of ciphertext $\mathbf{c}_1$ to a smaller number q_l is used to reduce the error term roughly by the ratio q_{l+1}/q_l, while preserving the correctness of decryption under the same secret key.

- ModulusSwitch$(\mathbf{c}_1, q_{l+1}, q_l)$: On input $\mathbf{c}_1 \in R_{q_{l+1}}^{n_1}$ and another smaller modulus q_l, output $\mathbf{c}_2 \in R_{q_l}^{n_1}$ which is the closest element to $(q_l/q_{l+1}) \cdot \mathbf{c}_1$.

2.5 Relinearization Algorithm in CDKS19

The CDKS19 scheme provides two special relinearization methods. They have similar structure patterns, so we only analyze the first method. The details are as follows. (See refer [18] for details).

1. **Parameter selection.** For a given security parameter λ, set the RLWE dimension be n, ciphertext modulus be q, polynomial distribution χ with small coefficients and error distribution ψ over R. Generate a random vector $\mathbf{a} \leftarrow U(R_q^d)$. For the party i, Sample the secret key $s_i \leftarrow \chi$. Sample an error vector $\mathbf{e} \leftarrow \psi^d$ and set the public key as $\mathbf{b}_i = -s_i \cdot \mathbf{a} + \mathbf{e}(\bmod q)$ in R_q^d.

2. **Evaluation key generation.**

- EvkGen(s_i): For the party i, input his secret key $s_i \leftarrow \chi$, generate the ciphertext $\mathbf{D}_i = [\mathbf{d}_{i,0}|\mathbf{d}_{i,1}|\mathbf{d}_{i,2}] \in R_q^{d \times 3}$ as the evaluation key.

(1) Sample $r_i \leftarrow \chi$;
(2) Sample $\mathbf{d}_{i,1} \leftarrow U(R_q^d)$ and $\mathbf{e}_1 \leftarrow \psi^d$, and set $\mathbf{d}_{i,0} = -s_i \cdot \mathbf{d}_{i,1} + \mathbf{e}_1 + r_i \cdot \mathbf{g}(\bmod q)$;
(3) Sample $\mathbf{e}_2 \leftarrow \psi^d$ and set $\mathbf{d}_{i,2} = r_i \cdot \mathbf{a} + \mathbf{e}_2 + s_i \cdot \mathbf{g}(\bmod q)$.

Note: the reason of introducing function $\mathbf{g}$ is to control the error growth, and the disadvantage is to increase the storage space.

(3) **Relinearization key generation.** For every party, they need to provide his public key and evaluation key to the cloud server. For the i-th party, he uses $\mathbf{b}_j$ of j–th party to generate his relinearization key as follows.

- Convert($\mathbf{D}_i, \mathbf{b}_j$): It takes as the input a pair of an uni-encryption $\mathbf{D}_i = [\mathbf{d}_{i,0}|\mathbf{d}_{i,1}|\mathbf{d}_{i,2}] \in R_q^{d\times 3}$ and a public key $\mathbf{b}_j \in R_q^d$ generated by (possibly different) parties i and j. Let $\mathbf{k}_{i,j,0}$ and $\mathbf{k}_{i,j,1}$ be the vectors in R_q^d such that $\mathbf{k}_{i,j,0}[\varsigma] = \langle \mathbf{g}^{-1}(\mathbf{b}_j[\varsigma]), \mathbf{d}_{i,0}\rangle$ and $\mathbf{k}_{i,j,1}[\varsigma] = \langle \mathbf{g}^{-1}(\mathbf{b}_j[\varsigma]), \mathbf{d}_{i,1}\rangle$ for $1 \leq \varsigma \leq d$, i.e., $[\mathbf{k}_{i,j,0}|\mathbf{k}_{i,j,1}] = \mathbf{M}_j \cdot [\mathbf{d}_{i,0}|\mathbf{d}_{i,1}]$ where $\mathbf{M}_j \in R_q^d$ is the matrix whose ς-th row is $\mathbf{g}^{-1}(\mathbf{b}_j[\varsigma]) \in R^d$. Let $\mathbf{k}_{i,j,2} = \mathbf{d}_{i,2}$ and return the relinearization key $\mathbf{K}_{i,j} = [\mathbf{k}_{i,j,0}|\mathbf{k}_{i,j,1}|\mathbf{k}_{i,j,2}] \in R_q^{d\times 3}$.

$$[\mathbf{k}_{i,j,0}|\mathbf{k}_{i,j,1}] = \begin{bmatrix} \mathbf{g}^{-1}(\mathbf{b}_j[1]) \\ \vdots \\ \mathbf{g}^{-1}(\mathbf{b}_j[d]) \end{bmatrix} [\mathbf{d}_{i,0}|\mathbf{d}_{i,1}] \in R_q^{d\times 2},\ [\mathbf{k}_{i,j,2}] = [\mathbf{d}_{i,2}] \in R_q^d$$

4. **Relinearization algorithm.** For the homomorphic multiplication of parties i and j, their ciphertexts follow the ciphertext expansion method in scheme LZY[+]19, e.g. $\mathbf{c}_i = (c_{i_0}|\, c_{i_1}| \cdots |c_{i_k}) \in R_{ql}^{k+1}$. Their joint secret keys are all $\mathbf{s} = (1, s_1, s_2, \cdots, s_k) \in R_q^{k+1}$. Decrypting their ciphertext product $\overline{\mathbf{ct}} = (\mathbf{c}_i \otimes \mathbf{c}_j) = \{c_{i,j}\}_{0\leq i,j\leq k}$, we can get

$$\langle \overline{\mathbf{ct}}, \mathbf{s}\otimes\mathbf{s}\rangle = c_{0,0} + \sum_{i=1}^{k}(c_{0,i} + c_{i,0})s_i$$

$$+ \sum_{i,j=1}^{k} \mathbf{g}^{-1}(c_{i,j}) \cdot \mathbf{K}_{i,j} \cdot (1, s_i, s_j) \in R_q^{k+1}$$

The function of the relinearization algorithm is to make $\langle \overline{\mathbf{ct}}, \mathbf{s}\otimes\mathbf{s}\rangle = \left\langle \overline{\mathbf{ct}'}, \mathbf{s}\right\rangle$ hold for the new ciphertext $\overline{\mathbf{ct}}' = (c_0', c_1', \cdots, c_k') \in R_q^{k+1}$ (corresponding to the joint secret key $\mathbf{s} = (1, s_1, s_2, \cdots, s_k) \in R_q^{k+1}$).

So, the relinearization algorithm is described as follows.

1. $c_0' \leftarrow c_{0,0}$;
2: for $1 \leq i \leq k$ do
   $c_i' \leftarrow c_{i,0} + c_{0,i} (\bmod q)$
3: for $1 \leq i, j \leq k$ do
   $(c_0', c_i', c_j') \leftarrow (c_0', c_i', c_j') + \mathbf{g}^{-1}(c_{i,j}) \cdot \mathbf{K}_{i,j} (\bmod q)$

Chen et al. [15] and Li et al. [17] designed multi-key variants of BGV [19] by generating a relinearization key based on the multi-key GSW scheme. However, it consists of $O(k^2)$ key-switching keys from $s_i \cdot s_j$ to the ordinary key each of which has $O(k)$ components. CDKS19 is an extension of these researches in the sense that our relinearization method and other optimization techniques can be applied to BGV as well. Compared with CCS19 [15] and LZY⁺19 [17], the memory (bit-size) complexity of evaluation key in CDKS19 reduces to $O(kn)$, and the computational costs of the homomorphic multiplication reduces to $O(k^2n)$.

3 Optimizations of the Relinearization

In the CDKS19 scheme, the main reason for increasing the dimension of ciphertext is the introduction of function **g**, which is used to reduce the error to avoid generating the wrong evaluation key. In this section, we optimize the relinearization algorithm of CDKS19 to improve the efficiency of homomorphic multiplication decryption.

3.1 Two Optimized Generation Algorithms of Evaluation Key

1. Method 1

(1) **Parameter reselection**. For a given security parameter λ, set the RLWE dimension be n and the modulus be q. The party set K contains k parties, for the party $i \in [K]$, uniformly sample the polynomial s_i at random with the coefficients $\{-1, 0, 1\}$ in R, generate his secret key $sk_i := (1, s_i) \in R_3^2$. Choose $\mathbf{a} \leftarrow U(R_q^d)$, $a \leftarrow U(R_q)$ and $e \leftarrow \psi$, $\mathbf{e} \leftarrow \psi^d$. Generate his public key $pk_i := (b_i, a) \in R_q^2$, where $b_i = -s_i \cdot a + e (\mathrm{mod}\, q)$, and his auxiliary key $\mathbf{p}_i := s_i\mathbf{a} + \mathbf{e} \in R_q^d$.

Output the $(sk_i, pk_i, \mathbf{p}_i)$ of party i. Therefore, the memory space of sk_i and pk_i is lower than that of the CDKS19 scheme.

(2) **Evaluation key generation**. The i-th party uses his auxiliary key $\mathbf{p}_i$ to generate the evaluation key.

- UniEnc$(\mu; \mathbf{p}_i) : \mathbf{D} = [\mathbf{d}_0 | d_1]$:
  1) Sample $r_i \leftarrow \chi$.
  2) Sample $\mathbf{e}_1 \leftarrow \psi^d$ and set $\mathbf{d}_0 = -\mathbf{p}_i + \mathbf{e}_1 + \mathbf{g} \cdot r_i (\mathrm{mod} q)$.
  3) Sample $e_2 \leftarrow \psi$ and set $d_1 = -r_i \cdot a + e_2 + \partial_\varsigma \mu (\bmod q)$, where $0 \leq \varsigma \leq d-1$.
- EvkGen(s): Given the secret s_i, return $\mathbf{D}_i \leftarrow$ UniEnc$(s_i; \mathbf{p}_i)$. We call $\mathbf{D}_i$ the evaluation key of party i, that is $ek_i := \mathbf{D}_i$.

Every parties provide their $(pk_i, \mathbf{p}_i, ek_i)$ to the cloud server for homomorphic computing.

Security. The evaluation key generation algorithm is IND-CPA secure under the RLWE assumption of parameter (n, q, χ, ψ). We prove it by showing that the distribution $\{a, (pk_i, \mathbf{p}_i), (\mathbf{d}_0, \mathbf{a}, d_1)\}$ is computationally indistinguishable from the uniform distribution over $R_q \times R_q^{d+2} \times R_q^{d \times 2+1}$ for an arbitrary $\mu \in R$.

Firstly, we change the public key $b_i \leftarrow U(R_q)$ and auxiliary key $\mathbf{p}_i \leftarrow U(R_q^d)$ according to the RLWE assumption.

Secondly, we change the partial evaluation key $\mathbf{d}_0 \leftarrow U(R_q^d)$.

Thirdly, since d_1 is RLWE assumption hard about $r_i \leftarrow \chi$, and is independent of the choice of s_i, we change $d_1 \leftarrow U(R_q)$. So, the new distribution $\{(pk_i, \mathbf{p}_i) \leftarrow U(R_q^{d+2})$,

$(\mathbf{d}_0, d_1) \leftarrow U(R_q^{d+1})\}$ is independent of the given plaintext μ, we conclude that the algorithm is IND-CPA secure.

(3) **Relinearization key generation**.

- Convert($\mathbf{D}_i, b_j$): Input an evaluation key $\mathbf{D}_i = [\mathbf{d}_{i,0}|d_{i,1}]$ and a public key $b_j \in R_q$ generated by (possibly different) parties i and j. Let $k_{i,j,0}$ and $k_{i,j,1}$ be the polynomials in R_q such that $k_{i,j,0} = \langle \mathbf{g}^{-1}(b_j), \mathbf{d}_{i,0} \rangle$ and $k_{i,j,1} = \langle \mathbf{g}^{-1}(b_j), \mathbf{a} \rangle$ for $0 \le \varsigma \le d-1$, i.e., $[k_{i,j,0}|k_{i,j,1}] = \mathbf{g}^{-1}(b_j) \cdot [\mathbf{d}_{i,0}|\mathbf{a}]$. Let $k_{i,j,2} = d_{i,1}$ and return the relinearization key $\mathbf{K}_{i,j} = [k_{i,j,0}|k_{i,j,1}|\, k_{i,j,2}] \in R_q^3$.

Correctness. A shared relinearization key consists of encryptions of s_i, s_j for all pairs $1 \le i, j \le k$. We first claim that, if $\mathbf{D}_i$ is an uni-encryption of $s_i \in R$ encrypted by the i-th party and b_j is the public key of the j-th party, then the output $\mathbf{K}_{i,j} \leftarrow$ Convert($\mathbf{D}_i, b_j$) of the conversion algorithm is an encryption of $s_i s_j \in R$ with respect to the secret $(1, s_i, s_j)$. It is derived from the following formulas:

$$
\begin{aligned}
\mathbf{K}_{i,j} \cdot (1, s_i, s_j) &= \mathbf{g}^{-1}(b_j)[\mathbf{d}_0|\mathbf{a}]\begin{bmatrix} 1 \\ s_i \end{bmatrix} + s_j k_{i,j,1} \\
&= \mathbf{g}^{-1}(b_j) \cdot \left[(\mathbf{g} \cdot r_i - s_i \cdot \mathbf{a} + \mathbf{e}_1) + s_i \cdot \mathbf{a}\right] - r_i b_j + s_j e_2 + \partial_\varsigma \cdot s_i s_j \\
&= \left[\mathbf{g}^{-1}(b_j)\right]_{1\times d} \cdot [\mathbf{e}_1]_{d\times 1} + b_j r_i + r_i e_j - r_i b_j + s_j e_2 + \partial_\varsigma \cdot s_i s_j \\
&= \mathbf{g}^{-1}(b_j) \cdot \mathbf{e}_i + s_j e_j + r_i e_j + \partial_\varsigma \cdot s_i s_j \\
&= \partial_\varsigma \cdot s_i s_j + e_{\text{small}}
\end{aligned}
$$

2. Method 2

This method is based on the encryption mode of CKKS17 scheme to improve the generation algorithm of evaluation key. In the first method, the traditional BGV-type encryption is used, so the form of decryption is $\langle \mathbf{c}, sk \rangle = m + te(\text{mod} q)$. The function $\mathbf{g}$ used in the relinearization algorithm is to reduce the error. However, if we get a decryption form like $\langle \mathbf{c}, sk \rangle = Pm + te(\text{mod} P \cdot q)$, we can make the error P times smaller than that of the original form so that the using times of function $\mathbf{g}$ can be decreased.

(1) **Parameters reselection**. For a given security parameter λ, set the RLWE dimension be n, the modulus be q, and an integer $P = P(\lambda, q)$. For the party $i \in [K]$, uniformly sample the polynomial s_i at random with the coefficients $\{-1, 0, 1\}$ in R, generate his secret key $sk_i := (1, s_i) \in R_3^2$. Choose $a \leftarrow U(R_q)$, $\mathbf{d} \leftarrow U(R_{P\cdot q}^d)$ and $e \leftarrow \psi$, $\mathbf{e} \leftarrow \psi^d$. Generate his public key $pk_i := (b_i, a) \in R_q^2$, where $b_i = -s_i \cdot a + e(\text{mod} q)$, and his auxiliary key $\mathbf{p}_i := s_i \mathbf{d} + \mathbf{e} \in R_{P\cdot q}^d$.

(2) **Evaluation key generation.** The i-th party uses his auxiliary key $\mathbf{p}_i$ to generate the evaluation key.

- UniEnc($\mu; \mathbf{p}_i$) : $\mathbf{D} = [\mathbf{d}_0|d_1]$:
  1) Sample $r_i \leftarrow \chi$.
  2) Sample $\mathbf{e}_1 \leftarrow \psi^d$, and set $\mathbf{d}_0 = -\mathbf{p}_i + \mathbf{e}_1 + \mathbf{g} \cdot r_i(\text{mod} P \cdot q)$.
  3) Sample $e_2 \leftarrow \psi$ and set $d_1 = -r_i \cdot a + e_2 + P \cdot \mu(\text{mod} P \cdot q)$.

- EvkGen(s): Given the secret s_i, return $ek_i := \mathbf{D}_i \leftarrow \text{UniEnc}(s_i; \mathbf{p}_i)$. Every parties provide their $(pk_i, \mathbf{p}_i, ek_i)$ to the cloud server for homomorphic computing.

Security. As the proof of method 1, the evaluation key generation algorithm in method 2 is also IND-CPA secure under the RLWE assumption of parameter (n, q, χ, P, ψ).

(3) **Relinearization key generation.**

- Convert($\mathbf{D}_i, b_j$): Input the evaluation key $\mathbf{D}_i = [\mathbf{d}_{i,0}|d_{i,1}]$ and a public key $b_j \in R_q$ generated by (possibly different) parties i and j. Let $k_{i,j,0}$ and $k_{i,j,1}$ be the polynomials in R_q such that $k_{i,j,0} = \langle \mathbf{g}^{-1}(b_j), \mathbf{d}_{i,0} \rangle$ and $k_{i,j,1} = \langle \mathbf{g}^{-1}(b_j), \mathbf{d} \rangle$ for $1 \leq \varsigma \leq d$, i.e., $[k_{i,j,0}|k_{i,j,1}] = \mathbf{g}^{-1}(b_j) \cdot [\mathbf{d}_{i,0}|\mathbf{d}]$. Let $k_{i,j,2} = d_{i,1}$ and return the relinearization key $\mathbf{K}_{i,j} = [k_{i,j,0}|k_{i,j,1}|\, k_{i,j,2}] \in R^3_{P \cdot q}$.

Correctness. A shared relinearization key consists of encryptions of s_i, s_j for all pairs $1 \leq i, j \leq k$. We first claim that, if $\mathbf{D}_i$ is an uni-encryption of $s_i \in R$ encrypted by the i-th party and b_j is the public key of the j-th party, then the output $\mathbf{K}_{i,j} \leftarrow \text{Convert}(\mathbf{D}_i, b_j)$ of the conversion algorithm is an encryption of $s_i s_j \in R$ with respect to the secret $(1, s_i, s_j)$. It is derived from the following formulas:

$$\begin{aligned}
\mathbf{K}_{i,j} \cdot \left(1, s_i, s_j\right) &= \mathbf{g}^{-1}(b_j)\left[\mathbf{d}_{i,0}|\mathbf{d}\right]\begin{bmatrix} 1 \\ s_i \end{bmatrix} + s_j d_{i,1} \\
&= \mathbf{g}^{-1}(b_j) \cdot (\mathbf{e}_1 + r_i \cdot \mathbf{g}) - r_i b_j + s_j e_2 + P \cdot s_i \cdot s_j \\
&= \mathbf{g}^{-1}(b_j) \cdot \mathbf{e}_1 + r_i e_j + s_j e_2 + P \cdot s_i \cdot s_j \\
&= P \cdot s_i \cdot s_j + e'_{\text{small}}
\end{aligned}$$

3.2 The Optimized Relinearization

The goal of relinearization is to re-linearize the tensor product of ciphertexts. So in this section, we design the optimized relinearization algorithm following the algorithm framework of CDKS19.

Let K be an ordered set containing all indexes of users that the ciphertext corresponding to, and we assume that the indexes are arranged from small to large and K has no duplicate elements, thus we can describe a ciphertext as a tuple $ct = \{\mathbf{c}, K, l\}$. For the ciphertext product $\overline{\mathbf{ct}} = (\mathbf{c}_i \otimes \mathbf{c}_j) = \{c_{i,j}\}_{0 \leq i,j \leq k} \in R_q^{(k+1)\times(k+1)}$, we would re-linearize it to $\overline{\mathbf{ct}}' = (c'_i)_{0 \leq i \leq k} \in R_q^{k+1}$.

- Relin($\overline{\mathbf{ct}}$; $\{(\mathbf{D}_i, b_j)\}_{1 \leq i \leq k}$): Given an extended ciphertext $\overline{\mathbf{ct}} = (c_{i,j})_{0 \leq i,j \leq k}$ and k pairs of evaluation/public keys $\{(\mathbf{D}_i, b_j)\}_{1 \leq i \leq k}$, generate a ciphertext $\overline{\mathbf{ct}}' \in R_q^{k+1}$ as follows:

1. Compute $\mathbf{K}_{i,j} \leftarrow \text{Convert}(\mathbf{D}_i, b_j)$ for all $0 \leq i, j \leq k$ and set the relinearization key as $\overline{rlk} = \{\mathbf{K}_{i,j}\}_{0 \leq i,j \leq k}$.
2. Run the following Algorithm to relinearize $\overline{\mathbf{ct}}$.

Input: $\overline{\mathbf{ct}} = (c_{i,j})_{0 \le i,j \le k}$, $\overline{rlk} = \{\mathbf{K}_{i,j}\}_{0 \le i,j \le k}$.
Output: $\overline{\mathbf{ct}'} = (c'_i)_{0 \le i \le k} \in R_q^{k+1}$.
1: $c'_0 \leftarrow c_{0,0}$
2: for $1 \le i \le k$ do
3: $c'_i \leftarrow c_{i,0} + c_{0,i} \pmod q$
4: end for
5: for $1 \le i, j \le k$ do
6: Method 1:
$(c'_0, c'_i, c'_j) \leftarrow (c'_0, c'_i, c'_j) + \sum_{\varsigma=0}^{d-1} (c_{i,j})_\varsigma \cdot \mathbf{K}_{i,j} \pmod q$,where $(c_{i,j})_\varsigma$ is a polynomial with the coefficients of 0 or 1.
Method 2:
$(c'_0, c'_i, c'_j) \leftarrow (c'_0, c'_i, c'_j) + P^{-1} \cdot c_{i,j} \cdot \mathbf{K}_{i,j} \pmod{P \cdot q}$.
7: end for

Correctness

For the method 1. Since the evaluation key $\mathbf{D}_i \leftarrow \text{UniEnc}(s_i; s_i)$ of the i-th party is an uni-encryption of $\mu_i = s_i$, we obtain that $\mathbf{K}_{i,j} \cdot (1, s_i, s_j) \approx \partial_\varsigma \cdot s_i s_j (\text{mod} q)$. From the definition of $\overline{\mathbf{ct}'} \in R_q^{k+1}$ and the joint secret key $\mathbf{s} = (1, s_1, s_2, \cdots, s_k) \in R_q^{k+1}$, we get

$$\begin{aligned} &< \overline{\mathbf{ct}'}, \mathbf{s} >= c'_0 + \textstyle\sum_{i=1}^k c'_i \cdot s_i \\ &= c_{0,0} + \textstyle\sum_{i=1}^k (c_{i,0} + c_{0,i})_i s_i \\ &+ \textstyle\sum_{i,j=1}^k \sum_{\varsigma=0}^{d-1} (c_{i,j})_\varsigma \cdot \mathbf{K}_{i,j} \cdot (1, s_i, s_j) \ (\text{mod} q) \\ &\approx c_{0,0} + \textstyle\sum_{i=1}^k (c_{i,0} + c_{0,i})_i s_i + \sum_{i,j=1}^k c_{i,j} \cdot s_i s_j \\ &= < \overline{\mathbf{ct}}, \mathbf{s} \otimes \mathbf{s} > (\text{mod} q) \end{aligned}$$

For the method 2. The evaluation key is $\mathbf{D}_i \leftarrow \text{UniEnc}(s_i; s_i)$, and $\mu_i = s_i$. We obtain that $\mathbf{K}_{i,j} \cdot (1, s_i, s_j) \approx P \cdot s_i s_j (\text{mod} q)$. So we get

$$\begin{aligned} &< \overline{\mathbf{ct}'}, \mathbf{s} > = c'_0 + \textstyle\sum_{i=1}^k c'_i \cdot s_i \\ &= c_{0,0} + \textstyle\sum_{i=1}^k (c_{i,0} + c_{0,i})_i s_i \\ &+ P^{-1} \textstyle\sum_{i,j=1}^k c_{i,j} \cdot \mathbf{K}_{i,j} \cdot (1, s_i, s_j) \,(\text{mod} P)\ (\text{mod} q) \\ &\approx c_{0,0} + \textstyle\sum_{i=1}^k (c_{i,0} + c_{0,i})_i s_i + \sum_{i,j=1}^k c_{i,j} \cdot s_i s_j \\ &= < \overline{\mathbf{ct}}, \mathbf{s} \otimes \mathbf{s} > (\text{mod } q) \end{aligned}$$

3.3 Analysis of the Relinearization

The previous analysis showed that the security of our optimized algorithm is the same as CDKS. Here we compare the memory space and time complexity between our improved relinearization algorithm and that of CDKS19 to analyze the advantages of our optimized algorithm. See Table 1 for details.

It can be seen from Table 1 that compared to the CDKS19 scheme, the memory space of the evaluation key and relinearization key is reduced about times. Because the public keys can be generated off-line, the slight changes of the public key in Table 1 have

Table 1. Memory space and time complexity of the relinearization algorithm

Relinearization types		Space			Time complexity
		Public key	Evaluation key	Relinearization key	
CDKS19		$2kdn\lceil\log q\rceil$	$k(d\times 3)n\lceil\log q\rceil$	$3dkn\lceil\log q\rceil$	$O(d^3n)$
Our method	Method 1	$(kd+2k)n\lceil\log q\rceil$	$k(d+1)n\lceil\log q\rceil$	$3kn\lceil\log q\rceil$	$O(d^2n)$
	Method 2	$kdn\lceil\log Pq\rceil+2kdn\lceil\log q\rceil$	$k(d+1)n\lceil\log Pq\rceil$	$3kn\lceil\log Pq\rceil$	$O(dn)$

neglected impact on the computation efficiency. Moreover, the time complexity cost in the relinearization is smaller than that in CDKS19. So our relinearization algorithm is more efficient than CDKS19.

4 New Construction of BGV-Type MKFHE Scheme

Our relinearization algorithm can be used as well as CDKS. It can also be designed as a leveled MKFHE scheme. Firstly, the ciphertext of different levels is expanded. Then the optimized relinearization algorithm is used to re-linearize the tensor product of ciphertexts. Finally the modulus-switching is carried out to convert the result ciphertext into the next level.

4.1 The Ciphertext Extension

In this subsection, we detail the ciphertext extension algorithm MKFHE.CTExt in Refs. [16, 17], which converts a BGV-type ciphertext to a larger dimensional ciphertext corresponding to a common larger dimensional joint secret key. In fact, the joint secret key is a concatenation of secret keys from a party set.

For the security parameter λ, given a bound K on the number of parties, a bound L on the circuit depth with L decreasing modulus $q_L \gg q_{L-1} \gg \cdots \gg q_0$ for each level and a small integer p coprime with all q_l (or an integer $P = P(\lambda, q)$). For the i-th party $(l = 0, \ldots, L)$, output his key pairs $(sk_i, pk_{l,i}, ek_{l,i})$ and the ciphertext $\mathbf{c}_{l,i} \in R_q^2$.

Let K be an ordered set containing all indexes of parties that the ciphertexts corresponding to, and we assume that the indexes are arranged from small to large and K has no duplicate elements, thus we can describe a ciphertext as a tuple $ct = \{\mathbf{c}_l, K, l\}$, where $\mathbf{c}_l = (c_{l,i_0} | c_{l,i_1} | \cdots | c_{l,i_k}) \in R_{q_l}^{k+1}$ corresponding joint secret keys $\mathbf{s}_K = (1, s_{i_1}, ..., s_{i_k}) \in R_3^{k+1}$. Note that $\mathbf{s}_K = (1, s_{i_1}, ..., s_{i_k}) \in R_3^{k+1}$ satisfies all levels of ciphertext decryption. That is $\mu \leftarrow <\mathbf{c}_l, \mathbf{s}_K> (\text{mod} q_l \bmod p)$. So when the party set updates, the ciphertext of party i also changes.

- MKFHE.CTExt($\mathbf{c}_l, K'$): On input a ciphertext tuple $ct = \{\mathbf{c}_l \in R_{q_l}^{k+1}, K = \{i_1, ..., i_k\}, l\}$ corresponding to k parties and another party set $K' = \{j_1, ..., j_{k'}\}$ for $K \in K'$, output an extended tuple $ct' = \{\overline{\mathbf{c}}_l \in R_{q_l}^{k'+1}, K' = \{i_1, ..., i_{k'}\}, l\}$. The extending algorithm is as follows:

1. Divide the ciphertext $\mathbf{c}_l$ into $k+1$ sequential sub-vectors indexed by $K = \{i_1, ..., i_k\}$ (except for the first sub-vector), i.e.,

$\mathbf{c}_l = (c_{l,i_0}|c_{l,i_1}|\cdots|c_{l,i_k}) \in R_{q_l}^{k+1}$, where the corresponding secret key is $\mathbf{s}_K = (1, s_{i_1}, ..., s_{i_k}) \in R_3^{k+1}$.

2. The extended ciphertext $\bar{\mathbf{c}}_l$ consists of $k'+1$ sequential sub-vectors, which can be indexed by $K' = \{j_1, ..., j_{k'}\}$, i.e., $\bar{\mathbf{c}} = (c'_{j_0}|c'_{j_1}|\cdots|c'_{j_{k'}}) \in R_{q_l}^{k'+1}$.

If index j in K' is also included in K, we set $c'_j = c_j$,otherwise we set $c'_j = 0$ (except for the first sub-vector). The corresponding secret key for decryption is $\mathbf{s}_{K'}=(1, s_{j_1}, ..., s_{j_{k'}}) \in R_3^{k'+1}$.

It's easy to verify that $\langle \mathbf{c}, \mathbf{s}_{K,l} \rangle = \langle \bar{\mathbf{c}}, \mathbf{s}_{K'} \rangle \bmod q_l$.

4.2 Detail Construction

1. **Structure of Our Scheme**

The specific structure of our leveled MKFHE scheme is shown in Fig. 1. The scheme architecture is mainly divided into two parts: key generation and homomorphic computation of ciphertexts. Take two parties as an example. For the key generation part: they select public parameters from the cloud server to generate their key pairs, and then upload them. The cloud server uses the parameters and the uploaded keys to generate the relinearization key. For the homomorphic computation part: all parties upload their ciphertexts to the cloud server. Firstly, the ciphertexts are expanded by the ciphertext extension algorithm. Then, homomorphic multiplication is performed on the ciphertexts, and use the relinearization key to re-linearize the tensor product of ciphertexts. Finally, output the computed ciphertext which can be decrypted as the resulting ciphertext or be switched as a new input ciphertext in the next level.

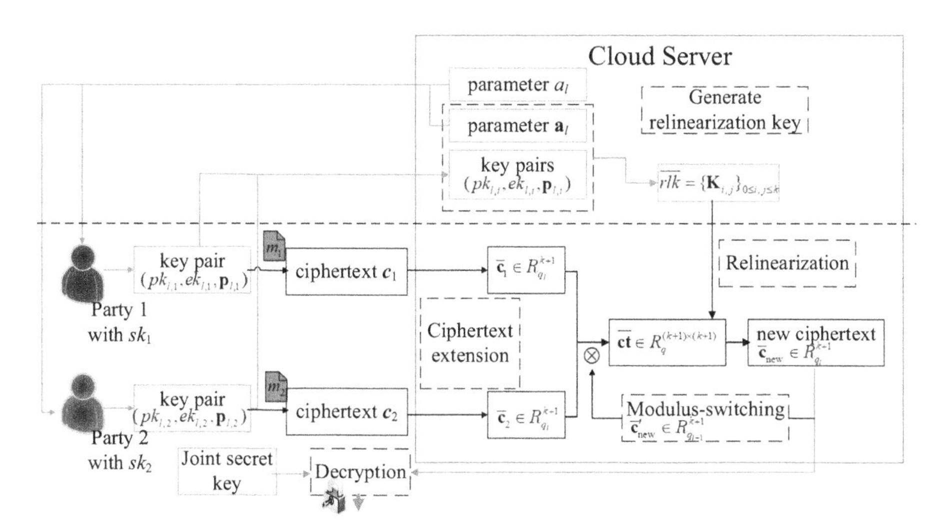

Fig. 1. The homomorphic multiplication structure of our leveled MKFHE

2. **Detail Construction**

According to the optimized relinearization, the general leveled BGV-type MKFHE is designed as follows.

• MKFHE.Setup($1^\lambda, 1^K, 1^L$): For the security parameter λ, given a bound K on the number of keys, a bound L on the circuit depth with decreasing modulus $q_L \gg q_{L-1} \gg \cdots \gg q_0$ (where $q_0 \gg 3$) for each level and a small integer p coprime with all q_l (or an integer $P = P(\lambda, q_l)$). We work over rings $R = \mathbb{Z}[X]/\Phi_m$ and $R_{q_l} = R/q_l R$ defined above. Let $\psi = \psi(\lambda)$ be a B-bound error distribution over R whose coefficients are in the range $[-B, B]$. Let $\chi = \chi(\lambda)$ be a distribution over R whose coefficients are in the range $(-B, B)$. Choose $L + 1$ random public vectors $\mathbf{a}_l \leftarrow U(R^d_{q_l})$ and polynomials $a_l \leftarrow U(R_{q_l})$ for $l \in \{0, \ldots, L\}$. All the following algorithms implicitly take the public parameter $pp = (R, B, \chi, \psi, \{q_l, \mathbf{a}_l, a_l\}_{l\in\{0,\ldots,L\}}, p, P)$ as input.

• MKFHE.KeyGen(pp): Input the public parameter pp, and generate the keys of circuit depth l for the i-th party.

1. Sample the polynomial s_i at random with the coefficients $\{-1, 0, 1\}$ in R, and set the secret key $sk_{l,i} := (1, s_i) \in R_3^2$.

2. Sample $e_{l,i} \leftarrow \psi$, and set the public key $pk_{l,i} := (b_{l,i}, a_l) \in R^2_{q_l}$, where $b_{l,i} = -a_l s_i + pe_{l,i} \bmod q_l$.

3. Sample $\mathbf{e}_{l,i} \leftarrow \psi^d$, and generate the auxiliary key $\mathbf{p}_{l,i} = s_i \mathbf{a}_l + \mathbf{e}_{l,i} \in R^d_{q_l}$.

4. Sample $r_{l,i} \leftarrow \chi$, generate the evaluation key $ek_{l,i} := \mathbf{D}_{l,i}$.

Output the key pairs of party i $(sk_i, pk_{l,i}, \mathbf{p}_{l,i})$.

• MKFHE.Enc($pk_{L,i}, \mu_i$): Input the plaintext $\mu_j \in R_p$ and the public key $pk_{L,i}$. Choose $r_i, e, e' \leftarrow \chi$, compute the ciphertext $\mathbf{c}_{L,i} = (c^L_{i,0}, c^L_{i,1}) = (r_i b_{L,i} + pe + \mu_i, ra_{L,i} + pe') \in R^2_{q_L}$.

• MKFHE.CTExt($\mathbf{c}_{L,i}, K$): Input the i-th party's ciphertext $\mathbf{c}_{L,i} \in R^2_{q_L}$, and output the ciphertext tuple $ct = \{\mathbf{c}_{L,i}, \{i\}, L\} \in R^{k+1}_{q_L}$.

• MKFHE.Dec($\mathbf{sk}_K, ct = \{\mathbf{c}_{L,i}, \{i\}, L\}$): Input the tuple $ct = (\mathbf{c}_{L,i}, K, l)$ and the joint secret keys $\bar{\mathbf{s}} = (1, s_1, ..., s_k) \in R_3^{k+1}$, output $\mu \leftarrow \langle ct, \bar{\mathbf{s}}\rangle \bmod q_l \bmod p$.

• MKFHE.Eval($(pk_{l,i}, ek_{l,i}, ct_i)_K, \mathcal{C}$): Let S be a subset of K, where $1 \leq |S| \leq k$. For another party j, performing the MKFHE.CTExt($\mathbf{c}_{l,i}, K$) and relinearization key generation program $\overline{rlk_{l,S}} \leftarrow$ Convert($\mathbf{D}_S, b_{l,j}$). The homomorphism operation is as follows:

1. MKFHE.EvalAdd($\overline{rlk_{l,S}}, \bar{\mathbf{c}}_{l,i}, \bar{\mathbf{c}}_{l,j}$): Input the ciphertexts $\bar{\mathbf{c}}_{l,i}, \bar{\mathbf{c}}_{l,j} \in R^{k+1}_{q_l}$ at the same level-l.

1) Compute $\bar{\mathbf{c}}_{l,\text{add}} := \bar{\mathbf{c}}_{l,i} + \bar{\mathbf{c}}_{l,j} \bmod q_l$, corresponding the joint secret key $\bar{\mathbf{s}} \in R_3^{k+1}$.

2) Compute $\bar{\mathbf{c}}'_{l-1,\text{add}} :=$ ModulusSwitch($\bar{\mathbf{c}}_{l,\text{add}}, q_l, q_{l-1}$).

2. MKFHE.EvalMult($\overline{rlk_{l,S}}, \bar{\mathbf{c}}_{l,i}, \bar{\mathbf{c}}_{l,j}$): Input the ciphertexts $\bar{\mathbf{c}}_{l,i}, \bar{\mathbf{c}}_{l,j} \in R^{k+1}_{q_l}$ at the same level-l.

1) Compute $\bar{\mathbf{c}}_{l,\text{mult}} := \bar{\mathbf{c}}_{l,i} \otimes \bar{\mathbf{c}}_{l,j} \bmod q_l$, corresponding the joint secret key $\hat{\mathbf{s}} = \bar{\mathbf{s}} \otimes \bar{\mathbf{s}} \in R_3^{(k+1)^2}$.

2) Perform $\bar{\mathbf{c}}'_{l,\text{mult}} =$ Relin($\bar{\mathbf{c}}_{l,\text{mult}}$; $(\mathbf{D}_{l,i}, b_{l,j})$), corresponding the joint secret key $\bar{\mathbf{s}} \in R_3^{k+1}$.

3) Compute $\bar{\mathbf{c}}'_{l-1,\text{mult}} \triangleq$ ModulusSwitch($\bar{\mathbf{c}}'_{l,\text{mult}}, q_l, q_{l-1}$).

4.3 Analysis of the Scheme

1. **Security Analysis**
Our optimized relinearization algorithm uses the single-key BGV-based encryption algorithm like CDKS19, as mentioned earlier, the security of optimized relinearization algorithm is based on RLWE assumption. Like most BGV-type MKFHE schemes [CZW17, LZY+19 et al.], our scheme security also rely on cyclic security assumption. And the modulus switching does not affect security. Therefore, our MKFHE is IND-CPA secure under the RLWE assumption of parameter pp^{RLWE}.

2. **Correctness**
In our scheme, the joint secret key $\bar{\mathbf{s}} \in R_3^{k+1}$ synthesized by all parties' secret key can satisfy all L levels ciphertext decryption. We take two ciphertexts' homomorphic multiplication as an example to analyze the correctness of our MKFHE scheme in the most complex case. That is, when the ciphertexts of parties i and j are homomorphiclly multiplied in level l, and there is a new party t participating in homomorphic multiplication in level $l-1$. For parties i and j, we get the product ciphertext $\bar{\mathbf{c}}'_{l-1,\text{mult}} = \text{ModulusSwitch}(\bar{\mathbf{c}}'_{l,\text{mult}}, q_l, q_{l-1}) \in R_{q_{l-1}}^{k+1}$ in level $(l-1)$. Set the new party set $K' \supset K$. The ciphertext of new party t is $\mathbf{c}_{l-1,t} \in R_{q_{l-1}}^2$. So the multiplication in level $(l-1)$ can be described as follows.

(1) Re-perform the ciphertext expansion algorithm $\bar{\mathbf{c}}'_{l-1,\text{M}} \in R_{q_{l-1}}^{k'+1} \leftarrow$ MKFHE.CTExt$(\bar{\mathbf{c}}'_{l-1,\text{mult}}, K')$ and $\bar{\mathbf{c}}_{l-1,t} \in R_{q_{l-1}}^{k'+1} \leftarrow$ MKFHE.CTExt$(\mathbf{c}_{l,t}, K')$, which corresponds to the new joint secret key $\bar{\mathbf{s}}' \in R_3^{k'+1}$.

(2) Use Method 1 to generate a new relinearization key of subset Y, where Y contains the parties i and j. According to the auxiliary keys of parties i and j, cloud server carries out a new auxiliary key $\mathbf{pb}_Y$ of the subset Y, where $\mathbf{pb}_Y = s_Y \mathbf{a}_{l-1} + \mathbf{e}_{l-1} \in R_{q_{l-1}}^d$ such that $s_Y = s_i s_j$. Perform the program UniEnc$(s_Y; \mathbf{pb}_Y) \rightarrow \mathbf{D}_Y = [\mathbf{h}_0 | h_1]$:

1) Sample $r_{l-1} \leftarrow \chi$.

2) Sample $\bar{\mathbf{e}}_{l-1,1} \leftarrow \psi^d$, and set $\mathbf{h}_0 = -\mathbf{pb}_Y + \bar{\mathbf{e}}_{l-1,1} + r_{l-1} \cdot \mathbf{g}(\text{mod} q_{l-1})$.

3) Sample $\bar{e}_{l-1,2} \leftarrow \psi$, and set $h_1 = -r \cdot a_{l-1} + \bar{e}_{l-1,2} + s_Y(\text{mod} q_{l-1})$.

Then, we get the relinearization key of subset Y is $\mathbf{K}_{Y,t} = [k_{Y,t,0} | k_{Y,t,1} | k_{Y,t,2}] \in R_{q_{l-1}}^3$, where $k_{Y,t,0} = \langle \mathbf{g}^{-1}(b_{l-1,t}), \mathbf{h}_0 \rangle$, $k_{Y,t,1} = \langle \mathbf{g}^{-1}(b_{l-1,t}), \mathbf{a}_{l-1} \rangle$ and $k_{Y,t,2} = h_1$.

(3) For the tensor product $\mathbf{ct}'' = \bar{\mathbf{c}}'_{l-1,\text{M}} \otimes \bar{\mathbf{c}}_{l-1,t}$, we can get the following result after relinearization.

$$
\begin{aligned}
&< \mathbf{ct}'', \bar{\mathbf{s}}' > (\text{mod} q_{l-1}) = c''_0 + \textstyle\sum_{\tau=1}^{k} c''_\tau \cdot s_\tau \\
&= c'_{0,0} + \textstyle\sum_{\tau=1}^{k'-1} (c'_{\tau,0} + c_{0,\tau}) s_\tau \\
&+ \textstyle\sum_{|Y|,t=1}^{k'-1} \sum_{\zeta=0}^{d-1} \left(c'_{|Y|,t} \right)_\zeta \cdot \mathbf{K}_{Y,t} \cdot (1, s_Y, s_t) \ (\text{mod} q_{l-1}) \\
&\approx c'_{0,0} + \textstyle\sum_{\tau=1}^{k'-1} (c'_{\tau,0} + c_{0,\tau}) s_\tau + \sum_{|Y|,t=1}^{k'-1} c'_{|Y|,t} s_Y s_t \\
&= < \bar{\mathbf{c}}'_{l-1,\text{M}} \otimes \bar{\mathbf{c}}_{l-1,t}, \bar{\mathbf{s}}' \otimes \bar{\mathbf{s}}' > (\text{mod} q_{l-1}) \\
&= \left(< \bar{\mathbf{c}}'_{l,\text{mult}}, \bar{\mathbf{s}} > (\text{mod} q_l) \right) \cdot \left(< \bar{\mathbf{c}}_{l-1,t}, \bar{\mathbf{s}}' > (\text{mod} q_{l-1}) \right) \\
&\approx \mu_i \mu_j \cdot \mu_t
\end{aligned}
$$

Therefore, in the homomorphic operation process, when the party set updates, all the parties do not need to regenerate their keys and ciphertexts, so our scheme satisfies multi-hop.

Table 2. The time complexity and error size between our scheme and CDKS19 in cloud server.

Relinearization types		Time complexity		Error size
		Relinearization key generation	Homomorphic multiplication decryption	
CDKS19		$O(kd^2n)$	$O(k^2d^3n)$	$O\left(k^2n^2d^2\lceil\log q_l\rceil B^2\right)$
Our scheme	Method 1	$O(kdn)$	$O(k^2d^2n)$	$O\left(k^2n^2d^2\lceil\log q_l\rceil B^2\right)$
	Method 2	$O(kdn)$	$O(k^2dn)$	$O\left(((q_l\lceil\log Pq_l\rceil)/2P)k^2n^2dB^2\right)$

Note. Here, we only analyze the result of once homomorphic multiplication decryption by two parties. See Appendix A and B for a specific computation.

3. **Efficiency Analysis**

Our MKFHE scheme realized the leveled fully homomorphic computation without using the key-switching technology. As mentioned in Subsect. 3.3, our optimized relinearization algorithms are more efficient than that of CDKS19. So, we analyze the time complexity of our scheme in the cloud server here to show the advantages. See Table 2 for details.

As shown in Table 2, compared with the CDKS19 scheme, our MKFHE scheme takes less time to generate the relinearization key and decrypt the multiplication of ciphertexts, and the error size is also decreased. Therefore, our construction is more efficient.

5 Conclusion

In this paper, two optimized relinearization algorithms are proposed and applied to the design of the leveled BGV-type MKFHE schemes. Compared to the prior MKFHE schemes, our scheme uses the relinearization algorithm instead of the key-switching technology to re-linearize the ciphertext more efficiently. In the next research content, we will focus on how to break through the restriction of the number of parties on the homomorphic computation efficiency.

Acknowledgments. This work was supported by National Key R&D Program of China (Grant No. 2017YFB080 2000), National Natural Science Foundation of China (Grant Nos. U1636114, 61872384, 61872289), National Cryptography Development Fund of China (Grant No. MMJJ20170112).

Appendix

A. Time Complexity

In this appendix, we calculate the time complexity in Tables 1 and 2.

We define the time complexity as the scalar operation (addition or multiplication) of a polynomial, denoted as Δ. For example, the time complexity of the product of two n-dimensional polynomials $a, b \in R_q$ is defined as Δn. Also, our optimized relinearization algorithm allows parties to generate their evaluation key $\mathbf{D}_i$ offline, so we do not calculate the time complexity of $\mathbf{D}_i$ any more.

A1. Time Complexity Calculation in Table 1

(1) For the CDKS19 scheme, the cloud server needs to perform $\left(2d^3 + d + con\right)$ (where con is a constant term) times polynomial multiplications to complete once relinearization, so the time complexity of once relinearization is about $O(d^3n)$.

In the same way, we can calculate the time complexity of our optimized relinearization algorithms.

(2) For the Method 1, the cloud server needs to perform $\left(2d^2 + d + con\right)$ times polynomial multiplications to complete once relinearization, so the time complexity of once relinearization is about $O(d^2n)$.

(3) For the Method 2, the cloud server needs to perform $(2d + con)$ times polynomial multiplications to complete once relinearization, so the time complexity of once relinearization is about $O(dn)$.

A2. Time Complexity Calculation in Table 2

(1) For the CDKS19 scheme

Every two parties need to perform $\left(2d^2 + con\right)$ times polynomial multiplications to generate a relinearization key. For k parties, at least $\lfloor k/2 \rfloor$ relinearization keys need to be generated. So, the time complexity of generating all the relinearization keys is about $O(d^2kn)$.

If the two parties decrypt the homomorphic multiplication of their ciphertexts successfully, the cloud server needs to perform $\left(2k^2d^3 + k^2d + k + con\right)$ times polynomial multiplications, so the time complexity is about $O(k^2d^3n)$.

(2) For the Method 1

Every two parties need to perform $(2d + con)$ times polynomial multiplications to generate a relinearization key. So, the time complexity of generating all the relinearization keys is about $O(kdn)$.

If the two parties decrypt the homomorphic multiplication of their ciphertexts successfully, the cloud server needs to perform $\left(2k^2d^2 + k^2d + k + con\right)$ times polynomial multiplications, so the time complexity is about $O(k^2d^2n)$.

(3) For the Method 2

Every two parties need to perform $(2d + con)$ times polynomial multiplications to generate a relinearization key. So, the time complexity of generating all the relinearization keys is about $O(kdn)$.

If the two parties decrypt the homomorphic multiplication of their ciphertexts successfully, the cloud server needs to perform $\left(2k^2d + 2k^2d + k + con\right)$ times polynomial multiplications, so the time complexity is about $O(k^2d^2n)$.

B. Error Analysis

Set the bound of ψ and χ is B, so for the $a, b \in \psi$, such that $||a \cdot b||_\infty \leq nB^2$.

1. For the Method 1

As shown in Method 1 of Subsect.3.1, the error size of $\mathbf{K}_{i,j} \cdot \left(1, s_i, s_j\right)$ is as follows.

$$||e_{\text{small}}||_\infty = ||\mathbf{g}^{-1}(b_j)\mathbf{e}_1 + s_j e_2 + r_i e_j||_\infty \leq n\lceil \log q_l \rceil (d+1)B + n\lceil \log q_l \rceil B^2;$$

After relinearization and decryption, we can get the following results.

$$\begin{aligned} &< \overline{\mathbf{ct}'}, \overline{\mathbf{s}} >= c_0' + \textstyle\sum_{i=1}^k c_i' \cdot s_i \\ &= c_{0,0} + \textstyle\sum_{i=1}^k (c_{i,0} + c_{0,i})_i s_i + \sum_{i,j=1}^k \sum_{\varsigma=0}^{d-1} \left(c_{i,j}\right)_\varsigma \cdot \mathbf{K}_{i,j} \cdot \left(1, s_i, s_j\right) \pmod{q_l} \\ &= c_{0,0} + \textstyle\sum_{i=1}^k (c_{i,0} + c_{0,i})_i s_i + \sum_{i,j=1}^k c_{i,j} \cdot s_i s_j + e_{\text{mult}} \end{aligned}$$

Therefore, after successful decryption, the final error size is as follows.

$$\begin{aligned} &||e_{\text{mult}}||_\infty = || \textstyle\sum_{i,j=1}^k \sum_{\varsigma=0}^{d-1} \left(c_{i,j}\right)_\varsigma \cdot e_{\text{small}} ||_\infty \\ &= k^2 nd\left(n\lceil \log q_l \rceil (d+1)B + n\lceil \log q_l \rceil B^2\right) \leq O\left(k^2 n^2 d^2 \lceil \log q_l \rceil B^2\right) \end{aligned};$$

2. For the Method 2

As shown in Method 2 of Subsect.3.1, we have

$$||e'_{\text{small}}||_\infty = ||\left(\mathbf{g}^{-1}(b_j)\mathbf{e}_1 + r_i e_j + s_j e_2\right)||_\infty = n\lceil \log P \cdot q_l \rceil (dB + B^2 + B)$$

After relinearization and decryption, we can get the following results.

$$\begin{aligned} &||e'_{\text{mult}}||_\infty = || \textstyle\sum_{i,j=1}^k P^{-1} \cdot c_{i,j} e'_{\text{small}} ||_\infty \\ &= ((q_l \lceil \log P \cdot q_l \rceil)/2P) k^2 n^2 (dB + B^2 + B) \leq O\left(((q_l \lceil \log P \cdot q_l \rceil)/2P) k^2 n^2 dB^2\right) \end{aligned}$$

Note. we usually choose the $P \succ q_l$, that is $P/q_l \approx 1$. So

$$O\left(((q_l \lceil \log P \cdot q_l \rceil)/2P) k^2 n^2 dB^2\right) \approx O\left(k^2 n^2 d \lceil \log q_l \rceil B^2\right).$$

3. For the CDKS19

As shown in Subsect. 2.5, the CDKS19 scheme completes once relinearization and decryption generating the error as follows.

$$||e_{\text{CDKS}}||_\infty = k^2 nd\left(n\lceil \log q_l \rceil (d+1)B + n\lceil \log q_l \rceil B^2\right) \leq O\left(k^2 n^2 d^2 \lceil \log q_l \rceil B^2\right).$$

References

1. López-Alt, A., Tromer, E., Vaikuntanathan, V.: On-the-fly multiparty computation on the cloud via multi-key fully homomorphic encryption. In: Proceedings of the Forty-Fourth Annual ACM Symposium on Theory of Computing, pp. 1219–1234. ACM (2012)
2. Qaosar, M., Zaman, A., Ssique, M.A., et al.: Privacy-preserving secure computation of skyline query in distributed multi-party databases. Information **10**(3), 119–135 (2019)
3. Xiong, J.B., Zhao, M.F., Bhuiyan, M., et al.: An AI-enabled three-party game framework for guaranteed data privacy in mobile edge crowdsensing of IoT. IEEE Trans. Industr. Inf. **17**(2), 922–933 (2021)
4. Hoffstein, J., Pipher, J., Silverman, J.H.: NTRU: a ringbased public key cryptosystem. In: International Symposium on AlgorithmicNumber Theory, pp. 267–288 (1998)
5. Doröz, Y., Hu, Y., Sunar, B.: Homomorphic AES evaluation using the modified LTV scheme. Des. Codes Crypt. **80**(2), 333–358 (2016)
6. Chongchitmate, W., Ostrovsky, R.: Circuit-private multi-key FHE. In: Fehr, S. (ed.) Public-Key Cryptography – PKC 2017. Lecture Notes in Computer Science, vol. 10175. Springer, Heidelberg (2017). https://doi.org/10.1007/978-3-662-54388-7_9
7. Che, X.L., Zhou, T.P., Li, N.B., et al.: Modified multi-key fully homomorphic encryption based on NTRU cryptosystem without key-switching. Tsinghua Sci. Technol. **25**(5), 564–578 (2020)
8. Clear, M., McGoldrick, C.: Multi-identity and multi-key leveled FHE from learning with errors. In: Gennaro, R., Robshaw, M. (eds.) Advances in Cryptology – CRYPTO 2015. Lecture Notes in Computer Science, vol. 9216. Springer, Heidelberg (2015). https://doi.org/10.1007/978-3-662-48000-7_31
9. Mukherjee, P., Wichs, D.: Two round multiparty computation via multi-key FHE. In: Fischlin, M., Coron, J.-S. (eds.) EUROCRYPT 2016. LNCS, vol. 9666, pp. 735–763. Springer, Heidelberg (2016). https://doi.org/10.1007/978-3-662-49896-5_26
10. Peikert, C., Shiehian, S.: Multi-key FHE from LWE, revisited. In: Hirt, M., Smith, A. (eds.) TCC 2016. LNCS, vol. 9986, pp. 217–238. Springer, Heidelberg (2016). https://doi.org/10.1007/978-3-662-53644-5_9
11. Brakerski, Z., Perlman, R.: Lattice-based fully dynamic multi-key FHE with short ciphertexts. In: Robshaw, M., Katz, J. (eds.) Advances in Cryptology – CRYPTO 2016. Lecture Notes in Computer Science, vol. 9814. Springer, Heidelberg (2016). https://doi.org/10.1007/978-3-662-53018-4_8.
12. Chillotti, I., Gama, N., Georgieva, M., Izabachène, M.: Faster fully homomorphic encryption: bootstrapping in less than 0.1 seconds. In: Cheon, J.H., Takagi, T. (eds.) ASIACRYPT 2016. LNCS, vol. 10031, pp. 3–33. Springer, Heidelberg (2016). https://doi.org/10.1007/978-3-662-53887-6_1
13. Gentry, C., Sahai, A., Waters, B.: Homomorphic encryption from learning with errors: conceptually-simpler, asymptotically-faster, attribute-based. In: Canetti, R., Garay, J.A. (eds.) Advances in Cryptology – CRYPTO 2013. Lecture Notes in Computer Science, vol. 8042. Springer, Heidelberg (2013). https://doi.org/10.1007/978-3-642-40041-4_5
14. Chillotti, I., Gama, N., Georgieva, M., Izabachène, M.: Faster packed homomorphic operations and efficient circuit bootstrapping for TFHE. In: Takagi, T., Peyrin, T. (eds.) ASIACRYPT 2017. LNCS, vol. 10624, pp. 377–408. Springer, Cham (2017). https://doi.org/10.1007/978-3-319-70694-8_14
15. Chen, H., Chillotti, I., Song, Y.: Multi-key homomorphic encryption from TFHE. In: Galbraith, S., Moriai, S. (eds.) Advances in Cryptology – ASIACRYPT 2019. ASIACRYPT 2019. Lecture Notes in Computer Science, vol. 11922. Springer, Cham (2019). https://doi.org/10.1007/978-3-030-34621-8_16

16. Chen, L., Zhang, Z., Wang, X.: Batched Multi-hop multi-key FHE from ring-LWE with compact ciphertext extension. In: Kalai, Y., Reyzin, L. (eds.) Theory of Cryptography. TCC 2017. Lecture Notes in Computer Science, vol. 10678. Springer, Cham (2017). https://doi.org/10.1007/978-3-319-70503-3_20
17. Li, N.B., Zhou, T.P., Yang, X.Y., et al.: Efficient multi-key FHE with short extended ciphertexts and directed decryption protocol. IEEE Access **7**, 56724–56732 (2019)
18. Chen, H., Dai, W., Kim, M., et al.: Efficient multi-key homomorphic encryption with packed ciphertexts with application to oblivious neural network inference. In: Proceedings of the 2019 ACM SIGSAC Conference on Computer and Communications Security, pp. 395–412. ACM, London (2019)
19. Brakerski, Z., Gentry, C., Vaikuntanathan, V.: (Leveled) Fully homomorphic encryption without bootstrapping. In: Proceedings of the 3rd Innovations in Theoretical Computer Science Conference, Cambridge, MA, USA, pp. 309–325 (2012)
20. Alperin-Sheriff, J., Peikert, C.: Faster bootstrapping with polynomial error. In: Garay, J.A., Gennaro, R. (eds.) Advances in Cryptology – CRYPTO 2014. CRYPTO 2014. Lecture Notes in Computer Science, vol. 8616. Springer, Heidelberg (2014). https://doi.org/10.1007/978-3-662-44371-2_17
21. Cheon, J.H., Kim, A., Kim, M., Song, Y.: Homomorphic encryption for arithmetic of approximate numbers. In: Takagi, T., Peyrin, T. (eds.) ASIACRYPT 2017. LNCS, vol. 10624, pp. 409–437. Springer, Cham (2017). https://doi.org/10.1007/978-3-319-70694-8_15
22. Li, B.Y., Micciancio, D.: On the security of homomorphic encryption on approximate. Cryptology ePrint Archive, Report 2020/1533 (2020). https://eprint.iacr.org/2020/1533
23. Cheon, J.H., Hong, S., Kim, D.: Remark on the security of CKKS scheme in practice. Cryptology ePrint Archive: Search Results 2020/1581 (2020). https://eprint.iacr.org/2020/1581
24. Lyubashevsky, V., Peikert, C., Regev, O.: On ideal lattices and learning with errors over rings. J. ACM (JACM) **60**(6), 43 (2013)
25. Lyubashevsky, V., Peikert, C., Regev, O.: A toolkit for ring-LWE cryptography. In: Johansson, T., Nguyen, P.Q. (eds.) EUROCRYPT 2013. LNCS, vol. 7881, pp. 35–54. Springer, Heidelberg (2013). https://doi.org/10.1007/978-3-642-38348-9_3
26. Brakerski, Z.: Fully homomorphic encryption without modulus switching from classical GapSVP. In: Safavi-Naini, R., Canetti, R. (eds.) CRYPTO 2012. LNCS, vol. 7417, pp. 868–886. Springer, Heidelberg (2012). https://doi.org/10.1007/978-3-642-32009-5_50
27. Halevi, S., Polyakov, Y., Shoup, V.: An improved RNS variant of the BFV homomorphic encryption scheme. In: Matsui, M. (ed.) Topics in Cryptology – CT-RSA 2019. CT-RSA 2019. Lecture Notes in Computer Science, vol. 11405. Springer, Cham (2019). https://doi.org/10.1007/978-3-030-12612-4_5
28. Brakerski, Z., Vaikuntanathan, V.: Efficient fully homomorphic encryption from (standard) LWE. In: Proceedings of Annual Symposium on Foundations of Computer Science, Los Alamitos, CA, USA, pp. 97–106 (2011)

Improved Attribute Proxy Re-encryption Scheme

Wang Zhan[1], Yuling Chen[1,4](✉), Wei Ren[3], Xiaojun Ren[4], and Yujun Liu[2]

[1] State Key Laboratory of Public Big Data, College of Computer Science and Technology, Guizhou University, Guiyang 550025, China
ylchen3@gzu.edu.cn

[2] School of Computer Science, China University of Geosciences, Wuhan 430074, China

[3] Blockchain Laboratory of Agricultural Vegetables, Weifang University of Science and Technology, Shouguang 262700, China

[4] Technical Center of Beijing Customs, Beijing, China

Abstract. With the development of cloud computing, storing and sharing medical data in the cloud is envisioned as a promising method for supporting a large scale of users. However, if the sensitive information contained in the medical data (e.g., prescribed drug) is leaked, the user privacy will be damaged. In this paper, we propose an attribute-based Proxy Re-encryption scheme based on data split. Security of data storage and the matching between doctors and patients are tackled by CP-ABE and Proxy Re-encryption. To protect patients' sensitive data, improve the security of files, and reduce computing overhead, we adopt medical file splitting technology and attribute key tree to update keys. The data is stored by two cloud servers. Even if either server is attacked, the clear text cannot be recovered without key information. Besides, mixed encryption is used to improve the operation efficiency.

Keywords: Data sharing · Cloud storage · Attribution-based encryption · Proxy Re-encryption · Data split

1 Introduction

With the rapid development of Cloud computing technology, more and more users choose to store and share their files in the Cloud, which is more convenient and efficient than traditional storage. At present, most of Cloud storage services are provided by third-party to help users manage information and store files. Now medical data sharing has become more and more popular. However, if the patient's medical records are leaked by the platform administrator or stolen by hackers, it will inevitably bring the risk of privacy disclosure to doctors and patients. Therefore, in order to ensure the safe storage of patients' medical data and facilitate the communication between doctors and patients, reducing the pressure on the server is the research hotspot now [1, 2].

In different environments, many scholars have proposed CP-ABE schemes with different functions and backgrounds. Blaze [3] and others first proposed the concept of

J. Xiong et al. (Eds.): MobiMedia 2021, LNICST 394, pp. 343–353, 2021.
https://doi.org/10.1007/978-3-030-89814-4_25

Attribute encryption, which provides a new form of data sharing. Sahai [4–6] and others put propose an identity-based encryption scheme.

In 2012, Seo [7] and others proposed the Proxy Re-encryption scheme (ABPRE), which combines the original attribute encryption scheme (ABE) with the Proxy Re-encryption, so that the sender can decrypt the data by satisfying the attributes when the sender is offline.

The contributions of our work in this paper are shown as follow:

1) A new scheme based attribute Proxy Re-encryption scheme is proposed. After file split, the data is divided into Body Data and Sensitive data, and the Body Data is encrypted symmetrically and stored in the Data Cloud. Symmetric encryption is used to encrypt the Body Data to update the key and reduce the calculation cost. The CP-ABE and Proxy Re-encryption are used to be responsible for the operation and matching, and then the cloud server can not get the complete file.
2) Owing to patients' different requirements, it is necessary to update the key of medical records. But the key update of Asymmetric encryption is inefficient. The Body Data is encrypted symmetrically, the key and ciphertext are stored separately, which is more efficient. This paper designs an attribute key tree for attribute encryption to achieve the purpose of key update, we adopt the attribute key generation tree to generate the keys of doctors and other roles.
3) Using the characteristics of Proxy Re-encryption and secondary encryption, users can expand the matching requirements (e.g. patients can not solve the disease through traditional interrogation). According to the patients' new requirements, a new shared structure is introduced by patients, and remote person consultation can be realized.

2 Related Work

Ibraimi et al. [8] proposed an access control scheme with Proxy Re-encryption technology, but the cost of calculating Proxy Re-encryption key in the user revocation process remains to be solved. Liu [9] proposed that in the cloud computing environment, the cloud service is composed of multiple servers, and the user's data is often stored in multiple servers, and the user's encryption operation may not be executed by all servers. Liang K et al. [10] proposed the attribute-based Proxy Re-encryption (CP-ABPRE), which extends the traditional Proxy Re-encryption (PRE) by allowing semi trusted agents to convert the ciphertext under the access policy into the text with the same Plain Text under another access policy (i.e., attribute-based re-encryption). Tiwari et al. [11] provide a flexible encrypted access control mechanism for data security access. A Proxy Re-encryption scheme based on ciphertext policy attribute is proposed. Niu et al. [12] proposed an improved Proxy Re-encryption sharing scheme, which improved the Proxy Re-encryption to store the medical records in the cloud server. Zhang et al. [13] proposed a sharing scheme of cloud storage combined with block chain based on attribute agent re-encryption. Luo et al. [14] used the cross domain multi authorization center to share the key, and used the key separation technology to realize the data privacy protection. However, the above scheme can not be used to update the key, and there is a great security risk. This scheme proposes a storage scheme of medical records based on attribute Proxy

Re-encryption. The medical records are divided into two parts, the Body Data and the Sensitive Data. Compared with the existing scheme, it reduces the operation efficiency and improves the security.

3 Preliminaries

3.1 System Role

The scheme mainly include six roles: Patient, Hospital, Doctor, Key Distribution Center, Data Cloud and Proxy Cloud, Process of our split attribute Proxy Re-encryption, which is shown in Fig. 1.

1) Patient: Responsible for submitting attributes to the key distribution center and managing key pairs.
2) Hospital (HS): The hospital is responsible for generating patient medical records for split and encryption.
3) Doctor: Responsible for receiving medical records and submitting their own attributes to the key distribution center.
4) Key Distribution Center (KDC): Responsible for accepting the attributes of doctors and patients, and generating all keys.
5) Data Cloud (DC): A third-party cloud service provider with huge storage space and the ability to store large-scale data.
6) Proxy Cloud (PC): A third-party cloud service provider with powerful computing capabilities, mainly responsible for Proxy computing.

4 Protocol and Algorithm Design

4.1 Scheme Design

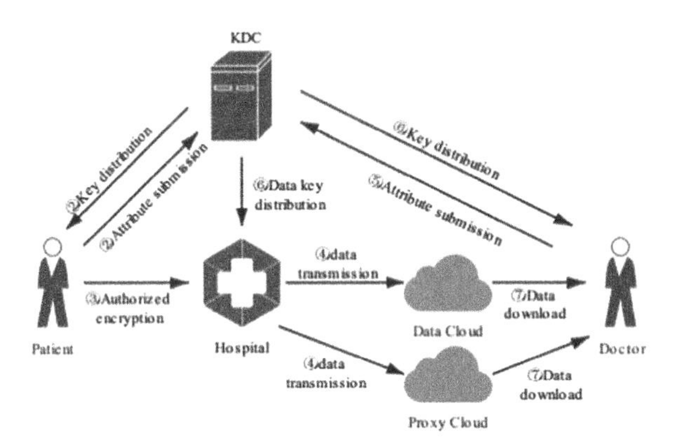

Fig. 1. Technological process of our split attribute Proxy Re-encryption.

The variable symbols used in the specific scheme, which is shown in Table 1.

Table 1. Symbolic variable

Symbol	Meaning
U	Attribute set
PK, SK	Public key and private key
GP	Public parameters
MSK	Master key
k_*	Data key
(M^*, ρ^*)	Shared structure
$rk_{A \to B}$	Proxy Re-encryption key
CT, m	Ciphertext, Plain Text
I	Key information

4.2 Algorithm Component

There are eight algorithms in this scheme:

$$setup1^k, U \to GP, MSK_*, PK \tag{1}$$

System Initialization Algorithm: Input security parameters and attribute sets of 1^k and U respectively, output GP as public parameters, MSK and PK as system master key and system public key according to security parameters and attribute set.

$$\begin{aligned} &Keygen_1(PK, MSK, S_A) \to PK_A, SK_A \\ &Keygen_2(PK, MSK, S_B) \to PK_{\boldsymbol{B}}, SK_B \\ &Keygen_3(MSK_*, Tree) \to k_* \end{aligned} \tag{2}$$

Key Generation Algorithm: Input system public key PK, system master key MSK and user submitted attribute S_A,S_B. Output public key and private key PK_A,PK_B and SK_A, SK_B. Master key MSK and attribute key $Tree$ to generate Data key k_*.

$$DataSegm \to m_1, m_2 \tag{3}$$

Data Split: The Clear text m is divided into two parts, including Sensitive data m_1, body data m_2.

$$ReKeygen_1\left(GP, SK_A, \left(M', \rho'\right), PK_B\right) \to rk_{A \to B} \tag{4}$$

Re-encryption Key Generation: Input public parameter GP, user private key SK_A. And the shared structure $\left(M^{'}, \rho^{'}\right)$ and public key PK_B, Output Generate Proxy Re-encryption key $rk_{A \to B}$.

$$\begin{aligned} &Encrypt_1(m_1, k_*) \to CT_A^{'} \\ &Encrypt_2(GP, m_2, k_*, I, M, \rho, PK) \to CT_A \end{aligned} \tag{5}$$

Information Encryption Algorithm 1: Input Body data m_1, Data key k_*. Output the ciphertext $CT_A{}^{'}$.

Information Encryption Algorithm 2: Input Public parameter GP, Sensitive data m_2, shared structure (m, ρ),the public key PK and data key k_* And key information I. Output ciphertext CT_A.

$$ReEncrypt(rk_{A \to B}, CT_A, \left(M^{'}, \rho^{'}\right), PK_B, SK_A) \to CT_B \tag{6}$$

Ciphertext Re-encryption Algorithm: Judge whether Bob is the system contract user, if Bob is the system contract user, Input ciphertext CT_A. Bob public key PK_B, the patient's private key SK_A. Share structure $\left(M^{'}, \rho^{'}\right)$, and Proxy Re-encryption key $rk_{A \to B}$ generate re-encrypted ciphertext CT_B.

$$ReDecryptCT_B, SK_B \to m_2, k_* \tag{7}$$

Re-encryption and Decryption Algorithm: The system checks whether Bob conforms the shared structure$\left(M^{'}, \rho^{'}\right)$, If Bob $\left(M^{'}, \rho^{'}\right)$ consistent with Bob, Bob can use his private key SK_B to Decrypt CT_B get sensitive data m_2 and data key k_*.

$$\begin{aligned} &Decrypt_1(CT_A, PK, SK_A) \to m_2 \\ &Decrypt_2\left(CT_A^{'}, k_*, GP\right) \to m_1 \end{aligned} \tag{8}$$

Ciphertext Decryption algorithm: If the submitted attribute conforms to the shared structure, Bob obtains the private key SK through the key Distribution Center (KDC). According to the PK, SK get the sensitive datam_2. Bob asks cloud for $CT_A{}^{'}$, using the data key k_*, And decrypt$CT_A{}^{'}$, get the datam_1.

4.3 Proposed Procedures

1) **Initialization stage**
   Run algorithm 1, randomly select groups G and G_T and generators $g, g_1 \in G$,α, $a \in Z^*{}_p$ taking Bilinear mapping $e : G \times G \to G_T$, Generate the system public parameter GP and Hash function to represent the identity of the Hash function ID, the role of the basic attributes of the Hash function H_1. The Hash function of doctor's basic information H_2. Hash function of message H_3.

$$\begin{aligned} &GP = \left(p, g, G, G_T, e, g_1, g^a, H_1, H_2, H_3, ID, e(g, g)^{\alpha}\right) \\ &ID : \{0, 1\}^* \to G, H_1, H_2, H_3 : G_T \to Z_p^* \\ &PK = (e(g, g)^a, g, g_1, g^a), \; MSK = \left(g^{\alpha}, a\right) \end{aligned} \tag{9}$$

Patients applied to the KDC and randomly selected $t \in Z^*{}_P$. Run algorithm 2 to generate the unique public and private key pairs of patients (doctors) (Fig. 2).

$$SK = (K = g^{at} g^{\alpha}, L = g^{t}, K_x = ID(x)^{t}, t \in S_A) \tag{10}$$

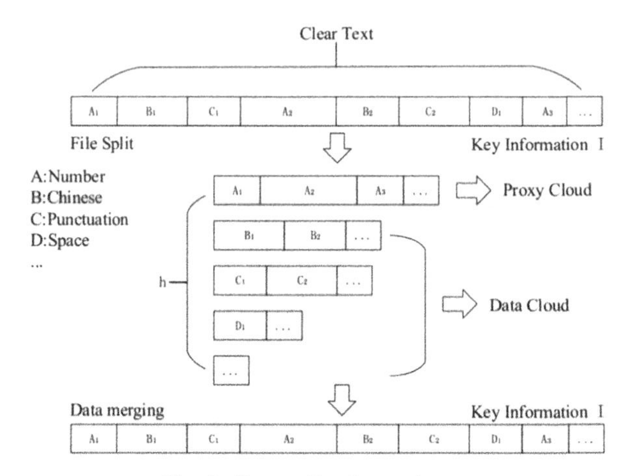

Fig. 2. Data split of our scheme.

2) **Encryption stage**
Run the algorithm 5 to encrypt and store it in the cloud. The data key is generated by the Key Distribution Center (KDC).The structure of the attribute based key tree is shown in the Figs. 3 and 4.

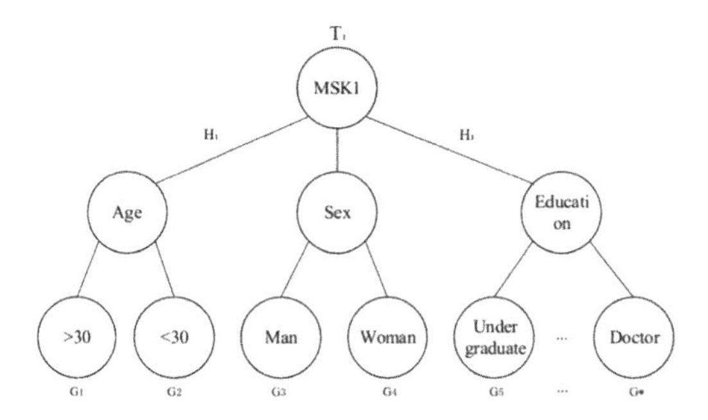

Fig. 3. The attribute based key tree of our scheme. (part 1)

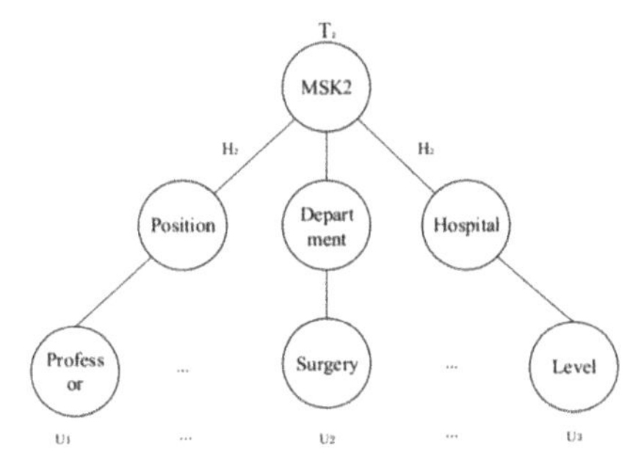

Fig. 4. The attribute based key tree of our scheme. (part 2)

At the same time, each node represents a different attribute group, and the leaf node represents each user's attribute. Each user has a unique identifier *ID*. For example, a 40 years old man with degree is a professor of surgery in a specialized hospital. After using the key to encrypt the segmented data segment, each user has his own k k_*. The key k is:

$$k = H_1\left(K_{G_1} \| K_{G_3} \| K_{G_5}\right) \| H_2\left(K_{U_1} \| K_{U_2} \| K_{U_3}\right) \tag{11}$$

When the user no longer authorized the doctor access rights or authorized to the next doctor, the key was changed, and the previously authorized user could not access the file normally. Proxy Cloud receives data m_1. Run the algorithm 5 to encrypt the data, in which the access shared structure set by the patient's M, ρ . Only when the attribute access applied by the doctor meets the shared structure can ciphertext be obtained CT_B. The encrypted ciphertext can be expressed as follows:

$$\begin{aligned} &CT_A = ((m, \rho), A_1, A_2, A_3, (B_1, C_1) \cdots (B_i, C_i)) \\ &A_1 = (m_2 + I + k_*) \cdot e(g, g)^{\alpha \cdot s}, A_2 = g^s, A_3 = g_1^s \\ &B_1 = \left(g^{\alpha}\right)^{\gamma_1} \cdot ID(\rho(1))^{-r_1}, \cdots, \\ &B_i = \left(g^{\alpha}\right)^{\gamma_i} \cdot ID(\rho(1))^{-r_i} \\ &C_1 = g^{r_1}, \cdots, C_i = g^{r_i} \end{aligned} \tag{12}$$

Where M is the matrix of $L \times N$, ρ is the mapping of M rows to attributes, and $\{\rho(i)|1 \leq i \leq l\}$ denotes the attributes in the access structure (m, ρ). S is the secret to be shared. $s, y_2, y_3 \cdots y_n \in Z^*{}_P$. For $i = 1$ to l, set $\gamma_1 = vM_i$,M_i is the vector corresponding to row i of matrix M, $v = (s, y_2, y_3 \cdots y_n)$,$r_1 \cdots r_i \in Z^*{}_P$ [15].

3) **Proxy key generation and Re-encryption stage**
At this time, it is necessary to determine whether the doctor attribute conform the new attribute set by the patients' $\left(M', \rho'\right)$. If the access structure is conformed, run algorithm 2.2 to generate the doctor's Public key and Private key. Among them, the Proxy Cloud randomly selects $\theta \in Z^*{}_P$. so that all doctors who conform the shared structure can obtain the same data key and realize consultation. The hospital (HS)

calculation of $rk_{A \to B}$.

$$\begin{aligned} & rk_{A \to B} = (rk_1, rk_2, rk_3, rk_4, R_X) \\ & rk_1 = K^{H_3(\delta)} g_1^{\theta} = g^{\alpha} g^{at} g_1^{\theta}, \\ & rk_2 = g^{\theta}, rk_3 = L^{H_3(\delta)}, \\ & rk_4 = C'_{(m', \rho')}, R_X = K_X^{H_3(\delta)} \end{aligned} \tag{13}$$

Judge whether the doctor's attribute conform the shared structure $\left(M', \rho'\right)$. Random selection $\delta \in G_T$.

$$\begin{aligned} & CT_B = (\left(m^I, \rho^I\right), A'_1, A'_2, A'_3, (B'_1, C'_1) \\ & \cdots \left(B'_i, C'_i\right), A_4, rk_4) \\ & A'_1 = \delta \cdot e(g, g)^{\alpha \cdot s'}, A'_2 = g^{s'}, A'_3 = g_1^{s'} \\ & B'_1 = (g^{\alpha})^{\gamma'_1} \cdot ID(\rho(1))^{-r'_1}, \cdots, \\ & B_i = (g^{\alpha})^{\gamma'_i} \cdot ID(\rho(1))^{-r'_i} \\ & C'_1 = g^{r'_1}, \cdots, C_i = g^{r'_i} \end{aligned} \tag{14}$$

When s conform the shared structure of $\left(M', \rho'\right)$, there exists a set of constants $\{\omega_i \in Z^*{}_P\}_{i \in l}$, make $\sum_{i \in l} w_i \gamma_i = s$, Where $\{\gamma_i\}$ is the secret of S.

$$A_4 = \frac{\frac{e(A_2, rk_1)}{e(A_3, rk_2)}}{\prod_{i \in l}(e(B_i, rk_3) e(C_i, R_{p(i)}))^{\omega_i}} \tag{15}$$

4) **Decryption stage**
When the user communicates with the doctor, the doctor submits the corresponding attributes to the key distribution system and conform the shared structure, the algorithm 8 is used to decrypt the encrypted ciphertext. The solution process is as follows.

$$\begin{aligned} & \frac{\frac{A_1}{e(A_2, K)}}{\left(\prod_{i \in l}\left(e(B_i, L) e\left(C_i, R_{p(i)}\right)\right)^{\omega_i}\right)} \\ & = \frac{(m_2 + I + k_*) \cdot e(g, g)^{\alpha \cdot s}}{e(g, g)^{\alpha \cdot s}} \\ & = m_2 + I + k_* \\ & D_{k_*}(CT_*) = m_1 \\ & m_2 + m_1 = m \end{aligned} \tag{16}$$

When the doctor obtains segment m_2. Data key k_*. After that, Body data m_1 and sensitive data m_2 to get clear text M. When the user is not satisfied with the doctor's diagnosis, or the doctor can not make an accurate judgment, the solution process is as follows.

$$\delta = \frac{\frac{A_1'}{e\left(A_2', K'\right)}}{\left(\prod_{i \in I}\left(e\left(B_i', L'\right)e\left(C_i', R_{p(i)}'\right)\right)^{\omega_i'}\right)} \tag{17}$$

$$(m_2 + I + k_*) = \frac{A_1}{(A_4)^{\frac{1}{H_3(\delta)}}}$$

5 Experiments and Analysis

5.1 Performance Analysis

In this paper, we simulate our experiment in a computer with Intel i5–6500 CPU @ 3.20 Hz and 8 GB memory windows, and test the algorithm efficiency with Python.

Then, we compared the efficiency and ciphertext length of attribute with Luo's and Zhang's scheme. Suppose that S represents the number of attributes and L_G, L_{G_T}, L_{Z_P} represent the length of G, G_P, Z_P respectively. We mainly discuss four components of our proposed scheme: system public key, system master key, user private key and ciphertext length (Table 2).

Table 2. Complexity comparison

Scheme	Luo's	Zhang's	Our's
PK	L_{G_T}	$(S+3)L_G + L_{G_T}$	$L_{G_T} + 3L_G$
MSK	L_G	$L_G + L_{Z_P}$	$L_G + L_{Z_P}$
SK	$(3S+2)L_G$	$(2S+2)L_G$	$(S+2)L_G$
CT	$SL_G + L_{G_T}$	$(2S+2)L_G + L_{G_T}$	$(S+1)L_G + L_{G_T}$

E is the exponentiation time, B is the bilinear mapping operation, S is the attribute, E_k is the time encrypted (Table 3).

Table 3. Efficiency comparison

Schemes	Luo's	Zhang's	Our's
Initialization	$E(G) + E(G_T) + B$	$(S+3)E(G) + E(G_T) + B$	$2E(G) + E(G_T) + B + hH(m_1)$
Key generation	$(2+S)E(G)$	$(3+S)E(G)$	$(2+S)E(G) + (h-1)E_k$
Encryption	$(10S+4)E(G) + 5E(G_T) + 9B$	$(3S+3)E(G) + 3E(G_T) + 7B$	$\frac{(6S+)E(G)+9B}{h} + \frac{h-1}{h}E_k$
Decryption	$SB + \gamma E(G)$	$(4S+6)E(G) + 5B$	$\frac{SB+\gamma E(G)}{h} + \frac{h-1}{h}E_k$

Figure 5 shows that the number of attributes has no impact on our proposed scheme in the initialization phase, compared with Luo's scheme, our proposed scheme performs file and data segment integrity hash verification in the initialization phase. Figure 6 shows that in the key generation stage, when the number of attributes increases, the computing time increases. In the key generation stage, the key update based on attribute will be executed, and the time consumed is different from other schemes. Figure 7 shows that the encryption time has some impacts with the number of attributes. The operation time of sensitive data after CP-ABE and Proxy Re-encryption is 1/h of the original file size, the total time of main data encryption, and the encryption efficiency of this scheme is improved by 34.24% compared. Figure 8 shows the time cost in decryption phase. And, in the decryption phase, our proposed scheme needs to decrypt the main data, and therefore it is more than others when the number of attributes is low.

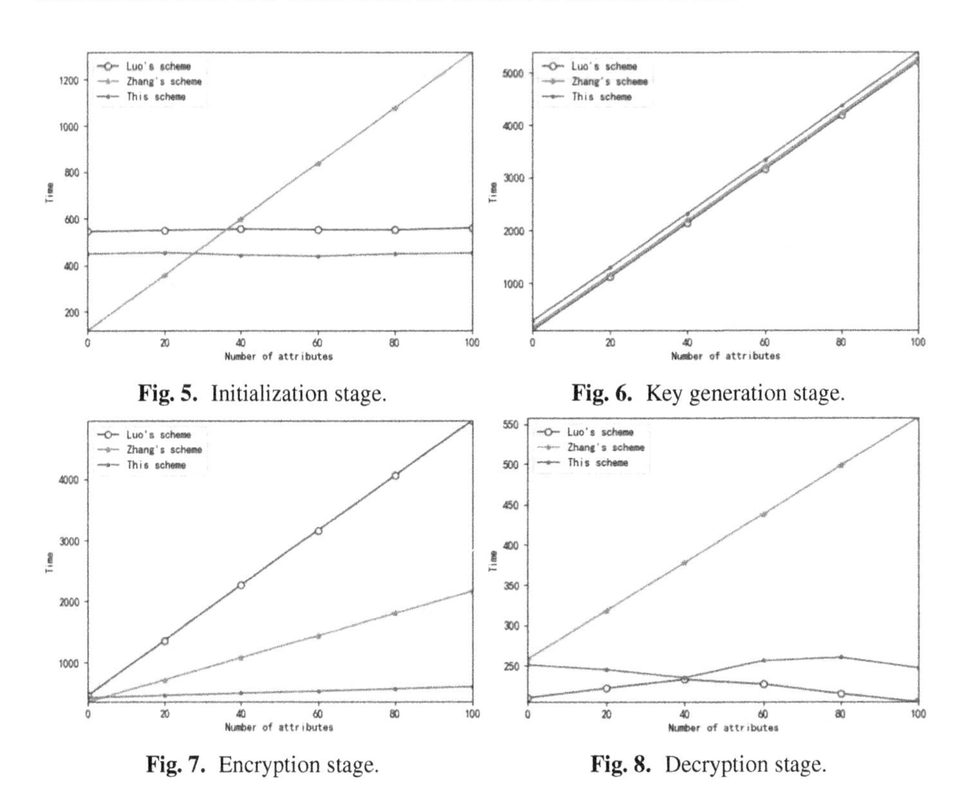

Fig. 5. Initialization stage.

Fig. 6. Key generation stage.

Fig. 7. Encryption stage.

Fig. 8. Decryption stage.

6 Conclusion

This scheme divides the data into different clouds for storage, reduces the server pressure, and solves the efficiency problem to a certain extent. The attribute key tree is used to realize the access control of one key at a time. All these make it possible for the safe storage of medical data. Nowadays, the application of cloud storage is constantly updated, and various technologies are constantly implemented. But the high complexity of asymmetric encryption can not be widely used.

Acknowledgment. This work is supported by the National Quality Infrastructure project that is Key R&D Program of China under Grant 2018YFF0212106. and supported in part by the National Natural Science Foundation of China under Grant 61962009. In part by the Major Science and Technological Special Project of Guizhou Province under Grant 20183001. and in part by the Open Funding of Guizhou Provincial Key Laboratory of Public Big Data under Grant 2018BDKFJJ013.

References

1. Feng, C.-S., Qin, Z.-G., Yuan, D.: Techniques of secure storage for cloud data. Chinese J. Comput. **38**(01), 150–163 (2015)
2. Fang, B., Jia, Y., Li, A., Jiang, R.: Privacy preservation in big data:a survey. Big Data Res. **2**(01), 1–18 (2016)
3. Blaze, M., Bleumer, G., Strauss, M.: Divertible protocols and atomic proxy cryptography. In: Nyberg, K. (eds.) Advances in Cryptology — EUROCRYPT 1998. EUROCRYPT 1998. LNCS, vol. 1403, pp. 127–144. Springer, Heidelberg (1998). https://doi.org/10.1007/BFb0054122
4. Sahai, A., Waters, B.: Fuzzy identity-based encryption. In: Cramer, R. (ed.) EUROCRYPT 2005. LNCS, vol. 3494, pp. 457–473. Springer, Heidelberg (2005). https://doi.org/10.1007/11426639_27
5. Bethencourt, J., Sahai, A., Waters, B.: Ciphertext-policy attribute-based encryption. In: 2007 IEEE Symposium on Security and Privacy (SP 2007), Berkeley, CA, pp. 321–334 (2007). https://doi.org/10.1109/SP.2007.11
6. Goyal, V., Pandey, O., Sahai, A., et al.: Attribute-based encryption for fine-grained access control of encrypted data. In: Proceedings of the 13th ACM Conference on Computer and Communications Security, pp. 89–98 (2006)
7. Seo, H.J., Kim, H.W.: Attribute-based proxy re-encryption with a constant number of pairing operations. J. Inf. Commun. Converg. Eng. **10**(1), 53–60 (2012)
8. Ibraimi, L., Petkovic, M., Nikova, S., Hartel, P., Jonker, W.: Mediated ciphertext-policy attribute-based encryption and its application. In: Youm, H.Y., Yung, M. (eds.) WISA 2009. LNCS, vol. 5932, pp. 309–323. Springer, Heidelberg (2009). https://doi.org/10.1007/978-3-642-10838-9_23
9. Liu, Q., Tan, C.C., Wu, J., Wang, G.: Reliable re-encryption in unreliable clouds. In: 2011 IEEE Global Telecommunications Conference - GLOBECOM 2011, Houston, TX, USA, pp. 1–5 (2011). https://doi.org/10.1109/GLOCOM.2011.6133609
10. Liang, K., Fang, L., Susilo, W., Wong, D.S.: A ciphertext-policy attribute-based proxy re-encryption with chosen-ciphertext security. In: 2013 5th International Conference on Intelligent Networking and Collaborative Systems, Xi'an, 2013, pp. 552–559 (2013). https://doi.org/10.1109/INCoS.2013.103
11. Tiwari, D., Gangadharan, G.R.: SecCloudSharing: secure data sharing in public cloud using ciphertext-policy attribute-based Proxy Re-encryption with revocation. Int. J. Commun. Syst. **31**(5), e3494 (2018)
12. Niu, S., Liu, W., Chen, L., Du, X.: An electronic medical record data sharing scheme based on Proxy Re-encryption. Comput Eng. 1–10 (2020)
13. Zhang, X., Sun, L.: Attribute proxy re-encryption for ciphertext storage sharing scheme on blockchain. J. Syst. Simul. **32**(06), 1009–1020 (2020)
14. Luo, E., Wang, G., Chen, S., Pinial, K.: Privacy preserving friend discovery cross domain scheme using re-encryption in mobile social networks. J. Commun. **38**(10), 81–93 (2017)
15. Liu, M., Liu, S., Wang, Y., Wang, J., Li, Y.,, Cao, H.: Optimizing the decryption efficiency in LSSS matrix-based attribute-based encryption without given policy. Chinese J. Electron. **43**(6), 1065–1072 (2015)

De-anonymization Attack Method of Mobility Trajectory Data Based on Semantic Trajectory Pattern

Wenshuai Zhang, Weidong Yang(✉), Haojun Zhang, and Zhenqiang Xu

Henan University of Technology, Zhengzhou 450001, China
{yangweidong,zhj,xuzhenqiang}@haut.edu.cn

Abstract. Anonymizing trajectory data based on pseudonyms is a common privacy protection method in data publishing scenarios. The so-called de-anonymization attack associates the anonymized trajectory data with the real identity of mobile object to further obtain the private information. The trajectory of a mobile object contains detailed spatio-temporal and semantic information. For anonymously released trajectory data, we propose a de-anonymization attack method based on semantic trajectory patterns, which uses a semantic trajectory pattern acquisition algorithm to obtain the frequent semantic trajectory pattern set of each mobile object, which is used as trajectory features to construct its mobility profiles, further design the corresponding similarity measure. Experiments on real trajectory datasets show that the method proposed in this paper can obtain a relatively high de-anonymization success rate.

Keywords: Privacy protection · De-anonymization attack · Frequent pattern mining

1 Introduction

Nowadays, with the application and development of mobile sensing, various positioning technologies and location services, the collection of spatio-temporal data has become largely convenient, and location service providers can easily collect the location trajectory data of billions of mobile objects. These data can be used to understand human behavior and to develop various services in various fields, such as vehicle congestion monitoring [1], path planning [2], friend recommendation [3], and travel mode analysis [4].

However, the mobility trajectory data set also contains a large amount of private information of mobile objects, such as residence, work location, physical condition, behavioral habits and other sensitive information. If these data are released or shared without protecting, it will seriously threaten the privacy of mobile objects. The literature [5] points out that continuous trajectory data exposure will be easier for attackers to obtain their behavior habits or interests. In order to protect the privacy of mobile objects' trajectory data, appropriate privacy-preserving mechanisms are usually used to

J. Xiong et al. (Eds.): MobiMedia 2021, LNICST 394, pp. 354–366, 2021.
https://doi.org/10.1007/978-3-030-89814-4_26

anonymize the original trajectory dataset before publishing it. These common privacy-preserving mechanisms can be divided into three main categories: 1) Modify the original trajectory (for example, by generalizing the location to the area to reduce the probability of an attacker identifying a mobile object) to protect privacy. 2) Add noise (for example, in the privacy protection mechanism based on differential privacy [6], by adding appropriate noise to the trajectory, the sensitive data is distorted while ensuring that the processed data can still maintain certain statistical properties). Reduce the trajectory accuracy, so as to achieve the purpose of privacy protection. 3) Use pseudonyms [7] to replace the real identity of the mobile object, and the real identity cannot be associated with the pseudonym in any way.

Among the above privacy protection mechanisms, the first two types of privacy protection mechanisms can have a better privacy protection effect on the trajectory data of mobile objects, but because the accuracy of the trajectory in time and space is reduced, the availability and completeness of the trajectory data set is greatly affected. The pseudonym technology has the advantages of not changing the original trajectory, easy to implement and maximum data availability, and is still one of the commonly used privacy protection methods.

In fact, even if the mobility trajectory is anonymized using pseudonym techniques, the attacker is still able to link the anonymous trajectory to the corresponding real identity with high probability. At present, the existing methods of de-anonymization attacks on mobility trajectory can be divided into two categories. One type is to extract movement features from location data and perform de-anonymization attacks through movement feature similarity. For example, Gambs et al. [8] proposed to construct mobile object behavior profiles for de-anonymization attacks by extracting the frequently occurring location points in the trajectory. This approach only considers single-dimensional location data, which makes the constructed mobility profiles not more accurately reflect the user behavior patterns embedded in the trajectories and affects the success rate of de-anonymization attacks. Another type of de-anonymization attack is to exploit the implied social relationships. This requires the assumption that the attacker can know the more complete real social relations, which is often difficult for the attacker to obtain the more complete real social relations in practical applications.

Researchers point out [9] that each person's mobile trajectory has its own inherent behavior pattern and does not change dramatically in the short term. At the same time, trajectories contain rich semantic information (e.g., semantic knowledge of their geographical location, behavioral preferences at a certain place, etc.) that can better reflect and represent the behavioral characteristics of different mobile objects.

Therefore, in order to characterize the mobility profile of different mobile objects more accurately and improve the success rate of de-anonymization. From the perspective of privacy attackers, this paper proposes a mobility trajectory de-anonymization attack method (TP-attack) based on semantic trajectory patterns, which combines the transfer time of mobile objects and the semantic trajectory characteristics, obtain the semantic trajectory pattern and design the corresponding similarity measurement to achieve anonymous trajectory de-anonymization attacks.

The main contributions of this paper are highlighted as follows:

- We propose a novel method to de-anonymization mobility trajectory. Considering the semantic information in the trajectory, we obtain the set of frequent semantic trajectory patterns in the trajectory as "fingerprint" to distinguish different individuals and build a mobility profile of mobile objects.
- Design a new similarity measure to compare the mobility profiles of mobile objects, so as to identify the real identity of the attacker's anonymous trajectory from the anonymous trajectory dataset based on the attacker's existing real trajectory information, and realize the anonymous trajectory de-anonymization attack.
- We perform experiments based on two real datasets. Results show that the set of frequent semantic trajectory patterns describes user behavior more accurately than others, Meanwhile, the experimental results show that the proposed method can obtain a high success rate of de-anonymization.

The rest of this paper is organized as follows. Section 2 introduces related work, Sect. 3 gives basic definitions and problem descriptions, Sect. 4 describes how to obtain semantic trajectory patterns, Sect. 5 introduces the TP-attack method, and Sect. 6 conducts related experiments and analyzes the experimental results, and Sect. 7 makes a conclusion with this paper.

2 Related Works

Protecting the anonymity of personal mobility is notoriously difficult due to sparsity [10] and hence mobility data are often vulnerable to deanonymization attacks. Many studies on trajectory privacy show that even if a person's data is anonymized, they still have unique patterns that may be exploited by malicious attackers who background knowledge. At present, many researchers have conducted in-depth research on the de-anonymization attack methods for mobility trajectories. This section divides them into two categories: One type is to extract movement features from location data and perform de-anonymization attacks through movement feature similarity that is, by extracting characteristic positions from the mobility trajectory (such as places frequently visited by mobile objects, stopping points, etc.) to express the behavior characteristics of mobile objects. Another type of de-anonymization attack is to exploit the implied social relationships.

Mulder et al. [11] proposed a de-anonymization attack method, by establishing a Markov model for each individual in the training set, and then using the model to maximize the likelihood to perform a de-anonymization attack on the individuals in the test set. However, this method only considers location information and does not consider the influence of time at all, and it cannot reflect the differences of different individuals.

Zhong et al. [12] proposed a novel attack method. First, perform characteristics analysis on anonymous mobility trajectories, use an optimized word frequency-inverse document frequency method to construct characteristic vectors, use partial trajectory fragments to analyze the anonymous data set of mobility trajectories, and combine characteristic similarity to analyze and match. Finally, from the anonymous trajectory data, the mobility trajectory with the highest similarity to the collected trajectory is analyzed to achieve the purpose of de-anonymization. The above methods based on trajectory characteristic positions have high computing efficiency, but do not consider the relevance in the time dimension, resulting in the accuracy of de-anonymization that needs to be improved.

Chris Y. T. Ma et al. [13] proposed a side information (i.e. location snapshots) based de-anonymization attack, where an attacker, obtaining a number of location snapshots of its victims, can recognize the trace of the victims from a set of anonymous traces. The authors use Bayesian inference to break the unlinkability between location snapshots and anonymized trace. H. Wang et al. used two mobile social network datasets as side information to evaluate the performance of de-anonymization attacks using external information. A Gaussian and Markov based algorithm is adapted to deal with spatiotemporal mismatches in different datasets [14]. H. Li et al. [15] measured the similarity between the disclosed locations in the MSN applications and the real mobility pattern, in terms of coverage rate and relative entropy, and presented an attack to infer MSN users' demographics from the disclosed locations by checking their similar point of interests.

3 Definitions and Attack Model

3.1 Definitions

The source of the mobility trajectory dataset is mainly the dataset generated by emerging Internet applications such as the Internet of Vehicles and mobile social networks. It has the following description:

Definition 1 (Stay Region). A stay region refers to a geographic location area where a mobile object stays within an area for a given time range. The semantic label of each stay area is the semantic information corresponding to the location point with the highest importance score of the stay area, symbolized as SR_i. In addition, the transfer time between two consecutive stay regions is $\alpha_i = t_{i+1\_arv} - t_{i\_lev}$. Where $t_{i_{arv}}$ represents the time when the first point p_1 reaches the stay region, and $t_{i_{lev}}$ is the time when the last point p_n leaves the stay region. In this way, a set of semantic stay points with transfer time for each mobile object can be obtained.

Definition 2 (Semantic Trajectory). A semantic trajectory is a sequence of n stay regions with time information and semantic annotations, denoted as $SR = (< SR_1, \alpha_1 >, \cdots, < SR_i, \alpha_i >)$, the length of the semantic trajectory is the number of semantic stay points.

Definition 3 (Semantic Trajectory Pattern). The trajectory pattern represents the regular movement of a mobile object. In this paper, it is usually denoted as sequences of semantic stay point with transfer time annotated, denoted as $SR_1 \overset{\alpha_1}{\rightarrow} \cdots \overset{\alpha_i}{\rightarrow} SR_i$.

Definition 4 (Frequent Semantic Trajectory Pattern Set). Given a set of semantic stay point sequences G, the minimum support θ, the frequent semantic trajectory pattern set of G is formalized as $FT_{\theta}^{G} = \{F_i | support^{G}(D_i) \geq \theta\}$ where the minimum support represents the percentage of trajectories containing D_i in the set G.

3.2 Attack Model

In order to protect the privacy of the trajectory of mobile objects, the mobility trajectory dataset needs to be anonymized before it is released. For any trajectory T_i, a pseudonym $w_i \in$ W is used to replace the real identity $v_i \in$ V of the creator of each trajectory, and each moving object corresponds to a unique pseudonym, so that the attacker cannot associate his real identity with his real position, thus protecting the mobile Object trajectory privacy.

The attacker has access to a set of anonymous trajectory dataset H, which includes the anonymous mobile trajectory of one or more attack targets, and the attacker can obtain several mobility trajectory fragments of the attack target at any time period in the future through observation or other methods. We call these trajectory fragments as background knowledge dataset K, in which user identities are known. The attacker's goal is to find out the true identity of the target's trajectory from the anonymous trajectory dataset by analyzing the trajectory characteristics of different mobile objects.

4 Semantic Trajectory Mode Acquisition

Frequent semantic trajectory patterns appearing in the trajectory of mobile objects reflect personal habits and behavior patterns, and can be used as quasi-identifiers to distinguish different individuals [16]. By extracting the stay region in the mobility trajectory and making it semantic, the semantic trajectory pattern that characterizes the life and behavior of the mobile object is further obtained, which is used as the trajectory feature to construct its mobility profile.

4.1 Extraction of Stay Region

When a mobile object generates activity in a certain geographic location, the speed will be lower than the average speed of the entire trajectory. The speed-based stay region recognition algorithm in [17] is used to cluster the stay points belonging to the same stay region to obtain the stay region sets.

4.2 Semantic Stay Region

Considering that mobile objects will have different behavioral information in the stay region for the same longitude and latitude, the semantics of the stay region is important to reveal the behavioral patterns. The literature [18] uses a predefined category of locations to represent the semantics of a stay region. However, each mobile object has different interest preferences in the stay region, and it is not very accurate to represent the semantics of mobile object activities in the stay region by this predefined approach. In fact, the semantics of an individual's important location can be derived from a person's long-term trajectory data. In this section, the semantic of stay region can be expressed by selecting the position with the highest importance score in the stay region.

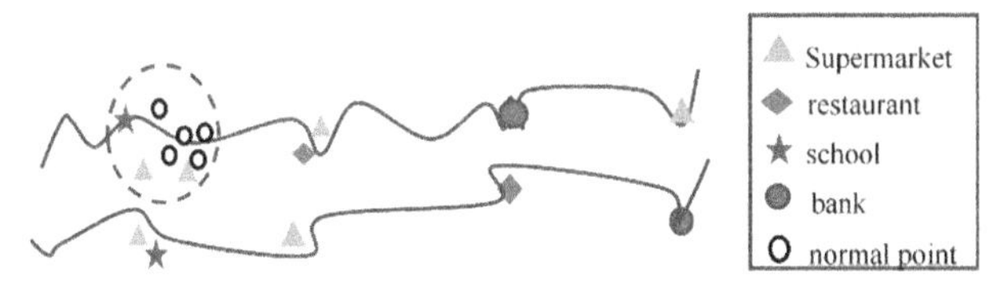

Fig. 1. Examples of stay region

Suppose that the position of the dash coil in Fig. 1 is a stay region of mobile object, which contains spatio-temporal points and POI. After identifying the stay region, calculate the importance score of each location in the area, select the highest score value as the stay point, and then call the Google API interface to semantically the stay point, so that the location data Correlate with semantic information to obtain a set of semantic stay region sequences. The importance score is defined as:

$$\text{score}(p_i) = \frac{sigp_i}{Dp_i} \tag{1}$$

where $sig(p_i) = \frac{|p_i|}{|sr|} \cdot \frac{|T_{pi}|}{|T|}$ as the importance of the location. $D(p_i)$ represents the distance between the position p_i and the nearest POI. $|p_i|$ indicates the number of times the mobile object visited the location, $|sr|$ indicates the number of positions in the stay region. $|T_{p_i}|$ indicates the number of traces containing the position. $|T|$ represents the total number of traces of mobile object u.

It is known that the POI (supermarket) location points in the stay region have a higher importance score, so the supermarket is used as the stay point in the stay region. In this way, a set of semantic stay points with transfer time for each mobile object can be obtained.

4.3 Semantic Trajectory Pattern Acquisition Algorithm

After obtaining the set of semantic stay region sequences, PrefixSpan [19] algorithm is improved, combined with the transfer time between stay regions, to obtain the semantic trajectory pattern with a support value less than θ, thereby obtaining the frequent trajectory pattern set of each mobile object. The specific process is shown in Algorithm 1:

Algorithm 1: Semantic trajectory pattern acquisition algorithm

Input: Semantic stay region sequence set R, minimum support θ, time thresholds τ
Output: frequent semantic trajectory pattern set Q

1: k ← 0
2: $P_0 \leftarrow \{T \times \{<>\}\}$
3: while $P_k \neq \emptyset$ do
4: for all P∈ P_k do
5: lastSR ← getlastStayRegion(P. prefix)
6: for all i ∈ P do
7: if support(P, i, τ) ≥ θ then
8: interval← ExtractIntervals(lastSR, i, P)
9: P_{k+1}. prefix ← link(prefix, i)
10: D_i ←GTP(P_{k+1}. prefix, $intervals$)
11: Output(D_i)
12: $P_{k+1} \leftarrow generateProjection(P, i)$
13: end if
14: end for
15: end for
16: k++
17: end while

Algorithm 1 gives the procedure of the semantic trajectory pattern acquisition algorithm. In order to obtain frequent sequences of stay regions with transfer times, the PrefixSpan algorithm is improved on by first taking the given set of semantic stay region sequences as the initial values of the projection database, and then finding the set of all frequent stay regions in the projection database as the initial set of prefixes. In each recursive process, the last set of stay regions of the current projection database prefix is first obtained, and then the set of stay regions with support satisfying the minimum threshold is found. The function in lines 8–11 functions to obtain the frequent time interval between two consecutive stay regions and combine the set of frequent stay regions with the current prefixes to generate a frequent semantic trajectory pattern with time interval. A projection database of this set of frequent stay regions is also obtained.

5 De-anonymization Attack Methods

5.1 An Approach Overview

From the de-anonymization attack model, it is clear that the attacker aims to identify the true identity corresponding to the attack target's trajectory from the anonymous trajectory dataset based on the trajectories in the background knowledge dataset. This section gives a strategy for an attacker to perform a de-anonymization attack. This is shown in Fig. 2.

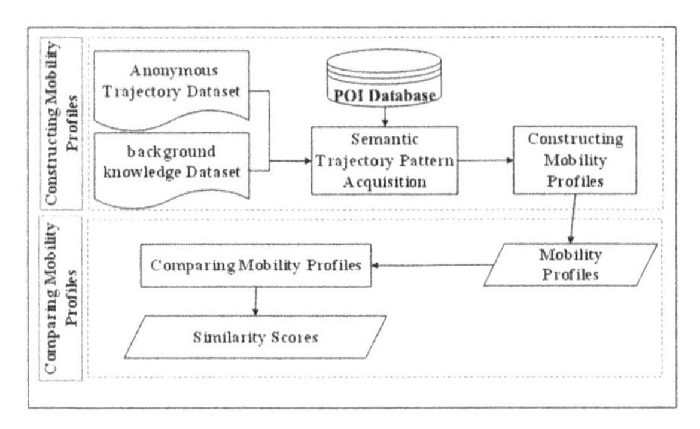

Fig. 2. The framework of our method.

The de-anonymization attack of mobile trajectory data can be divided into two phases:

- Constructing mobility profiles. The attacker constructs a mobility profile of the mobile object for several trajectories of the mobile object in the background knowledge dataset and the anonymous trajectory dataset, respectively.
- Comparing mobility profiles. A suitable similarity metric is designed to compare the similarity of the user mobility profiles constructed in the first phase so that the trajectories of the attack targets can be identified from the anonymous trajectory set.

5.2 Constructing Mobility Profiles

Mobility profiles describe the daily behavior patterns of mobile objects, which reflect the patterns of commonly visited locations. However, the longer the semantic mobility is, the greater the number of subsequences it generates. In order to avoid the error problem caused by repeated calculation, the maximum pattern set [20] is used as the mobility profiles of the mobile object. In this case, the maximal pattern set is composed of trajectory patterns that are not subsequences of any other pattern. The maximum trajectory pattern set is defined as:

Definition 5: (Maximum pattern set). Given mobile object u's frequent trajectory pattern set FT_{θ}^{G}, where Q and P are the trajectory patterns of mobile object u, then the maximum trajectory pattern set of user u is defined as:

$$\mathrm{M}\left(FT_{\theta}^{G}\right) = \left\{P \in FT_{\theta}^{G} \middle| \nexists Q \in FT_{\theta}^{G} (P \sqsubseteq Q)\right\} \tag{2}$$

Then for the anonymous dataset G of the anonymous mobile object u and the auxiliary data set $G^{'}$ of the real mobile object $u^{'}$ to construct the mobility profiles:

$$\mathrm{M}\left(FT_{\theta}^{G}\right) = (D_1, D_2, \cdots, D_i) \tag{3}$$

$$M\left(FT_{\theta}^{G^{'}}\right) = (V_1, V_2, \cdots, V_j) \tag{4}$$

Where D_i and V_j are the frequent patterns of u and the frequent patterns of $u^{'}$ obtained by frequent trajectory pattern acquisition algorithm, respectively.

5.3 Comparing Mobility Profiles

The more similar two mobile objects u and $u^{'}$ are, the more similar the corresponding mobility profile is. The essence of measuring the mobility profile of two mobile objects is to measure the similarity between the set of trajectory patterns. This section measures the distance between two maximal trajectory patterns on the basis of DTW [21]. Traditional distance calculation methods such as Euclidean distance are very sensitive to even small mismatches and require that the two time series are of equal length. In contrast, DTW is more suitable for our scenario and overcomes these drawbacks well.

Definition 6: (DTW): Give u_i and $u_i^{'}$, along with their corresponding maximal trajectory pattern $D_i = SR_1 \stackrel{\alpha_1}{\rightarrow} \cdots \stackrel{\alpha_n}{\rightarrow} SR_n$ and $V_j = SR_1 \stackrel{\alpha_1}{\rightarrow} \cdots \stackrel{\alpha_m}{\rightarrow} SR_m$, DTW aims to calculate the minimum distance between each semantic stay region SR_i and SR_j, which requires constructing an $n \times m$ Distance Matrix and finding a path $\varphi = \{\varphi(1), \cdots, \varphi(T)\}$, making the path through the elements minimizes the distance of SR_i and SR_j. We represent each element $\varphi(k) \in \varphi$ as the distance between stay region SR_i and SR_j. Where $\varphi(k) = (\varphi_{D_i}(k), \varphi_{V_j}(k))$, k = 1,2,···,T.

Obtaining the optimal Warping Curve Φ, the distance between SR_i and SR_j is minimized by:

$$DTW(D_i, V_j) = \min_{\varphi} \sum\nolimits_{k=1}^{T} \varphi(k) \tag{5}$$

Different mobile objects have different transfer times in the continuous stay region, reflecting different intentions. For any two maximum trajectory patterns D_i and V_j, we use the average of the overlapping time ratios between all consecutive staying regions in all the longest common subsequences as the time weight DOF (D_i, V_j) to distinguish different mobile objects.

$$\mathrm{DOF}(D_i, V_j) = \frac{\sum_{S \in lcs(D_i, V_j)} \sum_{i=1}^{len(S)-1} ot_S^{u,u^{'}}(i)}{|lcs(D_i, V_j)| \cdot (LCS(D_i, V_j) - 1)} \tag{6}$$

where $ot_S^{u,u^{'}}(i) = \frac{\sum_{1 \le i \le m} \alpha_{i\_max}^{'} - \alpha_{i\_min}^{'}}{\sum_{1 \le i \le k} \alpha_{i\_max} - \alpha_{i\_min}}$ represent the overlap time ratio between any two consecutive SR_{i-1} and SR_i in the longest common subsequence between u and $u^{'}$, $\sum_{1 \le i \le m} \alpha_{i\_max}^{'} - \alpha_{i\_min}^{'}$ represents the overlapping transition time between any two consecutive SR_{i-1} and SR_i, $\sum_{1 \le i \le k} \alpha_{i\_max} - \alpha_{i\_min}$ represents the all the occurring transition time between any two consecutive SR_{i-1} and SR_i. Therefore, the similarity between the maximum trajectory patterns D_i and V_j is calculated as:

$$\mathrm{SIM}(D_i, V_j) = DTW(D_i, V_j) \cdot DOF(D_i, V_j) \tag{7}$$

Given two mobile objects u and $u^{'}$, we only consider the trajectory pattern with the highest similarity value of the maximum trajectory pattern, denoted as $SIM_{MAX}(D_i, V_j)$

to measure between the maximum pattern sets, and obtain the mobility profiles similarity score, which is calculated as follows:

$$sim\left(u|u^{'}\right) = \frac{\sum_{D_i \in M(FT_\theta^G)} \sum_{V_j \in M(FT_\theta^{G'})} SIM_{MAX}(D_i, V_j) \cdot \mu(D_i, V_j)}{\sum_{D_i \in M(FT_\theta^G)} \sum_{V_j \in M(FT_\theta^{G'})} \mu(D_i, V_j)} \tag{8}$$

where $\mu\left(D_i, V_j\right) = \frac{support_u(D_i) + support_{u'}(D_i)}{2}$ denoted as a weight function constructed using the support value of the trajectory pattern [20].

6 Evaluation

6.1 Experimental Setup

The trajectory datasets used in the experiment are the Geolife dataset [22] and the CabSpotting dataset [23]. The specific information of the data set is shown in Table 1.

Table 1. Description of datasets.

Datasets	Mobile objects	Localization
Geolife	42	Beijing
CabSpotting	536	San Francisco

The experiment selected the most frequent 30-day trajectory data of mobile objects from these two datasets, and regarded the mobility of the entire time period as a mobility trajectory. In the experiment described in the article, we use the trajectory data of the mobile objects in the first 15 days of these two datasets as the anonymous trajectory data set H, that is, randomly renumber the trajectory of each mobile object in the previous 15 days to anonymize the trajectory data. The trajectory data of mobile objects in the next 15 days is the background knowledge data set K (trajectory collection of known user identities) for the de-anonymization attack stage.

Specifically, by de-anonymizing the anonymous trajectory dataset, Measuring the similarity of mobility profiles between the trajectory in the anonymous trajectory dataset H and the trajectory in the dataset K. The anonymous trajectory with the highest similarity to the trajectory in K is regarded as the same mobile object trajectory, and then compare the trajectory Whether the number of the mobile object of is consistent with the number of the known trajectory, if it is the same, the result of de-anonymization of the trajectory is accurate. Finally, the effect of the de-anonymization attack is measured by the following formula (9):

$$\mathrm{H}_{accuracy} = \frac{n}{N} \tag{9}$$

where n represents the number of trajectories that match correctly, and N represents the total number of trajectories in the anonymous dataset.

6.2 Competitors

In order to verify the effectiveness of the TP-attack method, we compare it with the following two mobility trajectory de-anonymization methods for de-anonymization success rates: 1) POI-attack [24]: The attacker extracts the trajectory from the anonymous trajectory feature POI, using POI collection to build mobility profile. Compare with the background knowledge you have to identify the true identity corresponding to the anonymous trajectory. 2) MMC-attack [8]: The attacker extracts the trajectory from the anonymous trajectory feature POI, constructing Markov chain model and compare the similarity of two Markov chains.

6.3 Experiment Analysis

In this section, we first studied the influence of the amount of background knowledge the attacker has on the success rate of de-anonymization. The experiment uses the two datasets mentioned in Table 1, and randomly selects $d(1 \leq d \leq 15)$ days of trajectory data from K to form the attacker's known dataset. Then use the TP-attack, POI-attack and MMC-attack methods to de-anonymize them. The X-axis and Y-axis represent the background knowledge of the attacker and the accuracy of trajectory matching. Figures 3 and 4 respectively show the influence of the amount of background knowledge the attacker has on the success rate of de-anonymization on the two datasets of Geolife and CabSpotting.

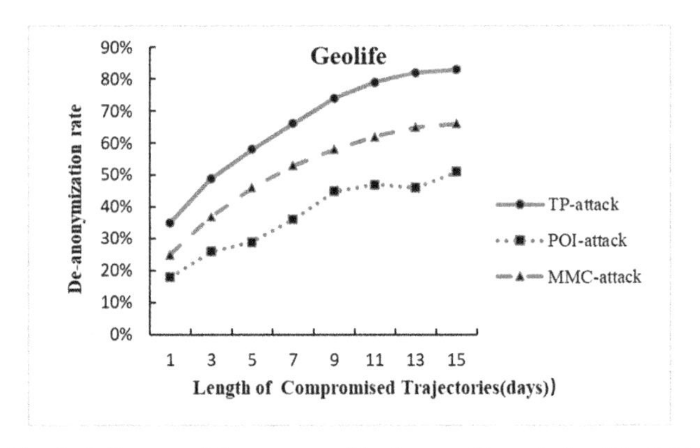

Fig. 3. Influence of the amount of background knowledge (Geolife)

By gradually increasing the attacker's background knowledge, the success rate of de-anonymity attacks is significantly increased; but with the increase of $d(d \geq 7)$, the rise gradually tends to be flat. In contrast, the success rate of TP-attack is better than the POI-attack and MMC-attack methods. The POI-attack method does not consider the time sequence; the MMC-attack method also does not consider the time factor, and the Markov chain is calculated multiple times, and the method of measuring the similarity of the Markov chain will have a certain impact on the success rate of de-anonymization. From Figs. 3 and 4, it can be seen that the more background knowledge an attacker has, the easier it is to carry out a de-anonymization attack.

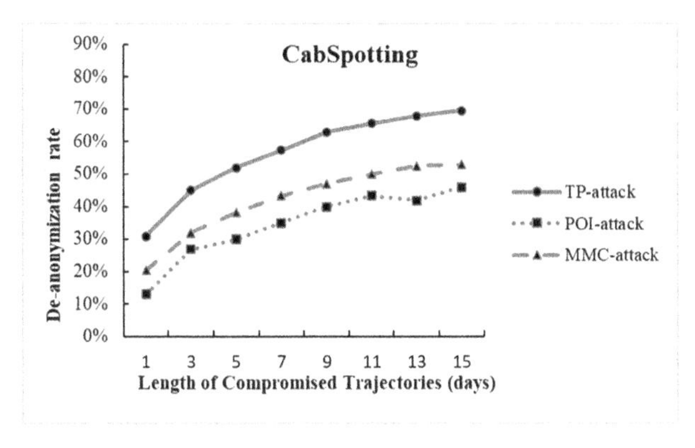

Fig. 4. Influence of the amount of background knowledge (CabSpotting)

7 Conclusion

In this paper, we propose a de-anonymization attack method based on the semantic trajectory pattern. By obtaining the semantic trajectory pattern of the mobile object, and using the transfer time of the mobile object between the stay region as the weight of the similarity measure, it can be more accurately characterize the trajectory patterns of different mobile objects, a unique mobility profiles of the mobile objects can be constructed for similarity measurement. The experimental results show that as the attacker has more background knowledge, the easier it is for the attacker to de-anonymization attack. At the same time, by comparing other mobility trajectory de-anonymization attack methods, the method based on the semantic trajectory pattern can obtain a higher success rate. Trajectory privacy protection technology is constantly improving, and in the next step, we will conduct research on de-anonymization attacks against other privacy protection technologies and explore the problems in trajectory privacy protection technology.

Acknowledgement. This work was supported by National Natural Science Foundation of China (61772173); Program for the Innovative Talents of the Higher Education Institutions of Henan Province (19HASTIT027); Open fund of Key Laboratory of Grain Information Processing and Control (under Grant No. KFJJ-2018105).

References

1. Kamble, S.J., Kounte, M.R.: Machine learning approach on traffic congestion monitoring system in internet of vehicles. Procedia Comput. Sci. **171**, 2235–2241 (2020)
2. Aggarwal, S., Kumar, N.: Path planning techniques for unmanned aerial vehicles: a review, solutions, and challenges. Comput. Commun. **149**, 270–299 (2020)
3. Uplavikar, N.M., Vaidya, J., Lin, D., Jiang, W.: Privacy-preserving friend recommendation in an integrated social environment. In: Kanhere, S., Patil, V.T., Sural, S., Gaur, M.S. (eds.) ICISS 2020. LNCS, vol. 12553, pp. 117–136. Springer, Cham (2020). https://doi.org/10.1007/978-3-030-65610-2_8

4. Zhao, X., Yan, X., Yu, A., et al.: Prediction and behavioral analysis of travel mode choice: a comparison of machine learning and logit models. Travel Behav. Soc. **20**, 22–35 (2020)
5. Gao, Q., Zhang, F.L., Wang, R.J., Zhou, F.: Trajectory big data: a review of key technologies in data processing. J. Softw. **28**(4), 959–992 (2017)
6. Andrés, M.E., Bordenabe, N., Chatzikokolakis, K., Palamidessi, C.: Geo-indistinguishability: differential privacy for location-based systems. In: Proceedings of the 2013 ACM SIGSAC Conference on Computer & Communications Security, pp. 901–914 (2013)
7. Pfitzmann, A., Köhntopp, M.: Anonymity, unobservability, and pseudonymity—a proposal for terminology. In: Designing Privacy Enhancing Technologies, pp. 1–9 (2001)
8. Gambs, S., Killijian, M.O., del Prado Cortez, M.N.: De-anonymization attack on geolocated data. J. Comput. Syst. Sci. **80**(8), 1597–1614 (2014)
9. Chang, S., Li, C., Zhu, H., Lu, T., Li, Q.: Revealing privacy vulnerabilities of anonymous trajectories. IEEE Trans. Veh. Technol. **67**(12), 12061–12071 (2018)
10. Aggarwal, C.C., Philip, S.Y.: A general survey of privacy-preserving data mining models and algorithms. In: Aggarwal, C.C., Yu, P.S. (eds.) Privacy-Preserving Data Mining, vol. 34, pp. 11–52. Springer, Boston (2008). https://doi.org/10.1007/978-0-387-70992-5_2
11. De Mulder, Y., Danezis, G., Batina, L., Preneel, B.: Identification via location-profiling in GSM networks. In: Proceedings of the 7th ACM Workshop on Privacy in the Electronic Society, pp. 23–32 (2008)
12. Zhong, J., Chang, S., Liu, X., Song, H.: De-anonymization attack method for mobile trace data. Computer Engineering (2016)
13. Ma, C.Y., Yau, D.K., Yip, N.K., Rao, N.S.: Privacy vulnerability of published anonymous mobility traces. In: Proceedings of the Sixteenth Annual International Conference on Mobile Computing and Networking, pp. 185–196 (2010)
14. Wang, H., Gao, C., Li, Y., Wang, G., Jin, D., Sun, J.: De-anonymization of mobility trajectories: dissecting the gaps between theory and practice. In: The 25th Annual Network & Distributed System Security Symposium (NDSS 2018) (2018)
15. Li, H., Zhu, H., Du, S., Liang, X., Shen, X.: Privacy leakage of location sharing in mobile social networks: Attacks and defense. IEEE Trans. Dependable Secure Comput. **15**(4), 646–660 (2016)
16. Sun, L., Yu, K.: Research on big data analysis model of library user behavior based on Internet of Things. Comput. Eng. Softw **40**(6), 113–118 (2019)
17. Lin, Z., Zeng, Q., Duan, H., Liu, C., Lu, F.: A semantic user distance metric using GPS trajectory data. IEEE Access **7**, 30185–30196 (2019)
18. Mazumdar, P., Patra, B.K., Lock, R., Korra, S.B.: An approach to compute user similarity for GPS applications. Knowl.-Based Syst. **113**, 125–142 (2016)
19. Cai, G., Lee, K., Lee, I.: Mining semantic trajectory patterns from geo-tagged data. J. Comput. Sci. Technol. **33**(4), 849–862 (2018)
20. Chen, X., Pang, J., Xue, R.: Constructing and comparing user mobility profiles for location-based services. In: Proceedings of the 28th Annual ACM Symposium on Applied Computing, pp. 261–266 (2013)
21. Kate, R.J.: Using dynamic time warping distances as features for improved time series classification. Data Min. Knowl. Disc. **30**(2), 283–312 (2015). https://doi.org/10.1007/s10618-015-0418-x
22. Zheng, Y., Xie, X., Ma, W.Y.: Geolife: a collaborative social networking service among user, location and trajectory. IEEE Data Eng. Bull. **33**(2), 32–39 (2010)
23. Piorkowski, M., Sarafijanovic-Djukic, N., Grossglauser, M.: CRAWDAD data set epfl/mobility. 24 February 2009 Downloaded from (2009)
24. Primault, V., Mokhtar, S. B., Lauradoux, C., Brunie, L.: Differentially private location privacy in practice. In: Third Workshop on Mobile Security Technologies (MoST) (2014)

Friendship Protection: A Trust-Based Shamir Secret Sharing Anti-collusion Attack Strategy for Friend Search Engines

Junfeng Tian and Yue Li(✉)

School of Cyberspace Security and Computer, Hebei University, Baoding 071000, China

Abstract. Online social networks (OSNs) provide users with applications to interact with friends or strangers. Among these applications, the friend search engine allows users to query other users' personal friend lists. However, if there is no suitable protection strategy, the application is likely to compromise the user's privacy. Some researchers have proposed privacy protection schemes to protect users from attacks that are initiated by independent attackers, but few researchers have conducted research on collusion attacks initiated by multiple malicious requestors. In this paper, we propose a resistance strategy against collusion attacks that are initiated by multiple malicious requestors in OSNs, introduce trust metrics, and limit users' ability to query through the Shamir secret sharing system (t, n) threshold function in the friend search engine to protect the user's friendships from collusion attacks by multiple attackers. The effectiveness of the proposed anti-collusion attack strategy is verified via synthetic and realistic social network datasets. Research on collusion attack strategies will help us design a safer friend search engine for OSNs.

Keywords: Friend search · Collusion attack · Threshold function

1 Introduction

Online social networks (OSNs) have increasingly become an indispensable social activity in people's lives. OSNs provide users with various applications to interact with family, friends and even strangers. Social networks provide users with a variety of services, such as interactive dating and online shopping. Users can interact, chat, and trade via social networks. As the number of social network users continues to increase and social networks continue to improve sociality, there are increasingly more attacks on users' privacy. Similar to mobile social networks, which are a social network category, the widespread nature of OSNs and their lack of effective privacy-preserving architecture make them a target for several attacks by adversaries [1]. Therefore, protecting the security of users' private information is important.

Social networks tend to display as many friend lists of users as possible, and a new social network application, the friend search engine, serves existing users in the social network and can be employed to attract potential users to join the network. The search

J. Xiong et al. (Eds.): MobiMedia 2021, LNICST 394, pp. 367–385, 2021.
https://doi.org/10.1007/978-3-030-89814-4_27

engine allows ordinary users to query a list of friends of individual users so that users can find friends of friends. People may join a social network because their friends have joined the social network, thereby attracting more users, which improves the sociality of social networks and expands the use of social networks. However, with the increasing number of social networks and the increasing number of users in social networks, some researchers have observed that friend search engines may expose more friendships than users are willing to show. The friends that users are not willing to show comprise private data of users, referred to as friendship privacy. The privacy-aware friend display scheme [2] can not only successfully protect the friendship privacy of users but also improve the sociality of OSNs. However, researchers studying collusion attacks in OSNs considered a collusion attack that was launched by multiple malicious requestors in coordination with each other and a collusion attack that can destroy the privacy of users' friendships and cause users to expose more friendships than they are willing to share [3]. And the data sharing framework can resolve potential data leakage [4]. We focus on the design of anti-collusion attack strategies that are aimed at the privacy of users' friendships in OSNs. The major contributions of this paper are presented as follows:

First, we analyse the collusion attack method in the friend search engine [3], in which multiple malicious requesters coordinate with each other to initiate queries to different but related users by designing a query sequence, thereby destroying the privacy of the friendships of the target user.

Second, we propose methods that can resist collusion attacks on friend search engines in OSNs. For resistance, we design a strategy for these collusion attacks, introduce trust metrics [5] to restrict access to requestors in friend search engines, and use the Shamir Secret Sharing (SSS) system (t, n) threshold function [6] to limit queries. In this paper, a trust-based (t, n) threshold anti-collusion attack strategy is proposed to prevent collusion attacks launched by multiple malicious requestors who coordinate the query sequence and query targets.

Third, to evaluate the effectiveness of the proposed anti-collusion attack strategy, we implement the strategy in a synthetic dataset and three large-scale real-world datasets. The experimental results show that the proposed anti-collusion attack strategy works effectively on large-scale datasets. By comparing the probability of a successful attack by a malicious inquirer using the proposed anti-collusion attack strategy with the probability of a successful attack by a malicious inquirer without the anti-collusion attack strategy, we determine that under the same attack conditions, the anti-collusion attack strategy in this work can reduce the probability of a successful attack, thereby protecting users' friendship privacy.

2 Related Works

2.1 Attacks Against Friendship Privacy

The number of users in OSNs continues to grow. Tens of thousands of users search for new friends and establish new contacts every day. Therefore, the privacy problem in friend search engines has attracted the attention of many researchers. Attacks against the privacy of friendships in OSNs can be divided into two categories, that is, attacks initiated by independent attackers and by colluding attackers.

Regarding independent attacks, research on modeling malicious attacks in OSNs shows that malicious individuals use the actual trust relationship between users and their family and friends to spread malware via OSNs [7]. By changing the display of malicious posts and personal information and hiding him/herself to avoid detection, an attacker in a chameleon attack, which is a new type of deception based on OSNs, is able to destroy users' privacy [8]. Studies have also shown that when the topology of OSNs does not contain cycles, malicious entities will violate users' privacy via active attacks if the network structure is not carefully designed [9]. Due to the rapid development of convolutional neural networks in recent years, applying them to social networks can result in very effective reasoning attacks and make high-precision predictions about private data [10]. In addition, by using the friend search application programming interface (API) to randomly grab data in OSNs, independent attackers can collect users' friendships. For some location-based social networks (LBSNs), despite the privacy protection strategies of location confusion and relative location, attackers can still perform a series of attacks by running LBSN apps and easily inferring the user's location [11].

Collusion attacks involve multiple malicious entities with the aim of launching a malicious attack through the coordination of multiple malicious entities to obtain more private information than is obtained in independent attacks. When users publish personal information in OSNs, attackers can launch inference attacks based on non-sensitive attributes and social relationships. An attacker utilizes users' profiles and social relationships in a collective manner to predict the sensitive information of related victims in a social network dataset that has been released [12]. Multiple malicious entities can be fake accounts that are created by a single attacker or different real attackers [13–15]. The router and users can maliciously collude to perform a collusion name guessing attack to compromise people's privacy [16]. Compared with independent attacks, collusion attacks are more complex and often exploit system vulnerabilities that independent attacks cannot detect. There is a complex collusion attack strategy, in which multiple malicious users coordinate their queries, share the query results, and dynamically adjust their queries based on the system's feedback to other malicious requestors [4].

2.2 Preservation of Friendship Privacy

The personalized privacy measurement algorithm can calculate the user's privacy level, thereby protecting user privacy data [17]. To protect the privacy of users' friendships in OSNs, some researchers have proposed a trust-based privacy protection friend recommendation scheme [18]. By comprehensively considering user interests and topological characteristics, a constraint traversal method is used to identify a strong trust path from trustees to trustees [19]. Research can alleviate cascading failures in trust relationships and be beneficial for the honest and free expression of opinions and experiences without users' privacy being compromised or trust relationships being affected. This research is built upon the dynamic modeling of cascading failures due to the occurrence of a trust crisis within a unidirectional or bidirectional social network [20]. The threat list and privacy protection mechanism illustrate the security requirements that OSNs should satisfy [21]. In response to privacy leakage caused by location information, researchers have proposed an environment-based system-level privacy protection solution that aims to automatically learn users' privacy preferences in different environments and provide users with transparent privacy control [22]. As noted by [23], some researchers have conducted comprehensive surveys on credibility in OSNs, such as the credibility of information for users and the evaluation of the trust level. In addition, a series of studies have proposed an unsupervised trust inference algorithm that is based on collaborative filtering in weighted social networks and a fast and robust trust inference algorithm [24, 25] to strengthen the security of social networks via trust inference.

However, researchers rarely consider privacy leakage problems caused by the friend search service provided by OSNs. Research on these problems can address the privacy needs of users' friends while ensuring the sociality of OSNs. The solution adopted by most OSNs is to allow each individual user to choose to completely display or completely hide their entire friend list. Moreover, OSNs often default their users to expose their entire friend list, of which most users are unaware [26]. It is conceivable that this setting aims to increase the sociality of the OSN. If users set their friend lists to completely hidden to protect the privacy of their friendships, this setting will substantially affect the sociality of OSNs. There are also some OSNs that set the users' friend list display to "show only a fixed number." For example, on Facebook, the number of friends displayed is set to 8, which limits the flexibility of users in changing their personal settings. However, some researchers have discovered that randomly displaying eight friends is sufficient for third parties to obtain data to estimate friend lists [27]. Moreover, regarding the different privacy settings of users, consider the following example: if A and B are friends, even if user A hides his or her friend list and the requestor cannot query the friend list of A, if user B is set to display his or her friend list, when the requestor queries the friend list of B, the friendships of B and A will be displayed and destroy A's privacy. This problem is referred to as the "mutual effect" [2].

To better protect the privacy of users' friendships in OSNs, a privacy protection strategy in the friend search engine [2] was shown to successfully resist attacks initiated by independent attackers. However, the strategy was unable to defend against collusion attacks initiated by multiple malicious attackers. Subsequently, an advanced collusion attack strategy coordinated by multiple malicious requestors [3] showed that multiple malicious requestors with limited knowledge of OSNs can successfully destroy users' privacy settings in the friend search engine. Another study [28] implemented web applications to detect malicious behavior, such as collusion attacks in the friend search engine. However, few researchers have investigated how to resist collusion attacks initiated by malicious attackers in friend search engines.

In this paper, we propose an anti-collusion attack strategy to fill these research gaps. This strategy distinguishes trusted users from untrusted users based on the credibility among users in OSNs and uses the (t, n) threshold function to limit the querying of requestors in the friend search engine to resist malicious attacks initiated by colluding attackers in OSNs.

3 Collusion Attack Strategy

3.1 Definition

In friend search engines, to strengthen the protection of the user's friendships, a certain number of friendships, such as k, will be displayed when responding to a query request.

These k friends are defined as the most influential friends of the users in the OSN. Assume that node N_a exists in the OSN with direct friends $N_{a.i}$ and that the set is $F_a^k(i < k)$. Requestor Q_1 wants to query N_a's friendships; two nodes, N_1 and N_2, exist, and $k = 1$. N_1 and N_2 are each user's most important friends.

Occupation. If requestor Q_1 queries node N_1, based on the friend search engine display strategy, the query result is $E(N_1, N_2)$. At this time, N_1 has shown his or her most important friend N_2, and N_1 is occupied.

Passive Display. Requestor Q_1 queries the important friend list of N_1. Based on the friend search engine strategy, the query result is $E(N_1, N_2)$. The most important friend who exposes N_2 is N_1, and N_2 is referred to as a passive display.

3.2 Attack Model

Maximum Number of Friends Displayed. Due to the different personal preferences of users in OSNs, their privacy settings will also be different. The maximum number of friends displayed, k, may also be different. This strategy assumes that all nodes have the same k value.

Attackers' Prior Knowledge. Typically, the success of a malicious requestor's attack is closely related to his or her knowledge of OSNs. The attack success rate of malicious requestors who know more about OSNs is expected to be higher. This paper assumes that malicious requestors have limited knowledge of OSNs and are limited only to target nodes.

Attack Target. The goal of the malicious query is to violate the privacy of the target user in the OSN (i.e., to query the $k + 1$th friend of the target node). When the privacy of the target user is set to show a number of friends less than k, the privacy of the target user cannot be violated. Each malicious requestor's attack target is unique, and each collusion attack has only one victim node. Although malicious requestors may infringe on the privacy of other users during the query process, only when the privacy of the target node is destroyed is the collusion attack considered successful.

Attack Strategy. Colluding attackers in OSNs can query users' friendships via the friend search engine and query the relationship between users and friends by coordinating the query sequence and query targets. We take a simple scenario—the small-scale complete graph that contains the four nodes shown in Fig. 1—as an example to analyze the perpetrated collusion attacks.

The OSN has four user nodes, N_1, N_2, N_3 and N_4, and two malicious requestors, MR_1 and MR_2. Additionally, $k = 1$. Set node N_3 as the target node, and perform the following query.

First, MR_1 queries N_3. The first important friend N_1 of node N_3 can be queried. Then, MR_2 successively queries N_1 and N_3 by returning N_1's first important friend as N_2 and occupying N_1. Finally, query N_3. Simultaneously, N_3 will expose the $k + 1$ friend, N_4, and the privacy of N_3 will be violated.

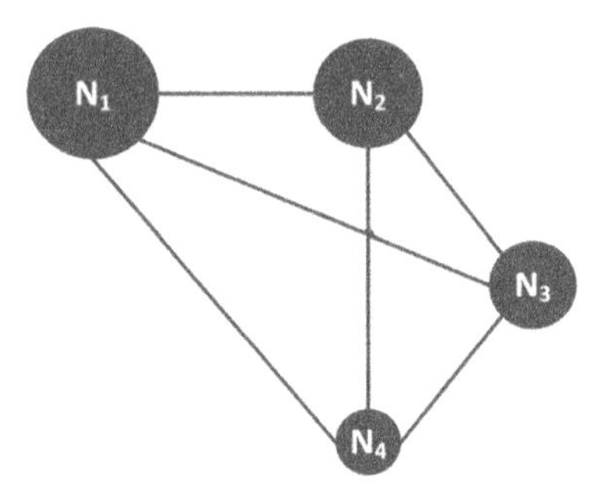

Fig. 1. Small-scale complete graph

4 Anti-collusion Attack Strategy

The adopted collusion attack strategy uses multiple malicious requesters to coordinate the query sequence and dynamically adjust the query target via the query results of other malicious requesters to query private friendships. However, if access control is given to the requestors in the friend search engine, collusion attacks can be resisted. In this section, we investigate the strategy to resist collusion attacks. In this work, the access control of requestors in the friend search engine is considered, credibility is employed as the restriction condition for requestor queries, and the SSS system (t, n) threshold function is utilized to control queries to the target user.

4.1 Credibility Calculations

Calculation of Direct Trust (DT_{ij})**.** For two user nodes that have historical interactions in the OSN, the credibility of the first user for the second user is referred to as direct trust. A user obtains the credibility evaluation of another user based on the historical performance of the user who has interacted with him or her. Therefore, the following factors are introduced when calculating direct trust.

Number of Interactions. The greater the number of interactions between two users is, the higher the trust between the users is.

Interaction Evaluation. After each interaction, the user gives a corresponding evaluation based on the process, results, and importance of the interaction event. The evaluation value of the *lth* interaction is recorded as $C_l \in [0, 1]$.

Interaction Time. Interaction evaluations that are similar to the current time better reflect the user's recent behavior. The closer the evaluation is to the current time, the greater is the impact on direct credibility.

Interaction Events. The weight of the event of the *lth* interaction between two users is denoted as W_l.

If node i and node j have interacted n times in the OSN, after the *lth* interaction is completed, node i evaluates node j to obtain evaluation value C_l and interaction event weight W_l. Subsequently, the lth interaction time t_l, importance of the lth interaction event W_l, evaluation value C_l of the interaction event of node i with node j, and influence of the number n of interactions between node i and node j on the evaluation value are considered. The calculation formula of direct trust is expressed as follows:

$$DT_{ij} = \alpha \cdot \frac{\sum_{l=1}^{n} \Phi(t_l) \cdot C_l \cdot W_l}{n} \tag{1}$$

where $\alpha = n/(n + 1)$ is a function of the number of interactions that is used to adjust the influence of the number of interactions on credibility. The user obtains a high degree of trust only when he or she obtains multiple satisfactory evaluation values. $\Phi\ (t_l) = exp\ (\lceil (t_n - t_l)/T \rceil)$ is the time decay coefficient, where t_n is the n th interaction time (i.e., current interaction time), t_l is the l th interaction time, and T is the time period. The evaluation of an interaction event that is more similar to the current interaction time has a greater impact on credibility. W_l and C_l are the weight of the interaction event between node i and node j and the evaluation value of node i for the event, respectively. This approach can prevent malicious requestors from interacting with the target user by using events with a low weight to gain the trust of the target user while deceiving the user during interaction events with high weights.

Recommended Trust (RT_{ij}). If node i wants to gain a comprehensive understanding of node j, node i needs to obtain the recommended trust for node j via intermediate node c, where node $c = \{c_1, c_2, c_3, \ldots, c_n\}$. The calculation of recommended trust is expressed as follows:

$$RT_{ij} = \sum\nolimits_{c=1}^{n} \left(DT_{ic} \cdot DT_{cj}\right) \tag{2}$$

where DT_{ic} is the direct trust of user i in user c, DT_{cj} is the direct trust of user c in user j, and the direct trust of user i in user c can be regarded as a recommendation for calculating the recommended trust weights.

Comprehensive Trust (OT_{ij}). The credibility of a user in the OSN must be integrated with his or her direct trust and the recommended trust of other users, which is referred to as comprehensive trust. The weights of direct trust and recommended trust are determined by experimental calculations. In real life, people are generally more inclined to believe their judgments, and the recommendations of others serve only as a reference. Thus, the calculation of comprehensive trust is expressed as follows:

$$OT_{ij} = u \cdot DT_{ij} + v \cdot RT_{ij} (u + v = 1, u > v) \tag{3}$$

where OT_{ij} is the direct trust of node i in node j, RT_{ij} is the recommended trust of node i in node j, and u and v are the weight coefficients of direct trust and recommended trust, respectively.

4.2 Shamir Secret Sharing System

The SSS system is a specific secret sharing scheme designed by Shamir based on language interpolation polynomial theory [29, 30]. This scheme clearly illustrates how to divide data D into n segments so that D can be easily reconstructed from t segments and so that even if all $t-1$ segments are mastered, D cannot be reconstructed.

In response to collusion attacks in OSNs, this paper uses the *SSS* (t, n) threshold function to control the querying of users' friendships. The (t, n) threshold secret sharing scheme consists of the following three stages.

System Parameter Setting. n is the number of all participants, t is the threshold, p is a large prime number, and $s \in Z_p$ is the secret to be shared.

Secret Distribution. The secret distributor D chooses a random t degree polynomial.

$$a(x) = s + a_1x^1 + a_2x^2 + a_3x^3 + \ldots \\ + a_{t-1}x^{t-1} \, mod \, p, \alpha_j \in_R Z_p \tag{4}$$

The condition $a(0) = s$ is satisfied. D sends $s_i = a(i)$ to participants P_i, $i = 1, 2..., n$.

Secret Reconstruction. Any number of participants can reconstruct the secret using their secret fragments. Let t participants who want to reconstruct the secret be P_i, $i = 1, 2..., t$, and let $A = |1, 2..., t|$.

λ_i is calculated based on the following formula:

$$\lambda_i = \prod\nolimits_{j \in A \backslash \{i\}} \frac{j}{j-i} \tag{5}$$

The original secret is restored based on the following formula:

$$s = \sum\nolimits_{i \in A} s_i \lambda_i \tag{6}$$

The security of the *SSS* depends on the assumption that the parties honestly perform the operations predetermined by the agreement. We consider reliable secret distributors and believe that the administrators of OSNs are honest in the strategy.

4.3 Friend Search Engine with the SSS System

In OSNs, users can access the friendships of other users by friend search engines. Multiple malicious requestors can share their query results with each other by coordinating the query target and query sequence, which causes the target user to expose more friends than the user is willing to display. A friend search engine that has introduced the trust metric and SSS can control the query of users. This control can guarantee that only users whose comprehensive trust reaches the trust threshold can successfully query the friendships of the target user.

Assume that secret distributor D is honest and that each anonymous requestor $P_i(i = 1, 2, ..., n)$ can obtain a correct secret fragment from D. The number of requestors is higher than the trust threshold for querying the friendships of the target user each time $n_A \geq 2$. The access control process of this solution is described as follows:

Obtain Comprehensive Trust. Requestors $P_i(i = 1, 2..., n)$ request querying the friendships of target user n_a, obtaining comprehensive trust T_{ai} of P_i, and sorting the results in descending order by value based on the interaction between target user n_a and requestor P_i in the OSN.

Classify the Query. Based on trust threshold *TR*, the requestors are divided into categories *A* and *B*. Category *A*: $T_{ai} \in [TR, 1]$ and category *B*: $T_{ai} \in [0, TR]$. The number of requestors in the two categories is denoted as n_A and n_B.

Confirm Threshold *t*. According to the definition of the (t, n) threshold function and the requirements of access control security, requestors who have not reached the system trust threshold cannot successfully query the target user's friendships. Since $T_{ai} < TR$, it is necessary to ensure that requestors in category *B* cannot successfully query the friendships of the target user. Thus, in each query process, $t = n_B + 1$.

Secret Distribution. The secret distributor *D* chooses a random *t* degree polynomial $a(x) = s + a_1x^1 + a_2x^2 + a_3x^3 + ... + a_{t-1}x^{t-1} mod\, p, \alpha_j \in_R Z_p, a(0) = S.D$ sends $s_i = a(i)$ to participants $P_i, i = 1, 2..., n$.

Secret Reconstruction. n_B requestors in category *B*, who are arranged in descending order of comprehensive trust, submit the secret fragments s_i obtained in reverse order, and n_A requestors and n_B requestors are divided into n_A groups for secret reconstruction.

Assume that requestor $P_i(i = 1, 2..., n)$, who queries the friendships of the target user, is arranged in descending order based on the comprehensive trust of the target usern_a. Category *A* is $P_1, P_2, P_3, ..., P_m$, and category *B* is $P_{m+1}, P_{m+2}, ..., P_n$. Threshold $t = n_B + 1$. Category *A* can be divided into *m* groups to reconstruct secrets (Table 1).

Table 1. Groups to reconstruct secrets

Group number	Group member
1	$P_1, P_n, P_{n-1}, \ldots, P_{m+1}$
2	$P_2, P_n, P_{n-1}, \ldots, P_{m+1}$
3	$P_3, P_n, P_{n-1}, \ldots, P_{m+1}$
...	......
a	$P_a, P_n, P_{n-1}, \ldots, P_{m+1}$

The comprehensive trust of the first requestor among the *m* groups of requestors who participate in the secret reconstruction is greater than the trust threshold set by the target user (i.e., only users trusted by the target user can successfully query the target's friendships). During each secret reconstruction process, the users $P_{m+1}, P_{m+2}, ..., P_n$, who have not reached the comprehensive trust level threshold, must submit their secret fragments $s_{m+1}, s_{m+2}, ..., s_n$ obtained from *D*. Users $P_1, P_2, P_3, ..., P_m$ will submit $s_{m+1}, s_{m+2}, ..., s_n$. The secret fragment $s_i(i \in [1, m])$ is secretly reconstructed. The threshold $t = n_B + 1$ can ensure that even if $P_{m+1}, P_{m+2}, ..., P_n$ constitute the group of submitted secret fragments, the secret cannot be successfully reconstructed.

5 Experiment

In this section, we experimentally verify the effectiveness of the proposed anti-collusion attack strategy. Our experimental research includes synthetic datasets to verify the validity of the credibility calculations and three large-scale real-world datasets to verify the security of the anti-collusion attack strategy.

5.1 Datasets

We generate random numbers that satisfy the previously described conditions of the credibility calculation method, including data on 1000 groups of user interactions, and verify the correctness of the trust calculations. In addition, we use three real-world social network datasets to verify the security of the anti-collusion attack strategy.

Synthetic Dataset. A random probability function is used to fit users' interactions in OSNs. The setting standards for the time interval of interactions between users and the weights of the interaction events are different for each OSN. We select the interaction data within the time interval ($\Phi(t_l) = 0.367879$) among users in the synthetic dataset. The number of interactions is set to 50; the weights of the interaction events take values in the range [1, 20]; and the interaction evaluation takes values in the range (0, 1] as an example to verify the rationality of the trust calculations, that is, $W_l \in [1, 20]$, $C_l \in (0, 1]$, and $n \in [1, 50]$. The trust between users may exceed 1 and should be normalized.

Facebook Dataset [31]. The data from Facebook.com capture the friendships among users, which can be modeled as undirected graphs.

Slashdot Dataset [32]. Slashdot is a technology-related news website and a specific user community, where users can submit and edit news about the current main technology. In 2002, Slashdot launched the Slashdot Zoo function, which enables users to mark each other as friends or enemies. The network establishes links between two friends or enemies among Slashdot users. Therefore, the data in this dataset are directional. This article uses 2009 Slashdot data, and the Slashdot dataset is converted to an undirected graph to reflect users' friendships. Regardless of the direction of the connection between two nodes in the network, an edge is created in the undirected graph for these two nodes.

Gowalla Dataset [33]. Gowalla is a location-based social networking site, where users share their location by signing in. The friendships collected from Gowalla are undirected. The complete dataset consists of 19,591 nodes and 950,327 edges. Due to data size limitations, this program selects only a portion of the data for testing.

We list the main attributes of each dataset in Table 2. The synthetic dataset is used to verify the rationality of the credibility calculations, and the remaining three datasets are used to verify the security of the proposed anti-collusion attack strategy.

Table 2. Social network dataset property

Dataset	Synthetic dataset	Facebook	Slashdot	Gowalla
Vertices	1000	63731	82168	196591
Edges	8997	817090	948464	582533
Average degree	—	25.773	12.273	9.668

5.2 Comparison Plan

As the collusion attack problem in friend search engines is more advanced than other attack problems, we have not obtained any relevant solutions to resist these collusion attacks. Therefore, we applied this strategy to the collusion attack and compared it with the original collusion attack. For the collusion attack strategy [3], we applied the proposed anti-collusion attack strategy. It is assumed that colluding attackers can successfully destroy the friendship privacy of users in every query if (t, n) threshold function access control is not adopted (i.e., probability that the colluding attackers successfully destroy the user's privacy is 1). The cases of using and not using (t, n) threshold function access control is compared to analyze the performance of the anti-collusion attack strategy proposed in this paper.

5.3 Performance Analysis

In this section, we analyze the rationality of the trust calculations and the security of the anti-collusion attack strategy using (t, n) threshold function access control.

Credibility Calculation Rationality. Based on random numbers, the values of direct trust and recommended trust are calculated by formula (1) and formula (2), and the value of the user's comprehensive trust is calculated by formula (3). We selected 1000 sets of data to prove the correctness of the trust calculations. The results are shown in Fig. 2.

Figure 2 shows that the results of the comprehensive trust calculations are normally distributed. In addition, they are consistent with realistic expectations.

Security Analysis. To improve the security and usability of the friend search engine, we assume that OSN administrators can be fully trusted in regard to the friend search engine. When the number of requestors is less than the number of query requests, the administrators can help the requestors complete the query. In this section, we conduct a security analysis based on four aspects: limit rate, number of users whose privacy has been violated, number of colluding attackers and success probability of collusion attacks.

Limit Rate (LR). The LR of the system is defined as the ratio of the number of users in category B to the number of all users, that is, the proportion of users who cannot successfully query in the friend search engine among all requestors. Based on formula (3) $OT_{ij} = u \cdot DT_{ij} + v \cdot RT_{ij} (u + v = 1, u > v)$, where $DT_{ij}, RT_{ij} \in [0, 1]$. The direct

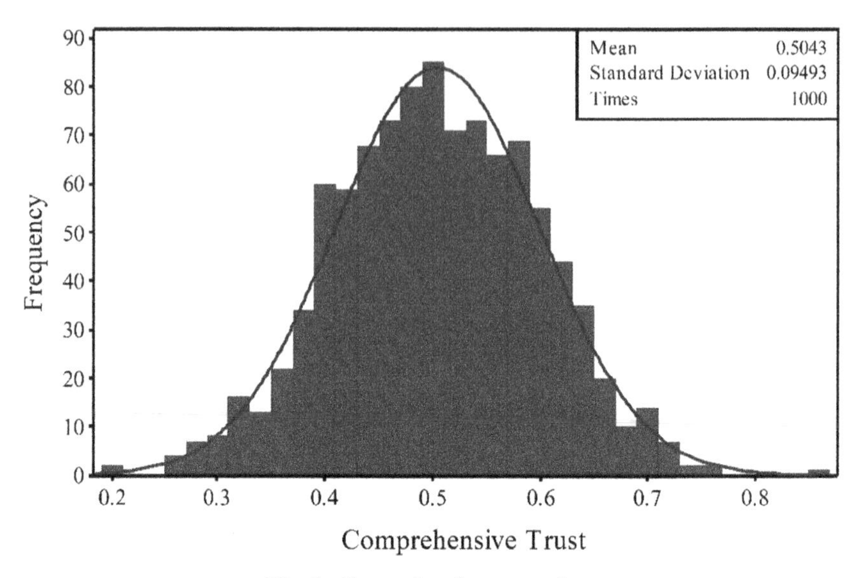

Fig. 2. Comprehensive trust values

trust weight coefficient u is set to 0.6, and the trust threshold is set to 0.5, 0.6, 0.7, 0.8, and 0.9. A total of 1000 experiments are conducted to verify the LR of the proposed strategy.

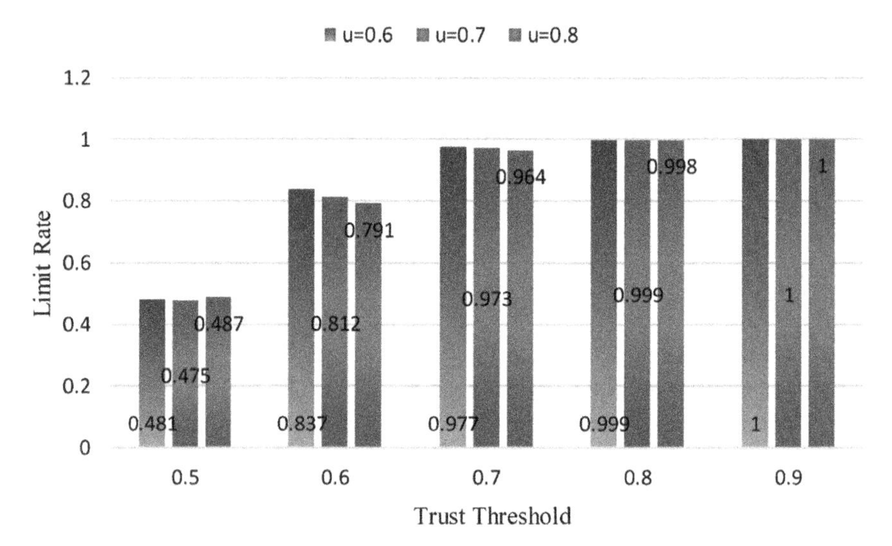

Fig. 3. Limit rate under the trust threshold with $u = 0.6$, $u = 0.7$ and $u = 0.8$

Figure 3 shows the LR and trust threshold results of this strategy. The value of the direct trust weight coefficient u are 0.6. When the trust threshold is 0.5, the LR of the strategy is approximately 40%. When the trust threshold is 0.6, the LR increases to 80%.

At 0.7, the LR increases to almost 100%; therefore, when the trust threshold is 0.7, almost no user reaches the trust threshold, and the friend search engine will not allow any querying. When the trust threshold is 0.6, 80% of users in the OSN cannot reach the threshold; thus, the number of requestors in the friend search engine is limited, and the safety of the friend search engine is increased.

Number of Users Whose Privacy Has Been Violated. Consider the trust threshold of 0.5 as an example. Sixty percent of users can make normal queries. In the worst case of the friend search engine query, the number of malicious requestors is not limited, and malicious requestors can destroy the privacy of the target user via a one-time collusion attack at the first layer. The probability of successfully destroying the user's privacy is $P(N_0|_{l=1}) = 1 - \prod_{i=1}^{k} \min\left(\frac{k}{d}, 1\right)$, where k is the maximum number of friends allowed to be displayed by the user and d is the degree of the node. The attack can destroy the privacy of 80% of the nodes in OSNs [3].

In a one-time collusion attack, the maximum number of malicious requestors is $n_A - 1$, and the collusion attack performs at least two queries. Thus, the probability of one collusion attack that destroys the target user's privacy at the first layer is $\left(\frac{n_A - 1}{n_A}\right)^2 \cdot P(N_0|_{l=1})$. When the trust threshold is set to 0.5 (lowest threshold), 40% of users' queries will be restricted. In this case, the anti-collusion attack strategy can reduce the number of users whose privacy is breached by at least 47.9%. Accordingly, the number of users whose privacy is violated decreases. By comparing the Facebook, Gowalla and Slashdot datasets, we obtain the results shown in Fig. 4.

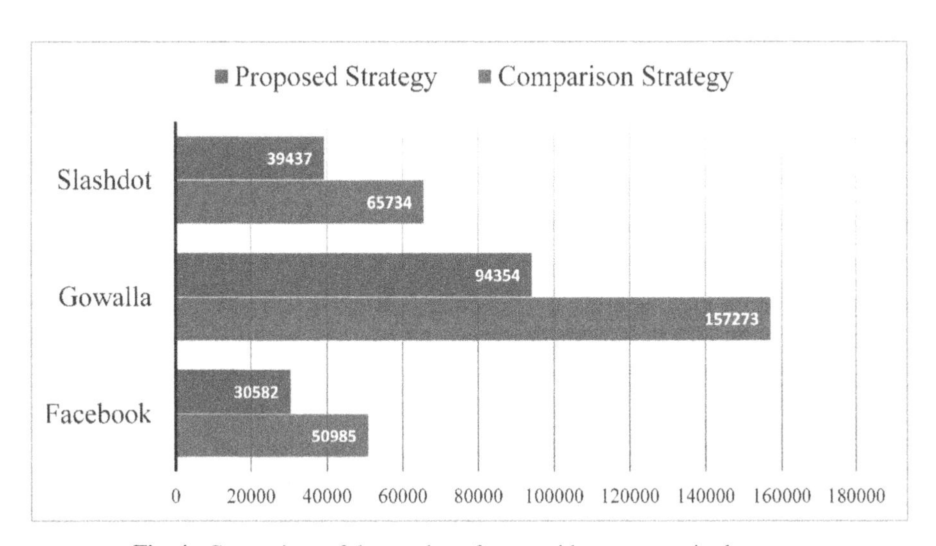

Fig. 4. Comparison of the number of users with a compromised strategy

Due to the limitation of the trust threshold, the number of users whose privacy is breached is significantly reduced. The number of users whose privacy is breached in the Facebook and Slashdot datasets is reduced by approximately 20,000, while the number of users whose privacy is breached in the Gowalla dataset is reduced by approximately 60,000. In the three datasets, the number of users whose privacy has been violated will be reduced by at least 40%. The proposed strategy greatly reduces the number of users whose privacy is violated, which improves the privacy security of users in OSNs.

Number of Colluding Attackers. Based on the (t, n) threshold function, in the query process of the friend search engine, n inquirers are required to participate in the query, and at least t requestors are required to perform secret reconstruction. Therefore, in a single query process, to ensure that malicious requestors can successfully query, it is necessary to ensure that t requestors are malicious requestors and that the comprehensive trust is higher than the trust threshold. In the best situation, two malicious requestors can destroy the privacy of the target user by making two queries. The total number of attackers required is $2n$, while in the comparison strategy, the number of inquirers required is only 2. Therefore, when the value of n set by the system is larger, more malicious attackers will be needed.

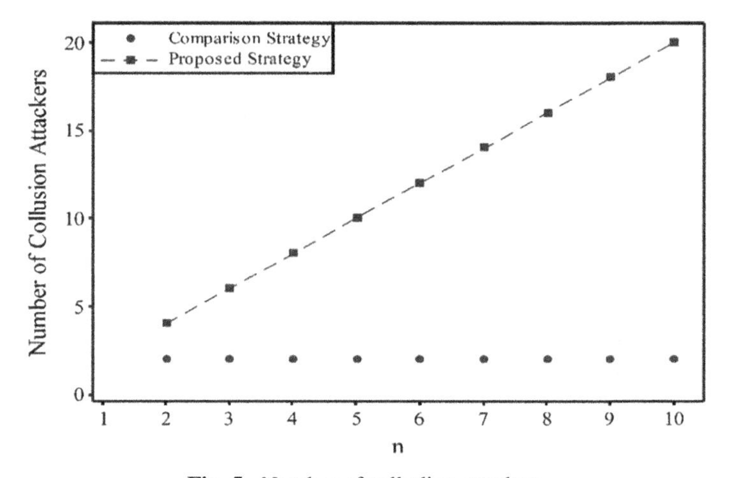

Fig. 5. Number of colluding attackers

Figure 5 shows that the number of colluding attackers varies with the number of queries n. The number of attackers in the proposed strategy is twice that of the comparison strategy. Under the same conditions, the colluding attackers will need more entities or accounts to make queries with the proposed strategy.

Probability of a Successful Collusion Attack. Assume that malicious requestors who have not interacted with the target user in the OSN want to query the target's friendships. First, long-term excellent interactions with the target are needed to obtain the trust of the target. A successful collusion attack requires multiple malicious requestors to cooperate to coordinate their query order and target, and each malicious requestor can successfully query the friend list of the query target. Therefore, multiple malicious requestors need to maintain excellent interactions with users in the OSN, which will require colluding malicious requestors to spend a substantial amount of time disguising their intentions to obtain the trust of the target user.

Consider the following attack [3] as an example of a successful collusion attack.

MR_1: Query N_0—> Retrieve $E(N_0, N_{0.1})$, $E(N_0, N_{0.2})$, $E(N_0, N_{0.3})$
MR_2: Query $N_{0.2}$—> Retrieve $E(N_{0.2}, N_{0.2.1})$, $E(N_{0.2}, N_0)$, $E(N_{0.2}, N_{0.2.3})$
MR_3:
Query $N_{0.2.1}$—> Retrieve $E(N_{0.2.1}, N_{0.2.1.1})$, $E(N_{0.2.1}, N_{0.2})$, $E(N_{0.2.1}, N_{0.2.1.2})$
Query $N_{0.2.2}$—> Retrieve $E(N_{0.2.1}, N_{0.2.1.1})$, $E(N_{0.2.1}, N_{0.2.2.2})$, $E(N_{0.2.1}, N_{0.2.2.3})$
// $N_{0.2.2}$ is "occupied"
Query $N_{0.2}$—> Retrieve $E(N_{0.2}, N_{0.2.1})$, $E(N_{0.2}, N_0)$, $E(N_{0.2}, N_{0.2.3})$
// find new friends:$N_{0.2.3}$
MR_4:
Query $N_{0.2.3}$—> Retrieve $E(N_{0.2.3}, N_{0.2.3.1})$, $E(N_{0.2.3}, N_{0.2})$, $E(N_{0.2.3}, N_{0.2.3.2})$
// $N_{0.2.3}$ can be passive show $N_{0.2}$
Query $N_{0.2}$—> Retrieve $E(N_{0.2}, N_{0.2.1})$, $E(N_{0.2}, N_{0.2.2})$, $E(N_{0.2}, N_{0.2.3})$
// $N_{0.2}$ is occupied
Query N_0—> Retrieve $E(N_0, N_{0.1})$, $E(N_0, N_{0.3})$, $E(N_0, N_{0.4})$
This collusion attack successfully destroyed N_0's privacy.

The collusion attack was coordinated by four malicious requestors. MR_1 makes the first request, and MR_2 determines the target to be queried based on the query results of MR_1. MR_3 queries based on the query result of MR_2. Thus, user $N_{0.2.2}$ will be "occupied," and the new friend $N_{0.2.3}$ of user $N_{0.2}$ can be queried. MR_4 makes a query based on the query result of MR_3 and obtains the $k + 1th$ friend of N_0, i.e., fourth friend $N_{0.4}$. The privacy of the friendships of user N_0 is destroyed.

Under (t, n) threshold function access control, four malicious requestors, i.e., MR_1, MR_2, MR_3, and MR_4, want to complete this query. First, they need to obtain the high trust of the target nodes, i.e., N_0, $N_{0.2}$, $N_{0.2.1}$, $N_{0.2.2}$, and $N_{0.2.3}$, and all four malicious requestors must have long-term excellent interactions with the target. If a malicious requestor cannot obtain the trust of the target, then $T_{ij} < T_t$, and the previously described attack cannot be successfully carried out. Therefore, a successful malicious attack by colluding attackers requires that all malicious requestors reach the trust threshold.

If malicious requestors already exist in the OSN and have interacted with the target user, this strategy restricts requestors whose trust level is below the trust threshold. A requestor cannot query the target user's friend list under (t, n) threshold function access control. Therefore, when the trust threshold is 0.5, 40% of users who do not reach the trust threshold will not be able to query. As described in the second part of this section, for the collusion attack strategy in [3], if (t, n) threshold function access control is not adopted, the probability that colluding attackers will successfully destroy a user's privacy is 1 for each query. In the (t, n) threshold secret sharing anti-collusion attack strategy combined with trust, the comprehensive trust of the requestors who can successfully query the friendships of the target user must be higher than the trust threshold, that is, malicious requestors need to be in category A. Next, we take the trust threshold of 0.5 as an example to discuss the probability that colluding attackers will successfully destroy the privacy of a user's friendships under (t, n) threshold function access control.

If there is a collusion attack, the worst case is that there are enough colluding attackers, and the privacy of the target user is destroyed by just two queries. During a single query, the maximum number of malicious requestors is $n_A - 1$. The maximum probability of malicious requestors who make two requests is $[0.6^{(n_A-1)} \cdot (\frac{n_A-1}{n_A})]^2$.

In Facebook, Gowalla and Slashdot, we observe that regardless of whether a popular node or an unpopular node is considered, the number of malicious requestors required to conduct a successful collusion attack can reach 10,000, which is the best case of a successful collusion attack in the three datasets. Therefore, in the case of $n_A \geq 2$, when there are at most $n_A - 1$ malicious requestors, the probability of a successful collusion attack $p \leq 0.09$.

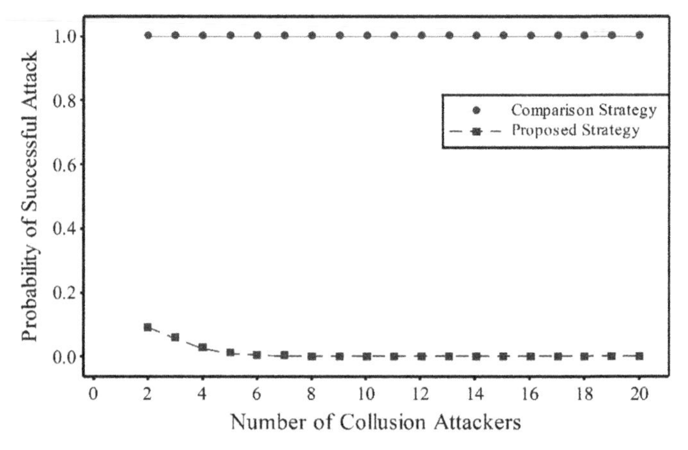

Fig. 6. Comparison of the probabilities of a successful collusion attack based on the number of colluding attackers

Figure 6 shows that when the number of malicious requestors is 2, the anti-collusion attack strategy based on (t, n) threshold secret sharing can reduce the probability of a successful collusion attack from 1 to 0.09. When the number of malicious requestors increases to 18, the anti-collusion attack strategy reduces the probability of a successful collusion attack to 0. When the system trust threshold is higher, it is more difficult for malicious requestors to conduct collusion attacks.

Therefore, the trust-based SSS anti-collusion attack strategy proposed in this work can substantially reduce the number of users whose privacy is compromised by means of credibility calculations, the trust threshold and the (t, n) threshold function. This strategy restricts user queries based on trust and uses the (t, n) threshold function of the SSS system for access control. This strategy can also reduce the probability of successful collusion attacks, which has a significant effect on resisting collusion attacks and can protect the friendship privacy of users in OSNs.

6 Conclusion

In this work, we propose an anti-collusion attack strategy based on trust and the SSS system (t, n) threshold. By analyzing the behaviors of users in OSNs, a method for calculating the credibility between two users is proposed. Comprehensive trust is calculated based on direct trust and recommended trust, and the comprehensive trust degree is employed as a control condition for querying in the friend search engine. The (t, n) threshold function is used for access control in the friend search engine. For users that already exist in the OSN, only requestors who satisfy the trust threshold can successfully query the friendships of the target user. The experimental results show that the proposed strategy can ensure that general users in the friend search engine can query normally and can greatly reduce not only the number of users whose privacy is destroyed but also the probability of successful collusion attacks initiated by malicious requestors by coordinating the query order and sharing the query results.

Theoretically, this work simplifies the complex privacy protection of a user's friendships to the user's access control strategy in the friend search engine. This research starts by theoretically analyzing the calculation of trust between two users and applies the (t, n) threshold function to control querying in the friend search engine to protect the privacy of the users' friendships.

Overall, the proposed strategy can successfully decrease the probability of collusion attacks in friend search engines. Specifically, attacking the same number of users requires more attackers, and the number of users who violate the same number of attackers is greatly reduced.

References

1. Maradapu, A., Venkata, V., Li, Y.: A survey on privacy issues in mobile social networks. IEEE Access **8**, 130906–130921 (2020)
2. Na, L.: Privacy-aware display strategy in friend search. In: 2014 IEEE International Conference on Communications (ICC), pp. 945–950 (2014)
3. Yuhong, L., Na, L.: Retrieving hidden friends: a collusion privacy attack against online friend search engine. IEEE Trans. Inf. Forensics Secur. **14**(4), 833–847 (2019)
4. Xiong, J., et al.: A personalized privacy protection framework for mobile crowdsensing in IIoT. IEEE Trans. Ind. Inf. **16**(6), 4231–4241 (2020)
5. Tian, J., Du, R., Cai, H.: Trusted Computing and Trust Management. Science Press, Beijing (2014)
6. Qiu, W., Huang, Z., Li, X.: Basics of Cryptographic Protocol. Higher Education Press, Beijing (2009)
7. Amusan, O., Thompson, A., Aderinola, T., Alese, B.: Modelling malicious attack in social networks. Netw. Commun. Technol. **5**(1), 37 (2020)
8. Elyashar, A., Uziel, S., Paradise, A., Puzis, R.: The chameleon attack: manipulating content display in online social media. In: Proceedings of the Web Conference 2020, vol. 2, pp. 848–859 (2020)
9. DasGupta, B., Mobasheri, N., Yero, I.G.: On analyzing and evaluating privacy measures for social networks under active attack. Inf. Sci. **473**, 87–100 (2019)
10. Mei, B., Xiao, Y., Li, R., Li, H., Cheng, X., Sun, Y.: Image and attribute based convolutional neural network inference attacks in social networks. IEEE Trans. Netw. Sci. Eng. **7**(2), 869–879 (2020)

11. Muyuan, L., Haojin, Z., Zhaoyu, G., Si, C., Le, Y., Shangqian, H.: All your location are belong to us: breaking mobile social networks for automated user location tracking. In: Proceedings of the International Symposium on Mobile Ad Hoc Networking & Computing, pp. 43–52(2014)
12. Zhipeng, C., Zaobo, H., Xin, G., Yingshu, L.: Collective data-sanitization for preventing sensitive information inference attacks in social networks. IEEE Trans. Dependable Secur. Comput. **15**(4), 577–590 (2018)
13. Binghui, W., Jinyuan, J., Le, Z., Gong, N.Z.: Structure-based Sybil detection in social networks via local rule-based propagation. IEEE Trans. Netw. Sci. Eng. **14**(8), 523–537 (2018)
14. Qingqing, Z., Guo, C.: An efficient victim prediction for Sybil detection in online social network. IEEE Access **8**, 123228–123237 (2020)
15. Tianyu, G., Jin, Y., Wenjun, P., Luyu, J., Yihao, S., Fangchuan, L.: A content-based method for Sybil detection in online social networks via deep learning. IEEE Access **8**, 38753–38766 (2020)
16. Xingwen, Z., Hui, L.: Privacy preserving data-sharing scheme in content-centric networks against collusion name guessing attacks. IEEE Access **5**, 23182–23189 (2017)
17. Chen, Z., Tian, Y., Peng, C.: An incentive-compatible rational secret sharing scheme using blockchain and smart contract. Sci. China Inf. Sci. **64**(10), 1–21 (2021). https://doi.org/10.1007/s11432-019-2858-8
18. Linke, G., Chi, Z., Yuguang, F.: A trust-based privacy-preserving friend recommendation scheme for online social networks. IEEE Trans. Dependable Secur. Comput. **12**(4), 413–427 (2015)
19. Mao, C., Xu, C., He, Q.: A cost-effective algorithm for inferring the trust between two individuals in social networks. Knowl.-Based Syst. **164**, 122–138 (2019)
20. Hamzelou, N., Ashtiani, M.: A mitigation strategy for the prevention of cascading trust failures in social networks. Future Gener. Comput. Syst. **94**, 564–586 (2019)
21. Fogues, R., Such, J.M., Espinosa, A., Garcia-Fornes, A.: Open challenges in relationship-based privacy mechanisms for social network services. Int. J. Hum. Comput. Interact. **31**(5), 350–370 (2015)
22. Huaxin, L., Haojin, Z., Suguo, D., Xiaohui, L., Xuemin, S.S.: Privacy leakage of location sharing in mobile social networks: attacks and defense. IEEE Trans. Dependable Secur. Comput. **15**(4), 646–660 (2018)
23. Alrubaian, M., Al-Qurishi, M., Alamri, A., Al-Rakhami, M., Hassan, M.M., Fortino, G.: Credibility in online social networks: a survey. IEEE Access **7**(c), 2828–2855 (2019)
24. Akilal, K., Slimani, H., Omar, M.: A robust trust inference algorithm in weighted signed social networks based on collaborative filtering and agreement as a similarity metric. Netw. Comput. Appl. **126**(March 2018), 123–132 (2019)
25. Akilal, K., Slimani, H., Omar, M.: A very fast and robust trust inference algorithm in weighted signed social networks using controversy, eclecticism, and reciprocity. Comput. Secur. **83**, 68–78 (2019)
26. Friending Facebook: 13 million US Facebook users don't change privacy settings. http://www.zdnet.com/article/13-million-us-facebook-users-dontchange-privacy-settings/. Accessed 16 Mar 2019
27. Bonneau J, Anderson J, Anderson R, Stajano F.: Eight friends are enough: social graph approximation via public listings. In: Proceedings of the 2nd ACM EuroSys Workshop on Social Network Systems SNS 2009, pp. 13–18 (2009)
28. Malka, S.S., Li, N., Doddapaneni, V.M.: A web application for studying collusion attacks through friend search engine. In: Proceedings - International Computing Software Application Conference, pp. 388–393 (2016)
29. Shamir, A.: How to share a secret. Commun. ACM **22**, 612–613 (1979)
30. Dawson, E., Donovan, D.: The breadth of Shamir's secret-sharing scheme. Comput. Secur. **13**(1), 69–78 (1994)

31. Viswanath, B., Mislove, A., Cha, M., Gummadi, K.P.: On the evolution of user interaction in Facebook. In: SIGCOMM 2009 – Proceedings of the 2009 SIGCOMM Conference on Co-Located Workshops, Proceedings of 2nd ACM Work Online Social Networks, WOSN 2009, pp. 37–42 (2009)
32. Leskovec, J., Lang, K., Dasgupta, A., Mahoney, M.: Community structure in large networks: natural cluster sizes and the absence of large well-defined clusters. Internet Math. **6**(1), 29–123 (2009)
33. Cho, E., Myers, S.A., Leskovec, J.: Friendship and mobility: user movement in location-based social networks. In: The 17th ACM SIGKDD International Conference on Knowledge Discovery and Data Mining, pp. 1082–1090. ACM (2011)

Privacy Computing Technology

A Rational Delegating Computation Protocol Based on Smart Contract

Juan Ma[1], Yuling Chen[1](✉), Guoxu Liu[2], Hongliang Zhu[3], and Changgen Peng[1]

[1] State Key Laboratory of Public Big Data, College of Computer Science and Technology, Guizhou University, Guiyang, China
ylchen3@gzu.edu.cn

[2] Blockchain Laboratory of Agricultural Vegetables, Weifang University of Science and Technology, Shouguang, China

[3] Engineering Lab for Cloud Security and Information Security Center of BUPT, Beijing, China

Abstract. The delegating computation has become an irreversible trend, together comes the pressing need for fairness and efficiency issues. To solve this problem, we leverage game theory to propose a smart contract-based solution. Firstly, according to the behavioral preferences of the participants, we design an incentive contract to describe the motivation of the participants. Secondly, to satisfy the fairness of the rational delegating computation, we propose a delegating computation protocol based on the smart contract. Specifically, rational participants are to gain the maximum utility and reach the Nash equilibrium in the protocol. Besides, we design a reputation mechanism with a reputation certificate, which measures the reputation from multiple dimensions. And the reputation is used to assure the client's trust in the computing party to improve the efficiency of the protocol. Finally, the analysis results show that the proposed protocol solves the complex traditional verification problem. Meanwhile, we prove the fairness and correctness of the protocol.

Keywords: Rational delegating computation · Smart contract · Nash equilibrium · Incentive contract · Reputation mechanism

1 Introduction

The delegating computation is that some of these devices are computationally weak due to various resource constraints. As a consequence, there are tasks, which potentially could enlarge a device's range of application, that are beyond its reach. A solution is to delegate computations that are too expensive for one device, to other devices which are more powerful or numerous and connected to the same network [1]. The traditional delegating computation protocols generally assume that participants need to be honest or malicious. In fact, a lot of participants are rational in the execution process, rational participants always choose a strategy to maximize their utilities. Besides, owing to the high cost of computation and communication complexity in the verification process of delegating computation, the efficiency of the protocol will be reduced.

J. Xiong et al. (Eds.): MobiMedia 2021, LNICST 394, pp. 389–398, 2021.
https://doi.org/10.1007/978-3-030-89814-4_28

Combining the game theory and traditional delegating computation, the rational delegating computation protocol is a part of rational cryptography. From the perspective of the participant's self-interest, the protocol utilizes the utility functions to ensure the correctness and reliability of the calculation results. In recent years, many scholars have conducted researches on rational delegating computation. Li et al. combined game theory and fully homomorphic encryption technology to construct a rational delegating computation protocol, which guarantees the interests of both participants [2]. Tian et al. constructed a rational delegating computation protocol based on Yao's garbled circuit and fully homomorphic encryption technology [3]. But in the process of delegating computation, if the malicious participants don't follow the rules in the protocol, the interests of honest participants will be reduced. Therefore, how to ensure the fairness of the protocol, which need to be considered.

Recently, the fairness of delegating computation is one of the hot topics in current research, and the existing researches utilize a third-party (e.g., bank [4], semi-trusted third-party [5, 6], trusted third-party [7, 8]) to overcome these issues. However, in the computation process, with a third-party, potential security problems will inevitably occur, e.g., unreasonable Nash equilibrium, privacy leakage, and low efficiency. To eliminate the drawbacks, many researchers adopt smart contract to realize the Peer-to-Peer transaction between the clients and the computing parties. Zhou et al. combined the game theory with traditional delegating computation and proposed a three-party game rational delegating computation protocol based on smart contracts [9]. Dong et al. combined game theory and smart contract to design a reasonable prisoner contract, collusion contract, and betrayal contract [10]. Song et al. researched the application of game theory in blockchain and proposed that the application of smart contract in delegating computation effectively prevent the collision problem of computing parties [11]. The payment scheme based on smart contract technology guarantees the fairness of participants. However, in the process of delegating computation, different tasks or utility have different impacts on the strategies of the computing parties, leading the clients to get an incorrect result. Therefore, how to select a reliable computing party is an urgent problem.

In the process of delegating computation, the reputation mechanism is designed to improve the reputation of honest participants and reduce the reputation of malicious participants. Jiang et al. proposed a rational delegating computation based on reputation and contract theory, which ensures that the client selects a reliable computing party [12]. Li et al. proposed a fair payment protocol based on Bitcoin time commitment, which ensures the fairness of participant's payment by using Bitcoin time commitment technology [13]. However, in the process of selecting the computing party, the reputation of the computing party will be updated every round. Therefore, how to efficiently view the latest reputation of the computing party is a key issue that needs to be resolved.

To solve the aforementioned problems, we propose a rational delegating computation protocol based on smart contract, which realizes the optimal utility of all rational participants, and guarantees the correctness of the calculation results and the fairness of the payment process. Besides, we design a reputation mechanism for measuring the reputation from different dimensions to select the high-quality computing parties. The main contributions are as follows:

1. According to the behavioural preferences of the participants, we design an incentive contract to motivate the participants to choose the strategy honestly, which reaches a reasonable Nash equilibrium result.
2. Based on smart contract, we propose a rational delegating computation protocol, which realizes the fairness of rational delegating computation. And we design a reputation mechanism for the client to choose high-quality computing parties. In addition, we prove the fairness and correctness of the protocol.

The rest of the paper is structured as follows. Section 2 introduces concepts such as game theory, Nash equilibrium, and smart contract. Section 3 proposes an incentive contract based on a smart contract. Section 4 establishes a reputation mechanism that is convenient for the client to choose the computing party. Section 5 proves the fairness and correctness of the protocol. Section 6 calculates the cost of the contract and analyzes the performance of the protocol. Section 7 summarizes the paper and the layout of future research.

2 Preliminaries

2.1 Game Theory

Definition 1 (Standard Game): The standard form of a n-player game is composed of three elements: player set P, strategy space S and utility function u, denoted as $G = \{P, S, u\}$, where $P = \{P1, ..., Pn\}$, $S = \{S1, ..., Sn\}$, $u = \{u1, ..., un\}$. Any specific strategy $si \in Si$ indicates that strategy si is the key element of the strategy set Si, and utility function $ui : S \rightarrow R$ (R represents the real number space) denotes the profits of the players i under different strategy profiles [14].

Definition 2 (Nash Equilibrium): A strategy profile $s^* = \{s1^*, ..., sn^*\}$ is a Nash Equilibrium of game $G = \{P, S, u\}$, if $u_i(si^*, s - i^*) \geq u_i(sj^*, s - j^*)$ holds for each player $Pi(i = 1, ..., n)$ and all $sj \in Si$. Obviously, if player $i \neq j$ complies with the strategy si^*, then the player j will not deviate from the strategy sj^*, as it will not benefit at all. In principle, there may be multiple Nash equilibrium in a game [15].

2.2 Smart Contract

Definition 3 (Smart contract): *Smart contracts* are in the blockchain environment, allowing the definition and execution of contracts signed on the blockchain. It is the automated execution of the contract, and its essence is a piece of code written on the blockchain. Smart contract is the basis of the program-ability of blockchain, and each node does not rely on a third-party to automatically execute the contract. Broadly speaking, a smart contract is a set of rules encoded in a programming language. Once the execution requirements of the code are met, the script will be automatically executed to realize the operation, and this process does not require the participation of a trusted third-party [16].

2.3 System Model

The system model considered in our construction comprises one client denoted by Cd and multiple computing parties denoted by $\{C1, C2, ..., Cn\}$.

As illustrated in Fig. 1, the system model consists of eight steps. Firstly, the client broadcasts the task to the computing party, and the interested computing party returns the response request. Next, the client will select two computing parties based on the reputation to perform outsourcing tasks and sign a contract with them. According to the content of the signed contract, the client and the computing party respectively deposit the deposit stipulated in the contract into the smart contract. Then, the computing party performs the task and sends the calculation result to the client. The client judges whether the computing party has executed the task correctly according to the calculation result, and makes the corresponding response and payment. Finally, according to the interactive behavior of the computing party, the client updates and uploads the reputation of the computing party.

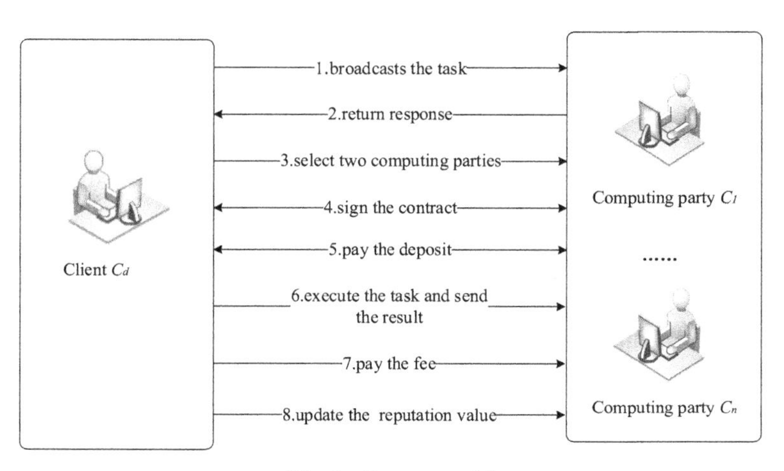

Fig. 1. System model

3 Incentive Contract

3.1 Deposit Variables

Since the incentive contract is introduced, we need to define more deposit variables. They are all non-negative.

- g: the client agrees to pay to a computing party for computing the task.
- v: the computing party's cost for computing the task.
- r: the deposit to invoke the *TTP*.
- c: the deposit a computing party pays to the client in order to get the task.
- $2g + r$: the deposit of a client in the smart contract.

The following relations is obvious:
(1) $g - v > 0$ (2) $r - 2c > 0$

3.2 Contract Content

We will introduce the specific content of Incentive contract, and we analyze each step in detail by studying the utility function of rational players. The contract is as follows.

Contract content

**Step1(Publish task):* The client C_d publishes the task.

**Step1.1:* C_d broadcasts the task X to all computing parties;

**Step1.2:* If C decides to accept task will return response request P.

**Step2:* C_d selects $C_i (i \in \{1, 2\})$ based on the reputation.

**Step3(Sign contract):* C_d, C_1 and C_2 must sign the *contract* to start it, otherwise *contract* terminates.

**Step4(Pay deposit):* C_d, C_1 and C_2 pay the deposit $(2g + r, c, c)$; otherwise *contract* terminates.

**Step5(Verification the result).* The client checks the results returned by the computing party.

**Step5.1:* If both C_1, C_2 deliver the results within the specified time, and the results are equal. Transferring C_d's deposit g to each C_i, returning C_i's deposit c.

**Step5.2:* Otherwise, the TTP is invoked and one of the following steps is performed;

**Step5.2.1:* If both C_1, C_2 are one party honest (return the correct result C_h) and one party malicious (return the wrong result C_f or return the random result C_r). Transferring C_d's deposit g to each C_h and C_f's (or C_r's) deposit c to C_d, and pays fee r to TTP;

**Step5.2.2:* If both C_1, C_2 return wrong result. Transferring deposit c to C_d, returning C_d's deposit $2g$, and pays fee r to TTP;

**Step5.2.3:* If both C_1, C_2 return random result. Transferring C_i deposit c to C_d, returning C_d's deposit $2g$, and pays fee r to TTP;

**Step5.3:* For each C_i, if C_i fail to deliver the result within the specified time. Transferring deposit c to C_d, returning C_d's deposit $2g$.

**Step6:* If the C_d deviates from the protocol, transferring C_d's deposit g to each C_i, returning each C_i's deposit c.

4 The Proposed Reputation Scheme

In this part, the reputation mechanism mainly includes five stages. The first stage, the content of the reputation certificate is defined. The second stage, the client selects the appropriate computing party to compute the task according to the reputation. The third stage, the client evaluates the current reputation based on the computing party's honesty. The fourth stage, the client computes the global reputation. In the fifth step, the computing party updates the reputation certificate.

4.1 Reputation Certificate

In the delegating computation process, the client selects the appropriate computing party according to its task requirements. In a reputation mechanism with a reputation certificate, the computing party has a high-quality reputation as the basis for obtaining the task. Generally, the reputation of the computing party is higher, the probability of being selected is greater. The reputation certificate format is shown in Table 1.

Table 1. Reputation certificate.

Symbol	Describe
$IDcd$	ID of the client
IDc	ID of the computing party
Rn	Number of rounds
$Sigcd$	Signature of the client
$Sigc$	Signature of the computing party
$HRVc$	The historical reputation of the computing party

4.2 Search of Historical Reputation

After the client publishes the task, the computing party needs to return a response and generates the latest reputation certificate to get the task. For the request returned by the computing party, the client selects two appropriate computing party based on the reputation. The format of the new reputation certificate is as follows.

$$CRCCn + 1 = IDC \| Rn + 1 \| HRVn \| Sign(IDC \| Rn + 1 \| HRVn)$$

4.3 Evaluation of Current Reputation

Usually, the computing party's strategies are affected by different tasks or utilities, which cause them to have different reputations. Reputation is whether the computing party's behavior is honest or not after each round of tasks. In general, the computing party has two reputation statuses: honest or malicious. Specifically, When the computing party's reputation value in round $t+1$ is higher than that in round t, i.e. $Lrept+1 \geq Lrept$, we consider the computing party to be honest. On the contrary, When the computing party's reputation value in round $t+1$ is lower than round t, i.e. $Lrept+1 \leq Lrept$, we consider the computing party to be malicious. The evaluation of reputation is as follows Eq. (1).

$$Lrept+1 = \alpha i, t\left[\frac{H}{S} * D(r) * A(r) * T(r)\right] \tag{1}$$

Let $\alpha i, t$ be the client Cd interactive evaluation to the computing party Ci in round t, here, $\alpha i, t \in \{0, 1, -1\}$. In order to reward the computing party who performed well in the execution of the task and punish the computing party who failed to complete the task as required, we propose the setting as shown in Eq. (2).

$$\begin{cases} \alpha i, t = 1, Lrepi, t = \frac{H}{S} * D(r) * A(r) * T(r) \\ \alpha i, t = 0, Lrepi, t = 0 \\ \alpha i, t = -1, Lrepi, t = -\left[\frac{H}{S} * D(r) * A(r) * T(r)\right] \end{cases} \tag{2}$$

If and only if there is no any interaction between the client Cd and the computing party Ci, $Lrepi, t = 0$. Obviously, for any Ci, when $t = 0$, $\alpha i, t = 0$, then $Lrepi, 0 = 0$. If and only if the computing party Ci honestly performs the delegation task in round t, $\alpha i, t = 1$, then $Lrepi, t = \frac{H}{S} * D(r) * A(r) * T(r)$. Conversely, if and only if the computing party Ci is behaves dishonestly in round t, $\alpha i, t = -1$, then $Lrepi, t = -\left[\frac{H}{S} * D(r) * A(r) * T(r)\right]$.

Where, $Lrepi, t$ satisfies $-1 \leq Lrepi, t \leq 1$, $Lrepi, t$ is the reputation of the current round, H represents the number of honest calculations by the computing party,S represents the total number of calculations by the computing party, $D(r)$ represents the complexity coefficient of the delegating computation and satisfies $0 \leq D(r) \leq 1$, $A(r)$ represents the benefit coefficient of the delegating computation and satisfies $0 \leq A(r) \leq 1$, $T(r)$ represents the time to submit the result of the calculation.

(1) When $Lrepi, t = -1$, it means that Ci is completely malicious in the process of delegating computation of round t;
(2) When $Lrepi, t = 0$, it means that there is no interaction between Cd and Ci in the process of delegating computation of round t;
(3) When $Lrepi, t = 1$, it means that Ci is completely honest in the delegating computation of round t.

4.4 Computation of Global Reputation

In this stage, after the client evaluates the reputation of the current round, the client computes the global reputation according to the historical reputation and the reputation of the current round. The computation of global reputation is as follows Eq. (3).

$$Grep_{n+1} = Grep_n + Grep_{n+1} \tag{3}$$

$Grepn+1$ represents the global reputation of the computing party, and $Grepn$ represents the global reputation of the computing party in the last round.

4.5 Update of Reputation Certificate

Finally, the client signs the global reputation of the computing party and uploads it to the blockchain. Then, the computing party updates the content of its reputation certificate, which facilitates the generation of the latest reputation certificate for the next round of tasks. The format of the latest reputation certificate is as follows.

$$CRCC, n+1 = CRCC - Cd, n+1 \| Sign(CRCC - Cd, n+1)$$

5 Protocol Proof

In this section, we prove the agreement in detail from the two aspects of fairness and correctness of the protocol. More details are given in Sects. 5.1 and 5.2.

5.1 Fairness Proof

Theorem 1. The rational delegating computation protocol based on smart contract is fair.

Proof. To ensure the fairness of the protocol, at the initial stage of the protocol, the client and the computing party respectively pay a deposit of to the smart contract. In the payment process of delegating computation, there are four cases as follows:

Case 1: When the computing party is honest, the smart contract transfers the deposit g to the computing party's account;

Case 2: When the computing party is malicious, the smart contract confiscates the malicious computing party's deposit c and transfers deposit c to the client's account;

Case 3: When the client is malicious, the smart contract confiscates the deposit $2g$ and transfers deposit g to the two computing parties' account respectively;

Case 4: When the client is honest, the client can get the correct calculation result.

In order to avoid the situation that the third-party is dishonest or bought off, based on the smart contract technology, we construct a rational delegating computation protocol to achieve fairness.

5.2 Correctness Proof

Theorem 2. The rational delegating computation protocol based on smart contract is correct.

Proof. There are three cases for the proof as following:

(1) When $C1$ chooses the honest strategy, if $C2$ chooses the honest strategy, the utility is $u(g-v)$; else if $C2$ chooses the random values strategy, the utility is $u(-c)$; else $C2$ chooses the malicious strategy, the utility is $u(-c-v)$; which is $u(g-v) > u(-c) > u(-c-v)$.

(2) When $C1$ chooses the malicious strategy, if $C2$ chooses the honest strategy, the utility is $u(g+c-v)$; else if $C2$ chooses the random values strategy, the utility is $u(-c)$; else $C2$ chooses the malicious strategy, the utility is $u(-c-v)$; which is $u(g+c-v) > u(-c) > u(-c-v)$.

(3) When $C1$ chooses the random value strategy, if $C2$ chooses the honest strategy, the utility is $u(g + c - v)$; else if $C2$ chooses the random values strategy, the utility is $u(-c)$; else $C2$ chooses the malicious strategy, the utility is $u(-c - v)$; which is $u(g + c - v) > u(-c) > u(-c - v)$.

Similarly, $C2$ also has the above three cases. According to analysis, only when participants choose the honesty strategy, maximizing their utility and reach Nash equilibrium.

6 Conclusion

Combining game theory with smart contract, in this paper, we propose a rational delegating computation protocol based on smart contract. More specifically, we analyze the strategies and utilities of participants, then we utilize smart contract to substitute the third-party to ensure the fairness of protocol. Also, we design a reputation scheme, which can measure the reputation from different dimensions to ensure the client chooses a reliable computing party. Our future work will focus on the collusion problem between computing parties in delegating computation.

Acknowledgement. This work was supported by the National Natural Science Foundation of China under Grant No. 61962009, Major Scientific and Technological Special Project of Guizhou Province under Grant No. 20183001, the foundation of Guizhou Provincial Key Laboratory of Public Big Data under Grant No. 2018BDKFJJ008.

References

1. Xue, R., Wu, Y., Liu, M.H., Zhang, L.F., Zhang, R.: Progress in verifiable computation. Scientia Sinica Inf. **45**(11), 1370–1388 (2015)
2. Li, Q.X., Tian, Y.L., Wang, Z.: Rational delegation computation protocol based on fully homomorphic encryption. Acta Electron. Sin. **47**(2), 216–220 (2019)
3. Tian, Y.L., Li, Q.X., et al.: Proven secure rational delegated computing protocol. J. Commun. **40**(07), 135–143 (2019)
4. Carbunar, B, Tripunitara, M.: Fair payments for outsourced computations. IEEE Trans. Parallel Distrib. Syst. **23**(2), 1–9 (2010)
5. Xu, G., Amariucai, G.T., Guan, Y.: Delegation of computation with verification outsourcing: curious verifiers. IEEE Trans. Parallel Distrib. Syst. **28**(3), 717–730 (2017)
6. Huang, H., Chen, X.F., Wu, Q.H., Huang, X.Y., Shen, J.: Bitcoin-Based fair payments for outsourcing computations of fog devices. Future Gener. Comput. Syst. **78**(2), 850–858 (2016)
7. Chen, X.F., Li, J., Susilo, W.: Efficient fair conditional payments for outsourcing computations. IEEE Trans. Inf. Forensics Secur. **7**(6), 1687–1694 (2012)
8. Yin, X., Tian, Y.L., Wang, H.L.: Fair and rational delegati on computation protocol. Ruan Jian Xue Bao/J. Softw. **29**(7), 1953–1962 (2018) (in Chinese).
9. Zhou, Q.X., Li, Q.X., Fan, M.M.: Anti-collusion delegation computation protocol of three-party game based on smart contract. Comput. Eng. **46**(8), 124–131, 138 (2020)
10. Dong, C., Wang, Y., Aldweesh, A., et al.: Betrayal, distrust, and rationality: smart counter-collusion contracts for verifiable cloud computing, pp. 211–227. ACM (2017)

11. Song, L.H., Li, T., Wang, Y.l.: Application of game theory in blockchain. Chinese J. Crypt. **6**(01), 100–111 (2019)
12. Jiang, X, Tian Y. Rational delegation of computation based on reputation and contract theory in the UC framework. In: Yu, S., Mueller, P., Qian, J. (eds.) Security and Privacy in Digital Economy, SPDE 2020. Communications in Computer and Information Science, vol. 1268. Springer, Singapore (2020). https://doi.org/10.1007/978-981-15-9129-7_23
13. Li, T., Tian, Y.L., et al.: A fair payment scheme based on blockchain in delegated computing. J. Commun. **41**(03), 80–90 (2020)
14. Tian, Y.L., Guo, J., Wu, Y., et al.: Towards attack and defense views of rational delegation of computation. IEEE Access **7**, 44037–44049 (2019)
15. Osboren, M.: An Introduction to Game Theory, pp. 151–232. Oxford University Press, New York (2004)
16. Liu, F.M., Chen, Y.T.: A review of blockchain technology research. J. Shandong Normal Univ. **35**(03), 299–311 (2020)

Privacy-Preserving Subset Aggregation with Local Differential Privacy in Fog-Based IoT

Lele Zheng(✉), Tao Zhang, Ruiyang Qin, Yulong Shen, and Xutong Mu

School of Computer Science and Technology, Xidian University, Xi'an 710071, China
{llzheng,xtmu}@stu.xidian.edu.cn, taozhang@xidian.edu.cn,
ylshen@mail.xidian.edu.cn

Abstract. As a typical novel IoT(Internet of Things) architecture, fog-based IoT is promising to decrease the overhead of processing and movement of large-scale data by deploying storage and computing resources to network edges. However, since the edge-deployed fog nodes cannot be fully trustable, fog-based IoT suffers some security and privacy challenges. This paper proposes a novel privacy-preserving scheme that can implement data aggregation from a subset of devices in fog-based IoT. Firstly, our scheme identifies the subset to be aggregated by computing the Jaccard similarity of attribute vectors of the query users and IoT devices, where the local differential privacy is employed to protect the attribute vectors. In addition, we use local differential privacy truth discovery to protect the data of IoT devices and improve the accuracy of the aggregation result. Finally, experiments show that our scheme is efficient and highly available by comparing it with state-of-the-art works. Theoretical analyses demonstrate that our proposed scheme has excellent performance on both computational costs and communication costs.

Keywords: Fog-based IoT · Local differential privacy · Truth discovery

1 Introduction

With the rapid development of the IoT, more and more physical devices (such as smartphones, vehicles, etc.) can freely connect to the Internet and generate various valuable data for different applications. Many emerging IoT applications rely

Supported by the National Key R&D Program of China (Grant 2018YFB2100400), the Natural Science Basic Research Program of Shaanxi (Program No. 2019ZDLGY13-03-01, 2020CGXNG-002, 2021ZDLGY07-05), the Fundamental Research Funds for the Central Universities (Grant No. JB210306), the Fundamental Research Funds for the Central Universities and the Innovation Fund of Xidian University(Grant No. YJS2103).

J. Xiong et al. (Eds.): MobiMedia 2021, LNICST 394, pp. 399–412, 2021.
https://doi.org/10.1007/978-3-030-89814-4_29

on generating large amounts of data to provide users with better services [20], but extensive computing, communication, and storage resources are required. To cope with such IoT applications' proliferation, fog computing has become a promising supplement to cloud computing by extending the network functions from the cloud to network edges.

However, there are still many unsolved problems in fog–based IoT architecture, in which security and privacy issues occupy one of the most critical positions [3,16]. The data submitted by the IoT device may contain sensitive information about the device (such as the location of the vehicle, the brand of the mobile phone, etc.). Once malicious attackers obtain these data, the user's privacy will be leaked. Therefore, the data of IoT devices should be protected before they are uploaded. Besides, compared with traditional IoT, fog-based IoT has a more complex network environment and architecture. The fog node needs to collect data from multiple data sources and provide novel aggregation services (selective data aggregation) for different applications. Fog computing can select some relevant nodes for data collection according to users' demands to offer personalized services. However, malicious attackers can infer the user's preferences from attributes of the IoT devices selected by the user. Thus, IoT devices' attributes should be protected while safeguarding the data because they also implicitly leak privacy, especially when the data source comes from a smartphone [10]. Therefore, it is urgent for us to protect IoT devices' attributes and data in fog-based IoT.

Our Contributions. Motivated by the above statements, in this paper, we investigate a novel privacy-preserving subset aggregation scheme. Using Jaccard similarity estimation and truth discovery subject to local differential privacy, our scheme can protect user's query vectors, IoT devices' attribute vectors, and values of IoT devices' data. The contributions of this paper can be summarized as follows:

1. We propose a novel secure subset aggregation scheme in fog-based IoT, where a Jaccard similarity estimation method that satisfies local differential privacy is utilized to select a desirable subset. Our scheme can protect the attribute of IoT devices and efficiently select nodes that meet the conditions.
2. We use local differential privacy technology to ensure IoT device data security and use the truth discovery method that satisfies local differential privacy to improve the aggregated results' accuracy.
3. Experiments claim our proposed scheme can effectively estimate the similarity and aggregate data. The aggregation result has high utility while we protect the privacy of both users and IoT devices. With the growth of the number of attributes and IoT devices, our scheme's time overhead increases very slowly.

The rest of the paper is organized as follows. Section 2 presents the related work. Section 3 introduces the system model and design goals. The details of our scheme are presented in Sect. 4. The analysis and experimental evaluations are shown in Sect. 5. Finally, Sect. 6 concludes the paper.

2 Related Work

The security of data aggregation has been studied for a long time. There are many privacy-preserving data aggregation schemes [2], such as multi-dimensional aggregation, fault-tolerant aggregation, and differential privacy aggregation. Moreover, Lu et al. [11] proposed a lightweight privacy protection data aggregation scheme for fog computing to enhance the Internet of Things, which focuses on saving communication bandwidth and many of the security as mentioned earlier features. Cheng et al. [5] proposed reliable and privacy-protected selective data aggregation based on fog-based IoT. HASSAN et al. [12] proposed a privacy protection subset aggregation scheme called PPSA in fog-enhanced IoT scenarios, which enables query users to obtain the sum of data from a subset of IoT devices. However, the enormous computational overhead in these schemes reduced the availability of them.

Similarity estimation is used in many fields, such as social networks, recommendation systems. The most commonly used methods are inner product, euclidean distance, cosine similarity, Jaccard similarity. However, conventional methods can not protect the privacy of two vectors. To solve this problem, Lu et al. [11] proposed a scheme based on homomorphic encryption, which used homomorphic encryption to calculate two vectors' inner product. The Euclidean distance based on homomorphic encryption has been presented by Zhang et al. [21]. Homomorphic encryption and zero-knowledge proof have been used to compute the cosine similarity in [19]. [6] computed Jaccard similarity between encrypted data by homomorphic encryption. These schemes have massive computation and communication costs, so the Jaccard similarity estimation schemes satisfied local differential privacy has been proposed. [1] proposed an LDP Jaccard similarity estimation scheme based on the Laplace mechanism, and [18] proposed a scheme based on the exponential mechanism.

Truth discovery is a technique that can improve the accuracy of the aggregation result, usually used in crowdsensing [8,13]. Truth discovery updates the weight based on the distance and result of the participant's data so that the impact of low-quality data is reduced. However, these truth discovery schemes can not protect the privacy of participants. Some works use encryption or secure multi-party computation techniques to protect privacy [14,15]. Li et al. [9] proposed a local differential privacy truth discovery scheme, which can effectively compute the truth value and protect users' privacy.

3 System Model and Design Goal

3.1 System Model

Figure 1 shows our system model. We divide it into four layers: IoT-device layer, Fog layer, Cloud layer, and Query User layer.

IoT-Device Layer: A set of IoT devices $I = \{I_1, I_2, \cdots, I_N\}$ are deployed at the IoT-device layer. Each device $I_i \in I$ is equipped with sensing, calculation,

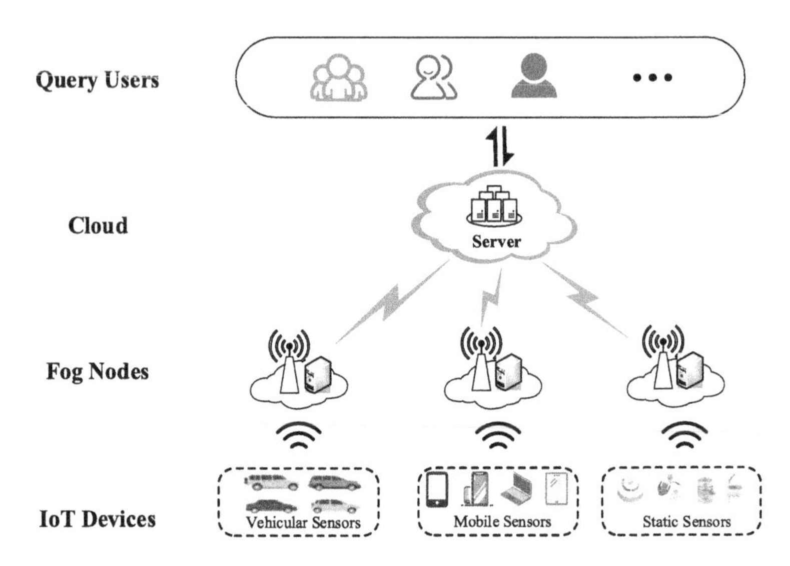

Fig. 1. System model

and communication modules, which allows them to collect, disturb, and upload data to the fog layer. In our model, IoT devices periodically collect data $D_i =< atr_i, x_i >$, where $atr_i \in \{0,1\}^m$ is the IoT devices' attributes and $x_i \in R$ is the value of the data. They respectively use privacy-preserving 1-bit minhash to perturb the attributes of IoT devices and use Laplace noise to perturb the values of the data. Finally, IoT devices upload their perturbed data $\widehat{D_i} =< \widehat{art}_i, \widehat{x}_i >$ to fog servers without disclosing any sensitive information.

Fog Layer: In the fog layer, fog servers with computing capabilities are deployed in each area. After receiving the user's query $\widehat{U}$ sent via the cloud layer, fog servers request IoT devices to upload their perturbed attributes $\widehat{art}_i$. In order to achieve efficiently privacy-preserving subset aggregation, fog servers use local differential privacy Jaccard similarity estimation to select IoT devices. Finally, the fog server will collect the value $\widehat{x}_i$, where $i \in I'$ from IoT devices, and upload them to the cloud layer.

Cloud Layer: In our model, the cloud layer is composed of servers, which are responsible for sending user queries and data aggregation from the fog layer. Here we use the local differential privacy truth discovery algorithm to improve the data's accuracy and send the obtained data to the user. All data and queries are disturbed, so there will be no privacy leakage.

Query User Layer: The user layer is mainly responsible for disturbing user queries and sending queries to the cloud layer. In our model, the query user selects a subset of IoT devices $I' \subseteq I$ to obtain the aggregate sum value of their prepared data D. User has an attribute vector $U \in \{0,1\}^m$ means the user wants to aggregate the data of the IoT devices which satisfy the demand U. That is to say, the user wants to query the data of the devices in a subset $I' \subseteq I$, and each I_i in the I' satisfies the similarity $S(U, atr_i) > \tau$, and τ is a threshold. The user also uses the private 1-bit minhash to perturb the attribute to protect the private information.

3.2 Security Model

In our security model, all entities, deployed at both device and fog layers, are assumed to be honest-but-curious participants, i.e., they are obliged to follow the protocols faithfully, but may be curious about the query information during data preparation and throughout the query processing steps. For example: 1) the fog node may try to identify the IoT devices Ii with exact data xi; 2) each IoT device may be curious about other IoT devices' data; 3) the user may be curious about each IoT device Ii's prepared data or, at least, prepared data in each IoT device in I. Note that the honest-but-curious assumption would be guaranteed in practice since the service providers should protect their own reputation and financial interests. Finally, there would be no collusion between entities such as sharing their results.

3.3 Design Goal

Considering the above system model, our goal is to design a privacy-preserving subset aggregation scheme in fog-based IoT. Precisely, the following three objectives should be achieved:

Privacy-Preserving. In our scheme, the user uses the private 1-bit minhash to perturb the attribute to protect the private information. IoT devices use private 1-bit minhash to perturb the attribute of the data and use Laplace noise to perturb the data's value.

Efficient. With fog servers deployed at the fog layer, the cloud server can communicate with IoT devices efficiently through fog servers. Our local differential Jaccard similarity estimation has low computational overhead compared with other schemes using homomorphic encryption. And IoT devices can offload the calculation task of Jaccard similartiy estimation to fog servers to get higher efficiency.

High Utility. In our scheme, the cloud server uses a truth discovery algorithm that satisfies local differential privacy to aggregate the data of IoT devices. Truth discovery improves the accuracy of the aggregation result by decreasing the weight of low-quality data so that users can get high utility results.

4 Our Proposed Scheme

4.1 Workflow

(1) System Initialization. In the system initialization, we assume the cloud server will bootstrap the whole system. The cloud server will set the number of private minhash functions to K, chose K different hash functions, and set the privacy budget to ϵ.

(2) Query Generation at The User's Side. The user has an attribute vector $U \in \{0,1\}^m$ meaning that the user wants to satisfy the demand U. That is to say, the user wants to query the aggregation result of the data of IoT devices which are in a subset $I' \subseteq I$, and each I_i in the I' satisfies the similarity $S(U, atr_i) > \tau$, where τ is a threshold.

The user also uses the noisy 1-bit minhash to perturb the query vector to protect the private information as $\hat{U} = h(U)$, where $h(\cdot)$ is a noisy minhash function. Then the user sends the query $\hat{U}$ to fog servers from different areas through the cloud server.

(3) Similarity Estimation at Fog Server. In this step, each fog server receives the perturbed query $\hat{U}$ from the user and sends the request of uploading attribute vectors to IoT devices. After receiving the request, IoT devices will use noisy 1-bit minhash to perturb their attributes atr_i, get the perturbed attributes $\widehat{atr_i} = h(atr_i)$, then upload them to the fog server. Fog server executes the similarity estimation to compute the similarity between each $\widehat{art_i}$ and $\hat{U}$, and selects the data whose similarity is greater than τ using the local differential privacy Jaccard silmilarity.

Finally, the fog server will collect the value of IoT devices' data whose attribute satisfies the user's query and upload the values $\{\hat{x}_{i \in I'}\}$ to the cloud server.

(4) Data Aggregation at Cloud Server. After receiving the data from fog servers, the cloud server uses local differential privacy truth discovery to increase the accuracy of the result of the aggregation. Finally, the cloud server sends the result to the user.

4.2 Local Differential Privacy Jaccard Similarity Estimation

In our model, users need the data that satisfies themselves, so we need to estimate the similarity of the user's query vector and IoT devices' attribute vectors. There are many ways to compute the similarity of two vectors, such as inner product, cosine similarity, Jaccard similarity. In fog-based IoT, we should protect users and IoT devices' privacy, so those privacy protection schemes of inner product and cosine similarity always use homomorphic encryption, which gives a heavy computation burden to users and IoT devices. Minhash is a standard method to estimate the similarity of two sets, and it can reduce the costs of communication and protect privacy to some extent. Local differential privacy Jaccard similarity protects privacy effectively by adding Laplace noise at the vector after minhash. Meanwhile, it can estimate the similarity effectively. We estimate the similarity as shown in Algorithm 1.

Minhash. Our scheme uses minhash algorithm to estimate the similarity of two vectors. A minhash function is defined as that, $h : \{0,1\}^m \rightarrow [m] := \{1, ..., m\}$ which is a random shuffle of a vector. The hash value of $x \in \{0,1\}^m$ is the position of the first 1 in x after the shuffle. Obviously, we can know the property of minhash that for any pair $x, y \in \{0,1\}^m$, we have $Pr[h(x) = h(y)] = J(x, y)$ where

Algorithm 1. Local Differential Privacy Jaccard Similarity Estimation

Require: user's query *vector* $< U >$, IoT device attributes *vector* $< I >$, number of hash functions k.
Ensure: similarity *est*.
1: $dist \leftarrow 0$
2: **for** $\{i = 0; i < k; i++\}$ **do**
3: $\quad dist \leftarrow dist + (U_i - I_i)^2$
4: **end for**
5: $est \leftarrow 1 - 2/k * dist + 8 * (\Delta f/\varepsilon)$
6: return *est*

the probability depends on the random choice of h. Then randomly choosing K different hash functions gets result $(h_1(x), \cdots, h_K(x)) \in [m]^K$. By linearity of expectation, the value $\frac{1}{K}\sum_{i=1}^{K}[h_i(x) = h_i(y)]$ is an unbiased estimator of $J(x, y)$.

1-bit Minhash. Riazi et al. [17] described a secure hash function construction based on mapping the produced minhash value to a random bit, building on the idea of b-bit minhash from [7]. Formally, given a random minhash function $h_{min} : \{0,1\}^m \rightarrow [m]$ and a random hash function $h_r : [m] \rightarrow \{0,1\}$, let $h(x) = h_r(h_{min}(x))$. For two vectors x, y, it is well known that $Pr(h(x) = h(y)) = (1 + J(x,y))/2$. Repeating the construction K times, we estimate $J(x, y)$ as $\frac{2}{K}\sum_{i=1}^{K}[h_i(x) = h_i(y)] - 1$.

Lemma 1. Let $x, y \in \{0,1\}^m$, and every vector has at least $\tau \geq 1$ attribute(s), we say that two vectors are neighboring if they differ in at most α positions, so that $J(x,y) \geq 1 - \alpha/\tau$. Let $h_1, \cdots, h_K$ be K random 1-bit minhash functions. Let $x^* = (h_1(x), \cdots, h_K(x)), y^* = (h_1(y), \cdots, h_K(y))$. Let $\delta > 0$. With probability at least $1 - \delta$, the number of different positions between x^* and y^* is at most $(K\alpha)/(2\tau) + \sqrt{3ln(1/\delta)(K\alpha)/(2\tau)}$.

Proof. Let the $X = \sum_{i=1}^{K} X_i$, where $i \in [K]$ and $X_i = [h_i(x) \neq h_i(y)]$ is a random variable. Since all X_i are independent and $Pr(X_i = 1) = \frac{1-J(x,y)}{2} \leq \alpha/(2\tau)$, we can get $E[x] = K\alpha/(2\tau)$. Then using Chernoff bound $Pr(X > (1+\beta)E[x]) \leq exp(\frac{-\beta^2 E[x]}{3})$, and let $\beta = \sqrt{3ln(1/\delta)/E[x]}$.

We can get $Pr\left(X > (K\alpha)/(2\tau) + \sqrt{3ln(1/\delta)(K\alpha)/(2\tau)}\right) \leq 1/\delta$, is that $Pr\left(X \leq (K\alpha)/(2\tau) + \sqrt{3ln(1/\delta)(K\alpha)/(2\tau)}\right) \geq 1 - 1/\delta$, so Lemma 1 is proved.

Noisy MinHash. Let K, α and τ be integers, and let $\varepsilon > 0$ and $\delta > 0$ be the privacy budget. By Lemma 1, we can get the sensitivity $\Delta f = (K\alpha)/(2\tau) + \sqrt{3ln(1/\delta)(K\alpha)/(2\tau)}$. According to [4], the user and each IoT node with $x \in \{0,1\}^m$ can add Laplace noise $N_{x,i} \sim Lap(\Delta f/\varepsilon)$ to get vector $\hat{x} = (h_1(x) + N_{x,1}, \cdots, h_K(x) + N_{x,K})$ with K 1-bit minhash functions $h_1, \cdots, h_K$, so getting (ε, δ)-differential privacy.

Algorithm 2. Local differential privacy Truth discovery

Require: noisy data from IoT devices $Vector < X >$.
Ensure: final truth $Result$.
1: Randomly initialize the truth $Result$
2: **repeat**
3: **for** $\{i = 0; i < n; i++\}$ **do**
4: Update the weight ws based on current estimated ground truth using Equation $w_i = \omega(\frac{d(x_i, Result)}{\sum_{i=1}^{n} d(x_i, Result)})$
5: **end for**
6: Update the truth Result based on current weights using Equation $Result = \frac{\sum_{i=1}^{n} w_i * x_i}{\sum_{i=1}^{n} w_i}$
7: **until** Convergence criterion is satisfied
8: return $Result$

Similarity Estimation. Given $\hat{x}$ and $\hat{y}$ from R^K, their similarity can be estimated as:

$$\hat{J}(\hat{x}, \hat{y}) = 1 - \frac{2}{K} \sum_{i=1}^{K} (\hat{x}_i - \hat{y}_i)^2 + 8(\Delta f/\varepsilon)$$

And $\hat{J}(\hat{x}, \hat{y})$ is an unbiased estimator for $J(x, y)$.

Using linearity of expectation, we prove as follows:

We use E_d to express the expectation of the distance of two vectors x and y. We can easily get $E_d = E[\sum_{i=1}^{K} (\hat{x}_i - \hat{y}_i)^2]$, because of the linearity of expectation, $E_d = \sum_{i=1}^{K} E[(\hat{x}_i - \hat{y}_i)^2]$. Every hash function has the same expectation, we can use h_1 to be the representative, then we expand E_d, get $E_d = KE[((h_1(x) - h_1(y)) + (N_{x,1} - N_{y,1}))^2] = KE[(h_1(x) - h_1(y))^2 + 2(h_1(x) - h_1(y)(N_{x,1} - N_{y,1}) + (N_{x,1} - N_{y,1})^2]$. Both $N_{x,i}, N_{y,i}$ are independently chosen, so $E[N_{x,i}] = E[N_{y,i}] = 0$, get $E_d = K(E[(h_1(x) - h_1(y))^2] + E[(N_{x,1} - N_{y,1})^2])$. According to the property of expectation, we know $E[(N_{x,1} - N_{y,1})^2] = Var[N_{x,1} - N_{y,1}] + (E[N_{x,1} - N_{y,1}])^2$, and we know $Var[N_{x,1}] = 2(\Delta/\varepsilon)^2$, then we can get $E_d = KE[(h_1(x) - h_1(y))^2] + 2KVar[N_{x,1}] = KE[(h_1(x) - h_1(y))^2] + 4K(\Delta f/\varepsilon)^2$. Because $(h_1(x) - h_1(y))^2$ just has two possible values, so $E_d = K(0^2 Pr[h_1(x) = h_1(y)] + 1^2 Pr[h_1(x) \neq h_1(y)]) + 4K(\Delta f/\varepsilon)^2$. $Pr[h_1(x) = h_1(y)] = J(x, y), Pr[h_1(x) \neq h_1(y)] = 1 - J(x, y)$, that is $E_d = K/2(1 - J(x, y)) + 4K(\Delta f/\varepsilon)^2$.

From what has been discussed above, we can get:

$$J(x, y) = 1 - \frac{2}{K} E[\sum_{i=1}^{K} (\hat{x}_i - \hat{y}_i)^2] + 8(\Delta f/\varepsilon)^2$$

4.3 Local Differential Privacy Truth Discovery

In the local differential privacy data aggregation model, all data are perturbed by IoT devices. The privacy of IoT devices has been protected, but the noise of the data may be significant. The traditional aggregation scheme treats all data equally, making some low-quality data affect aggregation results' accuracy.

By using truth discovery, the data with high weight will contribute more to aggregation results.

Our truth discovery is used on the perturbed data. The server will receive the data which have been perturbed, then aggregate the perturbed data $\{\hat{x}_{i \in I}\}$ from all IoT devices by conducting local differential privacy truth discovery to obtain the final output. The local differential privacy truth discovery can improve the utility of data and protect IoT devices' privacy. The method is shown in Algorithm 2.

Data Perturbation. Firstly, each IoT device perturbs its data by adding Laplace noise $\hat{x_i} = x_i + N_x \sim Lap(\Delta f/\epsilon)$, then uploads it to the server. We denote the perturbation mechanism as M, the original data as x_i. So the M satisfies ϵ-Local Differential Privacy. For any subset $S \in R$, and two different records x_1 and x_2, the following inequality holds:

$$Pr\{M(x_1) \in S\} \leq e^{\epsilon} Pr\{M(x_2) \in S\}$$

This definition compares the probability of observing the perturbed value of two different records x_1 and x_2 in the same range. In this step, the weights of IoT devices are inferred based on the current aggregated results. An IoT device will have a high weight if it provides information close to the aggregated results. Typically, the weights of IoT devices are calculated as follow:

$$w_i = \omega(\frac{d(x_i, result)}{\sum_{i=1}^{n} d(x_i, result)})$$

Where distance function $d(x, y)$ is a function that measures the distance between the value provided by the IoT device and the aggregated result, and weight computation function $\omega(\cdot)$ is a monotonically decreasing function. If the distance is small, the IoT device's data will get a high weight. In our scheme, weight computation function $\omega(\cdot)$ is $-log(\cdot)$.

Aggregation. In the aggregation step, the weights of IoT devices are fixed. We compute aggregated results as follow:

$$result = \frac{\sum_{i=1}^{n} w_i * x_i}{\sum_{i=1}^{n} w_i}$$

where w_i is the weight of i-th IoT device, and the x_i is the perturbed value of the i-th IoT device data. In this weighted aggregation scheme, the final result relies on those IoT devices which have high weights.

5 Evaluation

This section evaluates our scheme's computational costs and compares them with [12], which compute the similarity of two vectors with homomorphic encryption. The environment of the experiment is Intel core i7-4790 CPU @ 3.60GHz with windows 10 operating system.

We run our experiments 1000 times for different parameter values. We show the average execution result for each function in Fig. 2 with variables of the number of attributes m, the number of hash functions K, the number of IoT devices n, and the round of truth discovery t.

5.1 Experiment

1) In Fig. 2(a), we show the time cost of the generation of user's query and IoT devices' attribute vectors for different m and K. Because the times of computation are just related to the number of the attributes and hash functions, not related to what the attributes are, we randomly select some attributes to compute the time cost, and the computational cost will increase when K or m becomes larger.

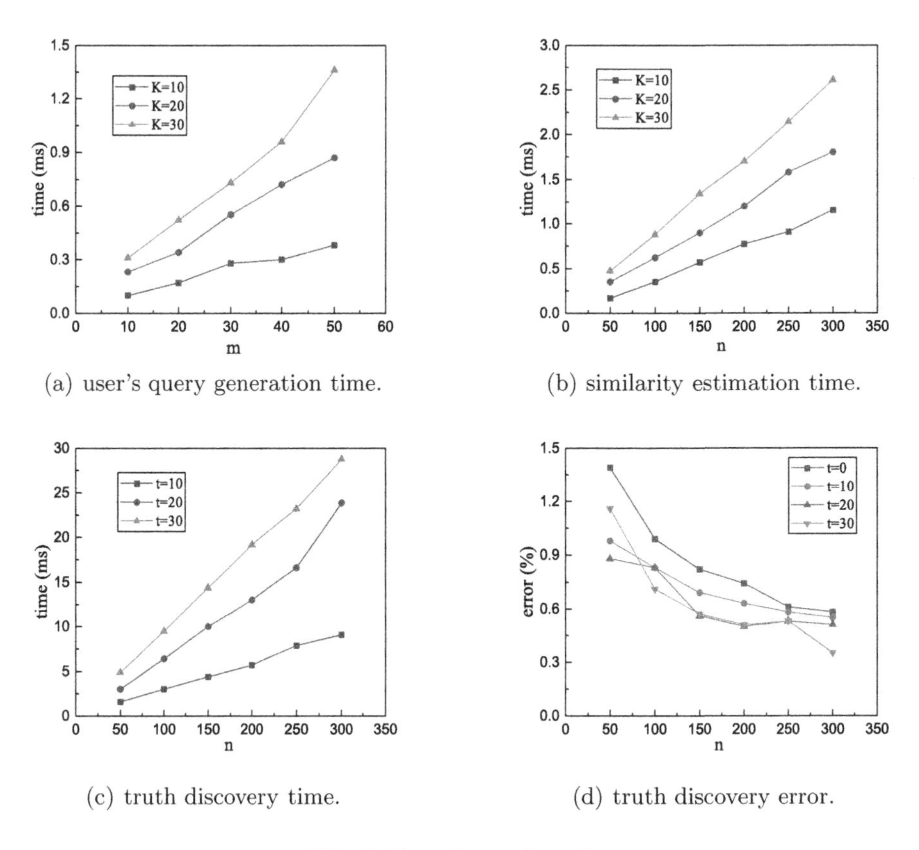

Fig. 2. Experimental result

2) In Fig. 2(b), we show the time cost of similarity estimation of two attribute vectors for different n and K. As same as 1), there is no effect on computation for which attributes two vectors have. The computational cost will increase rapidly as k increases.

3) In Fig. 2(c), we show the time cost of truth discovery of the noisy values for different n and t. We randomly generate some temperature data and add Laplace noise to them. When t becomes larger, the calculation cost will also increase.

4) In Fig. 2(d), we show the mean absolute percentage error (MAPE) of truth discover for different n and t. We compare the aggregation results of noisy data and raw data and compute the percentage of the raw data aggregation result, which difference between noisy results and raw result accounts. Through comparing with the method which does not use truth discover(when $t = 0$), we show that the error can be reduced by using truth discover.

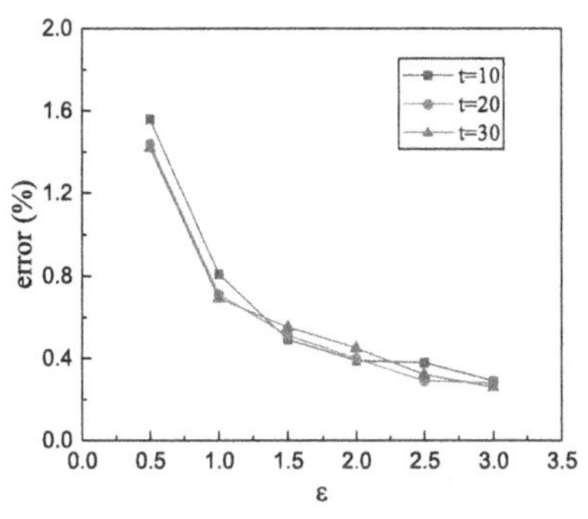

(a) MAPE on truth discovery, varying ϵ from 0.5 to 3.0, fixing $n = 100$.

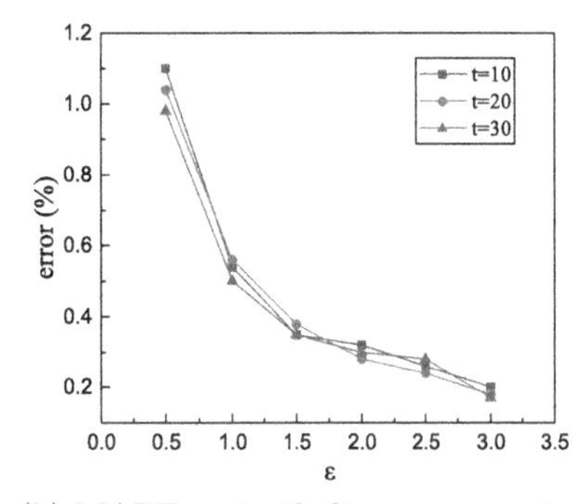

(b) MAPE on truth discovery, varying ϵ from 0.5 to 3.0, fixing $n = 200$.

Fig. 3. Experimental result

5) In Figs. 3(a) and 3(b), we show the mean absolute percentage error of truth discover for different ϵ and t when the $n = 100$ and $n = 200$. Our experimental results demonstrate that the error will reduce with increasing privacy budget ϵ.

6) In Table 1, we compare our user's query generation's time cost with user's query generation in [12] called PPSA. We let our $K = 20$ and 30 because their scheme's time cost is too high, so we let their key generation parameter $\kappa = 512$ and 1024. The number of attributes of our scheme and PPSA is m.

7) In Table 2, we compare our similarity estimation step's time cost and the matching step with PPSA. We set the same parameters, let our $K = 20$ and 30, and let their key generation parameter $\kappa = 512$ and 1024. Let $m = 20$, and n is the number of IoT devices. From the table, we can find out that our scheme's costs increase very slowly, but homomorphic encryption increases rapidly.

Table 1. User's query generation time (ms)

m	Algorithm			
	Our scheme		PPSA	
	K = 20	K = 30	κ = 512	κ = 1024
5	0.13	0.20	46	230
10	0.26	0.30	58	402
15	0.28	0.40	79	569
20	0.32	0.52	102	710
25	0.43	0.65	123	873
30	0.46	0.71	145	1003

Table 2. Subset selection time (ms)

n	Algorithm			
	Our scheme		PPSA	
	K = 20	K = 30	κ = 512	κ = 1024
50	0.27	0.42	714	4807
100	0.57	0.81	1405	9428
150	0.85	1.17	2057	14184
200	1.10	1.57	2746	18893
250	1.40	1.95	3403	23828
300	1.67	2.36	4095	28672

5.2 Analysis

In this section, we will analyze the computational costs and the communication costs of our scheme.

Computational Costs. In the user's query generation step, the user just needs to compute the noisy attribute vectors. A minhash function's time complexity is $O(m)$, where m is the number of attributes. The user needs to compute K hash values, so the time complexity of generating the step is $O(mK)$. In the matching step, the fog servers need to compute the similarity between the user's query vectors and the IoT devices' vectors. If there are n IoT devices, the time complexity is $O(nm)$. In the truth discovery step, the cloud server needs to aggregate all results which satisfy the user's query, if we run t rounds truth discovery, the time complexity is $O(tn)$.

Communication Costs. IoT devices need to send their attribute vectors and noisy data, so each IoT device's communication cost is $O(K)$, where K is the number of hash functions. Every fog server will receive the attribute vectors and noisy data from IoT devices, so the communication cost is $O(n_iK)$, where n_i is the number of IoT devices for i-th fog server. After similarity estimating, each fog server will upload the data to the cloud server. For each fog server, the most communication cost of this step is $O(n_i)$, so the cloud server's communication cost is $O(\sum n_i)$.

6 Conclusion

This paper proposed a scheme to select the particular subset privately and effectively by using Jaccard similarity estimation that satisfies local differential privacy. In addition, our scheme used local differential privacy truth discovery to increase the accuracy of the aggregation result. Using local differential privacy jaccard similarity estimation and truth discovery, our scheme can protect user's query vectors, IoT devices' attribute vectors, and IoT devices' data values. Thus there is no private information being disclosed in our model. Moreover, experimental and analysis results show that our scheme has excellent performance in efficiency and accuracy.

References

1. Aumüller, M., Bourgeat, A., Schmurr, J.: Differentially private sketches for Jaccard similarity estimation. In: Satoh, S., et al. (eds.) SISAP 2020. LNCS, vol. 12440, pp. 18–32. Springer, Cham (2020). https://doi.org/10.1007/978-3-030-60936-8_2
2. Bao, H., Lu, R.: A new differentially private data aggregation with fault tolerance for smart grid communications. IEEE Internet Things J. **2**(3), 248–258 (2017)
3. Chiang, M., Ha, S., Chih-Lin, I., Risso, F., Zhang, T.: Clarifying fog computing and networking: 10 questions and answers. IEEE Commun. Mag. **55**(4), 18–20 (2017)
4. Dwork, C., Roth, A.: The Algorithmic Foundations of Differential Privacy (2013)
5. Huang, C., Liu, D., Ni, J., Lu, R., Shen, X.: Reliable and privacy-preserving selective data aggregation for fog-based IOT. In: 2018 IEEE International Conference on Communications (ICC) (2018)
6. Le, T.T.N., Phuong, T.V.X.: Privacy preserving Jaccard similarity by cloud-assisted for classification. Wirel. Personal Commun. (5) (2020)
7. Li, P., König, A.: B-bit minwise hashing, October 2009
8. Li, Y., Gao, J., Lee, P.P.C., Su, L., He, C., He, C., Yang, F., Fan, W.: A weighted crowdsourcing approach for network quality measurement in cellular data networks. IEEE Trans. Mob. Comput. **16**(2), 300–313 (2017)
9. Li, Y., et al.: Towards differentially private truth discovery for crowd sensing systems, October 2018
10. Lu, K.: Checking more and alerting less: detecting privacy leakages via enhanced data-flow analysis and peer voting (2015)
11. Lu, R., Heung, K., Lashkari, A.H., Ghorbani, A.A.: A lightweight privacy-preserving data aggregation scheme for fog computing-enhanced IoT. IEEE Access **PP**, 1 (2017)
12. Mahdikhani, H., Mahdavifar, S., Lu, R., Zhu, H., Ghorbani, A.A.: Achieving privacy-preserving subset aggregation in fog-enhanced IoT. IEEE Access **7**, 184438–184447 (2019)
13. Meng, C., et al.: Truth discovery on crowd sensing of correlated entities, pp. 169–182 (2015)

14. Miao, C., et al.: Cloud-enabled privacy-preserving truth discovery in crowd sensing systems. In: Proceedings of the 13th ACM Conference on Embedded Networked Sensor Systems, pp. 183–196, SenSys 2015, Association for Computing Machinery, New York, NY, USA (2015). https://doi.org/10.1145/2809695.2809719
15. Miao, C., Su, L., Jiang, W., Li, Y., Tian, M.: A lightweight privacy-preserving truth discovery framework for mobile crowd sensing systems. In: IEEE INFOCOM 2017 - IEEE Conference on Computer Communications (2017)
16. Ni, J., Zhang, K., Lin, X., Shen, X.S.: Securing fog computing for internet of things applications: challenges and solutions. IEEE Commun. Surv. Tutorials **20**(99), 601–628 (2018)
17. Riazi, M.S., Chen, B., Shrivastava, A., Wallach, D., Koushanfar, F.: Sub-linear privacy-preserving near-neighbor search (2016)
18. Yan, Z., Wu, Q., Ren, M., Liu, J., Liu, S., Qiu, S.: Locally private Jaccard similarity estimation. Concurr. Comput. Pract. Exper. **31**(24), e4889 (2019)
19. Yang, D., Xu, B., Yang, B., Wang, J.: Secure cosine similarity computation with malicious adversaries. In: Chaki, N., Meghanathan, N., Nagamalai, D. (eds.) Computer Networks and Communications (NetCom). LNEE, vol. 131, pp. 529–536. Springer, New York (2013). https://doi.org/10.1007/978-1-4614-6154-8_52
20. Yu, S., Wang, G., Liu, X., Niu, J.: Security and privacy in the age of the smart internet of things: an overview from a networking perspective. IEEE Commun. Mag. **56**, 14–18 (2018). https://doi.org/10.1109/MCOM.2018.1701204
21. Zhang, J., Hu, S., Jiang, Z.L.: Privacy-preserving similarity computation in cloud-based mobile social networks. IEEE Access **PP**(99), 1 (2020)

A Survey on Spatial Keyword Search over Encrypted Data

Zenebe Yetneberk(✉)

School of Cyber Engineering, Xidian University, Xi'an, China

Abstract. Many real-time applications use spatial keyword queries, which provide location information and text descriptions of Points Of Interest (POI). Since a promising research subject can pose such a procedure, performing spatial keyword queries over encrypted data is difficult. Several schemes for addressing secure spatial and keyword queries have been proposed, but previous surveys have only summarised and tested secure query algorithms. They still lack a privacy-preserving overall review of the spatial keyword query. This paper outlines a secure spatial keyword query's three main components: secure spatial query, secure textual query, and secure spatial-textual query. Some evaluation criteria have been carefully selected to aid in the evaluation of existing spatial keywords. Following that, we expand on and evaluate recent related research's benefits and disadvantages using the suggested criteria. Finally, we see some unanswered problems that researchers can use to perform more analyses and studies.

Keywords: Privacy preservation · Secure queries · Spatial keywords query · Privacy-preserving query

1 Introduction

Many researchers have been paying attention to spatial keyword queries in recent years due to the widespread usage of location-based services. By outsourcing their spatial textual data, including indexes, to a cloud service provider (CSP), data owners can competently support the online spatial keyword query process. Users who want to run queries on their data can submit requests to CSP, which will handle them quickly.

However, those outsourced services may incur some privacy leakage problems since the spatial-textual data is sensitive, and the cloud server may not be completely trustworthy. Besides, collecting spatial-textual objects consumes both human and financial resources, considered business secrets for competitors and any unauthorized parties. Moreover, in any case, if data users' spatial keyword queries are illegally obtained, potential attackers could use that data to eavesdrop on the privacy information [17].

J. Xiong et al. (Eds.): MobiMedia 2021, LNICST 394, pp. 413–431, 2021.
https://doi.org/10.1007/978-3-030-89814-4_30

There are already some surveys on spatial keyword queries. Lisi *et al.* [8] proposed a benchmark comparing the performance of spatial keyword query algorithms. Eldawy and Mokbel [15] classified the present work in this field into three distinct aspects, namely, approach, design, and components. Chen *et al.* [7] explained the existing studies on the query of multi-modal road network location-based data and categorized the prevalent work. Qi *et al.* [27] introduced the key ideas for underlying safe region techniques within which no query update in a very given region ranged a few times and illustrated how they were applied in several continuous spatial query algorithms to provide various query types. However, none of them has considered secure and privacy-preserving problems of spatial keyword queries, as we compare them in the Table 1. Thus, we encourage to conduct a systematic survey that summarizes recent state-of-the-art spatial, text, and spatial keyword search solutions for protection and privacy.

Table 1. Comparison of our survey with other survey

Covered topics	[7]	[27]	[22]	[30]	Our survey
Give a comprehensive review	N	N	N	N	Y
Proposed a set of evaluation criteria	Y	N	N	Y	Y
Summarize the pros and cons	N	N	N	N	Y
Analyze the performance	N	N	N	Y	Y
Propose some open issues and future direction	Y	Y	Y	N	Y

This survey will classify and thoroughly review privacy-preserving spatial keyword queries by grouping them into three main parts: spatial, textual, and spatial keyword queries. We study each query type one by one for analyzing their pros and cons. To instruct our analysis on current works' success, we propose a set of evaluation criteria that assist our judgment on potential investigation patterns.

According to our survey, some open research issues and important research directions that merit additional research efforts have been identified. The following are some of the most important contributions to our work:

- An extensive review of privacy problems on secure spatial, textual, and spatial keyword query technologies.
- Recommending a set of assessment criteria, and seriously overview existing work as well as analyze the strengths and weaknesses of these work based on our proposed criteria.
- Based on a detailed analysis, we propose some open issues and forecast future research trends.

2 Background

This section introduces basic concepts related to the spatial query, text query, and spatial-textual query. Spatial-keyword queries have three main types [13].

- Boolean Range Query (BRQ), $Q(R, doc)$: This query returns spatial objects containing all the keywords in the set *doc* and the range R, where a spatial range is R, and a set of keywords is *doc*.
- Boolean kNN Query (BkQ), $Q(loc, doc, k)$: It returned a set of k objects, each containing all the *doc* keywords ranked according to its spatial proximity to the location of query *loc*.
- Top-k Query (TkQ), Q(loc; doc; k): Based on their combined textual and spatial significance scores, this results in a collection of ranked k objects, where the *doc* is a set of keys where the *doc* is a set of keywords in the query, *loc* is the location of query, and k is the parameter in the top-k query that represents the number of return objects.

Indexing Techniques: An efficient indexing structure is vital to managing big data in the cloud server, specifically in Location-Based Services (LBS) providers. These auxiliary structures accelerate the database transactions and query process by storing the database's indexed columns separately for fast look-ups. An efficient index needs to be customized to the database. Here, we are going to discuss the popular index structures, which can be found in different types of databases.

2.1 Textual Index

This index is used for databases containing only the most widely used indexes, such as bitmaps and inverted files.

- Bitmap: It stores data as an array of bits. Each bit indicates whether the object contains the keyword. Bitmaps use logical operations on the bits storing the data to answer a query. A bitmap is fast and suitable for attributes whose values frequently repeat [38].
- Inverted Files: This data structure stores the mapping from some content to its location. The inverted file includes mapping keywords to files containing the keyword. Index files are popular in large-scale search engines as they allow for a speedy full-text search [29].

2.2 Spatial Indexing

The spatial index scheme can be classified into three categories: R-tree, grid, and space-filling curve indices.

- R-tree: R-tree [19] or its variant (*e.g.* the R*-tree [1]) is included in this index. Most geo-textual indices belonging to this category use the inverted files for text indexing, as illustrated in Fig. 1. The R-tree-based indices loosely combine the R-tree and the inverted files. It makes it easy to arrange spatial, and text data [41] independently. On the other hand, most existing indices strongly combine the R-tree with a text index.

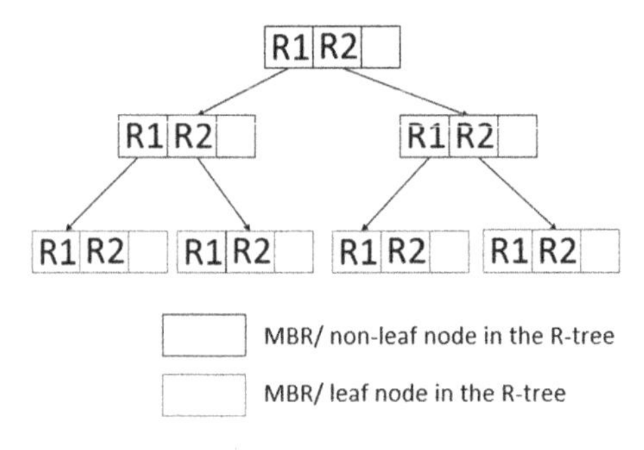

Fig. 1. R-tree

- Grid: This grid-based index entails allocation of relevant objectives to their position in the grid. It combines the index of a grid with a text index (*e.g.* the inverted file). Indices for text and grid can be prearranged either be organized separately [33] or combined tightly [21].
- Space-filling Curve: It is a curve whose range includes 2-dimensional space in its entirety. The index combines the inverted files with Z-curved based index [12] as illustrated in Fig. 2. A Hilbert curve-based index [9] is also included under this indexing category.

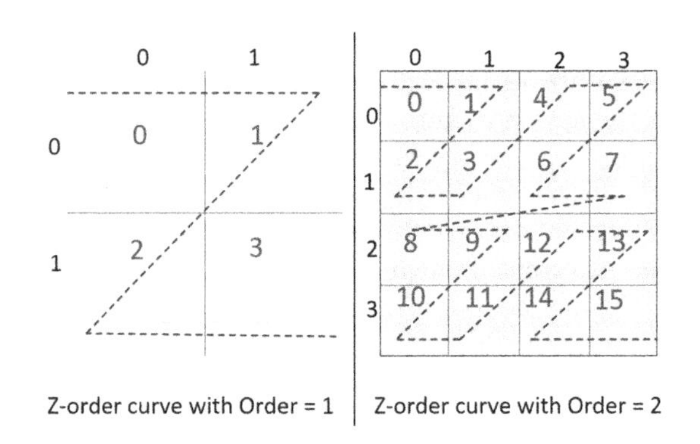

Fig. 2. Z-order curve with order 1 and order 2

2.3 Combination Indexing Schemes

Spatial and text indexing is associated with these geo-textual indices. Based on how the spatial and text indices are combined, the index can be categorized. It includes a text-first loose and tight combination.

A text-first index loses its combination index by using the text-first index. For the top-level index, the inverted file and organized posting are contained in a spatial structure in each inverted list. It can be R-tree, a space-filling curve, or a grid. On the other hand, the top-level spatial-first index is a spatial structure whose leaf nodes contain inverted files or bitmaps for the objects' text in the nodes.

Conversely, both the text and spatial index are tightly correlated with the tight combination index. A scheme to concurrently prune the search space with query processing. There are indexes of two kinds. In each inverted list, one incorporates the spatial information (*e.g.*, [12]) and the other integrates a text description into each spatial index node (*e.g.*, [13]).

Geo-Textual indexing: These are the 12 most common types of geo-textual indices for a set of spatial web objects [8]. The geo-textual indices are grouped into three features: the text index, the spatial index, and a hybrid of the two indices.

3 Evaluation and Categorizations Criteria for Secure Queries

In this unit, to categorize and compare the current system's benefits and drawbacks, we suggest a set of categorization requirements and evaluation criteria.

3.1 Categorization Criteria

We classify some related literature into three basic categories based on the query method; such as privacy-preserving spatial query, privacy-preserving textual query, and privacy-preserving spatial keyword query.

3.2 Evaluation Criteria

- Threat Model (TM): The principal research works commonly consider two different models. Some researchers contemplate the CSP as "Curious but Honest (CH)." The cloud service provider will monitor protocols that have been developed. Nevertheless, it will expose both data consumers' and owners' sensitive data. On the other hand, the CSP is considered by some researchers to be "Malicious (M)" *i.e.*, in which the CSP perhaps will return incorrect results for retrieval. Therefore, subsequent research studies seek to show recovery effects.
- Plain Text Attack (PTA)
  - *Chosen-Plaintext Attack (CPA)*: Let the cloud service provider in the dataset derive cyphertext for selected objects' plaintext and try to recover the cyphertext.
  - *Known-Plaintext Attack (KPA)*: Considering CSP, pairs of objects in the dataset can achieve plaintext-cyphertext. It seeks to overcome the secret key by using the data of these pairs to decode additional cyphertext.

3.3 Security Requirements (SR)

- Authenticity (Au): Authenticated query processing helps clients check if the query results returned by the CSP are accurate.
- Confidentiality (Cn): Since security information is essential, the data needed can only be accessed by legitimate data users. Thus, data information should not be revealed in the transmission process, and its confidentiality should be assured.
- Integrity (*In*): This refers to the user's query content in the entire process, where the cloud server is unable to get the actual query content or infer the query through the cyphertext.

3.4 Privacy Requirements (PR)

Privacy is the capability of a person or group of people to hide information about themselves. We categorize the privacy type into the following features:

- Query Privacy (QP): In the whole process, this applies to the user's query content. The cloud server is unable to get the actual query content or infer the query over the cyphertext.
- Data Privacy (DP): It refers to the data stored in the dataset. Under the process, the cloud server cannot get the actual data sets over cyphertext.
- Result Privacy (RP): This word applies to the content of the cloud server. We should consider the cloud server's resulting privacy related to the query's exact results that should not be revealed to it. From the user's point of view, they can obtain real data that satisfies the query's condition.
- Path Patterns Privacy (PPP): When executing the query, it refers to the index's traversal path. The CSP cannot reveal the actual traversal route.
- Access Pattern Privacy (APP): It refers to a relationship that defines matching points in a particular query containing the points. The cloud server cannot access this relationship between the query and the received results.

3.5 Performance Requirements (PR)

- Correctness (Cr): Correctness of data analysis is an essential aspect of performance. The query results should be correct and match the data user requests since some data analytics attacks may lead to incorrect analytical results and serious consequences.
- Scalability (Sc): When the number of objects increases, it relates to how the device reacts and allows us to assess its capacity for a broad network.
- Mobility (Mo): Mobility refers to the users' status, whether at the static or continuous positions. Static refers to the user's motion (rest at a fixed point), and dynamic refers to the user moving from one given place to another within a short period.

- Efficiency (Ef): Secure data collection and analysis should be effective and efficient since the service provider's energy power consumption needs to be minimized. Linear and non-linear are the two levels of efficiency in computational complexity measurement.
- Other requirements: Time Stamp (TS), which refers to the exact time that a query executes, and Place Description (PD), which gives some additional information about searched for places.

4 Literature Review

We analyze the published work in this survey only by looking for papers from the ACM Digital Library, Springer, IEEE Xplore Digital Library. We use search keywords, including privacy-preserving keyword query, spatial query, spatial keyword query, secure keyword, spatial and spatial keyword query. We are classifying some of the related references into three basic categories based on the query methods. Including privacy-preserving spatial query, privacy-preserving textual, and privacy-preserving spatial keyword search.

4.1 Secure Textual Query

It deals with multi-keyword Search, which has better flexibility and efficiency than the single keyword search [24], a cryptographic primitive Hierarchical Predicate Encryption (HPE) uses attribute hierarchy for simple range queries.

Boolean Search

a) Privacy-Preserving keyword Search (PPKS). A multi-round protocol between CSP and data users is a single keyword search technique. The keyword index links an individual keyword with its associated files. The heuristic pseudo-random function's primary importance is to encrypt a dictionary-based keyword index for individual files [6]. The virtues of $PPKS$ systems operate in multiple file formats, including compressed files, multimedia files, Etc. The keyword index should first be developed with priority.

b) Secure Privacy-Preserving Keyword Search (SPKS). The system helps the CSP decrypt the details and return the keyword-containing file [25]. $SPKS$ enables users to reduce communication and computational overheads, providing information and query privacy for the users. To efficiently search over encrypted data, it implements six algorithms.

Table 2. Summarazation and comparison of secure textual query

Keyword Search Category	Keyword Search Technique	Merits	Demerits
Boolean Keyword Search	PPKS [6]	—Due to pseudorandom function, it has better security	—Not applicable for multiple keyword search
	SPKS [25]	—Provide query and data privacy for the users —Has less communication and computational cost	—The system is not resiliently chosen and is known to be a plane text attack
	FKS [10,23]	—Utilized the multiway tree to improve search efficiency	—Give unsorted (ranked) search result —Search semantics is not considered
	APKS [24]	—Has better flexibility and efficiency compared to single keyword search	—Do not prevent keyword attack
	APKS+ [24]	—Prevents dictionary keyword attack —Accomplishes index and query privacy	—Not all the attributes are hierarchical
	ABE [20]	—Afford the best quality search on encrypted data —Fast accessing method	—Non-efficiency —Non-existence of attribute revocation mechanism
Ranked Keyword Search	RSSE [35]	—Provide effective protocol —Has strong security promise as compared to SSE schemes —Efficient support of relevance score dynamics	—Has a slight relevance score information leakage in contrast to keyword privacy
	K-gram [40]	—Effective and secure method in encrypted environment	—Can not support a multiway tree structure —Relatively need high storage space

c) Fuzzy Keyword Search (FKS). This search technique enhances the system's usability when searching for inputs that exactly match. The editing distance is used to quantify keyword similarity and construct fuzzy keyword sets. The data user can check the correctness and completeness of the search results. There are two approaches: wildcard and straight forward for dealing with the edit distance [10,23].

d) Authorized Private Keyword Search (APKS). A multi-keyword search that has greater flexibility and efficiency compared to a single keyword search [24]. Hierarchical Predicate Encryption (HPE) is a primitive cryptographic attribute hierarchy used in a fine-grained authorization system for a simple range of queries. Each user obtains a Local Trusted Authority (LTA) search authorization. During the encryption and decryption processes, $APKS+$ provides a hidden key to conceal the information from the attackers. Thus, $APKS+$ prevents attacks by dictionary keywords, protects the index and privacy of queries.

e) Attribute-Based Encryption (ABE). A novel cryptography solution that implements access control policies [20]. It asserts that one upload includes several policies that $PEKS$ and HVE do not endorse. Attribute-Based Encryption uses an access policy when searching for Boolean expressions for encrypted information. ABE affords better search quality on encrypted data and fast accessing methods.

f) Predicate Privacy Preserving in Public Key Encryption (PPP_PEKS). *PPP*_PEKS uses a randomization scheme to search for keywords to the best of *PEKS*. The technique of randomization will randomize keywords. Thus, the trapdoors do not provide any meaningful information [42]. A pair of secret keys exchange with the data owner and the recipient, which is not logical for many users. Two mechanisms were specifically present to accept guessing attacks: $PEKSrand - BG$ for brute-force guessing and $PEKSrand - SG$ for statistical guessing, as shown in Fig. 3. Ranked Search The ranked keyword search has a better performance than the Boolean Search by minimizing the major drawback. There is also support for indirect mapping between keywords and trapdoors. So, total computation, communication, and storage overheads are sensible in *PEKS*. Note: -Boolean searches on searchable Encryption have two significant drawbacks.

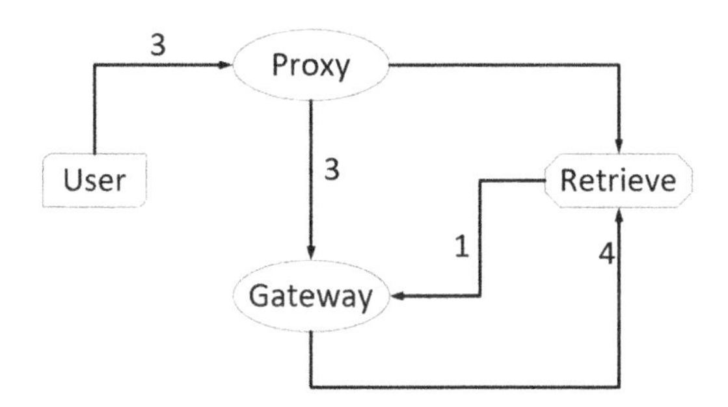

Fig. 3. Working principle of PEKSrand

- Retrieving all files leads to network traffic.
- The user wants to decrypt every file that contains the keyword queried.

a) Keyword Search-Based on Ranking over Encrypted Data. The Ranked Searchable Symmetric Encryption (RSSE) framework backing rank search is a cryptographic primitive build on the *SSE*. It comprises four algorithms, specifically *KeyGen*, *BuildIndex*, *TrapdoorGen*, *SearchIndex*. Accordingly, a well-organized *RSSE* framework adopts the Order Preserving Symmetric Encryption (OPSE) scheme and supports deterministic properties [35].

b) *K*-gram based fuzzy keyword Ranked Search (*K*-gam) Searchable encryption methods allow data users to query encrypted data based on keywords securely. These approaches only support precise keyword search and fail to perform spelling mistakes/morphological variants of words [40].

4.2 Secure Spatial Query

Nowadays, there are many searchable encryptions proposed based on a spatial query. Here we classify some papers based on the searchable encryption method supporting either kNN query or range query, and we will discuss their merits and demerits. By doing this, we can carefully select the technique for practical implementations that keep encrypting information retrieve.

Table 3. Summarazation and comparison of secure spatial query

Spatial query category	Spatial query schemes	Merits	Demerits
Secure kNN query processing methods	VD-kNN [3]	—Process encrypted Voronoi diagrams and return an exact result —It is more secure and accurate	—For k >1, this may place a heavy load on the data owner
	B-kNN [3]	—To minimise overheads, it uses the notion of query squares	—It is inefficient to query the process by itself —It has a high computational cost
	T-kNN [3]	—It can support any k range value —Decrease the data owner load —This method is less expensive compared to the VD-kNN method with higher accuracy	—It may not return true kNN at all times
Secure range query processing methods	*h*RQ [34]	—*h*RQ approach is secure and efficient —Protect ordering information	—It may introduce false positives
k-Nearest Neighbor Classification (Relational data)	PPkNN [36]	—It protects data confidentiality, the input query of the user, and hides the pattern of data access —To provide an efficient solution for classification problem	—It does not solve the DMED (Data Mining encrypted data) problem —It is not used to encrypted highly sensitive information
Secure k-Nearest Neighbor Query (Outsourced environments)	SkNN [4]	—It protects the queries the database	—Encrypted data are not secure

Secure kNN Query Processing Methods

a) Voronoi Diagram kNN Method for Secure NN queries (VD-kNN). A system works by handing out encrypted Voronoi diagrams and returning reliable results to protect the closest neighborhood queries [11].

b) Basic k Nearest Neighbor (B-kNN). This system summarises an individual pair of encrypted data points, and the CSP must govern according to Secure Distance Comparison Methods (SDCM). This scheme sums up an individual pair of encrypted data points, and the CSP must govern according to Secure Distance Comparison Methods (SDCM). CSP should determine the data point That is closer to the encrypted query point [11].

c) Triangulation-Based kNN (T-kNN). This method deals with any value of k by processing encrypted Delaunay triangulation [11] and reducing the data owner's load. The technique is an approximation for k>1 and achieves high precision in practice.

d) Privacy-Preserving *k*-Nearest Neighbor (PP*k*NN). PPkNN is a safe k-NN classifier based on encrypted, semantically secure data. A scheme where Alice does not engage in any calculations until the encrypted data outsource to the CSP [11]. The protocols used to generate a PPkNN scheme are Secure Multiplication (SM), Secure Squared Euclidean Distance (SSED), Secure Bit Decomposition (SBD), Secure Minimum, and Secure Bit-OR (SBOR). The PPkNN protocols protect the confidentiality of records, input requests, and obscure access to data.

e) Secure Processing of k-Nearest Neighbor Query over Encrypted Data (S*k*NN). The SkNN method is implemented to securely understand k-nearest data tuples to Q using the encrypted T database within CSP. Nothing is known to CSP about the actual contents of the t database and the query record Q [16]. The effective SkNN protocol satisfies the following conditions.

- Secure the patterns of CSP data access.
- Computing the k-Nearest Neighbors query Q correctly.
- Incurring low overhead computation of the end-user.

Yousef *et al.* [32] proposed A secure kNN system that preserves information confidentiality, user input queries, and data access patterns.

Secure Range Query Processing Method

(a) Mutable Order-Preserving Encryption (*m*OPE). This method tolerates safe range query evaluation and is the only known Order-Preserving Encoding System (mOPE) verifiably secure to date [26]. In [3,4], the mOPE is distinct from the previous techniques. In a client-server environment, the mOPE scheme works. The client has a hidden key to the asymmetric cryptography method, such as AES, and the data setoff (cyphertexts) can be stored on the CSP in increasing order of plaintexts.

b) Half-Space Range Query (*h*RQ). This scheme can accomplish polyhedral queries on encrypted data. $\hat{R}$-trees are essential for encrypting and outsourcing the index to the database [36]. The $\hat{R}$-trees index is preferable because there is a lower leakage of information than bucketization schemes such as a k-means, BNL-tree, and kD-tree [38].

4.3 Secure Spatial Keyword Query

Privacy-Preserving Top-K Spatial Keyword Queries (PkSKQ). In terms of spatial proximity and textual significance, This scheme considers k objects ideal for the query. The privacy-preserving top-k query, where an improved version of $ASPE$ encrypts spatial and textual data, is facilitated by a secure index. An enhanced variant of $ASPE$ is $ASPEN$ with Noise [32].

Table 4. Summarazation and comparison of secure spatial keyword query

Spatial keyword query category	Spatial keyword Query Schemes	Merits	Demerits
Privacy-preserving Top-k spatial Keyword queries	PkSKQ [32]	—Enable search over encrypted tree index —Secure pruning techniques based on keywords —Improved spatial-textual data query output on broad scales —Valid and secure scheme	—If CSP is colluding with the data users, the data users can reveal the secure structure —The use of pattern hiding techniques such as Private Information Retrieval (PIR) [39] and Oblivious RAM [18,31] is ineffective
Privacy-preserving boolean spatial keyword queries	PPBSKQ [14]	—Efficient and secure query —Improved Index (dimension expansion and space reduction)	—Transmission channel insecurity —Challenge to get reduced space

Keyword-Based Secure Pruning. The primary goal is to prune the nodes that do not contain any of the queried keywords. A powerful and storage-saving technique for storing information about an object's existence in a dataset is a Bloom filter [2].

Privacy-Preserving Boolean Spatial Keyword Queries (PPBSKQ). This scheme is a standard query for spatial keywords that takes both the spatial range and the keyword into account. All the query keywords must be understood by each resulting candidate and located with a definite query range [14]. The first

Table 5. Comparison of existing geo-textual indices

RF	Algorithm	Threat model		Plain-Text Attack		Security			Privacy					Performance				Other	
		CH	M	CPA	KPA	Au	Cn	In	QP	DP	RP	PPP	APP	Cr	Sc	Mo	Ef	TS	PD
[14]	PPBSKQ	✓		✓	✓	Δ	✓	✗	✓	✓	✓	✓	✗	✗	✓	static	linear	✗	✗
[25]	SPKS	Δ	Δ	✗	✗	Δ	✓	✓	✗	✓	✓	✓	✓	✓	Δ	static	linear	✗	✗
[10]	FKS	✓		✓	✓	Δ	✓	✗	✓	✓	✓	✗	✗	✗	Δ	static	nonlinear	✗	✗
[23]	FKS	✓		✓	✓	Δ	✓	✗	✓	✓	✓	✗	✗	✗	Δ	static	nonlinear	✗	✗
[24]	APKS	✓		✗	✗	Δ	✓	✗	✗	✓	✗	✗	✗	✗	✓	static	nonlinear	✗	✗
	APKS+	✓			✓	Δ	✓	✓	✓	✓	✓	✓	✓	✓	✓	static	Δ	✗	✗
[20]	ABE	Δ		Δ	Δ	Δ	✓	✓	✓	✓	✓	✓	✓	✓	Δ	Δ	Δ	✗	✓
[35]	RSSE	✓		✓	✓	Δ	✓	✓	✓	✓	✓	✓	✓	✓	✓	both	nonlinear	✗	✗
[40]	K-gram	✗	✗	Δ	✓	Δ	✓	✗	✓	✓	✓	✗	✓	✗	✓	static	Δ	✗	✗
[11]	VD-kNN	✓		✓	✓	Δ	✓	✓	✓	✓	✓	✓	✓	✓	✓	static	linear	✗	✗
	B-kNN	✓		✓	✓	Δ	✓	✓	✓	✓	✓	✓	✓	✓	✓	static	linear	✗	✗
	T-kNN	✓		✓	✓	Δ	✓	✓	✓	✓	✓	✓	✓	✓	✓	static	linear	✗	✗
[28]	SkNN	✓		✓	✓	Δ	✓	✓	✓	✓	✓	✓	✗	✓	Δ	static	nonlinear	✗	✗
[16]	PPkNN	✓		✓	✓	Δ	✓	✓	✓	✓	✓	✓	✓	✓	Δ	static	nonlinear	✗	✗
[36]	hRQ	✓		✗	✗	Δ	✓	✓	✓	✓	✗	✓	✓	✓	✗	static	nonlinear	✗	✗
[32]	PkSKQ	✓		✓	✓	Δ	✓	✗	✓	✓	✓	✓	✗	✗	✓	static	linear	✗	✗

Notes. ✓: Denotes that the method satisfies the criterion;
✗: Denotes that the method does not satisfies the criterion;
Δ: Denotes that the method does not care about the criterion or, (the method meets the criterion but not clarified)

difficulty is the creation of consolidated keyword data. In the meantime, under the widely accepted known background thread model [5,34], it can support query processing. A new spatial-textual Bloom filter encoding approach transforms spatial and text information into vectors to find this issue out. Based on ASPE [37], the mapped information in the Bloom filter encoder can be encrypted.

5 Comparison, Analysis, and Summary

Privacy-preserving refers to both query and data privacy. The queried keywords are one way secure and will not be revealed to the CSP since each keyword is encrypted with a secure key. Location privacy is also achieved by using improved encryption methods known to the data owner and end-user only. Keywords are secure, similar to query keywords. The document location is still not hidden from the CSP as the secure index is built on the encrypted documents based on their locations. Therefore, the system ensures the privacy preservation of both the query and the data, but the CSP can know the querys' accurate location.

In this survey, we summarize three basic privacy-preserving queries and evaluate them. The performance evaluation result of all reviewed papers under a group of secure keyword query (see Table 2), secure spatial query (see Table 3), and secure spatial keyword query in Table 4. We summarize the comparison of different methods against the evaluation parameters in Table 5.

Our survey shows that rank-based data retrieval has better information security, fast search access, and does not outflow data to untrusted authorities, and is sufficient for encrypted data searching. The protected kNN query protocol over encrypted data preserves data confidentiality, user requests privacy, and hides patterns of access to information. A privacy-preserving kNN classification is proposed to provide a novel solution to the PPkNN classifier issue over encrypted data.

In addition to privacy promises, a privacy-preserving range query scheme effectively concerns overhead storage, the processing time of queries, and overhead communication. The cloud's overhead storage should be low since most of the information stored in the cloud is typically huge. As many applications need real-time queries, the database processing time desired is negligible. Other than encrypted data objects, the communication overhead denotes the information transmitted between the data owner and the CSP. The data transferred between the data users and the CSP, rather than the query's exact results, is also referenced. Because of bandwidth restrictions and additional time requirements, the overhead is low for uploading and downloading.

Access pattern privacy and path pattern privacy, hiding data access patterns from the CSP, are not considered adequately in most privacy-preserving spatial and textual queries. Secure privacy-preserving keywords and authorized private keyword searches are also do not consider in a detailed manner. Another important parameter is to support scalability since most of the methods give good results for a few data. However, in reality, there are a vast amount of data in could server to process. Thus, these methods are not applicable for real-time applications.

Threat Model. The server follows the mandatory protocol for communication and correctly implements the algorithms needed. It may, however, attempt to obtain information about the data and the content of the queries with the help of domain knowledge. Most of the methods are built based on this curious but honest model, except for SPKS, ABE, and k-gram techniques. The critical issue occurs if it is not trusted by either the data user or the data owner.

Authorization and Access Control. During our survey, not almost all methods consider this privacy parameter, which controls every request to a server and determines the granted or denied request based on specific rules.

Computation and Communication Costs. In our survey, we have found that some of the methods are linear. Which includes SPKS, VD-kNN, B-kNN, T-kNN, PkSKQ, and PPBSKQ, and some others are groups under the category of nonlinear (FKS, APKS, RSSE, hRQ, PPkNN, SkNN. The rest of the methods are unknown whether they include or not.

Other Attributes. It includes attributes other than text and geographic locations, including timestamp (actual time location), place description, and some comments to find places. These attributes are essential and common in recent social media usage (Twitter) has high significance. Except for ABE and hRQ, all others have not these attributes.

6 Open Research Issues and Future Research Directions

Based on the proposed evaluation criteria, we have come across some open issues with spatial keyword search over encrypted data outlined in Section VIA. Furthermore, In Section VIB, we are trying to suggest a list of future directions for study.

6.1 Open Issues

Our serious survey has found some open research issues in privacy-preserving spatial, keyword, and spatial keyword queries, which need future exploration.

First, all the existing research work in our survey on secure and privacy-preserving spatial queries are based on the static objects in which users who initiate a query are at a fixed position. Besides, existing work assumes that the data to be searched is not updated within some time interval. However, in our everyday lives, we move around outside to use our smartphones and other mobile devices to receive information anywhere and check for locations. At this point, we need a secure and privacy-preserving dynamic spatial keyword, which is still an open issue.

Second, one of the vital privacy parameters is authorization or access control. All existing work research on the spatial keyword queries is based on a curious but honest server. They have not considered a complete dishonest threat model.

Third, there still a lack of sound solutions for having additional attributes like timestamp and place description in the existing secure queries. These attributes seem very simple but have a high significance in a practical situation as used on Twitter. Nevertheless, in our survey, except for ABE and hRQ methods, other methods do not include these attributes.

6.2 Future Direction

We propose some future research directions to implement a usable and secure authentication system with privacy preservation queries.

- Study on a secure and privacy-preserving keyword, spatial and spatial keyword queries systems are in our day-to-day activities. In the face of such unsafe and dynamic cyberspace, some open issues must be resolve as soon as possible. Present-day, the commonly used mobile application for find location has become common as it improves the querying time and system performance by guaranteeing its security and user privacy. How to shield the user's data is a central research topic. Specifically, when the user-sensitive information like medical records stored in a CSP or daily routing recorded in the CSP is not fully trusted.
- Secure and privacy-preserving spatial keyword queries on dynamic spatial keyword objects are not well exploited. A series of factors could affect security and system performance. Designing usable, fast, secure, and privacy-preserving systems is an important topic, particularly when security and privacy become hot issues. Besides, for the system to operate efficiently and accurately for customers, a suitable searching algorithm plays a key role. Advanced algorithms should further examine to support efficiency, security, and privacy. At the same time help to forecast some open issues and imminent research tips. We strongly suggest that improving the security and privacy of existing secure search schemes should be highlighted in future research.
- The cost of authentication is a source-restricted mobile device that should be considered. Most mobile devices (for instance, smart bracelets and mobile phones) have limited electricity, computation capability, and storage space. It makes them applicable in a native search service, an online business directory, a GPS navigation system, and other applications due to an increasing overview of geo-positioning technologies and geolocation services. Thus, it is essential to consider secure and privacy-preserving methods and algorithms implemented without quality degradation in all these limiting parameters.
- The Continuous k Nearest Neighbors (CkNN) and Continuous Range (CR) queries are specific types of Continuous Spatial Queries (CSQs). In both cases, a protected region-based query processing techniques can be managed using secure regions to find more kinds of queries. In the past few years, privacy-preserving spatial keyword queries have attracted substantial interest. Therefore, CSQ has implemented some sympathetic extension, which has implications for privacy-preserving continuous top-k spatial keyword queries and privacy-preserving continuous range spatial keyword queries.
- Most current research effort considering the curious but honest threat model of the cloud server. The server will obey the schemes planned but will attempt

to reveal the data users' and owners' sensitive data. A performance assessment of the curious but honest or dishonest threat model is not addressed well. Future research should improve the quality of model performance evaluation. To prove the effectiveness of privacy-preserving spatial keyword search method, more holistic model evaluation should be conducted.
- When the data owner sends the data, index, and search algorithm to the CSP, data, query, and location privacy should be maintained. The cloud service provider will process the query that the data user sends. In reality, for spatial keyword queries, a lightweight and powerful privacy-preserving scheme is highly predictable. A concrete application demands these severe issues are answered. In this situation, cryptography-based schemes may not be valid in the spatial keyword search. A trivial and well-organized solution is highly expected. Therefore, future research work should give proper attention to the effect of scalability in the system.

Therefore, future research work should give proper attention to the effect of scalability in the system.

7 Conclusion

We have reviewed the recent developments in privacy-preserving spatial, keyword, and spatial keyword queries. This paper pointed out the basic secure privacy-preserving queries and further proposed a sequence of evaluation criteria for assessing current works' performance. We presented a comparative evaluation of the topical outcomes by dividing the existing privacy-preserving searching methods into three categories based on the point of interest they are designed. We found that some of the designs of the current systems are based on honest but curious models. We build some open problems and predict forthcoming research advice based on our survey. We have confidence in taming the security and privacy of existing secure search schemes getting proper attention in the future study.

This paper marks the beginning of a new age of protected privacy research. Many previous surveys, for example, concentrated on secure spatial and secure keyword queries separately, rather than offering a detailed overview of privacy-preserving spatial keyword queries. The study of a stable, privacy-preserving spatial keyword query in both static and dynamic situations is still in its early stages, and there are many important research questions to be answered, such as:

- Our day-to-day activities, the widely used mobile application for finding a location, have become popular and capable of boosting querying time and device efficiency by maintaining user privacy and protection.
- Secure and privacy-preserving spatial keyword queries on dynamic spatial keyword objects are not well exploited.
- The cost of authentication in a source-restricted mobile device should be taken into account.

– The consistency of model performance assessment should be improved in future research.

We expect the importance of questions like these to grow with increasing commercial interest in a secure and privacy-preserving keyword, spatial and spatial keyword queries system.

Acknowledgements. This work was supported by the National Natural Science Foundation of China (No. 62072361), the Fundamental Research Funds for the Central Universities (No. JB211505)

References

1. Beckmann, N., Kriegel, H.P., Schneider, R., Seeger, B.: The r*-tree: an efficient and robust access method for points and rectangles. In: Proceedings of the 1990 ACM SIGMOD International Conference on Management of Data, pp. 322–331, SIGMOD 1990. Association for Computing Machinery, New York, NY, USA (1990). https://doi.org/10.1145/93597.98741
2. Bloom, B.H.: Space/time trade-offs in hash coding with allowable errors. Commun. ACM **13**(7), 422–426 (1970). https://doi.org/10.1145/362686.362692
3. Boldyreva, A., Chenette, N., Lee, Y., O'Neill, A.: Order-preserving symmetric encryption. In: Joux, A. (ed.) EUROCRYPT 2009. LNCS, vol. 5479, pp. 224–241. Springer, Heidelberg (2009). https://doi.org/10.1007/978-3-642-01001-9_13
4. Boldyreva, A., Chenette, N., O'Neill, A.: Order-preserving encryption revisited: improved security analysis and alternative solutions. In: Rogaway, P. (ed.) CRYPTO 2011. LNCS, vol. 6841, pp. 578–595. Springer, Heidelberg (2011). https://doi.org/10.1007/978-3-642-22792-9_33
5. Cao, N., Wang, C., Li, M., Ren, K., Lou, W.: Privacy-preserving multi-keyword ranked search over encrypted cloud data. IEEE Trans. Parallel Distrib. Syst. **25**(1), 222–233 (2014). https://doi.org/10.1109/TPDS.2013.45
6. Chang, Y.-C., Mitzenmacher, M.: Privacy preserving keyword searches on remote encrypted data. In: Ioannidis, J., Keromytis, A., Yung, M. (eds.) ACNS 2005. LNCS, vol. 3531, pp. 442–455. Springer, Heidelberg (2005). https://doi.org/10.1007/11496137_30
7. Chen, L., Shang, S., Yang, C., Li, J.: Spatial keyword search: a survey. GeoInformatica **24**(1), 85–106 (2019). https://doi.org/10.1007/s10707-019-00373-y
8. Chen, L., Cong, G., Jensen, C.S., Wu, D.: Spatial keyword query processing: An experimental evaluation. Proc. VLDB Endow. **6**(3), 217–228 (2013). https://doi.org/10.14778/2535569.2448955
9. Chen, Y.Y., Suel, T., Markowetz, A.: Efficient query processing in geographic web search engines. In: Proceedings of the 2006 ACM SIGMOD International Conference on Management of Data, pp. 277–288, SIGMOD 2006, Association for Computing Machinery, New York, NY, USA (2006). https://doi.org/10.1145/1142473.1142505
10. Cheng, L., Jin, Z., Wen, O., Zhang, H.: A novel privacy preserving keyword searching for cloud storage. In: 2013 Eleventh Annual Conference on Privacy, Security and Trust, pp. 77–81 (2013). https://doi.org/10.1109/PST.2013.6596039
11. Choi, S., Ghinita, G., Lim, H., Bertino, E.: Secure KNN query processing in untrusted cloud environments. IEEE Trans. Knowl. Data Eng. **26**(11), 2818–2831 (2014). https://doi.org/10.1109/TKDE.2014.2302434

12. Christoforaki, M., He, J., Dimopoulos, C., Markowetz, A., Suel, T.: Text vs. space: efficient geo-search query processing. In: Proceedings of the 20th ACM International Conference on Information and Knowledge Management, pp. 423–432, CIKM 2011. Association for Computing Machinery, New York, NY, USA (2011). https://doi.org/10.1145/2063576.2063641
13. Cong, G., Jensen, C.S., Wu, D.: Efficient retrieval of the top-k most relevant spatial web objects. Proc. VLDB Endow. **2**(1), 337–348 (2009). https://doi.org/10.14778/1687627.1687666
14. Cui, N., Li, J., Yang, X., Wang, B., Reynolds, M., Xiang, Y.: When geo-text meets security: privacy-preserving Boolean spatial keyword queries. In: 2019 IEEE 35th International Conference on Data Engineering (ICDE), pp. 1046–1057 (2019). https://doi.org/10.1109/ICDE.2019.00097
15. Eldawy, A., Mokbel, M.F.: The era of big spatial data. Proc. VLDB Endow. **10**(12), 1992–1995 (2017). https://doi.org/10.14778/3137765.3137828
16. Elmehdwi, Y., Samanthula, B.K., Jiang, W.: Secure k-nearest neighbor query over encrypted data in outsourced environments. In: 2014 IEEE 30th International Conference on Data Engineering, pp. 664–675 (2014). https://doi.org/10.1109/ICDE.2014.6816690
17. Ghinita, G.: Privacy for location-based services. Synthesis Lect. Inf. Secur. Privacy Trust **4**(1), 1–85 (2013)
18. Goldreich, O., Ostrovsky, R.: Software protection and simulation on oblivious rams. J. ACM **43**(3), 431–473 (1996). https://doi.org/10.1145/233551.233553
19. Guttman, A.: R-trees: A dynamic index structure for spatial searching. In: Proceedings of the 1984 ACM SIGMOD International Conference on Management of Data, pp. 47–57, SIGMOD 1984. Association for Computing Machinery, New York, NY, USA (1984). https://doi.org/10.1145/602259.602266
20. Hohenberger, S., Waters, B.: Attribute-based encryption with fast decryption. In: Kurosawa, K., Hanaoka, G. (eds.) Public-Key Cryptography - PKC 2013, pp. 162–179. Springer, Heidelberg (2013)
21. Khodaei, A., Shahabi, C., Li, C.: Hybrid indexing and seamless ranking of spatial and textual features of web documents. In: Bringas, P.G., Hameurlain, A., Quirchmayr, G. (eds.) DEXA 2010. LNCS, vol. 6261, pp. 450–466. Springer, Heidelberg (2010). https://doi.org/10.1007/978-3-642-15364-8_37
22. Krumm, J.: A survey of computational location privacy. Pers. Ubiquit. Comput. **13**(6), 391–399 (2009)
23. Li, J., Wang, Q., Wang, C., Cao, N., Ren, K., Lou, W.: Fuzzy keyword search over encrypted data in cloud computing. In: 2010 Proceedings IEEE INFOCOM, pp. 1–5 (2010). https://doi.org/10.1109/INFCOM.2010.5462196
24. Li, M., Yu, S., Cao, N., Lou, W.: Authorized private keyword search over encrypted data in cloud computing. In: Proceedings of the 2011 31st International Conference on Distributed Computing Systems, pp. 383–392, ICDCS 2011. IEEE Computer Society, USA (2011). https://doi.org/10.1109/ICDCS.2011.55
25. Liu, Q., Wang, G., Wu, J.: Secure and privacy preserving keyword searching for cloud storage services. J. Netw. Comput. Appl. **35**(3), 927–933 (2012)
26. Popa, R.A., Li, F.H., Zeldovich, N.: An ideal-security protocol for order-preserving encoding. In: 2013 IEEE Symposium on Security and Privacy, pp. 463–477 (2013). https://doi.org/10.1109/SP.2013.38
27. Qi, J., Zhang, R., Jensen, C.S., Ramamohanarao, K., HE, J.: Continuous spatial query processing: a survey of safe region based techniques. ACM Comput. Surv. **51**(3), 1–39 (2018). https://doi.org/10.1145/3193835

28. Samanthula, B.K., Elmehdwi, Y., Jiang, W.: K-nearest neighbor classification over semantically secure encrypted relational data. IEEE Trans. Knowl. Data Eng. **27**(5), 1261–1273 (2015). https://doi.org/10.1109/TKDE.2014.2364027
29. Schreter, I., Gottipati, C., Legler, T.: Inverted indexing, 15 May 2018. US Patent 9,971,770
30. Shen, J., Liu, D., Shen, J., Tan, H., He, D.: Privacy preserving search schemes over encrypted cloud data: a comparative survey. In: 2015 First International Conference on Computational Intelligence Theory, Systems and Applications (CCITSA), pp. 197–202 (2015). https://doi.org/10.1109/CCITSA.2015.46
31. Stefanov, E., et al.: Path ORAM: an extremely simple oblivious RAM protocol. In: Proceedings of the 2013 ACM SIGSAC Conference on Computer and Communications Security, pp. 299–310, CCS 2013. Association for Computing Machinery, New York, NY, USA (2013). https://doi.org/10.1145/2508859.2516660
32. Su, S., Teng, Y., Cheng, X., Xiao, K., Li, G., Chen, J.: Privacy-preserving top-k spatial keyword queries in untrusted cloud environments. IEEE Trans. Serv. Comput. **11**(5), 796–809 (2018). https://doi.org/10.1109/TSC.2015.2481900
33. Vaid, S., Jones, C.B., Joho, H., Sanderson, M.: Spatio-textual indexing for geographical search on the web. In: Bauzer Medeiros, C., Egenhofer, M.J., Bertino, E. (eds.) SSTD 2005. LNCS, vol. 3633, pp. 218–235. Springer, Heidelberg (2005). https://doi.org/10.1007/11535331_13
34. Wang, B., Yu, S., Lou, W., Hou, Y.T.: Privacy-preserving multi-keyword fuzzy search over encrypted data in the cloud. In: IEEE INFOCOM 2014 - IEEE Conference on Computer Communications, pp. 2112–2120 (2014). https://doi.org/10.1109/INFOCOM.2014.6848153
35. Wang, C., Cao, N., Ren, K., Lou, W.: Enabling secure and efficient ranked keyword search over outsourced cloud data. IEEE Trans. Parallel Distrib. Syst. **23**(8), 1467–1479 (2012). https://doi.org/10.1109/TPDS.2011.282
36. Wang, P., Ravishankar, C.V.: Secure and efficient range queries on outsourced databases using rp-trees. In: 2013 IEEE 29th International Conference on Data Engineering (ICDE), pp. 314–325 (2013). https://doi.org/10.1109/ICDE.2013.6544835
37. Wong, W.K., Cheung, D.W.l., Kao, B., Mamoulis, N.: Secure KNN computation on encrypted databases. In: Proceedings of the 2009 ACM SIGMOD International Conference on Management of Data, pp. 139–152, SIGMOD 2009, Association for Computing Machinery, New York, NY, USA (2009). https://doi.org/10.1145/1559845.1559862
38. Wu, M.C., Buchmann, A.P.: Encoded bitmap indexing for data warehouses. In: Proceedings 14th International Conference on Data Engineering, pp. 220–230. IEEE (1998)
39. Yi, X., Paulet, R., Bertino, E.: Private Information Retrieval, 1st edn. Morgan & Claypool Publishers, New York (2013)
40. Zhou, W., Liu, L., Jing, H., Zhang, C., Wang, S.Y.S.: K-gram based fuzzy keyword search over encrypted cloud computing. J. Softw. Eng. Appl. **06**(01), 29–32 (2013)
41. Zhou, Y., Xie, X., Wang, C., Gong, Y., Ma, W.Y.: Hybrid index structures for location-based web search. In: Proceedings of the 14th ACM International Conference on Information and Knowledge Management, pp. 155–162 (2005)
42. Zhu, B., Zhu, B., Ren, K.: PEKSRand: providing predicate privacy in public-key encryption with keyword search. In: 2011 IEEE International Conference on Communications (ICC), pp. 1–6 (2011). https://doi.org/10.1109/icc.2011.5962452

Real-Time Stream Statistics via Local Differential Privacy in Mobile Crowdsensing

Teng Wang[1(✉)] and Zhi Hu[2]

[1] Xi'an University of Posts and Telecommunications, Xi'an 710121, China
wangteng@xupt.edu.cn
[2] Northwest University, Xi'an 710127, China

Abstract. Mobile crowdsensing has enabled the collection and analysis of stream data. However, the direct processing of gigantic stream data will seriously compromise users' privacy since those stream data involve numerous sensitive information. To address the challenges of the vulnerabilities of untrusted crowdsensing servers and low data utility, we propose an effective real-time stream statistics mechanism that can not only achieve strong privacy guarantees, but also ensure high data utility. We firstly apply local perturbation on each user's stream data on the client side, which achieves ω-event ϵ-local differential privacy for each user at each timestamp. Then, we propose a retroactive grouping-based noise smoothing strategy that adaptively exploits the time correlations of stream data and smooths excessive noises, thus improving data utility. Finally, experimental results on real-world datasets show the strong effectiveness of our mechanism in terms of improving data utility.

Keywords: Stream data · Local differential privacy · Time correlation · Data utility

1 Introduction

Mobile crowdsensing has greatly promoted and facilitated the big data collection, analysis and utilization [15,20]. Without deploying thousands of static sensors, a large scale mobile crowdsensing system can be easily formed with portable smart mobile devices, such as mobile phones, smart watch/bands, etc. In such a system, each participant will joint one or more monitoring tasks and continuously contribute her/his personal stream data for statistics. Therefore, mobile crowdsensing has been widely adopted for many data-driven applications, such as traffic flow control, population density monitoring, health management [2,28,29].

However, the continual collection and analysis of stream data seriously violates the privacy of each participant and causes severe impacts [18,30]. Because

J. Xiong et al. (Eds.): MobiMedia 2021, LNICST 394, pp. 432–445, 2021.
https://doi.org/10.1007/978-3-030-89814-4_31

such stream data involve in individuals' identity, location, health status, or other sensitive information. What's worse, the privacy leaks will gradually accumulate as time continues to grow [6], which will produce negative consequences.

Differential privacy (DP) [12], as a rigorous privacy paradigm, has been widely adopted to provide users with privacy protection in real-time monitoring systems [10,11]. Many existing studies [5,7,14,17,22] have applied DP on stream data publishing when facing on a trustworthy server, which cannot be directly used for stream data processing in mobile crowdsensing.

As a distributed variant of DP, local differential privacy (LDP) [25] is enable to provide privacy guarantees for each user locally and is independent of any assumptions on servers. Based on LDP, Erlingsson et al. [13] proposed to provide longitudinal privacy protection for user's successive data. Ding et al. [8] designed an LDP-compliant method for continual data collection and estimation. Moreover, Joseph et al. [16] also focused on collecting up-to-date statistics over time under LDP. When considering the dynamic and time correlations of stream data, the privacy-preserving mechanism for real-time stream data processing is still in the early stage of research. In nature, the privacy-preserving concerns of real-time statistics mainly cover two-fold challenges in mobile crowdsensing:

1) **Non-local privacy.** Existing studies on real-time stream statistics mainly using centralized differential privacy that assumes a trusted server. But in mobile crowdsensing systems, the servers may be untrustworthy and interested in users' sensitive data.
2) **Low data utility.** In nature, the stream data are highly sparse and dynamic, and accompanied by complex time correlations. In its simplest way of LDP deployment, the direct perturbation on stream data is vulnerable to low data utility.

In this paper, to address the above challenges and further improve data utility, we propose a local differential privacy (LDP)-based stream data collection and statistics mechanism which incorporates a retroactive grouping strategy to guide noise smoothing in an adaptive way. In summary, we make the following contributions.

- We propose to perturb stream data of each user on the client side by applying LDP, which prevents privacy leakage from the data source when considering untrusted servers in crowdsensing. Based on the randomized response technique, the local perturbation is performed on each user's stream data to achieve ω-event ϵ-LDP at each timestamp.
- We design a retroactive grouping-based noise smoothing strategy to reduce noise error in stream aggregation and estimation. The retroactive grouping can adaptively exploit the time correlations of the dynamic streams, thus making noise smoothing effective and improving data utility.
- We conduct extensive experiments on two real-world datasets. The results demonstrate that our mechanism can improve data utility while ensuring strong privacy protections.

The remainder of the paper is organized as follows. Section 2 provides a literature review. Section 3 outlines the preliminaries and problem statement. Section 4 introduces our proposed mechanism. Section 5 presents the evaluation results. Finally, Sect. 6 concludes the paper.

2 Related Works

The privacy-preserving mechanisms for real-time stream data processing have been widely studied under differential privacy (DP). The design principle of existing schemes consists of transformation, modeling, sampling, grouping, and clustering.

Based on the compressibility of time-series data, Xiao et al. [27] proposed to transform the frequency domain of the raw data into a wavelet coefficient matrix and then add Laplace noise to the coefficients to achieve DP. As for large-scale spatio-temporal data, Acs et al. [4] combined sampling, clustering, Fourier perturbation, and smoothing to improve the utility of the published data. Besides, both [19] and [3] adopted discrete Fourier transformation technique under differential privacy protection.

As for dynamic stream data, Fan et al. [14] proposed FAST which adaptively samples data points and adds noise based on Kalman filter and PID feedback error. FAST can capture the dynamic changes of stream data by an adaptive sampling and filtering mechanism, thus reducing the total noise. Moreover, Wang et al. [21] proposed to use an unscented Kalman filter to publish time-series data in non-linear scenes. To publish infinite streams, Kellaris et al. [17] proposed ω-event privacy model which can protect any event sequences within any window of ω timestamps. Correspondingly, they also introduced two privacy budget allocation mechanisms BD and BA that ensure better data utility than uniform distribution mechanisms. Wang et al. [22,23] proposed RescueDP which integrates adaptive sampling and budget allocation, dynamic grouping, perturbation, and filtering to publish multiple infinite streams with high data utility. Chen et al. [7] proposed PeGaSus, which adopts a perturber-grouper-smoother framework to smooth excessive noises, thus improving data utility greatly.

However, the above mechanisms adopt DP to achieve privacy protection under the assumption of a trusted server, which cannot prevent insider attacks from inferring the privacy of users. To address this, local differential privacy (LDP) [9] can be accepted to ensure data privacy for distributed crowdsensing tasks. In this direction, LDP is widely used to alleviate the privacy concerns for many data collection and analytic tasks, such as frequency and mean value estimation, frequent items mining, marginal release, empirical risk minimization (ERM), deep learning, recommendation system, etc. [25].

So far, most studies focus on data collection or analysis with LDP only for one-time computation. As for evolving data and repeated collection, Erlingsson et al. [13] proposed the RAPPOR mechanism to publish binary attributes with LDP, which tries to achieve longitudinal privacy protection for each user. Besides, Ding et al. [8] developed LDP protocols for repeated collection and computation of counter data such as daily APP usage statistics. Moreover, Joseph et al. [16] proposed the THRESH mechanism under LDP for collecting up-to-date statistics over time, which only updates the global estimation when it becomes sufficiently inaccurate, thus improving data utility. Nonetheless, the LDP-compliant mechanism for evolving stream data is still in its infancy.

3 Preliminaries and Models

In this section, we introduce some background knowledge of local differential privacy, formalize the system/stream model, and present the problem definition.

3.1 Local Differential Privacy

Local differential privacy has been widely adopted to achieve privacy protection in distributed crowdsensing [25].

Definition 1 (ϵ-Local Differential Privacy (ϵ-LDP) [9]). *A randomized algorithm $\mathcal{A}$ satisfies ϵ-LDP if and only if for any pairs of input values v, v' in the domain of $\mathcal{A}$, and for any possible output $y \in Y$, it holds that*

$$\mathbb{P}[\mathcal{A}(v) = y] \leq e^{\epsilon} \cdot \mathbb{P}[\mathcal{A}(v') = y], \tag{1}$$

where $\mathbb{P}[\cdot]$ denotes probability and ϵ is the privacy budget. A smaller ϵ means stronger privacy protection, and vice versa.

Moreover, in the context of continuous stream processing, we consider ω-event privacy [17] as our privacy model since it can protect event sequences within any window of ω timestamps. The definition of ω-event privacy can be found in [17]. Under ω-event privacy, the sequential composition is a key character of LDP, which is defined as follows.

Theorem 1 (Sequential Composition). *Let $\mathcal{A}_i(v)$ be an ϵ_i-LDP algorithm on an input value v, and $\mathcal{A}(v)$ is the sequential composition of $\mathcal{A}_1(v), \cdots, \mathcal{A}_k(v)$. Then algorithm $\mathcal{A}(v)$ satisfies $\sum_{i=1}^{k} \epsilon_i$-LDP.*

3.2 Model Formalization

System Model. We consider a large distributed mobile crowdsensing system that consists of abundant local sensing nodes (i.e., mobile smart devices) and the central server. Each node randomly participates in a sensing task and uploads the sensing state to the central server in real-time. The central server will aggregate the data stream and conduct statistics. We don't make any assumptions about the credibility of the server. That is, the central server can be *honest* or *honest-but-curious*.

In such a system, suppose there are N sensor nodes (i.e., N participating users) that monitor and report on the same sensing task in real time. At each time t, let $r_i(t)$ denote the real state value of the user u_i under the current sensing task. If $r_i(t) = 1$, it means the user u_i holds the target state at time t. If $r_i(t) = 0$, it means the user u_i is not in the target state at time t.

Stream Model. Since the system model is distributed, the stream model is also distributed. Each user u_i holds an infinite source stream dataset D_i. Each tuple (u, s, t) in D_i is an atomic record denoting the user u was in state s at time t. Based on system model, $r_i(t)$ denotes the user u_i's stream data at time t for a given state s. Therefore, for each user u_i, her/his infinite stream data X_i can be represented as $X_i = \{r_i(1), r_i(2), \cdots, r_i(t), \cdots\}$.

3.3 Problem Definition

Let $X_i = \{r_i(1), r_i(2), \cdots, r_i(t), \cdots\}$ denote the real infinite stream of user u_i. Let $Z_i = \{z_i(1), z_i(2), \cdots, z_i(t), \cdots\}$ denote the reported noisy stream data of user u_i. Let $\widehat{C} = \{\hat{c}(1), \hat{c}(2), \cdots, \hat{c}(t), \cdots\}$ denote the statistics of the infinite stream that is estimated in the server side. Based on system model and stream model, the problem in this paper can be formalized as: *collecting the stream data of each user with ω-event ϵ-local differential privacy and estimating the stream statistics while guaranteeing a good data utility.*

4 Our Solution

In this section, we first show the design rationales and an overview of our mechanism. Then, we detail our solution for real-time stream statistics with local differential privacy.

4.1 Design Rationales and Overview

In mobile crowdsensing, the privacy-preserving of real-time stream statistics should cover two-fold concerns:

1) *Local privacy protection.* The users participate in the mobile crowdsensing task in a distributed way. Thus, the stream data of each user can be distributedly perturbed on the user side (i.e., client side) and then sent to the central server. To this end, we apply the randomized response technique to achieve local differential privacy for each user.
2) *Time correlation exploitation.* The direct perturbation on each user's stream data will destroy the time correlation of stream data, leading to a low data utility. Thus, we propose to learn the underlying time correlations of the stream under LDP and smooth the excessive noise to improve data utility.

Based on the above considerations, we proposed a local privacy-preserving mechanism for real-time stream statistics in mobile crowdsensing, as shown in Fig. 1. Our mechanism mainly consists of two components, that is, 1) local perturbation on the client side and 2) aggregation and estimation with error reduction on the server side. The details of each component will be introduced in the next sections.

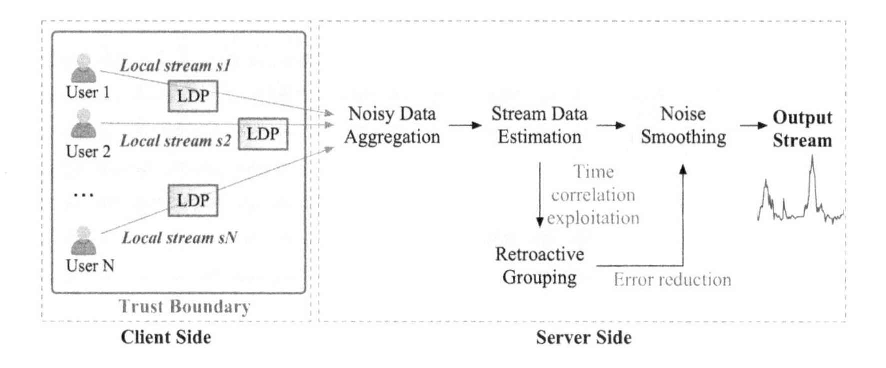

Fig. 1. A high-level overview of our mechanism

4.2 Local Perturbation on the Client Side

In the context of continuous stream processing, event-level [11], user-level [10], and ω-event [17] privacy models are three common paradigms. Since ω-event privacy can protect the stream data within any window of ω timestamps, we adopt it to protect stream in real-time settings in this paper. Besides, concerning the untrusted server in crowdsensing, the local perturbation of our mechanism will achieve ω-event ϵ-LDP for each user's stream data at each timestamp.

For any ω timestamps, the stream data of user u_i can be represented as an ω-bit vector B_i. At time t, the ω-bit vector of user u_i can be denoted as $B_i^t = \{r_i(t-\omega+1), \cdots, r_i(t-1), r_i(t)\}$. Then, our purpose can always achieve ω-event ϵ-LDP for ω-bit vector B_i^t at time t. To this end, we apply randomized response technique on stream data $r_i(t)$ at time t with privacy budget ϵ/ω. The specific perturbation rule is as follows.

$$z_i(t) = \begin{cases} r_i(t), & \text{with probability } f \\ 1 - r_i(t), & \text{with probability } 1-f \end{cases} \tag{2}$$

where f determines the privacy level and $z_i(t)$ is the noisy stream data of real stream data $r_i(t)$.

To guarantee ω-event ϵ-LDP, we set f as $\frac{e^{\epsilon/\omega}}{e^{\epsilon/\omega}+1}$, which will ensure the local privacy for the stream data within any window of ω timestamps. The Algorithm 1 shows the pseudocode of local perturbation. As we can see, after perturbation, each user's noisy data $z_i(t)$ will be sent to the server.

Theorem 2. *The local perturbation in Algorithm 1 satisfies ω-event ϵ-LDP. That is, Algorithm 1 provides ω-event ϵ-LDP for each user.*

Proof. Based on the definition of LDP, for any pairs of input stream data $r(t)$ and $r'(t)$ at time t, the ratio of probabilities for different outputs will satisfy

$$\frac{\mathbb{P}[\mathcal{A}(r(t))]}{\mathbb{P}[\mathcal{A}(r'(t))]} \leq \frac{f}{1-f} = \frac{\frac{e^{\frac{\epsilon}{\omega}}}{e^{\frac{\epsilon}{\omega}}+1}}{1-\frac{e^{\frac{\epsilon}{\omega}}}{e^{\frac{\epsilon}{\omega}}+1}} = e^{\frac{\epsilon}{\omega}} \tag{3}$$

Algorithm 1: Local Perturbation with ω-event ϵ-LDP on the Client Side

Input: $\mathbf{X} = [X_1, X_2, \cdots, X_N]^\top$: real infinite stream data, where each $X_i = \{r_i(1), r_i(2), \cdots, r_i(t), \cdots\}$,
ϵ: privacy budget,
ω: sliding window.
Output: $\mathbf{Z} = [Z_1, Z_2, \cdots, Z_N]^\top$: sanitized infinite stream data, where each $Z_i = \{z_i(1), z_i(2), \cdots, z_i(t), \cdots\}$.

1 **for** *each timestamp t* **do**
2   **for** *each user* u_i $(i = \{1, 2, \cdots, N\})$ **do**
3   Obtain the stream data $r_i(t)$ of user u_i;
4   Perturb $r_i(t)$ to $z_i(t)$ according to Eqn. (2);
5   Send $z_i(t)$ to server;
6 **Return Z**;

Thus, each user achieves $\frac{\epsilon}{\omega}$-LDP at each time t.

In addition, within any ω timestamps, it always holds that $\sum_{t-\omega+1}^{t} \frac{\epsilon}{\omega} = \epsilon$. Hence, Algorithm 1 provides ω-event ϵ-LDP for each user.

4.3 Aggregation and Estimation with Error Reduction on the Server Side

After local perturbation, the server will receive the noisy stream data $\mathbf{Z}(t)$ at each time t, where $\mathbf{Z}(t) = [z_1(t), z_2(t), \cdots, z_N(t)]^\top$. From $\mathbf{Z}(t)$, we can compute the received count of stream data, denoted as $\widehat{N}_t$. Let $x(t)$ be the unbiased estimation of stream data at time t. Based on the perturbation rule (i.e., Eq. (2)), it holds that

$$\frac{e^{\frac{\epsilon}{\omega}}}{e^{\frac{\epsilon}{\omega}} + 1} \cdot x(t) + \frac{1}{e^{\frac{\epsilon}{\omega}} + 1} \cdot (N - x(t)) = \widehat{N}_t \tag{4}$$

Therefore, the unbiased estimation of stream data can be computed as

$$x(t) = \left(\widehat{N}_t - \frac{N}{e^{\frac{\epsilon}{\omega}} + 1}\right) \cdot \frac{e^{\frac{\epsilon}{\omega}} + 1}{e^{\frac{\epsilon}{\omega}} - 1} \tag{5}$$

Theorem 3. *The variance of the unbiased estimation of stream data is* $\frac{N e^{\frac{\epsilon}{\omega}}}{(e^{\frac{\epsilon}{\omega}} - 1)^2}$ *at each time* t.

Proof. The specific proof refers to [26].

However, adding noise at each time t will incur high perturbation errors due to the sparsity and time correlations of stream data. Direct perturbation will break the time correlations among stream data, leading to a low data utility. Therefore, from the perspective of time correlation exploitation, we propose a retroactive grouping strategy to reduce total noise and improve the data utility.

The retroactive grouping strategy aims to divide the timestamps into groups based on the stream received so far. The grouping rule is that the stream data in the same group have a small deviation value from their average. Specifically, this paper adopts deviation function to compute the deviation value between the stream data and the current group. Let $\mathcal{G}_t$ be the group at time t, and $X[\mathcal{G}_t]$ be the set of corresponding stream values in group $\mathcal{G}_t$. The deviation function is formalized as

$$f(X[\mathcal{G}_t]) = \sum_{j\in\mathcal{G}_t} \left| x(j) - \frac{\sum_{j\in\mathcal{G}_t} x(j)}{|\mathcal{G}_t|} \right| \tag{6}$$

where $|\mathcal{G}_t|$ is the size of group.

The deviation value essentially reflects the absolute difference of a set of data from their average. Thus, based on deviation value, we can easily evaluate the quality of each potential group and conduct grouping at each timestamp.

The specific process of stream aggregation and estimation with error reduction on the server side is shown in Algorithm 2. We can see that the unbiased estimation of stream data will be computed firstly at each timestamp t, as shown in lines 2–3.

Algorithm 2: Stream Aggregation and Estimation with Error Reduction on the Server Side

Input: $\mathbf{Z} = [Z_1, Z_2, \cdots, Z_N]^\top$: the reported noisy stream data of N users.
Output: $\widehat{C} = \{\hat{c}(1), \hat{c}(2), \cdots, \hat{c}(t), \cdots\}$: the estimated count of infinite stream data.

1 **for** *each timestamp t* **do**
2   Estimate the received count $\widehat{N}_t$ of stream data;
3   Compute the unbiased estimation of stream data as $x(t) = \left(\widehat{N}_t - \frac{N}{e^{\epsilon/\omega}+1}\right) \cdot \frac{e^{\epsilon/\omega}+1}{e^{\epsilon/\omega}-1}$;
4   **if** $t = 1$ **then**
5     $\mathcal{G}_t = \{t\}$, and set $\mathcal{G}_{state} = open$;
6   **if** $\mathcal{G}_{state} = open$ **then**
7   Compute group deviation value $v_t = f(X[\mathcal{G}_{t-1} \cup t])$;
8   **if** $v_t < \theta_t$ **then**
9       $\mathcal{G}_t = \mathcal{G}_{t-1} \cup \{t\}$, and set $\mathcal{G}_{state} = open$;
10   **else**
11       $\mathcal{G}_t = \{t\}$, and set $\mathcal{G}_{state} = close$;
12   **else**
13     $\mathcal{G}_t = \{t\}$, and set $\mathcal{G}_{state} = open$;
14   Smooth noise as $\hat{c}(t) = \text{median}\{x(j) | j \in \mathcal{G}_t\}$;
15 **Return** $\widehat{C} = \{\hat{c}(1), \hat{c}(2), \cdots, \hat{c}(t), \cdots\}$

Next, lines 4–13 show the procedure of the retroactive grouping strategy. As we can see, the first timestamp is directly taken as a group and the group state is open, that is $\mathcal{G}_t = \{t\}$ and $\mathcal{G}_{state} = open$. Here, the open state of a group means that a new timestamp can be added to it. For each of the successive timestamps, depending on the group state, there will be two grouping operations.

1) If group state is open, we then compute group deviation value v_t as $v_t = f(X[\mathcal{G}_{t-1} \cup t])$ based on Eq. (6). If the deviation value v_t is smaller than threshold θ_t, the group will be updated by adding the current timestamp t into it and the group state is still open, that is $\mathcal{G}_t = \mathcal{G}_{t-1} \cup \{t\}$ and $\mathcal{G}_{state} = open$, as shown in lines 8–9. Otherwise, the current timestamp t is taken as a new group and the group state is close, that is $\mathcal{G}_t = \{t\}$ and $\mathcal{G}_{state} = close$, as shown in lines 10–11.
2) If group state is close, the current timestamp t will be directly taken as a new group and the group state is open, that is $\mathcal{G}_t = \{t\}$ and $\mathcal{G}_{state} = open$, as shown in lines 12–13.

Based on the noisy stream count and time group result, we can smooth the excessive noise at each time t and obtain the final estimated count of stream, that is $\hat{c}(t) = \text{median}\{x(j)|j \in \mathcal{G}_t\}$, as shown in the $14th$ line. Here, we adopt median smoothing in this paper.

From Algorithm 2, it can be observed that the threshold θ_t is a key parameter and plays an important role in grouping. An appropriate threshold determines the effectiveness of the retroactive grouping strategy. Intuitively, due to the dynamics of stream data, the threshold that selected according to the changing trend of the stream data will be a good threshold. Therefore, we no longer predefine a fixed threshold and instead update the threshold based on the observed feedback errors at the current timestamp. In this paper, we update the threshold based on the feedback error like [24].

5 Experiments

This section presents the performance evaluations of our proposed framework for real-time stream statistics with ω-event ϵ-LDP.

5.1 Evaluation Setup

Datasets. Our experiments are conducted on two real-world datasets.

- Retail [1] is a retail market basket dataset, which contains 16,470 unique items. We take the Retail dataset as stream data by taking the size of items as the length of the stream.
- Kosarak [1] is a webpage click-stream dataset, which was collected from a Hungarian online news portal. Kosarak dataset contains around one million users and 41,270 categories. We pre-process the Kosarak dataset into stream data by taking the size of click categories as the length of the stream.

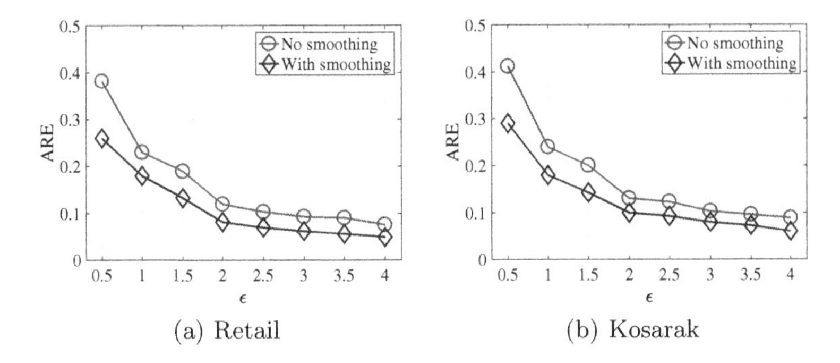

Fig. 2. Utility evaluation of retroactive grouping-based noise smoothing when ϵ changes ($\omega = 20$)

Metrics. We use the metric of Average Relative Error (ARE) to evaluate the data utility of our proposed mechanism. In formal, ARE is defined as

$$\mathrm{ARE}(C, \widehat{C}) = \frac{1}{T} \sum_{t=1}^{T} \frac{c(t) - \hat{c}(t)}{\max\{c(t), \delta\}} \tag{7}$$

where C and $\widehat{C}$ are the real and noisy stream statistics, respectively. The parameter δ is used to mitigate the effect of zero value or excessively small value.

Experimental Environment. We simulate a crowdsensing system for real-time stream statistics. The system has an aggregation server. Each user acts as a participant node and perturbs her/his stream using Algorithm 1 before sending data to the server. The privacy-preserved stream data are aggregated in the server end to compute the stream statistics. All algorithms and experiments are implemented using Python 2.7, running on a Windows 10 PC with CPU i7-10700, 16 GB RAM.

5.2 Experimental Results

In what follows, we present the evaluation results of our mechanism from different aspects varying from different privacy parameters. In all experiments, we consider privacy budget $\epsilon \in \{0.5, 1, 1.5, 2, 2.5, 3, 3.5, 4\}$ and sliding window $\omega \in \{10, 20, 30, 40, 50, 60, 70, 80\}$. Each experiment is conducted 100 times and the average result is reported.

1) Performance evaluation of retroactive grouping-based noise smoothing.

Figure 2 gives the ARE of our mechanism on Retail and Kosarak datasets when using noise smoothing or not. On the whole, the AREs decrease constantly with an increase of ϵ since the privacy protection level becomes lower when ϵ increases. Besides, as we can see, the ARE under noise smoothing is always smaller than that without noise smoothing. This is because the noise smoothing

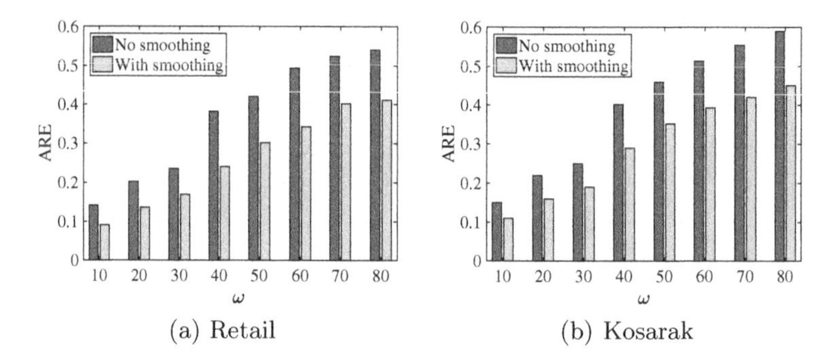

Fig. 3. Utility evaluation of retroactive grouping-based noise smoothing when ω changes ($\epsilon = 1$)

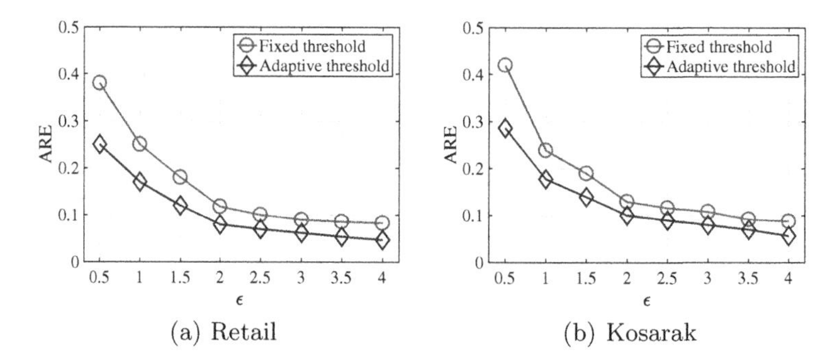

Fig. 4. Utility evaluation of threshold selection when ϵ changes ($\omega = 20$)

can reduce the total noise based on the consideration of time correlations of stream data. Therefore, it demonstrates that the retroactive grouping strategy can learn the stream changing trends, thus making noise smoothing effective.

Figure 3 shows the evaluation results of retroactive grouping-based noise smoothing when ω varies from 10 to 80. As we can see, the AREs increase constantly with the increase of ω since the privacy budget at each timestamp becomes smaller when ω increases. A smaller privacy budget leads to a larger variance. Nonetheless, it can be observed that the ARE is certainly reduced when applying noise smoothing. Thus, this demonstrates again that the retroactive grouping-based noise smoothing can greatly improve the data utility when dealing with dynamic stream data.

2) Performance evaluation of adaptive threshold selection.

Figure 4 gives the evaluation results of threshold selection varying ϵ from 0.5 to 4. At first, the AREs on both datasets decrease with the increase of ϵ since a larger ϵ means a lower privacy protection level, thus holding a higher data utility. Moreover, we can see that the adaptive threshold selection in our mechanism gives better accuracy in all cases than that using a fixed threshold, which is as expected. The reason here is that an adaptive threshold essentially

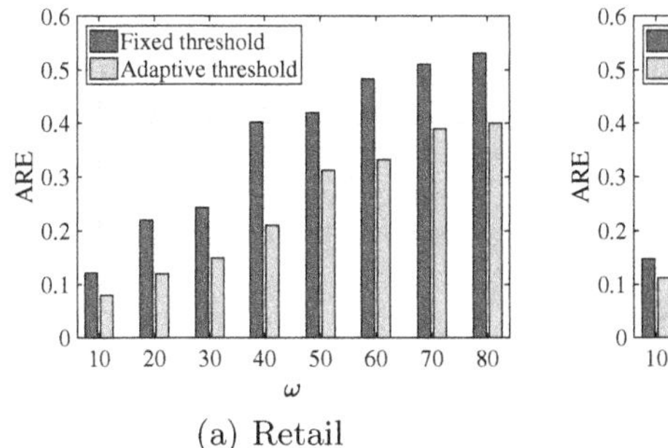

(a) Retail

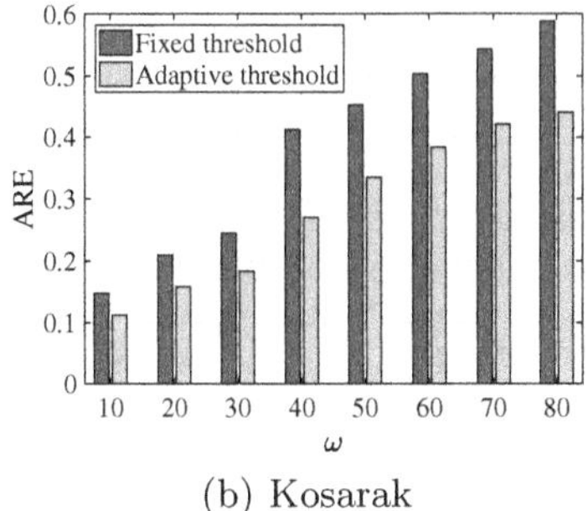

(b) Kosarak

Fig. 5. Utility evaluation of threshold selection when ω changes ($\epsilon = 1$)

reflects the changing trends of the dynamic stream data, which will make the time grouping more accurate, thus improving the data utility.

Figure 5 presents the AREs on two datasets varying ω from 10 to 80. Overall the performance gets worse when the size of the sliding window (i.e., ω) increases. A large ω can ensure the privacy of each user within a long time slot, thus reducing the accuracy. Moreover, the AREs are effectively reduced in both figures when using an adaptive threshold. This proves again that the utility improvement using an adaptive threshold is much significant than using a fixed threshold, which is as expected.

6 Conclusion

In this paper, we propose a privacy-preserving mechanism with ω-event ϵ-local differential privacy (LDP) for real-time stream statistics in mobile crowdsensing. Our mechanism applies LDP to introduce perturbation on the client side, which provides strong privacy guarantees for each user. Concerning the dynamics and time correlations of stream data, we leverage a retroactive grouping strategy to learn the time correlations and then smooth the excessive noise. All experiments have shown the effectiveness of our proposed mechanism in terms of improving data utility.

Acknowledgments. This work was supported in part by the National Natural Science Foundation of China (No. 62102311) and in part by the Scientific Research Program Funded by Shaanxi Provincial Education Department (Program No. 21JK0913).

References

1. Frequent itemset mining dataset repository. http://fimi.ua.ac.be/data/
2. Abdelhameed, S.A., Moussa, S.M., Khalifa, M.E.: Restricted sensitive attributes-based sequential anonymization (RSA-SA) approach for privacy-preserving data stream publishing. Knowl. Based Syst. **164**, 1–20 (2019)

3. Acs, G., Castelluccia, C., Chen, R.: Differentially private histogram publishing through lossy compression. In: Proceedings of IEEE ICDM, pp. 1–10 (2012)
4. Acs, G., Castelluccia, C.: A case study: privacy preserving release of spatio-temporal density in Paris. In: Proceedings of ACM SIGKDD, pp. 1679–1688 (2014)
5. Al-Hussaeni, K., Fung, B.C., Iqbal, F., Dagher, G.G., Park, E.G.: Safepath: differentially-private publishing of passenger trajectories in transportation systems. Comput. Netw. **143**, 126–139 (2018)
6. Cao, Y., Yoshikawa, M., Xiao, Y., Xiong, L.: Quantifying differential privacy under temporal correlations. In: Proceedings of IEEE ICDE, pp. 821–832 (2017)
7. Chen, Y., Machanavajjhala, A., Hay, M., Miklau, G.: PeGaSus: data-adaptive differentially private stream processing. In: Proceedings of ACM CCS, pp. 1375–1388 (2017)
8. Ding, B., Kulkarni, J., Yekhanin, S.: Collecting telemetry data privately. In: Advances in Neural Information Processing Systems, pp. 3571–3580 (2017)
9. Duchi, J.C., Jordan, M.I., Wainwright, M.J.: Local privacy and statistical minimax rates. In: IEEE Annual Symposium on Foundations of Computer Science, pp. 429–438 (2013)
10. Dwork, C.: Differential privacy in new settings. In: Proceedings of ACM-SIAM SODA, pp. 174–183 (2010)
11. Dwork, C., Naor, M., Pitassi, T., Rothblum, G.N.: Differential privacy under continual observation. In: ACM Symposium on Theory of Computing, pp. 715–724 (2010)
12. Dwork, C., Roth, A., et al.: The algorithmic foundations of differential privacy. Found. Trends Theor. Comput. Sci. **9**(3–4), 211–407 (2014)
13. Erlingsson, Ú., Pihur, V., Korolova, A.: Rappor: randomized aggregatable privacy-preserving ordinal response. In: Proceedings of ACM SIGSAC CCS, pp. 1054–1067 (2014)
14. Fan, L., Xiong, L.: An adaptive approach to real-time aggregate monitoring with differential privacy. IEEE Trans. Knowl. Data Eng. **26**(9), 2094–2106 (2014)
15. Guo, B., et al.: Mobile crowd sensing and computing: the review of an emerging human-powered sensing paradigm. ACM Comput. Surv. (CSUR) **48**(1), 1–31 (2015)
16. Joseph, M., Roth, A., Ullman, J., Waggoner, B.: Local differential privacy for evolving data. In: Advances in Neural Information Processing Systems, pp. 2381–2390 (2018)
17. Kellaris, G., Papadopoulos, S., Xiao, X., Papadias, D.: Differentially private event sequences over infinite streams. VLDB Endow. **7**(12), 1155–1166 (2014)
18. Li, M., et al.: All your location are belong to us: breaking mobile social networks for automated user location tracking. In: Proceedings of ACM MobiHoc, pp. 43–52 (2014)
19. Rastogi, V., Nath, S.: Differentially private aggregation of distributed time-series with transformation and encryption. In: Proceedings of ACM SIGMOD, pp. 735–746 (2010)
20. Sun, G., Sun, S., Yu, H., Guizani, M.: Toward incentivizing fog-based privacy-preserving mobile crowdsensing in the internet of vehicles. IEEE Internet Things J. **7**(5), 4128–4142 (2020)
21. Wang, J., Zhu, R., Liu, S.: A differentially private unscented Kalman filter for streaming data in IoT. IEEE Access **6**, 6487–6495 (2018)
22. Wang, Q., Zhang, Y., Lu, X., Wang, Z., Qin, Z., Ren, K.: Real-time and spatio-temporal crowd-sourced social network data publishing with differential privacy. IEEE Trans. Dependable Secure Comput. **15**(4), 591–606 (2016)

23. Wang, Q., Zhang, Y., Lu, X., Wang, Z., Qin, Z., Ren, K.: RescueDP: real-time spatio-temporal crowd-sourced data publishing with differential privacy. In: Proceedings of IEEE INFOCOM, pp. 1–9 (2016)
24. Wang, T., Yang, X., Ren, X., Zhao, J., Lam, K.Y.: Adaptive differentially private data stream publishing in spatio-temporal monitoring of IoT. In: IEEE 38th International Performance Computing and Communications Conference (IPCCC), pp. 1–8 (2019)
25. Wang, T., Zhang, X., Jingyu, F., Xinyu, Y.: A comprehensive survey on local differential privacy toward data statistics and analysis. Sensors **20**(24), 1–47 (2020)
26. Wang, T., Blocki, J., Li, N., Jha, S.: Locally differentially private protocols for frequency estimation. In: USENIX Security Symposium, pp. 729–745 (2017)
27. Xiao, X., Wang, G., Gehrke, J.: Differential privacy via wavelet transforms. IEEE Trans. Knowl. Data Eng. **23**(8), 1200–1214 (2011)
28. Yargic, A., Bilge, A.: Privacy-preserving multi-criteria collaborative filtering. Inf. Process. Manag. **56**(3), 994–1009 (2019)
29. Zhang, X., Hamm, J., Reiter, M.K., Zhang, Y.: Statistical privacy for streaming traffic. In: Network and Distributed System Security Symposium (NDSS), pp. 1–15 (2019)
30. Zheng, X., Cai, Z., Li, Y.: Data linkage in smart internet of things systems: a consideration from a privacy perspective. IEEE Commun. Mag. **56**(9), 55–61 (2018)

Privacy-Preserving Property Prediction for New Drugs with MPNN

Jiaye Xue[1], Xinying Liao[1], Ximeng Liu[1,2], and Wenzhong Guo[1,2(✉)]

[1] College of Mathematics and Computer Science, Fuzhou University, Fuzhou 350108, China
guowenzhong@fzu.edu.cn

[2] Fujian Provincial Key Laboratory of Information Security of Network Systems, Fuzhou University, Fuzhou 350108, China

Abstract. Message passing neural network (MPNN) is one of the excellent deep learning models for drug discovery and development, drug laboratory usually outsource the MPNN model to cloud servers to save the research and development cost. However, drug-related data privacy has become a noticeable hindrance for outsourcing cooperation in drug discovery. In this paper, we propose a lightweight privacy-preserving message passing neural network framework (SecMPNN) for property prediction in new drugs. To implement SecMPNN, we design multiple protocols to perform the three stages of MPNN, namely message function, update function, and readout function. The above new-designed secure protocols enable SecMPNN to adapt to the different numbers of participating servers and different lengths of encryption requirements. Moreover, the accuracy, efficiency, and security of SecMPNN are demonstrated through comprehensive theoretical analysis and a large number of experiments. The experimental results show the communication efficiency in multiplication and comparison increases 27.78% and 58.75%, the computation error decreases to 4.64%.

Keywords: Privacy-preserving · Message passing neural network · Drug discovery

1 Introduction

Message passing neural network (MPNN) for learning molecular fingerprints has been considered to be an ideal approach to predict the properties of new drugs. Compared with other molecular prediction models, MPNN has better predictive performance and interpretability. Drug laboratory can use the detection results of MPNN to learn the properties of new drugs and make more effective improvements. However, it is quite staggering for storage capacity and the computing power requirements for training such an MPNN model for predicting the properties of new drugs. For a single drug, the MPNN model was trained with batch

J. Xiong et al. (Eds.): MobiMedia 2021, LNICST 394, pp. 446–461, 2021.
https://doi.org/10.1007/978-3-030-89814-4_32

size 20 for 3 million steps (540 epochs), and the molecular data can over 130000 samples [1]. Therefore, compared with building their own servers, the drug laboratory prefers to build the MPNN model using cloud computing technology and outsource their new drugs to cloud servers. Besides, the process from drug discovery to market costs, on average, well over $1 billion and can span 12 years or more, due to high attrition rates, rarely can one progress to market in less than ten years. Neither researcher nor laboratory wants these new drug discoveries to be revealed to some hostile adversaries. Therefore, new drug information is a valuable commercial resource for the drug laboratory, establishing an lightweight and secure privacy-preserving framework for MPNN to predict the properties of new drugs is necessary.

To address the above problems, we use the bit decomposition method to design multiple secure n-party protocols to perform the MPNN framework (SecMPNN) for predicting the properties of new drugs. The main contributions of this work are listed as follows.

- We propose SecMPNN allows drug laboratory to share new drug data securely and use a high-performance MPNN framework to give assistance to drug research. In the framework, the drug laboratory need not be anxious about their data leaks to the adversary.
- Several secure n-party sub-protocols use to finish the multiplication, comparison in the secure computation of the message function, the update function, and the readout function of MPNN without revealing the original data. Compared with traditional methods, our protocols adapt to the different numbers of participating servers and different lengths of encryption requirements.
- Through comprehensive analysis, the correctness and security of SecMPNN are proved. The experimental results show the communication efficiency in multiplication and comparison increases 27.78% and 58.75%, the computation error decreases to 4.64%.

2 Related Work

Although quantum mechanics allows us to calculate the properties of drugs in principle, the equations caused by the laws of physics are too difficult to solve accurately. As a result, scientists have developed a series of approaches to approximate quantum mechanics, with different tradeoffs between speed and accuracy, such as density functional theory (DFT) with a variety of functionals [2]. Although DFT has been widely used, its speed is still too slow to be used in large-scale systems. In order to improve the accuracy of DFT, Mohassel et al. [15] used neural networks to approximate a particularly troublesome term in DFT called the exchange-correlation potential. However, their method can not improve the efficiency of DFT and relies on a large number of special atom descriptors.

To the best of our knowledge, SecureML [5] seems to be the first privacy protection training and prediction work based on secure multiparty computing technology. However, they are inefficient and can only support very simple networks and few hidden layers. Recently, various hybrid protocol frameworks have been proposed, like MiniONN [6] and DeepSecure [7]. However, both MiniONN and DeepSecure are computationally intensive cryptographic primitives, which are not easy to extend and can not make full use of the efficient data structure of MPNN.

3 Preliminaries

3.1 Message Passing Neural Network (MPNN)

MPNN is a sequence of connected layers that converted from an input layer to output layer. Each layer is consists of a group of neurons. Two kinds of common connected layer: fully connected layer and activation layer, are mainly used in MPNN. In [1], MPNN can be regarded as three parts: message function, update function, and readout function. We will briefly introduce these three parts as follow:

- **Message function**: The information h_v^t of each point v at time t in graph G is combined by $A_{e_{vw}}$ and its neighbor point information h_w^t, $w \in N(v)$ at time t, where $N(v)$ is the set of neighbors of point v. In detail, $A_{e_{vw}}$ represents a neural network layer as follows the pattern: $Input \rightarrow [FC \rightarrow ReLU] \times r \rightarrow FC$ where $Input$ is the attribute of edge e_{vw} and r denotes the number of repetition. This process can be express as $m_v^{t+1} = \sum_{w \in N(v)} A_{e_{vw}} h_w^t$.
- **Update function**: In this function, h_v^t update from time t to time $t+1$ according to $h_v^{t+1} = GRU(h_v^t, m_v^{t+1})$, while GRU is the gated recurrent unit. We set its specific process is as follows: $z_v^{t+1} = Sigmoid(W_{mz}m_v^{t+1} + b_{mz} + W_{hz}h_v^t + b_{hz})$, $r_v^{t+1} = Sigmoid(W_{mr}m_v^{t+1} + b_{mr} + W_{hr}h_v^t + b_{hr})$, $n_v^{t+1} = tanh(W_{mn}m_v^{t+1} + b_{mn} + r_v^{t+1} \odot (W_{hn}h_v^t + b_{hn}))$, $h_v^{t+1} = (1 - z_v^{t+1}) \odot n_v^{t+1} + z_v^{t+1} \odot h_v^t$.
- **Readout function**: After T time step, the graph information is obtained from the current stable node state and initial point state through the readout function: $R = \sum_{v \in V} Sigmoid(i(h_v^T, h_v^0)) \odot (j(h_v^T))$. In the above equation, i and j both are the neural networks, V is the point set in graph G. W and b is the weight and bias in different neural networks. $\odot$ denotes elementwise multiplication.

Finally, prediction results R of drug properties are obtained.

3.2 Basic Definitions

We describe the basic definitions of the data format, data split and secure n-party protocol in SecMPNN.

Data Format: In SecMPNN, we convert float number to integer number by multiplying the number 10^p, and then delete the remaining decimal places, where p is the number of decimal places. It means for a floating-point number x, we compute $\bar{x} = \lfloor x \cdot 10^p \rfloor$, where $\lfloor \cdot \rfloor$ denotes the round-down operation. For simplicity, we will omit the following overbar if there is no confusion. We use the binary complement representation of numbers to perform bit operations. The weight of the most significant bit (MSB) is a negative number of the corresponding power of 2 and the weight of other bits is a power of two. It means l-bit signed integer x can be expressed into the form $x^{(l-1)}x^{(l-2)}\dots x^{(0)}$ with $x^{(l-1)}$ being the MSB and $x = -x^{(l-1)} \cdot 2^{l-1} + \sum_{j=0}^{l-2} x^{(j)} \cdot 2^j$. The l is used for the performance benchmarks. However, it is stressed that all the protocols presented here work for any choice of l.

Data Split: We assume that there are n edge servers $S_i(i \in N_{n-1})$ participating in the work, $N_{n-1} = \{0, 1, \dots, n-1\}$. Given a number a is randomly split into n shared values as $a = \sum_{i=0}^{n-1} a_i$, where the $a_i(i \in N_{n-1})$ are called the shared values of a and will be stored in S_i, $[a] = \{a_i | i \in N_{n-1}\}$ represent the set of a_i.

Secure n-Party Protocol: All security n party agreements in this article meet the following formal definitions. Suppose $\mathcal{P}(\mathcal{I}, \mathcal{S})$ is one secure n-party protocol. Given random shared values of inputs $\mathcal{I} = \{[a], [b], \dots\}$ and n edge servers $\mathcal{S} = \{S_i | i \in N_{n-1}\}$, $\mathcal{P}$ outputs $\{f_i | i \in N_{n-1}\}$, where $\{f_i | i \in N_{n-1}\}$ are n random shared values of the computation result from $\{S_i | i \in N_{n-1}\}$ respectively. One needs to compute $f = \sum_{i=0}^{n-1} f_i$ to get the final calculation result f.

3.3 Basic Secure n-Party Protocols

The sub-protocols are introduced below which are operated among n edge serves.

Reveal: The secure n-party protocol should allow revealing the value a to all servers while covering up shared values. It can be expressed as $a = Reveal([a])$.

Secure Addition or Subtraction (SecAdd): In this protocol, servers compute $f(a, b) = a \pm b$. Since $a \pm b = \sum_{i=0}^{n-1} a_i \pm \sum_{i=0}^{n-1} b_i = \sum_{i=0}^{n-1}(a_i \pm b_i)$, it is easy to find out that $S_i(i \in N_{n-1})$ can perform the secure addition and subtraction locally without interaction with each other. Obviously, outputs satisfy $\sum_{i=0}^{n-1} f_i = a \pm b$.

Reshare: The input of this protocol is $[a]$ and output is shared value $[b]$ such that $b = a$, all shared values b_i are uniformly distributed, a_i and b_j are independent for $i, j \in N_{n-1}$. Firstly, S_i generates random $r_{i,(i+1)\%n} \in \mathbb{Z}_{2^l}$ and then $r_{i,(i+1)\%n}$ is sent from S_i to S_j. Finally, S_i computes $b_i = a_i + r_{i,(i+1)\%n} - r_{(i-1)\%n,i}$.

Secure Bit-Wise Addition Protocol (BitAdd): It is a protocol based on the *ripple-carry adder* (RCA) and carry is calculated by iterating from the least significant bit (LSB) to MSB. Given two same length of bit strings $[a^{(l-1)}]...[a^{(0)}]$ and $[b^{(l-1)}]...[b^{(0)}]$, it can be formulated as: $d^{(j)} = a^{(j)} \oplus b^{(j)} \oplus c^{(j)}$, $c^{(j+1)} = (a^{(j)}b^{(j)}) \oplus ((a^{(j)} \oplus b^{(j)})c^{(j)})$.

Secure Bit Comparison Protocol (BitCompare): It mentioned in [10] and it is trival to expand to three or more participating servers. The input is shared value of two l-bit original binary without sign bit as $[a^{(l-1)}] \dots [a^{(0)}]$ and $[b^{(l-1)}] \dots [b^{(0)}]$, the output is the shared value of result for comparison $[f]$ where $f = 1$ if and only if $[a^{(l-1)}] \dots [a^{(0)}] > [b^{(l-1)}] \dots [b^{(0)}]$.

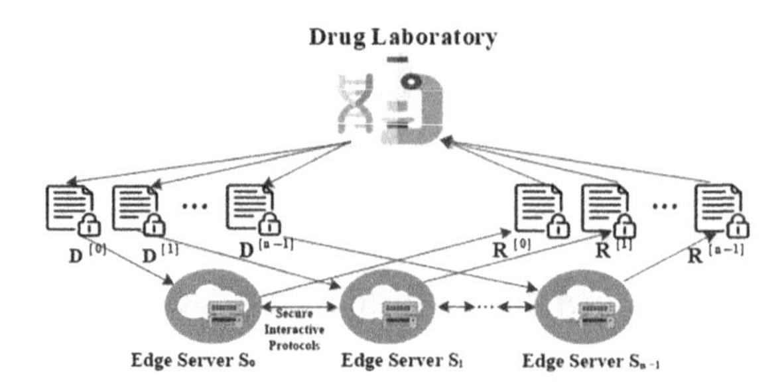

Fig. 1. System model.

4 System Architecture

4.1 System Model

There are two types of participants comprising SecMPNN shown in Fig. 1, namely the drug laboratory and the edge servers $S = \{S_i | i \in N_{n-1}\}$.

- Drug laboratory would like to use privacy-preserving property prediction for new drugs with MPNN model and is unwilling to share their new drugs data with others. In SecMPNN, drug laboratory randomly split the data D into $\{D^{[i]} | i \in N_{n-1}\}$. The encrypted data $D^{[i]} (i \in N_{n-1})$ are delivered to i^{th} edge servers.
- $S_i (i \in N_{n-1})$ are n cloud servers responsible for providing storage and computing power. In SecMPNN, They complete the calculation task of MPNN without knowing the plaintext of new drugs. The final outputs of servers $S_i (i \in N_{n-1})$ as $R^{[i]}$ are sent to drug laboratory using secure communication channels. The drug laboratory can obtain the final results by computing $R{=}\sum_{i=0}^{n-1} R^{[i]}$.

4.2 Attack Model

In SecMPNN, we use the model of *honest-but-curious*. In the model, $S = \{S_i | i \in N_{n-1}\}$ are all *honest-but-curious* parties, they complete each steps of protocols, and curious about the data which benefit themselves belonging to others that. In addition, we assume a simulator ζ can obtain the real view of the secure n-party protocol and generate random values. For the real view, ζ tries to generate a simulated view in polynomial time. A probabilistic polynomial algorithm can be found to distinguish the real view from the simulated view by adversary $\mathcal{A}$ is regarded as a successful attack. Also, we hypothesize that uniformly random values can be generated by edge servers, and edge servers cannot be simultaneously corrupted or collude with each other.

Algorithm 1. *SecMul*

Input: Shared values $[a]$ and $[b]$.
Output: Shared value $[f']$ such that $f' = ab$.
1: $[a'] \leftarrow Reshare([a])$.
2: $[b'] \leftarrow Reshare([b])$.
3: S_i send a'_i to servers $S_{(i+n-1)\%n}, ..., S_{(i+n-\lfloor (n-1)/2 \rfloor)\%n}$.
4: S_i send b'_i to servers $S_{(i+n-1)\%n}, ..., S_{(i+n-\lfloor n/2 \rfloor)\%n}$.
5: S_i computes $f_i \leftarrow a'_i \cdot b'_i + \sum_{j=1}^{\lfloor n/2 \rfloor} a'_i \cdot b'_{(i+j)\%n} + \sum_{j=1}^{\lfloor (n-1)/2 \rfloor} b'_i \cdot a'_{(i+j)\%n}$.
6: $[f'] \leftarrow Reshare([f])$.
7: **return** $[f']$.

5 Building Blocks: Secure n-Party Protocol

5.1 Secure n-Party Multiplication Protocol

The secure n-party multiplication protocol we designed can greatly reduce the number of rounds of communication and calculation. If we have two values a and b shared additively as $a = \sum_{i=0}^{n-1} a_i$ and $b = \sum_{j=0}^{n-1} b_j$, their product is $a \cdot b = \sum_{i=0}^{n-1} \sum_{j=0}^{n-1} a_i \cdot b_j$. The addends of the form $a_i \cdot b_i$ can be computed locally by server S_i. In order to find an addend of the form $a_i \cdot b_j$ $(i \neq j)$, the shared values a_i can be sent from S_i to S_j (or the shared values b_j from S_j to S_i). Knowing the shared values a_i and a_j, server S_j is still unable to get any information concerning a, but in order to obtain universal composability, all the shared values a_i and b_j still need to be reshared. Otherwise, the server may infer the original data from the previous information.

As illustrated in Algorithm 1, we introduce the details of secure n-party multiplication protocol under three or more participating servers.

5.2 Secure n-Party Comparison Protocol

The secure n-party comparison protocol is the most basic and important framework component. Suppose we have two numbers a, b to compute $f(a, b) = (a > b)$ and $f \in \{0, 1\}$, where $f = 1$ if and only if $a > b$. The completion of the secure n-party comparison protocol relies on the following security sub-protocols.

RandomBit: We now describe a protocol *RandomBit* in Algorithm 2 for securely generating shared values of a uniformly random bit. The protocol has no inputs, and the output is shared value $[r]$ of a uniformly random $r \in \{0, 1\}$. *RandomBit* protocol is based on the fact that squaring a non-zero element is a 2-to-1 mapping and given $A = \sqrt{a^2}$ one has no idea if the pre-image was a or -a. We will let the "sign" of such an a be our random bit.

RandomSolvedBits: The input of this protocol is none and output are shared values of random number $r \in \mathbb{Z}_{2^l}$ as $[r]$ and the shared values of complement $[r^{(l-1)}] \ldots [r^{(0)}]$. Servers carry out *RandomBit* for l times as $[r^{(l-1)}] \ldots [r^{(0)}]$ and compute $[r] = \sum_{i=0}^{l-2} 2^i \cdot [r^{(i)}] - [r^{(l-1)}] \cdot 2^{(l-1)}$ locally.

Algorithm 2. *RandomBit*

Output: Shared value $[r]$ while $r \in \{0, 1\}$.

1: **repeat**
2: S_i generate random $a_i \leftarrow \mathbb{Z}_{2^l}$.
3: $[a^2] = [a][a]$.
4: $A^2 = \text{Reveal}([a^2])$.
5: **until** $A^2 \neq 0$.
6: $A \leftarrow \sqrt{A^2}$.
7: Select random integer $j \in N_{n-1}$.
8: S_i generate random $b_i \leftarrow \mathbb{Z}_{2^l}$ while $i \neq j$.
9: S_i compute $x_i \leftarrow b_i * A - a_i$ and send x_i to S_j while $i \neq j$.
10: S_j compute $b_j \leftarrow (a_j - \sum_{i \in N_{n-1}}^{i \neq j} x_i)/A$.
11: $[c] \leftarrow [b] + 1$.
12: S_i compute $y_i \leftarrow c_i \% 2$ and send y_i to S_j while $i \neq j$.
13: S_i compute $r_i \leftarrow (c_i - y_i)/2$ while $i \neq j$.
14: S_j compute $r_j \leftarrow (c_j + \sum_{i \in N_{n-1}}^{i \neq j} y_i)/2$.
15: **return** $[r]$.

Algorithm 3. *Bits*

Input: Shared value $[a]$.
Output: The shared values of complement of $[a^{(l-1)}]...[a^{(0)}]$.

1: $[r], [r^{(l-1)}]...[r^{(0)}] \leftarrow \textit{RandomSolvedBits}()$.
2: $[c] \leftarrow [a] - [r]$.
3: $C \leftarrow \textit{Reveal}([c])$.
4: Generate shared values of complement $[C^{(l_C)}]...[C^{(0)}]$ of $C(l_C \geq l-1)$.
5: Add shared values with a prefix of zero $[0]$ from $[r^{(l-1)}]...[r^{(0)}]$ to $[r^{(l_C)}]...[r^{(0)}]$.
6: $[a^{(l_C+1)}]...[a^{(0)}] \longleftarrow \textit{BitAdd}([C^{(l_C)}]...[C^{(0)}], [r^{(l_C)}]...[r^{(0)}])$.
7: Change $[a^{(l_C+1)}]...[a^{(0)}]$ to a fixed length as $[a^{(l-1)}]...[a^{(0)}]$.
8: **return** $[a^{(l-1)}]...[a^{(0)}]$.

Bits: It is a protocol which given shared value $[a]$ securely computes the shared values of complement of $[a]$ as $[a^{(l-1)}], \ldots, [a^{(0)}]$ to realize the bit decomposition function. The input $[a]$ cover up by computing $[c] = [a] - [r]$, after writing the complement of c, use *BitAdd* protocol to add the shared values of complement of c and the shared values of complement of r to get the shared values of complement of a.

Finally, we apply the previous sub-protocol to compare two shared values. The details are shown in Algorithm 4. Note that, after getting two implementations by *Bits*, we compare the complement as the meaning of original code. Because the sign bit calculates as the highest power binary weight, it is easy to find that the result is right if a and b has the same sign and will be wrong if a and b has a different sign, we can XOR the sign bits of a and b to ensure the correctness. $\oplus$ is XOR operation for bits and it can be formulated as $[a] \oplus [b] = [a] + [b] - 2[a][b]$.

Algorithm 4. *Compare*

Input: Shared values $[a]$ and $[b]$.
Output: Shared value $[d]$ where $d = 1$ if $a > b$ and $d = 0$ otherwise.

1: $[a^{(l-1)}]...[a^{(0)}] \leftarrow Bits([a])$.
2: $[b^{(l-1)}]...[b^{(0)}] \leftarrow Bits([b])$.
3: $[c] \leftarrow BitCompare([a^{(l-1)}]...[a^{(0)}], [b^{(l-1)}]...[b^{(0)}])$.
4: $[d] \leftarrow [c] \oplus ([a^{(l-1)}] \oplus [b^{(l-1)}])$.
5: **return** $[d]$.

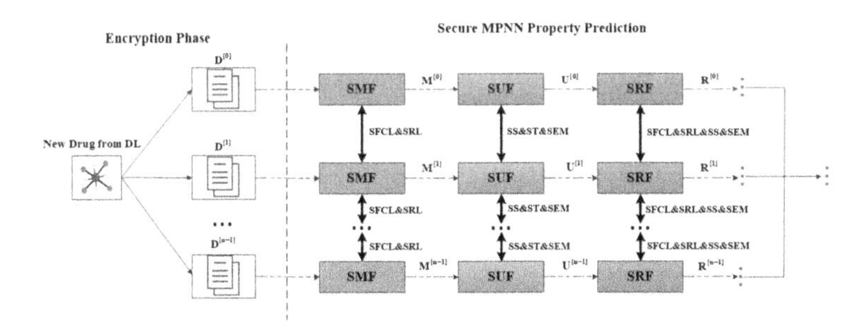

Fig. 2. Privacy-preserving MPNN for property prediction of new drugs

6 Privacy-Preserving Property Prediction of New Drugs

We provide the details of SecMPNN in this section, and discuss the feasibility of extending SecMPNN to three or more party settings. We use $x^{[i]}$ to denote the share values x, distributed in corresponding edge servers. We give an overview for SecMPNN and shown in Fig. 2. It is composed of three stages, namely secure message function (SMF), secure update function (SUF), and secure readout function (SRF). In SecMPNN, new drug data will be firstly split into random shared values and uploaded to the edge servers. Then, the following secure functions are performed securely.

6.1 Secure Message Function

By using SMF to extract new drug node features, it can be ensured that no plaintext information is leaked to the edge server. The input of SMF is the shared feature of chemical bond in drug molecules with a fixed size as $\{D^{[i]}|i \in N_{n-1}\}$, which are uploaded by the drug laboratory. SMF consists of two kinds of layers: secure fully connected layer (SFCL) and secure ReLU layer (SRL). In SFCL, specifically, for the jth hidden neuron in this layer, $S_i(i \in N_{n-1})$ will compute one component of the activation: $y_j^{[i]} = \sum_k w_{j,k} x_k^{[i]} + b_j^{[i]}$. We use x_k for the activation of the kth neuron in the previous layer. In addition, we use $w_{j,k}$ to denote the weights and b_j to denote the bias of the jth neuron in the current layer. In SRL, we utilize $s = Compare([0], [x])$ protocol to compare input $[x]$ and 0 to determine the sign of x. If $s = 0$, the result of SRL is $[x]$; otherwise, the result is [0]. There are 3 SFCLs and 2 SRLs deployed to implement SMF. After SMF, edge servers output $M = \{M^{[i]}|i \in N_{n-1}\}$ and send to SUF.

6.2 Secure Update Function

One part comprises the SUF, namely the secure gated recurrent unit (SGRU). The goal of the secure update function is to update each node feature without leaking any new drug feature information. Given the node feature output M by SMF, SUF set the secure update gates z and secure reset gates r by secure logistic *Sigmoid* function (SS), use secure logistic *tanh* function (ST) and secure element-wise multiplication (SEM) compute two gates hidden status. Note that, SS and ST can be approximated with piecewise polynomials:

$$f(x) = \begin{cases} P_0(x), & x_0 \leq x \leq x_1 \\ P_1(x), & x_1 \leq x \leq x_2 \\ & \dots \\ P_{k-1}(x), & x_{k-1} \leq x \leq x_k \end{cases} \quad (1)$$

Where $P(x)$ is an p-degree polynomial, $P(x) = a_0 + a_1x + \ldots + a_px^p$, $a_i(i \in Np)$ are the public weights. The higher the degree of the polynomial, the better the approximation effect that can be obtained. In order to better protect the privacy of the elements x, we use *SecMul* protocol to compute x^p and figure out that the polynomials can be solved securely by applying the *Compare* protocol. For SEM, we only need to use *SecMul* protocol to compute the corresponding element securely and no other complex operations are required. After SUF, edge servers output $U = \{U^{[i]}|i \in N_{n-1}\}$ and send to SRF.

6.3 Secure Readout Function

The destination of SRF is to complete extraction of property prediction from the feature graph. In SRF, two fully connected networks use SFCL and SRL, they can execute as securely as the SMF mentioned above. We utilize SS and SEM to connect the results of two fully connected networks, they can execute as securely as SUF mentioned above. There are 6 SFCLs, 3 SRLs, 1 SS and 1 SEM deployed to implement SMF. Finally, the shared values of output R will send to the drug laboratory securely. After SRF, edge servers output $R = \{R^{[i]}|i \in N_{n-1}\}$ and send to drug laboratory.

7 Theoretical Analysis

7.1 Correctness

Firstly, in the SMF, the SFCL actually performs linear dot product. For the SRL, the relationship between data x and 0 is determined by *Compare* protocol. If x is greater than 0, the shared value of x is retained; otherwise, the shared value of

x of all servers will become the shared value of 0. Secondly, in the SUF, SS and ST are fitted by additive function under the condition of ensuring the accuracy, and there are dot product operation of SFCL and SEM which can express as $\sum_{i=0}^{n-1} y_i = \sum_{i=0}^{n-1} Wx_i = W \sum_{i=0}^{n-1} x_i = Wx = y$, these operations also meet the additive conditions. Finally, in the SRF, SS and SEM also exist in this function.

7.2 Security

Through the general composability framework [11], We prove the security of our protocol. In our *honest-but-curious* model, adversaries were allowed to destroy up to one of the n servers. To prove that the protocol is secure enough to prove that, given its input and output, the corrupted party's viewpoint is simulatable. Specifically, we use the following definitions and lemmas.

Definition 1. *We say that if there is a probabilistic polynomial time simulator ζ protocol is secure, the simulator can generate a view of opponent $\mathcal{A}$ in the real world, and the view is computationally indistinguishable from its real view.*

Lemma 1. *A protocol is perfectly simulatable if all its sub-protocols are perfectly simulatable.*

Lemma 2. *If a random element r is uniformly distributed on $\mathbb{Z}_n$ and independent from any variable $x \in \mathbb{Z}_n$, then $r \pm x$ is also uniformly random and independent from x.*

Lemma 3. *The protocol Reshare, BitAdd, Reveal, BitCompare is secure in the honest-but-curious model.*

We advise the reader to [9] and [8] for the proofs of Lemma 1 and Lemma 2, [10] and [8] for the proof of Lemma 3. Since performing locally can be perfectly simulated, we mainly prove the security for the sub-protocols in our framework that need interactions among servers $S_i (i \in N_{n-1})$ in the following.

Theorem 1. *The protocol SecMul is secure in the honest-but-curious model.*

Proof. For $S_i (i \in N_{n-1})$, the view in the protocol execution will be $\mathcal{V}_i = (a'_i, \ldots, a'_{(i+\lfloor (n-1)/2 \rfloor)\%n}, b'_i, \ldots, b'_{(i+\lfloor n/2 \rfloor)\%n})$. These values are obtained through the *Reshare* protocol. Therefore, $\mathcal{V}_i$ is simulatable by the simulator ζ. Besides, the output of S_i will be $\mathcal{O}_i = (f_i)$, where $f_i = a'_i \cdot b'_i + \sum_{j=1}^{\lfloor n/2 \rfloor} a'_i \cdot b'_{(i+j)\%n} + \sum_{j=1}^{\lfloor (n-1)/2 \rfloor} b'_i \cdot a'_{(i+j)\%n}$. Since the operations are performed locally by S_i, $\mathcal{O}_i$ is also simulatable by the simulator ζ.

Theorem 2. *The protocol RandomBit, RandomSolvedBits, Bits, Compare are secure in the honest-but-curious model.*

Proof. For $S_i (i \in N_{n-1}, i \neq j)$, the view in the protocol execution will be $\mathcal{V}_i = (A, a_i, x_i, b_i, c_i, y_i)$. For S_j, the view will be $\mathcal{V}_j = (A, a_j, [x], b_j, c_j, y_i)$. A and $[a]$ are obtained through the *Reshare* protocol and *Reveal* protocol. $[b], [c], [y]$ are random number. Besides, the output of S_i will be $\mathcal{O}_i = (r_i)$ where $r_i = (c_i - y_i)/2$, the output of S_j will be $\mathcal{O}_j = (r_j)$ where $r_j = (c_j + \sum_{i \in N_{n-1}}^{i \neq j} y_i)/2$. According to Lemma 2, these values are uniformly random. The proof of security about *RandomSolvedBits*, *Bits* and *Compare* are similar as above. It is trivial to see that they are secure in the *honest-but-curious* model.

Table 1. Comparison of the protocol complexities (Here, l is the bit-width, and $m = log_2 l$.)

Approach	$SecMul$		$Compare$		Number of servers
	Rounds	Comm(bits)	Rounds	Comm(bits)	
[8]	1	$15l$	$m+3$	$5m^2 + 12(l+1)m$	3
[3]	1	$3lm$	44	$205l + 188lm$	3
[4]	1	$3lm$	15	$279l + 5$	3
Ours	1	$(3n + \lfloor n/2 \rfloor + \lfloor (n-1)/2 \rfloor) * l$	$l+1$	$66nl$	n

Theorem 3. *The interactive protocol of secure message function, secure update function, and secure readout function in SecMPNN are secure in the semi-honest model.*

Proof. $\mathcal{A}$ eavesdrops on the transmission channels among the n edge servers and records the messages about the interactive protocols inputs into an input tape $tape_{in}$ and outputs into an output tape $tape_{out}$. According to the definitions of the interactive protocols, $\mathcal{A}$ have $tape_{in} = \mathcal{V}_{SecAdd} \cup \mathcal{V}_{SecMul} \cup \mathcal{V}_{Compare}$ and $tape_{out} = \mathcal{O}_{SecAdd} \cup \mathcal{O}_{SecMul} \cup \mathcal{O}_{Compare}$. Here, the elements belonging to the same sub-protocol are pushed into the same view. Based on Lemma 1, $tape_{in}$ and $tape_{out}$ are simulatable. It is trival to deduce SecMPNN are secure in the *honest-but-curious* model.

7.3 Efficiency

As shown in Table 1, the total communications of $SecMul$ and $Compare$ in our framework are far less than those in the traditional work. While $l = 32$ and $n = 3$, the communication efficiency in multiplication and comparison increases 27.78% and 58.75%.

8 Experiment

In this section, we first present the experimental results of our framework about the secure n-party comparison protocol. Then, we evaluate the performance and security of SecMPNN. The experiments are conducted through QM9 data, which can be obtained publicly by [12]. In the experiment, we take the SMILES [13] string encoding of each drug as input, which is then converted into fingerprints using RDKit [14]. After that, we encrypt the fingerprints, and then send them to multiple edge servers. By adjusting the number of servers and the length of encrypted bits, we test the performance of SecMPNN under different conditions. Each server is equipped with an Intel(R) Core(TM) i5-9500 CPU @3.00 GHz and 8.00 GB of RAM.

8.1 Performance Evaluation of the Sub-protocols in SecMPNN

In our framework, since our secure n-party comparison protocol is based on bit decomposition, when evaluating the performance of the comparison protocol, we use the bit-width$l$ of the input data and the number of servers$n$ as variables, and test under different server numbers and bit widths. As shown in Fig. 4 (1)–(2), the runtime and the communication overhead both go up in the bit-width l and the number of servers n. However, they are in a matter of "milliseconds" and "Kilobyte".

8.2 Performance Evaluation of SecMPNN

We compared SecMPNN with existing other frameworks [16–18,20,21]. In [16], it proposed a cooperative learning framework and applied it to a specific region between two networks. It can extract features from the source network to other specific networks, but the optimization of the source network is ignored. In [17], it designed a secure model to encrypted data, however, it is difficult to apply it in practice because of its large communication and time overhead. In [18], it proposes a private protocol that can reduce the transmission in multiple servers. In [20,21], they achieved encryption with lightweight data, but the assumed trusted third party is difficult to find in practice. We put a summary of the comparison in Table 2.

Table 2. Summary of Comparative analysis

Function	[16]	[17]	[18]	[20]	[21]	Ours
F_1	NN	RF	LM	NN	LM	NN
F_2	×	HE	MPC	MPC	MPC	MPC
F_3	×	√	√	×	×	√
F_4	×	√	√	×	×	√
F_5	√	×	√	√	√	√

Notes. F_1: machine learning model: neural network (NN) or random forest (RF) or linear model (LM). F_2: Encrypted methods: homomorphic encryption (HE) or multi-party computation (MPC). F_3: without trusted third party. F_4: secure bit length. F_5: supporting lightweight.

Efficiency. Firstly, we test the cost of the drug laboratory for new drugs encryption. We can see from Fig. 4 (3) that the time required for encryption shows a linear increase with the increase in the number of servers and the increase in bit width. However, the time cost of the encryption and decryption stages in our framework is almost negligible. Secondly, we test the performance of our privacy-preserving MPNN property prediction framework that is performed by the different number of participating servers. Figure 4 (4)–(6) shows communication overhead of different functions for processing one instance with the networks as Fig. 3. The overall time of SecMPNN is mainly determined by the network condition. In the test with an encryption length of $l = 64$, each servers only consume about 33 Gigabyte of communication overhead to obtain the property prediction. This is mainly because we don't rely on any heavy cryptographic primitives.

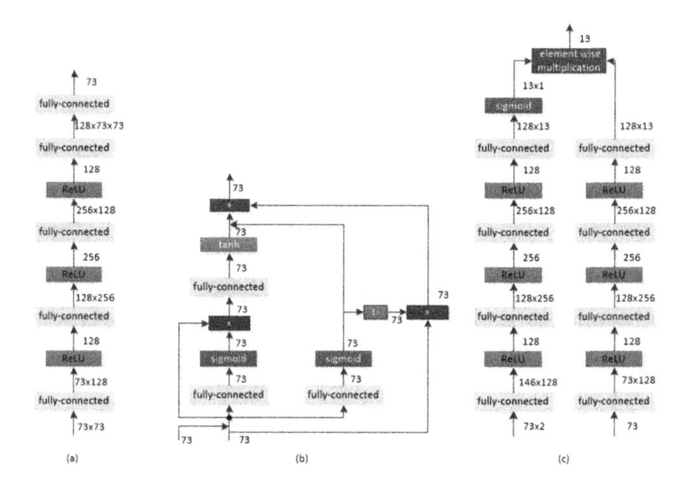

Fig. 3. MPNN architectures: (a) message function, (b) update function, and (c) readout function

Accuracy. In our SecMPNN, after the prediction of 50 drug formulas in the validation set, the average calculation error between the predicted value and the standard value is 4.64% and the accuracy is 95.36%. The calculation error mainly comes from the truncation error caused by the same data format as the integer, and the approximated error caused by approximating with piecewise polynomials of the nonlinear activation layer. We compare with the similar framework proposed in [19]. In [19], the accuracy of privacy-preserving and verifiable federated learning framework is 87.74%. It can be found that our SecMPNN framework can predict the properties of new drugs with high accuracy.

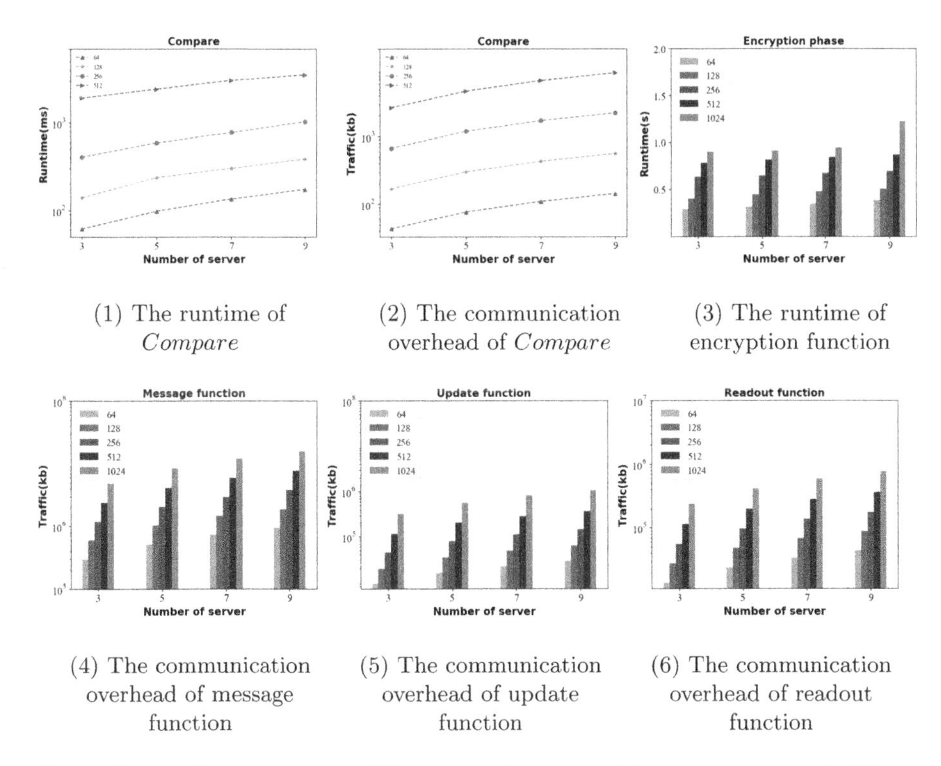

(1) The runtime of *Compare*

(2) The communication overhead of *Compare*

(3) The runtime of encryption function

(4) The communication overhead of message function

(5) The communication overhead of update function

(6) The communication overhead of readout function

Fig. 4. The performances with different number of participating servers and different length of encryption requirements.

9 Conclusions

In this paper, we proposed a novel lightweight framework for new drugs prediction as SecMPNN. The property prediction could be performed among edge servers securely. We first divide the drug data into shared values randomly, and then send them to the edge server. Using the bit decomposition method, a series of secure n-party protocols corresponding to different functions of MPNN are designed. Through empirical experiments and theoretical analysis, the security, effectiveness and accuracy of the framework are verified. In the future, we will further study more complex protocols to deal with more complex practical problems.

Acknowledgment. This work is supported by the National Natural Science Foundation of China (Grant No. U1705262, No. 62072109, No. U1804263), and the Natural Science Foundation of Fujian Province (Grant No. 2018J07005).

References

1. Gilmer, J., Schoenholz, S.S., Riley, P.F., Vinyals, O., Dahl, G.E.: Neural message passing for quantum chemistry. In: International Conference on Machine Learning, pp. 1263–1272. PMLR (July 2017)
2. Beck, A.D.: Density-functional thermochemistry. III. The role of exact exchange. J. Chem. Phys. **98**(7), 5648–5656 (1993)
3. Damgård, I., Fitzi, M., Kiltz, E., Nielsen, J.B., Toft, T.: Unconditionally secure constant-rounds multi-party computation for equality, comparison, bits and exponentiation. In: Halevi, S., Rabin, T. (eds.) TCC 2006. LNCS, vol. 3876, pp. 285–304. Springer, Heidelberg (2006). https://doi.org/10.1007/11681878_15
4. Nishide, T., Ohta, K.: Multiparty computation for interval, equality, and comparison without bit-decomposition protocol. In: Okamoto, T., Wang, X. (eds.) PKC 2007. LNCS, vol. 4450, pp. 343–360. Springer, Heidelberg (2007). https://doi.org/10.1007/978-3-540-71677-8_23
5. Mohassel, P., Zhang, Y.: SecureML: a system for scalable privacy-preserving machine learning. In: 2017 IEEE Symposium on Security and Privacy (SP), pp. 19–38. IEEE (May 2017)
6. Liu, J., Juuti, M., Lu, Y., Asokan, N.: Oblivious neural network predictions via miniONN transformations. In: Proceedings of the 2017 ACM SIGSAC Conference on Computer and Communications Security, pp. 619–631 (October 2017)
7. Rouhani, B.D., Riazi, M.S., Koushanfar, F.: DeepSecure: scalable provably-secure deep learning. In: Proceedings of the 55th Annual Design Automation Conference, pp. 1–6 (June 2018)
8. Bogdanov, D., Niitsoo, M., Toft, T., Willemson, J.: High-performance secure multi-party computation for data mining applications. Int. J. Inf. Secur. **11**(6), 403–418 (2012)
9. Bogdanov, D., Laur, S., Willemson, J.: Sharemind: a framework for fast privacy-preserving computations. In: Jajodia, S., Lopez, J. (eds.) ESORICS 2008. LNCS, vol. 5283, pp. 192–206. Springer, Heidelberg (2008). https://doi.org/10.1007/978-3-540-88313-5_13
10. Cramer, R., Damgård, I.B.: Secure Multiparty Computation. Cambridge University Press, Cambridge (2015)
11. Canetti, R.: Universally composable security: a new paradigm for cryptographic protocols. In: Proceedings 42nd IEEE Symposium on Foundations of Computer Science, pp. 136–145. IEEE (October 2001)
12. Ramakrishnan, R., Dral, P.O., Rupp, M., Von Lilienfeld, O.A.: Quantum chemistry structures and properties of 134 kilo molecules. Sci. Data **1**(1), 1–7 (2014)
13. Weininger, D.: SMILES, a chemical language and information system. 1. Introduction to methodology and encoding rules. J. Chem. Inf. Comput. Sci. **28**(1), 31–36 (1988)
14. RDKit: Open-source cheminformatics. www.rdkit.org. Accessed 11 Apr 2013
15. Hu, L., Wang, X., Wong, L., Chen, G.: Combined first-principles calculation and neural-network correction approach for heat of formation. J. Chem. Phys. **119**(22), 11501–11507 (2003)
16. Wu, S., Zhong, J., Cao, W., Li, R., Yu, Z., Wong, H.S.: Improving domain-specific classification by collaborative learning with adaptation networks. In: Proceedings of the AAAI Conference on Artificial Intelligence, vol. 33, no. 01, pp. 5450–5457 (July 2019)

17. Ma, Z., Ma, J., Miao, Y., Liu, X.: Privacy-preserving and high-accurate outsourced disease predictor on random forest. Inf. Sci. **496**, 225–241 (2019)
18. Zheng, W., Popa, R.A., Gonzalez, J.E., Stoica, I.: Helen: maliciously secure coopetitive learning for linear models. In: 2019 IEEE Symposium on Security and Privacy (SP), pp. 724–738. IEEE (May 2019)
19. Xu, G., Li, H., Liu, S., Yang, K., Lin, X.: Verifynet: secure and verifiable federated learning. IEEE Trans. Inf. Forensics Secur. **15**, 911–926 (2019)
20. Xia, Z., Gu, Q., Xiong, L., Zhou, W., Weng, J.: Privacy-preserving image retrieval based on additive secret sharing. arXiv preprint arXiv:2009.06893 (2020)
21. Xia, Z., Gu, Q., Zhou, W., Xiong, L., Weng, J.: Secure computation on additive shares. arXiv preprint arXiv:2009.13153 (2020)

Privacy-Preserving Computation Tookit on Floating-Point Numbers

Zekai Chen[1], Zhiwei Zheng[1], Ximeng Liu[1,2], and Wenzhong Guo[1,2(✉)]

[1] College of of Mathematics and Computer Science, Fuzhou University, Fuzhou 350108, China
guowenzhong@fzu.edu.cn

[2] Fujian Provincial Key Laboratory of Information Security of Network Systems, Fuzhou University, Fuzhou 350108, China

Abstract. Computation outsourcing using virtual environment is getting more and more prevalent in cloud computing, which several parties want to run a joint application and preserves the privacy of input data in secure computation protocols. However, it is still a challenging task to improve the efficiency and speed of secure floating point calculations in computation outsourcing, which has efficient secure integer calculations. Therefore, in this paper, we propose a framework built-up with a privacy-preserving computation toolkit with floating-point numbers (FPN), called PCTF. To achieve the above goal, we provide efficient toolkit to ensure their own data that FPN operations can be securely handled by homomorphic encryption algorithm. Moreover, we provide simulation results to experimentally evaluate the performance of the accuracy and the efficiency of PCTF, which will slowdown with 10x time consumption per in the secure floating-point addition and secure floating-point multiplication. Existing FPN division is constantly approaching result of division, or obtaining the quotient and remainder of division, in terms of precision, it is impossible to guarantee the precise range stably. However, our PCTF has higher precision in secure floating-point division, and the precision can be guaranteed at least 10^{-17}.

Keywords: Privacy computation · Homomorphic encryption · Multiple keys · Secure computation

1 Introduction

Nowadays, the cloud computing paradigm [1] revolutionizing the organizations' way of operating their data particularly in the way they store, access and process data. On one hand, according to IDC's data age 2025 [2] report, the annual data generated in the world will grow from 45 ZB to 175 ZB from 2019 to 2025. Local users can not afford such a large amount of data computing, more and more enterprise customers' applications such as Internet of Things (IoT) [3], self-driving car [4], and scientific research [5] need to be deployed in the data center or

J. Xiong et al. (Eds.): MobiMedia 2021, LNICST 394, pp. 462–476, 2021.
https://doi.org/10.1007/978-3-030-89814-4_33

cloud. On the other hand, cloud computing [6], with its cost-efficiency, flexibility, and offload of administrative overhead, is used by the majority of enterprise users. Organizations are prone to delegate their computational operations in addition to their data to the cloud.

Despite the tremendous advantages that the cloud offers, privacy and security issues in the cloud are preventing companies to utilize those advantages. When data are highly sensitive, e.g., the medical [6,7]. To protect the data privacy [8,9], data owners typically encrypt their data before outsourcing to the cloud. Homomorphic encryption [10] is a promising solution to avoid privacy leakage risks while protecting data confidentiality. Figure 1 plots a typical multiparty network with several different participants that own restricted storage ability and limited computing power. They jointly transmit data to the cloud server for data calculation. However, there exist challenges in the following two aspects. First, cloud service providers often store data belonging to multiple participants on the same server [11] to reduce operational costs. Therefore, different participants should be distributed with an individual key [12], to avoid multi-tenancy related attacks. How to achieve multiple keys computation without compromising the privacy of the outsourced data becomes a challenging issue. Second, cryptosystems are generally designed to protect only integer values. Computations on integers envivnately affect the accuracy of data and decision making results, which may even lead to wrong diagnosis in patients' ill [7] and wrong decision in self-driving [4].

In this paper, we present a **P**rivacy-Preserving **C**omputation **T**ookit on **F**loating-Point Numbers. The toolkit is constructed using secure computation based on homomorphic encryption and provide perfect or statistical privacy in the semi-honest model. The desirable features of PCTF include the following:

- **Secure FPN Computation:** We build a privacy-preserving outsourced computation of toolkit FPN operations with multiple keys. The toolkit offers multiplication, addition, subtraction, division with FPN. Secure multiplication addition, subtraction of FPN are trivial extensions of the integer operations. Meanwhile, we present new protocols for the division of FPN.
- **Multiple Users:** It has good expandability, including convenience and cost-effectiveness in supporting multiple secret keys in computing. Different data providers encrypt their data with their own public keys to the cloud server. The cloud server carries out secure computation in the case of multiple keys.
- **Efficiency and Accuracy:** Our calculation toolkit equips with secure and efficient encrypting FPN. Extensive experiments indicate that our scheme has high efficiency and lower accuracy loss, especially in the division.

The rest of this paper is organized as follows. In Sect. 2, we introduce the related work of secure multiparty computation protocols. In Sect. 3, we describe the preliminaries required in the understanding of our proposed FPN operations. In Sect. 4, we formalize the system model as well as the attacker model. Then we give a detailed introduction about privacy-preserving FPN computation protocols in Sect. 5. The security analysis and performance evaluation are presented in Sects. 6 and 7, respectively. Section 8 concludes this paper.

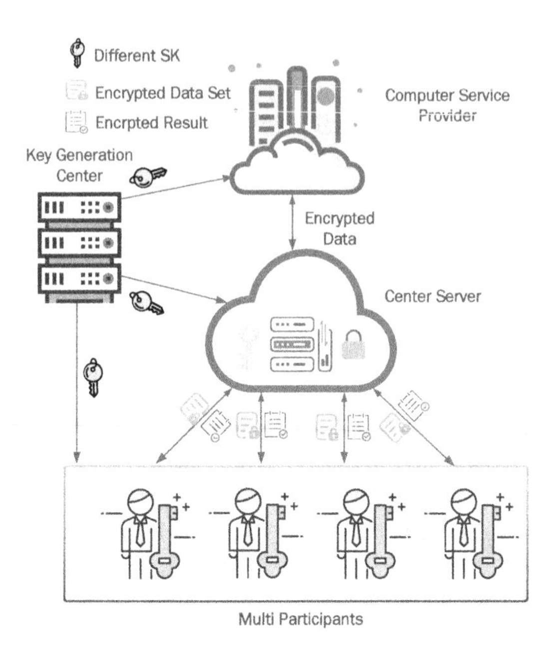

Fig. 1. System model

2 Related Work

Earlier work on privacy-preserving machine learning [13,14] has been proposed to provide privacy preservation during the training phase, but these schemes lacked implementation. Secure integer protocols have been studied in numerous researches. e.g., garbled circuit based protocol [15], homomorphic encryption based protocols [16–21] and secret sharing based protocols [21–24]. However, garbled circuit suffer from very high computation and communication complexities. Secret sharing needs all the servers to jointly compute some functions interactively. It requires multiple servers to store certain redundant data and requires pairwise secure channels between servers, so it is may difficult to implement in a number of real-world scenarios. Catrina et al. [20] presents a family of protocols for multiparty computation with rational numbers using floating-point representation. This approach offers more efficient solutions for secure computation than other usual representations, but it becomes quite complex when the divisor is secret. Veugen et al. [19] presents a secure integer division protocol that outputs an approximate quotient based on homomorphic encryption, such approximate protocols could be only useful for secure clustering and statistical analysis. Existing secure computation [17] with FPN base on homomorphic encryption can be realized at huge computation cost and communication overhead, and security division protocol is not implemented. Liu et al. [18] solves secure integer division by getting quotient and remainder, but it does not have precision, and it is impossible to obtain the complete value after division.

3 Preliminary

In this section, we introduce some terminologies and basic knowledge. First, some important notations used in this paper are provided in Table 1.

3.1 Distributed Two Trapdoors Public-Key Cryptosystem

DT-PKC splits a strong private key into different shares. In addition, the weak decryption algorithm should support distributed decryption to solve the authorization problem in the multi-key environment, follows the idea in [18], and works as follows:

- **raw_encryption**: Given a message $m \in \mathbb{Z}_N$, we compute $N = p \times q$ and $\lambda = \frac{(p-1)(q-1)}{2}$. We can choose a random number $r \in \mathbb{Z}_N$, and the ciphertext is generated as:

$$[\![m]\!]_{pk} = g^m r^N \bmod N^2 = (1 + m \times N) r^N \bmod N^2 \tag{1}$$

- **raw_decryption**: Given a message $[\![m]\!]_{pk}$, $N = p * q$ and $\lambda = \frac{(p-1)(q-1)}{2}$, using the decryption algorithm $D_{SK}(\cdot)$ and the corresponding private key $SK = \lambda$, the algorithm can compute to the plaintext. Since gcd $(\lambda, N) = 1$, the ciphertext is generated as:

$$[\![m]\!]^\lambda \bmod N^2 = r^{\lambda N}(1 + mN\lambda) \bmod N^2 = (1 + \mathrm{m}N\lambda) \tag{2}$$

$$m = L\left([\![m]\!]^\lambda \bmod N^2\right) \lambda^{-1} \bmod N \tag{3}$$

Table 1. Notations

Symbol	Definition
pk_u, sk_u	A pair of public-private keys
SK_1, SK_2	Partial decrypted private keys
$[\![m]\!]_{pk_u}$	A ciphertext encrypted with pk_u
$\lVert \cdot \rVert$	The data length of an arbitrary variable
κ	The security parameter variable
p, q	p and q are two big primes, $\lVert p \rVert = \lVert q \rVert = \kappa$
N	N is a big integer satisfying $N = p \times q$
$\mathbb{Z}_N$	An integer field of N
$\langle m \rangle$	Encrypted object of m with FPN

3.2 Privacy-Preserving Multiple Keys Integer Protocol

In particular, we assume there exits of semi-honest cloud server CP, CSP, [18] and their own public key pk_1, pk_2 from parties ($user_1$, $user_2$). Based on DT-PKC, the strong private key can be split into the different partial strong private key. Moreover, CP has partial strong private key SK_1, and CSP has partial strong private SK_2. All of the below protocols are considered under CP and CSP:

- **Secure Addition** (SAD): Consider CP with private input $([\![a]\!]_{pk_1}, [\![b]\!]_{pk_2})$, partial strong private key SK_1, and CSP with the partial strong private SK_2. The goal of the secure addition is to return the encryption of $[\![a+b]\!]_{pk}$ to CP, and then the $[\![a+b]\!]_{pk}$ will be returned to users. The basic idea of the SAD protocol is based on the following property which holds for any given $a, b \in \mathbb{Z}_N$, with different public key.
  - CP: generate $r_a, r_b \in \mathbb{Z}_N$, obtain $[\![r_a+r_b]\!]_{pk}$ and compute $D_{SK_1}([\![a+ra]\!]_{pk})$, $D_{SK_1}([\![b+r_b]\!]_{pk})$ to CSP.
  - CSP: decrypt r_a, r_b by SK_2, compute $[\![a+r_a+b+r_b]\!]_{pk}$ to CP.
  - CP: obtain $[\![a+b]\!]_{pk} = [\![a+r_a+b+r_b]\!]_{pk} \cdot [\![-r_a-r_b]\!]_{pk}$.
- **Secure Multiplication** (SM): In this protocol, the goal of the secure multiplication is to return the encryption of $[\![a \times b]\!]_{pk}$ to CP, and then the $[\![a \times b]\!]_{pk}$ will be returned to users. The basic idea of the SM protocol is based on the following property which holds for any given $a, b \in \mathbb{Z}_N$, with the different public key.
  - CP: generate $r_a, r_b \in \mathbb{Z}_N$, obtain $[\![a+r_a]\!]_{pk}, [\![b+r_b]\!]_{pk}$ and compute $D_{SK_1}([\![a+ra]\!]_{pk})$, $D_{SK_1}([\![b+r_b]\!]_{pk})$ to CSP.
  - CSP: decrypt $(a+r_a)(b+r_b)$ by SK_2, compute $[\![h]\!]_{pk} = [\![a \times b + a \times r_b + b \times r_a + r_a \times r_b]\!]_{pk}$ to CP.
  - CP: obtain $[\![a \times b]\!]_{pk} = [\![h]\!]_{pk} \cdot [\![-a \times r_b]\!]_{pk} \cdot [\![-b \times r_a]\!]_{pk} \cdot [\![-r_a \times r_b]\!]_{pk}$.
- **Secure Exponent Calculation** (SEXP): Given a plaintext number a and an encrypted integer number $[\![b]\!]_{pk}$ $(a, b \in \mathbb{Z}_N)$, the SEXP protocol will provide the encrypted data $[\![U]\!]_{pk}$, s.t., $[\![U]\!]_{pk} = a^{[\![b]\!]_{pk}}$. The SEXP protocol is described as follows: $a^{[\![b]\!]_{pk}} = (a^{[\![b+r]\!]_{pk}})^{a^{-r}} = a^{[\![b+r-r]\!]_{pk}}$.
- **Secure Greater Than** (SGT): Based on DT-PKC, the goal of the secure greater than is to return the encryption of $[\![u^*]\!]$ to CP, and then the $[\![u^*]\!]_{pk}$ will be returned to users. The basic idea of the SGT protocol is that we flip a coin to computer $[\![a]\!]_{pk} \cdot [\![2 \times b + 1]\!]_{pk}^{N-1}$ or $[\![2 \times b + 1]\!]_{pk} \cdot [\![a]\!]_{pk}^{N-1}$ to get result of relationship between a and b. If $u^* = 0$, it shows $a \geq b$; and If $u^* = 1$, it shows $a < b$.
- **Secure Maximum** (SMAX): Consider CP with input$([\![a]\!]_{pk_1}, [\![b]\!]_{pk_2}, [\![c]\!]_{pk_3})$. The goal of the secure maximum is to return the encryption of the maximum number between input to CP, and then the result will be returned to users. We can call SGT to computer $[\![u]\!]^* \leftarrow$ SGT $([\![a]\!]_{pk_1}, [\![y]\!]_{pk_2})$ to get $[\![u^*]\!]_{pk}$, and then $[\![A]\!]_{pk}$ and $[\![L]\!]_{pk}$ can be calculated by $[\![A]\!]_{pk} = [\![(1-u^*) \times a + u^* \times b]\!]_{pk}$ and $[\![L]\!]_{pk} = [\![(1-u^*) \times b + u^* \times a]\!]_{pk}$, s.t., $A \geq I$. Through multiple use of SGT, we will find the encryption of the maximum number.

3.3 Floating-Point Number

In this paper, we present FPN as $u = (s_u, b, e_u)$. Thus, we focus on the IEEE 754 standard for oating-point arithmetic. Therefore, IEEE numbers are stored using a kind of scientic notation. Floats are approximated using a binary fraction with the numerator using the first 53 bits starting with the most signicant bit and with the denominator as a power of two.

Under these circumstances, mantissa can be expressed as FPN. We consider the FPN is the set of number, which can be expressed as fractional number like $\frac{a}{b}$ $(a, b \in \mathbb{Z}_N)$. Meanwhile, b can be set for approximated 2^n. Therefore, we can get the numerator of the integer by multiplying by the denominator:

Algorithm 1: Floating-Point Number Encoding Protocol (FPE)

Input: f of FPN
Output: (s_f, b, e_f)
1 $base = 16$
2 $s, e < -\text{Frexp}(f)$
3 $e_f = \frac{(e - FLOAT\_BIT)}{log_2(base)}$
4 $s_f = round(\text{Fraction}(f) \times \text{Fraction}(base)^{-e_f}$
5 **return** $(s_f, b = base, e_f)$

In this protocol, we use Frexp function in python, which is used to calculate mantissa and exponent of FPN. Meanwhile, we use Fraction function in python, which can be used to calculate the numerator and denominator of FPN. According to the maximum number of FPN that python can represent, we can convert the mantissa into an integer. Meanwhile, we can see the mantissa from FPN convert integer number in Table 2.

Table 2. Frexp number table

Number	Mantissa	Base	Exponent
232342.3353423	0.88631567131919865	2	18
232342.3353423	15966443708343096	16	−9

Based on Algorithm 1, we easily get decoding of encoding-floating. Mantissa is defined as s. Base is defined as b. Exponent is defined as e. Thus, such a format is a number for which there exists at least one representation triplet (s, b, e):

$$f = s \times b^e \tag{4}$$

4 System Model and Privacy Requirement

In this section, we system model and e privacy requirement of PCTF.

4.1 System Model

In this session, we first present PCTF model that is comprised of four kinds of entities, namely a center server (CP), computer service provider (CSP), multi-participant (MP), and key generation center (KGC).

- **CP**: A CP stores and manages data outsourced from MP. CP also stores all the intermediate and final results in encrypted form. Furthermore, a CP is able to perform certain calculations over encrypted data.

- **CSP**: A CSP is responsible for assisting CP to complete the complex computations, which is able to partial decrypt ciphertexts sent by the CP, perform certain calculations over the partial decrypted data, and then reencrypt the calculated results.
- **MP**: Multi participants provide the encrypted data with the their own public keys. The goal of MP is to request a CP to perform some calculations over the encrypted data under multiple keys.
- **KGC**: The trusted KGC is tasked with the distribution and management of both public and private keys in the system. The KGC can split the strong private key into SK_1, SK_2 to CP and CSP, respectively.

4.2 Threat Model

In our threat model, MP, CP, and CSP are semi-honest parties, which strictly follow the protocol, but are also interested to learn data belonging to other parties. Therefore, we introduce an adversary $\mathcal{A}$ in our model. The goal of $\mathcal{A}$ is to decrypt the challenge MP's ciphertext with the following capabilities:

- $\mathcal{A}$ may eavesdrop on all communication links to obtain encrypted data.
- $\mathcal{A}$ may compromise the CSP to guess plaintext values of all ciphertexts sent from the CP by executing an interactive protocol.
- $\mathcal{A}$ may compromise the CP to guess plaintext values of all ciphertexts outsourced from the challenge MP, and all the ciphertexts sent from the CSP by executing an interactive protocol.
- $\mathcal{A}$ may compromise MP, with the exception of the challenge MP, to get access to their decryption capabilities, and try to guess all plaintexts belonging to the challenge MP.

The adversary $\mathcal{A}$, however, is restricted from compromising (1) both the CSP and the CP concurrently, and (2) the challenge MP. We remark that such restrictions are typical in adversary models used in cryptographic protocols.

5 Privacy-Preserving FPN Computation Protocols

In this section, we propose a set of generic protocols, which is used in cloud computation. All of the below protocols are under semi-honest parties. In particular, the FPN can be convert to FPE with Algorithm 1.

In order to realize PCTF, we can define the relative functions as follow:

- $\mathsf{FPE}(m)$: Function is defined by converting FPN to FPE, which outputs a result that $m = (s_m, b, t_m)$.
- $\mathsf{Encrypt}(s_m, e_m)$: Function encrypts the s_m and e_m by raw_encryption function, which outputs Encryption Object $\langle m \rangle = ([\![s_m]\!]_{pk}, b, [\![t_m]\!]_{pk})$.
- $\mathsf{Decrypt}(([\![s_m]\!]_{pk}, b, [\![t_m]\!]_{pk}))$: Given encrypted object, we choose raw_decryption to decrypt $[\![s_m]\!]_{pk}$ and $[\![t_m]\!]_{pk}$ separately. We also use FPE to assemble s_m and t_m into FPN as output.
- $\mathsf{SforCP}([\![m]\!]_{pk})$: Given encrypted integer number, we can choose random $r \in \mathbb{Z}_N$, and get $[\![r]\!]_{pk}$ to send $[\![m+r]\!]_{pk}$ from CP to CSP.
  - CP: generate $r \in \mathbb{Z}_N$, compute $D_{SK_1}([\![m+r]\!]_{pk})$ to CSP.
  - CSP: obtain $D_{SK_1}([\![m+r]\!]_{pk})$ compute $D_{SK_2}(D_{SK_1}([\![m+r]\!]_{pk}))$ to CP.
  - CP: obtain m.

5.1 Secure FPN Addition

Given two encrypted FPNs $\langle u\rangle = ([\![s_u]\!]_{pk}, b, [\![e_u]\!]_{pk})$ and $\langle v\rangle = ([\![s_v]\!]_{pk}, b, [\![e_v]\!]_{pk})$, the goal of the SFAD protocol is to securely compute two encrypted FPNs addition. The overall steps of SFAD protocol are shown as follows:

Algorithm 2: Secure Floating Addition Protocol (SFAD)

Input: Two encrypted FPNs $\langle u\rangle$ and $\langle v\rangle$
Output: Encrypted FPN $\langle T\rangle$

1 Both CP and CSP jointly calculate
2 $[\![G]\!]_{pk} \leftarrow \mathsf{SGT}([\![e_u]\!]_{pk_1}, [\![e_v]\!]_{pk_2})$
3 **if** *SforCP*($[\![G]\!]_{pk}$) **then**
4 $\quad [\![e_T]\!]_{pk} \leftarrow \mathsf{SAD}([\![e_v]\!]_{pk_2}, [\![O]\!]_{pk})$
5 $\quad [\![t]\!]_{pk} \leftarrow \mathsf{SAD}([\![e_u]\!]_{pk_1}, [\![e_v]\!]_{pk_2}^{N-1})$
6 $\quad [\![p]\!]_{pk} \leftarrow \mathsf{SM}([\![s_u]\!]_{pk_1}, \mathsf{SEXP}(b, [\![t]\!]_{pk}))$
7 $\quad [\![s_T]\!]_{pk} \leftarrow \mathsf{SAD}([\![s_v]\!]_{pk_2}, [\![p]\!]_{pk})$
8 **else**
9 $\quad [\![e_T]\!]_{pk} \leftarrow \mathsf{SAD}([\![e_u]\!]_{pk_1}, [\![O]\!]_{pk})$
10 $\quad [\![t]\!]_{pk} \leftarrow \mathsf{SAD}([\![e_v]\!]_{pk_2}, [\![e_u]\!]_{pk_1}^{N-1})$
11 $\quad [\![p]\!]_{pk} \leftarrow \mathsf{SM}([\![s_v]\!]_{pk_2}, \mathsf{SEXP}(b, [\![t]\!]_{pk}))$
12 $\quad [\![s_T]\!]_{pk} \leftarrow \mathsf{SAD}([\![s_u]\!]_{pk_1}, [\![p]\!]_{pk})$
13 $\langle T\rangle = ([\![s_T]\!]_{pk}, b, [\![e_T]\!]_{pk})$
14 **return** $\langle T\rangle$

During this protocol, using the SAD, SM, SEXP, SforCP protocols, it easily aligns two encrypted number of exponent part. Meanwhile, SAD might be used to add two encrypted numbers of mantissa to get the result.

5.2 Secure FPN Multiplication

Given two encrypted FPNs $\langle u\rangle = ([\![s_u]\!]_{pk}, b, [\![e_u]\!]_{pk})$ and $\langle v\rangle = ([\![s_v]\!]_{pk}, b, [\![e_v]\!]_{pk})$, the goal of the SFM protocol is to securely compute two encrypted FPNs multiplication. The overall steps of SFM protocol are shown as follows:

Algorithm 3: Secure Floating Multiplication Protocol (SFM)

Input: Two encrypted FPNs $\langle u\rangle$ and $\langle v\rangle$
Output: Encrypted FPN $\langle T\rangle$

1 Both CP and CSP jointly calculate
2 $[\![s_T]\!]_{pk} \leftarrow \mathsf{SM}([\![s_u]\!]_{pk1}, [\![s_v]\!]_{pk2})$
3 $[\![e_T]\!]_{pk} \leftarrow \mathsf{SAD}([\![e_u]\!]_{pk1}, |[\![e_v]\!]_{pk2})$
4 $\langle T\rangle = ([\![s_T]\!]_{pk}, b, [\![e_T]\!]_{pk})$
5 **return** $\langle T\rangle$

Base on SM and SAD protocol, we multiply the encrypted mantissa part by SM protocol and add the encrypted exponent part by SAD protocol.

5.3 Secure Integer Digits

Here, given encrypted integer $[\![n]\!]_{pk}$, we propose the secure digits protocol to obtain result $[\![d]\!]_{pk}$ to get digit of n.

Algorithm 4: Secure Digits Protocol (SDG)

Input: Encrypted integer $[\![n]\!]_{pk}$
Output: Encrypted digit $[\![d]\!]_{pk}$

1 In CP:
2 Random select $r_a \in \mathbb{Z}_N$, then Compute $[\![r_a]\!]_{pk}$
3 $[\![n + r_a]\!]_{pk} \leftarrow \mathsf{SM}([\![n]\!]_{pk_1}, [\![r_a]\!]_{pk})$
4 $[\![t]\!]_{pk} = D_{sk_1}([\![n \times r_a]\!]_{pk})$
5 Send $[\![t]\!]_{pk}, [\![r_a]\!]_{pk}$ to CSP
6 In CSP:
7 Receive $[\![t]\!]_{pk}$ from CP
8 $t = D_{sk_2}([\![t]\!]_{pk})$
9 $t_n = Math.Log_{10}t$ in plaintext
10 $t_{d_n} = int(round(tnum - int(tnum)) \times 100))$
11 Send $[\![t_{d_n}]\!]_{pk}, [\![int(t_n)]\!]_{pk}$ to CP
12 In CP:
13 Receive $[\![t_{d_n}]\!]_{pk}, [\![int(t_n)]\!]_{pk}$ from CSP
14 $t_r = Math.Log_{10}r_a$ in plaintext
15 $t_{d_r} = int(round(tnum - int(tnum)) \times 100))$
16 $[\![t_s]\!]_{pk} = [\![t_{dn}]\!]_{pk} \cdot [\![t_{dr}]\!]_{pk}^{N-1}$
17 $[\![lt]\!]_{pk} \leftarrow \mathsf{SGT}([\![t_s]\!]_{pk}, [\![50]\!]_{pk})$
18 $[\![d]\!]_{pk} = [\![int(t_n)]\!]_{pk} \cdot [\![int(t_r)]\!]_{pk}^{N-1} \cdot [\![lt]\!]_{pk} \cdot [\![1]\!]_{pk}$
19 **return** $[\![d]\!]_{pk}$

In this protocol, the goal of the SDG protocol is to securely compute the decimal digits of the encrypted number. During this protocol, we assume that CP can coordinate with CSP. The basic idea of the SDG protocol is based on the following property which holds for any given number:

$$\mathrm{d} = \lceil \log_{10} n \rceil + 1 \tag{5}$$

where all the arithmetic operations are performed under $\mathbb{Z}_N$. The overall steps in SDG are shown in Algorithm 4, we can suppose the d_s to get $\log_{10} n$ whether to add one after rounding up. The following equation is d_s:

$$\mathrm{d}_s = \lceil \log_{10}(\mathrm{n} \times r_a) - \lfloor \log_{10}(\mathrm{n} \times r_a) \rfloor - \log_{10} r_a + \lfloor \log_{10} r_a \rfloor \times 100 \rceil \tag{6}$$

$$[\![t_s]\!]_{pk} = [\![\lfloor \log_{10}(\mathrm{n} \times ra) \rfloor]\!]_{pk} \cdot [\![\lfloor \log_{10} ra \rceil]\!]_{pk}^{N-1} \tag{7}$$

$$[\![\mathrm{d}]\!]_{pk} = [\![t_s]\!]_{pk} \cdot SGT\left([\![d_s d_s d_s d_s]\!]_{pk}, [\![50]\!]_{pk}\right) \cdot [\![1]\!]_{pk} \tag{8}$$

5.4 Secure Integer Modular Calculation

Here, the goal of the SMOD protocol is to compute given the two encrypted integers $[\![f_1]\!]_{pk_1}$, $[\![f_2]\!]_{pk_2}$ division' quotient and remainder.

Algorithm 5: Secure Modular Calculation Protocol (SMOD)

Input: Two encrypted integers $[\![f_1]\!]_{pk_1}$, $[\![f_2]\!]_{pk_2}$
Output: Encrypted quotient $[\![q]\!]_{pk}$ and encrypted remainder $[\![r]\!]_{pk}$
1 Both CP and CSP jointly calculate
2 $[\![G_1]\!]_{pk} \leftarrow \mathsf{SGT}([\![0]\!]_{pk}, [\![f_1]\!]_{pk_1})$ and $[\![G_2]\!]_{pk} \leftarrow \mathsf{SGT}([\![0]\!]_{pk}, [\![f_2]\!]_{pk_2})$
3 $[\![sign]\!]_{pk}, [\![q]\!]_{pk} = [\![1]\!]_{pk}, [\![0]\!]_{pk}$
4 **if** *SforCP*$([G_1]_{pk})$ **then**
5 $\quad [\![f_1]\!]_{pk_1} = [\![f_1]\!]_{pk_1}^{N-1}$ and $[\![sign]\!]_{pk} = [\![sign]\!]_{pk}^{N-1}$
6 **if** *SforCP*$([G_2]_{pk})$ **then**
7 $\quad [\![f_2]\!]_{pk_2} = [\![f_2]\!]_{pk_2}^{N-1}$ and $[\![sign]\!]_{pk} = [\![sign]\!]_{pk}^{N-1}$
8 $[\![d_1]\!]_{pk} \leftarrow \mathsf{SDG}([\![f_1]\!]_{pk_1})$ and $[\![d_2]\!]_{pk} \leftarrow \mathsf{SDG}([\![f_2]\!]_{pk_2})$
9 $[\![d_s]\!]_{pk} = [\![d_1]\!]_{pk} \cdot [\![d_2]\!]_{pk}^{N-1}$
10 **for** $i \leftarrow 1$ **to** *SforCP*$([\![d_s]\!]_{pk})$ **do**
11 $\quad [\![p]\!]_{pk}, \langle t\rangle = [\![0]\!]_{pk}, \langle u\rangle$
12 $\quad$ **for** $j \leftarrow 1$ **to** 9 **do**
13 $\quad\quad [\![t_s]\!]_{pk} \leftarrow \mathsf{SM}([\![f_1]\!]_{pk_1}, [\![f_2]\!]_{pk_2}^{N-j-10\times i})$
14 $\quad\quad [\![lt]\!]_{pk} \leftarrow \mathsf{SGT}([\![t_s]\!]_{pk}, [\![0]\!]_{pk})$
15 $\quad\quad [\![t_p]\!]_{pk} \leftarrow \mathsf{SM}([\![lt]\!]_{pk}, [\![i]\!]_{pk})$
16 $\quad\quad [\![p]\!]_{pk} \leftarrow \mathsf{SMAX}([\![p]\!]_{pk}, [\![t_p]\!]_{pk})$
17 $\quad [\![f_t]\!]_{pk} \leftarrow \mathsf{SM}([\![f_t]\!]_{pk}, [\![p]\!]_{pk}^{N-1}) \cdot [\![f_t]\!]_{pk}$
18 $\quad ([\![q]\!]_{pk}, [\![r]\!]_{pk}) \leftarrow (\mathsf{SM}(([\![q]\!]_{pk}^{N-10})^{N-1} \cdot [\![p]\!]_{pk}, [\![sign]\!]_{pk}), [\![f_t]\!]_{pk})$
19 **return** $([\![q]\!]_{pk}, [\![r]\!]_{pk})$

To start with, CP computes decimal digits of $[\![f_1]\!]_{pk_1}$, $[\![f_2]\!]_{pk_2}$ to get times of the loop body is executed. Then, using the SGT, SMAX protocols, CP computes best number with the help of CSP, for $1 \leq i \leq 9$ to get $[\![q]\!]_{pk}$ and $[\![r]\!]_{pk}$.

5.5 Secure FPN Division

Algorithm 6: Secure Floating Division Protocol (SFDIV)

Input: Two encrypted FPNs $\langle u\rangle$ and $\langle v\rangle$
Output: Encrypted FPN $\langle T\rangle$
1 Both CP and CSP jointly calculate
2 $([\![q_1]\!]_{pk}, [\![r_1]\!]_{pk}) \leftarrow \mathrm{SMOD}([\![s_u]\!]_{pk_1}, [\![s_v]\!]_{pk_2})$
3 $([\![q_2]\!]_{pk}, [\![r_2]\!]_{pk}) \leftarrow \mathrm{SMOD}([\![r_1]\!]_{pk}^{pow(10,17)}, [\![s_v]\!]_{pk_2})$
4 $q_2 \leftarrow \mathsf{SforCP}([\![q_2]\!]_{pk})$, then $\langle t\rangle \leftarrow \mathsf{Encrypt}(q_2 * 10^{-17})$
5 $[\![e]\!]_{pk} \leftarrow \mathsf{SM}([\![e_u]\!]_{pk_1}, [\![e_v]\!]_{pk_2}^{N-1})$, then $[\![e_t]\!]_{pk} = [\![e]\!]_{pk} \cdot [\![e_t]\!]_{pk}$
6 $\langle T\rangle \leftarrow \mathsf{SFAD}(\langle t\rangle, ([\![q_1]\!]_{pk}, b = 16, [\![e_u]\!]_{pk_1}))$
7 **return** $\langle T\rangle$

In this protocol, the basic idea is based on the following property which gets for q_1, q_2, r_1 is the remainder of $\frac{u}{v}$ and r_2 is the remainder of $\frac{r_1 \times 10^{17}}{v} \in \mathbb{Z}_N$:

$$\frac{u}{v} = q_1 + \frac{r_1}{v} = q_1 + \frac{r_1 \times 10^{17}}{v} \approx q_1 + q_2 \times 10^{-17} \tag{9}$$

The overall steps involved in the SFDIV protocol are shown in Algorithm 6. In order to get higher precision, using SMOD protocol, CP computes $[\![q_1]\!]_{pk}$, $[\![r_1]\!]_{pk}$ of $\frac{[\![s_u]\!]_{pk_1}}{[\![s_v]\!]_{pk_2}}$, and then $[\![r_1]\!]_{pk}$ multiplies 10^{17} to do similarly steps to get $[\![q_2]\!]_{pk}$ and $[\![r_2]\!]_{pk}$. Observe that value of r_2 is very small, ignoring does not affect the accuracy. We can compute $\langle t \rangle = \mathsf{FPE}(r_1 * 10^{-17})$, $[\![e]\!]_{pk} = \mathsf{SM}([\![e_u]\!]_{pk_1}, [\![e_v]\!]_{pk_2}^{N-1})$.

$$x^* = ([\![q_1]\!]_{pk}, 16, [\![e_u]\!]_{pk}), \quad y^* = ([\![s_t]\!]_{pk}, 16, [\![e]\!]_{pk} \cdot [\![e_t]\!]_{pk}) \tag{10}$$

$$\langle x \rangle = \text{ Encrypt } (x^*), \langle y \rangle = \text{ Encrypt } (y^*) \tag{11}$$

$$\langle \frac{f_1}{f_2} \rangle = SFAD(\langle x \rangle, \langle y \rangle) \tag{12}$$

6 Security Analysis

In this section, we will analyze the security guarantees of the proposed protocols, which are based on DT-PKC.

Definition 1. *We assume that the adversary $\mathcal{A}$, and real world interacts with a simulator Sim to complete the process in the ideal world. We consider that FPN of $\langle x \rangle$, $\langle y \rangle$ are the input, Π is the corresponding protocol, and c denotes the computationally indistinguishable. If PCTF system is secure, which can be represented as* $\{\text{IDEAL}_{\Pi,\text{sim}}(\langle x \rangle, \langle y \rangle)\} \overset{c}{\approx} \{\text{REAL}_{\Pi,\text{A}}(\langle x \rangle, \langle y \rangle)\}$.

Lemma 1. *The protocol of SAD, SM, SGT, SMAX, and system of DT-PKC are secure in the honest-but curious model [17].*

Lemma 2. *Since the comparison result is that MP can decide the number to use to enter the corresponding branch together, we consider that the result of SGT is only safe for CP to know.*

Lemma 3. *FPN $f = (s_f, b_f, e_f)$. If only a part of it is leaked to others, the others cannot deduce the complete data based on the obtained part, so the data cannot be leaked to others.*

Theorem 1. *The SFM protocol is secure in the presence of semi-honest (non-colluding) adversaries $\mathcal{A} = (\mathcal{A}_{MP}, \mathcal{A}_{CP}, \mathcal{A}_{CSP})$.*

Proof. Sim_{MP} receives $\langle x \rangle$, $\langle y \rangle$ as input, which consist of s_x, t_x, s_y, t_y, and then simulates that $\mathcal{A}_{MP}$ obtains $[\![s_x]\!]_{pk}$, $[\![t_x]\!]_{pk}$, $[\![s_y]\!]_{pk}$, $[\![t_y]\!]_{pk}$. Sim_{CP} and Sim_{CSP} simulate that $\mathcal{A}_{CP}$ and $\mathcal{A}_{CSP}$ calculate $\mathsf{SAD}([\![s_x]\!]_{pk},[\![t_x]\!]_{pk})$ and $\mathsf{SM}([\![s_y]\!]_{pk},[\![t_y]\!]_{pk})$. According to Lemma 1, it is trivial to see that the view of $\mathcal{A}_{MP}$, $\mathcal{A}_{CP}$, $\mathcal{A}_{CSP}$ consists of the encrypted data and execute SAD and SM in security. Correspondingly, the SDG protocol is also secure in the semi-honest model. Thus they are indistinguishable in the real and the ideal executions.

Theorem 2. *The* SFAD *protocol is secure in the presence of semi-honest (non-colluding) adversaries* $\mathcal{A} = (\mathcal{A}_{MP}, \mathcal{A}_{CP}, \mathcal{A}_{CSP})$.

Proof. Before performing SAD and SM, the views of CP and CSP will be $View_1 = (\langle x \rangle, \langle y \rangle, SG_1, SG_2)$ and $View_2 = (\langle x \rangle, \langle y \rangle, [\![SG_1]\!]_{pk}, [\![SG_2]\!]_{pk})$, respectively. $\mathcal{A}_{CP}$ will obtain SG_1, SG_2. Due to Lemma 2, result of SGT is safe for CP to know. It's trivial to see they are simulated by Sim_{MP}, Sim_{CP}, Sim_{CSP}. Since the security of the protocol SAD, SM execution have been proven above, it's trivial to prove the security of the whole protocol SFAD as well. Correspondingly, the SMOD protocol is also secure in the semi-honest model.

Theorem 3. *The* SFDIV *protocol is secure in the presence of semi-honest (non-colluding) adversaries* $\mathcal{A} = (\mathcal{A}_{MP}, \mathcal{A}_{CP}, \mathcal{A}_{CSP})$.

Proof. Due to Lemma 1, Lemma 2, Lemma 3, result of SGT and the digit of FPN' s_x or digit of FPN' s_y are safe for CP to know. An execution of SMAX, SM, SAD, SFAD will be $View_1^j = (\langle x_1 \rangle^j, [\![q_1^i]\!]_{pk}^j, \langle y_1 \rangle^j, [\![r_1]\!]_{pk}^j, [\![res_1]\!]_{pk}^j)$, $View_2^j = (\langle x_2 \rangle^j, [\![q_2^i]\!]_{pk}^j, \langle y_2 \rangle^j, [\![r_2]\!]_{pk}^j, [\![res_2]\!]_{pk}^j)$, and $1 \leq j \leq 9$. Since res_2^j of $View_2^j$ is intermediate result of part of FPN, according to Lemma 3, Lemma 1, Theorem 2, $\mathcal{A}_{MP}$, $\mathcal{A}_{CP}$, $\mathcal{A}_{CSP}$ will be computationally indistinguishable. Thus, SFDIV protocol is secure.

7 Experiment Analysis

In this section, we conduct real-world experiments to show the performance of the proposed for evaluating our PCTF. In PCTF, we implement SFAD, SFM, SFDIV, and SDG in our PCTF by Python. The Python program runs on a PC (Intel(R) Core(TM) i5-9500 CPU @ 3.00 GHz and 8.00 GB RAM) to simulate the multi-users with the different public keys to compute tasks on the cloud server. Meanwhile, we test the performance of PCTF the compared with [17], as shown in Fig. 2. The accuracy and efficiency are as follows:

- Accuracy: To implement multi-keyword floating division which has not been addressed in [17] and [18], we propose SFDIV to address the issues. In our SFDIV protocol, we approximate the fractional part of the FPN. Since the remainder obtained by the first division, the remainder obtained by the expansion after the division is very small, and the precision of floating point numbers that can be represented by Python can be ignored. Therefore, we consider the loss of SFDIV is vary small and the computation accuracy is very high. Meanwhile, in our PCTF, we can promise the precision of SFDIV at least in 10^{-17}. The SFM and SFAD in encrypting computation, it has no loss of accuracy.
- Efficiency: We evaluate the efficiency of our proposed PCTF from criteria: time cost overhead and scalability. The evaluation results show whether our model is efficient and practical. During this process, we can see (a), (b), (c) in Fig. 2 that our SFAD and SFM have less computation cost and less joint computing time, comparing with [18]. But in our SFDIV, we can see (d) and (g) in Fig. 2 that we can know that as the accuracy increases, the calculation

time required for the division also increases. When the divisor is encrypted and the dividend is an integer or encoding number in (d), (e), (f), (g), (h), (i), we can get the computation cost only in CP. This reduces the risk of ciphertext leakage.

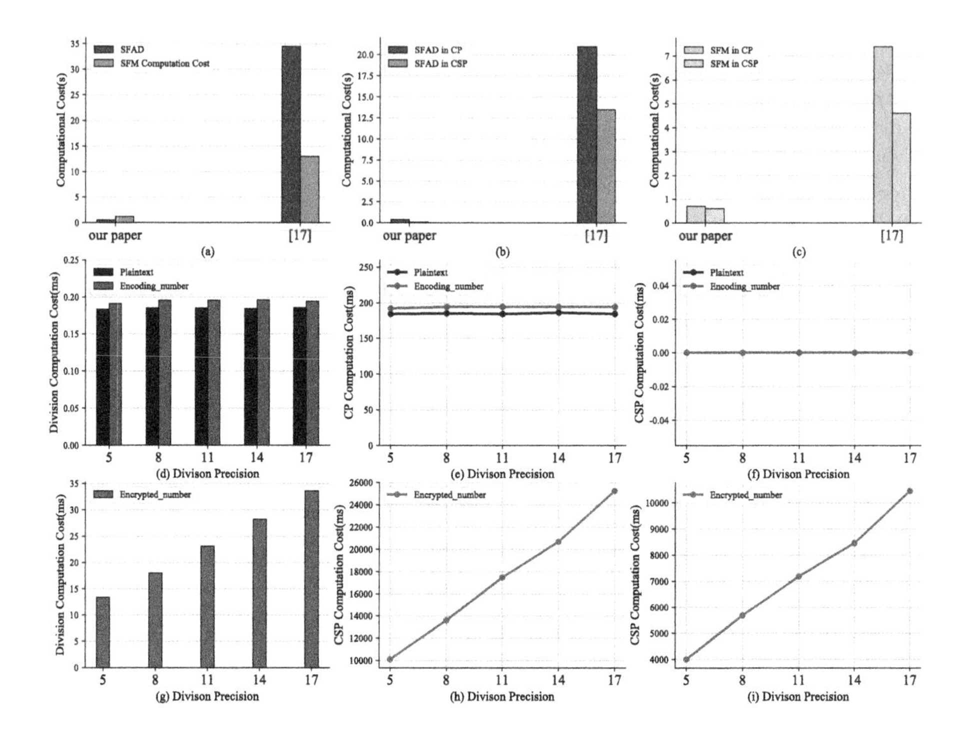

Fig. 2. Floating computational cost.

Table 3. The time cost of security protocol

Protocol	CP compute(s).	CSP(s) compute.	Total time(s)
SFAD	0.4323	0.1501	0.5823
SFM	0.8234	0.4139	1.2375
SFDIV	24.6281	9.9341	34.5623

As shown in Table 3, we calculate the FPN division in SFDIV, SFAD, and SFM. From the Fig. 2, we can see that our division time increases with the number of digits increasing, and it only takes 13 s when the accuracy is 10^{-5}, which is 4 times faster than the 50 s required in reference [18] times. With the increase of division accuracy, we can see that the time of the algorithm increases linearly. When dealing with some high-precision problems, our safe division can still maintain high accuracy. After 20 iterations of secure division, our error is only kept at 10^{-15}. This shows that multiple iterations do not affect the performance of the SFDIV.

8 Conclusion

In this paper, we proposed PCTF, a framework for computation tookit on FPN, which allows users to outsource encrypted data to a cloud service provider for storing and processing. We build a privacy-preserving computation tookit of FPN. The toolkit offers multiplication, addition, subtraction, division with FPN. SFM, SFAD are trivial extensions of the integer operations. Meanwhile, We present new protocols for the division of FPN. Our evaluations demonstrated that our framework compared with other FPN protocols effectively reduces computational and communication. Protocols are provably secure in the cloud service with semi-honest behavior.

In the future work, we will focus on the optimization of communication overhead and the expansion of machine learning methods. Although the total time cost of the calculation process is not high in the current scheme, the communication time in CP is still worth optimizing. In addition, to cope with the expansion of machine learning algorithms in different scenarios, the future work will focus on applying in a specific application domain.

Acknowledgement. This work is supported by the National Natural Science Foundation of China (Grant No. U1705262, No. 62072109, No. U1804263), and the Natural Science Foundation of Fujian Province (Grant No. 2018J07005)

References

1. Mell, P., Grance, T.: The NIST definition of cloud computing (draft). NIST Special Publication 800-145 (2011)
2. Seagate Cor: Data age 2025 - the Evolution of Data to Life-Critical (2018). https://www.seagate.com/cn/zh/our-story/data-age-2025/. Accessed 2020
3. Chamberlin, B.: IoT (Internet of Things) Will go Nowhere Without Cloud Computing and Big Data Analytics. http://ibmcai.com/2014/11/20/iot-internet-of-things-will-go-nowhere-without-cloud-computing-and-big-data-analytics/
4. Yeshodara, N.S., Nagojappa, N.S., Kishore, N.: Cloud based self driving cars. In: 2014 IEEE International Conference on Cloud Computing in Emerging Markets (CCEM), Bangalore, India, pp. 1–7 (2014). https://doi.org/10.1109/CCEM.2014.7015485
5. Quick, D., Martini, B., Choo, R.: Cloud Storage Forensics. Elsevier, Amsterdam (2013)
6. Mohassel, P., Zhang, Y.: SecureML: a system for scalable privacy-preserving machine learning. In: IEEE Symposium on Security and Privacy (SP), IEEE 2017, pp. 19–38 (2017)
7. Wang, X., Ma, J., Miao, Y., Liu, X., Yang, R.: Privacy-preserving diverse keyword search and online pre-diagnosis in cloud computing. IEEE Trans. Serv. Comput. (2019)

8. Sun, W., Yu, S., Lou, W., Hou, Y.T., Li, H.: Protecting your right: attribute-based keyword search with fine-grained owner-enforced search authorization in the cloud. In: Proceedings of INFOCOM, pp. 226–234. IEEE (2014)
9. Bidi Ying, D.M., Mouftah, H.T.: Sink privacy protection with minimum network traffic in WSNs. Ad Hoc Sens. Wirel. Netw. **25**(1–2), 69–87 (2015)
10. Paillier, P.: Public-key cryptosystems based on composite degree residuosity classes. In: Stern, J. (ed.) EUROCRYPT 1999. LNCS, vol. 1592, pp. 223–238. Springer, Heidelberg (1999). https://doi.org/10.1007/3-540-48910-X_16
11. Cloud Computing. https://en.wikipedia.org/wiki. Accessed 2019
12. López-Alt, A., Tromer, E., Vaikuntanathan, V.: On-the-fly multiparty computation on the cloud via multikey fully homomorphic encryption. In: Proceedings of 44th Annual ACM Symposium on Theory Computing, pp. 1219–1234 (2012)
13. Lindell, Y., Pinkas, B.: Privacy preserving data mining. In: Bellare, M. (ed.) CRYPTO 2000. LNCS, vol. 1880, pp. 36–54. Springer, Heidelberg (2000). https://doi.org/10.1007/3-540-44598-6_3
14. Sanil, A.P., Karr, A.F., Lin, X., Reiter, J.P.: Privacy preserving regression modelling via distributed computation. In: Proceedings of the ACM SIGKDD International Conference on Knowledge Discovery and Data Mining (KDD 2004), pp. 677–682. ACM (2004)
15. Lazzeretti, R., Barni, M.: Division between encrypted integers by means of garbled circuits. In: WIFS 2011, pp. 1–6 (2011)
16. Dahl, M., Ning, C., Toft, T.: On secure two-party integer division. In: Keromytis, A.D. (ed.) FC 2012. LNCS, vol. 7397, pp. 164–178. Springer, Heidelberg (2012). https://doi.org/10.1007/978-3-642-32946-3_13
17. Liu, X., Choo, K.-K.R., Deng, R.H., Lu, R., Weng, J.: Efficient and privacy-preserving outsourced calculation of rational numbers. IEEE Trans. Depend. Secur. Comput. **15**, 27–39 (2016)
18. Liu, X., Deng, R.H., Choo, K.-K.R., Weng, J.: An efficient privacy-preserving outsourced calculation toolkit with multiple keys. IEEE Trans. Inf. Forensics Secur. **11**(11), 2401–2414 (2016)
19. Veugen, T.: Encrypted integer division and secure comparison. IJACT **3**(2), 166–180 (2014)
20. Catrina, O., Saxena, A.: Secure computation with fixed-point numbers. In: Sion, R. (ed.) FC 2010. LNCS, vol. 6052, pp. 35–50. Springer, Heidelberg (2010). https://doi.org/10.1007/978-3-642-14577-3_6
21. Kiltz, E., Leander, G., Malone-Lee, J.: Secure computation of the mean and related statistics. In: Kilian, J. (ed.) TCC 2005. LNCS, vol. 3378, pp. 283–302. Springer, Heidelberg (2005). https://doi.org/10.1007/978-3-540-30576-7_16
22. Bogdanov, D., Niitsoo, M., Toft, T., et al.: High-performance secure multi-party computation for data mining applications. Int. J. Inf. Secur. **11**(6), 403–418 (2012)
23. Bunn, P., Ostrovsky, R.: Secure two-party k-means clustering. In: Proceedings of the 2007 ACM Conference on Computer and Communications Security, CCS 2007, Alexandria, Virginia, USA, October 28–31, 2007. ACM (2007)
24. Catrina, O., Dragulin, C.: Multiparty computation of fixed-point multiplication and reciprocal. In: DEXA 2009, no. 1, pp. 107–111 (2009)

Vendor-Based Privacy-Preserving POI Recommendation Network

Longyin Cui[1(✉)], Xiwei Wang[2], and Jun Zhang[1]

[1] University of Kentucky, Lexington, KY, USA
lcu225@uky.edu, jzhang@cs.uky.edu
[2] Northeastern Illinois University, Chicago, IL, USA
xwang9@neiu.edu

Abstract. Point-of-interest (POI) recommendation services are growing in popularity due to the choice overloading and overwhelming information in modern life. However, frequent data leakage and hacking attacks are reducing people's confidence. The awareness of privacy issues is multiplying among both the customers and service providers. This paper proposes a localized POI recommendation scheme combined with clustering techniques and introduces the concept of "virtual users" to protect user privacy without sacrificing too much accuracy.

Keywords: Recommender system · Privacy-preserving · Virtual user · Recommendation network

1 Introduction

The demand for Point of Interest (POI) recommendation services is proliferating. Location-based Social Networks (LBSN) providers, such as Yelp and Google Local have effectively increased their market shares. According to Yelp's Q4'19 report, there are 36 million unique mobile app users bringing in revenues over one billion dollars in 2019. Moreover, the total number of user reviews it collected since 2004 has surpassed 205 million [19]. From the Newzoo's Global Mobile Market Report 2020: (1), there are 3.5 billion smartphone users worldwide by the end of 2020; (2), in 2019, about 56% of the global website traffic was generated by mobile devices [13]. Figure 1 shows the primary structure and components of a typical LBSN, in which the records of check-in activities are usually used to generate recommendations.

Conventionally, the data collected is saved and stored in a central server. Such centralized recommender systems are vulnerable when facing data breach issues, and the cost or penalty is substantial. The Capital One data breach, which caused approximately 500 million dollars financial damage on top of other indirect costs [9]. The case study on this incident shows that, nowadays, companies worldwide are not yet adequately adapted to securing their cloud computing environments [14].

J. Xiong et al. (Eds.): MobiMedia 2021, LNICST 394, pp. 477–490, 2021.
https://doi.org/10.1007/978-3-030-89814-4_34

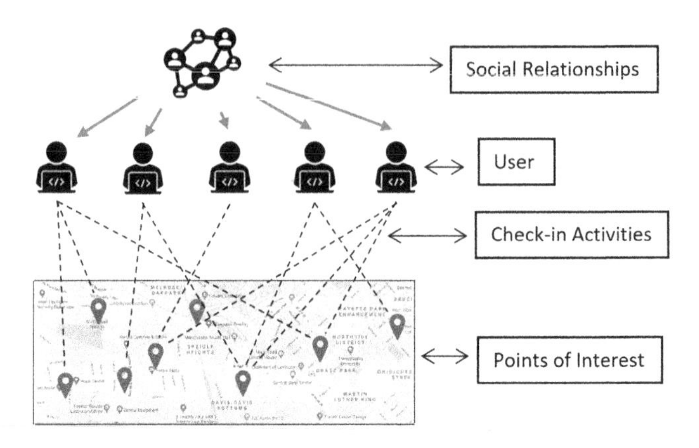

Fig. 1. Location-based social networks (LBSN).

In addition to improving compliance controls, another solution is to push the data processing task to the user end, securing their privacy by eliminating the need for a central server [3,12,17,20]. However, data itself is an essential resource, and the path of giving up storing user information to avoid legal fallouts could be a blind alley. Moreover, these frameworks still require private data, such as social network data or real-time GPS locations. Fetching such data is risky, even under the presence of a privacy disclaimer. In 2019, the Federal Trade Commission issued a 5 billion dollar fine on Facebook due to its violation on consumers' privacy [5].

In this paper, We propose a vendor-based recommendation network scheme. Instead of maintaining a central server or carrying out all computing activities on user sides, we localize the recommendation tasks on vendors for each small area. The vendor can be any Point-of-Interest, such as restaurants, gas stations, and grocery stores. Challenges arise when small business owners decide to build their recommender systems, such as *cold start problem* and *sparsity problem*. Due to the lack of correlation in the high dimensional space, predictions directly made on a sparse feedback matrix often suffer from low accuracy. We alleviated this problem by introducing '***virtual users***', a concept elaborated in later sections that can implicitly reflect real users' preferences.

Both recommendation providers and consumers benefit from this scheme. For service providers, their users are more likely to use the services because local businesses are more trusted than large corporations [16]. For users, their data can be analyzed more efficiently and the risk of having all data hacked at once is remarkably reduced. The customers' chosen vendors also reflect users' active visiting areas making the system naturally location-aware.

Our main contributions are summarized as follows:

- We propose a localized POI recommender system framework, where the computing tasks and data storage of a central server are distributed to smaller areas.

- We introduce the idea of "virtual users", a concept that enables localized recommender systems to collaborate with each other resolving the sparsity problem.
- We conduct experiments on real-world datasets, demonstrating the importance of geographical restriction and the effectiveness of our framework.

The rest of this paper is organized as follows: the background and related works are introduced in Sect. 2. The problem description and our proposed solution are discussed in Sect. 3. Next, in Sect. 4, the experiments are carried out, and the results are analyzed. Section 5 gives the conclusion and future work.

2 Background

2.1 Centralized Recommendations

The major difference between a centralized RS and a distributed or decentralized RS is how data is attained and processed. The data is stored on a single server for a centralized RS where new recommendations are generated immediately after data pre-processing. There are many ways to implement centralized POI recommendation models. To investigate the tradeoff between privacy preservation and recommendation accuracy, we selected several straightforward models to demonstrate the proposed framework. With that being said, the Recommender System Network (RSN) we propose is compatible with various methods.

For example, for a classic Matrix Factorization (MF) model, a regression technique is realized to collaboratively learn the latent factors of users and items (i.e., POIs) [15]. While users' feedback is reflected by their ratings, the latent factors indicate each user or item's hidden characteristics. A general MF based recommendation model can be represented by the following optimization problem as shown in Eqs. (1) and (2), where $\boldsymbol{r}_{ui}$ denotes the known rating given by user $\boldsymbol{u}$ to item $\boldsymbol{i}$, and $\hat{\boldsymbol{r}}_{ui}$ denotes the predicted rating. Vectors $\boldsymbol{p}_u$ and $\boldsymbol{q}_i$ represent the user and item latent factors, respectively.

$$\min_{p_u, q_i} \sum_{r_{ui} \in R_{train}} (r_{ui} - \hat{r}_{ui})^2 + \lambda(\|q_i\|^2 + \|p_u\|^2) \tag{1}$$

$$\hat{r}_{ui} = q_i^T p_u \tag{2}$$

In the well-known biased MF model [11,15], the predicted rating, however, is formalized as follows:

$$\hat{r}_{ui} = \mu + b_u + b_i + q_i^T p_u \tag{3}$$

where $\boldsymbol{\mu}$, $\boldsymbol{b}_u$, and $\boldsymbol{b}_i$ represent the global mean, the user bias, and the item bias, respectively. $\boldsymbol{R}_{train}$ is the set of observed ratings. Accordingly, the objective function is then updated as Eq. (4). The notations and the parameters will be discussed in detail in later sections.

$$\min_{p_u, q_i} \sum_{r_{ui} \in R_{train}} (r_{ui} - \hat{r}_{ui})^2 + \lambda(b_u^2 + b_i^2 + \|q_i\|^2 + \|p_u\|^2) \quad (4)$$

We involve Biased MF heavily in our experiment due to its excellent combination of simplicity and reliability.

2.2 Decentralized Recommendations

To convert a centralized RS into a decentralized RS, service providers need to push the data storage and processing tasks to the users' end. Either the user data can be distributed efficiently, or a secure protocol such as a safe peer-to-peer structure is provided to allow information exchange. In a decentralized RS, every user keeps a fraction of the training data and is responsible for generating their own recommendations locally. Some researchers managed to shift the learning process to the users' end to resolve privacy concerns [3,17,18].

However, there are inevitable vulnerabilities in these models. For example, when users exchange ratings directly, a malicious user is able to gather other users' ratings by giving positive feedback to all locations. Alternatively, when only latent factors are exchanged, a malicious user can tell that another user visited a specific place if they share a similar latent factor associated with the same location. Each of the researchers made their breakthroughs and have solved different problems, but many of them remain.

3 Model and Methodology

3.1 Preliminaries

In a centralized or traditional recommender system, suppose we use $\boldsymbol{u}$ to denote a user (customer) and $\boldsymbol{i}$ an item (POI), then $\boldsymbol{U}$ and $\boldsymbol{I}$ are the user and item sets, where we have $\boldsymbol{u} \in \boldsymbol{U}$ and $\boldsymbol{i} \in \boldsymbol{I}$. $\boldsymbol{m}$ and $\boldsymbol{n}$ represent the sizes of $\boldsymbol{U}$ and $\boldsymbol{I}$, respectively. A rating $\boldsymbol{r}_{ui}$ indicates the preference of user $\boldsymbol{u}$ over item/POI $\boldsymbol{i}$. In our dataset, each rating $\boldsymbol{r}_{ui} \in [1, 5]$, where 1 indicates least favored and 5 most favored. As it was introduced in the previous section, we use $\hat{\boldsymbol{r}}_{ui}$ for predicted ratings and $\boldsymbol{r}_{ui}$ for their observed counterparts. Aside from the objective function in Eq. (4), if we denote the rating matrix by $\boldsymbol{R}$, then we have the following formula:

$$R_{m \times n} \approx P_{m \times k} \cdot Q_{n \times k}^T \quad (5)$$

where $\boldsymbol{k}$ is the number of latent factors that are retained, $\boldsymbol{p}_u$ and $\boldsymbol{q}_i$ are column vectors of the two matrices, respectively. For the MF models, unless specified, we use $\boldsymbol{P} \in \mathbb{R}^{m \times k}$ to denote user latent factor matrix, and $\boldsymbol{Q} \in \mathbb{R}^{n \times k}$ to denote the item latent factor matrix. Furthermore, we define $\boldsymbol{T}_r$ as the training set and $\boldsymbol{T}_e$ as the test set. Typically, all MF methods require learning user and item's latent factors by regressing over the known user-item ratings from the pre-processed training dataset [2]. Because of this, both Eqs. (4) and (5) aim to find the optimal $\boldsymbol{P}$ and $\boldsymbol{Q}$ that minimizes $\left\|R - P \times Q^T\right\|$. Finally, we denote the constant in the regularizing terms in Equations such as (1) and (4) by $\boldsymbol{\lambda}$. Both $\boldsymbol{k}$ and $\boldsymbol{\lambda}$ are adjusted and tuned using cross-validation. We use Stochastic Gradient Descent (SGD) to solve the least squares optimization.

In our proposed RSN framework, we break down the centralized RS into multiple local entities. Each area of the city has an independent RS which maintains its own users, and is considered as a local group, denoted by $\boldsymbol{g}_i$ ($\boldsymbol{g}_1 \cup \boldsymbol{g}_2 \cdots \cup \boldsymbol{g}_n = \boldsymbol{U}$). The set of all groups is represented by $\boldsymbol{G}$. In addition to the physical user (real customers) set $\boldsymbol{U}$, we introduce a virtual user (generated fake customers) set $\boldsymbol{V}$. If we define the matrix that stores the virtual users' ratings as $\boldsymbol{R}_v$ and the real users $\boldsymbol{R}_r$, then we have:

$$R_{train} = \begin{bmatrix} R_r \\ R_v \end{bmatrix} \tag{6}$$

In practice, users are encouraged to choose a nearby vendor they trust to receive recommendations. In order to simulate the real scenario in the experiment, users in the dataset are clustered beforehand. The clustering is based on the Pearson Correlation Coefficient (PCC) of users' ratings. Specifically, for any pair of users $\boldsymbol{a}$ and $\boldsymbol{b}$, the similarity between the two is defined as:

$$S_{ab\_PCC} = \frac{\sum_{i \in I_{ab}} (r_{ai} - \mu_a) \cdot (r_{bi} - \mu_b)}{\sqrt{\sum_{i \in I_{ab}} (r_{ai} - \mu_a)^2} \cdot \sqrt{\sum_{i \in I_{ab}} (r_{bi} - \mu_b)^2}} \tag{7}$$

where $\boldsymbol{\mu}_a$ and $\boldsymbol{\mu}_b$ are the average ratings of users $\boldsymbol{a}$ and $\boldsymbol{b}$, and $\boldsymbol{I}_{ab}$ is the item set that $\boldsymbol{a}$ and $\boldsymbol{b}$ both rated. After the affinity matrix is constructed, we then perform the kernel k-means clustering to minimize their in-cluster variance. The PCC is chosen since it has the best performance with respect to mean absolute error (MAE) in neighborhood based RS models [4]. Once all the users are clustered, the cluster centroids are treated as virtual users and sent to all the other RSs.

3.2 Problem Description

We chose two separate cities and their nearby districts to evaluate our model. Most users are only active in a particular town and remain in specific places. This phenomenon is usually referred to as "location aggregation" [3]. For example, Fig. 2 shows the points are aggregated where each point is a visiting record. The x-axis and y-axis represent the user and POI IDs, respectively.

Users select local businesses they trust to share their data before getting recommendation services. Accordingly, we split all the users into different groups to simulate real user activities. This step in simulation is not required in practice since users choose their trusted vendors spontaneously. The paper estimates a user's active location (i.e., the latitude and longitude) by their previously visited POIs (Eq. 8 and 9):

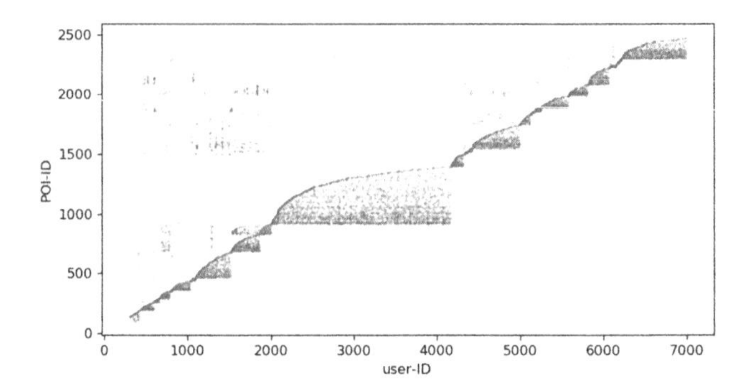

Fig. 2. Yelp dataset user visiting behavior analysis. Each point in the figure represents a check-in record.

$$Lat_u = \frac{1}{|I_u|} \sum_{i \in I_u} Lat_i \tag{8}$$

$$Lon_u = \frac{1}{|I_u|} \sum_{i \in I_u} Lon_i \tag{9}$$

where $\boldsymbol{I}_u$ denotes all the POIs a user visited.

However, the side effect of such action is what it makes the already sparse POI rating matrix even sparser. To increase the number of ratings that can be used to train each model, we provide each local RS with virtual users' ratings. A local RS generates a certain number of virtual users by clustering the existing users. We denote the cluster as $\boldsymbol{c}$ and the user set of that cluster as $\boldsymbol{U}_c$.

$$r_{vi} = \frac{\sum_{i \in I_{uv}, u \in U_c} r_{ui}}{|U_c|} \tag{10}$$

Equation 10 shows how a virtual user rating is estimated. For a virtual user $\boldsymbol{v}$, its rating toward location $\boldsymbol{i}$ is approximated by computing the mean value of the ratings left by users in the same cluster ($\boldsymbol{u} \in \boldsymbol{U}_c$) who visited the same location ($\boldsymbol{i} \in \boldsymbol{I}_{uv}$). The virtual users, whose ratings are shared by all RSs in a RSN, summarize the preferences of physical users.

3.3 Generating Recommendations

When acquiring recommendations, a user downloads the two factorized matrices $\boldsymbol{P}$ and $\boldsymbol{Q}$ as defined in Eq. (5), from the local recommender system. This way, the dimensions of the original rating matrix are indirectly reduced, decreasing the download time. The original ratings are also slightly perturbed, making it harder to backtrace the user's checking-in records. The user then reconstructs the rating matrix $\hat{\boldsymbol{R}}$ on his or her personal device. By searching and finding the most similar user, according to (7), the user can acquire all the predicted ratings for any unvisited POIs. Users can choose whether to share the personal information they hold.

The information downloaded is not only anonymous but also a combination of real users and virtual users. The reconstructed matrix $\hat{\boldsymbol{R}}$ is different from the rating matrix stored on the local server since the number of virtual users is dynamic, and all ratings are perturbed.

3.4 The Algorithm

Since the simulation of users choosing trustworthy vendors and generating virtual users play an essential role in our experiment, we organize the work and show it in *Algorithm 1.*

Algorithm 1: Preprocessing User Ratings

---

aInput: all ratings from R, all POIs' location information (longitude and latitude)
Output: n groups of processed training sets $\{Tr_1, Tr_2, ... Tr_n\}$and test sets$\{Te_1,Te_2, ... Te_n\}$

```
1  for u = 1 to U do
2  |  Calculate each user's longitude Long_u and latitude Lat_u according to (8)
   |  and (9)
3  |  Eliminate users outside target city
4  |  Perform k-means clustering based on Euclidean distance among users
5  end
6  for g = 1 to G do
7  |  Split R_g into training Tr_g set and test set Te_g
8  |  for user u in Tr_g do
9  |  |  Calculate the similarities to all other users according to (7)
10 |  end
11 |  Complete affinity matrix for the current group.
12 |  Perform the kernel k-means clustering to generate virtual user ratings Rv_g
   |  according to (10)
13 |  Append Rv_g to Rv
14 end
15 for g = 1 to G do
16 |  Update Tr_g by appending Rv according to (6)
17 end
18 return Tr and Te
```

Each vendor-based RS possesses a training set $\boldsymbol{Tr}_g$ in practice, learns the factorized matrices according to (4) and (5), and then make the matrices $\boldsymbol{P}$ and $\boldsymbol{Q}$ ready for download for its users. The final recommendations are generated on every user's personal device which further decreases the workload for each vendor-based RS.

3.5 System Update and Maintenance

While the data is static in our simulation, the real-world users continuously move and change their active visiting areas. Furthermore, for privacy concerns, there should not exist a link between cluster centroids and physical users, which prevents the updating process from using the same users. Therefore, each time a clustering is completed, its components, centroids, and the number of clusters will be different from the previous one. To make the cluster centroids better represent real users' personal preferences, two mechanisms are implemented:

- The clustering needs to be regularly performed to generate new virtual users.
- The old virtual users need to be turned inactive after the clustering becomes obsolete.

In real-world scenarios, each virtual user is attached with a timestamp. Once it reduces to zero, the virtual user expires and is then removed. When a virtual user is created, the local RS will broadcast it to all the RSs in the same network. The recipients will decide if the information is useful, depending on the overlap between virtual users' visited locations and the item set of the current RS.

4 Experiments

4.1 Datasets

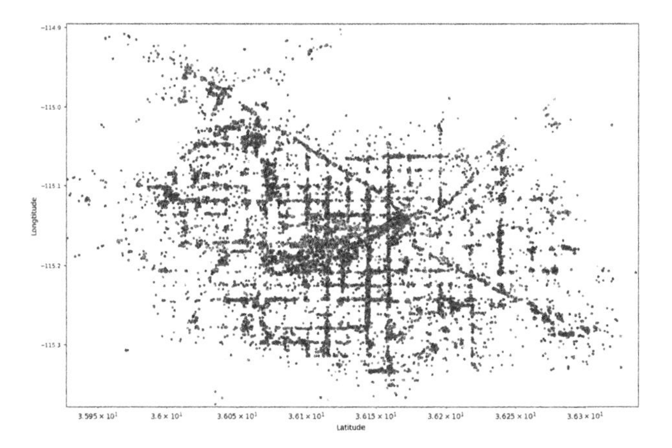

Fig. 3. The user visiting location plots (Las Vegas). Each point represents a user's visiting at real-world geographical position.

We use two subsets of the Yelp business review dataset [1]. The first set was collected in the Urbana-Champaign area, and the second was from the city of Las Vegas and its surrounding areas. The ratings' type is explicit rating (from 1 star to 5 stars) collected by Yelp between January 2007 and December 2017. We removed the users with too few ratings and repeated ratings.

In our test, all local RSs are in the same city or metropolitan area. However, in real-world scenario, if the RSs share any identical items or have overlapped item sets, communication can be established, and RSs in the same RNS can then enhance each other.

During pre-processing, we adopted multiple ways to test the appropriate number of RSs in a network. As discussed in previous sections, we need to group the users based on their visiting locations to simulate the real-world scenario. Figure 3 shows the users' visited POIs, which almost reflects the streets' shape in Las Vegas and nearby areas. However, when we attempt to guess users' real locations by plotting their Euclidean centers of all the visited places, the results did not illustrate apparent segregation. Clustering methods, including k-means, spectral, and density-based spatial, were all tested, and eventually, we chose k-means for its simplicity and straightforwardness. When comparing different clustering methods, we evaluate both the results of accuracy and the balance of user numbers among each area.

Table 1. Datasets statistics

Dataset area	Numer of users	Number of items	Number of ratings
Urbana-Champaign	2737	1502	22654
Las Vegas	31540	30374	802900

After pre-processing, the details of the datasets are listed in Table 1. We sort all the ratings in chronological order and split them by the ratio of 0.2 with the first 80% of ratings for training and 20% for testing.

4.2 Evaluation Metrics

We adopted two metrics to evaluate the model performance, the Root Mean Square Error (RMSE) and Mean Absolute Error (MAE). Although accuracy is not always the best metric to evaluate recommender systems [10], minor accuracy improvements, measured by RMSE or MAE, can still pose significant impacts on the quality of top-k recommendations [8,15]. Equations (10) and (11) show the details of the definition:

$$RMSE = \sqrt{\frac{\sum_{u,i \in Te}(r_{ui} - \hat{r}_{ui})^2}{m}} \tag{11}$$

$$MAE = \frac{\sum_{u,i \in Te}|r_{ui} - \hat{r}_{ui}|}{m} \tag{12}$$

4.3 Results and Discussion

In our experiment, we compare our results with three existing models:

- The MF model, promoted by one of the Netflix winners Simon Funk [6]. We chose the one that has integrated the baseline model proposed in [15]. The two latent matrices in (5) are factorized and learned using the objective function in (4). We used a well-built version to represent a typical centralized RS [7].
- The DMF model, a decentralized scheme that only allows users to exchange gradient loss among neighbors during training [3]. Like most decentralized RSs, there is no data stored on the server. All personal information is kept on users' devices.
- The baseline model, of which the prediction function is defined by Eq. 3 but without the last term. A predicted rating is merely calculated by adding the global mean, column bias, and row bias, and there are no iterative updates involved.

We opt for straightforward recommendation methods over complex models. In our experiment, the core model can be replaced or combined with other schemes. Each local RS can use different algorithms to generate recommendations.

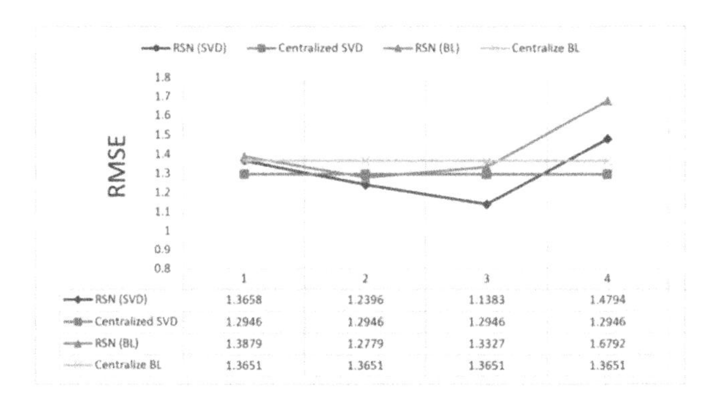

Fig. 4. Local RSs accuracy results (Urbana-Champaign). Four models' RMSE results for each local RS in Urbana-Champaign area.

In the Urbana-Champaign dataset, users were divided into four sections based on their frequently visited locations. In contrast, the Las Vegas metropolitan area has been categorized into ten smaller regions based on the same criterion. The results of every local RS are shown in Fig. 4 and Fig. 5. The different number of groups is due to the different number of users and the cities' scale. On one hand, if we keep the user size too small for an area, the number of users that can be clustered would be too small, leading to insufficient clusters. On the other hand, if this size is too large, each cluster will have too many users, causing the centroids to be too general to reflect physical users' preferences and interests.

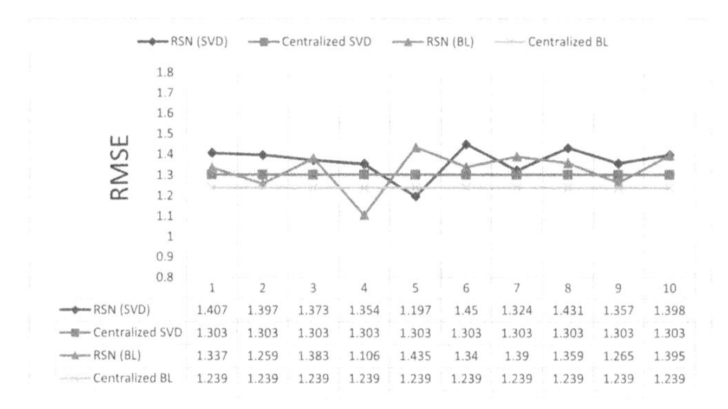

	1	2	3	4	5	6	7	8	9	10
RSN (SVD)	1.407	1.397	1.373	1.354	1.197	1.45	1.324	1.431	1.357	1.398
Centralized SVD	1.303	1.303	1.303	1.303	1.303	1.303	1.303	1.303	1.303	1.303
RSN (BL)	1.337	1.259	1.383	1.106	1.435	1.34	1.39	1.359	1.265	1.395
Centralized BL	1.239	1.239	1.239	1.239	1.239	1.239	1.239	1.239	1.239	1.239

Fig. 5. Local RSs accuracy results (Las Vegas). Four models' RMSE results for each local RS in Las Vegas area.

In Fig. 4 and Fig. 5, for local RSs in RSN, their performance oscillates up and down on the curves formed by centralized RSs. In most cases, their accuracy is only slightly better than each local RS in an RSN. Occasionally, for some specific areas, such as areas 2 and 3 in Fig. 4 or areas 4 and 5 in Fig. 5, RSs in RSN produced higher accuracy than the centralized RSs. In practice, each local RS in an RSN can virtually work with any model and does not have to uniformly use the same method, so theoretically our proposed model has the potential to outperform a centralized RS. The reason is similar to why a hybrid model performs consistently better than a pure model.

One thing to point out is the way we calculate the average RMSE and MAE. We assume that each region has a local RS to generate its recommendations using real ratings and virtual ratings. It is necessary to estimate the performance of the RSN using all local RSs' average MSE and RMSE. For MAE, the average is the exact mean value of all the MAEs from every local RS. For the RMSE, however, the average RMSE is estimated by calculating the MSE first, and then compute the average RMSE by taking the square root of the mean value of MSE.

As far as hyper-parameters, in Fig. 5, where every RS in the RSN uses biased MF as the default model, the number of latent factors $\boldsymbol{k}(40)$ and learning rate $\boldsymbol{\lambda}(0.1)$ were the same as the centralized MF model. In the DMF model, we used the same value for $\boldsymbol{k}$ and set the regularizer to 0.01 and the learning rates to 0.05. We probed each model with $\boldsymbol{k} \in \{5, 40\}$, the learning rate $\boldsymbol{\lambda} \in \{0.01, 0.5\}$, and the regularizer between $\{0.001, 10\}$.

Table 2 shows the results for different models. It is apparent that all MF models performed better on the Urbana-Champaign dataset. There could be two reasons. First, the Urbana-Champaign dataset is more compact, meaning a small area with relatively sufficient users and POIs to analyze their preferences. Although the Las Vegas dataset is from a densely populated city, it is still too sparse geographically. In fact, this dataset includes visiting records from the city of Las Vegas, North Las Vegas, Spring Valley, Paradise, and all small towns nearby. Second, as shown in Fig. 6, the percentage of new businesses in the Las

Table 2. Datasets statistics

Urbana-Champaign					
Model	MF	RSN(MF)	Baseline	RSN(Baseline)	DMF
RMSE	1.2946	1.3121	1.3650	1.4279	1.4984
MAE	1.0307	1.0568	1.0469	1.0940	1.2018
Las Vegas					
Model	MF	RSN(MF)	Baseline	RSN(Baseline)	DMF
RMSE	1.3034	1.3704	1.2394	1.32996	1.4360
MAE	1.0012	1.1015	0.9452	1.04457	1.1241

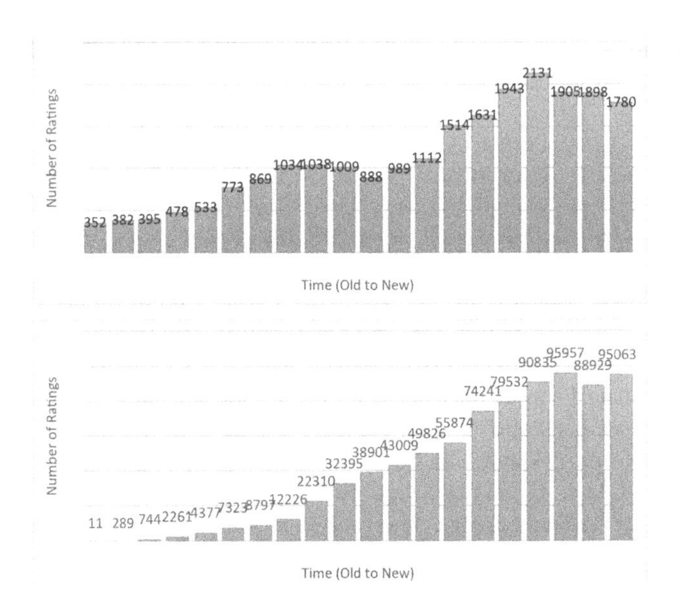

Fig. 6. Rating distributions. (Red: Urbana-Champaign, Blue: Las Vegas). The ratings are ordered from old to new. (Color figure online)

Vegas area is higher than that in Urbana-Champaign. As mentioned previously, we ordered the ratings chronologically, enabling the models to use old data to predict new ratings for simulating real-world scenarios.

In our first dataset, compared to the centralized Biased MF model, the accuracy tradeoff for our RSN model is very small if not trivial (as low as 0.0175 in RMSE and 0.0261 in MAE). The tradeoff is more significant when the recommendation method is changed from Biased MF to the Baseline, but it is still smaller than 0.1. This test result is under the circumstances that all local RSs in our RSN uniformly use the same method and much fewer ratings (1/4 of the total ratings in Urbana-Champaign dataset, 1/10 of the total ratings in Las Vegas dataset). With different recommendation methods implemented, the RSN can achieve better performance. Since each local RS maintains a much smaller set of users to generate recommendations, it reduces the training time. Also, with the help of "virtual users," it scored similar accuracy as a centralized RS. In contrast to a completely decentralized RS such as DMF, the RSN sacrifices much less accuracy for privacy preservation.

In this experiment, we estimated users' preferred vendors who hold their personal history using the locations of their most frequently visited stores. This although is not the most accurate way, is the best option in our assessment due to the limited information. We are positive that in practice, the overall performance of the proposed RSN framework will be better. As a summary, the RSN can lower the privacy risk and boost user confidence with minimal loss of prediction accuracy. Each local RS has a light workload and fast speed of convergence. Moreover, should there be any data breach, it is easier to investigate and control the damage.

5 Conclusion and Future Work

Conventional centralized RSs face high risks in protecting users' privacy because they carry all personal information in the same system. On the other hand, even though decentralized models can eliminate the risks of data breaches, they almost give up all further opportunities for data analysis or mining.

In this paper, we proposed a recommender system network scheme that takes advantage of centralized RSs and decentralized RSs. By introducing virtual users, we can lower the data leakage risk while still generate accurate recommendations. The scheme focuses on both the collaboration among users and among the RSs hosted by small local businesses that people trust.

Future work includes integrating a distributed neural network into our RSN, RS-to-RS communication, and exploring other ways to create virtual or synthetic users, i.e., users who are not physical but can reflect real users' interests and preferences

Acknowledgement. We would like to show our gratitude to the colleagues from the Department of Computer Science at the University of Kentucky and Northeastern Illinois University whose insights and expertise have inspired us. This research was supported by a Committee on Organized Research grant from Northeastern Illinois University.

References

1. Yelp dataset. https://www.yelp.com/dataset
2. Campos, P.G., Díez, F., Cantador, I.: Time-aware recommender systems: a comprehensive survey and analysis of existing evaluation protocols. User Model. User-Adapt. Interact. **24**, 67–119 (2013). https://doi.org/10.1007/s11257-012-9136-x
3. Chen, C., Liu, Z., Zhao, P., Zhou, J., Li, X.: Privacy preserving point of interest recommendation using decentralized matrix factorization (2020)

4. Desrosiers, C., Karypis, G.: A comprehensive survey of neighborhood-based recommendation methods. In: Ricci, F., Rokach, L., Shapira, B., Kantor, P.B. (eds.) Recommender Systems Handbook, pp. 107–144. Springer, Boston, MA (2011). https://doi.org/10.1007/978-0-387-85820-3_4
5. Federal_Trade_Commission: FTC imposes $5 billion penalty and sweeping new privacy restrictions on Facebook. Press release 24 (2019)
6. Funk, S.: Netflix update: try this at home (2006)
7. Hug, N.: Surprise: a python library for recommender systems. J. Open Source Softw. **5**, 2174 (2020). https://doi.org/10.21105/joss.02174
8. Koren, Y.: Factor in the neighbors: scalable and accurate collaborative filtering. ACM Trans. Knowl. Discov. Data (TKDD) **4**(1), 1–24 (2010)
9. Lu, J.: Assessing the cost, legal fallout of capital one data breach (2019)
10. McNee, S.M., Riedl, J., Konstan, J.A.: Being accurate is not enough: how accuracy metrics have hurt recommender systems. In: CHI 2006 Extended Abstracts on Human Factors in Computing Systems, pp. 1097–1101 (2006)
11. Mnih, A., Salakhutdinov, R.R.: Probabilistic matrix factorization. In: Advances in Neural Information Processing Systems, vol. 20, pp. 1257–1264 (2007)
12. Nedic, A., Ozdaglar, A.: Distributed subgradient methods for multiagent optimization. IEEE Trans. Autom. Control **54**, 48–61 (2009)
13. Newzoo: Newzoo global mobile market report 2020. https://newzoo.com/insights/trend-reports/newzoo-global-mobile-market-report-2020-free-version/
14. Novaes Neto, N., Madnick, S., de Paula, M.G., Malara Borges, N., et al.: A case study of the capital one data breach. Stuart E. and Moraes G. de Paula, Anchises and Malara Borges, Natasha, A Case Study of the Capital One Data Breach (January 1, 2020) (2020)
15. Koren, Y.: Factorization meets the neighborhood: a multifaceted collaborative filtering model. In: Proceedings of the 14th ACM SIGKDD International Conference on Knowledge Discovery and Data Mining (2008)
16. SBA_Office_of_Advocacy: Small business profile (2016). https://www.sba.gov/sites/default/files/advocacy/United_States.pdf
17. Wang, X., Nguyen, M., Carr, J., Cui, L., Lim, K.: A group preference based privacy preserving POI recommender system (2020)
18. Yan, F., Sundaram, S., Vishwanathan, S., Qi, Y.: Distributed autonomous online learning: regrets and intrinsic privacy preserving properties. IEEE Trans. Knowl. Data Eng. **25**, 2483–2493 (2012)
19. Yelp: Yelp - company - fast facts (2020). https://www.yelp-press.com/company/fast-facts/default.aspx
20. Yun, H., Yu, H., Hsieh, C., Vishwanathan, S., Dhillon, I.: Nomad: nonlocking, stochastic multimachine algorithm for asynchronous and decentralized matrix completion (2013)

Cyberspace Security and Access Control

Fine-Grained Collaborative Access Control for IPFS with Efficient Revocation

Yifei Li(✉), Yinghui Zhang, and Qixuan Xing

School of Cyberspace Security, Xi'an University of Posts and Telecommunications, Xi'an 710121, China

Abstract. Access control is important for IPFS and collusion resistance is a basic requirement in traditional attribute-based encryption (ABE). However, in many emergency situations, conditional collaboration between multiple users is needed to decrypt ciphertext data. In addition, in order to make access control mechanisms suitable for the change of user attributes, it is necessary to construct ABE schemes that support attribute revocation. In this paper, a fine-grained collaborative decryption scheme that supports attribute revocation in IPFS storage (GORE-ABE) is proposed. In this scheme, the tree access policy is adopted, and the users are divided into different groups. The premise for successful decryption is that the users participating in the collaboration are in the same group and that the user attribute set satisfies the tree policy. The analysis results show that the GORE-ABE scheme is secure and efficient.

Keywords: Attribute-based encryption · IPFS · Collaborative decryption · Attribute revocation · Ciphertext update

1 Introduction

As time goes on, more people store massive data on cloud storage [1]. However, centralized cloud storage servers are prone to failure. Such as, the service provider interrupts service or removes the user's files on the grounds of violating the regulations [2]. In addition, cloud storage fees are getting higher and higher. Inter-Planetary File System (IPFS) has the characteristics of decentralization, semi-trust and automatic deletion of duplicate data [3]. Therefore, before uploading data to IPFS, users must use encryption algorithm to process their own private data [4–9]. Attribute-based encryption (ABE) is a traditional and widely used technology [10–13]. Although traditional ABE can realize collusion resistance, it requires conditional collaborative decryption by multiple users in many emergency situations. In view of this, Li et al. [14] proposed the concept of collaborative decryption of multiple users within a group. Specifically, users are grouped and only users in the same group can cooperate to decrypt, however, this scheme only implements coarse-grained access control policy [14]. Xue et al. [15] proposed group-oriented fine-grained collaborative operation scheme, but this scheme lacks necessary mechanisms such as attribute revocation and hence is not practical.

J. Xiong et al. (Eds.): MobiMedia 2021, LNICST 394, pp. 493–500, 2021.
https://doi.org/10.1007/978-3-030-89814-4_35

In summary, a scheme that only supports coarse-grained collaborative access control means that the scheme allows collaborative decryption of all user attributes within the same group. In a scheme that does not support fine-grained revocation, when a user logs out of the system but still has the decryption capability, this is undoubtedly a huge security hole [16, 17]. Therefore, it is vital to construct a fine-grained collaborative decryption scheme that supports efficient revocation (GORE-ABE).

Our Contribution. In this paper, a fine-grained collaborative decryption scheme that supports attribute revocation in IPFS storage (GORE-ABE) is proposed. Specifically, the GORE-ABE scheme uses ABE to realize fine-grained collaborative decryption of multiple users within a group. This scheme transfers the storage of mass data to IPFS server and avoids the occurrence of single point of failure. The scheme also implements efficient attribute revocation and ciphertext update. Next, the security and performance of the GORE-ABE are analyzed.

2 Preliminaries

2.1 Bilinear Maps

Two multiplicative cyclic groups G and G_T (g is the generator) are defined. A map $e : G \times G \rightarrow G_T$. The e is considered bilinear if it satisfies the following three requirements. Specific conditions are as follows: 1) Bilinearity; 2) Non-degeneracy; 3) Computability.

2.2 Application Scenario of GORE-ABE

The GORE-ABE scheme not only encrypts the privacy of users, but also realizes the decryption mechanism of single user and multi-user cooperation in emergency, which can be applied to the following scenarios. 1) Single decryption scenario: As in the traditional ABE scheme, a single user with sufficient access rights can decrypt. 2) Collaboration scenario: when a single user cannot decrypt, it can cooperate with other users with different attributes in the same group to decrypt. The process of collaborative decryption is as follows. First, the requester U_p tries to decrypt, and if the decryption fails, it will send a request to the group for cooperative decryption (broadcasting its own translation key $\mathbb{T}_p$). Translation key $\mathbb{T}_p$ are composed of group generators g, group identifiers θ_τ and random numbers (σ_p and $\mathbb{C}$) and can be used to assist in collaborative access scenarios. Second, a user U_p' in the same group (the collaborator U_p') responds to the request and calculates an intermediate ciphertext $e(g,g)^{\sigma_p' q_z(0)}$ with his private key SK_p'. Thirdly, U_p' uses the received translation key U_p and its own translation key $\mathbb{T}_p'$ and intermediate ciphertext $e(g,g)^{\sigma_p' q_z(0)}$ to compute the intermediate ciphertext $e(g,g)^{q_z(0)\cdot\sigma_p}$ and pass it back to the requester U_p. Finally, the requestor U_p receives the intermediate ciphertext $e(g,g)^{q_z(0)\cdot\sigma_p}$ and tries to decrypt it again. If it fails to decrypt, the requestor continues to request other users in the same group to join the collaboration until it can decrypt it. If it can decrypt it, the subsequent decryption operations will continue.

3 Construction of GORE-ABE

Since traditional ABE schemes [e.g. 9, 11, 12] do not support collaborative access control, and Xue et al. [15], which supports collaborative access control, cannot achieve fine-grained revocation at the user attribute level. Therefore, GORE-ABE scheme is proposed, and its system model is shown in Fig. 1, including four entities: IPFS storage server, Data Owners, Data Users and Authority Center (CA). The specific scheme is as follows.

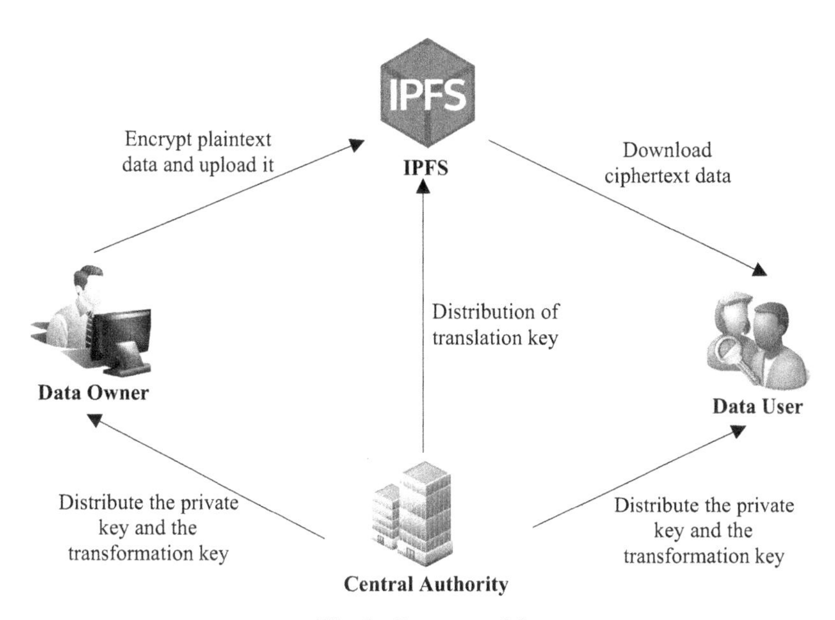

Fig. 1. System model

3.1 Setup

First, for each user group τ, CA defines the group secret key with random numbers ($\mathbb{A}, \mathbb{B}, \mathbb{C} \in \mathbb{Z}_{\mathbb{P}}$) and then defines map $e : G \times G \rightarrow G_T$. $\mathbb{H}$ is a hash function. Next, multiplicative cyclic groups G and G_T (g is generator) are picked by CA. Finally, In the scheme given ω project, selecting the secret key for each group $\theta_\tau \in \mathbb{Z}_{\mathbb{P}}$ and releasing unique identifier to each user U_p.

$$\text{Public key } PK = \left\langle \mathcal{D} = g^{\mathbb{B}}, \mathcal{F} = g^{\mathbb{C}}, J = g^{\frac{1}{\mathbb{B}}}, \mathcal{K} = g^{\frac{1}{\mathbb{C}}}, e(g,g)^{\mathbb{A}}, g, \mathbb{H}, G_T, G \right\rangle,$$
$$\text{Master key } MK = \left\langle g^{\mathbb{A}}, g^{\theta_1}, \ldots, g^{\theta_\tau}, \mathbb{B}, \mathbb{C} \right\rangle$$

3.2 KeyGen

The attribute set of user U_p is $S_p = \{a_{p,1}, \ldots, a_{p,\omega_p}\}$ ($a_{p,q}$ for S_p in the collection of the first q attributes, ω_p for S_p attributes in the number). First, CA use group key θ_τ marking to the user. Then, the CA to users randomly selected $\sigma_p \in \mathbb{Z}_{\mathbb{P}}$, CA randomly select users attributes for $\sigma_{a_{p,q}} \in \mathbb{Z}_{\mathbb{P}}$ and attribute index $Q_{a_{p,q}} \in \mathbb{Z}_{\mathbb{P}}$, $1 \leq q \leq \omega_p$. Finally, the KeyGen outputs the translation key $\mathbb{T}_p$ [15] used for collaborative decryption and private key SK_p.

$$SK_p = \left\langle M_p = g^{\frac{\mathbb{A}+\theta_{\mathcal{I}}}{\mathbb{B}}}, M_{p,q} = g^{\sigma_p}\mathbb{H}(a_{p,q})^{\sigma_{p,q}}, M'_{p,q} = g^{\frac{\sigma_{p,q}}{Q_{a_{p,q}}}}, 1 \leq q \leq \omega_p \right\rangle$$

$$\mathbb{T}_p = g^{\frac{\theta_{\mathcal{I}}+\sigma_p}{\mathbb{C}}}$$

3.3 Encryption

For GORE-ABE scheme, mixed cryptography is used. The translation nodes [18] is set on the tree access policy, and only on the transformation node can multi-user collaboration decryption be allowed. $\mathbb{S}$ is the secret value in the tree $\mathcal{T}$.

Let $\mathbb{Y}$ and $\mathbb{X}$ be leaf node sets and non-leaf node sets in $\mathcal{T}$ respectively, and then get the ciphertext CT_μ.

$$CT_\mu = \left\langle \begin{array}{l} C = \mathcal{D}^{\mathbb{S}}, \overline{C} = \mathcal{F}^{\mathbb{S}}, \mathcal{T}, \forall x \in \mathbb{X} : \hat{C}_x = \mathcal{F}^{q_x(0)}, \\ \forall y \in \mathbb{Y} : C_y = g^{q_y(0)}, C'_y = \mathbb{H}(att(y))^{q_y(0) \cdot Q_{a_{p,q}}}, \\ C' = \mu \cdot e(g,g)^{\mathbb{A}\mathbb{S}} \end{array} \right\rangle$$

3.4 Decryption

The requester U_p runs the decryption algorithm. If it can be decrypted, it is the traditional ABE decryption case alone. If it cannot be decrypted, it sends out the collaboration request signal to other users (collaborators) in the group to realize the multi-user cooperation decryption in an emergency U_p invokes the recursive decryption algorithm *DecNode*, inputs the user attribute set γ for decryption operation, and stores all the calculated node secret values in B_x.

- If x is a leaf node (attribute is p):

$$P_x = DecNode(CT_\mu, \gamma, u_p, x) = \frac{e(M_{p,q}, C_x)}{e\left(M'_{p,q}, C'_x\right)} = \frac{e\left(g^{\sigma_p}\mathbb{H}(a_{p,q})^{\sigma_{p,q}}, g^{q_x(0)}\right)}{e\left(g^{\sigma_p}, \mathbb{H}(a_{p,q})^{q_x(0)}\right)}$$
$$= e(g,g)^{\sigma_p q_x(0)}, att(x) = a_{p,q} \in S_p; \text{ otherwise, } P_x = \perp.$$

- If x is a no-leaf node:

 a) When z not translation nodes (not collaborate decryption), U_p try to decrypt separately (z is child node of x and attribute is p):

$$P_z = DecNode(CT_\mu, \gamma, u_p, z) = \frac{e(M_{p,q}, C_z)}{e\left(M'_{p,q}, C'_z\right)} = e(g,g)^{\sigma_p q_z(0)},$$
$$att(z) = a_{p,q} \in S_p; \text{ otherwise, } P_z = \perp.$$

b) When z is translation nodes, GORE-ABE scheme can achieve multi-user collaboration decryption: The collaborator decrypts the secret value:

$$P'_z = DecNode(CT_\mu, \gamma, u_p, z) = \frac{e(M_{p,q}, C_z)}{e(M'_{p,q}, C'_z)} = e(g,g)^{\sigma'_p q_z(0)},$$

Then user U'_ps translation key $\mathbb{T}_p$ and own (collaborator's) translation key $\mathbb{T}'_p$ to convert P'_z into P_z containing U'_ps identity and sends it to U_p. P_z are as follow.

$$P_z = P'_z \cdot e(\hat{C}_z, \frac{\mathbb{T}_p}{\mathbb{T}'_p}) = e(g,g)^{\sigma'_p q_z(0)} \cdot e(g^{\mathbb{C}} g^{q_z(0)}, g^{\frac{\theta_\tau + \sigma_p}{c}} \Big/ g^{\frac{\theta_\tau + \sigma'_p}{c}})$$
$$= e(g,g)^{q_z(0) \cdot \sigma_p}$$

- U_p merge user attribute private key set and continue decryption operation using Lagrange interpolation formula:

$$P_x = \prod_{z \in B_x} P_z^{\Delta_{index(z),\{index(z)\}}(0)} = e(g,g)^{\sigma_p q_z(0)}, \text{ among them: } \Delta_{p,S(x)} = \prod_{q \in S, q \neq p} \frac{x-q}{p-q}.$$

Then, U_p runs the recursive algorithm to decrypt the root node r as follows:

$$P_r = DecNode(CT_\mu, \gamma, r, u_i) = e(g,g)^{\sigma_p \cdot \mathbb{S}}.$$

Continue decrypting, output:

$$P = \frac{e(\overline{C}, \mathbb{T}_p)}{P_r} = \frac{e\left(g^{\mathbb{C} \cdot q_r(0)}, g^{\frac{\theta_\tau}{\mathbb{C}}} \cdot g^{\frac{\sigma_p}{\mathbb{C}}}\right)}{e(g,g)^{\sigma_p \cdot q_r(0)}} = e(g,g)^{\theta_\tau \cdot \mathbb{S}},$$

U_p download the ciphertext from IPFS and decrypt the symmetric key μ:

$$\mu = \frac{P \cdot C'}{e(C, M_p)} = \frac{e(g,g)^{\theta_\tau \cdot \mathbb{S}} \cdot \mu \cdot e(g,g)^{\mathbb{S}\mathbb{A}}}{e\left(g^{\mathbb{B}\mathbb{S}}, g^{\frac{\mathbb{A} + \theta_\tau}{\mathbb{B}}}\right)}.$$

- Finally, U_p uses μ to decrypt $Enc_\mu(M)$ to recover the plaintext M.

3.5 KeyUpdate

If an attribute ε is revoked, the corresponding index $Qa_{p,q}$ and ciphertext CT_μ are refactored. CA randomly chosen $Q'_{a_{p,q}} \in \mathbb{Z}_\mathbb{P}$ (It's different from $Qa_{p,q}$) and sent it to contain attributes ε for all users. After receiving it, the relevant users update their private keys: $M'_{p,q} = g^{\frac{Qa_{p,q}}{Q'a_{p,q}}}$. However, users without ε are not affected and $\mathbb{T}_p = g^{\frac{\theta_\tau + \sigma_p}{\mathbb{C}}}$ does not need to be updated.

CA sends the new attribute index $Q'_{a_{p,q}} \in \mathbb{Z}_\mathbb{P}$ to IPFS, which updates the ciphertext by introducing random number $\varphi(\varphi \in \mathbb{Z}_\mathbb{P})$. The updated ciphertext:

$$CT'_\mu = \left\langle \begin{array}{l} \forall x \in \mathbb{X} : \hat{C}_x = \mathcal{F}^{q_x(0)+\varphi}, C = \mathcal{D}^{(\mathbb{S}+\varphi)}, \overline{C} = \mathcal{F}^{\mathbb{S}+\varphi}, \mathcal{T}, \\ \forall y \in \mathbb{Y} : C_y = g^{q_y(0)+\varphi}, C'_y = \mathbb{H}(att(y))^{(q_y(0)+\varphi) \cdot Qa_{p,q}}, \\ C' = \mu \cdot e(g,g)^{\mathbb{A}(\mathbb{S}+\varphi)} \end{array} \right\rangle$$

4 Security Analysis

The failure-prone nature of cryptography has led to a widespread belief that these systems should be designed and analyzed in a formal way. Next, we consider the following security features.

Data Confidentiality: IND games are often used to describe the semantic security of cryptographic schemes. In such a game, the algorithm is said to be provably security if the two messages in the safe game are indistinguishable to the opponent [19]. GORE-ABE scheme roughly is the same as the data confidentiality proof in reference [12, 15], so the security of GORE-ABE scheme is the same as that of reference [12, 15], that is, it realizes the provable security under the general group model.

Anti-User Collusion: GORE-ABE scheme uses parameters such as group identifier, user random number and user attribute random number to resist user collusion.

Controlled Collaboration Within a Group: Only users in the same group can cooperate to decrypt, and all other actions are considered as collusion attacks.

Private Key Confidentiality: Translation key $\mathbb{T}_p$ [18] are composed of group generators g, group identifiers θ_τ and random numbers (θ_τ and $\mathbb{C}$) and can be used to assist in collaborative access scenarios without users revealing their private keys SK_p.

Secure Revocation of User Attribute: GORE-ABE scheme can effectively resist user collusion.

5 Performance Evaluation

The performance of GORE-ABE scheme is analyzed by comparing it with the scheme proposed by Xue et al. [15]. In short, the GORE-ABE scheme maintains high efficiency on the basis of realizing attribute revocation.

5.1 Computational Complexity

The computation overhead of addition and multiplication operation is very small, so the main operations in the algorithm include exponentiations and pairings. Although a small amount of exponentiations is added in GORE-ABE scheme, the computational overhead is not increased and the algorithm is more practical. In addition, the ciphertext update algorithm is executed by IPFS.

For convenience, expressed in P_a bilinear mapping, expressed in E_x index operation, the number of user attributes in W_n said with W_c said the number of attributes associated with the access structure expressed in W_y number of user attributes for decryption in W_l from the root node to leaf node between the number of leaf nodes, expressed in T_r convert the number of nodes, with $T_r^{'}$ for decryption of the conversion of the node number. In Table 1, we give the theoretical computation overhead above two schemes.

Table 1. Computation overhead.

Phase	Xue et al. scheme	Ours
Setup	$(5+\omega)E_x + P_a$	$(5+\omega)E_x + P_a$
KeyGen	$(2W_n + 3)E_x$	$(2W_n + 3)E_x$
Encryption	$(2W_c + 3 + T_r)E_x$	$(3W_c + 3 + T_r)E_x$
Decryption	$(2W_y + 2 + T_r^{'})P_a + (W_l + W_y)E_x$	$(2W_y + 2 + T_r^{'})P_a + (W_l + W_y)E_x$
KeyUpdate	N/A	E_x

5.2 Computational Complexity

The storage costs, communication costs, encryption and decryption time, ciphertext size and other parameters of GORE-ABE scheme are consistent with those of Xue et al. [15].

6 Conclusion

In this paper, we propose a fine-grained collaborative operation scheme that allows multiple users within a group (GORE-ABE). This scheme transfers the storage of mass data to IPFS and avoids single point of failure. In addition, the scheme also implements the user's attribute revocation and ciphertext update, which further enhances the practicability of the scheme. The GORE-ABE scheme is secure and has a broad prospect in practical application.

Acknowledgment. This work is supported by the National Natural Science Foundation of China (62072369, 62072371, 61772418), the Innovation Capability Support Program of Shaanxi (2020KJXX-052), the Shaanxi Special Support Program Youth Top-notch Talent Program, the Key Research and Development Program of Shaanxi (2019KW-053, 2020ZDLGY08-04, 2021ZDLGY06-02), and Sichuan Science and Technology Program under Grant 2017GZDZX0002.

References

1. Liang, Y., Cheng, H., Chen, W.: Building energy consumption data index method in cloud computing environment. Int. J. Perform. Eng. **16**(5), 747–756 (2020)
2. Wu, J., Ping, L., Ge, X., et al.: Cloud storage as the infrastructure of cloud computing. In: 2010 International Conference on Intelligent Computing and Cognitive Informatics, pp. 380–383. IEEE (2010)
3. Benet, J.: Ipfs-content addressed, versioned, p2p file system. arXiv preprint arXiv:1407.3561 (2014)
4. Xiong, J., Bi, R., Zhao, M., et al.: Edge-assisted privacy-preserving raw data sharing framework for connected autonomous vehicles. IEEE Wirel. Commun. **27**(3), 24–30 (2020). https://doi.org/10.1109/MWC.001.1900463

5. Tian, Y., Wang, Z., Xiong, J., et al.: A blockchain-based secure key management scheme with trustworthiness in DWSNs. IEEE Trans. Industr. Inf. **16**(9), 6193–6202 (2020). https://doi.org/10.1109/TII.2020.2965975
6. Xiong, J., Ma, R., Chen, L., et al.: A personalized privacy protection framework for mobile crowdsensing in IIoT. IEEE Trans. Industr. Inf. **16**(6), 4231–4241 (2020)
7. Tian, Y., Li, Q., Hu, J., Lin, H.: Secure limitation analysis of public-key cryptography for smart card settings. World Wide Web **23**(2), 1423–1440 (2019). https://doi.org/10.1007/s11280-019-00715-8
8. Chen, Z., Tian, Y., Peng, C.: An incentive-compatible rational secret sharing scheme using blockchain and smart contract. Sci. China Inf. Sci. **64**(10), 1–21 (2021). https://doi.org/10.1007/s11432-019-2858-8
9. Xiong, J., Chen, X., Yang, Q., et al.: A task-oriented user selection incentive mechanism in edge-aided mobile crowdsensing. IEEE Trans. Netw. Sci. Eng. **7**(4), 2347–2360 (2020)
10. Sahai, A., Waters, B.: Fuzzy identity-based encryption. In: Cramer, R. (eds.) Advances in Cryptology – EUROCRYPT 2005. EUROCRYPT 2005. Lecture Notes in Computer Science, vol. 3494. Springer, Heidelberg. https://doi.org/10.1007/11426639_27
11. Zhang, Y., Chen, X., Li, J., et al.: Attribute-based data sharing with flexible and direct revocation in cloud computing. KSII Trans. Internet Inf. Syst. **8**(11), 4028–4049 (2014)
12. Bethencourt, J., Sahai, A., Waters, B.: Ciphertext-policy attribute-based encryption. In: 2007 IEEE Symposium on Security and Privacy (SP 2007), pp. 321–334. IEEE (2007)
13. Zhang, Y., Zheng, D., Chen, X., et al.: Efficient attribute-based data sharing in mobile clouds. Pervasive Mob. Comput. **28**, 135–149 (2016)
14. Li, M., Huang, X., Liu, J.K., Xu, L.: GO-ABE: group-oriented attribute-based encryption. In: Au, M.H., Carminati, B., Kuo, C.C.J. (eds.) Network and System Security. NSS 2015. Lecture Notes in Computer Science, vol. 8792. Springer, Cham (2014). https://doi.org/10.1007/978-3-319-11698-3_20
15. Xue, Y., Xue, K., Gai, N., et al.: An attribute-based controlled collaborative access control scheme for public cloud storage. IEEE Trans. Inf. Forensics Secur. **14**(11), 2927–2942 (2019)
16. Zhang, Y., Deng, R.H., Xu, S., et al.: Attribute-based encryption for cloud computing access control: a survey. ACM Comput. Surv. (CSUR) **53**(4), 1–41 (2020)
17. Hur, J., Noh, D.K.: Attribute-based access control with efficient revocation in data outsourcing systems. IEEE Trans. Parallel Distrib. Syst. **22**(7), 1214–1221 (2010)
18. Bobba, R., Khurana, H., Prabhakaran, M.: Attribute-sets: a practically motivated enhancement to attribute-based encryption. In: Backes, M., Ning, P. (eds.) Computer Security – ESORICS 2009. ESORICS 2009. Lecture Notes in Computer Science, vol. 5789. Springer, Heidelberg (2009). https://doi.org/10.1007/978-3-642-04444-1_36
19. Pointcheval, D.: Provable security for public key schemes. In: Contemporary Cryptology. Advanced Courses in Mathematics - CRM Barcelona (Centre de Recerca Matemàtica). Birkhäuser Basel (2005). https://doi.org/10.1007/3-7643-7394-6_4

Towards Efficient Fine-Grained Access Control and Data Privacy Protection for Smart Home

Shuaiyong Shen[1(✉)], Yang Yang[1,2], Zuobin Ying[3], and Ximeng Liu[1,2]

[1] College of Mathematics and Computer Science, Fuzhou University, Fuzhou 350108, China
n190320059@fzu.edu.com

[2] Fujian Provincial Key Laboratory of Information Security of Network Systems, Fuzhou University, Fuzhou 350108, China

[3] School of Electrical and Electronic Engineering, Nanyang Technological University, Nanyang 639798, Singapore
james.ying@ntu.edu.sg

Abstract. Fine-grained access control become a research spotlight in the scenario of the smart home with Internet-of-Things (IoTs). However, most of the existing works could only realize ciphertext outsourcing with an unchangeable encryption key in the cloud. That is because the trivial solution needs great communication overhead. To overcome this challenge, we propose an updatable encryption scheme named SM9-UE for the IoT smart home scenario. The scheme in constructed on the basis of SM9, a Chinese official cryptography standard. SM9-UE realizes secure and lightweight ciphertext updating through using the token generated by the data owner, which is information independent with the plaintext. We give the formal security definition and prove it to be IND-UPD secure. Theoretical comparisons demonstrate that our scheme is efficient in terms of computation and storage. Experimental results also indicate that SM9-UE can be practically applied to IoT smart home.

Keywords: Internet of Things · Smart home · Updatable encryption · SM9 · Key rotation

1 Introduction

With the rapid development of modern information technology, Internet-of-Things (IoTs) technology has made our life more convenient and comfortable, such as monitoring and control [15]. It is estimated that more than 50 billion devices will be connected to the Internet by 2020, according to the report [12]. So far, as a response to the trend of our times, various mainstream companies, such as Samsung, Apple and Google [18], has launched their own framework for connecting diverse devices. Smart home is one of the most exciting IoT application scenarios that can revolutionlize our lives. For instance, those devices that

J. Xiong et al. (Eds.): MobiMedia 2021, LNICST 394, pp. 501–515, 2021.
https://doi.org/10.1007/978-3-030-89814-4_36

need to be controlled, such as air conditioners, TVs, doors and windows, can be conveniently closed outdoors by your cell phone.

Although smart home paints a beautiful blueprint for our daily lives, it also brings us serious security problems [13]. There are many ways to attack [11, 21], making the user's private data unsealable, easily leaked or tampered with by illegal elements, resulting in great economic losses to uses. At present, the situation of software access control determines whether users' private information can be safely maintained, many researchers have pointed out that some software can lead to users' data leakage due to over-privileged in the smart home. For instance, a smart doorlock allows the user to lock the door remotely through an App when he forgot to lock the door when leaving home. However, Apps often try to get more privileges than they acturally needs, such as access to the user's location information, access to the user's contact information, access to the battery status of the door lock, and so on. Whereas, users may not intend to grant these privileges at present. [10,19,20].

Both Lee *et al.* [16] and Yan *et al.* [23] have put forward schemes to solve this problem of over-privileged with better protection of user's data privacy. To enable more fine-grained access control, both schemes are functonally-based units of action. Lee *et al.* finish the least privilege and availability of device functionality, it sacrifices time overhead due to the access control list is adopted by access procedure. To overcome these deficiency, Yan *et al.* propose a new scheme with a cryptographic technique. It uses the IBE encryption scheme to encrypt the user data, and takes function ID as the public key to realize secure cloud outsourcing. However, when private data is stored permanently in an untrusted cloud, robust key management should periodically change its own secret key to keep the data secure: updating encrypted data from an old key to a fresh one. For example, the Payment Card Industry Data Security Standard (PCI DSS) emphasizes that credit card data must be stored in an encrypted form requiring key rotation [22]. So, a new scheme, which makes the secret key update periodically, is urgently needed. To address this security issue, we propose an SM9-UE scheme. Our contributions can be summarized as follows:

- ***Fine-grained access control.*** The proposed SM9-UE scheme takes the function ID of the device as the basic unit and implements fine-grained access control with solving the over-privilege problem.
- ***Key rotation.*** We propose a novel *ciphertext-independent* updatable encryption based on SM9 that can realize rotating the encryption key periodically, as well as alter current ciphertexts from the old key to the fresh one without costly computation and communication overhead.
- ***Security and practicability.*** We give the formal security definition and prove the SM9-UE is adaptive update indistinguishablility (IND-UPD) secure. Theoretical analysis and experimental results demonstrate that SM9-UE can meet the security needs in the IoT smart home.

1.1 Organization

The rest of our paper is organized as follows. In Sect. 2, we introduce preliminaries about this work. In Sect. 3, we present our system model and attack model.

Table 1. Summary of notations

Notation	Definition
$x\|\|y$	The concatenation of x and y, where x and y are bit strings or byte strings
$\oplus$	The bitwise XOR operator that operates on two bit strings of the same length
hid	Identifier of the encryption private key generating function
$H_v(\cdot)$	A cryptographic hash function
$H_1(\cdot), H_2(\cdot)$	Cryptographic functions derived from the cryptographic hash function
KDF($\cdot$)	The key derivation function
N	The order of the cyclic groups $\mathbb{G}_1, \mathbb{G}_2$ and $\mathbb{G}_T$, which is a prime number greater than 2^{191}
$BITS(m)$	Count the bit length of a bit string m
Δ_e	An update token under epoch e
$[u]g$	The u multiple of the element g in the additive groups $\mathbb{G}_1$ or $\mathbb{G}_2$
g^u	g to the power of u, where g is an element in the multiplicative group $\mathbb{G}_T$ and u is a positive integer
l	The length of secret key

Section 4 gives the concrete construction of the SM9-UE scheme. In Sect. 5, we give the formal security definition and prove its security. The performance of the proposed scheme is evaluated in Sect. 6. In Sect. 7, we draw a conclusion remark.

2 Preliminaries

In this section, we outline the definition of Updatable Encryption and SM9 Cryptographic Scheme as well as introduce some of the math involved. We follow the syntax of prior works in [6–8] and summarize the key notations used in this paper in Table 1.

2.1 Bilinear Group

$\mathbb{G}_1, \mathbb{G}_2$ are additive groups and $\mathbb{G}_T$ is a multiplicative group. All three groups have prime order N. g_1 and g_2 are the generators of $\mathbb{G}_1$ and $\mathbb{G}_2$, respectively. At the same time, there exist a homomorphism ψ from $\mathbb{G}_2$ to $\mathbb{G}_1$ such that $\psi(g_2) = g_1$. Bilinear pairing $\hat{e}$ is a map of $\mathbb{G}_1 \times \mathbb{G}_2 \to \mathbb{G}_T$ satisfying the following conditions:

- **Bilinearity**: for any $g_1 \in \mathbb{G}_1$, $g_2 \in \mathbb{G}_2$, $a, b \in \mathbb{Z}$, we have $\hat{e}\left(g_1^a, g_2^b\right) = \hat{e}\left(g_1, g_2\right)^{ab}$.
- **Non – degeneracy**: $\hat{e}\left(g_1, g_2\right) \neq 1_{\mathbb{G}_T}$.
- **Computability**: for any $g_1 \in \mathbb{G}_1, g_2 \in \mathbb{G}_2$, there exists an efficient algorithm to compute $\hat{e}\left(g_1^a, g_2^b\right)$.

2.2 Hard Problem

The security of our updatable encryption based SM9 construction is built upon following hard problems.

Problem 1. (Bilinear Inverse Diffie-Hellman Problem, BIDH [6,9]) For any $a, b \in \mathbb{Z}_N^*$, it is hard to compute the result of $\hat{e}(g_1, g_2)^{b/a}$.

Problem 2. (Decisional Bilinear Inverse Diffie-Hellman Problem, DBIDH [6,9]) For $a, b, r \in \mathbb{Z}_N^*$, it is very hard to distinguish $(g_1, g_2, g_1^a, g_2^b, \hat{e}(g_1, g_2)^{b/a})$ from $(g_1, g_2, g_1^a, g_2^b, \hat{e}(g_1, g_2)^r)$.

Problem 3. (τ-Bilinear Inverse Diffie-Hellman Problem, τ-BDHI [6,9]) For an integer τ and $\alpha \in \mathbb{Z}_N^*$, given $(g_1, g_2, g_i^\alpha, g_i^{(\alpha^2)}, \ldots, g_i^{(\alpha^\tau)})$ for some value $i \in \{1, 2\}$, computing $\hat{e}(g_1, g_2)^{1/\alpha}$ is hard.

Problem 4. (τ-Gap-Bilinear Inverse Diffie-Hellman Problem, τ-Gap-BIDH [6,9]) For integer τ and $\alpha \in \mathbb{Z}_N^*$, given$(g_1, g_2, g_i^\alpha, g_i^{(\alpha^2)}, \ldots, g_i^{(\alpha^\tau)})$ for some value $i \in \{1, 2\}$ and the DBIDH algorithm, it is hard to compute $\hat{e}(g_1, g_2)^{1/\alpha}$.

2.3 Updatable Encryption

We follow the syntax of prior work [17] and describe the definition of Updatable Encryption (UE).

Definition 1. *An updatable encryption scheme UE for message space $\mathcal{M}$ as a tuple of algorithms* $\{$**UE.Setup**, **UE.Next**, **UE.Enc**, **UE.Dec**, **UE.Upd**$\}$ *as follows.*

- **UE.Setup**: This algorithm takes a security parameter λ as input for getting a secret key k_0.
- **UE.Next**: This algorithm takes a secret key k_e for epoch e as input, it outputs a new secret key k_{e+1} and an update token Δ_{e+1} for each epoch $e + 1$.
- **UE.Enc**: According to the input of a message $m \in \mathcal{M}$ and key k_e for epoch e, this algorithm outputs a ciphertext C.
- **UE.Dec**: The deterministic algorithm is also run by client. In terms of input of a ciphertext C_e and k_e of some epoch e return $\{m'/\perp\} \leftarrow$ **UE.Dec** (k_e, C_e).
- **UE.Upd**: In terms of input of a ciphertext C_e from epoch e and the update token Δ_{e+1}, it returns the updated ciphertext $C_{e+1} \leftarrow$ **UE.Upd**(Δ_{e+1}, C_e).

2.4 SM9 Identity-Based Encryption (SM9-IBE)

SM9 Cryptographic Scheme is a standard for identification and cryptography adopted by China [9]. It is mainly composed of four aspects, namely Digital signature algorithm, Key exchange protocol, Key encapsulation mechanism (KEM), public key encryption algorithm. SM9-IBE is actually a hybrid scheme which consists of KEM and a Symmetric Encryption (SE) algorithm. It contains four operations: **Setup, KeyGen, Encrypt and Decrypt** as follows:

- **Setup**$(\lambda) \rightarrow (MPK, MSK)$. Given the security parameter λ, it outputs a master public key MPK and a master secret key MSK.
- **Extract**$(MSK, ID) \rightarrow d_{ID}$. Given the master secret key MSK and an identity ID, it outputs a personal decryption key d_{ID}.
- **Encrypt**$(MPK, ID, M) \rightarrow C$. Given the master public key MPK and an identity ID, it outputs a ciphertext C.
- **Decrypt**$(d_{ID}, C) \rightarrow M'$. Given a private key d_{ID} for identity ID and ciphertext colllection $C = (C_1, C_2, C_3)$. It outputs a message M'.

3 System Model and Attack Model

we now present some entities involved in our model and the attack model.

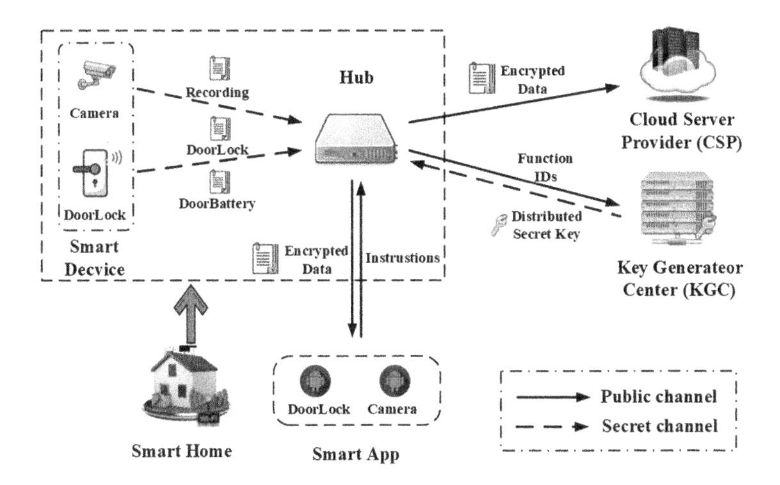

Fig. 1. System model

3.1 System Model

Our system model consists of five entities: Hub, Smart Device, Cloud Service Provider (CSP), Key Generation Center (KGC) and Smart App.

- **Hub** plays an important role since diversity devices may correspond to different communication protocols. It not only helps smart devices connect to the home network, but also serves as a transit point for the data transfer between Smart App and Smart Device. Additionally, this entity replaces the user to encrypt the data generated by the smart device and forward it to the cloud.
- **Smart Device**, such as camera and doorlock, is a data generator that can provide complete and secure data storage on a cloud system through Hub anytime, anywhere. Besides, complex operations is hard to realize on this entity side due to limitations in computing power, storage space, and bandwidth.

- **CSP** is the manager of the cloud server and also an honest but curious third party, which provides quite a few services such as data storage, transmission and timely updating encrypted data after the secret key has been updated.
- **KGC** is assumed to be absolutely trusted third party, who is responsible for the selection of system parameters, generation of the encryption master keys, and generate of user's encryption secret keys according to the its identity. More precisely, it needs to not only protect the secret key but also initializes the relevant system parameters.
- **Smart App**, such as camera app and doorlock app, installed on the user's mobile phone for providing an operation interface of smart devices. If the permissions of Smart App is too large, it is very likely that the sensitive data of the users will be stolen by the malicious app while using it to manage the smart devices.

3.2 Overview

As depicted in Fig. 1, the process of the proposed system model is specialized as follows.

- *System setup and secret key distribution.* At first, KGC selects an appropriate security parameter λ and then takes advantage of Setup algorithm in SM9-UE to generate the system's master public key MPK and master secret key MSK based on this parameter. Then, it generates the corresponding secret keys based on the series of function identity ID, afterwards it receives a certain epoch e sent by Hub and forwards those secret keys back to Hub, starting with $e = 0$. All secrets keys that are distributed must take place on a secure channel.
- *Update key distribution.* When moving from epoch e to epoch $e + 1$, Smart App first destroy by melting or burning the secret key assigned by Hub, Hub will send a series of function identity to KGC, and then KGC will update the corresponding secret key and generates an update token Δ_{e+1} using the KeyUpd algorithm in SM9-UE scheme, and then KGC will send the updated secret key and the corresponding update token to Hub. After Hub receive these, it thoroughly remove all previous secret key and update token.
- *Private the data encryption and upload.* Before storing the user's private data generated by diversity Smart Device on the CSP, Hub collect these data M and run the encrypt algorithm of SM9-UE scheme to encrypt them with the function identity ID and a secret key. Finally, Hub upload these ciphertext with corresponding function identity to the CSP.
- *Secure access to privacy data.* After the Smart App are installed on the phone user must carefully and prudently define the Smart devices function ID that can be accessed in each epoch. Then, Hub will send the corresponding secret key according to the defined function ID to the Smart App. Next, when Smart App needs private data, it will request to the Hub according to the corresponding function ID, Hub will download ciphertext data from CSP based on these function ID, and then send it to the Smart App. Smart App can decrypt it through using the secret key assigned by Hub.

- *Secure ciphertext update.* When moving from epoch e to epoch $e+1$, after CSP received the update token Δ_{e+1} sent by Hub, it first deletes update token Δ_e and then revokes an algorithm CTUpd to update all previously stored ciphertexts.

3.3 Attack Model

Our system will confront the following possible attacks: 1) The app would use malicous logic to obtain privacy while it is installed on the phone; 2) Vulnerable app design flaws that can be exploited by malicious attackers to upgrade their privileges, access unauthorized functions, and steal sensitive data, when they are installed on mobile phones; 3) Vulnerable app may be used by attackers to steal keys, so as to decrypt ciphertext from semi-trusted cloud and obtain sensitive data of user.

4 SM9-UE Scheme

We present a SM9-UE construction scheme, which is *ciphertext-independent* updatable encryption based on SM9-IBE algorithm.
Setup$(\lambda) \rightarrow (MSK, MPK)$. On input security parameter λ, the operation run as follows:

- Select three groups $\mathbb{G}_1$, $\mathbb{G}_2$ and $\mathbb{G}_{\mathbb{T}}$ of prime order N and a bilinear pairing map $\hat{e} : \mathbb{G}_1 \times \mathbb{G}_2 \rightarrow \mathbb{G}_T$. Pick generator $g_1 \in \mathbb{G}_1$, $g_2 \in \mathbb{G}_2$ randomly.
- Pick a random number $MSK \in \mathbb{Z}_N^*$ as master secret key of system and the master public key MPK can be computed as $MPK \leftarrow [MSK]\ g_1 \in \mathbb{G}_1$.
- Select the function identifier hid that use one byte to represent.
- So the public parameters is denoted as $params = \{\mathbb{G}_1, \mathbb{G}_2, \mathbb{G}_{\mathbb{T}}, \hat{e}, g_1, g_2, hid, N\}$ and output MSK and MPK.

KeyGen$(MSK, ID) \rightarrow k_0$. To generate a secret key pair k_e for an identity ID, some operation are performed as follows:

- Create $k_{0_1} \leftarrow$ SM9-IBE.Extract (MSK, ID).
- Create $k_{0_2} \leftarrow$ SE.KeyGen (λ).
- It outputs a secret key pair $k_0 \leftarrow (k_{0_1}, k_{0_2})$.

KeyUpd$(k_e) \rightarrow (\Delta_{e+1}, k_{e+1})$. To generate a new secret key pair taking input as a key pair k_e for epoch e, some operation will be executed as follows:

- Parse $k_e = (k_{0_1}, k_{e_2})$.
- Create $k_{e+1_2} \leftarrow$ SE.KeyGen (λ).
- It outputs an update token $\Delta_{e+1} \leftarrow (k_{e_2}, k_{e+1_2})$ and a new secret key pair $k_{e+1} \leftarrow (k_{0_1}, k_{e+1_2})$.

Enc$(MPK, ID, k_{e_2}, m) \rightarrow C$. Given a master public key MPK, an identity ID and a key k_{e_2} as outer key for epoch e, some operation are executed as follows:

- Compute $s = [H_1(ID||hid, N)]\ g_1 + MPK \in \mathbb{G}_1$.
- Pick a random element $r \in Z_N^*$.
- Create $C_1 \leftarrow [r]s$.
- Compute $w = \hat{e}\ (MPK, g_2)^r \in \mathbb{G}_T$.
- If the *sequence cipher* algorithm is used to encrypt data:
  i) Compute $k_1||k_2 = \text{KDF}(C_1||w||ID, l)$ where $l = BITS(m) + v$.
  ii) Create $C_2 \leftarrow k_1 \oplus m$ if k_1 is not all zero, otherwise redo the above.
- If the *block cipher* algorithm is used to encrypt data:
  i) Compute $k_1||k_2 = \text{KDF}(C_1||w||ID, l)$.
  ii) Compute $C_2 = \text{SE.Enc}(k_1, m)$, if k_1 is not all zero, otherwise redo the above.
- It finally outputs $C \leftarrow \text{SE.Enc}(k_{e_2}, C_1||C_2||H_v(C_2||k_2))$.

Dec$(k_e, C_e) \to m'$. Given a secret key pair k_e and a ciphertext C_e for epoch e, it will carry out some action as follows:

- Parse $k_e = (k_{0_1}, k_{e_2})$.
- $C'_e \leftarrow \text{SE.Dec}(k_{e_2}, C_e)$ and parse $C'_e = (C_1, C_2, C_3)$.
- Output *error* if $C^1 \notin \mathbb{G}_1$, otherwise go on.
- Compute $w' = \hat{e}\ (C_1, k_{0_1})$.
- If the data is encrypted by the *sequence cipher*:
  i) Compute $k_1||k_2 = \text{KDF}(C_1||w'||ID, l)$ where $l' = BITS(M) + v$.
  ii) Create $m' \leftarrow k_1 \oplus C_2$. If k_1 is not all zero, otherwise outputs *error*.
- If the data is encrypted by the *block cipher*:
  i) Compute $k_1||k_2 = \text{KDF}(C_1||w'||ID, l)$.
  ii) Create $m' \leftarrow \text{SE.Dec}(k_1, C_2)$.
- Finally, it outputs m' if $C_3 = H_v(C_2||k_2)$, otherwise outputs *error*.

CTUpd $(\Delta_{e+1}, C_e) \to C_{e+1}$. Given a update token Δ_{e+1} and ciphertext C_e from epoch e, it returns the updated ciphertext. Some operations will be performed as follows:

- Parse $\Delta_{e+1} = (k_1, k_2)$.
- Compute $C' \leftarrow \text{SE.Dec}(k_1, C_e)$
- It outputs $C_{e+1} \leftarrow \text{SE.Enc}(k_2, C')$.

5 Security Analysis

5.1 Security Model

Next, we homoplastically present the definition of Adaptive Update Indistinguishability (IND-UPD), which defined by *Lehmann et al.* in [17]. The IND-UPD notion ensures that an updated ciphertext obtained from the CTUpd algorithm does not reveal any information about the previous ciphertext, even when $\mathcal{A}$ adaptively compromises a number of keys and tokens before and after the challenge epoch. Then, the security is defined with the following game:

Init: The adversary $\mathcal{A}$ outputs an identity ID^* where it wishes to be challenged.
Setup: The challenger runs the Setup algorithm taking security parameters λ as input. Adversary $\mathcal{A}$ gets all system *params* except master secret key MSK.
Phase 1: The adversary $\mathcal{A}$ adaptively issues queries to following oracles:

- $\mathcal{O}_{\text{Enc}}(m, ID^*)$: The challenger sends a ciphertext C_e for epoch e to $\mathcal{A}$, where C_e is generated with the algorithm Enc as $C_e \leftarrow \text{Enc}(k_e, m)$.
- $\mathcal{O}_{\text{KeyUpd}}(k_e)$: The challenger run KeyUpd algorithm to update secret key k_e as $k_{e+1} \leftarrow \text{KeyUpd}(k_e)$. It output k_{e+1} to $\mathcal{A}$.
- $\mathcal{O}_{\text{CTUpd}}(C_e)$: On input ciphertext C_e for epoch e, run algorithm CTUpd as $C_{e+1} \leftarrow \text{CTUpd}(\Delta_{e+1}, C_e)$ and output a new ciphertext C_{e+1} for epoch $e+1$ to $\mathcal{A}$.

Challenge: Once the adversary $\mathcal{A}$ decides that **Phase 1** is over, it outputs two ciphertext C_0 and C_1 on which it wishes to challenged. The challenger picks a random bit $b \in \{0, 1\}$ and sets the challenge updated ciphertext $\tilde{C} = \text{CTUpd}(C_b)$. It sends $\tilde{C}$ as the challenge to the adversary $\mathcal{A}$.
Phase 2: The adversary $\mathcal{A}$ queries $q_{m+1}, \ldots, q_n$ where query q_i is one of:

- $\mathcal{O}_{\text{Enc}}(m)$: The challenger responds as in **Phase 1**.
- $\mathcal{O}_{\text{KeyUpd}}(k_e)$: The challenger responds as in **Phase 1**.
- $\mathcal{O}_{\text{CTUpd}}(C_e)$: The challenger responds as in **Phase 1**.

Guess: Finally, the adversary $\mathcal{A}$ outputs a guess $b' \in \{0, 1\}$. $\mathcal{A}$ wins if $b = b'$.

We refer to such an adversary $\mathcal{A}$ as an IND-UPD adversary. We define the advantage of the adversary $\mathcal{A}$ in attacking the scheme $\mathcal{E}$ as

$$Adv_{\mathcal{E},\mathcal{A}} = \left| \Pr\left[b = b'\right] - \frac{1}{2} \right|$$

Definition 2. *We say an updatable encryption scheme $\mathcal{E}$ is* IND-UPD *secure if for all probabilistic polynomial-time adversaries $\mathcal{A}$, it holds that $Adv_{\mathcal{E},\mathcal{A}} < \epsilon(\lambda)$ for some negligible function ϵ.*

5.2 Security Proof

In this section, the security of our proposed scheme will be analyzed precisely under the attack model proposed in Sect. 3.3. In our attack model, **CSP** is considered as an semi-honest entity while **Hub** and **KGC** is accounted as absolutely trusted party. In addition, the **Smart App** may be malicious and attack our system model through the following means to obtain sensitive data of users. We make an assay of how does our scheme defend against these attacks.

Theorem 1. *The proposed* SM9-UE *scheme achieves* IND-UPD *security with respect to Definition 2, if* SM9-IBE *and* SE *scheme is* IND-CPA *secure.*

Proof. Suppose $\mathcal{A}$ has advantage ϵ in attacking the SM9-UE system, then we construct a simulator $\mathcal{B}$ to break the IND-CPA security of the SM9-IBE and SE scheme with a non-negligible advantage. It provides some oracles to $\mathcal{B}$:

- $\mathcal{O}^{\text{SM9-IBE}}_{\text{Enc}}(m, ID_i)$: The challenger run algorithm SM9-IBE.Enc(m, ID_i) and output a ciphertext C to $\mathcal{B}$.
- $\mathcal{O}^{\text{SE}}_{\text{Enc}}(m)$: The challenger run algorithm SE.Enc(m) and output a ciphertext C to $\mathcal{B}$.

The simulator $\mathcal{B}$ works by interacting with the challenger and the adversary $\mathcal{A}$ as follows:

Initialization. The game begins with $\mathcal{A}$ first outputting an identity ID to $\mathcal{B}$. $\mathcal{B}$ initially generates keys $k^i \leftarrow$ SM9-IBE.Extract(ID), sets $e \leftarrow 0$, and guesses the challenge epoch e' uniformly random from $\{0, \ldots \tilde{e}\}$, where $\tilde{e}$ is an upper bound on the number of epochs for $\mathcal{A}$.
Setup. To generate system parameters, challenger run SM9-IBE.Setup algorithm and output $params = \{\mathbb{G}_1, \mathbb{G}_2, \mathbb{G}_T, \hat{e}, g_1, g_2, hid, N\}$ to $\mathcal{B}$.
Phase 1. $\mathcal{A}$ adaptively issues quires to following oracles:

- $\mathcal{O}_{\text{Enc}}(m, ID)$: Compute $C^i \leftarrow$ SM9-IBE.Enc(m, ID), and $\mathcal{B}$ queries C^i to its own oracle $\mathcal{O}^{\text{SE}}_{\text{Enc}}(C^i, k^o_e)$ to obtain a ciphertext C_e. Then $\mathcal{B}$ forwards C_e to $\mathcal{A}$.
- $\mathcal{O}_{\text{KeyUpd}}(k_e)$: Generate key $k^o_{e+1} \leftarrow$ SE.KeyGen(λ) and set $e \leftarrow e + 1$.
- $\mathcal{O}_{\text{CTUpd}}(C_e)$: $\mathcal{B}$ obtains C_{e+1} by querying C^i to its own $\mathcal{O}^{\text{SE}}_{\text{Enc}}(C^i, k^o_{e+1})$ and return C_{e+1} to $\mathcal{A}$.

Challenge. When $\mathcal{A}$ decides that **Phase 1** is over, it outputs two ciphertext C_0, C_1 for an challenge epoch e'. $\mathcal{B}$ forwards C_0, C_1 to challenger. The challenger pick a random bit $b \in \{0, 1\}$ and set challenge updated ciphertext $\tilde{C} =$ CTUpd(C_b). Finally $\mathcal{B}$ send it to $\mathcal{A}$.
Phase 2. $\mathcal{A}$ issues more queries to following some oracles:

- $\mathcal{O}_{\text{Enc}}(m_i, ID)$ where $e \neq e'$: $\mathcal{B}$ responds as in **Phase 1**.
- $\mathcal{O}_{\text{KeyUpd}}(k_e)$ where $e \neq e'$: $\mathcal{B}$ responds as in **Phase 1**.
- $\mathcal{O}_{\text{CTUpd}}(C_e)$ where $e \neq e'$: $\mathcal{B}$ responds as in **Phase 1**.

Guess. $\mathcal{A}$ outputs a guess $b' \in \{0, 1\}$. $\mathcal{B}$ forward b' to challenger.

As shown above, $\mathcal{A}$ obtain the challenge ciphertext $\tilde{C}$ in IND-CPA game for SE and SM9-UE scheme. If $\mathcal{A}$ successfully guesses which ciphertext is updated in the challenge epoch, $\mathcal{B}$ also outputs the right guess. Therefore, if $\mathcal{A}$ can break proposed SM9-UE scheme with probability $\epsilon(\lambda)$, $\mathcal{B}$ can break SM9-IBE and SE with the same probability. This completes the proof of Theorem 1.

6 Performance Evaluation

In this section, we give thorough and accurate performance analysis.

Table 2. Comparison of computational efficiency

Scheme	Security [4]	Ciphertext independent	Enc	Dec	TokenGen	CTUpd [5]
BLMR [2]	detIND-ENC	✔	2 Exp	2 Exp	2 Exp	$2n$ Exp
BLMR+ [17]	weakIND-UE	✔	2 Exp	2 Exp	2 Exp	$2n$ Exp
RISE [17]	randIND-UE	✔	2 Exp	2 Exp	1 Exp	$2n$ Exp
NYUE [14]	rand-UPD	✔	(60 Exp, 70 Exp)	22 e	2 Exp	(60 Exp, 70 Exp)
E&M [14]	detIND-UPD	✗	3 Exp	3 Exp	3 Exp	3 Exp
ReCrypt [5]	UP-REENC	✗	2 Exp	2 Exp	$2n$ Exp	$2n$ Exp
NYUAE [14]	rand-UPD	✗	(110 Exp, 90 Exp)	29 e	3 Exp	(110 Exp, 90 Exp)
SHINE0 [3]	detIND-UE	✗	1 Exp	1 Exp	1 Exp	1 Exp
mirrorSHINE [3]	detIND-UE	✗	2 Exp	2 Exp	1 Exp	2 Exp
OCBSHINE [3]	detIND-UE	✗	1 Exp	1 Exp	1 Exp	1 Exp
SM9-UE(Ours)	IND-UPD	✔	1 e	1 e	—	—

[4]The notions IND-ENC, randIND-UPD and detIND-UPD are from [17]. The notions UP-IND and UP-REENC are from [3,14,17]. All notions build upon the definitions given by EPRS[4].
[5]Exp and e denote a module exponentiation and a pairing computation, respectively. n denote the number of ciphertexts.

Table 3. Comparison of scheme in IoT

Scheme	Fine-grained access	ACL	Data security	Against over-privilege	Key rotation
FACT [16]	✔	✔	Unsafe	✔	✗
IoT-FBAC [23]	✔	✗	Safe	✔	✗
SM9-UE (Ours)	✔	✗	Safe	✔	✔

6.1 Theoretical Analysis

We analyzed the proposed scheme in theory by comparing it with previous works in terms of communication and storage cost, computation efficiency, which are summarized in Table 2 and Table 3.

As shown in Table 2, we summarize the security and computing efficiency of existing Updatable Encryption (UE) schemes and compare them with ours. Firstly, our scheme is *ciphertext-independent* updatable encryption, which greatly reduces the computation and communication overhead, since the data owner does not download part of the ciphertext from the cloud to make a token. Besides, as our proposed scheme is based on an elliptic curve discrete logarithm problem and bilinear pairings to enhance its security, encryption, and decryption require a pairing computation. In the CTUpd algorithm, the computational cost of our solution up to the SE.Enc algorithm and doesn't need any module exponentiation computation.

In Table 3, we present the comparison between our scheme with other existing IoT schemes from five aspects: fine-grained access, against over-privileged resistance, ACL, data security, and key rotation. To fine-grained access, Lee *et al.*

take the function ID as the object unit to control access and against the drawback of over-privileged in [16]. However, their scheme encountered a bottleneck with the increase in devices due to the introduction of an Access Control List (ACL). Both Yan *et al.* and our solution does not require ACL and safely stored encrypted data on the cloud. Nevertheless, Yan *et al.* is incapable dealing with the problem of secret key rotating except for download all ciphertext from the cloud, re-encrypt it and then upload it, which can incur very high communication and computing overhead. Our solution of the SM9-UE scheme enables the CSP to update the ciphertext independently and securely, which reduces the computational pressure of the data owner and makes the data from the smart device reach safely.

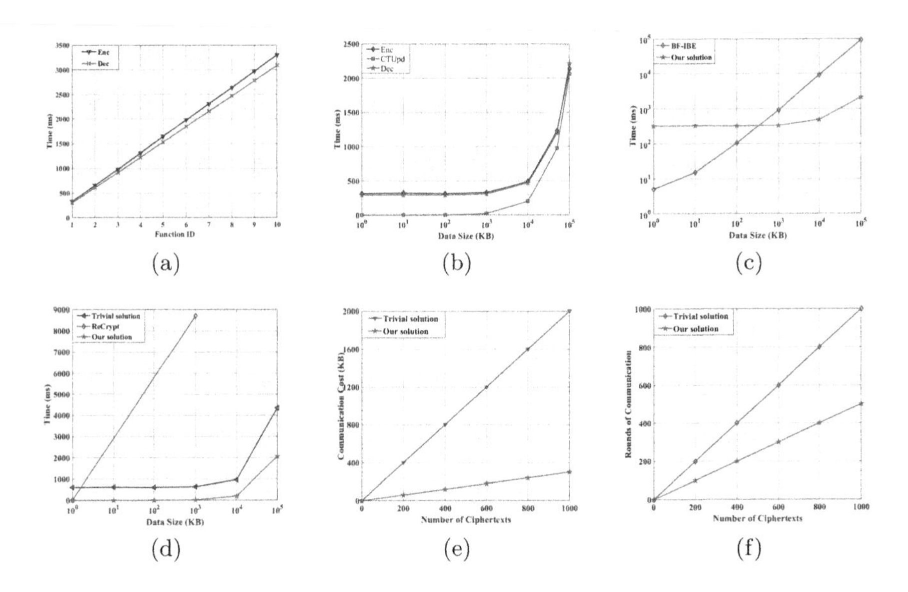

Fig. 2. SM9-UE scheme performance. (Color figure online)

6.2 Implementation and Evaluation

All SM9-UE constructing algorithms are implemented by using C programs. To speed up the simulation we performed, we are based on the the framework proposed by Guan's GmSSL[1] by using a 4.00 GHz Intel®Core™ i5-9500 CPU and 4 GB RAM under the Ubuntu 18.04.5 LTS operating system.

To simulate our scheme, we instantiate and implement the proposed scheme following the standard (Gm/T [9]) and the values of some parameters definition in [8]. The hash function H_v and Symmetric Encryption (SE) is implemented by using SM3 hash algorithm and SM4 block cipher algorithm [8] respectively. The block size is 128 bits. we take the average value of each algorithm executed 100 times as the final result in order to make the experimental results more accurate.

[1] https://github.com/GmSSL.

Time comsumption of Enc and Dec under the different number of function IDs is shown in Fig. 2(a). Let the number of function IDs range from 1 to 10 and the data size is set 1 MB. In the course of encryption phase (the blue line), the greater the number of the function IDs is, the more time that it will cost. When the number of function IDs reaches 10, encryption time only needs about 3.3 s. This demonstrate that our solution is practical and effective. After an App is installed, Hub assigns it the corresponding secret key based on the authorization given by the user. During the decryption phase (the red line), App uses the corresponding secret key to decrypt the ciphertext to obtain data. The decryption time also increases with the number of function IDs. Look at the whole picture, the encryption time is slightly higher than the decryption time and the gap grows as the number of function IDs.

To accurately demonstrate the performance of our scheme, we present the experiment result under different data sizes from 1 KB to 100 MB in Fig. 2(b). In the course of encryption phase, Hub encrypts the data generated by smart device under different function ID. The encryption time is relatively stable around 320 ms and does not float much when the data size is from 1 KB to 1 MB. As data size increases from 1 MB to 10 MB, the encryption time starts to grow slowly. When the data size changes from 1 MB to 100 MB, the encryption time increases significantly. The green line shows how the decryption time varies with different data sizes. This process executed by an App. When an App receives the ciphertext downloaded from the cloud by the Hub, it decrypts ciphertext with the secret key given by the Hub to obtain the data. Judging from the tendency of the decryption time with the size of the data, it is almost the same as the encryption time. The time required for encryption and decryption is actually similar for the same size data. During the ciphertext update phase (the blue line), CSP independently update ciphertext after Hub updates the secret key and receiving the token. It takes very little time to update the ciphertext when the encrypted data size is less than 1 MB. The ciphertext update time increases with the data size from 1 MB and getting closer and will even outlast encryption and decryption time.

In Fig. 2(c), we compare the Enc algorithms between our solution and BF-IBE [1] under different data sizes from 1 KB to 100 MB. Although our scheme performs slightly worse than the another when the data size is less than 1 MB, that is because our scheme requires a pair of computation cost during Enc algorithm. After the data size is super 1 MB, the greater the data size is, the better the performance of our solution than the another.

Next, we show a comparison of ciphertext update with a trivial solution. As we all know, a trivial solution requires downloading all ciphertext from the cloud, re-encrypting it and then upload it for the ciphertext update process. It is amazing that our solution only needs to generate a token locally without other operations. In Fig. 2(d), communication overhead required for data upload and download is ignored by us. It only compared the time required to update the ciphertext algorithm, as the data size gradually increases. For traivial solution (the blue line) in terms of his time with the increase in data size increases and significantly more than our solution. ReCrypt scheme (the green line) proposed by [5] has obvious advantage than trivial solution under tiny data size. However, as the data size gets larger, more time is required. It is estimated that 10 MB of data will require about 90 ms, which is beyond the scope of our diagram. Figure 2(e) and Fig. 2(f) shows a comparison of communication overhead under ours solution and trivial solution. The data size is set to 1 KB and the number of ciphertext ranges from 0 to 1000. Obviously, the communication overhead and communication rounds of our solution are lower than trivial solution.

7 Conclusion and Further Problem

In this paper, for the sake of preventing software over-privilege access and reducing the computing overhead of each entity accordingly, we propose a solution for updatable encryption based on SM9 in the IoT smart home, which is referred to as SM9-UE. It not only allows users to realize fine-grained access control, but also periodically rotate secret keys to enhance the security of the private data. Besides, when the secret key is rotated, the cloud service provider can independently update ciphertext while ensuring that plaintext is not retrieved. The reason why our way greatly reduces communication and computation overhead is user does not need to download the ciphertext locally and then re-encrypt and upload it. Then, we present a formal security definition and prove that our solution achieves IND-UPD secure. Finally, both the theoretical analysis and the experimental results show that SM9-UE is efficient and practical. Future works includes extending our scheme to other IoTs scenarios such as smart logistics and smart medical, designing new schemes to promote computation efficiency and further security enhancement.

References

1. Boneh, D., Franklin, M.: Identity-based encryption from the weil pairing. SIAM J. Comput. **32**(3), 586–615 (2003)
2. Boneh, D., Lewi, K., Montgomery, H., Raghunathan, A.: Key homomorphic PRFs and their applications. In: Canetti, R., Garay, J.A. (eds.) CRYPTO 2013. LNCS, vol. 8042, pp. 410–428. Springer, Heidelberg (2013). https://doi.org/10.1007/978-3-642-40041-4_23
3. Boyd, C., Davies, G.T., Gjøsteen, K., Jiang, Y.: Fast and secure updatable encryption. Technical Representation Cryptology ePrint Archive, Report 2019/1457, 2019. https://eprint.iacr.org (2020)
4. Davidson, A., Deo, A., Lee, E., Martin, K.: Strong post-compromise secure proxy re-encryption. In: Jang-Jaccard, J., Guo, F. (eds.) ACISP 2019. LNCS, vol. 11547, pp. 58–77. Springer, Cham (2019). https://doi.org/10.1007/978-3-030-21548-4_4
5. Everspaugh, A., Paterson, K., Ristenpart, T., Scott, S.: Key rotation for authenticated encryption. In: Katz, J., Shacham, H. (eds.) CRYPTO 2017. LNCS, vol. 10403, pp. 98–129. Springer, Cham (2017). https://doi.org/10.1007/978-3-319-63697-9_4

6. GM/T: Sm9 identity-based cryptographic algorithms part 1: General. [EB/OL] (2020). http://www.gmbz.org.cn/main/postDetail.html?id=20180322410400
7. GM/T: Sm9 identity-based cryptographic algorithms part 4: Key encapsulation mechanism and public key encryption algorithm. [EB/OL] (2020). http://www.gmbz.org.cn/main/postDetail.html?id=20180322410400
8. GM/T: Sm9 identity-based cryptographic algorithms part 5: Parameter definition. [EB/OL] (2020). http://www.gmbz.org.cn/main/postDetail.html?id=20180322410400
9. GM/T0044-2016: Sm9 identity-based cryptographic algorithms. [EB/OL] (2020). http://www.gmbz.org.cn/main/postDetail.html?id=20180322410400
10. Grace, M.C., Zhou, W., Jiang, X., Sadeghi, A.R.: Unsafe exposure analysis of mobile in-app advertisements. In: Proceedings of the Fifth ACM Conference on Security and Privacy in Wireless and Mobile Networks, pp. 101–112 (2012)
11. Huang, Z., Lai, J., Chen, W., Li, T., Xiang, Y.: Data security against receiver corruptions: soa security for receivers from simulatable dems. Inf. Sci. **471**, 201–215 (2019)
12. IDC, I.D.C.: Idc market in a minute: Internet of Things. [EB/OL] (2020). http://www.idc.com/downloads/idc_market_in_a_minute_iot_infographic.pdf
13. Khan, M.A., Salah, K.: Iot security: review, blockchain solutions, and open challenges. Future Generat. Comput. Syst. **82**, 395–411 (2018)
14. Klooß, M., Lehmann, A., Rupp, A.: (R)CCA secure updatable encryption with integrity protection. In: Ishai, Y., Rijmen, V. (eds.) EUROCRYPT 2019. LNCS, vol. 11476, pp. 68–99. Springer, Cham (2019). https://doi.org/10.1007/978-3-030-17653-2_3
15. Lee, I., Lee, K.: The internet of things (iot): applications, investments, and challenges for enterprises. Busin. Horiz. **58**(4), 431–440 (2015)
16. Lee, S., Choi, J., Kim, J., Cho, B., Lee, S., Kim, H., Kim, J.: Fact: Functionality-centric access control system for IOT programming frameworks. In: Proceedings of the 22nd ACM on Symposium on Access Control Models and Technologies, pp. 43–54 (2017)
17. Lehmann, A., Tackmann, B.: Updatable encryption with post-compromise security. In: Nielsen, J., Rijmen, V. (eds.) EUROCRYPT 2018. LNCS, vol. 10822, pp. 685–716. Springer, Cham (2018). https://doi.org/10.1007/978-3-319-78372-7_22
18. Neagle, C.: A guide to the confusing internet of things standards world network-world. Network World (2014)
19. Pearce, P., Felt, A.P., Nunez, G., Wagner, D.: Addroid: privilege separation for applications and advertisers in android. In: Proceedings of the 7th ACM Symposium on Information, Computer and Communications Security, pp. 71–72 (2012)
20. Ronen, E., Shamir, A.: Extended functionality attacks on IOT devices: the case of smart lights. In: 2016 IEEE European Symposium on Security and Privacy (EuroS&P), pp. 3–12. IEEE (2016)
21. Stergiou, C., Psannis, K.E., Kim, B.G., Gupta, B.: Secure integration of IOT and cloud computing. Future Gener. Comput. Syst. **78**, 964–975 (2018)
22. v3.2, P.D.: Pci security standards council: requirements and security assessment proce- dures. [EB/OL]. https://www.pcisecuritystandards.org/ Accessed 2020
23. Yan, H., Wang, Y., Jia, C., Li, J., Xiang, Y., Pedrycz, W.: Iot-fbac: function-based access control scheme using identity-based encryption in IOT. Future Gener. Comput. Syst. **95**, 344–353 (2019)

Scale Variable Dynamic Stream Anomaly Detection with Multi-chain Queue Structure

Fei Wu[1](✉), Ting Li[1], Kongming Guo[2], and Wei Zhou[2]

[1] State Grid Fujian Electric Power Company, Fuzhou 350001, China
[2] State Grid Info-Telecom Great Power Science and Technology Co., Ltd., Fuzhou 350003, China

Abstract. Most existing data stream anomaly detection algorithms do not involve in the tackling of multi-scale characteristic of stream data, a multi-chain queue based data storage structure that is especially suitable for the analysis of multi-scale stream data is designed, and a corresponding algorithm to identify the multi-scale stream anomaly is proposed. The algorithm employs an iteration strategy and takes 3θ as the criteria for discriminating the anomaly, to minimize each anomaly's effect to its neighbors, to detect simultaneously the anomalies in the data sequences that are at the same time and the anomalies in different times. Meanwhile, the increase of a new data sample and the delete of an obsolete observation data are implemented effectively through the operation of a queue, Hence, a better result of stream data anomaly mining is obtained with the changing of mining scale. Finally, through the experiments on a real stream dataset, the proposed algorithm is shown to be capable of finding out some true anomalies in different scales with a higher accuracy rate, when compared with the traditional sliding window based algorithms and the machine learning based algorithms.

Keywords: Anomaly detection · Stream data · Multi-chain queue · Multi-scale

1 Introduction

Streaming data is one of the common data types in the cloud environment of the Internet of things. Exploring the algorithm, significance and application of stream anomaly detection is a hot research topic in the field of data mining in recent years [1]. On the basis of traditional outlier detection, researchers have proposed many excellent outlier detection algorithms, such as distance based method [2], sliding window based method [3] and so on. However, for the multi-scale characteristics of network flow data, there are few related studies in the literature, and there is no report on the results of multi-scale flow anomaly detection.

Hence, this paper bases on the storage structure, similarity measurement and scale aggregation strategy of multi-scale network flow data, and designs a multi-chain queue based data storage structure to conduct dynamic stream anomaly detection with the changing of scale. It is found that the algorithm is capable of finding out some true anomalies at the same time and in different times during the dynamic changing of mining scales.

J. Xiong et al. (Eds.): MobiMedia 2021, LNICST 394, pp. 516–525, 2021.
https://doi.org/10.1007/978-3-030-89814-4_37

2 Related Work

Stream anomaly detection algorithm and its application is a hot research field of data mining in recent ten years. In terms of algorithm mechanism, related research mainly includes methods based on statistics and sliding window, methods based on data mining and methods based on artificial intelligence.

In the area of sliding window, Kontaki et al. [2] studied anomaly detection based on distance, and proposed an anomaly recognition method for continuous monitoring data stream based on sliding window. Zhang et al. [3] proposed an angle based subspace anomaly detection method based on sliding window. Lin et al. [4] studied anomaly detection of data flow in sensor networks based on sliding window and optimized clustering. Qiu et al. [5] proposed a stream data anomaly detection method based on long-term memory (LSTM) network and sliding window, aiming at the characteristics of large amount of stream data and rapid production. It can not only predict data, but also update and adjust the network in real time while learning. Yu et al. [6] proposed a data flow anomaly detection algorithm based on angle variance to solve the problem of high-dimensional spatial sparsity, and demonstrated its application in elevator fault detection.

In the aspect of data mining, Salehi et al. [7] studied the incremental local outlier detection algorithm to save memory, and proposed an efficient flow anomaly detection algorithm MiLOF. Gao et al. [8] studied the incremental stream data outlier mining based on cube. Kim et al. [9] used binary classification strategy for statistical testing of anomalies, thus realizing anomaly pattern detection of data flow. Zhu et al. [10] proposed an improved approximate average KNN outlier detection scheme based on grid for IOT flow data, which uses grid to filter most normal data, and also makes the data anomaly test simple. Ma et al. [11] proposed a network abnormal traffic recognition method based on bag of words model clustering to solve the problems of low recognition accuracy of existing flow anomaly detection methods and the need to determine the threshold for rapid recognition.

In the field of artificial intelligence, Bouguelia et al. [12] proposed an incremental unsupervised flow anomaly detection algorithm GNG-A with the object of data flow undergoing different types of changes and the goal of algorithm adaptive update maintenance. Based on neural network and Bayesian optimization, Alnafessah et al. [13] carried out hybrid anomaly detection of data stream. Xu et al. [14] proposed an abnormal network data mining algorithm based on fuzzy neural network for the influence of fuzzy weighted disturbance on network data clustering center.

Although researchers have carried out a lot of fruitful research in the aspects of algorithm strategy, adaptive ability, parallelizability, real-time, high-dimensional complexity, concept drift and so on, and also put forward some theoretical or practical methods, there is no report on the related results of multi-scale flow anomaly detection in the literature.

To sum up, many researchers have carried out a lot of meaningful research on the strategy, adaptive ability, parallelism of flow anomaly detection algorithm, as well as the real-time performance, high-dimensional sparsity, concept drift of flow data, and also put forward some very valuable methods. However, there are few related reports on the research of multi-scale flow anomaly, so the paper focuses on multi-scale flow anomaly detection, especially the design and implementation of flow anomaly detection algorithm are explored.

3 Scale Variable Dynamic Stream Anomaly Detection

In order to achieve efficient multi-scale anomaly detection of dynamic network flow data, one of the core tasks is to design the storage structure of flow data and control the storage space of flow data within a certain range. On the other hand, to design a scientific multi-scale aggregation scheme and its corresponding definition of similarity measurement. Therefore, the multi-scale dynamic flow anomaly detection algorithm based on composite chain queue is introduced from the aspects of data storage structure, multi-scale aggregation function, similarity measurement and core algorithm.

3.1 Multi-chain Queue Structure

In view of the dynamic multi-scale characteristics of network flow data, the flow data storage structure of composite chain queue type shown in Fig. 1 is designed. Stream data is stored in different chain queue sequences according to scale conditions that is noted as s, which can only be processed once and the sequence is guaranteed. At the same time, it is suitable for multi-scale stream anomaly analysis.

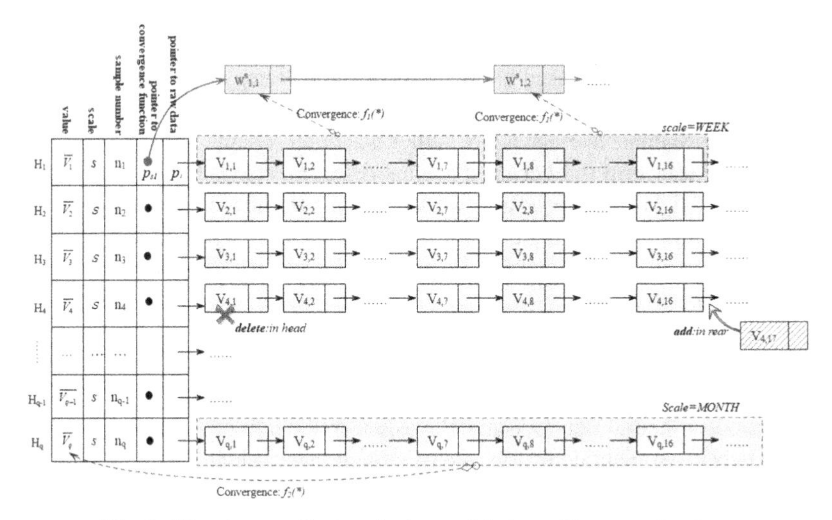

Fig. 1. The storage of network stream data in multi-queue structure.

In the chain queue structure, when a new stream data sample is added, the sample is stored in the rear of the specific chain queue according to the time $H_x(x = 1, 2, ..., q)$ and the scale condition s, such as $V_{4,1}$ in Fig. 1; while an observation is deprecated, the oldest sample in the head of the chain queue is deleted, such as $V_{4,17}$ in Fig. 1. In addition, the pointer p_x in the chain header node points to the list nodes, where the raw stream data observation values $V_{x,y}$ are stored, while x is the time of data acquisition, for 24 h in a day, $x = 24$, and is the index number of the data sample.$V_{x,y}$. in series … In order to distinguish the direction of data flow, data flow..$V_{x,y}$.. is subdivided into input flow and output flow, that is $V_{x,y} = (IS, OS)_{V(x,y)}$, denotes the input flow data value and

OS denotes the output flow data value. Furthermore, the pointer p_{sx} in the chain header node points to a series of data value $w^s_{x,z}$, which are aggregated from the raw stream data $V_{x,y}$ under the scale condition s specified in the header node of the chain queue. The aggregation function f_1 is usually the average value, while.z. is the serial number of the data sample for the moment x under the scale condition s, $z \in \{1, 2, 3, ...\}$. The aggregation data also distinguishes the input stream from the output stream.

3.2 Multiscale Convergence Function

When the network flow data is converged according to the scale condition s, the aggregation function f_1 used is shown in Eq. (1), that is, the k_{th} observation value. $w^s_{x,k}$. at the moment x with the scale condition s is generated by the function f_1. Similarly, the aggregation value $\overline{V_x}$ with the scale condition s is converged from $w^s_{x,k}$, by the convergence function f_2 (as shown in Eq. (2)), that is, $\overline{V_x}$ is generated by the convergence of the observation sequence $w^s_{x,k}$, $k = 1, 2, 3, ...$

$$w^s_{x,k} = f_1(x_k) = \frac{1}{\|s\|} \cdot \sum_{i=1}^{s} v_{x,\,(k-1)s+i} \tag{1}$$

$$\overline{V_x} = f_2(w^s_x) = \frac{1}{|n_x|} \cdot \sum_{k=1}^{n_x} w^s_{x,k} \tag{2}$$

3.3 Dissimilarity Measurement

As the core element of anomaly detection, dissimilarity measurement between data samples is defined by weighted Euclidean distance for stream data including incoming and outgoing flow, as shown in Eq. (3).

$$diff(V_1, V_2) = \omega \cdot (IS_{V_1} - IS_{V_2}) + (1 - \omega) \cdot (OS_{V_1} - OS_{V_2}) \tag{3}$$

Wherein, ω is the weight of incoming and outgoing flow. For the server under attack, it is better to take a higher ω value; for the server carrying out program instruction broadcast or data transmission, it is better to take a lower ω value.

3.4 Stream Anomaly Detection Algorithm

In the multi-chain queue stream data storage structure as shown in Fig. 1, when the scale condition s is changed, the stream data sequence $w^s_{x,z}$ $(x = 1, 2, 3, ..., q)$ with scale s is obtained through the aggregation function f_1 on the basis of raw stream data observation $V_{x,y}$, then, the overall stream value $\overline{V_x}$ for the moment x is computed through the aggregation function f_2 on the basis of $w^s_{x,z}$. Obviously, the stream data sequence $w^s_{x,z}$ is the original stream data observation sequence, when s is the original granularity of stream data. Therefore, the proposed stream anomaly detection algorithm should carry out anomaly detection on each series of stream data sequence $w^s_{x,z}$ $(z = 1, 2, 3, ...)$, as well as on the sequence of overall stream value $\overline{V_x}$ ($x = 1, 2, 3, ..., q$), to find out the outlier in each series and in each time moment. Finally, when the scale condition s is changed, each data sequence $w^s_{x,z}$ and the overall stream value $\overline{V_x}$ need to be updated and re-aggregated, and the stream anomaly detection need to be carried out again.

For the stream data series $P = \{P_1, P_2, \cdots, P_n\}$, the algorithm employs the 3θ criterion (θ is the standard deviation of the stream data observation in a sequence) to judge if an observation is abnormal. Moreover, the algorithm uses an iterative strategy to reduce the influence of a significant deviation outlier on the other data observation, that is, only the most abnormal data is detected in each cycle, and the program loops until all anomalies are detected.

According to the strategy of the algorithm, the pseudo-code for the iterative stream anomaly detection algorithm is as follows.

```
Set findIreOutlier( P = {P1, P2, L , Pn} , threshold=3θ){
 (1) S = ∅;
 (2) while( TRUE ){
   (2.1) C_S = ∅ ;
   (2.2) P̄ = compute_Mean(P) , θ = compute_Std(P) ;
   (2.3) for(i=1; i<=n; i++){
          value = diff(Pi, P̄);
          if(value > 3θ)  C_S = C_S ∪ Pi;
   (2.4) if( C_S = ∅ ) break;
          else{ P_O = find_max{C_S}; S = S ∪ P_O; P_O = P̄ }
   }
 (3) return S ;
}
```

Step (1) initializes the final anomaly set S to be empty $\emptyset$; Step (2) loops circulates until no more abnormality was found, while sub-step (2.1) initializes candidate anomaly set C_S to be empty $\emptyset$, sub-step (2.2) calculates the mean $\overline{P}$ and standard deviation θ of data series P with the function $\overline{P} = compute\_Mean(P)$ and $\theta = compute\_Std(P)$ respectively, sub-step (2.3) computes the dissimilarity $diff(P_i, \overline{P})$ between P_i and for each data sample P_i by using Eq. (3). If $diff(P_i, \overline{P})$ is greater than 3θ, P_i is regarded as a candidate anomaly and is added to.C_S.. Finally, sub-step (2.4) obtains the data object P_O with the largest deviation value (that is $P_O = \max\{C_S\}$), to be the anomaly found in current cycle, and P_O is added to S, and replayed with the $\overline{P}$ to eliminate the impact of the anomaly P_O on the other data.

3.5 Efficiency Analysis

For the data sequence with n objects, step (2.1) is finished in constant time, step (2.2) is to compute the mean and standard deviation whose time complexity is $O(n)$, step (2.3) is to calculate the deviation of each data sample whose time complexity is $O(n)$, and step (2.4) is to select the data with the largest deviation from the candidate anomalies, and the time complexity is not more than $O(n)$. Therefore, for the data sequence P with m exceptions, the time complexity of the algorithm is $O(n \cdot m)$.

For the multi-chain queue based multi-scale stream anomaly detection algorithm shown in Fig. 1, the time complexity of the stream anomaly is $O(q \cdot m)$, where m is the number of anomalies and q is the series number of chain queues. Among which, the time complexity for the detection of series $w^s_{x,z}$ $(H_x,\ x = 1, 2, 3, \ldots, q)$ with scale condition s is $O(t_x \cdot m_x)$, where t_x is the scale of the data sequence $w^s_{x,z}$ and m_x is the number of anomalies of $w^s_{x,z}$. Hence, the time complexity of anomaly detection is $O(t_x \cdot m_x \cdot q)$ for a total number of q data sequence. Therefore, the overall time complexity of multi-scale stream anomaly detection algorithm is $O(q \cdot m) + O(t_x \cdot m_x \cdot q)$.

4 Experiments

4.1 Data Information

The experimental is conducted on a PC with Windows 7, Intel Core i5-3570 CPU @ 3.4 GHz × 2, 8 GB memory. The object is the number of data packets obtained by four application servers for network supervision. The time span is from January 2010 to February 2011. The total number of data is 27234, including ID, machine number and eight characteristics of data capture, such as year, month, day, hour, minute, second and flow rate. The number of valid data of four servers A1, A2, A3 and A4 is 8077807581002857 respectively.

The five characteristics for the network flow data obtained from the four servers are shown in Table 1.

Table 1. Five characteristics of the data set

	Min	Q1	Q2	Mean	Q3	Max
A1	0	0	0	1.956	4.539	12.788
A2	0	5.616	9.234	104.559	12.969	52718.629
A3	0	1.384	1.451	397.335	1203.25	1699.000
A4	0	11.775	13.344	12.436	14.252	18.354

It can be seen from Table 1 that the median (Q2), the third quantile (Q3) and maximum (max) of servers A1, A2 and A3 are significantly different, which means that there is a large fluctuation in network flow and the possibility of abnormality is high; while the characteristic value of server A4 is relatively flat and there is no obvious fluctuation. Therefore, in order to highlight the value and advantages of multi-scale flow anomaly detection, the multi-scale anomaly detection of server A4 is introduced and the results are analyzed.

4.2 Result and Discussion

According to the 3θ criterion, the data from server A4 is iteratively detected on the "hour" scale, and the results are shown in Fig. 2. A total of 12 iterations were carried out, and 329 anomalies were detected (green labeled points shown in Fig. 2). The upper and lower thresholds and the number of anomalies of each round of detection are shown in Fig. 3.

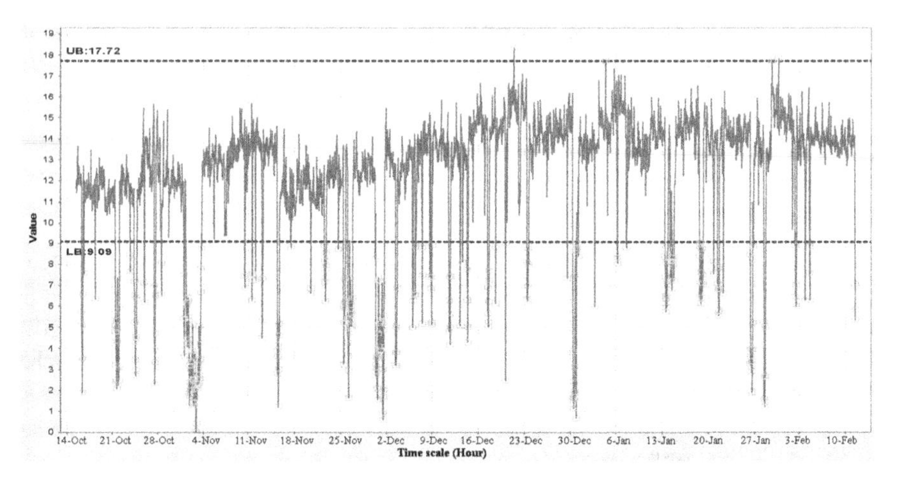

Fig. 2. Anomaly detection result with s = Hour, IR = 12.

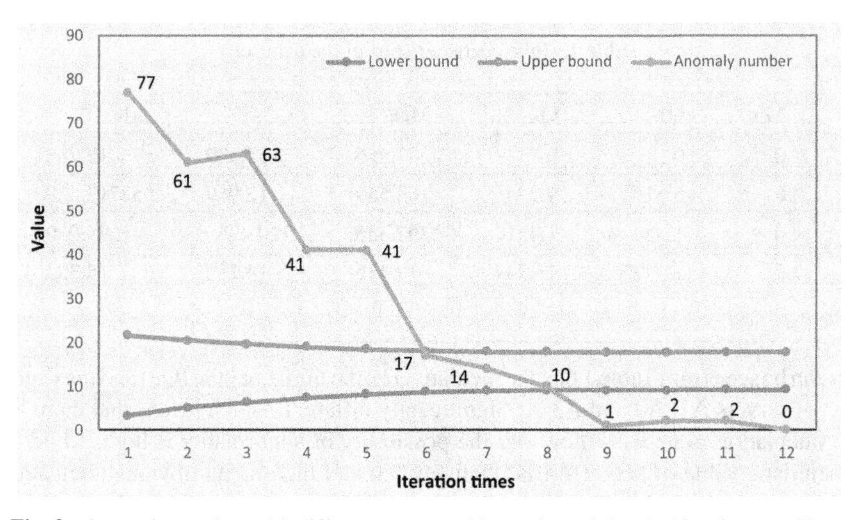

Fig. 3. Anomaly number with different upper and lower bound thresholds when s = Hour.

On the "day" scale, the iterative anomaly detection is performed for each time (24:00) sequence according to "day". The number of iterations and the number of anomalies are shown in Fig. 4. Compared with the "hour" and "day" scale, the number of anomalies (128) is significantly reduced, and the number of iterations is also greatly reduced, which is in line with the high granularity contraction characteristics of flow anomalies.

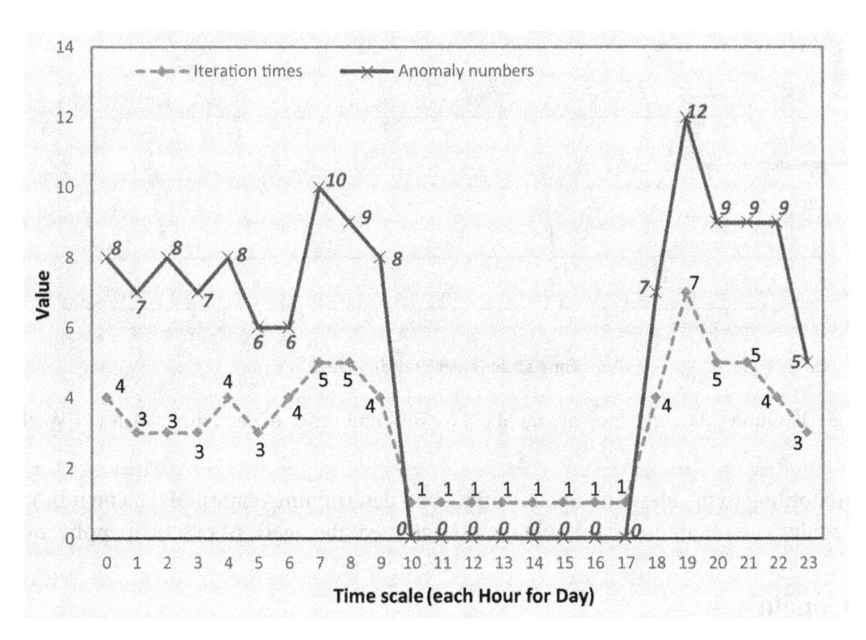

Fig. 4. Iteration times and anomaly numbers for different series of each time when s = Day.

On the "day" scale, iterative anomaly detection is performed on the corresponding sequence of each time (24 h) by "week". The number of iterations and the total number of anomalies are shown in Fig. 5.

Comparing with Fig. 4 and Fig. 5, it can be seen that with the rolling up of time scale, some non-significant anomalies are diluted, and the number of anomalies and the number of iterative detection are correspondingly reduced, such as 1–3 and 20–23 periods. In addition, under different scale conditions, although the number of anomalies is different, but the trend is roughly the same, so that people can basically determine whether there is an anomaly at a certain time or period, for example, there is an anomaly at 8 and 18, and there is less possibility of an anomaly at 11–17.

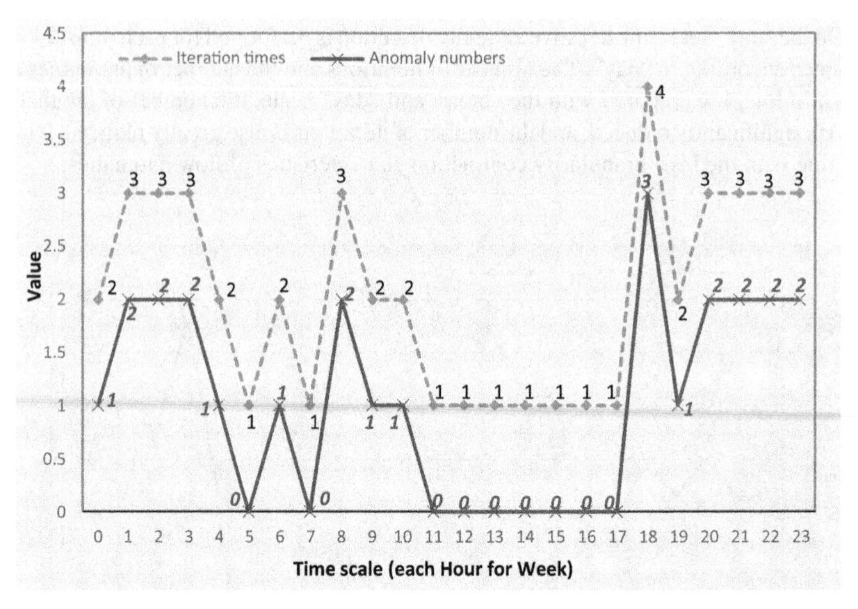

Fig. 5. Iteration times and anomaly numbers for different series of each time when s = Week.

According to the above analysis, mining and determining abnormal data on multiple time scales can obtain more reliable results and meet the needs of practical application.

5 Conclusions

This paper designs a multi-chain queue structure based algorithm for the detection of multi-scale dynamic stream anomaly. The algorithm can detect simultaneously the anomalies in the data sequence at a certain time and at each time under a given scale, and maintain the ability of anomaly detection when the scale conditions change, to realize the detection of multi-scale dynamic flow anomalies. The algorithm guarantees the sequence of stream data through the composite chain queue, and only processes it once, while the increase of new data and the elimination of historical data are efficiently realized through the queue in and out operations. In addition, the algorithm controls the memory usage by limiting the size of the composite chain queue storage structure, so as to ensure the ability of the algorithm to process massive stream data. Experimental research and efficiency analysis show that the proposed algorithm can detect anomalies in different scales, and then obtain more meaningful anomalies in multi-scale.

Further research work includes the control of total amount flow data storage in queued, and the anomaly measurement of high-dimensional flow data, and the reduction of false detection rate.

Acknowledgement. The work was supported in part by the Natural Science Foundation of Fujian Province, China (No. 2020H0043) and the Science and Technology Project of State Grid Fujian Electric Power Company (No. 5213001800LF).

References

1. Wang, H., Bah, M.J., Hammad, M.: Progress in outlier detection techniques: a survey. IEEE Access **7**, 107964–108000 (2019)
2. Kontaki, M., Gounaris, A., Papadopoulos, A.N., Tsichlas, K., Manolopoulos, Y.: Efficient and flexible algorithms for monitoring distance-based outliers over data streams. Inf. Syst. **55**, 37–53 (2016)
3. Zhang, L., Lin, J., Karim, R.: Sliding window-based fault detection from high-dimensional data streams. IEEE Trans. Syst. Man Cybern. Syst. **47**(2), 289–303 (2017)
4. Lin, L., Su, J.: Anomaly detection method for sensor network data streams based on sliding window sampling and optimized clustering. Safety Sci. **118**, 70–75 (2019)
5. Qiu, Y., Chang, X., Qiu, Q., Peng, C., Su, S.: Stream data anomaly detection method based on long short-term memory network and sliding window. J. Comput. Appl. **40**(05), 1335–1339 (2020)
6. Yu, L., Li, Y., Zhu, S.: Anomaly detection algorithm based on high-dimensional data stream. Comput. Eng. **44**(01), 51–55 (2018)
7. Salehi, M., Leckie, C., Bezdek, J.C., Vaithianathan, T., Zhang, X.: Fast memory efficient local outlier detection in data streams. IEEE Trans. Knowl. Data Eng. **28**(12), 3246–3260 (2016)
8. Gao, J., Ji, W., Zhang, L., Li, A., Wang, Y., et al.: Cube-based incremental outlier detection for streaming computing. Inf. Sci. **517**, 361–376 (2020)
9. Kim, T., Park, C.H.: Anomaly pattern detection for streaming data. Expert Syst. Appl. **149**, 1–16 (2020)
10. Zhu, R., Ji, X., Yu, D., Tan, Z., Zhao, L., et al.: KNN-based approximate outlier detection algorithm over IoT streaming data. IEEE Access **8**, 42749–42759 (2020)
11. Ma, L., Wan, L., Ma, S., Yang, T.: Abnormal traffic identification method based on bag of words model clustering. Comput. Eng. **43**(05), 204–209 (2017)
12. Bouguelia, M.-R., Nowaczyk, S., Payberah, A.H.: An adaptive algorithm for anomaly and novelty detection in evolving data streams. Data Min. Knowl. Discov. **32**(6), 1597–1633 (2018)
13. Alnafessah, A., Casale, G.: TRACK-plus: optimizing artificial neural networks for hybrid anomaly detection in data streaming systems. IEEE Access **8**, 146613–146626 (2020)
14. Xu, L., Wang, J.: Data mining algorithm of abnormal network based on fuzzy neural network. Comput. Sci. **46**(04), 73–76 (2019)

Network Security Situational Awareness Model Based on Threat Intelligence

Hongbin Zhang[1,2], Yan Yin[1(✉)], Dongmei Zhao[2], Bin Liu[3,4], and Hongbin Gao[1]

[1] School of Information Science and Engineering, Hebei University of Science and Technology, Shijiazhuang 050018, China
2019114017@stu.hebust.edu.cn

[2] Hebei Key Laboratory of Network and Information Security, Hebei Normal University, Shijiazhuang 050024, Hebei, China

[3] School of Economics and Management, Hebei University of Science and Technology, Shijiazhuang 050000, China

[4] Research Center of Big Data and Social Computing, Hebei University of Science, Shijiazhuang 05000, China

Abstract. In order to deal with the problems that the increasing scale of the network in the real environment leads to the continuous high incidence of network attacks, the threat intelligence was applied to situational awareness, and the situational awareness model based on random game was constructed. Threat perception of the target system was performed by comparing the similarity between the exogenous threat intelligence and the internal security events of the system. At the same time, internal threat intelligence was generated based on the threat information inside the system. In this process, game theory was used to quantify the current network security situation of the system, evaluate the security status of the network. Finally, the prediction of the network security situation was realized. The experimental results show that the network security situation awareness method based on threat intelligence can reflect the changes in the network security situation and predict attack behaviors accurately.

Keywords: Situation awareness · Threat intelligence · Game theory · Network security · Nash equilibrium

1 Introduction

As the scale of the network and the number of users continue to increase, the network develops toward large-scale, multi-service and big data, and the architecture of the network becomes complex. In this context, network attacks are numerous and evolving, such as computer viruses, malware, and information leakage. Many large industries and enterprises actively advocate, built and applied situational awareness systems to deal with the severe challenges faced by network security. Network security situational awareness is an effective means to ensure network security. It has become the focus of network security research to use situation awareness to discover potential threats and respond. At present,

J. Xiong et al. (Eds.): MobiMedia 2021, LNICST 394, pp. 526–536, 2021.
https://doi.org/10.1007/978-3-030-89814-4_38

most of the proposed network security situation awareness technologies and methods are based on small-scale networks. With the continuous expansion of network scales and appearance of new advanced attack technologies such as Advanced Persistent Threat (APT) [1], the accuracy of current situation awareness technology and the maneuverability reduced greatly. APT utilizes advanced attack means to carry out long-term, sustained network attacks on a specified target, which are highly stealthy and harmful [2].

In recent years, the emergence of Cyber Threat Intelligence (CTI) has brought new ideas to the research of situation awareness and has become a new direction in the field of situation awareness. CTI describes attack behavior, provides contextual data about network attacks, and guides network attacks and defenses. APTs can be detected and prevented by using threat intelligence to collect large amounts of data and analyze malicious behavior through effective data sharing and ensuring the security and quality of information exchange [3].

However, the current research on CTI is still in the initial stage, and the relevant research results are few. The reasonable and standardized use of threat intelligence for situational awareness is a key issue that needs to be addressed urgently. To address this issue, this paper proposed a network security situation awareness model based on threat intelligence, which collects situation elements of network asset status, risk status, and log warnings. And then the collected data information was filtered, cleaned and correlated with exogenous threat intelligence. Finally, the processed data was used to quantify and predict the situation through the game process between attack and defense. Meanwhile, the internal threat information is compared with the situational prediction results to generate insider threat intelligence.

2 Related Work

Situational awareness refers to extracting the elements of the system in a given specific time and space, understanding their meaning and predicting their likely impact. ENDSLEY [4] divided situational awareness into three levels: situational element extraction, situational understanding and situational projection. Regarding the research on network security situational awareness (NSSA), BASS [5] first proposed the concept of network situational awareness, that is, in a large-scale network environment, the elements that can cause changes in the network situation are acquired, understood, evaluated, displayed, and the future forecast of development trends.

In the area of research on situational awareness models, literature [6] proposed a security situation awareness game model based on Stochastic C-Petri Net for the complex problems of IoT defense. The model dynamically considers the confrontation between offense and defense, finds potential offensive behaviors, and makes effective defenses. Literature [7] proposed a situational awareness model based on topological vulnerability analysis to calculate network security situational values and achieve situational awareness by acquiring and analyzing network security situational elements. Literature [8] proposed a Markov Multi-stage Transferable Belief Model for eXfiltration APT attacks, which combined the kill chain model with the attack tree to conduct situational awareness for complex multi-stage attacks. In addition, common situational awareness models also include Markov chain model [9], Bayesian network model [10], and support vector machine (SVM) model [11].

However, the current research on situational awareness is mostly based on small-scale networks. In the context of large-scale networks, situational awareness technology needs to introduce advanced technologies to achieve more comprehensive security analysis and situation prediction. Since threat intelligence is mainly obtained by using the collection method of big data, it can provide the most complete and up-to-date security incident data, and greatly improve the detection ability of new and advanced dangers in network security situational awareness work. Literature [12] studied the APT attack chain and selected DNS traffic as the original data for the overall APT detection. By using a variety of different detection features, this article combines the latest threat intelligence and big data technology, which has a certain significance for APT attack detection. Literature [13] designs a security threat intelligence sharing system with threat intelligence as the entry point. The threat intelligence data shared by the third party is used to evaluate the security situation of the power grid security and detect abnormal behaviors in time. Using threat intelligence to realize intrusion intention identification greatly improves the capability of system security situational awareness.

By analyzing existing situational awareness models, a situational awareness model based on threat intelligence is proposed, which is divided into three parts: situational perception, situational comprehension and situational projection. The situational awareness model is shown in Fig. 1.

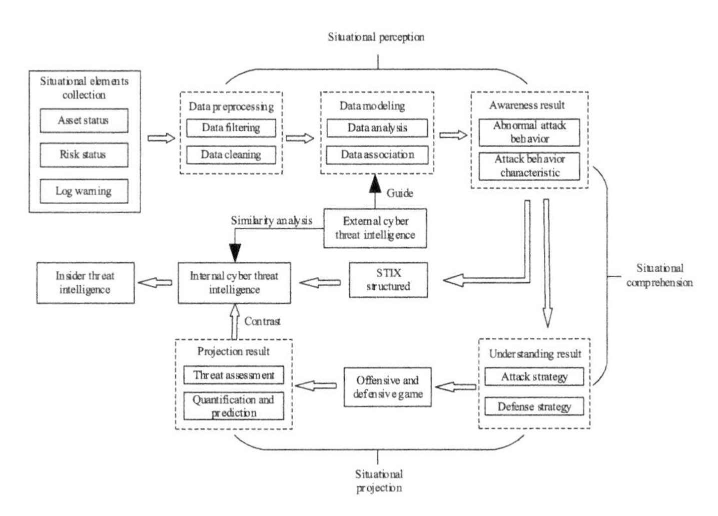

Fig. 1. Situational awareness model.

In the situation perception part, the internal security incidents of the system are analyzed and understood to detect the internal threats through the guidance of external threat intelligence. In the part of situation projection, the security situation value of the target system is quantified through the game process between the attacker and the defender, thus assessing the security state of the network. The Nash equilibrium existence theorem is used to predict the possible attack strategies of the attackers, and to urge the defenders to take appropriate measures in advance to protect the network from attacks.

3 Situation Perception Method Based on Similarity

This section proposes a threat information detection method based on the comparison of external threat intelligence and security event similarity. By comparing the similarity between them, it can judge whether the internal security incident is threatening or not, and discover the threat information inside the system. Finally, the situation prediction results are fed back to the internal threat information to determine the internal threat intelligence.

3.1 Definition of Threat Intelligence

Threat intelligence is defined by Gartner as: Threat intelligence is evidence knowledge about IT and information assets facing existing or brewing threats, including implementable context, mechanisms, indications, meaning, and executable recommendations. This paper divides threat intelligence into internal threat intelligence and external threat intelligence according to the source of threat intelligence.

External Cyber Threat Intelligence (ECTI). It is usually derived from the open source intelligence (OSINT) provided by the intelligence provider.

Internal Cyber Threat Intelligence (ICTI). Threat intelligence data generated by enterprises or organizations for the protection of internal information assets and business processes.

Table 1. The objects and their properties

Object	Threat properties	Description
Indicator	Name	The name used to identify the threat indicator
	Labels	Used to specify the type of indicator
	Pattern	Indicator detection mode
Attack Pattern	Description	Describe the details and context of the attack pattern
	Name	The name used to identify the attack pattern
Tool	Labels	Type of tool described
	Name	The name used to identify the tool
Vulnerability	CVE-id	Vulnerability identifier
Observed Data	Objects	The observed data in security events
	Frist-observed	The time window for observed data begins
	Last-observed	The time window for observed data ends

Before the preprocessing of threat intelligence data, it is necessary to unify the format of internal and external threat intelligence. In this paper, we structure ECTI with STIX

2.0 and select five types of objects, namely Indicator, Attack Pattern, Tool, Vulnerability, and Observed Data, as analysis elements. Each of these objects also contains numbers of threat attributes, these objects and attributes are used as the basis for ICTI collection to discover attack behaviors. The following Table 1 gives the description of the threat objects.

3.2 Situational Perception Method for Threats

The threat perception process is as follows: 1) Processing of attack data from the target system to identify the purpose, tools and impact of the attack; 2) Use STIX2.0 to extract the above-mentioned related objects and threat attributes to generate Security Incidents (SI); 3) Calculate the weight of SI to obtain information with threat value, which will be defined as internal threat information.

The security incident (SI) is 5 tuples corresponding to the above 5 objects, namely

$$SI = (Ind, AP, Tool, Vul, OD)$$

Where the weight of objects is represented by ω_i, the weights of all objects satisfy $\sum_{i=1}^{5} \omega_i = 1$. The weight of threat properties is represented by ω_{ij}. Weight value satisfy $\sum_{i=1,j=1}^{i,j} \omega_{ij} = 1$.

Calculate the frequency of occurrence of each attribute in each tuple in ECTI by Eq. (1).

$$P_{kj} = \frac{x_{kj}}{\sum_{k=1,j=1}^{k,=3,j=n} x_{kj}}, \mathrm{j} = 1, 2, 3, \ldots, \mathrm{n} \tag{1}$$

In the formula, x_{kj} represents the number of times that threat attributes appear in ECTI.

According to literature [14], Eq. (2) is used to obtain the relative superior value r_{ij}, and the weight $\omega_{ij}^{'}$ of each attribute in ECTI is calculated by Eq. (3):

$$r_{ij} = \frac{p_{ij} - \hat{j} p_{ij}}{\overset{\vee}{j} p_{ij} - \hat{j} p_{ij}} \tag{2}$$

In the formula, $\wedge$, $\vee$ denote small and large characters.

$$\omega_{ij}^{'} = \frac{\sum_{j=1}^{j=n} r_{kj}}{\sum r_{ij}} \tag{3}$$

Through the above calculation, the weight of the Indicator object can be obtained as $\omega_1^{'} = (\omega_{11}^{'}, \omega_{12}^{'}, \omega_{13}^{'})$. In the same way, the properties weights $\omega_{ij}^{'}$ of other tuples $(AP, Tool, Vul, OD)$ in SI can be obtained.

Through the collection and sorting of the internal security data, the weight ω_i of each tuple in the internal SI can be obtained, as well as the weight ω_{ij} of the properties under each tuple.

By comparing the difference between ω'_{ij} and ω_{ij}, the maximum difference is defined as Similarity.*sim* can be expressed by Eq. (4):

$$sim = \max\left\{ \left| \omega_{ij} - \omega'_{ij} \right| \middle| \omega_{ij} \in SI, \omega'_{ij} \in ECTI \right\} \tag{4}$$

Set the threshold value ε. When Sim $\leq$ ε, then SI can be defined as threat information for the source within the target system. After obtaining the threat information from the system, a game model is established for the offensive and defensive strategy, and the situation is predicted. After the prediction result is obtained, the threat information and the prediction result are compared and analyzed to determine whether they are consistent. If they are consistent, define the internal threat information as ICTI.

4 Offensive and Defensive Game Model

This paper uses stochastic game to model the process of network attack and defense, and the attack and defense process is described as a Network Security Awareness Model based on Stochastic Game (NSAM-SG).

4.1 Quantification Method of Security Situation Value

In the offensive and defensive game, no matter what strategy is adopted, costs and benefits will be generated. The difference between cost and benefit is defined as utility, and the utility is used to quantify the value of network security situation.

(1) **Attacker utility.** $U(\tau)_a$ is equal to the difference between the benefit and cost:

$$U(\tau)_a = profit_a - cost(\tau)_a + cost(\tau)_d \tag{5}$$

(2) **Defender utility.** According to the antagonistic relationship between the two players in the game, $U(\tau)_d$ is

$$U(\tau)_d = profit_d - cost(\tau)_d + cost(\tau)_a \tag{6}$$

(3) **Attack cost.** The cost of attack is directly proportional to the threat level (*TL*) of its corresponding threat intelligence. So

$$cost(\tau)_a = TL \tag{7}$$

According to literature [15], this paper divides threat levels into 5 categories, namely Root, User, Data, DoS and Probe, corresponding to 10, 5, 3, 2,0.5.

(4) **Defense cost.** Combined with the classification of defense categories in literature [16], defense costs are divided into four categories according to the complexity of defense measures: no defense measures, monitoring protection measures, prevent preventing measures and repair protection measures, the defense costs $cost(\tau)_d$ are 0,4,8,10 respectively.

(5) **Attacker profit.** Attack profit is related to the success rate β of threat attack and the harm *Impact* caused by Vulnerability. According to the CVSS rating standard, *Impact* is

$$Impact = \lambda \cdot VL \cdot [1 - (1 - Con_a) \cdot (1 - Int_a) \cdot (1 - Ava_a)] \tag{8}$$

Among them, λ is the correction factor; VL is the degree of difficulty of exploiting vulnerability, divided into four levels: critical, high, medium, and low. $CIA = (Con_a, Int_a, Ava_a)$ indicate the confidentiality, integrity, and availability hazards caused by the vulnerability to the system. Therefore, the attacker profit is

$$profit_a = \beta \cdot Impact \tag{9}$$

(6) **Defender profit.** The benefit of the defender is numerically equal to the benefit caused by the attack. Therefore, the defender profit is

$$profit_d = -profit_a \tag{10}$$

In conclusion, the security situation value S of the target network is

$$S = U(\tau)_d - U(\tau)_a \tag{11}$$

When $S > 0$, the current network is in a safe state, the larger the $|S|$, the safer the network. When $S < 0$, the current network is in a dangerous state.

4.2 Attack Prediction

In a game, Nash equilibrium means that the strategy chosen by either party to the game is optimal with respect to the strategy choices of the others. In the process of network offense and defense, both the attacker and the defender hope to reap the greatest benefits with the least cost of attack or defense. In this case, the attacker and defender will choose the best strategy based on each other's strategies. According to Nash equilibrium, NSAM-SG must have an equilibrium point. Therefore, we use the Nash equilibrium to predict aggressive behavior.

Suppose the attacker and the defender choose the attack strategy and the defense strategy according to the probability vectors $P_a = (x_1, x_2, x_3, \cdots, x_m)$ and $P_d = (y_1, y_2, y_3, \cdots, y_n)$ respectively.

The benefit expectations of attackers and defenders are defined as E_a, E_d:

$$E_a = \sum_{i=1}^{m} \sum_{j=1}^{n} x_i y_j U_a\left(T_a^i, T_d^j\right) \tag{12}$$

$$E_d = \sum_{i=1}^{m} \sum_{j=1}^{n} x_i y_j U_d\left(T_a^i, T_d^j\right) \tag{13}$$

The equilibrium utility expectations E_a and E_d of the mixed strategy $\left(x_i^*, y_j^*\right)$ are optimal, where $\left(x_i^*, y_j^*\right)$ satisfies:

$$\begin{cases} \forall x_i, \sum_{i=1}^{m} \sum_{j=1}^{n} x_i^* y_j^* U_a\left(T_a^i, T_d^j\right) \geq \sum_{i=1}^{m} \sum_{j=1}^{n} x_i y_j^* U_a\left(T_a^i, T_d^j\right) \\ \forall y_j, \sum_{i=1}^{m} \sum_{j=1}^{n} x_i^* y_j^* U_a\left(T_a^i, T_d^j\right) \geq \sum_{i=1}^{m} \sum_{j=1}^{n} x_i y_j^* U_a\left(T_a^i, T_d^j\right) \end{cases} \tag{14}$$

Among: The mixed strategy $x^{*}=(x_1^{*}, x_2^{*}, \cdots, x_m^{*})$ is the best choice strategy for the attacker, while the optimal defense strategy is $y^{*}=(x_1^{*}, y_2^{*}, \cdots, y_n^{*})$.

5 Experimental Results and Analysis

In the experiment, the CICIDS2017 intrusion detection data set was used to verify the ICTI generation method based on similarity and Nash equilibrium prediction method based on the NSAM-SG model. The experimental topology is shown in Fig. 2.

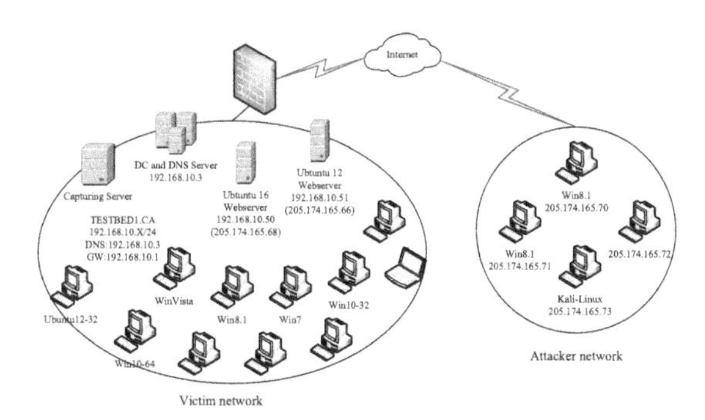

Fig. 2. Network topology.

This data set [17] contains benign traffic and the latest attack traffic. The data capturing period started at Monday 2017/7/3 and ended at Friday 2017/7/7. The implemented attacks include Brute Force FTP, Brute Force SSH, DoS, Heartbleed, Web Attack, Infiltration, Botnet and DDoS.

The experiment selects CAPEC-24, CAPEC-47, CAPEC-185, and CAPEC-122 as ECTI to guide the detection of insider threats. Analyze the attack behaviors in the data set and make it into attack information in STIX2.0 format, and then analyze the similarity between the attack information to determine whether the information is endogenous threat information.

The attribute weight of each tuple in ECTI can be calculated as follows:

$\omega_1' = (0.25, 0.42, 0.33)$, $\omega_2' = (0.58, 0.42)$, $\omega_3' = (0.33, 0.67)$, $\omega_3' = (0.33, 0.67)$, $\omega_4' = (1)$, $\omega_5' = (0.35, 0.325, 0.325)$.

The weight of each tuple in the SI are:

$\omega_1 = (0.5, 0.33, 0.17)$, $\omega_2 = (0.44, 0.56)$, $\omega_2 = (0.44, 0.56)$, $\omega_2 = (0.44, 0.56)$, $\omega_3 = (0.43, 0.57)$, $\omega_4 = (1)$, $\omega_5 = (0.5, 0.25, 0.25)$.

The similarity between SI and ECTI is $\mathrm{Sim} = (0.25)$. Set the fuzzy threshold $\varepsilon = 0.5$. Since $\mathrm{Sim} \leq \varepsilon$, the SI has threat value and can be used as the threat information in the system, which is then compared and analyzed with the result of the situation prediction.

According to the predicted results, the most likely attack strategy of the attacker is to use the DoS vulnerability to launch an attack. Therefore, the Tool, Vulnerability object and its properties in SI can be defined as the internal threat intelligence of the target system.

According to NSAM-SG model, the security situation value of data capture cycle is quantified. Through calculation, the security situation value of the entire network environment is obtained, and the situation change curve is shown in Fig. 3.

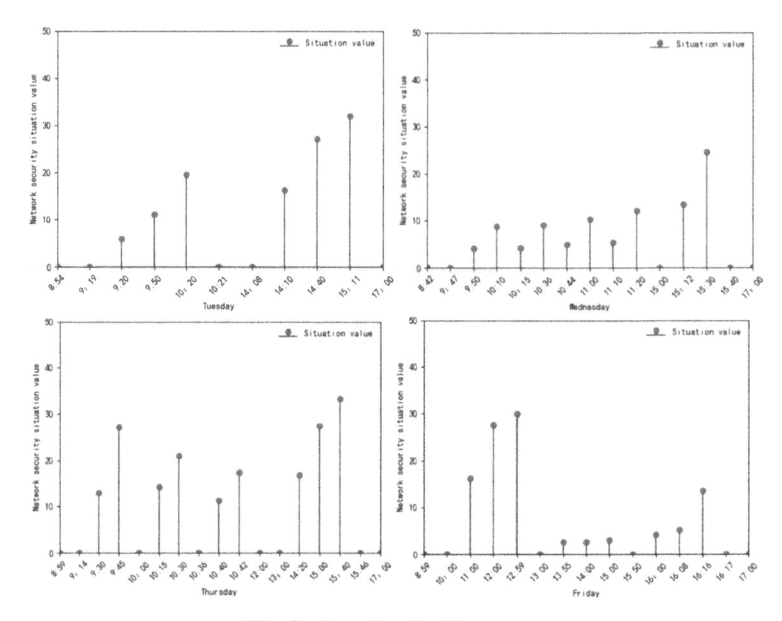

Fig. 3. Security situation curve.

It can be seen from the figure that as the attack progresses, the situation value shows an upward trend, so the danger faced by the entire network further deepens. Different attack modes lead to different system situation values. When the network is attacked by port scanning, the network has the lowest security posture value and the network is least affected. While the network is attacked by SSH brute-force or penetration attack, the network has a higher security posture value and the network is more severely impacted.

Finally, according to Brute force, DoS, SQL injection, XSS and Heartbleed vulnerabilities in the CICIDS2017 data set, as well as the corresponding defense measures, the payoff matrix P of both attacking and defending parties is calculated as:

$$P = \begin{bmatrix} 4.58,-4.58 & 6.41,-6.41 & 14.27,-14.27 & 18.89,-18.89 & 34.98,-34.98 \\ 8.15,-8.15 & 4.58,-4.58 & 23.93,-23.93 & 26.44,-26.44 & 27.98,-27.98 \\ 20.51,-20.51 & 1.83,-1.83 & 27.35,-27.35 & 24.55,-24.55 & 29.74,-29.74 \\ 29.38,-29.38 & 16.32,-16.32 & 4.59,-4.59 & 12.84,-12.84 & 34.98,-34.98 \\ 37.78,-37.78 & 18.65,-18.65 & 27.34,-27.34 & 4.12,-4.12 & 31.48,-31.48 \\ 17.09,-17.09 & 17.49,-17.49 & 12.23,-12.23 & 3.60,-3.60 & 34.98,-34.98 \\ 20.38,20.38 & 20.99,-20.99 & 37.78,-37.78 & 30.22,-30.22 & 2.50,-2.50 \end{bmatrix}$$

Calculating the Nash Equilibrium to obtain a mixed strategy probability distribution for both sides:$x^* = (0, 0.71, 0, 0.12, 0.17)$, $y^* = (0, 0, 0, 0.13, 0.22, 0, 0.47)$.

According to the prediction, the attacker's most likely strategy is to exploit the DoS vulnerability.

According to literature [18], DoS/DDoS attacks accounted for 36% of all samples in CICIDS2017. So, the attacker will most likely to launch the DoS attack next. Therefore, this paper proposes the method of Nash equilibrium for situation prediction is feasible.

6 Conclusion

This paper proposes a network situational awareness model based on threat intelligence, which uses threat intelligence as the basis for situational awareness to conduct situational perception and discover internal threat. By modeling the offensive and defensive game, the network security status is evaluated. Finally, the Nash equilibrium is used to predict the attack behavior. Through experimental analysis, NSAM-SG is well placed to detect unknown attacks and to accurately assess and predict the security situational of the network. The next work is to apply the proposed method to the real network environment and improve the shortcomings of the experiment. It will also introduce predicted attack strategy probabilities in the ICTI generation process to further standardize the generation of ICTI.

Acknowledgements. This research was supported by the National Natural Science Foundation of China under Grant No.61672206, No.61572170, S&T Program of Hebei under Grant No.18210109D, No.20310701D, No.16210312D, High-level Talents Subsidy Project in Hebei Province under Grant No.A2016002015, Technological Innovation Fund Project of Technological Small and Medium-sized Enterprises of Shijiazhuang under Grant No.9SCX01006, S&T research and development Program of Shijiazhuang under Grant No.191130591A.

References

1. Zhang, Q., Li, H., Hu, J.: A study on security framework against advanced persistent threat. In: IEEE International Conference on Electronics Information and Emergency Communication, pp. 128–131. IEEE, Macau (2017)
2. Cinar, C., Alkan, M., Dorterler, M., Dogru, I.A.: A study on advanced persistent threat. In: 2018 3rd International Conference on Computer Science and Engineering (UBMK) 2018, pp. 116–121. IEEE, Sarajevo (2018)
3. Li, Y., Dai, W., Bai, J., Gan, X., Wang, J., Wang, X.: An intelligence-driven security-aware defense mechanism for advanced persistent threats. IEEE Trans. Inf. Forensics Secur. **14**(01), 646–661 (2019)
4. Endsley, M.R.: Toward a theory of situation awareness in dynamic systems. Hum. Factors **37**(1), 32–64 (1995)
5. Bass, T.: Intrusion detection systems and multisensor data fusion: creating cyberspace situational awareness. Commun. ACM **43**(4), 99–105 (2000)
6. He, F., Zhang, Y., Liu, H.: SCPN-based game model for security situational awareness in the Intenet of Things. In: 2018 IEEE Conference on Communications and Network Security (CNS), pp. 1–5. IEEE, Beijing (2018)
7. Li, T.F., Li, Q., Yu, X.: Network security situation awareness model based on topology vulnerability analysis. Comput. Appl. **38**(S2), 157–163+169 (2018)

8. Ioannou, G., Louvieris, P., Clewley, N.: A markov multi-phase transferable belief model for cyber situational awareness. IEEE Access **7**, 39305–39320 (2019)
9. Salfinger, A.: Framing situation prediction as a sequence prediction problem: a situation evolution model based on continuous-time markov chains. In: 22nd International Conference on Information Fusion (FUSION). IEEE, Ottawa (2019)
10. Lin, P., Chen, Y.: Dynamic network security situation prediction based on bayesian attack graph and big data. In: 2018 IEEE 4th Information Technology and Mechatronics Engineering Conference (ITOEC). IEEE, Chongqing (2018)
11. He, Y.M.: Assessment model of network security situation based on K Nearest Neighbor and Support Vector Machine. Comput. Eng. Appl. **49**(09), 81–84 (2013)
12. Li, J.T.: APT Detection research based on DNS traffic and threat intelligence. Shanghai Jiaotong Univ. (2016)
13. Li, W.J., Jin, Q.Q., Guo, J.: Research on Security Situation Awareness and Intrusion Intention Recognition Based on Threat Intelligence Sharing. Comput. Modern. **2017**(03), 65–70 (2017)
14. Zhang, H.B., Yi, Y.Z., Wang, J.S., Cao, N., Duan, Q.: Network security situation awareness framework based on threat intelligence. Comput. Mater. Continua **56**(3), 381–399 (2018)
15. Lippmann, R.P., Fried, D.J., Zissman, M.A.: Evaluating intrusion detection systems: the 1998 DARPA Off-line intrusion detection evaluation. Comput. Netw. **34**(4), 579–595 (2000)
16. Xi, R.R., Yun, X.C., Zhang, Y.Z., Hao, Z.Y.: An improved quantitative evaluation method for network security. Chin. J. Comput. **38**(04), 749–758 (2015)
17. Canadian Institute for Cybersecurity. Intrusion Detection Evaluation Dataset (CIC-IDS2017) (2021). http://www.unb.ca/cic/datasets/ids-html
18. Zhao, D.: Research and implementation of construction and detection methods of virtual attack and real attack chains for feint attacks. Beijing Jiaotong Univ. (2019)

Research Progress and Future Trend Analysis of Network Security Situational Awareness

Junwei Zhang[1(✉)], Huamin Feng[2], Biao Liu[2], Ge Ge[3], and Jing Liu[4]

[1] School of Cyber Engineering, Xidian University, Xian 710126, China
zhangjunwei@stu.xidian.edu.cn
[2] Beijing Electronic Science and Technology Institute, Beijing 100070, China
{fenghm,liubiao}@besti.edu.cn
[3] National Administration of State Secrets Protection, Beijing 100031, China
[4] School of Cyberspace Security, Beijing University of Posts and Telecommunications, Beijing 100876, China
liudingjing@bupt.edu.cn

Abstract. With the continuous expansion of the network scale in recent years, network security problems have become increasingly prominent, and network security incidents have emerged one after another. Network security situation awareness is an essential part of network security defense that allows cybersecurity operators to cope with the complexity of today's networks and threat landscape. In this paper, we thoroughly review and systematize the origin and the models of network security situational awareness and the evolution of its definition, and then we give its definition. Additionally, we introduced the key technologies in this field from the three functional modules of network security situation extraction, network security situation assessment, and network security situation prediction, and analyzed their advantages and disadvantages. Last but not least, we explicitly propose four possible research directions that the researchers in network security can work on in the future.

Keywords: Network security · Situation awareness · Situation assessment · Situation prediction · Artificial intelligence

1 Introduction

Cyberspace has become the fifth national security domain outside the sea, land, air, and sky. Cyberspace security has become an important part of national security. However, with the rapid development of the scale of cyberspace, the problem of network security is becoming increasingly serious. Globally, there have been many serious cybersecurity incidents in recent years. such as the extortion outbreak in May 2017, "WannaCry", by encrypting data information in the system, make originally the data is not available, the opportunity to extort money a lot, "WannaCry" virus spread across nearly 150 countries and regions, including education, transportation, medical, energy networks, many industries are major attack by the virus. In April 2020, EDP, a Portuguese multinational

J. Xiong et al. (Eds.): MobiMedia 2021, LNICST 394, pp. 537–549, 2021.
https://doi.org/10.1007/978-3-030-89814-4_39

energy company, was attacked by ransomware. After being attacked, the attacker claimed to have obtained 10TB of EDP's sensitive data files and finally extorted a ransom of 1,580 bitcoins (equivalent to about 9.9 million euros).

For common vicious network security incidents, most network administrators are hindsight, that is, the event caused a certain impact was noticed, therefore, how to do active monitoring and active defense before the threat comes, try to avoid or reduce the occurrence of network security incidents, network security managers are very urgent needs [1]. In this paper, the origin and definition of network security situational awareness are summarized, and the technical methods of functional modules are introduced, analyzed, and compared. The research trend of network security situational awareness in the next few years and the challenges that researchers may face are proposed.

2 Definition and Development of Network Security Situational Awareness

In 1988, Endsley [2] proposed the concept of situational awareness for the first time at the International Human Factor Annual Conference, that is, "to recognize and understand environmental factors within a certain time and space, and to predict the future development trend".

Endsley's definition of situational awareness [3] has been widely accepted and applied to a variety of functional areas. He understands situational awareness as a state of knowledge and distinguishes it from the process used to achieve such a state by dividing it into three levels: situational element extraction, situational understanding, and situational prediction, as shown in Fig. 1.

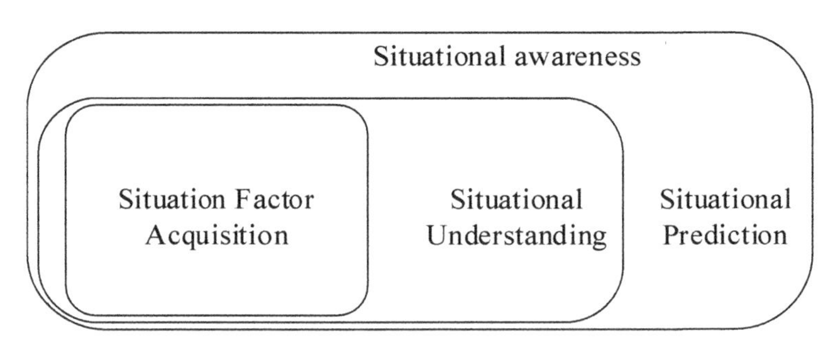

Fig. 1. Endsley situational awareness model.

First is the situation factor extraction, the main work of this level is to obtain the necessary data; the second level, situational understanding, is to analyze the data obtained from the first level. Finally, the situation prediction is made [4]. The data analysis results obtained at the second level are used to predict the situation in the short term in the future. This is also the earliest situational awareness model, which is the foundation of the network security situational awareness model.

A more classic situational awareness model is the data fusion model given by THE U.S. military JDL (Joint Directors of Laboratories), also known as the JDL model [5].

In this model, situational awareness is divided into five stages. In this model, situational awareness is divided into five stages. Followed by data preprocessing, event extraction, situation assessment, impact assessment, resource management, process control, and optimization. The main task is to monitor and evaluate the entire data fusion process in real-time, and integrate information at various levels to optimize related resources [6].

In 1999, the United States air force communications and information center Tim Bass, put forward the network space situational awareness (cyberspace situational awareness, CSA) concept [7], for the first time the concept of situational awareness is academia fusion in the field of network space safety, can effectively improve the cognition of managers to protect the network aims to shorten the time of the network security management decisions and provide the corresponding decision.

The situational awareness in the network applications mainly revolves around safety, Tim Bass in 2000 intrusion detection framework based on multisensory (see Fig. 2) [8]. The model is the prototype of network security situational awareness, reasoning framework consists of intrusion detection, intruder identity recognition, intrusion behavior, situation assessment, and threat assessment, etc.

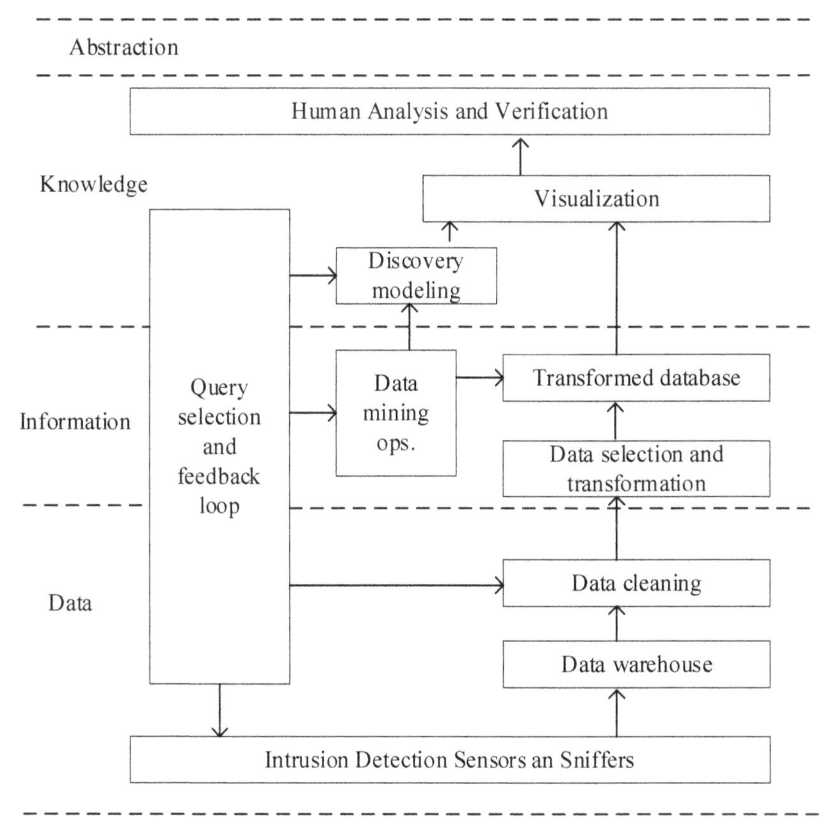

Fig. 2. Intrusion detection data fusion model.

In 2006, Wang Hui Qiang [9] discussed the concept of cybersecurity situational awareness, thinking that it refers to "in a large-scale network environment, acquiring, understanding, displaying, and predicting future development trends of security elements that can cause changes in the network situation", this definition is a Chinese translation defined by Endsley.

In 2007, Lai Ji Bao, Wang Hui Qiang et al. [10] proposed a network security situational awareness model based on Netflow. Using Netflow technology can well realize network security situation awareness, discover potential threats and vulnerabilities promptly, and present them to decision-makers in a visual manner, to achieve the purpose of comprehensive monitoring of the entire network. At the same time, because the system is dealing with massive amounts of data and information, performance optimization issues need to be further studied.

In 2009, Wei Yong et al. [11] proposed a network security situational awareness model based on information fusion. Introduce the improved DS evidence theory to fuse information from multiple data sources, and then use vulnerability information and service information to calculate the network security situation through situation element fusion and node situation fusion, and perform time-series analysis to achieve a quantitative analysis of the network security situation and trend forecasting.

In 2011, Jia Yan et al. [12] proposed a security situational awareness model for large-scale networks because of the characteristics of massive, multi-mode, and multi-granularity data in large-scale networks.

In 2014, Franke U [13] regarded network situational awareness as a subset of situational awareness, that is to say, network situational awareness is a part of situational awareness, which refers to the "network" environment. But that definition is a bit too vague and doesn't specify whether it's situational awareness for security.

In 2017, Gong Jian [14] put forward the network security situational awareness is the cognitive process of the network system security status, including from the system to measure the raw data fusion processing step by step and the background of the implementation of the system state and activity of semantic extraction, identify the existence of all kinds of network activity and the intention of abnormal activity, thus obtained according to the characterization of the network security situation and the trend of network system impacts normal behavior.

In 2019, Jia Yan and others [15] proposed the definition of network security situation awareness as the detection, extraction, understanding, evaluation, and future prediction of security elements that affect the network situation in a large-scale network environment.

With the improvement of application security requirements and technical difficulties, in recent years, academic research on network security situational awareness has become more and more common and in-depth. However, at present, a unified and comprehensive definition of network security situational awareness has not yet been formed, and most of them are correct. A detailed explanation of Endsley's definition of situational awareness. In this article, network security situation awareness is defined as the extraction of the characteristic elements that affect the network security situation in a complex network environment, and the necessary fusion and classification of the extracted characteristic elements, and then the use of technical methods for evaluation and analysis, and finally a series of complex processes for predicting the network security situation in the future based on the evaluation results.

3 Key Technologies of Network Security Situational Awareness

Although there are still some problems in the division of several stages of network security situational awareness by different researchers and the understanding of the relationship between different stages, most researchers divide network security situational awareness into three functional modules of situation element extraction, situation assessment, and situation prediction. This chapter introduces the key technologies of network security situational awareness in turn according to the classification of functional modules.

3.1 Key Technologies of Network Security Situation Feature Elements Extraction

Network security posture characteristic element extraction in the underlying network security situational awareness, is the foundation of network security situational awareness and security features elements mainly include static configuration of network information and dynamic information and include the information of network topology, the former vulnerability information, and status information, etc., the latter refers to the various protective measures of log collection and analysis techniques for the threat of information, etc.

When the researchers collected information, the foreign researchers mostly from a single factor analysis, specific elements of the corresponding specific data information to assess the security situation of specific, such as Jajodia [16] and Wang [17] and others study is only gathering network vulnerability information, evaluation by collecting the information of the network vulnerability, Ning [18, 19] only gathering network alarm information, analyze the status of the alarm information to evaluate the network threat; Barford [20] et al. used the data and information about the attack collected by Honeynet to evaluate the attack situation of the network. The common point of these studies is that they all collect, analyze and study a specific network element, and only obtain single situation information, which cannot obtain comprehensive information and then analyze the overall situation, and cannot adapt to the complex and changeable network environment.

Domestic researchers, on the other hand, from multi-source data information acquisition, starting from multiple layers, multiple Angle comprehensive assessment of network security situation, such as Wang Juan [21] is put forward based on the index system of network security situational awareness, extraction of multi-source information security data, according to the requirements of hierarchy, information source and the difference between structures, the layered index model, the extracted 25 candidate index, the index information assessment of network security situation; Wang et al. [22] proposed a botnet detection technology based on information fusion to effectively integrate the complex network security information of different sensors in time and space dimensions to improve the perception ability of botnet attacks. There is a lot of research is geared to the needs of the extraction of multi-source heterogeneous information network security work [23–26], Chang Yiheng and others proposed a security situation element extraction meth-od based on probabilistic neural network, which solved the problem of low efficiency and low accuracy of situation element extraction in a complex network environment. Multisource and redundant data interference for safety information [27]. Duan Yongcheng proposed a network security situation factor extraction method based on information gain random forest, which greatly improves the accuracy of situation factor extraction [28]. These studies use different technologies to collect and collate multi-source security data.

In conclusion, the foreign researchers mostly focus on the single factor extraction and analysis of the domestic researchers tend to the extraction of multi-source elements, because the characteristic of the network security situational awareness is a fusion of a variety of network information to consider the overall situation of network security, therefore, the elements of a multi-source extraction are inevitable, but the multi-source data and information not only reduces the extraction efficiency, desultorily data also brings to the information fusion and redundant processing difficulty, at present the optimization of the extraction method has a lot of space.

3.2 Key Technologies for Network Security Situation Assessment

Network security situation assessment is the core part of network security situation perception. Based on the fusion of all kinds of security equipment data and according to the needs of network security assessment, an assessment value of the current network security situation is obtained through formal reasoning calculation with the help of some mathematical model. In short, network security situation assessment is a mapping from situation factor to situation result value [29]. The network security situation assessment methods can be divided into three categories: mathematical model-based, knowledge-based reasoning, and pattern recognition.

Situation assessment method based on the mathematical model is the most common and most common method of analytic hierarchy process, the domestic has the value of the discloser is Chen Xiu Zhen and others [30] in 2006 put forward the hierarchical network system security threat situation of the quantitative evaluation model, this model can be divided into network system from top to bottom, host, service, and attack/holes 4 levels, as shown in Fig. 3, taking the evaluation of overall "after" top-down, local first strategy, and the model is based on IDS mass alarm information and network performance indicators, and the importance of the service, the host itself and the organizational structure of the network system combining.

The model exists some deficiencies: only IDS alert information in its assessment method a safe source of information in the actual network system deployment, such as firewall, system log safety factors are indispensable, if not the information included in the calculation, it loses the network security situation assessment technology can comprehensively reflect the advantage of the network security situation. Therefore, in 2009, 2010, and 2012 respectively, Lai Ji Bao [31], Zhang Yong [32], and Meng Jin [33] all improved the above hierarchical model, making the effect of more sources of hierarchical evaluation more accurate. In 2015, Jia et al. [34] proposed a layered framework for network security situation assessment. The framework can reflect the

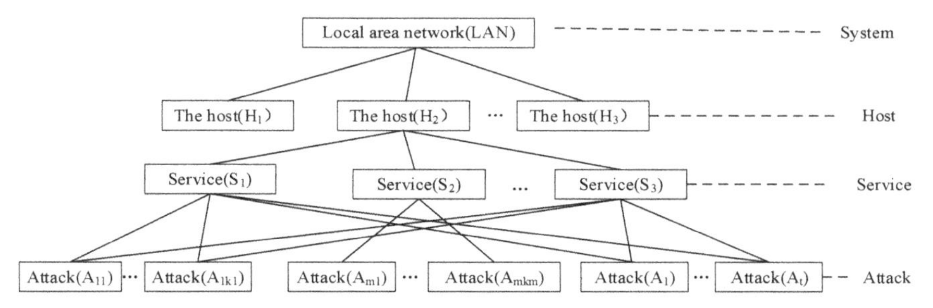

Fig. 3. Hierarchical network system security threat situation quantitative assessment model.

security status of information systems, but the disadvantage is that the framework is suitable for offline analysis, but it is not well suited for real-time analysis.

The main characteristic of the knowledge-based inference method is to rely on expert knowledge and experience in the process of constructing an evaluation model and then analyze network security situations through logical reasoning. More common are fuzzy logic reasoning, Bayesian reasoning, evidence theory, and so on.

Xie et al. [35] used a Bayesian network to model the uncertain factors in the network, calculate the probability of a successful attack, and evaluate the severity of the attack in real-time. Aguilar et al. [36] is a combination of fuzzy logic and neural network technology, base on the cognitive map presents the FCM (fuzzy cognitive map, the fuzzy cognitive map), the concept of using it to get important assets in a network dependence of damage assessment, fuzzy reasoning too difficult, however, and the figure of storage cost is big, not suitable for large, complex network environment. Boyer et al. [37] designed a situation assessment framework based on DS evidence theory to quantify network security situation. Li et al. [38] introduced a Bayesian network-based evidence network to carry out network security situation assessment, and the main idea is to carry out a similar reasoning assessment under the framework of evidence theory and with the full probability formula of Bayesian network as a reference. Yang Hao et al. [39] obtained the network vulnerability situation value by integrating vulnerability data and alarm data through DS evidence theory. The network security situation assessment method based on knowledge reasoning has a certain artificial intelligence, but the difficulty of obtaining inference rules and prior knowledge is the bottleneck of this method. Although the advantage of the evidence theory is that the required prior data is easy to obtain and can integrate different expert knowledge and data source information, it is also undesirable to have too high computational complexity when evidence conflicts.

The pattern recognition method establishes a situation template through machine learning and divides situations through pattern matching and mapping. More advanced than knowledge reasoning, it does not rely too much on expert knowledge and experience. The main methods include the grey correlation method, rough set theory, and cluster analysis method. Many researchers [40–45] have adopted grey correlation analysis, rough set theory, and cluster analysis to carry out network security situation assessment and achieved good results. The evaluation method of pattern recognition has the advantages of high efficiency, large processing capacity, and not relying too much on expert knowledge. The disadvantage is that the stage of pattern extraction is difficult to face more complex features, thus affecting the evaluation efficiency.

3.3 Key Technologies for Network Security Situation Prediction

The ultimate purpose of an evaluation is to predict and use historical data information to provide a management basis for future network security, which is the transformation from passive to active network security management. Network security situation prediction is the highest level of the whole situation perception and plays an important role in the defense of network security [46]. At present, research on network security situation prediction methods can be roughly divided into three categories: machine learning, Markov model [47], and gray theory.

Thanks to the improvement of hardware computing speed, machine learning methods based on neural networks and deep learning have developed rapidly in recent years. In the field of network security situational awareness, the automatic perception and self-learning mechanism of machines can be established to fit the thinking ability and analysis, and judgment ability of experts, to predict complex network security events more flexibly [48]. Lin et al. proposed a network security situation prediction based on BP neural network, and Tang [49] proposed a network security situation prediction method based on dynamic covariance BP neural network. The disadvantage of the BP neural network is its slow convergence speed, ease to fall into the local optimal solution, and ease to oscillate in the learning process. In addition to BP neural network, Zhang et al. [50] established a parametric optimized wavelet neural network security situation prediction model by using an improved niche genetic algorithm to improve the prediction accuracy of the network security situation. Feng et al. [51] proposed a network security situation prediction method based on cyclic neural networks. Ren Wei et al. [52, 53] proposed a situation prediction method based on RBF neural network by taking advantage of the characteristics of network security situation values with nonlinear time series and the advantages of neural network in dealing with chaotic and nonlinear data. Compared with the neural network, support vector machine (SVM) has a faster convergence speed, Hu, et al. [54] proposed a model of network security situation prediction based on graphs and SVM, and put forward by Lu and others [55] network security situation prediction based on support vector machine (SVM), is to use different methods to determine the optimal parameters of support vector machine (SVM), improve the prediction precision and shorten the training time.

Wang et al. [56] proposed a network security situation prediction method based on a fuzzy Markov chain and established a unified information base based on multi-source log data mining technology. Wen Zhi Cheng [57] proposed a prediction method based on the hidden Markov model. Liang et al. [58] proposed an algorithm based on weighted HMM to predict the security of mobile networks.

Lai Ji Bao [59] proposed network security situation prediction based on simple weighting and gray theory and established a prediction model based on gray theory. Zhang et al. [60] also carried out network security situation prediction by moving the grey correlation model and grey prediction algorithm. Deng Yong Jie et al. [61] proposed to combine neural networks with gray theory to predict network security situation, which also obtained good results.

Each forecasting technique has its advantages and limitations. Machine learning has excellent self-learning and adaptive capabilities, which can provide high convergence speed and strong fault tolerance. However, sufficient training data is needed to obtain parameters, and it is difficult to build neurons with self-learning and adaptive capabilities. For the Markov model, although it can perform various time-series predictions, it still needs a set of training data. In addition, it is almost impossible to recognize all possible states and their transitions, especially in complex networks. Grey theory can provide a small sample of data in the short-term prediction, thus providing better prediction without any training.

4 Summary and Prospect

According to the above analysis and summary, network security situational awareness started late, and many technologies are still immature and need to be further optimized and strengthened. The following is a discussion of the development trend of network security situational awareness:

First, big data analysis and processing technology. The extraction and preprocessing of network security situation elements are the most basic part of network security situation perception. The reality is that the network environment is becoming more and more complex, and the data types and formats are growing exponentially. The massive security information cannot be directly used as the analysis object of network security situation perception. Therefore, the application of big data analysis and processing technology in the extraction of network security situation elements will be the most important research in the future.

Second, the deep integration of artificial intelligence technology and network security situational awareness. The fourth part of the article introduces in detail the key technology of network security situational awareness of each function module, it is not hard to see, artificial intelligence, machine learning, researchers have become important methods in the aspect of network security situational awareness, but there are obvious flaws, artificial intelligence technology is in the rapid development phase, a new generation of artificial intelligence technology with the depth of the situational awareness can bring new vitality for the field, to solve the problems of the situation awareness at all levels to provide new methods and inspiration.

Thirdly, the visualization research of network security situational awareness. The ultimate goal of scientific research is applied. The application of cybersecurity situational awareness cannot only have some data. It needs a more direct way to express the deeper meaning of these data. Therefore, visualization is an indispensable part. However, in the process of reading related documents, it is found that there are little researches on visualization, so the visualization of network security situational awareness is also an important direction for future research.

Fourth, new problems arising from the expansion of the application scope. With the rapid development of big data and 5G, industrial control network is deeply integrated with the Internet. The new network pattern will inevitably bring new network security problems, and the application of network security situational awareness in complex network scenes will also be the focus of future research.

To summarize, network security situational awareness of research in the phase of development, there are a lot of not forming the theory of issues that need to be perfect, there are many key technologies that need to optimize modified, a new pattern of the network brings new security issues, new application scenario requires new methods of technology, network security situational awareness will exert its advantages to provide security for network security, national security escort.

Acknowledgments. This work was supported by the "High-precision" Discipline Construction Project of Beijing Universities (No. 20210071Z0403).

References

1. Gutzwiller, R., Dykstra, J., Payne, B.: Gaps and opportunities in situational awareness for cybersecurity. Digital Threats: Res. Pract. **1**(3), 1–6 (2020)
2. Endsley, M.R.: Design and evaluation for situation awareness enhancement. Hum. Fact. Soc. Annu. Meet. **32**, 97–101 (1988)
3. Endsley, M.R.: Situation awareness global assessment technique (SAGAT). In: Proceedings of the IEEE 1988 National Aerospace and Electronics Conference, Dayton, OH, USA, pp. 789–795. IEEE (1988)
4. Husák, M., Jirsík, T., Yang, S.J.: SoK: contemporary issues and challenges to enable cyber situational awareness for network security. In: Proceedings of the 15th International Conference on Availability, Reliability and Security, Virtual Event, Ireland, pp. 2. Association for Computing Machinery (2020)
5. Giacobe, N.: Application of the JDL data fusion process model for cyber security. Proc. SPIE Int. Soc. Opt. Eng. **7710**, 5 (2010)
6. Ao, Z.G.: Cyberspace operations situational awareness. In: Cyberspace operations: mechanism and planning, pp. 691–699. Publishing House of Electronics Industry, Beijing (2018)
7. Bass, T.: A glimpse into the future of ID. Login:: the magazine of USENIX & SAGE, vol. 24, pp. 40–45 (1999)
8. Bass, T.: Intrusion detection systems and multisensor data fusion. Commun. ACM **43**(4), 99–105 (2000)
9. Wang, H.: Survey of network situation awareness system. Comp. Sci. **33**, 5–10 (2006)
10. Lai, J., Wang, H., Jin, S.: Research on network security situational awareness system based on NetFlow. Comp. Appl. Res. **24**(08), 173–175 (2007)
11. Wei, Y., Lian, Y., Feng, D.: Network security situation assessment model based on information fusion. Comp. Res. Develop. **46**(3), 353–362 (2009)
12. Jia, Y., Wang, X., Han, W., Li, A., Cheng, W.: YHSSAS: security situational awareness system for large-scale networks. Comp. Sci. **38**(002), 4–8 (2011)
13. Franke, U., Brynielsson, J.: Cyber situational awareness – a systematic review of the literature - ScienceDirect. Comp. Secur. **46**, 18–31 (2014)
14. Gong, J., Zang, X., Su, Q., Hu, X., Xu, J.: Overview of network security situational awareness. J. Softw. **28**(4), 1010–1026 (2017)
15. Jia, Y., Han, W., Yang, H.: Research status and development trend of network security situational awareness. J. Guangzhou Univ. **3**, 1–10 (2019)
16. Jajodia, S.: Topological Analysis of Network Attack Vulnerability. ACM (2006)
17. Wang, L., Singhal, A., Jajodia, S.: Toward Measuring Network Security Using Attack Graphs, vol. 49. ACM (2007)

18. Pan, N.: Techniques and tools for analyzing intrusion alerts. ACM Trans. Inf. Syst. Secur. **7**(2), 274–318 (2004)
19. Pan, N., Xu, D.: Alert correlation through triggering events and common resources. In: Proceedings of the Computer Security Applications Conference, 2004, 20th Annual, pp. 360–369. IEEE Computer Society (2004)
20. Barford, P., Yan, C., Goyal, A., Li, Z., Paxson, V., Yegneswaran, V.: Employing honeynets for network situational awareness. Adv. Inf. Secur. **46**(1), 71–102 (2010)
21. Wang, J., Zhang, F., Fu, C., Chen, L.: Research on index system in network situational awareness. Comput. Appl. **27**(008), 1907–1909, 1912 (2007)
22. Hailong, W., Gong, Z.: Heterogeneous multi-sensor information fusion model for botnet detection. In: Proceedings of the 2010 International Conference on Intelligent Computation Technology and Automation, pp. 428–431 (2010)
23. Liu, X., Wang, H., Cao, B.: Network security situation awareness model based on multi-source fusion. J. PLA Univ. Sci. Technol. (2012)
24. Wu, H., Hu, A., Song, Y., Bu, N., Jia, X.: A new intrusion detection feature extraction method based on complex network theory. In: Proceedings of the 2012 Fourth International Conference on Multimedia Information Networking and Security, pp. 852–856 (2012)
25. Tsang, C., Kwong, S.: Multi-agent intrusion detection system in industrial network using ant colony clustering approach and unsupervised feature extraction. In: Proceedings of the 2005 IEEE International Conference on Industrial Technology, pp. 51–56 (2005)
26. Lai, J., Wang, H., Zheng, F., Feng, G.: Network security situation element extraction method based on DSIMC and EWDS. Comput. Sci. **37**(011), 64–69 (2010)
27. Chang, Y., Ma, Z., Li, X., Gong, D.: Security situation element extraction based on probabilistic neural network. Cybersp. Secur. **11**(128(10)), 60–65 (2020)
28. Duan, Y., Li, X., Yang, X., Yang, L.: Network Security Situation Factor Extraction Based on Random Forest of Information Gain (2019)
29. Zhang, J.: Research on some key technologies of network security situation assessment. Doctor's degree, National University of Defense Technology (2013)
30. Chen, X., Zheng, Q., Guan, X., Lin, C.: Hierarchical network security threat situation quantitative assessment method. J. Softw. **17**(004), 885–897 (2006)
31. Lai, J.: Research on several key technologies of network security situational awareness based on heterogeneous sensors. Doctor's degree, Harbin Engineering University (2009)
32. Zhang, Y.: Research and system implementation of network security situational awareness model. Doctor's degree, University of Science and Technology of China (2010)
33. Meng, J.: Research on key technologies of network security situation assessment and forecast. Doctor's degree, Nanjing University of Science and Technology (2012)
34. Jia, Y., Wu, H., Jiang, D.: A hierarchical framework of security situation assessment for information system. In: Proceedings of the 2015 International Conference on Cyber-Enabled Distributed Computing and Knowledge Discovery, pp. 23–28 (2015)
35. Peng, X., Li, J.H., Ou, X., Peng, L., Levy, R.: Using Bayesian networks for cyber security analysis. In: Proceedings of the 2010 IEEE/IFIP International Conference on Dependable Systems and Networks, DSN 2010, Chicago, USA. IEEE (2010)
36. Szwed, P., Skrzynski, P.: A new lightweight method for security risk assessment based on fuzzy cognitive maps. Int. J. Appl. Math. Comp. Sci. **24**(1), 213–225 (2014)
37. Boyer, S., Dain, O., Cunningham, R.: Stellar: a fusion system for scenario construction and security risk assessment. In: Proceedings of the Third IEEE International Workshop on Information Assurance (IWIA 2005), pp. 105–116 (2005)
38. Li, X., Deng, X., Jiang, W.: A novel method of network security situation assessment based on evidential network. In: Chen, X., Yan, H., Yan, Q., Zhang, X. (eds.) ML4CS 2020. LNCS, vol. 12486, pp. 530–539. Springer, Cham (2020). https://doi.org/10.1007/978-3-030-62223-7_46

39. Yang, H., Xie, X., Li, Z., Zhang, L.: Simulation of network security situation estimation model under multiple intrusion environment. Comp. Simulat. **033**(006), 270–273 (2016)
40. Zhao, G., Wang, H., Wang, J.: Research on survivability situation assessment of network based on grey relational analysis. Small Microcomp. Syst. **10**, 1861–1864 (2006)
41. Wang, C.: Assessment of network security situation based on grey relational analysis and support vector machine. Appl. Res. Comp. (2013)
42. Zhuo, Y., He, M., Gong, Z.: Rough set analysis model for network situation assessment. Comp. Eng. Sci. **34**(3), 1–5 (2012)
43. Li, X., Li, X., Zhao, Z.: Combining deep learning with rough set analysis: a model of cyberspace situational awareness. In: Proceedings of the 2016 6th International Conference on Electronics Information and Emergency Communication (ICEIEC), pp. 182–185 (2016)
44. Xiao, C., Qiao, Y., He, H., Li, J.: Multi-level fuzzy situation assessment based on optimal clustering criteria. Comp. Appl. Res. **30**(4), 1011–1014 (2013)
45. Wen, Z., Chen, Z., Tang, J.: Network security situation assessment method based on cluster analysis. J. Shanghai Jiaotong Univ. (Chin. Ed.) **50**(9), 1407–1414 (2016)
46. Leau, Y.-B., Manickam, S.: Network security situation prediction: a review and discussion. In: Intan, R., Chi, C.-H., Palit, H.N., Santoso, L.W. (eds.) ICSIIT 2015. CCIS, vol. 516, pp. 424–435. Springer, Heidelberg (2015). https://doi.org/10.1007/978-3-662-46742-8_39
47. Ioannou, G., Louvieris, P., Clewley, N.: A Markov multi-phase transferable belief model for cyber situational awareness. IEEE Access **7**, 39305–39320 (2019)
48. Lin, Z., Chen, G., Guo, W., Liu, Y.: PSO-BPNN-based prediction of network security situation. In: Proceedings of the 2008 3rd International Conference on Innovative Computing Information and Control, p. 37 (2008)
49. Tang, C., Yi, X., Qiang, B., Xin, W., Zhang, R.: Security situation prediction based on dynamic BP neural with covariance. Procedia Eng. **15**, 3313–3317 (2011)
50. Zhang, H., Huang, Q., Li, F., Zhu, J.: A network security situation prediction model based on wavelet neural network with optimized parameters. Digital Commun. Netw. 139–144 (2016)
51. Feng, W., Fan, Y., Wu, Y.: A new method for the prediction of network security situations based on recurrent neural network with gated recurrent unit. Int. J. Intell. Comput. Cybernet. **13**(1), 25–39 (2018)
52. Ren, W., Jiang, W., Jiang, X., Sun, Y.: Network security situation prediction method based on RBF neural network. Comp. Eng. Appl. **42**(31), 136–138, 144 (2006)
53. Jiang, Y., Li, C., Yu, L., Bao, B.: On network security situation prediction based on RBF neural network. In: Proceedings of the 36th China Control Conference (2017)
54. Hu, J., Ma, D., Liu, C., Shi, Z., Yan, H., Hu, C.: Network security situation prediction based on MR-SVM. IEEE Access **7**, 130937–130945 (2019)
55. Lu, H., Zhang, G., Shen, Y.: Cyber security situation prediction model based on GWO-SVM. In: Barolli, L., Xhafa, F., Hussain, O.K. (eds.) IMIS 2019. AISC, vol. 994, pp. 162–171. Springer, Cham (2020). https://doi.org/10.1007/978-3-030-22263-5_16
56. Wang, Y., Li, W., Liu, Y.: A forecast method for network security situation based on fuzzy Markov chain. In: Huang, Y.-M., Chao, H.-C., Deng, D.-J., Park, J.J.H. (eds.) Advanced Technologies, Embedded and Multimedia for Human-centric Computing. LNEE, vol. 260, pp. 953–962. Springer, Dordrecht (2014). https://doi.org/10.1007/978-94-007-7262-5_108
57. Wen, Z., Chen, Z.: Network security situation prediction method based on hidden Markov model. J. Cent. South Univ. **46**(10), 137–143 (2015)
58. Liang, W., Long, J., Chen, Z.: A security situation prediction algorithm based on HMM in mobile network. Wirel. Commun. Mob. Comput. **2018**, 5380481 (2018)
59. Lai, J., Wang, H., Liang, W., Zhu, L.: Study of network security situation awareness model based on simple additive weight and grey theory. In: Proceedings of the 2006 International Conference on Computational Intelligence and Security, pp. 1545–1548 (2006)

60. Zhang, F., Wang, J., Qin, Z.: Using gray model for the evaluation index and forecast of network security situation. In: Proceedings of the 2009 International Conference on Communications, Circuits and Systems, pp. 309–313 (2009)
61. Deng, Y., Wen, Z., Jiang, X.: Network security situation prediction method based on grey. Theory **2**, 69–73 (2015)

An Improved DDoS Attack Detection Model Based on Unsupervised Learning in Smart Grid

Zhili Ma[1,2], Hongzhong Ma[1], Xiang Gao[2], Jiyang Gai[3], Xuejun Zhang[3](✉), Fucun He[3], and Jinxiong Zhao[1]

[1] State Grid Gansu Electric Power Research Institute, Lanzhou 730070, Gansu, China
[2] State Grid Gansu Electric Power Company, Lanzhou 730000, Gansu, China
[3] School of Electronic and Information Engineering, Lanzhou Jiaotong University, Lanzhou 730070, Gansu, China
xuejunzhang@mail.lzjtu.cn

Abstract. The bidirectional communication system in smart grid is vulnerable to distributed denial of service (DDoS) attacks, due to its characteristics of complex system structure and difficult to control. The multiple nodes in the smart grid system will be compromised when the DDoS attack happen, thus resulting in the denial of legitimate services to users and disruption of the normal operation in power grid system. In order to defense such attack, some detection methods have been proposed in recent years. However, most of the existing detection methods have the characteristics of low detection accuracy and high false positive rate. In this paper, we proposed a novel DDoS attack detection method which only uses unlabeled abnormal network traffic data to build the detection model. Our method firstly uses Balanced Iterative Reducing and Clustering Using Hierarchies algorithm (BIRCH) to pre-cluster the abnormal network traffic data, and then explores autoencoder (AE) to build the detection model in an unsupervised manner based on the clustering subsets. In order to verify the performance of our method, we perform experiments on KDDCUP99 dataset and compare our method with existing classical anomaly detection methods. Results show that the proposed method has higher detection accuracy for abnormal traffic detection.

Keywords: Smart grid · DDoS attack detection · Autoencoder · BIRCH algorithm · Unsupervised learning

Supported by the NSFC project (grant no. 61762058, and no. 61861024), the Science and Technology project of Gansu Province (grant no. 20JR5RA404) and the Science and Technology project of State Grid Gansu Electric Power Research Institute (grant no. 52272219100P).

J. Xiong et al. (Eds.): MobiMedia 2021, LNICST 394, pp. 550–562, 2021.
https://doi.org/10.1007/978-3-030-89814-4_40

1 Introduction

As the most widely distributed and complex power Internet of Things (IoT) system, smart grid combines the existing power network and the modern information network to provide users with more novel services, such as bidirectional communication system, remote controlling of smart home appliances, updating of consumer behavior and stability tracking of the power grid. These services effectively respond to the energy demand and distribution of power services. However, more and more heterogeneous terminals are generated while improving the efficiency of the smart grid, which results in a wider exposure of the grid systems. So the vulnerability of smart grids to security threats has also increased. In particular, the bidirectional communication and software-oriented nature of the smart grid makes it highly vulnerable to cyberattacks, which not only affect the normal operation of the grid system, but also disturb social stability and cause property damage. Distributed Denial of Services (DDoS) attack is one of the major security threats to smart grid, in which attackers always use multiple puppet machines to attack targets at the same time. The DDoS attack seriously affects the continuity and availability of smart grid. Thus, proactive defense, one of the most promising methods to enhance power system network security [22], is becoming more and more important, which calculates the ultimate benefits of hackers and defenders under different conditions based on the constructed model, then predicts possible attack behaviors and evaluate the best defense strategy for the power system. Therefore, it is critical to proactively and accurately detect multiple nodes in smart grid system at the same time for guaranteeing the safe operation of the grid system.

Recently, the anomaly-based detection methods are the most common ways to solve problems in intrusion detection systems, such as the machine learning-based detection methods [12] which have made ideal achievements in network security, privacy protection and so on. Machine learning-based detection methods usually include supervised learning-based methods and unsupervised learning-based methods. The attack detection methods based on supervised learning is essentially a kind of classification method, which can achieve expected detection accuracy when being provided well-labeled datasets. These methods require a large number of labeled samples which are difficult to obtain or high cost of acquisition.

Based on the existing research, this paper proposes a DDoS attack detection method based on unsupervised learning to detect abnormal traffic of the communication network in smart grid. In this paper, we firstly use BIRCH algorithm [11] to pre-cluster abnormal network traffic in an unsupervised way to obtain different patterns of clustering subsets. Then, The resulted clustering subsets are respectively input into the AE for model training, through the process of "encoding-decoding" to reconstruct the input data and obtain the average reconstruction error (the training loss of the model). Then the obtained reconstruction error is used as detection threshold to detect the normal traffic and attack traffic in the network.

The main contributions of our work are summarized as follows:

1) In this paper, we use autoencoder to train the normal traffic and abnormal traffic, and build the threshold-based abnormal traffic detection models respectively. Experimental results show that the detection accuracy using abnormal traffic to build detection model is higher than that of using normal traffic to build detection model. Therefore, it is innovative to build the anomaly detection model only using anomaly traffic data.

2) Because the training data plays an important role on the detection performance of the threshold-based anomaly detection method. Thus, we use clustering algorithm BIRCH to pre-cluster the original abnormal traffic data in a way of unsupervised learning, and the data with similar patterns can be clustered to get different clustering subsets. Experiments in this paper show that the detection model built with pre-clustered datasets can achieve higher detection accuracy.

The rest of our paper is organized as follows. Section 2 mainly introduces the machine learning based methods for DDoS attacks detection in the communication network of smart grid. Section 3 introduces the structure of the architecture of the proposed abnormal traffic detection model in this paper. Section 4 mainly presents the experimental results and analysis of the proposed method on different datasets and the comparison with previous methods. Section 5 concludes our work and discussion of future research.

2 Related Work

In this section, we make a brief introduction of some existing DDoS attack detection methods in smart grid, such as the mainstream machine learning-based detection methods and threshold-based network anomaly detection methods. Most of the previous machine learning-based methods used shallow learning or the combination of linear and nonlinear to achieve ideal detection results. For example, Rohan et al. [5] built detection models based on machine learning algorithm K-nearest neighbors (KNN), Random Forests (RF) and Decision tree (DT) to detect abnormal network traffic in IoT environments. Aamir et al. [2] used supervised machine learning techniques to classify DDoS attacks, including random forests (RF), k-neighbors (KNN) and support vector machines (SVM). Zekriet al. [21] classified the DDoS attacks using the Naive-Bayes and Decision-Tree in cloud computing environments in real time. The threshold-based network traffic detection method achieves the data training by discovering the correlation between the data features, and reconstructs the data using the method of mapping the data to the optimal subspace. In the process of forming a subspace, the data with large reconstruction error are identified as abnormal samples which are different from normal samples. The classical Principal Component Analysis (PCA) method is a threshold-based anomaly detection algorithm. For example, Anukool Lakhina et al. [9] used PCA algorithms to separate the high-dimensional space of network traffic data into non-intersecting subspaces corresponding to normal and abnormal network conditions, and performed anomaly detection on that subspaces. Paffenroth et al. [13] used robust principal component analysis (RPCA) to detect anomalies, the method achieved low FPR on individual packets and further supported the hypothesis that the low dimensional subspace computed by RPCA is more representative of normal data. However, in the actual network environment, the majority of features of traffic data are nonlinear, and the PCA algorithm which has a process of linear transformation cannot capture the nonlinear relationship between features. Yang et al. [19] and Zhu et al. [23] have used reconstruction error in different ways to detect anomalies in network traffic. The Particle swarm optimization (PSO-BP) algorithm [20] based on BP and the OS-ELM [10] algorithm based on online extreme learning machine use reconstruction error as threshold for anomaly detection. Wei et al. [6] propose a novel DDoS attack real-time defense mechanism based on deep Q-learning network (DQN), which dynamically adjusts the service resource according to the current operating state of the system and ensures the response rate of normal service requests. PARK et al. [14] use stack noise reduction auto-encoder technology to achieve feature dimensional reduction and capture nonlinear correlation between features, and the results show that the anomaly detection method based on auto-encoder is superior than other traditional methods. Yang et al. [18] input normal traffic into AE, using minimal reconstruction error as the detection threshold to detect whether the traffic data is normal, but which cannot provide the specific type of attack, which is very important for defenders to adopt defense measures. Chen et al. [3] introduce a detection method based on unsupervised outliers, which shows that it achieves good detection accuracy. But their method integrates multi-autoencoders to a single model to detect anomaly samples, hence, the process of model training and anomaly detection has a large time overhead.

Through the analysis of the existing DDoS attack detection methods based on unsupervised learning in communication network, it can be found that the AE has been used for network anomaly detection research. But most studies used AE to learn the high-dimensional characteristics of training data and get its low-dimensional features, then using supervised learning algorithms to achieve the detection of abnormal traffic or normal network traffic data to build an anomaly detection model based on AE. In this paper, we perform experiments and find that the detection model based on abnormal network traffic has better detection performance than that obtained based on normal traffic data. Thus we propose an anomaly detection method built based on abnormal network traffic, and perform experiments on benchmark network intrusion detection dataset KDDCUP99. Results show that the model trained with the clustering subsets can achieve better performance in terms of detection accuracy.

3 Proposed Method

In this section, we present the architecture of our DDoS attack detection method as shown in Fig. 1.

As can be seen from Fig. 1, our method is an unsupervised learning-based anomaly detection framework and mainly consists of data preprocess, detection model training and detection phase. Firstly, the abnormal network traffic datasets need to be normalized and standardized, then the processed data will be input into the BIRCH algorithm and achieve the unsupervised pre-clustering

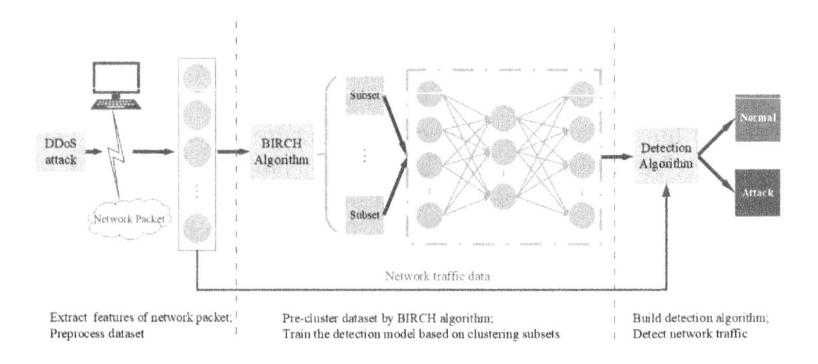

Fig. 1. Architecture of the proposed detection model in this paper.

of data using cluster feature tree (CFT). In the phase of model training, the clustering subsets are input into AE for the model training, and the minimized reconstruction error of the train data is used as the detection threshold for subsequent anomaly detection. In the detection phase, we input the test data into the trained model to get its output, then we calculate the reconstruction error between the output and its input. If the reconstruction error is higher than the preset detection threshold, the test data is marked as normal traffic data. Otherwise, it is marked as abnormal traffic data.

3.1 Data Pre-processing

In this paper, we use the open benchmark intrusion detection datasets KDD-CUP99 to perform experiments. Because the original dataset contains multiple types of data features, it directly affects the calculation of clustering features and the generation of clustering feature trees. Therefore, in our proposed method, we firstly use Min-Max techniques [8] to normalize the numerical data in the dataset and transform the raw data linearly, then we map the normalized data to [0, 1]. The conversion function is expressed as equation (1):

$$x\prime = \frac{x - min(x)}{max(x) - min(x)} \tag{1}$$

3.2 Pre-classify Dataset Using BIRCH Algorithm

As we know, the network traffic data generated in different environments will produce different distribution features. Thus, the inherent change of the features of data will inevitably affect the performance of the detection model. That is, the threshold obtained by using unbalanced network traffic data with complex data patterns and distribution features could affect detection accuracy rate for detecting attack traffic. Therefore, we use the BIRCH algorithm to pre-cluster the original dataset firstly, by which the data with similar patterns will be clustered into clustering subsets. As a result, the clustering subsets are used as train data to build detection models, which can avoid the influence of unbalanced training data on the performance of the detection model.

The important characteristic of the BIRCH algorithm is that it can accomplish high-quality clustering of large datasets with limited memory resources. In addition, BIRCH algorithm uses cluster features (CF) to summarize a cluster and cluster feature trees (CF-trees) to represent clustered hierarchies, which enables clustering methods achieve higher speed and greater scalability for operating to large databases. At the same time, this method is also very effective for incremental dynamic clustering. Therefore, it is always used in many fields in recent years.

Clustering features [16] can not only effectively reduce the storage space of data, but also efficiently calculate all the indicators in the BIRCH algorithm that form clustering decisions.

Given a cluster that contains N d-dimensional data: $\{X_i\}$, in which i = 1 ,2, ..., N, then the following indicators can be defined:

$$D = \sqrt{\frac{\sum_{i}^{N} \sum_{j}^{N} (X_i - X_0)^2}{N(N-1)}} \tag{2}$$

In which X_0 is the center of the cluster, and the D is the average distance between the two clusters.

The BIRCH algorithm consists of four stages [17]. At the stage 1, all samples are read in turn, and a CF tree is created in memory to contain as much attribute information as possible. Stage 2 is an optional process in which memory occupancy is compressed to the desired range by scanning leaf nodes to reconstruct a smaller cluster feature tree. At the stage 3, all CF tuples are clustered using hierarchical agglomerative clustering algorithm, and a better clustering tree is obtained. In this part, formula (2) is used to calculate an accurate distance measure. Stage 4 is also an optional process which uses the center point generated by stage 3 as the seed, then scanning the original dataset again and mapping the data points to its nearest seed for better clustering results.

Based on the above theory, this paper uses BIRCH algorithm to pre-classify the dataset. By calculating the number of clusters using Davies-Bouldin, the pre-processed dataset is clustered into subsets. And the subsets are labeled for subsequent model training.

3.3 Determine Detection Threshold and Build Detection Model

Autoencoder [15] is an artificial neural network algorithm. The method proposed in this paper mainly utilizes the AE algorithm to learn the data features of the clustering subsets, and reconstructs the input data. And uses the processes of "encoding" and "decoding" to get the minimized reconstruction error of the input data. As a result, we can distinguish the normal traffic and abnormal traffic by using the obtained reconstruction error as the detection threshold.

As shown in Fig. 2. The AE model consists of encoding layer, hidden layer and decoding layer. The function of the encoding layer aims to encode the input

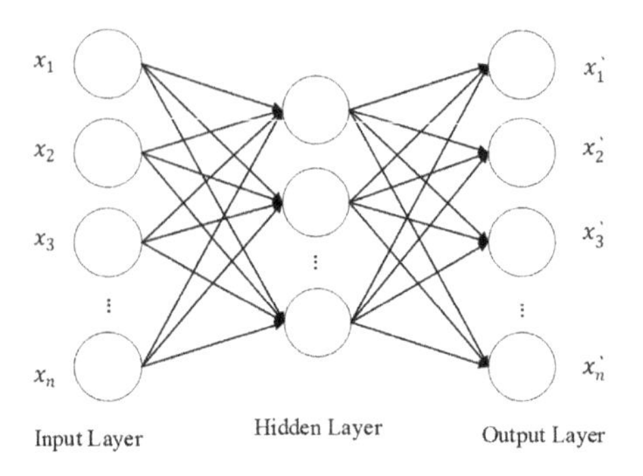

Fig. 2. The structure of the autoencoder.

data. The hidden layer aims to learn each feature of the input data. And the decoding layer aims to reconstruct the input data.

The encoding process is used to reduce the dimension of a high-dimensional features, compressing a given input data into a specified dimension which is equal to the number of neurons in the hidden layer, and the mapping of the input data x to the hidden layer is represented by h, which is expressed as (3):

$$h_i = f(x) = s(\sum_{j=1}^{d} W_{ij}^{input} x_j + b_i^{input}) \tag{3}$$

where x is the input vector, W is the weight matrix of the coding layer, b is the bias matrix, and s is a nonlinear activation function, which generally are sigmoid or Relu function, by contrast, the rule function has a better effect [7], so we use Relu function as the activation function in this experiment. It is expressed as (4):

$$f(x) = max(0, \ x) \tag{4}$$

decoding layer mainly aims to reconstruct the input data, and decode the low-dimensional data of the hidden layer to the size of the original vector space, which is regarded as an inverse process of encoding. The mapping function is (5):

$$h_i\prime = g(h) = s(\sum_{j=1}^{d} W_{ij}^{hiddern} h_j + b_i^{hidden}) \tag{5}$$

where s is the activation function of the decoder, and we use Relu as the activation function in this experiment.

The process of training the model is to calculate the minimized average reconstruction error between input data $\{x_1, x_2, ..., x_n\}$ and the reconstructed data $\{x_1\prime, x_2\prime, ..., x_n\prime\}$. Therefore, we use mean square error to calculate the error between the test sample and its reconstructed output in this paper. It is defined as (6):

$$\theta = L(x_i, x_i\prime) = \frac{1}{n}\sum_{i=1}^{n}(x_i - x_i\prime)^2 \tag{6}$$

The proposed method uses AE to learn the most important features of input data, and the reconstruction error is obtained through the process of "encoding-decoding". Then the reconstruction error is used as the detection threshold for subsequent anomaly detection. In the test process, the test data will be input into the trained model. Then we calculate the minimized reconstruction error between the data and its original input by mean square error function. If the reconstruction error is higher than the predetermined detection threshold, the test data is marked as normal traffic data. Otherwise, it is marked as abnormal traffic data.

4 Experimental Analysis and Results

In this section, we perform experiment on benchmark intrusion detection datasets KDDCUP99 to verify the performance of our method, and compare the results with the recently machine learning-based DDoS detection methods [2,5] including RF (Random Forest), NB (Naive-Bayes), DT (Decision-Tree), and LR (Logistic-Regression). The follows are results and details about all experiments.

4.1 Introduction of the Dataset

The KDDCUP99 dataset [1], a published security audit dataset from Columbia University's IDS Labs, contains 488,734 training data for 23 different types of attacks, each data has 38 features, including source IP, source port, destination IP, destination port, transaction protocol, status, duration and attack category. Since the experiments in this paper are based on normal traffic and abnormal traffic. Therefore, we select part of data labeled with normal and Dos from KDDCUP99 dataset as datasets for follow experiments.

4.2 The Results of the Experiment

In order to verify that the detection model built by abnormal traffic data can achieve better detect performance. We use the un-clustered traffic dataset labeled with Dos and the dataset labeled with Normal as the training data to build the detection model. As shown in Fig. 3, through experiment statistics we found that the minimum reconstruction error of normal traffic data is higher than the preset detection threshold, which means that if the reconstruction error of the test sample is higher than the preset detection threshold, the traffic data is marked as normal traffic data. Otherwise, it is marked as abnormal traffic.
At the same time, in order to verify the overall performance of the model, the Accuracy, Precision, Recall and F-score are used as evaluation indicators, and these values can be obtained by the follows: $Accuracy = \frac{TP+FN}{TP+TN+FN+FP}$, $Precision = \frac{TP}{FP+TP}$, $Recall = \frac{TP}{TP+FN}$, $F-score = 2 \cdot \frac{Precision \cdot Recall}{Precision+Recall}$, in which $TN(TrueNegative)$is used for normal traffic to be detected as normal traffic. FN (False Negative) is used for normal traffic to be detected as abnormal traffic. TP (True Positive) is used for abnormal traffic to be detected as abnormal traffic.

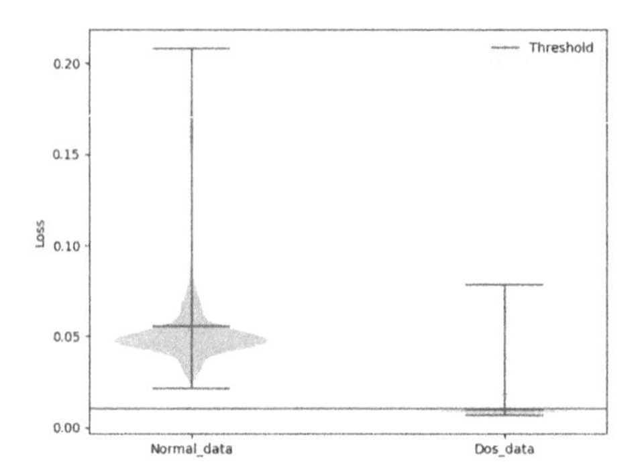

Fig. 3. Distribution of reconstruction errors for normal and Dos data

Table 1. Performance of detection models based on normal traffic and abnormal traffic.

Dataset	Accuracy (%)	Precision (%)	Recall (%)	F-score (%)
Dos	97.128	100	98.218	99.212
Normal	83.846	97.602	98.925	95.318

FP (False Positive) is used for abnormal traffic to be detected as normal traffic. In order to ensure the accuracy of the experimental results, for each method, we carry out ten experiments and take the average results of the ten experimental as the final results. As shown in Table 1, the results show that the detection model trained based on abnormal traffic data is more efficient, and the detection Accuracy, Precision, Recall and F-score are 97.128%, 100%, 98.218% and 99.212% respectively, the corresponding value obtained by the model based on normal traffic data are 83.846%, 97.602%, 98.925% and 95.318%. Especially the detection accuracy was 13.282% higher than that of obtained by the detection model based on normal traffic data. Therefore, it is obvious that the detection model obtained by the abnormal traffic data has higher detection performance.

Therefore, the detection model in this paper is built based on abnormal traffic data. In order to get the optimal clustering subsets, we determine the best number of cluster using DBI (Davies-Bouldin index). The index is also known as classification fitness indicator which is used to evaluate the merits of clustering algorithms [4]. Typically, the smaller the indicator, the better the clustering effect. The indicator is defined as:

$$DBI = \frac{1}{N}\sum_{i=1}^{N} max(\frac{s_i + s_j}{d_{ij}}),\ i \neq j \tag{7}$$

where s_i is the average distance from each point in cluster i to the cluster center, d_{ij} is the distance between the center of cluster i and the center of cluster j, and N is the number of clusters.

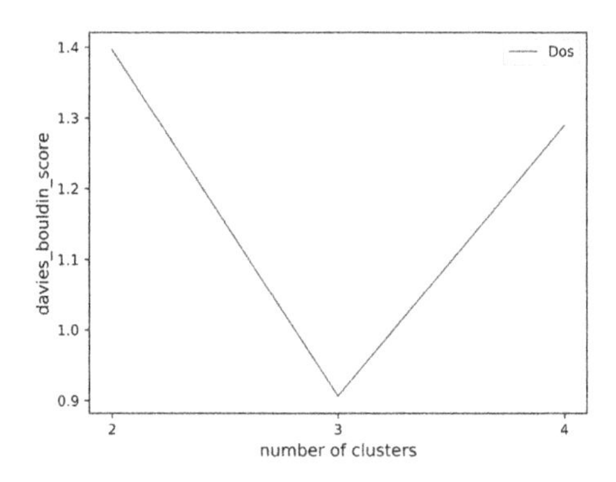

Fig. 4. Davies-Bouldin-Score

As shown in Fig. 4, we take different number of clusters to get the Davies-Bouldin score corresponding to them. We find that when the number of clusters was 3, the Davies-Bouldin score was the smallest (0.906), which brought the best clustering effect.

Table 2. Clustering results for the Dos dataset in KDDCUP99.

Dataset (Dos)	Data-size
Dos	102309
ID 0	100057
ID 1	2036
ID 2	216

The results are shown in Table 2. We use BIRCH algorithm to pre-cluster the original Dos traffic dataset in KDDCUP99, which categorizes original dataset from different data patterns and gets three subsets of clusters with different sample sizes, and the subsets were labeled ID 0 (100057 samples), ID 1 (2036 samples) and ID 2 (216 samples) respectively. Considering the comprehensive consideration, our method uses ID 0 as the train data to build the detection model.

In order to verify the overall performance of this method, this paper compared the proposed method with the recently machine learning-based DDoS attack detection methods on the KDDCUP99 dataset. In order to ensure the accuracy of the experimental results, based on each method, we carry out ten experiments and take the average results of the ten experimental as the final results. Table 3 shows the comparison of the indicators for each method, we can find the Accuracy, Precision and F-score of the proposed method are the highest, at 99.360%, 100% and 99.679% respectively. And it is the NB that obtain the highest Recall value of 99.994%, which is 0.666% higher than that of our

Table 3. The experimental results compared with the methods based on supervised learning.

Indicators	NB	RF	DT	LR	Proposed
Accuracy (%)	93.188	94.647	92.775	96.766	99.360
Precision (%)	93.007	94.942	92.953	95.949	100
Recall (%)	99.994	98.587	95.619	99.894	99.328
F-score (%)	95.776	96.648	94.018	97.671	99.679

method. Overall, our method is superior to the comparison method in terms of detection Accuracy, Precision and F-score.

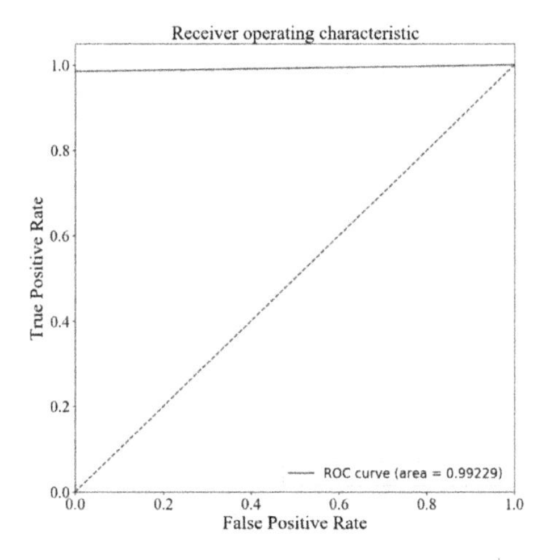

Fig. 5. ROC curve and AUC area

In order to visually evaluate the detection accuracy of the proposed method, we evaluate the classification effect of the method by ROC curve and AUC indicator. Based on the prediction results of the proposed algorithm, we calculate the values of TPR (True Positive Rate) and FPR (False Positive Rate) to form a ROC graph. The AUC is defined as the area under the ROC curve, with a range of values between 0.5 and 1. The reason to use AUC as the evaluation criterion is because in many cases the ROC curve cannot clearly explain which classifier is better. As a value, the larger of the corresponding AUC value, the better of the classifier effect. From Fig. 5, it can be known that the AUC of the proposed method in this paper reached to 0.99229, which shows that the method has a better classification effect.

Experiments show that the detection accuracy of abnormal traffic is obviously better than that of comparison method. This is because we use BIRCH cluster algorithm to pre-classify network traffic data with similar distribution features and obtain good clustered subsets. Based on the clustering subsets, we design a stacked autoencoder to train the detection model, which can learn the high-dimensional features of the pre-clustered subsets and get more suitable detection threshold. Thus, we achieve better detection performance.

5 Summary

This paper proposed an improved DDoS attack detection model based on unsupervised learning, the goal of which is to achieve more accurate and efficient DDoS attack detection in communication network of smart grid. Our method uses BIRCH algorithm to cluster network traffic data with similar features and utilizes greater cluster subset as training data. Then, we train the detection model using stacked AE in an unsupervised way, and use the average reconstruction error of train data (the training loss of the model) detection threshold. We perform experiments based on KDDCUP99 dataset, and compare the proposed method with recently machine learning-based DDoS detection methods, the results show that the proposed method is superior to the comparison method in detection performance.

In addition, the method proposed in this paper need to be further improved to adapt more extensive and complex smart grid environment. 1) In order to improve the generalization capability of the model, we will consider whether the BIRCH-AE based method is suitable for DDoS attack detection with various rates in different environment. 2) In order to enhance the robustness of the model, we consider building a real-time and adaptive attack detection framework to achieve efficient real-time monitoring of DDoS attacks in communication network of smart grid.

References

1. Dataset (1999). http://kdd.ics.uci.edu/databases/kddcup99/kddcup99.html
2. Aamir, M., Zaidi, S.: Clustering based semi-supervised machine learning for DDoS attack classification. J. King Saud Univ. Comput. Inf. Sci. **33**(4), 436–446 (2019)
3. Chen, J., Sathe, S., Aggarwal, C., Turaga, D.: Outlier detection with autoencoder ensembles. In: Proceedings of the 2017 SIAM International Conference on Data Mining (2017)
4. Davies, D.L., Bouldin, D.W.: A cluster separation measure. IEEE Trans. Pattern Anal. Mach. Intell. PAMI-*1*(2), 224–227 (1979)
5. Doshi, R., Apthorpe, N., Feamster, N.: Machine learning DDoS detection for consumer Internet of Things devices, pp. 29–35 (2018)
6. Feng, W., Wu, Y.: DDoS attack real-time defense mechanism using deep q-learning network. Int. J. Performability Eng. **16**(9), 1362–1373 (2020)
7. Glorot, X., Bordes, A., Bengio, Y.: Deep sparse rectifier neural networks. J. Mach. Learn. Res. **15**, 315–323 (2011)

8. Ihsan, Z., Idris, M.Y., Abdullah, A.H.: Attribute normalization techniques and performance of intrusion classifiers: a comparative analysis. Life Sci. J. **10**(4), 2568–2576 (2013)
9. Lakhina, A., Crovella, M., Diot, C.: Diagnosing network-wide traffic anomalies. Comput. Commun. Rev. **34**(4), 219–230 (2004)
10. Li, Y., Qiu, R., Jing, S., Li, D.: Intrusion detection system using online sequence extreme learning machine (OS-ELM) in advanced metering infrastructure of smart grid. PloS ONE **13**(2), e0192216 (2018)
11. Minxuan, Y.: Research and implementation of data mining system based on improved clustering algorithm. Ph.D. thesis, University of Electronic Science and Technology (2012)
12. Moustafa, N., Hu, J., Slay, J.: A holistic review of network anomaly detection systems: a comprehensive survey. J. Netw. Comput. Appl. **128**, 33–55 (2019)
13. Paffenroth, R., Kay, K., Servi, L.: Robust PCA for anomaly detection in cyber networks (2018)
14. Park, S., Seo, S., Kim, J.: Network intrusion detection using stacked denoising autoencoder. Adv. Sci. Lett. **23**(10), 9907–9911 (2017)
15. Spatiotemporal, E., Related, S., Context, A.I.: Unsupervised feature learning for audio classification using convolutional deep belief networks (2009)
16. Tao, Cui, X.: The research of high efficient data mining algorithms for massive data sets. Appl. Mech. Mater. **556–562**, 3901–3904 (2014)
17. Tian, Z., Ramakrishnan, R., Livny, M.: Birch: an efficient data clustering method for very large. ACM Sigmod Rec. **25**(2), 103–114 (1996)
18. Yang, K., Zhang, J., Xu, Y., Chao, J.: DDoS attacks detection with autoencoder. In: NOMS 2020–2020 IEEE/IFIP Network Operations and Management Symposium (2020)
19. Yang, S., Zhang, R., Nie, F., Li, X.: Unsupervised feature selection based on reconstruction error minimization. In: ICASSP 2019–2019 IEEE International Conference on Acoustics, Speech and Signal Processing (ICASSP) (2019)
20. Yu, S., Xu, L.: Research of intrusion detection based on PSO-BP algorithm. J. Shazhou Prof. Inst. Technol. (2018)
21. Zekri, M., Kafhali, S.E., Aboutabit, N., Saadi, Y.: DDoS attack detection using machine learning techniques in cloud computing environments. In: 2017 3rd International Conference of Cloud Computing Technologies and Applications (CloudTech) (2017)
22. Zhao, J., Zhang, X., Di, F., Guo, S., Mu, D.: Exploring the optimum proactive defense strategy for the power systems from an attack perspective. Secur. Commun. Netw. **2021**(6), 1–14 (2021)
23. Zhu, Q.H., Yang, Y.B.: Subspace clustering via seeking neighbors with minimum reconstruction error. Pattern Recogn. Lett. **115**(NOV.1), 66–73 (2017)

Intrusion Detection System Based on Improved Artificial Immune Algorithm

Jilin Wang(✉), Zhongdong Wu, and Guohua Wang

School of Electronic and Information Engineering, Lanzhou Jiaotong University, Lanzhou 730070, China
wangjilin515@163.com, {WUZHD,wangguohua}@mail.lzjtu.cn

Abstract. Artificial immunity is widely used in the field of intrusion detection by simulating the accurate identification function of biological immune system to foreign intrusions, among which negative selection algorithm is the most widely used. However, due to the large amount of network data and high dimensionality, it often leads to problems such as low detection accuracy. In this paper, the method of combining principal component analysis (PCA) with genetic algorithm (GA) and negative selection algorithm improves the accuracy of intrusion detection. Among them, principal component analysis performs dimensionality reduction and feature extraction on intrusion data, and genetic algorithm is used to optimize the generation part of detector. The performance test was performed on the NSL-KDD standard test data set. The results show that this method improves the accuracy of intrusion detection and reduces the false alarm rate, which proves the effectiveness of the method.

Keywords: Intrusion detection · Artificial immune algorithm · Negative selection algorithm

1 Introduction

As society enters the Internet age, the application of the Internet has changed the social form and people's lives. While enjoying the convenience of the Internet, people are also taking the risk of privacy leakage. Due to the complex interests involved in the Internet, the diversity of participant goals, and network security issues also affect social development and threaten the security and stability of the world's politics, military, economy, and culture, the security of the Internet has become an important guarantee. Internet infringement and cybercrime are spreading across various industries in various fields. The methods and frequency of infringement are often unexpected, the consequences and extent of the infringement are even more shocking. In the first half of 2020 alone, in February the US natural gas pipeline company was attacked and forced to close the compression facility. In April, the industrial control facility of the Israeli water supply department was attacked by a cyber attack. In May, Venezuela's national power grid was attacked, causing a large-scale blackout across the country [1]. Existing statistics show

J. Xiong et al. (Eds.): MobiMedia 2021, LNICST 394, pp. 563–576, 2021.
https://doi.org/10.1007/978-3-030-89814-4_41

that in terms of economy alone, the loss caused by cyber infringement in the United States has reached more than US$17 billion per year. Other Western countries such as France have exceeded 10 billion francs, and Britain and Germany also have billions of US dollars [2]. All sorts of incidents show that the problems of network security has reached a point that cannot be ignored. After nearly two decades of development, intrusion detection technology has become an important technology and research direction in the security field, and has been applied in many industries such as military, finance, government, commerce, transportation, and electric power, and plays a key role.

Immunity is a physiological function of the human body, and it is the third line of defense of the human body. When foreign antigens invade the human body and destroy healthy tissues and cause the internal balance of the human body to be out of balance, the immune defense line quickly functions to eliminate antigens and maintain human health. It can effectively deal with a large number of different types of virus invasion. The immune system's solutions are distributed, flexible and adaptive. These features are exactly what the field of intrusion prevention expects. Artificial Immune System (AIS) is a comprehensive intelligent scientific research direction that integrates control science, computer science and life science [3]. As an intelligent system developed by learning from and using various principles and mechanisms of the biological immune system, the artificial immune system is an improvement of the modern network behavior detection system from the perspective of biological immunity, and finally makes the system have a greater similarity with the biological immune system It can solve complex and changeable network viruses well. Modern researchers have made substantial improvements on the basis of the artificial immune system and applied it to the field of intrusion detection.

Zhang Ling [4] designed an intrusion detection algorithm based on random forest and artificial immunity (RFAIID), and proposed an antibody forest model; the clone selection algorithm was used to obtain an excellent large sample antibody set to improve the adaptability of intrusion detection. However, there is still the problem that the redundant attributes in the algorithm affect the detection speed of the algorithm. Xin Zhuang [5] combined rough set theory with artificial immunity, deleted redundant attributes, improved the operating efficiency of the algorithm, improved the stability of the model, accelerated the convergence speed, and ensured the superiority of the antibody. Feng Xiang [6] added segmentation technology and key bits to the negative selection algorithm to avoid the matching loopholes caused by the constant matching probability and reduce the system missed detection rate; and the clone selection algorithm in the genetic algorithm and the improved negative selection algorithm are compared In combination, the dynamics and diversity of detector generation are improved. Liu Hailong [7] proposed a high-dimensional real-valued detector distribution optimization algorithm, and used the principal component analysis method and affinity calculation to solve the problem of low detector distribution in the high-dimensional real-valued space in immune intrusion detection. The detector distribution was optimized through affinity comparison.

Amira Sayed A. Aziz [8] use detectors generated by immune algorithms to detect abnormal activities in the network. Minkowski distance function and Euclidean distance test detection process. Adeni jiuwashola David [9] improved the NSA detector generation stage, used neural network technology to build a model, and developed a new model called NNET-NSA, which has a higher detection rate. Soodeh Hosseini [10] proposed a new combination of abnormal process detection technology. This technique unifies the negative selection algorithm and the classification algorithm. This method reduces the training time while improving the accuracy of the system. Based on the artificial immune system, a host-based abnormal process detection framework is established. Nguyen Thanh Vu [11] combines artificial immune system (AIS) and deep learning to classify benign and malignant documents. Use AIS to build a clone of the malware detector to improve the accuracy of the unknown virus detection rate, and then use the Deep Belief Network (DBN) to calculate and train the risk level of the file, and evaluate the performance of the system.

It can be seen from the above that inspired by the artificial immune system, many improved methods have been successfully applied to intrusion detection of computer network systems. However, there are some problems that hinder the wide application of this method [12]:

(1) Uneven distribution of attack types. The current types of attacks on the network include DOS, U2R, R2L, and Probe. Since the probability of each attack is different, there will be an uneven data distribution problem in the data set. In attack detection, the smaller the amount of data, the lower the probability of being detected.
(2) Intrusion detection speed problem. The training speed and detection time are not so ideal when dealing with large-scale data. So far, few intrusion detection systems have both accuracy and speed. How to improve the detection speed is also a problem that should be paid attention to in the field of intrusion detection.
(3) Large-scale data sets are easy to be missed. In high-speed switching networks, intrusion detection systems cannot detect all data packets well, and the accuracy of analysis is not high. Faced with the current increasing data dimensions and complex network behaviors, there are a large number of misjudgment warnings or a longer judgment time, which is likely to cause the problem of missing reports and missed detections.
(4) Poor flexibility. The fundamental reason why the system is frequently successfully attacked today is that the protection strategy of each system is passive and static, and it does not have autoimmune functions and flexibility, which causes the system to face risks [13]. Therefore, it is necessary for intrusion detection to have the ability of autonomous learning, which can be adjusted according to the changes of the environment, so as to optimize the performance effect.

Based on the above, the research of this paper starts from the perspective of the combination of artificial immune algorithm and intrusion detection technology, solves the current vulnerability attack problem of computer network, improves and perfects some problems in artificial immune algorithm, and makes it can play a greater role in the field of intrusion detection. In this article, the method of combining principal component analysis (PCA) with genetic algorithm (GA) and negative selection algorithm is adopted. We will build an architecture of an intrusion detection system based on an improved negative selection algorithm, and divide the architecture into three modules for detailed introduction.

The first section mainly introduces some major network security incidents currently encountered in the world, and then introduces the research results and existing problems in the research direction of network intrusion detection at home and abroad in recent years. On this basis, the article explains the significance. The second chapter introduces the definition of negative selection algorithm and the related formulas of genetic algorithm. The third chapter builds an intrusion detection system model of an improved artificial immune algorithm; makes a related flow chart, explains the overall structure of the system, analyzes the function of each module of the system, and the fourth chapter uses the NSL-KDD data set to test, verified the validity of the model. The fifth chapter summarizes the research work of this article, analyzes the problems existing in the experimental research, and prospects the follow-up work.

2 Definition of Related Algorithms

2.1 Negative Selection Algorithm

Inspired by the theory of immune model, researchers at home and abroad have proposed a variety of immune intelligent algorithms, represented by negative selection algorithms, clone selection algorithms, immune network algorithms, and so on [14]. Among them, in 1997, Forrest of the University of New Mexico in the United States proposed a negative selection algorithm [15]. She studied the process of human thymocytes to produce immune T cells, that is, by negating the lymphocytes that produce immune responses to the body, and divided the detection procedures into self-collection and non-self collection. By negating the detection program that matches the pattern with normal traffic, the collection of detection programs that do not match the normal traffic is used as a detector to filter network traffic. Negative selection algorithms are widely used in intrusion detection, and have high research value (Fig. 1).

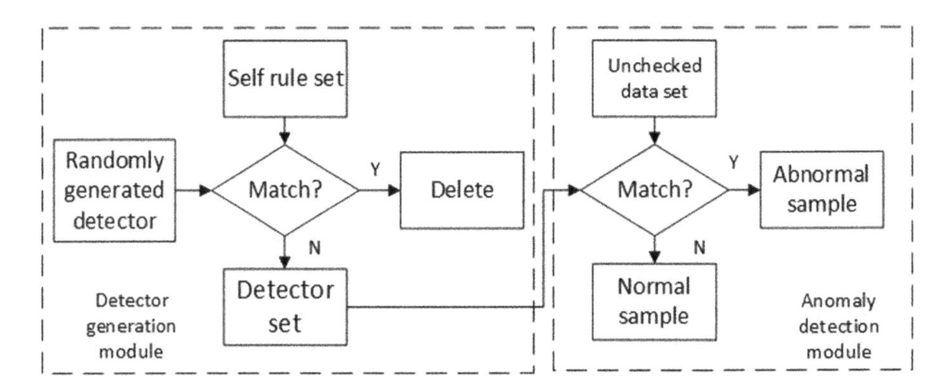

Fig. 1. Negative selection algorithm intrusion detection architecture.

Definition 1: Antigen. Antigen g $=$ $\{g|g = < X1, X2, X3, \cdots Xn >\}$ means all samples in the space.

Definition 2: Self-set and non-self-set. Self (S) represents normal samples in space; Nonself (N) represents abnormal samples in space.

$$\begin{aligned} S \cup N &= \mathrm{g} \\ \mathrm{S} \cap \mathrm{N} &= \phi \end{aligned} \tag{1}$$

Definition 3: Detector set. Mature detector set D, Represents a mature detector obtained by judging whether it matches the self-set S. D exists in the non-self region. Among them, rd represents the detector radius.

$$D = \{\mathrm{d}|d =< y1, y2, y3, \cdots, yn, rd >\} \tag{2}$$

Definition 4: Affinity calculation. Affinity is the firmness of the binding of an antibody binding site to an epitope. The tighter the binding, the less likely the antibody is to separate from the antigen. Affinity function (fit): The value of the degree of match between an antigen and an antibody is generally represented by a real number between 0 and 1. Use function fit(i,j) represents the affinity between antigen i and antibody j. Obviously, affinity is related to the degree of matching between antibody and antigen. The affinity can be expressed by the following formula:

$$fit(i,j) = \frac{1}{1 + tij} \tag{3}$$

In the formula, tij represents the distance between antigen i and antibody j. The distance function can be Hamming distance or Euclidean distance. The matching function is as follows:

$$d = \sum_{i=1}^{L} \delta \tag{4}$$

Hamming distance:

$$d(i,j) = \sqrt{|xi1 - xj1|^2 + \cdots + |xip - xjp|^2} \tag{5}$$

2.2 Genetic Algorithm

GA is an algorithm with learning function, which is a random search optimization algorithm. Its algorithm principle draws on the natural selection and genetic mechanism of nature in Darwin's theory of evolution. It was proposed by Holland in 1975 and published in the book "Adaptation in Natural and Artificial Systems". In this book [16], the GA system machine was comprehensively discussed for the first time and the corresponding data theory proof was given. This kind of algorithm is obviously different from traditional and optimal solution algorithms. It is no longer highly dependent on gradient information like the algorithm. Instead, it uses a group search strategy and makes the individuals between them complete the corresponding exchange and mutation. According to the fitness function, the calculation of the corresponding value is completed, and new individuals are formed by means of crossover, mutation and selection methods, that is, the problem solution is obtained, and the search space of the group is then realized to achieve the overall optimal solution. So this method can solve those non-linear problems, genetic algorithm puts forward a new idea and new method. Based on the theory of biological evolution, genetic algorithms are not directly related to specific problems. Instead, they receive the described objective function through a randomly generated population, and find the corresponding optimal solution with the help of genetic and natural selection models (Fig. 2).

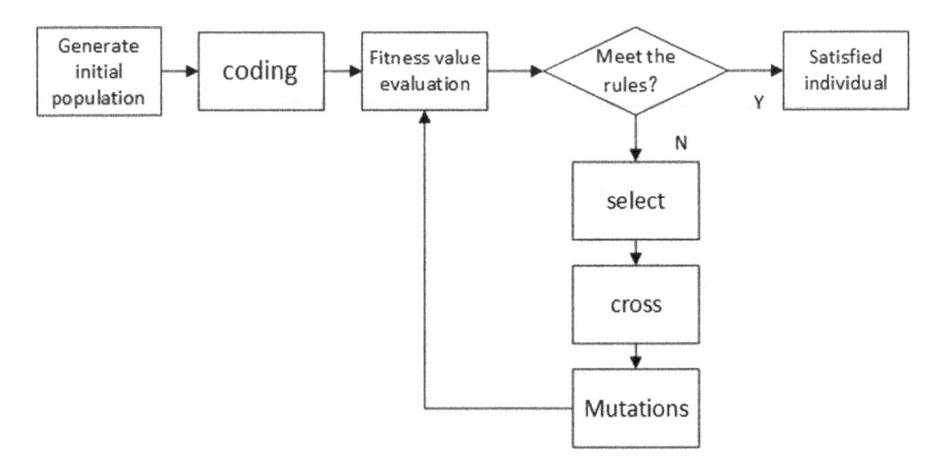

Fig. 2. Genetic algorithm flow chart

In the fitness function design link needs to strictly follow the rules of reasonableness, generality and specification. The quality of a fitness function should also meet the widest possible versatility, so that users do not need to change the fitness function when solving various problems. Generality requirements are higher requirements for fitness function design. In this research, the functions mentioned in reference [17] are used, as follows:

$$Fit(f(x)) = \left\{ \begin{array}{ll} 1 - 0.5x\left[\left|\frac{f(x)-b}{a}\right|\right]^{a}, & |f(x) - b| < a \\ \frac{1}{1+\left[\left|\frac{f(x)-b}{a}\right|^{\beta}\right]}, & |f(x) - b| \geq a \end{array} \right\} \tag{6}$$

Generally, under an ideal background, the value of b is $f\ \min(x) = y*$, if the current fitness value is 0.5, then a represents the distance from f(x) to $f\ \min(x)$.Since the applicable environment of the fitness function needs to be considered, it is necessary to set β to 2.

In the above formula, both a and b will be optimized for the next generation under the action of genetic algorithm. b is f(x)xi, At this time, the formula of a is:

$$a = \max X\left[0.5, \frac{|f\ \max(i) - f\ \max(i)*|}{30}\right] \tag{7}$$

3 Intrusion Detection System Based on Improved Negative Selection Algorithm

3.1 Data Set

The NSL-KDD data set is used in the selection of experimental data. The NSL-KDD data set is a network environment established by Lincoln Lab to simulate the US Air Force LAN. It collects 9-week TCPdump network connection and system audit data to simulate various user types, various network traffic and attack methods. Make it like a real network environment. This data is based on a large number of improvements on the traditional authoritative intrusion detection data set. NSL-KDD removes the redundant data in the KDD99 data set. All types of attack samples retain only one record, so that the data will not be repeated records and deviations. NSL-KDD data is widely used in intrusion detection research on computer networks, so it is feasible to conduct intrusion detection experiments based on the NSL-KDD data set. Each network connection in the data set is marked as normal (normal) or abnormal (attack), and the abnormal types are subdivided into 4 categories, a total of 39 attack types, of which 22 attack types appear in the training set, and 17 unknown attacks The type appears in the test set. The purpose of this design is to test the generalization ability of the classifier model. The ability to detect unknown attack types is an important indicator for evaluating the quality of an intrusion detection system [18].

3.2 Related Functional Modules

(1) Data Preprocessing Module. The first step is to preprocess the data set, and then divide the input data set into self-set and non-self-set. Also, because the characteristic attributes of the sample records are messy, the values are referenced after comparison according to 41 attribute standards, which will lead to inconsistent values, and the final test results will be disordered. Therefore, in order to improve the versatility of this experiment, we preprocess the collected data before testing the data. The specific method is: first standardize the collected data, use one-hot encoding to transform the discrete data in the data set in the data preprocessing part, and then use the principal component analysis algorithm to reduce the dimension of the features, reducing the complexity of the calculation and save time.

Convert the string to discrete numbers, use one-hot encoding to transform the features, convert the discrete type to a numeric type to normalize its value, and map the values to [0,1] to reduce error. After finding the maximum and minimum values of the attributes of each record, use the following formula to normalize them:

$$x' = \frac{x - \min i}{\max i - \min i} \tag{8}$$

After the values are normalized, the relationship between the original data is retained, and the errors caused by the large value difference of each attribute are also eliminated. And to ensure that the program converges faster at runtime (Table 1).

Table 1. Conversion of coding features.

Before mapping	After mapping
TCP	1, 0, 0
UDP	0, 1, 0
UDP	0, 0, 1

Principal component analysis (PCA) is a commonly used method of feature dimensionality reduction and feature extraction. Through orthogonal transformation, a group of potentially correlated variables is converted into a group of linearly uncorrelated variables. The converted group of variables is called principal component. Principal component analysis analyzes the data model, reduces the dimensionality of the data set while ensuring the minimum loss of information, and projects the feature space onto a smaller subspace to better describe the data.

The steps for preprocessing the data using principal component analysis are as follows:

Step1: Standardize the data set.

Step2: The eigenvectors and eigenvalues are extracted from the covariance matrix. The calculation formula of the covariance matrix is as follows:

$$Cvjk = \frac{1}{n-1}\sum_{i=1}^{n}(xij - xj')(xik - xk') \tag{9}$$

Where the mean vector:

$$x' = \frac{1}{n}\sum_{k=1}^{n} xi \tag{10}$$

The covariance between the two features is calculated as follows:

$$CM = \frac{1}{\mathrm{n}-1}((X - x')^T(X - x')) \tag{11}$$

Step3: Arrange the eigenvalues in descending order, and select the k eigenvectors corresponding to the k eigenvalues, and k is the dimension of the new eigenspace.

Step4: The projection matrix W is constructed by the selected k eigenvectors.

Step5: Pass the original data set through the projection matrix W to generate a k-dimensional feature subspace, $Y = X * W$.

(2) Detector Generation Module. The creation and selection of the detector is based on the NSA algorithm, which generates a random detector, matches the generated detector with the self-set, if it matches, deletes the detector, if it does not match, then generates a mature detector set D. Because the genetic operator has the characteristics of selection, mutation, crossover, fitness, etc., the genetic algorithm is used to improve the traditional negative selection algorithm, generate an optimized and balanced subset, and optimize the generation and distribution of detectors. When the algorithm runs to the set genetic total algebra, the algorithm will stop running. The specific process is as follows:

Step1: In the GA algorithm, scientific coding will affect the performance of the algorithm and its population diversity to varying degrees. Compared with real number coding, binary has a higher search ability. Binary coding is used here. Binary coding is the structure of the original problem is transformed into the bit string structure of the chromosome.

Step2: Use the fitness function mentioned in the above formula (6) to calculate the corresponding values of different populations.

Step3: Select. The selection function means to select certain individuals from the parent population for inheritance. In this article, random competitive selection method is used to perform selection operations. According to the roulette gambling selection method, a pair of individuals are selected each time, and then the fitness values of the two individuals are compared. The one with the higher fitness will be selected. Repeat this process until the total number reaches the specified number.

Step4: Cross. The group obtained after the selection is processed in a uniform crossover method according to the predetermined crossover rate. For randomly selected individuals X1, X2, two intersection points are determined by a random method, and then three integers of 0, 1, and 2 are randomly generated. When the random number is 0, the front part of X1, X2 crosses; when the random number is 1, the middle part of X1, X2 crosses, and when the random number is 2, the back part of X1, X2 crosses.

Step5: Mutations. The variogram is used to change the value of one or more bits in the chromosome. The probability of mutation is generally 0.02–0.03. The variogram aims to improve the fitness of the chromosome by introducing new characteristics.

Step6: Repeat steps 1 through 5 above until the set maximum genetic algebra is completed.

For the detection device, its main function is to detect foreign intrusions, which is similar to lymphocytes in the immune system. The optimized and balanced subset is obtained through the above steps, which is defined as a detector, and the detector matches with the self set. If it matches, the detector is deleted, and if it does not match, a mature detector set is formed. The implementation process of this mature detection is shown in Fig. 3.

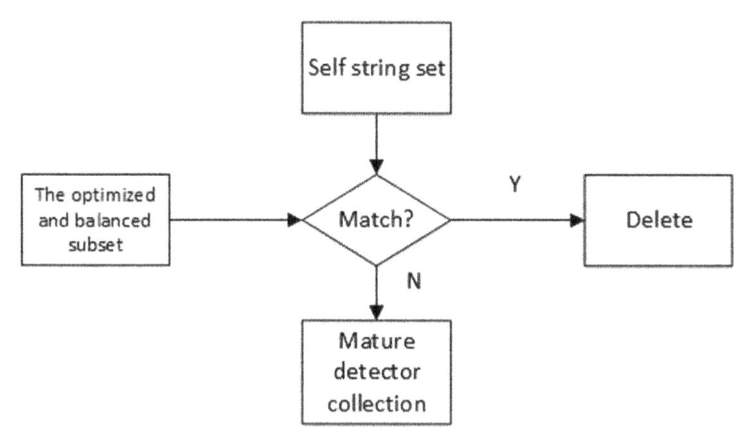

Fig. 3. The production process of the mature detector.

(3) Intrusion Detection Module. The mature detector is sent to the detection module, and the match between the sample to be tested and the generated detector is calculated and processed. If it matches, the intrusion data is found, and the relevant processing is performed immediately. Euclidean distance is used here, the match between detector d and antigen g is defined as:

$$mdg = 1/\left(1 + \sqrt{\sum (di - gi)^2}\right) \tag{12}$$

Calculate the Euclidean distance between the data to be detected and the self-set in the space to determine whether a match occurs. By setting a threshold r, compare the size of r and l to determine whether they match. When $r \leq l$, the sample is far away from the detector, and there is no intersection or mismatch. If the result is the opposite, then match. When generating the detector, for its radius, a fixed radius can be artificially set.

3.3 The Whole Frame

See Fig. 4.

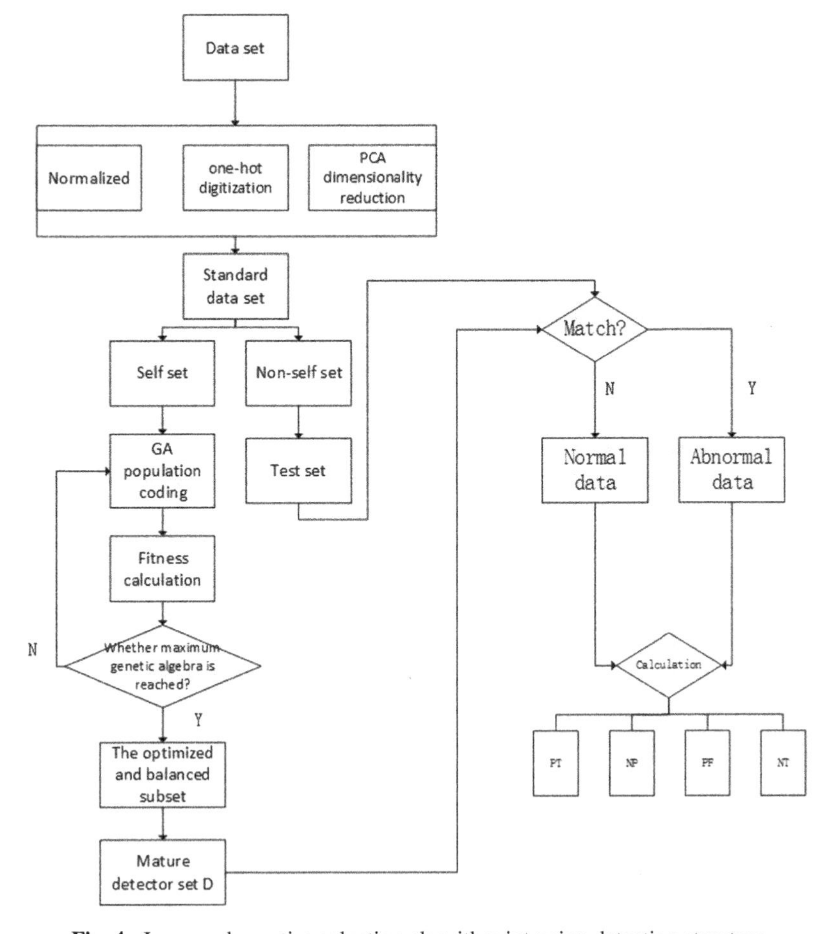

Fig. 4. Improved negative selection algorithm intrusion detection structure

4 Experiment

In order to verify this algorithm, the authoritative NSL-KDD data set is selected, which is a more classic data set in the current intrusion detection field. Each record data covers normal and abnormal features, and the latter can be subdivided into four types of attacks. The evaluation indicators used in the experiment are Accuracy (ACC), Detection rate (DR) and False positive rate (FAR). The meanings and calculation expressions of each indicator are shown in Table 2. Among them, TP represents the number of samples that actually classify abnormal samples as abnormal, TN represents the number of samples that actually classify normal samples as normal, FP represents the number of samples that actually classify samples that are actually normal as abnormal, and FN represents The samples that are actually abnormal are classified as the number of normal samples.

Table 2. Performance evaluation indicators.

	Meaning and computational expression
DR	Detection rate (DR) refers to the proportion of aggressive behaviors that are correctly classified in a concentrated antigen: $DR = \frac{TP}{TP+FN}$
FAR	False positive rate (FAR) refers to the ratio of the number of all normal antigens misidentified as attack in the antigen set to the total number of normal antigens in the test set: $FAR = \frac{FP}{TN+FP}$
Accuracy	Classification accuracy rate (ACC) refers to the ratio of the number of correctly classified antigens in the antigen set to the total number of samples in the antigen set: $ACC = \frac{TP+TN}{TP+TN+FP+FN}$

Now use the improved negative selection algorithm proposed in Chapter 3 to conduct experiments, and select a part of the NSL-KDD data set, which contains 20,000 normal and intrusive values. Before calculation, set the corresponding parameters as follows (Table 3):

Table 3. Experimental parameter settings.

Data	Number of mature detectors	Generations	The self set	Non-self test set	Non-self training set
20000	1000	200	93	6637	13270

Set the number of test sets as 1000. When the value of detector radius Rd is determined and the range of radius variation is maintained from 0.2 to 0.8, the accuracy and false positive rate of the algorithm are tested to see if they are still within the acceptable range. In each case, the experiment was run 20 times, and the average value was taken. The results are shown in the following Table 4:

Table 4. Values of detection rate and false alarm rate under different self-radius

rd	DR (%)	FAR (%)
0.2	89.25	9.3
0.3	94.97	6.7
0.4	95.83	2.3
0.5	97.98	1.1
0.6	96.51	1.9
0.7	94.97	1.3
0.8	92.41	2.1

It can be seen from the above table that when the self-radius is small, the space occupied by the normal sample is relatively small, and the non-self space occupied by the corresponding detector is relatively large. Therefore, during intrusion detection, most abnormal behavior samples can be detected, but the self-region is too small, causing some new normal samples to fall in the non-self-region, making the false alarm rate higher. With the increase of the self-body radius, the self-body area becomes larger and larger, making the non-self-body area occupied by the detector smaller and smaller, which will correspondingly cause a decrease in detection rate and false alarm rate. On the whole, when the self-body radius rd = 0.5, the overall performance is the best.

Table 5. Comparison of experimental results

The experimental method	DR (%)	FAR (%)	ACC (%)
Improved negative selection algorithm	96.43	0.97	97.88
The original negative selection algorithm	88.92	1.01	90.12
Single-SVM	89.42	3.62	95.32
BP + GA	82.18	2.74	89.75

From the results obtained in Table 5, it can be seen that the accuracy and detection rate of the intrusion detection behavior of the improved negative selection algorithm is much higher than that of the original negative selection algorithm, and the false alarm rate is also lower. Compared with other methods, the performance of this algorithm is also significantly better than several other algorithms. Through multiple sets of comparative experiments, using the NSL-KDD data set to test, the results show the effectiveness of the method.

5 Conclusion

Nowadays, intrusion detection systems are widely used as a line of defense for security protection. Artificial immune algorithms simulate the characteristics of biological immune systems against foreign intrusions, which have huge application space in intrusion detection. However, due to the large amount of network traffic and unobvious features, the correct rate of intrusion detection is often not high, and the detection effect is poor. Therefore, this article improves the Negative Selection Algorithm (NSA) from two aspects of feature extraction and detector generation, and discusses the application of the method of combining PCA and GA with negative selection algorithm in intrusion detection. When facing a large number of high-dimensional features of network data, PCA is used as a feature extraction method to achieve the purpose of dimensionality reduction. The GA is used to optimize the detector generation. However, this algorithm also has some shortcomings. For example, it consumes a lot of time during detection and is easy to fall into local optimal problems. These aspects need to be taken seriously in the future.

Acknowledgments. This work was supported by State Grid Gansu Electric Power Research Institute Project No. 520012 Intelligent Recognition and Defense of Intrusion Behavior of Electric Power Information and Physical Fusion System in Cyber Attack Environment.

References

1. Bi, R., Chen, Q., Chen, L., Xiong, J., Wu, D.: A privacy-preserving personalized service framework through Bayesian game in social IoT. Wireless Commun. Mobile Comput. **2020**, 1–14 (2020)
2. Xiong, J., Ma, R., Chen, L., et al.: A personalized privacy protection framework for mobile c2rowdsensing in IIoT. IEEE Trans. Industr. Inf. **16**(6), 4231–4241 (2020)
3. Cooper, E.L.: Evolution of immune systems from self/not self to danger to artificial immune systems (AIS). Phys. Life Rev. **7**(1), 55–78 (2010)
4. Zhang, L., Zhang, J., Sang, Y., et al.: Intrusion detection algorithm based on random forest and artificial immunity. Comput. Eng. **46**(8), 146–152 (2020)
5. Xin, Z., Wan, L.: Iintegrated intrusion detection model based on artificial immunity. Comput. Eng. Design **40**(10), 2799–2804 (2019)
6. Feng, X., Ma, M., Zhao, T., Yu, H.: Intrusion detection system based on hybrid immune algorithm. Comput. Sci. **41**(12), 43–48 (2014)
7. Liu, H., Zhang, F., Xi, L.: High dimensional real-valued detector distribution optimization algorithm for intrusion detection based on immunity. J. Tsinghua Univ. (Sci. Tech.) **52**(10), 1415–1419 (2012)
8. Amira Sayed, A., et al.: Artificial immune system inspired intrusion detection system using genetic algorithm. Informatica: Int. J. Comput. Inf. 347–357 (2012)
9. David, A.O., Joseph, U.J.: A novel immune inspaired concept with neural network for intrusion detection in cybersecurity. Int. J. Appl. Inf. Syst. **12**(30), 13–17 (2020)
10. Hosseini, S., Seilani, H.: Anomaly process detection using negative selection algorithm and classification techniques. Evol. Syst. (2019). https://doi.org/10.1007/s12530-019-09317-1
11. Thanh Nguyen, V., Hoang Dung, L., Dinh Le, T.: A combination of artificial immune system and deep learning for virus detection. Int. J. Appl. Eng. Res. **13**(22), 15622–15628 (2018)

12. Zhang, Y., Wang, L., Sun, W., et al.: Distributed intrusion detection system in a multi-layer network architecture of smart grids. IEEE Trans. Smart Grid **2**(4), 796–808 (2012)
13. Zhao, J., Zhang, X., Di, F., et al.: Exploring the optimum proactive defense strategy for the power systems from an attack perspective. Secur. Commun. Netw. **6**, 1–14 (2021)
14. Yang, H., Li, T.: Intrusion detection based on T cell receptor principle. Int. J. Performabil. Eng. **15**(9), 2407–2413 (2019)
15. Forrest, S., Perelson, A.S., Allen, L., et al.: Self-nonself discrimination in a computer. Comput. Soc. Symp. Res. Secur. Privacy 202–212 (1994)
16. Holland, J.H.: Adaptation in natural and artificial systems. Ann. Arbor. (1975)
17. Liu, Y.: Research on fitness function in genetic algorithm. J. Lanzhou Polytech. College **3**, 1–4 (2006)
18. Weikai, W., Lihong, R., Lei, C., et al.: Intrusion detection and security calculation in industrial cloud storage based on an improved dynamic immune algorithm. Inf. Sci. 43–557 (2018)
19. Aldhaheri, S., Alghazzawi, D., Li, C., Barnawi, A., Bandar, A.: Alzahrani. Artifificial immune systems approaches to secure the internet of things: a systematic review of the literature and recommendations for future research. J. Netw. Comput. Appl. 157 (2020)
20. Al-Qatf, M., Lasheng, Y., Al-Habib, M.: Deep learning approach combining sparse autoencoder with SVM for network intrusion detection. IEEE Access **6**, 52843–52856 (2018)

A Smart Access Control Mechanism Based on User Preference in Online Social Networks

Fangfang Shan[1,2(✉)], Peiyu Ji[1], Fuyang Li[1], and Weiguang Liu[1]

[1] Zhongyuan University of Technology, Zhengzhou 450000, China
6129@zut.edu.cn
[2] Zhengzhou University, Zhengzhou 450000, China

Abstract. Data privacy protection is crucial in the era of big data, and although access control mechanisms can effectively prevent privacy leakage, existing access control mechanisms of social networks rarely consider users' personal privacy preferences in the process of generating access control policies, so they cannot provide personalized services to users. We proposed an intelligent access control mechanism based on users' privacy preferences by extracting their privacy preference values through a quantifiable analysis mechanism, and then using the values and some key user social resource information as feature vectors. The experiments show that this mechanism can automatically generate appropriate access control policies to meet the potential privacy needs of different users, so as to better protect the privacy of social network data.

Keywords: Personal preference · Access control mechanism · Online social network

1 Introduction

Online social network can provide digital users with social interaction and information sharing, but it has privacy security problems. The overwhelming amount of user information can be collected by enemies using illegal means and correlated to deduce some more private user information, threatening the security of users' personal and property. However, few users are aware of the serious harm that can be caused by privacy breaches. Therefore, it is crucial to study the access control mechanism of social networks for protecting users' private data.

In general, most of the current social network access control is based on relationships, cryptographic algorithms, game theory and face recognition technologies. Pang et al. [1] proposed an access control scheme for Facebook social networks, which implements access control on resources according to the relationship between users. Cheng et al. [2] used regular expression to define access control policy, so that user-user relationship, user-resource relationship and resource-resource relationship can control access requester access resources. Backes et al. [3] proposed a new social relationship reasoning mechanism, which can predict the social relationship between two people without

J. Xiong et al. (Eds.): MobiMedia 2021, LNICST 394, pp. 577–590, 2021.
https://doi.org/10.1007/978-3-030-89814-4_42

any prior knowledge. Shan et al. [4] proposed a method to control the access rights of resources in social networks by using different relationships to correspond to different access rights. Voloch et al. [5] proposed a new role and trust based access control model to evaluate each user's trust by several standards. Users with specific roles and appropriate permissions can access some instances of data if they do not reach a sufficient level of trust. These roles and trust assessments provide more accurate and feasible information sharing decisions and better control of privacy in social networks. Yousra et al. [6] proposed a community-centered broker-aware access control (CBAC) model, which uses important concepts from social network analysis (SNA), namely broker, one-to-many relationship, temporary relationship and emerging access control models such as attribute-based and trust-based access control and decentralized strategies. Xu et al. [7] proposed a trust-based access control mechanism Trust2Privacy to protect the privacy of users after releasing information, which can effectively realize the conversion from trust to privacy. Zhu et al. [8] proposed an online social network rumor propagation model with forced silence function. Aljably et al. [9] proposed a privacy protection model, which uses limited local differential privacy (LDP) to save the composite copy of the collected data, so as to purify the user information collected from social networks. The model further uses reconstructed data to classify user activities and detect differences. Alemany et al. [10] proposed two soft-paternalism mechanisms that provide information to the user about the privacy risk of publishing information on a social network. That privacy risk is based on a complex privacy metric. The results show that there are significant differences in teenagers' behaviors towards better privacy practices.

Shan et al. [11] proposed a social network forwarding control mechanism based on game theory on the basis of analyzing the benefits of both forwarding parties. On the basis of analyzing the benefits of different game strategies selected by the forwarder and the publisher, they compared the historical data of forwarding operations with the threshold set by the publisher, and gave the final decision whether to allow forwarding. Can effectively prevent the forwarder's dishonest forwarding behavior. Wu et al. [12] verified the constructed privacy random game model and the new privacy risk measurement criterion, and solved the strategy by reinforcement learning algorithm, and obtained effective personal access control strategy.

In order to protect the privacy information of users in multi-user photos, Xu et al. [13] proposed a multi-user photo privacy protection model based on face recognition technology. Bernstein et al. [14] further proposed a screen shooting interference system to prevent camera privacy leakage. Li et al. [15] proposed an access control model CoAC for cyberspace, which can effectively prevent the security problems caused by the separation of data ownership and management rights, and the secondary/multiple forwarding of information. Marinescu et al. [16] proposed an access control method that can automatically learn the authorization rules to form a model, judge the newly launched access request, and prevent any access request that may attempt to use the loopholes in the authorization logic of social networks. In addition, privacy protection access control schemes [17] applied to information-centric networks and access control schemes for health information [18] also emerged.

Zhang PanPan et al. [19] proposed a game metric model based on privacy preference for the equilibrium problem between privacy protection and service quality. Zhang

Chao et al. [20] proposed a privacy-preserving social network information recommendation method based on information dissemination model. Lei [21] et al. proposed a hierarchical management scheme for friend matching using attributes to promote secure friend discovery in MSN. The scheme involves the establishment of several attribute centers, which perform fine-grained management according to various user attributes. Alshareef [22] et al. proposed a new collaborative access control framework that takes into account the relationship between multiple users' viewing and sharing items, and ultimately resolves conflicts in user privacy settings. G Liu et al. [23] model trust by proposing the three-valued subjective logic (3VSL) model. 3VSL properly models the unertainties that exist in trust, thus is able to compute trust in arbitrary graphs. Based on the 3VSL model, they further design the AssessTrust (AT) algorithm to accurately compute the trust between any two users connected in an OSN. And it is verified that 3VSL can accurately simulate the trust relationship between any pair of indirectly connected users in Advogato and PGP.

The existing access control mechanism for social privacy protection has formulated access strategies for users' privacy needs in different environments, but it ignores the impact of users' personal preferences on access control strategies. To solve this problem, this paper proposes a user preference analysis mechanism, and constructs an information sensitivity model by using the amount of privacy information. On the basis of this model, the influence of user personality differences on the degree of privacy is studied. By using the method of arctangent and privacy measurement, combined with information entropy, weighted information entropy and other technologies, the user preference analysis mechanism for social networks is finally constructed. This mechanism is used to quantify the user preference value as one of the feature vectors of SVM for the analysis of access control strategy. Experiments show that the access control strategy with the personal preference analysis mechanism significantly improves the protection of privacy information of different users for different social objects. More reasonable and effective solution to social platform user privacy resources leak.

2 Basic Knowledge

2.1 SVM

The earliest origin of SVM is in pattern recognition, a classifier algorithm developed from generalized portrait algorithm, which is an algorithm for binary classification of the data to be processed in a supervised learning manner, a generalized fast and reliable linear classifier, and also supports nonlinear classification, which is better for solving the problem of small samples.

The sample set $S = \{(x_i, y_i); i = 1, \ldots, m\}$ of the given training, where $x_i \in R^n$ represents the n-dimensional input vector, and y_i is the marker for each vector. SVM works by constructing a hyperplane (w, b).

$$w^T x_i + b = 0 \tag{1}$$

The distance between any point x and the hyperplane is:

$$\gamma = \frac{|w^T x + b|}{||w||} \tag{2}$$

After the duality problem and the soft spacing

$$\min_{w,\,b} \frac{1}{2} w^T w + C \sum_{i=1}^{N} \xi_i \tag{3}$$

$$y_i \left(w^T x_i + b \right) \geq 1 - \xi_i \tag{4}$$

$$\xi_i \geq 0, i = 1, 2, \ldots, N \tag{5}$$

$y_i \left(w^T x_i + b \right) \geq 1 - \xi_i$ is a quadratic programming problem, which becomes a maximization problem after duality:

$$\max_{\alpha} \sum_{i=1}^{N} \alpha_i - \frac{1}{2} \sum_{i=1}^{N} \alpha_i \alpha_j y_i y_j K\left(x_i, x_j\right) \tag{6}$$

$$s.t. \sum_{i=1}^{N} \alpha_i y_i = 0;\ 0 \leq \alpha_i \leq C, i = 1, 2, \ldots N \tag{7}$$

2.2 Information Entropy

Shannon proposed the concept of "information entropy" to solve the problem of quantitative measurement of information. From the perspective of information transmission, information entropy can represent the value of information. Simply put, the lower the probability of an event happening, the more information it can give when it happens. The calculation formula of information entropy is as follows:

$$H(x) = -\sum P(x_i) \log(2, P(x_i)) \tag{8}$$

3 Smart Access Control of User Preferences Based on SVM

3.1 Quantitative Analysis of User Preferences

The factors affecting user access control privileges are not only uniform or generalized sensitive information, but also influenced by user personality traits, social circles, specific content of posted resource information, etc. In order to more accurately formulate access control policies for network users, it is required to add the analysis of different users' personalized preferences based on the traditional social access control architecture. Therefore, this paper proposes a user preference analysis mechanism, and its technical route is shown in Fig. 1.

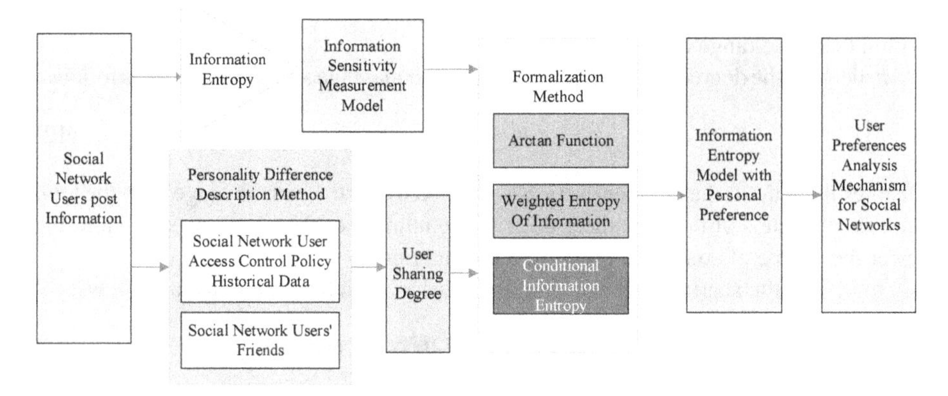

Fig. 1. A mechanism for analyzing user preferences.

In social networks, the information uploaded by users to social networks contains different degrees of privacy information, which requires different access control strategies. Firstly, information entropy is used to describe the privacy information contained in the information, to determine the degree of privacy of the information, and to construct the basic information sensitivity measurement model.

Specifically, $x_i \in X$ denotes the ith message posted by a social network user, $p(x_i)$ denotes the proportion of the amount of private information contained in the ith message to the amount of all the user's private information, and the source of the information posted by the user in the social network is denoted as below.

$$\begin{pmatrix} X \\ P(x) \end{pmatrix} = \begin{pmatrix} x_1 & x_2 & \ldots & x_i & \ldots & x_n \\ p(x_1) & p(x_2) & \ldots & p(x_i) & \ldots & p(x_n) \end{pmatrix} \tag{9}$$

Where $0 \le p(x_i) \le 1$, $\sum_{i=1}^{n} p(x_i) = 1$.

The information source entropy of the released information is denoted as:

$$H(x) = -\sum_{i=1}^{n} p(x_i)\log_2 p(x_i) \tag{10}$$

$H(x)$ represents the average private amount of information released by users in the social network, where N represents the total amount of information released by users.

Specifically, $x_i \in X$ denotes the ith message posted by the user in the social network, $f \in N$ denotes the number of friends the user has in the social network, f_a^i denotes the number of allowed visible friends set when the user uploads the ith message, and $w(f)$ is the user's social breadth function, calculated as follows.

$$w(f) = 2/(\pi \arctan f) \tag{11}$$

Where $f \in N, 0 \le f < +\infty$. This function satisfies the following properties: (1) the social breadth of the user $w(f) \in (0, 1)$; (2) social breadth increases with the number of users' friends in the social network. Therefore, function $w(f)$ monotonically increases within the value range of f.

h_i denotes the degree of confidentiality of the ith message and is calculated as follows:

$$h_i = f_a^i / f \tag{12}$$

Denotes the ratio of the number of friends allowed to view the ith message xi to the total number of friends of the user, the greater of the number of friends allowed to view, the lower the degree of confidentiality of the information.

s_i denotes the sharing degree of the ith message, which is calculated as follows:

$$s_i(f, f_a^i) = h_i \times w(f) = (2f_a^i)/(\pi f \arctan f) \tag{13}$$

The above equation is used to describe the number of friends f of a user and the influence of the number of friends f_a^i allowed to be seen in the access control policy on the sharing degree of information x_i. The more friends a user has and the more friends they are allowed to view, the higher the sharing degree of information xi, and vice versa.

The sharing space of information released by users is as follows:

$$\begin{pmatrix} X \\ S(X) \end{pmatrix} = \begin{pmatrix} x_1x_2 \ldots x_i \ldots x_n \\ s_1s_2 \ldots s_i \ldots s_n \end{pmatrix}, \quad 0 < s_i < 1, i = 1, 2, \ldots n \tag{14}$$

Define source entropy with personal preferences:

$$H_s(x) = -\sum_{i=1}^{n} s_i p(x_i) \log_2 p(x_i) \tag{15}$$

$H_s(x)$ describes the influence of different user preferences on the privacy degree of information published in social networks by sharing degree S_i (i = 1, 2…N), and realize the analysis and measurement of user preferences.

3.2 Access Control Mechanism of Social Network Based on User Preference

An access control mechanism model based on user preferences is shown in Fig. 2, the model includes:

User: Used to describe a registered user in a social network platform, user set U, $U \in \{u_1, u_2, \ldots, u_n\}$. Each user can upload user resources through the social platform and access other user resources under the conditions of permission. User resources include text, images, videos, etc. And different access control requests can be made according to different user resources.

User resources: Used to describe the user information generated by the user in using the social platform, including the user's autonomously uploaded resources and the access control policy records for the resources and the user's social network relationship graph.

All user's resource: It is used to describe the user information based on all registered users in the social platform in the process of use. Including all user resources and their access control policy records.

Preference analysis mechanism: Quantitative analysis of user's personal preferences and obtaining user preference values by analyzing user's access control policy records and user resources. It is one of the evaluation elements of the access control evaluation module.

Access control module: Generate resource access control policies with user preferences by analyzing user resources, user preference values, all user resources and their access control policy records.

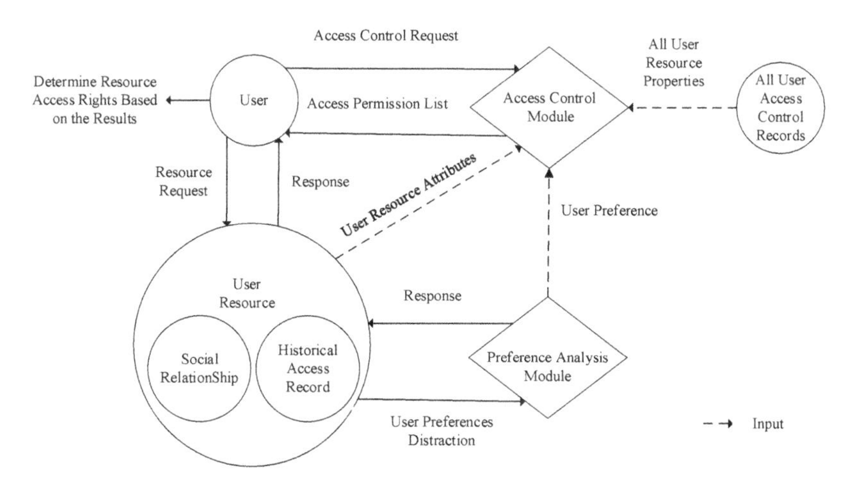

Fig. 2. An access effect model based on user preferences.

The basic steps are as follows:

When users publish user resources, they first make resource request and access control request for user resources and access control module respectively.

The user resource responds after receiving the request, and the user's historical access control records and social network relationships are analyzed to form user access control attributes.

User resources are taken as input to conduct user preference analysis and obtain user preference values.

Analyze all user resources and their access control policy records, and get all user resource attributes.

When the access control evaluation module receives the user resource attribute, all user resource attribute and user preference value, it generates the user resource access control policy, and returns the access permission list to the user after responding to the user's access control request.

The user determines the resource access control rights through the result of the permission list.

3.3 Access Control Policy Generation Method

We preprocessed the data of users' historical resource records, friend relationship graphs and user preferences obtained through preference analysis mechanism in the social network platform, and divided the processed data into training set and test sets, so as to avoid the over-fitting problem. After that, the machine learning model will be generated by the training set through the machine learning algorithm, and the prediction model will be generated through the test set. Finally, the resources to be processed will be input as the model, so as to get the access rights list of the user's resources this time. Training and validation of access control model is shown in Fig. 3.

User resource information is divided into high, medium and low levels (represented by H, M and L respectively). Personal privacy preference is divided into high, medium and low levels (represented by H, M and L, respectively). The degree of privacy is divided into five grades: very high, high, medium, low and very low (represented by VH, H, M, L and VL, respectively). The higher the amount of privacy information a user requests for access control resources, the higher the corresponding privacy degree will be, the higher the personal privacy preference will be, and the corresponding privacy degree level will also improve. The details are shown in Table 1.

4 Analysis of Experiment and Application Examples

In this experiment, the prediction model of SVM is used to carry out the experiment. The author conducted research in this field during his doctoral study and published relevant papers in this field. From the author's previous research results, we can conclude that SVM algorithm is the best choice for this mechanism.

4.1 Data Set

The author's research group has developed its own information sharing system with functions such as instant messaging and information release in social networks. The

Table 1. The relationship between the degree of privacy and Privacy information/Personal Privacy preference.

Privacy information	Personal privacy preference	Degree of privacy
H	H	VH
H	M	H
H	L	M
M	H	H
M	M	M
M	L	L
L	H	M
L	M	L
L	L	VL

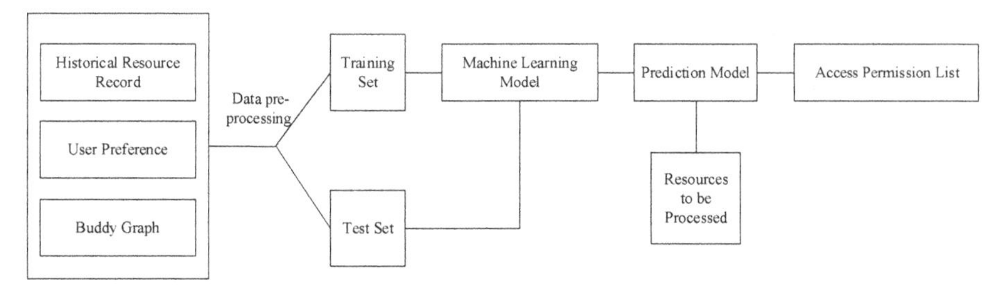

Fig. 3. Training and validation of access control model.

system is open to use in the institute. Registered users can add friends, upload resources and set access control policies. After several months of operation, the data of 60 users in the system were selected as the research object. The study subjects had an average of 200 friends in the system. Then, 10 generated resources of each research object are extracted as decision data, and the amount of decision data of the research object is equal to the number of its friends, with an average of 200. A total of 120,000 pieces of basic data were generated by 60 research subjects, which constituted the data set of the experiment in this paper. Each basic data is preprocessed to obtain a five-dimensional vector eigenvalue and a Boolean target value, which correspond to the training data. 120,000 training data after processing constitute the training sample set.

4.2 Comparison of Models

In this experiment, SVM algorithm is used to build the model. In order to verify the effectiveness of the personal preference analysis mechanism proposed by the author, the SVM algorithm is used to construct two models, one model with the personal preference analysis mechanism proposed by the author, the other model without the personal preference analysis mechanism. Compare the performance of the two models. When comparing results, the index of model performance is accuracy.

$$Accuracy = \frac{TP}{TP + FP} \tag{16}$$

Where TP represents the number of people the system recommends to be visible and the user Settings are also visible. FP represents the number of people whose user Settings are not visible but whose system Settings are.

4.3 Development Platform and Environment

The experimental environment is as follows.

CPU: Dual core I7-3770, 3.4 GHz;

Memory is DDR 4B; The hard disk is 256;

The operating system is Windows 10.

The design language is Python 3.6.5 (64-bit).

In the process of programming, SVM model is realized by Sklearn 0.21.3 package.

4.4 Experimental Results Comparison

The data set of this experiment is trained by proposing solutions. When SVM is used for data training, RBF function is selected. Constant C is used to balance training error and γ is used by RBF kernel function in machine learning process. When optimizing parameters C and γ, the grid search algorithm is used to improve the accuracy of the algorithm. Grid search method is an exhaustive search method for specifying parameter values. The optimal learning algorithm can be obtained by optimizing the parameters of the evaluation function through cross validation.

After grid search optimization, the values of the optimal parameters C and γ are 2^{10} and 2^0. It can be clearly seen from Fig. 4 that when the number of folds of cross validation is 10, the SVM model with the personal preference analysis mechanism is the most accurate. With the increase of the number of folds, the accuracy will increase and decrease, but 10 folds cross validation is the best. However, the SVM model without the personal preference analysis mechanism, no matter how many folds, is not as good as the SVM model with the personal preference analysis mechanism.

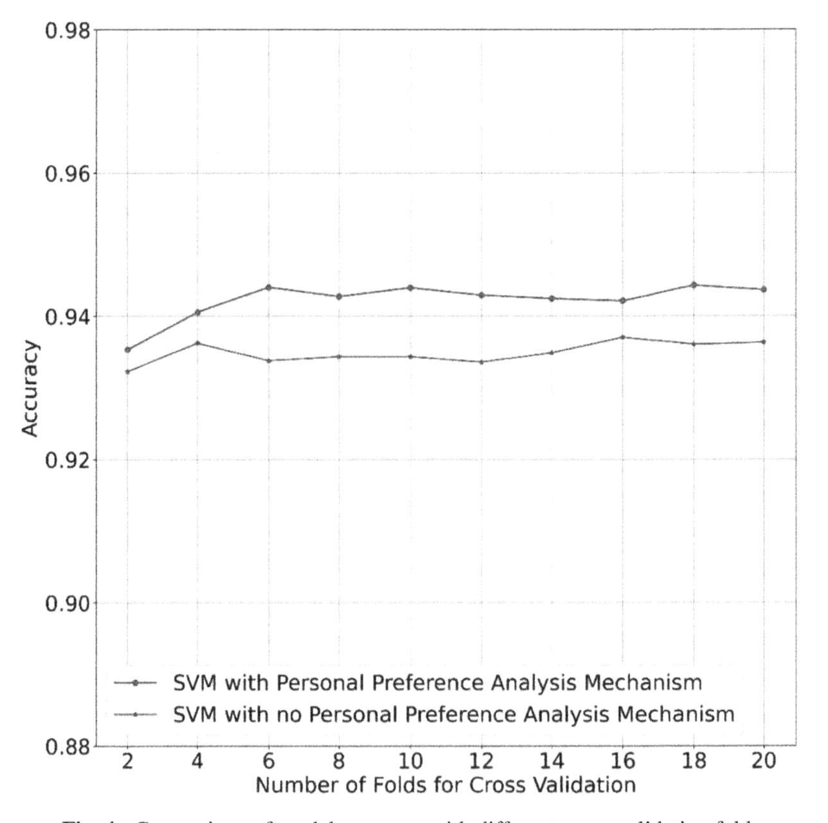

Fig. 4. Comparison of model accuracy with different cross-validation fold.

Figure 5 shows the AUC (Area Under Curve) image. What we know is that this model is optimal if the AUC value is closer to 1. It can be seen from Fig. 5 that the AUC value of the SVM model with the personal preference analysis mechanism added is greater than that of the SVM model without the personal preference analysis mechanism added. Through program calculation, the AUC value of the SVM model with the personal preference analysis mechanism is 0.9599, and that of the SVM model without the personal preference analysis mechanism is 0.9518. The difference between the two is 0.0081. According to Fig. 5, it can also be concluded that the SVM model with personal preference analysis mechanism is the optimal one.

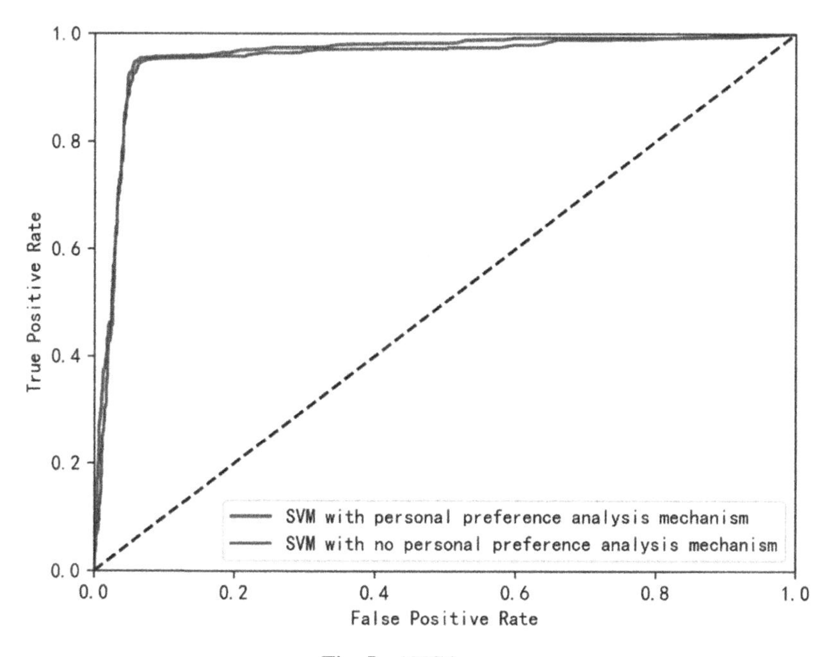

Fig. 5. AUC image.

5 Comparison

This section discusses several related works on access control schemes based on personal preferences, and compared with the schemes proposed in literature [7, 21, 22], the comparison results are shown in Table 2.

Table 2. Comparison of characteristics of different schemes.

	Literature [21]	Literature [22]	Literature [7]	The scheme of this paper
Quantify personal preferences	×	×	×	√
Privacy metric	×	√	√	√
User relationship	√	√	√	√
Sensitive level	√	×	√	√
Trust value	√	×	√	√
Policy individualization	√	√	×	√
History	√	√	√	√

The first column of Table 2 represents seven characteristics to be discussed in this section. The three columns in the middle are the characteristics of the other three programmes. The last column is the characteristics of the scheme in this article.

The scheme proposed in [21] a multi-user collaborative access control framework, which considers the relationship between multiple users and projects, solving conflicts in the privacy settings of the users involved. The method was experimented on an open source social network called Diaspora. However, this solution only solves the privacy Settings between users related to the project, users unrelated to the project are not taken into account. At the same time, users related to the project are difficult to define, and it is difficult to determine which users are related to the project and which are not. There is no privacy measurement and no integration of personal preferences into access control mechanisms.

Literature [22] proposed a scheme that only the friend data requestor whose attributes meet the access policy can decrypt the ciphertext and continue to conduct further communication. In this scheme, it was proposed that taking Weibo social network as an example, everybody has different attributes under different scenarios. However, when registering on Weibo, it is possible to initialize its own properties without setting the properties. Then according to the rules of the scheme, only friends with the access policy satisfied by the attribute can access the data, which is very unfair to a new user who has not set the attribute. At the same time, it also causes pressure to users. This scheme does not take into account the trust value between users and the sensitivity of privacy items, but only generates an access control policy after matching the encryption and decryption of attributes. The scheme in this paper will be more convenient to generate an access policy.

Literature [7] proposed Trust2Privacy, a trust-based access control mechanism to protect the personalized privacy of users after posting their information, which can effectively realize the transformation from trust to privacy. The scheme is similar to the scheme proposed in literature [21] and this paper, for example, the trust value between users and the relationship between users are considered. It is worth noting that the individual location information is taken into account in literature [7]. It is interesting and innovative to add location information to the privacy protection of social networks, but location can also expose a person's privacy. As for the transmission and protection of location, this literature does not involve more studies and put forward corresponding measures. Also, as we know, the current social media software needs to be authorized by the mobile device to obtain an individual's location, so I think the acquisition of location information is difficult for this mechanism. Similarly, literature [7] does not take into account the personal preferences of users, and the access control policies formulated are not more personalized.

6 Conclusion

This paper proposed a social network access control model based on user preferences, proposes a quantification of user privacy preferences by analyzing user historical access control records, and integrates relationship types, access control contents, and user personal preference information entropy as feature vectors, proposes access control policies for user resources in social network platforms by support vector machine algorithm, and generates corresponding access control permission list. The proposed method of quantifying user privacy preferences first constructs an information sensitivity model by the amount of user privacy information and the degree of information privacy, then determines the user sharing degree by personality difference description, and later uses a formal method to determine the information entropy of user personal preferences. Thus, the access control policy is made more consistent with users' personal privacy needs. Therefore, in the next step of the research, natural language processing is used to analyze user text resources and informative labeling and classification so as to protect user privacy in terms of public privacy information attributes, and privacy analysis of user audio and video resources to expand the scope of social network access control processing.

Acknowledgments. The research activities described in this paper have been conducted within "the key R & D and Promotion of Special Science and Technology Projects of Henan Province (212102310480)".

References

1. Pang, J., Yang, Z.: A new access control scheme for Facebook-style social networks. Comput. Secur. **54**(1), 44–59 (2015)
2. Cheng, Y., Park, J., Sandhu, R.: A user-to-user relationship-based access control model for online social networks. In: Cuppens-Boulahia, N., Cuppens, F., Garcia-Alfaro, J. (eds.) DBSec 2012. LNCS, vol. 7371, pp. 8–24. Springer, Heidelberg (2012). https://doi.org/10.1007/978-3-642-31540-4_2
3. Backes, M., Humbert, M., Pang, J, et al.: walk2friends: inferring social links from mobility profiles. ACM (2017)
4. Shan, F., Hui, L., Li, F., et al.: HAC: hybrid access control for online social networks. Secur. Commun. Netw. **2018**, 1–11 (2018)

5. Voloch, N., Nissim, P., Elmakies, M., Gudes, E.: A role and trust access control model for preserving privacy and image anonymization in social networks. In: Meng, W., Cofta, P., Jensen, C.D., Grandison, T. (eds.) IFIPTM 2019. IAICT, vol. 563, pp. 19–27. Springer, Cham (2019). https://doi.org/10.1007/978-3-030-33716-2_2
6. Yousra, A., Kamran, M.A., Basit, R., et al.: Community-centric brokerage-aware access control for online social networks. Future Gener. Comput. Syst. **109**, 469–478 (2018)
7. Xu, G., Liu, B., Jiao, L., et al.: Trust2Privacy: a novel fuzzy trust-to-privacy mechanism for mobile social networks. IEEE Wirel. Commun. **27**(3), 72–78 (2020)
8. Zhu, L., Wang, B.: Stability analysis of a SAIR rumor spreading model with control strategies in online social networks. Inf. Sci. **526**, 1–9 (2020)
9. Aljably, R., Yuan, T., Al-Rodhaan, M., et al.: Anomaly detection over differential preserved privacy in online social networks. PLoS ONE **14**(4), e0215856 (2019)
10. Alemany, J., Val, E.D., Alberola, J., et al.: Enhancing the privacy risk awareness of teenagers in online social networks through soft-paternalism mechanisms. Int. J. Hum. Comput. Stud. **129**, 27–40 (2019)
11. Shan, F., Li, H., Zhu, H.: A game theory-based forwarding control mechanism for social networks. J. Commun. **39**(003), 172–180 (2018)
12. Yu, W., Li, P.: SG-PAC: a stochastic game approach to generate personal privacy paradox access-control policies in social networks. Comput. Secur. **102**, 102157 (2020)
13. Xu, K., Guo, Y., Guo, L., et al.: My privacy my decision: control of photo sharing on online social networks. IEEE Trans. Dependable Secure Comput. **14**(2), 199–210 (2017)
14. Bernstein, M., Bakshy, E., Burke, M., et al.: Quantifying the invisible audience in social networks. In: Proceedings of Human Factors in Computing Systems, pp. 21–30 (2013)
15. Li, F.H., Wang, Y.C., Yin, L.H., et al.: A cyberspace-oriented access control model. J. Commun. **7**(5), 9–20 (2016)
16. Marinescu, P., Parry, C., Pomarole, M., et al.: IVD: automatic learning and enforcement of authorization rules in online social networks. In: S&P 2017, pp. 1094–1109 (2017)
17. Li, B., Huang, D., Wang, Z., et al.: Attribute-based access control for ICN naming scheme. IEEE Trans. Dependable Secure Comput. **15**(2), 194–206 (2018)
18. Yeh, L., Chiang, P., Tsai, Y., et al.: Cloud-based fine-grained health information access control framework for lightweight IoT devices with dynamic auditing and attribute revocation. IEEE Trans. Cloud Comput. **6**(2), 532–544 (2018)
19. Zhang, P.P., Peng, C.G., He, C.Y.: A privacy protection model based on privacy preference and its quantification method are presented. Comput. Sci. **45**(006), 130–134 (2018)
20. Zhang, C., Liang, Y., Fang, H.S.: Social network information recommendation method that supports privacy protection. J. Shandong Univ. (Nat. Sci. Ed.) **55**(03), 13–22 (2020)
21. Alshareef, H., Pardo, R., Schneider, G., et al.: A collaborative access control framework for online social networks. J. Logical Algebraic Meth. Program. **114**, 100562 (2020)
22. Lei, Z.A., El, B., Gw, C., et al.: Secure fine-grained friend-making scheme based on hierarchical management in mobile social networks. Inf. Sci. **554**, 15–32 (2021)
23. Liu, G., Yang, Q., Wang, H., et al.: Trust assessment in online social networks. IEEE Trans. Dependable Secure Comput. **18**(2), 994–1007 (2021)

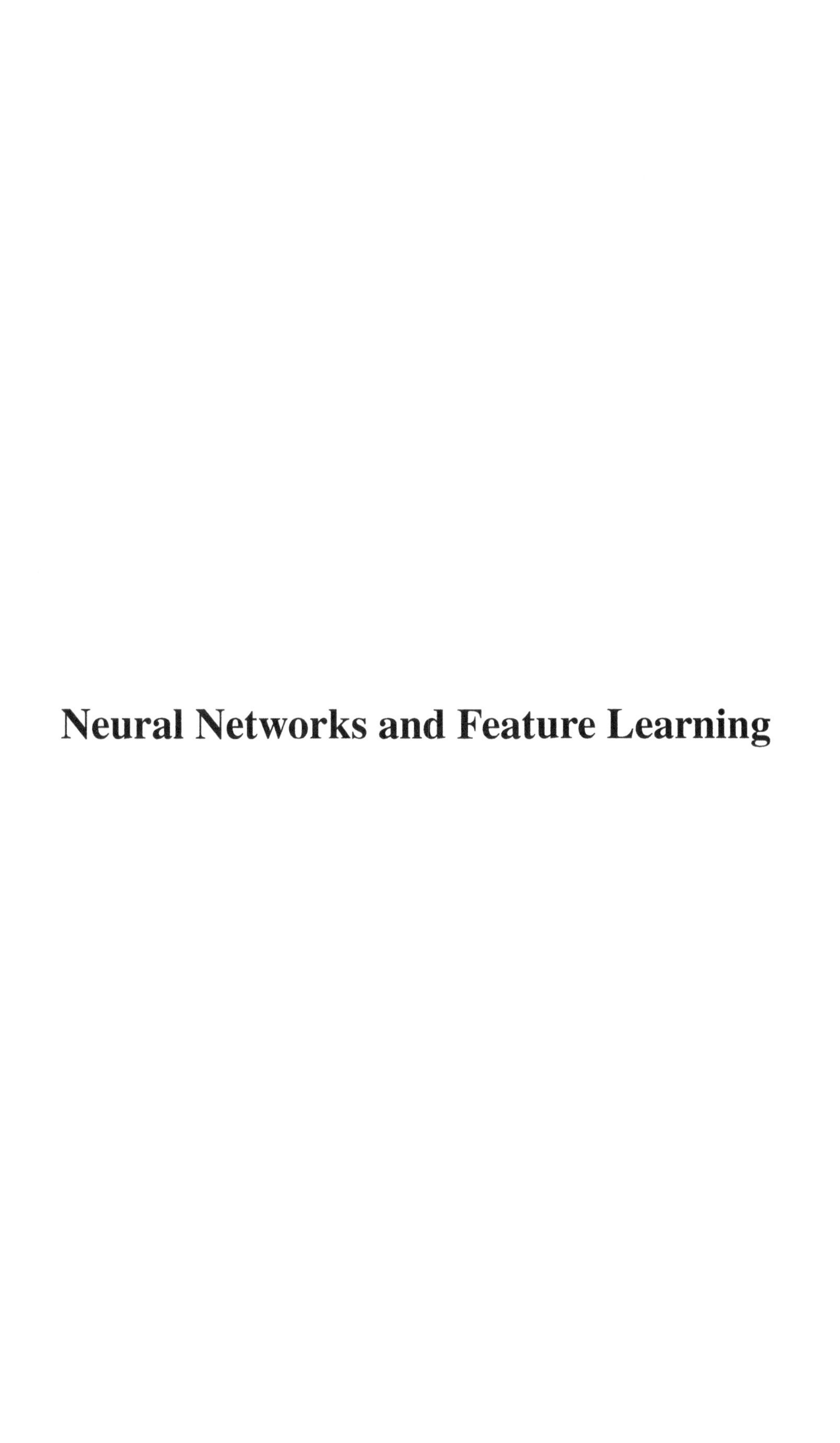

Neural Networks and Feature Learning

Multi-scale Two-Way Deblurring Network for Non-uniform Single Image Deblurring

Zhongzhe Cheng[1], Bing Luo[1], Li Xu[2](✉), Siwei Li[3], Kunshu Xiao[3], and Zheng Pei[1,2]

[1] The Center for Radio Administration Technology Development, School of Computer and Software Engineering, Xihua University, Chengdu 610039, China
[2] The School of Science, Xihua University, Chengdu 610039, China
[3] Xihua College of Xihua University, Chengdu 610039, China

Abstract. We propose a new and effective image deblurring network based on deep learning. The motivation of this work is based on traditional algorithms and deep learning which take an easy-to-difficult approach to image deblurring. In traditional algorithms, a rough blur kernel is obtained first, and then a precise blur kernel is gradually refined. In deep learning, the pyramid structure is adopted to restore clear images from easy to difficult. We hope to recover the clear image by two-way approximation. One network recovers the roughly clear image from the blurred image, and the other network recovers part of the structural information from the blank image, and finally the two networks are added together to obtain the clear image. Experiments show that since we decomposed the original deblurring task into two different tasks, the network performance has been effectively improved. Compared with other latest networks, our network can get clearer images.

Keywords: Multi-scale network · Two-way learning · Non-uniform deblurred

1 Introduction

Image deblurring is a traditional computer vision problem. Image blur is mainly formed by camera shake and object motion. It exists in various scenes in the world, such as natural images [8], human face images [9], text images [7], etc. The purpose of traditional deblurring algorithms is to obtain blur kernels and clear images from blurred images. In low-level vision, this is an ill-posed problem, because the same blurred image can be corresponded various pairs clear latent image and blur kernel [1–4]. At the same time, traditional algorithms have general effects on non-uniform blur. In real blurred scenes, image blur is often not affected by a single factor, which makes it difficult for traditional algorithms to model non-uniform blur, which affects the final deblurring effect.

J. Xiong et al. (Eds.): MobiMedia 2021, LNICST 394, pp. 593–601, 2021.
https://doi.org/10.1007/978-3-030-89814-4_43

With the development of deep learning, neural networks are gradually used in image deblurring. Because of the learning ability of neural network, it has good performance on the image with non-uniform blur, which can adaptively deblur each pixel. When the neural network was first used for deblurring, the researchers hoped to estimate the motion blur through the network and obtain the image blur carried by each pixel, which ultimately obtained the blur information of each position of the image [19,21,24]. In recent years, researchers have found that direct estimation of clear images is better than estimation of motion blur. These neural networks are roughly divided into two categories: multi-scale neural networks and generative adversarial networks. The multi-scale neural network deblurs the blurred images of multiple scales to achieve the effect of removing the blur from easy to difficult, which is similar to the pyramid structure in traditional algorithms. Generative adversarial network uses the generation of confrontation mechanism to get closer to the real clear image. Although the pixel difference between the deblurred image and the original image is larger, it is more in line with the human eye's perception of a real clear image.

Multi-scale neural network was first proposed by Nah et al. [10]. They construct a multi-scale network by analogy with the pyramid model in the traditional algorithm, and formed a complete de-blurring network by splicing the de-blurring results between different scales. However, the model has large parameters which results in slower network convergence and longer training time. Then, Tao et al. [11] proposed a multi-scale recurrent neural network based on [10], which greatly reduces model parameters by sharing parameters at different scales, which reduces training time. Zhang et al. [17] construct different parameter sharing and parameter independence method according to the role of each convolutional layer of the network in the network, which further improved the deblurring effect.

Generative adversarial networks are most commonly used for image generation, and then gradually applied to various computer vision tasks. Kupyn et al. [14] use generative adversarial network for image deblurring. Since the deblurred image generated according to the Mean-Square-Error(MSE) loss function does not necessarily in line with the human eye's definition of a clear image, the sharp edges of the image are still partially blurred. The generated confrontation network can use the discriminant model to make the generated image as close to the real image as possible. Although some image information will be lost, it is better than other neural networks in terms of image structure and realism. Then, Kupyn et al. [15] add the pyramid structure and the local-global adversarial loss function to improve the network performance on the basis of the original network.

We hope to propose a new network based on the original multi-scale network and integrate the optimization ideas of generating adversarial network into the network to improve performance. Our main contributions are as follows: (1) We propose a new deblurring network based on a multi-scale framework, which incorporates new optimization ideas and can obtain clearer results. (2) The network we proposed has two branches, which respectively obtain the final clear image from the two ways of the blurred image and the blank image. (3) We prove that the new network can obtain roughly clear image residual images, and the final deblurred image obtained is better than other networks through experiments.

2 Related Work

Multi-scale Network: The multi-scale network is similar to the pyramid framework in traditional algorithms, and it is based on the observation: after the blurred image is upsampled, the smaller the image size, the smaller the degree of blur. In other words, the multi-scale network first obtains a rough result by deblurring the low-scale image, and then refines the image through the large-scale network, and finally obtains a clear image. Nah et al. [10] construct a multi-scale deblurring network based on the above principles, but there are problems such as large model and parameters which result in some difficulty in training. Based on the work of [10], Tao et al. [11] use Recurrent Neural Network (RNN), which reduces the model size and the number of parameters. [11] adds connections between feature layers of different scales to obtain a better deblurring result. Gao et al. [12] find that the degree of blur at different scales is different, so using the same network for feature extraction will affect the network's extraction of clear image features. Gao et al. [12] adopt the parameter independence of the feature extraction layer to avoid it. At the same time, Gao et al. [12] also find that the network deblurring process after feature extraction is similar. Therefore, parameter sharing is adopted for part of the convolutional layer, which greatly reduces the number of parameters without reducing performance. However, the multi-scale networks described above are all based on a pyramid structure, and there are only differences in parameter sharing and independence between different networks. In terms of network structure, the three methods have no obvious differences. Cai et al. [25] adds the extreme channel prior to the multi-scale network framework at each level of the network, which improves the network performance by constraining the sparseness of the polar channel of the feature image. In general, the multi-scale network can restore better image content information, but the effect of image edge restoration is general, especially for sharp edges, which still contains some blur information.

Generative Adversarial Network: The main purpose of the generation adversarial network in the field of image deblurring is to restore sharp edges, so that the resulting clear image is more in line with human perception. Kupyn et al. [14] propose a new generative confrontation network. The generator network is composed of multiple residual blocks with the same structure, and the discriminator uses Wasserstein distance. The network loss function is constructed by three loss functions which contains MSE, confrontation loss and feature loss. Kupyn et al. [15] make improvements on the basis of the original network of [14]. The pyramid structure is integrated into the generation network, and the global-local discriminator is added to the discriminant network to further improve the network performance. Zhang et al. [17] propose a new optimization idea. Most of

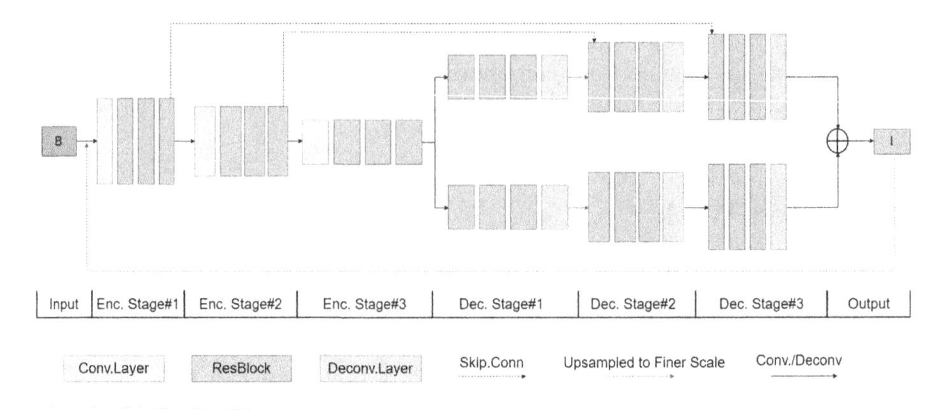

Fig. 1. Our proposed Multi-scale two-way learning network.

the discriminators that have been proposed hope that the network will discriminate clear images as 1 and blurred images as 0. Then, by continuously optimizing the generator, the deblurred image will gradually approach from 0 to 1. Zhang et al. [17] hope to move the deblurred image and the clear image closer to 0.5. This optimization result is similar to other methods. In the end, the deblurred image and the clear image will be close, and then the difficulty of optimizing 0 and 1 to 0.5 together is less than other methods, so the effect will be better. However, the above three methods are generative adversarial networks, so they all have a common problem: the discriminator will reduce the consistency of the deblurring result and the original image while optimizing the production plant generator, which leads to the difference in pixel values from the original image (Fig. 1).

3 Proposed Method

Most of the network's ideas for image deblurring are from easy to difficult, gradually removing image blur information. Whether it is the pyramid structure in the traditional algorithm, or the multi-scale network and residual learning in the neural network, the improvement of network performance often depends on artificially reducing the difficulty of network learning. We hope to build a network that one part restores the image based on the blurred image and the other part restores the image from blank image, which can complement each other in the process of restoring the image and finally get a clear image.

3.1 Multi-scale Two-Way Deblurring Network

As shown in Fig. 3, our proposed network is composed of multiple encoders and decoders. At the same time, the small-scale image is restored to obtain the deblurring results and then passed to the large-scale network after up-sampling. Each encoder is composed of a convolutional layer and three residual blocks, and

(a) Blurry Image (b) Nah et al. (c) Tao et al. (d) Ours

Fig. 2. Visual comparison on the dataset of GoPro dataset

the decoder is composed of a deconvolutional layer and three residual blocks. Different from [10,11] network structure, our network has two branches in the decoder part to recover the image content and the remaining information of the image respectively. In order to guide the two branches of the network to recover the corresponding image information, we use the connection between the feature maps to achieve the goal. The up part of the decoder inherits the feature map of the encoder, so the main image content is obtained first. The lower part of the decoder has no feature map skip connection, so the remaining image information is restored from blank image. Finally, we add results from two parts network can get the clear image.

The network we proposed can be expressed by the following formula:

$$I^i, I^i_{up}, I^i_{down} = Net(B^i, I^{i+1}; \theta) \tag{1}$$

where $I^i, I^i_{up}, I^i_{down}$ represents the output of the deblurred image, the up part and the low part of the decoder of the i-th network, and $I^i = I^i_{up} + I^i_{down}$; B^i represents the input of the blurred image; θ represents network parameters.

3.2 Loss Function

We use deblurred images and blurred images on various scales to calculate the mean square error as the loss function of the network. The general multi-scale

(a) Blurry Image (b) Down-net (c) Up-net (d) Finnal result

Fig. 3. Visual comparison of our proposed net

Table 1. Table reports the mean PSNR and SSIM obtained over the GoPro dataset [10]

Method	Nah et al. [10]	Tao et al. [11]	Kupyn et al. [14]	Kupyn et al. [15]	Ours w/o TWNet	Ours
PSNR	29.08	30.10	28.70	28.17	30.18	**30.56**
SSIM	0.914	0.924	0.958	0.925	0.926	**0.933**

loss function has the same weight on each scale, but our goal is to get the output on the largest scale, so we increase its weight, hoping that the network will prioritize the final output result. The loss function expression is as follows:

$$L = \sum_{i=1}^{S} \frac{\alpha_i}{T_i} \left\| F_i(B_i; \theta) - I_i \right\|_2^2 \tag{2}$$

where I_i, B_i represents the output of the deblurred image and blurred image of the i-th of scale; T_i represents the number of pixels; θ represents network parameters; α_i represents the weight of different scales.

4 Experiment

We implement our framework on the TensorFlow platform [18]. To be fair, all experiments are performed on the same dataset with the same configuration. For model training, we use Adam solver [26] with $\beta_1 = 0.9$, $\beta_2 = 0.999$ and $\epsilon = 10^{-8}$. The learning rate is initially set to 0.0001, exponentially decayed to 0 using power 0.3. We set the convolution kernel size 3×3, $\alpha_1 = \alpha_2 = 1$, $\alpha_3 = 5$. We randomly

Table 2. Table reports the mean PSNR and SSIM obtained over the Köhler dataset [5]

Method	Xu et al. [6]	Pan et al. [8]	Nah et al. [10]	Tao et al. [11]	Kupyn et al. [14]	Kupyn et al. [15]	Ours
PSNR	29.96	30.20	26.48	26.75	26.10	26.36	**26.80**
SSIM	0.876	0.888	0.807	0.837	0.816	0.820	**0.840**

crop 256×256 images from original paired clear and blurred images as training images. We use Xavier method [22] initialization parameters. Our experiments can converge after 4000 epochs.

Dataset Preparation: In order to create a large training dataset, methods based on early learning [19–21] synthesize blurred images by convolving clear images with real or generated uniform/uneven blur kernels. Due to the simplified image formation model, the synthesized data is still different from the data captured by the camera. Recently, researchers [10] propose a method to generate blurred images by averaging consecutive short exposure frames in videos taken by high-speed cameras (such as high-speed cameras). For example, GoPro Hero 4 Black, which can approximate long exposure blurry picture. These generated frames are more realistic because they can simulate complex camera shake and object motion, which are common in real photos.

In order to fairly compare the performance differences between different network frameworks, we train our network on the GoPro dataset, which contains 3214 image pairs. Like [10–13], we choose 2103 pairs as the training set and 1111 pairs as the test set.

Benchmark Dataset: We first conduct experiments on the test set of the GoPro dataset, which contains many complex blurs caused by camera shake and object motion. Table 1 shows our performance compared with other state-of-the-art methods. We choose Peak-Signal-to-Noise Ratios(PSNR) and Structural Similarity (SSIM) [23] as the evaluation criteria. It can be seen that the generative adversarial network [14,15] have significant advantages in restoring image structure (SSIM), but the result after deblurring is quite different from the original image pixel value (PSNR). At the same time, the multi-scale network is lower than our proposed method in both PSNR and SSIM. In addition, we give the experimental results of the network structure without two-way deblurring. Figure 2 shows the subjective effect of our and other methods. Figure 3 shows the output results of each branch of the bidirectional network. It can be seen that the up part of the network removes part of the image blur and the lower part of the network mainly focuses on the edge of the image. Information was supplemented.

Then, we also conducted experiments on the traditional dataset [5], this dataset consists of 4 images and 12 blur kernels, of which three blur kernels are larger in size and form larger blurs. It can be seen from the Table 2 that since we doesn't train network for large blurred images during the training process, the effect of large blurred images is general. Traditional algorithms directly model blurred images, so the effect of processing large blurred images is better. However, our network performance is still better than other neural networks, which shows the effectiveness of our proposed network.

5 Conclusion

Based on the multi-scale neural network, combined with the optimization idea of two-way approximation, we constructed a new multi-scale two-way deblurring network. This network, like other neural networks, has a significant effect on non-uniform deblurring. Compared with other state-of-the-art multi-scale networks, our network can better restore image edges and get better deblurring images.

Acknowledgement. This work was supported in part by National Natural Science Foundation of China (No. 61801398), The Young Scholars Reserve Talents program of Xihua University and The program for Vehicle Measurement, Control and Safety Key Laboratory of Sichuan Province (No. QCCK2019-005) and The Innovation and Entrepreneurship Project of Xihua Cup (No. 2021055) and The Talent plan of Xihua College of Xihua University (No. 020200107).

References

1. Shan, Q., Jia, J., Agarwala, A.: High-quality motion deblurring from a single image. ACM Trans. Graph. **27**, 1–10 (2008)
2. Cho, S., Lee, S.: Fast Motion Deblurring. ACM Trans. Graph. **28**, 1–8 (2009)
3. Fergus, R., Singh, B., Hertzmann, A., Roweis, S.T., Freeman, W.T.: Removing camera shake from a single photograph. ACM Trans. Graph. **25**, 787–794 (2006)
4. Xu, L., Jia, J.: Two-phase kernel estimation for robust motion deblurring. In: Daniilidis, K., Maragos, P., Paragios, N. (eds.) ECCV 2010. LNCS, vol. 6311, pp. 157–170. Springer, Heidelberg (2010). https://doi.org/10.1007/978-3-642-15549-9_12
5. Köhler, R., Hirsch, M., Mohler, B., Schölkopf, B., Harmeling, S.: Recording and playback of camera shake: benchmarking blind deconvolution with a real-world database. In: Fitzgibbon, A., Lazebnik, S., Perona, P., Sato, Y., Schmid, C. (eds.) ECCV 2012. LNCS, vol. 7578, pp. 27–40. Springer, Heidelberg (2012). https://doi.org/10.1007/978-3-642-33786-4_3
6. Xu, L., Zheng, S., Jia, J.: Unnatural l0 sparse representation for natural image deblurring. In: CVPR (2013)
7. J. Pan, Z. Hu, Z. Su, Yang, M.H.: L0-regularized intensity and gradient prior for deblurring text images and beyond. IEEE Trans. Pattern Anal. Mach. Intell. **39**(2), 342C355 (2017)
8. Pan, J., Sun, D., Pfister, H., Yang, M.H.: Blind image deblurring using dark channel prior. In: CVPR (2016)
9. Pan, J., Hu, Z., Su, Z., Yang, M.-H.: Deblurring face images with exemplars. In: Fleet, D., Pajdla, T., Schiele, B., Tuytelaars, T. (eds.) ECCV 2014. LNCS, vol. 8695, pp. 47–62. Springer, Cham (2014). https://doi.org/10.1007/978-3-319-10584-0_4
10. Nah, S., Kim, T.H., Lee, K.M.: Deep multi-scale convolutional neural network for dynamic scene deblurring. In: CVPR (2017)

11. Tao, X., Gao, H., Shen, X., Wang, J., Jia, J.: Scale-recurrent network for deep image deblurring. In: CVPR (2018)
12. Gao, H., Tao, X., Shen, X., Jia, J.: Dynamic scene deblurring with parameter selective sharing and nested skip connections. In: CVPR (2019)
13. Zhang, H., Dai, Y., Li, H., Koniusz, P.: Deep stacked hierarchical multi-patch network for image deblurring. In: CVPR (2019)
14. Kupyn, O., Budzan, V., Mykhailych, M., Mishkin, D., Matas, J.: DeblurGAN: blind motion deblurring using conditional adversarial networks. In: CVPR (2018)
15. Kupyn, O., Martyniuk, T., Wu, J., Wang, Z.: DeblurGAN-v2: deblurring (Ordersof-Magnitude) faster and better. In: ICCV (2019)
16. Aljadaany, R., Pal, D.K., Savvides, M.: Douglas-Rachford networks: learning both the image prior and data fidelity terms for blind image deconvolution. In: CVPR (2019)
17. Zhang, K., Luo, W., Zhong, Y., et al.: Deblurring by realistic blurring. arXiv (2020)
18. Abadi, M., et al.: TensorFlow: large-scale machine learning on heterogeneous systems (2015). Software available from tensorflow.org
19. Chakrabarti, A.: A neural approach to blind motion deblurring. In: Leibe, B., Matas, J., Sebe, N., Welling, M. (eds.) ECCV 2016. LNCS, vol. 9907, pp. 221–235. Springer, Cham (2016). https://doi.org/10.1007/978-3-319-46487-9_14
20. Schuler, C.J., Hirsch, M., Harmeling, S., Scholkopf, B.: Learning to deblur. TPAMI **38**(7), 1439–1451 (2016)
21. Sun, J., Cao, W., Xu, Z., Ponce, J.: Learning a convolutional neural network for non-uniform motion blur removal. In: CVPR (2015)
22. Glorot, X., Bengio, Y.: Understanding the difficulty of training deep feedforward neural networks. In: AISTATS, pp. 249–256 (2010)
23. Wang, Z., Bovik, A.C., Sheikh, H.R., Simoncelli, E.P.: Image quality assessment: from error visibility to structural similarity. IEEE Trans. Image Process. **13**(4), 600612 (2004)
24. Bahat, Y., Efrat, N., Irani, M.: Non-uniform Blind Deblurring by Reblurring. In: ICCV (2017)
25. Cai, J., Zuo, W., Zhang, L.: Dark and bright channel prior embedded network for dynamic scene deblurring. IEEE Trans. Image Process. **29**, 6885–6897 (2020)
26. Kingma, D.P., Ba, J.: Adam: a method for stochastic optimization. In: ICLR (2014)

BMP Color Images Steganographer Detection Based on Deep Learning

Shuaipeng Yang[1], Yang Yu[1(✉)], Xiaoming Liu[2], Hong Zhang[3], and Haoyu Wang[1]

[1] Beijing University of Posts and Telecommunications, Beijing, China
{2018140769,yangyu}@bupt.edu.cn
[2] CNCERT/CC Chaoyang District, Beijing, China
liuxm@cert.org.cn
[3] Tianjin Branch of CNCERT/CC Nankai District, Tianjin, China
zhangh@cert.org.cn

Abstract. A user who achieves covert communication by embedding secret information in the original image is called steganographer. Steganographer detection determines which user sent a secured image with a secret message. Existing steganographer detection algorithms take gray images as the main research content. To better adapt to the reality, we propose a WiserNet-based steganograph detection algorithm for the characteristics of BMP color images, and the process is divided into the following three steps: feature extraction through each channel convolution structure, prevent the conventional convolution structure destroy the correlation between the color image channel operation, reduce the number of the extraction of feature dimension. The use of a per-channel convolution structure makes it easier to extract color image features, and the low-dimensional feature vector reduces the time required for subsequent clustering algorithms, which improves the efficiency of steganographer detection. Simulation experiments are conducted for the classification of feature extractors, detection of different steganographic rates, and detection of different image scales. First, the steganalysis binary classification results of this algorithm are compared with similar algorithms, and the classification accuracy is 84.90% when the steganalysis rate is 0.4 BPC, which is 1.11% higher than Ye-Net and 0.83% higher than Xu-ResNet. Since there is very little published research on steganography detection of color images, four feature extractors, Ye-Net, Xu-ResNet, SRNet, and WiserNet, will be used in this experiment to replace the WiserNet-100 feature extractor in the steganography detection algorithm. The results show that the detection accuracy of the algorithm proposed in this paper reaches 93% when the embedding rate is 0.2 BPC, and the detection accuracy reaches 100% when the embedding rate is greater than 0.2 BPC. The steganographic detection accuracy reaches 84% when the graph scale is 60% and the steganographic rate is 0.2 BPC. In terms of detection time, the WNCISD-100 is 7.79 s, which is 50% less time-consuming compared to SRSD.

Keyword: WiserNet · BMP color image · Steganographer detection · DBSCAN

This work is supported by the National Key R&D Program of China under Grant No. 2016YFB0801004.

J. Xiong et al. (Eds.): MobiMedia 2021, LNICST 394, pp. 602–612, 2021.
https://doi.org/10.1007/978-3-030-89814-4_44

1 Introduction

Multimedia, such as texts, images, audio, and video, is considered to be one of the most effective mediums of communication and information sharing [1]. Image steganography ensures secret communication by embedding secret information in ordinary images [2, 3]. The most commonly used steganographic medium is the digital image [4]. As opposed to image steganography, image steganalysis is the art of revealing the secret information that is hidden by the steganographer in the images.

Currently, most image steganalysis techniques follow the pattern of separating suspicious images into overlay images or steganographic images. This problem is known as the steganographer detection problem. With the large amount of image data generated by social media and the development of steganographic algorithms, some users attempt to deliver secret information by image steganography through innocent users of social networks [5–7]. The detection performance of traditional image steganalysis can be significantly degraded. The effect of information theft forensics is not evident. Therefore, an increasing number of scholars have shifted the focus of image security detection from image steganalysis to locating the user who first sent the data carrying the hidden image, i.e., the steganographer, or criminals [8–10]. How to locate those users who send digital steganographic images carrying the hidden message and identify the steganographer from many common users is a major challenge for steganographer detection. Steganographer detection can effectively detect illegal users and avoid information being stolen,which we believe will play a key role in many important multimedia security applications in the future.

Steganographer detection aims to locate criminals among a large number of innocent users who may be carrying secret information using the steganography technique. The difficulty of this task is in the collection of useful evidence, that is, to detect secret messages generated by an unknown steganography method and the payloads embedded in suspicious images and to identify criminals based on the image features. Existing steganalysis methods are algorithms that rely on the binary classification of known data sets and payloads. And the detection performance decreases significantly in the case of unknown payloads. In the inference phase, the learned model is used to extract discriminative features, thus capturing the differences between illegal and innocent users. A series of experimental results show that the method performs well in both the spatial and the frequency domains even with low embedded payloads. The method has good robustness and offers the possibility to solve the payload mismatch problem.

In this paper, deep learning-based steganographer detection is studied for color images, and the WiserNet(Wider Separate-Then-Reunion Network) steganographer detection method for color images is proposed. The main findings of the paper are as follow:

This is the first study on BMP color image steganographer detection. The experimental results show that the model can identify the steganographer accurately in terms of various steganography rates.

In this paper, we introduce DBSCAN (Density-Based Spatial Clustering of Applications with Noise) for steganographer detection after feature extraction, which can effectively identify users with different steganography image ratios. The simulation and experiment show that this method has fair good robustness.

The proposed method is validated by simulation and experiments taking on a standard dataset, showing that our method achieves a low detection error rate on the spatial domain.

The rest of the paper is organized as follows: In Sect. 2, we give a detailed overview of the current framework of grayscale graph steganographer detection methods. In Sect. 3, we describe the proposed WiserNet-based color images steganographer detection framework in detail. In Sect. 4, we perform a series of comprehensive experiments to verify the performance of our proposed method. In Sect. 5, we summarize our proposed work and outline future work.

2 Related Work

This section will give a brief overview of the latest results on grayscale graph steganographer detection. So far, steganographer detection on color images has not been developed. Therefore, we introduce the steganographer approach based on BMP color images.

2.1 Color Images Steganalysis

The main battlefield for information hiding in spatial domain images is on grayscale images. However, the confrontation between color image steganography algorithms and color image steganalysis algorithms has also received increasing attention from researchers since most images in real life come with color. The mainstream grayscale image steganography algorithms, including the well-known S-UNIWARD [11], HILL [12], and MiPOD [13], employ the so-called minimizing additive embedding distortion framework. Later, Denemark and Fridrich [14] and Li [15] went a step further from additive distortion algorithms and constructed effective non-additive distortion algorithms to embed images using the correlation between adjacent pixels. Among them, Li [16] proposed the CMD(Clustering Modification Directions) steganography algorithm to achieve excellent steganography performance.

To illustrate, grayscale image steganography algorithms (such as S-UNIWARD and HILL) can be directly applied to color images. This is usually done by treating each color channel as separate grayscale images and embedding secret information into each color channel separately. The general practice is to embed bits of secret information independently into each color channel by treating each color channel as a separate grayscale image.

Subsequently, inspired by the CMD steganography algorithm, Tang [17] proposed a non-additive steganography algorithm for color images, CMD-C. The CMD-C color images steganography algorithm preserves not only the correlation of pixels within each color channel, but also the correlation between the three color channels, and uses these correlations for embedding. Therefore, the performance of resisting the steganalysis algorithms for color images is even better and obtains good color image steganography performance.

The currently dominant steganalysis algorithms are steganalysis detectors built using rich models with multi-dimensional features [18] and integrated classifiers [19]. A separation, followed by aggregation network, was proposed in the paper about WiserNet.

The authors considered the weighted summation operation in the conventional convolutional structure, i.e., the process of forming a linear combination of the input color channels, as a "linear complicity attack" [20]. It retained the strongly correlated content while weakening the irrelevant noise in the input, which was more favorable for the determination of the steganalysis results.

2.2 Steganographer Detection

Currently, far too little attention has been paid to the task of steganographer detection, which can be divided into two main categories based on the way of detecting the steganographer: steganographer detection based on clustering and steganographer detection based on anomaly detection.

In 2011, Ker et al. [21] first transformed steganographer detection into a clustering problem study. They first extract 274-dimensional PEV features for each image of the user, which was composed of 193-dimensional DCT coefficient features and 81-dimensional calibrated Markov features.

Subsequently, based on the extracted PEV features, Ker et al. used the MMD(maximum mean discrepancy) to calculate the distance between feature sets for each pair of users as a similarity metric between users. Finally, an aggregated hierarchical classification algorithm based on the similarity metric was used to distinguish steganographers from the many non-steganographer.

In 2012 and 2014, Ker et al. [22, 23] further improved their work by defining a steganographer as an anomaly among the communicating users and proposed to identify the steganographer using anomaly detection methods. Unlike the previous work, they used the method of local anomaly factor [24] to calculate the anomaly degree value of each user and rank them, and the user with the highest anomaly value would be identified as the steganographer.

In 2018 and 2019, after calculating the MMD distance of every two users, Zheng used the conjoint hierarchical clustering algorithm to detect the steganographer, and combined objects by establishing a hierarchical tree. Finally, all non-steganographer were grouped into one class, while the steganographers were separately classified into one class [20]. In 2019, Zheng enhanced feature extraction by multi-scale embedding and then selected the steganographer by Gaussian voting [27]. In 2020, Zheng detects steganographers based on LOF (local outlier factor) and selective strategy [28].

3 Method

The proposed framework for steganographer detection based on BMP color images is shown in Fig. 1. The framework mainly consists of three parts. In the first step, each color images of each user are extracted with a special cross using the trained WiserNet; second, based on the extracted feature vectors, the steganographic images and non-steganographic images are classified using the DBSCAN clustering method; in the third step, the steganographer is detected using the ranking determination method. Unlike the existing grayscale graph-based steganographic detection framework, the steganographer detection algorithm proposed here is further optimized based on WiserNet in the feature extraction step to cut the number of feature vectors to 100, which reduces the time consumption of clustering computation. The proposed method will be introduced in detail in the subsequent sections.

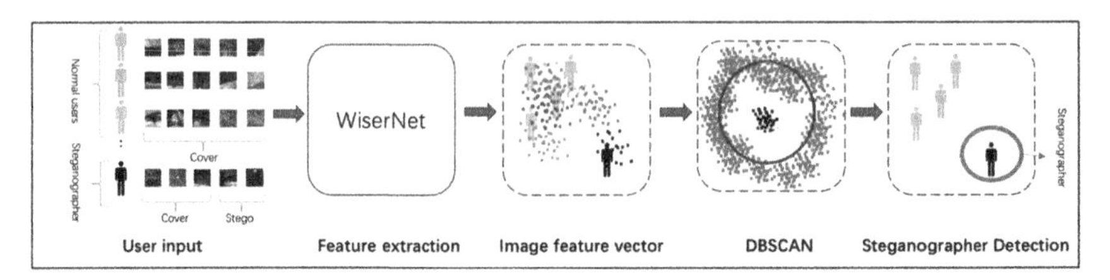

Fig. 1. Color image steganographer detection framework based on WiserNet.

3.1 Feature Extraction

The main object of this paper is true-color images, and we consider only the RGB true-color model, given a color image X of size M × N. It contains three color channels, namely, the red R channel, the green G channel, and the blue B channel. In this section, we do not consider the specific features of each color.

Thus, X can be expressed without loss of generality as $\{X_1, X_2, X_3\}$, where $X_i = (x_{i,pq})M \times N$, $x_{i,pq} \in \{0, 1, \cdots, 255\}$, $1 \leq i \leq 3$, $1 \leq p \leq M$, $1 \leq q \leq N$. In this paper, the feature extraction of color images is based on WiserNet. The first part of this network takes depthwise convolution, in which each input channel is convolved with a matrix of 30 convolution kernels to obtain the corresponding 30 independent output channels with 3 input channels and 30 convolution kernels, and 90 output feature maps will be obtained after passing the depthwise convolution structure. The second part performs the regular convolutional layer operation, which contains the BN (batch normalization) layer, ReLU (rectified linear unit) layer, and average pooling layer. The input feature image size of the regular convolution layer is reduced to 256 * 256, 128 * 128, and 32 * 32 in that order. The third part performs the pooling layer for processing, and the feature vector dimension is reduced by four times of fully connected network, that is, in descending order to 800, 400, 200, and 100. The final output is a 100-dimensional feature vector (Fig. 2).

3.2 DBSCAN Clustering Sorting to Detect Steganographer

After feature extraction to obtain a 100-dimensional feature vector, we take the DBSCAN algorithm based on density clustering to detect steganographer. Unlike the KMeans algorithm, This method does not require determining the number of clusters, but rather inferring the number of clusters based on the data, and can generate clusters for arbitrary shapes. DBSCAN is one of the classes that achieve the final clustering by the set of samples connected by the maximum density derived from the density reachability relation. The parameter (ϵ, MinPts) is used to describe the closeness of the neighborhood sample distribution. Where ϵ describes the neighborhood distance threshold of a sample

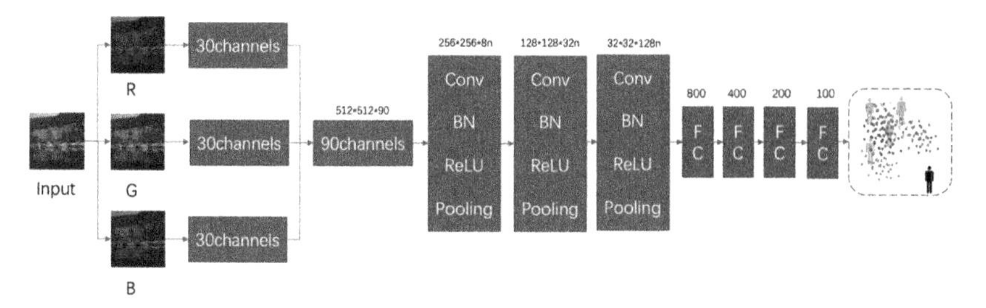

Fig. 2. Feature extraction framework based on WiserNet.

and MinPts describes the threshold of the number of samples in the neighborhood of a sample with distance. There can be one or more core objects inside the clusters of DBSCAN. If there is only one core object, all other non-core object samples in the cluster are in the ϵ-neighborhood of this core object; if there are multiple core objects, there must be one other core object in the ϵ-neighborhood of any one core object in the cluster, otherwise, these two core objects cannot be density reachable. The set of all samples in the ϵ-neighborhood of these core objects forms a DBSCAN clustering cluster.

4 Experiments

4.1 Data Set and Experimental Setup

All experiments in this paper were conducted on the BossBase (v1.01) dataset. Following the data set generation process used in the article [24], we obtained the BossBMP data set from 10,000 full-resolution original images of BossBase through subsampling operation.

BossBase dataset was also used in all experiments in this paper. In this chapter, the dataset generation process in the article [29] is used as a reference, and the original DNR format image is decontaminated using UFRaw software. The BMP format image is obtained by subsampling using ImageMagick, and then it is centered and cropped to 512 $\times$ 512 to obtain the BossBMP dataset. In the following illustration, the advanced CMD-C-Hill color image steganography algorithm is selected as the detection object, and the load is set to 0.1, 0.2, 0.3, 0.4, and 0.5 BPC (bits per color channel) for steganography operation, so as to obtain the carryover image corresponding to the carrier image.

4.2 Evaluation of the Effectiveness of Feature Extraction Methods

The proposed color image steganographer detection in this paper consists of two main parts, of which feature extraction based on WiserNet is the most critical technique. Therefore the effectiveness of the proposed feature extraction method for the steganalysis task is first tested before the performance evaluation of the steganographer detection task. Since the compared methods are based on the binary classification training models of steganograms and non-steganograms, in hopes of a fair comparison, the experiment processes the 100-dimensional features for binary classification results and compares them with the classical Ye's model, Xu's model, and SRNet networks (Tables 1 and 2).

Table 1. Performance evaluation parameters of feature extraction network.

Accuracy	Precision	Recall	F1
84.90%	82.04%	84.65%	82.80%

Table 2. Comparison of the performance and feature latitude of different color images steganography analysis methods.

Method	Classification precision (%)	Dimensional feature
Ye-Net [30]	83.79	144
Xu-Resnet [26]	84.97	128
SRNet [31]	87.06	512
WiserNet [34]	86.42	200
WiserNet-100	84.90	100

What stands out in the above tables is the marked improvement in the precision of our WiserNet-100 feature extraction network based on WiserNet. There is a 1.11% performance improvement compared with Ye-Net and equal to Xu-Resnet. It is also worth mentioning that there is a slight decrease in precision compared with the SRNet network, but SRNet outputs 512 feature latitude and WiserNet-100 has only 100 feature latitude. In specific scenarios, WiserNet-100 may reduce the time consumption of the whole experiment owing to the reduced latitude.

4.3 Effectiveness of Steganographer Detection

In practice, the steganographer may use different embedding rates to embed different hidden information in different images. Therefore, in our experiments, Ye-Net, Xu-Resnet, and SRNet are used as control groups for comparison regarding embedding rate.

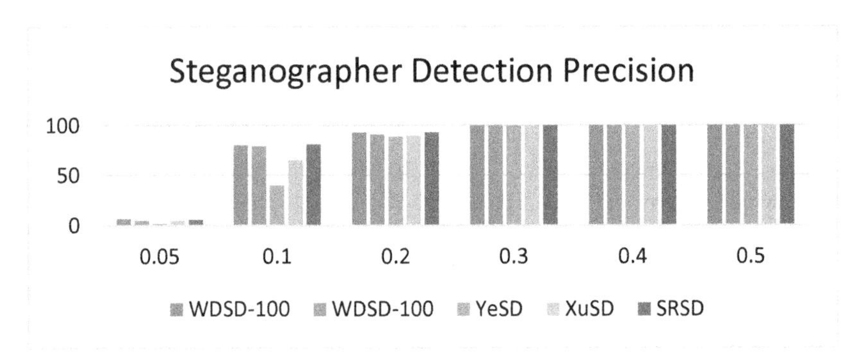

Fig. 3. Performance comparison of different steganographer detection methods with different embedding rates in the same steganography

The tested embedding probabilities contain 0.05, 0.1, 0.2, 0.3, 0.4, and 0.5. The above steganography algorithms use CMD-C-HILL to control the variables uniformly.

As can be seen from Fig. 3, the improved WiserNet-100 in this chapter has similar accuracy to the original WiserNet-200 in terms of Steganography analysis and detection. At 0.1 BPC, the detection accuracy of the improved WiserNet-100 is the same as that of WiserNet-200, and slightly higher than that of SRSD. At 0.2 BPC, the proposed detection algorithm maintains high accuracy, which is higher than traditional YESD and XUSD. After 0.3BPC, the detection rate of all the five detection algorithms reaches 100%.

Table 3. Comparison of the time consumption of different color image steganographer analysis methods

Method	Dimensional feature	Time(S)
Ye-Net	144	9.62
Xu-Resnet	128	9.10
SRNet	512	16.41
WiserNet	200	10.74
WiserNet-100	100	7.79

As can be seen from Table 3, when the performance of WiserNet-100 is the same as that of WiserNet-200, WiserNet-100 has a 2.95 s improvement in time consumption. The maximum detection time for SRSD with a higher feature dimension is 16.41Ss, which decreases as the feature dimension decreases. However, the detection time of WNCISD-100 is only 7.79S. In comparison, the time consumption is reduced to 50%. The reason for the similar accuracy of steganographer detection despite the difference in classification accuracy is that the experiments assume that there is only one steganographer, and the final identification of steganographers is done by tag sorting. Therefore, even though there is some difference in classification accuracy, a relatively high detection accuracy can still be achieved. Overall, the steganographer detection algorithm proposed in this chapter has an ideal university in specific scenarios.

4.4 Validity Assessment with Different User Graph Percentages

So far, all the above experiments are based on 100 users with 10 graphs each, of which 99 users with all cover images and the images of the steganographer are all steganographs, However, in practical life, it is likely that there is a user who sends a hidden message in the form of multiple messages and only a few steganographs at a time. Therefore, this experiment focuses on the percentage of steganographer who do not use steganography. The following results are obtained by repeating the experiment 100 times.

As can be seen from Fig. 4, the detection accuracy reaches 100% when the BPC is 0.5 and the figure share is greater than or equal to 40%; the detection accuracy reaches 89% when the BPC is 0.4 and the share is greater than or equal to 40% and reaches 100% detection when the figure share is 80%; the detection accuracy increases with the

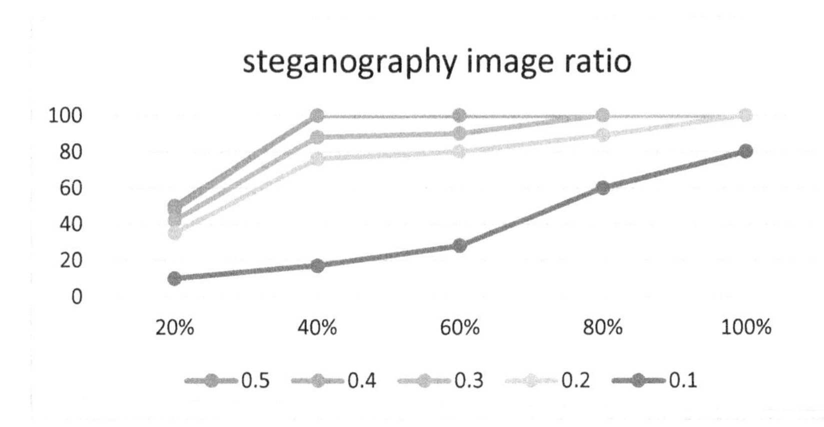

Fig. 4. WDSD performance comparison at different graph occupancy ratios and steganography rates.

increasing figure share and BPC. The performance is average when the BPC is 0.1, which is related to the low steganography rate resulting in the inconspicuous feature extraction. When the graph occupancy ratio is greater than or equal to 40%. The framework performs well in terms of detection performance.

5 Conclusion

This paper proposes a novel BMP color image steganographer detection method based on WiserNet. The first step is to extract the feature vectors of images from the WiserNet deep learning model through all three channels of BMP color images. The second step is to perform DBSCAN density clustering based on the extracted image feature vectors. The third step is to sort the clustering results and identify the steganographer.

Under the condition of the same detection accuracy, at present, relevant literature and case studies on color images are scarce. We use SRNET network with better feature extraction effect to carry out steganographer detection, which consumes 16.41 s. The algorithm proposed in this paper consumes 7.79 s, which reduces the time by half. In addition. This paper also verifies the performance of this method in complex realistic environments such as different embedding rates and different proportions of graphs, and the experiments show that this method achieves impressive performance in all the above cases.

In the future, the proposed color image steganographer detection will be applied to other image types, such as JPEG, so as to expand its application range.

References

1. Yang, J., Li, S.: Steganalysis of joint codeword quantization index modulation steganography based on codeword bayesian network. Neurocomputing **313**, 316–323 (2018)
2. Anderson, R.J., Petitcolas, F.A.P.: On the limits of steganography. IEEE J. Sel. Areas Commun. **16**(4), 474–481 (1998)

3. Islam, S., Gupta, P.: Effectof morphing on embedding capacity and embedding effiffifficiency. Neurocomputing **137**, 136–141 (2014)
4. Guo, SL., Yi, T.C., Wen, N.L.: A framework of enhancing images steganography with picture quality optimization and anti-steganalysis based on simulated annealing algorithm. IEEE Trans. Multimedia **12**(5), 345–357 (2010)
5. Zhang, X., Peng, F., Long, M.: Robust coverless images steganography based on DCT and LDA topic classification. IEEE Trans. Multimedia **20**(12), 3223–3238 (2018)
6. Zhou, H., Chen, K., Zhang, W., et al.: Comments on "steganography using reversible texture synthesis." IEEE Trans. Images Process. **26**(4), 1623 (2017)
7. Zheng, M., Zhong, S.-h., Wu, S., et al.: Steganographer detection via deep residual network. In: Proceedings of the IEEE International Conference on Multimedia and Expo, pp. 235–240 (2017).
8. Ker, A.D.: Batch steganography and the threshold game. In: Security, Steganography, and Watermarking of Multimedia Contents IX. SPIE, pp. 401–413 (2007)
9. Li, F., Kui, W., Lei, J., et al.: Steganalysis over large-scale social networks with high-order joint features and clustering ensembles. IEEE Trans. Inf. Forens. Secur. **11**(2), 344–357 (2017)
10. Kodovsky, J., Fridrich, J.: Ensemble classifiers for steganalysis of digital media. IEEE Trans. Inf. Forensics Secur. **7**(2), 432–444 (2012)
11. Goljan, M., Fridrich, J., Cogranne, R.: Rich model for steganalysis of color images. In: Proceedings IEEE International Workshop Information Forensics and Security (WIFS), pp. 185–190, December 2014
12. Goljan, M., Fridrich, J.: CFA-aware features for steganalysis of color images, Proc. SPIE **9409**, 94X090V-1–94090V-13 (2015)
13. Abdulrahman, H., Chaumont, M., Montesinos, P., et al.: Color images stegananalysis using correlations between RGB channels. In: Proceedings IEEE 10th International Conference Availability, Reliability and Security (ARES), pp. 448–454, August 2015
14. Abdulrahman, H., Chaumont, M., Montesinos, P., et al.: Color images steganalysis using RGB channel geometric transformation mea- sures. Secur. Commun. Netw. **9**(15), 2945–2956 (2016)
15. Abdulrahman, H., Chaumont, M., Montesinos, P., et al.: Color images steganalysis based on steerable Gaussian filters bank. In: Proceedings 4th ACM Information Hiding Multimedia Security Workshop (IH&MMSec), pp. 109–114 (2016)
16. Denemark, T., Sedighi, V., Holub, V., et al.: Selection-channel-aware rich model for steganalysis of digital images. In: Proceedings 6th IEEE International Workshop Information Forensic Security (WIFS), pp. 48–53, December 2014
17. Tang, W., Li, H., Luo, W., et al.: Adaptive steganalysis based on embedding probabilities of pixels. IEEE Trans. Inf. Forensics Secur. **11**(4), 734–745 (2016)
18. Boroumand, M., Fridrich, J.: Applications of explicit non-linear feature maps in steganalysis. IEEE Trans. Inf. Forensics Secur. **13**(4), 823–833 (2018)
19. Schmidhuber, J.: Deep learning in neural networks: an overview. Neural Netw. **61**, 85–117 (2015)
20. Tan, S., Li, B.: Stacked convolutional auto-encoders for steganalysis of digital images. In: 2014 Asia-Pacific Signal and Information Processing Association Annual Summit and Conference (APSIPA), pp. 1–4 (2014)
21. Qian, Y., Dong, J., Wang, W., et al.: Deep learning for steganalysis via convolutional neural networks. Proc. SPIE **9409** (2015). Art. no. 94090J
22. Pibre, L., Pasquet, J., Ienco, D., et al.: Deep learning is a good steganalysis tool when embedding key is reused for different images, even if there is a cover sourcemismatch. In: Proceedings Electron (2016)

23. Qian, Y., Dong, J., Wang, W., et al.: Learning and transferring representations for images steganalysis using convolutional neural network. In: Proceedings IEEE International Conference Images Processing (ICIP), pp. 2752–2756, September 2016
24. Zheng, M., Zhong, S.h.-h., Wu, S., et al.: Steganographer detection based on multiclass dilated residual networks. In: ACM International Conference on Multimedia Retrieval (ICMR) (2018)
25. Holub, V.: Content adaptive steganography-design and detection. PhD thesis. Department of Electricaland Computer Engineering, Binghamton University (2014)
26. Song, X., Liu, F., Yang, C., et al.: Steganalysis of adaptive JPEG steganography using 2D gabor filters. In: Proceedings of the 3rd ACM Workshop on Information Hiding and Multimedia Security -IH&MMSec 20'15. ACM Press (2015)
27. Xu, G., Wu, H.Z., Shi, Y.Q.: Structural design of convolutional neural networks for steganalysis. IEEE Signal Process. Lett. **23**(5), 708–712 (2016)
28. Ye, J., Ni, J., Yi, Y.: Deep learning hierarchical representations for image steganalysis. IEEE Trans. Inf. Forensics Secur. **12**(11), 2545–2557 (2017)
29. Suykens, J.A., Vandewalle, J.: Least squares support vector machine classifiers. Neural Process. Lett. **9**(3), 293–300 (1999)
30. Boroumand, M., Chen, M., Fridrich, J.: Deep residual network for steganalysis of digital images. IEEE Trans. Inf. Forensics Secur. **14**(5), 1181–1193 (2019)
31. Tang, W., Li, B., Luo, W., et al.: Clustering steganographic modification directions for color components. IEEE Sig. Process. Lett. **23** (2), 197–201 (2016)

UserRBPM: User Retweet Behavior Prediction with Graph Representation Learning

Huihui Guo, Li Yang(✉), and Zeyu Liu

Xidian University, Xi'an 710071, China
guohuihui@stu.xidian.edu.cn, yangli@xidian.edu.cn

Abstract. Social and information networks such as Facebook, Twitter, and Weibo have become the main social platforms for the public to share and exchange information, where we can easily access friends' activities and are in turn be influenced by them. Consequently, the analysis and modeling of user retweet behavior prediction have important application value in such aspects as information dissemination, public opinion monitoring, and product recommendation. Most of the existing solutions for user retweeting behavior prediction are usually based on network topology maps of information dissemination or design various hand-crafted rules to extract user-specific and network-specific features. However, these methods are very complex or heavily dependent on the knowledge of domain experts. Inspired by the successful use of neural networks in representation learning, we design a framework UserRBPM to explore potential driving factors and predictable signals in user retweet behavior. We use the graph embedding technology to extract the structural attributes of the ego-network, consider the drivers of social influence from the spatial and temporal levels, and use the graph convolutional network and the graph attention mechanism to learn its potential social representation and predictive signals. Experimental results show that our proposed UserRBPM framework can significantly improve prediction performance and express social influence better than traditional feature engineering-based approaches.

Keywords: Social networks · Retweet behavior prediction · Graph convolution · Graph attention · Representation learning

1 Introduction

Due to their convenient capability to share real-time information, social media sites (e.g., Weibo, Facebook, and Twitter) have grown rapidly in recent years. They have become the main platforms for the public to share and exchange information, and to a great extent meet the social needs of users. Under normal

---

Supported by the National Natural Science Foundation of China (62072359, 62072352), and the National Key Research and Development Project (2017YFB0801805).

J. Xiong et al. (Eds.): MobiMedia 2021, LNICST 394, pp. 613–632, 2021.
https://doi.org/10.1007/978-3-030-89814-4_45

circumstances, online social networks will record a large amount of information generated by people through interactive activities, including various user behavior data. User behaviors (also called actions) in online social networks contain posting messages, purchasing products, retweeting information, and establishing friendships, etc. By analyzing the distribution and causality of these behaviors, we can evaluate the influence between the initiator and the communicator of the behavior, as well, we can predict people's behaviors on social networks and deepen our understanding and understanding of human social behavior [18]. Till now, there is little doubt that the large amount of data generated by users' interaction provides an opportunity to study user behavior patterns, and the analysis and modeling of retweet behavior prediction have become a research hotspot. In addition to analyzing the retweeting behavior itself, retweeting can also help with a variety of tasks such as information spreading prediction [30], popularity prediction [38], and precision marketing [2].

Previous researches investigated the problem of user retweet behavior prediction from different points of view. On the first approach, some researchers build retweet behavior prediction models through network topology maps of information dissemination. Matsubara et al. [15] studied the dynamics of information diffusion in social media by extending an analysis model for information dissemination from the classic 'Susceptible-Infected' (SI) model. Wang et al. [26] proposed an improved SIR model, which used the mean field theory to study the dynamic behavior in uniform and heterogeneous network models. Their experiment showed that the existence of the network would influence information communication. This kind of research method studied retweeting behavior by modeling the propagation path of the message from a global perspective. The other approach is the machine learning method based on feature engineering. Liu et al. [13] proposed a retweeting behavior prediction model based on fuzzy theory and neural network algorithm, which can effectively predict the user's retweeting behavior and dynamically perceive the changes in hotspot topics. This research method relies on the knowledge of domain experts, and the process of feature selection may take a long time. However, in many online applications such as personalized recommendation [29,31] and advertising [2], it is critical to effectively analyze the social influence of each individual and further predict the retweeting behavior of users.

In this paper, we focus on user-level social influence. We aim to predict the action statuses of the target user according to the action statuses of her near neighbors and her local structural information. For example, in social networks, a person's behavior will be affected by her neighbors. As shown in Fig. 1, for the central user u, if some friends (red node) around her posted a microblog and other friends (white node) did not post it, whether the action statuses of user u will be affected by the surrounding friends and forward this tweet can be regarded as a user retweeting behavior prediction problem. The social influence hidden behind the retweeting behavior not only depends on the number of active users, but may also be related to the local network structure formed by "active" users. The problem mentioned above are common in practical applications, such as presidential elections [3], innovation adoption [21], and e-commerce [11]. Therefore, it has inspired many research work on user-level influence models, most of which [9,34] consider complicated handcrafted features, which require extensive knowledge of specific domains.

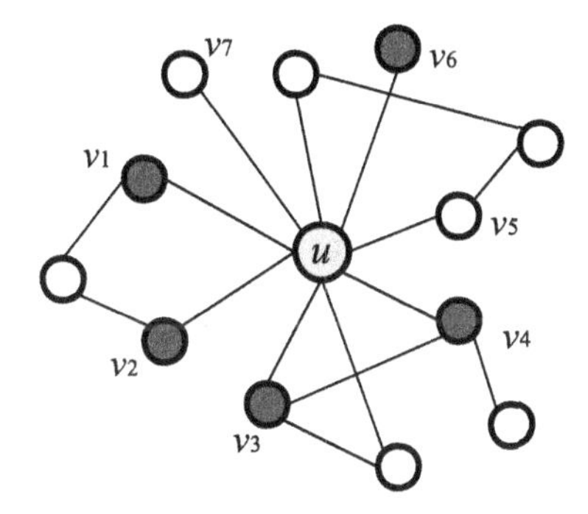

Fig. 1. A motivating example of user retweet behavior prediction. (Color figure online)

The recently proposed graph convolution networks(GCN) [4,12] is currently the best choice for graph data learning tasks. Inspired by the successful application of neural networks in representation learning [20,33], we designed an end-to-end framework UserRBPM to explore potential driving factors and predictive signals in user retweeting behaviors. We expect deep learning frameworks to have better expressive capability and prediction performance. The designed solution is to represent both influence driving factors and network structures into a latent space, and then use graph neural networks to effectively extract spatial features for learning, and further construct a user retweet behavior prediction model. We demonstrate the effectiveness and efficiency of our proposed framework on Weibo social networks. We compare UserRBPM with several conventional methods, and experiment results show that the UserRBPM framework can significantly improve the prediction performance.

The main contributions of this work can be summarized as follows:

- We designed an end-to-end learning framework UserRBPM to explore potential driving factors and predictive signals in user retweeting behaviors.
- We convert the retweeting behavior prediction into a binary graph classification, which is more operable and comprehensible.
- Experiment results demonstrate that the UserRBPM framework can achieve better prediction performance than existing methods.

Organization. The rest of this paper is organized as follows. Section 2 summarizes related work. Section 3 formulates the user retweet behavior prediction problem. We detail the proposed framework in Sect. 4. In Sect. 5 and Sect. 6, we conduct extensive experiments and analyze the results. Finally, we conclude our work in Sect. 7.

2 Related Work

In this section, we categorize and summarize prior work on user retweet behavior prediction and graph representation learning.

2.1 User Retweet Behavior Prediction

Many studies on user retweet behavior in social networks are based on the analysis and modeling of the dynamics in the process of information dissemination. Currently, researches on user behavior prediction in social networks take primarily two approaches. On the first approach, Yuan et al. [32] investigated the dynamics of friend relationships through online social interaction, and thus proposed a model to predict repliers or retweeters according to a particular tweet posted at a certain time in online social networks. Tang et al. [22] studied the conformity phenomenon of user behavior in social networks, and proposed a probabilistic model called Confluence to predict user behavior. This model can distinguish and quantify the effects of the different types of conformity. Zhang et al. [37] proposed three metrics: user enthusiasm, user engine, and user duration, to describe the user retweet behavior in the message spreading process, and studied the relationship between these three metrics and the influence obtained by the user retweet behavior.

The other approach is the machine learning method based on feature engineering, which solved the problem of user behavior analysis and prediction by manually formulating rules to extract the basic features of users and network structural features. Luo et al. [14] explored features: followers status, retweet history, followers interests, and followers active time with a learning-to-rank framework to discover who would retweet a tweet poster on Twitter. Zhang et al. [34] analyzed the influence of the number of active neighbors of a user on retweeting behavior, proposed two instantiation functions based on structural diversity and pairwise influence, and applied a classifier based on logistic regression to predict users' retweet behaviors. Jiang et al. [9] pointed out that the retweeting prediction is a sing-type setting problem. By analyzing the basic influence factors of retweet behavior in Weibo, the sing-type collaborative filtering method is used to measure users' personal preference and social influence for the purpose of predicting retweet behavior.

2.2 Graph Representation Learning

Graph representation learning has emerged as a powerful technique for solving real-world problems. Various downstream graph learning tasks have benefit from its recent developments, such as node classification [7], similarity search [35], and graph classification [19,36]. The primary challenge in this field is to find a way to represent or encode the structure of graphs so that it can be easily exploited by machine learning models. Traditional machine learning approaches relied on user-defined heuristics to extract features encoding structural information about a graph (e.g., degree statistics or kernel functions). However, recent years, have seen a surge in approaches for automatically learning to encode graph structure into low-dimensional embedding using techniques based on deep learning and nonlinear dimension reduction. Chen et al. [5] exploited graph attention networks(GAT) to learn user node representation by spreading information in heterogeneous graphs, and then leveraged limited labels of users to build end-to-end semi-supervised user profiling predictor. Zhang et al. [33] introduced the problem of heterogeneous graph representation learning and proposed a heterogeneous graph neural networks model HetGNN. Extensive experiments on various graph mining tasks, i.e., link prediction, recommendation, and node classification, demonstrated that HetGNN can outperform state-of-the-art methods.

3 Problem Formulation

In this section, we introduce necessary definitions and then formulate the problem of user retweet behavior prediction.

Definition 1. Ego network
The ego network model is one of the important tools for studying human social behavior and social networks. Compared with the global network version, the research version of the ego network pays more attention to individual users, and it is in line with the needs of personalized services in actual application systems. The research version of this paper can also be extended to other scenarios that include network relationships.

***r*-neighbors** Let $G = (V, E)$ denote a social network, where V is a set of users nodes and $E \subseteq V \times V$ is a set of relationships between users. We use $v_i \in V$ to represent a user and $e_{ij} \in E$ to represent a relationship between v_i and v_j. In this work, we consider undirected relationships. For a user u, its r-neighbors nodes are defined as $\Gamma_u^r = \{v : d(u, v) \leq r\}$, where $d(u, v)$ is the shortest path distance (in terms of the number of hops) between u and v in the network G, $r \geq 1$ is a tunable integer parameter to control the scale of the ego network.

***r*-ego network** The r-ego network of user u is the subnetwork induced by Γ_u^r, denoted by G_u^r.

Definition 2. Social action
In sociology, social action is an act which takes into account the actions and reactions of individuals. Users in social networks perform social actions, such as retweeting behaviors, citation behaviors. At each time stamp t, we observe a binary action status of user u, $s_u^t \in \{0, 1\}$, where $s_u^t = 1$ indicates user u has performed this action before or on the timestamp t, and $s_u^t = 0$ indicates that the user has not performed this action yet.

In this paper, our research motivation of user retweeting behavior prediction problem can be vividly illustrated by Fig. 1. For a user u in her 2-ego network (i.e., r = 2), if some users retweet a microblog m before or on the timestamp t, they are considered to be active. We can observe the action statuses of u's neighbors, such as $s_{v_1}^t = 1$, $s_{v_2}^t = 1$, and $s_{v_5}^t = 0$. Moreover, the set of active neighbors of user u is represented by $\psi_u^t = \{v_1, v_2, v_3, v_4, v_6\}$. As shown in Fig. 1, we study whether the action statuses of user u will be influenced by the surrounding friends and forward this microblog. Next, we will formalize the problem of user retweet behavior prediction.

Problem 1. User retweet behavior prediction [34]
User retweet behavior prediction models the probability of $u's$ action states conditioned on her r-ego network and the action states of her r-neighbors. More formally, given G_u^r and $S_u^t = \{s_v^t : v \in \Gamma_u^r \backslash \{u\}\}$, it can be concluded that the user retweet behavior prediction formula of user u after a given time interval Δt is as follows:

$$A_v = P\left(s_u^{t+\Delta t} \mid G_u^r, S_u^t\right) \tag{1}$$

Practically, A_v denotes the predicted social action status of user u. Suppose we have N instances, and each instance is a 3-tuple (u, a, t), where u is a user, a is a social action and t is a timestamp. For such a 3-tuple (u, a, t), we also know $u's$ r-ego network G_u^r, the action states of $u's$ r-neighbors S_u^t, and $u's$ future action states at $t + \Delta t$, i.e., $s_u^{t+\Delta t}$. We then formulate user retweet behavior prediction as a binary graph classification problem which can be solved by minimizing the following negative log likelihood objective w.r.t model parameter θ:

$$L(\theta) = -\sum_{i=1}^{n} \log\left(P_\theta\left(s_u^{t+\Delta t} \mid G_u^r, S_u^t\right)\right) \tag{2}$$

4 Model Framework

In this paper, we formally propose the UserRBPM to address the user retweet behavior prediction problem. The framework is based on graph neural networks to parameterize the probability in Eq. (2) and automatically detect the potential driving factors and predictive signals of user retweet behavior prediction. As shown in Fig. 2, UserRBPM is consisted of pre-trained network embedding layer, input layer, GCN/GAT layer and output layer.

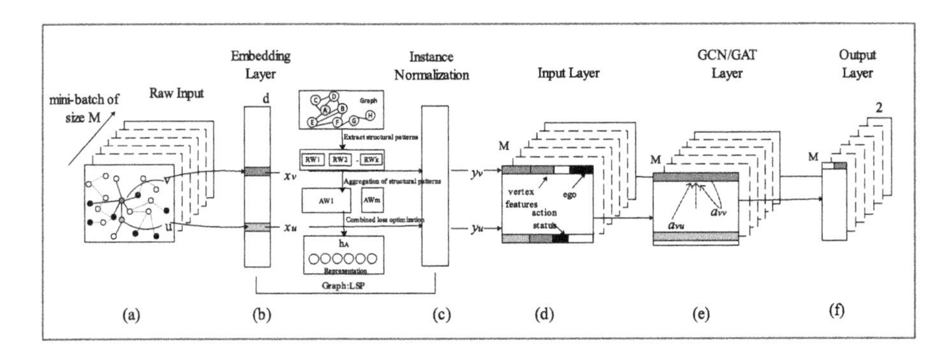

Fig. 2. Our proposed framework of UserRBPM (User Retweet Behavior Prediction Model).

4.1 Sampling Near Neighbors

Given a user u, the most straightforward way to extract its r-ego network is to perform a Breadth-First-Search (BFS) starting from user u. However, for different users, r-ego network scale (regarding the number of nodes) may vary greatly. Meanwhile, the size of user $u's$ r-ego network can be very large due to small-world property in social networks [27]. In real-world application scenarios, when sampling neighbor nodes of an ego user node, the problem that may arise is that each node has a different number of neighbors. Specifically, due to the small-world phenomenon in social networks, the size of user $u's$ r-ego network may be relatively very large or small. In addition, these different sizes of data are not suitable for most deep learning models.

In order to address the above problem, we select to perform random walk with restart (RWR) [23] from the original r-ego network to fix the sample size. Inspired by [1,24] which suggest that people are more susceptible to be influenced by active neighbors than inactive ones, we start a random walk on G_u^r from user u or its active neighbors. The walk iteratively travels to its neighborhood with a probability proportional to the weight of each edge. In addition, the walk returns back to the starting vertex u with a positive probability at each step. In this way, a fixed size of number of vertices can be collected, denoted by $\vec{\Gamma}_u^r$ with $\left|\vec{\Gamma}_u^r\right| = n$. We then regard the sub-network $\vec{G}_u^r$ induced by $\vec{\Gamma}_u^r$ as a proxy of the r-ego network G_u^r, and denote $\vec{S}_u^t = \{s_v^t : v \in \vec{\Gamma}_u^r \backslash \{u\}\}$ to be the action statuses of $u's$ sampled neighbors. When we use RWR, the starting node can be ego user or its active neighbors. The purpose of setting as described above is to make the starting node in the sequence obtained by walking as much as possible to keep in touch with surrounding neighbors, instead of being relatively single, so as to support the purpose of people being more susceptible to be influenced of active neighbors.

4.2 Graph Neural Networks Model

We design an effective graph neural networks model to incorporate both the structural properties in $\vec{G}_u^r$ and action statuses in $\vec{S}_u^t$, learned a hidden embedding vector for each ego user, then used to predict the action statuses of the ego user in the next time period $s_v^{t+\Delta t}$. As shown in Fig. 2, the graph neural networks model includes embedding layer, instance normalization layer, input layer, graph neural networks layer, and output layer.

Embedding Layer. For graph structure data such as social networks, we want to learn the users' social representation from users' relationship network data, that is, our main purpose is to discover network structural property and encode them into low-dimensional latent space. More formally, network embedding learns an embedding matrix $X \in R^{d \times |V|}$, with each column corresponding to the representation of a vertex (user) in the network G. In our scheme, we learn a low-dimensional dense real number vector $x_v \in R^d$ for each node v in the network, where $d \ll N$. The process of network representation learning can be unsupervised or semi-supervised.

In social networks, when considering the structural information, we can take the triadic closure, patterns characteristic of strong ties in social networks. As shown in Fig. 3, there will be such a case: The figure on the left contains a triadic closure. For the green node, it is equivalent to a different tree structure on the right (there is no triadic closure) after two neighborhood aggregations, which ignore the structural information of triadic closure. Therefore, there is a need for a method of graph representation learning that can adapt to different local structures.

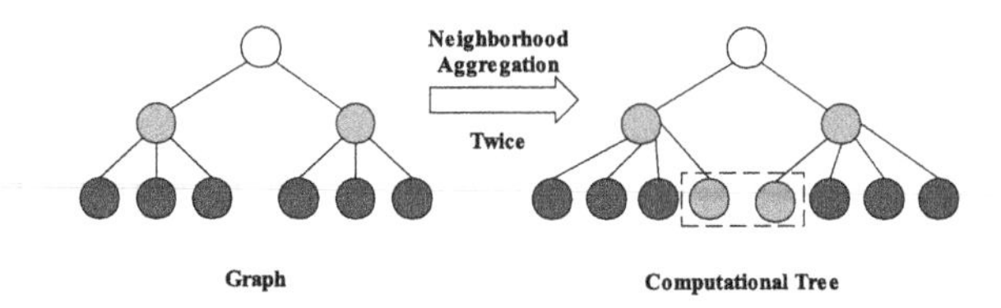

Fig. 3. Computational tree of a triadic closure graph. (Color figure online)

In our work, we utilize the GraLSP model [10] for graph representation learning, which explicitly incorporates local structural patterns into the neighborhood aggregation through random anonymous walks. Specifically, the framework captures the local structural patterns via random anonymous walks, and then these walk sequences are fed into the feature aggregation, where various mechanisms are designed to address the impact of structural features, including adaptive receptive radius, attention, and amplification. In addition, GraLSP can capture similarities between structures and are optimized jointly with near objectives of node. The process of GraLSP model for graph representation learning is shown in Fig. 2(b). In the case of making full use of the structural model, the GraLSP can outperform competitors in various prediction tasks in multiple datasets.

Instance Normalization Layer. In the training process of UserRBPM model, we applied Instance Normalization (IN) [25] to prevent overfitting, which is a regularization technique that loosens the model and allows for greater generalization. And for such tasks that focus on each sample, the information from each sample is very important. Therefore, we adopt such a technique in the task of retweet behavior prediction. After original data is normalized, the indicators are between [0, 1], which is suitable for comprehensive comparative analysis. Furthermore, it helps to speed up learning and also reduces overfitting.

Input Layer. As claimed in Fig. 2(d), the input layer constructs a feature vector for each user. The feature vector considered in our work consists of three parts: 1) the normalized low-dimensional embedding comes from the up-stream instance normalization layer; 2) two binary variables are also considered. The first variable represents the user's action statuses, and the second variable represents whether the user is an ego user; 3) the input layer also includes other personalized vertex features, such as spatial-level features (e.g., social roles) and temporal-level features (e.g., similarity, exposure, retweet rate.)

GCN Layer. The recently developed GCN [6] is a successful attempt to generalize the convolutional neural networks used in Euclidean space to graph structure data modeling. The GCN model naturally integrates the connection mode and feature attributes of graph structure data, and it is much better than many state-of-the-art methods on benchmarks. Graph Convolutional Network (GCN) is a semi-supervised learning algorithm for graph structure data, which can effectively extract spatial features for machine learning on such a network topology. Simultaneously, it can perform end-to-end learning of node feature information and structure information, which is one of the best choices for graph data learning tasks at present.

Graph Attention Networks. Essentially, both GCN and GAT are aggregation operations that aggregate the characteristics of neighbor nodes into the central node. GCN uses the Laplacian matrix to perform graph convolution operations, while GAT introduces the attention mechanism into GCN, which can add weight to the influence of neighboring nodes, thereby differentiating the influence of neighboring nodes. GAT assigns different weights to each node, paying attention to those nodes with greater effects, while ignoring some nodes with smaller effects. To a certain extent, the performance ability of GAT will be stronger, because the correlation between node features will be better integrated into the model.

Output Layer. In the output layer, each node corresponds to a two-dimensional representation, which is used to represent the user's behavior state (retweet/unretweet, cite/uncite, etc.). By comparing the representation of the ego user with groud truth, we then optimize the log-likelihood loss.

5 Experiment Setup

In this subsection, we first introduce the construction process and statistical characteristics of the dataset. Then, we present the existing representative methods and evaluation metrics. Finally, we introduce the implementation details of the UserRBPM framework.

5.1 Dataset Presentation and Processing

Presentation to Raw Datasets. We use real-world datasets to quantitatively and qualitatively evaluate the proposed UserRBPM framework. We used the Weibo dataset in the work of Tang et al. [17,34] and Wu et al. [28] also used the Weibo dataset in the work, and then we performed data preprocessing according to our research question. The microblogging network used in our research work is to crawl data from Sina Weibo. Particularly, when user u_1 follows user u_2, u_2's activities (such as tweet and retweet) will be visible to u_1. User u_1 can choose to tweet or retweet by user u_2. User u_1 is called the follower of user u_2 and user u_2 is called the followee of user u_1.

The Generation and Processing of Samples. In the problem of retweeting behavior prediction, since we can directly learn from the microblogs record which users have retweeted the microblogs, the extraction process of positive samples is relatively simple. Thus, for a user v who is affected by others, he performs a social action a at a certain timestamp t, and then we generate a positive sample. Compared with the extraction of positive samples, it is impossible to directly know from the microblogs records which users saw the message but did not retweet the microblogs. Therefore, the extraction method of negative samples is much more complicated.

For our research scenarios, there are two data imbalance problems. The first one comes from the number of active neighbors. As Zhang et al. [34] observed, structural features are significantly related to user retweeting behavior when the ego user has a relatively large number of active neighbors. For example, in the Weibo dataset, 80% of users have only one active neighbor and users with more than 3 active neighbors account for only 8.57%. Therefore, the model will be controlled by observation samples with few active neighbors. To illustrate the superiority of our proposed model in capturing local structural information, we established a balanced sub-dataset *Edata* (as shown in Table 1.) for fair data analysis and further training-test scheme. Specifically, we filter out samples in which the followers or followees did not have Weibo content. In addition, we only considered samples in which ego users have at least 3 active neighbors.

Table 1. Statistics of sub-dataset *Edata*.

Edataset	#Users	#Follow-relationships	#Oroginal-microblogs	#Retweet	#Ego Users
Weibo	1,500,290	20,297,550	274,150	15,755,810	151,300

The second problem is imbalanced labels. For instance, in our Weibo data set, the ratio between positive instances and negative instances is about 1:300. To address this problem, the most direct way is to select a relatively balanced dataset, that is, set the ratio of positive samples and negative samples to 1:3. In addition, we also used the global random downsampling method and microblog granularity-based down-sampling method to process imbalanced datasets. Among them, when we use the global random down-sampling method, the number of microblogs involved in the negative samples in the obtained dataset is small, and there is a case where only positive samples of the same microblog are not sampled to their corresponding negative samples. The down-sampling method based on microblog granularity can try its best to ensure that the number of positive and negative samples of the same microblog is also the same.

The Features of Our Design. We made detailed data observation and analyzed how the characteristics of users at the spatial and temporal levels influence retweeting behavior in addition to the structural attributes of social networks. To visualize the observation results, we design several statistical information, which respectively represent spatial-level features and temporal-level features. These characteristics can be regarded as user node features. In our work, the spatial-level features are specifically analyzed in terms of social roles. We studied the influence of social roles played by different users on the prediction performance of retweeting behavior. Inspired by the previous research work of Wu et al. [28], we divide users into three groups according to their network attributes: opinion leaders (OpnLdr), structural hole spanners (StrHole)) and ordinary users (OrdUsr). A detailed analysis of users' social roles and behaviors is shown in Table 2. For the temporal-level feature, we mainly analyzed the content of the messages posted by users. The features of our design are shown in Table 3.

Table 2. The statistics of social roles and relation statuses.

Social role	OrdUsr	OpnLdr	StrHole	Sum
Retweet Behavior	6,617,440(42%)	3,623,836(23%)	5,514,534(35%)	15,755,810
Original Post	68,537(25%)	123,367(45%)	82,245(30%)	274,150
Sum	1,125,217(75%)	150,029(10%)	225,043(15%)	1,500,290

Table 3. List of features used in our work.

Spatial-level features	Social role_Opinion leader (OpnLdr)
	Social role_Structure hole (StrHole)
	Social role_Ordinary users (OrdUsr)
Temporal-level features	The TF-IDF similarity between ego user and its followees' post content (Similarity)
	The number of microblogs posted by the followees (Exposure)
	The retweet rate of ego users to their followees (Retweet rate)
Handcrafted ego-network features [22]	The number/ratio of active neighbors
	Density of subnetwork induced by active neighbors
	Connected of components formed by active neighbors

5.2 Comparison Methods

In order to verify the effectiveness of our proposed framework, we compared the prediction performance of UserRBPM in this paper with existing representative methods. Firstly, we compared UserRBPM with previous retweeting behavior prediction methods which usually extract rule-based features. Secondly, by comparing the GraLSP method with other network embedding methods, it is verified that the local structure information plays a more important role in the prediction of forwarding behavior than the global information. The comparison method is as follows:

- Hand-crafted features + Logistic Regression(LR): We use logistic to train the classification model. The features we constructed manually include two categories: one is the user node features designed in our work, including spatial-level and temporal-level features; the other is the ego network features designed by Qiu et al. [17].

- Hand-crafted features + Support Vector Machine(SVM): We also use SVM as the classification model. The model use the same features as Logistic Regression.
- DeepWalk: DeepWalk [16] is a network embedding method that learns a social representation of a network by truncated random walks to obtain the structural information of each vertex.
- Node2vec: Node2vec [8] designs a biased random walks that can trade off between homophily and structural equivalence of the network.
- Our Proposed Method: In our proposed UserRBPM framework, we use GraLSP to extract the structural attributes of the r-ego network, design the user node features at the spatial-level and temporal-level, and finally apply GCN and GAT to learn latent predictive signals.

In order to quantitatively evaluate our proposed framework, we use the four popular metrics to evaluate the performance of retweeting behavior prediction. Specifically, we evaluate the performance of the UserRBPM in terms of Area Under Curve (AUC), Precision, Recall, and F1-Score.

5.3 Implementation Details

There are two stages for training our UserRBPM framework. In the first stage, we pretrain each module of UserRBPM, and in the second stage, we integrate the three modules of UserRBPM for fine-tuning.

Stage I: Pretrain of Each Module. For our framework, UserRBPM, we first perform a random walk with a restart probability of 0.8 and set the size of the sampled sub-network to be 30. For the embedding layer, the embedding dimension of the GraLSP model is set to three dimensions of 32, 64, and 128, and train GraLSP for 1000 epochs. Then we choose to use a three-layer GCN or GAT network structure, the first and second GCN/GAT layers both contain 128 hidden units, while the third layer (output layer) contains 2 hidden units for binary prediction. In particular, for UserRBPM with multi-head graph attention, both the first and second layers consist of $K = 8$ attention heads, and each attention head computes 16 hidden units (total $8 \times 16 = 128$ hidden units). The network is optimized by the Adam optimizer with the learning rate of 0.1, weight decay 5e-4, and dropout rate of 0.2. To evaluate the model performance and prevent information leakage, we performed five-fold cross-validation on our datasets. Specifically, we select 75% instances for training, 12.5% instances for validation, and 12.5% instances for test. In addition, the mini-batch size is set to be 1024 in our experiments.

Stage II: Global Fine-Tuning. In the global fine-tuning stage, if the dimension of embedding layer is set too large, the training process will be too slow, while a small setting will affect the performance of our model. After fine-tuning the model, we found that the model performance is relatively stable when the embedding dimension is set to 64. Then, we fix the parameters of pre-trained embedding module, and train the GCN/GAT layer with Adam optimizer for 1000 epochs, with the learning rate of 0.001. As the larger learning rate can make the model learn faster, thereby accelerating the convergence speed, but the performance of the model will be affected to some extent. Therefore, we set a relatively large learning rate at the beginning, and then gradually decrease with the training. Finally, we choose the best model by stopping using the loss on the validation sets as early as possible.

6 Experiment Results

6.1 Prediction Performance Analysis

Overall Performance Analysis. To verify the influence of the structural attributes of users' ego network and user nodes characteristics (extracted from the spatial and temporal level) on the prediction performance in social networks, as well as the interaction between features at different levels, we made the following comparison. As shown in Table 4, showing the prediction performance of different models.

Table 4. Prediction performance of different methods for retweeting behavior (%).

Methods	Precision	Recall	F1-score	AUC
Spatial-& Temporal-level& Handcrafted features + LR (ST& HC+LR)	69.74	71.58	70.65	77.27
Spatial-& Temporal-level& Handcrafted features + SVM (ST& HC+LR)	68.38	69.15	68.76	78.01
DeepWalk+ST+GAT	78.21	78.46	78.28	82.81
DeepWalk+ST& HC+GAT	79.68	80.24	79.96	82.75
Node2vec+ST+GAT	78.54	81.50	79.99	82.53
Node2vec+ST& HC+GAT	79.88	81.25	80.55	82.96
Our Method(UserRBPM)	81.97	82.58	82.27	83.21

Based on the analysis of four evaluation metrics used in our work, the performance of UserRBPM is better than the above-mentioned benchmark methods, which demonstrate that the effectiveness of our proposed framework. From the comparison among DeepWalk+ST&HC+GAT, Node2vec+ST&HC+GAT, and UserRBPM, we can observe that the GraLSP model we leverage in the embedding layer can indeed capture local structural patterns and significantly outperforms the first two methods in the experiment, confirming the GraLSP can indeed capture local structural patterns in retweeting behavior prediction. Meanwhile, from the comparison among ST&HC+LR, ST&HC+SVM, and UserRBPM, we notice that UserRBPM achieve an improvement of 13.59% in terms of precision. Such improvement verifies that the end-to-end learning framework UserRBPM can effectively detect potential driving factors and predictive signals in retweeting behavior prediction.

Comparing the first four methods with UserRBPM, it can be shown that the model which taking hand-crafted features as input hardly represent interaction effects, while network embedding technology can effectively extract high-dimensional structural attributes. Figure 4 shows that ST&HC+LR is notably better than HC+LR for retweeting behavior prediction. It reveals that users' spatial-level features and temporal-level features are the potential driving factors of retweeting behavior in social networks. Additionally, we observe that ST&HC+LR performs 4.42% better than HC+LR in terms of precision, verifying that the spatial-level and temporal-level features we designed have improved the prediction performance to a certain extent.

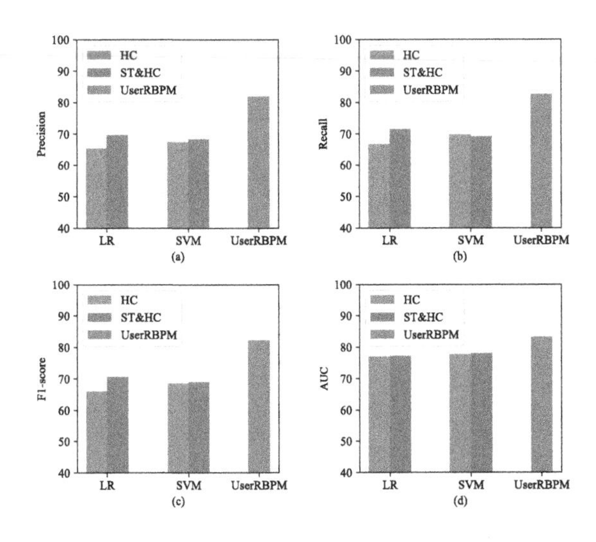

Fig. 4. Analysis results of different features.

Prediction Performance of Different Sampling Strategies. We use three sampling methods to obtain different training models. Among them, the directly sampling method (DSM) represents that we directly extract relatively balanced samples based on the ratio of the original positive and negative samples, that is, the ratio between positive and negative samples is set to 1:3. The number of positive samples and negative samples in completely random down-sampling (CRDM) and our down-sampling method (ODM) is the same. Experiment results are illustrated in Fig. 5. Compare to the completely random down-sampling method, the model trained with the samples obtained by our down-sampling method has better prediction performance. The better the prediction effect of the model obtained by the training data training, the more it shows that the dataset has universal significance and the learned model has a stronger generalization ability. In the original imbalanced datasets, the direct extraction of positive and negative samples with a ratio of 1:3 is simple, but the difference in the number of microblogs covered by the positive and negative samples is ignored. Therefore, the down-sampling method based on microblog granularity is more suitable for the user retweeting behavior prediction problem that we researched.

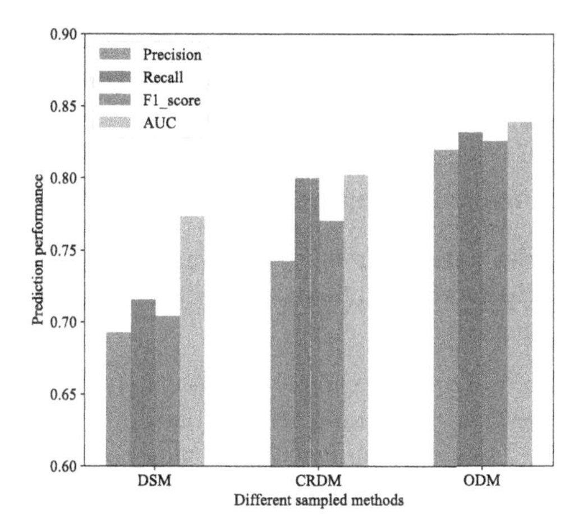

Fig. 5. Prediction performance of different sampling strategies.

Comparative Analysis of GCN and GAT. Table 5 is the prediction performance of two variants of graph deep learning, that is, the experimental results of using GCN and GAT to build models, respectively. In the application scenarios of our work, we observe that the performance of GCN in models constructed by different graph embedding technologies is generally worse than that of GAT. We attribute its disadvantage to the homophily assumption of GCN. This homophily exists in many real networks, but in our research scenario, different neighbor nodes may have different importance. Therefore, GAT is introduced to assign different weights to different neighboring nodes.

Table 5. Prediction performance of variants of UserRBPM (%).

Methods	Precision	Recall	F1-score	AUC
DeepWalk+ST& HC+GCN	77.49	79.28	78.37	74.89
DeepWalk+ST& HC+GAT	79.68	80.24	79.96	82.75
Node2vec+ST& HC+GCN	78.56	80.07	79.31	79.64
Node2vec+ST& HC+GAT	79.88	81.25	80.55	82.96
UserRBPM_GCN	80.82	80.58	80.70	82.23
UserRBPM_GAT	81.97	82.58	82.27	83.21

Besides, we wanted to avoid using hand-crafted features and make UserRBPM a pure end-to-end learning framework, so we compared the prediction performance with additional vertex features and no additional vertex features. Comparison results of prediction performance with/without vertex features, we observed that UserRBPM_GAT with hand-c rafted vertex features outperforms UserRBPM_GAT without hand-crafted vertex features by 1.48% in terms of precision, 0.81% in terms of recall, 1.15% in terms of F1-score, and 1.29% in terms of AUC. Experiment results demonstrate that, in addition to the pre-trained network embedding, we can still obtain comparable performance even without considering hand-crafted features.

6.2 Parameter Sensitivity Analysis

In addition, we consider parameter sensitivity in our work. We analyzed several hyper-parameters in the model and tested how different hyper-parameter choices affect the prediction performance.

Robustness Analysis. To verify the robustness of the UserRBPM framework, we changed the proportion of training set, validation set and test set and then redo the experiments. The results in Fig. 6 show that the model is effective under limited training data size. Even with small size of training set (20%–40%), our model can still have an acceptable and steady performance.

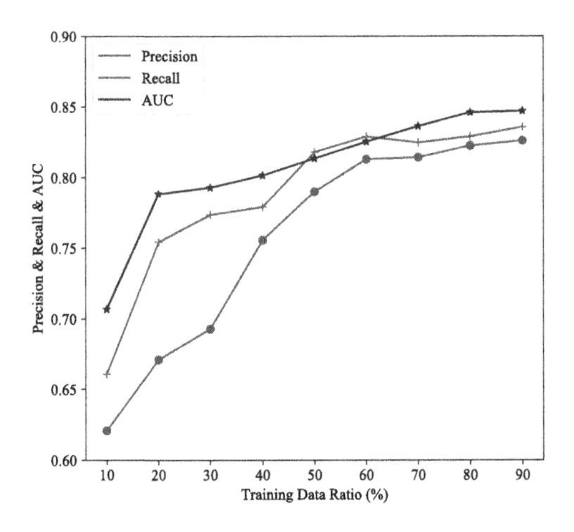

Fig. 6. Prediction performance with different training and test data size.

Effect of Instance Normalization. As mentioned in Sect. 4, this paper studied the technique used to accelerate model learning called *Instance Normalization (IN)*. This technique provides benefits to improve the classification performance. For instance, it can learn faster while maintaining or even increasing accuracy. Moreover, it also partially serves as a parameter tuning method. Therefore, we applied IN and obtained a boost in both performance and generalization. Figure 7 shows that the changes in the training loss of UserRBPM-GAT with/without IN layer during training. We can see that when there is an instance normalization layer, as the number of epochs increases, the training loss first drops rapidly and then remain stable. Instance normalization significantly avoids overfitting and makes the training process more stable.

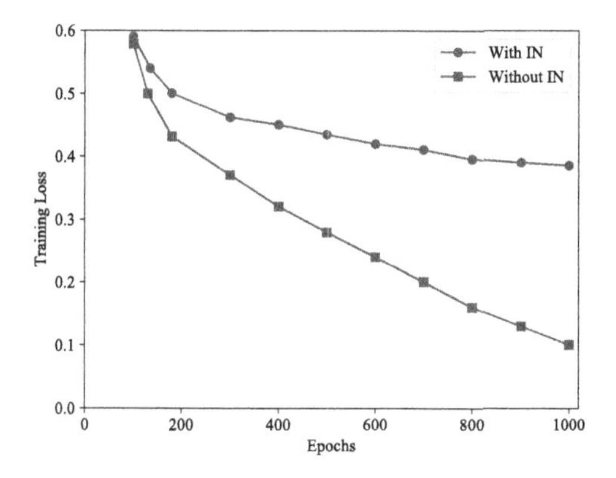

Fig. 7. The effect of instance normalization.

However, as shown in Fig. 8, we observe that the model without IN layer takes about 1011 s per epoch during the training process, while the model with IN layer takes about 1892 s per epoch. It was calculated that the model with IN layer increased the training time for each epoch by about 87% compared to the model without IN layer. Yet, we believe it is worthwhile to apply IN, as the additional training time is compensated with a faster learning rate (it requires less number of epochs to reach the same level of precision) and can ultimately achieve higher testing precision.

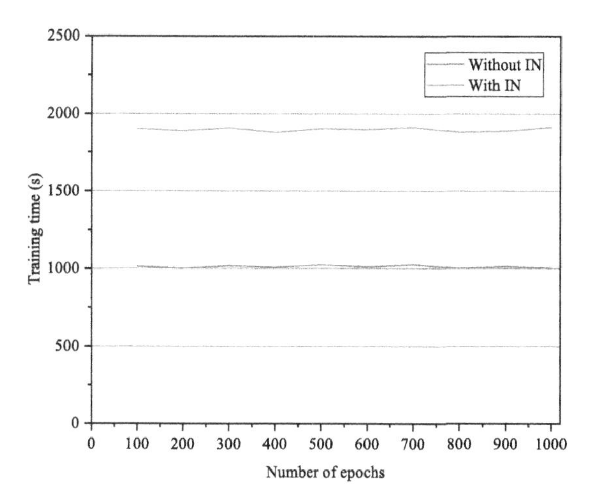

Fig. 8. Time overhead during each epoch.

7 Conclusion

In this work, we focus on user-level social influence in social networks and formulate the user retweet behavior prediction problem from a deep learning perspective. Unlike previous work that built a prediction model of retweet behavior based on network topology maps of information dissemination or conventional feature engineering-based approaches, we proposed UserRBPM framework to predict the action status of a user given the action statuses of her near neighbors and her local structural information. Experiments on a large-scale real-world dataset have shown that the UserRBPM significantly outperforms baselines with hand-crafted features in user retweet behavior prediction. This work explores the potential driving factors and predictable signals in user retweet behavior in hope that the deep learning framework has better expressive ability and prediction performance.

For future researches, the experimental dataset related to this research field still contains rich social dynamics that deserve further exploring. We can study user behavior in a semi-supervised manner, develop a generic solution based on heterogeneous graph learning, and then extend it to many network mining tasks, such as link prediction, social recommendation, similarity search, etc. Through such a learning scheme, we can leverage both unsupervised information and limited labels of users to build the predictor, and verify the effectiveness and rationality of user behavior analysis on real-world datasets.

References

1. Backstrom, L., Huttenlocher, D., Kleinberg, J., Lan, X.: Group formation in large social networks: membership, growth, and evolution. In: Proceedings of the 12th ACM SIGKDD International Conference on Knowledge Discovery and Data Mining, pp. 44–54 (2006)
2. Bakshy, E., Eckles, D., Yan, R., Rosenn, I.: Social influence in social advertising: evidence from field experiments. In: Proceedings of the 13th ACM Conference on Electronic Commerce, pp. 146–161 (2012)
3. Bond, R.M., et al.: A 61-million-person experiment in social influence and political mobilization. Nature **489**(7415), 295–298 (2012)
4. Chen, M., Wei, Z., Huang, Z., Ding, B., Li, Y.: Simple and deep graph convolutional networks. In: International Conference on Machine Learning, pp. 1725–1735. PMLR (2020)
5. Chen, W., et al.: Semi-supervised user profiling with heterogeneous graph attention networks. In: IJCAI. vol. 19, pp. 2116–2122 (2019)
6. Defferrard, M., Bresson, X., Vandergheynst, P.: Convolutional neural networks on graphs with fast localized spectral filtering. arXiv preprint arXiv:1606.09375 (2016)
7. Donnat, C., Zitnik, M., Hallac, D., Leskovec, J.: Learning structural node embeddings via diffusion wavelets. In: Proceedings of the 24th ACM SIGKDD International Conference on Knowledge Discovery & Data Mining, pp. 1320–1329 (2018)
8. Grover, A., Leskovec, J.: node2vec: Scalable feature learning for networks. In: Proceedings of the 22nd ACM SIGKDD International Conference on Knowledge Discovery and Data Mining, pp. 855–864 (2016)

9. Jiang, B., et al.: Retweeting behavior prediction based on one-class collaborative filtering in social networks. In: Proceedings of the 39th International ACM SIGIR conference on Research and Development in Information Retrieval, pp. 977–980 (2016)
10. Jin, Y., Song, G., Shi, C.: Gralsp: graph neural networks with local structural patterns. In: Proceedings of the AAAI Conference on Artificial Intelligence, vol. 34, pp. 4361–4368 (2020)
11. Kim, Y.A., Srivastava, J.: Impact of social influence in e-commerce decision making. In: Proceedings of the Ninth International Conference on Electronic Commerce, pp. 293–302 (2007)
12. Li, Q., Han, Z., Wu, X.M.: Deeper insights into graph convolutional networks for semi-supervised learning. In: Proceedings of the AAAI Conference on Artificial Intelligence, vol. 32 (2018)
13. Liu, Y., Zhao, J., Xiao, Y.: C-rbfnn: a user retweet behavior prediction method for hotspot topics based on improved rbf neural network. Neurocomputing **275**, 733–746 (2018)
14. Luo, Z., Osborne, M., Tang, J., Wang, T.: Who will retweet me? finding retweeters in twitter. In: Proceedings of the 36th International ACM SIGIR Conference on Research and Development in Information Retrieval, pp. 869–872 (2013)
15. Matsubara, Y., Sakurai, Y., Prakash, B.A., Li, L., Faloutsos, C.: Rise and fall patterns of information diffusion: model and implications. In: Proceedings of the 18th ACM SIGKDD International Conference on Knowledge Discovery and Data Mining, pp. 6–14 (2012)
16. Perozzi, B., Al-Rfou, R., Skiena, S.: Deepwalk: Online learning of social representations. In: Proceedings of the 20th ACM SIGKDD International Conference on Knowledge Discovery and Data Mining, pp. 701–710 (2014)
17. Qiu, J., Tang, J., Ma, H., Dong, Y., Wang, K., Tang, J.: Deepinf: social influence prediction with deep learning. In: Proceedings of the 24th ACM SIGKDD International Conference on Knowledge Discovery & Data Mining, pp. 2110–2119 (2018)
18. Riquelme, F., González-Cantergiani, P.: Measuring user influence on twitter: a survey. Inform. Process. Manag. **52**(5), 949–975 (2016)
19. Sun, F.Y., Hoffmann, J., Verma, V., Tang, J.: Infograph: Unsupervised and semi-supervised graph-level representation learning via mutual information maximization. arXiv preprint arXiv:1908.01000 (2019)
20. Sun, J., Wang, X., Xiong, N., Shao, J.: Learning sparse representation with variational auto-encoder for anomaly detection. IEEE Access **6**, 33353–33361 (2018)
21. Tang, J., Sun, J., Wang, C., Yang, Z.: Social influence analysis in large-scale networks. In: Proceedings of the 15th ACM SIGKDD International Conference on Knowledge Discovery and Data Mining, pp. 807–816 (2009)
22. Tang, J., Wu, S., Sun, J.: Confluence: conformity influence in large social networks. In: Proceedings of the 19th ACM SIGKDD International Conference on Knowledge Discovery and Data Mining, pp. 347–355 (2013)
23. Tong, H., Faloutsos, C., Pan, J.Y.: Fast random walk with restart and its applications. In: Sixth International Conference on Data Mining (ICDM'06), pp. 613–622. IEEE (2006)
24. Ugander, J., Backstrom, L., Marlow, C., Kleinberg, J.: Structural diversity in social contagion. Proc. Nat. Acad. Sci. **109**(16), 5962–5966 (2012)
25. Ulyanov, D., Vedaldi, A., Lempitsky, V.: Instance normalization: The missing ingredient for fast stylization. arXiv preprint arXiv:1607.08022 (2016)

26. Wang, J., Wang, Y.Q.: Sir rumor spreading model with network medium in complex social networks. Chinese Journal of Physics (2015)
27. Watts, D.J., Strogatz, S.H.: Collective dynamics of 'small-world' networks. Nature **393**(6684), 440–442 (1998)
28. Wu, H., Hu, Z., Jia, J., Bu, Y., He, X., Chua, T.S.: Mining unfollow behavior in large-scale online social networks via spatial-temporal interaction. In: Proceedings of the AAAI Conference on Artificial Intelligence, vol. 34, pp. 254–261 (2020)
29. Xiong, J., Chen, X., Yang, Q., Chen, L., Yao, Z.: A task-oriented user selection incentive mechanism in edge-aided mobile crowdsensing. IEEE Transactions on Network Science and Engineering (2019)
30. Yang, M.C., Lee, J.T., Lee, S.W., Rim, H.C.: Finding interesting posts in twitter based on retweet graph analysis. In: Proceedings of the 35th International ACM SIGIR Conference on Research and Development in Information Retrieval, pp. 1073–1074 (2012)
31. Yi, B., et al.: Deep matrix factorization with implicit feedback embedding for recommendation system. IEEE Trans. Ind. Inform. **15**(8), 4591–4601 (2019)
32. Yuan, N.J., Zhong, Y., Zhang, F., Xie, X., Lin, C.Y., Rui, Y.: Who will reply to/retweet this tweet? the dynamics of intimacy from online social interactions. In: Proceedings of the Ninth ACM International Conference on Web Search and Data Mining, pp. 3–12 (2016)
33. Zhang, C., Song, D., Huang, C., Swami, A., Chawla, N.V.: Heterogeneous graph neural network. In: Proceedings of the 25th ACM SIGKDD International Conference on Knowledge Discovery & Data Mining, pp. 793–803 (2019)
34. Zhang, J., Tang, J., Li, J., Liu, Y., Xing, C.: Who influenced you? predicting retweet via social influence locality. ACM Trans. Knowl. Discovery Data (TKDD) **9**(3), 1–26 (2015)
35. Zhang, J., Tang, J., Ma, C., Tong, H., Jing, Y., Li, J.: Panther: Fast top-k similarity search on large networks. In: Proceedings of the 21th ACM SIGKDD International Conference on Knowledge Discovery and Data Mining, pp. 1445–1454 (2015)
36. Zhang, M., Cui, Z., Neumann, M., Chen, Y.: An end-to-end deep learning architecture for graph classification. In: Proceedings of the AAAI Conference on Artificial Intelligence, vol. 32 (2018)
37. Zhang, X., Han, D.D., Yang, R., Zhang, Z.: Users' participation and social influence during information spreading on twitter. PloS One **12**(9), e0183290 (2017)
38. Zhao, Q., Erdogdu, M.A., He, H.Y., Rajaraman, A., Leskovec, J.: Seismic: a self-exciting point process model for predicting tweet popularity. In: Proceedings of the 21th ACM SIGKDD International Conference on Knowledge Discovery and Data Mining, pp. 1513–1522 (2015)

Mining Personal Health Satisfaction and Disease Index from Medical Examinations

Chunshan Li(✉), Xiao Guo, Dianhui Chu, Chongyun Xu, and Zhiying Tu

School of Computer Science and Technology, Harbin Institute of Technology, Weihai 264209, China
{lics,guox,chudh,cyxu,zytu}@hit.edu.cn

Abstract. People have regular medical examinations every year, the primary purpose of these is not only to discover a specific disease but to have a clear overview of their current health. Therefore, it is essential to be able to scientifically analyze and evaluate the results of medical examinations, and provide people with comprehensive feedback on physical health satisfaction (PHS) or a specific disease index (DI) feedback. PHS can reflect the current state of a person's body directly. In this paper, we propose a framework for calculating personal PHS and DI, by which the process of assessing people's health is indicated. We use the public dataset of the National Center for Health and Nutrition to conduct a scientific analysis of the population medical examinations data. Finally, based on a comprehensive experiment of 6166 participants' medical examinations data and some experiments with data from participants with heart disease, we verified the effectiveness of the proposed framework.

Keywords: Personal health satisfaction · Personal disease index · Health analysis

1 Introduction

In recent years, data mining has become more and more popular in the field of medical and health, and has been widely used in the areas of auxiliary medicine, disease prediction, online health assessment, etc., and has achieved remarkable results. Despite this, there are still many issues that need to be resolved urgently. As people's quality of life continues to improve, more and more people are paying attention to their health. Therefore, a large number of people go to a health check every year to accurately obtain the current physical health.

Health scoring systems have become increasingly popular over the years, such as Nuffield Health Score [1], but their scoring mechanism is based on pre-defined measurements of expert knowledge and experience. The SAPS acute physiology scoring system proposed by JR Le Gall et al. [2], the data set is derived from a large number of surgical and medical patient samples, and provides a method for converting the score into the probability of hospital death, the primary purpose of which is to study the current relationship between physical condition and mortality. As the number and dimensions

J. Xiong et al. (Eds.): MobiMedia 2021, LNICST 394, pp. 633–646, 2021.
https://doi.org/10.1007/978-3-030-89814-4_46

of health-related data grow, it becomes difficult to rely on expert knowledge. In 2014, Chen Ling et al. [3] implemented an automatic scoring system adapted to change in data attributes. The data set used personal health check data for the age group of 65 years and older in the past five years, and proposed the concept of personal health index (PHI), with the cause of death as a label. However, when people of other ages conduct physical examination analysis, it is difficult to obtain the cause of death of these people and the physical examination data for five consecutive years. Second, in their research, suffered some shortcoming when it came to pre-processing the data. In addition, people not only want to know the current physical condition through medical examination data, but also want to know that under current physical conditions, the probability of suffering from certain diseases, such as middle-aged and older people, will pay more attention.

Inspired by the above problems, in this paper, we have done the following four main tasks.

1. For the health checkup data of people of all ages, we propose the concept of personal health satisfaction (PHS) and construct a PHS calculation framework. Personal fitness satisfaction is used as a model training target. The patient can directly provide this attribute during the physical examination.
2. We train the model through a neural network algorithm to fill in some missing attributes and achieve impressive performance.
3. For the current physical condition and the probability of suffering from a certain disease, we propose the concept of disease index (DI) and use the PHS framework in question 1 to predict the disease index. This paper uses heart disease as an example to verify the effectiveness of the framework.
4. In addition, in order to make the framework reusable, we implement the algorithm and dataset based on an open-source platform (EasyML). For framework calculation process, each step result can be visualization using the EasyML.

In a word, we constructed a PHS computing framework that includes data filling, feature selection, model training, results prediction, and evaluation. Our job is the first to propose that the same set of frameworks can be used to predict personal health satisfaction and disease index.

2 Related Works

In recent years, a number of scoring systems have been introduced to assist in clinical decision making, such as APMCHE, SAPS, and MPM in patients in intensive care units [4]. At present, the most widely used and authoritative scoring method in the world, APACHE [5] (acute physiology and chronic health assessment) scoring system, American scholar Kuaus first proposed APACHE I in 1981 [6]. It consists of two parts. That is the acute physiology score (APS) and the pre-disease chronic health status (CPS) evaluation reflecting the severity of the acute disease. The former includes 34 physiological parameters. APACHE II [7] was launched in 1985, and the APACHE II score consists of three parts: 12 acute physiological variables, age, and chronic health status. The probability of death can be derived by using the disease category and the APACHE II score.

It can be seen that the use of 34 physiological parameters and 12 acute physiological variables in the selection of attributes are based on expert knowledge and experience. For disease prediction, Wilson et al. [8] studied coronary heart disease prediction, based on age, diabetes, smoking, JVC-V blood pressure categories, etc., to develop a gender-specific prediction equation to predict the risk of coronary heart disease, Palaniappan et al. [9] for heart disease prediction, the study developed a prototype of the Intelligent Heart Disease Prediction System (IHDPS) using data mining techniques, namely decision trees, naive Bayes and neural networks. Dangare [10] et al. developed a Heart Disease Prediction System (HDPS) using a neural network. The HDPS system predicts the likelihood of a patient having a heart attack. For prediction, the system uses 13 medical parameters such as gender, blood pressure, and cholesterol. Using medical profiles such as age, gender, blood pressure, and blood glucose can predict the likelihood of a patient suffering from heart disease, and it is easy to see that expert knowledge is also used when selecting attributes. Moreover, data mining in the health care field now uses more inpatient data, incidence rates [11–13], and mortality predictions [14]. The application of data analysis in large medical data sets is mostly based on population analysis of the distribution of some epidemics or high-risk diseases [15–18]. There are very few assessments based on personal health status and individual disease risk assessments, and there are many analyses of specific diseases that are used for classification, mostly in 2 categories, rather than judging the likelihood of illness. Chen Ling et al. [3] proposed the concept of personal health index, which is mainly aimed at the elderly over 65 years old. The framework is not applicable to people aged out this range. The method used in forecasting is a common algorithm in data mining. SVM and fail to compare with other methods.

3 PHS Calculation Framework

The PHS calculation framework (Fig. 1) mainly consists of five main parts: data preprocessing, feature selection, model training, result prediction, and PHS verification. The input is medical examinations dataset containing all attributes, and the output is personal health satisfaction or disease index ranging from 0 to 1.

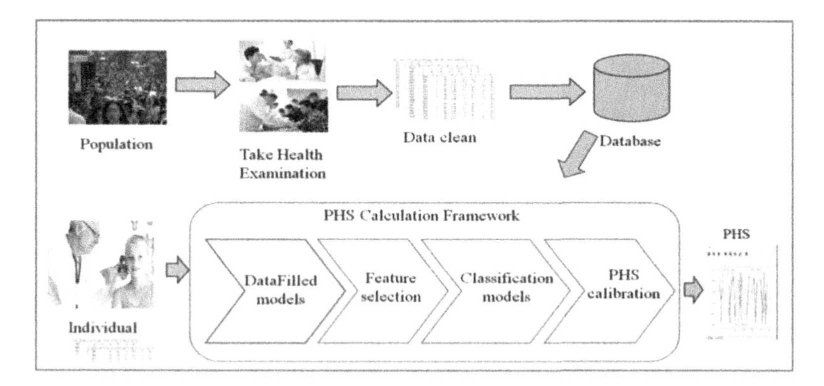

Fig. 1. An overview of the proposed PHS calculation framework

3.1 Data Pre-processing

The attribute values of the medical dataset include numeric values, serial numbers, and text. In the current work, only one text feature will be used as label: the current health satisfaction. The status, the recorded value is also the level information, "very good," "good," etc., we use numerical calculation for level recording. Firstly, we delete some unrecognized symbols and erroneous data. Secondly, with taking a look at missing values, there are many strategies to deal with missing values, a simple mean, regression, direct deletion..., After a few attempts, we kept all the available examples and focused on building model for each feature in order to infer the missing values. It is also the first part of the computational framework. We use simple forward neural networks to build those models. For this, we use three different methods for comparison experiments that are explained below. Finally, we normalize the values of the 30 attributes selected (described in detail in the next section).

We detail below the three neural network structures that we tried to use for reconstructing the missing values. The code is implemented with Tensorflow.

1) A Restricted Boltzman Machine (RBM) to reconstruct the input.

RBM [19] are widely used especially when there are a large amount of unlabeled data as a brick for a bigger network called deep belief network.

Here we use them as an order to mimic the reconstructed data distribution as we trained it with contrastive divergence procedure. We first define an energy function (Hopfield 1982):

$$\mathrm{E(v,h)} = -\sum\nolimits_{i\in visible} a_i v_i - \sum\nolimits_{j\in hidden} b_j h_j - \sum\nolimits_{i,j} v_j h_j w_{ij} \tag{1}$$

v_i, h_j binary states of visible unit i and hidden unit j.

a_i, b_j their biases and w_{ij} is the weight between them.

In a common RBM, the network assigns a probability to every possible pair of a visible and a hidden vector via this energy function:

$$\mathrm{p(v,h)} = \frac{1}{Z} e^{-E(v,h)} \tag{2}$$

$$\mathrm{Z} = \sum\nolimits_{v,h} e^{-E(v,h)} \tag{3}$$

$$\mathrm{p(v)} = \frac{1}{Z} \sum\nolimits_{h} e^{-E(v,h)} \tag{4}$$

It can be proven that when we derive the log probability, we eventually obtain such learning rule.

$$\Delta w_{ij} = lr(<v_i h_j>_{data} - <v_i h_j>_{model}) \tag{5}$$

Where lr is the learning rate.

Except that here our intermediary output is not stochastic but deterministic and their values are continuous. We still assume the learning rule is correct, and we can notice there is still convergence.

2) Auto-encoder

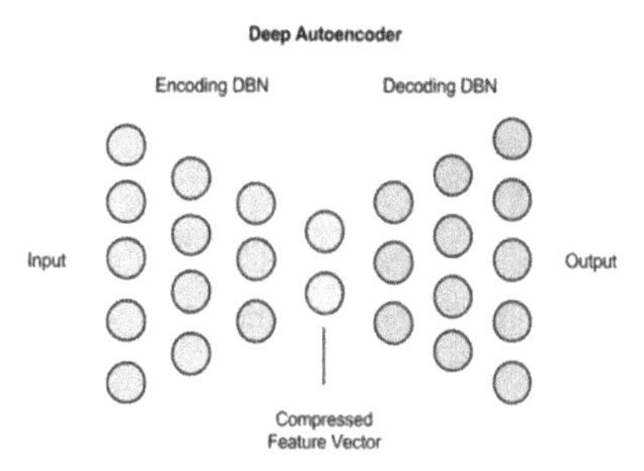

Fig. 2. Auto-encoder principle

In Fig. 2, AE [19] principle, we can notice some similitude with RBM, although this structure allows several hidden layers in both encoding and decoding part. Weights are this time not the same for both phase, it surely allow more freedom. And the objective function is not the same.

$$loss(\hat{y}, y) = \frac{1}{n}\sum_i |\hat{y}_i - y_i|^a \tag{6}$$

Where $\hat{y}$ is inferred value, y is true value.

There is some freedom to choose geometry or another penalty function, but the simple L2 norm does the job.

Though in the form, there is a bottleneck which can act as a regularizer, in our case, the data dimension is quite low. Consequently, this role is not so useful here.

Corrupt the input by adding some Gaussian noise as it is often the case to limit overfitting and learn more complex pattern has a limited influence, we blame the same reasons.

3) Simple Forward neural Network (FNN)

Once again, the structure is similar to the previous, but there is only one target the weights in the networks are now more specialized, though it makes the network even more sensible to overfitting.

$$loss(\hat{y}, y) = \frac{1}{n}\sum_i |\hat{y}_i - y_i|^a \tag{6}$$

4) few intermediary results

These figures (Fig. 3, Fig. 4) cannot be seen as proof of effectiveness. At this point, we cannot conclude. Their purpose is mostly to illustrate. Due to a large number of hyper-parameters to optimize, these results will change slightly.

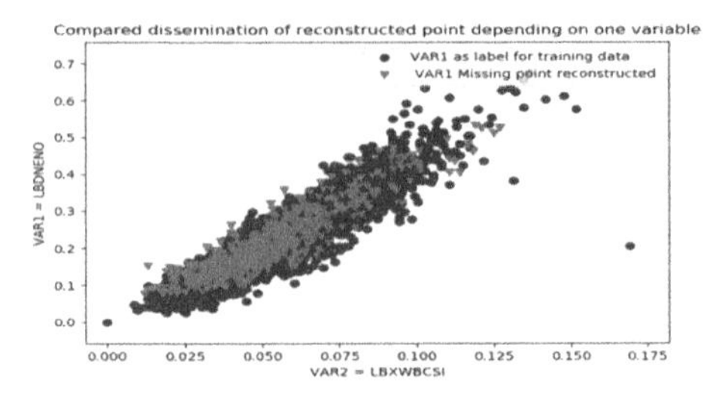

Fig. 3. FNN completion

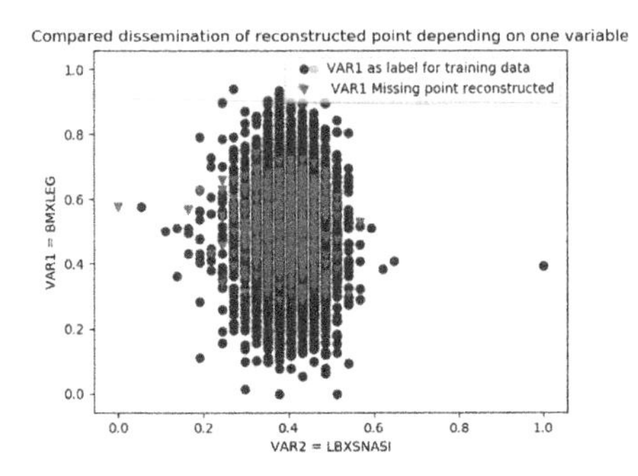

Fig. 4. FNN completion

Light and straightforward network, not too wide and not too deep seems to give a better result. It avoids overfitting, and it prevents activation function in the last layers to be saturated (especially for the simple neural network).

3.2 Feature Selection

There is a large number of attributes of the dataset. We will first go through a regularization phase. In the past, data mining in the field of medical research, attribute selection was mostly based on expert knowledge. In order to be able to pick out the attributes that have a greater impact on the results, we use the Ridge regression [21] algorithm to calculate each attribute and label value separately, and finally select 30 attributes that are better for the prediction.

Ridge regression solves some problems of ordinary least squares by penalizing the size of the coefficient. The minimum of the ridge coefficient is the sum of the squares of the residuals with penalty terms.

$$min||X_w - y||_2^2 + \alpha||w||_2^2 \tag{7}$$

Among them, $\alpha \geq 0$ is a complexity parameter of the amount of shrinkage: the larger the value of α, the larger the reduction.

We also performed a PCA that does not seem to improve the final results.

3.3 PHI and Prediction Models

In this sub-section, we first give the formal definition of PHI:

$\forall x \in \boldsymbol{R}^n$,
PHS: $x \rightarrow 1 - \text{ModelInfer}(x)$
$\boldsymbol{R}^n \rightarrow [0,1]$
x: the n-dimension vector that describe an individual;
n: the dimension that describe an individual;

Model Infer: the complementary function of PHS to 1.

After definition, we can will employ a proper predictor to calculate the PHI. We will ran several cross-validations with numerous learning algorithms. Are common algorithms such as SVM [22], TreeBoost [23], neural networks…, this one seems to outperformed other alternatives with slight differences depending on hyperparameters and kernels for SVM.

3.4 PHS Calibration

In this section, we wonder whether or not we need calibration (Fig. 5). Indeed, to give any individual an understandable and reliable index about its global health we need to be aware of the probability distribution over classes of our models, if the output, a continuous value, of these, is not linearly correlated to the belonging probability to levels, we would need to calibrate it.

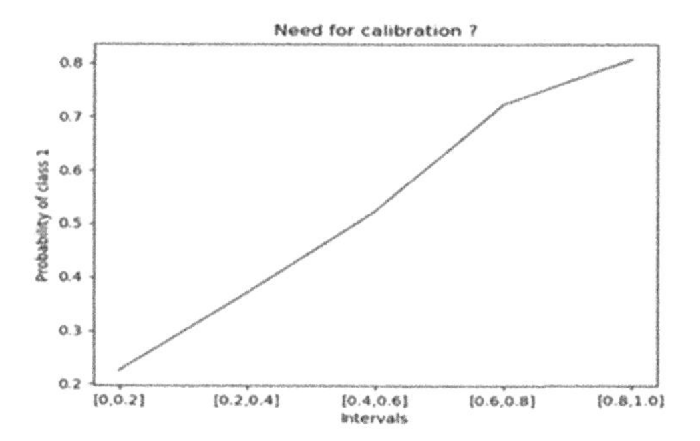

Fig. 5. Belonging probability to class depending on model output. Linear in this case

We can notice here the importance of penalty function among other parameters on the reliability curve. But a simple square error fulfills our expectation.

4 Experimental Evaluation

4.1 Dataset Description

The dataset we used has been released by the National Center for Health and Nutritional Surveys. The participants ranged from a few years to a few decades. The survey center randomly selected a total of 6,166 participants from all over the United States. The medical examination data is personal privacy, and the public dataset does not have any records that can identify the individual, and the individual is marked in the form of ID. Each person's medical examination data consists of four parts: Demographic, Examinations, Labs tests and Health questions, the specific content is shown in Table 1, due to space constraints, we only listed some attributes. And those are already selected one.

Table 1. Selected attributes by categories

Type	Attribute examples	Number
Demographic	Age, education, marital status, income,	13
Lab tests	Albumin, testosterone, blood lead,blood cadmium, Lymphocyte…	68
Examinations	Blood pressure, Weight, height, arm and leg length…	29
Health questions	Health condition(label), Mental condition(label)	2

4.2 Experiment Setup

For the following experiment, we used "Health condition" as the final target. In the data pre-processing stage, we first select 30 attributes by Ridge regression. In the dataset, there is a questionnaire on personal health satisfaction. The results are divided into 5 grades, 1–5, 1 means very satisfied, 5 means very dissatisfied. Looking at a first confusion matrix, the frontier between predicted classes is not clear, we can't get obvious prediction results. As a consequence, we have processed the label to keep only two classes (Table 2):

Table 2. Label change.

Former label	New label
1, 2	Healthy (0)
3, 4, 5	Not Healthy (1)

We ran cross-validation to select best learning algorithms including Aboost, TreeBoost, DecisionTree [24, 25], etc.

Algorithms are running in a Linux environment. All are written in Python with Scikit-learn package.

4.3 Experiment Results

(1) In the data pre-processing part, firstly, we perform a simple analysis of datasets, there are more than 4,000 records missing about 18 attributes, so the handling of missing values appear to be important.

Secondly, after regularization, as shown in Fig. 6, we can find that the missing values of two preponderant attributes reach 90%. In order to make the result more realistic, when filling in the missing values, the model is not built for these two attributes, only filled with '0', so in the filling model training, the attribute with the missing rate greater than 20% is directly filled with 0, less than 20% attribute will build a model for it.

We want to expose here the influence of the different methods of data blank completion we highlighted in the previous part on the final results (Table 3).

Table 3. Method results

Method	Score
FNN	0.72
RBM	0.70
AutoEncoder	0.72
KNN	0.71
Nothing	0.68

There is not significant difference between the different methods. We mostly blame the quality of the label for that though we decide to use FNN to deploy our code on the platform.

(2) In the feature selection part, we use the Ridge regression algorithm, after performing a grid experiment, the value of α is chosen to be 12, and the experimental result is shown in Fig. 6. An example of training accuracy with the neural network model.

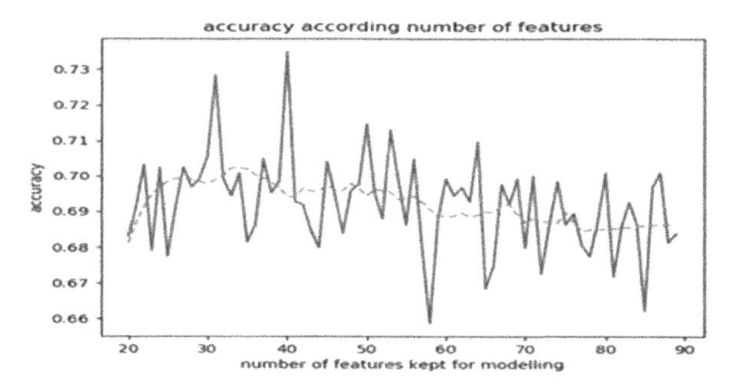

Fig. 6. Attribute number and accuracy

Where the abscissa is the number of selected attributes, and the ordinate is the accuracy of the result. The solid blue line is the specific accuracy obtained, and the yellow dotted line is a smoothed version that shows the trend. We can see that as the number of attributes increases, there is just a slight accuracy difference.

(3) Classification model selection: Firstly, Attempts have been made not to consider the label as they seem really subjective with common algorithms like k-means or Gaussian mixture models, the result in Fig. 7 (a&b). Even by contorting the geometry used, algorithms cannot distinguish that match Healthy/notHealthy group, and the process might be too dependent, too specific to the data.

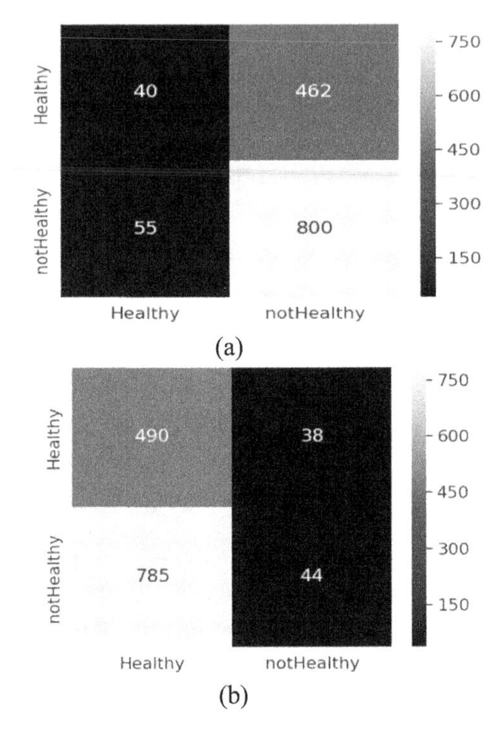

Fig. 7. (a) k-means. (b) Gaussian mixture model (GMM).

We performed cross-validation on quickly processed data normalization, data blank filled with 0 with some popular classification algorithms using 'Health condition' as a label. The comparison is shown in Fig. 8.

(4) (4)The influence of missing values filling algorithms. There is no significant difference between different methods though we decide to use FNN to deploy our code on the platform.
(5) PHS: The final prediction results of PHS, in Fig. 9, in order to make the display of the results more intuitive, we made a comparison between the actual value and the predicted value, and the abscissa in the figure. Indicates the 0-30th person, the ordinate is the prediction result range 0–1, the solid red line is the real value, only 0 and 1, two categories. the solid blue line is predicted value between 0–1.
(6) Calibration: as previously explained we need the output of our model to be linearly correlated to the belonging probability of the class.

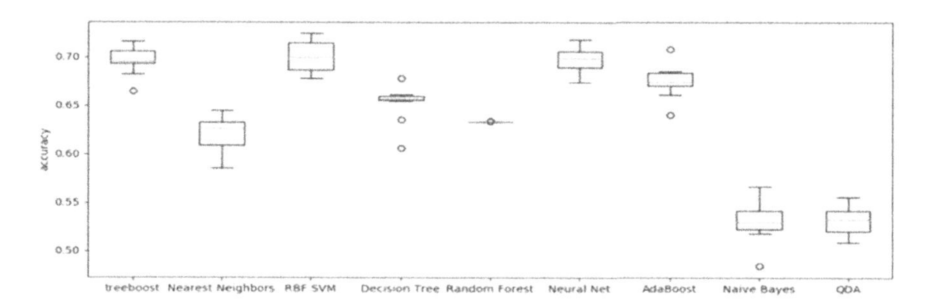

(a) Filled with "0", algorithm comparison

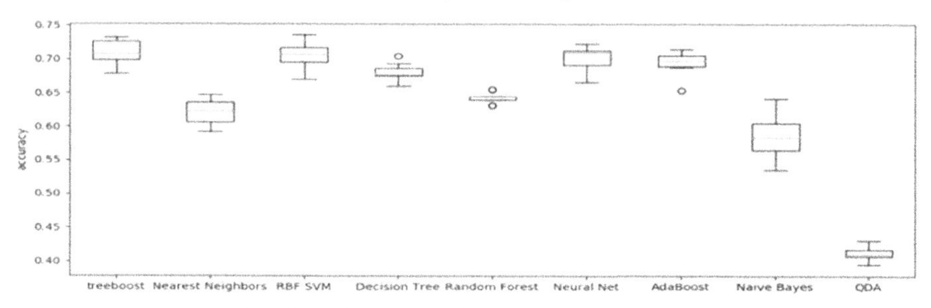

(b) Model filling, algorithm comparison

Fig. 8. Comparison of several classification algorithms

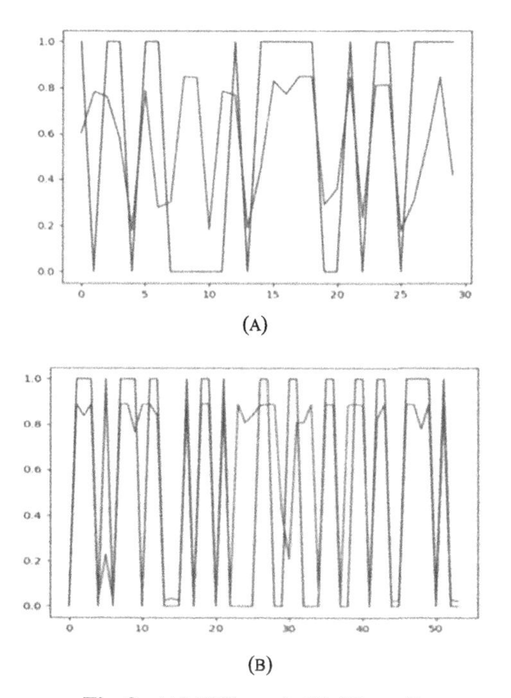

(A)

(B)

Fig. 9. (A) PHS result (B) DI result

We can notice that loss function(Fig. 10(a)) has a significant impact, but classical square loss function(Fig. 10(b)) fit our specification, calibration is in this case not necessary.

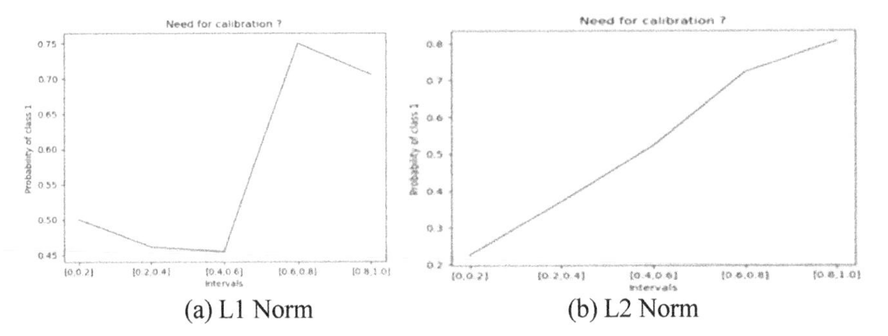

(a) L1 Norm (b) L2 Norm

Fig. 10. Class belonging probability according to the model output and the norm used

4.4 Data Processing Platform

In order to make PHS calculation framework and training model more reusable and scalable, we use an open-source machine learning platform that builds a Hadoop cluster that can support the distributed operation of the large dataset and make some modifications to the platform. The algorithm and data can be deployed on the platform through the interface. The advantage is that the algorithms can be moved on the page by dragging. The calculation result of each step of the whole process of the framework can be displayed in real-time so that we can better grasp the result. Moreover, any part of the framework can be optimized in the future, including algorithms and processes.

In Fig. 10, the left panel contains the algorithm, data, and tasks lists. The middle part shows the task running. The right panel let us see the detailed information about the task, real-time monitoring task, and running status. During the running process, the node is yellow, and the running success node is blue. After each algorithm runs successfully, you have access to the result directly from the output node.

5 Conclusion

In the era of big data, we can process a large amount of data. For our study case, we embarked in the healthcare field and tried to build a person's physical health satisfaction and a single disease index. Our work is constrained by the data we have access to indeed label is here highly subjective, although we managed to build a coherent index.

In other studies, PHS was proposed, but the label is based on the cause of death, to predict the current physical condition of the elderly, and it is to infer people's index with out of range. In addition, the missing data is not processed, and the missing data is deleted directly. We use the NN algorithm to train models filling with missing values.

When constructing the classification model, we did a lot of comparison experiments of classification algorithms, including before filling the data and filling the data.

A large number of experiments verify the validity and robustness of the framework. The best predictive PHS accuracy rate is 72.5%, and the disease index DI is 83%, which can better verify the validity and practicability of the model.

References

1. NuffieldHealthScore. http://info.nuffieldhealthscore.com/. Accessed 18 June 2014
2. Le Gall, J.R., Lemeshow, S., Saulnier, F.: A new simplified acute physiology score (SAPS II) based on a European/North American multicenter study. JAMA **270**(24), 2957–2963 (1993)
3. Chen, L., Li, X., Wang, S., et al.: Mining personal health index from annual geriatric medical examinations. In: 2014 IEEE International Conference on Data Mining, pp. 761–766. IEEE (2014)
4. Fisher, A., Burke, D.: Critical care scoring systems. In: Brown, S., Hartley, J., Hill, J., Scott, N., Williams, J. (eds.) Contemporary Coloproctology, pp. 513–528. Springer, London (2012). https://doi.org/10.1007/978-0-85729-889-8_35
5. Wong, D.T., Knaus, W.A.: Predicting outcome in critical care: the current status of the APACHE prognostic scoring system. Can. J. Anaesth. **38**(3), 374–383 (1991)
6. Knaus, W.A., Zimmerman, J.E., Wagner, D.P., et al.: APACHE-acute physiology and chronic health evaluation: a physiologically based classification system. Crit. Care Med. **9**(8), 591–597 (1981)
7. Knaus, W.A., Draper, E.A., Wagner, D.P., et al.: APACHE II: a severity of disease classification system. Crit. Care Med. **13**(10), 818–829 (1985)
8. Wilson, P.W.F., D'Agostino, R.B., Levy, D., et al.: Prediction of coronary heart disease using risk factor categories. Circulation **97**(18), 1837–1847 (1998)
9. Palaniappan, S., Awang, R.: Intelligent heart disease prediction system using data mining techniques. In: 2008 IEEE/ACS International Conference on Computer Systems and Applications, pp. 108–115. IEEE (2008)
10. Dangare, C., Apte, S.: A data mining approach for prediction of heart disease using neural networks. Int. J. Comput. Eng. Technol. (IJCET) **3**(3) (2012)
11. Esfandiary, N., Babavalian, M.R., Moghadam, A.-M.E., Tabar, V.K.: Knowledge discovery in medicine: current issue and future trend. Expert Syst. Appl. **41**(9) (2014)
12. Neuvirth, H., et al.: Toward personalized care management of patients at risk: the diabetes case study. In: SIGKDD. California, USA, pp. 395–403. ACM (2011)
13. Tran, T., Phung, D., Luo, W., Harvey, R., Berk, M., Venkatesh, S.: An integrated framework for suicide risk prediction. In: SIGKDD, Chicago, USA, pp. 1410–1418. ACM (2013)
14. Predicting mortality of ICU patients. http://physionet.org/challenge/2012/. Accessed 18 Feb 2014
15. Abbas, K., Mikler, A.R., Gatti, R.: Temporal analysis of infectious diseases: influenza. In: Proceedings of the 2005 ACM Symposium on Applied Computing, pp. 267–271. ACM (2005)
16. Koh, H.C., Tan, G.: Data mining applications in healthcare. J. Healthc. Inf. Manag. **19**(2), 65 (2011)
17. Obenshain, M.K.: Application of data mining techniques to healthcare data. Infect. Control Hosp. Epidemiol. **25**(8), 690–695 (2004)
18. Mellmann, A., Friedrich, A.W., Rosenkötter, N., et al.: Automated DNA sequence-based early warning system for the detection of methicillin-resistant Staphylococcus aureus outbreaks. PLoS Med. **3**(3), e33 (2006)

19. Hinton, G.E.: A practical guide to training restricted Boltzmann machines. In: Montavon, G., Orr, G.B., Müller, K.R. (eds.) Neural Networks: Tricks of the Trade. LNCS, vol. 7700, pp. 599–619. Springer, Heidelberg (2012). https://doi.org/10.1007/978-3-642-35289-8_32
20. Rifai, S., et al.: Higher order contractive auto-encoder. In: Gunopulos, D., Hofmann, T., Malerba, D., Vazirgiannis, M. (eds.) ECML PKDD 2011. LNCS, vol. 6912, pp. 645–660. Springer, Heidelberg (2011). https://doi.org/10.1007/978-3-642-23783-6_41
21. Khalaf, G., Shukur, G.: Choosing ridge parameter for regression problems, pp. 1177–1182 (2005)
22. Gupta, S., Kumar, D., Sharma, A.: Data mining classification techniques applied for breast cancer diagnosis and prognosis. Indian J. Comput. Sci. Eng. (IJCSE) **2**(2), 188–195 (2011)
23. Jianming, Z., Zhicai, Z., Keyang, C., et al.: Review on development of deep learning. J. Jiangsu Univ. Nat. Sci. Edit. **36**(2), 191–200 (2015)
24. Wu, X., Kumar, V., Quinlan, J.R., et al.: Top 10 algorithms in data mining. Knowl. Inf. Syst. **14**(1), 1–37 (2008)
25. Yang, S., Zou, L., Wang, Z., Yan, J., Wen, J.-R.: Efficiently answering technical questions-a knowledge graph approach. In: AAAI, pp. 3111–3118 (2017)

Visualizing Symbolic Music via Textualization: An Empirical Study on Chinese Traditional Folk Music

Liumei Zhang(✉) and Fanzhi Jiang

Xi'an Shiyou University, Xi'an 710065, Shannxi, China
zhangliumei@xsyu.edu.com, 19211060559@stumail.xsyu.edu.com

Abstract. The continuous integration of computer technology and art has promoted the development of digital music. In recent years, the massive music data has promoted the problems of music inquiry, music classification, music content understanding and so on, making a new subject develop continuously, that is music information retrieval. The current mainstream feature extraction of music is based on acoustics, such as pitch, timbre, loudness, zero-crossing rate, etc. Direct symbolic music feature extraction and music analysis are relatively rare. This paper aims to present an empirical study on text clustering ideas in natural language processing into the field of symbolic music style analysis. Firstly, Symbolic music is textualized and proposed by us as an inspiration. To be precise, textualization of symbolic music is converted into weighted structured data through tf-idf algorithm. In the following step, three different types of mainstream clustering algorithms, K-Means, OPTICS, and Birch are used to perform cluster analysis and comparison on the traditional Chinese folk music dataset we crawled, T-SNE algorithm is used to visualize the dimensionality reduction of high-dimensional data. Finally, a series of objective evaluation indicators of clustering are used to evaluate the three clustering algorithms. Through comprehensive evaluation of indicators, it is proved that the clustering algorithm has achieved an excellent clustering effect on the midi note dataset we extracted. As a result of the clustering, the professional music theory knowledge and the historical development characteristics of traditional Chinese folk music are comprehensively integrated, which reversely verifies that the 1300 midi music data sets have distinct modal characteristics of traditional Chinese folk music.

Keywords: Music information retrieval · Midi datasets · Traditional Chinese music · Text clustering · Feature extraction

Supported in part by the scholarship from National Natural Science Foundation of China (No. 61802301) and (No. 211817019), Shaanxi Natural Science Foundation of China (No. 2019JQ-056).

J. Xiong et al. (Eds.): MobiMedia 2021, LNICST 394, pp. 647–662, 2021.
https://doi.org/10.1007/978-3-030-89814-4_47

1 Introduction

The combination of music and technology has developed for a long time. As a branch of sound and music computing, music information retrieval has also experienced vigorous development in recent years [19]. The industry has conducted deep exploration of music information retrieval, which has derived many research directions. Music recommendation, for example, uses collected user information to process and infer users' specific explicit or hidden preferences for music to achieve accurate music recommendation [18]. There is also the calculation and identification of music emotion through acoustic characteristics [16]. They generally start from acoustics, and then use the proven Russell Ring two-dimensional emotional model to analyze music emotion. There is also research on content-based music style. At present, the research based on music content mainly takes audio signals as the research object. For example, the calculation method is used to analyze and understand the content of digital sound and music [1,6], as well as the research on music produced by instruments [7,8], as well as the theoretical limitations of various aspects of acoustics in the in-depth study of musical elements and musical structures [17,21]. As well as studies on information computation of human-generated music and songs [11,13].

Sergio Oramas [14] proposed a clustering scoring method for building music databases. Yu Qi [15] uses one-sided continuous matching similarity algorithm to cluster the feature database, and marks the cluster center, and then uses linear alignment matching (LAM) algorithm to accurately match each cluster center and its elements in the cluster, which ensures the accuracy of the retrieval. Changsheng Xu [21] proposed a new method to distinguish music styles. Firstly, the support vector machine was used to classify pure music and vocal music supervisedly and automatically. Secondly, the unique music features were extracted from the objective characteristics of the two kinds of music. Finally, the clustering method was used to reconstruct the music content, which used a large number of acoustic and energy features for feature extraction and analysis. Dong-Moon Kim [10] conducted short-time Fourier transform on waveform corpus to extract features from this corpus from the perspective of providing music customization, and then proposed a dynamic K-Means algorithm to recommend corpus fragments in music corpus from different schools and styles. Wei-HoTsai [17] different from the traditional clustering method, studied the unsupervised clustering of vocal music. It still used Fourier transform and other means to cut the vocal music, and then extracted the vocal characteristics with the waveform vocal music fragment. However, the scale of its vocal music corpus was small, which did not prove its universality. The research based on waveform music has common characteristics, that is, the research is difficult and the experimental accuracy is low. Rudi Cilibrasi [5] classified the corpus fragments based on the compression method. In order to visually display the information in the distance matrix, hierarchical clustering of the corpus was adopted. However, the test on large datasets was different from that on small datasets, and the results were not ideal.

Chinese traditional music refers to the music created by the Chinese using their own national inherent methods and taking their own national inherent forms with their own national inherent morphological characteristics. It includes not only ancient works that have been produced and circulated in history, but also contemporary works. It is an essential part of Chinese national music. Chinese traditional folk music has its inherent mode characteristics. Thousands of years ago, the law of five-degree intergrowth emerged simultaneously in the Chinese cultural region and the Greek cultural region. Although the law of three-point profit and loss and the law of five-degree intergrowth have small differences, they have very high similarities in the principle and method of the law of life [23]. As the ancestor of the Chinese music system, the records of the three-point profit and loss method first appeared in the Spring and Autumn Period's *Guanzi Diyuan*. At this time, the three-point profit and loss method is related to the records of Gong, Shang, Jiao, Zhi, Yu. After the *Lv's Spring and Autumn Annals -Rhythm* was completed, the three-point profit and loss law began to be linked to the rules on the length of twelve laws such as Huang Zhong and Lin Zhong [24]. On the basis of the five-tone scale, later gradually appeared other partial tones, evolved into seven national tone scale, but the main tone is still five-tone scale [20].

In this work, we first crawled the music corpus that established Chinese traditional music and extracted the corpus data in midi format for text representation of music symbols. Then, the tf-idf algorithm is used to weight the textualized note data and process them into vectors. In the next step, K-Means, OPTICS, and Birch clustering methods are used to cluster and compare the weighted data, and different clustering indicators are selected for evaluation. This paper uses the T-distributed Stochastic Neighbor Embedding (T-SNE) algorithm to reduce the dimension and visualize the clustering. According to the clustering results, combined with the music theory of traditional Chinese national music, this corpus has distinctive tonal characteristics of traditional Chinese national music. The main ideas of our research are as flowcharts Fig. 2 (Fig. 1).

Fig. 1. The Pentatonic scale of Gong Shang Jiao Zhi Yu.

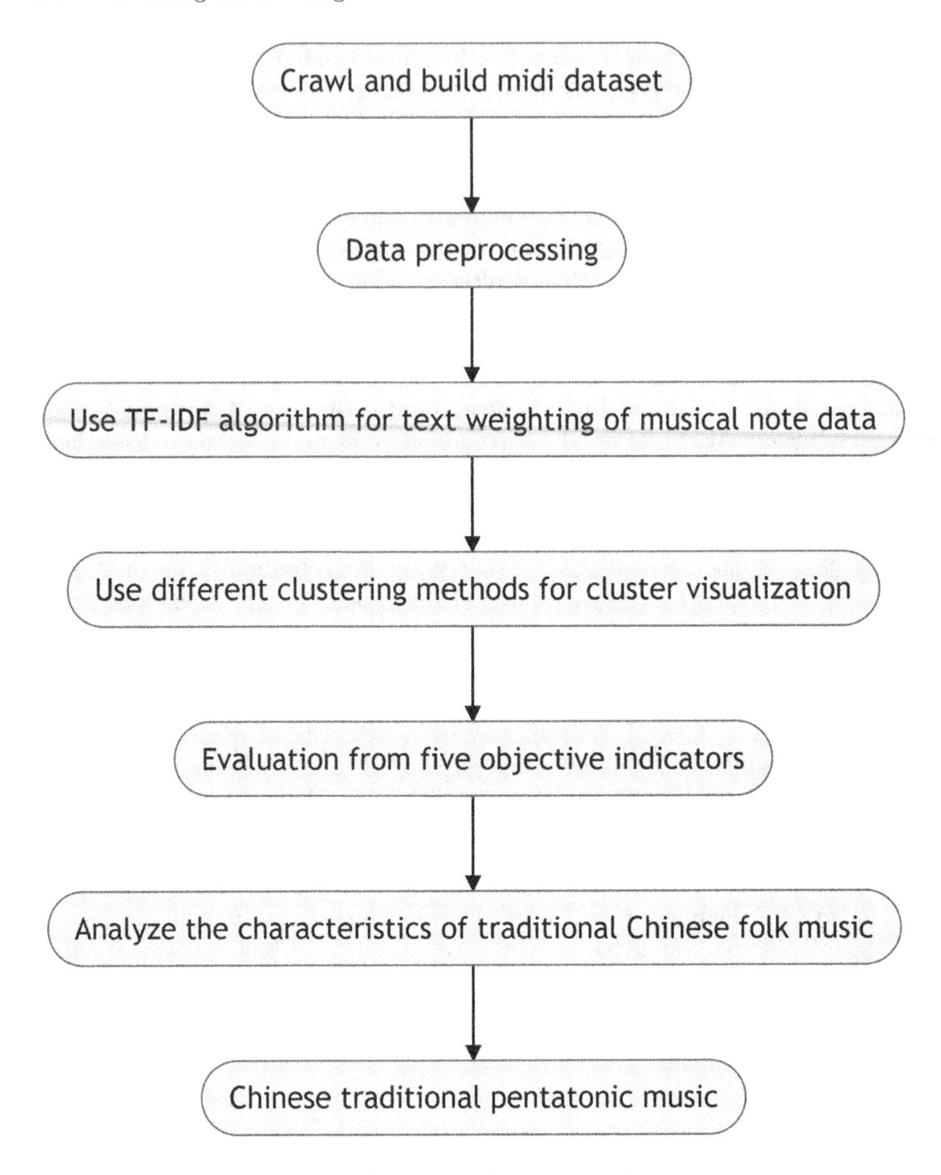

Fig. 2. Flow chart of main study idea.

2 Clustering Model

2.1 Basic Principles

In this part, we mainly describe the three clustering algorithms in depth from the background, motivation and calculation theory of the algorithm, paving the way for the processing and application of music data in the next part.

K-Means. Setting an initial value of k, representing k class clusters, aggregate all data points to the center of the closest class cluster, and keep iterating in accordance with this idea. Calculate the average value from the point to the center of the class cluster, and update the center of the class cluster until the iteration is stopped. This is the core idea of k-means algorithm.

Setting $Y = \{Y_1, Y_2, Y_3, ..., Y_n\}$. Each element in Y has m dimensions. Firstly, we initialize the center of the cluster according to the core idea of k-means and initialize k clustering centers $N_1, N_2, ..., N_n$, and then calculate the Euclidean distance from each element to each clustering center.

$$\operatorname{dis}(Y_i, N_j) = \sqrt{\sum_{t=1}^{m} (Y_{it} - N_{jt})^2} \tag{1}$$

The mean value of each dimension data of all elements in a specific region can be calculated as the center of the class cluster. The specific calculation equation is as follows:

$$N_l = \frac{\sum_{Y_i \in M_j} Y_i}{|M_l|} \tag{2}$$

The l cluster center is represented by N_l, where $1 \le l \le k$, $|M_l|$, the number of elements in the l cluster is represented by Y_i, and the i element in the l cluster is represented by Y_i, $1 \le i \le |M_l|$.

OPTICS. OPTICS is improved on the basis of the DBSCAN algorithm and is also a density-based clustering algorithm [2]. Choosing the appropriate two initial parameters and is the key to the DBSCAN algorithm, because it is very sensitive to the initial parameters, and whether the selection is appropriate will lead to huge differences in the calculation results. OPTICS came into being on this basis. Since the OPTICS algorithm is an improvement of the DBSCAN algorithm, some concepts are shared, such as: -Neighborhood, core object, direct density, density reachability, density connection, etc. The following is the definition related to OPTICS (assuming my sample Set is):

ϵ-Neighborhood: For $x_i \in X$, its ϵ-neighborhood contains the sub-sample set X whose distance between x_j and in the sample set is not greater than ϵ. ϵ-Neighborhood is a set, expressed as follows, the number of this set is recorded as $|N_\epsilon(x_j)|$.

$$N_\epsilon(x_j) = \{x_i \in X \mid \text{ distance } (x_i, x_j) \le \epsilon\} \tag{3}$$

Core object: For any sample $x_j \in X$, if its ϵ-neighbourhood corresponds to $N_\epsilon(x_j)$ at least contains $MinPts$ samples, that is, if $|N_\epsilon(x_j)| \ge MinPts$, then x_j is the core object.

Density direct: If x_i is located in the ϵ-neighborhood of x_j and x_j is the core object, it is said x_i to be direct by density x_j. The opposite is not necessarily true, that is, it cannot be said x_i to be directly reached by density X_j at this time, unless x_i is also the core object, that is, direct density does not satisfy symmetry.

Density is reachable: for x_i and x_j, if there is a sample sequence $p_1, p_2, ..., p_T$ that satisfies $p_1 = x_i$, $p_T = x_j$, and p_{t+1} is directly reached by the density p_t, it is said that the density is reachable. In other words, the density can be reached to meet the transitivity. At this time, the transferred samples in the sequence are all core objects, because only core objects can make other sample densities reachable directly. The density is reachable and does not satisfy the symmetry, which can be derived from the direct asymmetry of the density.

Density connected: For x_i and x_j, if there is a core object sample x_k, so that both x_i and x_j are reachable by the density x_k, then it is said that x_i and x_j the density are connected. The density connection relationship satisfies symmetry.

On the basis of the above definition of DBSCAN, OPTICS has introduced two definitions required by the algorithm:

Core distance: For a sample $x \in X$, for a given ϵ and $MinPts$, the smallest neighborhood radius that makes x a core point is called the core distance of x. It is mathematical expression is as follows, which $N_\epsilon^i(x)$ represents the node i closest to the node x in the set $N_\epsilon(x)$, such as $N_\epsilon^1(x)$ in $N_\epsilon(x)$ and Nearest node x:

$$cd(x) = \begin{cases} undefined\,|N_\epsilon(x)| < MinPts \\ d\left(x, N_\epsilon^{MinPts}(x)\right)|N_\epsilon(x)| >= MinPts \end{cases} \tag{4}$$

Reachability-distance: Let $x, y \in X$, for a given ϵ and $MinPts$, the reachability-distance of y about x is defined as:

$$rd(y, x) = \begin{cases} \text{undefined } |N_e(x)| < MinPts \\ \max\{cd(x), d(x, y)\}|N, (x)| >= MinPts \end{cases} \tag{5}$$

In particular, when x is the core point (the corresponding parameters are and), it can be understood $rd(y, x)$ according to the following formula:

$$rd(y, x) = \min\{\eta : y \in N_n(x) \& |N_\eta(x)| \geq \text{Min}\, Pts\} \tag{6}$$

Birch. The Birch algorithm builds a dendrogram called the cluster feature tree (CF tree). The CF tree can be constructed by scanning the data set in an incremental and dynamic manner. Therefore, it does not require the entire data set in advance [22].

It has two main stages: first scan the database to build a memory tree, and then apply the algorithm to the cluster leaf nodes. The CF tree is a highly balanced tree based on two parameters: branching factor B and threshold T. The CF tree is constructed when scanning data. When a data point is encountered, the CF tree will be traversed, starting from the root and selecting the nearest node at each level. If the closest leaf cluster of the current data point is finally determined, a test is performed to see whether the data point belongs to the candidate cluster or does not belong to the candidate cluster. Otherwise, a new cluster with a diameter larger than the given T will be created. Some other scans. It can also deal with noise effectively. However, when the clusters are not spherical, Birch may not work properly because it uses the concept of radius or diameter to control the boundaries of the clusters. In addition, it is sequence-sensitive and may generate different clusters for different sequence software input data.

3 Implementation

3.1 Dataset

This work is based on music information mining of symbolic music. Unlike waveform music, we have captured as many as 1,300 symbolic music of traditional Chinese people on the Internet and established a corpus of midi music. Since we didn 't make a detailed distinction when we caught it, its beats weren 't fixed, meaning their spectrum and beat numbers weren 't certain. But their average time is about ten seconds. py-midi and music21 toolkits are used to parse the music data set and rough pre-processing work. They separate the music according to the head and read the corpus with the music score object, and observe its main melody is in the piano track. So we extract the data of its piano track, and the basic scale spans the range of a group of small words to three groups of small words. Finally, a new data set based on text is established.

3.2 Data Preprocessing

Similar to text clustering, we first need to transform the music symbol information of this traditional music dataset document into mathematical information, so as to form high-dimensional space points and then calculate the similarity distance between the symbol element and the symbol element, so as to aggregate the symbol cluster. We use the classical statistical method if-idf algorithm to process the mathematical word vector of textualized note data, and give different weights. The specific algorithm of Ti-idf is as follows:

$$tf - idf = tf^{*}idf \tag{7}$$

$$f_{ij} = \frac{n_{i,j}}{\sum_k n_{k,j}} \tag{8}$$

In the formula, $n_{i,j}$ is expressed as times the word appears in the document d_j, and the denominator is expressed as the sum of the times of all words in the document d_j.

$$idf_i = \log \frac{|D|}{|\{j : t_i \in d_j\}|} \tag{9}$$

In the formula, $|D|$ is expressed as the total number of files in the corpus, and $|j : t_i \in d_j|$ represents the number of files containing words t_i (the number of files with $n_{i,j} \neq 0$). Tf-idf is used for the first mathematical processing of previously processed textualized music documents, including vectorizing notes data and giving specific weights based on the frequency of occurrence. Then, the structured data that can be used by the traditional clustering method are converted to music clustering. According to the music score object, the 19 extracted features can be observed across three octaves. In the piano, they are mainly distributed in the interval from a small word group to a small word group. The vectorized music data are the tensor of 87000 * 19.

3.3 Model Using

In order to prove the wide applicability of different clustering models in this dataset and the objectivity of clustering, we used three different clustering methods to cluster music data based on division, density and hierarchy. Among them, the k cluster determined by K-means clustering based on partition reaches the least square error, and the clustering results are dense, and the difference between classes is significant, as is shown in Algorithm 1. However, it may be difficult to select the k value or sensitive to noise points and outliers, resulting in large number of iterations and long time consuming. OPTICS does not explicitly generate data clustering, it only sorts the objects in the data object set and then calculates the sequence table, which contains a lot of information for extracting clustering, as is shown in Algorithm 2. It is not sensitive to the transformation of parameters in the process of clustering, but the clustering results are not as excellent as K-means and Birch. Birch has high efficiency in the calculation process, saves memory, and the clustering effect is better than OPTICS, as is shown in Algorithm 3. The time cost of the three clustering methods was statistically analyzed, as is shown in Table 1.

Table 1. Efficiency comparison of three clustering methods

Clustering Algorithm	K-Means	OPTICS	Birch
Clustering time (sec)	6.5	2.2	2.0

3.4 Algorithm

4 Objective Metrics for Evaluation

This paper evaluates the clustering effect from two kinds of metrics, and uses seven clustering methods. One of the measurement indexes is called internal evaluation method, which means to evaluate the algorithm by a single quantitative score, and the other is called external evaluation method, which compares the clustering results with the existing real classification results. Their classification is shown in Fig. 3.

Algorithm 1: K-means

Input: Noteset $D = \{x_1, x_2, x_3, ..., x_m\}$; Clusters k
Output: NoteCluster $C = C_1, C_2, ..., C_k$
1 k is selected Randomly from D:$\{\mu_1, \mu_2, \mu_3, ..., \mu_k\}$;
2 **repeat**
3 | $C = \varnothing(1 \leq i \leq k)$
4 **until** *something happens*;
5 **for** $j = 1, 2, 3, ..., m$ **do**
6 | Calculate the distance between sample x_j and each mean vector $\mu(1 \leq i \leq k) : d_{ji} = \| x_j - \mu_i \|_2$;
7 | Determine the cluster label of x_j according to the nearest mean vector:$\lambda = argmin_{i \in \{1,2,3,...,k\}} d_{ji}$;
8 | Corresponding cluster $C_{\lambda j} = C_{\lambda j} \cup \{x_j\}$ is divided into x_j ;
9 **end**
10 **for** $i = 1, 2, ..., k$ **do**
11 | Calculate the new mean vector $\mu_i^{'} = \frac{1}{|C_i|} \sum_{x \in C_i} x$;
12 | **if** $\mu_i^{'} \neq \mu_i$ **then**
13 | | Update μ_i to $\mu_i^{'}$;
14 | **else**
15 | | Keep the current mean unchanged ;
16 | **end**
17 **end**

The value of the ARI belongs to [−1, 1]. A higher value means that the more the clustering result matches the real situation. In a broad sense, ARI measures how well the two data distributions fit together. Homogeneity means that each cluster contains only members of a single class. Completeness refers to the extent to which all members of a given class are assigned to the same cluster. V measure score is symmetrical, it can be used to evaluate the consistency of two independent assignments on the same dataset. These indicators all use conditional entropy analysis to define some intuitive measures. The range of Mutual Information scores values is [−1, 1], and their larger values mean that the more the clustering results match the real situation. The range of Silhouette Coefficient is [−1, 1], and the closer the same category sample is, the farther away the sample of different categories is, the higher the score. A higher Carlinski-Harabasz score indicates that the model of the cluster is better. The index is the ratio of the difference between clusters and the degree of discreteness between clusters.

5 Experiment and Results

The experiments in this article are all carried out on computers equipped with Intel Core i7-9700 (3.00 GHz) CPU, 16 GB RAM and Microsoft Windows 10 operating system, and the development environment is Python 3.7. Figure 2

Algorithm 2: OPTICS

Input: Given parameter $\varepsilon, M, N_\varepsilon(i)$and $c_i, i = 1, 2, ..., N$.
Output: p $= \{p_i\}_{i=1}^N$

```
1  k = 1; v_i = 0, i = 1, 2, ..., N; ;
2  r_i = UNDEFINED, i = 1, 2, ..., N; I = 1, 2, ..., N; ;
3  while (I ≠ ∅) do
4      Get an element i from I ,and let I := I/i ;
5      if (v_i = 0) then
6          v_i = 1; ;
7          p_k = i, k = k + 1; ;
8          if (N_ε(i) ≥ M) then
9              //Insert the unvisited nodes in N_ε(i) into queueseedlist according
                 to the reachable distance ;
10             insertlist(N_ε(i), {v_t}_{t=1}^N, {r_t}_{t=1}^N, c_i, seedlist) ;
11             while (seedlistNOTEMPTY) do
12                 Get the first element j from seedlist ;
13                 v_j = 1 ;
14                 p_k = j, k = k + 1 ;
15                 if (|N_ε(j) ≥ M|) then
16                     break
17                 end
18                 //Insert the unvisited nodes in N_ε(i) into queueseedlist
                     according to the reachable distance ;
19                 insertlist(N_ε(i), {v_t}_{t=1}^N, {r_t}_{t=1}^N, c_i, seedlist) ;
20             end
21         end
22     end
23 end
```

Algorithm 3: Birch

Input: The dataset, threshold T, the maximum diameter (or radius) of a cluster R, and branching factor B.
Output: Compute CF points, where CF = (step of points in a cluster N , Linear sum of the points in the cluster LS, the square sum of N data SS).

1. (Load data into memory) An initial in-memory CF-tree is constructed with one scan of the data. Subsequent phases become fast, accurate and less order sensitive ;
2. (Condense data) Rebuild the CF-tree with a larger T ;
3. (Global clustering) Use the existing clustering algorithm on CF leaves ;
4. (Cluster refining) Do additional passes over the dataset and reassign data points to the closest centroid from step3 ;

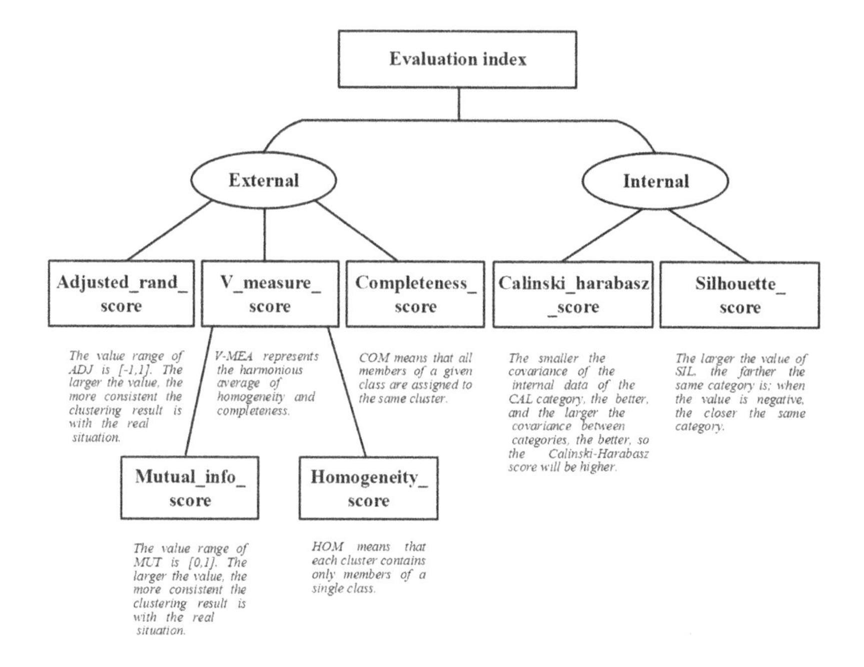

Fig. 3. Objective evaluation indicators for clustering effect

shows the main research route and ideas of this paper. Table 2 shows the comparison of the evaluation index results of K-Means, OPTICS and Birch.

Table 2. Comparison of objective indicators of clustering algorithm clustering effect

Algorithm	Silhouette	Calinski	Homogeneity	Completeness	V-measure	Adjusted rand	Mut information
K-Means	0.930	293355.316	0.934	1.000	0.966	0.949	0.966
OPTICS	0.908	165675.291	0.949	1.000	0.974	0.972	0.974
Birch	0.861	158523.624	0.859	1.000	0.924	0.878	0.924

T-distribution and Stochastic Neighbour Embedding algorithm is used for data dimension reduction and visualization of music data clustering. It is currently a very popular algorithm for dimension reduction of high-dimensional data. It is nonlinear and can adapt to the underlying data, support the optimization of parameters and confusion, and balance the local and global attention of data. We use it to reduce the data to three-dimensional data, and then visualize it. The elbow method is used to determine the optimal number of clusters. The reason why the elbow method is effective is based on the following observation: increasing the number of clusters helps to reduce the sum of intra-cluster variances of each cluster and calculate the sum of intra-cluster variances, the best value of k is 9, as is shown in Fig. 4.

After the clustering, after the clustering results, we use the reverse data to trace the clustered clusters, and trace back to the top seven largest number of note elements in each cluster, which corresponds to the five main voices and four partial voices of traditional Chinese pentatonic folk music: gong, Shang, Jiao, Zhi, Yu, Qingjiao, Bianzhi, Biangong, Run, as is shown in Fig. 5. The letter system is expressed as C, D, E, G, A, F, F#, B, Bb. C is used as the singing name of do, and the number of notes in various clusters is expressed as shown in the Table 3.

Table 3. Cluster tonic and other scattered notes in the cluster.

Cluster	Main note	First note	Second note	Third note	Fourth note	Fifth note	Sixth note
Cluster1	si fall	si	mi	sol	fa	re	do
Cluster2	la	sol	fa	mi	re	do	si
Cluster3	re	sol	fa	mi	do	si	la
Cluster4	mi	sol	fa	re	do	la	si
Cluster5	sol	fa sharp	fa	mi	re	do	la
Cluster6	fa sharp	do	sol	fa	mi	re	si
Cluster7	fa	fa sharp	sol	mi	re	do	si
Cluster8	si	sol	fa	mi	re	do	la
Cluster9	do	sol	fa	mi	re	si	la

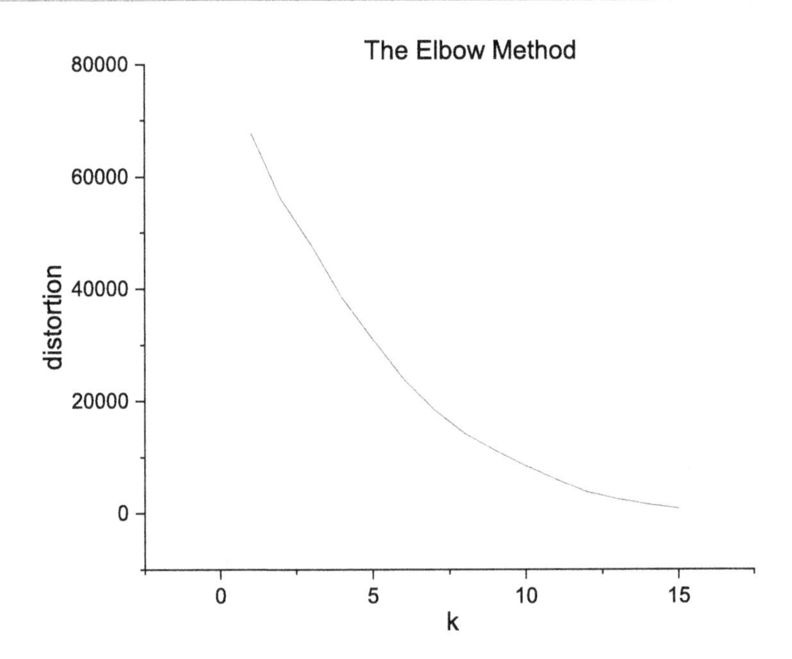

Fig. 4. The evaluation result of the error variance within the cluster to the K value.

Fig. 5. The pentatonic scale and the remaining four partial tones of the national seven-tone mode.

Chinese traditional pentatonic scale folk music mode, after a long development, mainly appeared Qingyue, Yanyue, and Yayue. There are three sources of Yayue in each dynasty. One is to inherit the palace music works of the Zhou Dynasty, the second is to reconstruct according to the music theory of the Zhou Dynasty, and the third is to make new music according to new sounds and customs. Its composition is Gong, Shang, Jiao, Bianzhi, Zhi, Yu, Biangong [25]. Qingyue refers to Qingshang music, also known as Qingshang music, is a traditional music rising in the Three Kingdoms, Jin Dynasty, Northern and Southern Dynasties and dominated in music life at that time. Its composition is the Gong, Shang, Jiao, Qingjiao, Zhi, Yu, and Biangong [12]. Yanyue is a very artistic song and dance music that provides entertainment and appreciation during the court dinner from Sui and Tang Dynasty to Song Dynasty. The palace yan music in Sui and Tang Dynasties reflects the highest achievement of music culture in this period. It originates from the accumulation of traditional music of Han nationality and the large-scale input of foreign music since Han and Wei Dynasties. Its composition is Gong, Shang, Jiao, Qingjiao, Zhi, Yu, Run [3]. Among them, cluster 1 has the structural characteristics of Yanyue mode, clusters 2, 3, 4, 8, and 9 have the structural characteristics of Qingyue mode, and clusters 5, 6, and 7 have the structural characteristics of Yayue mode. Figure 6, Fig. 7, and Fig. 8 show the clustering 3D visualization results under K-means, OPTICS, and Birch. It can be seen that the graph corresponds to the objective evaluation indicators in Table 2. K-Means and OPTICS show relatively better aggregation. The similar effect indicates that the note text data is more suitable for partition-based and density-based clustering methods.

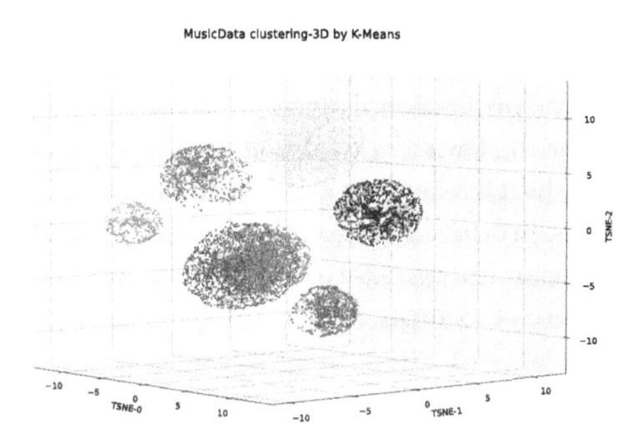

Fig. 6. Clustering 3D visualization results performed by K-means.

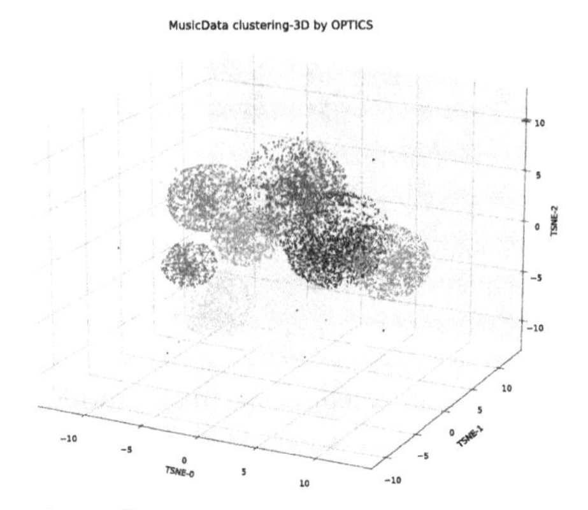

Fig. 7. Clustering 3D visualization results performed by OPTICS.

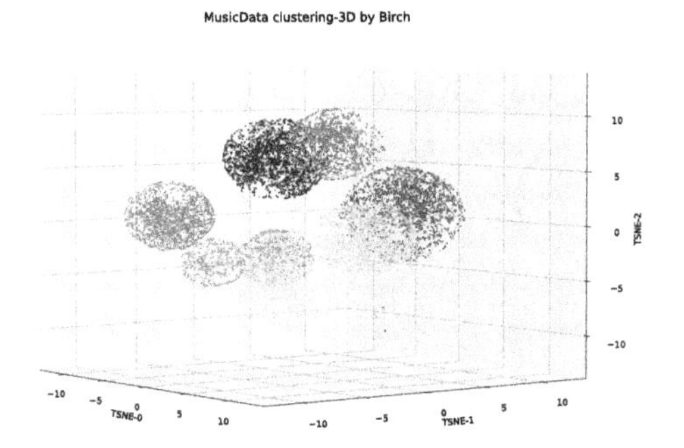

Fig. 8. Clustering 3D visualization results performed by Birch.

6 Conclusion

In this work, we first crawled and established a MIDI data set of traditional Chinese folk music, and proposed the idea of textualizing symbolic music data for cluster analysis. Then we combined professional knowledge of music theory with traditional Chinese folk music, cultural history conducted a textual clustering analysis on our own data set. According to the seven objective evaluation indicators of the clustering algorithm, K-Means and OPTICS produced relatively better results, combining professional music theory knowledge with traditional Chinese folk music. The historical development characteristics show that this dataset has distinct modal characteristics of traditional Chinese national music, and this experiment has produced an excellent combination of music grammar. We believe that the closer integration of music grammar and artificial intelligence is the future trend of computational music development. Regarding the next step, we think that we can extract more symbolic music characteristic texts and combine them with more professional music grammar for deeper exploration.

References

1. Akeroyd, M.A., Moore, B.C., Moore, G.A.: Melodyrecognition using three types of dichotic-pitch stimulus. J. Acoust. Soc. Am. **110**(3), 1498–1504 (2001)
2. Ankerst, M., Breunig, M.M., Kriegel, H.-P., Sander, J.: Optics: ordering points to identify the clustering structure. ACM SIGMOD Rec. **28**(2), 49–60 (1999)
3. Bing, L.: On Yanyue scale. Chin. Musicol. (2), 60–65 (1986)
4. Chongguang, L.: Fundamentals of Music Theory. People's Music Publishing House, Beijing (1962)
5. Cilibrasi, R., Wolf, R.d.: Algorithmic clustering of music based on string compression. Comput. Music J. **28**(4), 49–67 (2004)
6. Durey, A.S., Clements, M.A.: Features for melody spot-ting using hidden Markov models. In: 2002 IEEE International Conference on Acoustics, Speech, and Signal Processing, vol. 2, pp. II–1765. IEEE (2002)
7. Eronen, A.: Musical instrument recognition using ICA-based transform of features and discriminatively trained HMMs. In: Seventh International Symposium on Signal Processing and Its Applications, 2003. Proceedings, vol. 2, pp. 133–136. IEEE (2003)
8. Herrera, P., Amatriain, X., Batlle, E., Serra, X.: Towards instrument segmentation for music content description: a critical review of instrument classification techniques. In: International Symposium on Music Information Retrieval, vol. 9, p. 2 (2000)
9. Jain, A.K.: Data clustering: 50 years beyond k-means. Pattern Recognit. Lett. **31**(8), 651–666 (2010)
10. Kim, D., Kim, K.-s., Park, K.-H., Lee, J.-H., Lee, K.M.: A music recommendation system with a dynamic k-means clustering algorithm. In: Sixth International Conference on Machine Learning and Applications (ICMLA 2007), pp. 399–403. IEEE (2007)
11. Kim, Y.E., Whitman, B.: Singer identification in popular music recordings using voice coding features. In: Proceedings of the 3rd International Conference on Music Information Retrieval, vol. 13, p. 17 (2002)
12. Na, L.: On the difference between the European major scale and the national seven-tone unvoiced scale. GeHai (2), 78–80 (2010)
13. Liu, C.-C., Huang, C.-S.: A singer identification technique for content-based classification of MP3 music objects. In: Proceedings of the Eleventh International Conference on Information and Knowledge Management, pp. 438–445 (2002)
14. Oramas, S., Espinosa-Anke, L., Sordo, M., Saggion, H., Serra, X.: Information extraction for knowledge base construction in the music domain. Data Knowl. Eng. **106**, 70–83 (2016)
15. Qi, Y., Yongping, J., Du, X., Chuanze, L.: Application of a hierarchical clustering method in music retrieval. Comput. Eng. Appl. **47**(30), 113–115 (2011)
16. Roda, A., Canazza, S., De Poli, G.: Clustering affective qualities of classical music: Beyond the valence-arousalplane. IEEE Trans. Affect. Comput. **5**(4), 364–376 (2014)
17. Tsai, W.-H., Rodgers, D., Wang, H.-M.: Blind clustering of popular music recordings based on singer voice characteristics. Comput. Music J. **28**(3), 68–78 (2004)
18. Van den Oord, A., Dieleman, S., Schrauwen, B.: Deep content-based music recommendation. In: Advances in Neural Information Processing Systems, pp. 2643–2651 (2013)

19. Wei, L., Zijin, L., Yongwei, G.: Understanding digital music—summarization of music information retrieval technology. Fudan J. (Nat. Sci. Edit.) **57**(3), 271–313 (2018)
20. Xiaofeng, C.: Five-degree mutual generation and five-tone mode. Today Sci. Court (5), 64–64 (2006)
21. Xu, C., Maddage, N.C., Shao, X.: Automatic music classification and summarization. IEEE Trans. Speech Audio Process. **13**(3), 441–450 (2005)
22. Zhang, T., Ramakrishnan, R., Livny, M.: BIRCH: an efficient data clustering method for very large databases. ACM SIGMOD Rec. **25**(2), 103–114 (1996)
23. Xun, Z.: Interpretation of the five-degree interaction Law in Chinese music system. Contemp. Music (8) (2020)
24. Yijun, Z.: Analysis of the relationship between traditional Chinese musicology and ethnomusicology. Northern Music **02**(01), 39–40 (2020)
25. Zuxiang, Z.: Seven Tones of Yayue. Ph.D thesis (1987)

Activity Behavior Pattern Mining and Recognition

Ling Song[1], Hongxin Liu[1], Shunming Lyu[2], Yi Liu[1(✉)], Xiaofei Niu[1], Xinfeng Liu[1], and Mosu Xu[3]

[1] Shandong Jianzhu University, Jinan 250001, China
[2] State Grid Information and Telecommunication Branch, Beijing 100031, China
[3] The University of Melbourne, Melbourne, VIC 3010, Australia

Abstract. As human activity in mobile environments is facing with an ever-increasing range of data, therefore, a deeper understanding of the human activity behavior pattern and recognition is of important research significance. However, human activity behavior that consists of a series of complex spatiotemporal processes is hard to model. In this paper, we develop a platform to do pattern mining and recognition, the main work is as follows: (1) For comparing activity behavior, similarity matrix is computed based on activity intersection, temporal connections, spatial intersection, participant intersection and activity sequence comparison. (2) For calculating activity sequence similarity, an algorithm with $O(p(m - p))$ is proposed by line segment tree, greedy algorithm and dynamic programming. (3) Activity behavior pattern and socio-demographic pattern are derived by clustering analysis and mining. (4) Pattern is recognized under the inter-dependency relationship between activity behavior pattern and socio-demographic pattern.

Keywords: Activity behavior and socio-demographic pattern mining · Activity behavior and socio-demographic pattern recognition · Clustering analysis and mining · Activity sequence similarity · Activity behavior similarity

1 Introduction

As increasingly sensors and mobile devices are becoming smarter and more powerful, a remarkable consequence of these developments has been accumulated to more and more abundant human activity dataset, providing detailed information on the way people spend time. Activity behavior pattern mining derives a representative set of activity behavior patterns, and activity behavior recognition confirms what kind of activity patterns individuals belonged to, which could provide a priori knowledge in a wide variety intelligent system such as E-recommender system, social network analysis and mining, and urban transportation.

A large amount of related work has been done by previous researchers. We reviewed the following topics from various references: the influence of socio-demographic features, trajectory, spatiotemporal features, longest common subsequence (LCS) problem, and learning model.

J. Xiong et al. (Eds.): MobiMedia 2021, LNICST 394, pp. 663–678, 2021.
https://doi.org/10.1007/978-3-030-89814-4_48

Focusing on the influence of socio-demographic features, many studies concluded that human socio-demographic features could influence activity behavior significantly. In Lu and Pas's research, there existed relationships between socio-demographics and activity behavior [1]. According to socio-demographic features, after analyzing the multiple participations of different leisure activities, Kemperman and Timmermans determined the relevance of leisure activity that individual take part in and features of the living environment [2]. Focused on the activity patterns of individuals in dual-earner family, Bernardo et al. analyzed the function relationship of household socio-demographics and time use of activity [3].

With the aim of trajectory, on the assumption that individuals' real-world activity behavior is represented by their trajectory, Ying et al. proposed that the similarity between individuals is computed in terms of their maximal trajectory and potential friends are recommended based on social networks [4]. Zhang et al. defined a community as a group of individuals with similar activity and activity-travel trajectories [5]. Cai et al. focused on mining semantic trajectory pattern [6]. Outliers often reduce the performance of pattern mining and recognition, in order to reduce the outlier effect, Xiong et al. proposed a privacy and availability data clustering scheme (PADC), which enhances the selection of the initial center points and the distance calculation through detecting outliers during the clustering process [7].

Spatial and temporal information play an important role in feature extraction and representation for representing activity behavior. Banovic et al. converted the activity behavior logs into event sequences that show individuals' activity context, which are used to represent individuals' routine activity behavior [8]. For a deeper grasping of individuals' activity dynamics, Zhang et al. extracted individuals' spatial and temporal features from activities and/or trip events [5]. Activity feature extraction from video also capture both spatial and temporal information in computer vision for activity recognition [9]. You et al. presented a clustering algorithm with a trajectory similarity computation method by considering both semantic and geographic meaning [10]. In the work of Chakri et al., knowledge discovery is used to extracted space and temporal information from semantic trajectories [11]. With a kind of deep learning, convolution neural network is presented to catch spatiotemporal information in Ke et al. 's work [12].

The LCS has been widely used in bioinformatics and text comparison. Zhang et al. used the LCS of activity sequence to represent the similarity of two individuals [5]. The general conventional algorithm is dynamic programming with time complexity $O(mn)$. Many researchers have done a lot of work on optimized LCS on typical inputs and some kind of infrequent inputs. Based on the work of H. S. Stone, the algorithm run in $O(n\log n)$ on many inputs [13, 14]. Hirschberg designed two algorithms, in the general case, the running time runs in $O(pn + n \log n)$, and p is the length of the LCS. However, while p is near to m ($m < n$), the running time requires bounded by $O(p(m + 1 - p)\log n)$ [15]. When the expected length of an LCS is close to m, Nakatsu developed an algorithm in $O(n(m - p))$ [16]. Liu designed a parallel algorithm for LCS to make it more efficient [17].

Lots of studies have been done on how to build the learning model for activity behavior. Zhang and Jiang classified activities into the outdoor and the indoor activities, constructing a model of attendance prediction based on the Gradient Boosting Tree [18].

Hafezi et al. presented a pattern recognition model to identify time-use daily activity patterns by Fuzzy C-Means and regression tree based on household travel of diary time use survey [19]. Zhang et al. developed activity-travel model based on similarities of people's activity trajectories and spatiotemporal connection by social network analysis and community detection algorithm [5]. Banovic et al. forecasted human routine behavior by a Markov Decision Processes framework [8]. After given individual' time and location, Benetka et al. proposed a recommender system to return a list of ranked activities according to the probability of being done [20]. Activities and durations are predicted by computing their probabilities by two kinds of Long-Short Term Memory (LSTM) [21]. LSTM is also used to predict future activities and places [22]. In more applications, occupancy patterns are profiled by learning model based on time use survey [13, 23, 24]. Data sharing among connected and autonomous vehicles without any protection will cause private activity behavior leakage, Xiong et.al. constructed secure functions and implemented a privacy-preserving convolutional neural network (P-CNN) to ensure data security [25].

Although a large amount of work has been done in the previous studies, notably, the past studies seldomly focus on both activity behavior pattern and socio-demographic pattern mining. Therefore, we should make sense of the following problems:

(1) Similarity computation: on the one hand, the method needs to be related with comprehensively activities, location, and participants for computing similarity between activity behavior; on the other hand, an effective algorithm needs to be implemented in order to reduce computational complexity for computing similarity between activity sequences.
(2) Activity granularity: too fine and too coarse activity granularity makes it difficult to discover pattern characteristics. As different applications face different requirements, pattern mining and recognition system needs to adjust activity granularity flexibly.
(3) Visualization of clusters: for the purpose of obtaining the patterns, comprehensive analysis should be done based on the clustering results by visualization of activity sequence diagrams, scale charts and Probability Density Function (PDF) distribution graph of both activity behavior features and socio-demographic features.
(4) Pattern representation and recognition: Not only the representative set of activity behavior patterns but also socio-demographic patterns should be derived, bridging the relationship between them, which is further pattern recognized by activity or socio-demographic features.

2 Overview

The description of definitions and problem statements are firstly shown with the research outline following.

2.1 Definitions and Problem Statements

We give the following description about activity behavior.

Definition 1: Let $D = \{d_1, d_2,\ldots, d_{|D|}\}$ denote the set of individuals, where $|D|$ is the number of the individuals.

Definition 2: Let $A = \{a_1, a_2,\ldots, a_{|A|}\}$ represent the set of activity categories, where $|A|$ is the number of the activity categories, such as work and sport.

Definition 3: $\forall\, a_i$ ($a_i \in A$) has a starting time s_i and an ending time e_i, where s_i, $e_i \in \{0, 1,\ldots, t\}$, and t can be expressed in minutes from a certain time point.

Definition 4: Let $L = \{l_1, l_2,\ldots, l_{|L|}\}$ denote the set of location categories, where $|L|$ is the number of the activity categories, such as home and workplace.

Definition 5: Let $W = \{w_1, w_2,\ldots, w_{|W|}\}$ denote the set of participant categories, where $|W|$ is the number of the participant categories, such as parent and co-workers.

Definition 6: An activity action is defined as a quintuple $b(a, [s, e], l, w)$, where $a \in A$, s, e is the time that a begins and finish, $l \in L$ is the location of a takes place, and $w \subseteq W$ are the participants in the process of a.

Definition 7: For an individual d_i ($d_i \in D$), his/her activity behavior is described as a set of activity action, $B_i = \{b_{i1}(), b_{i2}(),\ldots, b_{|iB|}()\}$, where $b_{ij}()$ is an activity action, and $|iB|$ is the number of activities d_i involves in.

Definition 8: For the set D, the set of its corresponding activity behavior is described as $O = \{B_1, B_2,\ldots, B_{|D|}\}$.

Definition 9: For the set D, the set of its corresponding socio-demographic feature is described as $M = \{m_1, m_2,\ldots, m_{|D|}\}$, such as age, gender, education, occupation, and family income etc.

Based on the above definitions, the main problems of this paper are described as the followings.

Problem 1: Pattern mining of activity behavior is characterizing the cluster memberships of the set O, deriving clusters of homogeneous activity behavior pattern, $AP = \{P_1, P_2,\ldots, P_k\}$.

Problem 2: Pattern mining of socio-demographic feature is characterizing the cluster memberships of the set O, deriving clusters of homogeneous socio-demographic pattern, $SP = \{P'_1, P'_2, \ldots, P'_k\}$.

Problem 3: Given a set of socio-demographic or activity behavior features, pattern recognition problem can be stated to identify his activity behavior pattern and socio-demographic pattern he belongs to.

2.2 Research Outline

The framework of our model involves two modules: (1) Pattern mining module: aggregated similarity matrix is computed by comparing activity behavior and activity sequence between the individuals. Clusters are obtained by clustering algorithm based on the aggregated similarity matrix, and resulted in unique clusters of homogeneous daily activity behavior patterns by mining. Meanwhile, their corresponding socio-demographic patterns of homogeneous activity patterns are derived, implying inter-dependency between two kinds of patters. (2) Pattern recognition module: in view of the inter-dependency relationship between activity behavior patterns and socio-demographic patterns, pattern is recognized through socio-demographic or activity behavior features by the classifier.

3 Methodology

The process of pattern mining and recognition comprises four steps: (1) similarity adjacency matrix among individuals is calculated; (2) clusters are organized by clustering algorithm based on the similarity matrix; (3) representative patterns are obtained by characterizing the cluster memberships; (4) pattern are identified by classification algorithm.

3.1 Similarity Computation

Similarity comparison between two individuals is considered from the view of activity behavior and the view of activity sequence.

(1) Activity behavior similarity matrix

Activity behavior consists of a series of activity actions, so we calculate activity action similarity firstly. For two activity actions, k and l, they are represented as $b_k(a_k,\ [s,e]_k,\ l_k,\ w_k)$ and $b_l(a_l,\ [s,e]_l,\ l_l,\ w_l)$ respectively, the similarity between them, denoted as $sim_{act}[b_k,b_l]$, is defined as:

$$sim_{act}(b_k,b_l) = sim_a(a_k,a_l) \times sim_t([s,e]_k,[s,e]_l) \times sim_l(l_k,l_l) \times \frac{1}{P}\sum\nolimits_{p=1}^{P} sim_w\left(w_{k,p},w_{l,p}\right) \tag{1}$$

$sim_a(a_k,a_l)$ is the activity similarity function to compare a_k and a_l, $sim_l(l_k,l_l)$ is the location similarity function to compare l_k and l_l, $sim_w(w_{k,p},w_{l,p})$ is the participants similarity function to compare $w_{k,p}$ and $w_{l,p}$, $p = 1,2,\ldots P$, as there may be more than one participants in involved in the action. The above similarity function value is defined as 1 if the variables are the same, otherwise 0. The time similarity function $sim_t([s,e]_k, [s,e]_l)$ is defined as follows:

$$sim_t([s,e]_k,[s,e]_l) = \begin{matrix} \min(e_k,e_l) - \max(s_k,s_l),\ \min(e_k,e_l) > \max(s_k,s_l) \\ 0,\ \min(e_k,e_l) \le \max(s_k,s_l) \end{matrix} \tag{2}$$

For two individuals, $d_i, d_j \in D$, their daily activity behaviors are $B_i = \{b_{i1}(), b_{i2}(), \ldots, b_{im}()\}$ and $B_j = \{b_{j1}(), b_{j2}(), \ldots, b_{jn}()\}$, and m, n is the number of activities for d_i and d_j. The daily activity behavior similarity $Sim_{act}[B_i, B_j]$ is computed as the followings:

$$Sim_{act}(B_i, B_j) = \sum\nolimits_{k=1}^{m} \sum\nolimits_{l=1}^{n} sim_{act}(b_{ik}(), b_{jl}()) \tag{3}$$

In summary, activity behavior similarity matrix is defined as followings:

$$R_{act} = [Sim_{act}(B_i, B_j)]_{|D| \times |D|} \tag{4}$$

(2) Activity sequence similarity matrix

For computing activity sequence similarity, the time with adjacent time and similar activities is discretized into a time segment, and each segment is regarded as a weighted point. Thus, an individual's daily activities are described by two-dimensional representations of time and activities, and the comparison of two individual's activity sequences becomes three-dimensional data comparison. For reducing the time complexity, in view of the large degree of discretization of the time segment, we use line segment tree and greedy algorithm to reduce the dimension, and an algorithm with the complexity of $O(p(m-p))$ is proposed on the basis of previous studies of LCS problem [13–17].

For two individuals, d_i, $d_j \in D$, their activity sequences are described by S_i and S_j, notably, S_i and S_j actually refers to sequences of time segment after reducing the dimension. $LW_i[k]$ record the smallest j that S_i and S_j include a common subsequence of length k. We need to store list L to calculate LCS, so the whole space complexity is $O(n)$. Assuming that L has p values, then its space complexity is $O(p)$. When we calculate $LW_i[k+1]$ based on $LW_i[k]$ by dynamic programming, the list L is traversed only once, the process is $O(p)$. As the process needs to be done m times, where $m_1 \in \{0\}$, $m_2 \in \{0,1\}$, $m_3 \in \{0, 1, 2\}$,.... therefore, the worst time complexity is $1 + 2 + \ldots + p + (m-p)p = O((m-p)p)$ under the case of $m_p = p$. Thus, we compute activity sequence similarity $sim_{seq}[S_i, S_j]$. Activity sequence similarity matrix is defined as followings:

$$R_{seq} = [Sim_{seq}(S_i, S_j)]_{|D| \times |D|} \tag{5}$$

R_{act}, R_{seq} are normalized by Min-Max Normalization. We aggregate the above two adjacency matrixes to a composite matrix M, as shown in Eq. (6).

$$M = \omega_1 R_{act} + \omega_2 R_{seq}, \omega_1 + \omega_2 = 1 \tag{6}$$

3.2 Pattern Mining and Identification

In this paper, spectral clustering method is used to partition a set of N individuals into k clusters, Silhouette Coefficient (SC) analysis, Calinski and Harabasz (CH) score, and Davies-Bouldin index (DB) index are used to evaluate the clustering results.Activity time series analysis, statistical analysis of activity category, location category, and activity participant category, probability density function analysis of activity category, location category, and activity participant category are used to mine patterns, identifying the sets of representatives about activity behavior patterns. Statistical analysis of age, education, sex, metropolitan or not, race, occupation, family income, number of the family, number of children that is smaller than 18, working time, time with family, having an enterprise/farmer or not and labor force status etc. are used to mine patterns, identifying the sets of representative about socio-demographic patterns. Pattern is recognized by Random forest (RF).

4 Experiments and Analysis

The Multinational Time Use Study gathers more than a million diary days from over 70 national surveys [27]. The American Time Use Survey (ATUS) gives details of how, where, and with whom Americans take their time [28]. The experiment process consists of the followings: (1) Determining parameters of clustering; (2) Activity behavior pattern mining; (3) Socio-demographic pattern mining; (4) Pattern identification.

4.1 Data Preprocessing

We use ATUS of 2018 as our experiment dataset to explore individuals' daily life in 24 h (from 4:00am to 3:59am). Features about what, where, when and with whom are extracted. After preprocessing, we obtain a database of 13,133 individual samples. Notably, granularity of category in ATUS is too fine, so it's difficult to discover characteristics of pattern for some kind of application.

4.2 Determining Parameters of Clustering

In our experiments, granularity of categories can be merged flexibly. We merge some categories and select appropriate weights ω_1, ω_2 (Eq. (6)) and k (number of clusters) by SC, CH Score and DB Index. The parameters are set as followings:

(1) From the overall trend, with ω_1 increasing and ω_2 decreasing, the SC value has the trend of decreasing, CH Score has the trend of decreasing and DB Index has the trend of increasing, which means that ω_2 plays a more important role than ω_1. So, we set $\omega_1 = 0.1$, $\omega_2 = 0.9$, clustering result shows the best performance.
(2) It shows better clustering effect when k is smaller, but we hope to get more refined clusters that mine the diversity of patterns. Therefore, we need to look for a tradeoff. No matter what values ω_1, ω_2 are taken, $k = 7$ performs best under evaluating comprehensively with SC, CH Score and DB index, so we set $k = 7$.

4.3 Activity Behavior Pattern Mining

For discovery of activity behavior pattern, we operate more detailed data analysis and mining from the view of activity time series, activity category, location category and participant category based on the cluster results.

(1) Activity time series analysis

Figure 1 gives the activity time series of 7 clusters. One category corresponds to one color. X-axis is time (4:00am to 3:59am) and Y-axis is the number of individuals. We can observe that different clusters show obvious characteristics. Figure 2 shows probability density function (PDF) of number of individuals by time.

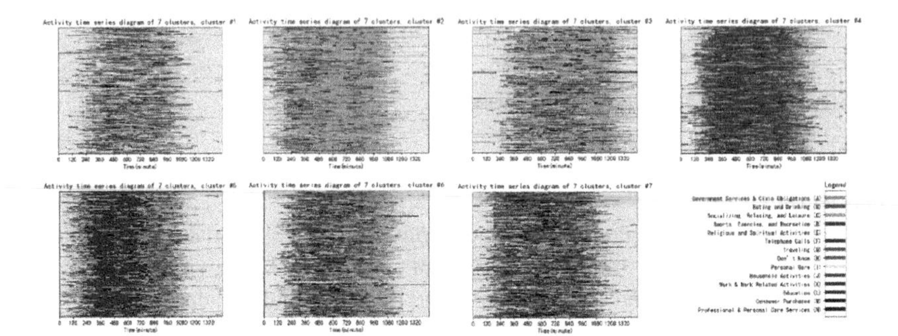

Fig. 1. Activity time series diagram with 14 categories (after activity merge).

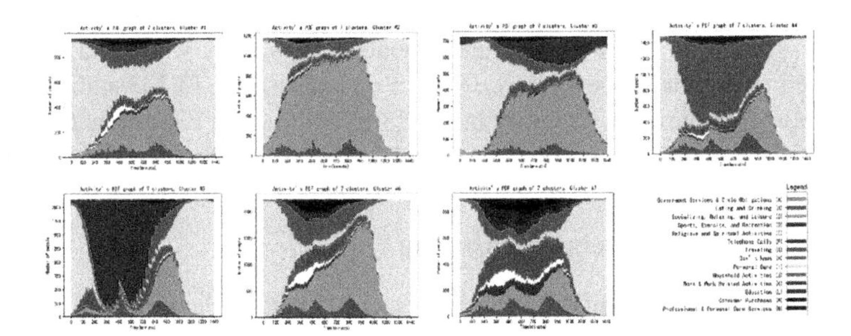

Fig. 2. Activity's PDF graph of clusters.

The typical activity characteristics of clusters as followings:

Cluster #1 has the largest proportion (59%) in I (Personal Care) with average 853 min. **Cluster #2** has the largest proportion (47%) in C (Socializing, Relaxing, and Leisure) with 682 average minutes. However, it takes up average 38min to work. **Cluster #3** has the third proportion (12%) in K (work & work related activities) with average 169 min. And the working hours are mainly distributed at 00:00 am, as Fig. 2 show. **Cluster #4** has the largest proportion (38%) is J (household activities) with average 473 min. **Cluster #5** has the largest proportion (37%) in K with average 534 min. And the working hours are mainly distributed at day. **Cluster #6** is the largest cluster with 23% of the individuals. It spends more time in B, D, E, G, L, M, N (including eating and drinking, sports, exercise, recreation, religious and spiritual activities, education, volunteer, travel, purchases of goods and services) than the other clusters, except #7. **Cluster #7**, compared with other clusters, the average time spend in B, D, E, G, L, M, N is the most. And it spends less time in C than #6.

(2) Activity location categories analysis
Cluster #1 has the largest proportion (58%) in J (including other Unspecified place, Blank, Other place, Don't know, Refused) with average 876 min. It spends average 463 min at A (Home or yard). **Cluster #2** has the largest proportion (52%) in A with average 766 min. **Cluster #3** has the third proportion (11%) in B(Workplace) with average 154 min, and the working places are mainly distributed at 00:00 am. **Cluster #4** has the second proportion (47%) in A with average 686 min. **Cluster #5** has the larger proportion (33%) in B with average 479 min. **Cluster #6** spends more time in D, E, F (including Restaurant or Bar, Place of worship, Store or Grocery) than the other clusters, except #7. **Cluster #7** takes least time at home, compared with other clusters, the average time in D, E, F, G, H, I (including Restaurant or Bar, Place of worship, Store or Grocery, School, Library, various mode of transportation, Gym, Outdoors) is the most.

(3) Activity participant categories analysis
Cluster #1 spends the least time with family (including Spouse, Unmarried partners, Own household children, Grandchildren, Parents, Brothers/sisters, Other related persons, Foster children, Own nonhousehold children < 18, Parents that not living in household, Other nonhousehold family members < 18, Other nonhousehold family members18 and older) and less time with other nonrelatives (including Housemates/roommates, Roomers/boarders, Other nonrelatives, Friends, Neighbors/acquaintances). **Cluster #2** takes the most time alone and least time with Co-workers (including Co-workers, people whom I supervise, customers, boss or manager). **Cluster #3** takes more time (average 127 min) with Co-workers. **Cluster #4** spends the most time (average 534 min) with family. **Cluster #5** spends the most time (average 212 min) with Co-workers. **Cluster #6** takes more time (average 505 min) with family and other nonrelatives. **Cluster #7** takes the least time alone, and takes the most time (average 573 min) with family and other nonrelatives.

4.4 Socio-demographic Pattern Mining

For discovery of socio-demographic pattern, we operate more detailed data analysis and mining from the view of age, sex, race, labor force status, occupation, family income, number of the family and underage children, time for work and family, having an enterprise/farmer or not, and Education level.

(1) Age and Sex
Cluster #3 is the youngest group, 59% of the people is between 20 and 59, and there are more male than female (53.7% vs. 46.3%). **Cluster #2** and **#6** are two old-aged groups. **Cluster #2** is the oldest group, 78% of the individuals are older than 50, and there are more male than female (55.3% vs. 44.7%). In **cluster #6**, the proportion from 40–79 years old takes up 66%, and there are more female than male (60.0% vs. 40.0%).

Cluster #1, **#4**, **#5** and **# 7** are the middle-aged groups. In **cluster #1**, the proportion from 30 to 69 years old takes up 62%, and there are more female than male (61.5% vs. 38.5%). **Cluster #4** is younger than **cluster #1**, the proportion from 30 to 69 years old takes up 76%, and there are more female than male (67.9% vs. 32.1%). In **Cluster #5,** the proportion from 30 to 69 years old takes up 85%, and there are more male than female (53.2% vs. 46.8%). In cluster #7, the proportion from 30 to 69 years old takes up 65%, and there are more female than male (54.5% vs. 45.5%).

(2) Race and Labor force status

Compared with the other clusters, **cluster #1, #2, and #3** have higher proportion of the blacks. In **cluster #1**, the black account for the largest proportion of all clusters (the white and the black, 67% vs. 24%). The proportion that is not labor force is higher, which takes up 48%. In **cluster #2**, the proportion of the white and the black is 79% vs. 18%. The proportion that is employed at work is the lowest, which takes up only 28%, on the contrary, the proportion that is not labor force is the highest, which takes up 68%. In **cluster #3**, the proportion of the white and the black is 71% vs. 22%. The proportion that employed at work and not in labor force is 54% and 38%, respectively.

Cluster #4, #5, #6, and #7 have higher proportion of the whites. In **cluster #4**, the proportion of the white is the largest, which takes up 86%, and the proportion of the black is the smallest, which takes up only 8%. The proportion that employed at work and not in labor force is 48% and 43%, respectively. In **cluster #5**, the proportion of the white takes up 82%, and the proportion of the black takes up 10%. The proportion that employed at work is 99%. In **cluster #6**, the proportion of the white takes up 83%, and the proportion of the black takes up 12%. The proportion that employed at work and not in labor force is 47% and 46%, respectively. In **cluster #7**, the proportion of the white and the black is 78% vs. 13%. The proportion that employed at work and not in labor force is 58% and 35%, respectively.

(3) Occupation and family income

Family income distribution and division is based on Pew classification [29]. **Cluster #1, #2, and #3** belong to the low family income groups. People who earn less than $40000 take up around one half, people who earn $40000-$99999 take up 32%–39%, and people who earn more than $100000 take up only 15%. People in these clusters are minority in the occupation of Science/Technology/Management. For **cluster #1 and #2**, there is no obvious difference in occupation distribution, however, for **cluster #3**, people that work in service and related occupation take up 22%.

Cluster #4 and #6 belong to the middle family income groups. People who earn less than $40000 take up around 34%–35%, people who earn $40000–$99999 take up37%–39%, and people who earn more than $100000 take up around 26%–29%. People engaged in Science/Technology/Management is the majority, which takes up 23%–26%; people engaged in service and related occupation take up 12%–15%; people engaged in manual occupation takes up 10%–13%, and people with unknown occupation takes up 47%–50%.

Cluster #5 and #7 belong to the high family income groups. People who earn less than $40000 take up around 24%–26%, people who earn $40000–$99999 take up 42%–43%, and people who earn more than 100000 take up 34%. Almost all people in **#5** have work. People engaged in Science/Technology/Management is the majority, which take up 45%, while people engaged in service and related occupation take up 27%, and people engaged in manual occupation take up 23%. People in **#7** engaged in Science/Technology/Management is the majority, which takes up 29%, people engaged in service and related occupation takes up 18%, people engaged in manual occupation takes up 11%, and people with unknown occupation takes up 38%.

(4) Number of the family and underage children

Cluster # 2 has the least family population. Single accounts for 42%, more than three people only account for 22.9%. At least one underage children account for only 18.5%.

Cluster #1, #3, and #6 are medium family. More than three people account for 35.8%–41.4%, and more than one underage child account for 29.4%–37.2%.

Cluster #4, #5, and #7 are large family. More than three people account for more than 50% and they have more underage children. **#4** has the largest number of family and underage children.

(5) Time for work and family

Cluster #2 and 3 has the longer working hours. **Cluster #2** has the longest working time,40.3% people work more than 700 min. Family time presents "Two ends big, middle small" state, that is, 50.8% people don't spend any time with family, and 24.9% people spend more than 600 min with family. In **# 3,** 41.3% work more than 500 min, and 80.5% take up less than 399 min with their family.

Cluster#1, #4, #5, #6, and #7 have the normal working hours ($\leq$500 min). In **#1,** 75.1% work less than 500 min, 10.2% people don't work, and 41.5% work less than 300 min. 67.3% people in **#4** work less than 500 min, 8.7% people don't work, and 39.1% work less than 300 min. They spend most of the time with the family and 51.7% spend more than 400 min with their family. 74.3% people in **#5** work less than 500 min, 1.3% people don't work, and 53.9% work less than 300 min. Although they are busy on work, they still spend more time with the family, in detail, 60.9% spend 1–399 min and 4.3% spend more than 400 min. 66.2% people in **#6** work less than 500 min, 8.3% people don't work, and 37.2% work less than 300 min. They spend more time with the family, with 34.7% spending more than 400 min with their family. 77.3% people in **#7** work less than 500 min, 11.7% people don't work, and 44.6% work less than 300 min. They spend more time with the family, about 36.8% spending more than 400 min with their family.

(6) Having an enterprise/farmer or not

Only less than 9% people in **cluster #1, #2, and #3** have an enterprise/farmer, however. More than 14% people in **cluster #4, #5, #6 and #7** have an enterprise/farmer.

(7) Education level

Cluster #1, #2, and #3 are three groups with the higher proportion of low education level and lower proportion of the high education level. People who are less than bachelor' degree take up about 75% and greater than or equal to bachelor' degree take up only 25%.

Cluster #4, #5, #6 and #7 are four groups with the higher proportion of the high education level and lower proportion of the low education level. People who are less than bachelor' degree take up only 51.0–57.3% and greater than or equal to bachelor' degree take up 41.4–49.0%.

4.5 Characteristics of Activity Behavior and Socio-demographic Pattern

From the above analysis, activity behavior pattern and socio-demographics pattern are derived, as Table 1 shown.

Table 1. Characteristics of activity behavior pattern and socio-demographic pattern

Pattern	Characteristics of activity behavior	Characteristics of socio-demographic
#1 10% (945)	The largest activity proportion in personal care, the least time to work, the largest location proportion is unspecified place, the least time with family and less time with other nonrelative	Middle-aged group, more female than male, with the higher proportion of the blacks, low family income, medium family population, the normal working hours, low education level
#2 (1179) 12%	The largest activity proportion in socializing, relaxing, and leisure, less time to work. the largest location proportion in home/yard, the most time alone and least time with co-workers	The oldest group, more male than female, with the higher proportion of the blacks, not labor force is the highest, low family income, the least family population, the employed people have the longest working hours, low education level
#3 (723) 8%	The working hours are mainly distributed after 00:00 am, the main location is the workplace, more time with co-workers	Youngest group, more male than female, with the higher proportion of the blacks, low family income, higher proportion in service and related occupation, medium family population, longer working time, less time with the family, low education level
#4 (1484) 15%	The largest proportion is household activities, caring for members, more time in home/yard, the most time with family	Middle-aged group, more female than male, with the higher proportion of the whites, middle family income, higher proportion in occupation of science, technology and management, large family population, the normal working hours, the most time with the family, with the higher proportion of having an enterprise/farmer, high education level
#5 (2136) 22%	The largest proportion is work related activities, the larger location proportion is workplace, the larger proportion with co-workers	Middle-aged group, more male than female, with the higher proportion of the whites, high family income, higher proportion in occupation of science, technology and management, large family population, the normal working hours, less time with the family, with the higher proportion of having an enterprise/farmer, high education level

(continued)

Table 1. (*continued*)

Pattern	Characteristics of activity behavior	Characteristics of socio-demographic
#6 (2229) 23%	More time in eating, sports, exercise, recreation, spiritual activities, education, travel, purchases of goods and various services, socializing, and leisure, more time in the place of restaurant, worship and store, more time with family and other nonrelatives	old-aged group, more female than male, with the higher proportion of the whites, middle family income, higher proportion in manual occupation, medium family population, the normal working hours, spend more time with the family, with the higher proportion of having an enterprise/farmer, high education level
#7 (897) 9%	The most time in eating, sports, exercise, recreation, spiritual activities, education, travel, purchases of goods and various services, socializing, and leisure, the least time at home, the most time in the place of restaurant, worship, store, library, school, varies mode of transportation and gym, least time alone and more time in family	Middle-aged group, more male than female, with the higher proportion of the whites, high family income, higher proportion in occupation of science, technology and management, large family population, the normal working hours, more time with the family, with the higher proportion of having an enterprise/farmer, high education level

From Table 1, both representative set of activity behavior pattern and socio-demographic pattern are derived, bridging the relationship between them, which is further pattern recognized by activity or socio-demographic features.

4.6 Pattern Identification

In this section, the pattern is recognized through socio-demographic or activity features by RF and parameters are set. For different application, we can select any kind of features as input for pattern recognition.

we use accuracy as evaluation criteria. Experiments are performed by 10 cross validation, and recognition accuracy is only 0.478. The reason lies in that some individuals in the clusters have no obvious characteristics, resulting the lower accuracy. For each cluster, we select 80% samples which squared Euclidean distance are closer to the centroid. With these samples, we do pattern recognition experiment as above, and recognition accuracy achieve 0.856.

Figure 3 and Fig. 4 are two test cases for pattern identification. The left of the figures is the input interface of pattern recognition. We input number of family, labor force status, number of underage children, relaxing time, family time, having an enterprise/farmer or not, metropolitan or not, race, age, sex, and family income. The right of the figures is the output interface of pattern recognition. The test case 1 is a typical working people and test case 2 is a typical housewife. We can see that largest probability that test case 1 is classified to the pattern #5 is 0.7270 and the largest probability that test case 2 is classified to the pattern #4 is 0.5994, which are consistent with analysis results in the Table 1.

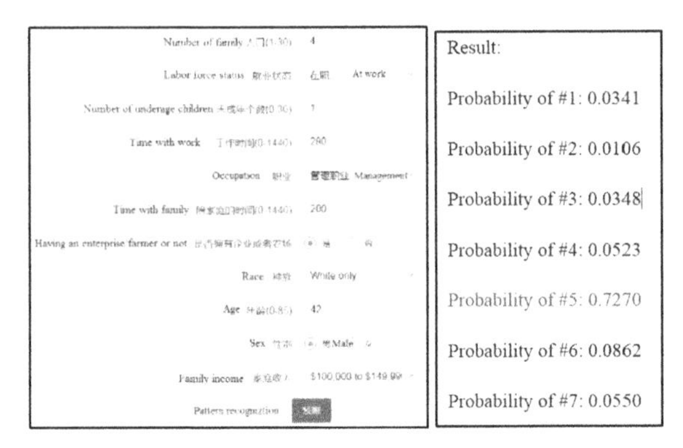

Fig. 3. The interface of pattern recognition (Test case 1).

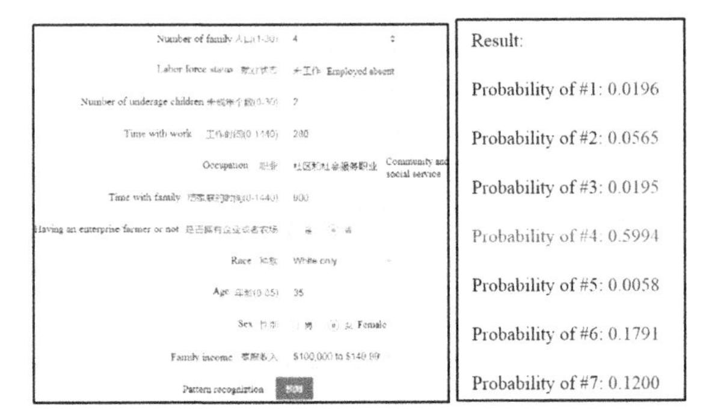

Fig. 4. The interface of pattern recognition (Test case 2).

5 Conclusion and Future Work

We address the human behavior pattern mining and recognition problems. This study contributes pattern mining and recognition by providing the linkage between activity behavior pattern and socio-demographic pattern based on the overall activity-based time use survey. In detail, we mine activity behavior patterns by deriving clusters of homogeneous daily activity behavior and activity sequence, where each pattern provides vital characteristics of activity behavior and socio-demographic characteristics. Furthermore, exploring more accurate activity behavior patterns play an important basic role in pattern recognition. Our proposed method can be applied to recommender system, activity schedule, social network analysis and mining, and urban planning.

Similarity computation between individuals is of primary importance in order to mine and recognize activity behavior pattern. The innovation is similarity computation between activity sequences. The time with adjacent time and similar activities is discretized into a time segment, as the degree of discretization is relatively large, the LCS algorithm with the complexity of $O(p(m-p))$ is proposed in this paper, which improves the efficiency of the algorithm. But it isn't suitable for the case when the degree of discretization is relatively low. In the future work, we will explore the semantic similarity computation for improving the understanding of semantic information. Furthermore, activity sequence pattern will be mined by network analysis and community detection algorithm.

Acknowledgments. This work is supported by the Major Science and Technology Innovation Project of Shandong Province under grant No. 2019JZZY010435, and National Natural Science Foundation of China under grant No. 51975332.

References

1. Ying, J.C., Lu, H.C., Lee, W.C., et al.: Mining user similarity from semantic trajectories. In: ACM SIGSPATIAL International Workshop on Location Based Social Networks. ACM (2010)
2. Zhang, W., Thill, J.C.: Detecting and visualizing cohesive activity-travel patterns: a network analysis approach. Comput. Environ. Urban Syst. **66**, 117–129 (2017)
3. Guochen, C., Kyungmi, L., Ickjai, L.: Mining semantic trajectory patterns from geo-tagged data. J. Comput. Sci. Technol. **33**(4), 849–862 (2018)
4. Banovic, N., Buzali, T., Chevalier, F., et al.: Modeling and understanding human routine behavior. In: CHI Conference on Human Factors in Computing Systems. ACM (2016)
5. Trong, N.P., Nguyen, H., Kazunori, K., Le Hoai, B.: A comprehensive survey on human activity prediction. In: Gervasi, O., et al. (eds.) ICCSA 2017. LNCS, vol. 10404, pp. 411–425. Springer, Cham (2017). https://doi.org/10.1007/978-3-319-62392-4_30
6. Lu, X., Pas, E.I.: Socio-demographics, activity participation and travel behavior. Transp. Res. Part A (Policy Pract.) **33**(1), 0–18(1999)
7. Xiong, J., et al.: Enhancing privacy and availability for data clustering in intelligent electrical service of IoT. IEEE Internet Things J. **6**(2), 1530–1540 (2019)
8. Kemperman, A., Timmermans, H.: Influence of socio-demographics and residential environment on leisure activity participation. Leis. Sci. **30**(4), 306–324 (2008)
9. Bernardo, C., Paleti, R., Hoklas, M., et al.: An empirical investigation into the time-use and activity patterns of dual-earner couples with and without young children. Transp. Res. Part A Policy Pract. **76**, 71–91 (2015)
10. You, W., Chenghu, Z., Tao, P.: Semantic-geographic trajectory pattern mining based on a new similarity measurement. ISPRS Int. J. Geo-Inform. **6**(7), 212 (2017)
11. Chakri, S., Raghay, S., Hadaj, S.E.: Semantic trajectory knowledge discovery: a promising way to extract meaningful patterns from spatiotemporal data. Int. J. Softw. Eng. Knowl. Eng. (2017)
12. Ke, Q., Bennamoun, M., An, S., Boussaid, F., Sohel, F.: Human interaction prediction using deep temporal features. In: Hua, G., Jégou, H. (eds.) ECCV 2016. LNCS, vol. 9914, pp. 403–414. Springer, Cham (2016). https://doi.org/10.1007/978-3-319-48881-3_28
13. Hunt, J.W., Szymanski, T.G.: A fast algorithm for computing longest common subsequences. Commun. ACM **20**(5), 350–353 (1977)
14. Aho, A.V., Hirschberg, D.S., Ullman, J.D.: Bounds on the complexity of the longest common subsequence problem. In: Symposium on Switching & Automata Theory. IEEE Computer Society (1974)
15. Hirschberg, D.S.: Algorithms for the longest common subsequence problem. J. ACM (JACM) (1977)
16. Nakatsu, N., Kambayashi, Y., Yajima, S.: A longest common subsequence algorithm suitable for similar text strings. Acta Informatica **18**(2), 171–179 (1982)

17. Liu, W., Chen, L.: Parallel longest common subsequence algorithm based on pruning technology. J. Comput. Appl. **26**(6), 1422–1424(2006)
18. Hafezi, M.H., Liu, L., Millward, H.: A time-use activity-pattern recognition model for activity-based travel demand modeling. Transportation **46**(4), 1369–1394 (2017). https://doi.org/10.1007/s11116-017-9840-9
19. Benetka, J.R., Krumm, J. Bennett, P.N.: Understanding context for tasks and activities, pp. 133–142 (2019). https://doi.org/10.1145/3295750.3298929
20. Krishna, K., Jain, D., Mehta, S.V., et al: An LSTM based system for prediction of human activities with durations. In: Proceedings of the ACM on Interactive, Mobile, Wearable and Ubiquitous Technologies, vol. 1, no. 4, pp.1–31 (2018)
21. Moon, G., Hamm, J.: A large-scale study in predictability of daily activities and places. In: The 8th EAI International Conference on Mobile Computing, Applications and Services. ICST, Institute for Computer Sciences, Social-Informatics and Telecommunications Engineering (2016)
22. Flett, G., Kelly, N.: An occupant-differentiated, higher-order Markov chain method for prediction of domestic occupancy. Energy Build. **125**, 219–230 (2016)
23. Diao, L., Sun, Y., Chen, Z., et al.: Modeling energy consumption in residential buildings: a bottom-up analysis based on occupant behavior pattern clustering and stochastic simulation. Energy Build. **147**, 47–66 (2017)
24. Barthelmes, V.M., Li, R., Andersen, R.K., et al.: Profiling occupant behaviour in danish dwellings using time use survey data. Energy Build., S0378778817342044-(2018)
25. Xiong, J., Bi, R., Zhao, M., Guo, J., Yang, Q.: Edge-assisted privacy-preserving raw data sharing framework for connected autonomous vehicles. IEEE Wirel. Commun. **27**(3), 24–30 (2020)
26. UCL Homepage. https://www.timeuse.org/mtus
27. BLS Homepage. https://www.bls.gov/tus/
28. U.S.Census Bureau: Highest Median Household Income on Record. http://www.census.gov/library/stories/2018/09/highest-median-household-income-on-record.html. Acessed 19 Stept (2019)
29. Zhang, J., Jiang, W., Zhang, J., Wu, J., Wang, G.: Exploring weather data to predict activity attendance in event-based social network: from the organizer's view. ACM Trans. Web **15**, 2 (2021). Article 10, 25 pages. https://doi.org/10.1145/34401341

ADS-B Signal Separation Via Complex Neural Network

Yue Yang[1], Haipeng Zhang[1], Haoran Zha[2](✉), and Ruiliang Song[1]

[1] Beijing R&D Center, The 54th Research Institute of CETC, Beijing 100010, China
[2] Harbin Engineering University, Harbin 150001, China
zhahaoran@hrbeu.edu.cn

Abstract. In the sphere of air surveillance, the Automatic Dependent Surveillance-Broadcast is a valuable method. However, the ADS-B system suffers from a considerable overlap issue, and has a significant influence on signal decoding, resulting in incorrect decoding or even data loss. A complicated neural network-based separation approach for ADS-B overlap signals is presented in this paper. Taking the two-signal overlap as the research object, the simulation data set is generated. After the Hilbert transform of the overlap ADS-B signal, it is input into the complex neural network, and finally the predicted waveform of the source signal is output to realize ADS-B signal separation. Experiments have shown that the technique is more efficient and has a lower error rate than previously proposed algorithms.

Keywords: Blind source separation · ADS-B · Complex neural network · Hilbert transform

1 Introduction

As an important air surveillance technology, automatic dependent surveillance broadcast (ADS-B) has the capability of tracking aircrafts in civil aviation. When the aircraft initiates a transmission of relevant information such as altitude, speed, latitude, and others, the data link makes it available to others. Other aircrafts decode and obtain information when receiving these data, so as to realize global surveillance [1, 2]. For the time being, the majority of ADS-B systems are based on the ground-based ground station to monitor the airspace.

ADS-B on land faces the problem of signal overlap. When multiple signals collide and overlap at the receiver, the demodulation module will result in incorrect decoding or even loss of critical information, reducing the security and reliability of the surveillance system. To avoid this situation, the overlap signals must be separated and the separated signals must be decoded to obtain the correct information [3]. The methods provided by the ICAO document standard can only receive one correct signal, limiting surveillance capacity. Nicolas Petrohilos advocated the Projection Algorithm (PA) and Extended Projection Algorithm (EPA), which contributed significantly to the field of ADS-B signal separation [4–7]. Wang [8] used the Alternating Direction Method of Multipliers to the

J. Xiong et al. (Eds.): MobiMedia 2021, LNICST 394, pp. 679–688, 2021.
https://doi.org/10.1007/978-3-030-89814-4_49

nonconvex blind adaptive beamforming issue. FastICA can also separate overlap ADS-B signals under certain conditions [9].

Advances in deep learning have led to its widespread application in the analysis of anomalies, channel identification, and speech recognition [10–14]. More recently, deep learning-based modulation signal categorization algorithms have increased [15–17]. Zhang [18] applied the lightweight deep neural network to the modulation recognition of electromagnetic signals. Adversarial attack is also one of the most concerned research directions in the field of artificial intelligence security [19]. At present, the application of deep learning in the field of blind source separation is mostly limited in the separation of speech signals. Wang [20] applied deep learning to single channel speech signal separation, and achieved good results. The application of deep learning in ADS-B field is mainly used to classify ADS-B signals. Yang [21] used deep learning to classify ADS-B signals and ACARS signals. At present, the common deep learning is based on real value, and some recent studies show that complex numbers have more abundant feature representation ability than real numbers. Trabelsi established the deep complex convolution neural network (CV-CNN) for the first time in 2017. Tu [22] validates the superior performance in automatic modulation classification achieved by the complex-valued networks. Soorya [23] utilizes neural networks with complex weights to learn fingerprints. In this paper, the simulation ADS-B signal data set is generated, and the complex neural network is built. The signal after Hilbert transform is used as the input to get the estimated source signal.

2 Additional Background Knowledge

2.1 Signal Structure

This paper deals with the ADS-B signal in 1090 MHz format. The length of an ADS-B signal is typically 120 μs or 64 μs, This work focuses on the long signal, which has two different elements: an 8 μs preamble and a 112 μs data block. The preamble section contains four pulses of 0.5 μs length at 0 μs, 1 μs, 3.5 μs, and 4.5 μs, respectively. The data block of ADS-B signal is modulated by Pulse Position Modulation (PPM). Figure 1 depicts the structure of an ADS-B signal.

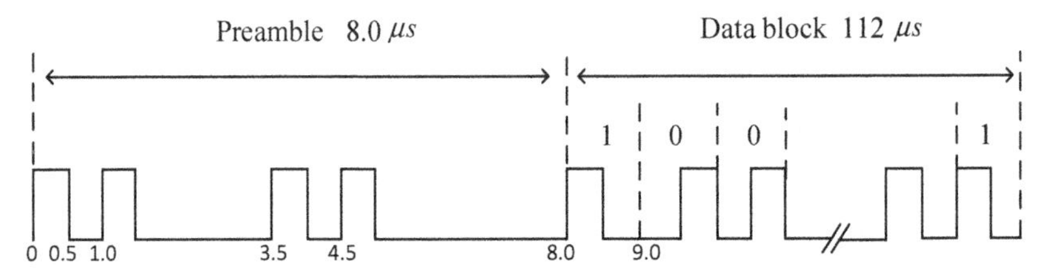

Fig. 1. ADS-B signal structure

2.2 Collision Pattern

Instead of requesting ahead of time, the ADS-B transmitter sends advisories and information depending on the pre-defined conditions and the state of the aircraft. Aircrafts spend the most of their time in the air in a constant condition. The aircraft is steady in the air, delivering 5.6 signals each second. The ADS-B signal is a bit sequence sent by each aircraft, and data from different planes will be delivered concurrently, resulting in signal overlap scenarios similar to the Aloha event. Because this collision mode accounts for the vast majority of occurrences, this study looks at what happens when two ADS-B signals collide.

3 Data Set Generation and Complex Neural Network

3.1 Data Set Generation

At present, it is hard to obtain effective measured data set of ADS-B overlap signal, so this paper has done a lot of related work on how to establish ADS-B signal data set. This study explores how to separate the overlap signals of two ADS-B signals, therefore generating all the overlap signals in the data set as a result of two ADS-B signals being seen to overlap in a given way under different situations.

Considering the difficulty and heavy workload of simulating the data sets in different signal time delay, different signal power difference and different SNR, the data set simulated in this paper is the overlap of two signals, and the time delay is 5 μs, 17 μs and 50 μs. The source signals' power differences are 2 dB, 3 dB, and 4 dB, respectively, the signal-to-noise ratio range is 5 dB–25 dB, and the interval step length is 5 dB. We first perform Hilbert transform on an ADS-B overlap signal, and then separate the real part from the imaginary part. The form of each signal stored in the data set can be regarded as the stitching of the real part and the imaginary part, in which the first half of each data is the real part, and the second half is the imaginary part. Among them, each data set is 90000 groups of signals, in which there are two source signals and one overlap signal. Figure 2 shows each signal in the dataset.

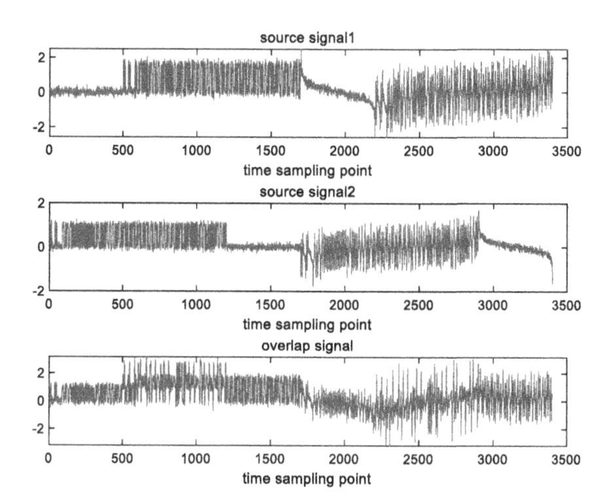

Fig. 2. Each group of signals in the dataset

3.2 Complex Neural Network

In order to realize the equivalent operation of 2D convolution in the complex domain, the complex filter matrix $\mathbf{h} = \mathbf{x} + i\mathbf{y}$ can be convoluted by the complex vector $\mathbf{W} = \mathbf{A} + i\mathbf{B}$.

$$\mathbf{W} * \mathbf{h} = (\mathbf{A} * \mathbf{x} - \mathbf{B} * \mathbf{y}) + i(\mathbf{B} * \mathbf{x} + \mathbf{A} * \mathbf{y}) \tag{1}$$

Figure 3 shows the schematic diagram of complex convolution operation, where M_I and M_R represent the imaginary part and real part characteristic graphs respectively, K_I and K_R represent the imaginary kernel and real kernel respectively.

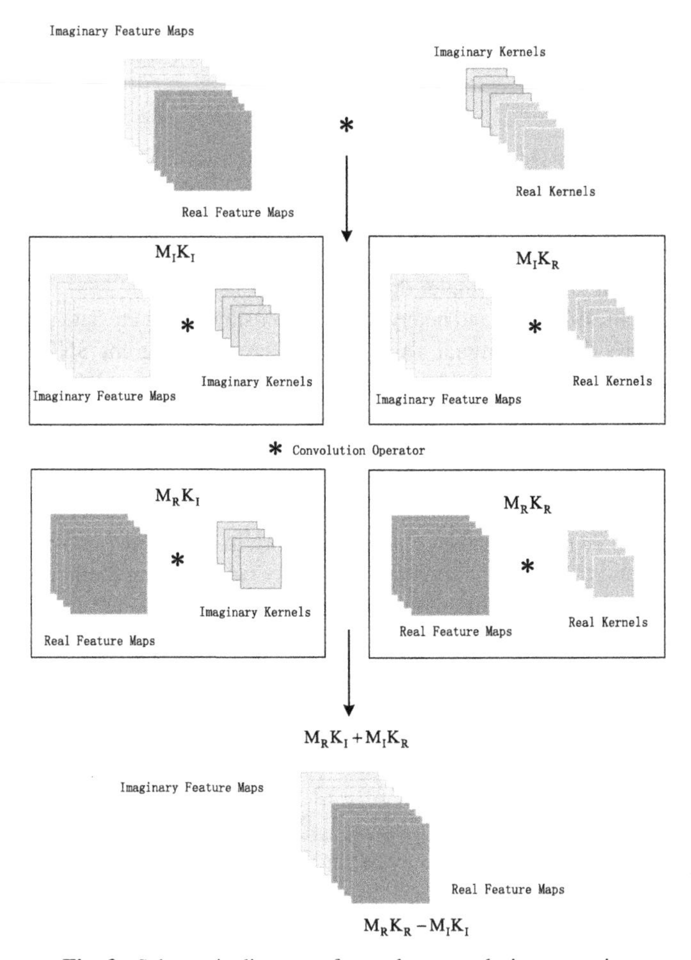

Fig. 3. Schematic diagram of complex convolution operation

In this paper, the real and imaginary parts of convolution operations are represented by matrix representation:

$$\begin{bmatrix} \Re(\mathbf{W} * \mathbf{h}) \\ \Im(\mathbf{W} * \mathbf{h}) \end{bmatrix} = \begin{bmatrix} \mathbf{A} & -\mathbf{B} \\ \mathbf{B} & \mathbf{A} \end{bmatrix} * \begin{bmatrix} \mathbf{x} \\ \mathbf{y} \end{bmatrix} \tag{2}$$

Batch normalization method can accelerate the learning speed of deep network, and batch normalization is very important for model optimization in some cases. However, because the standard formula of batch normalization is only applicable to real values, the batch normalization formula of complex values is adopted in this paper. Not only the input layer is normalized, but also the input of each middle layer is normalized.

In this paper, the problem is considered as a two-dimensional vector problem, which means that the data can be scaled according to the square root of the variance of the two principal components. This step can be completed by multiplying the 0-centered data $(\mathbf{x} - \mathbb{E}[\mathbf{x}])$ by the reciprocal of the square root of the 2×2 covariance matrix $\mathbf{V}$:

$$\widetilde{\mathbf{x}} = (\mathbf{V})^{-1/2}(\mathbf{x} - \mathbb{E}[\mathbf{x}]) \quad (3)$$

The normalization process can decorrelate the imaginary part and the real part of the element, which may avoid over fitting. This algorithm uses two parameters β and γ, in which the shift parameter β is a complex parameter and the scaling parameter γ is a 2×2 semi positive definite matrix.

Since the variance of real part and imaginary part of normalized input $\widetilde{\mathbf{x}}$ is 1, this chapter initializes γ_{rr} and γ_{ii} to $1/\sqrt{2}$, so that the variance modulus of normalized value is 1. Multiple batch normalization is defined as:

$$\mathrm{BN}(\widetilde{\mathbf{x}}) = \gamma\widetilde{\mathbf{x}} + \beta \quad (4)$$

In the training and testing process, the moving average with momentum is used to maintain the estimation of the normalized statistical data of the complex batch. The moving average values of V_{ri} and β are initialized to 0, the moving average values of V_{rr} and V_{ii} are initialized to $1/\sqrt{2}$, and the momentum of the moving average line is set to 0.9.

Proper weight initialization can avoid vanishing gradient. Therefore, the method of deriving complex weight parameters is as follows:

The complex weight has both polar and rectangular forms:

$$\mathbf{W} = |\mathbf{W}|e^{i\theta} = \Re\{\mathbf{W}\} + i\Im\{\mathbf{W}\} \quad (5)$$

where, θ and $|\mathbf{W}|$ is the phase and amplitude of $\mathbf{W}$.

4 Experiments

4.1 Experimental Setup

In the problem of data set selection, 5/6 of the data set is randomly selected as the training sample, 1/6 as the verification set and test sample for testing, and the data set is generated by random sampling.

In this experiment, the input of complex neural network is overlap signal waveform, and the output is two source signal waveforms predicted by complex neural network.

4.2 Evaluating Indicator

The evaluation indexes considered in this chapter are bit error rate and the similarity coefficient commonly used in traditional blind source separation. The definition of similarity coefficient is as follows:

$$\rho = \frac{\text{cov}(x, y)}{\sqrt{\text{cov}(x, x)\text{cov}(y, y)}} \tag{6}$$

where, $\text{cov}(x, y) = E\{[x - E(x)][y - E(y)]\}$ is the variance of x and y.

When the separated signal y approaches the source signal x, the similarity coefficient ρ approaches 1.

4.3 Experiment and Analysis

Experiment with the separation efficacy of overlap ADS-B signals by selecting two ADS-B signals at random, with a power difference of 3 dB between them, the time delay is 50 μs, and the SNR is 20 dB. Figure 4 shows the two original source signals as well as the overlap signals, and the input signal of neural network is shown in Fig. 5.

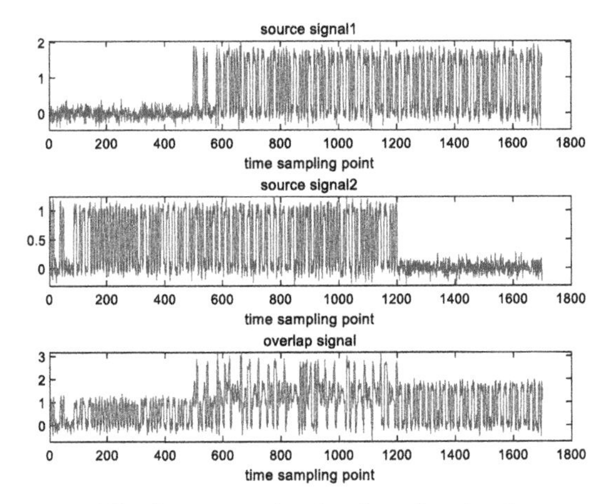

Fig. 4. Source signal and overlap signal

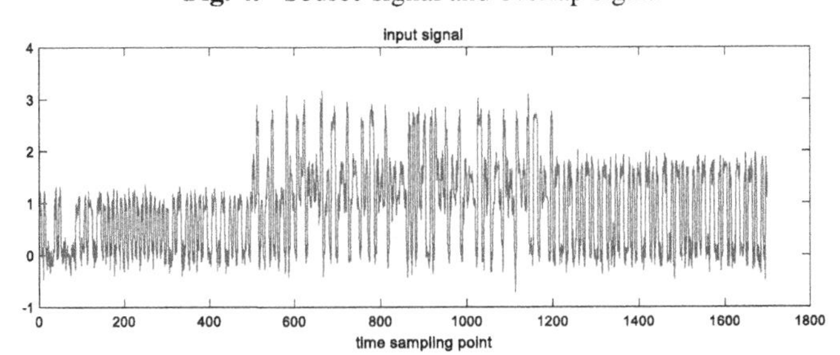

Fig. 5. Input signal of neural network

The output of the complex neural network is shown in Fig. 6. The real part of the intercepted output signal is shown in Fig. 7.

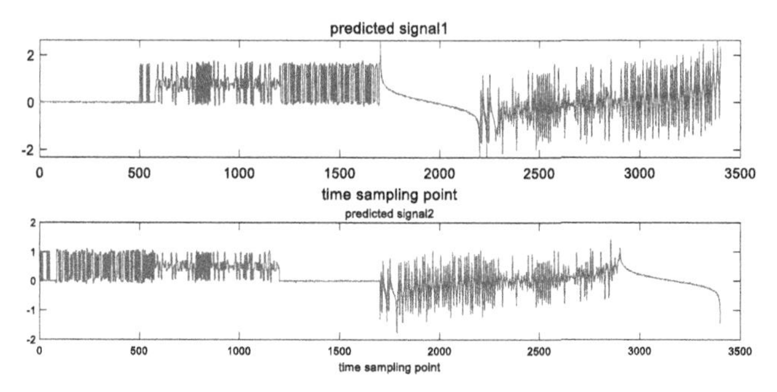

Fig. 6. Output signal of neural network

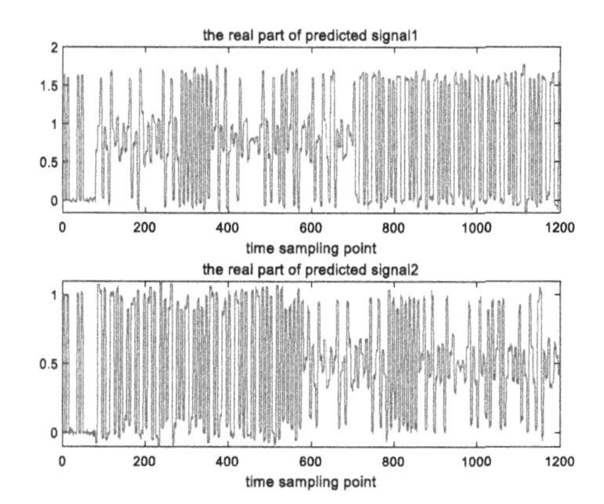

Fig. 7. Real part of output signal of neural network

As seen in Fig. 7, the actual sections of the two prediction signals generated by the complex neural network are comparable to the waveforms of the original signal in Fig. 6. It can be seen that complex neural network has the ability to separate ADS-B overlap signals.

The separation performance of the complex neural network is tested under the conditions that the two source signals have 2 dB, 3 dB, 4 dB power difference, the time latency between the two source signals is 5 μs, 17 μs and 50 μs, and the SNR is 5 dB, 10 dB, 15 dB, 20 dB and 25 dB.

When there is a 3 dB power differential between two signals and a 17 μs time delay, the curves of bit error rate of different algorithms vs SNR are presented in Fig. 8, and Table 1 shows the time consumption of different methods.

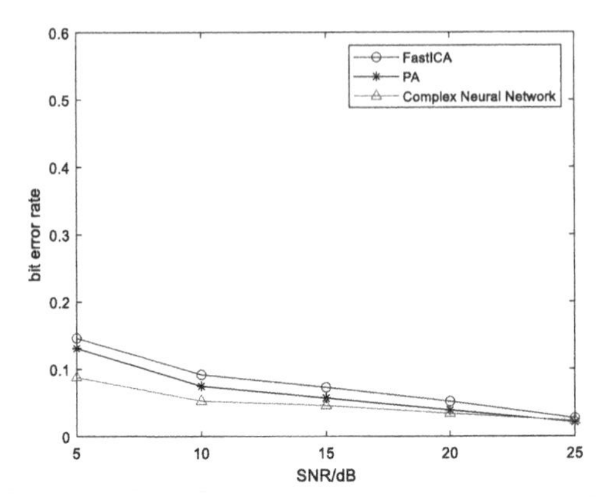

Fig. 8. Curve of bit error rate with SNR for different algorithms

Table 1. Separation time of different separation algorithms

Algorithm	Time (s)
FastICA	0.0419
PA	0.0828
Complex neural network	0.0014

As can be seen from Fig. 8, the complex neural network separation algorithm is better than traditional FastICA algorithm and PA algorithm. When the SNR is 5 dB and the power difference between the two ADS-B source signals is 3 dB, the BER is lower than 9%. It can be seen from Table 1 that the separation speed of the complex neural network reaches the millisecond level, which is significantly less than the other two algorithms.

When the two source ADS-B signals have different power differences, the curve of bit error rate changing with SNR is shown in Fig. 9, where the power differences are 2 dB, 3 dB and 4 dB; the SNR is 5 dB–25 dB. The similarity coefficient of separated waveform under different conditions is shown in Table 2.

This Fig. 9 shows that when the power difference between the source signals grows, the bit error rate drops. When the source signal has a power differential of 3 dB or 4 dB, the bit error rate is quite low. The greater the gap between the levels of the two sources, the less the influence on bit error rate is. Signal power difference sources increase in SNR, which causes the bit error rate of the same source signal power difference to drop simultaneously. When the power difference between the source signals is 4 dB and the SNR is 25 dB, the bit error rate is 1.8%. The similarity coefficient improves fast as the power difference between source signals and SNR grows. When the strength difference between source signals is 4dB, the similarity coefficient is 0.9314.

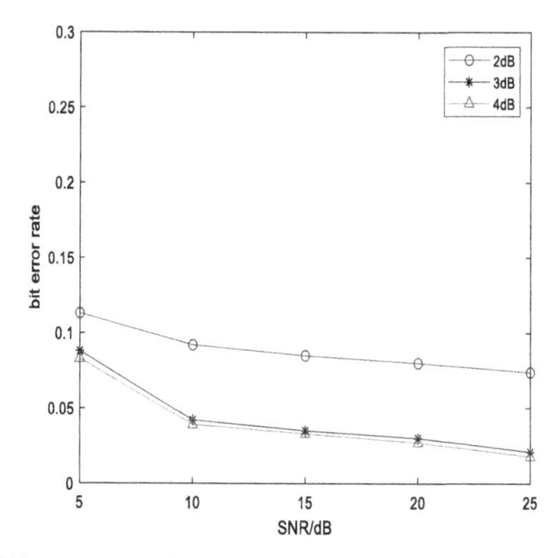

Fig. 9. Curve of bit error rate with SNR under different source signal power difference

Table 2. Similarity coefficient of separated waveform under different conditions

Power ratio	SNR				
	5 dB	10 dB	15 dB	20 dB	25 dB
2 dB	0.8053	0.8152	0.8474	0.8479	0.9203
3 dB	0.8782	0.9105	0.9127	0.9170	0.9230
4 dB	0.8860	0.9144	0.9181	0.9261	0.9314

5 Conclusion

This paper studies the application of deep learning in the separation of ADS-B overlap signals, and proposes a method of ADS-B signal separation based on complex neural network. The simulation data set is used in the experiment. The input of the complex neural network is the waveform of the overlap signal after Hilbert transform, and the output is the waveform of the two source signals predicted by the neural network after the Hilbert transform. The first half of each signal is the real part of the signal after Hilbert transform, and the second half is the imaginary part of the signal after Hilbert transform. Simulation results demonstrate that ADS-B signal separation algorithm based on complex neural network has better performance index, and bit error rate is lower than traditional algorithm. The time consumption reaches millisecond level, which can deal with large-scale signal overlap situation.

References

1. Strohmeier, M., Schafer, M., Lenders, V., Martinovic, I.: Realities and challenges of NexGen air traffic management: the case of ADSB. IEEE Commun. Mag. **52**(5), 111–118 (2014)

2. RTCA: Minimum operational performance standards for 1090 MHz extended squitter ADS-B, DO-260B, Washington, DC, pp. Appendix I (2009)
3. Werner, K., Bredemeyer, J., Delovski, T.: ADS-B over satellite: global air traffic surveillance from space. In: Tyrrhenian International Workshop on Digital Communications-Enhanced Surveillance of Aircraft and Vehicles, pp.47–52. IEEE (2014)
4. Petrochilos, N., van der Veen, A.-J.: Algorithms to separate overlapping secondary surveillance radar replies. In: 2004 IEEE International Conference on Acoustics, Speech, and Signal Processing, pp. 41–49. IEEE (2004)
5. Petrochilos, N., Piracci, E.G., Galati, G.N.: Separation of multiple secondary surveillance radar sources in a real environment for the near-far case. In: 2007 IEEE Antennas and Propagation Society International Symposium, pp. 3988–3991 (2007)
6. Petrochilos, N., Galati, G., Piracci, E.: Separation of SSR signals by array processing in multilateration systems. IEEE Tran. Aerosp. Eelctr. Syst. **45**(3), 965–982 (2009)
7. Petrochilos, N., van der Veen, A.-J.: Algebraic algorithms to separate overlapping secondary surveillance radar replies. IEEE Trans. Signal Process. **55**(7), 3746–3759 (2007)
8. Wang, W., Wu, R., Liang, J.: ADS-B signal separation based on blind adaptive beamforming. IEEE Trans. Veh. Technol. **68**(7), 6547–6556 (2019)
9. Zhang, Y., Li, W., Dou, Z.: Performance analysis of overlapping space-based ADS-B signal separation based on FastICA. In: 2019 IEEE Globecom Workshops (GC Wkshps), Waikoloa, HI, USA, pp. 1–6 (2019)
10. Zheng, D., You, F., Zhang, H., Gao, M., Jianzhi, Y.: Semantic segmentation method based on super-resolution. Int. J. Perform. Eng. **16**(5), 711–719 (2020)
11. Zhao, Y., et al.: Preserving minority structures in graph sampling. IEEE Trans. Vis. Comput. Graph. **27**(2), 1698–1708 (2021)
12. Lin, Y., Tu, Y., Dou, Z.: An improved neural network pruning technology for automatic modulation classification in edge devices. IEEE Trans. Veh. Technol. **69**(5), 5703–5706 (2020)
13. Gui, G., Liu, M., Tang, F., Kato, N., Adachi, F.: 6G: opening new horizons for integration of comfort, security, and intelligence. IEEE Wirel. Commun. **27**(5), 126–132 (2020)
14. Zhang, Z., Chang, J., Chai, M., Tang, N.: Specific emitter identification based on power amplifier. Int. J. Perform. Eng. **15**(3), 005–1013 (2019)
15. Lin, Y., Tu, Y., Dou, Z., Chen, L., Mao, S.: Contour Stella image and deep learning for signal recognition in the physical layer. IEEE Trans. Cogn. Commun. Netw. **7**(1), 34–46 (2021)
16. Jiabao, Y., Aiqun, H., Li, G., Peng, L.: A robust RF fingerprinting approach using multi-sampling convolutional neural network. IEEE Internet Things J. **6**(4), 6786–6799 (2019)
17. Lin, Y., Zhu, X., Zheng, Z., Dou, Z., Zhou, R.: The individual identification method of wireless device based on dimensionality reduction and machine learning. J. Supercomput. **75**(6), 3010–3027 (2017). https://doi.org/10.1007/s11227-017-2216-2
18. Zhang, S., Lin, Y., Tu, Y., et al.: Modulation recognition technology of electromagnetic signal based on lightweight deep neural network. J. Commun. **41**(11), 12–21 (2020)
19. Lin, Y., Zhao, H., Ma, X., Tu, Y., Wang, M.: Adversarial attacks in modulation recognition with convolutional neural networks. IEEE Trans. Reliab. **70**(1), 389–401 (2021)
20. Zhang, X.L., Wang, D.: A deep ensemble learning method for monaural speech separation. IEEE/ACM Trans. Audio Speech Lang. Process. **24**(5), 967–977 (2016)
21. Chen, S., Zheng, S., Yang, L., Yang, X.: Deep learning for large-scale real-world ACARS and ADS-B radio signal classification. IEEE Access **7**, 89256–89264 (2019)
22. Tu, Y., et al.: Complex-valued networks for automatic modulation classification. IEEE Trans. Veh. Technol. **69**(9), 10085–10089 (2020)
23. Gopalakrishnan, S., Cekic, M., Madhow, U.: Robust wireless fingerprinting via complex-valued neural networks. In: 2019 IEEE Global Communications Conference (GLOBECOM), Waikoloa, HI, USA, pp. 1–6 (2019)

A New Ensemble Pruning Method Based on Margin and Diversity

Zixiong Shen[1] and Xingcheng Liu[1,2,3](✉)

[1] School of Electronics and Information Technology, Sun Yat-sen University, Guangzhou 510006, China
isslxc@mail.sysu.edu.cn
[2] School of Information Science, Guangzhou Xinhua University, Guangzhou 510520, China
[3] Southern Marine Science and Engineering Guangdong Laboratory (Zhuhai), Zhuhai 519082, China

Abstract. Classification is one of the main tasks of machine learning, and ensemble learning has become a successful paradigm in the data classification field. This work aims to present a new method for pruning an ensemble classification model based on margin theory and ensemble diversity. Firstly, a new unsupervised form of instances margin metric is proposed, which does not need to consider the true class labels of the instances. This mechanism can improve the robustness of the algorithm against mislabeled noise instances. Then, the Jensen-Shannon (J-S) divergence between the classifiers is calculated based on the probability distribution of the class labels. Finally, all base classifiers are ordered with respect to a new criterion which combines the obtained margin values and the J-S divergence of base classifiers. Experiments show that the proposed method has a stable improvement on a significant proportion of benchmark datasets over existing ensemble pruning methods.

Keywords: Ensemble learning · Classification · Multiple classifier systems

1 Introduction

Ensemble learning is an important branch in the field of machine learning, which trains multiple base models explicitly or implicitly from data. As a mainstream machine learning paradigm, ensemble learning uses the strategy of "perturb and combine" to train multiple base classifiers, that is, randomly perturb the samples space or features space and randomly adjust the parameters of the base classifiers [1]. Individual classifiers may only focus on part of the information on

This work was supported by the Key Project of NSFC-Guangdong Province Joint Program (Grant No. U2001204), the National Natural Science Foundation of China (Grant Nos. 61873290 and 61972431), and the Science and Technology Program of Guangzhou, China (Grant No. 202002030470).

J. Xiong et al. (Eds.): MobiMedia 2021, LNICST 394, pp. 689–701, 2021.
https://doi.org/10.1007/978-3-030-89814-4_50

the decision boundary, which leads to certain limitations when making classification decisions. If the predicted information of multiple base classifiers can be integrated, a more reasonable classification result can be obtained. At present, many theoretical analyses and experiments have proved that integrating numerous base classifiers to train data can overcome the limitations of a single base classifier, improve the generalization performance of the original base classifiers, thus improve the classification accuracy [2].

However, training a large number of base classifiers requires additional storage resources, and the consumption of computing resources has also become a problem that can not be ignored in ensemble learning. Besides, it is not the fact that all classifiers used in the ensemble system can make the final classification results better. The probability of having highly similar base classifiers increases as the scale of ensemble model expands, and the accuracy of the entire ensemble classification system will also decrease with the increase of bad base classifiers. Zhou [3] has pruned the parallel ensemble methods in his research work and found that it can achieve better generalization performance with a smaller-scale ensemble. Some unnecessary base classifiers in the ensemble system are eliminated by a certain method, so that the generalization performance after pruning is better than the ensemble of all base classifiers before pruning. This is the so-called ensemble pruning, also called selective ensemble or ensemble selection [3]. The prerequisite for ensemble pruning is that all base classifiers have been generated, and no new base classifiers will be generated during the construction process. This is different from the classic serial ensemble learning method, which generates individual classifiers one by one during the training process, but ensemble pruning may discard any base classifier that has been generated [4].

The existing ensemble pruning methods can be mainly divided into three categories: sorting-based ensemble pruning, clustering-based ensemble pruning, and optimization-based ensemble pruning [5]. Sorting-based ensemble pruning method sorts individual classifiers in descending order based on a predetermined criterion, such as classification accuracy, diversity of the ensemble model. Next, the top-ranked base classifiers will be added to the final ensemble set [6,7]. The advantage of the sorting-based method is that it has lower computational complexity, but there is currently no unified sorting criterion. Clustering-based ensemble pruning method attempts to cluster similar base classifiers based on the generalization performance of the base classifiers [8]. Some representative base classifiers close to the cluster center can be used to fit the best decision boundary. Clustering-based methods are usually classified into two steps: firstly, divide all base classifiers into multiple clusters, which involves the problem, which clustering method is to be adopted. Secondly, select the appropriate base classifiers from the clusters, which involves the problem of the pruning strategy to be used [9]. Optimization-based ensemble pruning method transforms the ensemble pruning problem into an optimization problem, where, the final ensemble set is selected through optimizing the overall generalization capability [10]. Since searching for the optimal subset directly requires a lot of calculations, this method often resorts to optimization algorithms such as genetic algorithm, multi-objective optimization algorithm, and hill-climbing algorithm [11].

The main challenge of ensemble pruning is to design a practical algorithm that can reduce the ensemble scale without reducing the generalization performance [12]. At present, relevant research [13] has proved that the ensemble pruning method based on sorting is superior to the enumeration searching method, which directly selects the best subset in terms of classification accuracy and computational performance. In this paper, the concepts of instances margin [14] are applied to ensemble pruning, and the fact is that when ensemble pruning is performed, the instances with small margin values should be the main concern [15]. The performance of a base classifier is evaluated by those instances with small margin values. Besides, because the diversity of the classifier set is also an issue in the process of constructing an ensemble model, a new measuring criterion for pairwise difference is constructed to measure the diversity of the classifiers set. All the individual base classifiers are sorted by a predefined criteria, and the top-ranked base classifiers are incorporated into the final ensemble classifiers subset, so this method has higher computational efficiency than other state-of-the-art methods.

The rest of this paper is organized as follows. Section 2 presents the proposed ensemble pruning methodology. Section 3 gives the details of the experimental setup, results and comparative analysis. Discussions and concluding remarks are given in Sect. 4.

2 Proposed Method

In this section, a new ensemble pruning method based on margin and diversity named EPMD is proposed, which eliminates the useless base classifiers while improving the final accuracy of the ensemble learning framework. In order to solve the data classification problem effectively, the sample points near the classification decision boundary are more inclined to be focused on. Part of the original dataset is used as a training dataset to generate the base classifiers pool, and then a validation dataset is used to evaluate the generated base classifiers based on the proposed heuristic metrics. After obtaining the simplified subset of ensemble classifiers, classification tests are performed on the unused test dataset to obtain the final classification results.

2.1 Generate Base Classifier Pool

Suppose the initial dataset is a matrix of dimension $N \times n$: $D = \{\boldsymbol{x}_i, y_i)|i = 1, 2 \cdots, N\}$, including N samples $\boldsymbol{x}_i$ and N true class labels y_i, $y_i \in \{1, 2, \cdots, L\}$, that is, there are L classes in the original dataset. Each sample point $\boldsymbol{x}_i$ is a d-dimensional feature vector; $H = \{h_t|t = 1, 2, \cdots, T\}$ is a classifiers pool containing totally T base classifiers, each of which is equivalent to a mapping function of $\boldsymbol{x}_i : y_i^{'} = h_t(\boldsymbol{x}_i)$, and $y_i^{'}$ is the predicted class label.

Firstly, leave-m-out cross-validation is utilized to divide the initial data set into three equal parts, which are used as training dataset $D_{tr} \in \mathbb{R}^{N' \times n}$, validation dataset $D_{va} \in \mathbb{R}^{N' \times n}$ and test dataset $D_{te} \in \mathbb{R}^{N' \times n}$. The operation steps here are basically the same as the Bagging algorithm [16]. For the training dataset $D_{tr} \in \mathbb{R}^{N' \times n}$, Bootstrap [17] method is used to perform m random sampling with replacement. This work will be repeated until the number of samples in each sample set is the same as in the initial training dataset before sampling. After repeating T rounds of operation to obtain T sample sets $D_{tr\_t}(1 \leqslant t \leqslant T)$, the sampled training subsets are different from each other, and $|D_{tr\_t}| = |D_{tr}|$.

Next, Classification and Regression Tree (CART) [18] is utilized to train each training dataset D_{tr_t}, then obtain the ensemble classifiers $ES = \{h_1, h_2, \cdots, h_T\}$, and are added to the base classifier pool. The type of base classifiers used here is not unique, CART tree is utilized here because it is more sensitive to the perturbation of the input data, it is easier to produce diversified base classifiers and the computational complexity is not high.

2.2 Base Classifier Evaluation

Each base classifier of the ensemble system ES is used to classify the validation dataset samples, and the majority voting is used here to obtain the prediction results matrix of the validation dataset:

$$Mat = [\boldsymbol{R}_1, \boldsymbol{R}_2, \cdots, \boldsymbol{R}_t, \cdots, \boldsymbol{R}_T] \in \mathbb{R}^{N' \times T}, \tag{1}$$

where $\boldsymbol{R}_t = [C_t(\boldsymbol{x}_1), C_t(\boldsymbol{x}_2), \cdots, C_t(\boldsymbol{x}_i), \cdots, C_t(\boldsymbol{x}_{N'})]$ is the vector formed by the classification results of the t-th base classifier in the ensemble system ES.

According to the classification results matrix Mat, the votes number matrix $Vote \in \mathbb{R}^{N' \times L}$ of each data sample belonging to each class in the validation dataset calculated (that is, the number of all base classifiers that classify the data samples into a certain class). Sort the row elements of the votes number matrix Vote in descending order, and get the sorted votes number vector $\boldsymbol{v}(x_i) = [v_{c_1}, v_{c_2}, \cdots, v_{c_L}] \in \mathbb{R}^{1 \times L}$ for each data sample $\boldsymbol{x}_i$ in the validation dataset.

A new unsupervised form of instances margin metric is proposed here to eliminate useless weak classifiers:

$$\begin{aligned} margin(\boldsymbol{x}_i, y_i) =& \frac{1}{N} \cdot \frac{1}{\sum_{l=1}^{L}(v_{c_l})} \cdot \\ & \sqrt{(v_{c_1} - v_{c_2})^2 + (v_{c_2} - v_{c_3})^2 + \cdots + (v_{c_{L-1}} - v_{c_L})^2}. \end{aligned} \tag{2}$$

For a sample point $(\boldsymbol{x}_i, y_i)$ in the validation dataset, v_{c_1} represents the number of votes of the class with the most votes, that is, the vast majority of base classifiers in the ensemble system classify and predict the sample $(\boldsymbol{x}_i, y_i)$ into class c_1, and v_{c_2} represents the number of votes for the class with the second most votes, and so on, v_{c_L} represents the number of votes for the class label with the least number of votes.

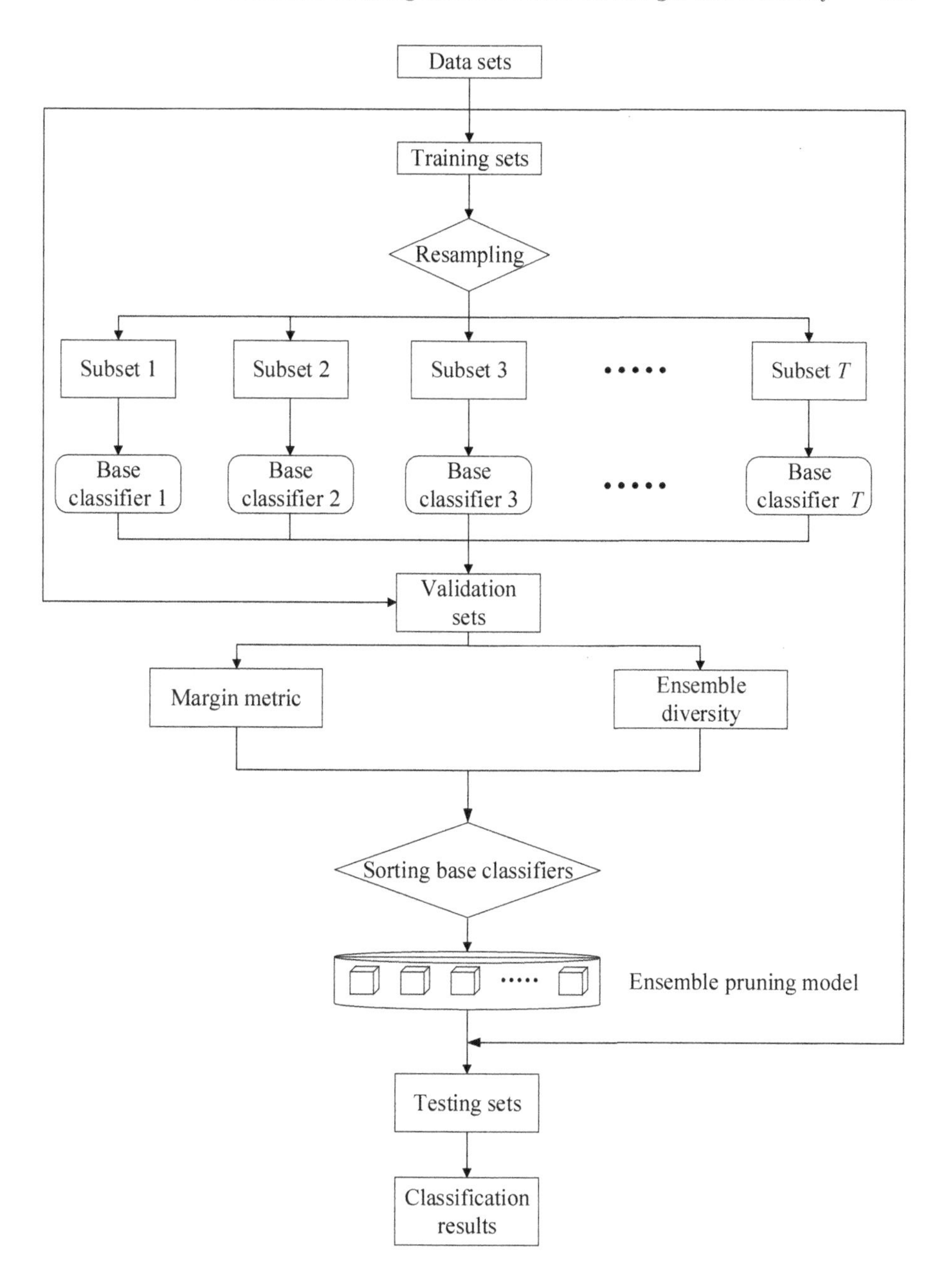

Fig. 1. Generation framework of the proposed method EPMD.

Next, the classification results matrix Mat is compared with the true class labels vector to find all the data points D_{va_t} that are correctly classified. For each classifier in the base classifiers pool, the number of validation dataset samples that are correctly classified and predicted is counted by formula (3):

$$N_R(h_t) = \sum_{i=1}^{N'} \Psi(C_t(\boldsymbol{x}_i), y_i), \tag{3}$$

where $I(true) = 1, I(false) = 0$.

Based on each classifier h_t in the base classifiers pool, the average margin value is calculated by formula (4):

$$\Phi(h_t) = \frac{1}{N_R(h_t)} \cdot \sum_{(\boldsymbol{x}_i, y_i) \in D_{va\_t}} margin(\boldsymbol{x}_i, y_i). \tag{4}$$

Considering that the existing selective ensemble learning algorithms based on margin values rarely involve the diversity between base classifiers, the Jensen-Shannon (J-S) divergence [19] is employed here from the perspective of information theory. For the classification results of each base classifier in the base classifiers pool, its probability distribution is calculated with respect to the class labels, and thereby the J-S divergence is obtained. The J-S divergence measures the degree of difference between the probability distributions of the classification results of different classifiers, eliminates base classifiers with low diversity, and improves the overall diversity of the ensemble system.

Let $p = \{p_1, p_2, ..., p_K\}$ and $q = \{q_1, q_2, ..., q_K\}$ be the two probability distributions on the random variable $\boldsymbol{X}$, where K is the number of discrete random variables. Then the J-S divergence between the probability distributions P and Q is defined as:

$$JS(\boldsymbol{p}, \boldsymbol{q}) = \frac{1}{2}\left[S\left((p), \frac{\boldsymbol{p}+\boldsymbol{q}}{2}\right) + S\left((q), \frac{\boldsymbol{p}+\boldsymbol{q}}{2}\right)\right], \tag{5}$$

where S is the Kullback-Leibler divergence (K-L) divergence between the two probability distributions:

$$S(\boldsymbol{p}, \boldsymbol{q}) = \sum_k p_k \cdot \log \frac{p_k}{q_k} \quad , \ (k = 1, 2, \cdots, K). \tag{6}$$

The J-S divergence can be obtained by the formulas (5) and (6), as shown in formula (7):

$$\begin{aligned} JS(\boldsymbol{p}, \boldsymbol{q}) &= H\left(\frac{\boldsymbol{p}+\boldsymbol{q}}{2}\right) - \frac{1}{2}H(\boldsymbol{p}) - \frac{1}{2}H(\boldsymbol{q}) \\ &= \frac{1}{2}\left[\sum_k p_k \cdot \log\left(\frac{2p_k}{p_k + q_k}\right) + \sum_k q_k \cdot \log\left(\frac{2q_k}{p_k + q_k}\right)\right]. \end{aligned} \tag{7}$$

The classification prediction result of a certain classifier h_t in the base classifier pool:

$$\boldsymbol{R}_t = [C_t(\boldsymbol{x}_1), C_t(\boldsymbol{x}_2), \cdots, C_t(\boldsymbol{x}_i), \cdots, C_t(\boldsymbol{x}_{N'})]. \tag{8}$$

Calculate the probability distribution of the class labels of the original dataset:

$$\boldsymbol{p}_t = (p_1, p_2, \cdots, p_l, \cdots, p_L)^T, \tag{9}$$

where p_l is the probability distribution about class label l:

$$p_l = \sum_{i=1}^{N'} I(C_t(\boldsymbol{x}_i) = l)/N'. \tag{10}$$

The J-S divergence between two different classifiers (Classifiers Jensen-Shannon divergence) is obtained by formula (7) as follows:

$$CJS(\boldsymbol{P_1}, \boldsymbol{P_2}) = \frac{1}{2}\Big[\sum_{l=1}^{L} p_l \cdot \log\Big(\frac{2p_l}{p_l + q_l}\Big) + \sum_{l=1}^{L} q_l \cdot \log\Big(\frac{2q_l}{p_l + q_l}\Big)\Big], \tag{11}$$

$$\boldsymbol{CJS} = \begin{bmatrix} CJS_{11} & CJS_{12} & \cdots & CJS_{1T} \\ CJS_{21} & CJS_{22} & \cdots & CJS_{2T} \\ \vdots & \vdots & \ddots & \vdots \\ CJS_{T1} & CJS_{T2} & \cdots & CJS_{TT} \end{bmatrix}. \tag{12}$$

When the J-S divergence between two different classifiers in the base classifier pool is larger, it indicates that the information difference between the probability distributions of the corresponding classification results is greater. Then the average degree of difference between the t-th base classifier and other base classifiers is defined as:

$$\widetilde{CJS_t} = \frac{\sum_{s=1, s\neq t}^{T} CJS_{st}}{T-1}. \tag{13}$$

The greater the average degree of difference between a certain base classifier and other base classifiers, the greater the diversity contribution that the base classifier makes to the ensemble system. In order to take average margin and diversity contribution of the base classifier into consideration simultaneously, a trade-off between margin and diversity (TMD) is defined as an objective function:

$$TMD(h_t) = \lambda \cdot (\frac{1}{\Phi(h_t)}) + (1-\lambda) \cdot \log(1 + e^{-\widetilde{CJS_t}}), \tag{14}$$

where $\lambda \in [0, 1]$ is a regularization factor, which is used to balance the importance of these two classifiers metrics $\Phi(h_t)$ and $\widetilde{CJS_t}$. By sorting all the classifiers in the base classifier pool in descending order according to the obtained value, a new base classifier sequence can be obtained: $ES' = \{h_1', h_2', \cdots, h_t', \cdots, h_T'\}$, which satisfies $TMD(h_{t-1}') \geqslant TMD(h_t')$, $0 \leqslant t \leqslant T$. The higher the ranking of the base classifier, the larger the value, and it is considered to have better generalization performance.

By selecting the first S base classifiers that can maximize the average classification accuracy of the ensemble system on the validation dataset, a selective ensemble classifiers subset is obtained as follows:

$$ES_{new} = \underset{S,(1 \leqslant S \leqslant T)}{arg\ max}\ accuracy(ES^{'}). \tag{15}$$

Finally, the ensemble pruning task has been modeled as an objective function value ordering problem as shown in (15), the pruned classifiers subset is used to predict the test dataset and then the final classification results are obtained. The flowchart of novel method for ensemble pruning proposed in this paper is shown in Fig. 1.

3 Experiments and Results

In this section, several experiments are presented to evaluate the performance of the proposed ensemble pruning method.

3.1 Datasets and Experimental Setup

Table 1. Summary of UCI repository datasets used in experiments

Datasets	# of samples	# of features	# of classes	Division ratio
Glass	214	9	6	70:70:70
Zoo	101	16	7	33:33:33
Air	359	64	3	120:120:119
Hayesroth	160	4	3	53:53:53
Appendicitis	106	7	2	35:35:35
M-of-N	1000	13	2	333:333:333
Car	1728	6	4	576:576:576
Ecoli	336	7	8	112:112:112
DNA test	1186	12	3	395:395:395
Tic-tac-toe	958	9	2	319:319:319
Seeds	210	7	3	70:70:70
Segment	2310	18	7	770:770:770
Tae	151	5	3	50:50:50
Vowel	528	10	11	176:176:176
Wdbc	569	30	2	189:189:189
Wpbc	198	25	2	66:66:66
Breast-w	699	10	2	233:233:233
X8D5K	1000	8	5	333:333:333
Penbased	10992	16	10	3600:3600:3600
Phoneme	5404	5	2	1800:1800:1800
Ringnorm	7400	20	2	2400:2400:2400
Spambase	4597	24	2	1532:1532:1532

Table 2. Average accuracy by the ensemble pruning methods and by complete bagging on test datasets

Datasets	Bagging (%)	SEMD (proposed) (%)	MDEP (%)	UMEP (%)	COMEP (%)
Glass	0.6595 (0.0625)	**0.6824** (0.0276)	0.6567 (0.0684)	0.6614 (0.0598)	0.6678 (0.0527)
Zoo	0.8220 (0.0936)	**0.8716** (0.0709)	0.8347 (0.0796)	0.8512 (0.0710)	0.8486 (0.0762)
Air	0.8282 (0.0490)	**0.8453** (0.0504)	0.8218 (0.0486)	0.8310 (0.0476)	0.8247 (0.0471)
Hayesroth	0.7294 (0.0881)	0.7709 (0.0707)	0.7377 (0.0783)	0.7666 (0.0712)	**0.7879** (0.0614)
Appendicitis	0.8314 (0.0625)	**0.8377** (0.0580)	0.8303 (0.0641)	0.8337 (0.0643)	0.8354 (0.0638)
M-of-N	0.9157 (0.0277)	**0.9484** (0.0218)	0.9230 (0.0265)	0.9274 (0.0263)	0.9421 (0.0227)
Car	0.9373 (0.0148)	**0.9442** (0.0139)	0.9392 (0.0139)	0.9401 (0.0152)	0.9420 (0.0131)
Ecoli	0.7945 (0.0410)	**0.7994** (0.0379)	0.7950 (0.0420)	0.7963 (0.0406)	0.7956 (0.0379)
DNA test	0.9057 (0.0168)	**0.9122** (0.0188)	0.9040 (0.0167)	0.9074 (0.0182)	0.9065 (0.0180)
Tic-tac-toe	0.8655 (0.0227)	**0.8948** (0.0193)	0.8696 (0.0245)	0.8663 (0.0253)	0.8867 (0.0209)
Seeds	0.8840 (0.0471)	**0.8907** (0.0428)	0.8831 (0.0475)	0.8840 (0.0464)	0.8859 (0.0471)
Segment	0.9524 (0.0100)	**0.9557** (0.0093)	0.9519 (0.0102)	0.9537 (0.0100)	0.9553 (0.0096)
Tae	0.4814 (0.0689)	0.4912 (0.0766)	0.4778 (0.0599)	0.4904 (0.0716)	**0.4914** (0.0706)
Vowel	0.7197 (0.0410)	**0.7362** (0.0373)	0.7124 (0.0435)	0.7165 (0.0412)	0.7150 (0.0390)
Wdbc	0.9385 (0.0192)	0.9379 (0.0214)	**0.9386** (0.0175)	0.9371 (0.0209)	0.9366 (0.0196)
Wpbc	0.7239 (0.0523)	**0.7345** (0.0459)	0.7244 (0.0482)	0.7279 (0.0500)	0.7274 (0.0542)
Breast-w	**0.9594** (0.0125)	0.9559 (0.0127)	0.9582 (0.0125)	0.9554 (0.0144)	0.9545 (0.0127)
X8D5K	0.9917 (0.0071)	**0.9937** (0.0066)	0.9932 (0.0059)	0.9931 (0.0068)	0.9929 (0.0063)
Penbased	0.9732 (0.0039)	**0.9755** (0.0035)	0.9728 (0.0041)	0.9729 (0.0042)	0.9736 (0.0036)
Phoneme	0.8695 (0.0081)	**0.8709** (0.0083)	0.8687 (0.0079)	0.8685 (0.0079)	0.8697 (0.0091)
Ringnorm	0.9461 (0.0067)	**0.9506** (0.0055)	0.9457 (0.0067)	0.9456 (0.0065)	0.9453 (0.0067)
Spambase	0.9288 (0.0073)	**0.9310** (0.0062)	0.9287 (0.0078)	0.9282 (0.0074)	0.9290 (0.0071)

The experiments are run on 22 randomly selected datasets from UCI Machine Learning repository [20] and Keel repository [21], these datasets are selected for comparative analysis since they are usually widely used in the similar ensemble learning methods. Table 1 gives a summary of these datasets. In the experiments, a dataset is randomly divided into three equal parts by cross-validation, which are training dataset, validation dataset and test dataset. CART tree [18] is utilized as the base classifier, which comes from the Classregtree classifier implemented in MATLAB 2016a. The initial ensemble classifiers scale is set to 200 base classifiers; the results of 100 repeated experiments are averaged to obtain the final classification accuracy. In order to ensure the fairness of the experiments, the division ratio of each experimental dataset remains the same when different classification methods are performed on the same dataset, which ensures that the training dataset, validation dataset, and test dataset are the same for each classification method.

3.2 Comparative Analysis of the Proposed Method to the State-of-the-Art Ensemble Learning Methods

Four algorithms are selected for comparison with the proposed method EPMD in the experiments, all of which are ensemble learning classification algorithms. Among them, Bagging [16] is the most classic ensemble learning algorithm without ensemble pruning; both UMEP [6] and MDEP [22] are ensemble pruning algorithms that use margin theory for selective ensemble; COMEP [23] is a selective ensemble algorithm that uses the normalized variation of information and the normalized mutual information to sort and select ensemble classifiers. In the experiments, for convenience, α in MDEP is set to 0.2 according to the original paper, and λ in the algorithm proposed by this paper and COMEP are both set to 0.2. For different data sets, different λ values will have slightly different results.

Average classification accuracies of five ensemble learning methods on 22 test datasets are given in Table 2, the results with better performance of the proposed method are highlighted in bold. For most datasets, the method proposed by this paper can show better classification performance compared with Bagging and other three ensemble pruning methods. In addition, this experiment also calculated the size of the classifiers subset after ensemble pruning, and compared the four selective ensemble classification algorithms. The running speed of the ensemble learning method mainly depends on the complexity and number of the base classifiers in the ensemble system; for algorithms that uniformly use the same base classifiers, minimizing the scale of the ensemble system can reduce the running time and storage overhead. After selective ensemble, the average number of classifiers in the ensemble classifier subset obtained by the four ensemble pruning methods based on sorting is shown in Table 3. Our proposed method is slightly higher than the COMEP method in the classifier scale after selective ensemble, but the overall gap is not big, and the ensemble scale is significantly smaller than the other two classification algorithms UMEP and MDEP. It reveals that using the method of our proposed method can significantly reduce the number of classifiers in the ensemble system and then reduce the computational cost.

The time complexity of the proposed method can be simply expressed as: $\mathcal{O}(T \times m \times log(m) \times n) + \mathcal{O}(T \times N^{'}) + \mathcal{O}(T \times log(T)) + \mathcal{O}(T)$, where m is the number of samples in each sampling subset, and the other symbols here have the same meaning as before. Since the number of samples, $N^{'}$, and the number of features, n, are both fixed values, the final time complexity can be approximately expressed as: $\mathcal{O}(T \times log(T))$. It can be seen that for the same data set, the running time of the algorithm depends to a large extent on the number of base classifiers. As the number of base classifiers in the ensemble model continues to increase, the running time consumed will also become longer.

Table 3. Average number of base classifiers selected by the ensemble pruning methods on test datasets

Datasets	SEMD (proposed)	MDEP	UMEP	COMEP
Glass	20.600	30.390	24.790	**11.980**
Zoo	15.810	19.120	14.080	**8.920**
Air	20.230	38.770	42.650	**19.390**
Hayesroth	16.810	23.600	16.410	**9.040**
Appendicitis	**10.460**	18.540	13.670	18.930
M-of-N	18.650	37.800	45.400	**13.930**
Car	**14.960**	40.040	37.540	16.020
Ecoli	12.050	21.260	16.170	**10.980**
DNA test	**16.950**	40.250	37.470	19.090
Tic-tac-toe	18.200	42.090	50.220	**15.430**
Seeds	**7.170**	16.200	12.660	7.710
Segment	**14.430**	41.320	31.560	15.420
Tae	21.670	22.180	18.120	**10.950**
Vowel	**23.590**	55.230	51.010	27.650
Wdbc	**10.940**	23.280	15.680	11.410
Wpbc	12.260	20.030	17.270	**9.400**
Breast-w	**8.180**	26.040	16.620	9.020
X8D5K	**6.990**	13.140	13.480	7.380
Penbased	**28.510**	66.820	67.790	35.350
Phoneme	**24.060**	65.750	62.640	25.250
Ringnorm	**28.540**	61.190	58.960	42.080
Spambase	**23.120**	50.830	54.380	23.650

4 Conclusion

In this paper, we proposed a novel ensemble pruning algorithm or selective ensemble algorithm based on margin theory and ensemble diversity (EPMD).

This method presents a new unsupervised form of average samples margin measurement criterion, it considers the margin distance from the unknown data samples to the classification decision boundary, and then characterize the overall performance of the classifiers in the ensemble system according to the samples margin values. In addition, the J-S divergence of the classification results is calculated with respect to the probability distribution of the class labels, and used to evaluate the diversity of the base classifiers in the ensemble system. All base classifiers are sorted according to the proposed measurement criteria, and a simplified ensemble classifiers subset is generated by selecting the subset of classifiers that can maximize the overall accuracy. Our idea comes from the fact that the combination of a large number of base classifiers is not always a perfect ensemble, and relatively few excellent classifiers with diversity are sufficient to obtain the best generalization performance. Different from the ensemble pruning methods that simply pursue the maximization of accuracy, the proposed method also obtains the complementarity between the base classifiers to some extent.

To evaluate the performance of the proposed ensemble pruning algorithm, we compare it with the classic Bagging algorithm and three state-of-the-art ordering-based methods. The experiments are performed on the 22 benchmark datasets from UCI repository and KEEL repository. The results show that our proposed method has varying degrees of advantages on most datasets. The proposed method has achieved higher classification accuracy than the comparison methods on 18 benchmark datasets. In addition, the size of the pruned classifiers set is smaller than that of any other comparison methods on 13 benchmark datasets. Therefore, the proposed method can achieve relatively good generalization performance with a relatively small ensemble scale, thereby achieving the purpose of ensemble pruning. Since the experiments are only performed on a single-type base classifier, more different types of base classifiers will be further explored along with considering the impact of hyperparameters on the classification results.

References

1. Jan, Z., Verma, B.: Multicluster class-balanced ensemble. IEEE Trans. Neural Netw. Learn. Syst. **32**(3), 1014–1025 (2021)
2. Zhu, Z., Wang, Z., Li, D., Zhu, Y., Wenli, D.: Geometric structural ensemble learning for imbalanced problems. IEEE Trans. Cybern. **50**(4), 1617–1629 (2020)
3. Zhou, Z.-H., Jianxin, W., Tang, W.: Ensembling neural networks: many could be better than all. Artif. Intell. **137**(1–2), 239–263 (2002)
4. Ali, M.A., Üçüncü, D., Ataş, P.K., Akyüz, S.Ö.: Classification of motor imagery task by using novel ensemble pruning approach. IEEE Trans. Fuzzy Syst. **28**(1), 85–91 (2020)
5. Tsoumakas, G., Partalas, I., Vlahavas, I.: An ensemble pruning primer. In: Okun, O., Valentini, G. (eds.) Applications of Supervised and Unsupervised Ensemble Methods. SCI, vol. 245, pp. 1–13. Springer, Heidelberg (2009) . https://doi.org/10.1007/978-3-642-03999-7_1
6. Guo, L., Boukir, S.: Margin-based ordered aggregation for ensemble pruning. Pattern Recognit. Lett. **34**(6), 603–609 (2013)

7. Zhang, C.-X., Zhang, J.-S., Yin, Q.-Y.: A ranking-based strategy to prune variable selection ensembles. Knowl. Based Syst. **125**, 13–25 (2017)
8. Lazarevic, A., Obradovic, Z.: Effective pruning of neural network classifier ensembles. In: International Joint Conference on Neural Networks. Proceedings (Cat. No. 01CH37222), IJCNN 2001, vol. 2, pp. 796–801 (2001)
9. Onan, A., Korukoğlu, S., Bulut, H.: A hybrid ensemble pruning approach based on consensus clustering and multi-objective evolutionary algorithm for sentiment classification. Inform. Process. Manag. **53**(4), 814–833 (2017)
10. Ykhlef, H., Bouchaffra, D.: An efficient ensemble pruning approach based on simple coalitional games. Inf. Fus. **34**, 28–42 (2017)
11. Qian, C., Yu, Y., Zhou, Z.-H.: Pareto ensemble pruning. In: Proceedings of the Twenty-Ninth AAAI Conference on Artificial Intelligence, AAAI 2015, pp. 2935–2941. AAAI Press (2015)
12. Zhu, X., Ni, Z., Ni, L., Jin, F., Cheng, M., Li, J.: Improved discrete artificial fish swarm algorithm combined with margin distance minimization for ensemble pruning. Comput. Ind. Eng. **128**, 32–46 (2019)
13. Martínez-Muñoz, G., Hernández-Lobato, D., Suárez, A.: An analysis of ensemble pruning techniques based on ordered aggregation. IEEE Trans. Pattern Anal. Mach. Intell. **31**(2), 245–259 (2009)
14. Schapire, R.E., Freund, Y., Barlett, P., Lee, W.S.: Boosting the margin: a new explanation for the effectiveness of voting methods. In: Proceedings of the Fourteenth International Conference on Machine Learning, ICML 1997, San Francisco, CA, USA, pp. 322–330 (1997). Morgan Kaufmann Publishers Inc
15. Feng, W., Dauphin, G., Huang, W., Quan, Y., Liao, W.: New margin-based subsampling iterative technique in modified random forests for classification. Knowl. Based Syst. **182**, 104845 (2019)
16. Breiman, L.: Bagging predictors. Mach. Learn. **24**(2), 123–140 (1996)
17. Efron, B.: Bootstrap methods: another look at the Jackknife. Ann. Stat. **7**(1), 1–26 (1979)
18. Breiman, L., Friedman, J., Stone, C.J., Olshen, R.A.: Classification and Regression Trees. CRC Press, Boca Raton (1984)
19. Lin, J.: Divergence measures based on the Shannon entropy. IEEE Trans. Inf. Theory **37**(1), 145–151 (1991)
20. Bache, K., Lichman, M.: UCI machine learning repository. UCI Machine Learning Repository University of California, Irvine, School of Information and Computer Sciences, December 2013
21. Alcala-fdez, J., et al.: Keel data-mining software tool: data set repository, integration of algorithms and experimental analysis framework. J. Mult. Valued Log. Soft Comput. **17**(2–3), 255–287 (2011)
22. Guo, H., Liu, H., Li, R., Changan, W., Guo, Y., Mingliang, X.: Margin & diversity based ordering ensemble pruning. Neurocomputing **275**, 237–246 (2018)
23. Bian, Y., Wang, Y., Yao, Y., Chen, H.: Ensemble pruning based on objection maximization with a general distributed framework. IEEE Trans. Neural Netw. Learn. Syst. **31**(9), 3766–3774 (2020)

Construct Forensic Evidence Networks from Information Fragments

Zhouzhou Li(✉), Xiaoming Liu, Mario Alberto Garcia, and Charles D. McAllister

Southeast Missouri State University, Cape Girardeau, MO 63701, USA
{zli2,xliu,mgarcia,cdmcallister}@semo.edu

Abstract. The advancement of modern technologies led to many challenges for digital forensic investigators, who now need to know every fragmentary, trivial, and isolated digital information to stitch an overall, solid evidence picture for each case. While the reality is they lack a mature model to help them accomplish their high-demanded works (such as data collection, protection, retrieval, recovery, analytic, archive). In this paper, we introduce a new concept, "Evidence Networks", and propose to build solid evidence networks from information pieces by reusing the existing successful network models. Additionally, the jigsaw puzzle model is used to analyze the practical problems that an evidence network may encounter.

Keywords: Digital forensics · Network model · Path cost · Correlation coefficient

1 Introduction

The word forensics means "to bring to court" [1]. It is the application of scientific principles to provide evidence in criminal cases. Its major tasks include how to collect crime scene evidence, prove the causes of accidents, test crime scene evidence in labs, etc. It is a cross-disciplinary science with close connections to crime scene investigation, drug analysis, genetics, physics, organic chemistry, criminal procedures, and the criminal justice system [2].

Digital Forensics (DF) is the modern day version of forensic science and deals with the retrieval, recovery, investigation, and use of material found in digital devices. Using the data collected from electronic devices, digital forensic investigators can assist in recovering lost or stolen data, trace it back to the source, and help create a detailed investigative report that can remedy any crime [3].

Supported by the Institute of Cybersecurity and the Department of Computer Science, Southeast Missouri State University.

J. Xiong et al. (Eds.): MobiMedia 2021, LNICST 394, pp. 702–714, 2021.
https://doi.org/10.1007/978-3-030-89814-4_51

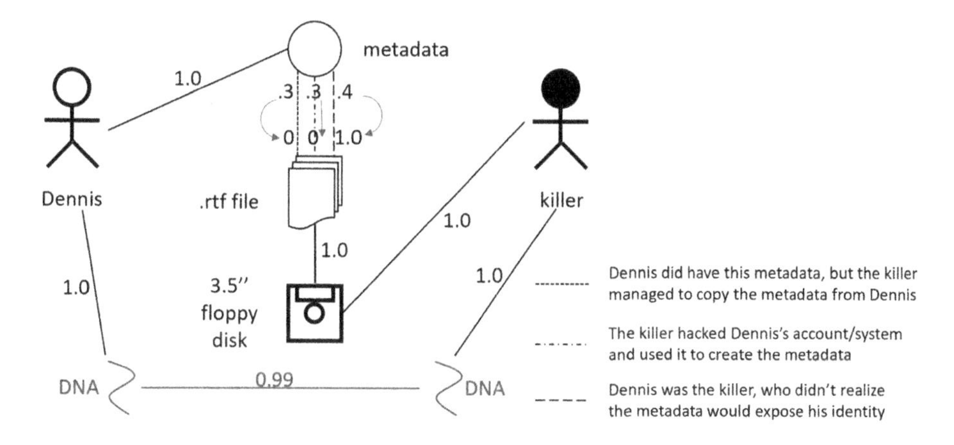

Fig. 1. Evidence network for the BTK case

In the famous BTK Killer case, the killer was responsible for the murder of ten people between 1974 and 1991, who was not identified for many years, until he sent a 1.44 MB floppy disk to the police (to mock the police's tracing capability). The digital forensic investigator for this case found a .rtf file on the disk and studied its metadata, which usually is not observable to non-technical users. The author's name and the associated organization left in the metadata provided solid evidence about the identity of the killer. With the guidance of this clue, the killer was eventually arrested and pleaded guilty. This is an good example of constructing solid evidence or clues from the fragmentary, trivial, and isolated digital data, where traditional forensics does not work.

Digital forensics has expanded from computer forensics to cover investigation of all devices capable of storing digital data [4]. With roots in the personal computing revolution of the late 1970s and early 1980s, the discipline evolved in a haphazard manner during the 1990s, and it was not until the early 21st century that national policies emerged.

With more and more technologies (including hardware and software) being introduced to the world, the forms of data storage and transmission changed substantially. The source data for investigation could be found in a PC, in a cell phone, cached in a small cell (base station or access point), or even distributed in multiple physical storage devices (servers) with each piece has no apparent relation to other pieces. Also, plenty data formats (software) for different types of information were invented. The emergence of Internet, Cloud Computing, and IoT, as well as their prevalence, intensifies the situation. For example, a $1.5 IoT device could be equipped with a multiple-core CPU, multiple wireless transceivers, a multiple-MB flash storage and a proprietary file system, which make it a valuable data source for forensic investigation. However, constructing a whole picture from all evidence pieces becomes more difficult.

Not only to deal with different hardware and software for data storage and transmission, a digital forensic investigator also needs to handle incomplete data.

A data user may not be an expert in techniques, who operates the data at higher layer, leaving the physical layer out of his/her attention. For example, a user keeps adding and deleting small files on his/her PC, which leaves many fragments on the hard disk drive. Some fragments may not be cleaned timely, so that they may still contain important and useful information for digital forensics, while without the attentions from the user(s). There is also a possibility that the criminals try to destroy, cover, or tamper the real data for a malicious purpose. A digital forensic investigator needs to grasp all kinds of tools/skills for data recovery.

In short, besides getting familiar with the modern hardware for data storage and transmission, the tools for data retrieval and analysis, constructing solid evidence from the fragmentary, covered, and tampered material is a big challenge to forensic investigators.

Our insight in the course of a digital forensic investigation is the forensics investigators can make full use of the outcomes of modern Computer Science to improve their work efficiency. If we treat all evidence pieces as network nodes, stitching them together is like finding a path from the source node to the destination node. The path cost will be decided by all the segments (hops). Therefore, the existing routing algorithm can be reused for digital forensics.

The remainder of the paper is organized as the follows. In the 'Review' section, we study the existing work for constructing evidence from isolated information. In the 'Proposal' section, we present the "Evidence Network" concept and re-use the existing network models (especially the router models) to describe, analog, and analyze the evidence networks. Also, in this section, we use the classical puzzle stitching problem to analog the problems we may encounter with the evidence network and provide our solution to those practical problems. To demo the counter-intuitive effects for false negative and positive cases, we put quantitative illustrations in the experiment section. In the 'Conclusion and Future Work' section, we conclude our work and indicate how to extend our work in future research.

2 Review

To our best knowledge, this work is the first to apply network models and graphic models for forensic evidence stitching.

In [5], the authors listed issues of DF. In [6], the authors explained the potential crisis behind today's golden age of digital forensic due to lacking a clear strategy for enabling research efforts that build upon one another. Especially, they pointed out that data management will be a big challenge to DF. "Even data that can be analyzed can wait weeks or months before review because of data management issues". They listed a few high watched areas for improvement: "the growing size of storage devices", "the increasing prevalence of embedded flash storage and the proliferation of hardware interfaces", "the proliferation of operating systems and file formats", using Cloud for remote processing and storage, and to split a single data structure into elements. All these can be summarized as a "isolated huge information islands" problem - each piece of information becomes trivial with comparison to the huge storage size, while it has unknown connections to the external.

To solve this problem, previous literature applied different approaches. However, they all have practical issue. Thus, this problem is still not addressed well.

With the success of Internet and Ciscos, network models become matured. They can connect isolated nodes to form a bigger network. The problem that network models solved has highly-abstract similarity with the DF problem we are concerned - just replace the keyword 'node' with 'evidence' in the previous sentence, we will find both have a similar structure.

The essence of a network model has its root in graph theory. Nodes and edges are the elements. In order to find a similar graphic model for our DF problem, we studied different models with focus on "Mosaic", "Panoramic", and "Jigsaw Puzzle" models.

In [7], the authors pointed out that "the format used to represent image data can be as critical in image processing as the algorithms applied to the data". That means, an algorithm cannot cover different formats. Data transforming is a general activity for image comparison and recognition. Then minor difference between the original and transformed images should be expected and not counted as an important factor.

In [8], the passive authentication techniques, operate in the absence of digital watermarks, signatures, or specialized hardware were discussed. This work revealed even there is no direct, obvious connection between two evidence pieces, there are indirect approaches to build/reconstruct their connections.

[9] and [10] introduced how to construct Mosaic. [11] introduced image alignment and image stitching algorithms. Both inspired our ideas for the DF problem.

[12] proposed to create full view panoramic mosaics from image sequences.[13] proposed to automatically create panoramic images based on local and overlap information. As a summary, they tried to utilize the temporal and spacial overlaps to build adjacent relations between pieces. This idea is hard to apply to evidence networks because evidence pieces may not have partial overlaps, but the "subset comparison" idea led us to the "Jigsaw Puzzle" model. [13] also explained their methods for eliminating visible shifts in brightness and removing moving objects which appeared in several image pieces, which gave some hints for us to solve the 'false positive' problem.

[14] revealed the connection between 'global' and 'local': apply global alignment (block adjustment) to the whole sequence of images while warps each image based on the results of pairwise local image registrations. It reminded us to think about the DF problem not just from the local comparison's perspective.

[15–21] proposed practical ideas and algorithms for file carving, which is a process of reassembling computer files from fragments in the absence of file system metadata. These papers used statistical methods or information theory (entropy) to categorize fragments into bins/bags to achieve high classification accuracy. Their methods can be re-used to permute evidences. However, a file fragment has a fixed size, which is not analog to a evidence because different evidences most likely are heterogeneous. Also, a file fragment only has one prior and one successor. While an evidence may have multiple connections to external. File carving theories and models cannot sufficiently describe high dimension evidence network.

3 Proposal

The forensic investigators can collect as many as possible evidence fragments from a criminal scene, the suspect's home, office, and other related places. However, they need a tool or model to piece the fragments into a whole picture of the crime that occurred.

Our proposal is to reuse the network models (especially, the router models) with necessary modifications to describe fragmentary evidences and the correlations between them. We aim to construct a network of evidences and figure out the high correlation path from one person/evidence to another person/evidence, which can guide further investigation.

As shown in Fig. 1, the BTK killer with unknown identity (black head) gave the police a floppy disk. The connection between the floppy disk and the killer has a 1.0 metric, which means a high correlation. The floppy disk contains a .rtf file. The connection between the .rtf file ad the floppy disk also has a 1.0 metric, which means another high correlation. It is worthy to note that (the content of) the .rtf file must be highly correlated to the case, otherwise we cannot give a 1.0 weight to this connection. To clarify this point, we can think about a Microsoft hidden file, which can be seen in almost every floppy disk. But, of course, Microsoft should not be put in the suspect list. The investigator found the name 'Dennis' (as well as his organization) in the metadata of the .rtf file, which causes the high correlation (metric = 1.0) between the metadata and 'Dennis'. However, there is an uncertainty about how the .rtf file can get such a metadata. We consider 3 possibilities here:

1. This 'Dennis' did have this metadata, but the killer managed to copy the metadata from this 'Dennis'
2. The killer hacked 'Dennis' account/system and used it to create the metadata.
3. 'Dennis' was the killer, who didn't realize the metadata would expose his identity.

To represent the uncertainty, we use 3 lines to connect the .rtf file and its metadata. Each line has an initial correlation/weight (0.3, 0.3, and 0.4), with the last one having somewhat higher weight than the others. These are rough estimates. Because at this moment, the investigators do not have detailed information about them, rough figures are fine. We can adjust them later to reflect any new update, just like when updating costs for network links.

Then from the killer to 'Dennis', we can find 3 (non-loop) paths with the following metrics:

- Path 1's metric = 1.0 * 1.0 * 0.3 * 1.0 = 0.3
- Path 2's metric = 1.0 * 1.0 * 0.3 * 1.0 = 0.3
- Path 3's metric = 1.0 * 1.0 * 0.4 * 1.0 = 0.4

The difference between traditional network models and this evidence network model is the way calculate the path 'cost'. Here we use the product of all connections' metrics, while traditional network models use the sum.

We can calculate a sum for all paths. In this case, it is 0.3 + 0.3 + 0.4 = 1, which means a strong correlation between the killer and this 'Dennis', no matter if they are the same person or not. The uncertainty is how they are correlated. Among all the paths, the most 'correlated' path has the metric '0.4', which is the best path so far. Therefore, the strategy would lead to investigate this path first.

In the BTK killer case, later, the investigator determined another path from the killer, who once left his DNA to a victim's nail, to this 'Dennis' - the investigator managed to get 'Dennis' DNA indirectly and found it matched the killer's.

Through this example, we can see how the constructed "Evidence Network" can provide guidance to the digital forensic investigations:

1. Even though there is uncertainty, a strong connection between the killer and 'Dennis' is identified (calculated). The metric for this purpose is the sum of all paths.
2. A best path (like a 'shortest' path in Internet) will be investigated first. To calculate the metric of a path, we use the product of every segment's correlation/weight. The idea is similar to a routing algorithm. The difference is the way we calculate the 'cost' of a path.
3. The correlation between two evidences/persons could be dynamic or initially rough. With some possibilities are proved true or false later, we can update the correlation metric. Just like the path cost fluctuating in Internet. In Fig. 1, (later) after we see the DNAs matching, we can update the correlations between the metadata and the .rtf file from [0.3, 0.3, 0.4] to [0, 0, 1.0].
4. The sum of all paths could be greater than 1.0. This implies that more than one solid evidence has been identified.

Therefore, a simple network topology can be used to describe the BTK killer scenario. We have the killer, the floppy disk, the .rtf file, the metadata, 'Dennis' in the topology. They are the nodes in the evidence network. Like in a real network, there are connections between the nodes, which represent the "correlations" between two nodes. By using the network model, we construct a path from the killer to 'Dennis'. And we use correlations to quantify the path and an overall correlation can be calculated by multiply all segments' correlations (we will discuss why the product rather than the sum is used to describe the overall path).

The BTK killer case is a simple example of 'Evidence Networks'. We need to consider more complicated examples to see if the idea still works.

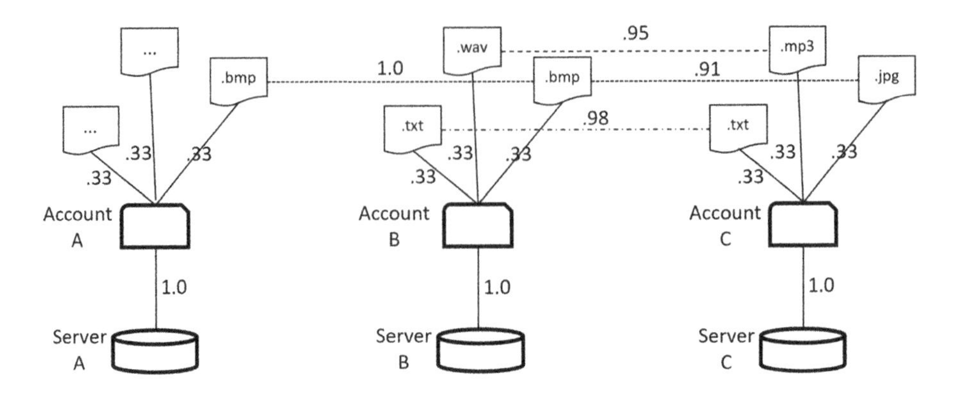

Fig. 2. Evidence network for associated accounts

Figure 2 shows an evidence network constructed to help decide whether two accounts are associated. Account B on Server B owns 3 files: a .bmp, a .txt, and a .wav. Account C on Server C also owns 3 files: a .jpg, a .txt, and a .mp3. The C's .mp3 file looks like a compressed version of B's .wav file because by applying a comparison algorithm we find they are similar. Also, the C's .jpg file looks like a compressed version of B's .bmp file because they are similar. The C's .txt file looks like a minor modified version of B's .txt file because they only have a few words different. Without considering Account A, we are pretty sure Account B and C are associated accounts. However, Account A has a .bmp file, which is the same as the .bmp file of B. Because it is a copy of the original .bmp file (not a compressed version), the two .bmp files have a 1.0 correlation, which is the highest correlation in the evidence network. Shall we grant A an associated account? What if we can only put two accounts in the associated account list? A+B or B+C, which is better? This is actually a global-local issue, or multi-dimensional issue.

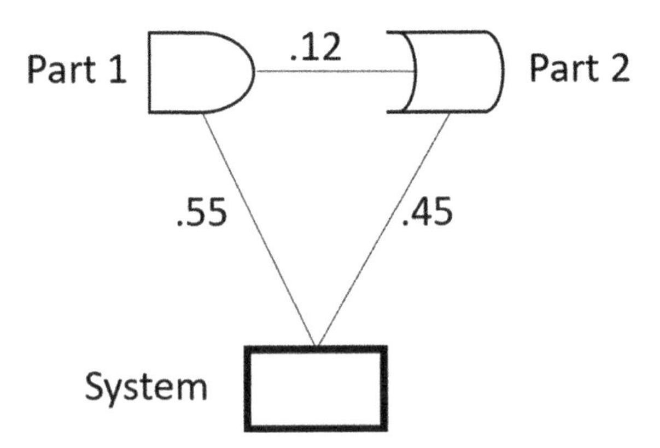

Fig. 3. A evidence network with complemented parts

Figure 3 gives another example of 'Evidence Networks'. Part 1 and Part 2 have a very weak correlation (metric = 0.12 if we apply an algorithm to calculate their correlation coefficients) because they are two different parts of a system. However, they should be combined together to form the system. Part 1 has a (metric = 0.55) connection to the system. Part 2 has (metric = 0.45) connection to the system. Using the system as the bridge, Part 1 can reach Part 2 with a 'cost' = 0.55 * 0.45 = 0.2475, which is higher than 0.12. The correlation between the two parts is higher than what we see directly. This is a kind of 'false negative' issue.

There may be other potential issues. We cannot enumerate all cases to derive the potential issues when we apply the 'Evidence Network' model for digital forensic investigation. Therefore, we need an intuitive and powerful tool to identify and analyze all the issues.

Fig. 4. A equal-size-piece puzzle made from [22]

Forming an Evidence Network is analog to stitching all puzzle pieces to from a complete picture. Both rely on local adjacent relations to construct a global graph. Both are expected to have the same issues. Therefore, we use a simple puzzle (Fig. 4) to intuitively explain the possible issues an Evidence Network may encounter.

The basic method to rebuild the original image is to find adjacent pieces for each piece. Comparing the edges of every pair of pieces can help identify the neighbors of a piece.

In Fig. 4, because each piece in the puzzle has the same shape, we can only exploit their color features to conclude their neighbor pieces. Only the four edges of a piece will be compared to the four edges of other pieces to find similar edges, which are candidate adjacent edges. The central area of each piece is ignored for this purpose. In computer graphics, edges are represented by tensors. We can calculate their Pearson Correlation Coefficients to evaluate their similarity. Pearson Correlation Coefficients are values between [−1.0, 1.0]. The bigger the absolute value is, the stronger is their similarity. In Fig. 4, two edges indicated by cutting line 2, 3, 4, or 7 have high correlation coefficients (more than 0.89), thus they are similar. Even we shuffle all the pieces, it is easy to determine their adjacent relations in a row through calculation.

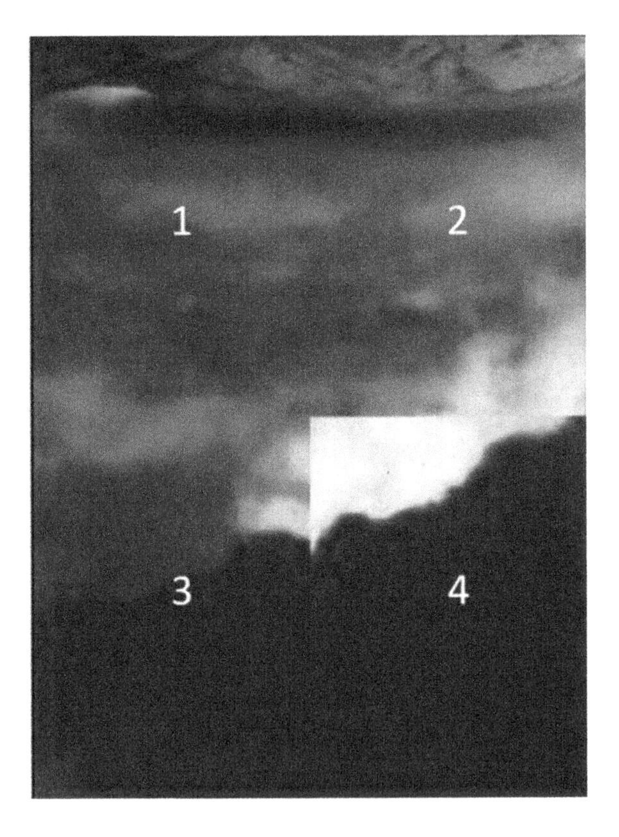

Fig. 5. False positive case

However, it is more difficult to determine their adjacent relations in a column. For example, the two edges indicated by cutting line 5 or 6 do not have high

Pearson correlation coefficients. This is because the cutting lines happen to be located around the natural color separation curves. These will be the false negative cases if we only rely on the Pearson correlation. Fortunately, each piece has four edges, i.e., four dimensions. Even when we cannot find a qualified adjacent piece in one dimension, we can try other dimensions. This 'false negative case' issue is equivalent to the 'Complemented Part' issue of the Evidence Networks - the cutting line is located at a natural separation, which makes two pieces/parts not similar at all. The solution is also given by the puzzle example: try other dimensions/paths to globally evaluate the adjacent relation of two pieces/parts.

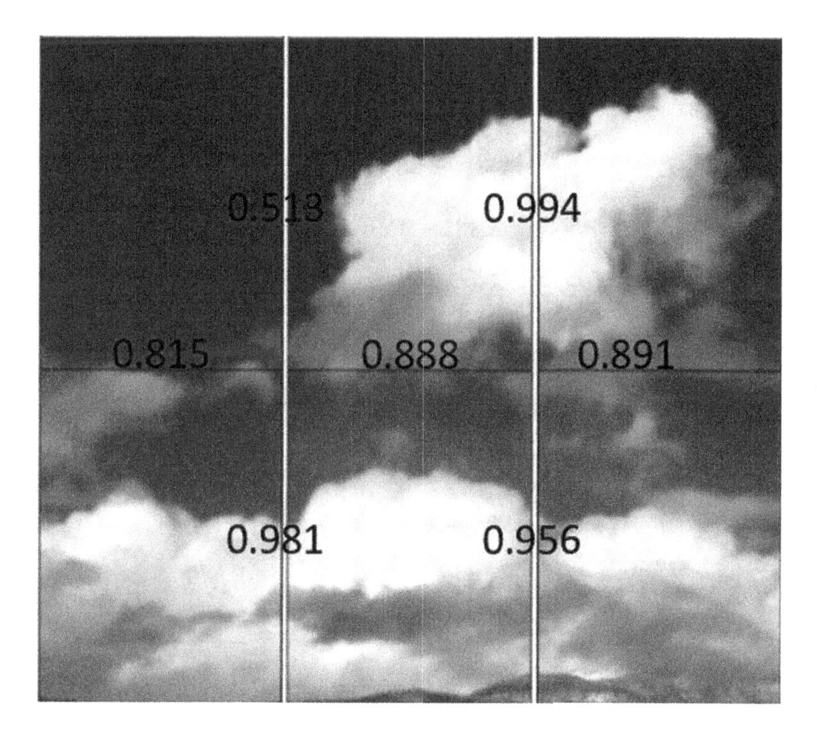

Fig. 6. False negative case with counterintuitive quantitative information

Figure 5 gives a 'False Positive' example of the puzzle stitching algorithm. It is obvious that piece 4 should not be there. But, if we compare the edges between piece 3 and 4, we will find they have a high Pearson correlation coefficient. That's the reason why the algorithm will put piece 4 there. Again, this reflects the calculation result from only one dimension. If we consider other edges and use a vote mechanism, piece 4 should be excluded from the candidates. This issue is equivalent to the 'Associated Accounts' issue in the Evidence Networks. Both have one and only one perfectly matching dimension. The solution is: if more than one candidates are found, global and multi-dimensional evaluation should be conducted.

4 Experiments

We calculated the correlation coefficients for each adjacent line in Fig. 4 and Fig. 5. Both Fig. 6 and Fig. 7 provide quantitative illustration of the false negative and positive cases, which are counter-intuitive though.

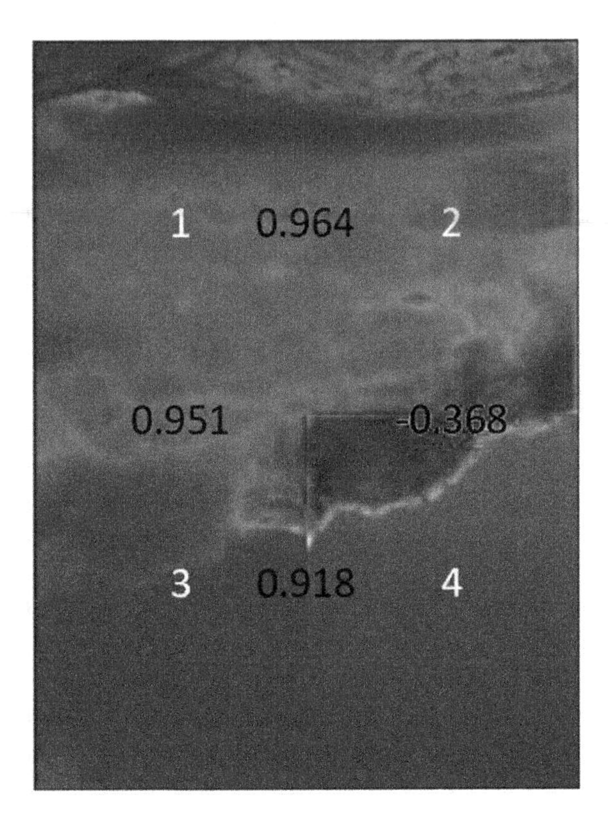

Fig. 7. False positive case with counterintuitive quantitative information

5 Conclusion and Future Work

In this paper, we reuse the well-developed network models and graphic models for constructing evidence networks for digital forensic investigation. Our evidence network model can reveal the hidden connection between two evidence fragment, the uncertainty of the connection, as well as the guidance to the promising investigation direction. Like a router model, the connection could dynamically vary. Our work also solved the 'false positive' and 'false negative' problems by giving an intuitive explanation from the similar 'Jigsaw Puzzle' model.

In the future, we plan to introduce noise to the models and apply filters to augment the models.

References

1. Hayes, D.R.: A Practical Guide to Computer Forensics Investigations. Pearson Education, London (2015)
2. https://study.com/articles/Criminology_vs_Criminalistics_Whats_the_Difference.html. Criminology Vs. Criminalistics: What's the Difference? Accessed 8 Jan 2021
3. https://cytelligence.com/resource/what-is-digital-forensics. What Is Digital Forensics? Accessed 8 Jan 2021
4. Reith, M., Carr, C., Gunsch, G.: An examination of digital forensic models. Int. J. Digit. Evid. **1**(3), 1–12 (2002)
5. Delp, E., Memon, N., Min, W.: Digital forensics. IEEE Signal Process. Mag. **26**(2), 14–15 (2009)
6. Garfinkel, S.L.: Digital forensics research: the next 10 years. Digit. Investig. **7**, S64–S73 (2010)
7. Adelson, E.H., Anderson, C.H., Bergen, J.R., Burt, P.J., Ogden, J.M.: Pyramid methods in image processing. RCA Eng. **29**(6), 33–41 (1984)
8. Farid, H.: Image forensics. Annu. Rev. Vis. Sci. (2019)
9. Hansen, M., Anandan, P., Dana, K., Van der Wal, G., Burt, P.: Real-time scene stabilization and mosaic construction. In: Proceedings of 1994 IEEE Workshop on Applications of Computer Vision, pp. 54–62. IEEE (1994)
10. Burt, P.J., Adelson, E.H.: A multiresolution spline with application to image mosaics. ACM Trans. Graph. (TOG) **2**(4), 217–236 (1983)
11. Szeliski, R.: Image alignment and stitching: a tutorial. Found. Trends® Comput. Graph. Vis. **2**(1), 1–104 (2006)
12. Szeliski, R., Shum, H.-Y.: Creating full view panoramic image mosaics and environment maps. In: Proceedings of the 24th Annual Conference on Computer Graphics and Interactive Techniques, pp. 251–258 (1997)
13. Uyttendaele, M., Eden, A., Skeliski, R.: Eliminating ghosting and exposure artifacts in image mosaics. In: Proceedings of the 2001 IEEE Computer Society Conference on Computer Vision and Pattern Recognition, CVPR 2001, vol. 2, pp. II–II. IEEE (2001)
14. Shum, H.-Y., Szeliski, R.: Systems and experiment paper: construction of panoramic image mosaics with global and local alignment. Int. J. Comput. Vis. **36**(2), 101–130 (2000)
15. Richard III, G.G., Roussev, V.: Scalpel: a frugal, high performance file carver. In: DFRWS (2005)
16. Memon, N., Pal, A.: Automated reassembly of file fragmented images using greedy algorithms. IEEE Trans. Image Process. **15**(2), 385–393 (2006)
17. Garfinkel, S.L.: Carving contiguous and fragmented files with fast object validation. Digit. Investig. **4**, 2–12 (2007)
18. Hand, S., Lin, Z., Guofei, G., Thuraisingham, B.: Bin-carver: automatic recovery of binary executable files. Digit. Investig. **9**, S108–S117 (2012)
19. Casey, E., Zoun, R.: Design tradeoffs for developing fragmented video carving tools. Digit. Investig. **11**, S30–S39 (2014)
20. Qiu, W., Zhu, R., Guo, J., Tang, X., Liu, B., Huang, Z.: A new approach to multimedia files carving. In: 2014 IEEE International Conference on Bioinformatics and Bioengineering, pp. 105–110. IEEE (2014)

21. van der Meer, V., et al.: File Fragmentation in the wild: a privacy-friendly approach. In: 2019 IEEE International Workshop on Information Forensics and Security (WIFS), pp. 1–6. IEEE (2019)
22. https://www.tmonews.com/2014/11/t-mobile-wideband-lte-touches-down-in-boise/

Matching Ontologies Through Siamese Neural Network

Xingsi Xue[1,2(✉)], Chao Jiang[1,2], and Hai Zhu[3]

[1] Intelligent Information Processing Research Center, Fujian University of Technology, Fuzhou 350118, Fujian, China
[2] School of Computer Science and Mathematics, Fujian University of Technology, Fuzhou 350118, Fujian, China
[3] School of Network Engineering, Zhoukou Normal University, Zhoukou 466001, Henan, China

Abstract. Ontology, the kernel technique of Semantic Web (SW), formally names the domain concepts and their relationships. However, as the ontologies are created and developed by different domain experts and communities, a concept may be named in various ways, bringing about the concept heterogeneity problem. To solve the problem, in this paper, a Siamese Neural Network (SNN)-based Ontology Matching Technique (OMT) is proposed, which is able to improve the matching efficiency by using a part of Reference Alignment (RA) to decrease the training time and improve the quality of matching results by using a logic reasoning approach to remove the conflict correspondences. The experimental results demonstrate that SNN-based OMT can determine high-quality alignment which outperforms the state-of-the-art OMTs.

Keywords: Ontology matching · Siamese neural networks · OAEI

1 Introduction

Over the past decades, Semantic Web (SW) technologies have been widely utilized, which provides great convenience for people to handle and link a variety of data [1,3,18,20]. Ontology, the kernel technique of SW, formally names the domain concepts and their relationships. However, as the ontologies are created and developed by different domain experts and communities, a concept may be named in various ways, bringing about the problem of concept heterogeneity. To address this heterogeneity problem, it is vital to determine correspondences between heterogeneous concepts, called the Ontology Matching (OM) [23].

Due to the complication of OM, it is arduous and impracticable to manually establish correspondences. Hence, diverse (semi)automatic OM Techniques (OMTs) have been proposed [8,12,19,24,26]. Machine learning (ML) is widely used in various fields [4–7,14–16]. In particular, ML-based OMTs are deemed as promising methods. Doan et al. [9] first presented an ML-based OMT that

J. Xiong et al. (Eds.): MobiMedia 2021, LNICST 394, pp. 715–724, 2021.
https://doi.org/10.1007/978-3-030-89814-4_52

similarity measures were described by a joint probability distribution of concepts concerned. Mao et al. [17] regarded the OM as a binary classification problem and used the Support Vector Machine (SVM) to solve it. khoudja et al. [13] integrated the state-of-the-art ontology matchers through the neural network to improve the results' quality. Bento et al. [2] used convolutional neural networks (CNN) to carry out the OM which shows good performance. Jiang et al. [11] proposed a Long Short-Term Memory Networks (LSTM)-based OMT to matching biomedical ontologies by using the semantic and structural information of concepts. However, the three neural network-based approaches need whole Reference Alignment (RA), which is unrealistic to obtain in the real matching scene, and three models' training is time-consuming, which could reduce the matching efficiency. In this paper, a Siamese Neural Network (SNN)-based OMT is proposed to further enhance the quality of alignments, which can predict the similarity value of two concepts by capturing the semantic feature. In particular, SNN-based OMT just utilizes a small part of RA, which is able to determine excellent alignments, and decrease the training time, and use a logic reasoning approach to remove the conflict correspondences to improve the quality of matching result.

2 Preliminary

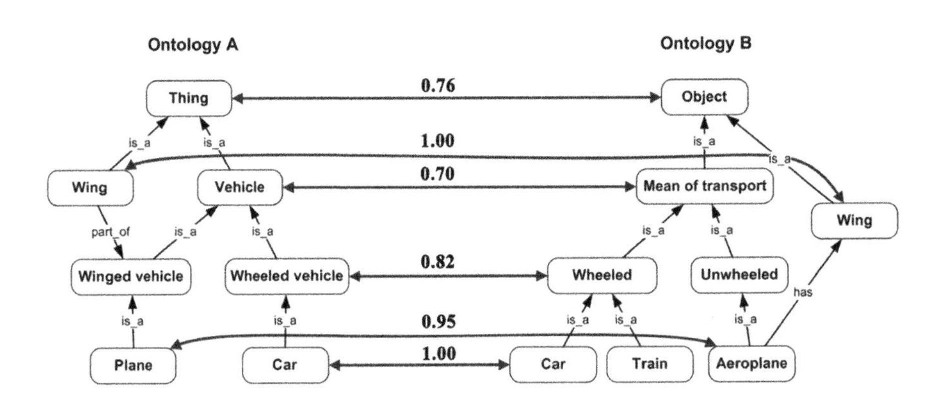

Fig. 1. Two ontologies and their alignment.

2.1 Ontology and Ontology Matching

Definition 1. An ontology is defined as a quadruple [21]

$$O = (IN, OP, DP, CL) \tag{1}$$

where IN is a series of individuals; OP is a series of object properties; DP is a series of data properties; CL is a series of classes. In addition, IN, OP, DP, and CL are the concepts. Figure 1 shows an example of two ontologies and their alignment. Two ontologies A and B contain a number of concepts in the left and right, respectively. In addition, the elements of rounded rectangles are classes, e.g. "car", while one-way arrows are object properties, e.g. "is_a".

Definition 2. An ontology alignment is a number of correspondences, and a correspondence is a quadruple [25]

$$corres = (c', c, s, t) \tag{2}$$

where c' and c are the concept from two to-be-matched ontologies; s is the similarity in [0, 1]; and t is the type of relation between c' and c. In Fig. 1, the double-sided arrows connect two concepts constituting the correspondences. For instance, "Plane" in ontology A and "Aeroplane" in ontology B are connected building a correspondence with 0.95 similarity. Besides, all the correspondences form an alignment, and RA is a golden alignment provided by the domain experts, which is used to test the performance of OMTs.

Definition 3. The process of ontology matching is a function [22]

$$A = \phi(O_1, O_2, RA, R, P) \tag{3}$$

where A is the final alignment; O_1 and O_2 are two to-be-aligned ontologies; RA is the reference alignment; R is the used resources; P is the used parameters. In the process of ontology matching, it is vital to determine the correspondences among heterogeneous concepts.

2.2 Performance Metrics

Generally, three metrics, i.e. precision, recall, and f-measure, are utilized to test the matching results' quality [10], which are expressed as follows:

$$P = \frac{correct\_matched\_correspondences}{all\_matched\_correspondences} \tag{4}$$

$$R = \frac{correct\_matched\_correspondences}{all\_correct\_correspondences} \tag{5}$$

$$F = 2 \times \frac{P \times R}{P + R} \tag{6}$$

where P and R respectively represent the accuracy and completeness of the results. P equals 1 indicating all matched correspondences are correct, while R equals 1 meaning that all correct correspondences are matched; F is the harmonic mean of P and R to balance them.

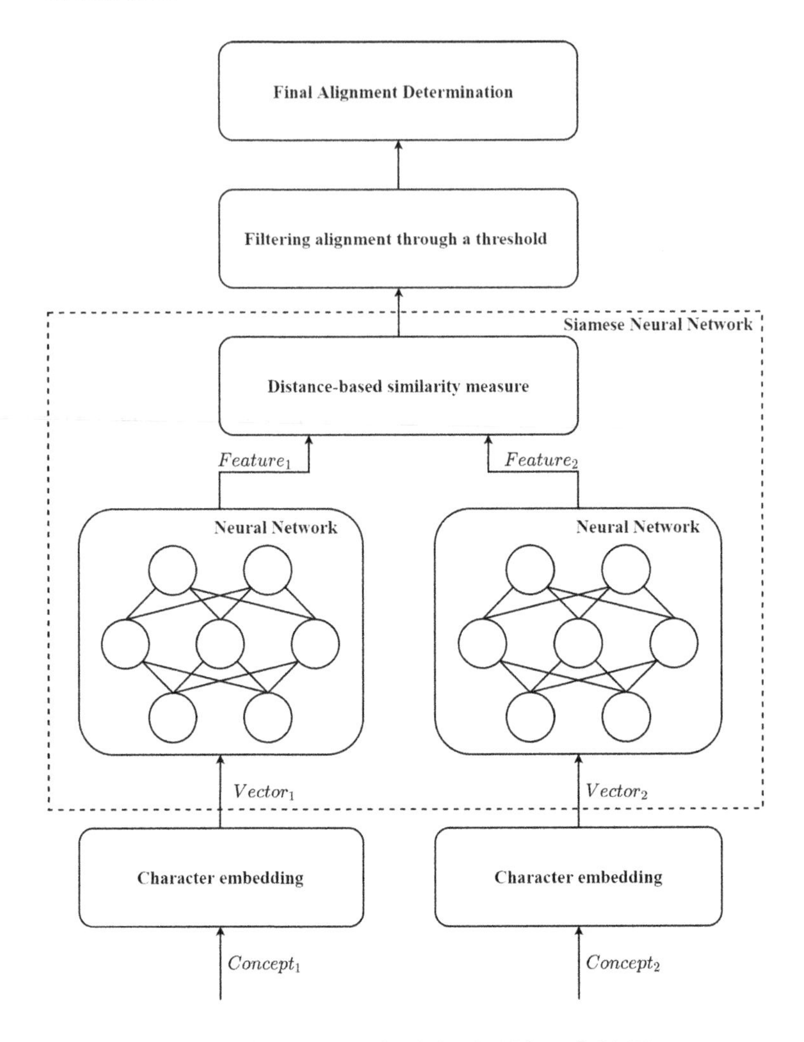

Fig. 2. The framework of the SNN-based OMT.

3 Methodology

In this study, an SNN-based OMT is proposed to solve the concept heterogeneity problem and the framework of SNN-based OMT is depicted in Fig. 2.

3.1 Data Set

The well-known Ontology Alignment Evaluation Initiative (OAEI) benchmark[1] is utilized to train the model and test the performance of our proposal. The succinct statement of benchmark testing cases is demonstrated in Table 1. To

[1] http://oaei.ontologymatching.org/2016/benchmarks/index.html.

enhance the efficiency of training process, a small part of RA is used that only a 2xx testing case is selected to train the model. In particular, given two to-be-matched ontologies O_1 and O_2, the correspondence $corres = (c', c)$ in RA is the positive sample, then the same number of negative samples are constituted by replacing c' with a concept c'' randomly selected from O_1. In addition, to ensure the quality of the final result, the trained testing case is not utilized to test.

Table 1. The succinct statement on benchmark testing cases.

ID	Succinct statement
1XX	Two same ontologies
2XX	Two ontologies with different lexical, linguistic or structural characters
3XX	The real ontologies

3.2 Siamese Neural Network

As seen in Fig. 2, the concepts' information is transformed as numeric vectors to feed the neural networks by using the character embeddings[2], whose possible character is a representation vector in 300 dimensions, and the value in each dimension is normalized in the interval [0, 1].

After, the trained SNN is utilized whose two networks are the same structure and weights. In this paper, the structure of two networks is Bi-directional Long Short Term Memory (Bi-LSTM), which is able to capture the semantic relationships and features of concept pairs. Then the distance-based similarity value of two concepts can be calculated through $Feature_1$ and $Feature_2$, which is defined as follows:

$$distance = \frac{||Feature_1 - Feature_2||}{||Feature_1|| + ||Feature_2||} \tag{7}$$

$$similarity = 1 - distance \tag{8}$$

where $||\cdot||$ represents the Euclidean norm, then the normalized distance is utilized to compute the similarity value, i.e. the smaller the distance, the greater the similarity value. In particular, the Adam optimizer and the contrastive loss are adopted and to optimize the two same networks:

$$L_{contrastive} = \sum_{i=0}^{N} \frac{y_i \times d_i + (1 - y_i) \times max\{margin - d_i, 0\}}{2N} \tag{9}$$

where N is the quantity of samples; $margin$ is a margin value; y_i is the sample type that y_i equals 1 representing the positive sample and 0 denoting the negative sample; d is the distance through the Eq. 7.

[2] https://github.com/minimaxir/char-embeddings.

3.3 The Alignment Determination

By means of the trained SNN, the $M \times N$ similarity matrix can be obtained that M and N are the concepts number of O_1 and O_2, respectively. Each element in the similarity matrix is the similarity value computed by using the Eqs. 7 and 8. In addition, the maximal value in each row or column will be chosen as the final correspondence. Furthermore, to ensure the precision of the final alignment, a threshold is adopted to filter the correspondences. After that, a logic reasoning approach is utilized that: (1) the correspondences are sorted by descending, (2) the correspondence with the greatest similarity is selected, (3) the rest correspondences are chosen one by one if it does not conflict with previous correspondence. To be specific, in Fig. 3, two correspondences (c_1', c_1) and (c_2', c_3) conflict that c_1' is the subclass of c_2' and c_3 is the subclass of c_1, then the correspondence with the lower similarity score will be discarded.

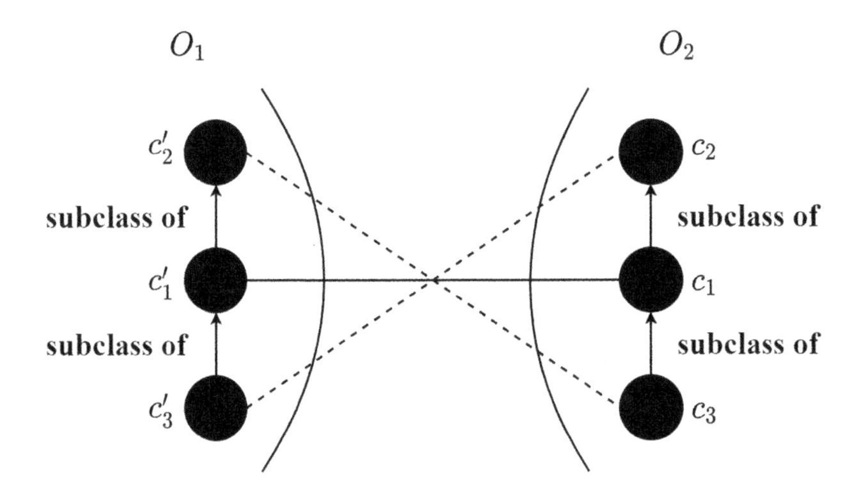

Fig. 3. An example of inconsistent correspondences.

4 Experiment

In the conducted experiment, the OAEI benchmark is used for evaluating the quality of alignments. Table 2 shows the comparison in terms of the average matching results' quality among SNN-based OMT and OAEI's OMTs, i.e. XMap, LogMap family, Pheno family, Lily, CroMatcher, and AML. As shown in Table 2, SNN-based OMT's recall, precision, f-measure, and f-measure per second are better than other competitors, which shows that SNN-based OMT outperforms the state-of-the-art OMTs.

In addition, Tables 3, 4, and 5 demonstrate the comparison in terms of R, P, and F, respectively. The bold text indicates the best OMT on the corresponding testing case and the symbols "+", "=", "-" represent the numbers that our proposal is superior, equal, and inferior to other OMTs, respectively. The compared

Table 2. Comparison among SNN-based OMT and OAEI's OMTs in terms of the average matching results' quality.

OMT	Recall	Precision	f-measure	Runtime (second)	f-measure per second
XMap	0.40	0.95	0.56	123	0.0045
PhenoMP	0.01	0.02	0.01	1833	0.0000
PhenoMM	0.01	0.03	0.01	1743	0.0000
PhenoMF	0.01	0.03	0.01	1632	0.0000
LogMapBio	0.24	0.48	0.32	54439	0.0000
LogMapLt	0.50	0.43	0.46	96	0.0048
LogMap	0.39	0.93	0.55	194	0.0028
Lily	0.83	0.97	0.89	2211	0.0004
CroMatcher	0.83	0.96	0.89	1100	0.0008
AML	0.24	1.00	0.38	120	0.0031
SNN-based OMT	0.88	0.97	0.92	108	0.0085

Table 3. Comparison among SNN-based OMT and OAEI's OMTs in terms of recall.

OMT	1XX	2XX	3XX
edna	**1.00**	0.56	0.82
aflood	**1.00**	0.74	0.81
AgrMaker	0.98	0.60	0.79
aroma	**1.00**	0.69	0.78
ASMOV	**1.00**	**0.85**	0.82
DSSim	**1.00**	0.62	0.67
GeRoMe	**1.00**	0.71	0.60
kosimap	0.99	0.57	0.50
Lily	**1.00**	0.86	0.81
MapPSO	**1.00**	0.73	0.29
RiMOM	**1.00**	0.81	0.82
SOBOM	0.97	0.46	0.55
TaxoMap	0.34	0.23	0.31
SNN-based OMT	**1.00**	0.81	**0.83**
+/=/-	4/9/0	12/0/1	13/0/0

OMT are edna, aflood, AgrMaker, aroma, ASMOV, DSSim, GeRoMe, kosimap, Lily, MapPSO, RiMOM, SOBOM, and TaxoMap. About testing cases 1XX, most of OMTs are able to obtain excellent alignments as the low heterogeneity of 1xx, that information of all concepts of ontologies have not been discarded and changed. Regarding 2XX, since the to-be-matched ontologies have a variety

Table 4. Comparison among SNN-based OMT and OAEI's OMTs in terms of precision.

OMT	1XX	2XX	3XX
edna	0.96	0.41	0.47
aflood	**1.00**	**0.98**	0.9
AgrMaker	0.98	0.98	0.92
aroma	**1.00**	**0.98**	0.85
ASMOV	**1.00**	0.96	0.81
DSSim	**1.00**	0.97	0.94
GeRoMe	**1.00**	0.92	0.68
kosimap	0.99	0.94	0.72
Lily	**1.00**	0.97	0.84
MapPSO	**1.00**	0.75	0.54
RiMOM	**1.00**	0.93	0.81
SOBOM	0.98	0.97	0.92
TaxoMap	**1.00**	0.9	0.77
SNN-based OMT	**1.00**	**0.98**	**0.94**
+/=/-	4/9/0	11/2/0	13/0/0

Table 5. Comparison among SNN-based OMT and OAEI's OMTs in terms of f-measure.

OMT	1XX	2XX	3XX
edna	0.98	0.47	0.59
aflood	**1.00**	0.84	0.85
AgrMaker	0.98	0.74	0.85
aroma	**1.00**	0.80	0.81
ASMOV	**1.00**	0.90	0.81
DSSim	**1.00**	0.75	0.78
GeRoMe	**1.00**	0.80	0.63
kosimap	0.99	0.7	0.59
Lily	**1.00**	**0.91**	0.82
MapPSO	**1.00**	0.74	0.37
RiMOM	**1.00**	0.86	0.81
SOBOM	0.97	0.62	0.68
TaxoMap	0.50	0.36	0.44
SNN-based OMT	**1.00**	0.88	**0.88**
+/=/-	5/8/0	11/0/2	13/0/0

of heterogeneity features, i.e. lexical, linguistic, or structure heterogeneity, it is arduous to determine the outstanding alignments. But SNN-based OMT is able to obtain a high quality of alignments. As the real-world testing cases 3XX, although they have more complex heterogeneous characteristics than 1XX and 2XX, our proposal is an effective approach to solve the real-world OM problem and performs much better than other OMTs. To sum up, SNN-based OMT is able to effectively and efficiently solve the OM problem.

5 Conclusion

To solve the concept heterogeneity problem, in this paper, an SNN-based OMT is proposed, which is able to improve the matching efficiency by using a small part of RA to decrease the training time and improve the quality of matching result by using a logic reasoning approach to remove the conflict correspondences. The experimental results demonstrate that SNN-based OMT can determine high-quality alignment which outperforms the state-of-the-art OMTs. In the real matching scene, it is unrealistic to gain the RA. In the future, we will focus on the improvement of the efficiency and effectiveness of SNN-based OMT without using the RA, and be interested in capturing the semantic information by using both character-level and word-level information of concepts.

Acknowledgement. This work is supported by the Natural Science Foundation of Fujian Province (No. 2020J01875) and the National Natural Science Foundation of China (Nos. 61801527 and 61103143).

References

1. Antoniou, G., Van Harmelen, F.: A Semantic Web Primer. MIT Press, Cambridge (2004)
2. Bento, A., Zouaq, A., Gagnon, M.: Ontology matching using convolutional neural networks. In: Proceedings of The 12th Language Resources and Evaluation Conference, pp. 5648–5653 (2020)
3. Berners-Lee, T., Hendler, J., Lassila, O.: The semantic web. Sci. Am. **284**(5), 34–43 (2001)
4. Chang, K.C., Chu, K.C., Wang, H.C., Lin, Y.C., Pan, J.S.: Energy saving technology of 5g base station based on internet of things collaborative control. IEEE Access **8**, 32935–32946 (2020)
5. Chen, C.H.: A cell probe-based method for vehicle speed estimation. IEICE Trans. Fundam. Electron. Commun. Comput. Sci. **103**(1), 265–267 (2020)
6. Chen, C.H., Song, F., Hwang, F.J., Wu, L.: A probability density function generator based on neural networks. Physica A Stat. Mech. Appl. **541**, 123344 (2020)
7. Chu, S.C., Dao, T.K., Pan, J.S., et al.: Identifying correctness data scheme for aggregating data in cluster heads of wireless sensor network based on Naive Bayes classification. EURASIP J. Wirel. Commun. Netw. **2020**(1), 1–15 (2020)
8. Da Silva, J., Revoredo, K., Baião, F.A., Euzenat, J.: Alin: improving interactive ontology matching by interactively revising mapping suggestions. Knowl. Eng. Rev. **35**, e1 (2020)

9. Doan, A., Madhavan, J., Domingos, P., Halevy, A.: Ontology matching: a machine learning approach. In: Staab, S., Studer, R. (eds.) Handbook on Ontologies. International Handbooks on Information Systems, pp. 385–403. Springer, Heidelberg (2004). https://doi.org/10.1007/978-3-540-24750-0_19
10. Euzenat, J., Shvaiko, P.: Ontology Matching, vol. 18. Springer, Heidelberg (2007). https://doi.org/10.1007/978-3-540-49612-0
11. Jiang, C., Xue, X.: Matching biomedical ontologies with long short-term memory networks. In: 2020 IEEE International Conference on Bioinformatics and Biomedicine (BIBM), pp. 2484–2489. IEEE (2020)
12. Jiang, C., Xue, X.: A uniform compact genetic algorithm for matching bibliographic ontologies. Appl. Intell., 1–16 (2021)
13. Ali Khoudja, M., Fareh, M., Bouarfa, H.: A new supervised learning based ontology matching approach using neural networks. In: Rocha, Á., Serrhini, M. (eds.) EMENA-ISTL 2018. SIST, vol. 111, pp. 542–551. Springer, Cham (2019). https://doi.org/10.1007/978-3-030-03577-8_59
14. Lin, J.C.W., Shao, Y., Djenouri, Y., Yun, U.: ASRNN: a recurrent neural network with an attention model for sequence labeling. Knowl. Based Syst. **212**, 106548 (2021)
15. Lin, J.C.W., Shao, Y., Zhou, Y., Pirouz, M., Chen, H.C.: A Bi-LSTM mention hypergraph model with encoding schema for mention extraction. Eng. Appl. Artif. Intell. **85**, 175–181 (2019)
16. Liu, H., Wang, Y., Fan, N.: A hybrid deep grouping algorithm for large scale global optimization. IEEE Trans. Evol. Comput. **24**(6), 1112–1124 (2020)
17. Mao, M., Peng, Y., Spring, M.: Ontology mapping: as a binary classification problem. Concurr. Comput. Pract. Exp. **23**(9), 1010–1025 (2011)
18. McIlraith, S.A., Son, T.C., Zeng, H.: Semantic web services. IEEE Intell. Syst. **16**(2), 46–53 (2001)
19. Patel, A., Jain, S.: A partition based framework for large scale ontology matching. Recent Patents Eng. **14**(3), 488–501 (2020)
20. Rhayem, A., Mhiri, M.B.A., Gargouri, F.: Semantic web technologies for the internet of things: systematic literature review. Internet Things, 100206 (2020)
21. Xue, X., Wang, Y.: Optimizing ontology alignments through a memetic algorithm using both matchfmeasure and unanimous improvement ratio. Artif. Intell. **223**, 65–81 (2015)
22. Xue, X., Wang, Y.: Using memetic algorithm for instance coreference resolution. IEEE Trans. Knowl. Data Eng. **28**(2), 580–591 (2015)
23. Xue, X., Wu, X., Jiang, C., Mao, G., Zhu, H.: Integrating sensor ontologies with global and local alignment extractions. Wirel. Commun. Mob. Comput. **2021** (2021)
24. Xue, X., Yang, C., Jiang, C., Tsai, P.W., Mao, G., Zhu, H.: Optimizing ontology alignment through linkage learning on entity correspondences. Complexity **2021** (2021)
25. Xue, X., Yao, X.: Interactive ontology matching based on partial reference alignment. Appl. Soft Comput. **72**, 355–370 (2018)
26. Xue, X., Zhang, J.: Matching large-scale biomedical ontologies with central concept based partitioning algorithm and adaptive compact evolutionary algorithm. Appl. Soft Comput., 107343 (2021)

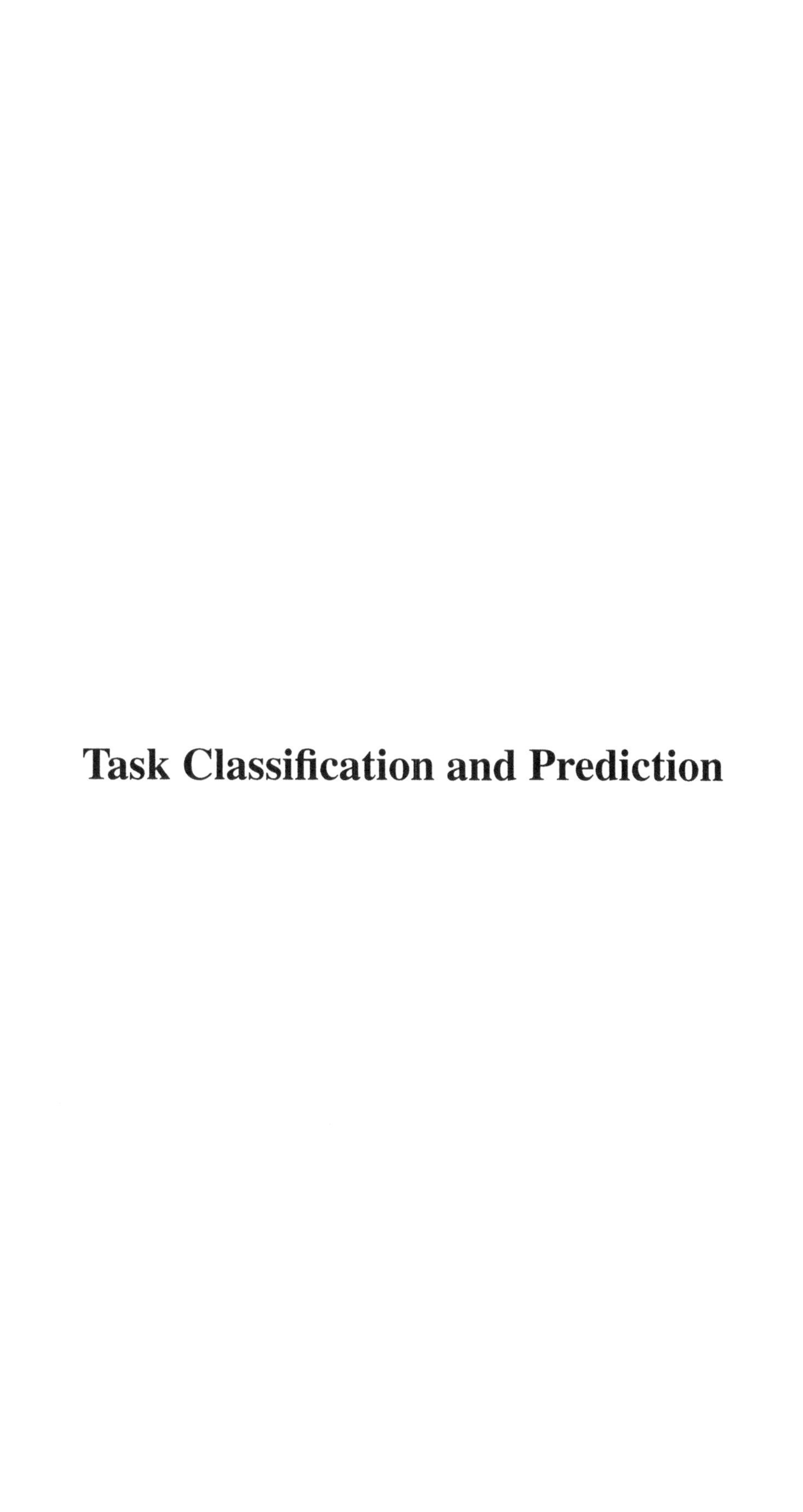

Task Classification and Prediction

Study on the Influence of Attention Mechanism in Large-Scale Sea Surface Temperature Prediction Based on Temporal Convolutional Network

Yuan Feng, Tianying Sun(✉), and Chen Li

College of Information Science and Engineering, Ocean University of China, Qingdao 266100, China
suntianying@stu.ouc.edu.cn

Abstract. The short term and small-scale sea surface temperature prediction using deep learning has achieved good results. The long-term sea surface temperature prediction technology in large-scale sea area is limited by the large and complex data. So how to use deep learning to select more valuable data and realize high precision of sea surface temperature prediction is an important problem. In this paper, attention mechanism and Temporal convolutional network (TCN) are used to predict the Indian Ocean 40°E–110°E and –25°S–25°N from 2015 to 2018 with 1° × 1° spatial resolution. The attention mechanism is used to distinguish the importance of the data, and the prediction models of full-feature (81 dimensions) and partial-feature (66 dimensions) are constructed. The experimental results show that the fitting degree of partial-feature models to sea surface temperature time series does not decrease significantly. The method proposed in this paper uses less data to ensure that the experimental accuracy does not decline significantly, and improves the long-term sea surface temperature prediction technology in large-scale sea area.

Keywords: Sea surface temperature · Attention mechanism · Temporal convolutional network

1 Introduction

Sea surface temperature is an important contributor to the health of regional marine ecosystems [1], and its changing trend may lead to the growth, reproduction and distribution of marine species. The rapid ocean warming trend will have a strong impact on marine fisheries [2]. The prediction of sea surface temperature has important guiding significance for large and medium-scale ocean physical phenomena. For example, the definition of the Indian Ocean Dipole (IOD) index is related to the abnormal changes of regional sea surface temperature. Large-scale annual sea surface temperature forecasts are helpful for climate monitoring, flood and drought risk warning and other aspects. The Indian Ocean will affect the surrounding areas, central South America, the southern

J. Xiong et al. (Eds.): MobiMedia 2021, LNICST 394, pp. 727–735, 2021.
https://doi.org/10.1007/978-3-030-89814-4_53

tip of Africa, southeastern Australia, Northeast Asia and other regions to have climate anomalies [3–6]. In this paper, the Indian Ocean is selected as the research object to make long-term forecast of sea surface temperature.

Dong et al. used the CFCC-LSTM neural network to predict the sea surface temperature 1 day, 7 days, and 30 days in advance on the Bohai Sea dataset (spatial resolution $0.05° \times 0.05°$). Experimental results show that the CFCC-LSTM model prediction error (RMSE) is 0.1466 °C, 0.2722 °C and 0.7260 °C [7]. Zhong et al. also conducted short-term predictions of sea surface temperature in a small area on the Bohai Sea dataset (spatial resolution $0.25° * 0.25°$), and used the LSTM layer to model the time series data of sea surface temperature. The experimental results show that LSTM The model predicts RMSE for 1, 7, and 30 days to be 0.0767 °C, 0.3844 °C, and 0.3928 °C [8]. Yang et al. improved the fully connected LSTM model and made 1, 7 and 30-day lead time predictions for the Bohai Sea and the East China Sea. The prediction results show that the longer the prediction lead time, the larger the prediction area and the lower the prediction accuracy [9]. Guan et al. used the entire China Sea and its adjacent sea areas as the study area (spatial resolution $0.05° \times 0.05°$), classified the sea area into 130 small areas through the SOM algorithm, and established LSTM models to predict sea surface temperature. The RMSE is 0.5 °C (one month in advance), 0.66 °C (12 months in advance). Building and training a model for each small area requires a lot of calculations, especially in large research areas with high spatial resolution [10]. The current research selects daily data with small time granularity and high spatial resolution to predict short-term sea surface temperature in a small area of sea, and only uses sea surface temperature for prediction. There are relatively few studies on long-period sea temperature forecasting in the ocean.

The study of large and medium scale physical ocean phenomena needs to deal with a large number of complicated ocean data. This paper uses the monthly ocean-atmosphere data of the past ten years and five months to predict the sea surface temperature of large-scale sea areas, 7 months in advance. The attention mechanism selects special features and compares the effects of partial-feature models and full-feature models. Experiments show that the data set used has low spatial resolution and large time granularity. The attention mechanism reduces the data set but the experimental accuracy does not decrease. The method this paper proposed is more suitable for the study of physical ocean phenomena in the ocean.

2 Data and Method

2.1 A Subsection Sample

This paper faces the interior of the Indian Ocean, which is the third largest ocean in the world (30°E–135°E, 30°N–66.5°S), located between Asia, Oceania, Africa and Antarctica. This paper uses the reanalysis data set provided by the National Center for Meteorological and Environmental Prediction (NCEP) with a spatial resolution of $1° \times 1°$, which includes atmospheric temperature, geopotential height, vertical velocity, water vapor, east-west wind speed, north-south wind speed, and undersea data such as temperature, Ocean current velocity in east-west and north-south directions. These monthly ocean-atmosphere data are available at https://psl.noaa.gov/data/gridded/.

In this paper, we select monthly data with longitude (40°E–110°E) and latitude (−25°S–25°N) from 1980 to 2018, and organize two sets of data sets to model and forecast separately. This paper selects the data for ten consecutive years and January to May in the eleventh year to predict the sea surface temperature from June to December in the eleventh year. The surface of the ocean is dynamically affected by waves, wind shear, and heat exchange, and the mixing of thermal expansion, ocean circulation and turbulence from the interior of the ocean will also produce dynamic effects [11]. Therefore, each month includes atmospheric, sea surface, and subsea parameter factors (81 in total). This paper selects atmospheric temperature, geopotential height, vertical speed, water vapor, east-west wind speed, north-south wind speed, at different heights (1000 850 500 300 hPa), a total of 24 atmospheric factors. The sea surface parameters include the sea height of the center point (SSH), the sea surface temperature of the center point (SST) and the sea surface temperature of 15 points around (17 factors). The subsea parameters include temperature at different sea depths (5, 15, 25, 35, 45, 55, 65, 75, 85, 95 m), currents in the east-west direction (U), currents in the north-south direction (V), and salinity (SSS) at the center point (40 factors).The features are numbered in order, and the model constructed by the above method of selecting features becomes a full feature model. The total number of features is 10125.

This paper uses the attention mechanism and the 2015 full-feature model trained by TCN to get the feature importance ranking. The attention mechanism can obtain the proportion of the impact of the full feature on the predicted value. The sum of all feature influence ratios is 1. In this paper, through experiments, the least important 15-dimensional features are discarded from the monthly data, in order, the vertical height is 400 m, the 35, 65 m underwater temperature, the salinity at the depth of 25, 55, and 95 m, currents in the east-west direction (U) at the depth of 15 and 65 m, currents in the north-south direction (V) at the depth of 5, 15, 25, 95 m and the sea surface temperature of the three points farthest from the center point. The total number of features after selection is 8250.

The training set is divided into two parts: feature and label. The model learns the relationship between features and label. This paper organizes the data in a sliding window. The test set is used to test the effect of the model. The training set required for the prediction model in 2015 (a total of 25 * 2533 pieces of data, also known as the number of samples, m) is the 2533 effective data points of the Indian Ocean from 1980 to 1989 and from January to May in 1990 as features, Sea surface temperature from June to December in 1990 is used as a marker; data from 1981–1990 and January to May in 1991 is used as a feature, and sea surface temperature from June to December in 1991 is used as a marker. The training set finally takes the data from 2004–2013 and January-June in 2014 and the corresponding sea surface temperature from July to December in 2014. The test set is the data from 2005–2014 and from January to June in 2015 and the corresponding sea surface temperature from July to December in 2015, a total of 2533 data.

2.2 Method

This paper uses the method of combining TCN architecture proposed in [12] and attention mechanism. TCN applies a variety of ideas such as residual connection, one-dimensional convolution, dilated convolution and causal convolution, so that the TCN structure has more advantages when dealing with long time series problems. In this paper, the training set is processed into a three-dimensional matrix of (m, timesteps, feature_nums) (full feature model feature_nums = 81, partial feature model feature_nums = 66), m is the number of samples, timesteps = 125 represents how many months we will use, feature_nums is the number of one month of data.

This paper sets the hyper parameters in the TCN model. The size of the convolution kernel is 8 and the number of the convolution kernels is 24, dilations = [1, 2, 4, 8, 16, 32, 64, 128, 256]. The size of the convolution kernel determines the value of each cell in the feature map is related to which areas of the input. When convolving the input matrix, the convolution kernel extracts the information of 8 time steps of the input matrix every time it slides. After experiments, the initial size of the convolution kernel is set to 8. In each residual block, the convolution kernel becomes larger according to the expansion parameter list, and the experimental result is the best. Dong always uses 7-day, 20-day, and 50-day historical sea temperature data as features to predict the future sea surface temperature of 1 day, 7 days, and 30 days. This paper uses the sea temperature data of the past ten years to make annual forecasts, which is more applicable to the research requirements of large and medium-sized ocean physical phenomena. Using dilated convolution can consider longer historical information, and is more suitable for predicting internal changes in the ocean than ordinary convolution. The number of convolution kernels determines the number of feature maps generated by the convolutional layer, and each feature map contains different information. There are too many feature maps, and some accidental changes in the sea temperature data set will be learned by the model. There are too few feature maps, and it is difficult to learn the relationship between features and sea surface temperature. After experiments, the model has the best effect when the number of convolution kernels is set to 24.

The attention mechanism can make the neural network have the ability to focus on a subset of its input (or features). Since we have a large number of input features, in the process of neural network learning, the attention mechanism can learn the relationship between these features and the output by itself, and increase the weight of some features that contribute more to the output to better find the relationship between features and output. We add an attention mechanism before putting the data into the TCN. That is, add a DENSE layer, and use softmax to assign a value between 0–1 to each feature, and then multiply the obtained weight matrix with the original input to obtain new 81-dimensional feature data, and then put it into TCN for learning. In the training process, the neural network will automatically learn the weight matrix, amplify the important features, and reduce the weight of the unimportant features to improve the ability of the model. This paper discusses the importance of features, adds attention mechanism to all-feature 81-dimensional data, and generates feature influence ratio histogram.

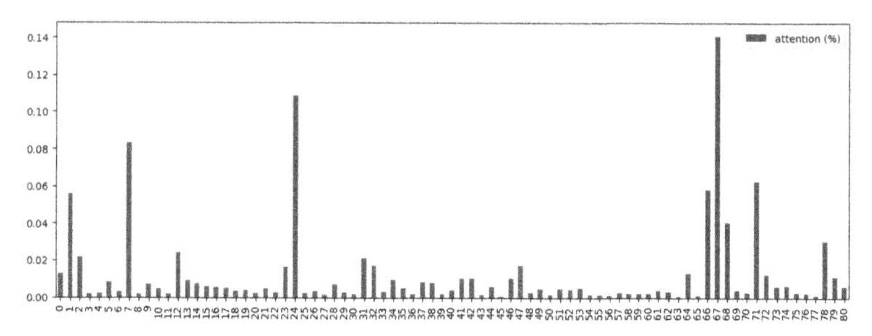

Fig. 1. The importance of whole features is marked by attention mechanism.

3 Result and Analysis

In this paper, a three-dimensional matrix (m, 125, feature_nums) is used to train the model, and 20% of the training set is randomly selected as the verification set, and the model parameters are adjusted to obtain the optimal model. This paper uses one evaluation mechanism-correlation, to compare the prediction accuracy of models trained with different features. The correlation degree represents the similarity between the predicted SSTs time series and the real SSTs time series. The higher the correlation degree, the higher the fit of the model to the true value and the higher the accuracy of the model. The long-term temperature changes in large-scale sea areas are affected by various factors, such as ocean currents, wind fields, and sea water velocity. The sea temperature changes are regular and periodic, and the ability of the model to learn the trend of sea temperature changes is measured by the similarity of the temperature time series.

Table 1 shows the average value of the correlations between all-feature and partial-feature models (8 in total) from 2015 to 2018. By studying the mean value of correlation in the sea area, the full-feature and part-feature models predict that the SSTs sequence fits the observed true value to a higher degree. It can be seen that the correlation degree of some feature models has decreased in 15 and 16 years, but the correlation degree has been improved compared with the full feature model in 17 and 18 years, which proves that the partial feature model proposed in this paper reduces the amount of data, the correlation degree does not decrease. In 2016, the reverse IOD phenomenon in the Indian Ocean showed that the average sea surface temperature in the west (50°E–70°E, 10°S–10°N) decreased, and the east (90°E–110°E, 10°S–0°) in the regional average sea surface temperature has risen. This abnormal sea surface temperature change is related to many factors such as ocean currents and wind fields. The wind field affects the seawater velocity, and the seawater flow affects the temperature field. The decline in correlation of some feature models in 2016 may be related to discarded features. The method of narrowing the data set through the attention mechanism is suitable for other normal years, but the fit for abnormally changed years is not enough.

Through the spatial distribution map of the correlation degree, the difference of the correlation degree in the area can be seen, so as to compare the effect of the full feature and the partial feature model. It can be seen from Fig. 1 that the correlation degree is higher overall from 2015 to 2016, and the correlation degree in 2017 and 2018 between

Table 1. Correlation comparison table of full feature and partial feature models.

Feature_nums	2015	2016	2017	2018
81dims	0.8806	0.9347	0.7717	0.8079
66dims	0.8660	0.8313	0.7856	0.8112

75°E–100°E, 7°S-–2°S, and there is a downward trend. From 2015 to 2016, the results of partial-feature models and full-feature models are close, and the error increase trend is not obvious. Partial- feature model performed better than the full-feature model in 2017, and the performance in 2018 was similar to the full-feature model. In general, the correlation between −10°S-−15°S and 87°E–100°E in the study area is lower than that of the surrounding sea area. This may be due to the complicated changes in sea temperature caused by the movement of regional ocean currents. It will make the model mistaken for noise, thus failing to learn the changes in sea temperature.

Figure 2 shows the comparison of the predicted SSTs time series and the real SSTs time series of two data points (−25°E, 63°S) and (−12°E, 73°S). The upper two figures are the prediction results of full-featured models, the two figures below are the prediction results of partial- feature model. The correlation degree is calculated from the 7-month real SSTs time series and the predicted SSTs time series, and the temperature fitting can be specifically observed through the curves of the real and predicted temperature values over the months. Observe the changes of the real temperature curve and the predicted curve at any data point, and you can measure the specific performance of the model on a single data point. It can be seen from the figure that the prediction results of partial-feature model and the full-feature model are close, and can roughly fit the trend of the true value of the data point. This paper uses the attention mechanism to select features. From the full-feature model to the partial feature, there is less data, but it does not affect the prediction result of the model. It can be seen from the figure that the error of the partial-feature model is basically close to the error of the full-feature model (Fig. 3).

Figure 2 Data points: July-December 2015 real sea surface temperature change curve and predicted sea surface temperature change curve

Through the comparison between the spatial distribution map of the correlation degree and the average correlation degree, the results show that this paper uses the attention mechanism to screen out more important features and reduce the total amount of data, but the correlation degree does not significantly decrease. Long-term SST information has far-reaching significance for some climate changes and the stability of ocean systems. The data required for this kind of research spans a long time and has complex features. Multi-source and multi-modal ocean-atmosphere data covers various factors that affect the temperature change of the ocean, but the data set is complex. It is important to distinguish the importance of features through attention mechanism. The method is less subjectively influenced by the researcher, and the prediction results rely on accurate calculation and objective analysis of the computer.

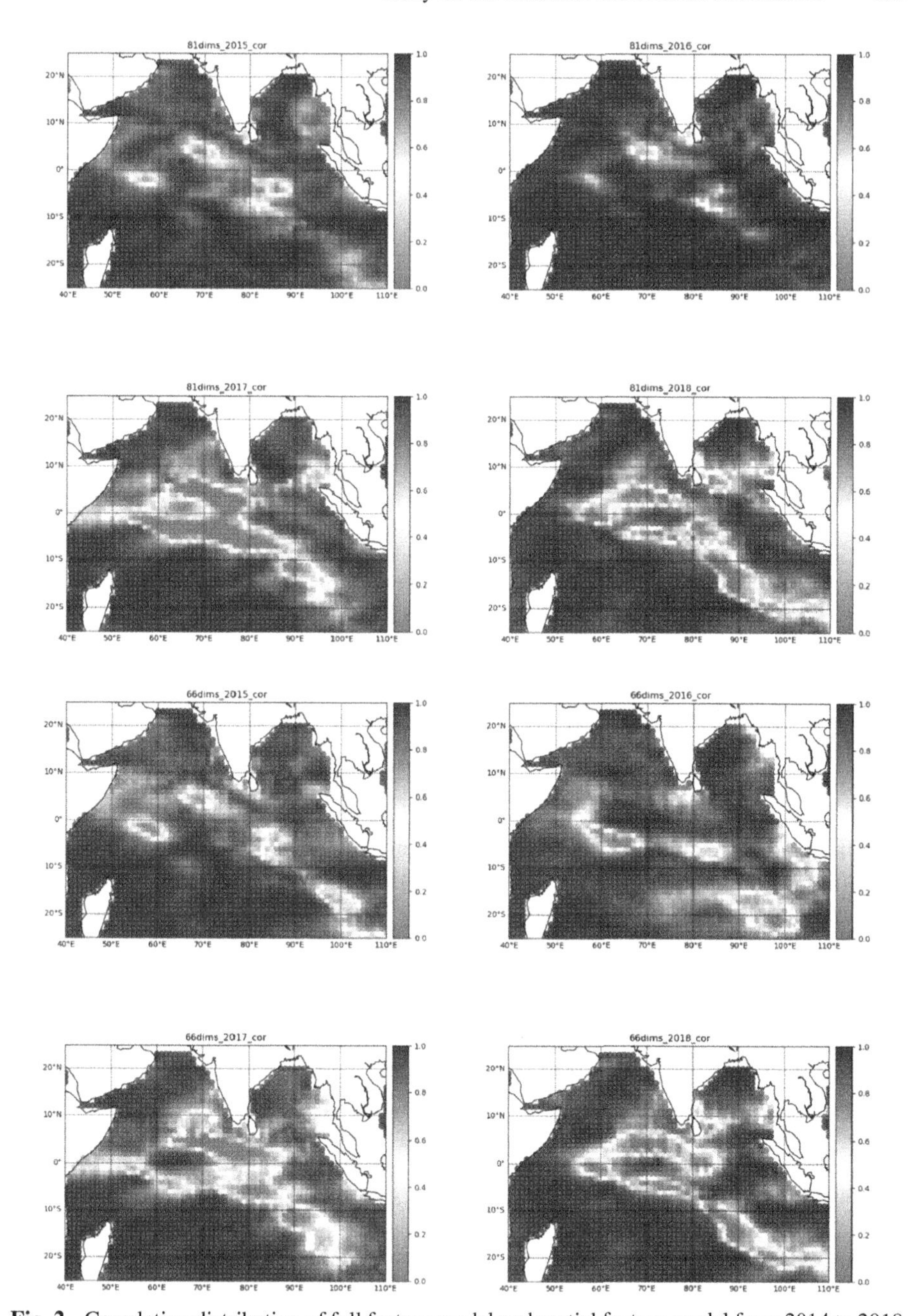

Fig. 2. Correlation distribution of full feature model and partial feature model from 2014 to 2018

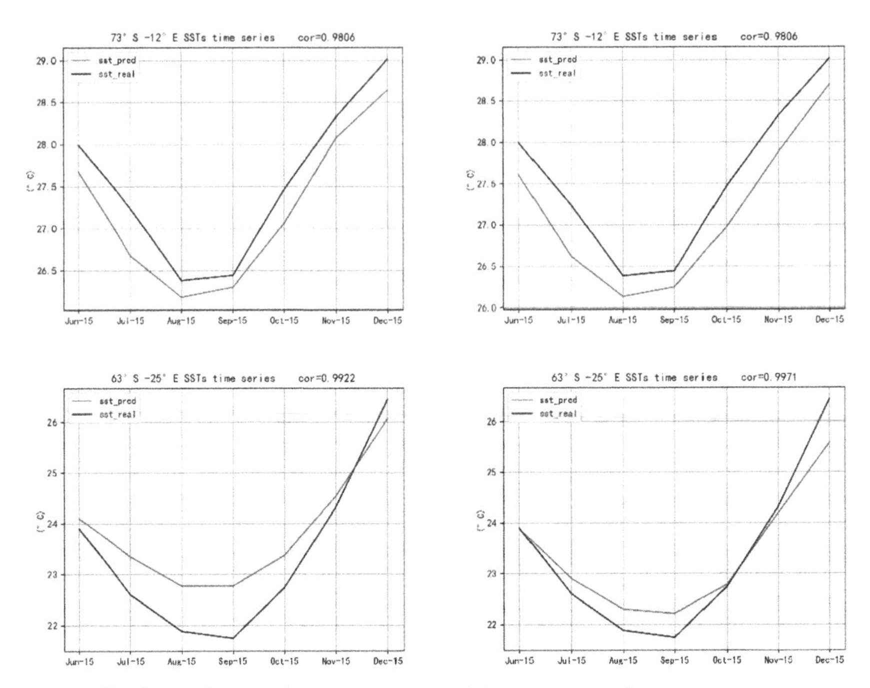

Fig. 3. Real sea surface temperature and forecast sea surface temperature

4 Summary

This thesis uses the attention mechanism to distinguish the importance of features, discards the least important 15 dimensions from all features, and compares the results of long-term sea surface temperature forecasting with all and some feature extraction schemes. The experimental results show that the TCN prediction model with some features maintains a stable high accuracy during the aging period. According to related research on sea surface temperature, the occurrence of large ocean climate phenomena is related to many factors. The full-feature model includes the regional influence of the wind field through the sea surface temperature of surrounding data points, and also considers the heat of ocean currents. The training data of full-feature model includes multi-factor and multi-level data, what need longer training time and higher hardware requirements. The result of sea surface temperature prediction used SSTs alone is not reliable enough, and the error in long-term sea surface temperature prediction is too large. Distinguishing the importance degree through the attention mechanism solves the problem of huge data set, and the accuracy is high. This paper uses the attention mechanism to complete the long-term sea surface temperature forecast of the large ocean basin structure in a partial feature manner. The model effect is stable and can better cope with the occurrence of abnormal sea temperature changes. It is useful for studying large-scale ocean physical phenomena.

This paper discusses the impact of reducing the size of the data set on the experimental results through TCN and attention mechanism. The results show that the data set is smaller and the accuracy is higher. Based on TCN system, historical ocean data were used to predict sea surface temperature 7 months in advance, and the attention mechanism was used to distinguish the importance of all features. Construct a new data set training model with partial features, and predict the sea surface temperature from July to December in 2015–2018. Through experiments, it is found that the experimental results of some feature models are better than the full feature model in 2016 and 2017. This method to reduce data is very effective in dealing with huge and detailed ocean data. This thesis uses attention machines to select features instead of relying on humans to distinguish. It has far-reaching significance for the combination of marine physics and deep learning.

References

1. Alexander, M.A., et al.: Projected sea surface temperatures over the 21st century: changes in the mean, variability and extremes for large marine ecosystem regions of Northern Oceans. Elem. Sci. Anthr. **6**, 9 (2018). https://doi.org/10.1525/elementa.191
2. Liang, C., Xian, W., Pauly, D.: Impacts of ocean warming on China's fisheries catches: an application of "mean temperature of the catch" concept. Front. Mar. Sci. **5**, 5 (2018). https://doi.org/10.3389/fmars.2018.00026
3. Ashok, K., Guan, Z., Saji, N.H., et al.: Individual and combined influences of ENSO and the Indian Ocean Dipole on the Indian summer monsoon. J. Clim. **17**(16), 3141–3155 (2004)
4. Ashok, K., Guan, Z., Yamagata, T.: Impact of the Indian Ocean Dipole on the relationship between the Indian monsoon rainfall and ENSO. Geophys. Res. Lett. **28**(23), 4499–4502 (2001)
5. Behera, S.K., Luo, J., Masson, S., et al.: Impact of the Indian Ocean Dipole on the East African short rains: a CGCM study. CLIVAR Exch. **27**, 43–45 (2003)
6. Chan, S.C., Behera, S.K., Yamagata, T.: Indian Ocean Dipole influence on South American rainfall. Geophys. Res. Lett. **35**(14), L14S12 (2008)
7. Yang, Y., Dong, J., Sun, X., Lima, E., Mu, Q., Wang, X.: A CFCC-LSTM model for sea surface temperature prediction. IEEE Geosci. Remote Sens. Lett. **15**(2), 207–211 (2018). https://doi.org/10.1109/LGRS.2017.2780843
8. Zhang, Q., Wang, H., Dong, J., Zhong, G., Sun, X.: Prediction of sea surface temperature using long short-term memory. IEEE Geosci. Remote Sens. Lett. **14**(10), 1745–1749 (2017). https://doi.org/10.1109/LGRS.2017.2733548
9. Yang, Y., Dong, J., Sun, X., Lima, E., Mu, Q., Wang, X.: A CFCC-LSTM model for sea surface temperature prediction. IEEE Geosci. Remote Sens. Lett. **15**, 207–211 (2018). https://doi.org/10.1109/lgrs.2017.2780843
10. Li, W., Guan, L., Qu, L., Guo, D.: Prediction of sea surface temperature in the china seas based on long short-term memory neural networks. Remote Sens. **12**(17), 2697 (2020). https://doi.org/10.3390/rs12172697
11. Wu, X., Yan, X.-H.: Estimation of subsurface temperature anomaly in the north Atlantic using a self-organizing map neural network. J. Atmos. Ocean. Technol. **29**(11), 1675–1688 (2012). https://doi.org/10.1175/jtech-d-12-00013.1
12. Bai, S., Kolter, J.Z., Koltun, V.: An empirical evaluation of generic convolutional and recurrent networks for sequence modeling. ArXiv e-prints, April 2018

Automatic Modulation Classification Using Convolutional Recurrent Attention Network

Yuzheng Yang[1], Sai Huang[1,2], Shuo Chang[1], Hua Lu[2(✉)], and Yuanyuan Yao[3]

[1] Key Laboratory of Universal Wireless Communications, Ministry of Education, Beijing University of Posts and Telecommunications, Beijing 10087, People's Republic of China
{yangyuzheng,huangsai,changshuo}@bupt.edu.cn

[2] GuangDong Communications and Networks Institute, Guangdong 510700, People's Republic of China
luhua@gdcni.cn

[3] School of Information and Communication Engineering, Beijing Information Science and Technology University, Beijing 100101, People's Republic of China
yyyao@bistu.edu.cn

Abstract. The development of wireless communication technology is much faster than the pace of its security. The interference such as adversarial attack degrades the accuracy and efficiency of communication environment. Automatic modulation classification (AMC) is viewed as an effective method to discover and identify the modulation mode of wireless signal corrupted by noise and interference. This paper proposes a novel modulation classification framework using Convolutional Recurrent Attention Network (CraNET) which is mainly composed of the convolution block and long short term memory network (LSTM) based attention block. Convolution block extracts the signal features in the feature extraction module. In the weighting module, the LSTM based attention block selectively weights the extracted features to weaken the content that has no contribution to the performance improvement. Extensive simulation verifies that the CraNET based modulation classification method performs higher accuracy and superior robustness than that of other existing methods.

Keywords: Automatic modulation classification · Attention block · Adversarial attack · Convolution block · Long short term memory network

---

This work was supported in part by the Guangdong Key Field *R&D* Program under Grant (2018B010124001), in part by the National Key Research and Development Program of China under Grants (2018YFF0301202, 2019YFB1804404, 2020YFB1807602), in part by the Beijing Natural Science Foundation under Grants (4202046, 19L2022), in part by the National Natural Science Foundation of China under Grant (61801052,61941102), and in part by the Industrial Internet Innovation and Development Project of Ministry of Industry and Information Technology under Grant (TC200H031).

J. Xiong et al. (Eds.): MobiMedia 2021, LNICST 394, pp. 736–747, 2021.
https://doi.org/10.1007/978-3-030-89814-4_54

1 Introduction

With the booming development of wireless communication technology [1], both the number and variety of wireless devices are increasing rapidly. At the same time, the way of wireless communication has become more diversified. However, rapid technological innovation has broken the stability of technological security. Diversified wireless devices and communication methods promote the development of interference technology such as adversarial attacks [2] to interfere with normal communication, which seriously degrades the accuracy and efficiency of communication environment, and causes unpredictable security risks. Compared with the traditional interference, this kind of interference is not noise, but is more invasive than noise and more difficult to distinguish the resistance interference. Automatic modulation classification (AMC), which is regarded as the basis link of spectrum sensing technology [3], identifies the modulation mode of wireless signal corrupted by noise and interference. AMC is mainly composed of three major technologies: data acquisition, feature extraction and modulation classification [4–6], and it is a common method in the field of anti active malicious interference such as jamming and spoofing.

AMC is divided into two main directions: likelihood based approach (LB) and feature extraction based approach (FB) [7]. LB classifiers first calculate the likelihood of the received signals under the hypotheses of different modulation schemes and then validate modulation schemes with the maximum likelihood (ML). The recognition accuracy of LB can achieve optimal performance in the Bayesian sense via minimizing the probability of misclassification. However, it will be degraded in the complex and changeable signal data, priori knowledge such as the probability distribution function of the signal also needs to be known in advance. FB method makes up the defect of LB method. FB method extracts some statistical information about the input signal for modulation classification [8–10]. FB method is a sub-optimal method, and it is difficult to deal with the situation of low SNR channel. However, it has the advantages of low computational complexity and requiring few prior knowledge about the received modulated signals. Thus, this paper aims at solving problems induced by FB method and improving the performance of FB method.

In order to achieve better performance, the neural network, which is robust to channel and noise uncertainties, is introduced into signal modulation classification. O'Shea *et al.* proposed a convolution neural network, which is used to identify the modulation formats [11]. However, no changes have been made to the structure of the traditional convolution neural network, which is firstly utilized for image classification. Therefore, it can not considers the sequential features of wireless signal, the recognition accuracy of this network is also limited. Sun *et al.* proposed a new approach to the automatic modulation classification of cochannel signals [12], which improved CNN based network structure and achieved certain results. However, the proposed network only considers the classification of the digital modulation signal, and the classification of the analog modulations is not considered. Mendis *et al.* [13] proposed an AMC framework which consists of spectrum correlation function (SCF) based feature characterization mechanism and deep belief network (DBN) based identification schem. However, the robustness of the network depends on the SCF image feature. Yao *et al.* [14] proposed an innovative joint model using CNN-LSTM network. In this model, QAM classification accuracy improved by Haar-wavelet Crest Searching. However, other modulation types' classification accuracy has not been improved due to the limitations of LSTM in processing long sequences. Tu Ya *et al.* [15] pointed out that DL model requires many training data to combat with over-fitting, then they extended Genera-tive Adversarial Networks (GANs) to the semi-supervised learning to overcome the problem. Yun Lin *et al.* [16] pointed out that waveforms in the physical layer may not be suitable for the prevalent classical DL models. They also evaluated the security problems caused by adversarial attacks to modulation recognition with CNN [17]. The results showed that CNN was highly vulnerable to adversarial attacks.

To solve the problems mentioned above, in this paper, we propose a CraNET based modulation classification method which can be divided into three parts: feature extraction module, weighting module and modulation classification module. CraNET's structure is mainly composed of convolution block and long short term memory network (LSTM) based attention block. Convolution block extracts the signal features in the feature extraction module. Different from LSTM which only outputs the last hidden state of LSTM layer, the proposed LSTM based attention block uses all hidden states to calculate the output in the weighting module. The contribution of each hidden state to the final output is enhanced or weakened by weighted summation. Moreover, the modulation classification decision is made in the modulation classification module using the softmax function. Extensive simulation verifies that the CraNET based modulation classification method performs higher accuracy and superior robustness than that of other existing methods.

The rest part of this paper is organized as follows: Sect. 2 introduces the signal model and modulation classification framework. Section 3 presents the CraNET structure, and introduces the process of classification. Simulation results are given in Sect. 4, which includes the comparison with other network performance. Section 5 summarizes the content of this paper.

2 Problem Statement

2.1 Signal Model

Since the normal communication will be shut down or transmit the signal on another carrier when it is interfered by the adversarial attack, the single antenna receiver receives the signal from one source at a given time. Hence the received baseband signal is given by:

$$r(n) = he^{(2\pi f_0 n+\theta_0)} s_k(n) + w(n), n = 1, 2, ..., N \tag{1}$$

where h is the Rayleigh channel coefficient, s_k is the complex baseband envelope of the received signal generated from the k-th modulation hypothesis $H_k \in$

$(H_1, H_2, ..., H_K)$, N denotes the total number of signal symbols, $w(n)$ is noise with its mean and variance are 0 and σ^2 respectively, and it is used as additive white Gaussian noise (AWGN). f_0 denotes the frequency offset and θ_0 denotes the phase offset. Since the average power of received signal is normalized, we define the SNR as:

$$\gamma_{SNR} = \frac{|h|^2}{\sigma_w^2} \tag{2}$$

2.2 CraNET Based Modulation Classification Framework

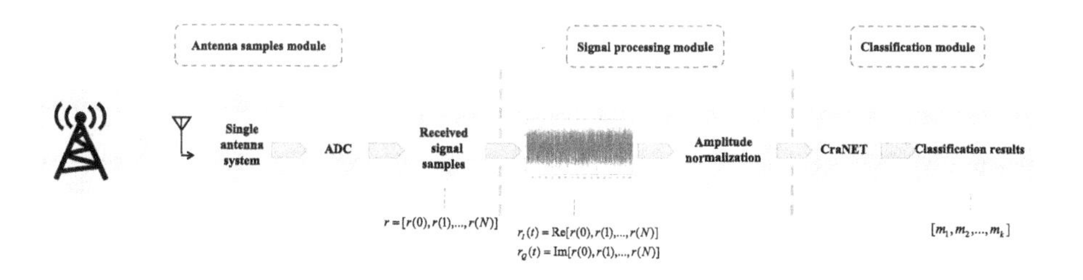

Fig. 1. CraNET based modulation classification framework

Figure 1 presents the CraNET based modulation classification framework, which consists of three modules: antenna samples module, signal processing module and classification module. The signal is received through the single-input and single-output (SISO) system, and the antenna samples module first receives the original signal by sampling, then converts the band-pass signal into baseband signal. Signal processing module divides the baseband signal into in-phase and quadrature (I/Q) channels and normalizes the signal. In the classification module, CraNET is first trained by the labeled data and is utilized to classify the modulation format of the unlabeled data. The in-phase and quadrature components are expressed as $I = Re[r(n)]$ and $Q = Im[r(n)]$. The complex representation of signal is given by:

$$r(n) = r_I(n) + r_Q(n) = \mathrm{Re}[r(n)] + j * \mathrm{Im}[r(n)] \tag{3}$$

Since the received signal may have great difference in amplitude, the signal amplitude is normalized. After the signal amplitude is normalized, a channel in complex signal is given by:

$$r_{I/Q}(n) = \frac{[r_{I/Q}(0), r_{I/Q}(1), ..., r_{I/Q}(N)]}{\max\{abs[r_{I/Q}(0), r_{I/Q}(1), ..., r_{I/Q}(N)]\}} \tag{4}$$

In the classification module, the well-trained classifier learns signal features by extracting I/Q features. Finally, the classifier gives the score value corresponding to each signal and selects the label corresponding to the maximum score as the output. The classification module outputs the classification results M according to the maximum probability value, and K is the number of a set of candidate modulations which is $M \in \{H_1, H_2, ..., H_K\}$.

3 CraNET Based Modulation Classification

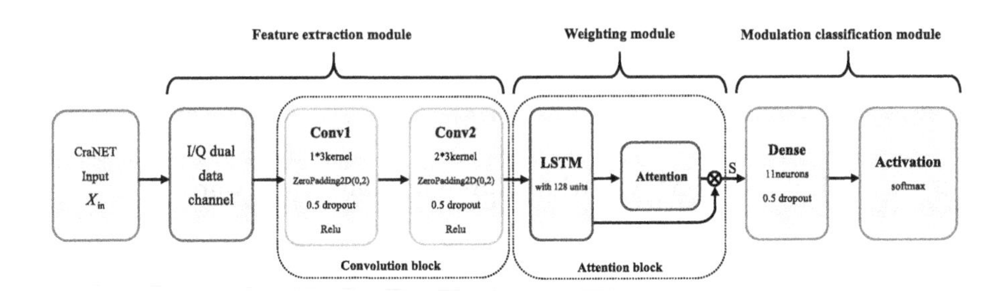

Fig. 2. CraNET based network structure

This section mainly introduces the CraNET structure, and then illustrates the classification process of the CraNET.

Figure 2 shows the overall network structure, which is divided into three parts: feature extraction module, weighting module and modulation classification module. The feature extraction module extracts the features of I/Q channels of the signal by convolution block which has two convolution layers. Then, these features are converted to hidden states and weighted by attention block. Finally it outputs s as the representation vector to the modulation classification module. The modulation classification module is composed of dense layer with softmax function, and the classification decision is finally conducted by using the representation vector. Since the convolutional neural network can find the intrinsic features of the data, it should be placed at the front of the network structure. Then the attention block can find the connection between the context information of the input sequence, and selectively focuses on the specific frame-related information of the input sequence. The proposed network takes full consideration of the data characteristics of the signal and combines the advantages of the attention mechanism.

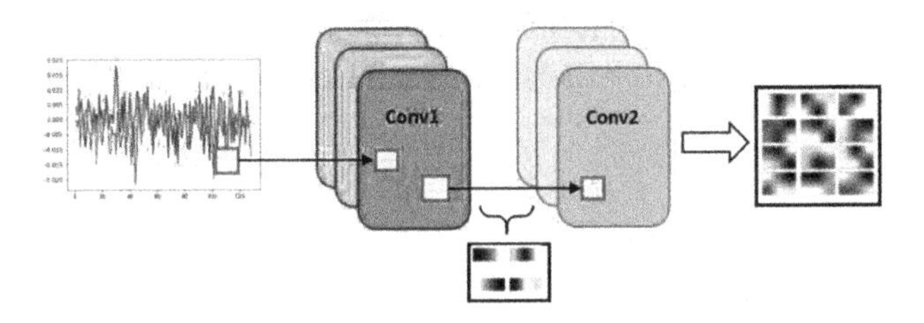

Fig. 3. CNN based network structure

3.1 CNN for Feature Extraction

Figure 3 shows the structure of CNN. Feature extraction is a crucial step in the modulation signal processing. In this module, both the analog signal and digital signal are considered for the experimental data set by the automatically features learning ability of CNN. CraNET first needs to find the intrinsic expression of the data through the convolutional network. The modulation recognition process in most wireless communication systems can be regarded as linear mixing, time shift, rotation, scale scaling and invariance with convolution. Convolutional neural networks have the characteristics of shift invariance in image processing problems [18,19], thus, the convolution layer is used in the first two layers of the whole classification model.

Table 1. Parameters of CNN

Input: I/Q signal matrix (Dimension:1*2*N)				
Layers	**Kernel Size**		**Padding**	**Stride Step**
Conv1	256@(1*3)	(0*2)	1	0.5
Conv2	80@(2*3)	(0*2)	1	0.5
Output: representation tensor (Dimension:80*2*N)				

Parameters of CNN as shown in Table 1. To balance the tradeoff between accuracy and efficiency, two convolution layers are used in the convolution block. The first convolution layer uses 256 convolution kernels of size 1 * 3, the second convolution layer uses 80 convolution kernels of size 2 * 3, and both layers use ReLU function as the activation function. Meanwhile, the weight $||W||_2$ of two norms are added to prevent the parameter value too large and overfitting in the process of parameter updating of convolution layer, and the dropout method is used to improve the generalization ability of the network. The first layer is used to get the edge and gradient detection of the input data in one dimension, and the second layer integrates the features extracted by the first layer through a larger convolution kernel. The convolution block converts the I/Q data into a representation tensor and passes it to the weighting module.

3.2 Attention Mechanism for Weighting

Figure 4 shows the calculation process of attention weight distribution. First, input the representation tensor $(x_1,...,x_T)$ into the LSTM layer after feature extraction, then the LSTM cell transforms the representation tensor into hidden states. Attention block takes the hidden states $h_1,...,h_T$ into the softmax function, and the attention distribution α_t is given out by the softmax function. The final output vector s is calculated by the weighted sum of all hidden states.

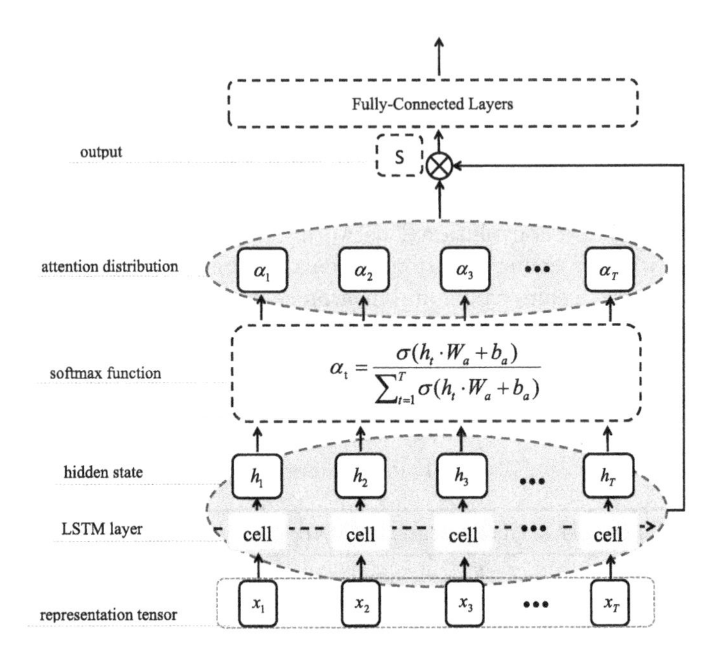

Fig. 4. Attention mechanism based network structure

The attention mechanism is based on LSTM network, and LSTM network is actually a RNN network. Denote the hidden states at the T time steps of an LSTM layer as $h_1,...,h_T$, where T is the number of time steps. The hidden state h_t is obtained by x_t and h_{t-1}. Meanwhile, T is the first dimension of the representation tensor $(x_1,...,x_T)$ and its value is 80. Unit is the hidden neuron of the gate structure in the LSTM cell, and its value is equal to the length of the output vector of the LSTM cell. To summarize the final output of LSTM layer, a feasible method is to concatenate all hidden states as one vecter and then output $H = (h_1,...,h_T)$. In this way, the extracted information from all time steps will be used.

Attention mechanism is introduced to measure the importance of all hidden states. The contribution of each hidden state to the final output is enhanced or weakened by weighted summation. In this paper, the proposed model uses a self-attention mechanism to adaptively derive the final output of LSTM layer using all hidden states. First, LSTM layer maps the input representation tensor $(x_1,...,x_T)$ to a sequence of hidden states $h_1,...,h_T$, then sends h_t to a softmax activation function through a shared time-distributed neural network layer with attention weight matrix W_a and bias b_a:

$$u_t = h_t \cdot W_a + b_a \tag{5}$$

The attention distribution α_t is calculated by sending the output u_t through a softmax activation layer:

$$\alpha_t = \frac{\sigma(u_t)}{\sum_{t=1}^{T} \sigma(u_t)} = \frac{\sigma(h_t \cdot W_a + b_a)}{\sum_{t=1}^{T} \sigma(h_t \cdot W_a + b_a)} \tag{6}$$

The final output vector s is the weighted sum of all hidden states H, its vector dimension is the same as h_t:

$$s = \sum_{t=1}^{T} \alpha_t h_t \tag{7}$$

By adding attention mechanism, the network achieves a better trade-off between increasing the amount of information that sent to the next layer and reducing the number of parameters that need to be trained in the next layer.

In the modulation classification module, the obtained s is fed into the fully connected layer as the final output of the attention block and the probability of all hypothetical modulation modes is calculated by the softmax function, the hypothetical modulation mode with the largest probability is selected as the final recognized modulation mode. Modulation signal labels $y \in \{1, 2, ..., K\}$ have K categories. When given a sample x, w_k is the weight vectors of the k-th class, the conditional probability of the k-th class for softmax regression prediction is:

$$p(y = k|x) = \frac{\exp(w_k^T x)}{\sum_{k=1}^{K} \exp(w_k^T x)} \tag{8}$$

Finally, the classification results are output by selecting the maximum statistics, which is given as follows:

$$k = \arg \max_{1 \leq k \leq K} p(y = k|x) \tag{9}$$

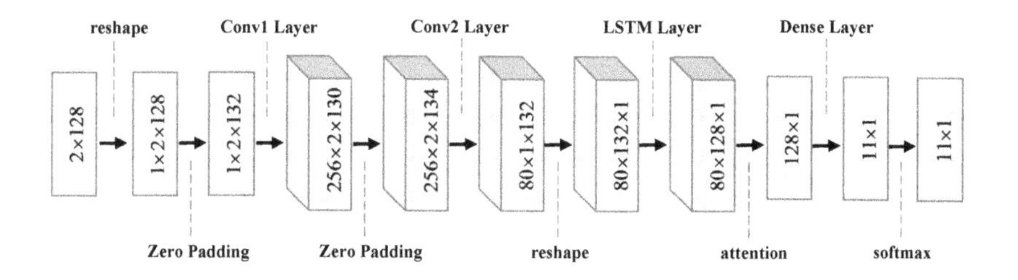

Fig. 5. Data-driven structure of CraNET

4 Simulation

In this section, the recognition and classification of signal features are carried out by constructing neural networks, and the training process of network learning is accelerated by tensorflow-gpu framework. The candidate modulation set M contains six kinds of modulation modes, which is given by M = *[AM-DSB, AM-SSB, BPSK, CPFSK, GFSK, PAM4]*. The signal data set is taken from RadioML [20] and SNR ranges from -20 dB to 18 dB with a step of 2 dB. Each sample vector has a size of 2×128. The Data-driven structure of CraNET is shown in Fig. 5. Adam optimization algorithm is used to replace the traditional random gradient descent algorithm. During the training, the loss value of the verification set is monitored by the Earlystopping in the callbacks, and the maximum number of tolerance is set to 8 times. The performance of the neural network used for modulation signal recognition is evaluated by classification accuracy to verify the performance and robustness of CraNET. The average classification accuracy is given by:

$$P_{CA} = \sum_{k=1}^{K} P(\hat{H} = H_k | H_k) P(H_k) \tag{10}$$

where K is the number of category, H_k is the hypothesis of k-th class.

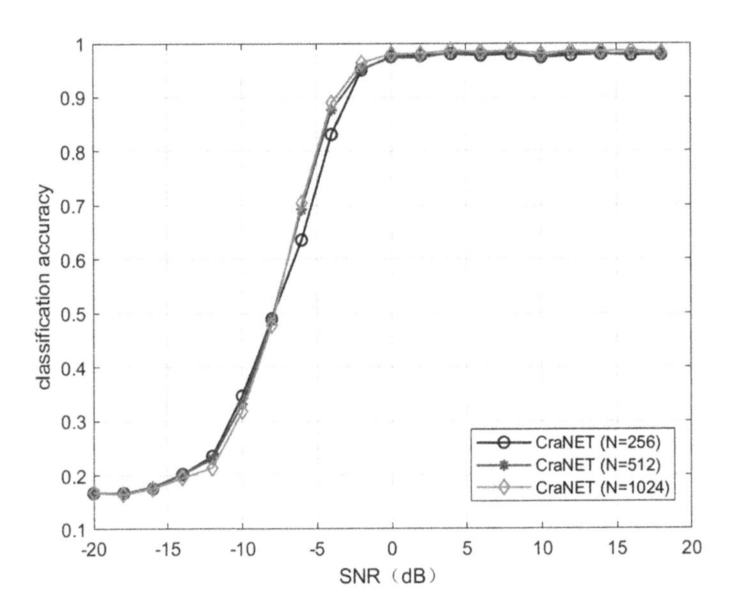

Fig. 6. The performance of CraNET with different batch size

Figure 6 shows the accuracy on different batch size set as 256, 512 and 1024 of CraNET, respectively. The batch size parameter determines the direction of gradient descent, and the direction of gradient descent is one of the influencing factors of classification accuracy. If the dataset is sufficient, the gradient calculated by training with small batch data is almost the same as that by training with large batch data. Thus, it can be found that the classification performance increases slightly with the batch size. It can be observed that SNR is close to 1.0 when it is above 0 dB, and CraNET has good performance with the accuracy more than 0.8 above −5 dB. Therefore, CraNET has better performance under normal SNR condition.

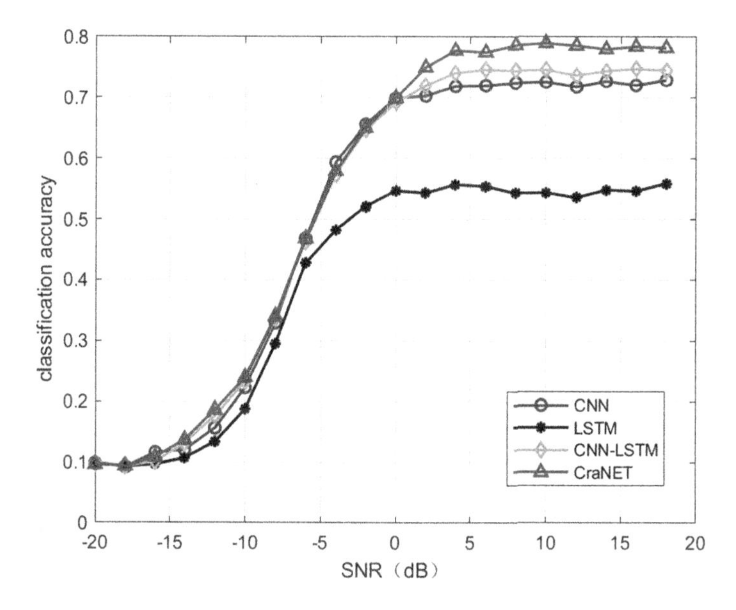

Fig. 7. The performance comparison with other AMC methods

Figure 7 compares the accuracy of CraNET with that of CNN, LSTM and CNN-LSTM network after training on the same data set. The data set contains eleven modulation modes of RadioML, which is given by M = *[8PSK, AM-DSB, AM-SSB, BPSK, CPFSK, GFSK, PAM4, QAM16, QAM64, QPSK, WBFM]*. Since the data set is generated through real channel simulation. The influence of various external factors is strengthened, and the situation is closer to the real modulation signal, which makes the accurate recognition rate decrease. Due to the increase of candidate types, the classification performance also declined. Through the curve in the graph, we can see that the accurate recognition rate of all classification network models is low when the SNR is lower than 0 dB, and the accurate recognition rate tends to a high constant value gradually when the SNR is higher than 0 dB. Overall, the accurate recognition rate of CraNET is always higher than that of other networks above −5 dB, and the highest accurate recognition rate of CraNET is about five percent higher than that of the CNN. Therefore, it is verified that CraNET has better performance than other networks.

5 Conclusions

This paper proposes a novel modulation classification framework using Convolutional Recurrent Attention Network which can be divided into three parts: feature extraction module, weighting module and modulation classification module. CraNET's structure is mainly composed of convolution block and long short term memory network (LSTM) based attention block. Convolution block extracts the signal features in the feature extraction module. Different from LSTM which only outputs the last hidden state of LSTM layer, the proposed LSTM based attention block uses all hidden states to calculate the output in the weighting module. The contribution of each hidden state to the final output is enhanced or weakened by weighted summation. Moreover, the modulation classification decision is made in the modulation classification module using the softmax function. CraNET is compared with other networks by Monte Carlo simulation to verify the superiority and robustness of CraNET.

References

1. Jain, A.K., Acharya, R., Jakhar, S., Mishra, T.: Fifth generation (5g) wireless technology "revolution in telecommunication". In: 2018 Second International Conference on Inventive Communication and Computational Technologies (ICICCT), pp. 1867–1872 (2018)
2. Sadeghi, M., Larsson, E.G.: Adversarial attacks on deep-learning based radio signal classification. IEEE Wirel. Commun. Lett. **8**(1), 213–216 (2019)
3. Dobre, O.A., Abdi, A., Bar-Ness, Y., Su, W.: Survey of automatic modulation classification techniques: classical approaches and new trends. IET Commun. **1**(2), 137–156 (2007)
4. Tian, X., Chen, C.: Modulation pattern recognition based on resnet50 neural network. In: 2019 IEEE 2nd International Conference on Information Communication and Signal Processing (ICICSP), pp. 34–38 (2019)
5. Fragkiadakis, A.G., Tragos, E.Z., Askoxylakis, I.G.: A survey on security threats and detection techniques in cognitive radio networks. IEEE Commun. Surv. Tutor. **15**(1), 428–445 (2013)
6. Gouldieff, V., Palicot, J., Daumont, S.: Blind modulation classification for cognitive satellite in the spectral coexistence context. IEEE Trans. Signal Process. **65**(12), 3204–3217 (2017)
7. Huang, S., Yao, Y., Wei, Z., Feng, Z., Zhang, P.: Automatic modulation classification of overlapped sources using multiple cumulants. IEEE Trans. Veh. Technol. **66**(7), 6089–6101 (2017)
8. Huan, C.-Y., Polydoros, A.: Likelihood methods for MPSK modulation classification. IEEE Trans. Commun. **43**(2/3/4), 1493–1504 (1995)
9. Chang, D., Shih, P.: Cumulants-based modulation classification technique in multipath fading channels. IET Commun. **9**(6), 828–835 (2015)
10. Karami, E., Dobre, O.A.: Identification of SM-OFDM and AL-OFDM signals based on their second-order cyclostationarity. IEEE Trans. Veh. Technol. **64**(3), 942–953 (2015)
11. O'Shea, T.J., Corgan, J., Charles Clancy, T.: Convolutional radio modulation recognition networks. arXiv preprint arXiv:1602.04105 (2016)

12. Sun, J., Wang, G., Lin, Z., Razul, S.G., Lai, X.: Automatic modulation classification of cochannel signals using deep learning. In: 2018 IEEE 23rd International Conference on Digital Signal Processing (DSP), pp. 1–5 (2018)
13. Mendis, G.J., Wei, J., Madanayake, A.: Deep learning-based automated modulation classification for cognitive radio. In: 2016 IEEE International Conference on Communication Systems (ICCS), pp. 1–6 (2016)
14. Yao, T., Chai, Y., Wang, S., Miao, X., Bu, X.: Radio signal automatic modulation classification based on deep learning and expert features. In: 2020 IEEE 4th Information Technology, Networking, Electronic and Automation Control Conference (ITNEC), vol. 1, pp. 1225–1230 (2020)
15. Tu, Y., Lin, Y., Wang, J., Kim, J.U.: Semi-supervised learning with generative adversarial networks on digital signal modulation classification. Comput. Mater. Continua **55**(2), 243–254 (2018)
16. Lin, Y., Tu, Y., Dou, Z., Chen, L., Mao, S.: Contour Stella image and deep learning for signal recognition in the physical layer. IEEE Trans. Cogn. Commun. Netw., 1 (2020)
17. Lin, Y., Zhao, H., Ma, X., Tu, Y., Wang, M.: Adversarial attacks in modulation recognition with convolutional neural networks. IEEE Trans. Reliab., 1–13 (2020)
18. Wang, W.G.L., Li, X.: Deep convolutional neural network and its application in image recognition of road safety projects. Int. J. Perform. Eng. **15**(8), 2182 (2019)
19. Songyin, B.X., Pan, D.W., En, F., Pan, F.: Traffic sign detection via efficient ROI detector and deep convolution neural network. Int. J. Perform. Eng. **16**(10), 1566 (2020)
20. O'Shea, T.J., West, N.: Radio machine learning dataset generation with gnu radio. In: Proceedings of the 6th GNU Radio Conference (2016)

Improved Speech Emotion Recognition Using LAM and CTC

Lingyuan Meng[1], Zhe Sun[1], Yang Liu[1(✉)], Zhen Zhao[1], and Yongwei Li[2]

[1] School of Information Science and Technology, Qingdao Universicity of Science and Technology, Qingdao 266061, China
yangliu@qust.edu.cn

[2] National Laboratory of Pattern Recognition, Institute of Automation, Chinese Academy of Sciences, Beijing 100089, China

Abstract. Time sequence based speech emotion recognition methods are difficult to distinguish between emotional and non-emotional frames of speech, and cannot calculate the amount of emotional information carried by emotional frames. In this paper, we propose a speech emotion recognition method using Local Attention Mechanism (LAM) and Connectionist Temporal Classification (CTC) to deal with these issues. First, we extract the Variational Gammatone Cepstral Coefficients (VGFCC) emotional feature from the speech as the input of LAM-CTC shared encoder. Second, CTC layer performs automatic hard alignment, which allows the network to have the largest activation value at the emotional key frame of the voice. LAM layer learns different degrees on the emotional auxiliary frame. Finally, BP neural network is used to integrate the decoding outputs of CTC layer and LAM layer to obtain emotion prediction results. Evaluation on IEMOCAP shows that the proposed model outperformed the state-of-the-art methods with a UAR of 68.5% and an WAR of 68.1% respectively.

Keywords: Speech emotion recognition · Attention · CTC · VGFCC · IEMOCAP

1 Introduction

As the fundamental research of affective computing, the study of speech emotion recognition has attracted great attention in recent years. Traditional machine learning approaches, such as Hidden Markov Model (HMM), Gaussian Mixture Model (GMM), Kernel Regression, Maximum Likelihood Classification (MLC), and Support Vector Machine (SVM) are widely adopted for emotion recognition using extracted features in previous works [1–3]. However, traditional speech emotion recognition methods are difficult to implement when encountering large training datasets.

Supported by The Natural Science Foundation of Shandong Province (No. ZR2020 QF007).

J. Xiong et al. (Eds.): MobiMedia 2021, LNICST 394, pp. 748–756, 2021.
https://doi.org/10.1007/978-3-030-89814-4_55

With the development of deep learning technology, deep neural networks (DNN) could not only easily handle large-scale training data, but also learn the deep level characteristics of speech emotion. For example, George trained a convolution recurrent neural network (CRNN) to perform continuous emotion prediction [4]. Zhang used spectrogram training based on full convolutional neural networks to convert the sequence transformation problem into image recognition problem [5]. Keren et al. used deep residual network (ResNet) to enhance speech to improve the performance of speech emotion recognition [6]. The intensity of speech emotions changes continuously over time, however, most DNN cannot distinguish between emotional and non-emotional frames in the speech, thus leading to performance degradation.

Graves et al. proposed the Connectionist temporal classification (CTC) algorithm [7], which can learn the error of the entire samples so that the network can automatically converge to the point where the emotional characteristics are most obvious [8,9]. However, the CTC algorithm only considers whether the current frame belongs to an emotional frame, while neglecting the difference of the amount of emotional information contained in each frame. The attention mechanism (AM [10] calculates the relative weights of the emotional features in the speech signal and each time domain, and selects the time domain signal with larger weight for recognition, so as to ensure that key information will not be lost [11]. Chen proposed a fusion model of AM and CTC in which the CTC module uses the emotional semantic coding sequence output by the decoder to calculate the loss [12], and the AM module encodes the emotional semantics output by the encoder. However, AM is global attention and does not consider the time order of the sequence.

In this paper, we propose a speech emotion recognition method based on Local Attention Mechanism (LAM) and CTC. First, we use SVM to extract the Variational Gammatone Cepstral Coefficients (VGFCC) features of the input speech, which are then used as the input of shared encoder; Second, the CTC layer performs back propagation using cross-entropy error for training and aligns the key emotional frames in the speech. Meanwhile, the LAM layer calculates the context relevance for the matching degree and the encoder output by attention mechanism algorithm, and different levels of information are extracted from different emotional frames for learning. Finally, the outputs of the CTC and LAM layers are integrated through BP neural network.

The main contributions of this paper include three aspects:

1) We apply local attention mechanism to reduce the interference of irrelevant speech frames on the current speech frame weight calculation;
2) By optimizing the decoder network structure and using supervised learning to merge the results of the CTC layer and LAM layer, the proposed model can fully learn the emotional features from speech;
3) Experimental results on benchmark dataset IEMOCAP demonstrate that our model gains an absolute improvement of 0.5%–0.8% over state-of-the-art strategies.

2 Proposed Method

The proposed framework as shown in Fig. 1 mainly consists: VGFCC feature extraction, CTC-Attention shared encoder, LAM layer, CTC layer, decoder and output layer.

2.1 VGFCC Feature Extraction

VGFCC considers the nonlinear and non-stationary characteristics of speech signals. First, set the maximum length of the voice $x(n)$ to 7.5 s, cut the longer voice into 7.5 s, and fill the shorter voice with zeros. Second, perform pre-emphasis, framing, and windowing on $x(n)$ (Hamming window) processing, and the sampling rate is set 16000 Hz. Third, $x(n)$ is decomposed into K intrinsic modal function (IMF) components and all IMF components are subjected to fast Fourier transform (FFT) to obtain the spectral amplitude. Next, all the spectrum amplitudes are modulus squared and summed to obtain the energy spectrum of the signal and the energy spectrum is filtered by the Gammatone filter. Finally, the discrete cosine transform is performed on the filtered result to obtain the VGFCC coefficient.

2.2 Encoding Module

CTC-Attention Shared Encoder: CTC layer and LAM layer share the same encoder which is a double hidden layer LSTM, since LSTM can selectively affect the state of the neural network at each moment through a special gate structure. The output of each LSTM unit is used as the input of the CTC layer and LAM layer.

LAM Layer: Following the Encoding model, a structured LAM network aggregates information from the LSTM hidden states H^{lstm} and produces a fixed-length vector **Conv** as the encoding of the emotion feature.

Given hidden states H^{lstm} as input, first determine the number of speech frames T, and then determine the size of the optional range K.

When T is less than K, the model will use the global attention mechanism. The network computes a vector of attentional weights c and the weighted sum of the hidden states h^{attn} as follows:

$$\mathrm{c} = \mathrm{softmax}\left(w_{s2} \tanh\left(W_{sl} H^{lstm^{T}}\right)\right) \tag{1}$$

$$h^{\mathrm{attn}} = \sum_{i=1}^{T} c H^{lstm}, \tag{2}$$

where W_{s1} and w_{s2} are trainable parameters.

When T is larger than K, the model uses a sliding window to limit the calculation range of the vector of attentional weights, and slides the window according to the current calculated position. As shown in Fig. 2, X_n is the hidden

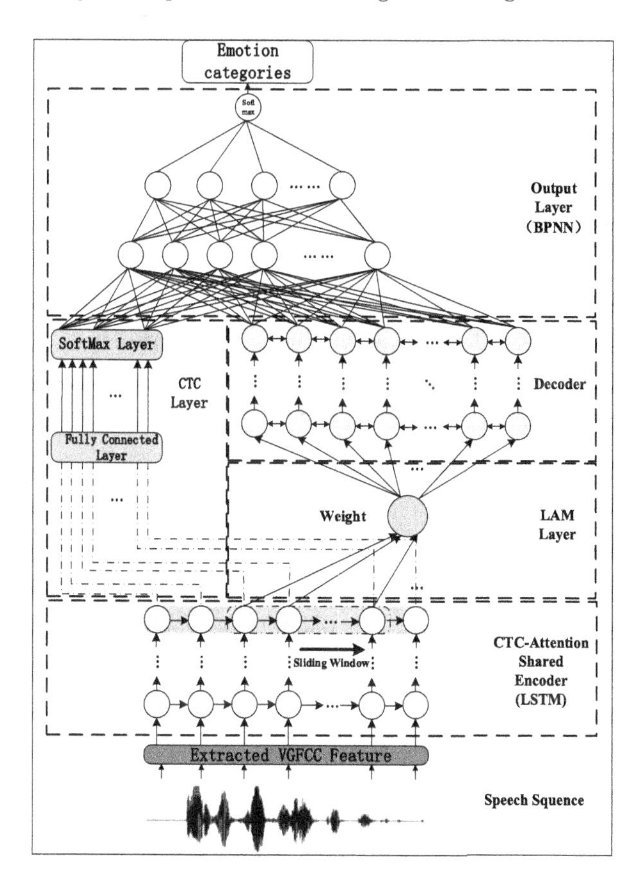

Fig. 1. Overall structure of the proposed framework.

layer output of the encoder neuron, and the current calculation position u is in the middle of the sliding window. If the lower limit of the sliding window is less than position 1 of X_1, start from position 1. If the upper limit of the sliding window is greater than the last position of the sequence, it ends with the last position. The length of the sliding window D is set to $K/2$. Note that the calculation formula of the weight vector c_{slide} and the hidden layer state weighted sum h^{attn} is:

$$c_{\text{slide}} = \text{softmax}\left(w_{s2} \tanh\left(W_{sl} H^{lstm^K}\right)\right) \quad (3)$$

$$h^{\text{attn}} = \sum_{i=u-D}^{u+D} c_{\text{slide}} H^{lstm} \quad (4)$$

CTC Layer: Define the intermediate label sequence $\pi = (\pi_1, \ldots \pi_T)$. Assumingis y' the expansion after adding the separator for y. Define a many-to-one mapping

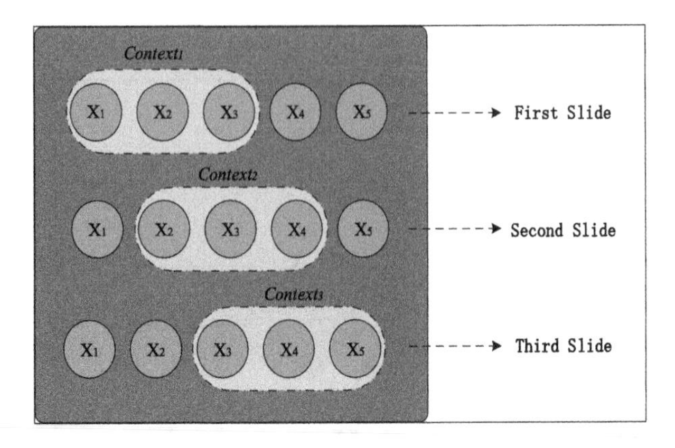

Fig. 2. The procedure of LAM.

as follow:

$$B : L' \rightarrow L^{\leq T} \tag{5}$$

where $L^{\leq T}$ is the output set of π of possible intermediate label sequences to obtain the probability of output label $P(y \mid x)$:

$$P(y \mid x) = \sum_{\pi \in B^{-1}(y')} p(\pi \mid x) \tag{6}$$

At each time step t of the input sequence x, the corresponding output pi_t must be calculated. Assuming that the output sequence between each time step is conditional and independent, $P(y \mid x)$ is the probability of a single label in the intermediate label sequence. (7) can be obtained as follow:

$$P(\pi \mid x) \approx \prod_{t=1}^{T} P\left(\pi_1 \mid \pi_1, \pi_2, \ldots, \pi_{t-1}, x\right), \forall \pi \in L' \tag{7}$$

The value of the negative log probability of the CTC loss function can be defined as follow:

$$L_{CTC}(S) = - \sum_{(x,y') \in S} \ln \sum_{\pi \in B^{-1}(y')} \prod_{t=1}^{T} q_t\left(\pi_t\right), \forall \pi \in L' \tag{8}$$

The CTC algorithm uses the HMM forward and backward algorithm to improve the calculation speed:

$$p_{CTC}(y \mid x) = \sum_{t=1}^{T} \sum_{u=1}^{|y'|} \frac{\alpha_t(u)\beta_t(u)}{q_t\left(\pi_t\right)} \tag{9}$$

where $\alpha_t(u)$ is the forward probability of the u-th label at time step t, and $\beta_t(u)$ is the backward probability of the u-th label at time step t.

2.3 Decoder

LSTM can only process the speech sequence in order, while the BLSTM could process the global speech sequence. Therefore, we select the BLSTM with double hidden layers as the decoder unit.

2.4 Output Layer

The output sequence of CTC and LAM is fused through double layers BP neural network. The gradient descent method is used for learning until the mean square error value reaches the threshold or the maximum number of iterations.

3 Experiment and Analysis

3.1 Database Description

We perform experiments on the public database, Interactive Emotional Dyadic Motion Capture (IEMOCAP) released by the School of Engineering of the University of Southern California. We use 12 h of IEMOCAP audio data which are divided into short sentences and labeled by three experts for discrete emotion categories. The labeled categories are "Angry", "Excited", "Happy", "Neutral" and "Sad". Since the emotions of "Excited" and "Happy" are very similar, both data are merged into the "Happy" category, and then these speech data are used as the emotion dataset for experiment which include 5531 utterances: "Happy" (1636), "Angry" (1103), "Sad" (1084), and "Neutral" (1708).

3.2 Experimental Setup

The experiment used Keras as the development framework and was completed on GeForce1080-Ti graphics card and Ubuntu 18.04 LST system. We convert the speech signal to 16 kHz and use 16bit to quantize the speech signal. We choose cross entropy as the cost function, Adadelta as the optimizer, and relu as the activation. The number of neurons selected for the encoder and decoder is 256, the learning rate is 0.01, the number of training cycles is 500, and the batch size model adjustment parameter is 64. 5531 pieces of speech emotion data are randomly divided into 80% training set and 20% testing set. The data set has a total of 4 emotion categories, and the proportion in the entire training/test set is still the same as the entire corpus, and the parameters are shown in the numbers above.

3.3 Comparative Experiments

In order to quantitatively evaluate the performance of the proposed model, the classification results of comparative experiments are provided in Table 1. The average training time, weighted average recall (WAR) and unweighted average recall (UAR) of the model are used as evaluation criterions.

Table 1. Results of comparative experiments. Note: Bold front denotes the best performance.

Model	UAR [%]	WAR [%]	Average training time [min]
LSTM	58.2	57.1	**76**
LSTM-CTC	63.7	62.9	109
LSTM-selfattention	66.1	65.8	114
AttRNN-RNN	67.6	67.5	105
LAM-CTC	**68.5**	**68.1**	85

We selected four classic models for comparison, including LSTM model [11], LSTM-selfattention model [11], LSTM-CTC model[12] and AttRNN-RNN model [12].

As shown in Table 1, the proposed model LAM-CTC is superior to other classic models in terms of UAR and WAR. Compared with the AttRNN-RNN model which performs best among the four models, the UAR and WAR of the proposed model increase by 0.9% and 0.6%, respectively. LAM effectively improves the tightness of the emotional context of certain long sentences, and resolve the problem that the global attention cannot perform emotional weight calculation for local speech, thereby ignoring the change of local emotions over time. In addition, when using BLSTM as the unit of the decoder, the current calculation vector contains more post information compared to only using RNN.

The average training time of proposed model is 85 min which only costs 78% of that of the LSTM-selfattention model and 84.7% of that of the AttRNN-RNN model. Since the global attention mechanism needs to calculate the weight of each frame at once when calculating the sample weight, LAM only needs to calculate the weight of the frame in the current window which slides backward overtime. At the same time, LAM calculates fewer parameters which could effectively reduce training time.

3.4 Ablation Experiments

The effectiveness of the LAM and CTC modules was verified through ablation experiments. The settings of the ablation experiments were as follows: (1) Replace LAM with global attention to verify the contribution of LAM to the performance (Attention-CTC). (2) Remove the CTC from the LAM-CTC model to verify the contribution of CTC to the performance (LAM). The comparison results of the ablation experiments are shown in Table 2.

As is shown in Table 2, the UAR and WAR of proposed model increase by 0.7% and 1.2% respectively compared with Attention-CTC because LAM could improve the tightness of the context of speech emotion. Compared with LAM, the UAR and WAR of the proposed model increase by 2.2% and 5%, respectively. By fusing the CTC module which could fully learn the key emotional frames of the speech, the proposed model can learn more about the emotional features in the

Table 2. Results of ablation experiments. Note: Bold front denotes the best performance.

Model	UAR [%]	WAR [%]	Average training time [min]
Attention-CTC	67.8	66.9	118
LAM	66.3	63.1	**67**
LAM-CTC	**68.5**	**68.1**	85

emotional speech. The training time of LAM-CTC only consumes 73.5% of that of the Attention-CTC model, while the LAM model only takes 67 min. Because LAM only needs to calculate the weight of the frame in the current window when calculating the context vector weight, and the attention window slides backward overtime. At the same time, LAM calculates fewer parameters, so using LAM can reduce training time. The results of ablation experiments show that both LAM and CTC could improve the performance of speech emotion recognition.

4 Conclusion

In this paper, we propose a deep learning model that integrates LAM and CTC for speech emotion recognition. Our method uses LAM to match the attention model according to the length of the speech sequence, and changes the calculation range of the context vector on the long sentence for modeling. CTC is utilized to perform alignment, which allows the network to have the largest activation value at the emotional key frame of the voice. The ablation experiment shows the effectiveness of reducing the interference of irrelevant speech frames on the current weight calculation. Compared with state-of-the-art models, the UAR and WAR achieves 68.5% and 68.1% with absolute increments more than 0.9% and 0.6%.

Acknowledgement. This work is supported by the Natural Science Foundation of Shandong Province (No. ZR2020QF007).

References

1. Schuller, B., Rigoll G., Lang, M.: Hidden Markov model-based speech emotion recognition. In: 2003 IEEE International Conference on Acoustics, Speech, and Signal Processing, 2003. Proceedings. (ICASSP 2003), pp. II–1 (2003). https://doi.org/10.1109/ICASSP.2003.1202279
2. Dong, F., Zhang, G., Huang, Y., Liu, H.: Speech emotion recognition based on multi-output GMM and SVM. In: 2010 Chinese Conference on Pattern Recognition (CCPR), pp. 1–4 (2010). https://doi.org/10.1109/CCPR.2010.5659255
3. Caihua, C.: Research on multi-modal mandarin speech emotion recognition based on SVM. In: 2019 IEEE International Conference on Power, Intelligent Computing and Systems (ICPICS), pp. 173–176 (2019). https://doi.org/10.1109/ICPICS47731.2019.8942545

4. Trigeorgis, G., et al.: Adieu features? End-to-end speech emotion recognition using a deep convolutional recurrent network. In: 2016 IEEE International Conference on Acoustics, Speech and Signal Processing (ICASSP), pp. 5200–5204 (2016). https://doi.org/10.1109/ICASSP.2016.7472669
5. Zhang, Y., Du, J., Wang, Z., Zhang, J., Tu, Y.: Attention based fully convolutional network for speech emotion recognition. In: 2018 Asia-Pacific Signal and Information Processing Association Annual Summit and Conference (APSIPA ASC), pp. 1771–1775 (2018). https://doi.org/10.23919/APSIPA.2018.8659587
6. Ariav, I., Cohen, I.: An end-to-end multimodal voice activity detection using WaveNet encoder and residual networks. IEEE J. Sel. Top. Signal Process. **13**(2), 265–274 (2019). https://doi.org/10.1109/JSTSP.2019.2901195
7. Graves, A., Metze, F.: A first attempt at polyphonic sound event detection using connectionist temporal classification. In: 2017 IEEE International Conference on Acoustics, Speech and Signal Processing (ICASSP), pp. 2986–2990 (2017). https://doi.org/10.1109/ICASSP.2017.7952704
8. Miao, Y., Gowayyed, M., Metze, F.: EESEN: end-to-end speech recognition using deep RNN models and WFST-based decoding. In: 2015 IEEE Workshop on Automatic Speech Recognition and Understanding (ASRU), pp. 167–174 (2015). https://doi.org/10.1109/ASRU.2015.7404790
9. Shan, C., et al.: Investigating end-to-end speech recognition for Mandarin-English code-switching. In: ICASSP 2019 - 2019 IEEE International Conference on Acoustics, Speech and Signal Processing (ICASSP), pp. 6056–6060 (2019) https://doi.org/10.1109/ICASSP.2019.8682850
10. Su, J., Zeng, J., Xie, J., Wen, H., Yin, Y., Liu, Y.: Exploring discriminative word-level domain contexts for multi-domain neural machine translation. IEEE Trans. Pattern Anal. Mach. Intell. **43**(5), 1530–1545 (2021). https://doi.org/10.1109/TPAMI.2019.2954406
11. Xie, Y., Liang, R., Liang, Z., Huang, C., Zou, C., Schuller, B.: Speech emotion classification using attention-based LSTM. IEEE/ACM Trans. Audio Speech Lang. Process. **27**(11), 1675–1685 (2019). https://doi.org/10.1109/TASLP.2019.2925934
12. Chen, X.: Research on Speech Emotion Recognition Method Based on Time Series Deep Learning Model. Harbin Institute of Technology (2018)

High Spatial Resolution Remote Sensing Classification with Lightweight CNN Using Dilated Convolution

Gang Zhang(✉), Wenmei Li, Heng Dong, and Guan Gui

College of Telecommunications and Information Engineering, Nanjing University of Posts and Telecommunications, Nanjing 210003, China
{1219012317,liwm,dongh,guiguan}@njupt.edu.cn

Abstract. High spatial resolution remote sensing (HSRRS) classification is one of the most promising topics in the field of remote sensing. Recently, convolutional neural network (CNN), as a method with superior performance, has achieved amazing results in the HSRRS classification tasks. However, the previously proposed CNN models generally have large model sizes, and thus require a tremendous amount of computing and consume a lot of memory. Furthermore, wearable electronic devices with limited hardware resources in actual application scenarios fail to meet the storage and computing requirements of these complex CNNs. To solve these problems, lightweight processing is important and practical. This paper proposes a new method called DC-LW-CNN, which applies Dilated Convolution (DC) and neuron pruning methods to maintain the classification performance of CNN and reduce resource consumption. The simulation results indicate that different DC-LW-CNNs have better performance effects in HSRRS classification task. Not only that, they also achieve this remarkable performance with smaller model size, less memory, and faster feedforward computing speed.

Keywords: High resolution remote sensing scene · Convolutional neural network · Dilated convolution · Neuron pruning

1 Introduction

HSRRS images have rich information about the spatial distribution of features, spectrum and texture. This characteristic information is helpful for image target recognition and classification with different application areas such as smart city, hyperspectral imaging, medical imaging, etc. [1–3]. Many deep learning-based methods have achieved remarkable results in digital images and wireless communication tasks due to their powerful feature extraction capabilities [2–6]. Among these deep learning methods, CNN method is able to extract middle-level and high-level information in images, which has certain advantages in image classification tasks. Current CNN methods have been applied in the field of HSRRS image scene classification and has excellent performance [7–9].

J. Xiong et al. (Eds.): MobiMedia 2021, LNICST 394, pp. 757–767, 2021.
https://doi.org/10.1007/978-3-030-89814-4_56

The key of HSRRS image classification is to extract feature information from HSRRS images. In order to improve the classification accuracy, some researchers use CNN models to focus on extracting feature channels and texture information [10–12]. However, CNNs under these methods require more neural network layers and processing steps. Furthmore, more processing steps will need more intermediate characteristic parameters and lead to more parameter calculations. The increase in model size and the decrease in operating efficiency are common. Therefore, it is of practical significance to lighten the neural network [13–16].

In this paper, we propose a lightweight processing method on the basis of maintaining the original CNN's performance in HSRRS image scene classification. Lightweight CNN requires less memory consumption and floating point operations (FLOPs). The main contribution of the paper includes two aspects:

- The DC-CNN is obtained by applying the DC method into CNN. This method expands the receptive field of the convolution operation so that more feature information of identical sub-object can be retained during the downsampling process of the feature map. The results show that this method improves the classification accuracy and lightens the model.
- On the basis of DC-CNN, neuron pruning is applied to prune redundant neurons and recover the model original performance after fine-tuning. The benefit of a smaller DC-LW-CNN is to reduce the memory consumption and FLOPs.

2 Related Works

The CNN method based on deep learning has been applied to HSRRS image scene classification due to its powerful feature learning ability. Considerable research efforts using CNN have gone into improving the intra-class diversity and inter-class difference of HSRSS images by extracting more feature information. R. Zhu et al. proposed an attention-based deep feature fusion (ADFF) framework using attention maps to perform deep feature fusion and defining a new cross-entropy loss function [10]. Z. Wang et al. used smooth expansion convolution and separable-shared operations to solve grid artifacts and improve the performance of network models in dense prediction tasks [17]. Jinbo Xiong et al. integrated securre functions into the VGG-6 model to construct a lightweight p-CNN [18].

However, these optimized CNNs generally depend on a large model size and high FLOPs. It means that the large CNN requires more memory consumption and runs slowly on hardware with limited resources. In other words, this will make it difficult to embed complex CNN model into portable devices with limited resources[13–16]. Therefore, it is necessary to apply lightweight processing to remove redundancy in the CNN. From the perspective of reducing network storage, FLOPs and reducing the complexity of the network model, general lightweight methods can be divided into five categories: knowledge distillation, compact convolution kernel design, parameter quantization, low rank decomposition and parameter pruning [13]. In some work on using parameter pruning,

H. Hu et al. pruned the redundant neurons to lighten VGG-6 by Apoz criterion to obtain an effective CNN architecture [19]. Z. Chen et al. used low-rank approximation to obtain the approximate network, and applied channel pruning to the approximate network in a global manner [20]. According to the principle of pruning redundancy judgment under different algorithms, the redundant filters are also pruned to construct a lightweight CNN [21,22].

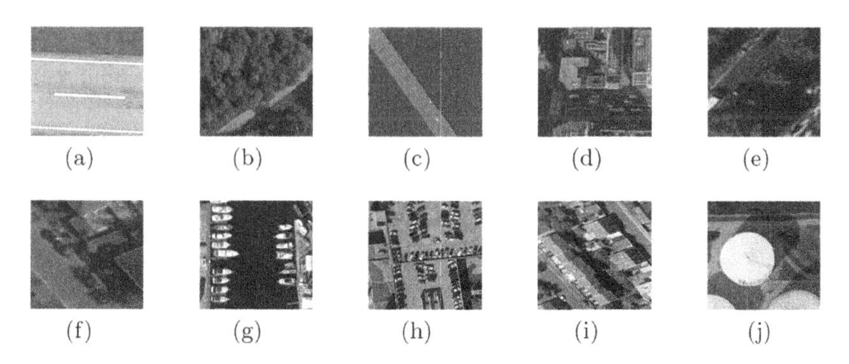

Fig. 1. Typical 10 examples of the HSRRS: (a) airport runway, (b) avenue, (c) bridge, (d) city building, (e) city green tree, (f) city road, (g) marina, (h) parking lot, (i) residents, (j) storage.

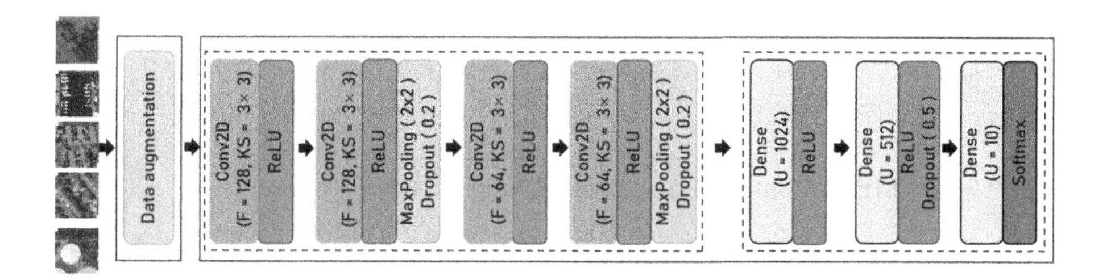

Fig. 2. Original CNN architecture for classifying HRRS images.

3 Theoretical Basis

In this section, we will introduce the structure of the original CNN in [7] and the performance of the original CNN in detail.

3.1 Structure of the Original CNN

The objects of classification are 10 types of HSRRS images (see Fig. 1). They are extracted from the University of California (UC) merced land use dataset

[23] and the remote sensing image classification benchmark (RSI-CB) dataset [24]. In this paper, the original CNN will be further optimized (see Fig. 2). It has 4 convolutional layers. The standard convolution kernel of 3×3 specification is mainly used to extract the local feature information in the HSRRS images. Following every two convolutional layers, a max-pooling layer is set to compress the input feature map. In order to improve the generalization ability of the CNN, we also set the dropout layer to make some feature detectors stop working. The convolutional layers are followed by 3 fully connected layers to integrate these feature information. Each neural network layer is followed by a nonlinear rectified linear unit (ReLU) activation function to enhance the expression ability of the model and fit the correct label better. Finally, the soft-max loss function is applied to calculate the probability distribution of the input sample images on each label. In the testing phase, the neural network model selects the labels with the highest probability as the output results.

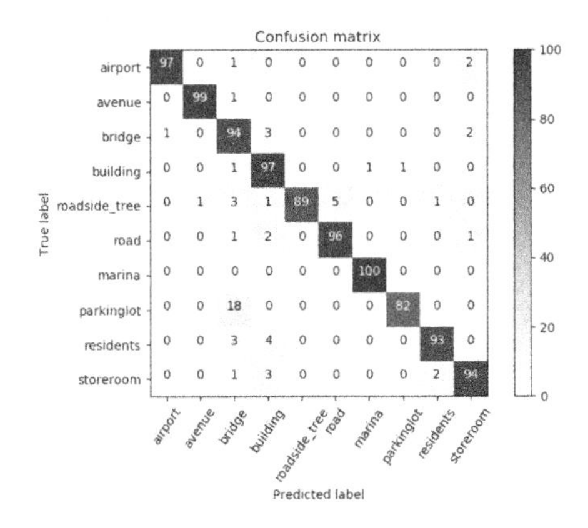

Fig. 3. Confusion matrix of the original CNN test results.

3.2 Performance of the Original CNN

Through training and testing the original CNN, the original CNN has a superior classification effect in the HSRRS image classification task. The confusion matrix during testing is shown in Fig. 3. Its average accuracy (AA) is 94.10%. It is worth noting that it also has the disadvantages of large number of parameters and slow calculation speed. Therefore, we aim to construct DC-LW-CNN through DC and neuron pruning methods to obtain a certain lightweight effect.

4 The Proposed DC-LW-CNN Method

In this section, we will specifically introduce DC-LW-CNN method. We use the DC to replace the standard convolution of the original CNN to obtain the DC-CNN (see Fig. 4). Subsequently, neuron pruning method is applied for further processing to obtain the final DC-LW-CNN.

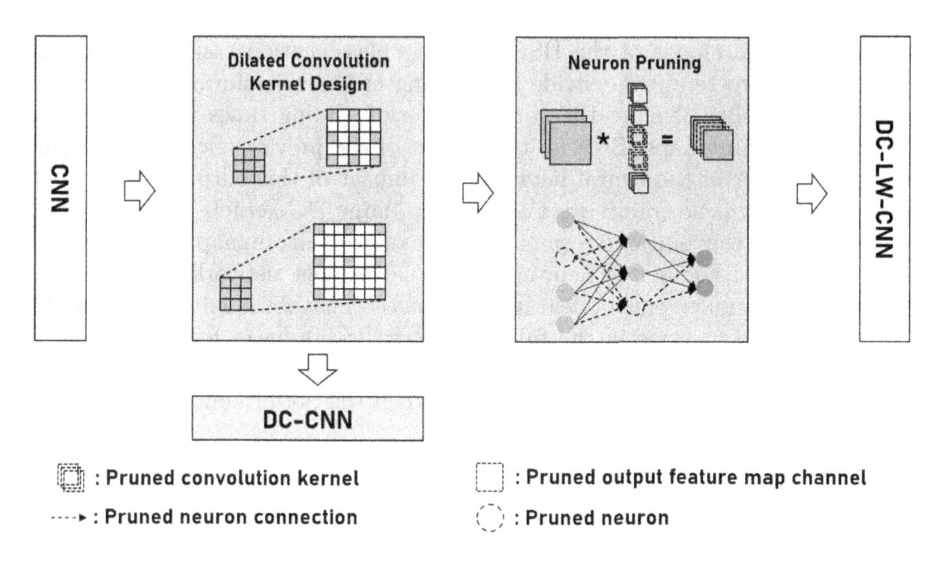

Fig. 4. Overview of DC-LW-CNN.

4.1 Dilated Convolution

When CNNs are used to deal with image classification problems, CNNs stacked with multi-layer convolutional pooling have achieved good feature extraction results. It is worth noting that the information of the original image mapped by a pixel on the output feature map determines the upper limit of information extraction. In order to extract more feature information, we can rely on more downsampling operations to increase the receptive field. However, it will make some objects difficult to be detected and result in a decrease in classification performance. At the same time, the multi-layer convolution operation is also accompanied by more resource consumption. In application scenarios with limited resources, it is not advisable to rely on more convolutional pooling operations to improve CNN feature extraction capabilities. DC, as a universally applicable convolution method, can increase the receptive field of convolution without introducing additional parameters and retain more feature information of feature maps. Furthermore, the positions of different objects in the HSRRS image are basically distributed in blocks, and a large receptive field is suitable. As shown in Fig. 4, when we set the 2-D dilation rate to r, the dilated convolution is constructed by inserting $r-1$ holes (zeros) between pixels in the standard convolutional kernel. The size of the convolution kernel changes from $K \times K$ to $K_d \times K_d$, where $K_d = K + (K-1) \times (r-1)$. Since the DC kernel is larger than the standard convolution kernel, the application of DC can increase the receptive field. Furthermore, DC ($r = 1$) can be equivalent to the standard convolution.

The HSRRS image contains many types of sub-objects, and the same sub-objects are relatively concentrated. These sub-objects form a larger area in the HSRRS image and contain more information. Therefore, we apply the DC convolution method to retain more feature information of the same sub-objects and ensure the performance of the HSRRS image classification task. At the same time, we hope to retain the stride size setting of the convolution operation to reduce the number of large DC core convolution sliding times and the size of the output feature maps. Because the output of the previous layer is the input of the next layer in the neural network, the impact of this output feature map size reduction will be transitive. Furthermore, large DC kernels will be applied in the next few convolutional layers. The size of the feature map will be further reduced and the number of neurons in the subsequent network layers will be reduced. The smaller output feature map of the convolutional layers and the fewer number of neurons in the fully connected layers mean fewer parameters and fewer parameter operations. As a result, the parameters, memory and the FLOPs of the entire CNN will be reduced when the performance of classifying HSRRS images is maintained.

4.2 Neuron Pruning

The purpose of neuron pruning here is to prune redundant convolution kernels in convolutional layers and neurons in fully connected layers. Neuron pruning is a method of obtaining redundant neurons by discriminating criteria, then removing these relatively unimportant neurons. The use of neuron pruning further simplifies the DC-CNN, allowing fewer parameters to participate in the calculation, thereby achieving a better lightweight effect.

The specifics of this method in this paper are: Firstly, we pre-train DC-CNN model that requires pruning. Secondly, since the output of the previous network layer will affect the processing of the subsequent network layers, we select the neural network model layer to be pruned in the order of the network layers by calculating the weight sum of the nodes in this layer $W_n^{(i)}$ and the average weight sum of all nodes in this layer $\bar{W}_n$. The $W_n^{(i)}$ and $\bar{W}_n$ can be described as

$$W_n^{(i)} = W(O_n^{(i)}) = \sum_{k\times k} w(O_n^{(i)}) \tag{1}$$

$$\bar{W}_n = \bar{W}(O_n) = \frac{\sum_i^M \sum_{k\times k} w(O_n^{(i)})}{M} \tag{2}$$

where $O_n^{(i)}$ is the i-th node in c-th layer, $w(\cdot)$ is the weight of the node, k is the size of the convolution kernel, M is the number of nodes in the layer. Then, according to the discrimination result, we prune the redundant neurons. If $W_n^{(i)} < \bar{W}_n$ is satisfied, we will prune $O_n^{(i)}$. Instead, we will save $O_n^{(i)}$. As shown in Fig. 4, when the redundant neurons of different types of neural network layers and their connections are pruned, the model will become sparse. Finally, we load the weights of the unpruned neurons and fine-tune the network model to restore its performance through a few epochs retraining. The specific pruning algorithm flow can be seen in Algorithm 1.

Algorithm 1. The proposed DC-LW-CNN method.

Input: 10 types of HSRRS images; The batch_size is set to 64 during training and 100 HSRRS images of each type are used for validating and testing;
Output: The DC-LW-CNN;
1: Set 50 epochs to train the original CNN; Turn on the early_stopping setting and monitor the validation set loss of CNN; The patience of early_stopping is set to 5;
2: Train the original CNN and save the weight of the original CNN;
3: Load weight and test the original CNN;
4: Train and test DC-CNN with r of 3, 5, 7 respectively and remain the other settings unchanged;
5: Load weight of DC-CNNs and neuron pruning;
6: **for** n-th layer in DC-CNN_layers **do**
7: **for** i-th node in n-th layer **do**
8: Calculate the $W_n^{(i)}$ and $\bar{W}_n$ according to the weight;
9: **end for**
10: **if** $W_n^{(i)} < \bar{W}_n$ **then**
11: Prune node $O_n^{(i)}$;
12: **else**
13: Save node $O_n^{(i)}$;
14: **end if**
15: Set 10 epochs to fine-tune DC-CNN;
16: **end for**
17: **return** DC-LW-CNN.

5 Simulation Results

In this section, we demonstrate the effectiveness of the DC method and neuron pruning method through simulation. For DC-CNNs and DC-LW-CNNs obtained by simulation, we evaluate and compare the performances of the changed model with the original CNN. The evaluation indicators are the four aspects of AA, parameters, memory and FLOPs. In these CNNs, we set the learning rate to 0.001. The optimizer is the RMS Prop optimizer and the loss function is the Categorical cross entropy loss function. Meanwhile, we monitor the loss value of the validation set to stop training early. Our device is equipped with GTX 1080Ti and deep learning library is Keras with Tensorflows as the backend.

5.1 Classification Performance

In this paper, we obtain the performance of different CNNs through the confusion matrix of true labels and predicted labels of HSRRS images. The AA of the original CNN is 94.10% (see Fig. 3). We apply DC with r of 3, 5 and 7 to four convolutional layers respectively. The confusion matrixs of DC-CNNs are shown in Fig. 5. The corresponding AA rates are 95.80%, 95.10% and 95.70% respectively. Figure 6 shows the classification performance of DC-LW-CNNs. The AA of DC-LW-CNNs are 95.50%, 95.30% and 95.40%.

In general, DC-CNNs and DC-LW-CNNs have better classification performance than the original CNN. It meets the precondition of lightweight processing. Comparing all the confusion matrices, we can find that some scenes are easily misclassified by the original CNN. For example, parking lot scene is easily misclassified as bridge scene. DC-CNNs have alleviated this problem to some extent. Among them, DC-CNN ($r = 3$) makes the classification accuracy of each scene reach 90% and above. As the size of the DC core becomes larger, the performance of other DC-CNNs classifying for each type of HSRRS images is not so stable. For DC-LW-CNNs, due to the removal of redundant neuron nodes, some feature information is missing, which leads to the degradation of AA performance and some new scene misclassification problems. However, they do not have the same degree of misjudgment performance as the original CNN.

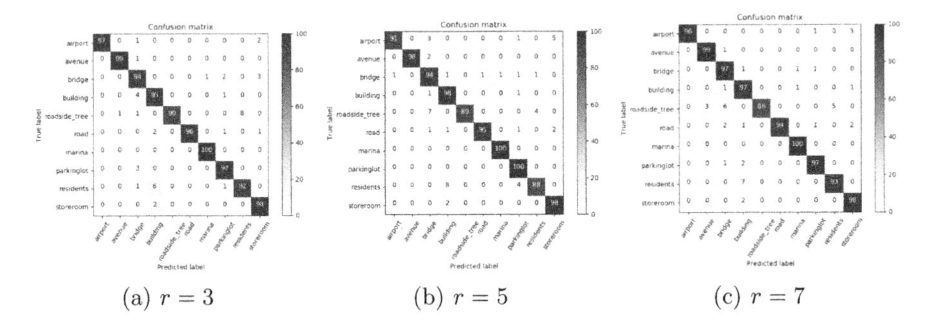

(a) $r = 3$ (b) $r = 5$ (c) $r = 7$

Fig. 5. Confusion matrix of the DC-CNNs test results (a)–(c) ($r = 3, 5, 7$).

5.2 Compression and Acceleration Performance

In the case of maintaining the better performance, the lightweight effect of the CNN is mainly reflected in compression and acceleration. We evaluate the effect of compression and acceleration through parameters, memory and FLOPs. We give the model architecture information of DC-LW-CNNs (see Table 1) and the

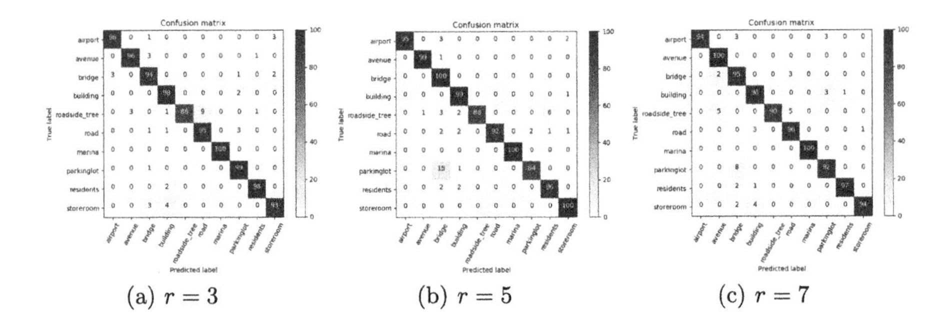

(a) $r = 3$ (b) $r = 5$ (c) $r = 7$

Fig. 6. Confusion matrix of the DC-LW-CNNs test results (a)–(c) ($r = 3, 5, 7$).

Table 1. Model architecture comparison.

Model	DConv1	DConv2	DConv3	DConv4	FC1	FC2	FC3
DC-LW-CNN ($r = 3$)	81	75	35	34	359	260	10
DC-LW-CNN ($r = 5$)	79	73	33	31	326	250	10
DC-LW-CNN ($r = 7$)	70	73	29	33	316	241	10

* "DConv" is the dilated convolution layer and "FC" is the fully-connected layer.

Table 2. Model performance comparison.

Methods	Parameters	Memory (MB)	FLOPs	AA (%)
CNN [3]	59,775,242	250.196	5,773,887,488	94.10
DC-CNN ($r = 3$)	48,568,586 (↓18.75%)	223.466 (↓10.68%)	5,371,037,696 (↓6.98%)	95.80 (↑1.81%)
DC-CNN ($r = 5$)	38,541,578 (↓35.52%)	183.510 (↓26.65%)	4,986,472,448 (↓13.64%)	95.10 (↑1.06%)
DC-CNN ($r = 7$)	29,694,218 (↓50.32%)	148.146 (↓40.78%)	4,620,191,744 (↓19.98%)	95.70 (↑1.70%)
DC-LW-CNN ($r = 3$)	9,086,165 (↓84.80%)	57.685 (↓76.94%)	1,957,829,994 (↓66.09%)	95.50 (↑1.49%)
DC-LW-CNN ($r = 5$)	5,990,782 (↓89.98%)	44.261 (↓82.31%)	1,722,254,136 (↓70.17%)	95.30 (↑1.28%)
DC-LW-CNN ($r = 7$)	4,753,632 (↓92.04%)	37.472 (↓85.02%)	1,430,955,568 (↓75.22%)	95.40 (↑1.38%)

related data information of different methods (see Table 2). DC-LW-CNNs have been pruned redundant neuron nodes and achieved certain lightweight effect. For them, the lightweight effect of the convolutional layer is obvious, and the nodes of each convolutional layer are pruned by about 50%. The larger the value of r we set, the better the lightweight effect. Simulation results show that the compression and acceleration performance of DC-LW-CNN ($r = 7$) is better. Its parameters are $\{4,753,632\}$. It occupies 37.472 MB memory. Its FLOPs is $\{1,430,955,568\}$. In terms of actual storage and operation of the CNN model, the memory is reduced by 85.02%, and the FLOPs is reduced by 75.22%.

6 Conclusion

In this paper, a DC-LW-CNN method using DC and neuron pruning is proposed to reduce the parameters and keep the performance the original CNN. The application of larger DC kernels expands the receptive field of the output feature maps and reduces the number of convolution operations. The neuron pruning method removes the redundancy of the neural network and realizes the compression and acceleration of the neural network. The simulation results show that the DC-LW-CNN ($r = 7$) has comparable performance in HSRRS image scene classification with 8% of the parameters, 15% of the memory and 25% of the FLOPs in the original CNN. At the same time, the classification accuracy of each type of scene has reached 90% or above, and the classification performance is more stable. However, the difference between some scenes comes from the feature information of small objects, and a large receiving field is not required to extract the features of some small objects. Therefore, DC-CNN ($r = 7$) does not have relatively stable classification performance. In future work, we plan to apply different DCs to different convolutional layers and combine the residual blocks to extract multi-scale information in HSRRS images to obtain a better and stable classification effect. At the same time, more different lightweight methods should be applied to the CNN model to obtain better lightweight effects.

References

1. Su, H., Yu, Y., Du, Q., Du, P.: Ensemble learning for hyperspectral image classification using tangent collaborative representation. IEEE Trans. Geosci. Remote Sens. **58**(6), 3778–3790 (2020)
2. Su, H., Zhao, B., Du, Q., Du, P.: Kernel collaborative representation with local correlation features for hyperspectral image classification. IEEE Trans. Geosci. Remote Sens. **57**(2), 1230–1241 (2019)
3. Yu, Y., Su, H.: Collaborative representation ensemble using bagging for hyperspectral image classification. In: IGARSS 2019 - 2019 IEEE International Geoscience and Remote Sensing Symposium, Japan, pp. 2738–2741 (2019)
4. Patil, N., Ingole, K., Mangala T. R.: Deep convolutional neural networks approach for classification of lung diseases using X-Rays: COVID-19, pneumonia, and tuberculosis. Int. J. Perform. Eng. **16**(9), 1332–1340 (2020)
5. Wang, Y., Gui, G., Ohtsuki, T., Adachi, F.: Multi-task learning for generalized automatic modulation classification under non-Gaussian noise with varying SNR conditions. IEEE Trans. Wirel. Commun. **20**, 3587–3596 (2021)
6. Zhang, Y., et al.: CV-3DCNN: complex-valued deep learning for CSI prediction in FDD massive MIMO systems. IEEE Wirel. Commun. Lett. **10**(2), 266–270 (2020)
7. Li, W., Liu, H., Wang, Y., Li, Z., Jia, Y., Gui, G.: Deep learning-based classification methods for remote sensing images in urban built-up areas. IEEE Access **7**, 36274–36284 (2019)
8. Du, P., Li, E., Xia, J., Samat, A., Bai, X.: Feature and model level fusion of pretrained CNN for remote sensing scene classification. IEEE J. Sel. Top. Appl. Earth Obs. Remote Sens. **12**(8), 2600–2611 (2019)
9. Cheng, G., Xie, X., Han, J., Guo, L., Xia, G.: Remote sensing image scene classification meets deep learning: challenges, methods, benchmarks, and opportunities. IEEE J. Sel. Top. Appl. Earth Obs. Remote Sens. **13**, 3735–3756 (2020)
10. Zhu, R., Yan, L., Mo, N., Liu, Y.: Attention-based deep feature fusion for the scene classification of high-resolution remote sensing images. Remote Sens. **11**(17), 1–24 (2019)
11. Li, F., Feng, R., Han W., Wang, L.: High-resolution remote sensing image scene classification via key filter bank based on convolutional neural network. IEEE Trans. Geosci. Remote Sens. **58**(11), 8077–8092 (2020)

12. Li, E., Samat, A., Du, P., Liu, W., Hu, J.: Improved bilinear CNN model for remote sensing scene classification. IEEE Geosci. Remote Sens. Lett. (2020)
13. Cheng, Y., Wang, D., Zhou, P., Zhang, T.: Model compression and acceleration for deep neural networks: The principles, progress, and challenges. IEEE Sig. Process. Mag. **35**(1), 126–136 (2018)
14. Sharma, A.K., Foroosh, H.: Slim-CNN: a light-weight CNN for face attribute prediction. In: 15th IEEE International Conference on Automatic Face and Gesture Recognition (FG 2020), pp. 329–335. Buenos Aires, Argentina (2020)
15. Wang, Y., et al.: Distributed learning for automatic modulation classification in edge devices. IEEE Wirel. Commun. Lett. **9**(12), 2177–2181 (2020)
16. Li, W., et al.: Classification of high-spatial-resolution remote sensing scenes method using transfer learning and deep convolutional neural network. IEEE J. Sel. Top. Appl. Earth Obs. Remote Sens. **13**(1), 1986–1995 (2020)
17. Wang, Z., Ji, S.: Smoothed dilated convolutions for improved dense prediction. In: Proceedings of the 24th ACM SIGKDD International Conference on Knowledge Discovery and Data Mining, pp. 2486–2495 (2018)
18. Xiong, J., Bi, R., Zhao, M.: Edge-assisted privacy-preserving raw data sharing framework for connected autonomous vehicles. IEEE Wirel. Commun. **27**(3), 24–30 (2020)
19. Hu, H., Peng, R., Tai, Y.-W., Tang, C.-K.: Network trimming: a data-driven neuron pruning approach towards efficient deep architectures (2016). https://arxiv.org/abs/1607.03250
20. Chen, Z., Chen, Z., Lin, J., Liu, S., Li, W.: Deep neural network acceleration based on low-rank approximated channel pruning. IEEE Trans. Circuits Syst. I: Regul. Pap. **67**(4), 1232–1244 (2020)
21. He, Y., Zhang, X., Sun, J.: Channel pruning for accelerating very deep neural networks. In: IEEE International Conference on Computer Vision (ICCV), Venice, Italy, pp. 1389–1397 (2017)
22. Lin, S., Ji, R., Li, Y., Deng, C., Li, X.: Toward compact convNets via structure-sparsity regularized filter pruning. In: IEEE International Conference on Image Processing (ICIP), vol. 31, no. 2, pp. 574–588 (2020)
23. Yang, Y., Newsam, S.: Bag-of-visual-words and spatial extensions for land-use classification. In: Proceedings of the 18th SIGSPATIAL International Conference on Advances in Geographic Information Systems, pp. 270–279 (2010)
24. Li, H., et al.: RSI-CB: a large scale remote sensing image classification benchmark via crowdsource data (2017). https://arxiv.org/abs/1705.10450

LSTM-Based Battlefield Electromagnetic Situation Prediction

Hengchang Zhang, Shengjie Zhao(✉), and Rongqing Zhang

School of Software Engineering, Tongji University, Shanghai 201804, China
{hengchang_zhang,shengjiezhao,rongqingz}@tongji.edu.cn

Abstract. In the modern battlefield, the complex electromagnetic environment has brought considerable challenges to assessing the electromagnetic situation. Traditional electromagnetic situation assessment methods are difficult to process massive high-dimensional data, making it challenging to predict the electromagnetic situation. In recent years, the in-depth development of deep learning has provided a breakthrough in the electromagnetic situation field. However, related research mainly focuses on threat assessment and electromagnetic situation complexity prediction. There are few papers on electromagnetic situation prediction using machine learning. In this paper, we propose an electromagnetic situation prediction method based on long short-term memory (LSTM). We first build an attack-defense model based on deep reinforcement learning to simulate the electromagnetic situation. Then we use LSTM to predict the development of the situation and improve the loss function to reduce the prediction error. Furthermore, we analyze the impact of different situation features on the final win rate and use a small amount of situation information to predict the win rate with high accuracy. The experimental results show that this method can effectively predict the electromagnetic situation, providing excellent decision support for commanders.

Keywords: LSTM · Electromagnetic situation · Situation prediction

1 Introduction

The battlefield electromagnetic situation refers to the state and the development trend of the confrontation between the two sides on the battlefield. With the wide application of spectrum equipment and electromagnetic technology, the battlefield electromagnetic environment has become increasingly complex, which poses a con-

This work is supported in part by the National Key Research and Development Project under Grant 2019YFB2102300 and 2019YFB2102301, in part by the National Natural Science Foundation of China under Grant 61936014 and 61901302, in part by the Fundamental Research Funds for the Central Universities (China), and is also funded by State Key Laboratory of Advanced Optical Communication Systems Networks, China.

J. Xiong et al. (Eds.): MobiMedia 2021, LNICST 394, pp. 768–781, 2021.
https://doi.org/10.1007/978-3-030-89814-4_57

siderable challenge to assessing and predicting the electromagnetic situation. Most existing situation assessment methods focus on providing commanders with the past and current electromagnetic conditions but cannot predict future development, so it is difficult to formulate strategies based on the development trends. Therefore, accurately predicting and evaluating the battlefield electromagnetic situation has become an urgent problem in the future battlefield.

Endsley [1] divided the situation assessment process into three stages: situation awareness, situation comprehension, and situation prediction. At present, the research on electromagnetic situation mainly focuses on the first two stages, namely electromagnetic situation awareness and electromagnetic situation comprehension. Cai *et al.* [2] analyzed the elements of the battlefield electromagnetic situation, and used a signal population distribution method to calculate the complexity of the battlefield electromagnetic environment. Shen *et al.* [3] designed a 3D electromagnetic situation visualization system based on geographic information system (GIS). Yuan *et al.* [4] proposed an index system to characterize the complexity of the electromagnetic environment, and predicted the complexity based on Bayesian networks.

Traditional electromagnetic situation assessment methods are complex and difficult to process massive high-dimensional data. Deep learning methods, as a comparison, have more powerful representation learning capabilities and can automatically extract various complex features from the original data. In recent years, various deep learning algorithms have been proposed for different types of problems, such as Convolutional Neural Networks [5], Recurrent Neural Networks [6], and Residual Networks [7], etc. These methods are widely used in image recognition, object detection, language modeling, and other fields [8]. In the employment of deep learning in electromagnetic situation investigation, related research focuses on electromagnetic threat prediction [9] and electromagnetic interference prediction [10], but there are few papers about the battlefield electromagnetic situation prediction.

In this paper, we propose an LSTM-based electromagnetic situation prediction method for the two-sided battlefield, which can achieve efficient prediction of various electromagnetic situation parameters with high accuracy. This method requires a large amount of data for model training. To solve the problem of insufficient data sources, we first establish an attack-defense model based on deep reinforcement learning. Then, we build an LSTM situation prediction model and improve the loss function to smooth the prediction results. We use the simulated data to train and test the model, and find that compared to MSE loss, the improved loss function can reduce the prediction error (RMSE) by about 10.35%. In addition, we analyze the impact of different situation features on the final win rate by analysis of variance and use a small amount of situation information to predict the win rate with high accuracy.

The rest of this paper is organized as follows: Sect. 2 proposes the attack-defense model based on deep reinforcement learning to generate data for situation prediction. Section 3 introduces the LSTM model we use. Section 4 introduces our experiments and the results on situation prediction. Section 5 concludes the paper.

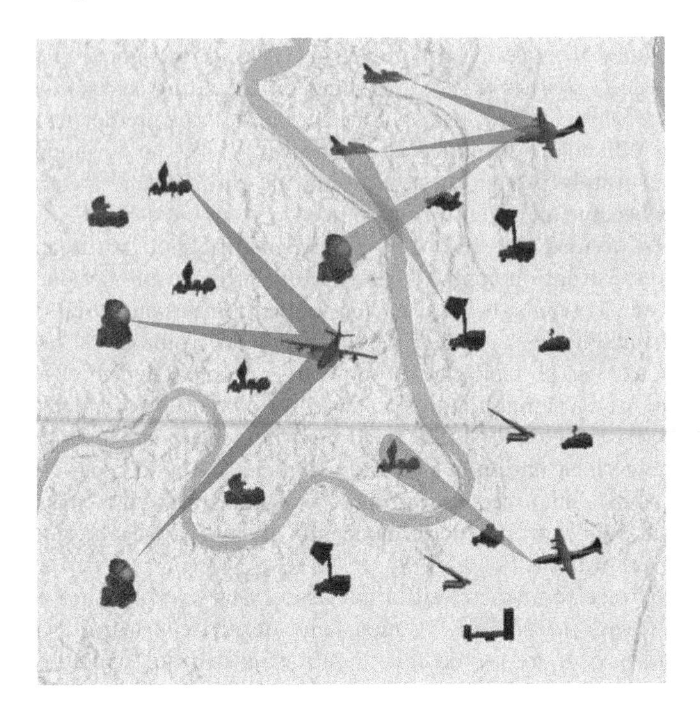

Fig. 1. The simulated battlefield.

2 Attack-Defense Model

The battlefield electromagnetic situation is obtained from the original battlefield data through data fusion methods. However, the battlefield data are confidential to all countries, and it is hard to get. Therefore, we build an attack-defense model, a type of wargaming [11], to generate simulated data. Wargaming has a long history. It is mainly used to simulate the battlefield environment, on which the commander can deploy their strategies and prove their ideas. The traditional wargaming model requires people to manually set the battlefield environment and manually manipulate the battlefield elements to advance the war. The cost of acquiring data in this way is unacceptable. In 2017, Deepmind proposed AlphaGo Zero [12], which is a Go program based on deep reinforcement learning. The program generates data through self-playing games and uses the data for training, thus becoming a Go expert from scratch. This inspires us to build a deep reinforcement learning model to compete with each other in the simulated battlefield environment and use the data to analyze and predict the battlefield situation.

2.1 Situation Description

Figure 1 is the scene diagram of the attack-defense model. In one game, the two sides participating in the game each has a certain number of missiles and other equipment. Use $s_t = [s_t^r, s_t^b]$ to represent the state of two sides at time-step t. s consists of two matrices with the same shape.

The shape of s^r or s^b is 3 rows and $M + W$ columns. Each column represents one type of equipment. The first row represents the equipment's quantity, the second represents the code of the equipment, and the third represents the total remaining value of the equipment in this column. The first M columns represent M types of missiles. There can be multiple missiles of each type, so the first row of the first M columns can take a number greater than 1. The last W columns represent equipment other than missiles. There is at most one piece of equipment in each column, so the first row of the last W columns can only be 0 or 1.

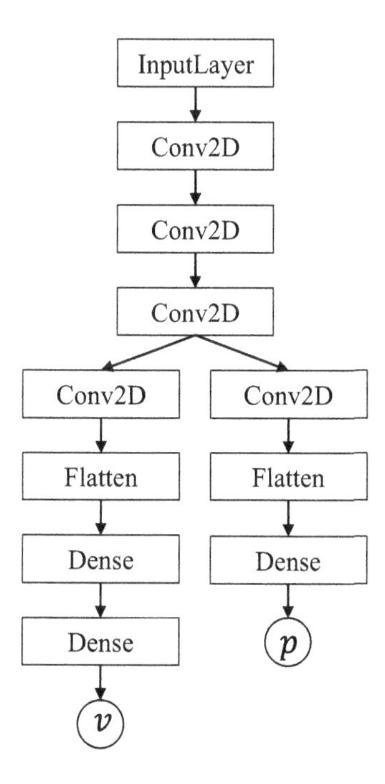

Fig. 2. Structure of the CNN used in the attack-defense model.

2.2 Decision Model

The game is turn-based. Both sides attack alternatively. In each time-step of the game, one side chooses one of its missiles to attack the other side's non-missile unit, so the decision can be denoted by an $M * W$ matrix. The probability of a successful attack is related to the accuracy of the missile acc_i and the opponent's overall defense value def, as in

$$prob = \frac{c \cdot acc_i}{c + def} \tag{1}$$

where c is a preset parameter. If a unit is successfully hit, the unit's value will be reduced by the missile's attack power. If the value of a unit is reduced to 0, then the unit will be removed.

The decision model's purpose is to make the appropriate decision π according to the current state s. We use a neural network guided MCTS [12] to search for the best move. Figure 2 describes the structure of the neural network $(p, v) = f_\theta(s)$.

2.3 Model Training Process

The neural network f_θ is trained based on the data generated from games of self-play.

At the initial state $i = 0$, the neural network's parameter θ_0 is randomly initialized. Then at each iteration $i > 0$, multiple games are played. At each time-step t in one game, an MCTS search $\pi_t = a_{\theta_{i-1}}(s_t)$ is executed under the guidance of the neural network. A game terminates at time-step T when one side's total value becomes 0 or when the game exceeds a maximum length. A sample (s_t, π_t, z_t) is produced at each time-step t, where $z_t = 1$ if the current player is the final winner of the game, and $z_t = -1$ if otherwise. The neural network is trained from the data (s, π, z) to minimize the difference between p and π, also v and z.

When the training of the neural network f_θ is completed, we then use the model to play more games, and use the data generated to analyze the situation.

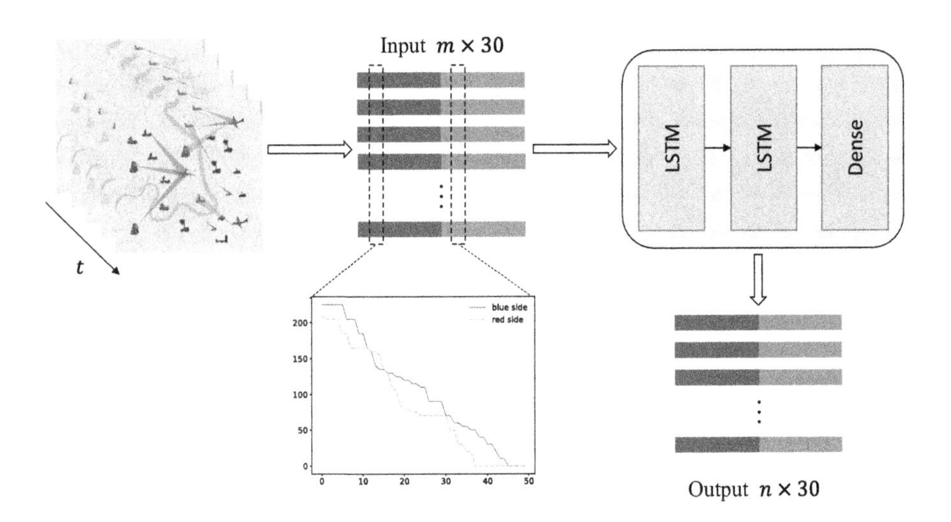

Fig. 3. Overall architecture of the LSTM.

3 LSTM-Based Prediction Method

In the previous section, we build an attack-defense model to simulate battlefield electromagnetic data. This section proposes the situation prediction method based on LSTM. First, extract the situation features from the original data, then preprocess the data, in addition, build an LSTM model to predict the situation. Figure 3 is the overall architecture of the proposed method, where m and n are the number of input time-steps and output time-steps, respectively.

3.1 Feature Extraction

The data generated above represents the number and type of battlefield elements. For a game with T time-steps, the dimension of the data collected is $[T, 2, 3, 100]$. The original data are sparse, and it is difficult to obtain the overall situation from it. We extract 15 features from the original data, as is listed in Table 3. Figure 4 shows 2 of the 15 features of both sides in a 48-time-step game.

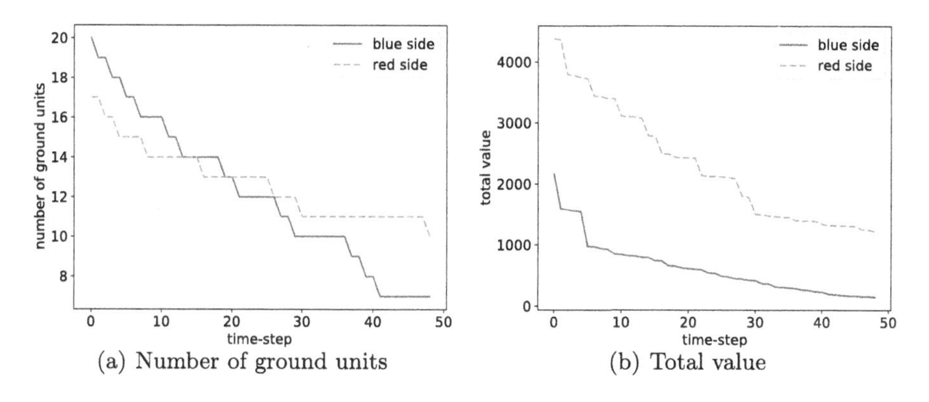

Fig. 4. 2 of 15 features in one game. (Color figure online)

As a result, the situation prediction problem is transformed into a time series prediction problem, which predicts the situation of both red and blue in the future m time-steps through the known situation of the red and blue sides of n time-steps.

3.2 Data Preprocessing

As shown in Fig. 4, the difference between the absolute values of different features is very large. For example, the number of ground units is generally dozens, but the value of total units can reach several thousand. The vast difference between different features will affect the training of the model. The feature with a larger value will occupy a larger weight in the initial stage of training, which makes it difficult for the feature with a relatively small value to play a role. Therefore, the raw data need to be normalized. We use the min-max normalization, namely:

$$x_i' = \frac{x_i - min(x)}{max(x) - min(x)}. \tag{2}$$

In this question, if the 30 features are normalized separately, the contrast between the two sides will be lost. Figure 4(b) represents the total value of the red and blue sides of 48 time-steps in one game. It can be seen that in the beginning, the total value of the red side is more than twice that of the blue side. If the two sides' features are normalized separately, after normalization, the total value is shown in Fig. 5(a). It can be seen that the total value of the two sides has both become 1 at the initial state.

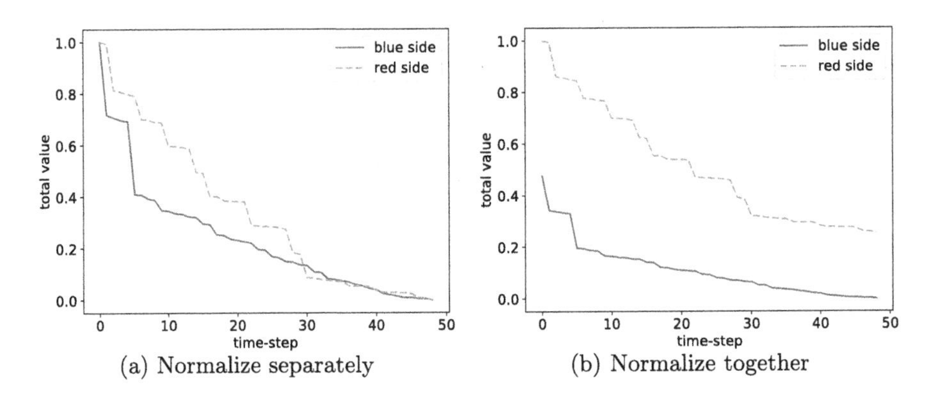

Fig. 5. Comparison of two normalization methods. (Color figure online)

To avoid this problem, we normalize the same feature of both sides simultaneously. The result is shown in Fig. 5(b). This method maps the total value to the $[0, 1]$ interval while preserving the comparison between the two sides, and the normalized data are suitable for training in a neural network.

After feature extraction, the $[T, 2, 3, 100]$ raw data become the $[T, 30]$ features. To fully utilize the data for training the neural network, we need to convert the time series data to pairs of input and output sequences using the sliding window method. When predicting n time-steps with m time-steps, use data of the former m time-steps as input and the data of the following n time-steps as output. In this case, for a game with T time-steps, the number of samples that can be generated is $T - m - n + 1$.

3.3 LSTM Model

In time series forecasting problems, RNN (Recurrent Neural Network) [13] is a commonly used method. In recent years, RNN has achieved great success in many problems such as speech recognition, machine translation, and image description. However, RNN is hard to handle the long-term dependencies [14]. To address this problem, LSTM [15] introduces the cell state to maintain long time memory and replaces the hidden layer neuron with a special LSTM structure with three gates, namely input gate, forget gate, and output gate. The input gate controls how much new information is added to the cell state. The forget gate decides how much cell state to forget. The output gate controls how much of the updated cell state to output.

We build a neural network with two layers of LSTM with 128 units and one fully connected layer. We use Relu as the activation function and Adam as the optimizer. In addition, we add a penalty term to the MSE loss function to smooth the prediction results and improve the prediction accuracy, as in

$$loss = \mathrm{MSE} + \sum_{i} \sum_{t} \mathrm{Relu}(d_{t+1}^{i} - d_{t}^{i}) \tag{3}$$

where d_t^i is the value of the i-th feature at time-step t. In Sect. 4, we will introduce the prediction results and the comparison of different loss functions.

4 Experimental Results

In the previous section, we extract 30 features from the self-play data and propose an LSTM model to predict the electromagnetic situation. This section uses the simulated data generated by the attack-defense model to evaluate the prediction method. We further analyze the features by ANOVA (Analysis of Variance) and predict the win rate by a small part of these features.

4.1 Situation Prediction

We use the data generated by self-play to train and test the LSTM model proposed in Sect. 3. A total of 800 games are generated. Each game contains up to 50 time-steps. We randomly select 80% of the data as the training set and use 20% of the data as the test set. The batch size is set to 16.

Table 1. Results of different loss function

Loss function	RMSE (10^{-2})	MAPE (%)
MSE	1.2986	4.72
MSE+penalty	1.1642	3.44

The result on the test set is shown in Table 1, the input time-step is 12, and the output time-step is 1. MAPE and RMSE are defined by

$$\mathrm{MAPE} = \frac{100\%}{n} \sum_{i=1}^{n} \frac{|\widetilde{y_i} - y_i|}{y_i} \tag{4}$$

$$\mathrm{RMSE} = \sqrt{\frac{1}{n}\sum_{i=1}^{n}(\widetilde{y_i} - y_i)^2} \tag{5}$$

where $\widetilde{y_i}$ and y_i indicate the predicted value and true value, respectively.

Figure 6 compares the result of the model with a pure MSE loss function and the model with an MSE combines penalty loss function. It can be seen that when the penalty is added, the prediction result is smoother and more close to the real situation.

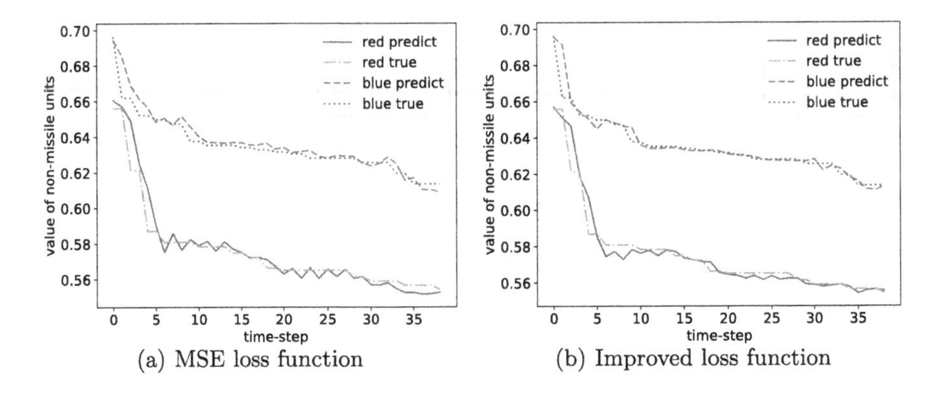

Fig. 6. Prediction results of different loss function.

In the attack-defense model, a move is decided only depends on the current state of both sides and has nothing to do with the path to the current state. In the situation prediction problem, however, more input time-steps can provide the neural network with more information about player preferences and equipment attributes, thereby improving prediction accuracy. We test a variety of combinations of input time-steps and output time-steps, and evaluate the results by RMSE. The results are listed in Table 2.

Table 2. RMSE of different input and output time-step combinations

RMSE (10^{-2})		Input Time-Step				
		1	2	3	4	5
Output Time-Step	1	1.5396	1.2524	1.2470	1.1832	1.1773
	2	1.5731	1.5277	1.4655	1.4327	1.4001
	3	1.8373	1.7904	1.7224	1.7101	1.6616
	4	2.0359	2.0386	1.9866	1.9206	1.8697
	5	2.3092	2.2801	2.2724	2.2617	2.1951

4.2 Win Rate Evaluation Based on Situation

Apart from the situation prediction, the evaluation of the current winning probability is also a critical subject. A proper assessment of the current situation is conducive to the formulation of the following action strategies. In the attack-defense model, the output v of the neural network $f_\theta(s)$ gives the win rate estimation, but the input is the entire current state s. We hope to evaluate the winning probability through the situation. On the one hand, it can reduce the dimension of input data and reduce data acquisition difficulty; on the other hand, it can simplify the model and improve prediction efficiency.

At the t-th time-step of one game, the state of both sides is $s_t = [s_t^r, s_t^b]$. From the aforementioned neural network, the estimated win rate of the current state v_t can be obtained by $(p_t, v_t) = f_\theta(s_t)$. Moreover, the features d_t can be obtained through the feature extraction process. We then take d_t as input and v_t as output to train a neural network to evaluate the win rate based on the situation.

Figure 7(a) is the estimation of the win rate v_t from the red and blue sides in one game. It can be seen that their estimations are not always the same. That is, for example, in some consecutive time-steps, both sides may "think" they are going to lose.

Therefore, we average the win rate estimations for two consecutive rounds. If the red side attacks at time-step t, the win rate estimation is denoted as w_t^r, then the estimation at the next time-step is denoted as w_{t+1}^b. Combining these two time-steps, the win rate of the red side at time-step t is

$$\widetilde{w}_t^r = \frac{w_t^r + 1}{w_t^r + w_{t+1}^b + 2} \tag{6}$$

and the win rate of the blue side is $\widetilde{w}_t^b = 1 - \widetilde{w}_t^r$. After the processing, the win rate of both sides in Fig. 7(a) is converted into Fig. 7(b). We can then take $(d_t, \widetilde{w}_t^r)$ as a sample.

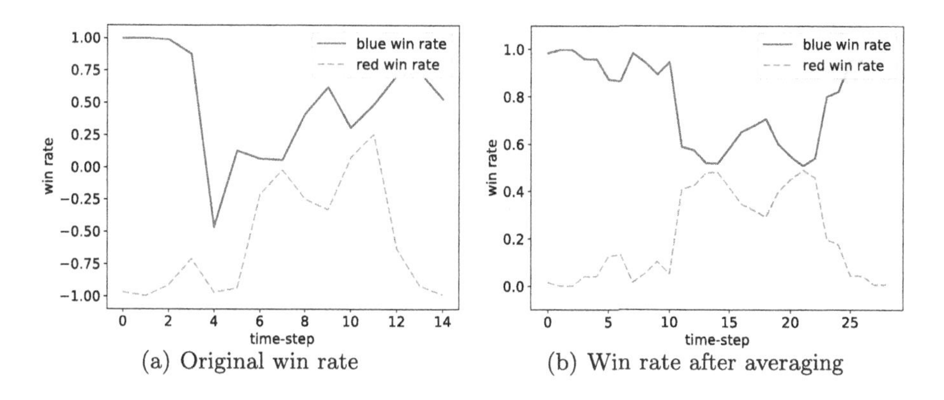

(a) Original win rate (b) Win rate after averaging

Fig. 7. Win rate in one game. (Color figure online)

We build a neural network with seven fully connected layers with dropouts, where the activation function is Relu, the output layer is Sigmoid to ensure the output is in $[0,1]$, the loss function is MSE, and the optimizer is Adam. The network's input is the 30 features of both sides d_t, and the output is the predicted win rate of the red side $\widetilde{w}_t^r$.

We use the data of 300 games, which contains a total of 12126 samples, to train and test the neural network. We use 9700 samples as the training set and others as the test set. The MAPE on the test set is 10.3%, and the MSE is 0.0107.

4.3 Analysis of Variance

In the previous part, we use all the 30 features to predict the win rate. However, it may be challenging to obtain all the situation features on the actual battlefield for various reasons. Therefore, it is necessary to study using fewer features to predict the win rate.

Analysis of variance is widely used to examine the influence of the independent variable A on the dependent variable Y. It partitions the sum of squares (SS) of the dependent variable Y into the sum of squares between groups (SS_A) and the sum of squares within groups (SS_E). SS_A represents the deviation of Y when A takes different values, and SS_E represents the random error of Y when A takes a fixed value. If A is a continuous variable, the values of A can be divided into several levels. ANOVA is a form of hypothesis testing, and the null hypothesis is that all values of factor A have the same effect on Y. If the null hypothesis is true, (7) can be proven, where MS is mean square, k is the total number of possible values A can take, n is the total number of cases, and $F_{k-1,n-k}$ is a F-distribution with $(k-1, n-k)$ degrees of freedom.

$$\frac{MS_A}{MS_E} \sim F_{k-1,n-k} \tag{7}$$

If $MS_A/MS_E > F_{k-1,n-k}(\alpha)$, reject the null hypothesis; otherwise, accept the null hypothesis, where a is the significance level.

The results of the ANOVA for the 30 features are shown in Table 3, which is sorted in descending order over the F value. A larger F value means the feature has a greater influence on the win rate.

We use the first i and last i features in Table 3 ($i = 1, 2, ..., 10$) as the input of the neural network to predict the win rate. The MSE between the predicted value and true value on the test set are shown in Fig. 8. It can be seen that the MSE of prediction using features with larger F-values is relatively smaller. Furthermore, as the number of input features increases, the difference of the MSE between the best and the worst tends to shrink. The MSE of the first ten features is 0.0117, which is very close to using all the 30 features, i.e., 0.0107.

Table 3. Results of ANOVA

Feature	F	PR(>F)
red: value of non-missile units	2863.0286	0
red: number of non-missile units	1675.8144	0
blue: value of non-missile units	1373.2211	7.9974E−285
red: value of anti-ground missiles	375.53513	2.0574E−82
red: value of anti-air missiles	343.7655	1.0750E−75
red: number of missiles	276.7351	1.8566E−61
blue: total value	188.9768	1.1120E−42
blue: number of missiles	182.6833	2.5074E−41
blue: number of non-missile units	138.3295	9.1909E−32
red: value of ground units	130.6649	4.1827E−30
blue: value of anti-ground missiles	117.6721	2.7310E−27
red: value of air units	115.0895	9.9193E−27
blue: value of missiles	76.1177	3.0200E−18
blue: value of anti-air missiles	59.5973	1.2552E−14
red: number of anti-air missiles	56.2550	6.8097E−14
blue: number of anti-ground missiles	55.1852	1.1705E−13
red: total value	55.0848	1.2315E−13
blue: number of anti-air missiles	39.3230	3.7141E−10
blue: value of ground units	36.1119	1.9166E−09
red: number of air units	34.0109	5.6202E−09
red: number of ground units	31.8621	1.6922E−08
red: total defense	31.6575	1.8797E−08
red: value of missiles	24.9382	6.0022E−07
blue: number of ground units	22.9143	1.7139E−06
red: number of all units	22.6500	1.966E−06
red: number of anti-ground missiles	12.1402	4.9519E−04
blue: number of all units	11.4244	7.2711E−04
blue: number of air units	5.0296	0.0249
blue: total defense	3.2456	0.0716
blue: value of air units	0.4646	0.4954

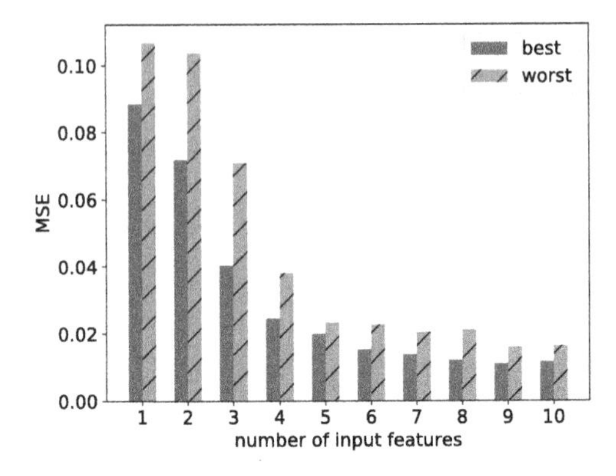

Fig. 8. MSE using different features for prediction.

5 Conclusion

In this paper, we first constructed an attack-defense model based on deep reinforcement learning and used it to generate a large amount of battle data. Next, the electromagnetic situation feature extraction method was proposed, 30 features were extracted from the original data to characterize the electromagnetic situation, and the electromagnetic situation prediction model based on LSTM was constructed. Third, we proposed a win rate evaluation model with situation features as input and win rate as output. We further used ANOVA to analyze the extracted features and showed that we could use a small number of the most relevant features to predict the win rate while maintaining high prediction accuracy.

References

1. Endsley, M.R.: Toward a theory of situation awareness in dynamic systems. Human Factors **37**(1), 32–64 (1995)
2. Cai, X., Song, J.: Analysis of complexity in battlefield electromagnetic environment. In: 2009 4th IEEE Conference on Industrial Electronics and Applications, pp. 2440–2442. IEEE (2009)
3. Shen, D., Jiang, B., Liu, T., et al.: Realizing of a battlefield electromagnetic situation system. In: 2017 36th Chinese Control Conference (CCC), pp. 10304–10309. IEEE (2017)
4. Yuan, H., Wenrui, D., Chunlei, L.: Assessment and prediction of complex electromagnetic environment based on Bayesian network. In: 2017 IEEE International Conference on Unmanned Systems (ICUS), pp. 120–125. IEEE (2017)
5. Krizhevsky, A., Sutskever, I., Hinton, G.E.: ImageNet classification with deep convolutional neural networks. Commun. ACM **60**(6), 84–90 (2017)
6. Jozefowicz, R., Vinyals, O., Schuster, M., et al.: Exploring the limits of language modeling. arXiv preprint arXiv:1602.02410 (2016)

7. He, K., Zhang, X., Ren, S., et al.: Deep residual learning for image recognition. In: Proceedings of the IEEE Conference on Computer Vision and Pattern Recognition, pp. 770–778 (2016)
8. Minar, M.R., Naher, J.: Recent advances in deep learning: an overview. arXiv preprint arXiv:1807.08169 (2018)
9. Wei, C., Qi, L., Wu, R., Lin, Y.: Electromagnetic spectrum threat prediction via deep learning. In: Liu, S., Yang, G. (eds.) ADHIP 2018. LNICST, vol .279, pp. 433–442. Springer, Cham (2019). https://doi.org/10.1007/978-3-030-19086-6_48
10. Shu, Y.F., Wei, X.C., Fan, J., et al.: An equivalent dipole model hybrid with artificial neural network for electromagnetic interference prediction. IEEE Trans. Microw. Theory Tech. **67**(5), 1790–1797 (2019)
11. Perla, P.P., McGrady, E.D.: Why wargaming works. Naval War Coll. Rev. **64**(3), 111–130 (2011)
12. Silver, D., Schrittwieser, J., Simonyan, K., et al.: Mastering the game of go without human knowledge. Nature **550**(7676), 354–359 (2017)
13. Cleeremans, A., Servan-Schreiber, D., McClelland, J.L.: Finite state automata and simple recurrent networks. Neural Comput. **1**(3), 372–381 (1989)
14. Bengio, Y., Simard, P., Frasconi, P.: Learning long-term dependencies with gradient descent is difficult. IEEE Trans. Neural Netw. **5**(2), 157–166 (1994)
15. Hochreiter, S., Schmidhuber, J.: Long short-term memory. Neural Comput. **9**(8), 1735–1780 (1997)

Automatic Modulation Classification Based on Multimodal Coordinated Integration Architecture and Feature Fusion

Xiao Zhang and Yun Lin(✉)

College of Information and Communication Engineering, Harbin Engineering University, Harbin 150001, China
{zxiao,linyun}@hrbeu.edu.cn

Abstract. With the rapid advancement of the 5G wireless communication technology, automatic modulation classification (AMC) is not only faced with more complex communication environment, but also needs to deal with more modulation styles, which increases the difficulty of modulation recognition invisibly. However, most deep learning (DL)-based AMC approaches currently merely use time domain or frequency domain monomodal information and ignore the complementarities between multimodal information. To address the issue, we exploit a signal statistical graph domain-I/Q waveform domain multimodal fusion (SIMF) method to achieve AMC based on AlexNet, complex-valued networks and multimodal technology. The extracted multimodal features from signal statistical graph domain and I/Q waveform domain are fused to obtain a richer joint feature representation and T-SNE algorithm is used to visualize the extracted feature. Moreover, coordinated integration architecture was adopted to achieve mutual collaboration and constraints between multiple modalities to maintaining the unique characteristics and exclusivity of each modal. Simulation results demonstrate the superior performance of our proposed SIMF method compared with unimodal model and feature fusion model.

Keywords: Automatic modulation classification · 5G wireless communications · Multimodal deep learning · Feature fusion

1 Introduction

Recently, from 2020 onwards, 5G wireless communication networks will be introduced worldwide and entered the commercial development stage [1]. Some countries have launched research and exploration on 6th generation mobile networks (6G). 6G wireless communication networks are supposed to provide global reach, improved spectral/energy/cost performance, increased information, and improved protection [2, 3]. In order to meet the standards of wireless communication technology, as an important technology that cannot be ignored in non-cooperative communication, AMC technology plays an important role. However, AMC not only faces an increasingly complex communication environment, but also needs to deal with more modulation patterns, which

J. Xiong et al. (Eds.): MobiMedia 2021, LNICST 394, pp. 782–795, 2021.
https://doi.org/10.1007/978-3-030-89814-4_58

invisibly increases the difficulty of modulation recognition. Therefore, it is important and urgent to explore a more efficient AMC method.

DL has sparked widespread interest as a major achievement in artificial intelligence. Recently, DL has also become a research hotspot in the field of AMC [4–10], and provides a new way to improve recognition performance by virtue of multiple advantages such as pattern recognition and feature expression. The basic idea of the AMC based on the image recognition network model is to transform the signal recognition problem into an image recognition problem. Peng *et al.* [11] convert signals into three-channel images and studies DL models for AMC. In the sluggish and flat fading channels, G. Jajoo *et al.* [12] proposed a novel method focused on constellation structure to identify PSK and QAM modulation of different orders. Lin et al. [13] suggested the contour stellar image (CSI) approach for transforming signal waveforms into statistically significant pictures, which can migrate deep statistical information from the initial wireless signal waveforms.

The sum of information encoded in the phase of a signal is adequate to retrieve the remainder of the information encoded in its magnitude. Currently, most of the researchers interpret I/Q data with an ordered pair of real-valued numbers, and numerous architectures have been developed for this data format. [14–16]. Cheng *et al.* [17] proved that the complex-valued networks has certain advantages over the real-valued networks in the AMC, because it has a richer representation ability and better generalization characteristics. Tu *et al.* [18] proposed to apply complex-valued networks for AMC and it is proved that the performance of complex-valued networks in AMC is better than that of real-valued networks.

But as presented above, the majority of current DL-based approaches for AMC use monomodal information from single-dimensional domain [19, 20]. The general shortcoming of these AMC methods is that they do not make reasonable use of the existing multimodal information and ignore the complementarity among them. To address this problem, multimodal technology [9, 21, 22] has been applied to the field of AMC. Zhang et al. [23] proposed a multi-modality fusion model and attempts to integrate different modality features for the first time to improve performance in the area of AMC. Wu et al. [24] propose a CNN-based AMC method of multi-feature fusion that converts signals into cyclic spectra (CS) and constellation diagram (CD) of image representations. To realize AMC based on deep residual networks, Qi et al. [25] exploit a waveform-spectrum multimodal fusion method.

In this paper, we proposed a signal statistical graph domain-I/Q waveform domain multimodal fusion (SIMF) method to achieve AMC based on AlexNet, complex-valued networks and multimodal technology. This paper focuses on the use of multimodal information technology in AMC. The following is a list of the paper's key contributions.

1. We extract multimodal information from the original modulated signal dataset. Specifically, the first modality is the CSI of the signal statistical graph domain, and the second modality is the signal I/Q waveform domain. The two modalities are fused by multimodal fusion technology. We adopted the T-SNE algorithm to visualize the extracted features.

2. We adopt coordinated integration architecture to achieve mutual collaboration and constraints between multiple modalities to maintaining the unique characteristics and exclusivity of each modal.
3. The modulated signal datasets of different sampling points are established to evaluate our model performance. The experimental results prove that the classification accuracy of our proposed SIMF method is better than other methods.

The following is the structure of this paper: Signal modal and data preprocessing are covered in Sect. 2. We introduce our scheme in detail in Sect. 3. Section 4 shows the simulation results. Finally, the conclusions are drawn in Sect. 5.

2 Signal Modal and Preprocessing

2.1 Signal Model

In a communication system, the general equation of the signal received by the receiver can be expressed as

$$r(t) = h(t) * s(t) + n(t) \tag{1}$$

where $n(t)$ represents the additive white Gaussian noise (AWGN) with zero mean, $h(t)$ denotes the channel impulse response, $s(t)$ is modulated signals, and $r(t)$ is the received signals, $*$ is the convolution operation. The received signal $r(t)$ is usually transformed into a discrete version $r[n]$ and represented by I/Q format data. It is comprised of the in-phase component r_I and the quadrature component r_Q. Thus, the discrete signal $r[n]$ can be described

$$r[n] = r_I[n] + \tilde{j} r_Q[n] \tag{2}$$

2.2 Signal Preprocessing

Most of the AMC methods are based on monomodal information from single depicting dimension and ignore the complementarities among the multimodal information of electromagnetic signal. In this article, we will study two modal representation approach of electromagnetic signal I/Q waveform domain and statistical graph domain. By taking advantage of the correlation between signal multimodal data and eliminating the redundancy among multimodal data we will learn a more accurate representation method of data features.

Signal I/Q Waveform Domain. The first modality is the I/Q vector which is obtained by the imaginary part and real part of the received signals. The i-th sample of signal I/Q vector is defined as

$$x_i^{I/Q} = \begin{bmatrix} x_i^I \\ x_i^Q \end{bmatrix} \tag{3}$$

where $x_i^I = \mathrm{Re}[r[n]], x_i^Q = \mathrm{Im}[r[n]]$.

Signal Statistical Graph Domain. In this paper, we transform the original signal data into CSI as the second modality. The transformation method is as follows. In view of the original signal CD in different regions have different characteristics of the sample point density, density of using rectangular window function in the CD sliding on the graph, the statistics window in different regions of points, number of sampling points divided by the whole CD figure get normalized dot density value, the final size normalized dot density values mapped to different color, make the original signal CD figure into new CSI. The CSI estimation method can be outlined as follows (Fig. 1):

$$\rho(i,j) = \frac{\sum_{i=x_1}^{x_2} \sum_{j=y_1}^{y_2} \mathrm{dots}(i,j)}{\sum_{x_1=W_0}^{W_1} \sum_{y_1=H_0}^{H_1} \sum_{i=x_1}^{x_2} \sum_{j=y_1}^{y_2} \mathrm{dots}(i,j)} \tag{4}$$

Where W_0, H_0 are top left coordinate of CSI; W_1, H_1 are the lower right corner of CSI; x_0, y_0 are the upper left corner coordinates of the density window function; x_1 and y_1 the lower right corner coordinates of the density window function. Figure 2 shows the CSI of the twelve signals at 8 dB.

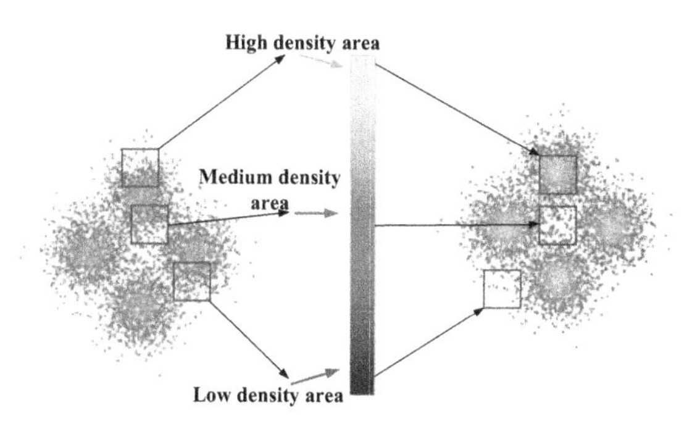

Fig. 1. CD converted to CSI

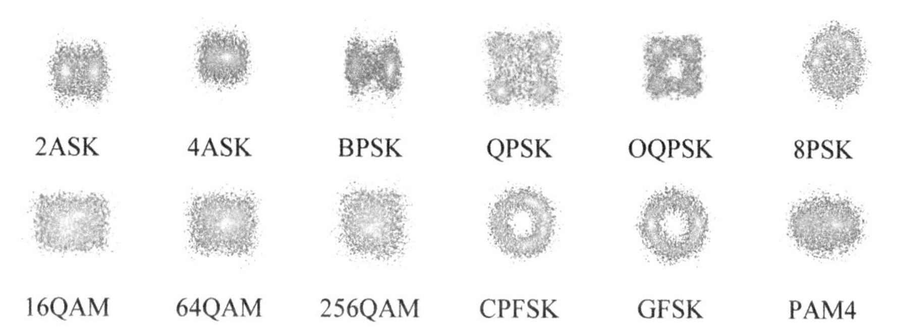

Fig. 2. Contour stellar images of twelve signals at 8 dB

CD as a binary image, it does not distinguish among pixels at a single sampling point and pixels at multiple sampling points. By contrast, colors and shapes of CSI can give us more specific information about the wireless signal they represent and provide finer grained features. It can retain the statistical characteristics of the signal even under the interference of noise.

3 The Proposed Modulation Classification Method

In this section, we'll go through the prototype and of the AlexNet networks as signal statistical graph domain features extraction and Complex-valued networks as I/Q waveform domain features extraction, then introduce the SIMF method we proposed.

3.1 Signal Statistical Graph Domain Features Extraction

Recently, DL has gradually become a hot topic in the teaching field of various subjects. Alex Krizhevesky et al. proposed a deeper, broader CNN model and won the most difficult visual object recognition challenge in the 2012 ImageNet Large Visual Recognition Challenge (ILSVRC). Therefore, AlexNet is employed as feature extractor to extract the acquirement of the powerful features.

The architecture of AlexNet is shown in Fig. 3. The first convolution layer uses Local Response Normalization (LRN) to perform convolution and maximum pooling, where 96 different acceptance filters of size 11 × 11 are used. The maximum pool operation is performed using a 3 × 3 filter with a step size of 2. The same action is performed in the second layer using 5 × 5 filters. 384, 384 and 296 feature maps were used in the third, fourth and fifth convolutional layers of 3 × 3. Finally, there are two fully connected layers (FC) and a softmax layer.

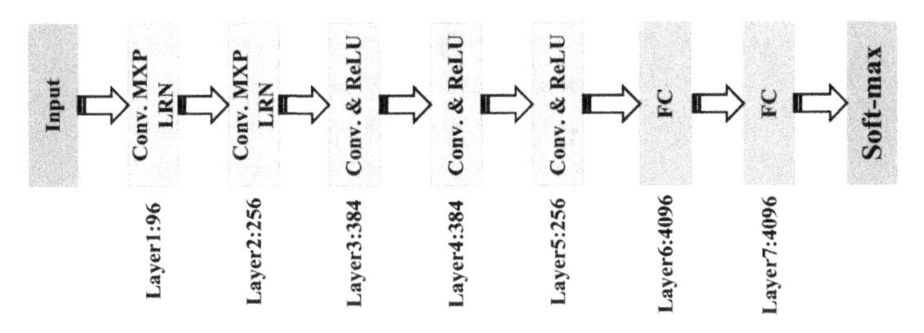

Fig. 3. Architecture of AlexNet

Transfer learning is adopted in our framework because it could accelerate and optimize the learning efficiency of models. We keep the parameters of the original AlexNet model and only set the number of neurons in the last two FC to 1024 to fit the size of our data set. We extract the deep features of CSI with AlexNet based on transfer learning. Figure 4 is the feature extraction process by AlexNet. In our work, the input of AlexNet networks are RGB image sets of CSI with size of 227 × 227 × 3 and the output of the last FC layers are 1024 deep CSI features we extracted.

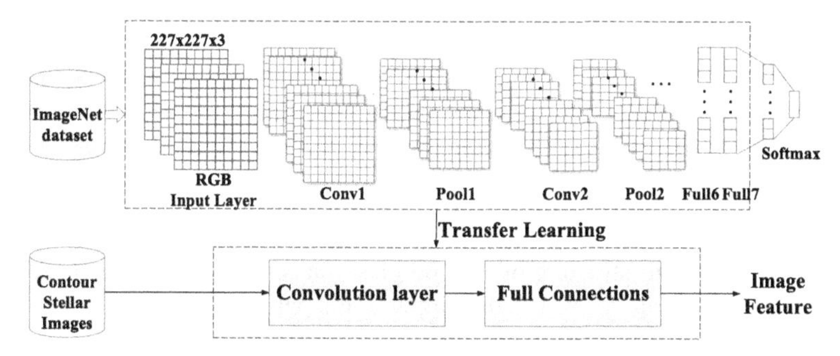

Fig. 4. The transfer process of the AlexNet

3.2 Signal I/Q Waveform Domain Features Extractions

Current mainstream deep learning technologies and architectures are all based on real number manipulation and representation, while some recent basic theoretical analysis shows that complex numbers have more expressive capabilities. In addition, electromagnetic signals mostly appear in the form of I/Q complex, and the phase information of electromagnetic signals can also represent the time-domain characteristics of electromagnetic signals. Therefore, the network with the expression and processing of complex forms can help to improve the electromagnetic signal recognition rate to some extent. Thus, we use complex-valued networks as signal I/Q waveform domain features extractions. Next, we will present the implementation of the complex-valued building blocks of the complex neural network.

Complex Convolutional Layer. In order to perform the equivalent operation equivalent to traditional real 2D convolution in complex number domain, in this scheme, the complex filter matrix $W = A + iB$ is convolved by the complex vector $\mathbf{h} = \mathbf{x} + i\mathbf{y}$. By using real numbers to simulate the operation of complex numbers, so that x and y are real matrices. Since the convolution operator is distributed, the vector h is convolved through the filter W, and the following can be obtained:

$$\mathbf{W} * \mathbf{h} = (\mathbf{A} * \mathbf{x} - \mathbf{B} * \mathbf{y}) + i(\mathbf{B} * \mathbf{x} + \mathbf{A} * \mathbf{y}) \tag{5}$$

Complex Batch Normalization. In deep learning models, batch normalization can accelerate the convergence speed of models and alleviate the problem of "gradient dispersion" in deep networks. The complex number normalization is specifically designed according to the complex number network, and its formula is as follows:

$$BN(\dot{x}) = \gamma \cdot \dot{x} + \beta \tag{6}$$

Where γ is variance, β is shift factor.

Complex Dense Layer. FC is used as a classifier in deep neural network. In order to make full use of the advantages of complex-valued networks, complex-FC is used at the end of complex-valued networks. Similar to the complex convolution kernel, the complex FC also carries out I/Q two-channel association information mining of electromagnetic data through the alternating product of the real part imaginary part of the weight of the complex full connection layer and the real part imaginary part of the signal. Let $W = A + iB$ represents complex dense vector weight, and $\vec{s} = x + iy$ represents complex-valued input. Similar to the complex convolutional operation, we can get:

$$\mathbf{W} \cdot \vec{s} = (\mathbf{A} \cdot x - \mathbf{B} \cdot y) + i(\mathbf{B} \cdot x + \mathbf{A} \cdot y) \tag{7}$$

In our work, the overall structure of the complex-valued network we used as features extractor is shown in Fig. 5. We use the output of the penultimate FC layer as the deep features of the signal I/Q waveform domain.

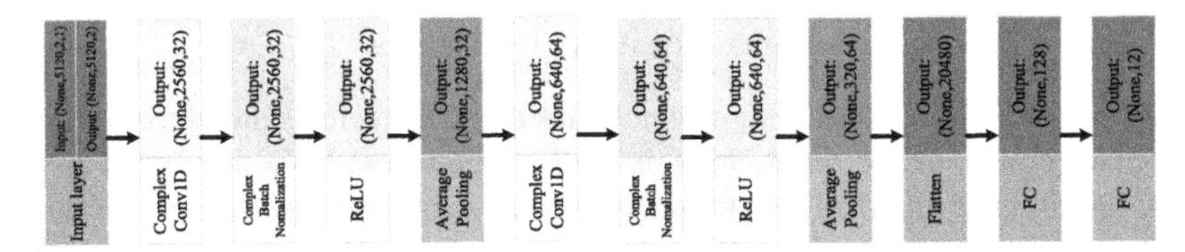

Fig. 5. The structure of Complex-valued model

3.3 The Proposed SIMF Method

In order to fuse the theories and methods of AMC, deep learning, complex-valued networks and multimodal deep learning, we proposed a framework based on the multimodal information fusion of signal statistical graph domain and I/Q waveform domain to achieve AMC. The fusion of multimodal information makes use of the complementarity of different modal information to realize the fusion of multi domain information to obtain more comprehensive and easily distinguishable features. Therefore, the final fusion feature representation is more discriminative than the single-mode representation, which can improve the classification accuracy and robustness of the model. Moreover, in order to make the separate single mode and the feature fusion multi-mode maintain the cooperative classification goal, we adopt the coordinated integration architecture to maximize the similarity of different models while maintaining the independent operation of each model.

Our proposed SIMF framework is shown in Fig. 6. Specifically, first, we convert the original signal into CSI. Then, AlexNet network is selected as the feature extractor of the signal statistical graph domain, and at the same time, the high-level features of the electromagnetic signal I/Q data are extracted using a complex-valued networks. The features of the two modals are fused by series splicing method. At last, the fusion features are input into the FC to map the classification results, thereby obtaining the multimodal model. Finally, the use of coordinated integration architecture to achieve mutual collaboration and constraints between multiple modalities is conducive to maintaining the unique characteristics and exclusivity of each modal. It takes into account the problems of modal coordination and feature fusion at the same time. Here, the fusion target of the two unimodal models of the signal statistical graph domain model and I /Q waveform

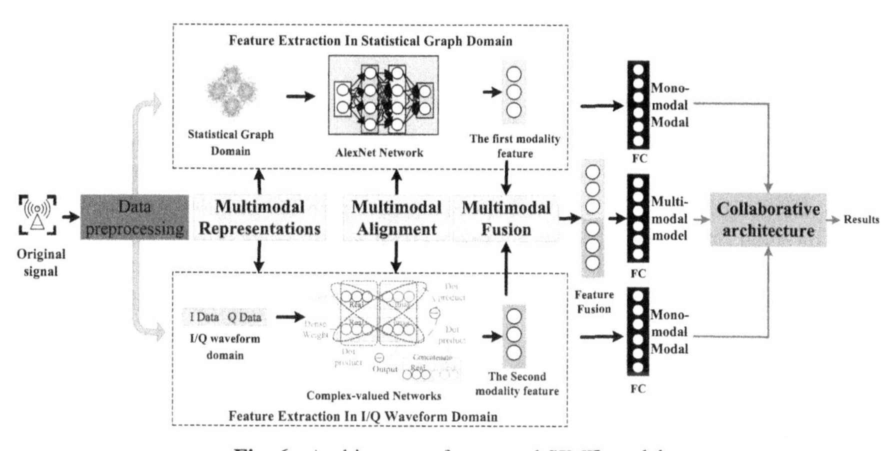

Fig. 6. Architecture of proposed SIMF model

domain and the feature fusion multimodal model are expressed by the same loss function, as shown in formula (9). In this process, the alignment of multimodal is ensured. In this way, the similarity structure between and within the modals is maintained, and at the same time the goal of mutual cooperation between the modals is achieved.

Next, we shall discuss the method of coordinated integration architecture in detail. Since the input of the model contains two modalities of the statistical graph domain and the I/Q waveform domain of same original signal data source, the penalty of the two predicted label distributions of two unimodal models must be taken into consideration. We use cross entropy to calculate the penalty of different modalities of predicted label distributions. Let $p_\theta(x_i)$ and $q_\theta(x_i)$ represent the output probability distributions of two unimodal models. The following is a description of cross entropy:

$$\text{Cross Entropy} = -\sum_{i=1}^{n} p_\theta(x_i)\ln(q_\theta(x_i)) \tag{8}$$

Let x_i^m (for $m \in \{1, 2\}$), $i \in \{1, 2, ...N\}$ denote the m-th modal information of the i-th example. x_i^c is the fusion features from the two unimodal modalities of the i-th example. $\Theta=\{\theta^c, \theta^m\}$ are obtained by training. N represents the number of training samples. In this way, the loss function of the feature fusion multimodal model and the two unimodal models can be represented by the same fusion target as follows:

$$LossFuction = -\frac{1}{N}\sum_{i=1}^{N} t_i \ln\left(p_{\theta^c}\left(x_i^c\right)\right) - \frac{1}{N}\sum_{m=1}^{2}\sum_{i=1}^{N} p_{\theta^c}(x_i^c)\ln\left(p_{\theta^m}\left(x_i^m\right)\right) \tag{9}$$

where $p_\theta(x_i)$ is probability distribution and is obtained by softmax function that is defined as:

$$p_\theta(x_i) = \text{softmax}(x_i) = \frac{1}{\sum_{k=1}^{K} e^{\theta_k^T x_i}}\left[e^{\theta_1^T x_i}, e^{\theta_2^T x_i}, \ldots, e^{\theta_k^T x_i}\right]^T \tag{10}$$

where K denotes the number of all classes, t_i is the true label probability distribution.

4 Experiments

4.1 Experimental Dataset

BPSK, QPSK, 16QAM, 64QAM, and 256QAM are the modulation types used in the traffic channel of 5G mobile communications, according to open protocol and specifications. Looking forward to dynamic spectrum access in 5G mobile communications, digital modulation recognition has more practical significance. Hence, in this paper, we use MATLAB to generate 12 digital modulated signals, including 2ASK, 4ASK, BPSK, QPSK, OPQSK, 8PSK, 16QAM, 64QAM, 246QAM, CPFSK, GFSK, PAM4. The noise environment considered in the experiment is additive white Gaussian noise (AWGN) with SNR from −10 dB to 8 dB and a stride of 2 dB. In this experiment, under each SNR, 1250 samples are generated for each modulated signal, among them are 1000 training samples and 250 test samples.

4.2 Results and Discussions

Figure 7 presents average accuracy of the proposed SIMF method, feature fusion method, AlexNet based unimodal model method and complex-valued networks based unimodal

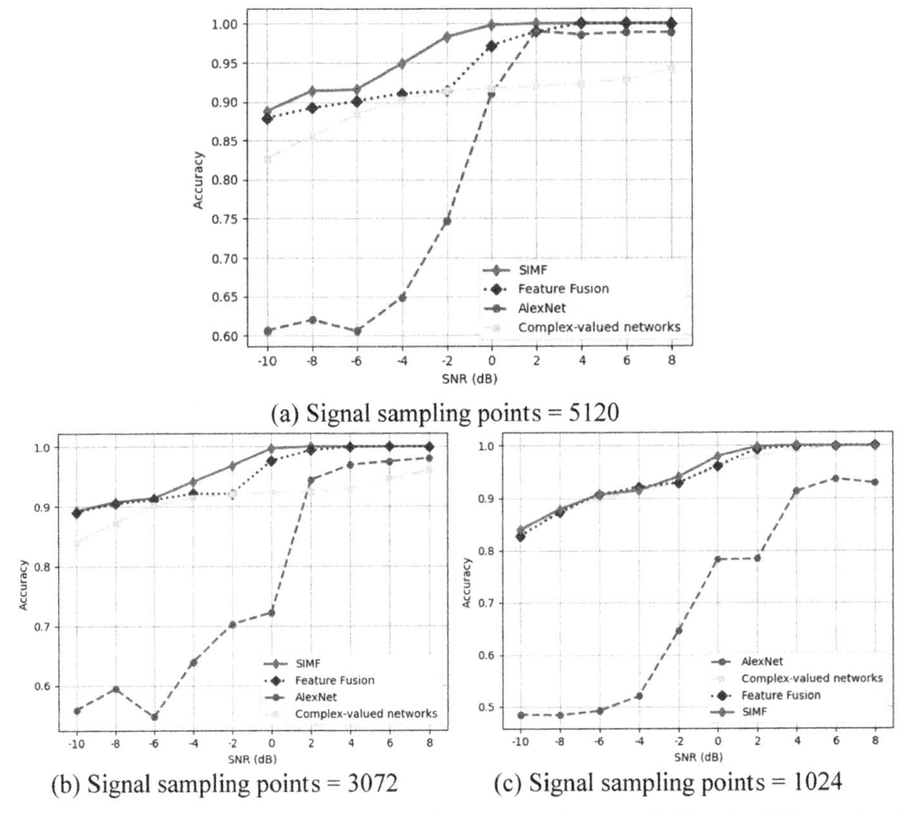

Fig. 7. Average classification accuracy of different methods versus SNR under different signal sampling points.

model method versus signal to noise ratio (SNR) under different signal sampling points. In the experimental simulation, we set the signal sampling points as 5120, 3072 and 1024 respectively to explore the influence of different signal sampling points on the performance of different methods.

From the figures, the average classification accuracy of the SIMF method proposed in this paper is significantly better than the other three methods under the overall SNR for different signal sampling points, especially when the SNR is greater than 0 dB, the accuracy reaches 100%. Moreover, the average accuracy of the multimodal model with feature fusion is significantly higher than that of the other two unimodal models. This indicates that the multimodal fusion method can realize the complementarity among the multimodal information, so as to obtain a more comprehensive joint feature to improve the accuracy and model robustness. On the other hand, as the number of signal sampling points decreases, we can see that the AlexNet based unimodal model is greatly affected by it. The smaller the number of signal sampling points, the lower the average accuracy of the AlexNet based unimodal model. In addition, the performance of AlexNet based unimodal model is greatly affected by SNR, because its average accuracy is significantly reduced with the decrease of SNR. Under the condition of large number of signal sampling points and high SNR, the performance of AlexNet based unimodal model is better than that of complex-valued networks unimodal model. In addition, the signal sampling number has little effect on the complex-valued networks unimodal model, and better classification accuracy can be achieved under low SNR. Because the multimodal model can combine the advantages of two unimodal models, the accuracy of the multimodal model in the overall SNR is significantly improved. When the number of signal sampling points is 1024, the performance of the proposed SIMF method and the multimodal model is not significantly improved due to the limitations of the AlexNet based unimodal model.

Figure 8 shows the confusion matrixes of different AMC methods with SNR being −2 dB. It can be seen from Fig. 8(d) that the main errors of the AlexNet unimodal model occur between 8PSK and 2ASK, 4ASK and PAM4, as well as among 16QAM, 64QAM and 256QAM. This can be explained by the CSI diagrams of the dataset, since their CSI diagrams are similar. From Fig. 8(c), it can be seen that the main error of the complex-valued network unimodal model occurs between 8PSK and QPSK. In comparison, because the feature fusion multimodal model can achieve the complementary advantages of the two unimodal models, the probability of correct classification is significantly improved. Most importantly, we can see from Fig. 8 (a) that our proposed SIMF method has an obvious advantage in classification correctness probability compared with the other three models due to its combination of feature fusion and coordinated integration architecture.

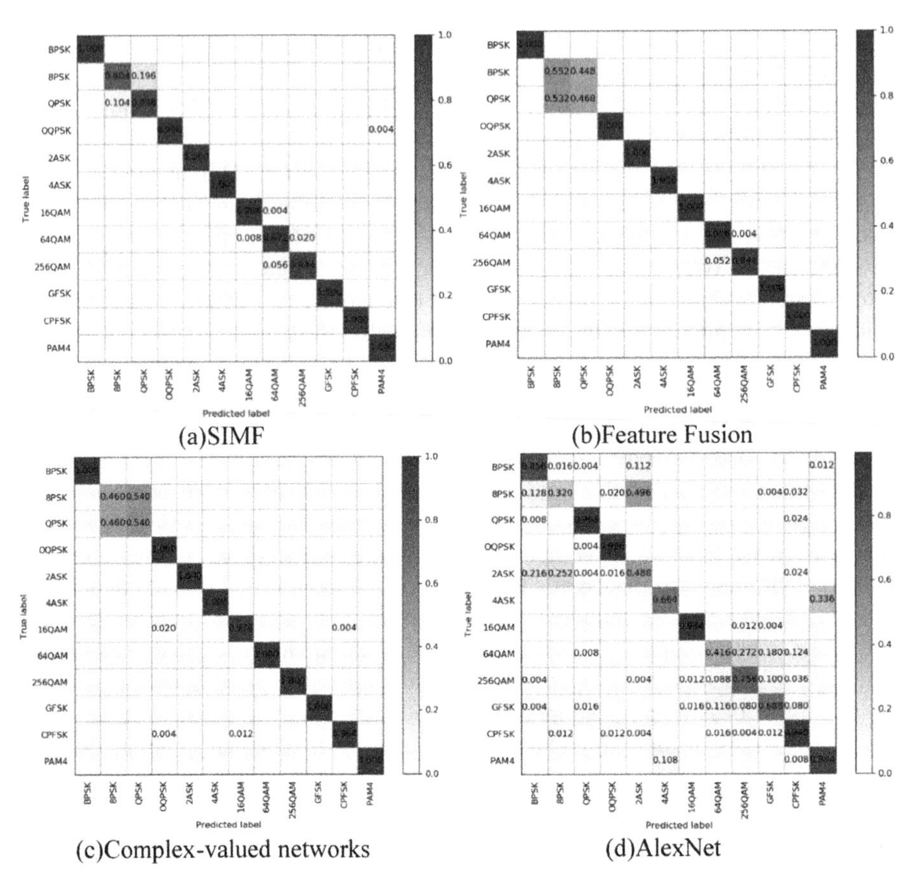

(a)SIMF (b)Feature Fusion

(c)Complex-valued networks (d)AlexNet

Fig. 8. Confusion matrix of different methods under SNR = −2 dB

To better understand the classification performance of each kind of signal under different methods, we analyze the correct classification probability of twelve modulation types differs with SNR in Fig. 9. From the Fig. 9 we can see that the AlexNet unimodal model has some error in the correct classification of each modulated signal when the SNR is lower than 2 dB. This is because the CSI of the signals is greatly affected by the SNR. However, the classification accuracy of the complex-valued networks unimodal model for 8PSK and QPSK signals is always low, because complex-valued networks are unable to acquire enough features to differentiate these categories when only I/Q element is used as input. In comparison, our proposed SIMF model can achieve 100% accuracy when the SNR is greater than 0 dB, thanks to the combination of feature fusion and coordinated integration architecture.

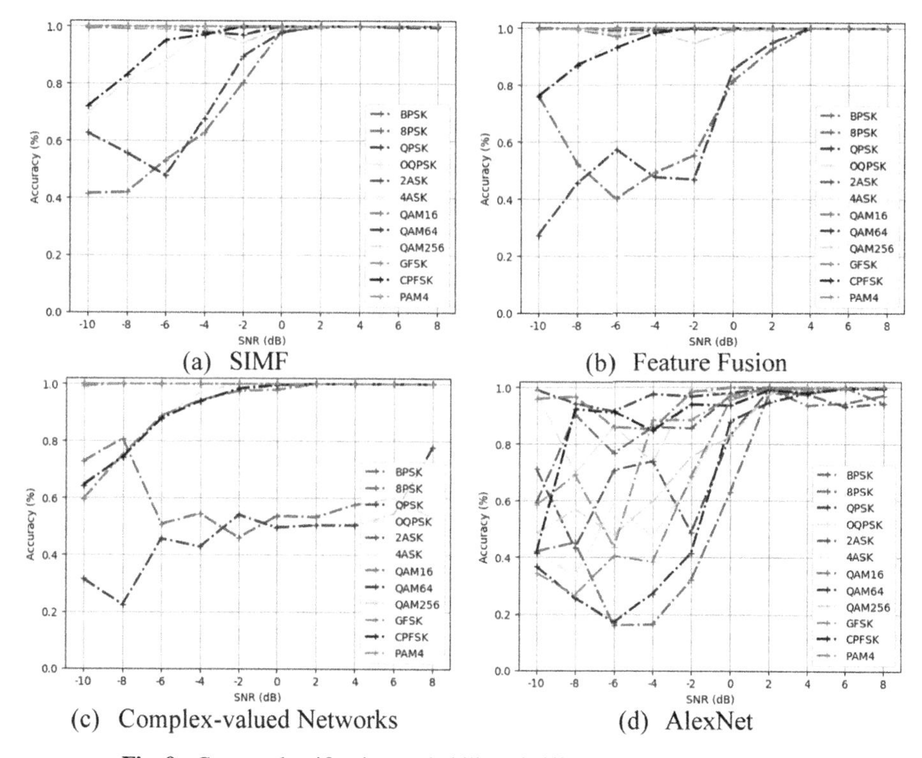

Fig. 9. Correct classification probability of different methods versus SNR

Figure 10 is a visualization of features extracted by different methods at SNR = 6 dB. The specific method is to extract the output of the penultimate FC of the networks, and use the T-SEN [26] method for dimensionality reduction and visual display. Each color represents a signal modulation type. It can be seen that the characteristic aliasing of each signal in the AlexNet unimodal model is serious and the distance between classes is short. However, both the complex-valued networks unimodal model and the feature fusion multimodal model have the aliasing of the two signal features (the signal features represented by orange and yellow). By comparison, the class spacing of each signal feature in the proposed SIMF method is relatively large and easy to distinguish, which reflects the advantages of the proposed method.

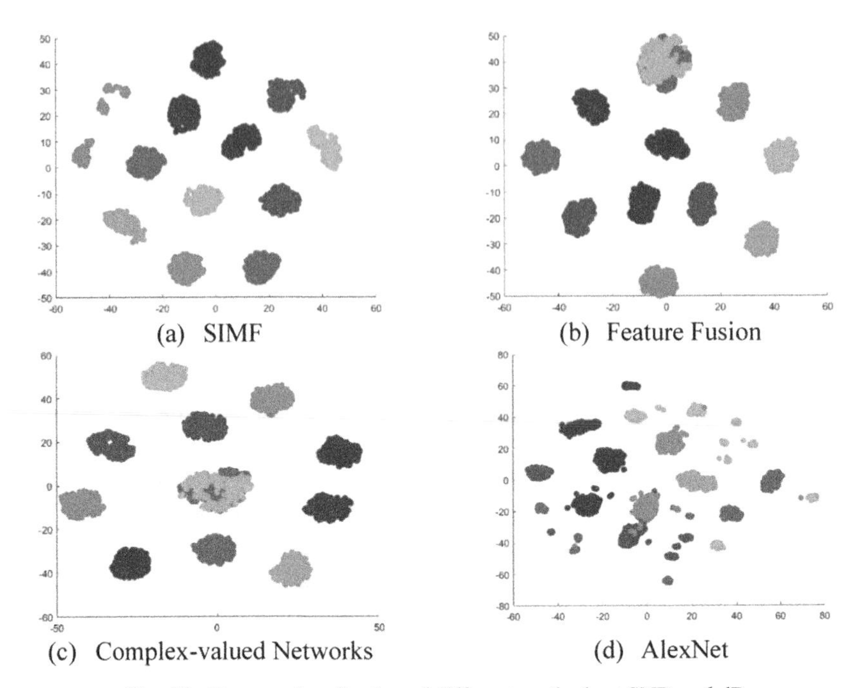

Fig. 10. Feature visualization of different methods at SNR = 6 dB

5 Conclusions

In order to take advantage of the complementarity between multimodal information and realize the conversion and fusion of multi-domain information, we proposed SIMF framework based on the multimodal information fusion of signal statistical graph domain and I/Q waveform domain to achieve AMC. We use series fusion to obtain a more informative joint feature representation of multimodal features. Furthermore, we use a coordinated integration architecture to achieve mutual cooperation and constraints between multiple modes, which is conducive to maintaining the unique characteristics and exclusivity of each modal. The final simulation results show that our proposed framework achieves superior performance than other unimodal models and multimodal models under the entire SNR. In further work, we will continue to explore other multimodal fusion methods to improve the robustness and performance of our model.

References

1. You, X., Zhang, C., Tan, X., et al.: AI for 5G: research directions and paradigms. Sci. China Inf. Sci. **62**(2), 1–13 (2019)
2. You, X., Wang, C.X., Huang, J., et al.: Towards 6G wireless communication networks: Vision, enabling technologies, and new paradigm shifts. Sci. China Inf. Sci. **64**(1), 1–74 (2021)
3. Zhang, P., Kai, N.I.U., Hui, T., et al.: Technology prospect of 6G mobile communications. J. Commun. **40**(1), 141 (2019)
4. Wang, Y., Wang, J., Zhang, W., et al.: Deep learning-based cooperative automatic modulation classification method for MIMO systems. IEEE Trans. Veh. Technol. **69**(4), 4575–4579 (2020)

5. Tu, Y., Lin, Y., Wang, J., et al.: Semi-supervised learning with generative adversarial networks on digital signal modulation classification. Comput. Mater. Continua **55**(2), 243–254 (2018)
6. Lin, Y., Zhu, X., Zheng, Z., et al.: The individual identification method of wireless device based on dimensionality reduction and machine learning. J. Supercomput. **75**(6), 3010–3027 (2019)
7. Lin, Y., Tu, Y., Dou, Z.: An improved neural network pruning technology for automatic modulation classification in edge devices. IEEE Trans. Veh. Technol. **69**(5), 5703–5706 (2020)
8. Yun, L., Haojun, Z., Xuefei, M., et al.: Adversarial attacks in modulation recognition with convolutional neural networks. IEEE Trans. Reliab. **70**(1), 389–401 (2021). https://doi.org/10.1109/TR.2020.3032744
9. Lin, J., Wei, M.: Network security situation prediction based on combining 3D-CNNs and Bi-GRUs. Int. J. Perform. Eng. **16**(12), 1875–1887 (2020)
10. Restuccia, F., Melodia, T.: Physical-Layer Deep Learning: Challenges and Applications to 5G and Beyond. arXiv preprint arXiv:2004.10113 (2020)
11. Peng, S., Jiang, H., Wang, H., et al.: Modulation classification based on signal constellation diagrams and deep learning. IEEE Trans. Neural Netw. Learn. Syst. **30**(3), 718–727 (2018)
12. Jajoo, G., Kumar, Y., Yadav, S.K.: Blind signal PSK/QAM recognition using clustering analysis of constellation signature in flat fading channel. IEEE Commun. Lett. **23**(10), 1853–1856 (2019)
13. Lin Yun, T., Ya, D.Z., et al.: Contour stella image and deep learning for signal recognition in the physical layer. IEEE Trans. Cogn. Commun. Netw. **7**(1), 34–46 (2021). https://doi.org/10.1109/TCCN.2020.3024610
14. Zhang, Z., Luo, H., Wang, C., et al.: Automatic modulation classification using CNN-LSTM based dual-stream structure. IEEE Trans. Veh. Technol. **69**(11), 13521–13531 (2020)
15. Xu, J., Luo, C., Parr, G., et al.: A spatiotemporal multi-channel learning framework for automatic modulation recognition. IEEE Wirel. Commun. Lett. **9**(10), 1629–1632 (2020)
16. O'Shea, T.J., Roy, T., Clancy, T.C.: Over-the-air deep learning based radio signal classification. IEEE J. Sel. Top. Sig. Process. **12**(1), 168–179 (2018)
17. Cheng, X., He, J., He, J., et al.: Cv-CapsNet: complex-valued capsule network. IEEE Access **7**, 85492–85499 (2019)
18. Tu, Y., Lin, Y., Hou, C., et al.: Complex-valued networks for automatic modulation classification. IEEE Trans. Veh. Technol. **69**(9), 10085–10089 (2020)
19. Meng, F., Chen, P., Wu, L., et al.: Automatic modulation classification: a deep learning enabled approach. IEEE Trans. Veh. Technol. **67**(11), 10760–10772 (2018)
20. Li, R., Li, L., Yang, S., et al.: Robust automated VHF modulation recognition based on deep convolutional neural networks. IEEE Commun. Lett. **22**(5), 946–949 (2018)
21. Baltrušaitis, T., Ahuja, C., Morency, L.P.: Multimodal machine learning: a survey and taxonomy. IEEE Trans. Pattern Anal. Mach. Intell. **41**(2), 423–443 (2018)
22. Zhang, C., Yang, Z., He, X., et al.: Multimodal intelligence: Representation learning, information fusion, and applications. IEEE J. Sel. Top. Sig. Process. **14**(3), 478–493 (2020)
23. Zhang, Z., Wang, C., Gan, C., et al.: Automatic modulation classification using convolutional neural network with features fusion of SPWVD and BJD. IEEE Trans. Sig. Inf. Process. Netw. **5**(3), 469–478 (2019)
24. Wu, H., Li, Y., Zhou, L., et al.: Convolutional neural network and multi-feature fusion for automatic modulation classification. Electron. Lett. **55**(16), 895–897 (2019)
25. Qi, P., Zhou, X., Zheng, S., et al.: Automatic Modulation Classification Based on Deep Residual Networks with Multimodal Information. IEEE Trans. Cogn. Commun. Netw. **7**, 21–33 (2020)
26. Van der Maaten, L., Hinton, G.: Visualizing data using t-SNE. J. Mach. Learn. Res. **9**(11), 1–48 (2008)

Image Classification with Transfer Learning and FastAI

Ujwal Gullapalli[1], Lei Chen[1(✉)], and Jinbo Xiong[2]

[1] Georgia Southern University, Statesboro, GA 30460, USA
LChen@georgiasouthern.edu
[2] Fujian Normal University, Fuzhou 350117, China

Abstract. Today deep learning has provided us with endless possibilities for solving problems in many domains. Diagnosing diseases, speech recognition, image classification, and targeted advertising are a few of its applications. Starting this process from scratch requires using large amounts of labeled data and significant cloud processing usage. Transfer learning is a deep learning technique that solves this problem by making use of a model that is pre-trained for a certain task and using it on a different task of a related problem. Therefore, the goal of the project is to utilize transfer learning and achieve near-perfect results using a limited amount of data and computation power. To demonstrate, an image classifier using FastAI that detects three types of birds with up to 94% accuracy is implemented. This approach can be applied to solve tasks that are limited by labeled data and would gain by knowledge learned from a related task.

Keywords: Transfer learning · Deep learning · Image classification

1 Introduction

Traditionally, deep learning models are used in isolation. To solve a given task, they are trained on millions of data points. They achieve good results for that particular task, but when a different task needs to be solved, the model would again need to be trained on a large dataset that is related to that specific problem. The previous model cannot be used to solve the new task as the model will be biased from its training data and it will not fit well. The critical issue is that most models to solve complex tasks would require a large amount of labeled data and computing power. Transfer learning can solve these issues by making use of the knowledge gained on previously learned tasks and applying it to new tasks. The transfer learning model will work effectively if the pre-trained model is well-generalized. For example, the state-of-the-art image classification models such as VGG-16, VGG-19, InceptionV3, XCeption, and Resnet50 are trained on the ImageNet database [1]. The database contains over 1.2 million images. These models are generalized and their accuracy is determined on a specific dataset.

J. Xiong et al. (Eds.): MobiMedia 2021, LNICST 394, pp. 796–806, 2021.
https://doi.org/10.1007/978-3-030-89814-4_59

When these pre-trained models are directly applied to a different task or a similar problem they will still suffer a significant loss in performance. Therefore, the primary motivation for the paper was to explore transfer learning by building an image classifier that classifies three types of birds with less data. The procedure and techniques used for building the classifier is inspired from Jeremy Howard's and Sylvain's work on Deep Learning [2]. The development environment for implementing transfer learning can be built locally in an IDE such as Pycharm, Spyder, Visual Studio, or by using a cloud service provider such as Google Colab, PaperSpace, AWS. In this case, Paperspace cloud service is used. It provides free and paid computer resources and uses Jupyter Notebooks as its development environment. To implement the model, FastAI, a deep learning library that runs on top of PyTorch is used. It is a library designed to be quick and easy to rapidly deploy state-of-the-art models to solve problems [3].

2 Related Works

Transfer learning has been around for a few decades, but it has received less interest compared to other areas of machine learning such as unsupervised and reinforcement learning. Andrew Ng, a renowned computer scientist, and a pioneer in AI, believes that apart from supervised learning, transfer learning will be the driver of commercial success for machine learning [4]. Transfer learning is being used across multiple domains. One of the key areas is learning from one language and applying that knowledge to another language. There has been interesting research in this area like zero-shot translation [5]. It uses a single Neural Machine Translation model to translate between multiple languages and has achieved state-of-the-art results.

The other interesting area being utilized is for simulation learning. The model is trained in simulation and the knowledge is transferred to the real-world robot. This is because the robots can be slow and expensive to train [6]. Similarly, self-driving technology is also utilizing simulation training and there are open-source resources where a self-driving car can be trained [7,8]. When it comes to image classification most of the pre-trained models are based on convolutional neural networks that are trained on the Imagenet database. [9] studied if the models that performed better on the Imagenet database also performed well on other vision tasks. They found when the models are used as feature extractors or for finetuning, there was a strong correlation between Imagenet accuracy and transfer accuracy. [10] looked at how transferable are the features in deep neural networks. One of their observation was that the performance benefits of transferring features decrease the more dissimilar source task and target task are.

Some of the challenges in transfer learning are figuring out the best features and measuring the transferability of a model. [11] proposed attentive feature distillation and selection (AFDS) technique adjusts the strength of transfer learning regularization and figures out the important features to transfer. They deployed AFDS on Resnet101 and were able to outperform all existing transfer learning methods. [12] proposed a new metric called LEEP (Log Expected Empirical Prediction) that can predict the performance and convergence speed of transfer and meta-transfer learning methods. It outperformed other metrics such as negative conditional entropy and H scores.

3 Methodology

To demonstrate transfer learning, we built an image classifier that classifies three different types of birds. We used a model pre-trained on an Imagenet and applied it to our task which is to classify three different types of birds. We used a process knows as fine-tuning, that mainly comprised of three steps:

- **Importing the Model:** A model that is trained on a benchmark dataset like Imagenet is imported. It can classify everyday objects, animals, birds, etc. The last layer of the pre-trained model that predicts various classes is removed and replaced with the output layer consisting of three classes of birds.
- **Freezing:** After modifying the output layer, the layers that have come from the pre-trained model were frozen. This means that frozen layer weights will not be updated when the model is trained. Only the modified layer is updated during training.
- **Training:** In final step, the model is trained till it achieves a lowest error rate without overfitting.

There are many types of transfer learning techniques. In some of the cases, the last few layers are removed instead of just the head. This is because, the pre-trained model initial layers can act as general feature extractors and last layers will be more geared towards the original dataset.

4 Building the Bird Classifier

The birds we are going to classify are three birds that are commonly found in the backyard in the state of Georgia. These are Northern Cardinal, Blue Jay, and Yellow-Rumped Warbler (Fig. 1). The image classifier consists of a GUI within the Jupyter notebook. It takes bird image as input and predicts its class.

4.1 Setup Environment

Importing required libraries: FastAI to implement the model and JMD image scraping library to get our data. The development was done in Jupyter Notebook environment.

Fig. 1. Types of birds to be classified. (Color figure online)

```
import fastbook
from fastbook import *
fastbook.setup_book()
from fastai.vision.widgets import *

from jmd_imagescraper.core import *
from pathlib import Path
from jmd_imagescraper.imagecleaner import *
```

4.2 Data Collection

Created the directories to store data.

```
bird=['Northern Cardinal','Blue Jay','Yellow Rumped warbler']
path = Path().cwd()/"bird"
```

The problem required images of three different birds: Blue Jay, Northern Cardinal, and Yellow-Rumped Warbler. A total of 150 images per bird type are collected from the DuckDuckGo search engine using the JMD image scraper (Fig. 2).

```
duckduckgo_search(path,"Northern Cardinal",
"Northern Cardinal",max_results=150)

duckduckgo_search(path,"Blue Jay",
"Blue Jay",max_results=150)

duckduckgo_search(path,"Yellow Rumped Warbler"
,"Yellow rumped warbler",max_results=150)
```

Fig. 2. Collected images from DuckDuckGo search engine. (Color figure online)

4.3 Model Creation

FastAI library provides a simple Datablock method to feed data into our model for training.

```
bird=DataBlock(
    blocks=(ImageBlock,CategoryBlock),
    get_items=get_image_files,
    splitter=RandomSplitter(valid_pct=0.2,seed=42),
    get_y=parent_label,
    item_tfms=Resize(128)
   )
```

The parameters used are as follows:

- blocks: ImageBlock is the input data that is used for the model and category block refers to the labels.
- get_items: get_image_files is used to get the image file names from the path.
- splitter: It is used to specify the method to split the validation set and the training set. Here, the data is split randomly and 20% is allocated to validation set. A seed is used to make sure that every time the program runs, the validation set stays the same.
- get_y: It is specified to locate labels of the data. The method parent_label is going to look at name of the parent in the path of the image file.
- item_tfms: The data from the internet can be of any size. Hence, image is resized into 128×128.

After creating the datablock, a dataloader is created. A dataloader is used to grab a bunch of images as a batch for GPU processing. This makes the GPU process much faster.

```
dls=bird.dataloaders(path)
```

Checking the images from the validation batch (Fig. 3).

```
dls.valid.show_batch(max_n=4,nrows=1)
```

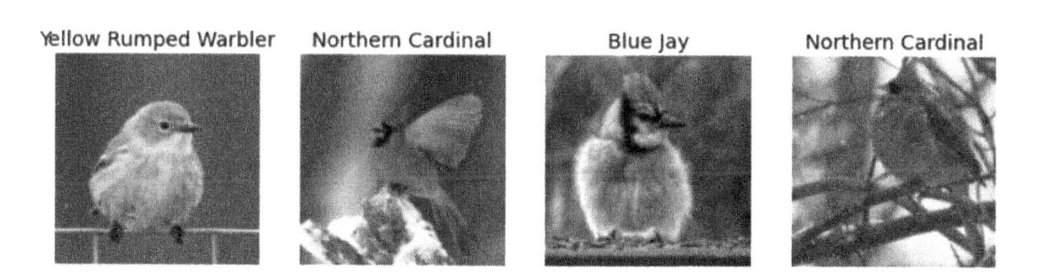

Fig. 3. Images from validation batch.

4.4 Data Transformation

The default resizing method crops the image to fit a square. This might lead to losing some of the information from the picture. There are other resize methods such as squishing, stretching, padding, etc. Squishing or stretching the image can lead to uneven shapes of the bird. This can affect the performance of the model. After trying different image transformation methods, the RandomResizedCrop method gave the best performance for this problem. It crops the random size of the original image and a random aspect ratio of the original aspect ratio is created. This also helps make the model less likely to overfit as it sees a different part of the image every time (Fig. 4).

```
bird = bird.new(
item_tfms=RandomResizedCrop(224, min_scale=0.4),
batch_tfms=aug_transforms())

dls = bird.dataloaders(path)

dls.train.show_batch(max_n=4, nrows=1, unique= True)
```

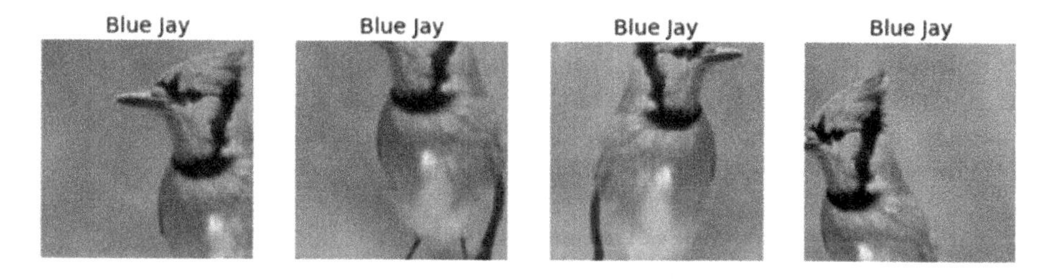

Fig. 4. Training set images after transformation.

4.5 Model Training

There are a lot of architectures such as VGG19, InceptionNET, and ResNet that can be used for this image classification problem. Here, a Convolution Neural Network using the Resnet architecture consisting of 18 layers is created. This model is pre-trained on Imagenet database and it can classify general images into 1000 classes. The metric used was error rate, which tells how frequently the model is making incorrect predictions.

```
learn=cnn_learner(dls, resnet18, metrics=error_rate)
```

The model is then trained for 4 epochs. An epoch specifies the number of times the image will be used as training data.

```
learn.fine_tune(4)
```

epoch	train_loss	valid_loss	error_rate	time
0	1.300787	0.381271	0.177778	00:04

epoch	train_loss	valid_loss	error_rate	time
0	0.208502	0.319453	0.144444	00:05
1	0.132655	0.240997	0.088889	00:05
2	0.087438	0.250987	0.077778	00:05
3	0.070439	0.332649	0.133333	00:05

Fig. 5. Training results for each epoch.

The results (Fig. 5) showed that after training for just 3 epochs the error rate was down to around 7% from 17%. This is for the images randomly downloaded from the internet. There is a chance that there might be false data and this can affect the model performance. FastAI provides a way to check the images the model is having trouble predicting. Checking the images with top 5 loses (Fig. 6).

```
interp.plot_top_losses(5, nrows=5)
```

The top 5 loses showed that the model is having trouble predicting the Yellow-Rumped Warbler. The first image in the top loss results is blurry and the second image is not relevant and they should be discarded from the dataset.

The data was cleaned with FastAI inbuilt cleaner widget (Fig. 7).

```
cleaner = ImageClassifierCleaner(learn)
cleaner
```

Fig. 6. Top 5 images the model is having trouble classifying. (Color figure online)

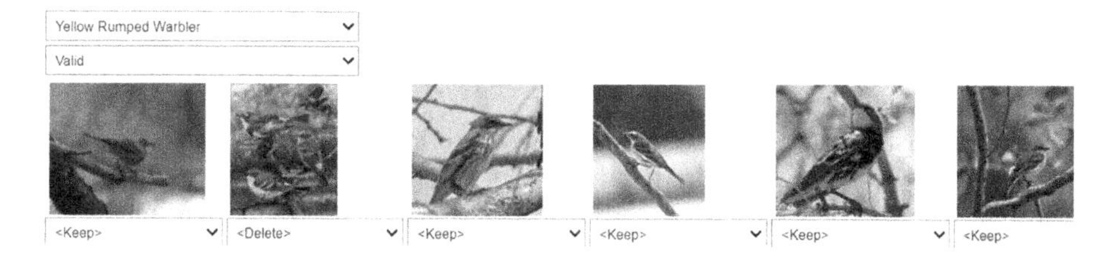

Fig. 7. FastAI inbuilt cleaner.

After cleaning the data and retraining for 2 epochs. The error rate was reduced to around 5% (Fig. 8).

epoch	train_loss	valid_loss	error_rate	time
0	0.180719	0.165416	0.057471	00:05
1	0.121468	0.165792	0.057471	00:05

Fig. 8. Retraining results

4.6 Results

The results were visualized using a confusion matrix (Fig. 9).

```
interp = ClassificationInterpretation.from_learner(
learn)
interp.plot_confusion_matrix()
```

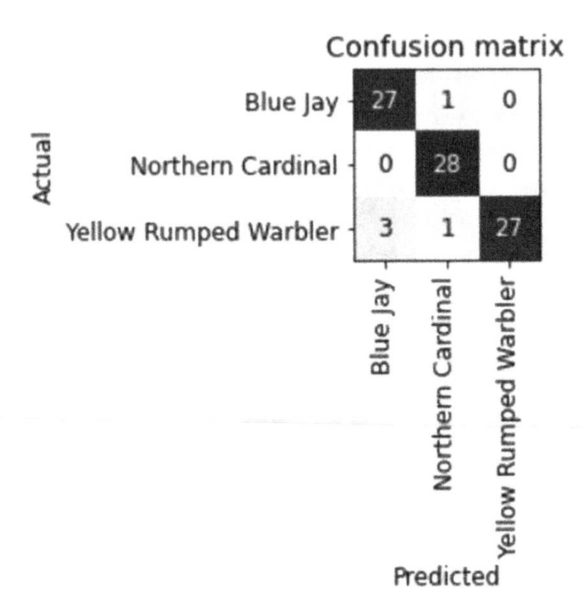

Fig. 9. Results visualized using confusion matrix. (Color figure online)

The confusion matrix showed that there were four Warblers misclassified as Blue Jays and Northern Cardinal and one Blue Jay was misclassified as Northern cardinal. We have a total of 5 incorrect predictions out of 87 making our error rate 0.057%.

4.7 Building a Classifier Widget

In the final step, a classifier widget within Jupyter notebook environment was built. It looks similar to an image upload GUI. After building the widget, the classifier was tested using random images from Google image search. The first step is to export the model, so that it can be reused for other tasks or it can be deployed in an application.

Here, the model is exported and loaded again to build the classifier widget.

```
learn.export()
learn_export = load_learner('export.pkl')
```

A function is defined to build the classifier widget.

```
btn_upload = widgets.FileUpload()
btn_run = widgets.Button(description='Classify')
out_pl = widgets.Output()
lbl_pred = widgets.Label()

def on_click_classify(change):
```

```
    img = PILImage.create(btn_upload.data[-1])
    out_pl.clear_output()
    with out_pl: display(img.to_thumb(128,128))
    pred,pred_idx,probs = learn_export.predict(img)
    lbl_pred.value = f'Prediction: {pred}; Probability: {probs[
                                        pred_idx]:.04f}'

btn_run.on_click(on_click_classify)
```

The widget is loaded.

```
VBox([widgets.Label('Upload Bird!'),
      btn_upload, btn_run, out_pl, lbl_pred]
```

A random image of a warbler from google image search is uploaded and tested.

The classifier predicted that the image is of a Yellow-Rumped Warbler with 99% probability (Fig. 10).

5 Summary

Based on the results, it can be summarized as the following:

- A large labeled dataset is not required to solve a task using transfer learning.
- The time taken to train the model is very little, which also reduces the computation costs.
- FastAI Library is very effective in implementing state-of-the-art models and transfer learning.

Fig. 10. Testing a random warbler image from Google. (Color figure online)

There are many techniques of transfer learning. The fine-tuning technique demonstrated here takes a pre-trained model for one given task and then tunes the model to make it perform on a similar task. This simple process can help us solve real-world problems. For example, we can build an Image detection system that tracks endangered birds or a warning system that detects a poisonous jellyfish in the water.

6 Conclusion

This research explored transfer learning and its importance. It also applied the transfer learning technique using a Resnet18 model for bird classification and achieved great results. This was done only using 450 images and it took less than one minute to train the model. Therefore, by leveraging the power of transfer learning many tasks that do not have large amounts of labeled data can possibly be solved, which is typically the constraint in a real-world setting.

References

1. Image Net Database. Image Net. http://www.image-net.org/. Accessed Apr 2021
2. Howard, J., Gugger, S.: Deep Learning for Coders with Fastai and PyTorch: AI Applications Without a PhD. Information, 1st edn., July 2020. https://www.oreilly.com/library/view/deep-learning-for/9781492045519/
3. Howard, J., Gugger, S.: Fastai: a layered API for deep learning. Information **11**(2), 108 (2020). https://doi.org/10.3390/info11020108
4. Andrew, N.G.: Deep Learning for Building AI Systems (2016). https://nips.cc/Conferences/2016/TutorialsIntros
5. Johnson, M., et al.: Google's multilingual neural machine translation system: enabling zero-shot translation. TACL **5**, 339–351 (2017). https://doi.org/10.1162/tacl_a_00065
6. Rusu, A.A., Vecerik, M., Rothörl, T., Heess, N., Pascanu, R., Hadsell, R.: Sim-to-Real Robot Learning from Pixels with Progressive Nets (2016). arXiv:1610.04286
7. OpenAI, Universe. Open AI Universe. https://github.com/openai/universe. Accessed Apr 2021
8. Udacity. Self-Driving-Car-Sim. Udacity. https://github.com/udacity/self-driving-car-sim. Accessed Apr 2021
9. Kornblith, S., Shlens, J., Le, Q.V.: Do better ImageNet models transfer better? In: 2019 IEEE/CVF Conference on Computer Vision and Pattern Recognition (CVPR). IEEE (2019)
10. Yosinski, J., Clune, J., Bengio, Y., Lipson, H.: How transferable are features in deep neural networks? In: Advances in Neural Information Processing Systems 27, pp. 3320–3328, December 2014. arXiv:1411.1792
11. Wang, K., Gao, X., Zhao, Y., Li, X., Dou, D., Xu, C.: Pay Attention to Features, Transfer Learn Faster CNNs (2019). openreview.net. https://openreview.net/forum?id=ryxyCeHtPB
12. Nguyen, C.V., Hassner, T., Seeger, M., Archambeau, C.: LEEP: A New Measure to Evaluate Transferability of Learned Representations (2020). arXiv:2002.12462

Object Recognition and Detection

RSS Fingerprint Based Signal Source Localization Using Singular Value Decomposition

Mingzhu Li[1] and Lei Cai[2(✉)]

[1] School of Electronics and Information Engineering, Hebei University of Technology, Tianjin 300401, China

[2] Guangdong Communications & Networks Institute, Guangdong 510700, China

cailei@gdcni.cn

Abstract. As the technique determines the position of a target device based on radio frequency (RF) fingerprint, received signal strength (RSS) fingerprint source localization technology is attracting increasing attention due to its numerous applications. In this paper, we propose a novel RSS Fingerprint Based Signal Source Localization algorithm. We use multi-dimensional interpolation to establish a fingerprint database, and singular value decomposition (SVD) is utilized to extract effective information from the fingerprint data. We divide the fingerprint database into multiple sub-fingerprint databases according to the location area and k-nearest neighbor (KNN) algorithm. Moreover, we adjust the fingerprints in each sub-fingerprint database to complete the offline training phase. In the online positioning phase, in order to improve the positioning efficiency, we use the k-dimensional (KD) tree algorithm to predict the fingerprint database using the fingerprint data received from the test point and determine the final position of the target source. Extensive measurements are carried out and it is proved that the proposed method is superior to existing ones.

Keywords: Radio frequency fingerprint · Signal source localization · Singular value decomposition

1 Introduction

With the continuous development of modern society, both daily life and public utilities are inseparable wireless communications [1]. Therefore, it is necessary to monitor, to locate and to standardize radio signals. Radio signal sources are affected by the reflection, refraction, diffraction and other radial propagation of obstacles. The current radio signal source positioning mainly relies on GPS systems or fixed monitoring stations or mobile monitoring vehicles [2], using Time of Arrival (TOA), Time Difference of Arrival (TDOA) or Angle of Arrival (AOA) and other ranging methods for positioning. However, these methods have higher requirements for hardware equipment and consume corresponding costs. The use of radio frequency (RF) fingerprint positioning

J. Xiong et al. (Eds.): MobiMedia 2021, LNICST 394, pp. 809–823, 2021.
https://doi.org/10.1007/978-3-030-89814-4_60

technology can obtain higher positioning accuracy in an environment with complex radio signal propagation characteristics [3], which can basically meet the needs. Moreover, this technology uses the received signal strength (RSS) fingerprint and the corresponding location coordinate information to calculate the positioning accuracy. It does not involve the location information of the access point (signal transmitter) and save working time [4].

The RSS was widely utilized as a feature in localization [5], as the RSS can be obtained easily. In the offline database building stage, the fingerprint signal collected in a large area are relatively sparse, and the RF signal are affected by attenuation and multipath effects during propagation. Therefore, the collected RSS fingerprints are directly used, which will cause a decrease in positioning accuracy. To solve this problem, some scholars have proposed preprocessing and clustering of fingerprint data. For instance, Fang [6] proposed the fingerprint feature extraction method of the signal projection, which projected the collected fingerprint signals to the relevant physical space. In literature experiment part, the principal components analysis (PCA) method was used to prove that this method can reduce the influence of the external environment on data and reduce the positioning error. When Youssef released the Horus positioning system [7], he proposed clustering to improve the accuracy of this positioning system. The specific method is to first divide the fingerprints into several areas after the establishment of the fingerprint database. And in the positioning stage, after obtaining the fingerprint vector of the points to be located, this method determines the point to be located in a certain area. Finally, in this area the accuracy of the position is determined. The method greatly reduces the positioning complexity in a large scale in environment and effectively improves the positioning accuracy.

And the research team led by Castor designed and completed a RF fingerprint positioning system called "Nibble" [8]. Nibble uses the signal noise ratio (SNR) of a RF signal reception as the location signal. In the literature [9], the author improved the accuracy of RF fingerprint by selecting the channel impulse response (CIR) as the fingerprint, and proved the superiority of fingerprint as the signal of physical layer. However, with underlying physical information they rely on specialized software and hardware to receive them [10]. The additional hardware leads to an increase in cost, which greatly limits the development of these fingerprints. Besides, with the rapid breakthrough of machine learning in various fields, more and more scholars are considering using machine learning to solve this problem, such as semi-supervised learning method [11, 12]. Researcher Pan proposed a semi-supervised learning database construction method [13], which uses part of the RSS fingerprints with marked locations and part of the RSS fingerprints without marked locations to build a fingerprint database through a manifold model, thereby reducing the workload.

Positioning is divided into receiving end positioning and signal transmitting source positioning. Since both species are based on radio signal strength, there is interoperability. Therefore, this paper draws on receiving end positioning, adopting RF fingerprint location technology to carry out signal source localization research and improve the offline database building stage and online positioning stage. We propose a novel RSS Fingerprint Based Signal Source Localization algorithm. First, in the offline training phase, we perform multi-dimensional interpolation on the RSS data collected from the field to obtain all the fingerprints in the positioning area. The SVD processing of the obtained data not only retains useful information but also reduces the impact of external factors on fingerprints. At this time, we complete the establishment of the fingerprint database. We divide the database into multiple intersecting sub-fingerprint databases according to the location area, and use the KNN algorithm to adjust the data in the sub-fingerprint databases. In the online positioning phase, we use the KD tree algorithm to predict the fingerprint database using the fingerprint data received from the test point and determine the final position of the target source by the improved weighted k- nearest neighbor (WKNN) algorithm.

2 RF Fingerprint Location Process

In this section, we introduce the process of RF fingerprint location technology, including the offline phase and the online phase.

RF fingerprint positioning is to establish the corresponding relationship between the fingerprint signal space and the geographic location space to achieve location determination [14]. This process is divided into two phases: the offline phase of fingerprint database construction and the online phase of location positioning, as shown in Fig. 1. In the first phase, we build a fingerprint database that stores the RF fingerprint signals and the corresponding location coordinates. In the next phase, the received signal characteristics of the text point to be located are matched with the fingerprints' in the database. After the matching fingerprint is obtained, the position coordinates can be determined.

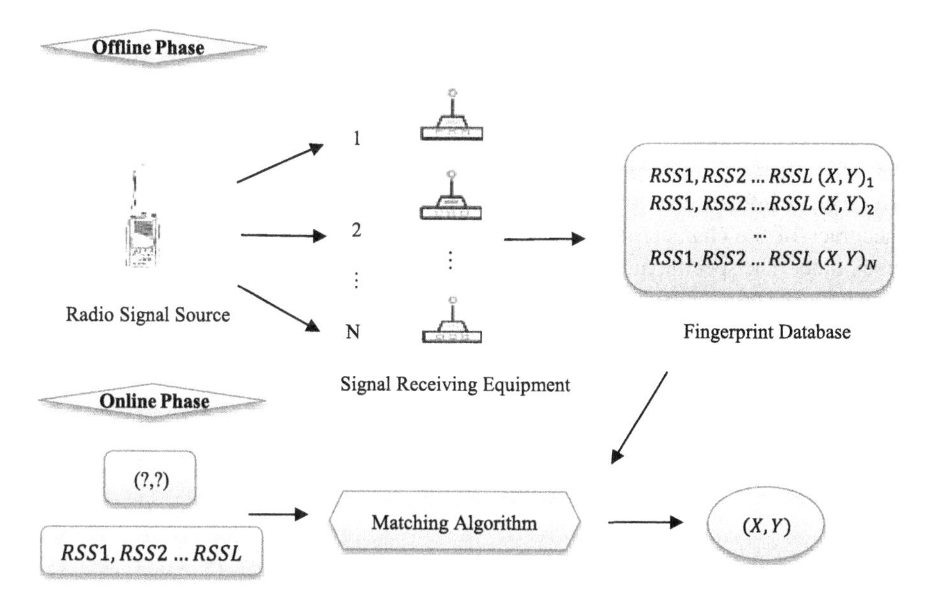

Fig. 1. Principle of RF fingerprinting positioning

The choice of matching algorithm for radio signal source RF fingerprint positioning should be reasonable because the choice of algorithm has a great influence on positioning accuracy. Commonly used location fingerprint location algorithms mainly include NN, KNN and WKNN method.

2.1 Nearest Neighbor

Suppose the characteristic value of the signal strength at reference points received by receiving devices in the offline phase is (1)

$$RSS_i = (RSS_{1i},\ RSS_{2i},\ RSS_{3i} \ldots RSS_{Li}) \tag{1}$$

Among them, L is the number of collection nodes, RSS_{ji} represents its signal strength of the reference point i received by the jth collection node, and the signal strength of each reference point received by the collection node is stored in the fingerprint database. For the measured signal strength characteristic value of the test point is (2)

$$RSS = (RSS_1,\ RSS_2,\ RSS_3 \ldots RSS_L) \tag{2}$$

For each reference point, the vector Euclidean distance between the test point and the reference point i with respect to the signal strength is obtained by formula (3) as

$$D_i = \sqrt{(RSS_1 - RSS_{1i})^2 + (RSS_2 - RSS_{2i})^2 + \ldots + (RSS_L - RSS_{Li})^2} \tag{3}$$

Take the estimated point coordinates corresponding to the minimum signal strength distance as the output coordinates of the positioning result.

2.2 *K*-Nearest Neighbor

The KNN method is an improvement of the NN. Considering that the NN method may have a great influence on the result due to the selection of reference points and the inaccuracy of the characteristic data measured by the test points, we consider adding some points. The specific method is: instead of taking the smallest distance, the distance obtained in the previous step is to take the k smallest distances from small to large in turn. For the coordinates corresponding to these k distances, the average value is calculated as the output, and the two-dimensional coordinate system the following is expressed as:

$$(x, y) = \frac{1}{k}\sum_{k=1}^{k}(x_k, y_k) \tag{4}$$

2.3 Weighted *K*-Nearest Neighbor

The WKNN method is an improvement of the KNN [15]. The k minimum distances selected by the KNN method contribute the same to the final positioning result, and the average value is taken. But the smaller the distance is and the greater the contribution to the coordinate of the result is. Therefore, weights are assigned to the k minimum distances. The positioning result in the two-dimensional coordinate system is expressed as:

$$(x, y) = \sum_{k=1}^{k} w_k (x_k, y_k) \tag{5}$$

Among them, the sum of w_k is 1.

$$w_k = \frac{1/D_k}{\sum_{k=1}^{k} 1/D_k} \tag{6}$$

3 Proposed Method

In the offline process, one problem is that the fingerprint signal collection is time-consuming and labor-intensive. The traditional fingerprint collection work is to first determine a large number of reference points in the area to be tested, then collect and store the RF signal reception intensity value for a period of time at each reference point in the database. When the collection area is relatively large, the establishment of database will consume huge manpower and material resources. Another problem is the preprocessing of fingerprint signals. Directly using the collected fingerprints for positioning will cause the dropping of the positioning accuracy. In the online phase, due to the large positioning space, the calculation complexity of positioning process is relatively high, the calculation amount is large, and the efficiency of determining the unit is low. In order to solve the above problems, this section proposes a novel RSS Fingerprint Based Signal Source Localization algorithm that introduces some new methods and processes in two phases.

3.1 Offline Training Phase

Fill in Fingerprint Database Data. In order to solve the problem of high database construction cost, we use multi-dimensional interpolation to reduce the number of sampling points and the cost of database construction. In the fingerprint collection stage, only a few numbers of fingerprints need to be interpolated to obtain a sufficient number of fingerprints.

Using MATLAB 2016a $F = scatteredInterpolant(x, y, v)$ to create a two-dimensional interpolation of $v = F(x, y)$, it can be understood as the RSS at a point (x, y) in the space. The basic idea is to first establish an interpolation class based on the existing training samples, and then evaluate the interpolation points. We can calculate the F value at a set of query points (for example (xq, yq)) to obtain the inserted value $vq = F(xq, yq)$. After multiple interpolations, the fingerprints received by each receiving device from the transmitting source can be obtained [16] (Fig 2).

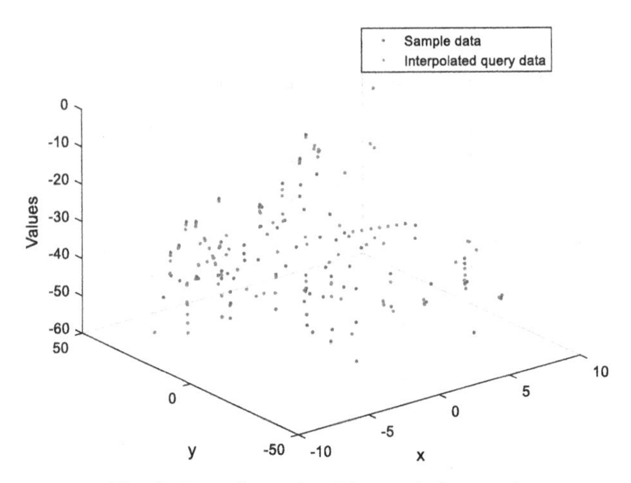

Fig. 2. Two-dimensional interpolation result

Extraction of Effective Fingerprint Information. In order to solve the problem of low positioning accuracy when directly using the collected RSS fingerprints for positioning, SVD is used to extract effective information from the complete data after the multi-dimensional interpolation.

The main application of matrix singular value decomposition in statistics is PCA, which is a data analysis method used to find "patterns" hidden in large amounts of data [17]. The collected fingerprint data is formed into matrix *A*. For the matrix $A_{m\times n}$, there are $U_{m\times m}$, $V_{n\times n}$, $S_{m\times n}$, satisfying $A = U * S * V'$. Except for the singular value of *A*, the elements on the diagonal of matrix *S* are all *0*, and these values on the diagonal are arranged from large to small. The singular value is the weight, and the larger the weight is, the greater the information is. In many cases, the sum of the top 10% or even 1% of the values accounts for more than 99% of the sum of all. Therefore, a singular value of the first *r* can be used to approximate the matrix.

$$C = \sqrt{(S_{11})^2 + (S_{22})^2 + \ldots + (S_{\mathrm{nn}})^2} \tag{7}$$

$$C_r\sqrt{\sum (S_{rr})^2} \quad r = 1, 2, \ldots, n \tag{8}$$

where $S_{11}, S_{22}, S_{33,\ldots,}S_{nn}$ are singular values, when $C_r/C \geq 80\%$, r can be obtained.

$$A_{m\times n} \approx U_{m\times r}\sum\nolimits_{r\times r} V^T_{r\times n} \tag{9}$$

where $\Sigma_{r\times r}$ is a diagonal matrix formed by $S_{11,}S_{22,}S_{33,\ldots,}S_{rr}$. $A_{m\times n}$ is the result after extracting valuable information.

Currently, the common data preprocessing methods include arithmetic average filtering method and median average filtering method. For the same experimental environment, the same collection node, reference point, test point layout, we use these methods to process the collected fingerprint data and establish fingerprint databases, and we perform SVD processing on the data on the basis of the above methods, and then compare it in the same online positioning algorithm (KNN). The test point data is positioned and analyzed in different fingerprint databases. The estimated position errors are obtained, as shown in Table 1.

From the data in the Table 1, it can be seen that when the fingerprint data are processed by two commonly used data preprocessing methods and then by SVD processing, the test points have a better positioning effect when online positioning is performed in the fingerprint database established than the one without SVD processing.

Table 1. The positioning errors (m) of the test points in the fingerprint databases established after the data are processed by the original data preprocessing methods and the SVD processing on the basis of these methods.

Serial numbers of test points	Arithmetic average filtering method	SVD processing afterwards	Median average filtering method	SVD processing afterwards
1	6.35	6.35	8.29	6.08
2	8.70	5.62	10.36	7.34
3	15.03	13.41	16.25	15.99
4	22.06	18.98	21.38	18.20
5	11.43	9.13	10.64	10.02
6	21.90	21.90	20.45	17.28
7	15.80	14.33	17.51	15.03
8	16.52	13.09	14.47	12.62
9	29.71	29.19	28.52	28.52
10	27.18	23.08	26.91	21.21

Establish Sub-fingerprint Database. After obtaining the complete fingerprint database, the database is divided into multiple intersecting sub-fingerprint databases according to the geographic location area, and the RSS fingerprint in each sub-fingerprint database is adjusted according to the KNN algorithm.

Taking one of the sub-fingerprint databases as an example, each reference point is used as a test point for online positioning in this database. The matching algorithm used is the KNN algorithm, and k is obtained by multiple cross-validation. At this time, we set the threshold which is the Euclidean distance between predicted positioning coordinates and actual coordinates. When the result of a certain test point is larger than the threshold, this point and the fingerprint information containing that will be deleted from database and transferred to the pending all the data in this database meet the requirements. After performing the above process on all data in sub-fingerprint databases, we filter out the repeated data from the pending area, and return to their previous databases to perform online positioning again. If the positioning result still does not meet the requirements, remove it from the fingerprint database. Finally, the information contained in the fingerprint database are the position coordinates of each reference point, the RSS fingerprint data and the location of the sub-fingerprint database. The above is a new process improved based on the offline phase of traditional RF fingerprint positioning, and the flowchart is shown in Fig. 3.

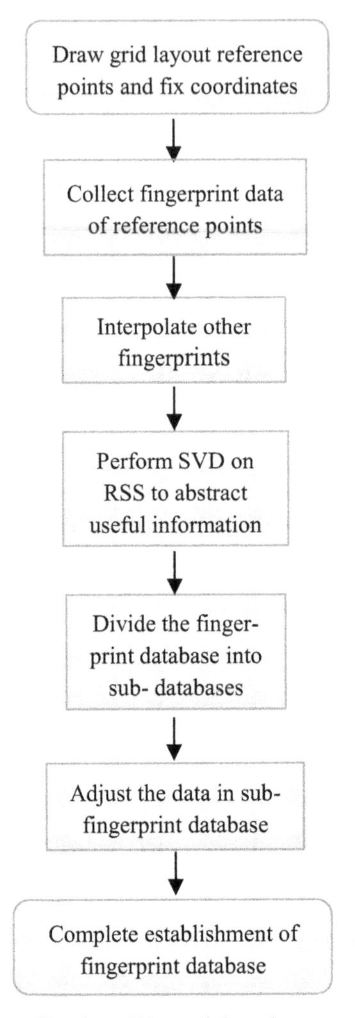

Fig. 3. Offline training phase

3.2 Online Positioning Stage

At this stage, we put the RSS fingerprint data collected from the test point into the fingerprint database for SVD processing to extract useful information, and use the KD tree algorithm to predict the location of the sub-fingerprint database. At last, we apply the improved WKNN algorithm for the predictive positioning.

Determine the Location of the Test Point. In order to improve the efficiency of searching, we consider using a special structure to store training data to reduce the number of distance calculations. Using the KD tree can save the search for most data points, thereby reducing the amount of search calculations.

The KD tree is a tree data structure that stores instance points in a k-dimensional space for quick retrieval [18]. It is also a binary tree, which represents a division of the k-dimensional space. Constructing the KD tree is equivalent to continuously dividing the space with a hyperplane perpendicular to the coordinate axis to form a series of k-dimensional hyper rectangular regions. Each node of the KD tree corresponds to a k-dimensional super rectangular area.

Using this method to determine the location of the sub-fingerprint database where the test point is located includes three steps. The first step is to build a tree, the second is to search for the nearest neighbor, and the last step is to predict. KD tree building uses the variance of the values of n features from the n-dimensional features of m samples to be calculated, where m corresponds to the number of reference points in the fingerprint database, and n corresponds to the number of receiving devices. We use the jth dimension feature n_j with the largest variance as the root node. For this feature, we choose the sample corresponding to the median n_{jv} of the value of n_j as the division point. For all samples with the value of the jth feature less than n_{jv}, we divide it into the left subtree, and for the jth dimension feature for samples with a value greater than or equal to n_{jv}, we divide it into the right subtree. For the left subtree and the right subtree, we use the same method as before to find the feature with the largest variance to change the node, and recursively generate the KD tree (Fig. 4).

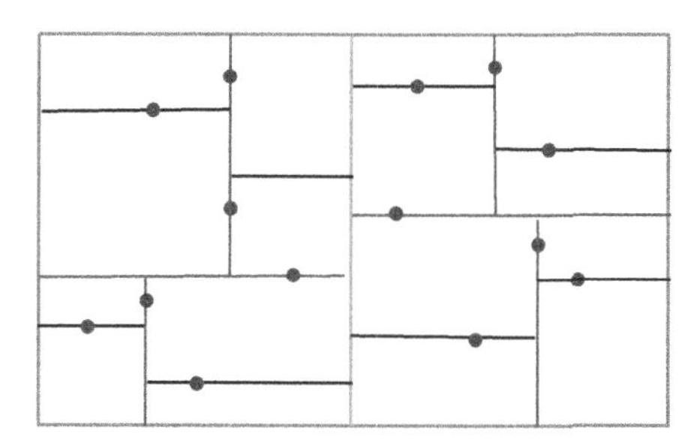

Fig. 4. Schematic diagram of KD tree space division for 2D data

For a test point, we first find the leaf node that contains the test point in the KD tree, take the test point as the center and the distance from the modified point to the sample instance of the leaf node as the radius to obtain a hypersphere. The nearest neighbor must be inside the hypersphere. Then return to the parent node of the leaf node, we should check whether the super rectangular body contained in another child node intersects the super sphere. If it intersects, we can go to this child node to find if there is a closer neighbor, and update the nearest neighbor. If they do not intersect, we directly return to the parent node and continue to search for the nearest neighbor in another subtree. When backtracking to the root node, the algorithm ends, and the nearest neighbor node saved at this time is the final nearest neighbor.

On the basis of KD tree search for nearest neighbors, we select the first nearest neighbor sample and set it as selected. In the second round, we ignore the selected samples and reselect the nearest neighbors. In this way, we run k times to get the k nearest neighbors of the target. According to the majority voting method, it is predicted that the category with the largest number of categories in neighbors can determine the location of the sub-fingerprint database where the test point is located.

Improved Positioning Algorithm. It is necessary to improve the vector Euclidean distance formula for calculating the RSS of the test point and the reference point to reduce the positioning error [19]. $RSS_L - RSS_{Li}$ is the absolute error, and the deviation from the true value is reflected in the same unit dimension, but due to the different magnitudes, the contribution to D_i may be inconsistent and cause errors. Therefore, the absolute error is converted into a relative error. The relative error is expressed as a percentage, it is a dimensionless value. Generally speaking, the relative error can better reflect the credibility of the data, so it is changed to the relative error to improve the above possible drawbacks. Therefore, the formula for obtaining D_i under the same intensity becomes:

$$D_i = \sqrt{\left(\frac{RSS_1 - RSS_{1i}}{RSS_{1i}}\right)^2 + \left(\frac{RSS_2 - RSS_{2i}}{RSS_{2i}}\right)^2 + \ldots + \left(\frac{RSS_L - RSS_{Li}}{RSS_{Li}}\right)^2} \quad (10)$$

The improved positioning method in the online phase uses the improved KNN (K = 4) and WKNN (K = 5) algorithms. The results are shown in the Figs. 5 and 6.

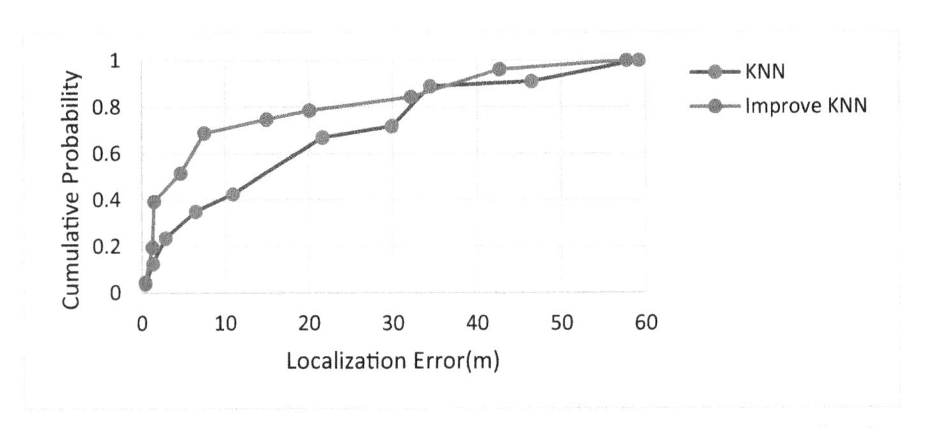

Fig. 5. Comparison of positioning results between KNN and improved KNN in the online phase of traditional method

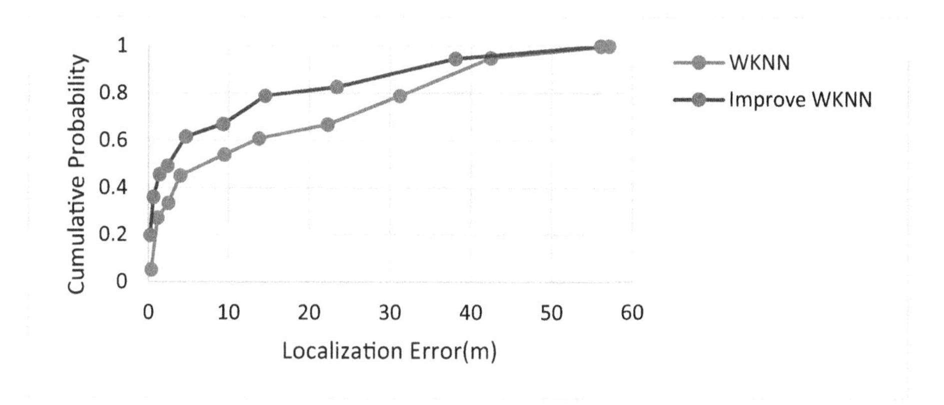

Fig. 6. Comparison of positioning results between WKNN and improved WKNN in the online phase of traditional method

4 Experiments Validation

4.1 Experiments Setup

We choose an open space with fewer obstacles outdoors, this area is about $175 \times 100\,\text{m}^2$. Then we divide a grid every 5 m in this area, and place the reference point at the center of the grid. As shown in the Fig. 7, a total of 154 black points are reference points, which are the positions of the signal source when the RSS fingerprint is collected. The red points are a total of 52 test points, and 5 receiving devices. A, B, C, D, E are arranged on the roof around the area. All receiving equipment are placed in a position that is visible and encloses the entire area to ensure that the signals emitted at each reference point can be received by the equipment. Radio signals are transmitted at each set reference point, and 5 receiving devices simultaneously receive the signals and record the RSS fingerprints.

The signal source is Motorola GP328Plus walkie-talkie, the radio frequency is 430.11 MHz. Each receiving device consists of a receiving antenna and a Tektronix RSA306b receiver. This receiver is connected to a laptop and display multiple RSS values received in the same time period. After that, the data received by each receiving device is processed, the maximum value and the minimum value are removed, and the remaining data is averaged. Finally, the 5 received field strength values are sorted into a set of fingerprint data. In order to save the time of querying the IP address and realize the rapid positioning process, a laptop is connected and transmits data through the LAN composed of 3 Huawei Q2S routers. The instrument connection of a group of receiving equipment is shown in the Fig. 10. The fingerprint information of other reference points that are not marked is derived from multi-dimensional interpolation.

After the completing of the fingerprints in the database, we use SVD to preprocess the data to extract effective information to reduce the impact of the external environment. At this time, let $C_r/C = 90\%$.

As shown in the Fig. 7, the area is divided into 7 intersecting areas and the points are included in the corresponding sub-fingerprint database. All points from the red line to the left are included in the sub-fingerprint database 1, from the purple line to the bottom all points are included in the database 2 and the points from the blue line to the right are included in the database 3. The points between the green lines are included in the database 4 and between the orange lines are included in the database 5. Besides, the points to the upper left of the orange and purple lines are included in database 6, and points to the upper right of the green and purple lines are included in the database 7.

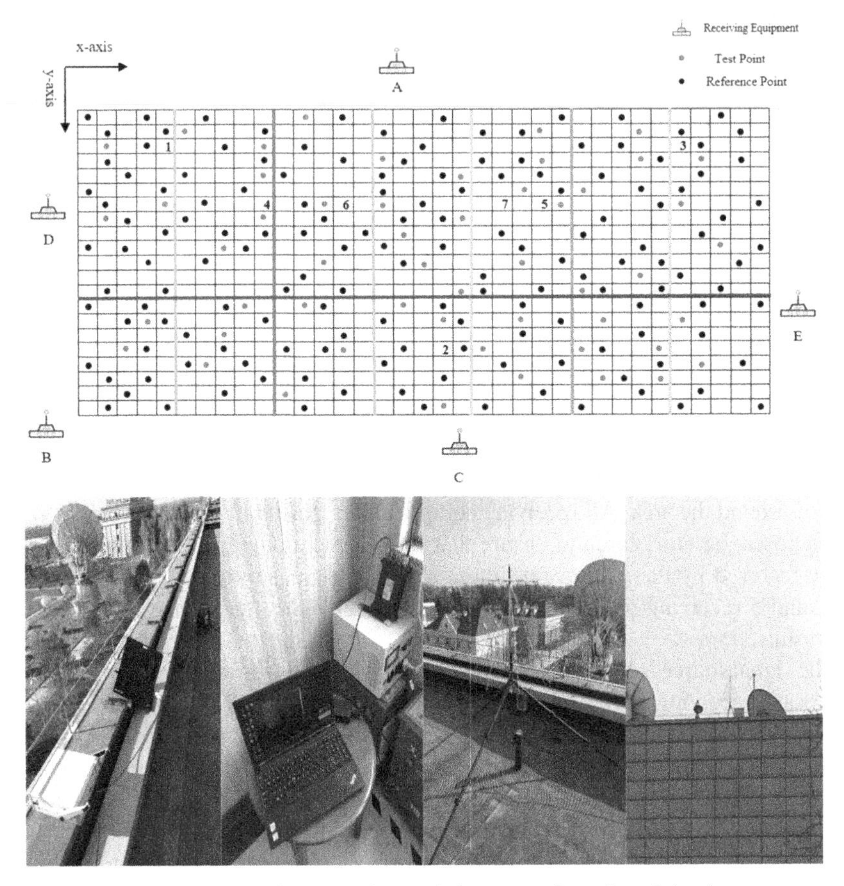

Fig. 7. The location of reference points and the connection of receiving instruments

According to the process of offline establishment of fingerprint database proposed in the previous section, the representative data of the fingerprint database are shown in the Table 2.

Table 2. Representative data in the fingerprint database.

	Coordinate(m)		RSS (dBm)					
Serial number	X	Y	A	B	C	D	E	Sub-fingerprint database
1	2.5	17.5	−38.43	−37.27	−37.21	−28.39	−38.87	1,6
2	2.5	22.5	−35.32	−41.26	−38.75	−31.67	−39.59	1
3	7.5	7.5	−37.53	−38.73	−37.32	−29.80	−39.25	1,6
4	7.5	12.5	−35.67	−40.58	−36.90	−27.89	−40.54	6
5	17.5	2.5	−27.98	−42.61	−40.67	−27.22	−45.27	1,6
6	17.5	7.5	−32.18	−43.02	−38.64	−28.36	−41.94	1
7	27.5	2.5	−32.55	−40.67	−40.56	−28.40	−41.34	1,4,6
8	37.5	27.5	−31.58	−40.03	−33.96	−24.27	−36.53	1,4,6
9	42.5	7.5	−33.14	−40.65	−45.00	−25.88	−37.78	4,6
10	62.5	22.5	−24.57	−45.10	−47.07	−32.49	−33.26	4,6

5 Experimental Results

Using the novel RSS Fingerprint Based Signal Source Localization algorithm proposed in this paper to predict positioning (the K in the KD tree and the improved WKNN algorithm are both obtained by cross-validation, $K = 6{,}5$) and compare to the previous methods. As shown in the Fig. 8, the new improved method has an error probability of

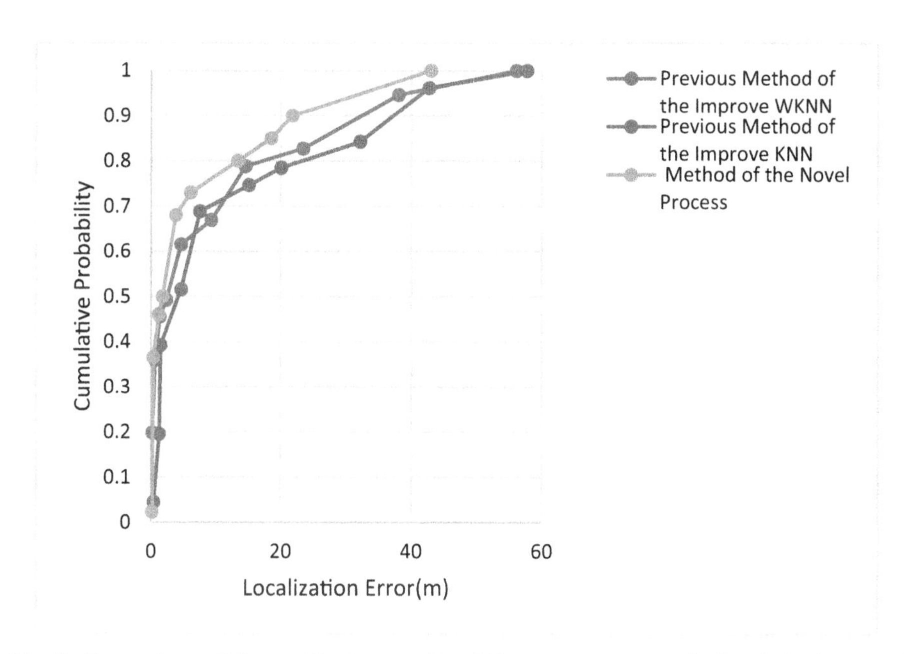

Fig. 8. Comparison of the positioning results of the new process method and the improved traditional methods

less than 13 m can reach 80%, less than 19 m can reach 85% and less than 22 m can reach 90%.

6 Conclusion

In short, the results of positioning using the novel RSS Fingerprint Based Signal Source Localization algorithm proposed in this paper are significantly better than the traditional positioning algorithm based on RF fingerprint technology. Using SVD to process the collected fingerprint data can reduce the impact of the external environment on the data. Besides, in the offline training phase, we use multi-dimensional interpolation to complete the fingerprints in the fingerprint database and KNN to adjust the fingerprints in the sub-fingerprint database. In the online positioning phase, we use the KD tree to determine the area of the test point and then use the improved WKNN algorithm to complete the positioning. Therefore, by using the combination with the improved positioning process that introduces other methods mentioned in this paper, the positioning accuracy can be greatly improved.

Acknowledgement. This work was supported in part by the Guangdong Key Field R&D Program under Grant (2018B010124001) and in part by the Beijing Natural Science Foundation under Grants (4202046). I want to thank these projects for supporting my research in terms of funding.

References

1. Huang, S., Yuan, Y.Y., Wei, Z.Q.: Automatic modulation classification of overlapped sources using multiple cumulants. IEEE Trans. Veh. Technol. **66**(7), 6089–6101 (2017)
2. Rose, C., Britt, J., Allen, J.: An integrated vehicle navigation system utilizing lane-detection and lateral position estimation systems in difficult environments for GPS. IEEE Trans. Intell. Transp. Syst. **15**(6), 2615–2616 (2015)
3. Rong, X.F., Na, Y.: Analysis and simulation of positioning error of RSSI position fingerprint. J. Xi'an Technol. Univ. **30**(6), 574–576 (2010)
4. Yang, H., Wei, Y.Y., Lin, B.: Positioning Technology. Publishing House of Electronics Industry, Beijing (2013)
5. Shen, Y.: Research on Access Point Selection Algorithm in Wireless Indoor Positioning Based on Fingerprint. Zhejiang University, Zhejiang (2014)
6. Fang, S.H., Lin, T.N., Lin, P.C.: Location fingerprinting in a decorrelated space. IEEE Trans. Knowl. Data Eng. **20**(5), 685–691 (2008)
7. Youssef, M., Agrawal, A.: Location-clustering techniques for WLAN location determination systems. Int. J. Comput. Appl. **28**(3), 278–284 (2006)
8. Gao, G.W.: Design and Implementation of Indoor Positioning System Based on WLAN and Zigbee. Beijing University of Posts and Telecommunications, Beijing (2011)
9. Shen, Y.: Research on Access Point Selection Algorithm in Wireless Indoor Positioning Based on Fingerprint. Zhejiang University, Zhejiang (2014)
10. Lin, Y., Tu, Y., Dou, Z.: Contour stella image and deep learning for signal recognition in the physical layer. IEEE Trans. Cogn. Commun. Networking **7**(1), 34–46 (2021)
11. Huang, S., Lin, C., Xu, W.: Identification of active attacks in Internet of Things: joint model- and data-driven automatic modulation classification approach. IEEE Internet Things J. **8**(3), 2051–2065 (2021)

12. Tu, Y., Lin, Y., Wang, J.: Semi-supervised learning with generative adversarial networks on digital signal modulation classification. CMC-Comput. Mater. Continua **55**(2), 243–254 (2018)
13. Lin, Y., Zhao, H., Ma, X.: Adversarial attacks in modulation recognition with convolutional neural networks. IEEE Trans. Reliab. **70**(1), 389–401 (2021)
14. Hu, A.M.: Research on Indoor Radio Frequency Fingerprint Location Technology Based on Machine Learning. Beijing Jiao Tong University, Beijing (2019)
15. Zhou, J.F., Liu, T., Zou, L.: Design of machine learning model for urban planning and management improvement. Int. J. Performability Eng. **16**(6), 958–967 (2020)
16. MATLAB documentation. https://ww2.mathworks.cn/help/matlab/ref/scatteredinterpolant. Accessed 22 Jan 2021
17. Wang, Z., Zhang, H.Q., Peng, L.: Medical data information extraction and classification method based on singular value decomposition. J. Chengdu Univ. Inf. Technol. **35**(5), 537–541 (2020)
18. KD tree search for nearest neighbors. https://www.cnblogs.com/pinard/p/6061661.html. Accessed 22 Jan 2021
19. Hui, H.: Research on Fingerprint Location Detection System Based on SA44B Indoor Radio Signal Source Location. Hebei University of Technology, Tianjin (2015)

Ship Detection in SAR Images Based on an Improved Detector with Rotational Boxes

Xiaowei Ding[1], Changbo Hou[1,2](✉), and Yongjian Xu[1]

[1] Harbin Engineering University, Harbin 150001, HLJ, China
houchangbo@hrbeu.edu.cn

[2] Key Laboratory of Advanced Marine Communication and Information Technology, Ministry of Industry and Information Technology, Harbin Engineering University, Harbin 150001, HLJ, China

Abstract. In the SAR ship data under complex backgrounds, especially in the coastal area, the horizontal bounding box detection algorithm makes a large number of coastal noise interference targets feature extraction and bounding box regression. In addition, the horizontal bounding box cannot well reflect the characteristics of large aspect ratio of ships. Therefore, this paper proposes an improved YOLOv3 detection algorithm based on the rotational bounding box, which increases the encoding method of the angle parameter, and generates the prediction bounding box at a fixed angle interval. Different angle intervals will have different effects. Focus loss function is used to solve the problem of positive and negative sample balance and difficult sample feature learning. The experimental results show that the average precision of the R-YOLOv3 algorithm based on the rotational bounding box on the SAR ship data set is 87.3%, which is a 13.5% gain compared with the classic YOLOv3, which reflects the high precision of the ship targets.

Keywords: SAR image · Ship target detection · Convolutional neural network

1 Introduction

Ship detection is one of the main technologies for maritime surveillance, which is extremely important for maintaining maritime security, monitoring maritime transportation and improving the capability of maritime defense and early warning [1]. At present, the data sources of ship detection are mainly optical sensors, infrared sensors and SAR sensors. Synthetic aperture radar (SAR) [2] is an active side microwave imaging sensor. Compared with infrared and optical passive sensors, it has the advantages of penetrating clouds and working day and night. It can collect large-area data anytime and anywhere under any weather conditions such as daytime, nighttime and foggy days, and generate high-resolution images by comprehensive utilization of signal processing, pulse compression and synthetic aperture principle. The traditional methods of target detection in SAR images can be divided into threshold-based methods [3], salient region-based methods [4], texture-based methods [5] and statistical analysis-based methods [6]. Among

J. Xiong et al. (Eds.): MobiMedia 2021, LNICST 394, pp. 824–836, 2021.
https://doi.org/10.1007/978-3-030-89814-4_61

these methods, Constant False Alarm Rate (CFAR) [7] and its variants are most widely used. In recent years, with more and more successful launch of SAR satellites, SAR image acquisition becomes more and more easy and high-resolution image data increases. The development of high-precision and high-efficiency target detection system has attracted much attention. Deep learning algorithm has a broad application prospect in ship target detection and image classification [8, 9] due to its strong autonomous learning ability [10] and feature representation ability [11]. At present, the algorithms based on convolutional neural network are Faster R-CNN [12], YOLOv1 [13], YOLOv3 [14], SSD [15], RetinaNet [16], etc. Subsequently, more and more improved algorithms are applied to SAR ship target detection. For example, in 2017, Kang et al. [17] proposed a region-based R-CNN target detection algorithm based on multi-scale feature fusion, which combines shallow and deep features and is conducive to eliminating false alarms. Li et al. [18] constructed the first public dataset SSDD in the field of ship target detection in SAR images in 2017, and proposed an improved detector based on Faster R-CNN based on multi-technology fusion. Then Kang et al. [19] proposed a new Faster R-CNN detection network by combining CFAR algorithm and Faster R-CNN algorithm. Faster R-CNN uses the sliding window generated by CFAR algorithm as a candidate region to detect small ship targets, and obtains better detection performance.

Ship detection will have recognition problems. Usually, the direction of ship targets in SAR images is diverse. Most of the ship detection in the offshore area will only receive the influence of sea clutter, and the detection task is simple. However, in the coastal area, the scene is complex, and the ship arrangement is very dense. The annotation method of horizontal box cannot distinguish the target from the target, as well as the target and the background. The background interference is serious, and it is easy to cause missed detection. It is difficult to distinguish small ship targets from speckle noise, which intensifies the difficulty of ship target detection. Therefore, based on YOLOv3, this paper proposes a ship target detection algorithm based on rotational boxes in SAR image. Firstly, R-YOLOv3 uses ResNet-50 as the backbone network of feature extraction, and improves the extraction of complex features by residual module [20]. Secondly, in order to improve the detection precision of small SAR ships, R-YOLOv3 draws on the idea of FPN [21] to perform multi-scale fusion and independent prediction of the extracted features. Then, in view of the characteristics of ship direction diversity and the problem of large background interference in coastal areas, the detection algorithm adds the encoding method of angle parameters, and generates the prediction boundary box according to the fixed angle interval, and improves the calculation method of rotational boxes cross-parallel ratio to solve the influence of angle change on the detection accuracy. Finally, the algorithm uses the focus loss function to reduce the weight of the samples that are easy to classify, so that the model focuses on learning the characteristics of the foreground objects with less difficult classification and improves the detection precision of the model for ships.

2 Methods

2.1 Overall Scheme of R-YOLOv3

The one-stage detector R-YOLOv3 proposed in this paper uses the idea of residual network structure and RPN multi-scale feature fusion to realize the boundary anchor box regression with angle information. The structure of our method is shown in Fig. 1.

Firstly, the SAR image to be detected is used as the input of the feature extraction network. The feature of the ship target is extracted through the ResNet50 backbone network, and the feature mapping of five layers with different sizes is obtained. The features of three different sizes extracted from the last layers of the feature extraction network, and are fused to achieve the regression of the border on the fused feature map. In the bounding box prediction stage, the box is predicted by the coordinate and angle information of the target. Considering the sensitivity of the intersection ratio to the angle change, a new intersection ratio strategy is adopted. The selection of loss function has an important influence on the performance of the algorithm. The Focal Loss function is used to solve the imbalance of positive and negative samples and the problem of difficult to classify samples, reduce the weight of easy to classify samples, and make the model focus on learning the characteristics of foreground objects with less difficult classification.

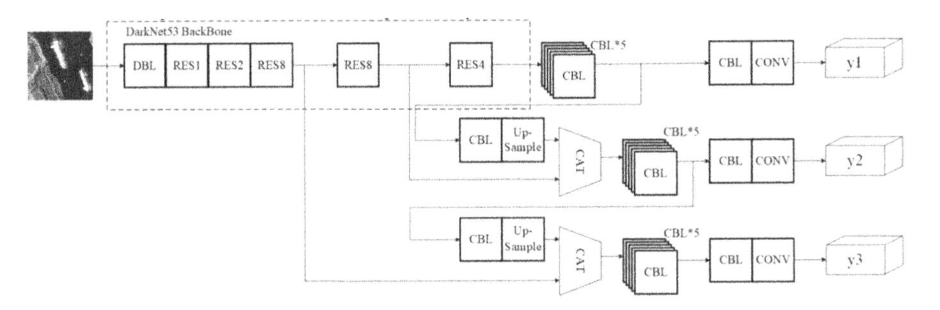

Fig. 1. Structure of R -YOLOv3 based on ResNet-50

2.2 Backbone Network for Feature Extraction

R-YOLOv3 uses ResNet-50 as the backbone network for extracting target features, and its structure is shown in Fig. 2. ResNet-50 is mainly composed of multiple residual units. Each residual unit contains the jump connection from input to output and the output obtained by three convolution operations. The convolution kernel sizes of the three convolution layers are 1×1, 3×3 and 1×1. This residual structure solves the problem of gradient disappearance and model detection accuracy reduction after network deepening. In order to improve the small target detection precision of the model, R-YOLOv3 draws on the idea of FPN feature fusion, and ResNet-50 has carried out five down-samplings. Therefore, the whole feature extraction process is divided into five stages ($conv1$, $conv2$, $conv3$, $conv4$, $conv5$). The output characteristic figure of the last layer of each stage is (C_1, C_2, C_3, C_4, C_5).Because the number of pixels of the target is reflected by the resolution information, the large-scale feature map extracted by the shallow network has high resolution, less semantic information, small network receptive field and accurate location information, which can better detect small targets. The semantic information extracted by the deep network is rich in low-resolution features, and it feels rough about the large position information. Small targets are easy to miss. R-YOLOv3 performs up-sampling and concat respectively on the extracted features at three different scales. The features at different scales are fused with each other, and finally are predicted on the feature maps at three scales. The sizes of the three feature scales output are 32, 16 and 8 times the down-sampling of the input image resolution, and the size of the feature map output is 10×10, 20×20 and 40×40.

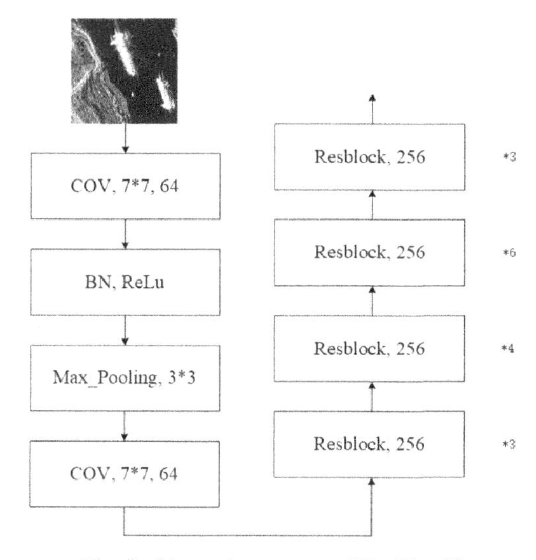

Fig. 2. Network structure of ResNet-50

2.3 Rotational Boundary Box Prediction

In the detection algorithm based on rotational box due to the increase of the angle parameter, the rotational anchor box generated at a certain anchor point is arranged at a certain angle interval, as shown in Fig. 3. Then, by calculating the intersection ratio between the angled anchor box and the ground truth, the anchor box with the largest intersection ratio is assigned to the ground truth. However, there will be a problem with this calculation method. At this time, the anchor box on the matching through the maximum crossover ratio is not the optimal boundary box. Assuming that two anchor boxes A and B, the length and width size of anchor box A is, the length and width size of anchor box B is, and the angle interval set by anchor box is 45°. The size of the real box C is, and its angle is 20°. At this point, when the angle value of anchor box A is 0°, the intersection ratio with the real box reaches the maximum, which is 0.40; When the angle of anchor box B is 0°, the intersection ratio with the ground truth reaches the maximum, which is 0.45. According to the above matching criteria, the anchor box B matches the real box C at the angle of 0°. However, under the same angle deviation, the size deviation between anchor box B and real box C is 10, while there is no deviation between anchor box A and real box. Therefore, anchor box A is more suitable for allocation to real box C.

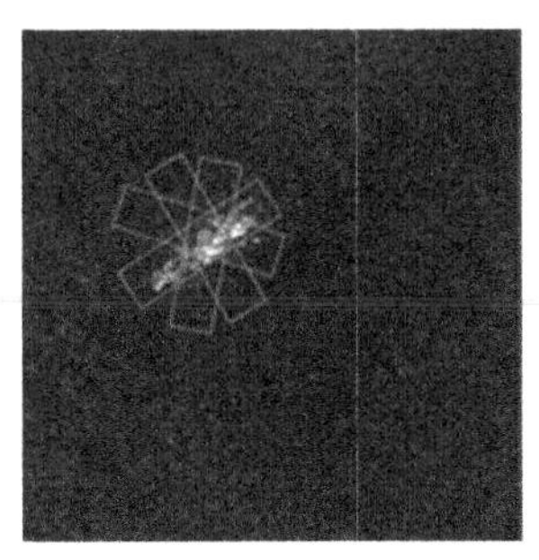

Fig. 3. Multi-angle rotational anchor box

Therefore, the matching result between the anchor box with angle and the ground truth is not completely optimal by directly using the above intersection and union ratio calculation.

$$ArIoU(A, B) = \frac{area\left(\hat{A} \cap B\right)}{area\left(\hat{A} \cup B\right)}|\cos(\theta_A - \theta_B)| \quad (1)$$

Among them, the first half of the formula is IoU between horizontal boxes, the second half is to measure the angle deviation between the two boxes, and the value range after the product is [0, 1].

R-YOLOv3 uses K-Means clustering algorithm to calculate the clustering results of the data set annotation box as the target prior box. The feature map of each scale matches three prior boxes, and nine prior boxes are obtained by three dimensional copolymerization classes. In the prediction boundary box stage, each cell on each feature map predicts three boundary boxes. The regression network learns the boundary box offset of the target according to the input characteristics. In addition to the four values of the center point coordinates, the angle parameter is also added. The predictive value conversion formula of rotational boundary box is defined as.

$$\begin{aligned} b_x &= \sigma(t_x) + c_x \\ b_y &= \sigma\left(t_y\right) + c_y \\ b_w &= p_w e^{t_w} \\ b_h &= p_h e^{t_h} \\ b_\theta &= (t_\theta + i)p_\theta \end{aligned} \quad (2)$$

$b_x, b_y, b_w, b_h, b_\theta$ is the center coordinate, width, height and angle of the prediction boundary box. c_x, c_y is the offset of the grid where the target center is located and the upper left corner of the feature map. i represents the angle interval of the anchor box.

2.4 Loss Function

In the R-YOLOv3 model, the size of the input image is 320 × 320.6300 prediction boxes will be generated. Most of them do not contain ship targets, and the distribution of positive and negative samples is unbalanced. In the training process, there will be a problem of uneven distribution of positive and negative samples, and the background contains a lot of noise interference. Compared with the foreground ship targets, it is very small and difficult to classify the samples. It is difficult to effectively train the model. Although the positive and negative sample imbalance problem is adjusted by setting the Ignore _ thread threshold and reducing the confidence of the boundary box that does not contain the target. However, the imbalance between positive and negative samples and the difficulty in classifying samples still exist. Therefore, the R-YOLOv3 algorithm further uses the Focal Loss function [22] to solve this problem, reduce the weight of the easy-to-class samples, make the model focus on learning the characteristics of the foreground objects with less difficult-to-class numbers, and improve the detection accuracy of the model for ships. The loss function of YOLOv3 algorithm is the sum of coordinate loss, confidence loss and classification loss function.

$$loss = bboxloss + confidenceloss + classloss \tag{3}$$

The class loss function used in YOLOv3 is the direct summation of the cross entropy loss function [23] of various training samples. It is defined as.

$$CE(p_t) = -\log_a(p_t) \tag{4}$$

Where, $p_t = \begin{cases} p & y = 1 \\ 1 - p & y = 0 \end{cases}$, p predicts the probability of the sample output category, y represents the label of the category.

In order to extract the features with more information from the model, the cross-entropy loss function is usually multiplied by a weight coefficient α inversely proportional to the probability of target existence, which weakens the contribution of a large number of negative samples to the model and increases the weight proportion of positive samples.

$$CE(p_t) = -\alpha_t \log_a(p_t) \tag{5}$$

$$\alpha_t = \begin{cases} \alpha & y = 1 \\ 1 - \alpha & y = 0 \end{cases} \tag{6}$$

Although the weight parameter solves the problem of imbalance between positive and negative samples in the training process, the ship targets in the distant waters in the ship data set are easy to detect, while the background noise in the coastal area is large, and it is difficult to distinguish the ship from the background. Therefore, in the Focal Loss calculation formula, a modulation parameter is introduced to reduce the weight of simple samples through this super parameter, so that the model focuses on the learning of difficult-to-class samples.

$$FL(p_t) = -(1 - p_t)^{\gamma} \log_a(p_t) \quad (7)$$

Where, $\gamma \geq 0$. It is the standard cross entropy loss function when $\gamma = 0$.

3 Experiments and Results

3.1 Dataset

The experimental data in this paper are SAR Ship Detection Dataset (SSDD). SSDD data set is the first open dataset for SAR ship target detection created by Naval Aeronautics and Astronautics University in 2017. The dataset has a total of 1160 images, containing 2456 ship targets, only including the ship category. The experimental data set mainly adopts the rotational box labeling method. Different from the horizontal box data that are represented by the upper left and the lower right, the parameters of the rotational box are the coordinates of four points. For the ship target near the coast, the rotational box divides the target and background pixels, reduces the interference of artificial targets in the coastal area on the ship target detection, and the length-width ratio of SAR ship target is larger than other targets. The annotation method of the rotation box well reflects the real shape and size of the ship, and the length-width ratio and size of the horizontal box and the real shape of the ship, as shown in Fig. 4. In this paper, the partition ratio of the experimental training set and the test set is 8: 2.

Fig. 4. Two target labeling methods

3.2 Evaluation Indicators

In the experiment of this article, three typical evaluation indicators in the target detection algorithm are used to evaluate the performance of different ship detection algorithms. They are accuracy, recall and average precision (AP).

(1) Accuracy. Also known as the precision rate, it is expressed as the probability of the true positive sample among all the samples predicted to be positive samples, reflecting the correctness of the detection target.

$$precision = \frac{TP}{TP + FP} \tag{8}$$

(2) Recall. Also known as recall rate, it is expressed as the probability that all positive samples that are correctly detected account for all positive samples.

$$recall = \frac{TP}{TP + FN} \tag{9}$$

(3) Average accuracy. Because both the precision rate and the recall rate have single-point value limitations and cannot reflect the complete performance of a detection model, the average precision is generally used to evaluate the pros and cons of the model. Use the accuracy rate of all detected images as the value of the ordinate and the recall rate as the value of the abscissa to draw the accuracy-recall rate curve (PR curve). The average accuracy is the area enclosed by the PR curve and the two coordinate axes.

$$AP = \int_0^1 P(R)dR \tag{10}$$

3.3 Detection Results

Since the size of the final extracted feature map of the R-YOLOv3 feature network is 32, 16 and 8 times of the down-sampling of the original image, the size of the input image should be an integer multiple of 32. The image size of the SSDD + dataset used in this section is not fixed. Therefore, first, the long side is cut into a square according to the short side of the image, and then the side length is filled with resizing to form a fixed size. This method can retain the original size of the ship as much as possible, especially the small target of the ship, reduce the deformation of the target and reduce the loss of feature information. Considering the distribution characteristics of the simple background and the complex background of the data set, the test set selects the images with the number 1 and 9 at the end of the image name, so as to ensure the uniform distribution of the target in the simple distant sea area and the target in the offshore complex background, and better evaluate the performance of the detection model. Before the training of the model, nine prior boxes were generated by K-Means clustering according to the true values of the tilt boundary box with angle information marked by the dataset, which were (5, 7), (7, 12), (8, 18), (9, 24) (11, 36), (15, 33), (17, 55), (27, 87) and (42, 142).

The picture size is 320×320. The batch size set by the training is 4, and the initial learning rate is 0.001. The SGD algorithm is used to optimize the model. The momentum coefficient is 0.9, and the weight attenuation rate during training is 0.0005. The IoU threshold set by the NMS is 0.5, and the confidence threshold is 0.01. The angle interval of the generated anchor box is 45°. All the samples in the training set achieve one-time training, which is called epoch. Each experiment trains 50 epochs.

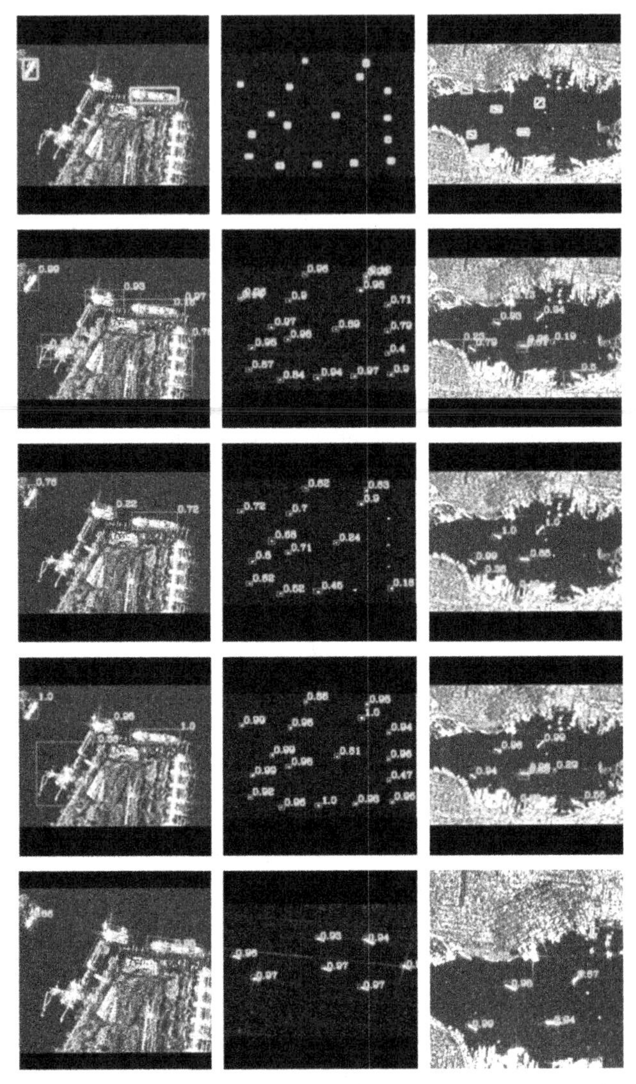

Fig. 5. SAR ship detection results. (a) Ground truth. (b) SSD. (c) Faster R-CNN. (d) Original YOLOv3. (e) R-YOLOv3

The detection visualization results of YOLOv3, SSD, Faster R-CNN and R-YOLOv3 are shown in Fig. 5. The observation results show that the three detection algorithms based on horizontal box have false detection, especially the interference of coastal background has a negative impact on the detection accuracy. In addition, Faster R-CNN has a good detection effect on large ships, but there is a missed detection of small ship targets. R-YOLOv3 separates the ship target from the coastal background, clusters the real target information to generate a more accurate priori box, provides more accurate target location information, reduces the risk of missed detection and false detection, and improves the detection accuracy of the model.

Table 1. Comparison of evaluation metrics of different methods

Model	Backbone	Precision	Recall	AP
SSD	VGG-16	62.8	76.2	70.7
Faster R-CNN	VGG-16	66.0	80.1	74.6
YOLOv3	DarkNet-53	64.5	78.0	73.8
R-YOLOv3	ResNet-50	80.8	90.1	87.3

The quantitative results of the four network models are shown in Table 1. It can be seen from the comparison of the results in the table that the detection algorithm of the rotating box not only retains the aspect ratio and shape characteristics of the ship target, but also reduces the interference of the redundant background near the coast and reduces the difficulty of detecting the ship target near the coast. R-YOLOv3 improves the accuracy of 13.5% compared with the algorithm YOLOv3 based on the horizontal box, and obtains 12.7% and 16.6% gains respectively compared with Faster R-CNN and SSD. The ships in SSDD data set have the characteristics of multi-direction. The grid on each feature map extracted by the R-YOLOv3 model predicts the boundary box at different angles, which will have different effects on the detection model. The detection results of different angle intervals are shown in Table 2. Since the ship ' s orientation angles are mostly near the SSDD dataset, the model has better performance when the angle interval is set. There are fewer horizontal and vertical ship targets in the data set, so when the angle parameter is set to be, the accuracy of the predicted boundary box information is low, and the detection performance of the model is poor.

Table 2. Detection result of different angle intervalls

Angle	Precision	Recall	AP
30	81.2	88.0	84.2
45	80.8	90.1	87.3
60	80.0	86	82.7
90	80.7	84.8	80.2

3.4 Detection Results of Different Backbone Networks

In order to test the advantages and disadvantages of different backbone networks and the influence on the detection performance of the model, this section completed the R-YOLOv3 detection model based on the extraction characteristics of different backbone networks. The extracted backbone networks mainly include DarkNet-53, MobileNetv2, ResNet-50 and ResNet-101.

In addition, the experiment also verifies the ability of the focus loss function to solve the imbalance problem of positive and negative samples and the classification problem of difficult samples.

Table 3. Comparison of evaluation metrics of different backbone networks

Backbone	Loss	Precision	Recall	AP
DarkNet-53	CE	70.7	84.3	81.7
DarkNet-53	Focal Loss	73.2	89.8	85.3
MobileNetv2	Focal Loss	63.1	87.0	81.8
ResNet-50	Focal Loss	80.8	90.1	87.3
ResNet-101	Focal Loss	83.7	90.8	87.8

According to the results shown in Table 3, it can be concluded that in the R-YOLOv3 network model based on the same backbone network DarkNet-53, the focus loss function reduces the probability of negative samples by improving the balance of positive and negative samples, and accelerates the convergence rate of the model. At the same time, it enhances the learning ability of the model for SAR ship samples with complex backgrounds, especially for offshore targets, so that the model is effectively trained. From the evaluation index, the AP value of the R-YOLOV3 detection model using the focus loss function is nearly 4% higher than that of the model based on the cross entropy function, and more accurate results are obtained in the detection accuracy. The recall rate of the detection model is greatly improved, and the leakage rate of the ship target in the complex background is reduced, indicating the effectiveness of the focus loss function. In the R-YOLOv3 network model based on different backbone network extraction features, compared with the lightweight network MobileNetv2, the multi-scale fusion detection structure of DarkNet-53 combines the low-level features and high-level features well, enhances the feature expression ability, improves the detection ability of the model for small ship targets, and obtains a 4.7% gain. The R-YOLOv3 framework based on the residual network ResNet-50 and ResNet-101 shows that the expansion of network depth makes the model easier to train and learn the target characteristics. ResNet50 and ResNet101 improve the detection performance by 2% and 2.5% respectively compared with the model based on DarkNet-53.

4 Conclusion

The characteristics of large length and width ratio and direction diversity of ships in the sar image make it possible to detect the rotating frame. This paper presents an improved YOLOv3 algorithm for SAR ship detection. In order to improve the accuracy of small target detection, ResNet-50 is selected as the feature extraction network of YOLOv3. In view of the large aspect ratio and background interference of SAR ships, the boundary box prediction with angle information is introduced. The improved rotational box cross-parallel ratio strategy is adopted to reduce the influence of angle parameters, and the Focal Loss function is used to solve the imbalance between positive and negative samples exacerbated by the anchor box and the difficulty of sample learning. The experimental results show that the average accuracy of the improved YOLOv3 algorithm based on the rotational box on the SAR ship data set has been improved to some extent, but there are still missing and false positives for the difficult samples in the coastal area. In the future, the detection algorithm needs to be improved to improve the detection precision.

Acknowledgments. This work is supported by the National Key Research and Development Program of China under Grant 2018AAA0102702; the National Natural Science Foundation of China (62001137); the Natural Science Foundation of Heilongjiang Province (JJ2019LH2398); the Fundamental Research Funds for the Central Universities (3072020CFT0801).

References

1. Liu, L., Ouyang, W., Wang, X., et al.: Deep learning for generic object detection: a survey. Int. J. Comput. Vis. **128**(2), 261–318 (2020)
2. Su, H., Wei, S., Liu, S., et al.: HQ-ISNet: high-quality instance segmentation for remote sensing imagery. Remote Sens. **12**(6), 989 (2020)
3. Yadav, S., Biswas, M.: Threshold-based clustering of SAR image using gaussian kernel and mean-shift methods (2019)
4. Zhang, Q., Wu, Y., Zhao, W., et al.: Multiple-scale salient-region detection of SAR image based on gamma distribution and local intensity variation. IEEE Geosci. Remote Sens. Lett. **11**(8), 1370–1374 (2014)
5. Ressel, R., Lehner, S.: Texture-based sea ice classification on TerraSAR-X imagery. In: Proceedings of the 22 IAHR International Symposium on ICE 2014 (IAHR-ICE 2014) (2014)
6. Eltoft, T., Doulgeris, A., Anfinsen, S.N.: Model-based statistical analysis of PolSAR data. In: IEEE International Geoscience & Remote Sensing Symposium. IEEE (2009)
7. Robey, F.C., Fuhrmann, D.R., Kelly, E.J., et al.: A CFAR adaptive matched filter detector. IEEE Trans. Aerosp. Electron. Syst. **28**(1), 208–216 (1992)
8. Wu, Q., Li, Y., Lin, Y., et al.: Weighted sparse image classification based on low rank representation. Comput. Mater. Continua **56**(1), 91–105 (2018)
9. Tu, Y., Lin, Y., Hou, C., et al.: Complex-valued networks for automatic modulation classification. IEEE Trans. Vehicular Technol. (99), 1 (2020)
10. Lin, Y., Tu, Y., Dou, Z.: An improved neural network pruning technology for automatic modulation classification in edge devices. IEEE Trans. Veh. Technol. **69**(5), 5703–5706 (2020)
11. Yu, J., Hu, A., Li, G., et al.: A robust RF fingerprinting approach using multi-sampling convolutional neural network. IEEE Internet Things J. (99), 1
12. Ren, S., He, K., Girshick, R., et al.: Faster R-CNN: towards real-time object detection with region proposal networks. IEEE Trans. Pattern Anal. Mach. Intell. **39**(6) (2015)
13. Qinggang, W., Xueming, Z.: Remote sensing object detection via an improved YOLO network. Int. J. Performability Eng. **16**(11), 1803 (2020)
14. Redmon, J., Farhadi, A.: YOLOv3: an incremental improvement. In: IEEE Conference on Computer Vision and Pattern Recognition, Utah: arXiv preprint: 1804. 0276 (2018)
15. Liu, W., Anguelov, D., Erhan, D., et al.: SSD: single shot multibox detector. In: European Conference on Computer Vision **6**, 21–27 (2016)

16. Lin, T.Y., Goyal, P., Girshick, R., et al.: Focal loss for dense object detection. IEEE Trans. Pattern Anal. Mach. Intell. **99**, 2999–3007 (2017)
17. Kang, M., Ji, K., Leng, X., et al.: Contextual region-based convolutional neural network with multilayer fusion for SAR Ship detection. Remote Sens. **9**(8), 860 (2017)
18. Li, J., Qu, C., Shao, J.: Ship detection in SAR images based on an improved faster R-CNN. Sar in Big Data Era: Models, Methods & Applications. IEEE (2017)
19. Kang, M., Leng, X., Lin, Z., et al.: A modified faster R-CNN based on CFAR algorithm for SAR ship detection. In: 2017 International Workshop on Remote Sensing with Intelligent Processing (RSIP). IEEE (2017)
20. He, K., Zhang, X., Ren, S., et al.: Deep residual learning for image recognition. In: IEEE Conference on Computer Vision & Pattern Recognition. IEEE Computer Society (2016)
21. Lin, T.Y., Dollar, P., Girshick, R., et al.: Feature pyramid networks for object detection. In: 2017 IEEE Conference on Computer Vision and Pattern Recognition (CVPR). IEEE Computer Society (2017)
22. Lin, T.Y., Goyal, P., Girshick, R., et al.: Focal loss for dense object detection. IEEE Trans. Pattern Anal. Mach. Intell. (99), 2999–3007 (2017)
23. Rampun, A., López-Linares, K., Morrow, P.J., et al. Breast pectoral muscle segmentation in mammograms using a modified holistically-nested edge detection network. Med. Image Anal. 57 (2019)

An Object Detection and Tracking Algorithm Combined with Semantic Information

Qingbo Ji[1,2], Hang Liu[1], Changbo Hou[1,2](✉), Qiang Zhang[1], and Hongwei Mo[3]

[1] College of Information and Communication Engineering, Harbin Engineering University, Harbin 150001, HLJ, China
houchangbo@hrbeu.edu.cn

[2] Key Laboratory of Advanced Marine Communication and Information Technology, Ministry of Industry and Information Technology, Harbin Engineering University, Harbin 150001, HLJ, China

[3] College of Intelligence Systems Science and Engineering, Harbin Engineering University, Harbin 150001, HLJ, China

Abstract. This paper proposes a novel algorithm for object detection and tracking combined with attribute recognition that can be used in embedded systems. The algorithm, which is based on a single-shot multibox detector (SSD) and kernelized correlation filter (KCF), can distinguish other objects similar to the tracking object, thereby solving the problem of confusion between similar objects. Different classification tasks in the multi-attribute recognition algorithm share the feature extraction module, which uses depthwise separable convolution and global pooling instead of standard convolution and fully connected (FC) layers, thereby improving the overall recognition accuracy and computational efficiency. Additionally, an attribute weight fine-tuning mechanism is added to improve the overall precision and ensure that different tasks are fully learned according to the degree of difficulty. Moreover, this algorithm reduces the size of the model without decreasing the accuracy, making it possible to be run on an embedded device. The results of experiments performed on OTB-100 demonstrate that a superior accuracy of 82.85% is achieved, and the precision and F1 indicator values reach 69.84% and 70.85%, respectively. The precision rate (PR) and success rate (SR) of the overall algorithm respectively reach 85.34% and 80.88%, which are higher than those achieved by the SSD algorithm. However, the size of the proposed attribute recognition algorithm is only 3.05 MB, which is about 1% of the size of other algorithms, indicating that the proposed algorithm not only improves the overall recognition accuracy, but also effectively reduces the model size.

This work is supported by the National Key Research and Development Program of China under Grant 2018AAA0102702; Natural Science Foundation of Heilongjiang Province (JJ2019LH2398); Fundamental Research Funds for the Central Universities (3072020CFT0801, 3072019CF0801 and 3072019CFM0802).

J. Xiong et al. (Eds.): MobiMedia 2021, LNICST 394, pp. 837–854, 2021.
https://doi.org/10.1007/978-3-030-89814-4_62

Keywords: Object detection and tracking · Attribute recognition · Similar object confusion · Shared feature extraction module

1 Introduction

As the development of artificial intelligence (AI) and embedded devices has become increasingly smarter and autonomous, object detection and tracking have become the key components of real applications, such as autopilot programs, intelligent monitoring, and smart robots. Embedded devices dominated by AI have consistently attracted the attention of researchers. However, limited by the computing and storage capacity of embedded hardware, it is difficult to directly deploy high-performance convolutional neural networks (CNNs) to devices. Therefore, when designing the algorithm network structure, the attribute recognition accuracy, the calculation amount, and the size of the network should all be thoroughly considered. Although the related research on tracking tasks has made great progress, the existing tracking algorithms still face difficulties in overcoming dense objects [1], cross-motion [2], and confusion between similar objects [3]. Moreover, the further technical development of computer vision has resulted in the desire to extract high-level semantic information, i.e., to identify the refined attributes of the target while tracking it. In existing research, detection tracking and attribute recognition algorithms have usually been studied separately, though the achievement of fine-grained attribute recognition results to make up for the defects of the tracking algorithm has not been considered. For example, considering frame sequences with multiple pedestrians, if the trajectories of the people in the image are crossed, a tracking algorithm usually uses the apparent characteristics of other pedestrians to update the model due to the confusion of other pedestrian targets, which eventually leads to tracking the wrong pedestrian. Thus, improving the performance of object detection and tracking via the use of semantic information is one of the goals of detection and tracking algorithms.
The contributions of the present work are summarized as follows:

A multi-attribute object recognition method is integrated into a detection and tracking algorithm. When multiple similar targets are detected, the attribute recognition results are used to screen specific tracking targets;

A detection method based on CNN has been introduced into the tracking model to extract the semantic and spatial features. Thus, the performance of the algorithm has been improved in complex background such as motion occlusion, scale variation;

To reduce the amount of calculation and ensure classification accuracy, the standard convolution in the feature extraction module is replaced with depthwise separable convolution, and global pooling or convolutional layers are used in the classifier instead of fully connected (FC) layers, which can avoid the geometric distortion caused by unifying the size of the input image when using FC layers.

2 Related Work

2.1 Attribute Recognition

Attribute recognition (AR) and classification have been widely studied in the field of computer vision to obtain more detailed information about an object. Related methods primarily depend on contextual information or side information [4] and can be applied to the recognition and classification of faces and pedestrian attributes [5], which plays an important role in the video surveillance field. In contrast to some low-level features, like the histogram of oriented gradients (HOG) [6] and the local binary pattern (LBP) [7], attributes can be considered as high-level semantic information [8]. Therefore, training a model by using attributes can improve its robustness.

There are two ways to achieve pedestrian attribute recognition (PAR), namely via traditional and deep learning methods. Traditionally, more convincing features could be obtained by manually labeling the features, training more effective classifiers, and associating the features with attributes. Compared with deep learning methods, traditional methods like HOG, scale-invariant feature transform (SIFT)[9], and the support vector machine (SVM) [10] are not very robust, because they all use manually-designed, low-level features.

Deep learning methods can be classified as either multi-label or multi-task learning methods, and have achieved appreciable results in the attribute recognition field.

Multi-label learning treats all classifications as one task, and uses an FC layer to classify attributes. The prediction results are not in a single category, but are a list of attributes. Zhu et al. [11] proposed a multi-label convolutional neural network (MLCNN), which predicts multiple attributes in the same framework; the authors divided a person in an image into several parts, which were then input into MLCNN via the detection of body parts. Finally, the softmax function was used to classify the attributes.

Regarding multi-task learning, the front of the neural network shares parameters, and the multiple FC layers in the back of the network are juxtaposed. Abdulnabi et al. [12] proposed a joint multi-task attribute classification algorithm based on CNN called multi-task CNN (MTCNN). The model takes the entire image as the input, shares features among different attributes, and reduces the loss by end-to-end training. As compared with traditional feature extraction methods, multi-task-based AR algorithms can extract deeper features and are more robust. In addition, the correlations between different attributes are used to share features, which improves the calculation efficiency and overall accuracy.

2.2 Neural Network Detection Framework

Object detection has been the basis of computer vision tasks, the goal of which is to extract the region of interest (ROI) from images or videos [13]. The existing object detection algorithms can be classified into two categories, namely region proposal-based two-stage detection and one-stage detection, based on the regression method used to predict the object location and category.

Two-stage algorithms mainly use a combination of region proposal with a CNN to locate and classify the objects. Girshick et al. have successfully proposed R-CNN [14], fast R-CNN [15], and faster R-CNN [16]. R-CNN uses a selective search to select the region, and its feature vector is extracted by a CNN. It then conducts classification via an SVM and finally uses a regressor for box regression. However, disadvantages of R-CNN are that the CNN and SVM are trained separately and each region proposal extracts feature vectors, which causes poor performance in real-time scenarios. Moreover, it takes too long to process each image, which necessitates more space and time while training.

Fast R-CNN and faster R-CNN have been proposed to overcome the shortcomings of R-CNN. Fast R-CNN extracts features of the entire image instead of an ROI. In contrast, faster R-CNN uses a region proposal network (RPN) instead of a selective search to select the candidate region, which improves the running speed of the algorithm. However, these algorithms still cannot meet real-time requirements.

Although one-stage methods do not perform as well as two-stage methods in terms of accuracy, they are superior in speed. Many representative algorithms have been proposed, including you only look once(YOLO) [17], the DenseBox and the single-shot multi-box detector (SSD) [18]. The DenseBox [19] network based on VGG-19 first utilizes an image pyramid and pays more attention to small and severely obscured targets. YOLO, an evolution of GoogleNet, is a single CNN that can predict multiple bounding boxes and classification possibilities. However, its shortcomings are obvious; compared with SSD, YOLO has a lower accuracy rate and larger errors for the recognition of small objects.

2.3 Object Tracking Methods

Recently, with the improvement of computing power and the expansion of labeled datasets, many object tracking methods like multi-domain network(MDNet) [20] and tree structure based CNN(TCNN) [21] have made great progress via deep learning techniques. However, deep learning algorithms are expensive in terms of computation and use online fine-tuning, and it is difficult for them to meet real-time performance requirements. Thus, researchers have made improvements to the tracking speed, such as with the Staple [22], generic object tracking using regression network(GOTURN) [23], and large margin object tracking with circulant feature maps(LMCF) [24] algorithms, among which GOTURN can achieve a tracking speed of 100 FPS. The existing object tracking methods can be classified into two categories, namely discriminative tracking algorithms and generative tracking algorithms.

Generative tracking algorithms [25] model the target area of the current frame and determine a similar area in the next frame as the prediction area. Representative methods include the mean-shift method [26], Kalman filter tracking, and the particle filter tracking algorithm [27]. Researchers have also proposed an adaptive object tracking method based on the mean-shift method, which solves the problem of tracking failure caused by constant changes in the target scale in videos. Moreover, the video processing speed can reach 125 FPS. However, because they do not consider the background information, these methods have low accuracy despite their fast processing speeds.

In contrast, discriminative object tracking algorithms borrow ideas from machine learning, and treat the tracking problem as two classification problems. In the current frame, the target and background areas are treated as two types of training samples, and the optimal discriminant function is trained via the machine learning method. In the process of tracking subsequent frames, the area with the smallest optimal discriminative function is the target area. The circulant structure of tracking-by-detection with kernels (CSK) algorithm [28] utilizes the cycle sample matrix as the dataset of the classifier, which greatly reduces the amount of calculation and radically increases the tracking speed. However, the CSK tracking algorithm is performed in the gray space, which lacks color information. Moreover, it can only track objects with a fixed size, and cannot robustly track objects in the case of changing scales. The kernel correlation filter (KCF) [29] has been proposed to improve the accuracy while ensuring speed, and uses the fast HOG instead of single-channel gray features. Compared with generative tracking algorithms, discriminative tracking algorithms perform better in terms of speed and accuracy. Moreover, the proposal of the KCF has solved the problem of the large amount of calculation; thus, it was chosen as the object tracking algorithm used in the present study.

3 Proposed Method

This paper proposes an object detection and tracking algorithm combined with attribute recognition that can be applied in embedded systems. Figure 1 presents the flow chart of the proposed method. We use the SSD to detect the target position in the first frame as the KCF initialization coordinates and use the attribute recognition network to identify and record the attribute characteristics of the target. To reduce the error accumulation of the KCF, SSD is utilized to obtain the location information of the object every N frames, and the obtained object area is used as the initial value of the KCF tracking algorithm to initial the model. Since the SSD target detection model may detect multiple targets of the same category or even multiple similar targets, multiple attributes of the target are obtained through the attribute recognition algorithm, and multiple targets in the detection results are compared with the attributes of the previously tracked targets to filter out the attributes. We filter out one or several targets with the closest attributes, and select the target with the closest Hamming distance between these targets and the tracking result of the KCF algorithm as the final tracking target and initialize the KCF algorithm. Both the template and the tracking result of KCF are encoded into binary codes and then calculated the distance to assist in identifying whether the target is well tracked.

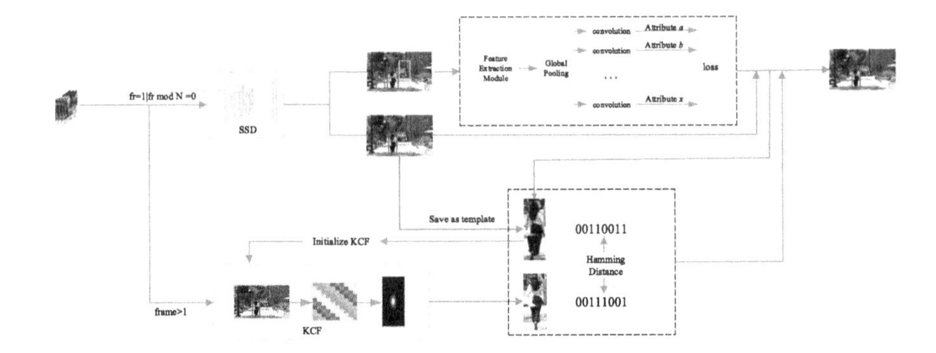

Fig. 1. Flowchart of the proposed object detection and tracking algorithm with attribute recognition.

3.1 Network Architecture

Because the fine-grained multi-task attribute algorithm places high requirements on the computing power and storage sources of the device, the application scenarios of the proposed method are mostly mobile devices with limited computing and storage capabilities, such as self-driving cars and unmanned aerial vehicles, which makes it difficult to deploy high-performance CNNs directly into the devices [30]. Therefore, when designing the algorithm network structure, not only the attribute recognition accuracy of the network, but also the calculation amount and size of the network, must be thoroughly considered.

Attribute Recognition Network Architecture. In natural scenarios, the attribute categories of an object are diverse. For example, considering attribute recognition for pedestrians, there are different attributes, such as gender, clothing style, hair length, etc. Regarding the task of car attribute recognition, the attributes include the color, make, model, and type. When using the deep learning method to solve different tasks, a one-to-one single-task learning mechanism is usually adopted, for which a network is designed separately for each task and different networks solve different tasks. Although the single-task learning mechanism is highly targeted, it ignores the internal connections between different tasks and lacks diversified information that can be borrowed from other attributes, thereby reducing the overall recognition accuracy. With the increase of the number of attributes, the calculation amount of the single-task learning mechanism increases linearly, and the overall calculation efficiency is extremely low. Because different attributes are related to each other, it is possible that the potential information of one task will help improve the recognition accuracy of the model on another task. For example, pedestrians with a male gender usually have short hair, and those with short skirt attributes in their dress style are usually female. Therefore, a multi-task-based recognition algorithm can identify the attributes of the target, the connections between the tasks can be used, and the shared feature representations between different tasks can be fully mined, thereby allowing the model to better summarize the overall attributes.

Multi-task-based attribute recognition can be classified as either hard-sharing or soft-sharing [31], and Fig. 2 presents the structures of the two types of sharing. In soft-sharing, each task has its own model parameters, and regularization is introduced to make the parameters between different tasks as similar as possible. In contrast, hard-sharing fully considers the sharing of the information source among different parallel networks. The common attributes of multi-tasks are learned while sharing the hidden layer, and the classification problems of different tasks are then completed via the specific task layer. If the number of tasks is large, the shared hidden layer of the model needs to extract as many common effective features as possible to meet the needs of different tasks. Thus, the hard-sharing method can also reduce the risk of model overfitting.

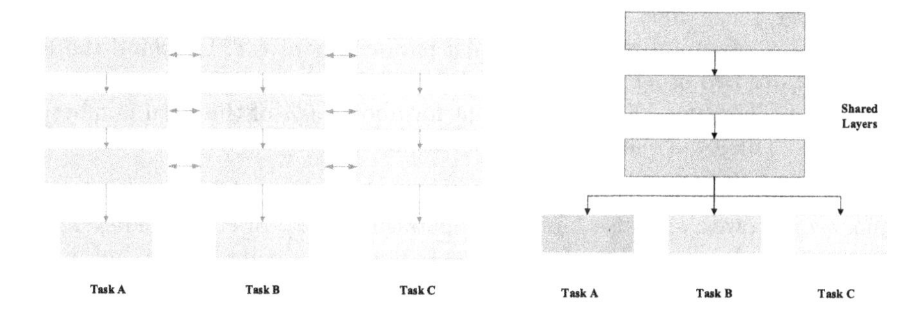

Fig. 2. Parameter sharing of multi-task-based attribute recognition. The left represents soft-sharing and the right represents hard-sharing module.

Based on the preceding analysis, the hard-sharing mode of multi-task recognition is used to design the attribute recognition network, as presented in Fig. 3. In the network, different attributes share a feature extraction module. A global pooling layer is used to reduce the number of parameters and plays a role in the FC layer to obtain global features. A convolutional layer is then used as a classifier for different attributes, and the loss of each classifier is weighted as the overall loss value.

Network of the Feature Extraction Module. In the proposed method, traditional convolution is replaced with a depthwise separable convolutional network to compress the storage and computation on the algorithm level, which significantly reduces the computing requirements and storage overhead on embedded systems. Usually, several FC layers are connected in the feature extraction module in the CNN, and the feature map is mapped into a fixed-length feature vector. In other words, each node of the FC layer is connected to all the nodes of the previous layer, and is used to synthesize the acquired features and acts as a classifier in the CNN. However, there is a flaw in the FC layer, namely the huge

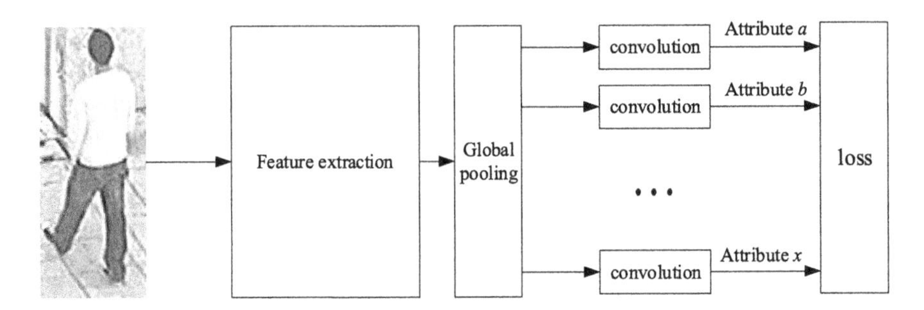

Fig. 3. The attribute recognition network.

number of parameters. Taking VGGNet [32] for example, the size of the feature map that the feature extraction module outputs to the FC layer 1 is 7×7×512, the FC layer 1 has 4096 neurons, and it has 102,70,468 parameters, accounting for 74% of the total number of model parameters of 138M. Behind the FC layer, there are two other FC layers with 16,777,216 and 4,096,000 parameters, respectively. The three FC layers account for about 90% of the total number of parameters. This brings about two problems, as follows.

(1) A large number of parameters results in the substantial consumption of memory and power, and also limits the operating speed. In embedded systems, not only does a large number of parameters occupy the limited space, but, more importantly, there are some differences between the on-chip memory and the off-chip memory (double-rate synchronous dynamic RAM for weights). The RAM of the embedded systems takes up valuable chip area, so the capacity is usually small (the capacity of the Xilinx Zynq UltraScale + MP Soc is only 32.4 MB), while the capacity of off-chip memory can be very large (the memory capacity of the ZCU developed board is 4 GB) with a low reading speed and high power consumption. To improve the operating speed and reduce the power consumption, access to the off-chip memory must be reduced, which means that the size of the network must be reduced as much as possible.

(2) A large number of parameters brings about overfitting and reduces the generalization performance. Due to the sampling error included in the training set, too many parameters will also fit the sampling error to the model while parameter fitting. The intuitive performance results in the model having a small loss in the training set, but a large loss in the test set.

Global pooling and a convolutional layer can be used instead of FC layers to solve these problems. Figure 4 presents the structures of the FC layer and global pooling layer. The FC layer expands the feature map into a vector and then conducts classification. For example, in VGGNet, the feature extraction module first obtains a 7×7×512 feature map and converts it into a 1×1×512 map through the FC layer, and then connects 1000 neurons to output a fractional vector. The global pooling layer converts the 7×7×512 map into a 1×1×512 map without any parameters. Max pooling or average pooling is then used to transform the feature map into a 1000-dimensional score vector, and the required parameters are $512 \times 1000 = 0.5$M. It is evident that the number of parameters for global pooling is much less than that of the FC layer. Because the number of global pooling parameters is greatly reduced, overfitting can be avoided, and there is almost no effect on the performance of the global pooling and FC layers. Additionally, if there exists an FC layer in the CNN, the input image must be converted into the same size, which will cause geometric distortion and other factors that will affect the accuracy. Regarding the convolutional layer, the parameters are shared in each channel, and the sizes of the kernel and input image are not related; therefore, the convolutional layer can accept images of different sizes.

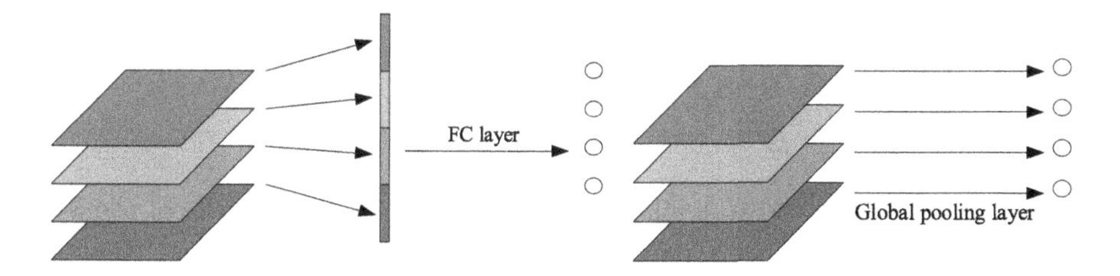

Fig. 4. The structures of the FC layer and global pooling layer.

3.2 Loss Function

The loss function is the end of, and the key to, the learning quality of a CNN, and is widely used to describe the differences between the predicted and true values. Usually, when processing a single-labeled multi-class classification problem, a CNN uses the softmax function as the activation function of the output layer. For example, the output layer of a CNN has k nodes, and the activate function is defined as follows:

$$f(z_j) = \frac{e^{z_j}}{\sum_k e^{z_k}} \tag{1}$$

where e^{z_j} is the probability for the j-th class, $f(z_j)$is the probability distribution, and z_j refers to the j-th neuron.

Via the softmax function, the probability value in the output vector is mapped to the interval $(0, 1)$, and each element in the vector is regarded as the confidence of the corresponding category. Considering the implementation of the CNN, the multi-attribute recognition of objects is a multi-label classification task for which the softmax activation function is not suitable. Regarding multi-label classification, the softmax function can be replaced with the sigmoid function to compute the loss. For the i-th sample $X_i \in \mathbb{R}$ in the sample space, and it passes the CNN to obtain an L-dimensional attribute predict vector $\hat{y_i} = [\hat{y}_{i1}, \hat{y}_{i2}, ..., \hat{y}_{iL}] \in \mathbb{R}^L$. It then predicts every element $\hat{y}_{im}$ in the vector $\hat{y}_i$, where $m \in [1, L]$ indicates the confidence of the corresponding attributre.

The true label of the sample $X_i \in \mathbb{R}$ is $y_i = [y_{i1}, y_{i2}, ..., y_{iL}] \in \mathbb{R}^L$. Each value of element y_{im} in the true label vector is shown in Eq. (2), where $m \in [1, L]$ indicates whether the attribute corresponding to the sample exists.

$$y_{im} = \begin{cases} 1, & \textit{if this attribute exists} \\ 0, & \textit{otherwise} \end{cases} \tag{2}$$

The DeepMAR algorithm trains the input image and multi-attribute labels, and uses the sigmoid cross-entropy loss to jointly consider multiple attributes. Aiming at the problem of the uneven distribution of multi-attribute label loss values, the cross-entropy loss function is improved as follows:

$$\begin{cases} L_{wce} = -\frac{1}{N}\sum_{i=1}^{N}\sum_{l=1}^{L} w_l(y_{il}log(\hat{p}_{il}) + (1 - y_{il})log(1 - \hat{p}_{il})) \\ w_l = e^{-\frac{p_l}{\sigma^2}} \end{cases} \tag{3}$$

where y_{il} refers to the true label, $\hat{p_{il}}$ refers to the predicted score of the l-th attribute for sample X_i, w_i refers to the weight of the l-th attribute, p_l refers to the propotion of positive samples of the l-th attribute in the training set, and σ refers to a hyperparameter. By improving the sigmoid cross-entropy loss function, attributes with a small proportion of positive samples can be given larger weight values; therefore, when the attribute is incorrectly classified as a negative sample by the CNN, a larger "penalty" will be given to improve the ability to learn when the samples are unbalanced.

However, using the same network structure to classify different attributes results in different classification difficulties. It is obviously inappropriate to determine the weight of loss via the sample ratio. Therefore, the weight penalty mechanism is introduced in consideration of the difficulty of classification. In other words, the network determines the weights of attributes according to the classification difficulties of different attributes, and sets a high loss weight for more difficult classification tasks. The difficulty of classifying different attributes is measured by the loss corresponding to each attribute in the training process.

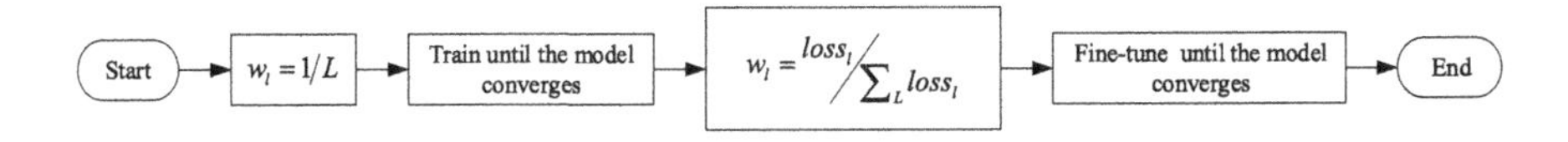

Fig. 5. The process of attribute weighting in the loss function.

In this process, all the attributes are first treated with equal degrees of importance (weight $w_l = \frac{1}{L}$). Next, the model is trained until the loss function remains stable, and the loss $loss_i$ corresponding to each attribute is recorded. The value w_l is then taken as the proportion of the overall loss. Finally, the model is fine-tuned until the loss function remains stable.

4 Experimental Results

4.1 Training Process

The PEdesTrian Attribute (PETA) dataset was used for the training and testing of the attribute recognition network. Each image in the dataset is labeled with 61 binary attributes and 4 multi-valued attributes. In this research, 15 attributes were selected for training and testing: "upper Plaid," "upper Thin-Stripes," "upper Black," "upper Brown," "upper Green," "upper Gray," "upper Red," "upper White," "upper Yellow," "lower Black," "lower Brown," "lower Gray," "lower White," "lower HotPants," and "lower Long Skirt." For the experiment, 1,000 images were randomly chosen for testing, and the others were used for training.

In the training process, different weights must be assigned for attribute recognition tasks of different difficulty levels (as shown in Fig. 1). Figure 6 presents the trend comparison of the loss functions of the weighted and unweighted tasks when the total number of training iterations was 221.4×10^3. First, the weights of all tasks were made equal, and training was stopped at the 100×10^3-th iteration. The weights were then saved in a file named "caffemodel," based on which a comparative experiment of continuing training and fine-tuning was conducted. In the continuing training process, all the task weights were kept equal, which means no weight was assigned. However, in the fine-tuning process, weights were respectively assigned to different tasks according to the degree of difficulty at the 100×10^3-th and 200×10^3-th iterations.

Figure 6 reveals that, due to the introduction of the weight penalty mechanism, larger loss weights were assigned to the more difficult identification tasks; therefore, the loss of the weighted tasks in the later stage of training became gradually less than that of the unweighted tasks. This proves that the proposed penalty mechanism has a better multi-task learning ability.

4.2 Experimental Results and Analysis

Attribute Recognition Experiment Results and Analysis. To quantitatively evaluate the effectiveness of the proposed algorithm structure and loss function design, different indexes in the PETA dataset were tested, and the performance was compared with those of four existing algorithms that exhibit good performance in multi-attribute recognition tasks. The ELF-em algorithm is a traditional discriminant method that uses integrated features based on color and texture features. The other three comparison algorithms are all based on a CNN. In addition, the size of the model was also taken into consideration to facilitate the implementation of the algorithm on embedded systems.

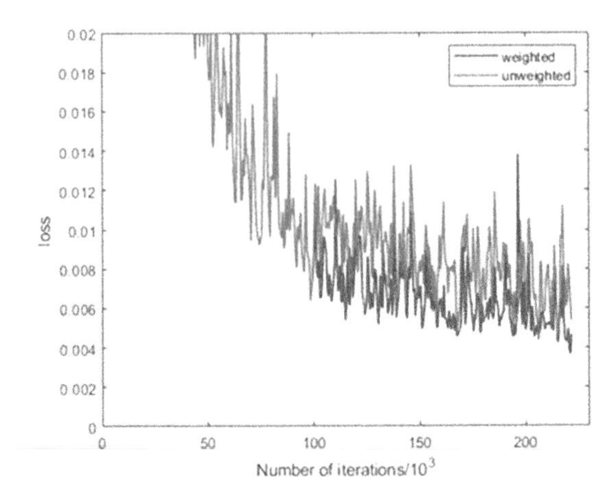

Fig. 6. Comparison between the loss values of the assigned and unassigned weights in the training phase.

Table 1. Evaluation results of attribute recognition by different methods. The comparison algorithms are all target attribute recognition algorithms, and target detection algorithms are not included.

Method	mAP	Accuracy	Precision	Recall	F1	Size of model
ELF-mm	75.21%	43.68%	49.45%	74.24%	59.36%	–
CNN-mm	76.65%	45.41%	51.33%	75.14%	61.00%	217.45
ACN	69.66%	62.61%	80.12%	72.26%	75.98%	232.57
DeepMAR	73.79%	82.60%	74.92%	76.21%	75.56%	232.57
Proposed	70.13%	82.85%	69.84%	71.91%	70.85%	3.05

As presented in Table 1, the accuracy of the proposed method was significantly higher than the accuracies of the other algorithms, which demonstrates that the loss function design of the algorithm and the weighting operation of different attributes in the training process significantly improved the accuracy, especially for the more difficult attributes. However, because different attributes are weighted according to the classification difficulty, the proposed algorithm did not perform as well under the condition of uneven sample data (the PETA dataset has fewer positive samples and more negative samples), which resulted in a poor performance in terms of the mAP, precision, recall, and F1 indicators.

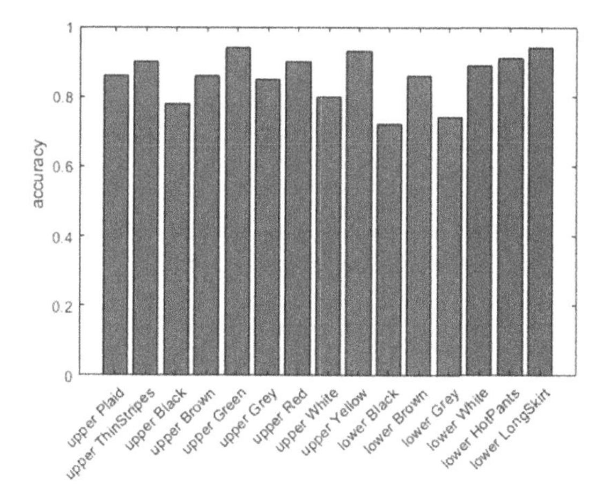

Fig. 7. Accuracy distribution map of each attribute.

In the training process of the model, weights for different tasks were redistributed according to the proportion of classification difficulty at the 100×10^3-th and 200×10^3-th iterations so that the classification tasks of different attributes were fully learned during the training process. In the final result, the accuracy of each attribute was more even and without obvious polarization, as shown in Fig. 7. The accuracies for "lower Black" and "lower Grey" were respectively 0.72 and 0.74, and slightly lower than the accuracies for the other attributes. This is because the two colors of black and grey present large changes in the images under different lighting and angles, which may easily cause misdetection. However, this did not severely affect the overall distribution due to the large loss weights assigned to these two attributes.

Regarding the size of the model, CNN-mm uses AlexNet as its backbone, while ACN and DeepMAR adopt CaffeNet. Many standard convolutions are used in these two backbones, and the size of the convolutional kernel is relatively large. Although the performances of various indicators are improved to a certain extent, the size of the model usually exceeds 200 MB. In the proposed method, the standard convolution is replaced with depthwise separable convolution, which effectively reduces the model size while ensuring greater accuracy; the final size of the model is only 3.05 MB.

Overall Algorithm Experiment Results and Analysis. There are many tracking algorithms that exhibit good real-time performance and robustness, including KCF, Struck, CSK, and Staple. To evaluate the performance of the proposed algorithm, these methods were tested with the SSD detection model on the OTB-100 and LaSOT datasets. To ensure the objectivity and fairness of the experimental results, the public datasets VOC2007 and VOC2012 were adopted for the SSD model training dataset of the proposed algorithm. Because these three datasets are quite different from the OTB-100 and LaSOT test sets,

13 video sequences in the OTB-100 test set ("BlurBody," "Couple," "Girl," "Gym," "Crows," "Human2," "Human6," "Human7," "Human8," "Human9," "Man," "Trellis", and "Woman") and 33 video sequences with a person category (20 test video sequences) in LaSOT were used for testing.

Table 2. Precision rate (PR) and success rate (SR) of different object detection and tracking algorithms. The best results are presented in bold.

	Proposed	SSD	Struck	CSK	Staple	KCF
PR(%)	**85.34**	75.28	48.98	37.51	93.10	70.51
SR(%)	**80.88**	74.75	46.53	27.49	73.17	60.23

The one-pass evaluation (OPE) method was chosen to test the KCF, Struck, CSK, and Staple tracking algorithms by manually setting the target position in the first frame. The SSD object detection and tracking algorithm, as well as the proposed algorithm, detect the object automatically without providing the initial object position. Table. 2 presents the comparison results.

As presented in Fig. 8, the precision and success results were plotted to verify the feasibility and effectiveness of the proposed algorithm. Compared with the SSD object detection model, the success rate was slightly improved and the precision was increased by nearly 10%. Moreover, compared with the Struck, CSK, and KCF algorithms, the precision and success rates of the proposed algorithm were greatly improved.

Intuitive comparison has also been made to test our detection and tracking method in some sorted video sequence. As Fig. 9 show, the yellow box represents the tracking result of proposed method, and the blue, green, red, purple boxes represent the KCF, CSK, Staple, Struck, respectively.

However, the precision and success rates of the proposed algorithm were not higher than those of the Staple algorithm. This is because the proposed algorithm requires the SSD model to detect the object position as the training base sample of the KCF in the initial frame and in tracking failure situations. Moreover, the datasets VOC2007 and VOC2012 used for training the SSD model are quite different from the test video. In other words, the "person" category in the video sequence of the test set mostly includes images from the perspective of a road monitor, whereas the "person" images in the training set are all images from the horizontal perspective. Moreover, the video sequence of the "bottle" category in the test set is longer than that in the training set.

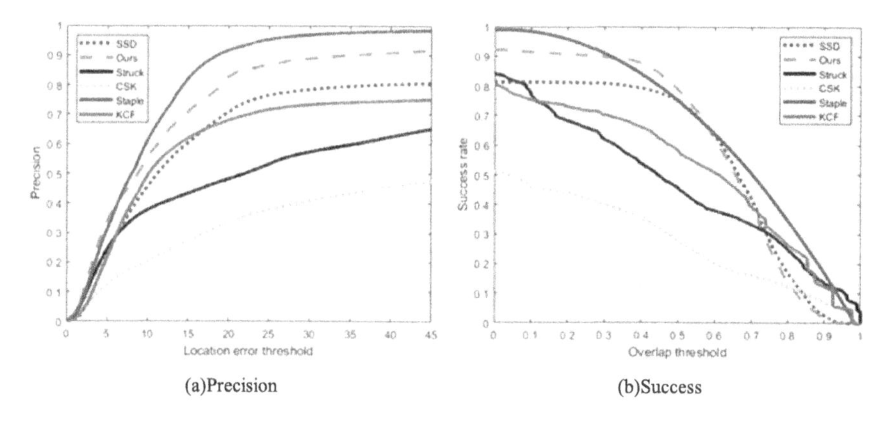

(a)Precision (b)Success

Fig. 8. Curves of OPE test results.

Fig. 9. Intuitive comparison of tracking results in different video sequences.

Due to the large difference between the training and test sets, the performance of SSD was not very good (the precision and success rates were 74.51% and 75.12%, respectively), which reduced the accuracy of the proposed algorithm. To verify this point of view, another set of experiments was conducted by adding approximately 200 pedestrian images from the perspective of a road monitor, which were not included in the test set, to the training set. The test results after retraining the SSD model are reported in Table 3, which demonstrates that the proposed method exhibited a great improvement, while the performance of SSD was slightly improved with the increase in the number of images in the training set. The precision rate of the proposed method was basically the same as that of the Staple algorithm, while the success rate was 13% higher.

Table 3. Test results after adding images into the training set.

Method	SSD		Proposed	
	Test results	Increase	Test results	Increase
PR%	78.32	3.04	92.29	6.95
SR%	75.83	1.08	86.55	5.67

5 Conclusion

In this work, an object detection and tracking method based on attribute recognition was proposed. This method utilizes the attribute difference of the object to distinguish multiple similar targets to avoid confusion with other objects during tracking. Additionally, to increase the overall recognition accuracy and computing efficiency, the performance is improved via the combine of SSD detection method and KCF tracking algorithm, assisted by shared feature extraction, and global pooling. Experimental results indicate that the recognition accuracy of the proposed method is higher than the accuracies of other algorithms, and the size of the model is significantly reduced. The motion blur, confusion between similar objects and scale variation are well overcome. In follow-up research, an attempt will be made to automatically adjust the weights in the weight penalty mechanism, instead of manually setting fixed parameters in various stages.

References

1. Wang, X., Xiao, T., Jiang, Y., Shao, S., Sun, J., Shen, C.: Repulsion loss: detecting pedestrians in a crowd. In: 2018 IEEE/CVF Conference on Computer Vision and Pattern Recognition (CVPR) (2018)
2. Feichtenhofer, C., Pinz, A., Zisserman, A.: Detect to track and track to detect (2017)
3. Hoiem, D., Chodpathumwan, Y., Dai, Q.: Diagnosing error in object detectors. In: Proceedings of the 12th European conference on Computer Vision - Volume Part III (2012)
4. Mao, L., Yan, Y., Xue, J.H., Wang, H.: Deep multi-task multi-label CNN for effective facial attribute classification. In: Proceedings of International Conference on Computer Vision and Pattern Recognition (2020)
5. Sarafianos, N., Xu, X., Kakadiaris, I.A.: Deep imbalanced attribute classification using visual attention aggregation (2018)

6. Zhou, W., Gao, S., Zhang, L., Lou, X.: Histogram of oriented gradients feature extraction from raw Bayer pattern images. IEEE Trans. Circuits Syst. II: Express Briefs (99), 1 (2020)
7. Zhao, G., Wang, X., Cheng, Y.: Hyperspectral image classification based on local binary pattern and broad learning system. Int. J. Remote Sens. **41**(24), 9393–9417 (2020)
8. Wang, X., Zheng, S., Yang, R., Luo, B., Tang, J.: Pedestrian attribute recognition: A survey (2019)
9. Song, Z., Zhang, J.: Image registration approach with scale-invariant feature transform algorithm and tangent-crossing-point feature. J. Electron. Imaging **29**(2), 1 (2020)
10. Yu, X., Wang, H.: Support vector machine classification model for color fastness to ironing of vat dyes. Textile Res. J. 004051752199236 (2021)
11. Zhu, J., Liao, S., Lei, Z., Li, S.Z.: Multi-label convolutional neural network based pedestrian attribute classification. Image Vis. Comput. **58**, 224–229 (2017)
12. Abdulnabi, A.H., Wang, G., Lu, J., Jia, K.: Multi-task CNN model for attribute prediction. IEEE Tran. Multimedia (2015)
13. Guan, H., Cheng, B.: Taking full advantage of convolutional network for robust visual tracking. Multimedia Tools Appl. **78**(8), 11011–11025 (2018)
14. Olmez, E., Akdogan, V., Korkmaz, M., Er, O.: Automatic segmentation of meniscus in multispectral MRI using regions with convolutional neural network (R-CNN). J. Digital Imaging **33**(30) (2020)
15. Girshick, R.: Fast R-CNN. Computer Science (2015)
16. Ren, S., He, K., Girshick, R., Sun, J.: Faster R-CNN: towards real-time object detection with region proposal networks. IEEE Trans. Pattern Anal. Mach. Intell. **39**(6), 1137–1149 (2017)
17. Redmon, J., Divvala, S., Girshick, R., Farhadi, A.: Unified, real-time object detection, You only look once (2015)
18. Chen, X., Yu, J., Wu, Z.: Temporally identity-aware SSD with attentional LSTM. IEEE Transactions on Cybernetics (2018)
19. Shan, Y.: ADAS and video surveillance analytics system using deep learning algorithms on FPGA. In: 2018 28th International Conference on Field Programmable Logic and Applications (FPL), pp. 465–466 (2018)
20. Wang, J., Li, A., Pang, Y.: Improved multi-domain convolutional neural networks method for vehicle tracking. Int. J. Artif. Intell. Tools **29**(07n08), 2040022 (2020)
21. Lee, J., Kim, S., Ko, B.C.: Online multiple object tracking using rule distillated siamese random forest. IEEE Access **8** (2020)
22. He, W., Li, H., Liu, W., Li, C., Guo, B.: rstaple: a robust complementary learning method for real-time object tracking. Appl. Sci. **10**(9), 3021 (2020)
23. Ahmad, M., Ahmed, I., Khan, F.A., Qayum, F., Aljuaid, H.: Convolutional neural network based person tracking using overhead views. Int. J. Distribut. Sensor Netw. **16** (2020)
24. Gao, L., Li, Y., Ning, J.: Maximum margin object tracking with weighted circulant feature maps. IET Comput. Vis. **13**(1), 71–78 (2019)
25. Chen, Y., Sheng, R.: Single-object tracking algorithm based on two-step spatiotemporal deep feature fusion in a complex surveillance scenario. Math. Probl. Eng. (2021)
26. Xiang, Z., Tao, T., Song, L., Dong, Z., Wang, H.: Object tracking algorithm for unmanned surface vehicle based on improved mean-shift method. Int. J. Adv. Rob. Syst. **17**(3), 172988142092529 (2020)

27. Tang, Y., Liu, Y., Huang, H., Liu, J., Xie, W.: A scale-adaptive particle filter tracking algorithm based on offline trained multi-domain deep network. IEEE Access (99), 1 (2020)
28. Henriques, J.F., Caseiro, R., Martins, P., Batista, J.: Exploiting the circulant structure of tracking-by-detection with kernels (2012)
29. Henriques, J.F., Caseiro, R., Martins, P., Batista, J.: High-speed tracking with kernelized correlation filters. IEEE Trans. Pattern Anal. Mach. Intell. **37**(3), 583–596 (2014)
30. Zhang, W., Zhang, Z., Zeadally, S., Chao, H.C., Leung, V.C.: A multiple-algorithm service model for energy-delay optimization in edge artificial intelligence. IEEE Trans. Ind. Inform. (99), 1 (2019)
31. Sun, T., et al.: Learning sparse sharing architectures for multiple tasks. In: Proceedings of the AAAI Conference on Artificial Intelligence, vol. 34, pp. 8936–8943 (2020)
32. Zamani, N.S.M., Zaki, W.M.D.W., Huddin, A.B., Hussain, A., Mutalib, H.A.: Automated pterygium detection using deep neural network. IEEE Access **8**, 191659–191672 (2020)

Chinese License Plate Recognition System Design Based on YOLOv4 and CRNN + CTC Algorithm

Le Zhou[1(✉)], Wenji Dai[1], Gang Zhang[1], Hua Lou[2], and Jie Yang[1]

[1] College of Telecommunications and Information Engineering, Nanjing University of Posts and Telecommunications, Nanjing 210003, China
{1319015203,jyang}@njupt.edu.cn

[2] Changzhou College of Information Technology, Changzhou 213164, China

Abstract. License plate recognition (LPR) is widely used in the intelligent transportation systems. Traditional recognition methods have many disadvantages with slow detection speed and low recognition accuracy. In order to solve these problems, this paper proposes an end-to-end LPR method, which is based on YOLOv4 and Convolutional Recurrent Neural Network (CRNN) with Connectionist Temporal Classification (CTC) algorithm, which can effectively improve the detection speed and accuracy. First, based on the excellent classification and detection performance of YOLOv4, it is applied to accurately locate the license plate of the input car image. Then, we use CRNN to recognize the character information imported in the license plate image and add the CTC algorithm to the CRNN network to achieve the alignment of the input and output formats of the character information. Experimental results show that the accuracy rate of license plate recognition detection reaches as high as 97%, and the detection speed is as low as around 30 FPS (Frames Per Second).

Keywords: License plate recognition · YOLOv4 · Convolutional recurrent neural network · Connectionist temporal classification

1 Introduction

In the Intelligent Transportation Systems, the license plate recognition (LPR) system is widely used. LPR analyzes and processes the captured vehicle images under the complex background, so as to obtain the position of the license plate, then automatically recognize the characters on the license plate, and finally output the license plate information automatically. Usually, the license plate is considered as an identity of each vehicle and it is unique. LPR uses this feature of the license plate to identify and count vehicles. In a modern transportation system, the recognition of license plates affects the development of smart transportation. It is also an important factor affecting transportation modernization.

J. Xiong et al. (Eds.): MobiMedia 2021, LNICST 394, pp. 855–865, 2021.
https://doi.org/10.1007/978-3-030-89814-4_63

Traditional LPR methods mainly consist of the following parts: image acquisition, license plate location, character segmentation, and character recognition. First, the position of the license plate is obtained by performing a series of processing on the target image. Second, a certain method is used to divide the characters appearing on the license plate. Third, individual characters are extracted and judged one by one, until the final recognition result is output. Because the license plate has different geometric characteristics, Niu *et al.* [1] binarized the image into black and white and then performed canny edge detection to lock the license plate position. A single character is projected vertically, and the segmented characters are recognized one by one using the template matching method. The experimental results show that the detection performance is well. A. H. Ashtari *et al.* [2] proposed a color-based classification method, which divides it into stable-sized blocks through conversion in the color space. For each block, each small block with the help of a designed filtering process is checked to determine whether it contains a license plate or a certain part of the license plate. Meeras *et al.* [3] proposed the use of three levels of preprocessing local binary pattern classifiers (3L-LBPs) and a large number of AdaBoost cascades to detect license plate regions and improve the speed of license plate detection. Khan *et al.* [4] put forward such a viewpoint that LPR is composed of the following parts: 1) select the brightness channel from the CIE-Lab color space; 2) perform binary segmentation on the selected channel, and then perform image refinement; 3) fuse directional gradient histogram (HOG) and geometric features, and then use a new entropy-based method to select appropriate features; and 4) use support vector machine (SVM) for feature classification, with good results.

In the above-mentioned traditional LPR methods, different angles and positions of the license plates, as well as the accuracy of the license plate character segmentation, have a great influence on the accuracy of the license plate character recognition. The current LPR methods based on the character segmentation methods cannot meet the needs of the practical applications. Therefore, colleagues have proposed an end-to-end LPR algorithm. The advantages of no need for character segmentation, direct input of a complete license plate image at the input, and direct output of recognition results at the output, make end-to-end LPR algorithms highly sought after.

Nowadays, deep learning has been widely used, such as in target detection and classification, safety surveillance [5, 6] and automatic modulation classification [7], as well as in the extraction of abstract and semantic features [8]. Hua *et al.*[9] used deep learning for human emotion recognition (HERO), and the detection accuracy was greatly improved compared with traditional methods. In the wireless signal modulation classification, the automatic modulation classification algorithm based on deep learning has significantly improved the performance and efficiency of the communication system [10], so it is widely used in the field of wireless communication. In the field of intelligent Internet of Things [11], Li *et al.* [8] applied deep learning to remote sensing image classification, which can not only significantly improve the classification accuracy but also enrich the application of deep learning in the field of intelligent Internet of Things. Of course, the license plate detection and recognition system is also an important part of the intelligent Internet of Things. Through studying advanced deep learning algorithms and applying them to the LPR system, not only helps to improve the accuracy of recognition, but also helps to improve the detection speed, and the operating efficiency of the entire transportation system can also be improved.

Lin et al. [12] proposed the LPR convolutional neural network to improve the character recognition rate of fuzzy images, which does not require character segmentation. Li et al. [13] used VGG to extract low-level CNN features. This method cannot solve the shortcoming of slow VGG network training. Therefore, the method of using a single deep neural network for license plate detection and recognition is not very well. Although the end-to-end recognition framework avoids character segmentation, and it increases the accuracy of LPR system. However, due to the complex combination of Chinese license plate characters and the diversity of shooting angles, the positioning and recognition speed of the license plate is slow and the accuracy is not satisfied.

Based on the deficiencies of the algorithms, this paper proposes a LPR algorithm based on YOLOv4 and CRNN-CTC. The algorithm firstly uses the YOLOv4 network to locate the original image on the license plate detection network and then extracts the image convolution features. With the help of the recurrent neural network (RNN) as the standard model of natural language processing, it can hand text context information very well. Secondly, convolutional features of the images are extracted through the CNN network, then extract the convolution feature sequence of the image. Finally, aiming at the problem that the training characters cannot be aligned, it is solved by introducing the CTC algorithm. In terms of detection speed, the method proposed in this paper is faster. At the same time, the experimental results show that the detection accuracy is also very well.

2 Our Proposed LPR Method

As an application example of character detection and recognition technology, automobile LPR system plays an important role in the construction of smart transportation. The whole system consists of the following parts: license plate location, license plate character recognition, and post-processing recognition and correction, as shown in Fig. 1. This paper puts forward the idea of combining the latest YOLOv4 detection algorithm and the improved CRNN recognition algorithm, which can quickly and accurately detect license plates and recognize license plate characters with excellent performance.

Fig. 1. The overall structure of our model

2.1 License Plate Detection Based on Yolov4

For LPR systems, some previous algorithms are based on sliding window search targets that cannot meet the needs. And some improved algorithms on this basis use selective search to find possible targets such as R-CNN and Faster-RCNN, but the final results are usually determined by CNN or other methods. For the different sizes of photos obtained from different shooting angles and the complexity of the environment, using a sliding window to detect license plates will be very time-consuming and has a high error rate. As shown in Fig. 2, this article uses YOLOv4 [14] to detect license plates. The AP value of the YOLOv4 network developed based on the YOLOv3 network increased by 10%, and at the same time, the corresponding FPS value increased by 12% [15].

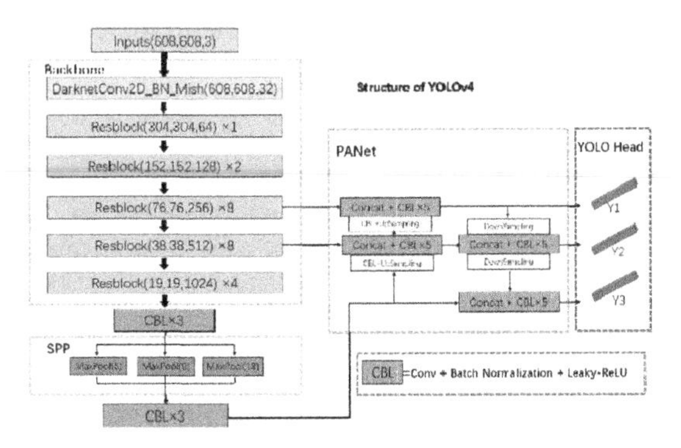

Fig. 1. YOLOv4 network structure.

The model structure of YOLOv4 is shown in Fig. 2. Compared with DarkNet53 of YOLOv3, YOLOv4 uses CSPDarkNet53. And the Neck part is composed of the Spatial Pyramid Pooling (SPPnet) and the Path Aggregation Network (PANet) in the deep convolutional network. The head part is composed of 3 YOLO headers. The CSPDarknet53 network in YOLOv4 is built based on the previous Darknet53 and CSPNet networks, that is, the 5 Resblock bodies in the figure above. The CSPDarknet53 network has many advantages. It can further reduce the amount of calculation and reduce cost. At the same time, the learning ability of the CNN network can be improved, and it also has reliable accuracy. The Darknet53 network with CSP structure is composed of 53 convolutional layers, the sizes of the convolutional layers are 1×1 and 3×3, respectively. In order to better extract the fusion of the target, SPPnet is inserted between the main network and the output layer. At the same time, each convolutional layer is connected to a batch normalization (BN) layer and a Mish activation layer, which can effectively extract target features. It can also further expand the acceptance range of backbone features and play a very important role in separating important context features. PANet is an improved network based on Mask R-CNN. Based on feature fusion, it introduces a bottom-up path augmentation structure. Through bottom-up path enhancement, it reduces the number of convolutional layers that need to pass through the information flow from high-level to low-level. At the same time, the information transmission path is shortened, and the low-level information is transmitted to the high-level, which ultimately makes the positioning information more accurate. And the introduction of adaptive feature pooling makes the extracted ROI features richer. The introduction of fully-connected fusion, which focuses on the overall Context information, and introduces a fully connected branch of the front background two classifications to obtain more accurate segmentation results. YOLOv4 extracts the middle layer in the feature utilization part, the middle and lower layers, and multiple feature layers at the bottom for target detection [16].

In the loss function part, unlike other YOLO models, YOLOv4 uses bounding box regression loss, object classification loss, and object confidence loss. When performing bounding box regression, traditional target detection models (such as YOLO V3), etc. directly set the MSE (mean square error) loss function referring to the center point coordinates of the real box and the prediction box and the width and height information, and then it also uses Intersection-over-Union (IoU) loss instead of MSE, but the performance is not very well. This research uses the latest loss function Complete-IoU (CIoU) [17] of YOLOv4. CIoU considers scale information of the overlap, center distance, and aspect ratio of the frame based on IoU. Such as (1)

$$\mathcal{L}_{CIoU} = 1 - IoU + \frac{\rho^2\left(\boldsymbol{b}, \boldsymbol{b}^{gt}\right)}{c^2} + \alpha\upsilon \tag{1}$$

In this loss function, $\boldsymbol{b}$ represents the center point of the anchor box, while $\boldsymbol{b}^{gt}$ represents the center point of the target box, and $\boldsymbol{\rho}$ represents the Euclidean distance between the two center points. $\boldsymbol{c}$ represents the diagonal distance of the smallest rectangle that can simultaneously cover the anchor box and the target box. Considering the constraint of aspect ratio consistency, CIoU loss adds the aspect ratio constraint $\boldsymbol{\alpha\upsilon}$ to the previous loss. Where $\boldsymbol{\alpha}$ is used as a trade-off parameter:

$$\alpha = \frac{\upsilon}{(1 - IoU) + \upsilon} \tag{2}$$

The parameter $\boldsymbol{\upsilon}$ is used to measure the consistency of the aspect ratio:

$$\upsilon = \frac{4}{\pi^2}\left(arctan\frac{w^{gt}}{h^{gt}} - arctan\frac{w}{h}\right)^2 \tag{3}$$

where $\boldsymbol{w^{gt}}$ represents the height of the real frame, and $\boldsymbol{h^{gt}}$ represents the width of the real frame, $\boldsymbol{w}$ represents the height of the prediction box, while $\boldsymbol{h}$ represents the width of the prediction box. If the height of the true frame and the predicted frame are similar, then $\boldsymbol{\upsilon} = 0$, the penalty term will not work. So intuitively, in order to better control the height and width of the predicted frame, a penalty item is added to this in order to make it closer to the height and width of the real frame. In this way, CIoU loss considers three important geometric factors for the target frame regression function: overlap area, center point distance, and aspect ratio. Therefore, when solving BBox regression problems, CIoU can achieve better convergence speed and accuracy.

2.2 CRNN Network

The Recurrent Neural Network (RNN) can capture contextual information in sequences. In particular, the two-way neural network can combine historical information and future information to predict the current instance, and it is more effective to use contextual information to perform continuous motion analysis in the time domain than to process each motion separately [18]. The CRNN structure is shown in Fig. 3 consisting of Convolutional Layers, Recurrent Layers, and Transcription Layers.

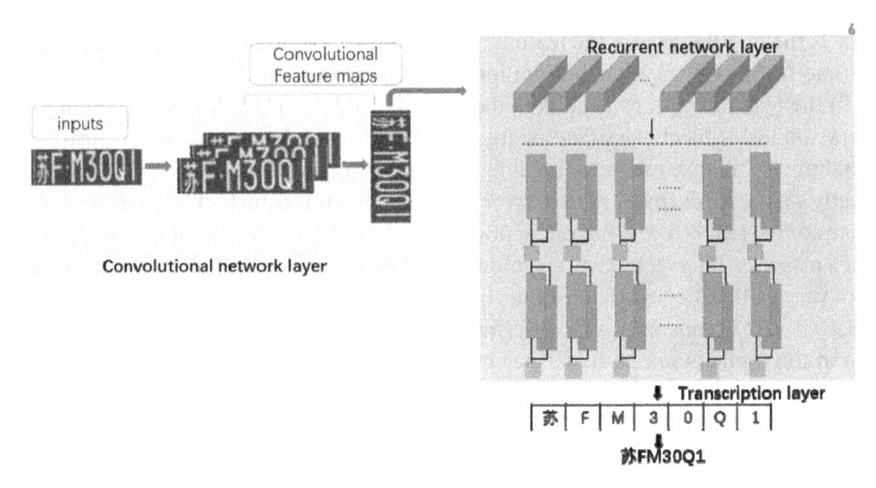

Fig. 2. CRNN Network Structure.

The convolutional layer here is a CNN without the fully connected layer, which is used to extract the Convolutional feature maps of the input image. All images need to be compressed before entering the convolutional layer and then formed into the same size. For the extraction of image feature vectors, the convolutional layer and the maximum pooling layer are mainly used. Then input the obtained feature vector to the loop layer. A deep two-way LSTM network forms a recurrent network layer, which is mainly used to extract features of text sequences. Each two-way memory network contains forward and backward propagation memory networks. The bottom network receives the sequence in the original order, while the top network receives the same input in reverse order. The two networks are not necessarily the same. Importantly, these two-way structures are stacked together, causing their output to be merged into the final prediction. For the predicted value obtained by the loop layer, the transcription layer converts it into a license plate label sequence and finally outputs it. As shown in Fig. 3, after the input image is compressed, it is sent to the convolutional network to extract image features and then converted into a convolution feature matrix. Then it is handed over to a deep two-way LSTM for character sequence feature extraction. Finally, after the RNN output is softmax, it is the character output.

2.3 Connectionist Temporal Classification (CTC)

For Recurrent Layers, if the common Softmax cross-entropy loss is used, each column output needs to correspond to a character element. Then during training, each sample picture needs to mark the position of each character in the picture, and then align it to each column of the Feature map through the CNN receptive field to obtain the label corresponding to the output of the column for training. In actual situations, it is very difficult to mark such alignment samples (in addition to marking characters, but also to mark the position of each character), and the workload is very large. Besides, because the number of characters in each sample is different, the font style is different, and the font size is different, the output of each column does not necessarily correspond to each character.

To solve this problem, we proposed the use of a connectionist temporal classification (CTC) algorithm. CTC is a loss function. It is based on the concept of a dynamic programming algorithm. It acts on the input and output links of training and only learns the mapping relationship between the input and output links [19, 20, 21]. Only focus on the output sequence we need, without considering the symbol correspondence in the input. It only cares about the convergence of the model to the set of expected sequences and does not care about the region where the symbol is generated. The model can still be trained without knowing the exact location of the symbol corresponding to the ground truth in the input image. Moreover, CTC plays an important role in solving such problems.

3 Experimental Results

3.1 Dataset Generation

In this section, we conducted experiments to verify the effectiveness of the proposed algorithm. We use NVIDIA 1080Ti for this experiment. The environment configurations are Linux Ubuntu 18.04, python 3.6 and pytorch1.3.0. In the experiment, the initial value of the learning rate is 0.001, and it decays exponentially when validation accuracy does not improve in a few previous epochs. The training batch cycle is 100, and each iteration of 100 rounds output a result. The test object is a vehicle license plate with white letters on a blue background of the Chinese mainland. The license plate consists of the Chinese characters representing 31 provinces, the combination of English letters A ~ Z (excluding I and O) and numbers 0–9, a total of 63 characters form a fixed-length 7-digit number license plate. The license plate photos were taken during the day and night, with a total of 20,000 pictures. To ensure that the effect of the test data evaluation is similar to the real scene model, the data set is divided into a training set and a validation set. The training data set is 15,000, the validation set is 1,000, and there are 4,000 test sets.

3.2 Evaluation Criterion and Comparisons

In this section, we compare the results obtained by our proposed method with other state-of-the-art methods through two key factors: the computational complexity of the model and the performance of the model. Due to the complexity of license plate detection, there is no unified established standard for evaluation, so we adopt the evaluation rules of general text detection, that is, we use the accuracy and speed of model detection to measure.

The recognition of license plates in images can be regarded as specific examples of text detection in natural scenes. In our work, the CIoU [17] is used to evaluate the index to evaluate the license plate, referring to formula (1). When the value is 1, the prediction box and manual comment box of the algorithm are completely overlapped, where IoU is represented by formula (4).

$$IoU = \frac{Y}{X} \times 100\% \tag{4}$$

Among them, Y represents the manually marked target box, and X represents the target box predicted by the algorithm.

In order to deepen the understanding of the entire recognition system, we cut out the pictures of license plate detection and divide the experimental results into two parts, license plate detection, and license plate recognition. The following figures (A)-(D) represent the additional measurement results under different angles and light. Picture (A) is the license plate detection result under the standard shooting angle in the daytime; Picture (B) is the license plate taken at a depression angle of 30 degrees during the day; Picture (C) is the license plate taken at the angle of 30 degrees in the daytime; Picture (D) is the license plate taken at a 30 degrees inclination at night.

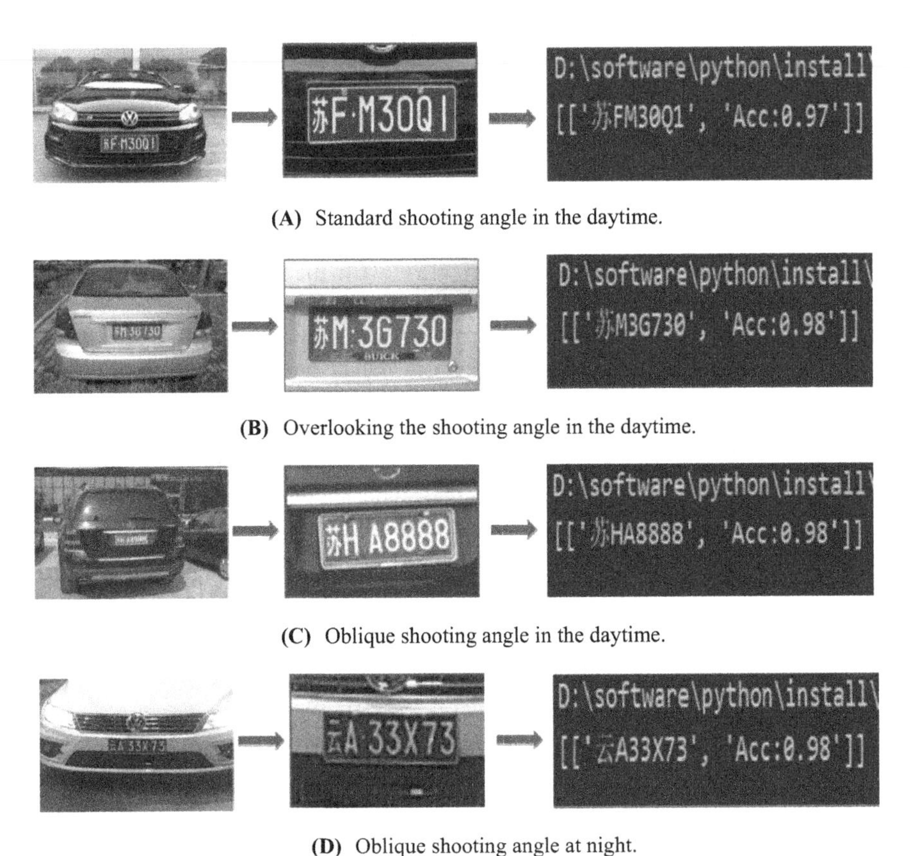

(A) Standard shooting angle in the daytime.

(B) Overlooking the shooting angle in the daytime.

(C) Oblique shooting angle in the daytime.

(D) Oblique shooting angle at night.

Fig. 3. Partial samples test results.

As can be seen from the above test results, the model proposed in this paper can well detect the target license plate, and the license plate information recognition is relatively accurate. At the same time, the location information of the license plate can be very accurately represented by the preset target box. During the day and night, the inclination angle of 0–30 degrees can achieve better detection results, and the accuracy rate can reach more than 97%. To further verify the network performance of this experiment, the reference [22] algorithm (YOLO), reference [23] algorithm (Faster R-CNN) and reference [24] algorithm (EasyPR) and the HyperLpr [17] recognition network are compared with the algorithm proposed in this paper. Among them, EasyPR is an open-source Chinese LPR system, developed based on SVM, and the detection accuracy is relatively high. HyperLpr uses the SSD algorithm. It can be seen from Table 1 that compared with EasyPR and Faster R-CNN LPR network, for algorithm detection speed, the method proposed in this article is slightly inferior, but the detection accuracy is greatly improved. Moreover, in the current practical application, 27FPS also meets the number of playback frames of most videos, which fully meets the actual needs of users.

Table 1. Comparison results of different algorithms.

Model	Accuracy (%)	FPS (Frame/second)
YOLO [22]	96.54	–
Faster R-CNN [23]	85.9	102
EasyPR [24]	93	90
HyperLpr [17]	97	15
The proposed algorithm	97	27

4 Conclusion

For the important issue of license plate detection in smart transportation construction, algorithms based on deep learning are always looking for the best detection model. Although various classification and recognition algorithms are emerging in an endless stream, the license plate recognition algorithm based on the YOLOv4 + CRNN-CTC network that we proposed still has certain advantages. Based on the outstanding performance of the YOLOv4 network in image classification and detection, we combine it with the CRNN-CTC network to locate the license plate in the input image and combine the convolutional recurrent neural network (CRNN) and the connection temporal classification (CTC) model to realize the license plate recognition. The recognition process of this method is an end-to-end recognition process and does not need to segment the characters of the license plates, which perfectly avoids the errors caused by the segmentation problem and affects the recognition accuracy. In order to evaluate the performance of this method, the license plates under different light and different inclination angles are used for testing. The experimental results prove the reliability and effectiveness of this method.

References

1. Liu, Y., Yan, J., Xiang, Y.: Research on license plate recognition algorithm based on ABC-Net. In: IEEE 3rd International Conference on Information Systems and Computer Aided Education (ICISCAE), pp. 465–469 (2020)
2. Ashtari, A.H., Nordin, M.J., Fathy, M.: An Iranian license plate recognition system based on color features. IEEE Trans. Intell. Transp. Syst. **15**(4), 1690–1705 (2014)
3. Al-Shemarry, M.S., Li, Y., Abdulla, S.: Ensemble of adaboost cascades of 3L-LBPs classifiers for license plates detection with low quality images. Expert Syst. Appl. **92**, 216–235 (2018)
4. Khan, M.A., Sharif, M., Javed, M.Y., Akram, T., Yasmin, M., Saba, T.: License number plate recognition system using entropy-based features selection approach with SVM. IET Image Process. **12**(2), 200–209 (2018)
5. Cao, W., et al.: CNN-based intelligent safety surveillance in green IoT applications. China Commu. **18**(1), 108–119 (2021)
6. Zhao, Y., Yin, Y., Gui, G.: Lightweight deep learning based intelligent edge surveillance techniques. IEEE Trans. Cogn. Commun. Netw. **6**(4), 1146–1154 (2020)
7. Wang, Y., Gui, G., Ohtsuki, T., Adachi, F.: Multi-task learning for generalized automatic modulation classification under non-Gaussian noise with varying SNR conditions. IEEE Trans. Wireless Commu. early access
8. Li, W., et al.: Classification of high-spatial-resolution remote sensing scenes method using transfer learning and deep convolutional neural network. IEEE J. Selected Topics Appl. Earth Observ. Remote Sens. **13**, 1986–1995 (2020)
9. Hua, W., Dai, F., Huang, L., Xiong, J., Gui, G.: HERO: human emotions recognition for realizing intelligent Internet of Things. IEEE Access **7**, 24321–24332 (2019)
10. Wang, Y., et al.: Distributed learning for automatic modulation classification in edge devices. IEEE Wireless Commun. Lett. **9**(12), 2177–2181 (2020)
11. Popoola, S.I., Adebisi, B., Hammoudeh, M., Gui, G., Gacanin, H.: Hybrid deep learning for botnet attack detection in the internet of things networks. IEEE Internet Things J.
12. Lin, C., Lin, Y., Liu, W.: An efficient license plate recognition system using convolution neural networks. In: 2018 IEEE International Conference on Applied System Invention (ICASI), pp. 224–227 (2018)
13. Li, H., Wang, P., Shen, C.: Toward end-to-end car license plate detection and recognition with deep neural networks. IEEE Trans. Intell. Transp. Syst. **20**(3), 1126–1136 (2019)
14. Bochkovskiy, A., Wang, C.Y., Liao, H.Y.M.: YOLOv4: optimal speed and accuracy of object detection. arXiv:2004.10934 (2020)
15. Wang, Y., Wang, L., Jiang, Y., Li, T.: Detection of self-build data set based on YOLOv4 network. In: 2020 IEEE 3rd International Conference on Information Systems and Computer Aided Education (ICISCAE), pp. 640–642 (2020)
16. Li, Y., et al.: A deep learning-based hybrid framework for object detection and recognition in autonomous driving. IEEE Access **8**, 194228–194239 (2020)
17. Qian, Y., et al.: Spot evasion attacks: adversarial examples for license plate recognition systems with convolutional neural networks. Comput. Secur. **95** (2020)
18. Hao, S., Miao, Z., Wang, J., Xu, W., Zhang, Q.: Labanotation generation based on bidirectional gated recurrent units with joint and line features. In: 2019 IEEE International Conference on Image Processing (ICIP), pp. 4265–4269 (2019)
19. Liu, H., Jin, S., Zhang, C.: Connectionist temporal classification with maximum entropy regularization. Adv. Neural Inf. Process. Syst. **31**, 831–841 (2018)
20. Feng, X., Yao, H., Zhang, S.: Focal CTC loss for chinese optical character recognition on unbalanced datasets. Complexity 2019 (2019)

21. Miao, Y., Gowayyed, M., Na, X., Ko, T., Metze, F., Waibel, A.: An empirical exploration of CTC acoustic models. In: 2016 IEEE International Conference on Acoustics, Speech and Signal Processing (ICASSP), pp. 2623–2627 (2016)
22. Redmon, J., Divvala, S., Girshick, R., Farhadi, A.: You only look once: unified, real-time object detection. In: 2016 IEEE Conference on Computer Vision and Pattern Recognition (CVPR), pp. 779–788 (2016)
23. Li, Y., Xu, G., Li, W.: FA: A fast method to attack real-time object detection systems. In:2020 IEEE/CIC International Conference on Communications in China (ICCC), pp. 1268–1273 (2020)
24. Xu, M., Du, X., Wang, D.: Super-resolution restoration of single vehicle image based on ESPCN-VISR model. In: IOP Conference Series Materials Science Engineering, vol. 790 (2020)

Detection and Localization Algorithm Based on Sparse Reconstruction

Zhao Tang[1] and Xingcheng Liu[1,2,3](✉)

[1] School of Electronics and Information Technology, Sun Yat-sen University, Guangzhou, China 510006
isslxc@mail.sysu.edu.cn

[2] School of Information Science, Guangzhou Xinhua University, Guangzhou, China 510520

[3] Southern Marine Science and Engineering Guangdong Laboratory (Zhuhai), Zhuhai 519082, China

Abstract. Wireless sensor networks (WSN) have received wide attention in many fields of applications. Secure localization is a critical issue in WSN. In the presence of malicious anchors, the traditional solution is to detect the malicious anchors, use the information collected from the normal anchors and then estimate the target location. The way of thinking and operating is reformed by modeling the behavior of the malicious anchors as perturbations. The secure localization is formulated as a sparse reconstruction problem. A gradient projection algorithm with variable step sizes is proposed to solve the sparse reconstruction. The proposed algorithm utilizes sparse reconstruction formulation for obtaining anchors information and identifying the malicious anchors by exploiting the sparsity of malicious anchors. The proposed algorithm is further modified to enhance the accuracy. The simulation results demonstrate that the proposed algorithm can effectively identify the cheating anchors and achieve great target anchors localization accuracy. The proposed algorithm performs better than any other algorithms of interest.

Keywords: Wireless Sensor Networks (WSN) · Malicious anchor detection · Sparse recovery · Gradient projection · Secure localization

1 Introduction

In the scenario of wireless sensor networks (WSN), a large quantities of wireless sensor nodes are anchored and deployed to collect information and process data [1]. In the coverage, WSN monitor the objects effectively and send considerable

This work was supported by the Key Project of NSFC-Guangdong Province Joint Program (Grant No. U2001204), the National Natural Science Foundation of China (Grant Nos. 61873290 and 61972431), the Science and Technology Program of Guangzhou, China (Grant No. 202002030470), and the Funding Project of Featured Major of Guangzhou Xinhua University (2021TZ002).

J. Xiong et al. (Eds.): MobiMedia 2021, LNICST 394, pp. 866–879, 2021.
https://doi.org/10.1007/978-3-030-89814-4_64

information to the observer [2]. WSN have been applied in various fields such as underwater exploration, environment monitoring, fire surveillance [3]. Due to the key roles of WSN and fragility of nodes anchored in the environment, node secure location is a significant issue [4].

Limited by the function of WSN and the vulnerability of the nodes anchored in the wild environment, the node secure localization significantly matters. The current node localization formulation in WSN is usually classified into two categories: range-based and range-free mechanisms [5]. The range-based algorithms adopt ranging technology to gather the distance information among nodes and unknown nodes, by means of Radio Signal Strength Indicator(RSSI) [6], Time-Difference of Arrival (TDoA), Angle of Arrival (AoA) [7]. The range-free algorithms make use of the connectivity of networks to gather information of the target anchors, by means of Distance Vector Hop(DV-Hop) [8], Approximate Point-in-Triangulation Test(APIT) [9] and so on.

Localization systems are vulnerable to attackers, who wish to invalidate the WSN' functionality. Therefore, it is of significance to focus on the accuracy and robustness of the localization. Secure localization algorithms proposed before are straightforward: the first step is to filter out cheating anchors based on the consistency of the anchors signal data set, the second one is to locate the target. An algorithm implementing Isolation Forest is proposed to filter out the malicious anchors [10]. MNDC and EMDC algorithms exploit cluster and evaluation of the consistency of RSSI and ToA measurement to detect the cheating anchors [11]. Gradient Descent (GD) method with a selective pruning stage for inconsistent measurements is used to achieve localization [12].

In this paper, we propose a detection algorithm by modeling malicious anchors' misbehavior into perturbations and reconstructing the sparse vector to detect the malicious anchors and then locating the target. The paper is organized as follows: Section 2 presents the network model and formulation. The proposed algorithm is developed in Sect. 3. The comparative experiments and simulation results are demonstrated in Sect. 4. The summary and conclusion are drawn in Sect. 5.

2 Network Model and Problem Formulations

2.1 Network Model

The network localization and the algorithm are considered in two-dimensional sensor networks where the measurement of distance is stable and available through ranging technology of TDoA. Each node provides location reference, including its location information and the measured distance. The notations used in this paper are listed in Table 1.

2.2 Problem Formulation

Especially, there is a WSN including anchor set $\{\mathbf{A}_1, \mathbf{A}_2, \ldots, \mathbf{A}_n\}$. To describe easily, we make one common node as the target anchor with the real location as $\boldsymbol{t}$=$[t_x, t_y]^T$. And thus the true distance between target anchors and others can

Table 1. Summary of notations

Notations	Meanings
n	Number of anchors
m	Number of malicious anchors
r	Number of reference anchors
k	Number of observations
$\boldsymbol{A}_i$	Location of the i-th anchor
$\boldsymbol{t}$	Location of the target node
d_i	Measured distance of the i-th anchor
n_i	Noise components of the i-th anchor
$\boldsymbol{p}$	Sparse vector
u_i	Attack components of the i-th anchor
ϕ_i	Positive part of u_i
ψ_i	Negative part of u_i
α^k	Step size of the k-th iteration
λ^k	Positive Scalars of the k-th iteration
P	Operation of projection
β, μ	Scalars for size election
σ_δ	Strength of the attacks

be represented $\|\boldsymbol{A}_i - \boldsymbol{t}\|, i = 1, 2, ..., n$. The cheating anchors report their fake measurement results, which can be simulated in Eq. 1.

$$d_i = \|\boldsymbol{A}_i - \boldsymbol{t}\| + n_i + u_i, i = 1, 2, ..., n. \tag{1}$$

where d_i is the distance in the presence of measurement errors and malicious anchors. n_i represents the random errors, which are given by $n_i \sim \mathcal{N}(0, \sigma^2)$. u_i simulates these misbehaviors attributed by malicious anchors, which are bounded by $\mathcal{N}(\mu_\delta, \sigma_\delta^2)$. μ_δ is the mean, σ_δ^2 is the variance.

The component of u_i is nonzero if i-th anchors is cheating, otherwise the value of u_i is zero or nearly zero. Since the set of u_i and the target location are unknown in advance, we arrange the set of u_i, t_x and t_y to sparse vector $\boldsymbol{p} = \{u_1, u_2, ...u_n, t_x, t_y\}$. The goal is converted to the recovery of $\boldsymbol{p}$.

3 Proposed Algorithm

Based on the above assumption, malicious anchors detection and target anchors localization are formulated into the sparse reconstruction problem. In this section, we proposed the algorithm using Basic Gradient Projection [13] for sparse reconstruction and sequential probability ratio testing to locate the target anchors and identify the cheating anchors. The algorithm includes the following steps:

1. Determine the initial target localization by recursive weighted least squares.
2. Perform the sparse reconstruction by Basic Gradient Projection.

The flow chart of the proposed algorithm is presented in Fig. 1. More specific description is presented as follows.

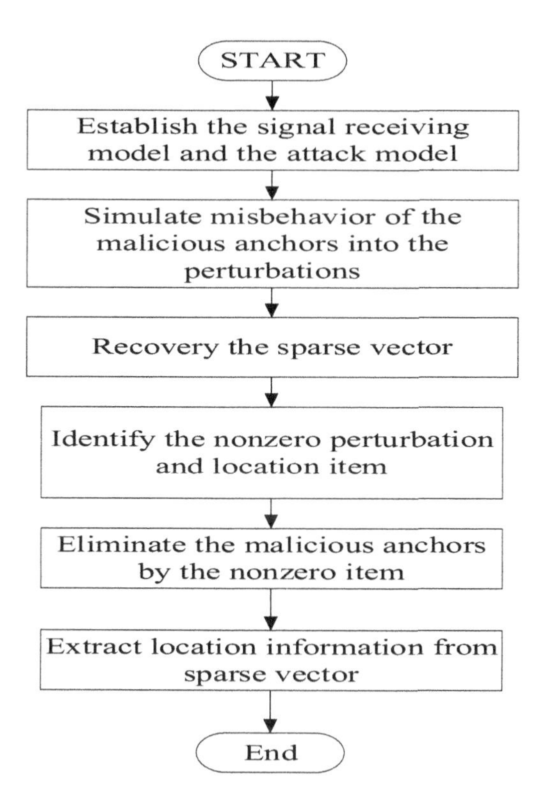

Fig. 1. Flow chart of the proposed algorithm

3.1 Determine the Initial Location

The following steps of the proposed algorithm benefit from starting points with an accurate value. Hence, we firstly proposed recursive weighted linear least squares for locating the initial location. The recursive weighted least squares are relatively independent.

In the received signal strength measurement, we can make use of the energy loss of signal to measure the distance between transmitter and receivers. The logarithmic distance path loss model [13] is Eq. 2.

$$P_R = P_{Ti} - 10a \log \frac{d_i}{d_0} + \varepsilon_i, i = 1, 2, ..., n. \tag{2}$$

where P_{Ti} presents the power of the i-th transmitter, and P_R represents the signal of power from the target anchor. a denotes path loss exponent. d_i denotes the distance between the i-th anchor and the target node, while d_0 is the referenced distance. ε_i is the measurement noise, which is bounded with $N(0,\sigma_i^2)$. After rewriting the above equal by dividing $10a$, we can get Eq. 3.

$$z_i = 10^{y_i}, i = 1, 2, ..., n. \tag{3}$$

where $z_i = 10^{\frac{P_{Ti}-P_R}{10a}}$, $y_i = lgd_i + \frac{\varepsilon_i}{10a}$. After UT and mathematical transform, we can get the mean and variance of z_i, which is denoted as Eq. 4, Eq. 4.

$$\overline{z_i} \approx \alpha_i \|\boldsymbol{A}_i - \mathbf{t}\|^2, \tag{4}$$

$$\sigma_{z_i}^2 \approx \beta_i \|\boldsymbol{A}_i - \mathbf{t}\|^4. \tag{5}$$

where $\alpha_i = \frac{2}{3} + \frac{1}{6}10^{\frac{\sqrt{3}\sigma_i^2}{5a}} + \frac{1}{6}10^{-\frac{\sqrt{3}\sigma_i^2}{5a}}$ and $\beta_i = \frac{2}{3}(1-\alpha_i)^2 + (10^{\frac{\sqrt{3}\sigma_i^2}{5a}} - \alpha_i)^2 + \frac{1}{6}(10^{-\frac{\sqrt{3}\sigma_i^2}{5a}} - \alpha_i)^2$ [13]. The formulate the linear system model is Eq. 6,

$$\mathbf{b} = \mathbf{At} + \mathbf{w}. \tag{6}$$

We select the r-th anchor node as the reference anchor node, $\mathbf{b}$ is the observed vector,

$$\mathbf{b} = \begin{bmatrix} \frac{z_1}{\alpha_1} - \frac{z_r}{\alpha_r} + (x_r^2 + x_r^2) - (x_1^2 + x_1^2) \\ \frac{z_2}{\alpha_2} - \frac{z_r}{\alpha_r} + (x_r^2 + x_r^2) - (x_2^2 + x_2^2) \\ \cdots \\ \frac{z_n}{\alpha_n} - \frac{z_r}{\alpha_r} + (x_r^2 + x_r^2) - (x_n^2 + x_n^2) \end{bmatrix}. \tag{7}$$

while $\mathbf{A}$ is coefficient matrix,

$$\mathbf{A} = 2\begin{bmatrix} x_r - x_1 & y_r - y_1 \\ x_r - x_2 & y_r - y_2 \\ \vdots & \vdots \\ x_r - x_n & y_r - y_n \end{bmatrix} \tag{8}$$

We denoted the covariance matrix as $\mathbf{c}(\mathbf{t})$. The solution of the weighted linear least squares for Eq. 6 is reduced to the minimization objective function $f(\mathbf{t})$, with the solution as Eq. 10.

$$f(\mathbf{t}) = (\mathbf{b} - \mathbf{At})^T * \mathbf{c}(\mathbf{t})^{-1} * (\mathbf{b} - \mathbf{At}). \tag{9}$$

$$\hat{\mathbf{t}} = [\mathbf{A}^T\mathbf{c}(\hat{\mathbf{t}})^{-1}\mathbf{A}]^{-1}\mathbf{A}^T\mathbf{c}(\hat{\mathbf{t}})^{-1}\mathbf{b}. \tag{10}$$

In order to calculate accurately, the recursive formula can be obtained by putting the solution into the covariance matrix as Eq. 11.

$$\hat{\mathbf{t}}^{\mathbf{k}} = [\mathbf{A}^T\mathbf{c}(\hat{\mathbf{t}}^{\mathbf{k-1}})^{-1}\mathbf{A}]^{-1}\mathbf{A}^t\mathbf{c}(\hat{\mathbf{t}}^{\mathbf{k-1}})^{-1}\mathbf{b} \tag{11}$$

We terminate with the solution $\hat{\mathbf{t}}^{\mathbf{k}}$ if the stopping criterion, $\|\hat{\mathbf{t}}^{\mathbf{k}} - \hat{\mathbf{t}}^{\mathbf{k-1}}\| \leq \gamma$, is satisfied.

3.2 Gradient Projection for Sparse Reconstruction

In this section, we assemble the unknown perturbation u_i, i=1, 2, ...,n and the target location $\mathbf{t}=[t_x,t_y]^T$ into unknown vector $\boldsymbol{p}=[u_1,u_2,...,u_n,t_x,t_y]$ with sparse features. The goal is converted to the recovery of $\boldsymbol{p}$. The problem can be transformed to the optimization problem:

$$\hat{\boldsymbol{p}} = \arg\min_{\boldsymbol{p}} \mathbf{G}(\boldsymbol{p}) = \arg\min_{\boldsymbol{p}} \sum (d_i - \|\mathbf{A}-\mathbf{t}\|_2 - u_i)^2 + \lambda\|\mathbf{u}\|_1 \tag{12}$$

To recover this vector, the Basic Gradient Projection algorithm is proposed to reconstruct the vector $\boldsymbol{p}$. Two vectors with positive value, $\boldsymbol{\phi}=[\phi_1,\phi_2,...,\phi_n]^T$ and $\boldsymbol{\psi}=[\psi_1,\psi_2,...,\psi_N]^T$, are introduced to split the $\boldsymbol{p}$ into positive and negative part. $\mathbf{G}(\boldsymbol{p})$ can be split into loss function part and regularization function part. Let the vector $\boldsymbol{\tau}=[\boldsymbol{\phi}^T,\boldsymbol{\psi}^T,\mathbf{t}^T]^T$ be the entire unknown vector in this process. Equation 12 can be rewritten as:

$$\hat{\boldsymbol{p}} = \arg\min_{\boldsymbol{p}} \mathbf{G}(\boldsymbol{p}) = \Sigma(d_i - \|\mathbf{A}-\mathbf{t}\|_2 - \phi_i + \omega_i)^2 + \lambda \cdot 1_N^T(\boldsymbol{\phi}+\boldsymbol{\psi}). \tag{13}$$

The process of iteration is:

$$\begin{cases} \boldsymbol{v}^k = P(\boldsymbol{\tau}^k - \alpha^k \nabla G(\boldsymbol{\tau}^k)) \\ \boldsymbol{\tau}^{k+1} = \boldsymbol{\tau}^k + \lambda^k(\boldsymbol{v}^k - \boldsymbol{\tau}^k) \end{cases} \tag{14}$$

where α^k is the variable step size, λ^k is a positive scalar. $P(\mathbf{z})$ denotes the operation of projecting $\mathbf{z}$, specially projecting onto the corresponding positive orthant along the negative gradient direction. Before the initial estimation, we can get the start point $\hat{\mathbf{t}}=[\hat{t_x},\hat{t_y}]$.

There are several step selection schemes. In our case, the iteration points produced tend to locate the boundary of the set. We choose the Armijo rule [13] along the projection arc, in which the value of λ^k is 1 for all k and α^k is the first number in the sequence of $\{1,\beta,\beta^2,...\}$ until the inequality. 15 meets,

$$\begin{aligned} &\mathbf{G}(P(\boldsymbol{\tau}^k - \alpha^k \nabla \mathbf{G}(\boldsymbol{\tau}^k))) \\ \le\ &\mathbf{G}(\boldsymbol{\tau}^k) - \mu \nabla \mathbf{G}(\boldsymbol{\tau}^k)^T(\boldsymbol{\tau}^k - P(\boldsymbol{\tau}^k - \alpha^k \nabla \mathbf{G}(\tau^k))), \end{aligned} \tag{15}$$

where $\beta \in (1,2)$ and $\mu \in (0,0.5)$. After fixing the value of α^k, we set $\boldsymbol{\tau}^k = P(\boldsymbol{\tau}^k - \alpha^k \nabla \mathbf{G}(\boldsymbol{\tau}^k))$. The iteration is terminated with the solution $\boldsymbol{\tau}^{k+1}$ by the stopping criterion.

4 Simulation Results

To evaluate the proposed algorithm, the mean localization Error(MLE) and the metrics set are introduced [14], including True Positive Rate(TPR) referring the proportion of correctly identifying malicious anchors; False Negative Rate(FNR) referring the possibility falsely identifying the cheating node as an honest node;

False Positive Rate (FPR) referring the possibility falsely identifying an honest node as a malicious one.

The evaluation of the proposed algorithms also includes comparison with other algorithms, Malicious Anchor Node Detection based on Isolation Forest(MANDIF) [10], using isolation forest and sequential probability ratio testing to detecting the malicious nodes; Malicious Nodes Detection using Clustering and Consistency (MNDC) and Enhanced Malicious Nodes Detection using Clustering and Consistency (EMDC) [11]; GD algorithm with fixed steps and variable steps [12]. Two kinds of GD algorithms are proposed, one is the fixed step size algorithm GD, the other is the variable step size algorithm GD. For better performance, we took the variable step size algorithm into the experiment. The rule of the change of the step size is $\gamma(i) = 15 - \frac{15(i-1)}{M}$, in which $\gamma(i)$ represents the step size of the i-th iteration, and M is the maximum number of iterations.

Table 2. Setting of Experiment I

n	m	β	μ	α_0	λ^k	K
30	1	0.5	0.1	1	1	1000

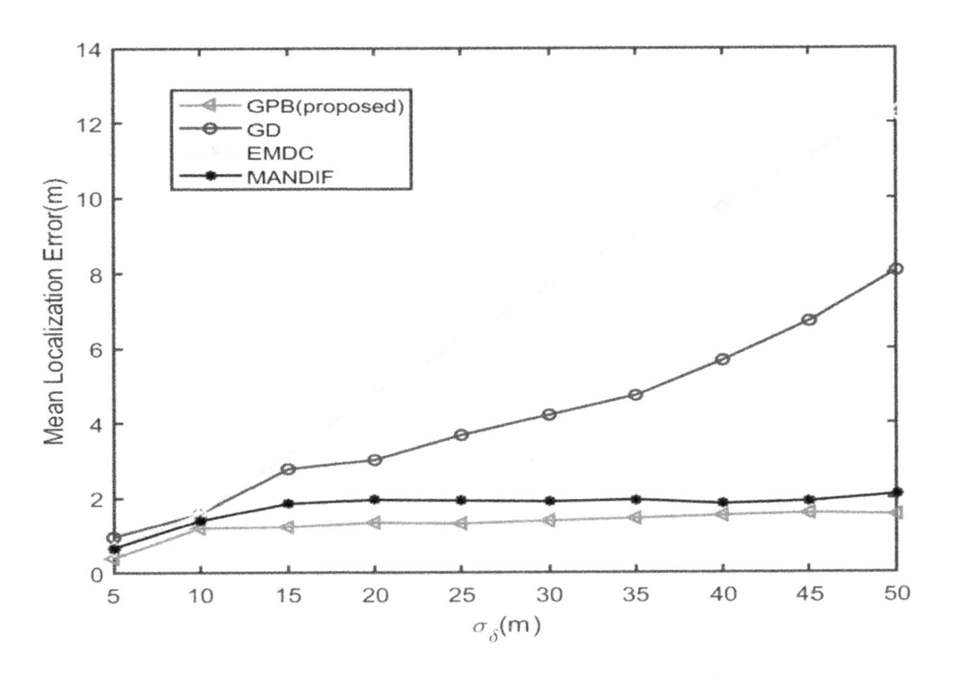

Fig. 2. Mean Localization Error (MLE) curves with σ_δ

In the simulation experiment, we deploy m anchors containing n malicious anchors randomly in the square field of $100m \times 100m$. The experiments below were repeated over 1000 times to obtain accurate and stable results. In Experiment I, we set the relatively simple environment where the malicious anchors are in a small scale. The corresponding parameters are summarized in Table 2. σ_δ is the stength of the attacks. The Mean Localization Error (MLE) with varying σ_δ from 5 to 50 is manifested as Fig. 2. It is can be depicted that while the EMDC and GD performance degrades as σ_δ increases, the MANDIF and GPB algorithms perform with stable and excellent features. Within the value range of σ_δ, the average value of MLE of EMDC, GD, MANDIF and GPB are 6.623, 3.948, 1.914 and 1.308 respectively. The proposed algorithm can decrease the localization error remarkably: by 80.3%, 66.9%, 31.7% compared to EMDC, GD and MANDIF respectively.

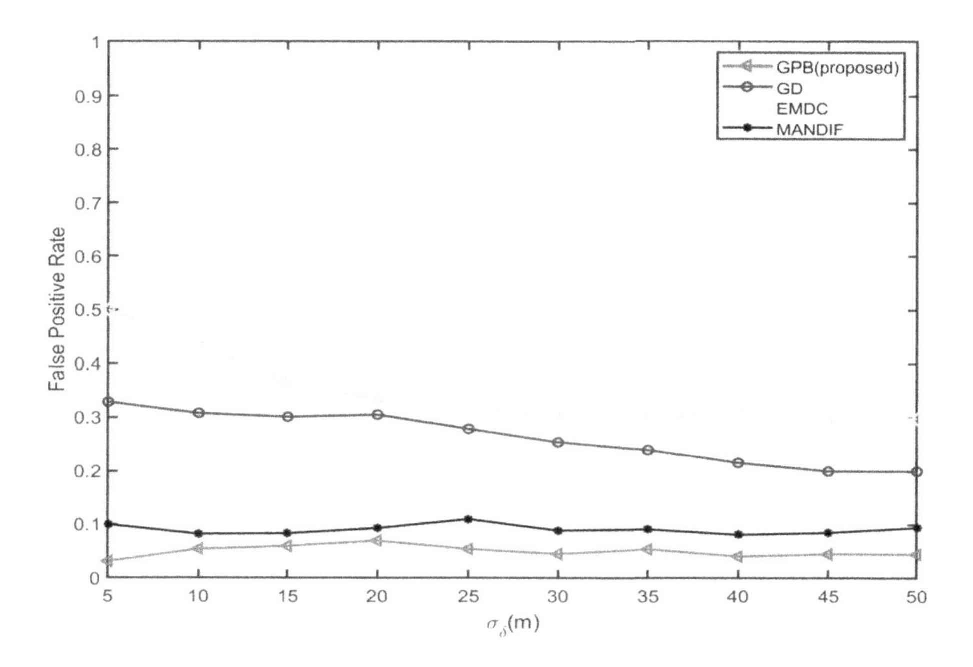

Fig. 3. FPR curves with σ_δ

The False Negative Ratio(FNR), True Positive Ratio(TPR), and False Negative Ratio(FNR) curves of comparison algorithms and the proposed algorithm with varying σ_δ are shown in Fig. 3, Fig. 4, Fig. 5 respectively. MNDC is proposed to achieve malicious anchors detection and secure localization. There is a very important premise condition to implement the method in EMDC: to guarantee the measurements of RSSI are not attacked while the measurements of ToA are under attack. This precondition can be difficult to guarantee in the practical scenario. The more violent the attack is, the larger the detection interval will be. As a result there will be a lower false detection rate. However, the FPR of GPB remains below 0.0699, The TPR remains above 0.9200, performing well in a violent-attack environment and soft-attack environment. The stable and excellent capability of the proposed algorithm comes from the fact that sparse recovering is not affected by the size of the value of a non-zero item.

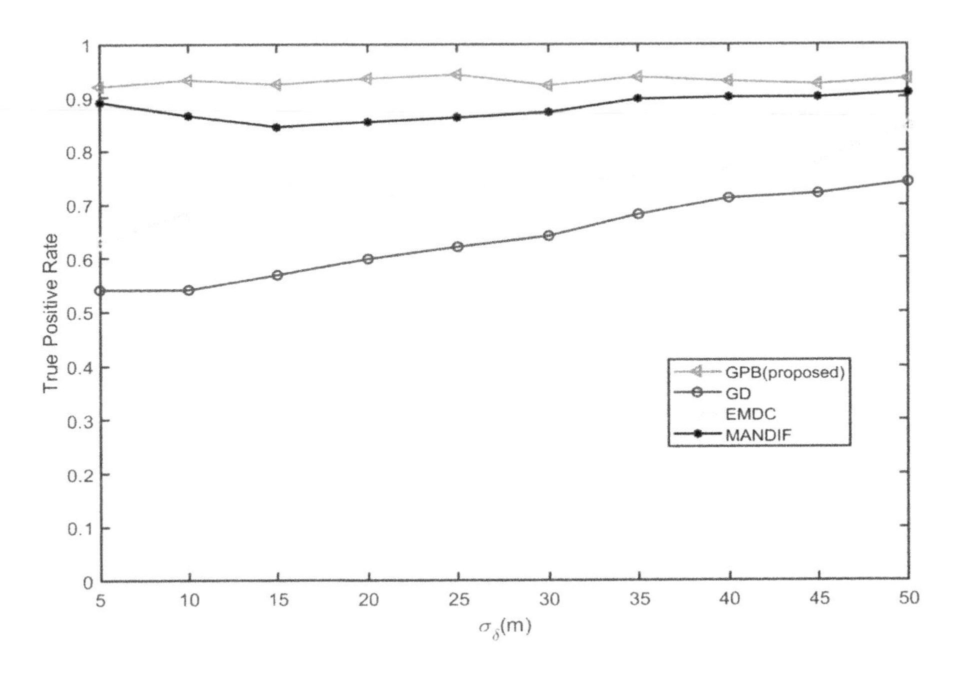

Fig. 4. TPR curves with σ_δ

Then, we consider a situation where the attack is more violent and the malicious anchor nodes occupy a larger proportion. Based on the theorem and the fact that: as long as the number of the malicious nodes $m \leq \frac{n-2}{2}$, the target node localization and all malicious anchors identification can be achieved at the same time. We set the number of the malicious nodes to take up 40% of the total nodes. Like Experiment I, m anchors including n malicious anchors were deployed randomly in the field of $100m \times 100m$. The corresponding parameters are summarized in Table 3.

Table 3. Setting of experiment II

n	m	β	μ	α_0	λ^k	K
30	9	0.5	0.1	1	1	1000

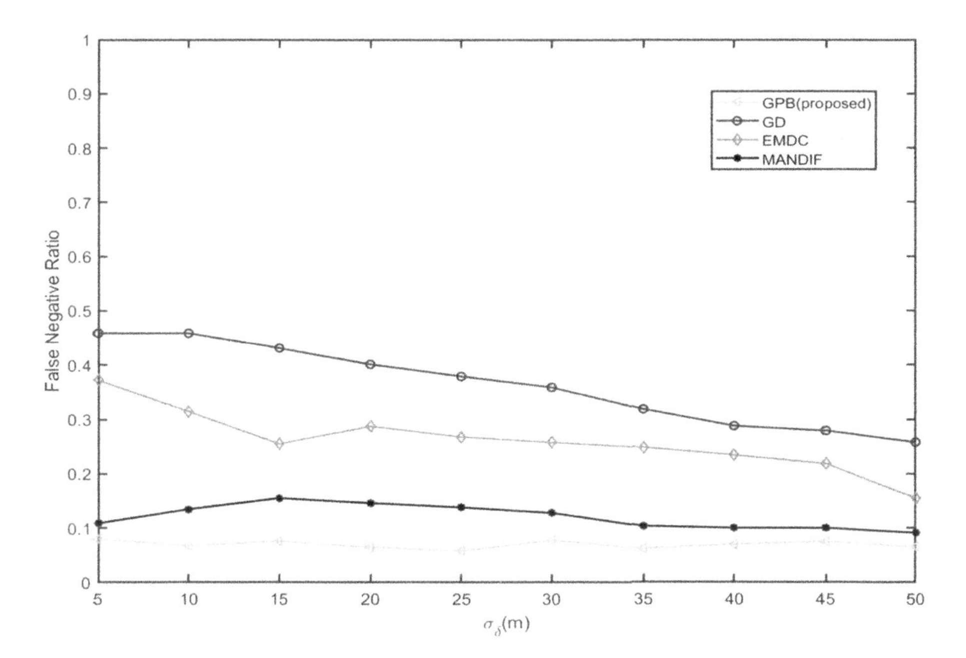

Fig. 5. FNR curves with σ_δ

The Mean Localization Error (MLE) manifested as Fig. 6. It is worth of denoting that, within the value range of σ_δ, the average value of localization error of EMDC, GD, MANDIF and GPB are 20.78, 18.84, 13.59 and 13.59. The positioning accuracy of other algorithms has declined, while the proposed one remains accurate, decreasing by 43.1%, 37.2%, 13.0% compared with EMDC, GD and MANDIF respectively.

The FPR, TPR and FNR curves of the proposed algorithm and other algorithms are shown in Fig. 7, Fig. 8, Fig. 9. The FPR of GPB keeps below 0.0812, The TPR remains above 0.9010, performing better than other algorithms.

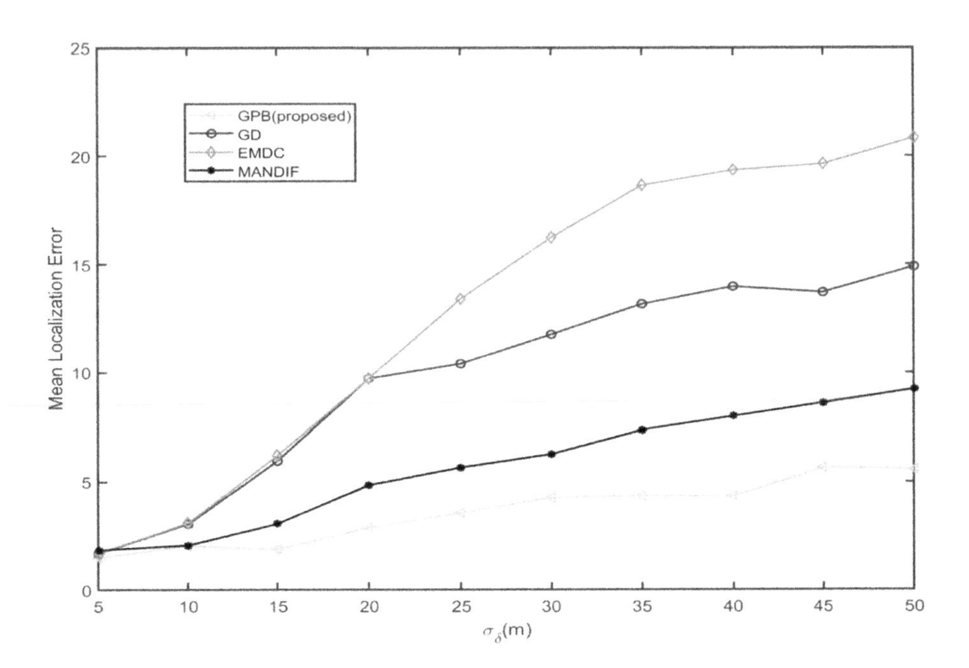

Fig. 6. Mean Localization Error (MLE) curves with σ_δ

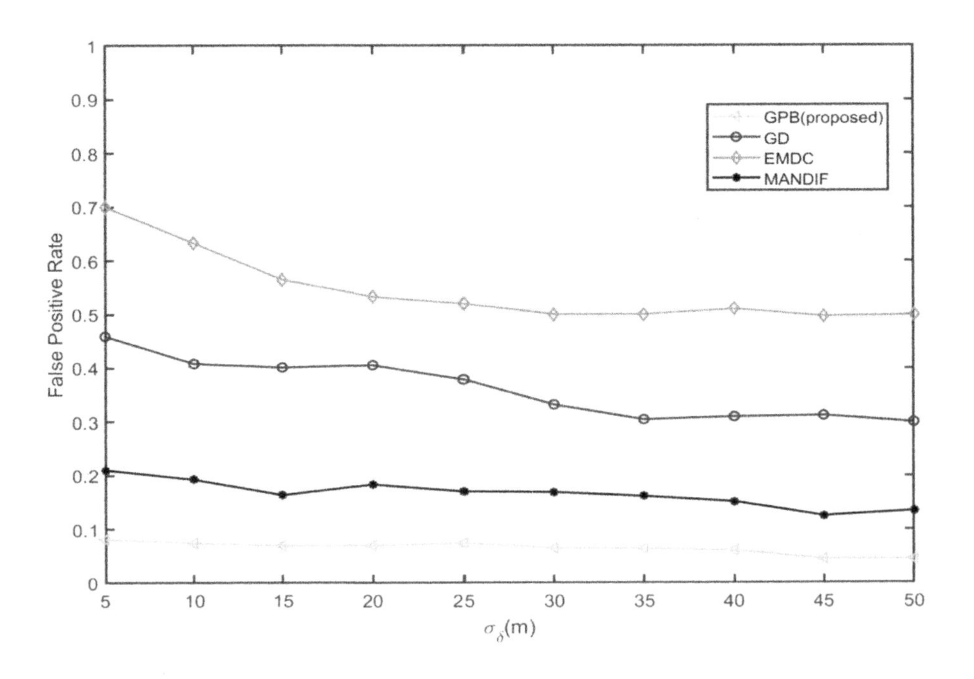

Fig. 7. FPR curves with σ_δ

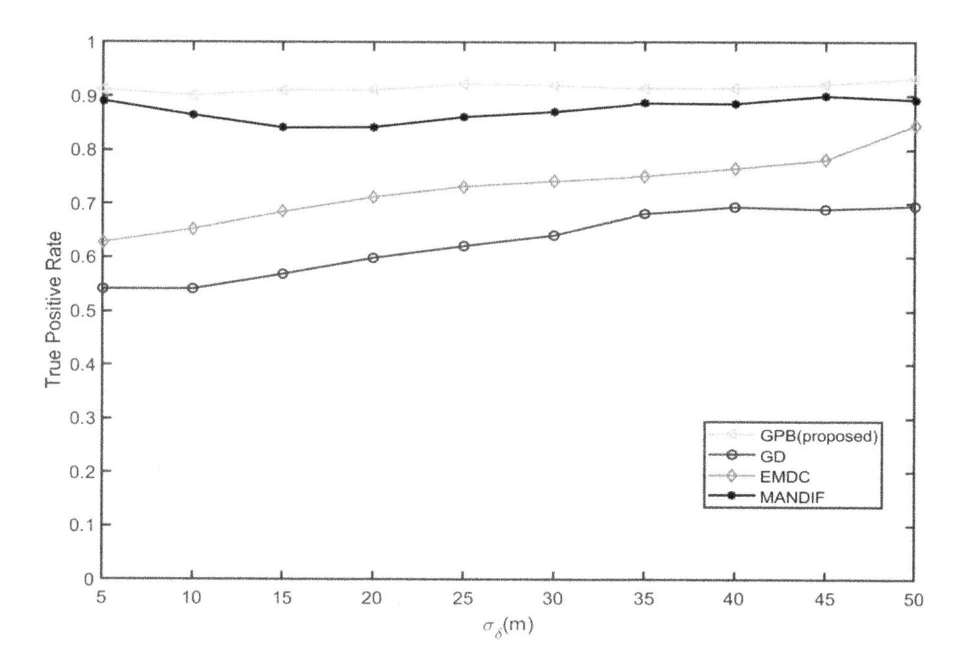

Fig. 8. TPR curves with σ_δ

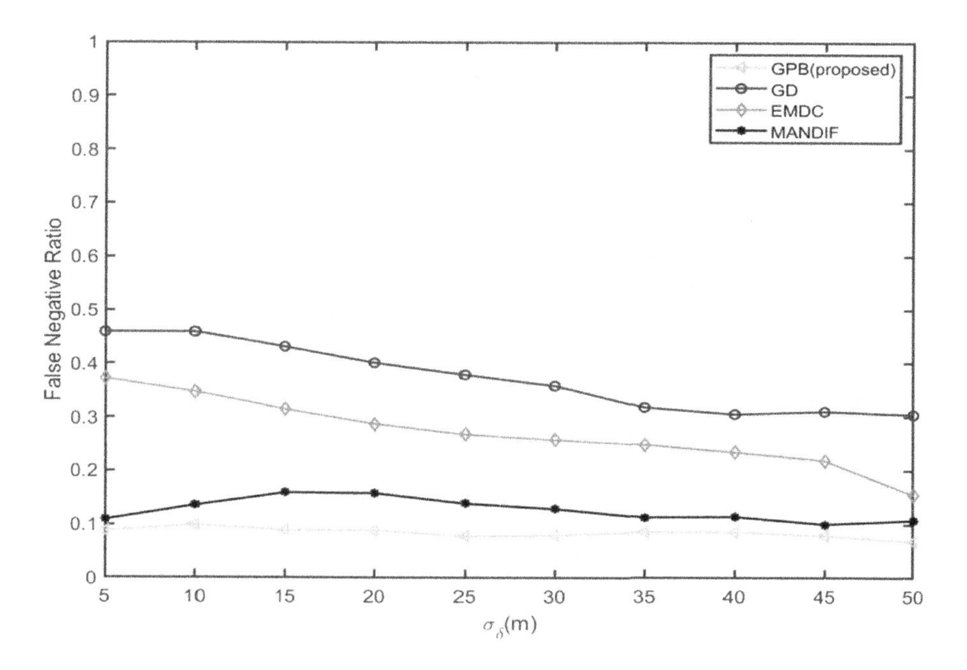

Fig. 9. FNR curves with σ_δ

5 Conclusions

In this paper, the localization problem is formulated as the sparse vector recovery problem. The Gradient Projection Basic algorithm(GPB) is proposed to identify the non-zero item and detect the malicious anchors. In the early stage of the proposed algorithm, the recursive weighted linear square is proposed to obtain the initially estimated position. The comparative experiments and simulation results demonstrate that the proposed algorithm can identify the malicious anchors and achieve successful localization with the probability above 0.9, which outperforms other algorithms of interest.

References

1. Xiong, J., Zhao, M., Bhuiyan, M.Z.A., Chen, L., Tian, Y.: An AI-enabled three-party game framework for guaranteed data privacy in mobile edge crowdsensing of IoT. IEEE Trans. Industr. Inf. **17**(2), 922–933 (2021)
2. Tian, Y., Wang, Z., Xiong, J., Ma, J.: A blockchain-based secure key management scheme with trustworthiness in DWSNs. IEEE Trans. Industr. Inf. **16**(9), 6193–6202 (2020)
3. Zhou, B., Chen, Q.: On the particle-assisted stochastic search mechanism in wireless cooperative localization. IEEE Trans. Wireless Commun. **15**(7), 4765–4777 (2016)
4. Jiang, W., Xu C., Pei, L., Yu, W. Multidimensional scaling-based TDOA localization scheme using an auxiliary line. IEEE Signal Process. Lett. **23**(4), 546–550 (2016)
5. Patwari, N., Ash, J.N., Kyperountas, S., Hero, A.O., Moses, R.L., Correal, N.S.: Locating the nodes: cooperative localization in wireless sensor networks. IEEE Signal Process. Mag. **22**(4), 54–69 (2005)
6. Luo, Q., Peng, Y., Li, J., Peng, X.: Rssi-based localization through uncertain data mapping for wireless sensor networks. IEEE Sens. J. **16**(9), 3155–3162 (2016)
7. Huang, B., Xie, L., Yang, Z.: Tdoa-based source localization with distance-dependent noises. IEEE Trans. Wireless Commun. **14**(1), 468–480 (2015)
8. Liu, X., Yin, J., Zhang, S., Ding, B., Guo, S., Wang, K.: Range-based localization for sparse 3-d sensor networks. IEEE Internet Things J. **6**(1), 753–764 (2019)
9. Liu, X., Xiong, N., Li, W., Xie, Y.: An optimization scheme of adaptive dynamic energy consumption based on joint network-channel coding in wireless sensor networks. IEEE Sens. J. **15**(9), 5158–5168 (2015)
10. Peng, J., Liu, X.: A malicious anchor detection algorithm based on isolation forest and sequential probability ratio testing (SPRT). In: Guo, S., Liu, K., Chen, C., Huang, H. (eds.) CWSN 2019. CCIS, vol. 1101, pp. 90–100. Springer, Singapore (2019). https://doi.org/10.1007/978-981-15-1785-3_7
11. Liu, X., Su, S., Han, F., Liu, Y., Pan, Z.: A range-based secure localization algorithm for wireless sensor networks. IEEE Sensors J. **19**(2), 785–796 (2019)
12. Garg, R., Avinash, L.V.: An efficient gradient descent approach to secure localization in resource constrained wireless sensor networks. IEEE Trans. Inf. Forensics Secur. (2012)

13. Hamidi, S., Shahbazpanahi, S.: Sparse signal recovery based imaging in the presence of mode conversion with application to non-destructive testing. IEEE Trans. Signal Process. 1 (2015)
14. Mukhopadhyay, B., Srirangarajan, S., Kar S.: Robust range-based secure localization in wireless sensor networks. In: 2018 IEEE Global Communications Conference (GLOBECOM) (2019)

Application of Yolov5 Algorithm in Identification of Transmission Line Insulators

Jinxiong Zhao[1(✉)], Jiaxiu Ma[2], Junwei Xin[3], and Rutai An[3]

[1] State Grid Gansu Electric Power Research Institute, Lanzhou 730070, China
[2] School of Information, Renmin University of China, Beijing 100872, China
[3] Lanzhou Longneng Technology Co., Ltd., Lanzhou 730050, China

Abstract. As an important infrastructure, the power system assumes a position that cannot be ignored in the national economy. The insulator in the transmission line is one of the main components of the power system. A complete and defect-free insulator is a prerequisite to ensure a good insulation between the current-carrying conductor and the ground. At present, it has become a mainstream practice to identify insulators through drones. However, due to the small number and single types of insulator data currently disclosed, the network does not have a large number of samples to learn more characteristics of insulators, which hinders the improvement of the accuracy of the network model to a certain extent. In this article, based on the existing 848 transmission line insulator data set, we train the yolov5 algorithm to generate a network with a recognition rate. The experimental results show that the mAP of the trained model is 11.41% higher than that of J-Method and 37.25% higher than the average of the other four methods mentioned by J-Method.

Keywords: Transmission line · Insulator · Yolov5 · Algorithm · Data enhancement

1 Introduction

As an important infrastructure, the power system assumes a position that cannot be ignored in the national economy [1]. The insulator in the transmission line is one of the main components of the power system. A complete and defect-free insulator is a prerequisite to ensure a good insulation between the current-carrying conductor and the ground. At present, it has become a mainstream practice to identify insulators through drones. Traditional manual identification of insulators in high-voltage transmission lines is extremely difficult, which is mainly manifested in the high identification cost and low identification efficiency and the inability to identify insulators in complex geographic environments.

J. Xiong et al. (Eds.): MobiMedia 2021, LNICST 394, pp. 880–887, 2021.
https://doi.org/10.1007/978-3-030-89814-4_65

In recent years, A series of intelligent autonomous identification algorithms have been produced, which can identify the defects of insulator equipment in time with the rapid development and continuous application of computer vision and image recognition technology. Liu et al. propose an improved algorithm based on Faster-RCNN for the presence of complex foreign bodies in the substation environment, which strengthens the detection of small targets [2]. However, there are certain sample scene differences in the identification of insulator defects, and the identification efficiency is very low. Jiang et al. combine the Faster R-CNN algorithm and the Soft-NMS algorithm to solve the identification of insulator defects in the interference environment, but only limit to a small increase in the average identification efficiency [3]. Based on the idea of segmentation network, Gao et al. propose a Mask R-CNN algorithm to improve the performance of the model [4], but this method is based on infrared image recognition. Aiming at the complexity of the transmission line image, Hou et al. integrate the AlexNet, VGG16 and Faster R-CNN network structures, but the redundancy of the network structure reduces the recognition efficiency [5].

In this paper, we use the yolov5 algorithm for sample training and insulator image defect recognition, which improves the accuracy and mAP value of algorithm recognition, and effectively solves the problem of inaccurate recognition of insulator defects in actual line inspection scenarios.

2 Insulator Data Enhancement

2.1 Introduction to U-Net Network

U-Net is proposed in 2015, which is an Encoder-Decoder structure and used to solve the problem of medical image segmentation. The U-Net network structure is shown in Fig. 1. On the left is the decoding process, which consists of convolution operation and pooling layer downsampling. On the right is the encoding process, which is restored to the original resolution after up-sampling. A skip-connection is added during the decoding and encoding process, and the entire network has a U-shaped symmetrical structure [6]. U-Net uses data enhancement methods such as translation, rotation, and elastic deformation during training to make up for the lack of data [7]. Since the number of transmission line insulators is too small to be used to train large-scale instance segmentation networks like Mask-RCNN, U-Net network is adopted [8]. Firstly, we label the insulator position on the image that taken by the transmission line for creating insulator Dataset. Secondly, we train the Dataset with U-Net network and generate a model.Thirdly, we use the model to predict other unlabeled insulator Dataset and segment the insulator masks. Finally, the segmenting insulator masks are overlaid on the image of ordinary transmission line, which to generate a new insulator Dataset for increasing the yolov5 training samples.

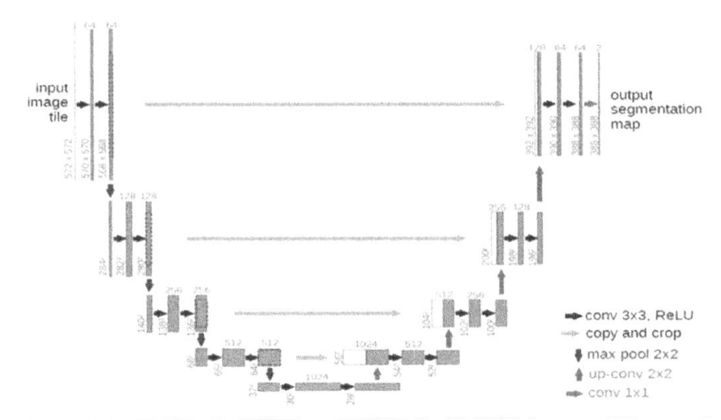

Fig. 1. U-Net network structure that comes from https://blog.csdn.net/fang_chuan/article/details/94965995

2.2 Data Enhancement

At present, the transmission line Insulator Dataset(In-D) is one of the frequently used datasets, with a total of 848 [9]. The data consists of two parts: normal insulators(Nor-Ins) and defective insulators(Def-Ins). Nor-Ins are 600 images of normal insulators captured by drones. Def-Ins are 248 insulator images, which are all generated through the following five steps of data enhancement:

Step 1: Use the TVSeg tool to segment the insulator from a small part of the original image, and the result of the segmentation is the mask image of the insulator;

Step 2: Use Affine Transform to randomly rotate, mirror, blur, add noise, Cutout, etc. to the original image and its insulator mask image to achieve data enhancement;

Step 3: Use the enhanced insulator mask image (including the coordinate position) to train the U-Net network;

Step 4: Use the trained U-Net network to segment the insulator mask image and the original background image (Background) of the original image;

Step 5: Generate different image data sets containing insulators.

After data preprocessing, 648 training sets and 200 test sets are finally generated. The test sets used in this paper are all images normally containing insulators taken by drones, the purpose is to ensure that the model is not over-fitting and can reflect the true accuracy rate.

3 Introduction to Yolov5 Algorithm

YOLOv5 is a single-stage target detection algorithm [10]. The algorithm adds some new improvements on the basis of YOLOv4, which greatly improves its speed and accuracy [11]. The architecture diagram of YOLOv5 is shown in Fig. 2 and the main improvement ideas are as follows:

(1) Input terminal: In the model training stage, some improvement ideas are proposed, mainly including Mosaic data enhancement, adaptive anchor frame calculation, and adaptive image scaling;

(2) Benchmark network: integrate some new ideas in other detection algorithms, mainly including Focus structure and CSP structure;
(3) Neck network: The target detection network often inserts some layers between BackBone and the final Head output layer, and Yolov5 just adds the FPN + PAN structure;
(4) Head output layer: The anchor frame mechanism of the output layer is the same as YOLOv4. The main improvements are the loss function GIOU_Loss during training and the DIOU_nms filtered by the prediction frame.

The YOLOv5 algorithm has 4 versions, and yolov5 is divided into four models according to size: yolov5s, yolov5m, yolov5l, and yolov5x [12].

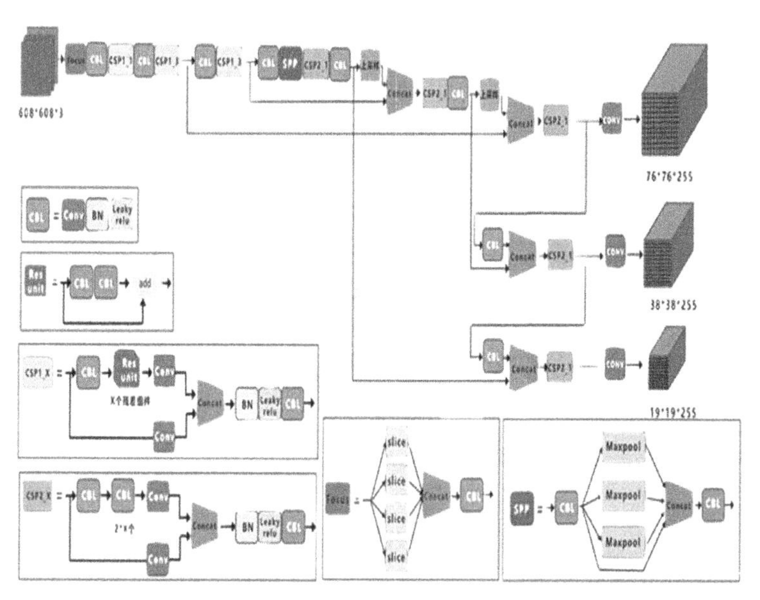

Fig. 2. YOLOv5 architecture diagram that comes from https://www.jianshu.com/p/4c348d78143e

4 Experiment

4.1 Experiment Preparation

The data set used in this experiment is In-D, the training data and the test data are 648 and 200, respectively, accounting for 76.42% and 23.58% of the total data. In the experiment, it is divided into 21 batches, and each batch size is 32 sheets. The size of each image is 640x640, with a total of 50 epochs before and after iteration. The running configuration environment of the experiment is shown in Table 1. The operating system is Ubuntu 18.04, the programming language is Python 3.6, and the training framework is Pytorch 1.7.1.

Table 1. The running configuration environment of the experiment

Experimental environment	Configuration
Operating system	Ubuntu 18.04
CPU	Intel(R) Xeon(R) Gold 6230 CPU @ 2.10 GHz
RAM /GB	188
Programming language	Python 3.6
Deep learning framework	Pytorch 1.7.1

The key metrics used in the experiment are precision rate (P), recall rate (TPR) and mAP value. The specific formula is as follows [13]:

$$\text{P(Precision)} = \text{TP} / (\text{TP} + \text{FP}) \tag{1}$$

$$\text{TPR} = \text{TP}/(\text{TP} + \text{FN}) \tag{2}$$

$$\text{mAP} = \frac{\sum_{i=1}^{K} AP_i}{K} \tag{3}$$

Where:

$$p_{\text{int }erp}(r) = \max_{r' \geq r} P(r')$$

$$AP = \sum_{i=1}^{\text{n - 1}} (r_{i+1} - r_i) p_{\text{int }erp}(r_{i+1})$$

$$r = TP/(TP + FN)$$

Where True Positive (TP) means that the true result is P and the predicted result is P, False Positive(FP) means that the true result is N and the predicted result is P, True Negative (TN) means that the true result is P and the predicted result is N, False Negative (FN) means that the true result is N and the predicted result is N, Recall(r) refers to the proportion of positive examples (TP + FN) correctly identified by the model to all positive examples in the Dataset, K is the number of categories., AP represent how good or bad the trained model is in the current category.

4.2 Experimental Result

The YOLOv5 algorithm was used in the experiment, and four models of YOLOv5s, YOLOv5m, YOLOv5l, and YOLOv5x were tried respectively. The network structure of these four models is the same, but the depth of the model and the number of convolution kernels are different, which results in different model sizes.

As shown in Fig. 3, the size of the four models YOLOv5s, YOLOv5m, YOLOv5l, and YOLOv5x are 14.4MB, 42.5MB, 93.7MB and 175.1MB, respectively. It can be clearly seen that the depth of YOLOv5x is 607 layers, and the capacity of its model is also the largest of the four models, 175.1MB, which is 2.15 times higher than the average size of the four models.

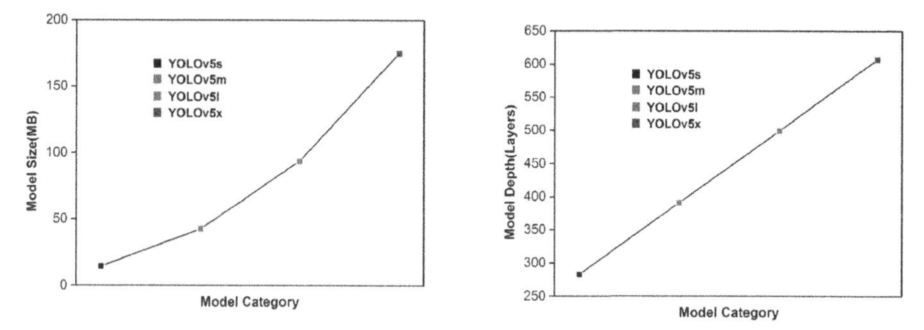

Fig. 3. The size and depth of the four models of YOLOv5s, YOLOv5m, YOLOv5l, and YOLOv5x

According to the three indicators previously set to measure the quality of the model, we test the four types of networks. As shown in Fig. 4a, YOLOv5l has the highest P index among the four models, 0.9263. YOLOv5l ranks second in the R index of the four models, which is only 0.03 lower than the highest R index (YOLOv5x), as shown in Fig. 4b. This data shows that YOLOv5l and YOLOv5x have very similar performance in terms of R index.

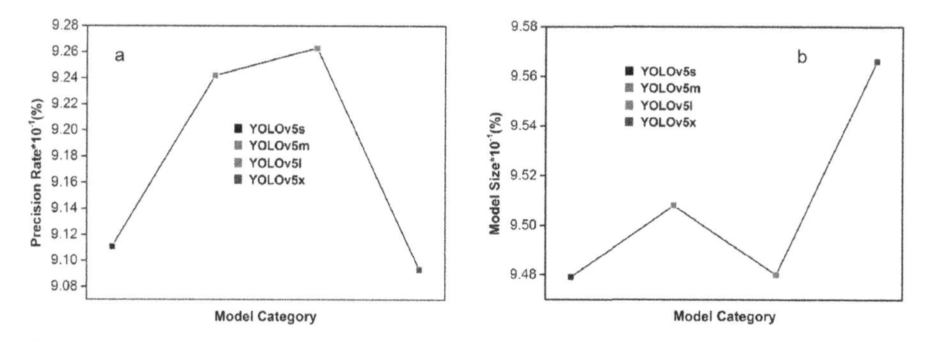

Fig. 4. a) The P and b) R index of the four models of YOLOv5s, YOLOv5m, YOLOv5l, and YOLOv5x.

Based on the index P and the index R, and according to the calculation rules of mAP, it is concluded that the mAP value of YOLOv5l is the highest (0.9727). This data shows that the overall performance of this model is the best. It is worth noting that although YOLOv5x is not the best of the four models, its mAP value is very close to that of YOLOv5l, and the difference between the two is only 0.03, as shown in Fig. 5a. Compared with the latest report of J-Method [14], YOLOv5l is 11.41% higher than its improved method and 37.25% higher than the average of the other four methods mentioned (Fig. 5b). That can be seen from the above data that YOLOv5l does have a great improvement in performance.

This experiment is only based on the actual business needs of the power system. The key considerations are the P index, R index and mAP value. Therefore, the calculation cost is not considered here. For operations that require particularly stringent accuracy in the power system business, it is recommended to use the YOLOv5l model. For businesses with general accuracy requirements and limited computing resources, it is recommended to use YOLOv5s and YOLOv5m with smaller model size.

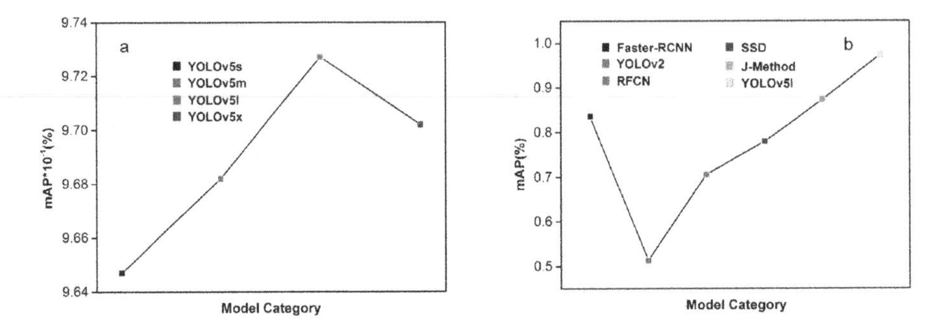

Fig. 5. a) The mAP value of the four models of YOLOv5s, YOLOv5m, YOLOv5l and YOLOv5x, b) Comparison of mAP value of YOLOv5l and J-Method.

5 Conclusion

In short, based on the existing In-D, the yolov5 algorithm is applied to the identification of insulators in transmission lines, which solves the problem of low recognition rate and low mAP value. The experimental results show that the mAP of the trained model is 11.41% higher than that of J-Method and 37.25% higher than the average of the other four methods mentioned by J-Method. That provides a new idea for the identification of insulators in transmission lines.

Acknowledgement. This research work is supported by Science and Technology Project of State Grid Gansu Electric Power Company (No. 52272219000E).

Conflicts of Interest. The authors declare that they have no conflicts of interest about this paper.

References

1. Zhao, J.X., Zhang, X., Di, F.Q., et al.: Exploring the optimum proactive defense strategy for the power systems from an attack perspective. Secur. Commun. Netw. 1–14 (2021)
2. Liu, L., Han, R., Han, Y.F., et al.: The application of the improved faster-RCNN target detection method in the detection of foreign body in substation suspension. Electrical Measure. Instrument. **58**(01), 142–146 (2021)

3. Jiang, S., Sun, Y., Yan, D.S.: Insulator identification of aerial photographic inspection images based on deep learning algorithm. J. Fuzhou Univ. (Nat. Sci. Edition) **01**, 58–64 (2021)
4. Gao, Y., Tian, L.F., Du, Q.L.: Detection of overheating defects of composite insulator based on mask R-CNN. China Electric Power **54**(01), 135–141 (2021)
5. Hou, C.P., Zhang, H.G., Zhangi, W., et al.: Identification method for spontaneous explosion defect of transmission line insulators. J. Electric Power Syst. Autom. **31**(06), 1–6 (2019)
6. Liu, J., Wang, J., Ruan, W., et al.: Diagnostic and gradation model of osteoporosis based on improved deep U-Net network. J. Med. Syst. **44**(1) (2020)
7. Li, Q.J., Fan, S.S., Chen, C.S.: An intelligent segmentation and diagnosis method for diabetic retinopathy based on improved U-NET network. J. Med. Syst. **43**(9), 1–9 (2019)
8. Ibtehaz, N., Rahman, M.S.: MultiResUNet: rethinking the U-Net architecture for multimodal biomedical image segmentation. Neural Netw. **121** (2019)
9. Mellnik, A.R., Lee, J.S., Richardella, A., et al.: Spin-transfer torque generated by a topological insulator. Nature (2014)
10. Li, S., Gu, X., Xu, X., et al.: Detection of concealed cracks from ground penetrating radar images based on deep learning algorithm. Constr. Build. Mater. **273**, 121949 (2021)
11. Yang, G., Feng, W., Jin, J., et al.: Face mask recognition system with YOLOV5 based on image recognition. In: 2020 IEEE 6th International Conference on Computer and Communications (ICCC). IEEE (2020)
12. Shu, L., Zhang, Z.J., Lei, B.: Research on a dense-Yolov5 algorithm for infrared target detection. World Sci. Res. J. **19**(01), 69–75 (2021)
13. Orovi, A., Ili, V., Uri, S., et al.: The real-time detection of traffic participants using YOLO algorithm. In: 2018 26th Telecommunications Forum (TELFOR). IEEE (2019)
14. Padilla, R., Passos, W.L., Dias, T., et al.: A comparative analysis of object detection metrics with a companion open-source toolkit. Electronics **10**(3), 279–306 (2021)

Author Index

Printed in the United States
by Baker & Taylor Publisher Services